4th International High Performance Buildings Conference at Purdue 2016

West Lafayette, Indiana, USA
11 - 14 July 2016

ISBN: 978-1-5108-2874-2

Printed from e-media with permission by:

Curran Associates, Inc.
57 Morehouse Lane
Red Hook, NY 12571

Printed by Curran Associates, Inc. (2016)

Ray W. Herrick Laboratories
School of Mechanical Engineering
Purdue University
177 S. Russell Street
West Lafayette, IN 47907-2099, USA.

Phone: +1 (765) 494-2132
Fax: +1 (765) 494-0787

rhlab@ecn.purdue.edu

Additional copies of this publication are available from:

Curran Associates, Inc.
57 Morehouse Lane
Red Hook, NY 12571 USA
Phone: 845-758-0400
Fax: 845-758-2634
Email: curran@proceedings.com
Web: www.proceedings.com

Ray W. Herrick Laboratories
Center for High Performance Buildings

Table of Content

B-01: Supervisory Control Optimization (IBO)

Time: Monday July 11, 2016: 1:00 PM - 3:00 PM — *Location*: 214 A&B
Session Chair: Jie Cai

B-02: Building Envelope Systems

Time: Monday July 11, 2016: 1:00 PM - 3:00 PM — *Location*: 214 C&D

Session Chair: Ying-Chieh Chan

B-03: Model Predictive Controls (IBO)

Time: Monday July 11, 2016: 3:30 PM - 5:30 PM — *Location*: 214 A&B
Session Chair: Robert Turney

B-04: Net Zero Energy Buildings

Time: Monday July 11, 2016: 3:30 PM - 5:30 PM — *Location*: 214 C&D
Session Chair: Matthew Miller

B-05: Building Data Analytics and Diagnostics (IBO)

Time: Tuesday July 12, 2016: 9:45 AM - 11:45 AM — *Location*: 214 A&B
Session Chair: John House

ID: 3361

Francesco Leonardi[1], Hayden Reeve[1], Timothy Wagner[1], Ziyou Xiong[1], June Park[2]
[1]United Technologies Research Center, United States of America; [2]Carnegie Mellon University, United States of America

Keywords: Point Mapping, Cost-Effective Deployment, Intelligent Building Applications

ID: 3674

Adam Regnier, Jin Wen
Drexel University, United States of America

Keywords: Fault, Diagnostics, Air Handling Unit, Bayesian Network, Noisy MAX

ID: 3223

Fadi M. Alsaleem[1], Michael Munroe[2], Mostafa RAFAIE Rafaie[1]
[1]Wichita State University, United States of America; [2]Emerson Climate Technologies,United States of America

Keywords: Filter Blockage, Indoor Air Flow, CART, Mann-Kendall

ID: 3029

Debashis Dey, Bing Dong, Zhaoxuan Li
University of Texas at San Antonio, United States of America

Keywords: AHU, FDD, Bayesian Network, Expert Rules

ID: 3182

Yunpeng Hu, Jiani Liu, Li Zhou, Yang Liu, Qingling Qiu
Wuhan Buziness University, China, People's Republic of

Keywords: Fault Detection, Chiller, Sensor, Principal Component Analysis, Undetectable Zone

ID: 3017

Fadi M. Alsaleem, Amro Quedan
Wichita State University, United States of America

Keywords: Mann-Kendall, Data Analysis, Low Refrigerant, Big Data

B-06: Facades, Lighting and Solar Radiation

Time: Tuesday July 12, 2016: 9:45 AM - 11:45 AM — *Location*: 214 C&D
Session Chair: Andrea Gasparella

Iason Konstantzos[1,2], Athanasios Tzempelikos[1,2], Nicole M. Murchison[3], Robert W. Proctor[3]

[1]School of Civil Engineering, Purdue University; [2]Ray W. Herrick Laboratories, School of Mechanical Engineering, Purdue University; [3]Psychological Sciences, Purdue University

Keywords: Daylight Glare, Roller Shades, DGP, Visual Comfort, Visible Sun

Giovanni Pernigotto[1], Alessandro Prada[2], Paolo Baggio[2], Andrea Gasparella[1], Ardeshir Mahdavi[3]

[1]Free University of Bozen-Bolzano, Faculty of Science and Technology, Bolzano, Italy; [2]University of Trento, Department of Civil Environmental and Mechanical Engineering, Trento, Italy; [3]Vienna University of Technology, Department of Building Physics and Building Ecology, Vienna, Austria

Keywords: Solar Irradiance Model, Diffuse Irradiance Models, Irradiance Models for Tilted Surfaces, Building Energy Simulation

Anna Maria Atzeri[1], Francesca Cappelletti[2], Andrea Gasparella[1], Athanasios Tzempelikos[3]

[1]Free University of Bolzano/Bozen, Faculty of Science and Technology, Bolzano, Italy; [2]University IUAV of Venezia, Dpt. of Design and Planning in Complex Environments, Venezia, Italy; [3]Purdue University School of Civil Engineering and Ray W. Herrick Labs, West Lafayette, Indiana (USA)

Keywords: Roller Shades, Thermal Comfort, Visual Comfort, Model Validation

Moyu Yan[1], Ming Qu[1], Steve Peng[2]

[1]Purdue University, Department of Civil Engineering, West Lafayette, IN, United States of America; [2]California State University, College of Business and Economics, East Bay, CA, United States of America

Keywords: Flat Plate Thermal Collector, Ellipse Tube Thermal Collector, Efficiency of Thermal Collector

Ying-Chieh Chan[1], Iason Konstantzos[1], Athanasios Tzempelikos[1], Matthew Miller[2]

[1]Purdue University; [2]Kawneer Company Inc.

Keywords: Interface, Exterior Solar Protection Devices, Shading System

B-07: Demand Response and Load Management (IBO)

Time: Tuesday July 12, 2016: 1:30 PM - 3:30 PM — *Location*: 214 A&B
Session Chair: William Hutzel

B-08: Building Analysis and Retrofits

Time: Tuesday July 12, 2016: 1:30 PM - 3:30 PM — *Location*: 214 C&D
Session Chair: Francesca Cappelletti

B-09: Control Oriented Modeling (IBO)

Time: Tuesday July 12, 2016: 4:00 PM - 6:00 PM — *Location*: 214 A&B
Session Chair: José Agustín Candanedo

B-SPC: Student Paper Competition

Time: Tuesday July 12, 2016: 4:00 PM - 6:00 PM — *Location*: 278
Session Chair: Panagiota Karava

B-10: Occupant Behavior and Predictive Models (IBO)

Time: Wednesday July 13, 2016: 9:45 AM - 12:00 PM — *Location*: 214 A&B

Session Chair: Athanasios Tzempelikos

B-11: Thermal Storage and Heat Pumps

Time: Wednesday July 13, 2016: 9:45 AM - 12:00 PM — *Location*: 214 C&D
Session Chair: Marco Baratieri

B-12: Building Simulation and Optimization

Time: Wednesday July 13, 2016: 1:00 PM - 3:00 PM — *Location*: 214 A&B
Session Chair: Hui Shen

B-13: Ventilation, Passive Cooling and IAQ

Time: Wednesday July 13, 2016: 3:30 PM - 5:30 PM — *Location*: 214 A&B
Session Chair: Brandon E. Boor

POLICY FOR PUBLISHING

POLICY FOR PUBLISHING CONFERENCE PAPERS IN ARCHIVAL JOURNALS

The organizing committee recommends that all authors who have papers that satisfy the criteria for archival journal articles submit them to such journals for publication. Because of the relatively small number of copies, publication in the proceedings of the International Compressor Engineering Conference at Purdue does not constitute prior publication as far as many journal editors are concerned. If in doubt, an inquiry should be made with the respective editor. In most cases, the limit of eight pages will require the creation of an expanded journal article version, which would qualify as a new publication anyway.

The criteria for archival papers are that they should contain original, quantitative, detailed scientific materials, which are placed in proper perspective relative to prior work and are supported by references to the appropriate literature. Thus, not all conference papers are suitable. Some describe present practice, or report on procedures that were applied first by someone else, or report design features which, although important to conference participants, have relatively short temporal interest.

The organizing committee requests that authors who choose to submit their papers to archival journals acknowledge that those papers were, in part, presented at the 2016 Conferences and send a copy of each submitted manuscript to the conference chairperson.

Purdue University sponsors this conference to provide a forum for the free exchange of ideas on the engineering state-of-the-art in the compressor field. The ideas expressed are those of the author(s) and do not necessarily represent the opinions or policies of the University or the cooperating organizations. The papers are not subject to the formal review procedures of the separate cooperating organizations as they would be if they were to be published by the cooperating societies. Instead, they are subject to the review procedures of the Organizing Committee.

CONTACT INFORMATION

Contact about the conferences and technical information:
Mrs. Kim Stockment
Conference Coordinator
Ray W. Herrick Laboratories
Purdue University, 177 S Russell St.
West Lafayette, IN 47907-2031 USA
Telephone: (765) 494-6078; Fax: (765) 494-0787
E-mail: hlconf16@purdue.edu
Website: www.engineering.edu/HerrickConf

Contact about the conference venue, hotel information and registration:
Mr. Ethan Kingery,
Conference Coordinator
Purdue University Conferences
128 Memorial Mall, Stewart Center, Room 116
West Lafayette, IN 47907-7220 USA
Telephone: (765) 496-3667; Fax: (765) 494-0567
E-mail: ekingery@purdue.edu
Website: www.conf.purdue.edu

FOREWORD

On behalf of the Organizing Committee, the Ray W. Herrick Laboratories, Purdue University, the Co-Sponsoring Organizations and the Endorsing Organizations, it is a great pleasure to present the Final Program of the 23rd International Compressor Engineering Conference, the 16th International Refrigeration and Air Conditioning Conference, and the 4th International High Performance Buildings Conference at Purdue 2016.

The Organizing Committee received more than 650 abstracts and accepted 476 papers for publication. Every effort was made to include papers of current engineering and scientific interest. In addition, an internal review was conducted of each paper. Nevertheless, the Organizing Committee takes no responsibility for the correctness or completeness of any papers published.

I would like to thank the authors for having chosen these conferences to present their work and for all their efforts in preparing and submitting papers. Thanks to them we present a conference program that is exciting and informative. I would also like to thank the session chairs and session co-chairs for their help in conducting the technical program. In addition, I would like to thank the members of the advisory committee, the Co-Sponsoring and Endorsing Organizations and their representatives for their support and helpful suggestions. In particular, I would like to thank the members of the Organizing Committee and the many graduate student helpers as well as Ms. Kim Stockment, our conference secretary, for their tremendous efforts in making our conferences a success.

We hope that you enjoy these conferences and return home with new ideas and professional contacts. The next Purdue Conferences are planned for July 16-19, 2018. We hope that we will see you again at that time.

Yours sincerely,

Eckhard A. Groll
General Chair

ORGANIZING COMMITTEE

General Conference Chair
Eckhard Groll

International Refrigeration and Air Conditioning Conference
Chair: James Braun
Co-Chair: Andrew Hjortland

International Compressor Engineering Conference
Chair: Travis Horton
Co-Chair: Harshad Inamdar

International High Performance Buildings Conference
Chair: Thanos Tzempelikos
Co-Chair: Jason Konstantzos

Short Course Chairs
Orkan Kurtulus – Compressor Short Course
Bill Murphy – Refrigeration Short Course
James Braun – 2016 IBO Workshops
Gregor Henze – 2016 IBO Workshops

Director, Herrick Laboratories
Patricia Davies

Student Organization Committee
Stephen Caskey
Domenique Lumpkin
Vatsal Shah

Student Paper Competition
Neera Jain

Conference Coordinator
Kim Stockment

ADVISORY BOARD

ADVISORY BOARD

SPEAKERS

MONDAY, JULY11 | 9:30 AM | LOEB THEATRE

Viraj Vithoontien

Program Leader, Montreal Protocol Program, World Bank

"Climate Action driving our future forward, key perspectives on Climate Change and Refrigeration and Air Conditioning linkages from the World Bank Montreal Protocol Program"
Climate Change considerations are transforming both our current and future development trajectory, largely driven by concerted actions taken at the international level. The 2016 International Refrigeration and Air Conditioning Conference is happening during a seminal year of key international decisions that are helping to define future development pathways. This session will present an overview of climate change issues at the international level and discuss the many inter-connections between this and the Refrigeration and Air Conditioning sectors today. The World Bank Montreal Protocol Program operates directly at this nexus, working with developing countries and their industries to help integrate climate change concerns into their industry and national and local development programs through a combination of technical advisory and financial support.

TUESDAY, JULY 12 | 8:30 AM | LOEB THEATRE

Ian Beausoleil-Morrison

Professor, Faculty of Engineering & Design, Carleton University

"Maximizing the use of solar energy to radically reduce the energy needs of housing?"
Heating, cooling, and ventilating the places we live in, and providing the hot water, lighting, and appliance services we need, consumes tremendous amounts of energy; this contributes significantly to environmental and energy security issues. For example, housing accounts for 30% of all electricity and 26% of all natural gas consumed in Canada, and produces 13.5% of the country's greenhouse gas (GHG) emissions. Although these figures vary from region to region, a similar situation exists in most OECD countries.

If we are to meet ambitious GHG emission reduction targets and reduce dependency on fossil fuels, innovative concepts are required for providing the energy services required by housing; incremental efficiency improvements will not be sufficient.

This talk will demonstrate new techniques aimed at maximizing the use of solar energy to radically reduce housing energy demands. This will include a hybrid active-passive system to increase passive solar gains whilst protecting against overheating. Concepts for the longterm storage of solar thermal energy at the single-house scale will also be explored.

TUESDAY, JULY 12 | 12:00 PM | PURDUE MEMORIAL UNION BALLROOMS
Manpreet Singh

Purdue University Food Science Department

"Refrigeration: A Key Strategy to Global Food Security"

Improved storage and transportation of food is an important aspect that can help in reducing food waste and is a key focus of several food security initiatives worldwide. Not only do transportation mechanisms need to be efficient, but also sustainable to be viable options for use by the industry. Refrigeration and advancing the technology to maintain a cold-chain can play a vital role in tackling the issue of 1.3bn tons of food wasted annually. Extreme weather calls for energy-intensive cooling technologies compared to milder climates, and unfortunately the compounding factor of developing and under-developed countries with weather extremes pose a bigger challenge. Additionally, the demand for chilled and frozen foods is limited due to access to effective refrigeration in developing countries. Therefore, prevention of food loss is a challenge that the food industry is faced with in addition to impact on quality (nutritional and microbial spoilage) and public health risks (due to pathogens).

WEDNESDAY, JULY13 | 8:30 AM | LOEB THEATRE
Drusilla Hufford

Environmental Protection Agency

"EPA's SNAP Program: Listing Safer Alternatives and the Climate Action Plan"

In the 1990 Clean Air Act Amendments, Congress tasked EPA with phasing out ozone depleting substances, and in addition, with the task of identifying safer substitutes. In response, EPA created the SNAP program, which in its over twenty years of existence has identified and listed more than 400 alternatives for use in major consumer and industrial uses. The President's June 2013 announcement of the Climate Action Plan specifically tasked SNAP with an important role in prioritizing listing of lower-global warming potential options, and in reviewing – and possibly prohibiting - options already listed as acceptable in light of the introduction of newer, safer alternatives. We will focus in this talk on EPA's activities under SNAP to respond to the CAP. In addition, we will focus and on how those activities support the broader engagement of the U.S. domestically and internationally in the effort to address rapid growth in the use of hydrofluorocarbons (HFCs). Also a focus of the CAP was the effort to amend the international treaty, the Montreal Protocol to phase down HFCs, and this talk will also update listeners on the status of this effort.

THURSDAY, JULY14 | 8:30 AM | LOEB THEATRE
Gene Fields

Vice President of Global Compressor Technologies, Johnson Controls

"The Past, Present, and Future of Scroll Compressors"

A colorful tour through the history of this great technology, including the key advancements since the introduction of the mechanism more than a century ago; as well as the many companies and people behind these things. The progress in manufacturing as well as design techniques made this story fascinating; and the Purdue Conference played a major role in the global evolution. From the complex mathematics of involute equations, displacements, volume ratios, and gas forces; to the various applications, including AC, refrigeration, homes/buildings, and automotive; the progress continues to inspire. The basics of simulation models will also be discussed, and how they led the way for very successful scroll products. Finally, future advancements with inverter technology, increased capacities, system integration, and expanders are discussed.

SPEAKER GUIDE

BREAKFAST FOR PRESENTERS, CHAIRPERSONS AND VICE CHAIRPERSONS

A complimentary lunch (on Monday) and breakfast (Tuesday – Thursday) is scheduled in the Purdue Memorial Union, West Faculty Lounge for presenting authors, session chairs & vice-chairs scheduled for each day. You can find your presentation time in the Conference Overview. **It is important to attend the speaker lunch or breakfasts if you are presenting a paper to meet your session chair and address any presentation questions you might have. Updated presentations can also be loaded at this time.** Instructions concerning audio/visual/projection systems and technical session organization will be addressed. **Please only attend the lunch or breakfast on the day in which you present.** If you have presentations scheduled for different days, you should attend the lunch or breakfasts **each day** so that you can meet your Session Chairperson and the other Presenters. If you are the Presenter in more than one session on the same day, please meet both Session Chairpersons. If you are a Presenter but not an author of a technical paper, you should be the person who attends this breakfast so that you may alert the Chairpersons that you are the Presenter.

PROGRAM FORMAT

*NEW Presentations are on a strict time schedule this year. **Each presenter gets ONLY 20 minutes total, for both presentation and question/answers.** Each presenter is allowed 15 minutes to present and 3-5 minutes for questions. Should there be a speaker who does not show up or there is a gap for whatever reason, **we will NOT move the speaker up, as has been done in the past.** This means some sessions will indeed end early, while others will run long.

PICTURES AND VIDEOTAPING DURING THE CONFERENCES & PICTURE PHONES

Conference attendees may **NOT** take pictures, videotapes, nor use picture phones at any presentation without the consent of the author or presenter. An official photographer will be taking pictures during the conferences. Attendees may **NOT** take pictures during the tours because the research is often highly sensitive.

PRESENTATIONS

If you have last minute changes to your presentation (different from what was uploaded into Conftool), please see your session co-chair during the daily breakfast Tuesday to Thursday or during the luncheon (Monday) to make arrangements to upload your new file.

FUTURE CONFERENCE DATES

July 16-19, 2018

SHORT COURSES

Short Courses are organized and presented by the faculty of Herrick Laboratories, in cooperation with our sponsoring organizations. Often, speakers from outside organizations and university serve as presenters for these courses. One CEU credit can be earned through attendance to any of the offered short courses.

The courses will meet from 8:00 a.m. until 5:00 p.m. on Sunday, July 10, 2016, and will meet concurrently.

OIL MANAGEMENT COMPRESSORS AND THEIR SYSTEMS

The short course focuses on oil management in hermetic, positive displacement compressors, such as scroll, rotary and reciprocating compressors, and their systems. It consists of eight 50-minute lectures and will provide ample time for discussion. In particular, the following topics will be part of the short course:

- Impact of lubricants on compressor and system performance and reliability.
- Thermodynamic and transport properties of lubricant and HFC/HFO/CO2/NH3/HC refrigerant mixtures, including miscibility, solubility, and viscosity.
- Impact of oil on heat transfer and pressure drop, and oil retention in heat exchangers.
- Review of components used for oil management in vapor compression systems, such as oil separators, accumulators, and filters.
- Oil management in simple and complex vapor compression systems, including supermarket applications using multiple compressors on a rack, air-source heat pump systems, VRF systems, systems with vapor injection, industrial CO2 and NH3 refrigeration systems.
- Performance of liquid-flooded compression with regeneration systems.

FINAL FRONTIERS IN VAPOR COMPRESSION CYCLE EFFICIENCY

The efficiency of vapor compression cycles used for refrigeration and comfort cooling/heating has steadily increased since the US Department of Energy initiated minimum efficiency requirements in the early 1980s. Some current offerings have SEER values that are more than double the initial DOE minimum SEER of 10.0. These efficiency improvements have been the result of better compressor designs, larger and more efficient heat exchangers, better expansion devices, and more efficient motor technology. While efficiencies have risen substantially, they are still just a fraction of the theoretical Carnot efficiency.

The short course speakers will evaluate the various components of the vapor compression cycle to identify all areas that still offer economic opportunities for even higher system efficiencies. Case studies using optimization methods will show how to "invent" new heat exchanger shapes that go beyond tube-fin and microchannel designs, as well as overall system optimization while considering system level performance metrics. In addition, an overview will be provided for other thermodynamic cycles that may hold promise to compete with, or even surpass, the traditional vapor compression cycle that is used in the majority of refrigeration or comfort cooling/heating applications.

2016 INTELLIGENT BUILDING OPERATIONS WORKSHOP

This will be the third (previously held in 2011 and 2013) in a series of workshops that brings together researchers and developers of intelligent building features and systems. The main goals of this one-day workshop are to

1. Review fundamentals of optimal control and automated diagnostics in intelligent buildings,
2. Understand the state-of-the-art in commercialized intelligent building technologies,
3. Identify existing gaps that should be addressed in future research and development.

The format of the workshop will include a review of relevant fundamentals, followed by brief case studies that highlight recent developments and demonstrations of model-predictive controls and automated diagnostics. Ample time will be allotted for open discussion to address what is necessary to move from research to widespread adoption of intelligent building technologies.

A single registration fee will cover participation in both the IBO Workshop and the 2016 Purdue Conferences. Special IBO paper sessions are incorporated within the High Performance Building and Refrigeration and Air Conditioning Conferences where current research critical for achieving scalable and cost effective intelligent building operations will be presented. Participants in the IBO Workshop are strongly encouraged to participate in these conferences.

SOCIAL EVENTS

MONDAY, JULY 11

Reception at the Lafayette Brewing Company

The reception is hosted by **UTC Carrier** and will be held from 6:00 to 8:00 p.m. Shuttle buses will depart from the Grant Street Parking Garage between 5:30 and 6:00 p.m. transporting attendees and guests to the reception. Attendees may also use CityBus and the Trolleys or walk to the Reception. There will be transportation available after the Reception. You may need to arrange shuttle service with your individual hotels. Some local restaurants and bars are open on Monday evenings in downtown Lafayette and the Purdue campus.

Driving Directions to the Lafayette Brewing Company: From Grant Street Parking Garage: Head south on Andrew Place toward W State Street. Turn right onto W State Street. Turn Right onto N Grant Street. Sharp right onto Northwestern Avenue. Use the two left lanes to turn left onto Wiggins St. Keep left to stay on Wiggins Street. Continue half mile onto Old U.S. 231. Turn right onto N 6th Street. Turn left onto Main street. Lafayette Brewing Company is on the left. Parking is available on the street, or in the lot directly south of the bar that will cost $1 as you exit.

TUESDAY, JULY 12

Luncheon – Purdue Memorial Union, North & South Ballrooms

The buffet luncheon will be held from 12:00 to 1:30 p.m. It is hosted by **Danfoss, Inc.** There will be a Keynote Address by Purdue University's Manpreet Singh of the Food Science Department.

WEDNESDAY, JULY 13

Steak BBQ – The Trails Banquet Facility, West Lafayette

The informal steak barbeque will be held from 6 p.m. to dusk and is hosted by **Emerson Climate Technologies.** Buses will begin departing from the Grant Street Parking Garage entrance across from the Union Club Hotel at 5:45 p.m. The last bus will leave at approximately 6:15 p.m. for the barbeque. There will be games, a beer wagon and a chance to visit outside. The buses will return attendees and guests to the Purdue Memorial Union, University Inn, Holiday Inn and the Hilton Garden Inn after the barbeque. You may also drive your own vehicle to this venue.

Driving Directions to The Trails: From Grant Street Parking Garage, exit on North Street, turn left. Turn right on Grant Street, turn right on Northwestern Avenue (stop light) immediately curve left on Wiggins Street. Follow US 231 (on right) down the hill to State Road 43 North. Turn left on State Road 43 North. Proceed 5-1/2 miles to Burnetts Road, turn right. The Trails will be on your right.

STUDENT EVENTS

Student Paper Competition: Winners of the Student Paper Competition from each conference will be announced and awarded at the Steak BBQ at the Trails. Plan to join us and congratulate these students on a job well done.

ASHRAE Student Meeting – Monday, July 11 from 3:30-5:30 pm, Stewart 279
Purdue Student Chapter of ASHRAE meeting with ASHRAE President, Tim Wentz. (Open to ALL student ASHRAE members, not just Purdue student members).

Student Mixer – Tuesday, July 12 from 8-10 pm, The Stacked Pickle
Gather with other students from around the world at The Stacked Pickle located at 516 Northwestern Avenue #1700, West Lafayette, IN 47906.

CONFERENCE OFFICE / HOSPITALITY ROOM

Conference Office is located in Stewart Center, Room 306

Monday	8:00 a.m. to 4:00 p.m.
Tuesday	8:00 a.m. to 4:00 p.m.
Wednesday	8:00 a.m. to 4:00 p.m.
Thursday	8:00 a.m. to 4:00 p.m.

E-Paper Center

The E-Paper Center is located in Room 302, Stewart Center. Attendees may print up to 10 papers via the computer stations set up in room 302. You may also use the e-stations to print airline boarding passes, check emails, etc. Please be considerate of other guests, and limit time on the computer to 15 minute intervals.

Internet Access

Wireless internet is available to all attendees via the Conference Division/ATT WIFI hotspot. Please choose ATT Wifi as your internet connection. When you open a browser, you will be automatically connected. Your will need to provide your email address and accept terms & conditions. If you have difficulties, please visit the Conference Office, Stewart Center, 306.

Display Center

There will be several tables available in the Conference Hospitality Room, Stewart Center, Room 306. Please check with Conference staff before displaying any material. No commercialism is allowed. There is also a Bulletin Board where messages can be posted for other attendees and the Organizing Committee will post any last minute schedule changes.

PRINTING, COPYING AND FAXING

Attendees may print from a CD/disk/memory stick, make copies, send faxes, and more at The BoilerCopy Maker in the Purdue Memorial Union, Second Floor. Charges vary for different services.

SMOKING POLICIES ON CAMPUS

Smoking is prohibited on the West Lafayette Campus, except in designated smoking areas. All smoking material shall be extinguished and disposed of in an appropriate receptacle at the perimeter of the campus. Designated smoking areas are published in maps around the Purdue campus. The closest smoking area to Stewart Center is located near State Road 26.

TEMPERATURES INSIDE & OUTSIDE DURING THE CONFERENCES

Outside summer temperatures in Indiana are usually between 75-95°F and very humid. Room temperatures in technical session rooms are often cool for our European visitors; come prepared!

TIME DURING CONFERENCES

The time zone in Indiana is Eastern Daylight Time. This is the same time as New York City in the summer.

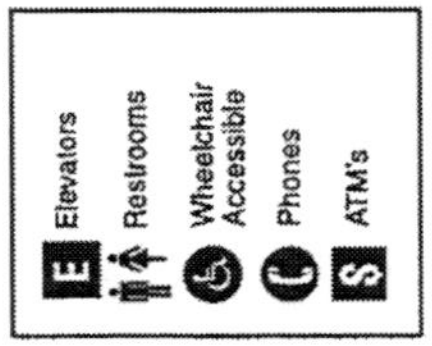

STEWART CENTER

FIRST FLOOR

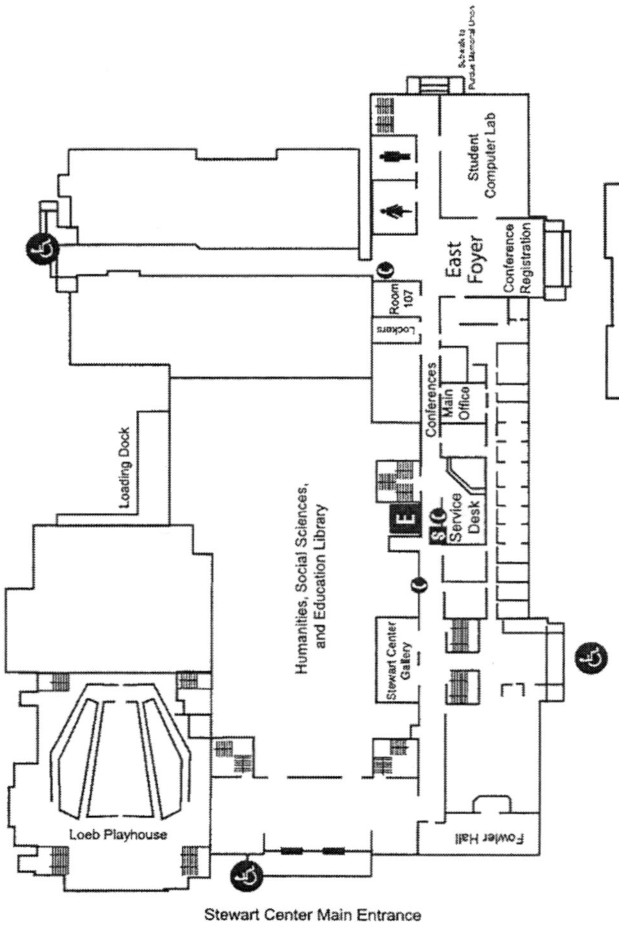

SECOND FLOOR

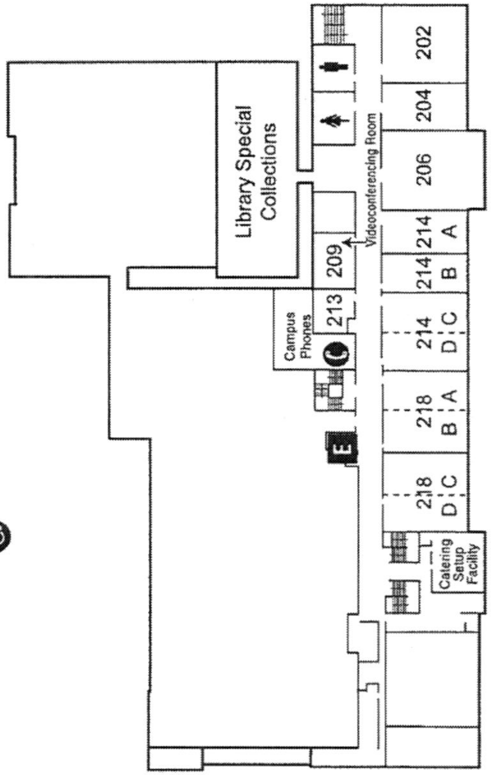

THIRD FLOOR

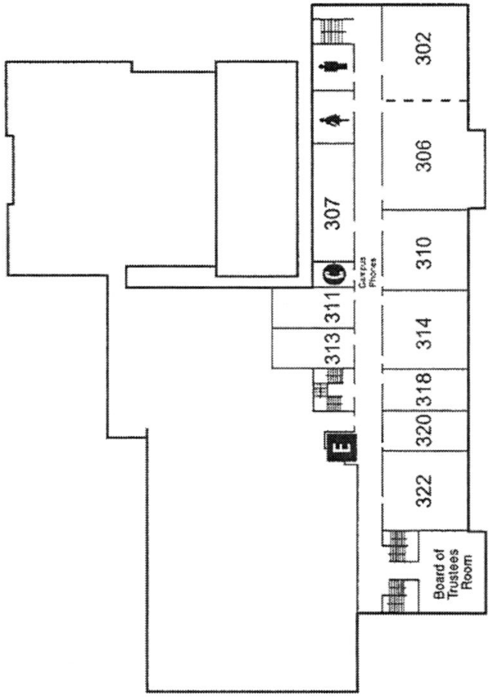

Carrier
turn to the experts
Danfoss
EMERSON
Climate Technologies
HIGHLY
KAWNEER
Parker
Rheem
The new degree of comfort.
Tecumseh

SPONSORS

A Multi-Agent Control Approach for Optimization of Central Cooling Plants

Rita JARAMILLO[1*], James BRAUN[2], Travis HORTON[3]

[1]*School of Mechanical Engineering, Purdue University, West Lafayette, IN
E-mail: rjaramil@purdue.edu

[2]Professor of Engineering, School of Mechanical Engineering
Purdue University, West Lafayette, IN
E-mail: jbraun@purdue.edu

[3] Assistant Professor of Civil Engineering with Courtesy Appointment in Mechanical Engineering
Purdue University, West Lafayette, IN
E-mail: wthorton@purdue.edu

* Corresponding Author

ABSTRACT

This paper presents an application of a multi-agent control approach for supervisory control of large central cooling plants. The starting point for this work was a multi-agent control simulation framework developed by Cai (2015). To adapt the framework to this problem, agents representing the performance of the different devices of the plant were developed and an optimization method capable of handling non-convex functions and discontinuous design spaces was developed and incorporated in the framework. A case study of an existing cooling plant was utilized to evaluate the approach in terms of optimality and computational resources. Simulations were carried out for different performance conditions to predict the performance of the plant under three different control strategies: 1) multi-agent control, 2) centralized optimization based on mathematical programming techniques and 3) a heuristic control strategy. The results showed that significant savings can be achieved through the implementation of multi-agent control. It is expected that, if each hardware component of the plant comes with an integrated agent that represents its behavior, then the proposed multi-agent framework could automatically generate the multi-agent structure and control algorithm after some relatively simple pre-configuration steps. This will reduce the site-specific engineering and will provide a more economic and easy to configure solution for central cooling systems.

INTRODUCTION

A large central cooling system consists of several chillers, cooling towers and pumps that supply chilled water to satisfy the cooling requirements of one or more buildings. Optimal supervisory control of such systems involves the determination of the mode of operation and set points that minimize operating costs while satisfying cooling and comfort requirements. The problem is complicated because of the presence of both continuous and discrete control variables. Most of the research related to optimal supervisory control of central cooling systems that has been conducted in the last three decades has focused on centralized control approaches, such as Braun et al. (1989a, b), Ahn and Mitchell (2001); Yao et al. (2004); Torzhkov et al. (2010), and Zhang and Turner (2012). Although these studies have demonstrated the effectiveness of optimal or near-optimal control in reducing operational costs, the results have not been widely implemented. Some limitations of centralized optimal control are the need for detailed information on the performance profiles of the cooling plant equipment in order to build a model for the optimization process, and, once implemented, the plant model and control sequences will need to be updated by experts every time a modification (such as the introduction of new equipment) is made to the plant.

A promising approach that addresses limitations of earlier approaches is the implementation of distributed multi-agent-based optimal control. The use of intelligent agents makes it possible to solve the optimization problem in a distributed manner by breaking a big complex problem into smaller, more manageable pieces that can be solved independently and in parallel by individual agents. The individual solutions can then be handled by a coordination agent that achieves some consensus. Since intelligent agents can solve individual problems to optimize performance without having total knowledge of the system, they would also add adaptive capability to the control system, i.e., the system could be more easily reconfigured to adapt to changes such as the introduction of new equipment. However, some drawbacks of this approach are the additional data transfer equipment required and the optimality traded off for reduced computations. Many approaches for distributed control and multi-agent systems have been proposed and demonstrated in a number of fields. However, a review of multi-agent control shows relatively few applications in the HVAC field (Treado, 2010; Sun et al, 2010; Kelly and Bushby, 2012). In these studies, results that document the performance of the control algorithms are very scarce and the proposed multi-agent control strategies were validated using simulations on small-scale HVAC systems with very simple models on the cooling plant side.

This paper presents an application of a multi-agent control approach for supervisory control of large central cooling plants. The work starts from a multi-agent control simulation framework developed by Cai (2015) for optimization-based supervisory control of distributed air-conditioning systems. In this setting, assuming that each hardware component of a system has an integrated agent that represents its behavior, then the framework can automatically generate the multi-agent structure and control algorithm after some relatively simple pre-configuration steps, reducing site-specific engineering. Although the proposed framework provides good flexibility in design of control topology it has some limitations and it might not be directly applied to some kinds of equipment: the distributed consensus-based algorithms utilized are conceived for convex functions and continuous design spaces. Therefore, they are not good at handling discrete variables such as multiple operating modes that are often present in HVAC systems. To adapt the framework to the current problem, agents representing the performance of the different devices of central cooling plants were developed and inserted in the framework and an optimization method capable of handling non-convex functions and discontinuous design spaces (Genetic Algorithm) was developed and incorporated in the framework. A case study of a cooling plant without significant storage was addressed to conduct a simulated demonstration of the approach for different operating conditions. The results in terms of optimality and computational resources were compared with two benchmarks: centralized optimization with conventional mathematical programming techniques and a heuristic control strategy.

1. MULTI-AGENT APPROACH

The multi-agent control approach utilized here consists of two main elements: a multi-agent simulation framework, which includes component agents representing the behavior of each physical component of the plant, and two optimization algorithms: a consensus-based distributed algorithm originally included in the framework called the alternating direction method of multipliers (ADMM) and a genetic algorithm (GA). The GA was developed and incorporated in the framework to provide an alternative for finding the global optimal operating point of a system in the presence of non-convex functions and discontinuous design spaces.

1.1. Component Agents

Previous work developed by Jaramillo et al. (2014) presented the development of a mathematical model of a cooling plant using MATLAB software. In this setting, each hardware component of the plant (i.e. chillers, cooling towers and pumps) was represented using semi-empirical models as a separate set of mathematical relationships with its own parameters, inputs and output variables. The parameters of the models were determined from regression of performance data. From there, agents representing each component behavior can be easily created. Each agent consists of a cost function (power consumption in this case) and a set of constraints. Once created, the agents can be incorporated into the multi-agent framework and interconnected according to the arrangement of the physical plant.

1.2. Multi-agent Framework

The multi-agent control framework developed by Cai et al. (2015) was built in MATLAB as a simulation prototype that can be replicated in other programming environments to support hardware implementation. The framework defines a general component agent structure as well as the flow connections between agents. A general component agent is written as a super class from which each component class can inherit the basic agent structure. The properties of the agent class consist of a collection of cost functions, and linear and non-linear equality and inequality constraints that characterize the behavior of a specific hardware component. The cost functions could be actual power consumption that needs to be minimized or some other performance metrics. Another important property of each component agent is the agent's group number. This parameter is used in the setup of the distributed optimization-based controllers: all the component agents with the same group number will be assigned to one local optimizer controller. The grouping might depend on physical distances among the devices, function, network structure or other criteria. A centralized controller will be synthesized if all the component agents are assigned the same group number. In this setting, the procedure to create a multi-agent control system is straight forward: assuming that all the component agents are available, one can simply drag and drop them in a project canvas and then specify the inter-agent connections, which are stream variables. Figure 1 illustrates the procedure to create a multi-agent control system for a central cooling plant.

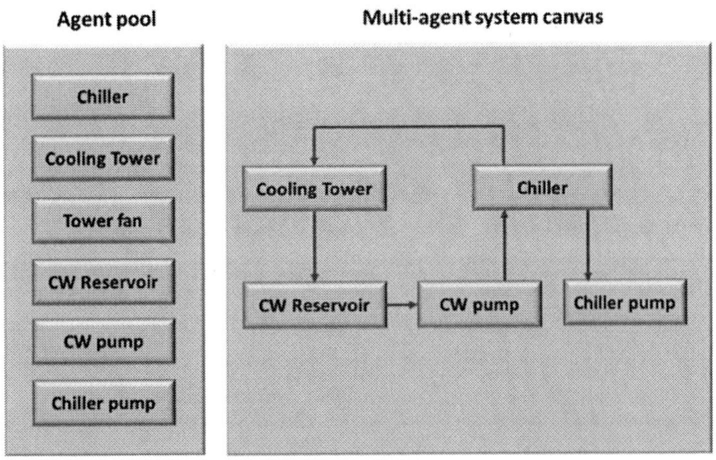

Figure 1. Procedure to Set-up a Multi-agent System. Adapted from Cai et al. (2015)

Once the component agents and their connections are defined, the framework will automatically compile the code and compose the optimization problem according to the specified configuration. The compilation process consists of several steps such as extraction of the cost functions of the different agents to construct the total cost function and the elimination of redundant equality constraints. These steps will be carried out for each group of components. Then, for each group i the composed optimization problem takes the form of Eq. 1.

$$\begin{aligned} & min_{X_i} f_i(X_i) \qquad i = 1, \dots, N \\ & subject\ to \quad A_i X_i \le B_i \\ & \qquad A_{eq,i} X_i = B_{eq,i} \\ & \qquad g_i(X_i) \le 0 \\ & \qquad h_i(X_i) = 0 \end{aligned} \tag{1}$$

In the above expression X_i is a vector of the local variables of sub-problem i and N is the number of subproblems. To complete the process of composing the distributed problem, consensus constraints need to be specified. These constraints enforce the local copies of the same variable to match between different agents. Taking the system in Figure 1 as an example, the water flow leaving the condenser pump is the same entering the chiller's condenser. If these two devices are assigned different group numbers, there will be two local variables corresponding to the same water flow properties. The minimization of the variable-speed pump power will favor lower condenser water flow while the minimization of the chiller power will be benefited by higher condenser water flow. So the minimization of power consumption of each group will drive two local variables representing the same physical quantity in opposite directions. Therefore, consensus constraints of the form expressed in Eq. 2 are necessary to enforce these local variables to converge to the same value.

$$X_i = F_i Z \qquad i = 1, \dots, N \tag{2}$$

where Z is a vector that contains the global variables of the problem and F_i is a matrix that picks out the variables of Z that correspond to X_i.

1.3. Multi-agent Optimization Algorithms

Distributed Optimization: Alternating Direction Method of Multipliers

The ADMM is an augmented Lagrangian method for solving distributed consensus problems that was introduced in the 1970s. A description of the method can be found in Summers and Lygeros (2012). The method was adapted to solve problems of the form expressed in Eq. 1 and Eq. 2. The augmented Lagrangian for these kind of problems is given in Eq. 3

$$L = \sum_{i=1}^{N} L_i(X_i, Z, Y_i) = \sum_{i=1}^{N}\left(f_i(x_i) + Y_i^T(X_i - F_i Z)\right) + \frac{\sigma}{2}\|X_i - F_i Z\|_2^2 \quad (3)$$

Where Y_i are vectors of the Lagrange multipliers and σ is a penalty parameter. The ADMM algorithm consists of the iterations

$$\begin{aligned} X_i^{k+1} &= argmin_{X_i} L_i(X_i^k, Z^k, Y_i^k) \quad s.t. \quad X_i \in \boldsymbol{C}_i \\ Z^{k+1} &= argmin_Z L_i(X_i^{k+1}, Z^k, Y_i^k) \\ Y_i^{k+1} &= Y_i^k + \sigma(X_i^{k+1} - F_i Z^{k+1}) \end{aligned} \quad (4)$$

The method alternatively minimizes X and Z, which allows the X_i minimizations to be done in parallel. In this particular form of the problem, the minimization of each variable of the vector Z reduces to an averaging of the equivalent local variables and Lagrange multipliers as expressed in Eq. 5.

$$(Z^{k+1})_j = \frac{1}{N_j}\sum_{i=1}^{N}\left((X_i^{k+1})_j + \frac{(Y_i^k)_j}{\sigma}\right) \quad (5)$$

With this distributed formulation, the original optimization problem is fragmented into several sub-problems with reduced dimensions and less constraints, which can be solved in parallel. A hardware implementation of this distributed decision making process is shown in Figure 2. The bottom layer corresponds to the sensing network that collects the required operating conditions. Above the sensing layer is a component agent layer that includes the agents representing the behavior of all devices. On top of the component agent layer, there is an optimizer agent layer, which is responsible for solving each sub-problem. Each optimizer agent calls the related component agents iteratively to optimize its corresponding cost function independently and in parallel with the other optimization agents. The consensus requirements among the local variables are enforced by a coordination layer that collects the local copies of all the variables, updates the dual variables accordingly and feeds them back to the optimizer agents to let them re-optimize with respect to the updated information. The iteration process continues until the termination criteria are met.

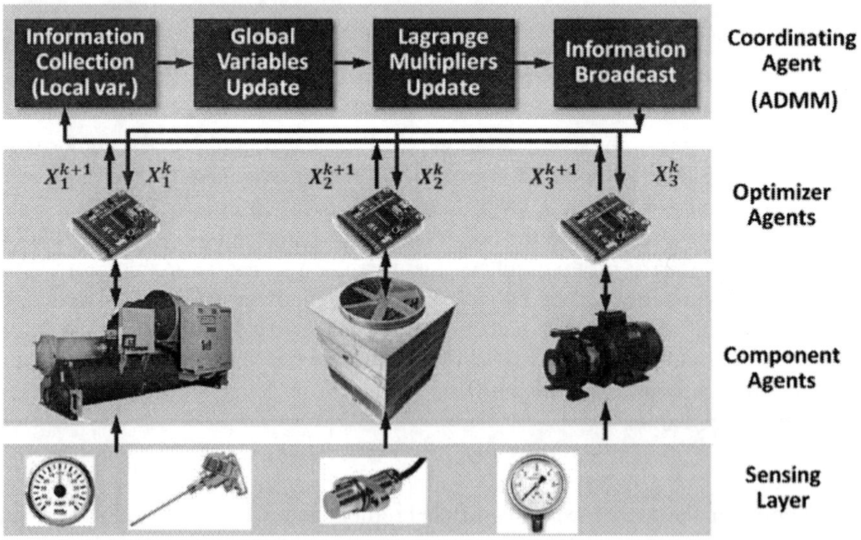

Figure 2. Hardware Implementation of the Multi-agent Controller with ADMM

Centralized, parallel optimization with Genetic Algorithm

Genetic algorithms (GA) are stochastic methods based on the principles of genetics and natural selection. A generic GA starts with a population of N_p candidate solutions randomly generated from the search space. An objective function is used to quantify the fitness of each individual and genetic operators of selection, crossover and mutation are applied to produce a new generation of individuals. The process of populating new generations continues until certain convergence criteria are satisfied. The algorithms for the genetic operators depend on the encoding method and the application. The GA conceived for optimization in the context of the multi-agent framework is real coded and has incorporated certain characteristics to make it suitable for solving large dimension problems in a more efficient way. Some of those features, based on the work presented by Zhu et al. (2014) are the generation of the initial population as a uniform array of points that covers the search space delimited by bounds and linear constraints and an elitist strategy that combines all the individuals from the previous generation and the new population, and selects the fittest individuals to constitute the new generation. One disadvantage of GAs is that their performance is not always satisfactory in the presence of equality constraints. To sort this difficulty, the proposed GA was combined with Broyden's method (a quasi-Newton method) to handle the non-linear equality constraints that arise from the behavior of the components.

2. CASE STUDY

The case-study considered in this work is a simplified version of the Northwest Chiller Plant at the main campus of Purdue University in West Lafayette, IN. The plant delivers chilled water through 37 Km of underground piping to partially meet the cooling requirements of more than 150 buildings on the campus (an average of 92.4 MMTon-hr per year). A schematic of the simplified plant, illustrating all the variables of interest is shown in Figure 3. The simplified case-study consists of three centrifugal chillers, each one with 2000 Ton nominal cooling capacity to produce chilled water for the campus, an evaporative counter-flow cooling tower with three cells, and three variable speed condenser water pump each one with 6000 gpm nominal flow rate. In the chilled water loop, the water returning from campus is cooled as it is circulated through the chiller and then returned to campus by high-pressure pumps. The chilled water pumps are considered part of the campus chilled water distribution systems and do not add to the costs of operation of the plant, therefore they will not be included in the model. In the condenser water loop, the chillers reject heat to the water that is circulated through the cooling tower cells and stored in the cold well, a common reservoir with capacity of 90,000 gal (341m^3). From there, the water is pumped again to the chillers by the condenser water pumps.

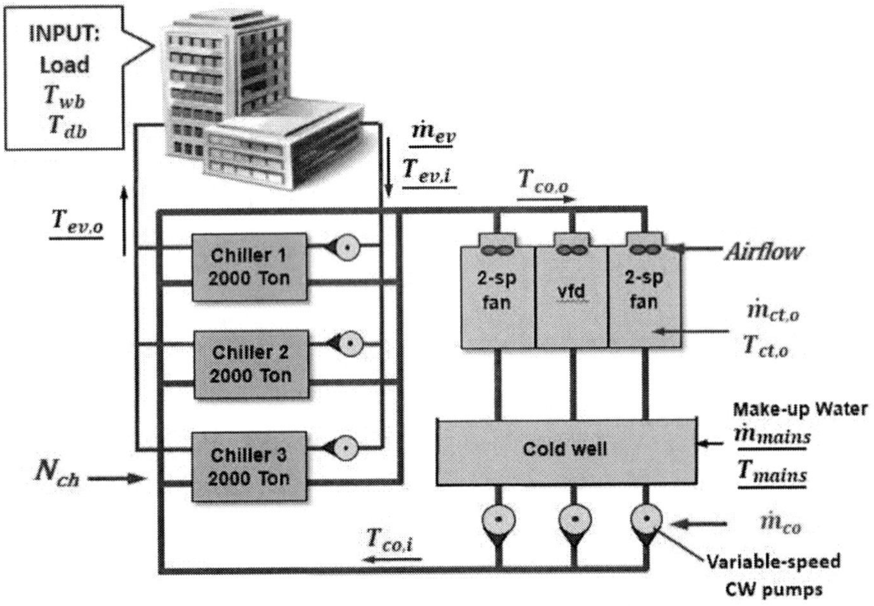

Figure 3. Schematic of the cooling plant

The baseline control for the plant involves the following heuristic strategies: (1) The chillers are sequenced based on the operators experience to meet the cooling load. (2) The condenser water pumps are sequenced with the chillers so that each chiller's condenser operates approximately with the nominal design flow. (3) There is feedback control of the tower fans to maintain a constant condenser water supply temperature set-point of 73°F, even though the minimum inlet water temperature for each chiller condenser is 55°F.

Optimal supervisory control of such a plant involves determining the values of the control variables that minimize the total power consumption at any time in response to uncontrolled variables (ambient dry-bulb and wet-bulb temperatures, and cooling load). Three independent control variables are considered in this problem: the total cooling tower air flow (*Airflow*), the total condenser water flow ($\dot{m}_{co}$) and the number of active chillers (N_{ch}) at any time. Other dependent optimization variables (in blue) appear as a result of the distributed formulation of the problem. The boundary conditions are the underlined variables in Figure 4. The chilled water supply set-point ($T_{ev,o}$) is fixed at 43°F, and the chilled water supply flow rate ($\dot{m}_{ev}$) which is fixed according to the nominal chiller's evaporator flow rate (3200 gpm per active chiller).

2.1. Optimization problem composition

Distributed formulation

If the component agents are grouped according to its kind, four groups are formed and a distributed formulation is synthetized by the framework as shown in Eq. 6 to Eq. 10. The functions in the expressions correspond to the cost functions (power consumption) and constraints related to the behavior of each device of the plant.

Group 1: Chillers

$$min_{\left[\dot{m}_{co}^{(1)},T_{co,i}^{(1)},T_{co,o}^{(1)}\right]}\{Chiller_{pow}(\dot{Q}_{ev},T_{ev,o},T_{co,o})\} \tag{6}$$

$$\text{Subject to: } T_{co,o}^{(1)} = Chiller_{Tco,o}\left(\underline{\dot{Q}_{ev}},\underline{T_{ev,o}},\dot{m}_{co}^{(1)},T_{co,i}^{(1)}\right)$$

Group 2: Cooling tower cells

$$min_{\left[Airflow^{(2)},\dot{m}_{co}^{(2)},T_{co,o}^{(2)},\dot{m}_{ct,o}^{(2)},T_{ct,o}^{(2)}\right]}\{Power_{ct}\} \tag{7}$$

$$\text{Subject to: } T_{ct,o}^{(2)} = CoolingTower_{T_{ct,o}}\left(\underline{T_{wb}},\underline{T_{db}},Airflow^{(2)},\dot{m}_{co}^{(2)},T_{co,o}^{(2)}\right)$$

$$\dot{m}_{ct,o}^{(2)} = CoolingTower_{\dot{m}_{ct,o}}\left(\underline{T_{wb}},\underline{T_{db}},Airflow^{(2)},\dot{m}_{co}^{(2)},T_{co,o}^{(2)}\right)$$

Group 3: Condenser water pumps

$$min_{\left[\dot{m}_{co}^{(3)}\right]}\{Power_{pump}\} \tag{8}$$

Group 4: Cold well

$$min_{\left[\dot{m}_{co}^{(4)},T_{co,i}^{(4)},\dot{m}_{ct,o}^{(4)},T_{ct,o}^{(4)}\right]}\{0\} \tag{9}$$

$$\text{Subject to: } T_{co,i}^{(4)} = Coldwell\left(\dot{m}_{co}^{(4)},\dot{m}_{ct,o}^{(4)},T_{ct,o}^{(4)},\underline{T_{mains}}\right)$$

The consensus constraints from this formulation are:

$$\begin{gathered}
Airflow = Airflow^{(2)}\\
\dot{m}_{co} = \dot{m}_{co}^{(1)} = \dot{m}_{co}^{(2)} = \dot{m}_{co}^{(3)} = \dot{m}_{co}^{(4)}\\
T_{co,i} = T_{co,i}^{(1)} = T_{co,i}^{(4)}\\
T_{co,o} = T_{co,o}^{(1)} = T_{co,o}^{(2)}\\
\dot{m}_{ct,o} = \dot{m}_{ct,o}^{(2)} = \dot{m}_{ct,o}^{(4)}\\
T_{ct,o} = T_{ct,o}^{(2)} = T_{ct,o}^{(4)}
\end{gathered} \tag{10}$$

In this distributed formulation, the original centralized problem is fragmented into four sub-problems with reduced dimensions and less constraints, which can be solved in parallel.

Genetic Algorithm Formulation

The GA is intended to be compatible with the multi-agent framework topology; therefore, the distributed formulation expressed in Eq. 6 to Eq. 9 can be also utilized to solve the problem with the GA. In this context, the cost function is the sum of the cost functions of each optimizer agent; therefore, the optimization problem is solved in a centralized manner by a coordinator agent, while the computation of the objective function and non-linear constraints is distributed and executed in parallel by the different optimizer agents.

3. OPTIMIZATION RESULTS

This section presents the results of the multi-agent approach applied to the case study described above. Four different operating conditions of the plant were considered as shown in Table 1. The optimization results are compared with two benchmarks: the heuristic control strategy previously described and centralized optimization with mathematical programming techniques. It is important to note that the problem considered here involves a discontinuous control variable (number of active chillers) and the cost function (power) is non-convex. In such cases, the convergence of the solution to the optimum point cannot be guaranteed. Therefore, performance maps of the plant were elaborated for each of the operating conditions considered to obtain the "true optimum" operating point. This optimum point was used as a baseline to assess the convergence of the different optimization methods. Table 1 presents the total power consumption obtained with the heuristic control strategy, and the power consumption and savings corresponding to the "true optimum" operating point. The savings were evaluated with respect to the heuristic strategy.

Table 1. Optimum operating point of the Cooling Plant and power savings for four Operating Conditions.

Operating conditions			Heuristics	"True optimum" operating point			
Test Nr	Load Ton	Wet bulb °F	Power kW	Active Chillers	Power kW	Savings kW	Savings %
1	2000	50	1145.46	2	1028.88	116.58	11.33
2	2000	80	1351.06	1	1302.61	48.45	3.72
3	4000	50	2324.22	3	1989.87	334.35	16.80
4	4000	80	2702.12	2	2612.62	89.50	3.43

The optimization problem was solved using three methods: 1) Centralized optimization using mathematical programming techniques (MATLAB function FMINCON); 2) Multi-Agent distributed optimization with ADMM, and 3) Multi-Agent optimization with a GA. Ten runs were made for each operating condition and method to account for the variability of the results which is caused by either the stochastic nature of the GA or the randomly generated initial guess used for the other methods. The average results for each operating condition are shown in Table 2. The results include average power savings (kW), average computational time (s), and the RMS error (kW) in the savings calculations relative to "true" savings from Table 1. The RMSE and computational time results can be more easily visualized in the bubble chart presented in Figure 4.

Table 2. Optimization results comparison

Operating conditions			FMINCON			ADMM			GA		
Test Nr.	Total Load Ton	Wet bulb °F	RMSE Power kW	Average Savings kW	Average time s	RMSE Power kW	Average Savings kW	Average time s	RMSE Power kW	Average Savings kW	Average time s
1	2000	50	15.04	101.55	6.43	0.02	101.5	28.25	0.87	115.7	6.46
2	2000	80	19.83	28.62	6.90	16.08	28.6	86.29	0.02	48.4	6.41
3	4000	50	7.61	326.74	4.44	7.88	326.7	43.55	0.28	334.1	8.30
4	4000	80	17.61	71.89	7.34	25.75	71.9	79.70	0.05	89.4	13.61

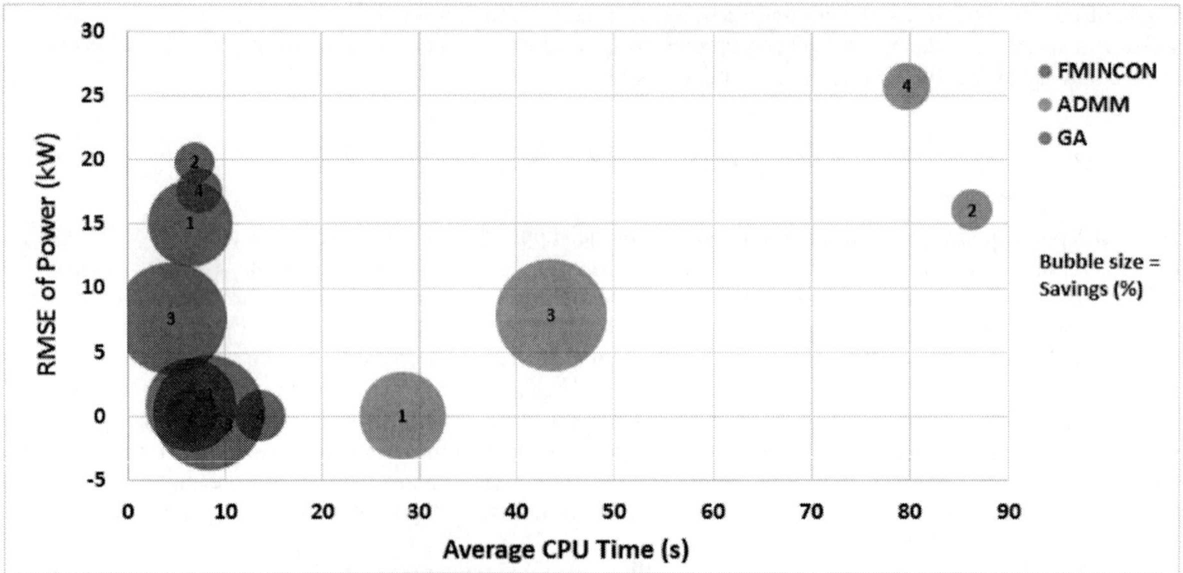

Figure 4. Optimization result comparison

It is important to note that, since the centralized optimization based on mathematical programming techniques (FMINCON) and the ADMM cannot handle discontinuous variables, the optimum number of active chillers for each operating condition had to be specified as an input for these methods. The GA, in contrast, can handle a mixture of integer and continuous variables and the number of active chillers was considered as another optimization variable.

The results show that the GA was able to find near-optimal solutions under all the operating conditions considered. This can be noted in Figure 4, where the RMSE of power (root mean square error in the power consumption compared to the true optimal) was much lower for the GA than the one obtained with the other methods. Further, the processing time for the GA was only 15% of the ADMM (average 8.7s compared to 59.5s). Given the non-convex shape of the cost function, the effectiveness to reach the optimum point and convergence speed of both the conventional centralized optimization and the ADMM method were highly dependent on the operating conditions considered and the goodness of the initial guess.

Comparison with the heuristic strategy, shows that significant power savings can be attained with all the optimization methods. The differences in the predicted average savings are not as significant as might have been expected, except for the second operating condition where the savings obtained with the ADMM and the central optimization with FMINCON are 60% of the savings obtained with the GA.

CONCLUSIONS

In this paper, a multi-agent control approach was applied to a central cooling system without significant storage, such that quasi-steady behavior could be assumed. The resulting non-convex optimization problem was solved for different operating conditions of the plant using a multi-agent simulation framework with two different algorithms: distributed optimization with the alternating direction of multipliers (ADMM) and centralized optimization with a genetic algorithm (GA). Comparison of the results with the "true optimum" point obtained from performance maps of the plant and centralized optimization using mathematical programming techniques showed that the GA was able to find near-optimal solutions for all the cases considered, while the effectiveness of other methods was highly dependent on the operating conditions and the goodness of the initial guess provided for the optimization. Other advantages of the GA are that it can handle a mixture of continuous and discontinuous variables and its convergence speed was much faster than the ADMM.

Even though the GA outperformed the other optimization methods considered in this case-study, the results might be different for a more complex scenario involving a higher number of optimization variables. Further, since the convergence of the ADMM is highly dependent on the goodness of the initial guess it can be expected that in a real setting, the initial guess provided by the optimum point of the previous time step will be closer to the actual optimum point, and the convergence of the ADMM method will be considerably improved. Future work will include an extensive simulation evaluation of the approach using a more complex model of the cooling system and one year of performance data.

NOMENCLATURE

Acronyms

ADMM	Alternating direction method of multipliers
GA	Genetic algorithm
HVAC	Heat, ventilation and air conditioning
RMSE	Root mean square error

Symbols

A	Matrix of coefficients of linear inequality constraints
A_{eq}	Matrix of coefficients of linear equality constraints
Airflow	Cooling tower total air flow
B	Vector of constants terms for linear inequality constraints
B_{eq}	Vector of constants terms for linear equality constraints
C	Feasible region of optimization variables
f	Cost function to be optimized
F_i	Matrix that assigns the local variable array Xi its corresponding global variables
g	Non-linear inequality constraint
h	Non-linear equality constraint
L	Augmented Lagrangian
$\dot{m}$	Mass flow rate
Power	Power consumption
$\dot{Q}$	Heat transfer rate
T	Temperature
T_{mains}	Cooling tower make-up water temperature
X	Local optimization variables
Y	Lagrange multipliers
Z	Global optimization variables

Subscripts/Superscripts

co	Chiller's condenser
ct	Cooling tower cell
cw	Cold well (condenser water reservoir)
db	Dry-bulb
ev	Chiller's evaporator
i	Inlet conditions
i	Subproblem number
k	Iteration number
N	Number of subproblems for multi-agent optimization
o	Outlet conditions
w	Water
wb	Wet-bulb

REFERENCES

Ahn, B. C., & Mitchell, J. W. (2001). Optimal control development for chilled water plants using a quadratic representation. Energy and Buildings, 33, 371-378.

Braun, J. E., Klein, S. A., Mitchell, J. W., & Beckman W. A. (1989a). Methodologies for optimal control of chilled water systems without storage. ASHRAE Transactions, 95(1), 652-662.

Braun, J. E., Klein, S. A., Mitchell, J. W., & Beckman, W. A. (1989b). Applications of Optimal Control to Chilled Water Systems without Storage. ASHRAE Transactions, 95(1), 663-674.

Cai, J., Kim, D., Jaramillo, R. C., Braun, J. E., & Hu, J. (2015). A general multi-agent control approach for building energy system optimization. Manuscript submitted for publication.

Jaramillo, R. C., Braun, J. E., & Horton, W. T. (2014). Application of Near-Optimal Tower Control and Free Cooling on the Condenser Water Side for Optimization of Central Cooling Systems. Third International High Performance Buildings Conference. West Lafayette, IN.

Kelly, G. E., & Bushby, S. T. (2012). Are intelligent agents the key to optimizing building HVAC system performance? HVAC& R Research, 18(4), 750-759.

MATLAB (Version R2013a) [Computer software]. Natick, MA: The MathWorks, Inc.

Summers, T. H., & Lygeros, J. (2012). Distributed model predictive consensus via the alternating direction method of multipliers. In 2012 50th Annual Allerton Conference on Communication, Control, and Computing (Allerton),, 79-84.

Sun, B., Luh, P. B., Jia, Q. S., Jiang, Z., Wang, F., & Song, C. (2010). An integrated control of shading blinds, natural ventilation, and HVAC systems for energy saving and human comfort. In Automation Science and Engineering (CASE), 2010 IEEE Conference on, 7-14.

Torzhkov, A., Sharma, P., Li, C., Toso, R., & Chakraborty, A. (2010). Chiller Plant Optimization: An Integrated Approach for Chiller Sequencing and Control. 49th IEEE Conference on Decision and Control. 15-1.

Treado, S. J. (2010). An Agent-Based Methodology for Optimizing Building HVAC System Performance, ASHRAE Transactions, 116(2), 124-134.

Yao, Y., Lian, Z., Hou, Z., & Zhou, X. (2004). Optimal operation of a large cooling system based on an empirical model. Applied Thermal Engineering, 24(16) pp. 2303–2321.

Zhang, Z., & Turner, W. (2012). Method for Estimating Energy Savings Potential of Chilled-Water Plant Retro-Commissioning. ASHRAE Transactions, 118(2).

Zhu, J., Zhou, D., Li, F., & Fu, T. (2014). Improved Real Coded Genetic Algorithm and Its Simulation. Journal of Software, 9(2), 389-397.

ACKNOWLEDGMENTS

The authors would like to acknowledge the collaboration provided by Purdue Physical Facilities, West Lafayette, IN.

Evaluation of Optimal Chiller Plant Control Algorithms in Model-Based Design Platform with Hardware-in-the-Loop

Pengfei LI[1]*, Shui YUAN[1], Keunmo KANG[1], Hayden REEVE[1]

[1] United Technologies Research Center,
East Hartford, CT, USA
(Phone: 860-610-7215, Fax: 860-557-8709, E-mail: lip1@utrc.utc.com)

* Corresponding Author

ABSTRACT

This paper presents a framework and results of using model-based design (MBD) methodology to evaluate the benefit and trade-offs of different chiller plant control algorithms for medium-sized commercial buildings including an optimization-based algorithm that can be deployed rapidly with little installation and commission effort. A high-fidelity dynamic simulation model for selected building types and climate zones were developed with Modelica and implemented in the hardware-in-the-loop (HiL) platform for controller hardware verification purpose. Baseline and optimization-based control algorithms were deployed in a building automation system (BAS)'s controller hardware with their performance monitored through web-based interface in real-time. Through extensive model-in-the-loop (MiL) analysis with 128 case studies that cover significant amount of chiller plant load scenarios, an average energy saving of 15% was achieved for the medium office building type and 10% for the large hotel building type in selected climate zones. A simple payback analysis was conducted and the commercialization requirement of less than 3 year payback period was met.
Keywords: *Model-Based Design, Low-Cost Optimal Control, Whole-Building Dynamic Modeling, Modelica*

1. INTRODUCTION

Chiller systems account for 31% of the total cooling electricity consumption of medium-sized commercial buildings within 25k-200k square feet. In the last decade, advanced controls such as model predictive control (MPC) has demonstrated energy savings that typically range from 5% to 15% (Bengea *et al.*, 2015; Li *et al.*, 2015). However, the installation and commissioning efforts to deploy MPC into existing building automation system (BAS) are often cost prohibitive and therefore undermine the energy saving benefit it brings into the game.

This paper presents the key accomplishments in developing the model-based design (MBD) framework that enables the energy evaluation of baseline and advanced chiller plant controls in model-in-the-loop (MiL) and hardware-in-the-loop (HiL) platforms. A key feature of this paper is the various chiller plant load scenarios considered in a high-fidelity whole-building dynamic modeling environment. Another key feature is the comparison of baseline chiller plant control (fixed-setpoints) to state-of-the-art advanced control methods such as ASHRAE 90.1 and heuristic-based algorithms as well as a low-cost optimal control developed internally at UTRC.

A high-fidelity integrated building HVAC and chiller plant dynamic model with equipment-level closed-loop controls was developed with Modelica by leveraging the work from DOE's previously funded projects for EnergyPlus (Crawley *et al.*, 2001) and Modelica Buildings Library (Wetter *et al.*, 2014) and UTRC's in-house model Library. The integrated model was demonstrated to be significantly faster than real-time with Dymola's variable-step solver and was shown to be numerically robust for a wide range of operating conditions including chiller plant start-up and shut-down as well as reversed water flow scenarios during transient operation.

The integrated building HVAC and chiller plant model was later successfully deployed in HiL platform coupling with real-world chiller plant controllers to assess baseline and low-cost optimal control operation and performance. The installation time and effort for baseline and low-cost optimal control deployment were found to be 8 hours and 14 hours, respectively. The baseline and low-cost optimal control operation has been monitored for the entire weekly profile and the integrated building HVAC and chiller plant system has progressed well for multiple chiller plant start up and shut down operation without any numerical issues and meanwhile generating reasonable results.

Finally, this research investigates an extensive set of 128 case studies (that exceed the project target of 54) that provide detailed understanding on how different climate zones, plant configurations, and building types may affect energy savings. Each case study is a weekly simulation using a whole-building dynamic HVAC system model coupled with many closed-loop PI controllers and supervisory controllers at the chiller plant level. Through extensive analysis, an average energy saving of 15% was achieved for the medium office building type and 10% for the large hotel building type in selected climate zones. A simple payback analysis was conducted and the commercial requirement of less than 3 year payback period was met.

2. WHOLE-BUILDING DYNAMIC MODELING PLATFORM AND INTEGRATION

This section provides details for the building, AHU, and chiller plant modeling, respectively. The building model was developed based on a translation of selected DOE EnergyPlus reference building models using a UTRC in-house model Library. The AHU model was developed based on LBNL's Modelica Buildings Library 2.1.0 and the chiller plant models were developed based on the UTRC in-house Library. The chiller plant controls Library was developed based on Modelica Standard Library 3.2.1.

2.1 Dynamic Modeling of Building Envelope & Zone and AHU Models

The building types of medium office and large hotel were selected due to their significant coverage for medium-size building footprint. The building models were developed with Modelica following the assumptions made in the corresponding EnergyPlus reference building models from DOE. The inputs to the building model are weather profiles (OAT, OARH, and solar), building occupancy, lighting, and equipment schedules, ground temperature, infiltration flow rate, as well as heating and cooling setpoints. Figure 1 illustrates the dynamic building modeling approach where EnergyPlus model's input (i.e., IDF file) and output files were used to identify the modeling assumptions and inputs need to be incorporated in the Modelica model. For air flow distribution, the variable air volume (VAV) models were modeled directly in the building model.

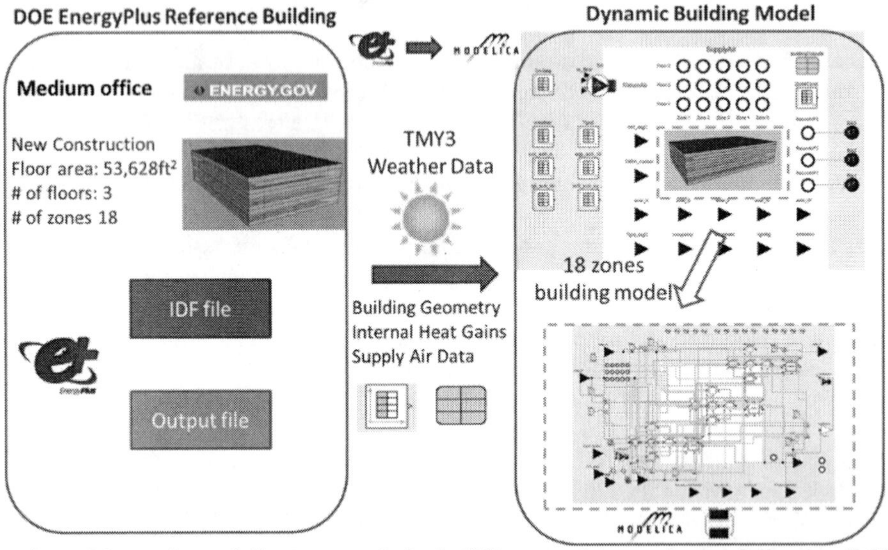

Figure 1: Illustration of dynamic modeling approach for building envelope and zone following DOE's EnergyPlus reference building (medium office)

The dynamic AHU model was developed based on LBNL's Modelica Buildings Library. The cooling coil model adopted handles both sensible and latent heat transfer with numerical discretization along the flow paths.

2.2 Chiller Plant Modeling

The chiller plant model includes dynamic models of chiller, cooling tower, pumps, and valves. These physics-based component models were adopted from UTRC's in-house Modelica Library and were validated with experimental data from UTRC's data. The chiller plant system model was built up by considering the core dynamics of chiller plant for controls evaluation. The chiller plant model consists of 2 variable-capacity chillers with the capacity determined based on building types and climate zones. Chilled-water temperature leaving the chillers is maintained at its setpoint by staging chillers and varying cooling capacity of each chiller. The cooling-water loop has 2 cooling towers with a variable-speed fan, which is used to maintain temperature of cooling water returned to chiller at its setpoint. The cooling water is circulated by two constant-flow pumps. The chilled-water loop has 2 configurations: primary-only and primary-secondary. The primary-only configuration has 2 variable-speed pumps to maintain the differential pressure. It has a bypass path connecting the pumps' outlet and return pipeline. There is a modulating value in the bypass path that is controlled to maintain a minimal flow of chilled water entering chillers. The primary-secondary has 2 water loops sharing a bypass path. The chilled water in the primary loop is circulated by 2 constant-speed pumps; the secondary loop uses 2 variable-speed pumps to maintain differential pressure between the supply and return pipelines at its setpoint. Figure 2(a) and 2(b) shows the schematic of the chiller plant model with primary-only and primary-secondary configurations.

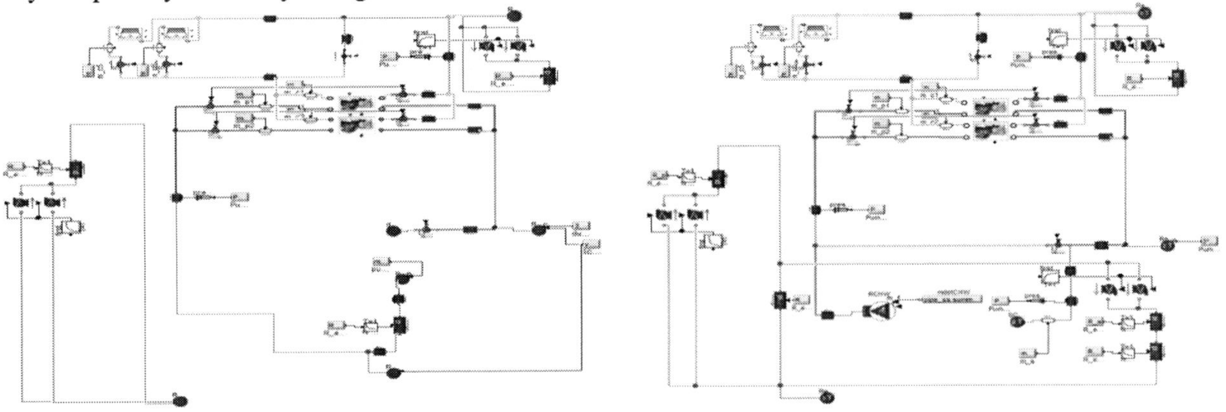

(a) Primary-only configuration (b) Primary-secondary configuration

Figure 2: Chiller plant model layout in Dymola

2.3 Chiller Plant Control Logics

A ChillerPlantControl Library was developed based on Modelica Standard Library 3.2.1. The Library includes baseline chiller plant control logics available from Automated Logic (ALC)'s WebCtrl® program. Figure 3 shows an overview of the Library that includes chiller staging, pump and fan PI control logics.

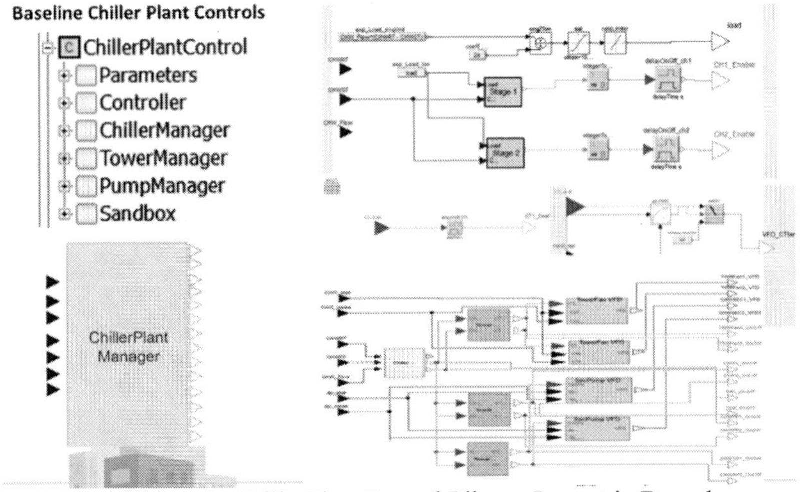

Figure 3: ChillerPlantControl Library Layout in Dymola

2.4. Integration of Building HVAC and Chiller Plant Model

In model-in-the-loop (MiL) platform, system-level coupling was tested incrementally before integrating all subsystem models together. To prepare the model for the HiL platform, each step of the subsystem integration was evaluated with fixed-step solver as well. Figure 4 schematically illustrates the coupling between building, AHU, and chiller plant model with closed-loop controls in each system and the corresponding model inputs and outputs.

At building level, individual zone temperature controller was implemented to adjust the zonal supply flow rate based on the temperature setpoints. Each AHU receives the information of required flow rate aggregated from each zone from the building model and provides consistent flow rate that would meet the flow requests to maintain the zone temperature within the setponts. At AHU level, there is a PI controller that measures the supply air temperature (SAT) as the feedback signal and modulates the chilled-water valve connected to the cooling coil model to maintain the SAT towards its setpoint. At chiller plant level, the AHU sides' pressure and temperature were connected to the supply and return ports of the chiller plant model so that the chiller plant's pump will provide sufficient pressure to deliver the required chilled-water flow rate to meet the SAT setpoint controls.

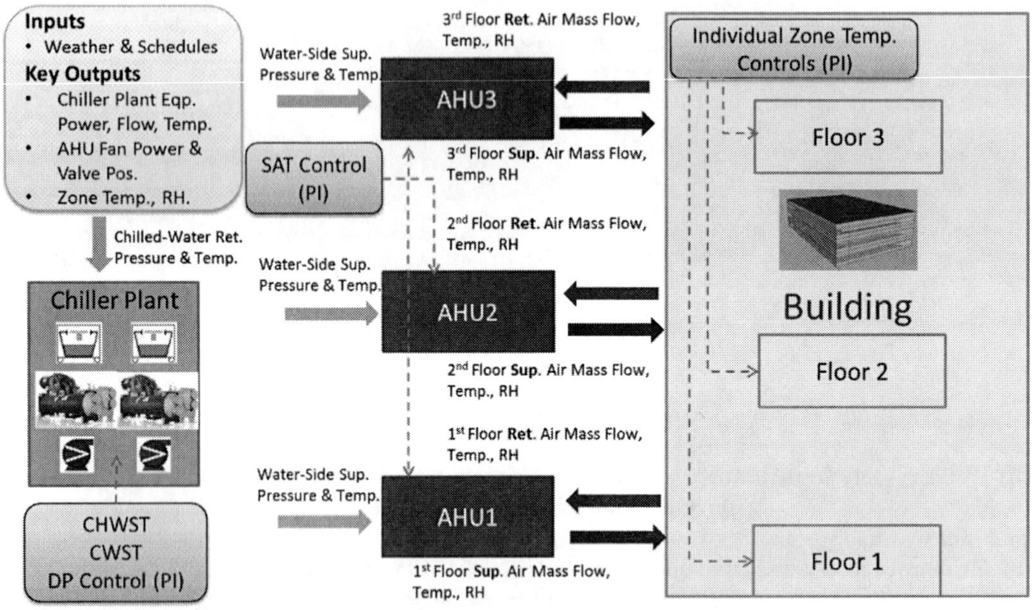

Figure 4: Schematic of model integration for building HVAC and chiller plant systems

2.5 Model Inputs and Outputs

Figure 5 shows the model inputs of outdoor air temperature (OAT), relative humidity (RH), occupancy schedule for different zones as well as zone temperature sepoint schedule using Chicago weather as the example for illustration purpose. Two weekly profiles were selected in the energy evaluation phase. The first weekly profile (July 16th to July 21st) represents a typical summer week that includes the summer design day selected by the EnergyPlus model. The second weekly profile (Oct. 2nd to 6th) represents the shoulder season week scenario.

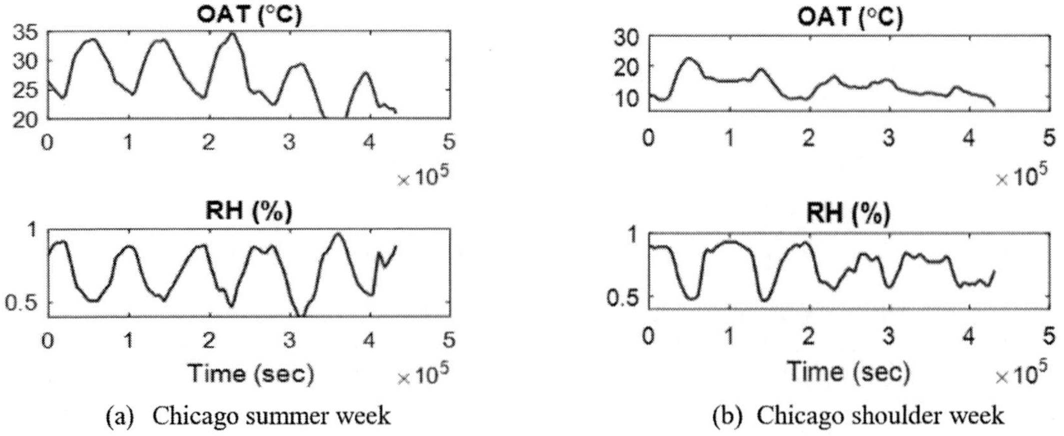

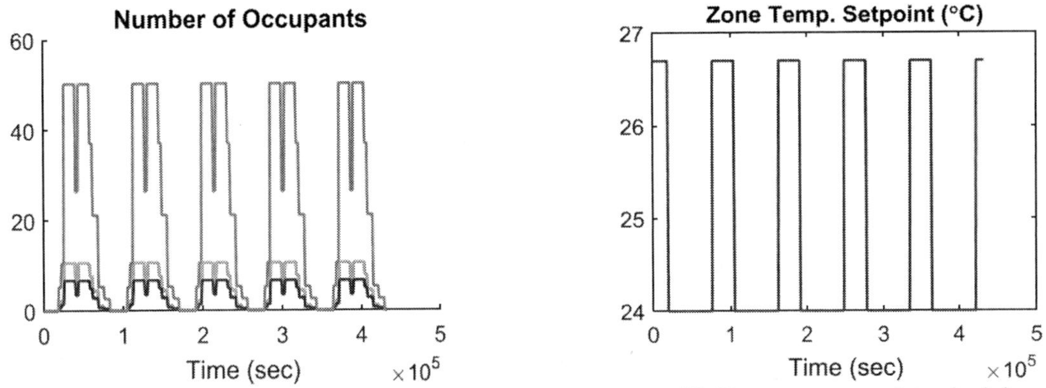

(c) Occupancy schedule (d) Zone temp. setpoint schedule

Figure 5: Selected inputs to the integrated building HVAC and chiller plant model

Figure 6(a) and 6(b) present example results of whole-building dynamic model for the chiller plant load and zone temperature profiles in medium office building, respectively. The key point to illustrate is that he integrated model has the capability to simulate the building behavior at different levels of fidelity.

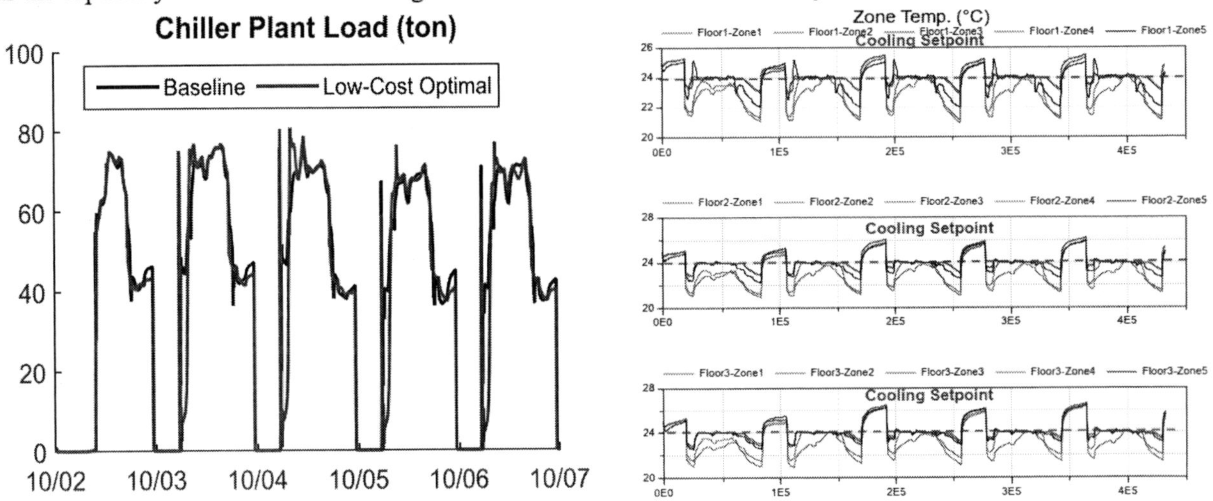

(a) chiller plant load profiles (b) Zone temperature profiles

Figure 6: Example simulation results of medium office during a weekly profile

3. HARDWARE-IN-THE-LOOP CONTROLLER VERIFICATION

Hardware-in-the-Loop (HiL) simulation is widely used in automotive and aerospace industries to verify control system software implementation as a part of their model based control development process. HiL allows verifying control requirements using a real controller with an emulated plant. HiL can be also useful for building HVAC system control development as real controllers can be evaluated in a more realistic environment than desktop numerical simulations. The HiL system developed for our study is depicted in Figure 7.

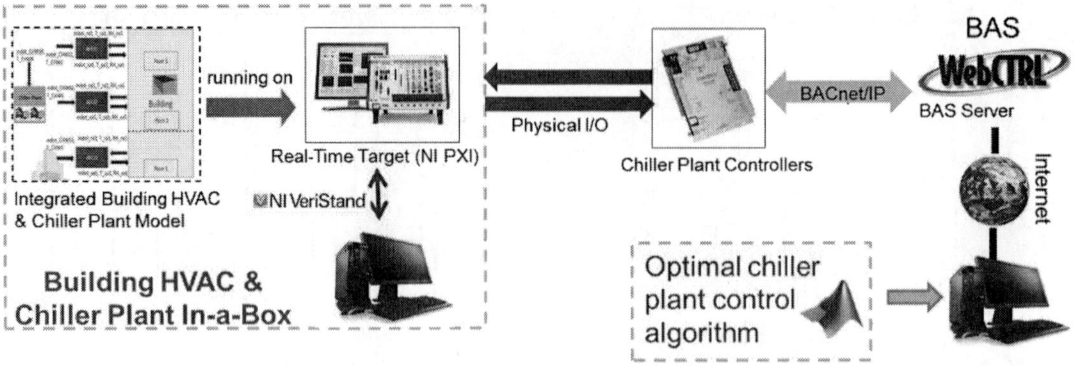

Figure 7: HiL system for building system control verification

With Matlab/Simulink support, the whole-building dynamic model discussed above is built as a real time capable Dynamic Link Library (DLL). The DLL runs on National Instruments (NI) PXI real time target. This is the same PXI hardware that is commonly used with NI LabView and Microsoft Windows. For HiL simulation, as the model is required to run in real-time, real time operating system is used instead of Windows. NI VeriStand is used in order to deploy model to the PXI system and manage simulations. Physical inputs and outputs (I/O) of the PXI system are connected to ALC controller I/O that typically involves analog and binary signals. The ALC controllers run control algorithms that are built in ALC Eikon logic builder. The control algorithm deployment to the controllers and management are done through the WebCtrl server that is connected to the ALC controllers via BACnet over IP network as well as operators via the Internet.

In new system deployment or retrofit applications, field engineers can use HiL simulation as a verification tool before deploying to a real job site. This enables to find and resolve issues in control algorithms before actual implementation so that actual commissioning of control system in the field is smoother with fewer issues. In research applications such as our study, HiL simulation can be used as a demonstration tool that is more realistic than desktop simulation. HiL simulation could even possibly replace real building demonstration when it is not available or feasible.

There are two steps needed to be carried out before running the baseline controller in HiL platform. The first step is model preparation. To enable HiL evaluation, the integrated model needs to be exported from Dymola to Simulink and then compiled as DLL file using NI VeriStand platform. The integrated model needs to be proven to work robustly with fixed-time step solver first in both Dymola and Simulink. The model preparation step usually takes several iterations since fixed-time solver has finer requirements for the dynamic models. The second step is the controller deployment, which happens after the model preparation phase and the associated time and efforts are documented below. Figure 8 shows the HiL setup at UTRC.

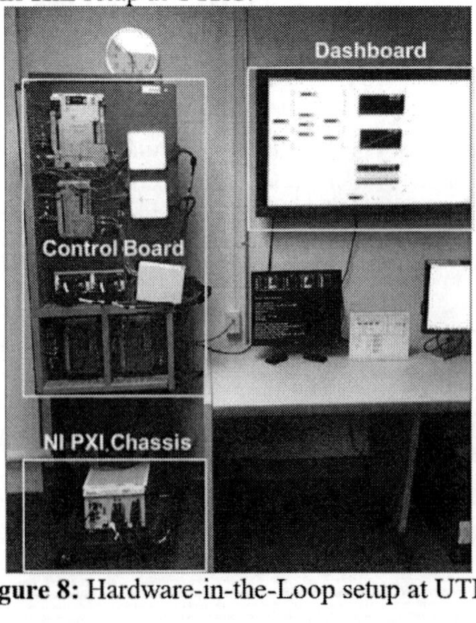

Figure 8: Hardware-in-the-Loop setup at UTRC

The baseline chiller plant control logics are available from Automated Logic (ALC)'s WebCtrl®. Figure 9 shows the chiller staging control logic.

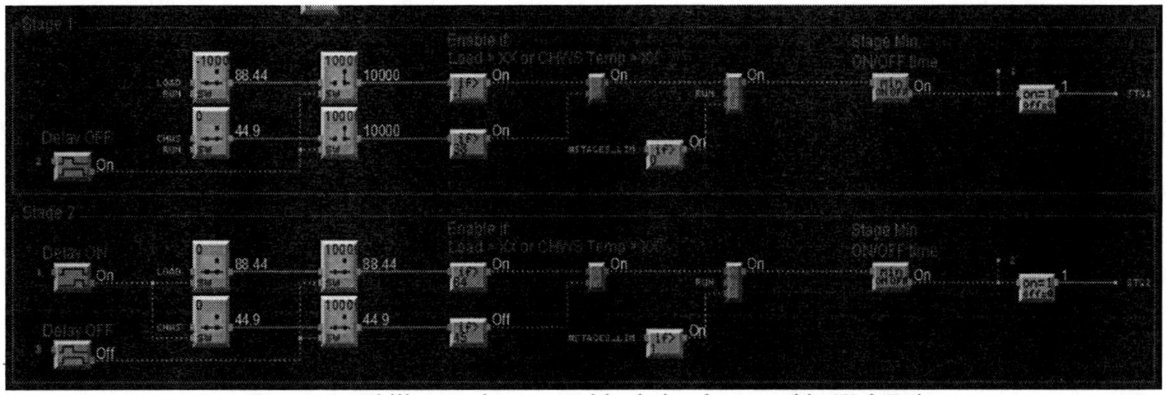

Figure 9: Chiller staging control logic implemented in WebCtrl

Table 1 shows the control setpoints selected for AHU and chiller plant systems. The optimal chiller plant control algorithm will manipulate the chilled-water supply temperature (CHWST) based on the estimation of air-side loads.

Table 1: Baseline and optimal control setpoints for AHU and chiller plant systems

Setpoints	Baseline	Optimal Control
AHU SAT setpoint	55F (12.78°C)	55F (12.78°C)
Zone setpoint (occ./unocc.)	75F/80F (23.9/26.7°C)	75F/80F (23.9/26.7°C)
CHWST setpoint	45F (7°C)	Optimized
CWST setpoint	85F	85F
Pressure diff. setpoint	30kPa	30kPa

Figure 10 shows the chilled-water temperature setpoint commanded by the low-cost optimal control during HiL testing.

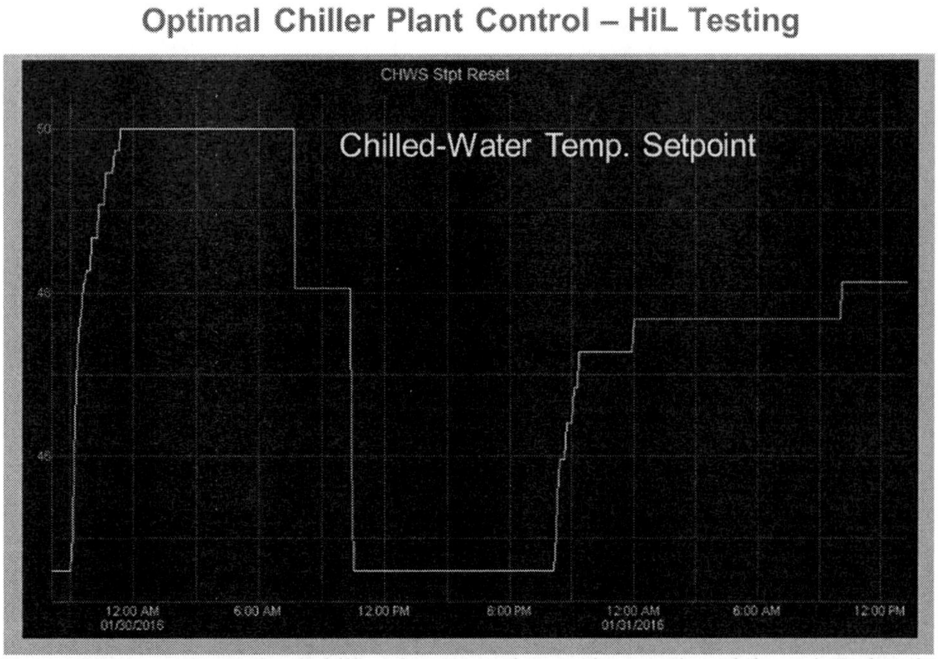

Figure 10: Low-cost optimal chiller plant control operation monitored through WebCtrl®

4. RESULTS AND ANALYSIS

This section presents the energy evaluation results by comparing different chiller plant controls and return of investment analysis for the low-cost optimal control.

4.1 Case Configurations for Case Studies

As described above, the baseline chiller plant control provides a constant chilled-water temperature setpoint of 7°C. The low-cost optimal control is realized by determining the maximum leaving chilled water temperature setpoint based on air-side load estimation. An online learning algorithm is employed to estimate the cooling coil parameters, which is used as a constraint in the optimization formulation to determine the degree of freedom to lift chilled water temperature setpoint. Table 2 provides a summary of the 4 case configurations exploited in our case studies. Each case configuration is represented by a high-fidelity whole-building HVAC system dynamics model that includes the chiller plant, AHUs, VAVs, zones, and the respective local PI controllers for each subsystem as well as supervisory controls at the chiller plant based on the Modelica platform.

Table 2: Summary of case configurations in the case studies

Case Configurations	**Case Configuration Definitions**
Case Configuration 1	Medium Office + Primary-Only Chiller Plant Configuration  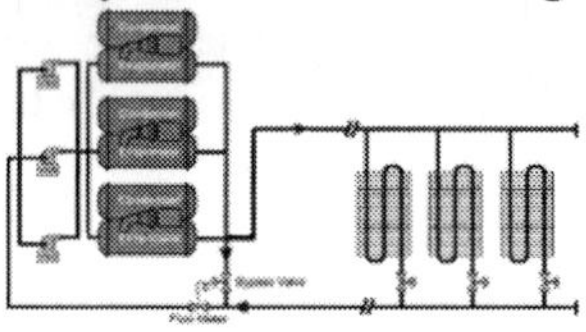
Case Configuration 2	Medium Office + Primary-Secondary Chiller Plant Configuration  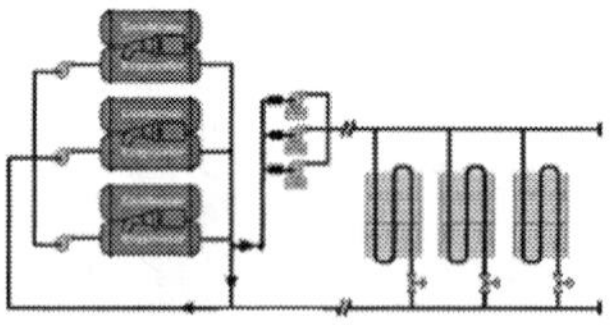
Case Configuration 3	Large Office + Primary-Only Chiller Plant Configuration  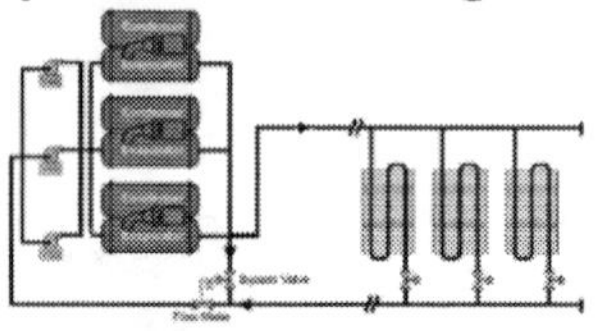
Case Configuration 4	Large Office + Primary-Secondary Chiller Plant Configuration  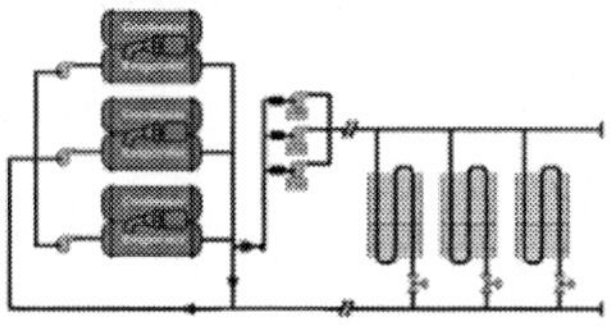

Table 3 lists all the weather profiles tested for each case configuration

Table 3: Weather profile scenarios for all test cases in a given case configuration

Test Cases	**Test Case Scenarios**
Test 1	Miami Summer
Test 2	Miami Shoulder
Test 3	Las Vegas Summer
Test 4	Las Vegas Shoulder
Test 5	Baltimore Summer
Test 6	Baltimore Shoulder
Test 7	Chicago Summer
Test 8	Chicago Shoulder

Figures 11(a) and 11(b) show the outdoor air temperature and relative humidity (RH), respectively.

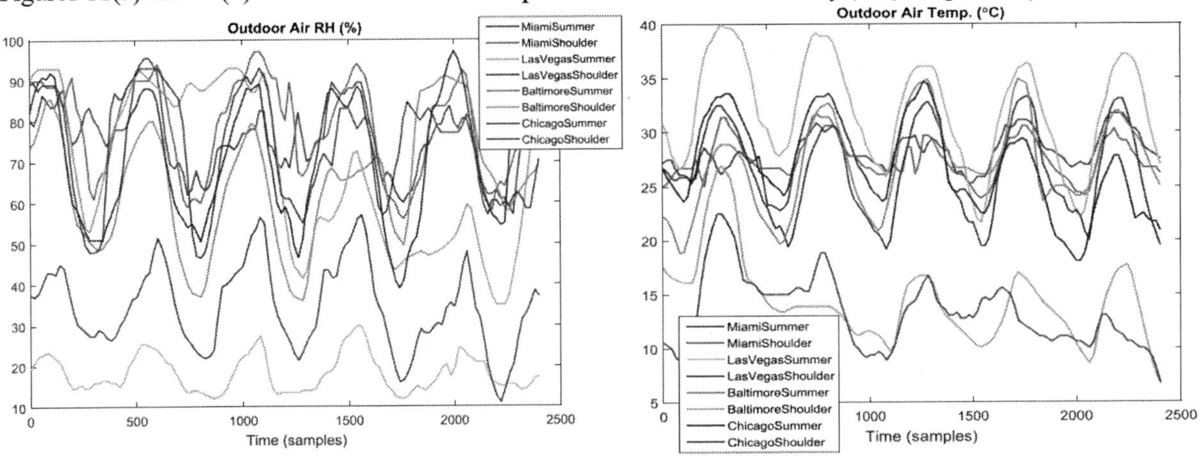

(a) Outdoor air temperature

(b) Outdoor air relative humidity (RH)

Figure 11: Outdoor air conditions of all test cases in Table 3

4.2. Comparison of Different Chiller Plant Control Algorithms with Baseline System

Figures 12 to 15 show the total chiller plant energy consumption and energy savings from the OAT-based reset, trim-respond, and the low-cost optimal control algorithms respectively for the 4 case configurations in Table 2. Table 4 shows a summary for the 4 different chiller plant control algorithms employed in this case study.

As can be observed from Figures 12 to 15, the OAT-based reset algorithm shows consistent savings across all the case configurations but is less effective in terms of achieving higher energy savings as seen from the low-cost optimal control. The trim-respond control algorithm yields comparable performance as the OAT based reset algorithm but the performance is less consistency in terms of overall energy savings achieved across all the case configurations. Note that the energy waste for trim-respond algorithm in the hotel primary secondary case configuration is mainly due to two reasons. The first reason is by trimming the CHWST setpoint up and down over time, the chiller staging will be affected and the cases with more energy consumption typically has more frequent staging behaviors of the 2nd chiller and therefore chiller 2 has more on time compared to the baseline. The second reason is by lifting the CHWST setpoint the pump will consume more power. In the case of hotel building (larger than the office), the trade-off between chiller powers and pump powers are more pronounced and therefore caused the fact that the increase of pumps' energy is more than the reduction of chillers' energy.

Table 4: Summary of chiller plant control logics employed in the case studies

Control Algorithms	**Descriptions**
1. Baseline Control	Constant chilled-water supply temperature (CHWST) setpoint of 7°C. Staging logic based on chiller plant load.
2.OAT-Based Reset (ASHRAE 90.1)	A linear schedule to reset CHWST setpoint based on outdoor air temperature (ASHRAE 90.1). Staging logic based on chiller plant load.
3.Heuristic-Based (Trim-Respond)	Trim-Respond logic resets CHWST setpoint based on the demand measured by AHU's chilled-water valve position. One request is generated when one chilled-water valve position becomes greater than a prescribed threshold (e.g., 90%). Staging logic based on chiller plant load.
4. Low-Cost Optimal	Maximize CHWST setpoints while performing real-time load estimation. Staging logic based on chiller plant load.

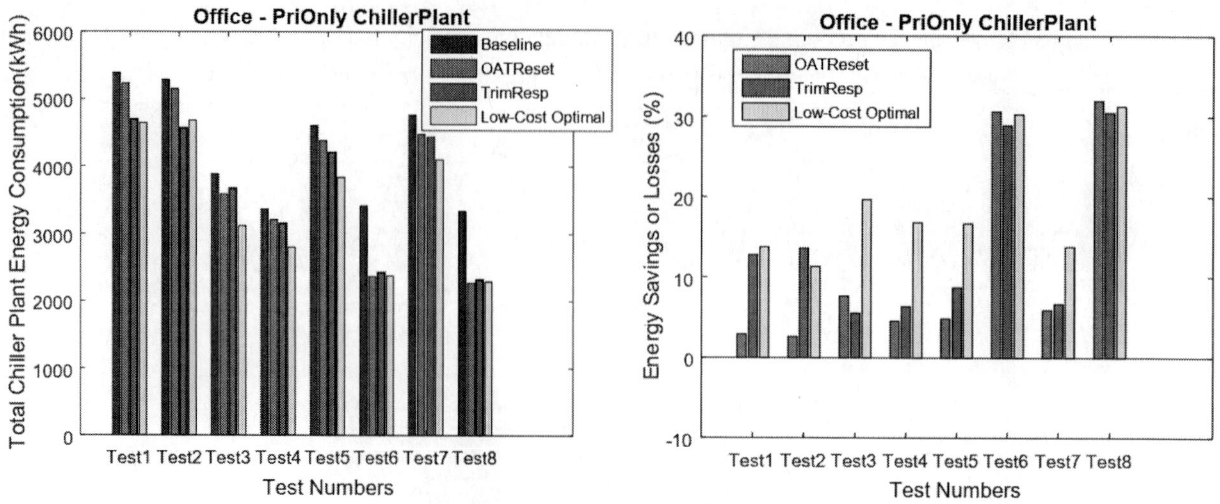

Figure 12: Comparison of energy consumption among baseline, OAT-reset, trim-respond, and low-cost optimal controls and respective energy savings to baseline (case configuration 1 – office & primary only chiller plant)

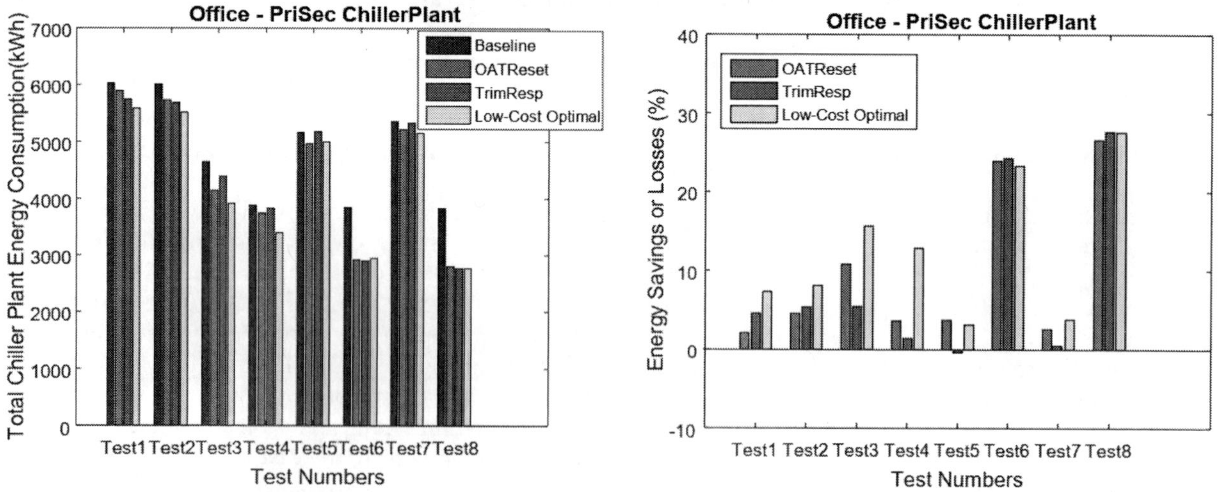

Figure 13: Comparison of energy consumption among baseline, OAT-reset, trim-respond, and low-cost optimal controls and respective energy savings to baseline (case configuration 2 – office & primary secondary chiller plant)

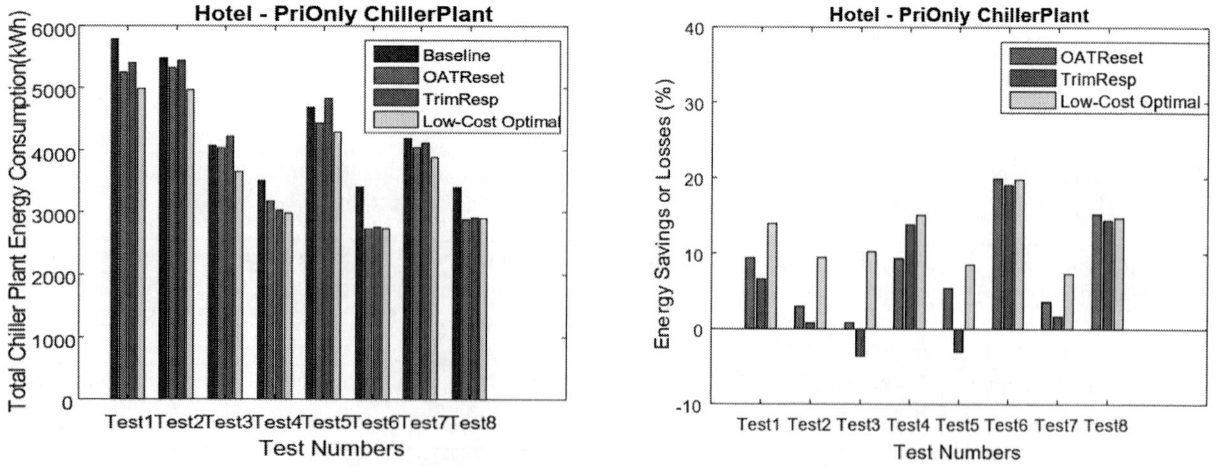

Figure 14: Comparison of energy consumption among baseline, OAT-reset, trim-respond, and low-cost optimal controls and respective energy savings to baseline (case configuration 3 – hotel & primary only chiller plant)

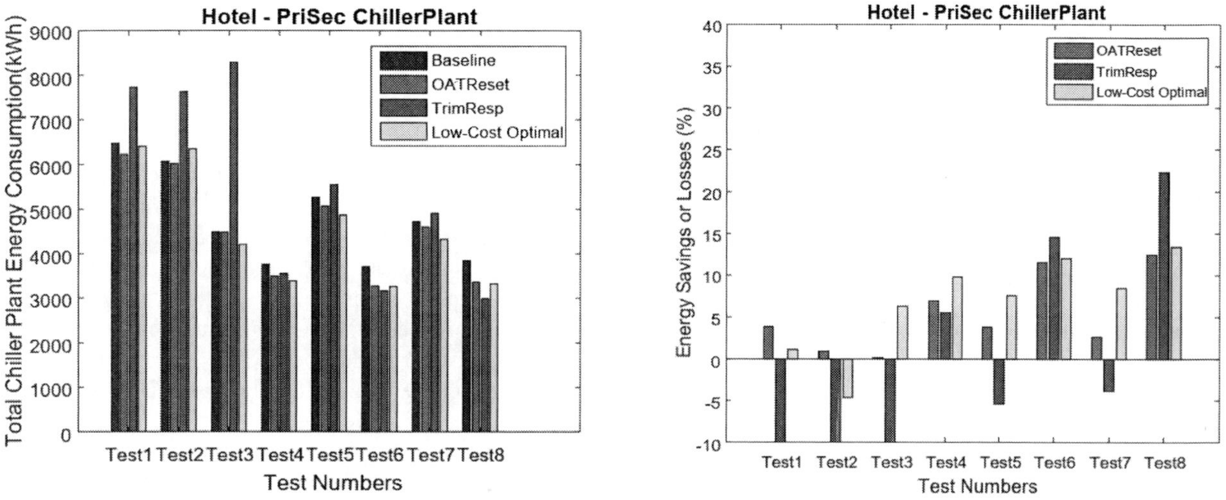

Figure 15: Comparison of energy consumption among baseline, OAT-reset, trim-respond, and low-cost optimal controls and respective energy savings to baseline (case configuration 4 – hotel & primary secondary chiller plant)

4.3 Return of Investment Analysis for Low-Cost Optimal Controls

This section presents a summary of the cost-benefit evaluation for the implementation of optimal chiller plant control algorithm. As demonstrated in the previous section, reduced energy consumption has been observed while the thermal comfort in the building zones has been maintained.

Figure 16 shows the average energy saving achieved by the low-cost optimal control across all the case configurations. The equation below was used to calculate the average energy savings for each configuration.

$$Average = 100\left(\sum_{test\ no.}^{1\ to\ 8} E_{total,base} - \sum_{test\ no.}^{1\ to\ 8} E_{total,optimal}\right) / \sum_{test\ no.}^{1\ to\ 8} E_{total,base}$$

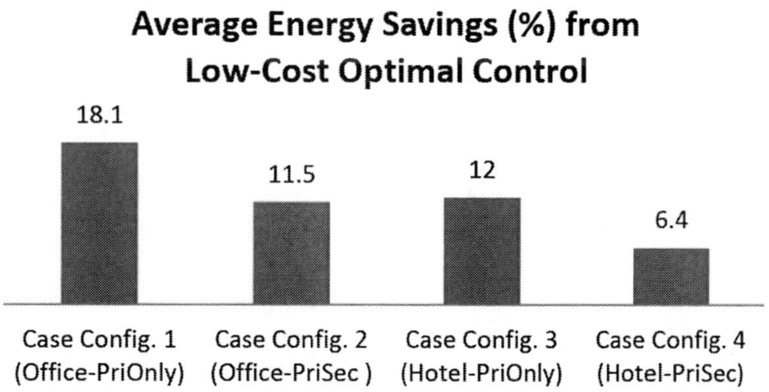

Figure 16: Average energy savings achieved by low-cost optimal control in all case configurations

The cost associated with the time required for a future commercial deployment of an optimal chiller plant control application, including customer engagement and site preparation, data mapping, and application installation and commissioning, is estimated to be approximately \$1150 (14 hours) based on an hourly rate of \$82 for HVAC contractor (CBEI, 2015).

An annual simulation of the medium office building model was conducted in EnergyPlus to determine the annual cooling energy consumption of the baseline system. As a result, the annual cooling capacity is determined to be 82.1 megawatt hour (Mwh) for the climate zone of Miami. Assuming an average chiller plant COP of 3 (CBEI, 2015), and a cost of electricity per kWh to be \$0.126 (CBEI, 2015), and ~12.5% energy consumption reduction in chiller plant operation, then 100% of the installation cost can be recovered in 3 years, which is less than the target of 3-year payback period and meets the commercial requirement identified for this project.

5. CONCLUSIONS

In summary, the scalable low-cost optimal chiller plant control algorithm has successfully demonstrated its effectiveness through an extensive set of 128 case studies covering a variety of chiller plant load variations with each case being a weekly simulation of whole-building dynamic HVAC system models with closed loop local controls and supervisory chiller plant controls. In particular, 4 case configurations were studied in details for both Office and Hotel sites and primary only and primary secondary chiller plant configurations. For each case configuration, the chiller plant control algorithms were evaluated in typical summer and shoulder weekly profiles across the climate zones of Miami, Las Vegas, Baltimore, and Chicago, respectively. In addition, the deployment of low-cost optimal controller hardware was verified using the hardware-in-the-loop (HiL) platform and the commissioning time and effort of baseline and low-cost optimal controller were evaluated through HiL as well.

A detailed analysis through model-in-the-loop (MiL) platform suggests a promising average energy saving of ~15% for medium office building across both primary only and primary secondary chiller plant configurations. For large hotel building, an average energy saving of ~10% is achieved for both primary only and primary secondary chiller plant configurations. Through simple payback analysis, the low-cost optimal chiller plant control can be paid back in less than 3 years which exceeds the commercial requirement identified in this project.

ACKNOWLEDGEMENT

This work is funded by Consortium for Building Energy Innovation (formally known as Energy Efficient Buildings Hub), sponsored by the Department of Energy under Award Number DE-EE0004261. The authors are grateful to James Braun and Timothy Wagner for their project management and technical consultation.

DISCLAIMER

REFERENCES

Bengea, S. C., Li, P., Sarkar, S., Vichik, S., Adetola, V., Kang, K., Lovett, T., Leonardi, F., & Kelman, A. D. (2015). Fault-tolerant optimal control of a building HVAC system. Science and Technology for the Built Environment, 21(6), 734-751.

Li, P., Vrabie, D., Li, D., Bengea, S. C., Mijanovic, S., & O'Neill, Z. D. (2015). Simulation and experimental demonstration of model predictive control in a building HVAC system. Science and Technology for the Built Environment, 21(6), 721-732.

Crawley, D. B., Lawrie, L. K., Winkelmann, F. C., Buhl, W. F., Huang, Y. J., Pedersen, C. O., ... & Glazer, J. (2001). EnergyPlus: creating a new-generation building energy simulation program. *Energy and Buildings*, 33(4), 319-331.

Wetter, M., Zuo, W., Nouidui, T. S., & Pang, X. (2014). Modelica Buildings Library. *Journal of Building Performance Simulation*, 7(4), 253-270.

Consortium for Building Energy Innovation (CBEI) Project 2.3 BP4 Final Deliverable (2015), Pre-commercial demonstration of cost-effective advanced HVAC controls and diagnostics for medium-sized commercial buildings, United Technologies Research Center.

Closed-Loop Scheduling for Cost Minimization in HVAC Central Plants

Michael J. RISBECK[1*], Christos T. MARAVELIAS[1], James B. RAWLINGS[1], Robert D. TURNEY[2]

[1] University of Wisconsin–Madison, Department of Chemical and Biological Engineering,
Madison, Wisconsin, USA
Email: risbeck@wisc.edu, christos.maravelias@wisc.edu, james.rawlings@wisc.edu

[2] Johnson Controls, Inc., Building Efficiency,
Milwaukee, Wisconsin, USA
Email: robert.d.turney@jci.com

* Corresponding Author

ABSTRACT

In this paper, we examine closed-loop operation of an HVAC central plant to demonstrate that closed-loop moving-horizon scheduling provides robustness to inaccurate forecasts, and that economic performance is not seriously impaired by shortened prediction horizons or inaccurate forecasts when feedback is employed. Using a general mixed-integer linear programming formulation for the scheduling problem, we show that optimization can be performed in real time. Furthermore, we demonstrate that closed-loop operation with a moderate prediction horizon is not significantly worse than a long-horizon implementation in the nominal case, and that closed-loop operation can correct for inaccurate long-term forecasts without significant cost increase. In addition, we show that terminal constraints can be employed to ensure recursive feasibility. The end result is that forecasts of demand need not be extremely accurate over long times, indicating that closed-loop scheduling can be implemented in new or existing central plants.

1. INTRODUCTION

Economically optimal operation of HVAC central plants is complicated by time-varying electricity cost and peak demand charges that cause nominal operation to be expensive. To achieve minimum utility cost, temporal load shifting via thermal energy storage (TES) is necessary (Henze, Felsmann, & Knabe, 2004). Efficient use of thermal storage requires advance scheduling for the central plant, which is challenging due to the presence of both discrete (i.e., on/off) and continuous (i.e., how much) decisions. Because buildings are subject to disturbances and because electricity prices may be volatile, the true economic cost of a given schedule may change significantly as time passes. In addition, model error for buildings or storage tanks can accumulate toward the end of the optimization horizon, invalidating the predictions of the schedule. Finally, units may break down unexpectedly, making a given schedule infeasible at future times. Thus, central plant schedules must be periodically updated to account for altered forecasts, disturbances, and other new information.

Although it is common to include a building model within the optimization problem (Mendoza-Serrano & Chmielewski, 2012; Y. Ma et al., 2012; J. Ma, Qin, & Salsbury, 2014) to compute the demand profile based on, e.g., predicted building temperature trajectory, in this work, we assume that demand is provided as a forecast that is treated as a fixed parameter within the optimization problem. We make this choice so as to avoid the issues of building model accuracy and to focus more completely on the central plant and the scheduling decisions that must be made. As a result of the central plant focus, each optimization more closely resembles a static scheduling problem than a dynamic optimal control problem. Thus, we refer to the architecture as "closed-loop scheduling," although many of the considerations in control problems (e.g., stability, state estimation, disturbance forecasting) are still relevant. In the following two subsections, we discuss some key considerations for production scheduling in general and for HVAC central plant scheduling.

1.1 Closed-Loop Scheduling

Traditionally, production scheduling has been considered as a static process; that is, a schedule is developed (via optimization or other method) and then followed until its end and possibly repeated periodically (Pantelides, 1994). The tacit assumptions are that both the production environment and production goals do not change throughout the scheduling horizon, and that decisions after the scheduling horizon are of no consequence. Unfortunately, many production environments are actually dynamic, and time does not split nicely into independent intervals. Thus, rescheduling often becomes necessary. While event-based rescheduling (whereby the schedule is recomputed only under a specific set of circumstances) can handle disturbances (Li & Ierapetritou, 2008), a closed-loop implementation can better address the effects of the moving production window (Subramanian, Maravelias, & Rawlings, 2012). In this scheme, a new schedule with a receding (but not shrinking) horizon is determined at each time point. From this schedule, the initial decisions (e.g., unit commitments) are implemented, and the plant is allowed to operate at the current levels for the remainder of the timestep. After this period, measurements are taken to determine the current plant state (e.g., storage levels), and demand forecasts are shifted and updated to include the new point at the end of the horizon. These updated forecasts are then used to generate a new schedule, and the process is repeated.

The two main questions are how to handle the uncertainty of forecasts or other parameters and how to perform the optimization. For parametric uncertainty, three possibilities are (i) to solve for a robust schedule that can meet all of the possible forecasts (Y. Ma et al., 2012), (ii) to solve a stochastic optimization that includes recourse actions (Y. Ma, Matuško, & Borrelli, 2015), or (iii) to simply solve the deterministic optimization using the best available information. However, the robust schedule may be overly conservative when quality forecasts are available, while the stochastic optimization can greatly increase the computational difficulty of the optimization problem due to the number of scenarios. For optimization, parametric optimization approaches have been proposed to shift the computational burden off line by determining the optimal schedule as a function of problem parameters such as equipment availability (Ryu, Dua, & Pistikopoulos, 2007). Assuming the parametric program can be solved, online computation is reduced to a simple lookup; however, due to exponential scaling in the number of parameters (and implicitly the prediction horizon), these methods are limited to small systems and short horizons. Instead, we propose to solve the deterministic formulation online at each timestep using a state-of-the-art solver. As will be demonstrated in the examples, advances in computing power and solution methods allow these problems to be solved quickly and in real time, eliminating what has historically been a major hurdle for online optimization.

1.2 Central Plant Optimization

There are a number of issues specific to central plants that complicate the scheduling process. First, utility consumption can vary nonlinearly with other decision variables. Traditional scheduling problems often assume that the resources consumed by a given task are either fixed or linearly related to batch size (Shah, Pantelides, & Sargent, 1993; Pantelides, 1994), which in the central plant problem essentially becomes an assumption of constant coefficient of performance (COP). While a constant COP assumption is common (Braun, 2007; Henze, Biffar, Kohn, & Becker, 2008), most equipment models are nonlinear (Lee, Liao, & Lu, 2012), so accurate cost prediction must consider variable COP as a function of load. In addition, TES is not 100% efficient, and thus requires a (possibly nonlinear) model to be embedded within the optimization problem (Touretzky & Baldea, 2014). Finally, demand charges are particularly challenging, as they penalize (typically monthly) peak electricity usage and can thus affect system cost over long time periods. They can be dealt with by retaining the past peak as the horizon advances (Cole, Edgar, & Novoselac, 2012) and by increasing their weighting near the end of the billing period (Risbeck, Maravelias, Rawlings, & Turney, 2015), although challenges still remain.

While many techniques can address a single chiller or a simple approximation (e.g., constant COP) of the central plant, we wish to explicitly address the combinatorial problem posed by making the optimal selection from among the available units. To reduce the number of discrete decisions, various approaches to central plant scheduling have been proposed, including near-optimal heuristics (Braun, 2007) and a predefined order for discrete operating modes (Y. Ma et al., 2012; Touretzky & Baldea, 2016). These methods often come at the cost of operational flexibility. As an alternative, the optimal configuration can be determined offline as a function of total central plant load (Henze et al., 2008). This relationship then serves as a surrogate for the central plant and eliminates the discrete on/off decisions. However, such a formulation makes it difficult to appropriately constrain equipment switching. Thus, we propose to simply include all on/off decisions within the optimization problem as discrete decision variables. As we will demonstrate, carefully constructed formulations can lead to quick solution times despite the combinatorial complexity.

2. SCHEDULING MODEL

To schedule equipment usage in the central plant, we formulate a mixed-integer linear programming (MILP) model to determine the cost-optimal loading for each piece of equipment over a finite prediction horizon. Full formulation details can be found in Risbeck et al. (2015). The optimization makes the following decisions:

- Which central plant units are active?
- For active units, what is their loading?
- How much water is being sent to/from active storage tanks?

Decisions are indexed in time by $t \in \mathbf{T}$ running from 1 to the (fixed) prediction horizon T. At each timestep, using the current forecast for utility costs and resource demand, the optimization is solved, and the optimal $t = 1$ decisions are implemented. In the following sections, we discuss the mathematical form of the model as well as some relevant forecasting considerations.

2.1 Mathematical Formulation

For the scheduling model, we employ a modified version of the formulation originally presented in Risbeck et al. (2015). For brevity, we describe only the key features. To facilitate generality, the formulation uses an abstract representation of the central plant by considering "generators" (units in the central plant) and "resources" (material flows in or out of the plant). Generators are indexed by $j \in \mathbf{J}$, and the category includes, e.g., conventional chillers and pumps, while resources are indexed by $k \in \mathbf{K}$ containing, e.g., chilled water and electricity. In the following discussion, Roman letters indicate decision variables, while Greek letters denote parameters.

The objective function is to minimize operating costs, calculated as

$$\min \sum_{k \in \mathbf{K}} \sum_{t \in \mathbf{T}} \rho_{kt} p_{kt} + \sum_{k \in \mathbf{K}} \rho_k^{\max} p_k^{\max}, \tag{1}$$

in which $p_{kt} \geq 0$ is the purchase of each resource, ρ_{kt} is a forecast of (time-varying) resource prices, $\rho_k^{\max}$ is the demand charge, and $p_k^{\max}$ is the maximum resource purchase, calculated by

$$p_k^{\max} \geq p_{kt} + \psi_{kt}, \quad k \in \mathbf{K}, t \in \mathbf{T}. \tag{2}$$

Here, $\psi_{kt} \leq 0$ is exogenous resource use margin (i.e., the difference between current exogenous use and peak exogenous use). Inclusion of this term is necessary because resources such as electricity may be used elsewhere in the overall system, and thus the optimizer should attribute cost only to resource use in excess of the nominal peak.

At each time t, demand ϕ_{kt} for each resource k must be met by the sum of production q_{jkt} in generators, charge or discharge y_{kt} of storage, and direct purchase p_{kt} as expressed in the following constraint:

$$\sum_{j \in \mathbf{J}} q_{jkt} + y_{kt} + p_{kt} \geq \phi_{kt}, \qquad k \in \mathbf{K}, t \in \mathbf{T}. \tag{3}$$

Depending on the particular resource, bounds may enforce that certain terms are zero, e.g., for chilled water, $p_{kt} = 0$ since chilled water cannot be purchased, and for electricity, $y_{kt} = 0$ unless there is a battery or other electricity storage available.

To schedule generators, binary variables u_{jt} are used to determine whether each generator is on or off. When a given generator is running ($u_{jt} = 1$), it often produces some resources ($q_{jkt} > 0$) while consuming others ($q_{jkt} < 0$). For example, a conventional chiller will produce chilled water while consuming electricity. Denoting these sets $\mathbf{K}_j^+$ and $\mathbf{K}_j^-$ respectively, these relationships are governed by equipment models that require constraints of the form

$$u_{jt} \xi_{jk}^{\min} \leq q_{jkt} \leq u_{jt} \xi_{jk}^{\max}, \qquad j \in \mathbf{J}, k \in \mathbf{K}_j^+, t \in \mathbf{T}, \tag{4}$$

$$q_{jkt} = u_{jt} \Xi_j(q_{j\mathbf{K}_j^+ t}), \qquad j \in \mathbf{J}, k \in \mathbf{K}_j^-, t \in \mathbf{T}, \tag{5}$$

in which the ξ_{jk} give equipment capacities and $\Xi_j(\cdot)$ is an equipment model. Note that (4) accounts for turndown when the generator is running, while (5) calculates resource consumption according to the generator model. In order to keep the optimization as an MILP, the generator models $\Xi_j(\cdot)$ are approximated as piecewise-linear and can be made arbitrarily accurate at the cost of additional discrete decision variables.

Other constraints in the formulation include a first-order linear model for storage tanks, dwell time enforcement to prevent rapid on/off switching, symmetry removal to improve solution times, and auxiliary equations for piecewise-linear models. Additional details can be found in Risbeck et al. (2015).

2.2 Forecasts

From the above constraints, we require four main types of parameters: resource price ρ_{kt}, exogenous margin ψ_{kt}, demand ϕ_{kt}, and equipment models $(\xi_{jk}^{\min}, \xi_{jk}^{\max}, \Xi_j(\cdot))$. We assume that the equipment models are readily available, e.g., empirical models based on manufacturer's or historical data (Lee et al., 2012), leaving three parameters yet to be determined. These remaining parameters are all forecasts that are subject to some error.

For electricity, time-varying prices are often published one day in advance (Albadi & El-Saadany, 2007), but if the prediction horizon is longer than 24 h, a longer forecast is required. For example, the accurate 24 h schedule could just be repeated, or historical averages can be used. Note that demand charge values are published in advance, and simpler rate structures (e.g., peak/off-peak pricing) may not be subject to any uncertainty. For central plants with heating equipment, natural gas is often sold at a constant price with no demand charge, and thus there is no forecasting uncertainty.

The more challenging forecasts to obtain are based on system demand and resource use. Both the demand and exogenous use forecasts require estimates of how much of a given resource will be consumed by the system. For heating and cooling loads, estimates can be made using linear models (Mendoza-Serrano & Chmielewski, 2012; Y. Ma et al., 2012; J. Ma et al., 2014) or detailed nonlinear models (US Department of Energy, 2016), but historical data can be used as well. For such models to make accurate predictions, weather data must be incorporated (Zavala, Constantinescu, Krause, & Anitescu, 2009). Exogenous electricity use typically requires some baseline data, although such measurements are often readily available.

3. COMPUTATIONAL EXAMPLES

We now present three simulations of different central plants being scheduled in closed loop. These simulations seek to answer the following questions:

- What effect does prediction horizon have on closed-loop performance?
- What is the effect of inaccuracy in demand forecasts?
- Can the optimization problems be solved in real time?
- Can terminal constraints improve robustness or reduce the necessary horizon?

For a simulation of length N with nominal prediction horizon T, the horizon of the subproblem at time t is taken as $\min(T, N-t)$ so that no time points after $t = N$ are considered by any optimization. This choice ensures a fair cost calculation so that the long horizons are not penalized for "planning ahead" at the end of the simulation. All timesteps are 1 h, and all optimizations are performed using Gurobi 6.5 on an Intel Core 2 2.66 GHz $\times$ 4 CPU with 8 GB of RAM.

3.1 Horizon Length

For simplicity, we start with a small system consisting of two conventional chillers and a storage tank as shown in Figure 1a. The system must meet time-varying demand subject to time-varying electricity prices and a demand charge. Using this system, we examine the effect of prediction horizon T on total electricity cost. We expect that, past a certain value, horizon length does not significantly affect electricity cost. In order to favor longer horizons, we assume that the storage tanks are completely efficient, which allows chilled water to be stored for long times without penalty. We employ a simulation horizon of $N = 240$, which gives 10 days of simulation.

The optimal $T = 240$ solution is shown in Figure 1. Because it includes all decisions considered throughout the simulation horizon, this simulation is the best possible schedule in terms of electricity cost. The main features we see

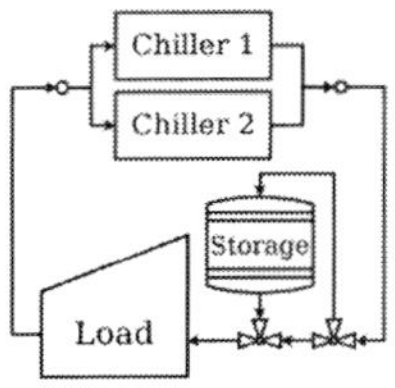

(a) Central plant equipment.

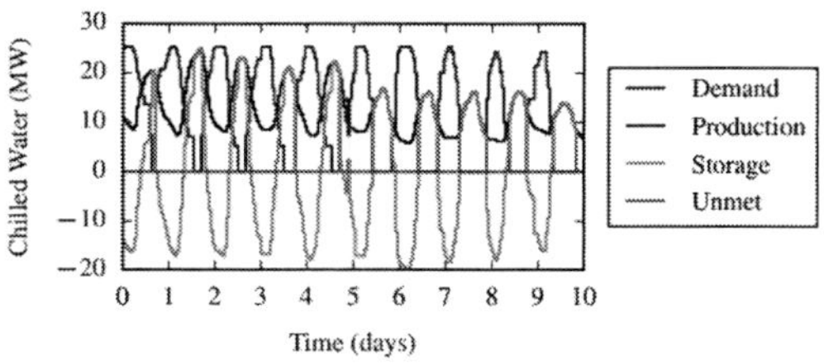

(b) Chilled water production. Negative values for "Storage" indicate charging, while positive values indicate discharging.

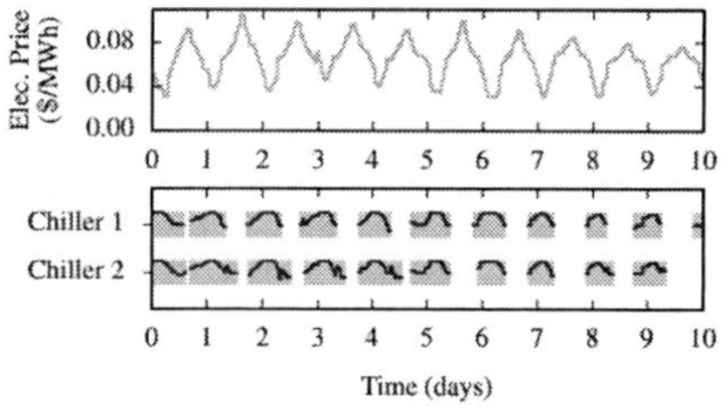

(c) Electricity price and equipment usage. Colored blocks show active units, while dark lines show loading between min and max capacity.

Figure 1: Optimal schedule for horizon length example.

Table 1: Closed-loop costs for varying horizons. Simulation length was $N = 240$. Costs relative to horizon $T = 240$.

Horizon	Use Charges	Demand Charges	Total Charges
168	−0.28%	+8.47%	+0.02%
96	−0.66%	+22.18%	+0.13%
48	−0.47%	+15.96%	+0.10%
24	−0.27%	+16.87%	+0.32%
12	+3.01%	+106.11%	+6.58%
6	+11.70%	+246.32%	+19.83%
2	+26.80%	+502.40%	+43.27%

are that the chillers are operated at full capacity during the night (when electricity is cheap), and they are largely shut off during the day (when electricity is expensive).

Using various smaller horizons, we repeat the closed-loop simulation. Total costs for each horizon are given in Table 1. From this table, we see that a long horizon is not necessary for low cost. Once the horizon is 24 h or longer, there is only a negligible effect on electricity cost, with the 24 h horizon less than 1% higher than the full 240 h horizon. We show the chilled water production for some of the horizons in Figure 2. From this figure, it is clear that the production profile for $T = 2$ is highly suboptimal and that there are large changes in load roughly every 2 h (which also corresponds to the minimum *on* dwell time for equipment). By contrast, the $T = 24$ solution is close to the $T = 240$ solution, indicating that the much shorter horizon would be acceptable for closed-loop operation. Note that the curve for $T = 240$ corresponds to the "Production" curve from Figure 1b.

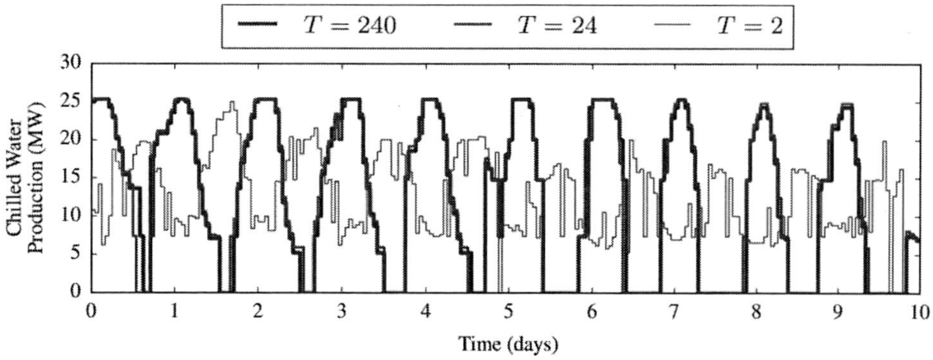

Figure 2: Chilled water production for various horizons. Horizon lengths are given in hours.

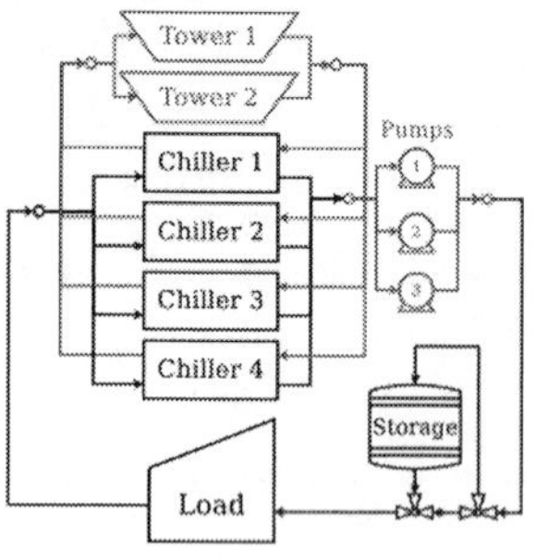

(a) Central plant equipment.

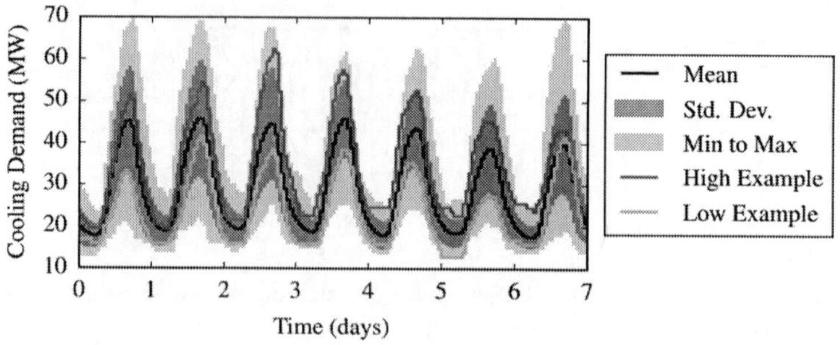

(b) Historical average demand and actual demand profile for second example. Data is averaged from May to September, 2010 in Stanford, CA.

Figure 3: System and demand for uncertain demand forecast example.

3.2 Uncertain Demand Forecasts

We turn next to examining the effect of uncertainty in the forecasts used by the model. Whereas the previous example assumed perfect forecasts were available (so as to not disadvantage a longer prediction horizon), in real systems, only an estimate of resource demand is available, and we wish to examine whether the prediction horizon becomes more or less significant in this case. In addition, we wish to determine whether closed-loop scheduling is a viable option without extremely accurate forecasts.

For this example, we use the more complex central plant shown in Figure 3a, which includes pumps and cooling towers. We simulate an approximate schedule as follows. First, we average historical data from May to September to determine the average cooling demand at each hour of the week. This average, along with two sample profiles, is shown in Figure 3b. Then, as a forecast of demand, we use a profile with 2 h of actual values followed by exponential decay from the actual to the average values. Thus, the forecast is accurate only at the beginning of the prediction horizon and quickly falls back to nominal average values. Certainly, this scheme would not be possible unless historical data is already available, but it is intended to represent the accuracy of a reasonable load-prediction methodology.

In the closed-loop simulation, each optimization is stopped when it has found the optimal solution or after 30 s, whichever comes first. If the time limit is reached, then the current incumbent solution is used. After the first timestep, the previous solution is shifted and provided as an initial guess to the optimizer. While this solution is not always feasible due to the updated forecast, the solver is often able to find a nearby feasible solution, which greatly helps solution times.

We first simulate the system using the "Low Example" from Figure 3b as the true demand profile and then then repeat the simulation using the "High Example." Costs for the system are shown in Table 2. The clairvoyant solution is the optimal schedule using true demand profile, which is the best the system could possibly do but requires perfect forecasts. For each of the three horizons and both demand profiles, we see that cost is higher than the clairvoyant solution (as expected) and further that the differences between the horizons are largely eliminated. For the low demand profile, each solution is less expensive than the average solution (which simply meets the Average demand from Figure 3b) even though it was the basis of the forecasts. Meanwhile, the conservative policy (which meets demand one standard deviation higher than the mean) is significantly more expensive and thus is not a good choice. For the high demand profile, the average profile is highly infeasible, but no infeasibility was observed in the closed-loop solution. Thus, as long as the demand forecast is accurate in the near term and reasonably representative for the remainder of the horizon, the closed-loop implementation can correct for forecasting errors without significant cost penalty.

Another important consideration is solution time. In a practical implementation, each optimization must be completed at least within the timestep, although ideally it should be much quicker. Indeed, in this example, solution times are fast as shown in Figure 4. Note that we omit the final $T-1$ times from each set, as these optimizations use smaller horizons and are thus easier. From this figure, we see that all of the $N=12$ and $N=24$ optimizations were solved to optimality within 30 s, which is certainly fast enough for practical implementation. The $N=48$ optimizations require more computational effort, although all instances were solved to within 0.5% of optimality within the time limit. We also see little difference between the high and low forecasts, indicating some insensitivity toward parameters.

Table 2: Costs for system with inaccurate demand forecast. Entries with a horizon indicate closed-loop results, while others are open-loop optimal. Costs are relative to optimal clairvoyant schedule.

(a) Costs for low demand example.

Horizon	Use Chg.	Dem. Chg.	Tot. Chg.
$N = 12$	+0.74%	+31.43%	+3.40%
$N = 24$	−0.56%	+42.39%	+3.16%
$N = 48$	−0.85%	+45.73%	+3.19%
Average	+27.79%	+27.51%	+27.76%
Conservative	+67.14%	+73.04%	+67.65%

(b) Costs for high demand example.

Horizon	Use Chg.	Dem. Chg.	Tot. Chg.
$N = 12$	+0.31%	+37.24%	+3.52%
$N = 24$	−0.17%	+32.68%	+2.68%
$N = 48$	−0.17%	+33.29%	+2.74%
Average	−18.66%	−19.00%	−18.69%
Conservative	+6.39%	+9.92%	+6.70%

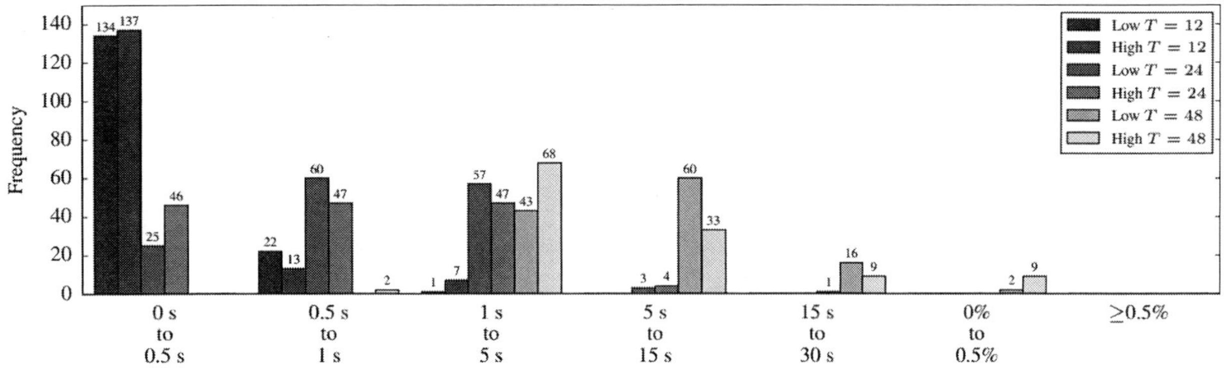

Figure 4: Solution times for uncertain demand example. Times indicate time needed to prove an optimal solution, while percentages are optimality gap after termination at 30 s. "High" and "Low" refer to demand forecast.

3.3 Terminal Constraints

While the previous examples have shown that a long prediction horizon is not necessary for low cost, it may still be desirable to obtain longer-term predictions, e.g., for diagnostics or higher-level planning. On such timescales, the discrete on/off decisions made by the full scheduling model are not as relevant, and instead the key choice is storage utilization. Thus, at the cost of less accurate predictions for resource consumption, computation can be made considerably easier by removing the discrete decision variables and piecewise-linear equipment models, replacing them with fully linear models (or other convex approximation). These changes yield a linear programming (LP) surrogate model that can be solved as a much more easily than the full scheduling model and thus can quickly provide a forecast for optimal storage use. Maximum equipment capacities are retained so that any surrogate schedule can be made feasible in the true by model relaxing dwell times and rounding up to minimum equipment capacities.

Although the surrogate model does not provide the discrete decisions needed to actually implement an equipment schedule, it can be used to enhance the solution quality of the full model by providing a reference trajectory for the storage tank. With the addition of a terminal constraint to the full model that requires the amount of storage to terminate on the reference trajectory, recursive feasibility (with respect to demand satisfaction) can be ensured. Because the surrogate model does not accurately calculate resource consumption rates, the storage trajectories it produces are likely to be suboptimal, but they are feasible. Thus, by using this trajectory as a terminal constraint, the full model is able to reoptimize the beginning of the trajectory for lower cost while still keeping enough storage to meet the future demand considered by the surrogate model.

As an example, we consider a larger system producing both hot and chilled water with the equipment shown in Figure 5a. We also use a demand profile within an extreme spike above the nominal capacity of the plant, thus requiring storage to be charged in advance. The optimal schedule is shown in Figures 5b and 5c. Notice that in the afternoon of the second day, the chillers are running at near-full capacity with significant withdrawal of storage. We then simulate the system in closed loop using both the "standard" approach used in the previous examples and also with the "surrogate" approach of this section that includes terminal constraints. Costs for these methods are given in Table 3. From this table, we see that while the longer-horizon standard approach does give the lowest (feasible) cost, the

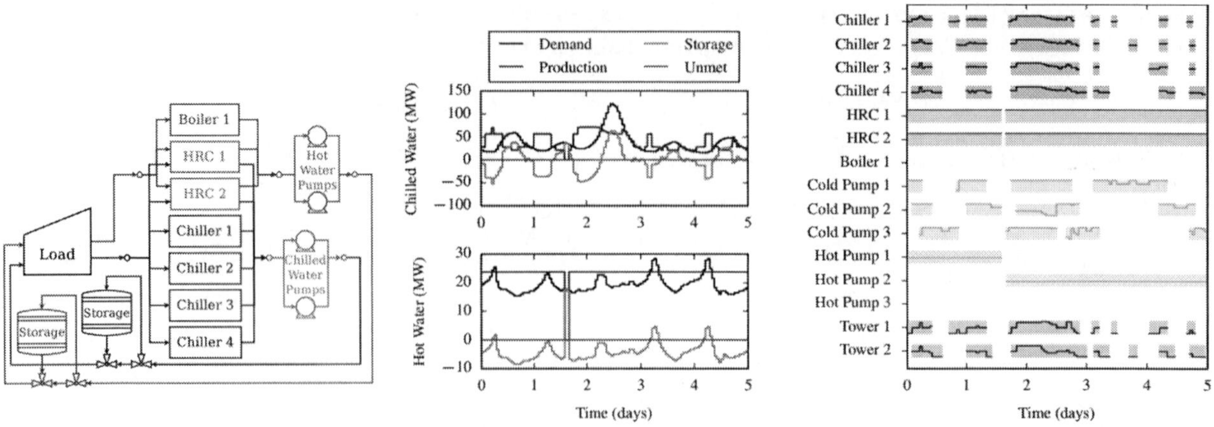

(a) Central plant equipment. Cooling towers not shown.

(b) Chilled and hot water production.

(c) Equipment usage.

Figure 5: System and optimal schedule for terminal constraint example.

Table 3: Costs for full system. Simulation length was $N = 120$. Costs are relative to horizon Standard $T = 120$. Surrogate entries use terminal constraints from $T = 120$ surrogate models, while standard entries do not.

Horizon	Use Charges	Demand Charges	Total Charges
Standard $T = 72$	+0.27%	+0.00%	+0.24%
Standard $T = 24$	+0.89%	+0.00%	+0.78%
Standard $T = 12$*	+0.17%	+26.00%	+3.45%
Surrogate $T = 24$	+0.69%	+0.00%	+0.60%
Surrogate $T = 12$	+1.19%	+0.31%	+1.08%

*Infeasibility during simulation.

shorter horizons with the surrogate model are competitive. In particular, while the standard $T = 12$ has the lowest economic cost, it fails to meet the requisite demand due to insufficiently charging the storage tank; by contrast, the surrogate method with $T = 12$ avoids infeasibility due to the added terminal constraint.

For this larger system, we would also like to examine solution times, which are shown in Figure 6. Here we notice that the surrogate method subproblems require slightly more time due to the added terminal constraint, but are essentially just as easy to solve as the standard subproblems without the terminal constraint. The longer $T = 72$ standard models are often not solved to optimality within the time limit, but they come close. Note that the $T = 120$ surrogate models used to derive the terminal constraints can be solved in roughly 1 s (not included in Figure 6), and so they do not appreciably affect overall computational burden. Indeed, all methods are sufficiently quick for practical use.

4. CONCLUSIONS

In this paper, we have presented an MILP-based scheduling model for HVAC central plants and demonstrated that closed-loop scheduling is viable for practical use. While long-term demand forecasts may be inaccurate, only immediate decisions of a given schedule are actually implemented, and the schedule is reoptimized at the next timestep using updated forecasts before taking further action. As a result, plant operation is most directly influenced by accurate information, while the more speculative choices made at the end of the prediction horizon are not utilized. This feature reduces the burden for demand forecasts, which need only be accurate in the near term. In addition, we have shown that extremely long prediction horizons are unnecessary, with little improvement in cost for horizons longer than 48 h. Central plants with large, well-insulated storage tanks can benefit from a longer prediction horizon, but in such cases, a simplified LP surrogate model can be used to determine the storage trajectory, which can then be used as a terminal constraint in the shorter-horizon detailed MILP optimization. For two realistically-sized central plants discussed in the

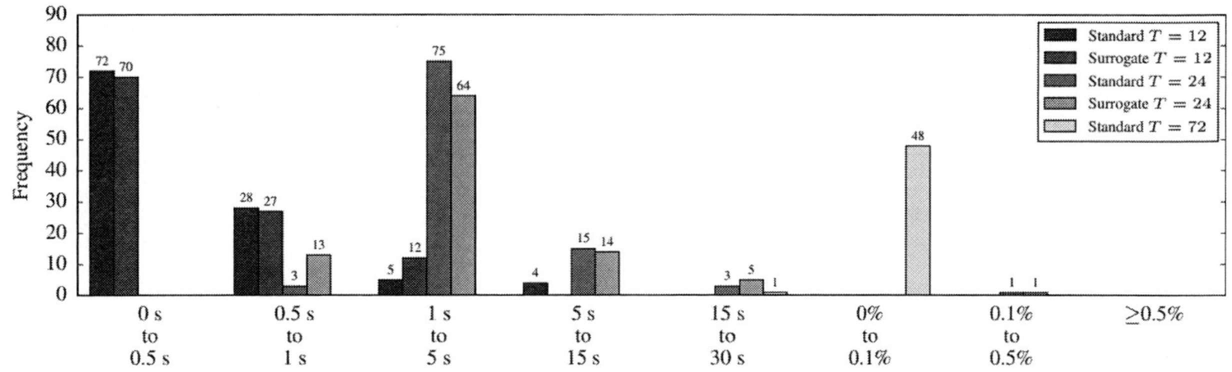

Figure 6: Solution times for terminal constraint example. Times are for solving the full model and do not include time spend solving the surrogate model to obtain terminal constraints. Bins are as in Figure 4.

examples, most subproblems could be solved to optimality within 30 s, while all could produce solutions no more than 0.5% away from optimality within that same time limit. Thus, with careful attention to problem formulation, real-time optimization is possible with standard commercial MILP solvers, leading to improved economic performance without relying on heuristic methods.

NOMENCLATURE

In the text, we use T for the length of the prediction horizon and N for the length of a simulation. All symbols used in the scheduling formulation are given below.

Sets and Subscripts

$j \in \mathbf{J}$	generators
$k \in \mathbf{K}$	resources
$\mathbf{K}_j^+ \subset \mathbf{K}$	resources produced by generator
$\mathbf{K}_j^- \subset \mathbf{K}$	resources consumed by generator
$t \in \mathbf{T}$	time points

Variables

p_{kt}	resource purchase
$p_k^{\max}$	peak resource purchase
q_{jkt}	resource production in generator
u_{jt}	generator binary on/off
y_{kt}	use of storage

Parameters

$\Xi_j(\cdot)$	model for generator consumption
$\xi_{jk}^{\min}, \xi_{jk}^{\max}$	min, max capacity for generator
ρ_{kt}	resource price
$\rho_k^{\max}$	demand charge
ϕ_{kt}	demand forecast
ψ_{kt}	exogenous margin

REFERENCES

Albadi, M. H., & El-Saadany, E. F. (2007). Demand response in electricity markets: An overview. In *2007 IEEE power engineering society general meeting* (pp. 1–5). IEEE.

Braun, J. E. (2007). A near-optimal control strategy for cool storage systems with dynamic electric rates. *HVAC&R Res.*, *13*(4), 557–580.

Cole, W. J., Edgar, T. F., & Novoselac, A. (2012). Use of model predictive control to enhance the flexibility of thermal energy storage cooling systems. In *American control conference (ACC), 2012* (pp. 2788–2793).

Henze, G. P., Biffar, B., Kohn, D., & Becker, M. P. (2008). Optimal design and operation of a thermal storage system for a chilled water plant serving pharmaceutical buildings. *Energ. Buildings*, *40*(6), 1004–1019.

Henze, G. P., Felsmann, C., & Knabe, G. (2004). Evaluation of optimal control for active and passive building thermal storage. *Int. J. Thermal Sci.*, *43*(2), 173–183.

Lee, T.-S., Liao, K.-Y., & Lu, W.-C. (2012). Evaluation of the suitability of empirically-based models for predicting energy performance of centrifugal water chillers with variable chilled water flow. *Appl. Energ.*, *93*, 583–595.

Li, Z., & Ierapetritou, M. (2008). Process scheduling under uncertainty: Review and challenges. *Comput. Chem. Eng.*, *32*(4-5), 715–727.

Ma, J., Qin, S. J., & Salsbury, T. (2014). Application of economic MPC to the energy and demand minimization of a commercial building. *J. Proc. Cont.*, *24*(8), 1282–1291.

Ma, Y., Borrelli, F., Hencey, B., Coffey, B., Bengea, S., & Haves, P. (2012). Model predictive control for the operation of building cooling systems. *IEEE Ctl. Sys. Tech.*, *20*(3), 796–803.

Ma, Y., Matuško, J., & Borrelli, F. (2015). Stochastic model predictive control for building HVAC systems: Complexity and conservatism. *IEEE Trans. Cont. Sys. Tech.*, *23*(1), 101–116.

Mendoza-Serrano, D. I., & Chmielewski, D. J. (2012). HVAC control using infinite-horizon economic MPC. In *Decision and control (CDC), 2012 IEEE 51st annual conference on* (pp. 6963–6968).

Pantelides, C. C. (1994). Unified frameworks for optimal process planning and scheduling. In *Proceedings on the second conference on foundations of computer aided operations* (pp. 253–274).

Risbeck, M. J., Maravelias, C. T., Rawlings, J. B., & Turney, R. D. (2015). Cost optimization of combined building heating/cooling equipment via mixed-integer linear programming. In *American control conference* (pp. 1689–1694). Chicago, IL.

Ryu, J.-h., Dua, V., & Pistikopoulos, E. N. (2007). Proactive scheduling under uncertainty: A parametric optimization approach. *Ind. Eng. Chem. Res.*, *46*(24), 8044–8049.

Shah, N., Pantelides, C., & Sargent, R. (1993). A general algorithm for short-term scheduling of batch operationsII. computational issues. *Comput. Chem. Eng.*, *17*(2), 229–244.

Subramanian, K., Maravelias, C. T., & Rawlings, J. B. (2012). A state-space model for chemical production scheduling. *Computers & Chemical Engineering*, *47*, 97–110.

Touretzky, C. R., & Baldea, M. (2014). Integrating scheduling and control for economic MPC of buildings with energy storage. *J. Proc. Cont.*, *24*(8), 1292–1300.

Touretzky, C. R., & Baldea, M. (2016). A hierarchical scheduling and control strategy for thermal energy storage systems. *Energ. Buildings*, *110*, 94–107.

US Department of Energy. (2016). *EnergyPlus 8.5.0.* Retrieved from `https://energyplus.net/`

Zavala, V. M., Constantinescu, E. M., Krause, T., & Anitescu, M. (2009). On-line economic optimization of energy systems using weather forecast information. *J. Proc. Cont.*, *19*(10), 1725–1736.

ACKNOWLEDGMENT

Sample data, equipment models, and funding provided by Johnson Controls, Inc.

A VOLTTRON^{TM} based implementation of Supervisory Control using Generalized Gossip for Building Energy Systems

Venkatesh Chinde*, Adam Kohl, Zhanhong Jiang, Atul Kelkar, Soumik Sarkar

Iowa State University, Mechanical Engineering Department,
Ames, IA, USA
{vchinde, adamkohl, zhjiang, akelkar, soumiks}@iastate.edu

* Corresponding Author

ABSTRACT

Building energy systems comprising of many subsystems with local information and heterogenous preferences demand the need for coordination in order to perform optimally. The performance required by a typical airside HVAC system involving a large number of zones are multifaceted, involves attainment of various objectives (such as optimal supply air temperature) which requires coordination among zones. The use of traditional centralized optimization involving a large number of variables is very difficult to solve in near real time. This paper presents a novel distributed optimization framework to achieve energy efficiency in large-scale buildings. The primary goals are to achieve scalability, robustness, flexibility and low-cost commissioning. The results are presented using the proposed distributed optimization framework based on a physical testbed in the Iowa Energy Center and demonstrates the advantages of the proposed methodology compared to a typical baseline strategy. The paper outlines a real-life implementation of the proposed framework based on the VOLTTRON^{TM} platform, recently developed by the Pacific Northwest National Laboratory (PNNL).

1. INTRODUCTION

Buildings are complex systems with a large variety of heterogeneous sub-systems involved in complex electro-mechanical interaction. Hence, energy usage optimization poses a difficult technical challenge that needs to be robust, scalable and easily deployable. In this context, a multi-agent based control strategy for optimal control of centralized air conditioning systems is proposed in (Cai, Kim, Putta, Braun, & Hu, 2015), where two distributed optimization algorithms have been formulated to minimize the energy consumption satisfying the constraints on the load. Recently, (Cai, Braun, Kim, & Hu, 2016) proposes a multi-agent control based demand response strategy for multi-zone buildings. In (Jiang, Chinde, Kohl, Sarkar, & Kelkar, 2016), authors have proposed a modular optimization framework using Generalized Gossip for building energy systems that naturally generalizes to the entire spectrum of "complete disagreement" to "complete agreement (consensus)" which will be useful for handling multiple energy resources and constraints. This paper implements and tests the effectiveness of the Generalized Gossip-based distributed optimization framework on the Iowa Energy Center's Energy Resource Station (ERS) testbed with the objective to reduce energy costs while maintaining zone comfort. These goals are accomplished by implementing the proposed framework within a multi-agent environment called VOLTTRON^{TM}.

VOLTTRON^{TM} an open source multi-agent system platform, developed by PNNL (Akyol et al., 2012; Lutes et al., 2014) which has several built-in features such as security, resource management etc., and leaving the developers to write agents and control their behavior. The rationale (Weiss, 1999) behind multi-agent systems (MAS) is well suited for diverse applications ranging from the internet to the building sector. These applications are inherently distributed, complex and solutions to such problems using centralized architecture pose limitations to the hardware or software involved. Multi-agent systems, as distributed systems have succeeded in solving complex problems while offering several desirable features such as speed, efficiency, scalability, and flexibility etc., one such complex problem is the case of buildings which is discussed in this paper.

There are several agent development platforms which support the MAS conceptual design such as JADE, ZEUS, VOLTTRON^{TM} etc. The description regarding these platforms and comparison studies in the context of microgrid applications is provided in (Kantamneni, Brown, Parker, & Weaver, 2015). It is noted that VOLTTRON™ is best employed by facilities and building managers, since it efficiently utilizes sensor and instrumentation data. The use of VOLTTRON™ platform has been demonstrated in several applications (Haack, Akyol, Carpenter, Tews, & Foglesong, 2013; Haack, Akyol, Tenney, et al., 2013) such as achieving the power consumption goal in a building while charging Electric Vehicle within user schedule, control of an electric bus for a local transit company and coordinating commercial building energy usage. In (Khamphanchai et al., 2014; Khamphanchai, Pipattanasomporn, Kuzlu, & Rahman, 2015) a low-cost, user-friendly building energy management platform has been built on the top of VOLTTRON to facilitate improved sensing and control of equipment in small and medium-sized buildings. The platform is aimed at improving energy efficiency and participate in demand response by controlling the loads in buildings. Overall, VOLTTRON™ provides a natural way of designing and implementing distributed control algorithms.

The organization of the paper is as follows. In section 2, we discuss the problem setup. In section 3, we discuss the generalized gossip based distributed optimization framework. In section 4, the overview and implementation aspects of VOLTTRON™ are discussed followed by conclusions presented in section 5.

2. PROBLEM SETUP

Building are complex energy systems composed of multiple subsystems which have different mathematical structures and evolve in different scales either in time or space. These components are interconnected due to which there are inherent dependencies between the local and the system-wide events. The major issue that persists in the building systems research is to increase the energy efficiency in buildings, the reason for this is two folds: the percentage share in the total energy consumption and the greenhouse gas emissions. These key issues are of paramount importance to meet the national energy demands and the environment challenges. Building efficiency needs to be considered as a means to provide occupant comfort and safe indoor environment and also building systems, in particular, has distributed nature, which demands the need for distributed control architectures and optimization. There are several disadvantages associated with the centralized optimization such as handling large scale systems with given time complexity and scalability issues often lead to the path for distributed optimization. The computational complexity in the distributed setup is addressed primarily based on the modularity in the modeling and control design phases. Distributed optimization problems arise in various fields of engineering where the set of agents coordinate to perform the task optimally.

Multi-agent coordination and control framework is adopted and tested in diverse applications and in the context of building systems application (Cai et al., 2015) provides a multi-agent control methodology for optimal control of centralized air conditioning systems. We have proposed a modular optimization framework in (Jiang et al., 2016), where the supervisory optimization scheme is completely decoupled from the data-driven micro-level modeling aspect leading to a significantly scalable and flexible architecture. The primary goals of the proposed framework are to achieve scalability, robustness, flexibility and low-cost commissioning. We present a formulation of the proposed methodology with regards to an illustrative example scenario for air-side heating, ventilation and air-conditioning (HVAC) system as presented in Fig. 1. In this energy supply-demand problem, individual zones become energy consumers that are served with conditioned air by an air handling unit (AHU).

2.1 Air-side HVAC (AHU-VAV) System

The general layout of a typical AHU-VAV HVAC system is shown in Fig. 1. While a central AHU provides conditioned air to each variable-air-volume (VAV) box, VAVs in turn supplies conditioned air to each zone. From a supervisory decision-making perspective, a few setpoints (e.g., supply air temperature (SAT) setpoint, mixed air temperature setpoint and duct static pressure setpoint) need to be determined for energy usage minimization while maintaining zone comfort levels. For simplicity, we consider SAT setpoint as the optimization variable to demonstrate the effectiveness of the proposed algorithm. Consider a situation where every zone in a building has the same comfort requirement and the same external/internal loads. In that case, a common SAT setpoint can be determined that satisfies the requirement of each zone. However, in reality due to the diversity of zones, AHU SAT is typically kept at a very low value (e.g., $55°F$) such that VAVs can reheat the supply air as needed before it enters the zones. Therefore, optimization can help decide a variable setpoint that reduces the excess energy use in this 'first cooling and then reheating' process. SAT setpoint can be further optimized based on the knowledge of outside air condition and zone thermal dynamics. For more details on the problem description and the assumptions refer to (Jiang et al., 2016).

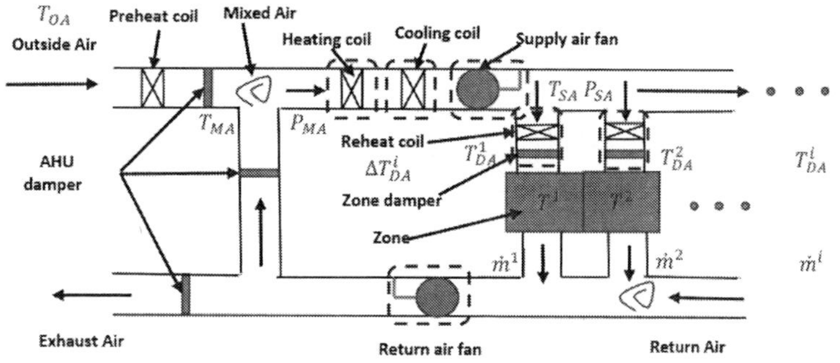

Figure 1: ***Typical layout of an AHU-VAV HVAC system***

3. DISTRIBUTED OPTIMIZATION USING GENERALIZED GOSSIP

This section describes a solution approach for the distributed building energy optimization problem formulated above. Most of the material in this section is taken from (Jiang et al., 2016). The approach uses a recently proposed generalized gossip-based algorithm. Theoretical contributions to the proposed framework can be found in (Jiang, Sarkar, & Kushal, 2015).

3.1 Background of Generalized Gossip protocol

Consider an undirected graph $\mathcal{G} = (\mathcal{V}, \mathcal{A})$ consisting of N agents, where $\mathcal{V} = \{1, 2, ..., N\}$ and $\mathcal{A} \subseteq \mathcal{V} \times \mathcal{V}$. If $(i, j) \in \mathcal{A}$, then agent i can communicate with agent j. Let the distributed building energy optimization problem be defined on the network as follows:

$$\begin{aligned} &\text{minimize} f(x) = \sum_{i=1}^{N} f^i(x) \\ &\text{subject to } x \in \mathbb{X} \end{aligned} \tag{1}$$

where $f^i : \mathbb{R}^M \longrightarrow \mathbb{R}$ are agent level objective functions (possibly convex or non-convex), $\mathbb{X}$ is a nonempty, closed, and compact subset of $\mathbb{R}^M$. x is a vector whose i^{th} component is represented by x^i.

The basic definitions (Boyd, Xiao, & Mutapcic, 2003; Johansson, Keviczky, Johansson, & H., December, 2008) and assumptions used in this paper are:

Definition 1: A vector $g \in \mathbb{R}^M$ is a subgradient of a convex function $f: \mathbb{R}^M \longrightarrow \mathbb{R}$ at a point $z \in \mathbb{R}^M$ if

$$f(y) \geq f(z) + g^T(y - z), \forall y \in \mathbb{R}^M \tag{2}$$

Definition 2: The set of all subgradients of a convex function of f at $z \in \mathbb{R}^M$ is called the subdifferential of f at z, and is denoted by $\partial f(z)$:

$$\partial f(z) = \{g \in \mathbb{R}^M | f(y) \geq f(z) + g^T(y - z), \forall y \in \mathbb{R}^M\} \tag{3}$$

Assumption 1 (Subgradient boundedness): There exists a scalar G for all $i = 1, ..., N$ such that

$$\|g^i(x)\|_2 \leq G, \forall g^i(x) \in \partial f^i(x), \forall x \in \mathbb{X} \tag{4}$$

Note, this assumption can be derived from the *Lipschitz continuity* relationship.

A vector notation of the update law for the proposed algorithm (derived from (Nedic & Ozdaglar, Janury, 2009) and (Sarkar, Mukherjee, & Ray, 2013)) for the optimization variable is as follows:

$$x(k+1) = (1 - \theta)\Pi(k)x(k) + \theta(x(k) - \nabla(k)). \tag{5}$$

where $\nabla(k)$ is the subgradient of f^i at $x^i(k)$ computed by agent i. $\Pi \in \mathbb{R}^N \times \mathbb{R}^N$ is the agent interaction matrix. θ is the user-defined control parameter.

3.2 Optimization algorithm overview

The proposed supervisory optimizer aims to determine optimal AHU supply air temperature based on information exchange among local zones. The crucial advantage of this framework is that local zones can use any local controllers and suitable modeling scheme. However, as long as they can compute subgradient for local cost function for energy optimization and achieving comfort, the supervisory control layer can run the generalized gossip protocol for global energy optimization.

In this context, each local zone needs modeling of thermal dynamics in order to compute subgradients for their local cost functions (refer (Jiang et al., 2016)). In this paper, although we use a data-driven ARX modeling scheme for zones, any other technique with same input-output conditions can be seamlessly accommodated. However, the local cost functions may be nonlinear and nonconvex and therefore, subgradients can be found using numerical differentiation. The modeling and optimization scheme also consider the local controllers for heating/cooling coil in AHU and dampers, reheat coils in VAV boxes. Currently, simple PI controllers are used for these local controllers (which is common for most of the HVAC equipment in commercial buildings (Bengea et al., 2015).

The basic workflow of the supervisory control framework is illustrated in Fig. 2. In this framework, an optimization interval is considered within which it is assumed that predictions from zone thermal modeling would be reliable, as well as the optimized AHU supply air temperature setpoint, would be appropriate. The supervisory decision-making process begins with subgradient-based optimization depending on the initial conditions. The building operation starts with optimized AHU supply air temperature setpoint. After expiry of the optimization interval, the supply air temperature setpoint is re-optimized based on current conditions. For validating the proposed algorithms scheme, a case study in

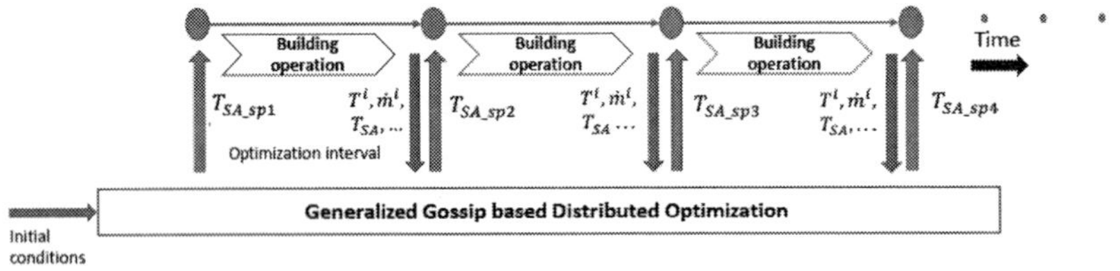

Figure 2: ***Workflow of the supervisory control framework***

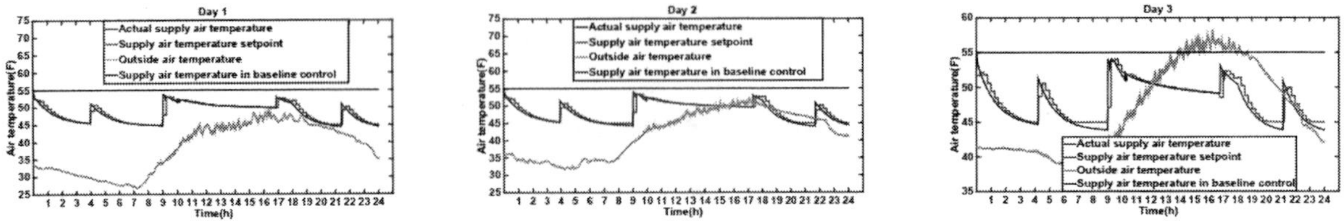

Figure 3: AHU supply air temperature under supervisory control and baseline control with different outside air temperatures.

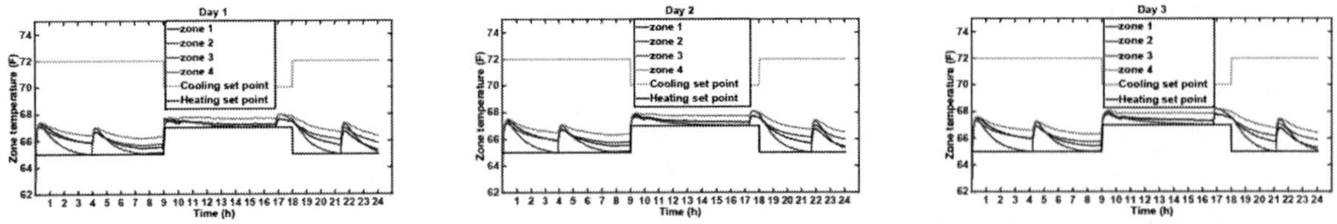

Figure 4: Zone temperature regulation during days with different outside air temperatures under supervisory control.

this paper is performed on a simulation (one AHU, four zones) based on the physical Energy Resource Station testbed in the Iowa Energy Center (Center, 2010). A typical baseline supply air temperature schedule is considered where the setpoint is kept constant at 55 °F. Moreover, the zone thermal modeling was performed using actual historical

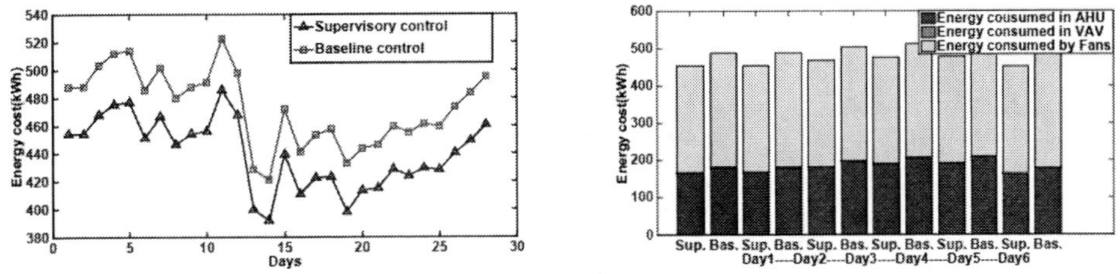

(a) (b)

Figure 5: (a) Energy cost in 28 test days in winter; and (b) Within 6 days energy consumed in AHU, VAV and by fans by supervisory control and baseline control.

data collected from the testbed during winter season. For validating the algorithm under different ambient conditions, the testing period in this case study is one month. As shown in Fig. 3, optimized supply air temperature setpoint varies under different ambient conditions and is different from the constant baseline condition. Zone temperature regulation performance for all 4 zones with different heating/cooling setpoints during unoccupied and occupied hours is shown in Fig. 4. The reason why local zone temperature control aims at approaching heating setpoint is because the testing was performed in the winter season in order to save energy. Figure 5(a) shows that in all 28 days, zone temperature by the proposed supervisory control scheme consumes less energy compared to baseline control. The energy consumption during 6 representative days by AHU heating/cooling coils, VAV reheat coils and AHU fans under baseline and supervisory control is shown in Fig. 5(b). The results demonstrate that in the supervisory framework cooling/heating energy consumed in AHU and the fan energy is reduced compared to baseline control, which validates the effectiveness of the proposed algorithm and control framework for HVAC systems.

4. VOLTTRON™ PLATFORM: OVERVIEW & IMPLEMENTATION

A new open source language-agnostic agent platform called VOLTTRON™ has been recently developed by PNNL (Lutes et al., 2014; Haack, Akyol, Carpenter, et al., 2013) for smart city applications with built-in security and resource management. Customized applications can be built on this platform for efficiently managing energy usage among appliances and devices, including HVAC systems, lighting, and electric vehicles. Key features include: (i) real-time data processing, (ii) automatic adjustment of data resolution and sampling frequency, (iii) data correlation from multiple domains, and (iv) support for distributed sensing, optimization and control applications. A software architecture for VOLTTRON™ based implementation of the proposed application is shown in Fig. 6(b). While data being used by applications live on the message bus, historical data can be stored in a cloud service and used for future on- or off-line uses. There are numerous database options of which, we consider SQLite as the VOLTTRON™ Historian which is very simple because of the automatic database creation during the launch. Furthermore, it provides resource guarantees for agents in the platform, including memory and processor utilization, authentication and authorization services, directory services for agent and resource location.

Fig. 6(a) shows the general layout of the ERS testbed. It consists of eight zones divided into two for each direction, East, South, West and North, respectively. The A test rooms (East-A,South-A, West-A, Interior-A) are served by air handling unit AHU-A, while B test rooms (East-B, South-B, West-B, Interior-B) are served by air handling unit AHU-B. The common and general service areas of the building are not typically used for research testing. All the test rooms within the ERS are intended to simulate a typical office space and each room has approximately 266 ft^2 of floor space. The building automation control system at the ERS is a modern commercial grade DDC system - Distech Controls' EC-Net *AX*™ Web-based Multi-protocol Building Automation and Energy Management Platform(Center, 2010). The system at our facility is a fully native BACnet system. The DDC systems is fully equipped with instrumentation and senosrs required for various building energy efficiency related research. There are total 1200 monitoring and control points with 600 of them collect data every minute. For the implementation purpose, we have considered A-test rooms served by air handling unit AHU-A.

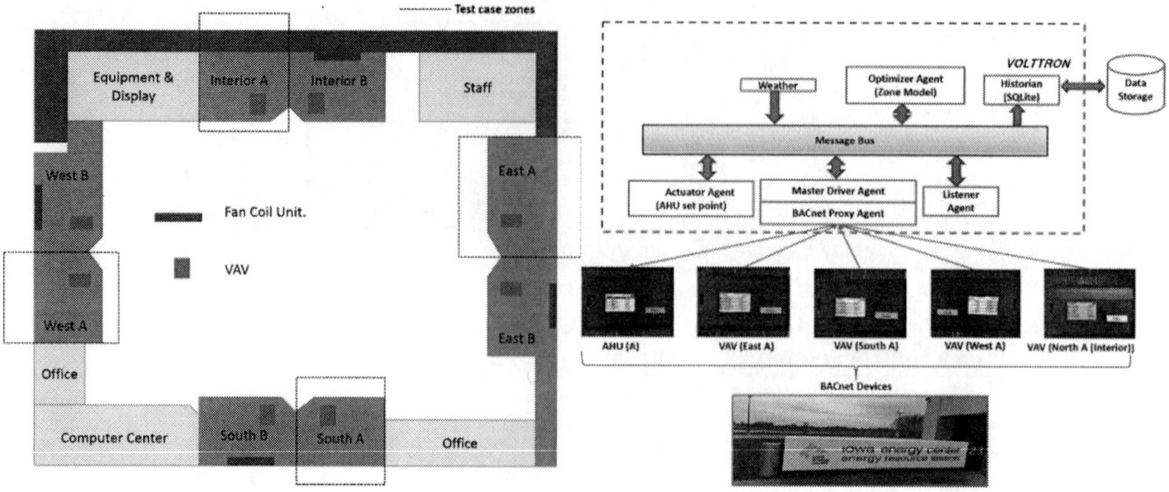

(a) (b)

Figure 6: (a) *ERS Test bed*; and (b) *VOLTTRON^{TM} based implementation of the proposed framework for supervisory control of Building HVAC system; the Message Bus provides a common platform for sharing information between applications, e.g., actuator scheduler, weather forecast service and optimization agent; historical data can be stored in a cloud service built on VOLTTRON^{TM} Historian (SQLite) or otherwise for future use.*

4.1 VOLTTRON^{TM} Agents: Description

The VOLTTRON^{TM} user guide provides a systematic procedure, templates, and examples for developing agents and to install the VOLTTRON^{TM} platform, by which one can develop simple agents for their own purpose. In the proposed setup we have built optimization agent on the top of VOLTTRON^{TM} which avails the existing benefits. The agents in VOLTTRON^{TM} are given as follows: (1) BACnet Proxy agent (2) Master Driver Agent (3) Weather Agent (3) Optimization agent (4) Weather agent (5) Listener Agent (6) Actuator Agent. The functionalities of individual agents are described below:

1. BACnet Proxy Agent: Device communication on a network happens through the virtual BACnet device and this agent is specifically used for communicating with the BACnet devices and managing the virtual BACnet device.
2. Master Driver Agent: Coordination among the drivers of devices is accomplished using this agent.
3. Weather agent: This agent interacts with remote applications such as online weather services like Weather Underground (www.wunderground.com). The weather information is used by the Zone Model to calculate the temperatures needed for computing the subgradients.
4. Optimization Agent: This agent serves the purpose of predicting the temperatures given the weather information and the parameters of the zone model identified using the data from the ERS test bed. The algorithm described in (Jiang et al., 2016) is implemented in Python to calculate the subgradients based on which the optimal AHU supply air temperature (SAT) is calculated.
5. Listener Agent: It logs all the activity on the message bus and is useful in testing the agent functionality in terms of publishing the required data. It can be used as a starting point for developing other agents.
6. Actuator Agent: This agent has the capability to assert control over the devices by accepting commands from the agents and scheduling times to issue commands to devices.

Message Bus in VOLTTRON^{TM} is a common place where data from all the agents and the devices are collected. All agents use publish/subscribe mechanism to communicate with the other agents. All the data published onto the message bus either by the devices or the agents will be collected by the historian and stores for retrieval or analysis purpose such as plotting etc.

4.2 Implementation

A detailed description of the steps from installing to configuring devices and writing agents can be found in (Lutes et al., 2014). We briefly provide the implementation steps in our setting below:

- Install the VOLTTRON^{TM} by following the steps given in user guide (Lutes et al., 2014) and move to the developer branch. ECLIPSE an integrated development environment (IDE) is used as a tool for agent development.
- The machine running the VOLTTRON^{TM} is physically connected through Ethernet network to the DISTECH network manager (router) to which all the individual devices of interest are connected. For the sake of convenience, we have shown only the five devices (1AHU and 4 VAVs) in Fig. (6(b)).
- Since the Host machine is Windows, we installed Virtual Machine (VM) to run Linux and VOLTTRON^{TM} and configured VM to use a bridged adapter to avoid problems with different IP address.
- The immediate step after installing VOLTTRON^{TM} is to find and configure the BACnet devices. The scripts such as bacnet_scan and grab_bacnet_config files which are there in the VOLTTRON^{TM} repository can be used to detect the devices and generates a registry configuration file in CSV format for the BACnet driver which consists of a list of point names. This points list can be changed based on user interest.
- Configure the BACnet Proxy Agent by filling the address field in the configuration file with the IP address of Linux machine and specify the subnet mask. Since the BACnet network is on a default port (47808) there is no need to mention the port number.
- Setup the Master Driver Agent which consists of list device configuration files. For each device, we need to create driver configuration file, CSV file and the entry in the Master Driver Agent configuration file.
- Import the VOLTTRON^{TM} into ECLIPSE and launch VOLTTRON^{TM}, BACnet Proxy Agent, Master Driver Agent and Listener Agent using the steps given in (Lutes et al., 2014). The output console of Listener Agent displays the device topics and messages.
- Develop the Optimization Agent by subscribing to the data from the message bus and implementing the algorithm scheme provided in section (3).
- Launching the Optimization Agent similar to other agents and it publishes the Supply Air Temperature (SAT) setpoint to message bus. Launching Historian agent stores the data to the SQLite database.
- Launch the Actuator Agent and schedule the timing to override the SAT setpoint of the AHU device.

In this context, we have developed an Optimization agent and a screenshot showing the supply air temperature in the output console is shown in Fig. 7.

5. CONCLUSIONS

This paper presents the implementation aspects of the Generalized Gossip-based distributed optimization for building energy systems using VOLTTRON^{TM}. Optimization agent has been developed in VOLTTRON^{TM} to realize the proposed framework which generates the optimal supply air temperature setpoint. The future work is aimed at increasing the scope of the optimization framework to entire air-side as well as water-side HVAC system and comparison of the energy cost using the proposed framework with the existing framework at ERS testbed.

REFERENCES

Akyol, B., Haack, J., Carpenter, B., Ciraci, S., Vlachopoulou, M., & Tews, C. (2012). Volttron: An agent execution platform for the electric power system. In *Third international workshop on agent technologies for energy systems valencia, spain.*

Bengea, S. C., Li, P., Sarkar, S., Vichik, S., Adetola, V., Kang, K., ... Kelman, A. D. (2015). Fault-tolerant optimal control of a building hvac system. *Science and Technology for the Built Environment*, *21*(6), 734–751.

Boyd, S., Xiao, L., & Mutapcic, A. (2003). Subgradient methods.

Cai, J., Braun, J. E., Kim, D., & Hu, J. (2016). A multi-agent control based demand response strategy for multi-zone buildings. In *American control conf.* (Accepted)

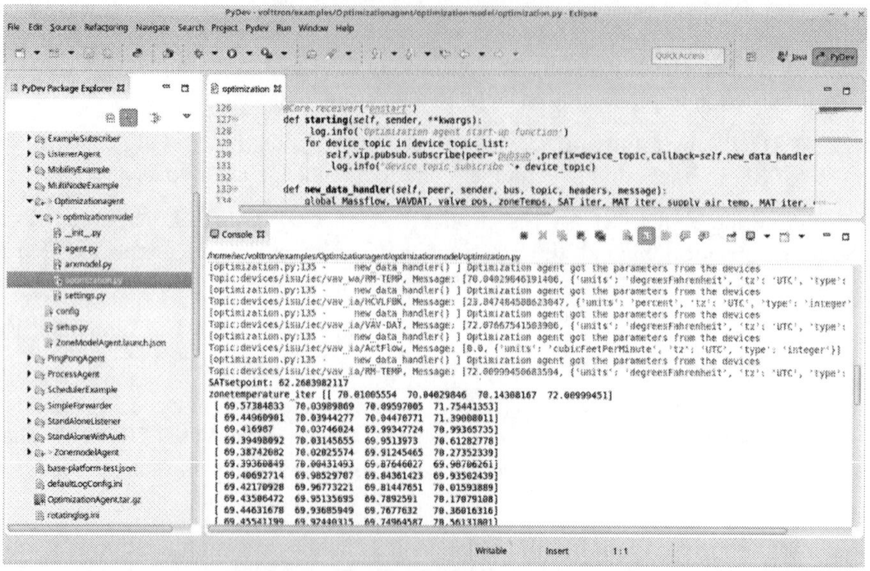

Figure 7: ***Opitmization Agent implementation in VOLTTRON***^{TM}

Cai, J., Kim, D., Putta, V. K., Braun, J. E., & Hu, J. (2015). Multi-agent control for centralized air conditioning systems serving multi-zone buildings. In *American control conference (acc), 2015* (pp. 986–993).

Center, I. E. (2010). Ers technical description. *Iowa Energy Center. Ankeny, I A.*.

Haack, J., Akyol, B., Carpenter, B., Tews, C., & Foglesong, L. (2013). Volttron: an agent platform for the smart grid. In *Proceedings of the 2013 international conference on autonomous agents and multi-agent systems* (pp. 1367–1368).

Haack, J., Akyol, B., Tenney, N., Carpenter, B., Pratt, R., & Carroll, T. (2013). Volttron: An agent platform for integrating electric vehicles and smart grid. In *Connected vehicles and expo (iccve), 2013 international conference on* (pp. 81–86).

Jiang, Z., Chinde, V., Kohl, A., Sarkar, S., & Kelkar, A. (2016). Scalable supervisory control of building energy systems using generalized gossip. In *American control conference (acc), 2016 to appear.*

Jiang, Z., Sarkar, S., & Kushal, M. (2015). On distributed optimization using generalized gossip. *in Proceedings of 54th Conference on Decision and Control, IEEE, Osaka.*

Johansson, B., Keviczky, T., Johansson, M., & H., J. K. (December, 2008). Subgradient methods and consensus algorithms for solving convex optimization problems. *in Proceedings of the 47th IEEE Conference on Decision and Control, Caucun, Mexico*, 4185-4190.

Kantamneni, A., Brown, L. E., Parker, G., & Weaver, W. W. (2015). Survey of multi-agent systems for microgrid control. *Engineering Applications of Artificial Intelligence, 45*, 192–203.

Khamphanchai, W., Pipattanasomporn, M., Kuzlu, M., & Rahman, S. (2015). An agent-based open source platform for building energy management. In *Smart grid technologies-asia (isgt asia), 2015 ieee innovative* (pp. 1–6).

Khamphanchai, W., Saha, A., Rathinavel, K., Kuzlu, M., Pipattanasomporn, M., Rahman, S., ... Haack, J. (2014). Conceptual architecture of building energy management open source software (bemoss). In *Innovative smart grid technologies conference europe (isgt-europe), 2014 ieee pes* (pp. 1–6).

Lutes, R. G., Haack, J., Katipamula, S., Monson, K., Akyol, B., Carpenter, B., & Tenney, N. (2014). *Volttron: User guide.* Report PNNL-23182. Richland, WA: Pacific Northwest National Laboratory.

Nedic, A., & Ozdaglar, A. (Janury, 2009). Distributed subgradient methods for multi-agent optimization. *Automatic Control, IEEE Transactions on, 54*, 48-61.

Sarkar, S., Mukherjee, K., & Ray, A. (2013). Distributed decision propagation in proximity networks. *International Journal of Control, 86*(6), 1118-1130.

Weiss, G. (1999). *Multiagent systems: a modern approach to distributed artificial intelligence.* MIT press.

ACKNOWLEDGMENT

The authors would like to thank Iowa Energy Center for supporting this work-Opportunity Grant No.OG-15-005.

Integrated Control of RTUs and Refrigeration Equipment in Convenience Stores

Donghun Kim[1]*, James E. Braun[1]
[1]Ray W. Herrick Laboratories, School of Mechanical Engineering, Purdue University
West Lafayette, IN, U.S.

kim1077@purdue.edu
jbraun@purdue.edu

* Corresponding Author

ABSTRACT

Convenience stores with refrigerated display cases are ubiquitous in much of the developed world. For such buildings it is common to utilize multiple rooftop units (RTU) to serve the retail area and multiple coolers or freezers to provide cooling for merchandise. The overall objective of this paper is to develop and assess an integrated control approach for typical convenience store applications using a virtual testbed. The integrated controls will coordinate the operation of multiple RTUs and refrigeration systems for the purposes of reducing energy consumption and electrical peak demand. For practical implementation, a model-based predictive control algorithm is developed and evaluated that requires no additional sensors. Convenience stores are everywhere and opportunities for optimizing controls would be significant.

1. INTRODUCTION

Retail stores with refrigerated display cases are ubiquitous in much of the developed world. An initial analysis of data from a (USA) retail directory indicates that there is approximately 3 billion square feet of buildings that incorporate both air conditioning and refrigeration equipment (big box super-centers, supermarkets, convenience stores, large drug stores, etc.) with approximate electrical energy consumption of 1283 trillion BTU per year (Michaels, 2003). This represents about 10% of the total electrical energy used in commercial buildings across the USA. A conventional control approach for these buildings relies on local feedback control, where each unit is cycled on and off using its own thermostat. Because a thermostat operates regardless of the overall building's behavior, the conventional control approach could result in unnecessary energy use and high electrical peak demand via poor coordination among the units. Therefore, significant energy savings and/or electrical peak demand reduction can be achieved by optimally coordinating both air conditioning and refrigeration equipment.

The goal of this paper is to develop and evaluate a practical control algorithm for on/off staging of multiple air conditioning and refrigeration units. A control approach, termed the Plug-and-Play (PnP) RTU Coordinator (Kim et. al. 2015a) which was previously developed to coordinate multiple rooftop units in an optimized manner, was generalized and extended for energy and demand savings in retail stores that include both air conditioning and refrigeration. In Section 2, the control algorithm is described. Section 3 describes a simulation testbed applied to a case study building. Results of controller evaluations for the store over the summer are provided in Section 4.

2. CONTROL APPROACH FOR RETAIL STORES

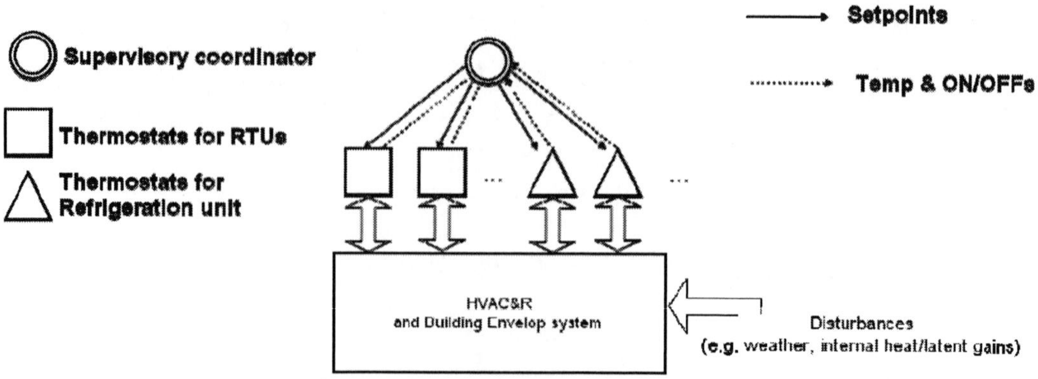

Figure 1: Unit coordinator elements and I/O for retail store applications

Since a conventional thermostat control approach does not consider overall building performance, it is natural to design a controller targeting reduction of energy consumption and peak demand in a centralized manner. Previously, a control algorithm coordinating an open space building served by multiple rooftop units (RTU) was developed and demonstrated. The control approach, termed the plug-and-play (PnP) RTU Coordinator (Kim et. al. 2015a), was designed to minimize the time required to configure the control strategy in order to enable a more cost effective control implementation for small/medium commercial building applications in which buildings are served by multiple RTUs. The control solution is not site-specific and provides reduced energy consumption and peak demand with low sensor requirements. Therefore, it is a strong candidate to provide energy/demand savings for retail store applications.

Due to its unique characteristic that it relies only on thermostat signals, i.e. temperatures and ON/OFF stage commands, the coordination algorithm can be directly applied to buildings equipped with both air handling and refrigeration units as depicted in Fig. 1. The RTU Coordinator has been slightly modified in order to apply to retail store buildings having refrigeration equipment. This is because the RTU Coordinator was designed for open space building applications while many retail store buildings, e.g. convenience stores, have separated zones served by different units. The algorithm details are explained as follows.

To formulate the control problem for optimal coordination between air conditioning and refrigeration units, let $p \in \mathcal{N}$ (a natural number) be the number of thermostats or equivalently the number of air conditioning and refrigeration units. The measured outputs are the thermostat temperatures, denoted as $y(k) \in \mathcal{R}^p$. The manipulated variables are the unit stages, denoted as $u(k) \in \mathcal{N}^p$. The control problem at a current time step k is:

$$\min_{u(j)\in\mathcal{N}^p,\delta\in\mathcal{R}^+,\Gamma_l\in\mathcal{R}^{p+},\Gamma_u\in\mathcal{R}^{p+}} \sum_{j=1}^{N_p}\sum_{i=1}^{p} P_i u_i(k+j) + d\cdot\delta + c_l^T\Gamma_l + c_u^T\Gamma_u \qquad (1)$$

$$T_l - \Gamma_l \le E(y(k+N_p)\,|\,\mathcal{G}_k) \le T_u + \Gamma_u$$

$$\sum_{i=1}^{p} P_i u_i(k+j) \le \delta \quad (\forall j \in \{1,\cdots,N_p\})$$

where P_i is the rated power for ith unit, and hence the first term in the control objective represents a scaled energy consumption over a predicted horizon, N_p. $E(y(k+N_p)\,|\,\mathcal{G}_k)$ is the optimal N_p-step prediction given data $\mathcal{G}_k = \{y(k-1), y(k-2),\cdots,u(k+N_p-1),u(k+N_p),\cdots,u(k-1),u(k-2),\cdots\}$. T_u, T_l ($\in \mathcal{R}^p$) are temperature upper and lower bounds, respectively. c_l, c_u $(\in \mathcal{R}^{p+})$ and d $(\in \mathcal{R}^+)$ are weights on optimization variables of Γ_l, Γ_u $(\in \mathcal{R}^{p+})$ and δ $(\in \mathcal{R}^+)$. From the last constraint, it can be seen that δ is an upper bound on each instantaneous electric demand over the prediction horizon. Therefore minimizing δ will naturally lower the electric peak demand over a prediction horizon. In addition, note that δ is dependent on all sequences of stages of p -units. Therefore, it is clear that the control problem supervises both air conditioning and refrigeration units. Γ_l and Γ_u can be seen as comfort violations from the first constraint of Eqn. (1). These variables are required to guarantee the existence of a solution for the optimization problem.

Note also that we only want to regulate the N_p -step ahead predicted temperatures within a temperature bound and not each of the predicted temperatures for less than the N_p steps. This is acceptable because our prediction horizon, N_p, is relatively short, e.g. 30 min to 1 hour. The N_p -step temperature regulation reduces the large number of inequality constraints that would be necessary if all of the predicted temperatures were constrained. The control problem of the RTU Coordinator can be obtained by eliminating δ, Γ_l and Γ_u in the objective function and constraints in Eqn. (1) (See Kim. et. al. 2015a). Therefore, the unit coordinator can be seen as a generalized version of the RTU Coordinator.

3. DEVELOPMENT OF CONVENIENCE STORE BUILDING MODEL

Figure 2: Google satellite view of the store

A simulation study was useful in evaluating the overall benefits of the unit coordinator compared with conventional control over a long period of time. This is because it was not possible to perform long-term side-by-side testing of the unit coordinator and conventional control for this site. A detailed model was developed and is briefly described in this section. This model was used as a virtual testbed to evaluate performance of the unit coordinator.

A convenience store (floor area 223.6 m^2, satellite view in Fig. 2) located in Acworth GA was identified as a case study building in collaboration with Oak Ridge National Laboratory and Emerson Climate Technologies. The store has four zones as shown in Fig. 3. The blue zone is the main customer service area served by two equivalent 3-ton rooftop units (RTU 1 and RTU 2). The green zone is a medium temperature walk-in refrigerated chamber served by two identical cooler units (< 2 ton) for beverages. The red zone is a low temperature (freezer) case having one unit (< 1 ton). The three refrigeration units have their own refrigeration equipment consisting of an outdoor air-cooled condensing unit and indoor evaporator coils. The pink zone is storage/office space served by one RTU (RTU 3).

Each unit is controlled by a separate thermostat shown as circles in the Fig. 3 (the thermostat locations for the storage and freezer are not shown).

A detailed model for the building was developed using a reduced-order coupled computational fluid dynamics (CFD) model (Kim et. al. 2015b). The model captures spatial temperature variations for open space areas, e.g. the main service and cooler zones in this building, in a computational efficient manner that allows long-term simulations. For a more detailed description, refer to Kim et. al. 2016. The model has been validated using measurements from the site. The occupancy profile was estimated by measuring percent door openings and was used to calculate occupant and lighting loads for occupied zones based on assumptions of maximum occupancy and nominal heat rates (ASHRAE, 2001).#

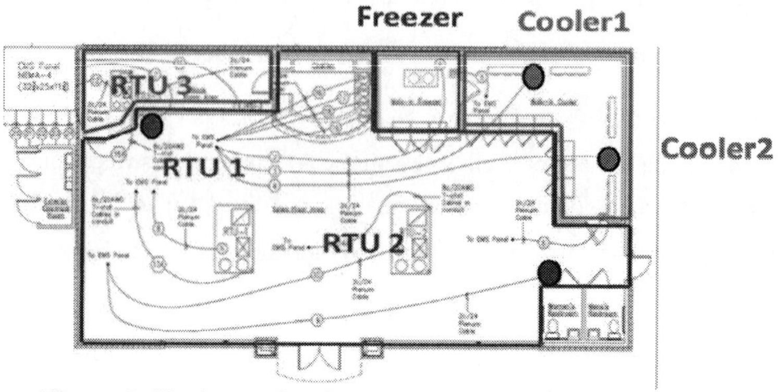

Figure 3: Zoning and unit numbering for a convenience store

4. RESULTS AND DISCUSSION

The simulation model described in Section 3 was executed in the Matlab Simulink environment with a stiff ordinary differential equation numerical solver, *ode15s*, that provides variable simulation time steps. The set-points for the conventional controls during an occupied period (6:00AM to 12:00AM) were set as (22°C, 4°C, -15°C) for RTUs, coolers and freezer, respectively. Deadbands of 1°C were used for all thermostats.

The unit coordinator controls predicted temperatures between 20 to 23°C for RTUs, between 2 to 5°C for coolers, and between -17 to -13°C for the freezer, respectively. An ARX(3,2) model structure with two weeks of historical I/O data were used to establish the controller model for the unit coordinator. A 30-min look ahead horizon was used for the controller in this study. For the penalty terms of δ , Γ_l and Γ_u , 100, 1000 and 1000 were assigned in this simulation study.

A 5-minute sampling time was applied for both the conventional and unit coordinator controllers. The sampling time for the conventional control is to account for anti-cycling time for units to avoid too frequent cycling. A four-month simulation, from May to Aug, was performed for each controller with TMY3 data.

In a setting with multiple RTU's with different efficiency ratings, it's possible to save energy by increased use of the more efficient units. Since the two RTUs for the main service area are identical and two coolers are also identical, the primary expected benefit of coordination for this case is demand reduction. Fig. 4-5 show sample comparisons between the conventional and unit coordination controls from the four-month simulation results. The left hand side of the figures indicates the thermostat temperatures and the right hand side indicates the unit staging for the controls. For the conventional case, all units cycle significantly during the day due to part-load conditions. There are some situations where 5 units are operating at the same time resulting in high peak power consumption. An example can be seen at around 2:00 PM in Fig. 4. The total HVAC power at this time is around 15 kW as shown in the left hand side of Fig. 6.

However, the Unit Coordinator algorithm predicts 30-min ahead and supervises unit stages that can bring the zone temperatures within the comfort bounds with the minimum peak power for the 30-min ahead prediction period. This coordination can be seen in Fig. 5. When there is a need for operating several air conditioning units at the same time, e.g. marked as red dashed lines in Fig. 5 where both RTU1 and RTU2 were turned on, the coordinator adjusts sequences of stages of refrigeration equipment in order to reduce electric peak demand. This reduces the peak demand from 15 kW to 9 kW as shown in Fig. 6. Of course, when the cooling load is close to full capacity, there is no opportunity for demand reduction. However, in practice, units are selected to meet a peak load that may occur for only a few days a year and units work under a part-load condition for most times. This means there is significant peak demand reduction potential in most cases with greater potential.

Due to balancing between reduction of peak demand and comfort violations (difference in thermostat temperatures from desired setpoints) of the unit coordinator as shown in Eqn. (1), comfort violations for the unit coordinator are slightly higher (around 1°C) than for the conventional thermostat algorithm. However, the magnitudes of the violations are in an acceptable range as shown in Fig. 5, since the unit coordinator also tends to minimize the comfort violations.

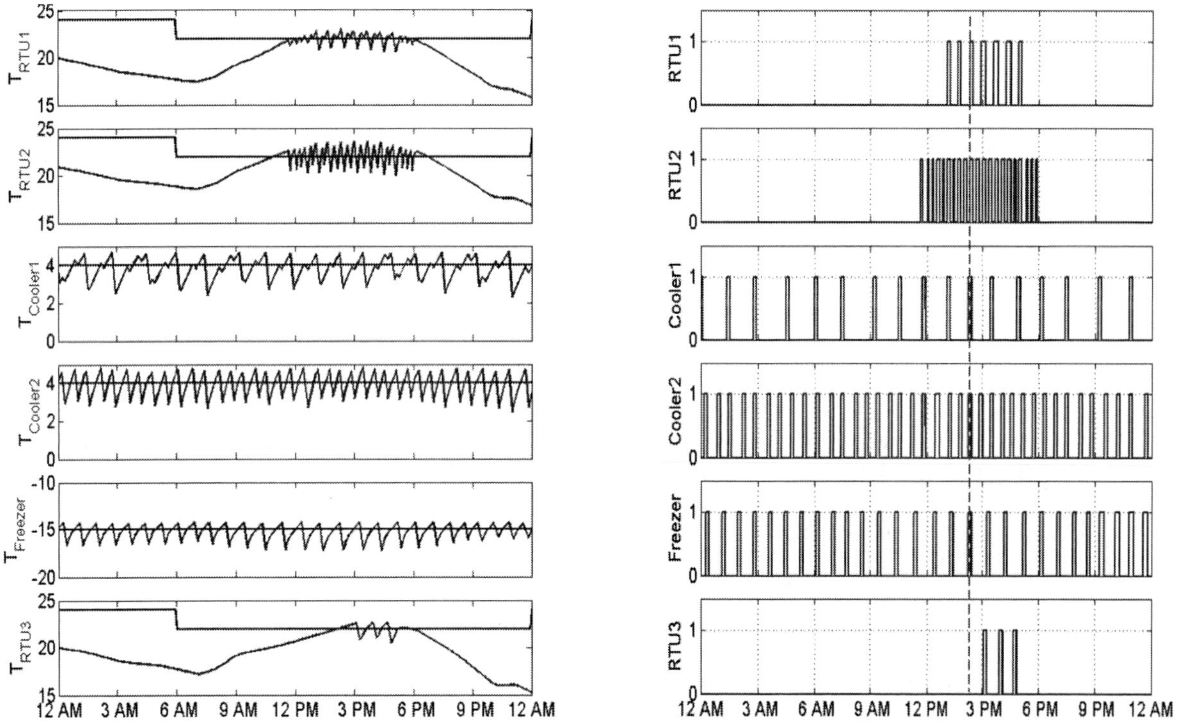

Figure 4: Representative RTU cycling and associated thermostat responses for conventional control at a convenience store

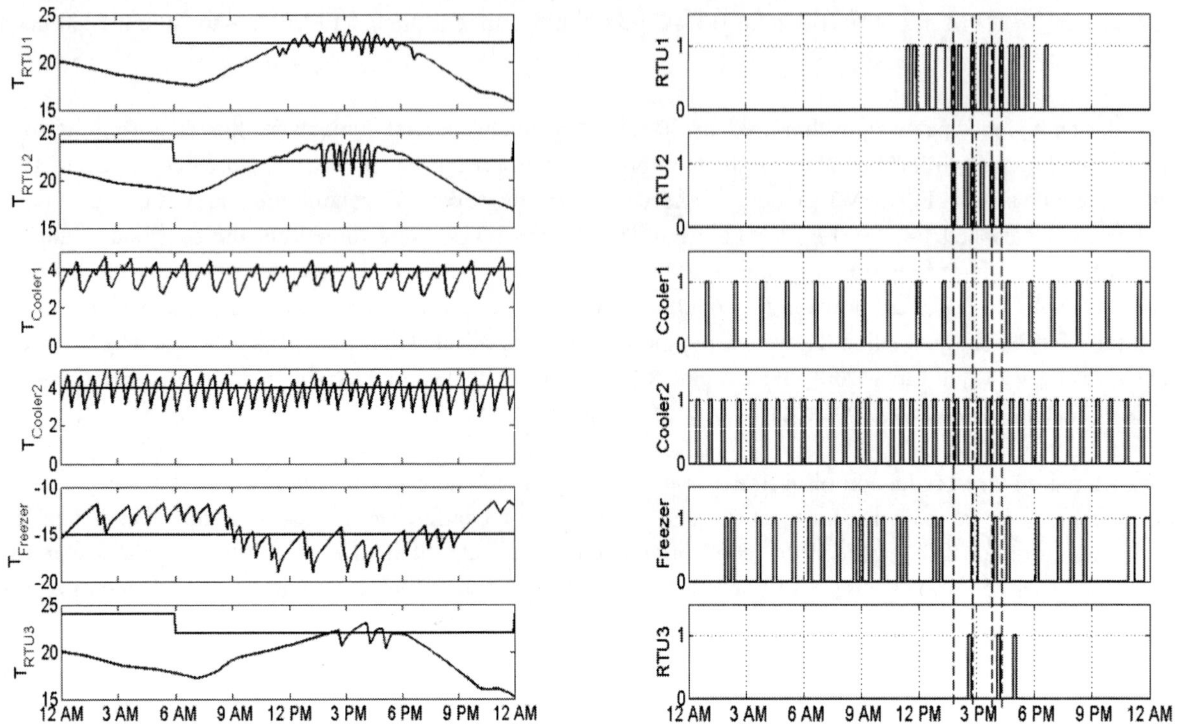

Figure 5: Representative unit cycling and associated thermostat responses for unit coordinator at a convenience store

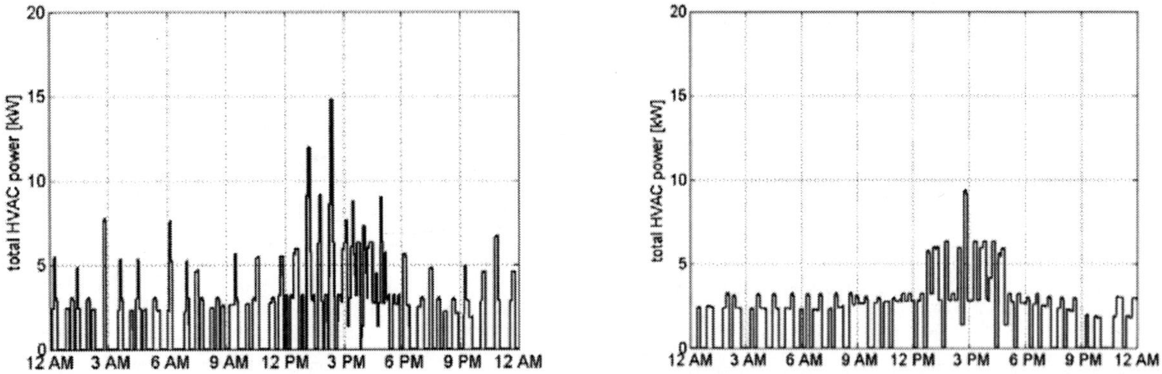

Figure 6: Profiles of total HVAC&R power for conventional (left) and unit coordinator (right) at a convenience store

Comparisons of energy usage and peak demand between the conventional and unit coordinator for the summer season are summarized in Table 1. Peak demands were calculated after performing a 15-min moving average on instantaneous power profiles for the simulation period. As expected, the energy savings are small due to lack of diversity in unit efficiencies. The primary energy savings are due to reduced cycling losses. However, the benefit of peak demand reduction is significant. The demand savings for the unit coordinator at the convenience store for the summer season are about <u>18 %</u>.

Table 1. Cooling Season Comparisons for a convenience store

	May		June		July		August		4-Month Totals	
	Conv	PnP	Conv	PnP	Conv	PnP	Conv	PnP	Conv	PnP
Energy (MWh)	2.3	2.2	3.5	3.2	3.9	3.5	3.9	3.6	13.6	12.5
Peak Power (kW)	13.0	11.8	14.8	11.7	14.8	11.7	14.3	11.7	56.9	46.9
Energy Savings (%)		4.3		8.6		10.2		10.2		8.1
Peak Demand Cost Reduction (%)		9.2		20.9		20.9		18.2		17.6

5. CONCLUSIONS

A practical control algorithm for coordinating both air handling and refrigeration equipment was developed and evaluated using a simulation testbed for a convenience store in terms of energy and demand savings. The simulations allowed evaluations of savings for the unit coordinator compared to conventional control over an entire cooling season. Based on these evaluations, typical cooling season energy and peak demand savings are expected to be about 8 % and 18 %, respectively. The controller was designed to minimize implementation costs in that it does not require additional sensors and is self-learning. It can be generally applied to retail stores served by multiple air handling and refrigeration units.

NOMENCLATURE

RTU	rooftop unit
y	Thermostat temperature
u	unit stage
p	the number of units
k	current time step
N_p	prediction horizon
P_i	rated power for i^{th} unit
T_u	temperature upper bound
T_l	temperature lower bound

REFERENCES

A. S. H. R. A. E. (2001). Fundamentals. *American Society of Heating, Refrigerating and Air Conditioning Engineers, Atlanta, 111.*

Kim, D., Braun, J. E., Cai, J., & Fugate, D. L. (2015a). Development and experimental demonstration of a plug-and-play multiple RTU coordination control algorithm for small/medium commercial buildings. *Energy and Buildings, 107*, 279-293.

Kim, D., Braun, J. E., Cliff, E. M., & Borggaard, J. T. (2015b). Development, validation and application of a coupled reduced-order CFD model for building control applications. *Building and Environment, 93*, 97-111.

Kim, D., Ramaraj, S., Braun, J.E., Ben Ayed, S., Borggaard, J.T., & Cliff, E. M. (2016). Milestone report Project 2.6 – Integrated Controls for Retail Stores with Refrigerated Cases, *Consortium for Building Energy Innovation*

Michaels, J., & Leckey, T. (2003). Commercial buildings energy consumption survey. *URL http://www. eia. doe. gov/emeu/cbecs.*

ACKNOWLEDGEMENT

The authors wish to acknowledge Dr. Eugene. M. Cliff, Dr. Jeff T. Borggaard and Dr. Samah B. Ayed from Virginia Tech, Sugirdhalakshmi Ramaraj from Purdue, and David L. Fugate from Oak Ridge National Laboratory. This work was supported by the Department of Energy through the Consortium for Building Energy Innovation (CBEI).

Model-based Optimal Control of a Building HVAC System

Fuxin Niu[1]*, Zheng O'Neill[1], Xiaohui Zhou[2], Defeng Qian[1]

[1]The University of Alabama,
Tuscaloosa, AL, USA
(Phone: 402-547-8907, Email: fniu@crimson.ua.edu)

[2]Iowa Energy Center,
Ankeny, IA, USA

* Corresponding Author

ABSTRACT

A model-based optimal control strategy is explored to minimize a building Heating, Ventilation and Air-Conditioning (HVAC) energy consumption in the heating mode. Energy performance models for individual component, including the Air Handler Unit (AHU) heating coil, the supply fan and the Variable Air Volume (VAV) terminal box reheat coil, are built through a data-driven method. Thermal response of the room air is established using a non-linear regression based identification approach. The AHU supply air temperature and the room air temperature are considered as the constrained condition. A platform of AMPL (A Modeling Language for Mathematical Programming) is used to for mathematical modeling and links to the optimization solvers. A case study using the data collected from the Energy Resource Station at the Iowa Energy Center was conducted using the proposed strategy. The comparison between the baseline and the simulation-based case with the proposed model-based optimal control indicated that a 22.1% savings potential of total energy consumption could be achieved.

1. INTRODUCTION

In the U.S., A Variable Air Volume (VAV) system is one of most commonly used air system for multiple-zone commercial buildings due to its capability to meet the varying heating and cooling loads of different building thermal zones. One of the key components of a VAV system is the terminal VAV box. There is an air damper and a reheat coil in the box. How to effectively and efficiently control the HVAC system with the VAV box plays a significant role to reduce energy consumption and maintain acceptable indoor environment in buildings.

Li et al. (2015) implemented an optimization-based model predictive control (MPC) algorithm for building HVAC systems and demonstrated its benefits through building energy consumption reductions as well as thermal comfort improvements. A MPC simulation framework was first presented with its associated performance benchmarked. The experimental results from the same building located at the Philadelphia Navy Yard were then presented. For the simulation study, it was estimated that the MPC could reduce the total electrical energy consumption by around 17.5%. For the subsequently experimental demonstration, the performance improvement of the MPC algorithm was estimated relative to baseline days with similar outdoor air temperature patterns during the cooling and shoulder seasons, and it was concluded that the MPC strategy reduced the total electrical energy consumption by more than 20% while improving thermal comfort in terms of zone air temperature. Cho and Liu (2008) developed and field-implemented optimal terminal VAV box control algorithms. The thermal conditions and energy consumption were compared between conventional and improved control algorithms using the measured data. The results showed that optimal VAV box control algorithms can stably maintain the room air temperature and reduce energy consumption compared with conventional control algorithms. Lu et al. (2005) presented the global optimization technologies for the HVAC systems. The objective function of a global optimization and constraints were developed based on physical models of the major HVAC components. A modified genetic algorithm was then used to solve the global optimization problem for minimizing the overall HVAC system energy consumption. Simulation studies for a centralized HVAC plant using the proposed optimal method showed that the optimization method improved the system performance significantly compared with traditional control strategies. Yang and Wang (2012) proposed an optimal control strategy to control the HVAC system for maintaining building's indoor environment with a high-energy efficiency. The control strategy utilized a swarm intelligence to determine the optimal amount of energy dispatched to individual equipment in the HVAC system. A case study was conducted to simulate the real time control process in a specified building environment. Compared to the Constant Air Volume (CAV) system and the non-optimized VAV system, the proposed optimal control strategy was capable of saving more energy in building operations under the same environmental condition. Kusiak et al. (2010) presented a data-driven approach to derive energy performance models. Data-mining algorithms were employed to select significant parameters and estimate individual HVAC component energy consumption. To minimize the total energy consumption, a single-objective optimization model was formulated and solved by the particle swarm optimization algorithm. The particle swarm optimization algorithm searches the near optimal solutions of the supply air temperature and static pressure setpoints in an air handling unit (AHU). The optimization results demonstrated a 7.66% savings. Fong et al. (2006) studied a simulation based Evolutionary Programming (EP) coupling approach, which incorporated the component-based simulation and EP optimization. The HVAC system under the investigation had an air-side system that contributes significantly to the overall energy consumption. From the optimization results, the proposed technique worked well in providing the optimum combination of the chilled water supply temperature and AHU supply air temperature for a cost-effective energy management throughout an entire year. Nassif et al. (2004) optimized the set points using a two-objective genetic algorithm for a supervisory control strategy. The optimization process was applied to an existing HVAC system using a detailed physical VAV model. The energy demand from the simulation case with the optimized control strategy is 19.5% less than that from the actual building with the non-optimized control strategy. The application of the two-objective optimization algorithm could help a better control in terms of minimal daily energy use and maximal daily thermal comfort in the building as compared to the one-objective optimization approach.

Modeling effort and required computation resource are bottlenecks for on-line implementations of MPC at a large scale. Other optimization based control methods utilized customized physics-based models that are complex and not scalable. The HVAC system operation may get the similar benefits by adapting optimization based control from the information captured in historical operation data. The goal of this preliminary study is focusing on how to utilize the short term historical data to get the current time optimal operation point. In this paper, model-based optimal control for a building HVAC system is proposed to minimize the total HVAC system energy consumption with constraints of occupants' thermal comfort. The individual component energy consumption, including an AHU, a fan and VAV boxes, was formulated. The fan energy consumption model was derived through a data-driven approach based on a polynomial regression algorithm. The optimal control approach was tested using measurements from the Energy Resource Station (ERS) Building at the Iowa Energy Center. We will first briefly introduce the ERS building, then the models for individual HVAC component and zone thermal dynamics, and optimization approach. Finally, we will talk about the savings potentials that were demonstrated through the simulation based study, limitations and future work.

2. BUILDING INTRODUCTION

The data used for data-driven energy performance models was collected at the ERS Building of the Iowa Energy Center in Ankeny, Iowa. The Iowa Energy Center established the ERS Building for the purposes of examining various energy-efficiency measures and demonstrating innovative HVAC concepts. The facility is divided into a general area and two test areas (A and B). Each test area includes four thermal zones served by one AHU. The basic description of the ERS facility is shown in Figure 1. Minimizing the total energy consumption from the HVAC system B is the goal of this case study. The HVAC system B is comprised of a central AHU and an overhead ducted air distribution that terminates with four room-level VAV terminal boxes. Each test room is equipped with a pressure-independent, single-duct VAV box. Each VAV box has both a hydronic and an electric reheat coil (Note: only hydronic reheat coils are used in this study). VAV boxes with reheat coils were traditionally controlled using the single maximum control logic. The supply airflow rate setpoint is reset from the zone maximum airflow rate setpoint when the space is at a full cooling stage proportionally down the zone minimum airflow rate when no cooling is required. This minimum airflow rate is maintained as the space temperature falls through the dead band into the heating mode. The hot water valve then modulates to maintain the space at the heating setpoint until it is fully open. The measured points included outdoor air temperature, return air temperature, AHU mixed air temperature, AHU supply air temperature and volumetric air flow rate, individual VAV box supply air temperature and volumetric air flow rate, individual room air temperature, AHU supply fan power consumption, inlet and outlet water temperatures and mass flow rate of AHU heating coil, and inlet and outlet water temperatures and mass flow rates of VAV box reheat coils. The data with one minute sampling frequency from Oct 30th 2013 to Jan 30th 2014 was used in this case study. The data are aggregated to 10 minutes interval for the training and testing of proposed data-driven models. Data sets used in this case study were collected from regular operation modes. No special function tests with excitations were performed.

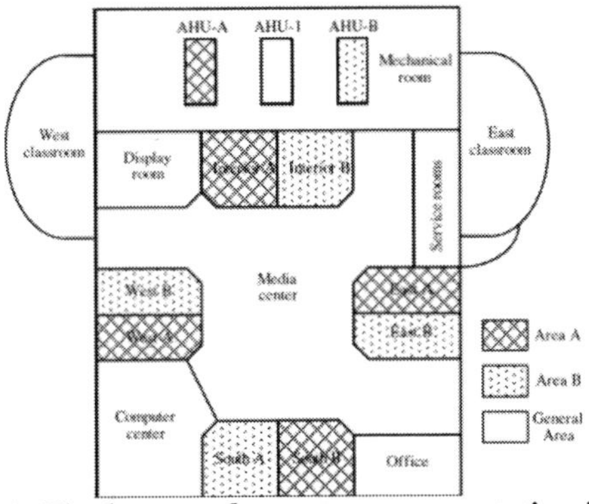

Figure 1: The layout of energy resource station in Iowa

3. MODELING AND OPTIMIZATION

The optimization objective is to minimize the total energy consumption of the HVAC system including AHU, fan and VAV box in the heating mode over the three months. First, data-driven energy performance model for the individual energy consumption component was formulated. The thermal comfort constraints for room air temperatures and AHU supply air temperature were included in the optimization problem formulation. The optimization problem was solved using a platform of AMPL (A Modeling Language for Mathematical Programming) (AMPL 2016). Control variables include AHU supply air temperature, VAV box supply air flow rates and supply air temperatures.

3.1 Data-driven Energy Performance Model

AHU model

The AHU mixed air temperature is formulated as follow. The return air temperature is calculated using the weighted average of air temperature from all four rooms.

$$T_{mix} = rT_{oat} + (1-r)T_{ra} \tag{1}$$

$$T_{ra} = \frac{\sum_{i=1}^{N_{vav}} (\dot{m}_{vav}^{i} \cdot T_{r}^{i})}{\sum_{i=1}^{N_{vav}} \dot{m}_{vav}^{i}} \tag{2}$$

Where

T_{mix} is the mixed air temperature (°C),

r is the fresh air ratio,

T_{oat} is the outdoor air temperature (°C),

T_{ra} is the return air temperature (°C),

N_{vav} is the number of VAV box, N_{vav} is 4 in this case study.

The AHU thermal power consumption from the heating coil is formulated as follow. The AHU mass flow rate is the sum of the individual VAV box supply air flow rate.

$$Q_{AHU} = \dot{m}_{s,AHU} c_p (T_{s,AHU} - T_{mix}) \quad (3)$$

$$\dot{m}_{s,AHU} = \sum_{i=1}^{N_{vav}} \dot{m}_{vav}^{i} \quad (4)$$

Where

Q_{AHU} is the AHU thermal power consumption (W),

$\dot{m}_{s,AHU}$ is AHU supply air mass flowrate (kg/s),

c_p is the air specific heat (J/(kg·°C)),

$T_{s,AHU}$ is the return air temperature (°C),

T_{mix} is the number of VAV box,

$\dot{m}_{vav}^{i}$ is the supply air mass flowrate at the ith VAV box (kg/s).

Fan model

Fan electrical power consumption was built based on the polynomial regression method:

$$P_{fan} = a_1 \dot{m}_{s,AHU}^{3} + a_2 \dot{m}_{s,AHU}^{2} + a_3 \dot{m}_{s,AHU} + a_4 \quad (5)$$

Where

P_{fan} is the fan power consumption (W),

a_1, a_2, a_3, a_4 are the coefficients of the fan power consumption curve.

Table 1 presents the identified coefficients of the proposed fan power curve. The comparisons between the regression model prediction and the actual measurements are shown in Figure 2. The R-square value of regression model is 0.87, and the root mean square error (RMSE) is 149.

Table 1: Coefficients of fan power curve

	Value
a_1	5371
a_2	-7396
a_3	3405
a_4	77

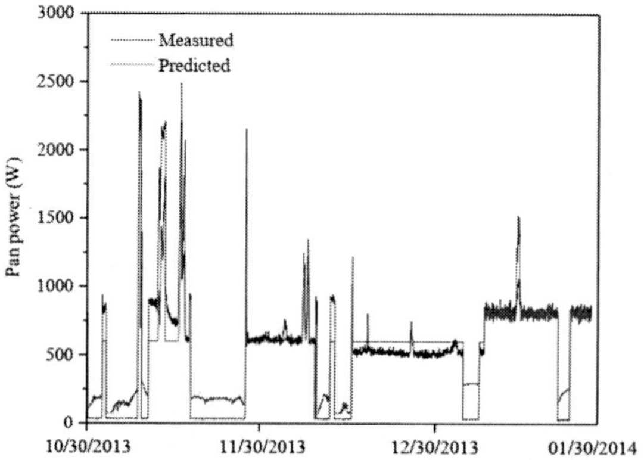

Figure 2: Fan electrical power consumption comparison

VAV box model

Reheat energy consumption by reheat coils of the VAV box at ith VAV box was modeled by:

$$Q_{reheat}^{i} = \dot{m}_{vav}^{i} c_p (T_{s,vav}^{i} - T_{s,AHU}) \quad (6)$$

Where

Q^{i}_{reheat} is the reheat thermal power consumption of the ith VAV box (W),

$T^{i}_{s,vav}$ is the supply air temperature at the ith VAV box (°C).

Room air thermal response model

A non-linear regression method based identification approach is applied to predict room air temperature with input variables of outdoor air temperature T_{oat}, VAV box supply air temperature $T_{s,vav}$ and VAV box supply air flow rate m_{vav}. For the interior room, we are assuming that the outdoor air temperature does not affect the interior room air temperature. Therefore, the room air temperature is mainly related with VAV box supply air temperature and supply air flow rate. The comparison of three identification methods, namely ARX model, State Space model (Niu et al. 2015), and the proposed non-linear regression model, are shown in Figure 3 using western room air temperature measurement data. For the non-linear regression method, air temperature identifications in four rooms were formulated using Equations (7) to (10). R-square and RMSE for these three data-driven methods are listed in Table 2. In this case study, the non-linear regression method is selected as the room air thermal response model due to the lowest RMSE.

$$T_r^{\text{east}} = 17.4828 + 0.2088T_{oat} + 0.5644T_{s,vav} \cdot m_{vav} \tag{7}$$

$$T_r^{\text{south}} = 19.3329 + 0.161T_{oat} + 0.2079T_{s,vav} \cdot m_{vav} \tag{8}$$

$$T_r^{\text{west}} = 17.7979 + 0.1826T_{oat} + 0.5063T_{s,vav} \cdot m_{vav} \tag{9}$$

$$T_{\text{r}}^{\text{interior}} = 20.5 + 0.1246T_{s,vav} \cdot m_{vav} \tag{10}$$

Where

T_r^{east} is the eastern room air temperature (°C),

T_r^{south} is the southern room air temperature (°C),

T_r^{west} is the western room air temperature (°C),

$T_{\text{r}}^{\text{interior}}$ is the interior room air temperature (°C).

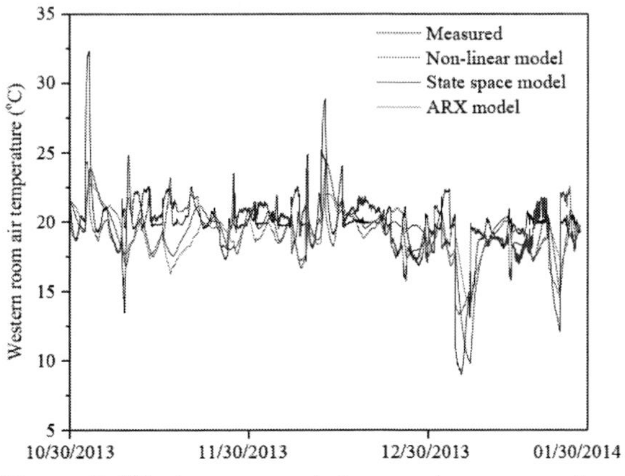

Figure 3: Western room air temperature comparison

Table 2: R-square and RMSE for three models included in Figure 3

	Non-linear regression method	ARX model	State Space model
R-square	0.53	0.06	0.23
RMSE	1.7	2.5	2.2

3.2 Optimization formulation

In this preliminary case study, the optimization objective function is to off-line minimize the total energy consumption of HVAC system including AHU, the fan and VAV boxes. The heating in the AHU heating coil and the VAV reheat coil is provided by a central gas-fired boiler. The coefficient of the boiler is assumed as 0.85. The control variables include the VAV box supply air flowrate $\dot{m}^i_{vav}$ (related to VAV box damper position), VAV box supply air temperature $T^i_{s,vav}$ (related to VAV box damper position and reheat valve position) and AHU supply air temperature $T_{s,AHU}$. It is expected that local AHU and VAV box controllers will take these optimized setpoints and decide the actuator actions accordingly.

$$\textit{Objective:}\ \text{Minimize}\quad Q_{total} = \frac{Q_{AHU}}{\eta} + P_{fan} + \frac{\sum_{i=1}^{N_{vav}} Q^i_{reheat}}{\eta} \tag{11}$$

The individual room air temperature and AHU supply air temperature are constrained in a fixed range. In addition, the heat transfer rates in the AHU heating coil and reheat coils cannot exceed their capacities.

$$\textit{Subject to:}\quad T_r^{\min} \le T_r^i(t) \le T_r^{\max} \tag{12}$$

$$T_{s,AHU}^{\min} \le T_{s,AHU} \le T_{s,AHU}^{\max} \tag{13}$$

$$\dot{m}_{s,AHU} c_p (T_{s,AHU} - T_{mix}) \le Q_{capa,AHU} \tag{14}$$

$$\dot{m}^i_{vav} c_p (T^i_{s,vav} - T_{s,AHU}) \le Q^i_{capa,reheat} \tag{15}$$

T_r in each room will be estimated using Equations (7) to (10). The minimum room air temperature $T_r^{\min}$ was set as 20 °C. The maximum room air temperature $T_r^{\max}$ was set as 22.2 °C. These maximum and minimum air temperatures were set as the exactly same with those in real operations at the ERS building.

3.2 Modeling Platform and Optimization Solver

For the proposed optimization-based control algorithm, AMPL was used to solve the optimization problem. AMPL is a high-level modeling language specifically tailored for optimization problem formulation with such features as automatic differentiation. AMPL include automatic differentiation tools and provides a convenient interface to state-of-the-art optimization solvers, including Interior Point OPTimizer (IPOPT) (Wachter and Biegler 2006). In this case study, the resulting optimization problem is solved using the IPOPT solver. All the model equations were defined in the AMPL platform first. After reading all the constant parameters and input variables, the AMPL calls the IPOPT solver to help to compute the optimal solution.

4. RESULTS AND DISCUSSIONS

Data from Oct 30th 2013 to Jan 30th 2014 was used to analyze the proposed model-based optimal control strategy. Figure 4 presents the room air temperature behaviors under the optimal control. The results indicate that the room air temperatures are more stable than the measurement data shown in Figure 3 (western room as the example). Majority room air temperatures fall in the range of 20 °C to 22.5 °C. The temperature variation in the southern room is the biggest due to the large disturbance such as solar radiation. Figure 5 and 6 show the optimized AHU supply air temperature and supply air flow rate (Note: the summation of optimized VAV box supply air flow rates). The AHU supply air temperature was narrowed between 13 °C and 16.5 °C. But the AHU supply air flow rates were increased for most of days. Whenever the outdoor air temperature was low, the model-based optimal supply air flowrate was high.

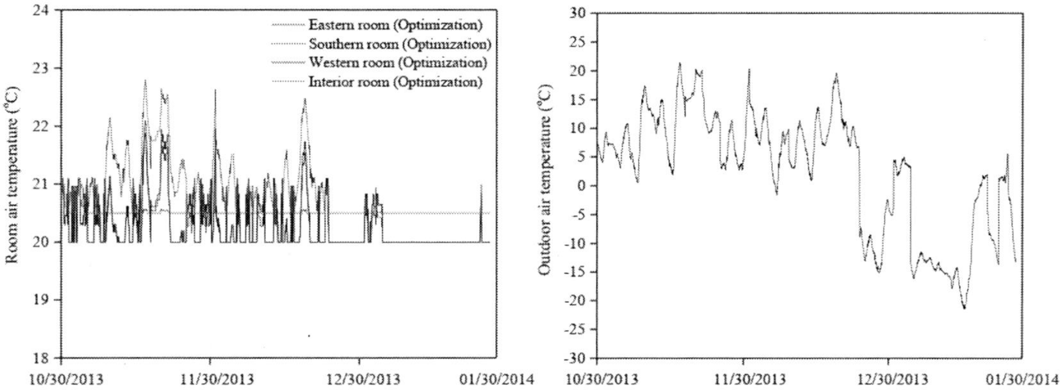

Figure 4: Room air temperatures from the case with optimal control and outdoor air temperature

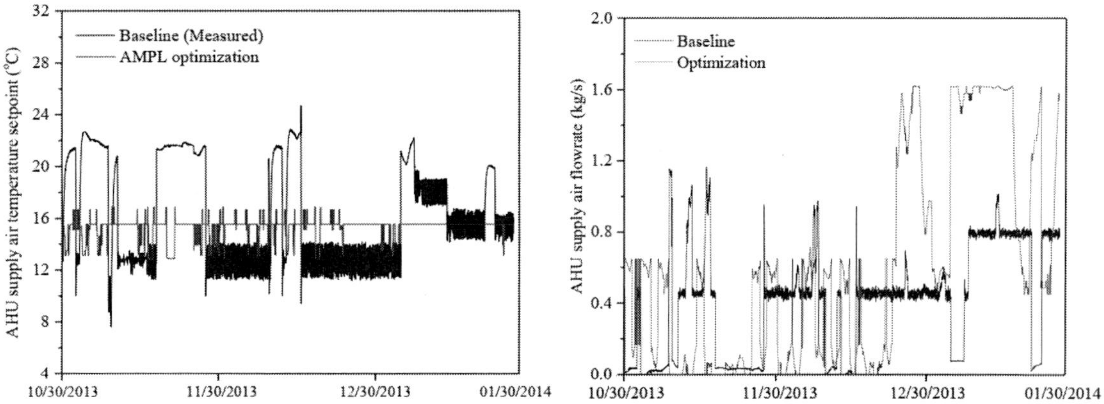

Figure 5: Optimized AHU supply air temperature and air flowrate

Figure 6 and 7 compare the VAV box supply air temperature and air flowrate between the baseline and the case with optimal control. The baseline VAV box supply air temperature is in the range of 15 °C to 35 °C. The model-based optimal VAV box supply air temperature was raised by 5 °C, which was in the range of 20 °C to 40 °C. The VAV box supply air flowrates are increased except for the interior room when the outdoor air temperature was low in January. Because the interior room air temperature is influenced less by the outdoor air temperature. For the model-based optimal control, the VAV supply air flow rate with the low outside air temperature condition is very high. The air mass flow rate of the southern room can get up to 0.7 kg/s.

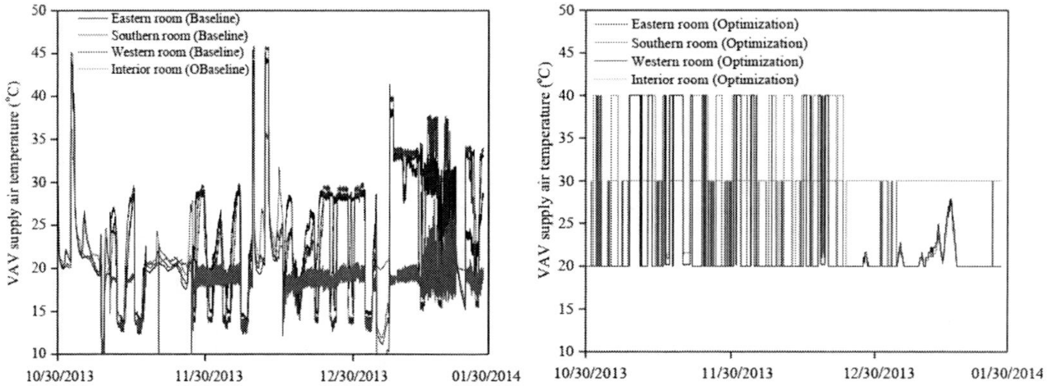

Figure 6: VAV box supply air temperature

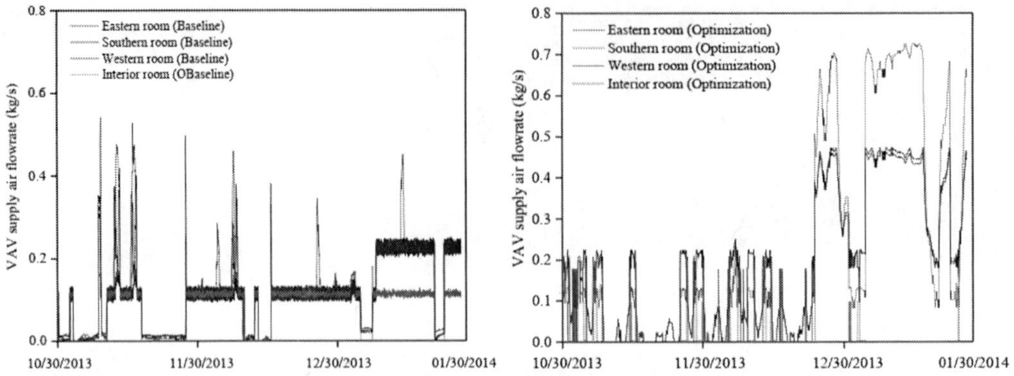

Figure 7: VAV box supply air flowrate

Figure 8 shows the energy consumption comparison between the baseline and the case with the proposed optimal control including the AHU thermal power consumption, the VAV box reheat thermal power consumption, the fan power consumption and the total power consumption. The AHU thermal power consumption of the model-based optimal control is lower than that from the baseline. The VAV box reheat thermal power consumption of the model-based optimal control is slightly lower than that of the baseline. However, the fan power consumption of the model-based optimal control is higher than that of the baseline. Finally, the total power consumption of the model-based optimal control is lower than that of the baseline. Figure 9 shows the quantified energy comparisons of the case with the proposed optimal control and the baseline. The total energy consumption can be saved by 22.1%. The VAV box reheat energy consumption is reduced by 208 kWh from 2,442 kWh to 2,234 kWh. The AHU thermal energy consumption is reduced by 962 kWh from 1792 kWh to 830 kWh. The fan energy consumption is increased by 165 kWh from 293 kWh to 458 kWh. The results indicate that the major energy consumption of the HVAC system is consumed by VAV reheat coil. By increasing the air flow rate, the AHU thermal power consumption and the VAV reheat coil energy consumption will be reduced. 22.1% of the energy saving can be achieved by applying the model-based optimal control.

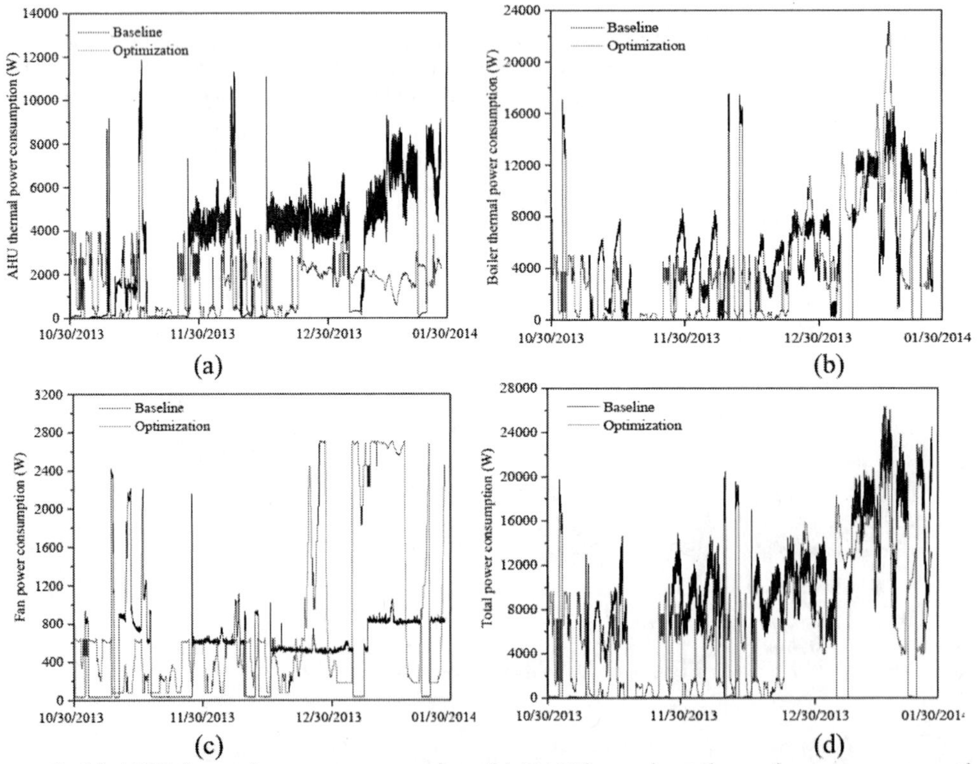

Figure 8: (a) AHU thermal power consumption; (b) VAV box reheat thermal power consumption; (c) Fan power consumption; (d) HVAC system total power consumption

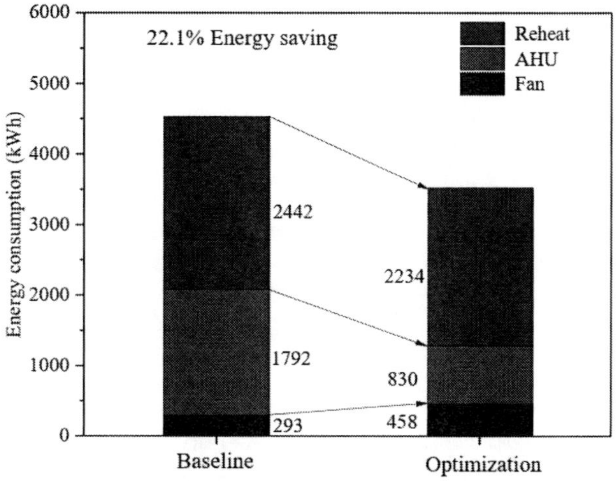

Figure 9: Energy consumption comparison

5. CONCLUSIONS AND FUTURE WORK

In this paper, a model-based optimal control is introduced by optimize the total HVAC system energy consumption. Data-driven energy performance models for individual component including the AHU, the fan and the VAV box reheat coil energy consumption, are built. The proposed optimal problem was solved using IPOPT based on the AMPL programming platform. Measurements from the real building were used for creations of data-driven models and as the baseline for the energy consumption comparison. The results of the model-based optimal control application in this simulation-based off-line study in the heating mode indicated that:

- The model-based optimal control greatly saves the total energy consumption.
- The room air temperature by the model-based optimal control is more stable.
- The energy saving of model-based optimal control is realized by increasing fan supply air flow rate to reduce the AHU energy consumption and VAV box reheat coil energy consumption.

Some future works are listed as follows:

- The data-driven model is based on a fixed historical data set. It is better to use a moving window to incorporate HVAC operation changes.
- Additional optimal control in the cooling mode will be conducted to analyze the annual energy performance.
- Demonstration and implementation of the proposed optimal control in the Building Energy Management System in a real building.

NOMENCLATURE

T	temperature	(°C)
r	ratio	(–)
N	number	(–)
Q	power consumption	(W)
m	air mass flow rate	(kg/s)
c	air specific heat	(J/(kg·°C))
a	coefficient	(–)

Subscript

mix	mix
oat	outdoor air temperature
ra	return air
vav	variable air volume

s	supply
AHU	air handle unit
fan	fan
reheat	reheat
r	room
min	minimum
max	maximum
east	eastern
south	southern
west	western
interior	interior

REFERENCES

AMPL. (2016). AMPL, a mathematical programming language. *http://ampl.com/*

Li, P., Vrabie, D., Li, D., Bengea, S. C., Mijanovic, S., & O'Neill, Z. D. (2015). Simulation and experimental demonstration of model predictive control in a building HVAC system. *Science and Technology for the Built Environment*, *21*(6), 721-732.

Cho, Y. H., & Liu, M. (2008, January). Optimal terminal box control algorithms for single duct air handling units. In *ASME 2008 2nd International Conference on Energy Sustainability collocated with the Heat Transfer, Fluids Engineering, and 3rd Energy Nanotechnology Conferences* (pp. 203-208). American Society of Mechanical Engineers.

Lu, L., Cai, W., Chai, Y. S., & Xie, L. (2005). Global optimization for overall HVAC systems—Part I problem formulation and analysis. *Energy conversion and management*, *46*(7), 999-1014.

Lu, L., Cai, W., Soh, Y. C., & Xie, L. (2005). Global optimization for overall HVAC systems—Part II problem solution and simulations. *Energy Conversion and Management*, *46*(7), 1015-1028.

Yang, R., & Wang, L. (2012, May). Optimal control strategy for HVAC system in building energy management. In *Transmission and Distribution Conference and Exposition (T&D), 2012 IEEE PES* (pp. 1-8). IEEE.

Kusiak, A., Li, M., & Tang, F. (2010). Modeling and optimization of HVAC energy consumption. *Applied Energy*, *87*(10), 3092-3102.

Fong, K. F., Hanby, V. I., & Chow, T. T. (2006). HVAC system optimization for energy management by evolutionary programming. *Energy and Buildings*, *38*(3), 220-231.

Nassif, N., Kajl, S., & Sabourin, R. (2004, June). Evolutionary algorithms for multi-objective optimization in HVAC system control strategy. In *Fuzzy Information, 2004. Processing NAFIPS'04. IEEE Annual Meeting of the* (Vol. 1, pp. 51-56). IEEE.

Wächter, A., & Biegler, L. T. (2006). On the implementation of an interior-point filter line-search algorithm for large-scale nonlinear programming. *Mathematical programming*, *106*(1), 25-57.

Niu, F., O'Neill, Z., Zuo, W., & Li, Y. (2015). Assessment of different data-driven algorithms for AHU energy consumption predictions.

ACKNOWLEDGEMENT

The authors would like to thank the Iowa Energy Center for supporting this work by providing data.

Comparison and Validation of Modelling Methods for Non-Homogenous Walls Incorporating Vacuum Insulation Panels

Brock CONLEY[1]*, Christopher BALDWIN[1], Cynthia A. CRUICKSHANK[1], Mark CARVER[2]

[1]Carleton University, Mechanical and Aerospace Engineering,
Ottawa, Ontario, Canada

[2]Natural Resources Canada, CanmetENERGY
Ottawa, Ontario, Canada

ABSTRACT

This paper compares the effective thermal resistance of 5 wall assemblies incorporating vacuum insulation panels (VIP) encased in extruded polystyrene (XPS), known as composite panels, through three different evaluation methods. The methods consisted of using steady-state and in-situ experimental testing as well as computer simulation. The computer modelling of the wall assemblies used simplifications, such as symmetry, and involves creating a thermal model for each unique 2D profile within the wall assembly independently. The results of the modelling method were compared to an experimentally calculated thermal resistance based on measured heat flux through a representative assembly under steady-state conditions in a guarded hot box and the measured heat fluxes under in-situ conditions at the Natural Resources Canada CanmetENERGY seasonal test facility. The feasibility of using a composite panels to address concerns of VIP constructability was determined to be successful with a strong dependence on the ratio of the VIP dimensions to the overall performance of the wall assembly design. An effective thermal resistance range of 8.44 m^2K/W to 9.07 m^2K/W was found during steady-state testing in the guarded hot box, and an agreement within 0.1% to 7.9% was found between steady-state and computer model results. In addition to the initial modelling, a 4.5% agreement was achieved with a new method using simplifications that are based on the coverage area of the unique cross section. The approach utilizes the current simplifications used for homogenous walls, and further develops the methodology to prove they are applicable to non-homogenous wall assemblies as well.

1. INTRODUCTION

As of 2012, space heating accounted for 62% of Canada's secondary residential energy use, the most significant category by a large margin (Natural Resources Canada, 2014). New strategies and policies to reduce overall energy consumption, with a focus on reducing space heating energy, have been implemented in the residential sector. Voluntary performances standards have been developed for new homes including R-2000 (Natural Resources Canada, 2015), LEED Canada for Homes (Canada Green Building Council, 2014) and Passive House (Canadian Passive House Institute, 2014). These standards add a series of performance criteria, in addition to conventional building code, in an effort to reduce a home's energy consumption and include a limit on energy and water consumption, and prescribe minimum levels of insulation, ventilation, etc. Generally, extensive modelling, proof of concept and/or builder training are required to obtain the energy efficiency designation.

A common method of maintaining a home within the constrained energy budget is to increase the overall air tightness and insulation in the dwelling above conventional construction standards through additional sealing and insulation. The approach often taken to increase the insulation value is to introduce additional levels of insulation. This practice however is not always possible or favourable. For example, adding thickness to the walls may increase the dwelling's footprint or reduce the useful floor space within the home. As a consequence, many studies are being performed on vacuum insulated panels (VIPs), which offer a high thermal resistance per unit thickness when compared to conventional materials. VIPs consist of a metallic enclosure and a vacuum maintained inside, effectively eliminating the conduction through the centre of panel, however a thermal bridge will occur along the edges (Mukhopadhyaya et. al, 2011; Baldwin et. al, 2015). Literature (Mukhopadhyaya et. al, 2014) has shown that a VIP can reach a thermal conductivity of 0.0034 $W/m{\cdot}K$, at the centre of panel, however a previous study found that a reduction of 32% of the thermal resistance at the panel perimeter exists (Conley and Cruickshank, 2016). There are concerns about whether the fragility and the non-homogenous nature of the panels will cause problems within residential dwellings and associated modelling.

The constructability of VIPs in building envelopes have been highlighted as a problem in publications (Mukhopadhyaya et. al, 2011). Since the identification of this problem, different techniques have been implemented in order to augment the constructability of VIPs and make them feasible within buildings. The most common approach has been to integrate VIPs within foam sheets and utilize them as a prefabricated panel with designated points for mechanical fastening. The motivation behind this technique is to mitigate the risk of puncture through improper placement of fasteners during the construction of the envelope and to keep the building practices of VIPs similar to the current state of standard materials.

To date, limited work has been conducted on modelling the non-homogenous nature of the panels in building applications. A weighted average method (Staube and Smegal, 2011) has been used to model wall envelopes. For standard wall construction that uses homogenous materials, the weighted average method utilizes the framing factor (FF), known as the amount of wall area that incorporates structural support, in order to reduce the size of experimental model. The method was used (Schiedel et. al, 2013) such that a repeatable section using a significantly smaller width and a fictitious stud size. This method reduces the overall size of the model and is a simplifying convention used by researchers and building science professionals. Currently, a standard of simplifying cross sections does not exist for wall envelope assemblies that contain multiple non-homogenous layers (e.g. walls incorporating VIPs).

2. EXPERIMENTAL APPROACH

In total, 5 different wall compositions and designs were evaluated at steady-state conditions within Carleton University's guarded hot box and one wall was evaluated at in-situ conditions using the Natural Resources Canada (NRCan) CanmetENERGY seasonal test facility. The purpose of the tests was to experimentally validate models that were built in THERM (THERM, 2015), a steady-state heat transfer simulation software used to determine the thermal resistance of wall assemblies.

2.1 Experimental Wall Assemblies

VIPs encased in foam or an insulation material, is a concept that has been studied (Erbenich and Knoll, 2011) as an approach to increase the use of VIP on construction sites. However, this method of protection was not nearly tested or metered as much as the current study. In the application of this study, it was important to allow sufficient space between and around the VIPs to insure a minimal chance of puncture by mechanical fastening during installation. By delivering the panels as a prefabricated piece, it would limit the changes to construction practices currently in place thereby improving the constructability of the novel material.

The composite panels prototyped under this study were designed as two 12.7 mm (0.5") thick sheets of extruded polystyrene (XPS) on either side of the VIP, acting as the shell, and additional 25.4 mm (1") thick XPS spacers around the VIPs. VIPs from two different manufacturers were used and had sizes of 863 mm by 559 mm (34" by 22") and 559 mm by 559 mm (22" by 22") from one manufacturer and 498 mm by 597 mm (19 3/8" by 23 1/2") panels from another manufacturer; an example of the wall composition can be found in Figure 1. The variation in the composite panel design was performed by changing the size of spacer between the VIPs, the orientation of the VIPs and the type of VIP within the panel. This led to the development of five different types of composite panels to

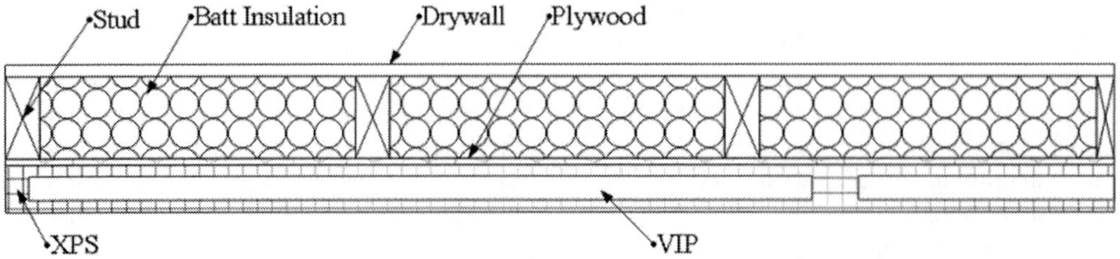

Figure 1: Horizontal cross section with material composition of composite panels labelled

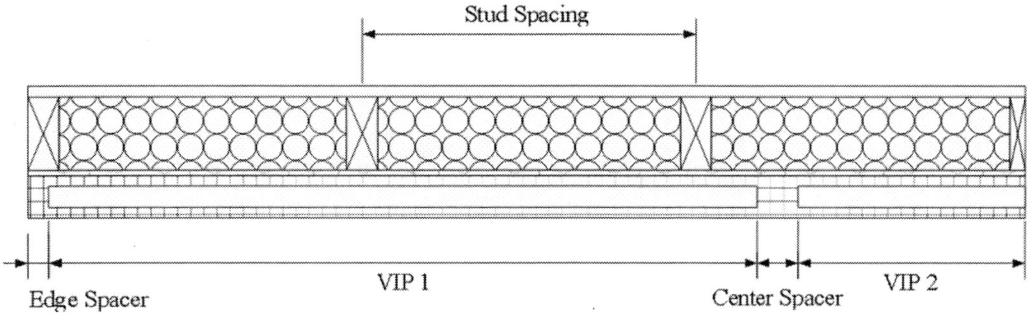

Figure 2: Horizontal cross section with dimensions labelled. Dimensions for each wall assembly found in Table 1

be tested at steady-state and modelled using THERM. Figure 2 is similar to Figure 1 but the composition labels were removed and the related dimensions (edge spacer, VIP spacer and two VIP sizes) were labelled instead. The associated dimensions were added to Table 1 for all five composite panel designs.

Table 1: Horizontal dimensions of the composite panels up to the middle of the wall

Test	Stud spacing, mm (in)	Edge spacer, mm (in)	Centre spacer, mm (in)	VIP 1, mm (in)	VIP 2, mm (in)
1	305 (12)	25 (1)	51 (2)*	559 (22)	559 (22)
2	406 (16)	25 (1)	51 (2)	864 (34)	279 (11)**
3	406 (16)	0 (0)	25 (1)	597 (23.5)	597 (23.5)
4	406 (16)	25 (1)	0 (0)	597 (23.5)	597 (23.5)
5	406 (16)	25 (1)	51 (2)*	559 (22)	559 (22)

*There are two centre spacers in half the composite panel.
**The centre of a 22" VIP aligns with the line of symmetry.

For each wall assembly, a large number of cross sections are present when the composite wall panels are utilized. Unlike conventional wall envelopes when layers of insulation are homogenous in the vertical direction with the exception of the stud layer, there are two additional compositions that are introduced based on the stud and VIP locations. The four unique cross sections in the assembly are stud-VIP, stud-no VIP, cavity-VIP and cavity-no VIP.

2.2 Guarded Hot Box

The guarded hot box test facility at Carleton University (Baldwin et. al, 2015) evaluates that overall thermal resistance of a wall assembly. The experimental set-up consists of three distinct controlled chambers: climate, metering and guard. The conditions within the chambers are set to simulate the outdoor and interior temperature conditions during the test period. As illustrated in Figure 3, the metering chamber is located within the guarded chamber in order to minimize or eliminate the heat loss through the metering chamber walls by maintaining a small temperature difference of approximately 0.1°C between the two chambers. Since the conduction through the chamber walls is driven by temperature difference, maintaining the chambers at roughly the same temperature throughout the test period limits heat loss and forces the heat input to the metering chamber through the specimen. The test facility follows ASTM C-1163 (ASTM, 2011) to determine whether steady-state conditions are met. The conditions are averaged over a 4 hour test period and the test continues until steady-state conditions are achieved. According to ASTM C-1163 this occurs when, for five consecutive test periods:

- the average specimen surface temperature in the metering chamber did not vary by more than ± 0.25°C.
- the average specimen surface temperature in the climate chamber did not vary by more than ± 0.25°C.
- the average temperature within the air curtain did not vary by greater than ± 0.25°C.
- the average energy input to the metering chamber did not vary by more than ± 1%.

After these requirements have been met, the collected data from 5 consecutive time periods can be averaged and analyzed to determine the overall thermal resistance, R, in m^2K/W of the assembly through Equation 1

$$R=\Delta T/(q/(t\cdot A)) \quad (1)$$

where ΔT is the temperature difference between the average hot and average cold side temperatures in °C, q is the energy input to the metering chamber in Wh, t is the time period of the test in hours and A is the metered area in m^2.

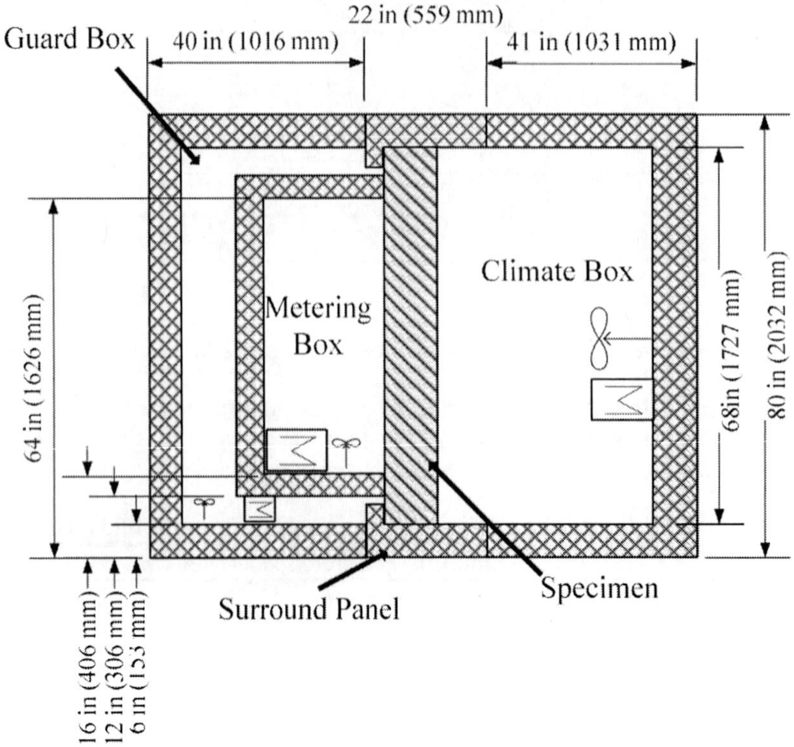

Figure 3: Schematic of Carleton University's guarded hot box

2.3 In-Situ Seasonal Test Facility

To compliment the steady-state test results of the guarded hot box facility and provide another experimental comparison, the initial design iteration was tested at in-situ conditions. During in-situ wall testing, the interior temperature and humidity are tightly controlled at steady conditions while the exterior conditions are determine by the naturally varying outdoor environment. In contrast to steady-state conditions, the wall assembly is under test conditions for a full seasonal cycle and allows for the moisture transport and hygrothermal performance to be monitored and evaluated. At the NRCan CanmetENERGY seasonal test facility, there is the opportunity to monitor three separate 2.45 m by 2.45 m (8' by 8') wall assemblies and evaluate the heat and moisture transport through the assemblies with the use of thermocouples, heat flux plates and other moisture monitoring devices. The measurements are taken at 10 minute interval. By using the heat flux and temperature difference at the specific cross sections of the overall wall assembly, the thermal resistance can be experimentally obtained.

In order to find the heat flux at the four unique cross sections of the panels, heat flux plates were installed at each point and thermocouples were used to determine the temperature difference across those points. Using Equation 2 it is possible to calculate the thermal resistance for the points and the effective R-value of the wall can be computed using the weighted averages method

$$R = \Delta T / q'' \tag{2}$$

where ΔT is the temperature difference across the material in °C, q'' is the heat flux measured in W/m^2 to find the thermal resistance, R in m^2K/W.

3. MODELLING APPROACH

An important aspect of wall assembly design is to develop calibrated thermal wall envelope models to easily assess the thermal performance. Within this study, models were developed to determine the effective thermal resistance and compare the results to the experimental data. The material properties and boundary conditions were taken from the THERM library or the manufacturers' material specifications sheets, when unavailable within THERM.

Previously, simplifications have been made in conventional wall assemblies by utilizing a fictitious stud size to represent the FF such that the geometry uses a larger, centred stud and properly portioned cavity area on one side, and a comparison between the actual and simplified geometries. It is generally acceptable to reduce the cross section using symmetry simplification and in some circumstances use the coverage area of material to create a singular cross section to input to the computer simulation. However, it is unknown whether these simplifications based on the framing factor and coverage area are valid when a wall assembly containing greater than two unique cross sections is modelled. Therefore, within this study a model was created for each of the five wall designs. Then a singular cross section was created based on the average coverage area of the four previously mentioned cross sections.

4. GUARDED HOT BOX RESULTS

In order to determine the thermal resistance of the assembly, the guarded hot box must reach steady state conditions outlined by ASTM C-1163 (ASTM, 2011). Figure 4 shows an example of the measured temperature profile through one of the cross sections over a 20 hour testing period. The temperature at each interface is denoted in the legend by the "outside-inside" materials, where "PLY" is plywood, "DRY" is drywall and "FIBER" is fiberglass insulation and the remaining acronyms have been defined. The graph shows that the temperature throughout the assembly and the interior surface remains essentially constant throughout the test. A cyclic temperature exists on the exterior wall surface due to the refrigeration unit cycling in the climate chamber. Another method to verify steady-state conditions, according to ASTM C-1163, is to determine if the heat transfer through the metering and guarded interface is less than 1% of the total energy input over the test period. The temperatures on the inside and outside of the metering chamber walls were instrumented with thermocouples, and since the effective thermal resistance and wall area are known, the heat loss from the chamber can be calculated. For example, when evaluating Test 3 the temperature difference was 0.01°C and the total heat transfer at the interface test was found to be 0.6 Wh, which was approximately 0.3% of the overall heat input during the test. Over the entire testing period of Test 3, 159.4 Wh of heat was added to maintain the steady-state conditions in the metering chamber over a 20 hour period. All of the energy flowed through the 1.83 m^2 metering area with an average temperature difference of 36.9°C across the

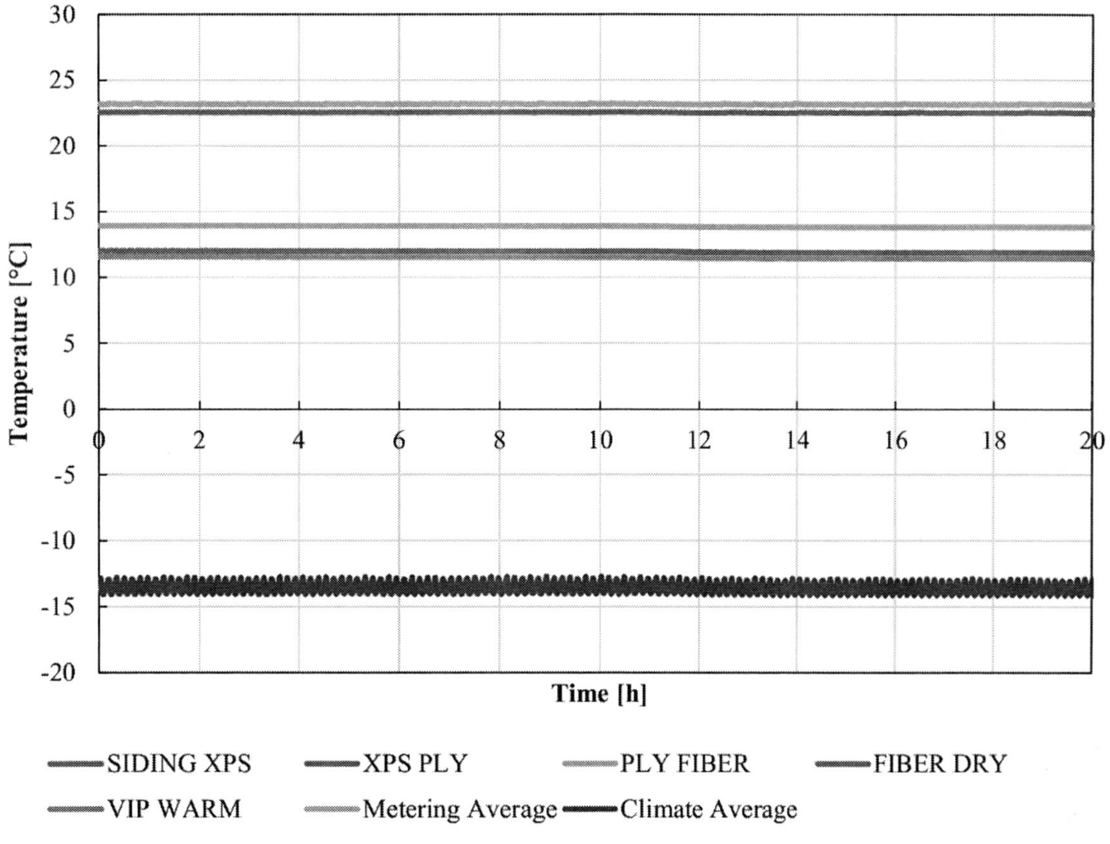

Figure 4: Steady-state temperature profile during guarded hot box test

assembly. Using Equation 1, the thermal resistance value for the wall design is 8.44 m^2K/W. In Table 2, the exterior and interior surface temperatures, the heat input to maintain steady-state conditions and the effective thermal resistance for the other four wall assembly designs are listed.

Table 2: Summary of steady-state test conditions and results from guarded hot box over 4 hour test period

Test	Climate (°C)	Metering (°C)	Heat Input (Wh)	Effective RSI (m^2K/W)
1	-13.3	23.0	139.9	8.97
2	-10.0	23.2	127.1	8.97
3	-13.7	23.1	159.4	8.44
4	-11.8	23.4	126.0	9.07
5	-11.8	23.3	126.0	8.97

5. IN-SITU RESULTS

The seasonal test facility is still in the initial year of operation and is able to test three experimental designs at a time. Therefore, at this point, the only wall design that has experimental in-situ data is the initial design iteration with the remainder of the wall designs to be tested in the coming years. The in-situ measurements were taken from January 19th 2016 to March 4th 2016 in Ottawa, Ontario, Canada. The indoor conditions were tightly controlled, while the exterior conditions comprised of the constantly changing outdoor environment. As such, Figure 5 shows the temperature profile with the first name indicating the interior facing material of the interface, for a point in line with the cavity and centre of VIP from the interior to exterior of the wall. On the exterior surface, it is obvious that the temperature is constantly changing, while the interior is remaining essentially constant with mild variance due to the heating unit cycling.

In accordance with ASTM C-1155 (ASTM, 2007), the thermal resistance of a wall assembly in-situ can be found when there is sufficient data points and a required convergence factor is met. Heat flux plates were placed at each

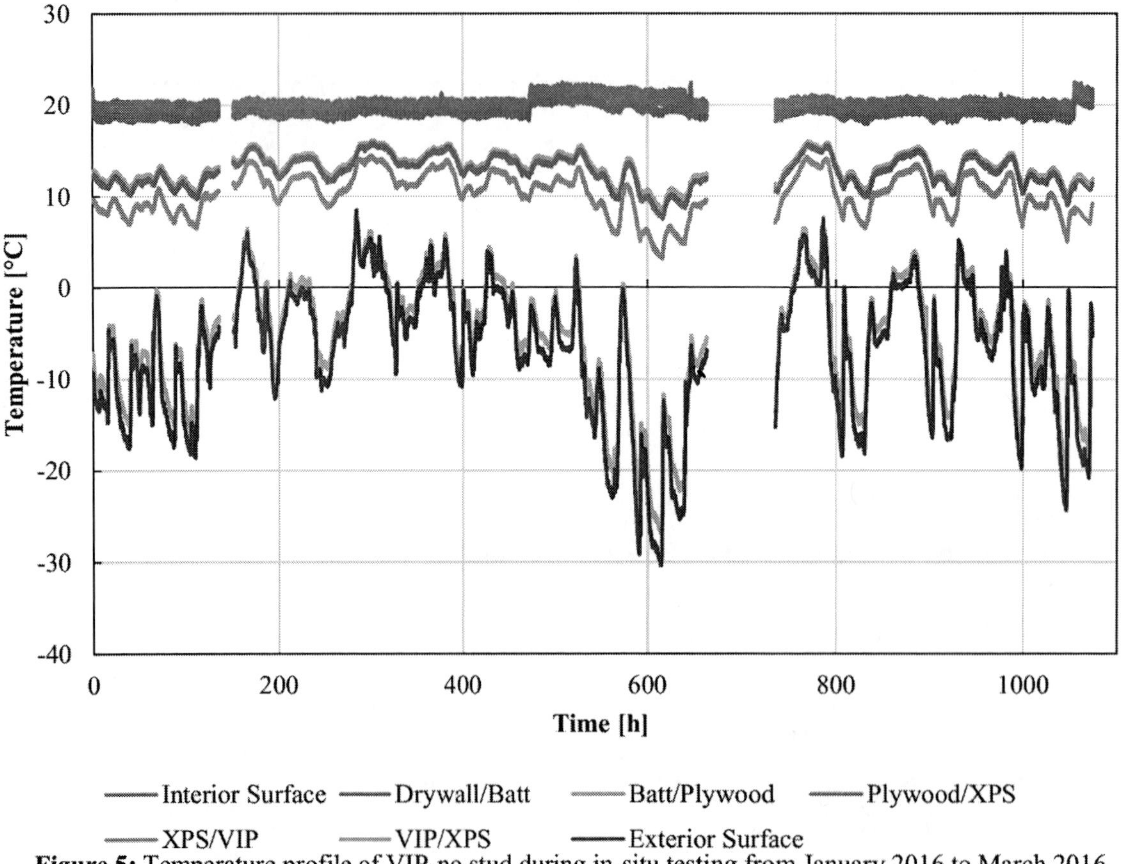

Figure 5: Temperature profile of VIP-no stud during in-situ testing from January 2016 to March 2016

unique cross section where a different heat flux would be occurring. Using the measured heat flux, the coverage area of that point specific point and the temperature difference an effective R-value can be found.Table 3 is a summary of the thermal resistance of each unique cross section based on the heat flux and temperature. The weighted RSI of each point is summed to find the effective thermal resistance assembly. Also included in the table is the thermal resistance of the centre of VIP, which was compared to the value from the manufacturer. If a large discrepancy exists, the experimental value will be used during simulations.

Table 3: Results from in-situ testing and effective thermal resistance

Cross Section	RSI (m^2K/W)	% Area	Weighted RSI (m^2K/W)
Stud-VIP	7.73	9.00%	0.70
Cavity-VIP	9.47	76.94%	7.29
Cavity-no VIP	4.68	9.34%	0.44
Stud-no VIP	2.91	4.72%	0.14
VIP	5.49		
		Effective RSI	8.56

6. MODELLING RESULTS

The modelling of the wall assemblies was split into two sections: accepted methods of simplifications, including symmetry and height ratios, and a new method presented where simplification are expanded to create a single representative cross section of the main wall.

6.1 Initial modelling

Each wall was modelled using a symmetrical simplification, such that only half of the wall geometry needed to be modelled in THERM. Since the walls were non-homogenous in the vertical and horizontal directions, two models needed to be used for each wall design; one that included VIPs and another that did not, due to encasing the perimeter of the VIP with XPS. This keeps the VIP from sliding out and aids in fastening the panel to the exterior wall. As mentioned prior, the THERM material library contained many of the necessary materials to replicate the main wall construction, however the VIPs did not exist and the thermal resistance per unit thickness was taken from the manufacturer's data sheet. A summary of the materials used are listed in Table 4. There are different VIP thermal conductivities listed for the different manufactures used during the study.

Table 4: Thermal conductivity values used during THERM simulations

Material Component	Thermal Conductivity (W/m·K)
Extruded Polystyrene (XPS)	0.026
Gypsum Wall Board	0.16
Lumber	0.14
Plywood (medium density)	0.17
Fiberglass Insulation	0.036
Vacuum insulation panels (VIPs)	0.0036 or 0.0043

In Figure 6, an example of the models for a VIP-present horizontal cross section is shown with the interior surface facing up. The boundary conditions on the left and right of the wall assembly are adiabatic, and the exterior and interior surfaces have "exterior winter", denoted by the blue surface, and "interior vertical surface", denoted by the red surface, conditions respectively. The set the temperature of the interior wall at 21.1°C and the exterior -17.8°C with film coefficients of 2.44 W/m^2K and 26 W/m^2K respectively. The boundary conditions used were within the

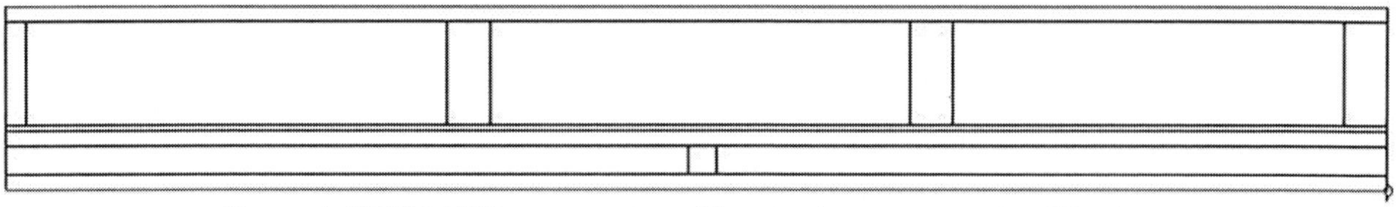

Figure 6: THERM VIP cross section of Test 3 using symmetry simplifications.

THERM library and were also used for other wall assembly models. For each cross section, the thermal resistance was evaluated based on the height ratio of cross section wall coverage to the overall wall height, and an effective thermal resistance value for the wall can be determined. In Table 5, the simulated thermal resistance values, height coverage and associated height ratio to find its contribution to the weighted and effective thermal resistance are presented for Test 1 and 2.

Table 5: Summary of THERM results using weighted average technique for Tests 1 and 2

	Component	RSI (m^2K/W)	Height (mm, in)	Height Ratio	Weighted RSI (m^2K/W)
Test 1	VIP	8.67	2286, 90	0.94	8.13
	No VIP	4.35	152, 6	0.06	0.27
				Effective RSI	8.40
Test 2	VIP	9.27	2235, 88	0.92	8.50
	No VIP	4.48	203, 8	0.08	0.37
				Effective RSI	8.87

6.2 Single representative cross section

The simplifications used in other publications for walls without a second non-homogenous layers solely account for the FF within the assembly. However, when the composite panels are introduced the amount of XPS and VIP also need to be taken into consideration. Another aspect that should be accounted for is the amount of coverage area where the VIP aligns with the stud. In order to deal with all these parameters and include them to create a single cross section in THERM, percentages of VIP coverage, XPS coverage, VIP stud coverage, XPS stud coverage and framing factor were determined and implemented onto a 610 mm (24") wide cross section. The cross section was organized, as seen in Figure 7, such that for the wall framing two fictitious stud sizes were used where the sum is based on the framing factor and the distribution is based on VIP-stud and XPS-stud coverages. The amount of VIP and XPS inside the composite panel is based on the overall layer composition. The coverage percentages and thermal resistance results for a representative assembly of Test 1 and Test 2 are summarized in Table 6.

Table 6: Geometry of single representative cross section

	Framing factor	VIP (mm)	XPS (mm)	VIP-stud (mm)	XPS-stud (mm)
Test 1	14%	523	85	39	46
Test 2	11%	523	85	44	23

7. COMPARISON OF METHODS

The experimental data from the in-situ and steady-state guarded hot box testing can provide an experimental validation for the computer simulation models and provide insight to whether the assumptions and simplifications that were made are accurate. Since the in-situ data is Test 1, 305 mm (12") on-centre (OC) framing with 863 mm by 559 mm (34" by 22") and 559 mm by 559 mm (22" by 22") VIP encased in XPS, the thermal resistance values were compared. The effective RSI values found from the in-situ and steady-state tests were 8.56 m^2K/W and 8.97 m^2K/W respectively and when compared to the THERM simulations, a percent difference of 1.9% and 6.5% was found. The results are summarized in Table 7. The advantages of using the new modelling method presented in the study is the

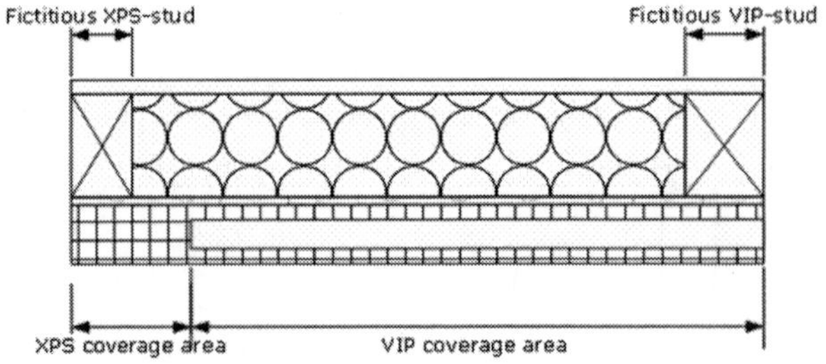

Figure 7: Single representative cross section using fictitious stud sizes

ability to simulate the main wall section using a single cross section. Even though more calculations of the coverage areas are required, the ability to edit, modify or change the dimensions in the new modelling method is improved. The benefits include the ability to utilize four dimensions in the composite panel and two fictitious stud sizes to construct a representative cross section that describes a complex, non-homogenous wall assembly.

Table 7: Percent difference of all evaluation methods for Test 1

	In-Situ	Steady-State	Single Cross Section	THERM
RSI (m^2K/W)	8.56	8.97	8.15	8.40
% Difference	1.9%	6.5%	4.2%	-

Furthermore, when the other steady-state results were compared to the simulation results, a large variation in differences existed. When Test 2 and 3 were analyzed the difference of effective thermal resistance was found to be 1.1% and 0.1% respectively, however the difference for Test 4 was found to be about 7.9%. The Test 4 wall design utilized a layer of VIP without XPS spacers between the VIPs to create the greatest effective thermal resistance. However, when this was applied within the simulation for a full scale wall assembly, XPS was required to fill the gaps caused by the geometry of the VIPs. The geometry caused approximately 165 mm (6.5") of XPS spacers to be present at the left and right composite panel edges, and a total of 445 mm (17.5") of XPS at the top and bottoms of composite panels. When excluding the no VIP section of the simulation, a thermal resistance becomes 9.25 m^2K/W, which reduced the percent difference to 2.0%. Therefore, if the geometry of the panels are not easily incorporated into a standard wall construction, such that a large amount of XPS is required to fill in the composite panel, a large degradation in thermal performance will exist. In Table 8, the effective thermal resistances from the steady-state testing and the computer simulations were summarized for the remaining wall designs.

Table 8: Percent difference of steady-state and computer modelling for each wall design

	Steady-State	THERM	% Difference
Test 1	8.97	8.40	6.5%
Test 2	8.97	8.87	1.1%
Test 3	8.44	8.43	0.1%
Test 4	9.07	8.38	7.9%
Test 5	8.97	8.62	3.9%

8. CONCLUSIONS

In conclusion, the steady-state and in-situ experimental testing and THERM modelling effectively evaluated the thermal resistance of five different wall envelope design utilizing VIPs encased in XPS composite panels. It was observed that VIP geometry can have a dramatic impact on the overall thermal resistance of the wall design. For the second set of manufactured VIPs, the VIP dimensions required more XPS than the initial set of panels that underwent testing and lower thermal performance was found due to the dimensions causing a larger fraction of XPS surrounding the perimeter of the panels.

The composite panels are a feasible solution to the significant constructability challenge that exists and must be addressed before VIP can be implemented into residential housing. When modelling the panels, either using symmetry or new method utilizing the ratio of coverage area of specific cross sections were found to be efficient at estimating the effective thermal resistances for Test 1 and 2. However, the benefits of being able to modify the representative cross section easily and the ability to use a single cross section to measure the thermal resistance are reasons to recommend using this method. In contrast to the first modelling method, it is much easier to compare the composition and coverage area of two separate wall designs using the new method since the geometries of the wall designs would be similar. The coverage areas of VIP and XPS and the associated material coverage at stud locations, from the fictitious sizes added to the model, can be clearly seen and compared in the cross section in Method 2, unlike Method 1 where it would require more information about the wall assembly, specifically height ratios of cross sections.

Furthermore, optimizing the dimensions of the VIPs within the assembly and the amount of XPS required to safely install the composite would be a valuable tool during the design of composite panels. The amount of XPS should be minimized since the effective thermal resistance of that cross section on 12" and 16" OC are 4.35 and 4.48 m^2K/W, nearly half of the effective thermal resistance when VIP are introduced to the cross section, but a minimum amount of XPS must be present as a tolerance for mechanical fastening to the dwelling. During this study, it was found that a 1.5" strip could be sufficient, however the necessary marking of where the XPS spacers are located on the surface of the composite panel is needed mitigate the risk of puncture.

REFERENCES

ASTM. (2007). *Standard Practice for Determining Thermal Resistance of Building Envelope Components for In-Situ data.* Conshocken, PA: ASTM International.

ASTM. (2011). *Standard Test Method for Thermal Performance of Building Materials and Envelop Assesmblies by Means of a Hot Box Apparatus.* West Conshohocken, PA: ASTM International.

Baldwin, C., Cruickshank, C. A., Schiedel, M., & Conley, B. (2015). Comparison of steady-state and in-situ testing of high thermal resistance walls incorporating vacuum insulation panels. *6th Internation Building Physics Conference.* Turino.

Canada Green Building Council. (2014). *Leed Canada Rating Systems.* Retrieved from http://www.cagbc.org/CAGBC/LEED/CommercGreenBuild/RatingSystems/CAGBC/Programs/LEED/CommercialInstitutional/RatingsSystems/LEED_Canada_Rating_S.aspx?hkey=5490b62b-b10f-45b7-9c41-2b5a299655b8

Canadian Passive House Institute. (2014). *Canadian Passive House Institute.* Retrieved from http://www.passivehouse.ca/

Conley, B., & Cruickshank, C. A. (2016). Evaluation of Thermal Bridges in Vacuum Insulation Panels Assemblies through Steady-State Testing in a Guarded Hot Box. *eSim 2016.* Hamilton.

Erbenich, G., & Knoll, S. (2011). New Modular VIP Systems (For Advanced Handling on Construction Site). *10th International Vacuum Symposium.* Ottawa.

Mukhopadhyaya, P., Kumaran, K., Ping, F., & Normandin, N. (2011). Use of vacuum insulation panel in building envelope construction: advantages and challenges. *13 th Canadian Conference on Building Science and Technology,* (pp. 1-10). Winnipeg.

Mukhopadhyaya, P., MacLean, D., Korn, J., van Reenen, D., & Melleti, S. (2014). Building applications and thermal performance of vacuum inuslation panels (VIPs) in Canadian subarctic climate. *Energy and Buildings, 85,* 672-680.

Natural Resources Canada. (2014). *National Energy Use Handbook: 1990 to 2011.* Ottawa: Office of Energy Efficiency, Natural Resources Canada.

Natural Resources Canada. (2015, 12 04). *The R-2000 Standard.* (Natural Resources Canada) Retrieved from http://www.nrcan.gc.ca/energy/efficiency/housing/new-homes/5087

Schiedel, M., Baldwin, C., & Cruickshank, C. A. (2013). In-situ experimental validation of THERM finite element analysis for a high R-value wall using vacuum insulation panels. *International Conference on Energy Sustainability.* Minneapolis.

Staube, J., & Smegal, J. (2011). *Builiding America Special Research Project: High-R Walls Case Study Analysis.* Sommerville, MA.: Building Science Corp.

THERM. (2015, March 10). (Lawrence Berkeley National Laboratory (LBNL)) Retrieved from http://windows.lbl.gov/software/therm/therm.html

ACKNOWLEDGEMENT

The authors would like to acknowledge the financial support of the Natural Science and Engineering Research Council (NSERC) of Canada and Natural Resources Canada (NRCan).

Experimental Study on Thermal Performance Improvement of Envelop Integrated with Phase Change Material in Air-conditioned Room

Yanru Li[1], Xi Meng[1], Suo Wang[1], Enshen Long[1,2,*]

[1]College of Architecture and Environment, Sichuan University,
Chengdu, Sichuan Province, China
Contact Information (86-28-85401015, 86-28-85401015, longes2@163.com)

[2]Institute of Disaster Management and Reconstruction, Sichuan University-Hong Kong PolyU,
Chengdu, Sichuan Province, China
Contact Information (86-28-85401015, 86-28-85401015, longes2@163.com)

* Corresponding Author

ABSTRACT

Compared with the traditional building envelope, the thermal mass of building envelope integrated with phase change material was increased greatly, which would reduce the building energy consumption, improve thermal comfort, and shift the peak electricity load. Due to latent heat energy storage when phase changing, the wall integrated with PCM can release heat storage or cold storage to maintain the indoor thermal environment for a period of time after closing the air conditioning. This work presents the results of an experimental study of thermal performances of wall integrated with PCM and without PCM when the air condition runs continuously and intermittently. A building with these two kinds of walls is chosen and the inner surface temperature and heat flow are measured. The building have a domestic heat pump as a cooling system. The results show that the PCM can reduce the inner surface temperature 1°C, and reduce the inner surface heat flow about 40% when the air condition runs continuously. When the air condition runs intermittently by the working schedule, the PCM can also reduce the inner surface temperature 1°C, and the cold storage releasing time of wall integrated with PCM is 2 hours longer than wall without PCM. The PCM can improved the thermal performance of building envelop significantly.

1. INTRODUCTION

Electricity consumption and indoor thermal environment of a building be affected by building envelope thermal performance. One of the approaches for cutting down building energy consumption and enhancing indoor thermal environment is to integrate phase change material (PCM) into building envelope. Compared with the traditional building envelope, the thermal mass of building envelope integrated with phase change material was increased greatly, which would reduce the building energy consumption, improve thermal comfort, and shift the peak electricity load. When the temperature increasing, PCM absorbs heat while changing from solid to liquid. And when the temperature decreasing, PCM releases heat while changing from liquid to solid. Thermal energy can be absorbed and stored in PCM through a melting process. And compared with sensible heat storage system, latent heat storage system with PCM has higher energy storage density while requiring smaller masses and volumes of material. Wang et al. (2013) evaluated the thermal performance of ultrathin envelope integrated with PCMs by aid of numerical simulation for PCM heat conductivity coefficient, phase change heat, phase-transition temperature and PCM layer position. Feng and Liang (2006) built experimental rooms with phase change wall and ordinary wall, and used household air-condition for cooling. Results showed that the indoor air temperature of room with phase change wall was 1 or 2°C lower than that of room with ordinary wall. And heat flux was reduced by the phase change wall. In artificial climate laboratory, Kuznik and Virgone (2009) did experiments about the indoor thermal environment of PCM room and ordinary room in summer, winter and the transition season. Experimental results showed that the phase change wall could reduce the indoor temperature fluctuations. Behzadi and M. Faril (2011) did computer simulation about PCMs impregnated in building materials. The results showed that the use of PCMs can effectively reduce the daily fluctuations of indoor air temperatures and maintains it at the desired comfort level for a longer period of time. Yan et al (2012) studied the thermal properties of the PCM wall formed by different methods, and results showed that the surface temperature and heat flow through the PCM walls prepared by different methods are lower than that of traditional wall. Lei et al (2016) studied the energy performance of building envelopes integrated with PCMs for cooling load reduction in tropical climate through simulations. The results showed that PCM can effectively reduce heat gains through building envelopes throughout the whole year. Sun et al (2014) made selection of optimum phase transition temperature for the PCMs for the various climates using indoor and outdoor air temperature.

Existing studies show that PCM can improve indoor thermal environment of building. However, the applications of PCM wall in continuous air conditioned room and intermittent air conditioned room are different. This work presents the results of an experimental study of thermal performances of wall integrated with PCM and without PCM when the air condition runs continuously and intermittently. A building with these two kinds of walls is chosen and the inner surface temperature and heat flux are measured.

2. EXPERIMENTAL SET-UP

2.1 composite PCM wall

This paper test the thermal performances of wall integrated with PCM and without PCM. Figure 1 shows the cross-sectional schematic view of walls. The size of walls are both 600mm×600mm×260mm. For the composite PCM wall, its envelope consists of one layer of concrete plaster of 10mm, one layer of solid bricks of 220mm, one layer of PCM of 20mm and one layer of concrete plaster of 10mm. For the reference wall, its envelope consists of one layer of concrete plaster of 10mm, one layer of solid bricks of 240mm and one layer of concrete plaster of 10mm.And the physical properties of each layer material are shown in Table 1. The phase-transition temperature range of PCM is from 18°C to 26°C, which has phase change latent heat of 178.5kJ/kg.

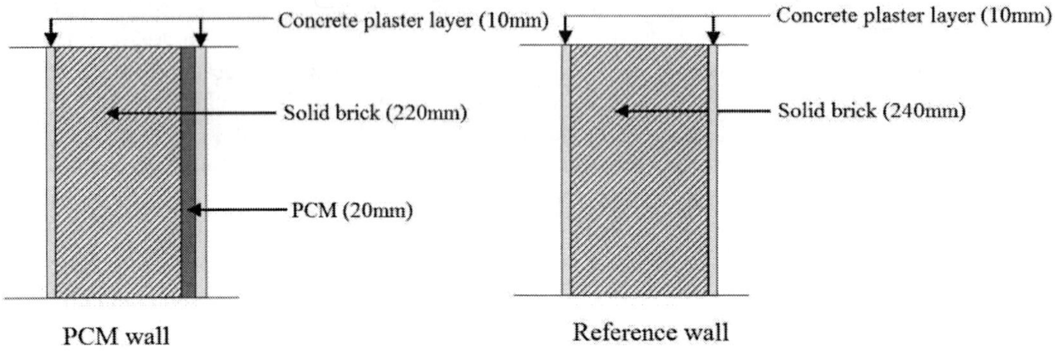

Figure 1: The cross-sectional schematic view of walls

Table 1: Physical properties of each layer material

Material	Density (kg/m³)	Thermal conductivity (W/(m·K))	Specific heat capacity (kJ/kg·K)
Mortar	1406	0.3505	1050
Solid brick	1536	0.7505	523
PCM	1300	0.25(phase change), 0.5(solid or liquid)	1785

2.2 Experimental methodology

To study the building envelope thermal properties improvement by integrated with PCM, a dynamic testing experimental building are built in Sichuan University. From Figure 2 and Figure 3 we can see the experimental building consists of two rooms with the size of 3.5m (length) × 3.0m (width) ×2.2m (height). The north wall of Room 1 is a composite wall with a variety of wall types, and its structure is shown in Figure 4. The size of wall unit is 600mm×600mm×260mm. For the interface of wall unit, there is 80mm EPS to reduce the heat transfer between wall blocks and guarantee a one-dimensional heat transfer in the center of the wall unit. The PCM wall and reference wall studied in this paper are in the Room 1's north wall. And the indoor and outdoor environment of this two kinds of walls can be think the same. In order to achieve the continuous and intermittent air conditioning situation, a wall-mounted air condition (KFR-35gw/HFJ+3) is installed in the Room 1's south wall.

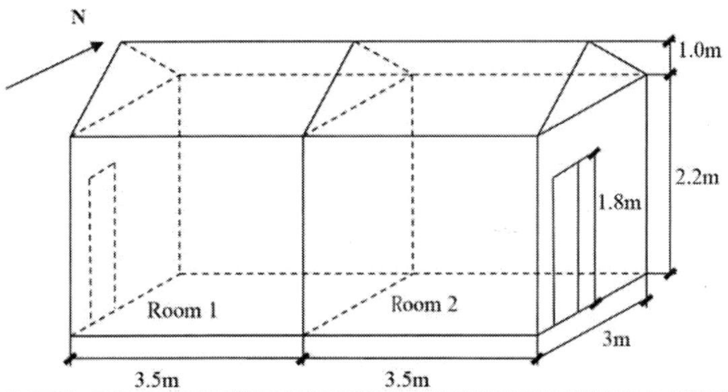

Figure 2: The experimental building

Figure 3: The appearance of the experimental building

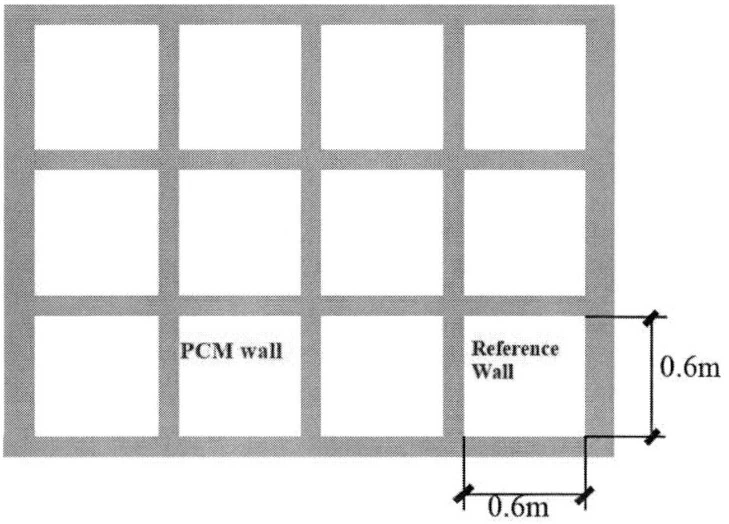

Figure 4: The structure of Room 1's north wall

The measured parameters in this experiment include indoor and outdoor air temperature, inner surface fluxes, inner surface temperatures and outside surface temperatures of PCM wall and reference wall, all of which are recorded by JTRG-II building thermal temperature automatic tester. To ensure the phase change process of PCM, the air condition temperature is set at 16°C during the experiment.

3. RESULT AND DISCUSSION

3.1 The continuous air conditioning situation

The experiment of continuous air conditioning situation is from 16:00 June 2nd 2015 to 11:00 June 7th 2015. Figure 5 presents the variation of indoor and outdoor air temperature from June 4th to June 5th. During the experiment of continuous air conditioning situation, the minimum outdoor air temperature is 19.25°C, while the maximum is 36.10°C. The outdoor air temperature between day and night is 12-16°C. Though the air conditioning temperature is set at 16°C during the experiment, the indoor air temperature is changing like sawtooth wave because the air condition lacking of frequency conversion and the power is larger. The compressor stops working when the indoor air temperature reaches the set temperature, and runs when the indoor air temperature is higher than the set temperature. The indoor air temperature in the continuous air conditioning situation changes between 13.4°C to 17°C, and the average temperature of 15.4°C. As the outdoor air temperature fluctuations, the indoor air temperature presents a cyclical fluctuations while the air conditioning set temperature is invariable. The indoor air temperature increases significantly during 16:00-20:00 every day. And the indoor air temperature fluctuations delay 1 hour the outdoor air temperature because of the thermal inertia of building envelope.

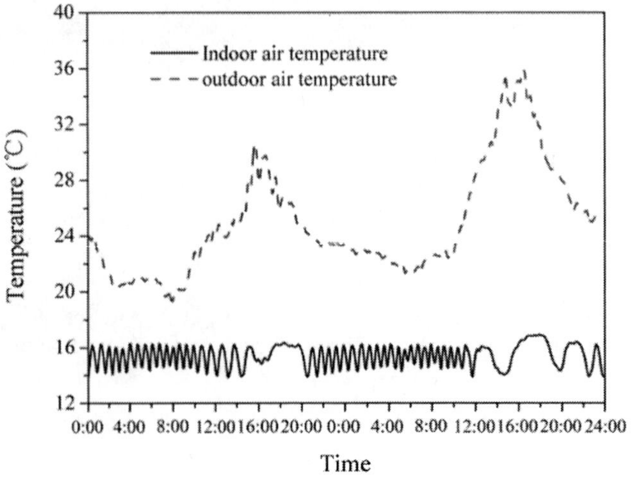

Figure 5: The variation of indoor and outdoor air temperature from June 4th to June 5th

Figure 6 presents the variation of inner surface temperature, and Figure 7 presents the variation of inner surface heat flux. During the experiment, the inner surface temperature of the PCM wall is lower than the reference wall. The average inner surface temperature of the PCM wall is 16.91°C, while it of the reference wall is 17.75°C. PCM can reduce the inner surface temperature about 1°C. In the situation of continuous air conditioning, the inner surface temperature fluctuation is different, and the inner surface temperature fluctuation of reference wall is large because its ability to resist the outdoor air temperature fluctuation is weak. Table 2 shows that the fluctuation amplitude of reference wall's inner surface temperature is 2.5°C, while it of PCM wall's is 1.9°C. With latent heat storage of PCM, the PCM wall can effectively decrease heat transfer through wall, and make heat attenuation and temperature delayed. Due to the test value of heat flux is negative, the inner surface heat flux of PCM wall is significantly lower than the reference wall, and the PCM wall has excellent energy saving potential. PCM can reduce the inner surface heat flux about 40%.

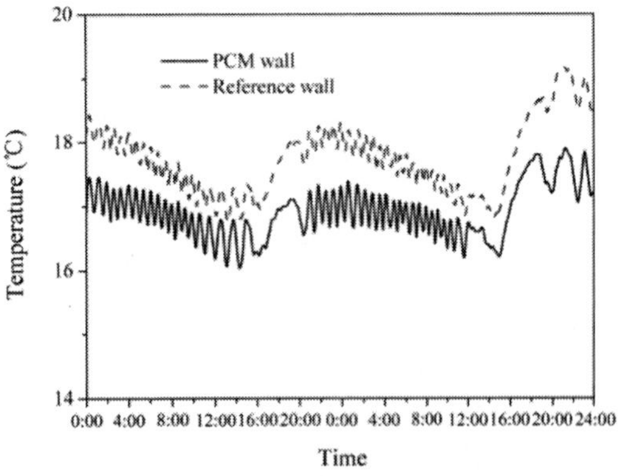

Figure 6: The inner surface temperature from June 4th to June 5th

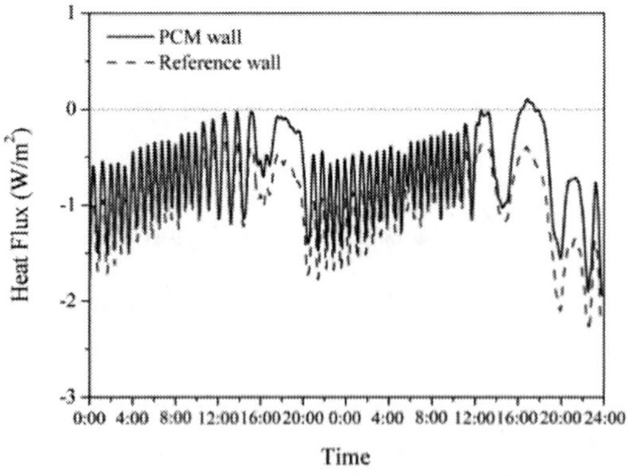

Figure 7: The inner surface heat flux from June 4th to June 5th

Table 2: The inner surface temperature fluctuations in continuous air conditioning situation

Wall type	Maximum value (°C)	Minimum value (°C)	Fluctuation amplitude (°C)
PCM wall	17.9	16.0	1.9
Reference wall	19.2	16.7	2.5

3.2 The intermittent air conditioning situation

The intermittent air conditioning refers to the residents regulate indoor thermal environment according to their own need. The intermittent operation mode is this paper refers to work schedule in office building, namely opening air condition during 8:00-18:00 and closing in other times.

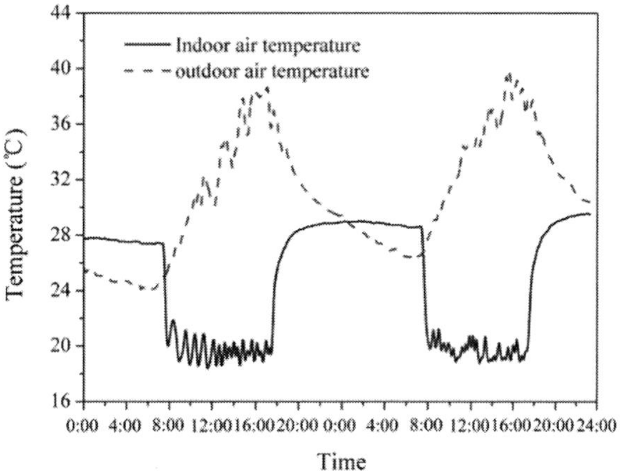

Figure 8: The variation of indoor and outdoor air temperature from July 31st to August 1st

The experiment of intermittent air conditioning situation is from July 31st to June 1st. Figure 8 presents the variation of indoor and outdoor air temperature during the experiment. The indoor air temperature changes significantly when the air conditioning operation mode changes. The indoor air temperature reduces quickly after opening the air condition, and goes to the first trough 20 minutes later. After 20 minutes, the indoor air temperature reduces slowly. The indoor air temperature is changing like sawtooth wave because the air condition lacking of frequency conversion and the power is larger. The indoor air temperature rises to normal in 1 hour after closing the air condition.

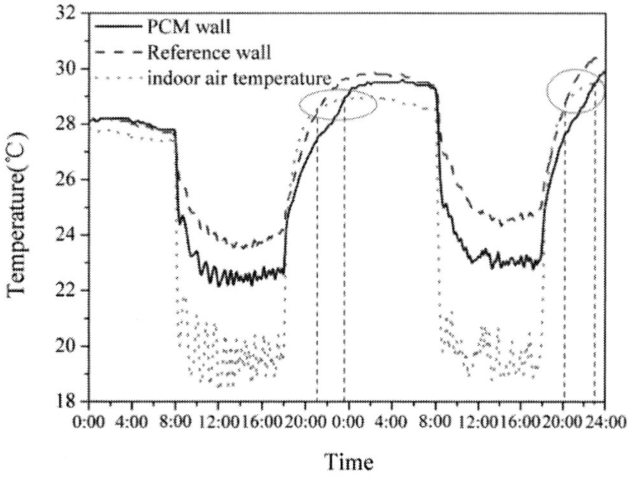

Figure 9: The inner surface temperature from July 31st to August 1st

It can be seen from Figure 9 that the inner surface temperatures reduce quickly and begin to cold storage after opening the air condition. The inner surface temperatures change quickly in the first 30 minutes since opening the air condition, and it begins to change slowly. The wall inner surface temperatures remain constant while the air condition has operated for 4 hours, which means the cold storage of walls is almost close to complete. The inner surface temperature of PCM wall is 1°C lower than the reference wall during the operating time of the air condition. After turning off the air condition at 18:00, the inner surface temperatures increase rapidly. With rapid heating rate, the indoor air temperature rise to normal indoor temperature after 1 hour since the air condition turned off, and then the temperature remains constant. The indoor air temperature rises to higher than the inner surfaces temperature after 30 minutes since closing the air condition, which means the walls begin to release the cold stored during the air conditioning time. The inner surface temperature of the reference wall rises higher than the air temperature at around 20:00-21:00, and cold releasing is completed. The inner surface temperature of the PCM wall rises higher than the air temperature at around 23:00-23:30, and cold releasing is completed. When the temperature decreasing after opening the air condition, PCM absorb cold while changing liquid to solid. And when temperature increasing after closing the air condition, PCM release cold while changing solid to liquid. Due to the latent heat storage of PCM when phase changing, the cold storage by PCM wall is larger than the reference wall, so the cold storage releasing time of the PCM wall is 2 hours longer than the reference wall. However, the wall integrated with PCM can release heat storage or cold storage to maintain the indoor thermal environment for a period of time after closing the air conditioning.

4. CONCLUSION

This paper studied the thermal performance improvement of building envelope by PCM when air condition runs continuously and intermittently. The results show that the PCM can reduce the inner surface temperature 1°C, and reduce the inner surface heat flow about 40% when the air condition runs continuously. When the air condition runs intermittently by the working schedule, the PCM can also reduce the inner surface temperature 1°C, and the cold storage releasing time of wall integrated with PCM is 2 hours longer than wall without PCM. The PCM can improved the thermal performance of building envelop significantly.

REFERENCES

Wang J., Long E.S., Qin W., and Xu L. 2013. Ultrathin envelope thermal performance improvement of prefab house by integrating with phase change material. Energy and Buildings 67:210-216.

Feng G.H., and Liang R.B. 2006. Research on cool storage time of a phase change wallboard room in the summer. Proceedings of the Sixth International Conference for Enhanced Building Operation, Shenzhen, China, 36:133-136.

Kuznik F., and Virgone J. 2009. Experimental assessment of a phase change material for wall building use. *Applied Energy* 86:2038-2046.

Behzadi S., and Farid M.M. 2011. Experimental and numerical investigations on the effect of using phase change materials for energy conservation in residential buildings. *HVAC&R Research* 11(3):366-376.

Yan Q.Y., Huo R., and Li L.S. 2012. Experimental study on the thermal properties of the phase change material wall formed by different methods. *Solar Energy* 86:309-3102.

Lei J.W., Yang J.L., and Yang E.H. 2016. Energy performance of building envelopes integrated with phase change materials for cooling load reduction in tropical Singapore. *Applied Energy* 162:207-217.

Sun X.Q., Zhang Q., Medina M.A., and Lee K.O. 2014. Energy and economic analysis of a building enclosure outfitted with a phase change material board (PCMB). *Energy Conversion and Management* 83:73-78.

ACKNOWLEDGEMENT

This project is funded by the National Natural Science Foundation of China (No.51478280 and No.51178282)

Analysis of the measurements reliability in dynamic test of the opaque envelope

Nadja BISHARA[1]*, Alessandro PRADA[2], Giovanni PERNIGOTTO[3], Marco BARATIERI[4] and Andrea GASPARELLA[5]

[1]Free University of Bozen-Bolzano, Faculty of Science and Technology, Bolzano, Italy
Tel: +39 0471 017685, Fax: +39 0471 017009, nadja.bisharaschroetter@natec.unibz.it

[2]University of Trento, Department of Civil Environmental and Mechanical Engineering, Trento, Italy
Tel: +39 0461 282653, Fax: +39 0461 282672, email: alessandro.prada@unitn.it

[3]Free University of Bozen-Bolzano, Faculty of Science and Technology, Bolzano, Italy
Tel: +39 0471 017632, Fax: +39 0471 017009, email: giovanni.pernigotto@unibz.it

[4]Free University of Bozen-Bolzano, Faculty of Science and Technology, Bolzano, Italy
Tel: +39 0471 017201, Fax: +39 0471 017009, email: marco.baratieri@unibz.it

[5]Free University of Bozen-Bolzano, Faculty of Science and Technology, Bolzano, Italy
Tel: +39 0471 017200, Fax: +39 0471 017009, email: andrea.gasparella@unibz.it

*Corresponding Author

ABSTRACT

The characterization of opaque building components' thermal behavior is essential in early design stages to compare different design and refurbishment alternatives. Evaluating the dynamic thermal response to an external solicitation is necessary for an effective design especially for the climates with important annual cooling demand. According to EN ISO 13786:2007, opaque elements can be characterized through some dynamic parameters (i.e., periodic thermal transmittance, decrement factor and time shift), which can be calculated starting from the wall materials' thermal properties. However, when either the material thermo-physical properties are unknown (e.g., in existing buildings) or the assumptions about 1-dimensional heat flux on which the method is based are not met (e.g., in platform-frame structures), proper accuracy of results is not ensured. Hence, for these cases and components, a direct measurement of dynamic parameters would be extremely useful.

Differently from the procedure to determine the steady-state thermal transmission properties, which is well established, there are no technical standards defined for the experimental measurement of dynamic thermal parameters so far. Since these experimental procedures are still under development, the extent to which several aspects limit accuracy and precision of the measurements are not yet cleared. In this framework, this paper presents both experimental and numerical tests, which have been performed in order to assess the impact of different sources of uncertainty on the estimation of EN ISO 13786 dynamic parameters by means of heat flux meter measurements on a timber wall specimen.

1. INTRODUCTION

The overall energy consumption of buildings is mainly due to annual heating and cooling demand, whose relative incidence can be different according to climate and use conditions. In south Europe cities, the cooling demand can be as much important as the heating one and, for the Mediterranean localities, it can be the main cause of building energy consumption. In the current European framework (e.g., Directive 31/2010/EU) and in the EU Member legislations (e.g., Italian Minister Decree 26/06/2015), energy efficiency for space cooling has to be pursued in the first place by means of both window shading systems and adequate opaque components design. As regards the latter, an efficient opaque component has to be able to store the heat absorbed from solar irradiation and to reduce and delay its impact on the indoor energy balance. For this reason, the law provides prescriptions about the thermal parameters characterizing the dynamic performance of the opaque envelope, relating the temperature excitations at the component's exterior surface to the heat flux responses at the interior one. According to EN ISO 13786:2007 (CEN, 2007a), these quantities are the periodic thermal transmittance, the time shift and the decrement factor. While periodic thermal transmittance and decrement factor express the ability of the component to damp the effects of unsteady solicitations, the time shift indicates the ability of delaying their impact on the internal surface heat balance. If a component's material properties are well known, the calculation of the dynamic thermal parameters can be easily performed according to the methods described by EN ISO 13786 itself. However, if the thermal material properties are not comprehensively known, as it is often in case for the existing building stock (Martín *et al.*, 2010), or in case of inhomogeneous components characterized by multi-dimensional heat flows (e.g., in platform-frame structures; Martín *et al.*, 2012), inaccurate results can be found. In those contexts, a direct measurement of dynamic thermal parameters can be necessary to overcome uncertainty, insufficient knowledge of the inputs or other issues related to the applicability of EN ISO 13786 method.

While measurement strategies for laboratory and in-situ determination of steady-state thermal properties are well-defined in technical standards, such as EN 1934:1998 (CEN, 1998), standardized procedures for an experimental determination of dynamic characteristics are still missing. Nevertheless, as it can be seen in the literature (e.g., Brown and Stephenson, 1993a and 1993b; Martín *et al.*, 2010; Martín *et al.*, 2012; Sala *et al.*, 2008; Ulgen 2002), the topic has been addressed by some researchers in the last decades, often with different points of view and targets, studying the possibility of exploiting and adapting hotbox apparatus to the measurements of EN ISO 13786 dynamic parameters. Brown and Stephenson (1993a and 1993b) experimentally investigated seven wall structures in order to evaluate their dynamic heat transmission characteristics. Other researchers compared measured and calculated dynamic parameters and found some discrepancies. Ulgen (2002) tested ten different walls and saw high differences, especially for the decrement factors. Sala *et al.* (2008) measured the dynamic thermal characteristics of differently insulated brick walls and observed significant deviations between measured and simulated response time. Martín *et al.* (2010) determined the response factors of a wall with unknown material properties, measuring the heat flux on both sides of a specimen with heat flux meters. Moreover, the authors simulated their experimental setup by means of a Computational Fluid Dynamics code, finding a good agreement between measured and simulated heat flows. A further research (Martín *et al.*, 2012) focused on the dynamic thermal response of structures including thermal bridges, revealing a difference between measured and numerically simulated amplitude of the heat flux.

According to what observed in the literature, methodologies based on hotboxes and hotboxes equipped with heat flow meters can be used for the assessment of dynamic thermal characteristics of opaque envelope components. However, the extent to which the results are affected by different sources of uncertainty is not fully clear and this makes difficult the development of in-situ dynamic measurement procedures based on heat flow meters. In this work, experimental and numerical simulations were combined in order to better understand the impact of uncertainty on the measurement of dynamic parameters. A modified hotbox and heat flow meter setup was used to test a timber wall construction under dynamic boundary conditions and a numerical model was developed and calibrated against the collected experimental data. With this model, factors affecting the measurement accuracy were analyzed and discussed.

2. METHOD

2.1 Experimental setup and procedure

Experimental investigations were performed with a modified hotbox apparatus at the building physics lab of the Free University of Bozen-Bolzano. A standard hotbox apparatus according to EN 1934:1998 consists of two climatic chambers separated by a test specimen, providing two different constant thermal conditions which generate a heat flow monitored by means of a heat flow meter (*HFM*) installed in one of the two chambers. This setup was modified by removing one chamber and installing a copper coil electrical heater very close to the test specimen's surface (Figure 1). The chamber has side lengths of 1.5 m and is equipped with PID-controlled cooling (i.e., an evaporator) and heating units (i.e., an electrical resistance). A radiant screen made of plywood is positioned within the climatic chamber, in front of the test specimen, in order to prevent radiant heat exchange with the chamber walls. A fan working constantly induces an air flow through the 3.9 cm broad gap between the screen and the wall specimen at 1.5 m s^{-1}. The electrical heater covers a wall surface area of 1.2 m by 1.2 m and generates a nominal power of 1500 W, operated cyclically for 2 h followed by a break of 22 h, in order to impose a periodic forcing solicitation of a side of the tested wall.

In this research work, we used as sample a timber wall (length 1.31 m, height 1.23 m, thickness 0.24 m) made of four parallel spruce wood layers and surrounded by insulating materials (i.e., 0.28 m of EPS plus 0.09 m of sheep wool) to minimize the lateral heat flow exchanged with the air of the lab. For the same reason, the chamber kept a constant internal temperature of 23 °C, equal to lab setpoint.

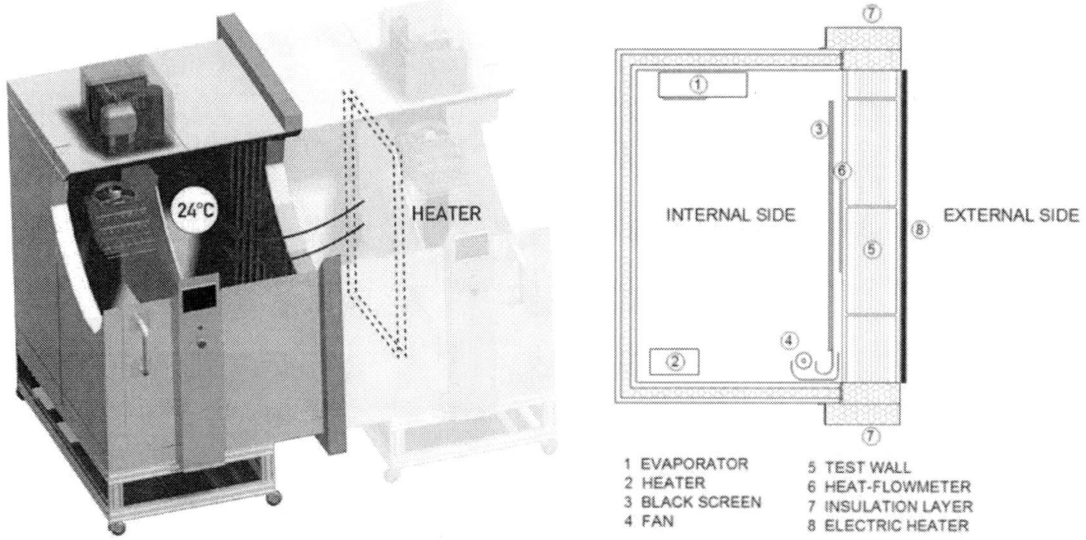

Figure 1: Experimental setup.

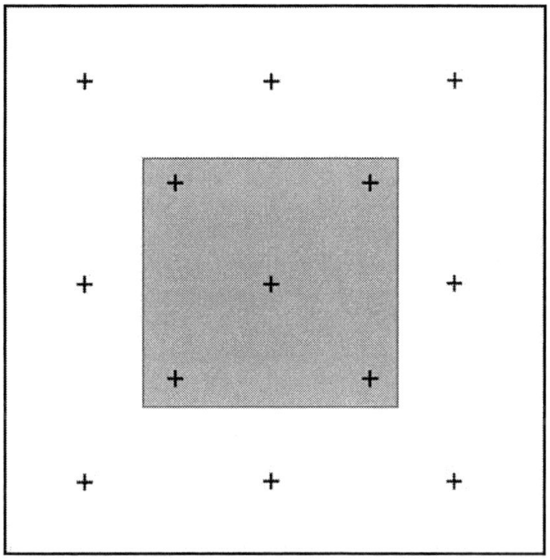

Figure 2: Schematic of the measurement setup (cross: thermocouples, grey area: *HFM*).

A set of 26 type T thermocouples measures surface temperatures at both sides of the timber wall in various positions, with one-minute time interval (Figure 2). The *HFM* used in this test, a square plate of 0.5 m by 0.5 m with an overall thickness of 0.012 m, was installed centrally with respect to the specimen's surface and consisted of a thermopile with 250 type T thermocouples. Further type T thermocouples measured the air temperature within the climatic chamber and within the laboratory environment, respectively.

2.2 Experimental evaluation of dynamic parameters

The external surface temperature and the internal surface heat flux are recorded for four days after the first three days of test. Then, the Fast Fourier Transform (*FFT*) algorithm (Press *et al.*, 2007) was applied to calculate the first 720 harmonics of external surface temperature and internal surface heat flux. For each of these days, the first term of the Fourier approximation (i.e. the fundamental one with a period of 24 h) was used to compute periodic thermal transmittance, decrement factor and time shift in accordance with EN ISO 13786:2007. Since direct measurements of surface heat flux and temperature values were collected, surface thermal resistances were neglected in these calculations. For each daily cycle, the periodic thermal transmittance Y_{ie} was computed as the ratio between the amplitude of the first harmonic of internal surface heat flux $\phi_{i,1}$ and the amplitude of the first harmonic of external surface temperature $\theta_{e,1}$:

$$Y_{i_e} = \frac{\Phi_{i,1}}{\theta_{e,1}} \quad (1)$$

The decrement factor *f* was computed by relating the periodic thermal transmittance to the steady state thermal conductance, which was found equal to 0.468 W m^{-2} K^{-1} during a preliminary steady state test:

$$f = \frac{Y_{i_e}}{C_s} \quad (2)$$

The time shift Δt_{ie} was computed using the phase displacements of the fundamental, respectively ϕ for the external temperature solicitation and ψ for the internal heat flux:

$$\theta_e \approx \overline{\theta}_e + \theta_{e,1} \cos(\omega t + \varphi) \quad (3)$$

$$\Phi_i \approx \overline{\Phi}_i + \Phi_{i,1} \cos(\omega t + \psi) \quad (4)$$

$$\Delta t_{i_e} = \frac{(\psi - \varphi)}{(2\pi)} \cdot 24 \quad (5)$$

2.3 Numerical Simulation Model

Additionally to the defined experimental model, numerical simulations were carried out with the software Delphin (Grunewald, 1997; Nicolai, 2008). Delphin 5 is a validated software (Nicolai, 2011) based on a finite control volume method, which facilitates the modelling of one- and two-dimensional heat and mass transfer processes. Aiming at an assessment of the measurement procedure and sensors reliability under dynamic conditions and at an investigation of the impact of the difference sources of uncertainty (e.g., lateral heat loss, installation of a guard ring, contact resistances), a two-dimensional numerical model was developed and calibrated against experimental data.

The model includes the timber wall specimen and an internally installed *HFM* (Figure 3). Assuming homogeneous air flow conditions on *HFM* and wall specimen, and exploiting symmetry, it was sufficient to consider a construction with half of its height merely. An equidistant grid with elements sized 1 mm by 1 mm was chosen as basic configuration, with the possibility of using elements larger up to 10 mm of height to facilitate balancing numerical errors and reducing simulation time. After some tests on some alternatives, a final simulation model with 13 952 cells was found in compliance with the criteria adopted for ensuring a normalized accuracy of residuals of 10^{-5}.

The measured surface temperatures were set as surface boundary conditions to the numerical model at the external side of the specimen, where the electrical heater was located (red dashed line in Figure 3). Top and bottom sides were defined as adiabatic, since they were well insulated and thus the heat flow through those surfaces could be assumed as negligible. Furthermore, this allowed robust findings from the adoption of 2-dimensional instead of 3-dimensional modelling as well. Differently from previous analyses (Pernigotto *et al.*, 2015), in this research the focus was put on uncertainty issues about heat transfer by conduction through the specimen and the *HFM*. Consequently, the model was limited to the solid domain and heat transfer by convection was dealt as a boundary condition to apply. Taking into consideration a model calibration, convective heat transfer coefficients were initially set according to the technical standard EN ISO 6946:2007 (CEN, 2007b): for the top of the external side of the specimen, a convective heat transfer coefficient of 7.7 W m^{-2} K^{-1} and a bulk temperature equal to the measured laboratory room temperature were set (blue line in Figure 3) while at the internal side, where forced convection was prevalent, we used a convective heat transfer coefficient of 10 W m^{-2} K^{-1}, calculated according to EN ISO 6946:2007, and the measured average air gap temperature.

Since the actual material properties of the spruce timber wall were unknown, a dry bulk density of 370 kg m^{-3}, a specific heat capacity (dry material) of 1350 J kg^{-1} K^{-1} and a thermal conductivity for a radial wood fiber direction of 0.098 W m^{-1} K^{-1} were assumed according to Kollmann (1982). Figure 4 captures an overview about the defined *HFM* geometric and thermal properties: the main layer, composed by rubber material embedding the thermopile, is positioned between an interface material and an aluminum layer, both with high thermal conductivity. *HFM* material properties were taken from manufacturer datasheets and, as regards the rubber layer, checked experimentally by means of Laser Flash Analysis.

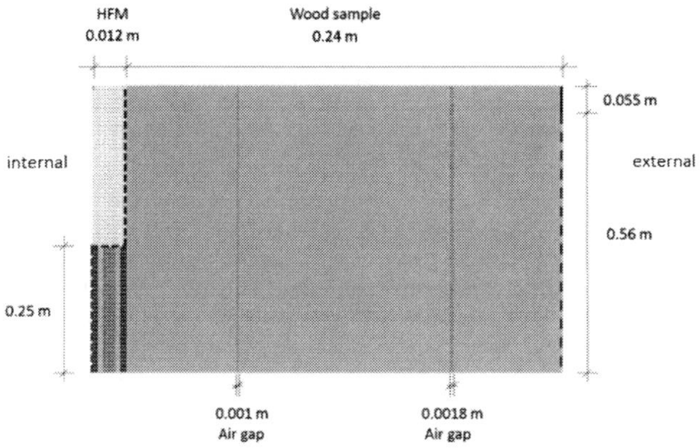

Figure 3: Structure of the numerical model.

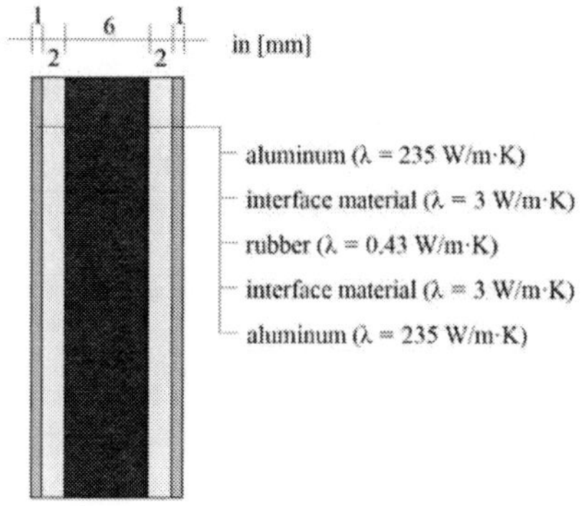

Figure 4: *HFM* geometric and thermal properties.

3. RESULTS AND DISCUSSION

3.1 Experimental results

Measured external surface temperature is shown in Figure 5 with a blue curve, reaching peak values of 52.5 °C, and minimum values of 23 °C. Filtered internal air-gap temperature, shown as red curve, averages to 23.6 °C, whereas it shows a periodically recurring unsteady behavior caused by the climatic chamber's air conditioning unit. Figure 6 shows the measured heat flux curves, both original and filtered according to second order Butterworth low-pass filter. Values move periodically in the interval between -0.5 W m^{-2} and 2.9 W m^{-2}, whereby the results show an increasing trend within the first days, due to thermal inertia effects of the test specimen.

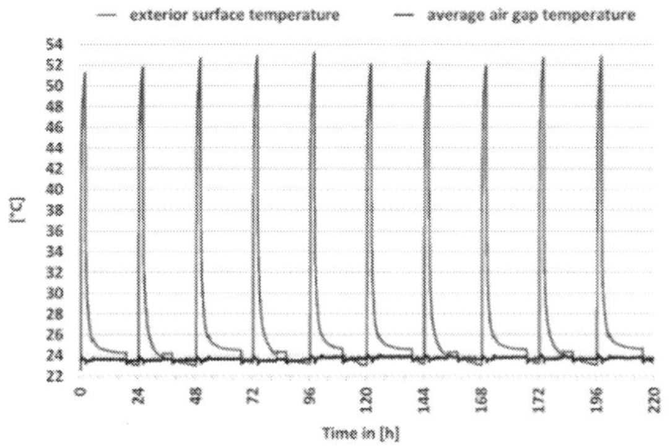

Figure 5: Measured temperature values.

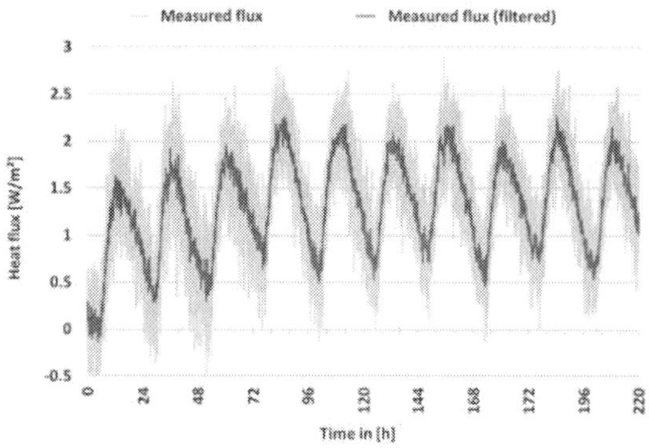

Figure 6: Measured heat flux values.

From the measured data, EN ISO 13786 dynamic parameters were calculated for each day of monitoring. Periodic thermal transmittance values averaged to 0.1285 W m^{-2} K^{-1}, decrement factor to 0.2745, and time shift to 11.67 h (Table 1).

Table 1: Dynamic thermal parameters computed from experimental measurements.

	Y_{ie} [W m^{-2} K^{-1}]	f [-]	Δt_{ie} [h]
Day 4	0.1263	0.2698	11.27
Day 5	0.1386	0.2962	11.83
Day 6	0.1301	0.2780	12.17
Day 7	0.1188	0.2540	11.42
Average	**0.1285**	**0.2745**	**11.67**

3.2 Numerical simulation results

Simulated heat flux values of a first numerical model (blue line in Figure 7) were compared to the measured results (orange curve in Figure 7): while phases match well, some disagreement can be seen between amplitudes. Furthermore, the simulated curve displays noticeable cyclic peaks, which are not so clearly observed in experimental data. As it can be seen in Figure 5 (red line), the laboratory room temperature conditions were not constant and this affected both measured and simulated heat fluxes. However, since the measured heat flux is very noisy (light orange curve in Figure 7), after the filtering process the impact of unsteady environment conditions can be hardly detected.

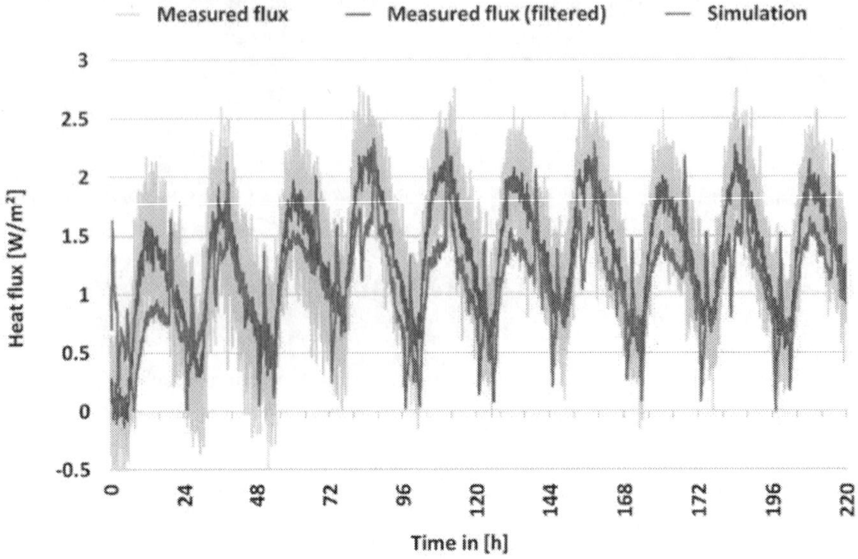

Figure 7: Measured and simulated heat flux.

3.3 Calibration of the numerical model

The preliminary numerical model achieved a good agreement only with measured phases and so a calibration process was performed against measured heat fluxes. After some preliminary sensitivity analyses, which highlighted the different impact of material properties and surface heat transfer coefficients, the calibration procedure was split into two stages.

In the first step, density and thermal conductivity of the timber wall material were adjusted, resulting in a better fit of amplitudes. Results are shown in Figure 8, for the whole measurement cycle (Figure 8a), and more in detail for the fifth and sixth day (Figure 8b). Adjusted density and thermal conductivity of spruce wood are 520 kg m^{-3}, and 0.141 W m^{-1} K^{-1}, respectively.

The second calibration phase was on the convective heat transfer coefficient at the internal surface, initially determined according to an approximated correlation. An increased constant value of 14 W m^{-2} K^{-1} led to well corresponding measured and calculated heat flux values, as it can be seen in Figure 9a and 9b.

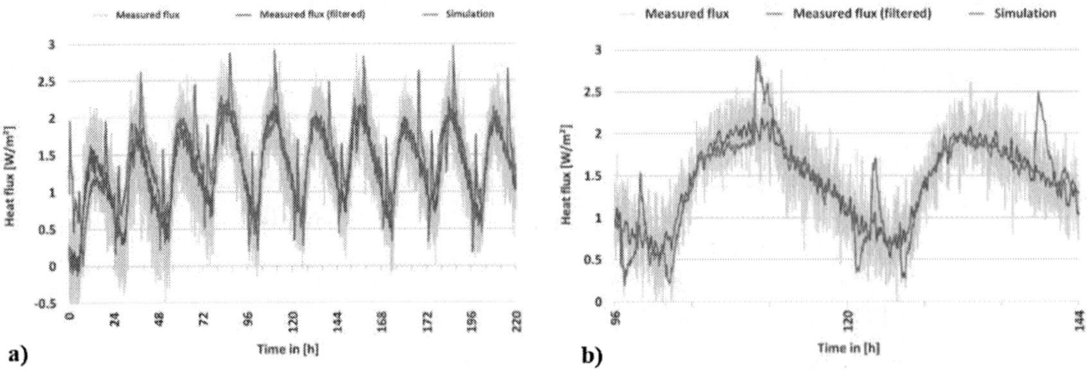

Figure 8: Results of a first calibrated numerical model - adjusted spruce wood properties.

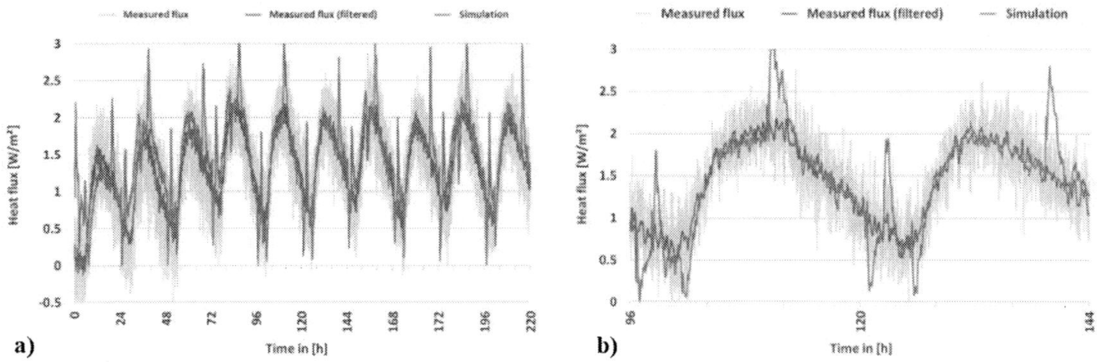

Figure 9: Results of a second calibrated numerical model - adjusted convective heat transfer coefficient.

For all models, dynamic thermal parameters were computed, basing on the first harmonics from the *FFT* analysis (Table 2). While results from a first numerical model led to high deviations from the measured average periodic thermal transmittance (-21.82 %), the calibrated numerical model presented a reduced error of 9.32 %. Moreover, deviation of the numerically achieved time shift from the measured one was very low (0.59 %).

Table 2: Dynamic thermal parameters from numerical models compared to experimental data.

	Y_{ie} [W m^{-2} K^{-1}]	Deviations from measured values	t_{ie} [h]	Deviations from measured values
Measured values	0.1285	-	11.67	-
Preliminary numerical model	0.1004	-21.82 %	11.93	2.16 %
Calibrated numerical model	0.1404	9.32 %	11.74	0.59 %

3.4 Analysis of boundary conditions

The calibrated numerical model was used to analyze the impact of different boundary conditions on the results. First, the incidence of convective conditions at the internal surface of the wall specimen was discussed. Then, we investigated the impact of some alternative experimental configurations on one-dimensional heat flux attainment (size of the *HFM* and heater, or presence of a guard ring). Finally, other possible sources of measurement errors, such as air gaps between *HFM* and specimen because of poor installation issues were studied.

Also in these analyses, the EN ISO 13786 dynamic parameters were calculated and compared to those obtained from the calibrated model (Table 3).

Table 3: Impact of guard ring and HFM poor installation on dynamic thermal parameters.

	Y_{ie} [W m^{-2} K^{-1}]	Deviations from calibrated model	t_{ie} [h]	Deviations from calibrated model
Numerical model including guard ring	0.1400	-0.34 %	11.78	0.28 %
Numerical model with air gap between *HFM* and wood	0.1288	-8.27 %	12.01	2.27 %

3.4.1 Convective boundary conditions: Convective processes induced by forced convection through over the *HFM* surface may have essential influence on results obtained by the sensor readings (Pernigotto *et al.*, 2015). So far, all simulated heat fluxes refer to the contact area between *HFM* and timber wall. Varying the position though the *HFM*, it can be appreciated a different level of noisiness in the results. For example, in Figure 10, the heat fluxes simulated at the center of the *HFM* and at the surface exposed to the air gap are compared to those at the interface with the specimen: the noisiness increases with the proximity to the air gap. In particular, the heat fluxes simulated at the *HFM* surface exposed to the air gap are affected by a noise similar to the one observed in the experimental ones. Indeed, because of the *HFM* installation on the planar surface of the specimen, there is a variation on the air gap thickness which can cause some turbulences and emphasize some unsteadiness in the behavior of the fan. According to Flanders (1985, 1994), this can cause an error to the *HFM* measurement which can be reduced, for example, by installing a tape through over the *HFM* and its edges (1985). The efficacy of this solution will be tested and improved both experimentally and numerically in further developments of the current setup, in order to optimize the measurement quality by minimizing the noise effects.

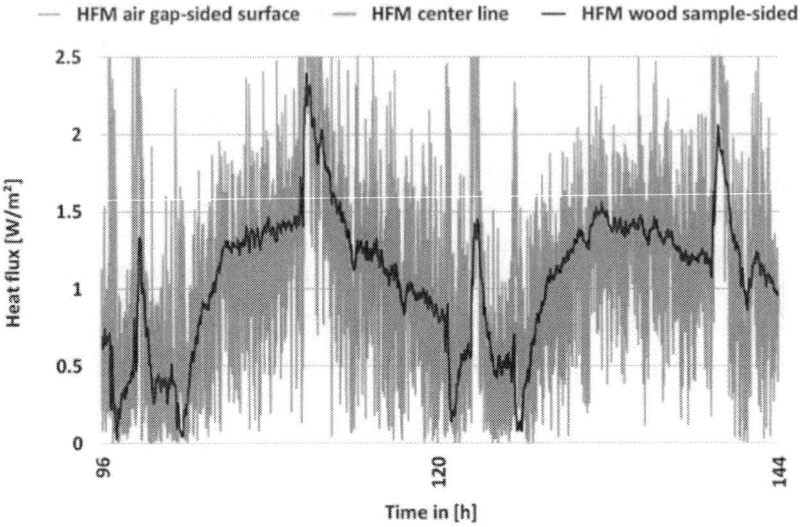

Figure 10: Different output positions for heat flux.

3.4.2 Conditions with potential impact on one-dimensional heat flux formation: The next analyses focus on the assumption of one-dimensional heat flux through the test specimen. In Figure 11, heat propagation within the timber wall is shown in temperature fields of different time steps. Due to different sizing of sample and electrical resistance, heat distribution across the sample is inhomogeneous (Figure 11a - 11c). Starting at time 103 h (Figure 11a), the external surface is excited by the electrical resistance. The influence of inhomogeneous temperature boundary conditions is clearly visible at this stage, and is even more emphasized after one hour (i.e., time 104 h in Figure 11b). After 10 hours, highest mean temperature is reached within the specimen (i.e., time 113 h in Figure 11c). By comparison, a variant with homogeneous temperature boundary condition results in a one-dimensional heat flux formation (Figure 11d). As it can be seen, the same size for specimen and electrical resistance would be of benefit but the impact of multi-dimensional heat flux on the *HFM* is negligible.

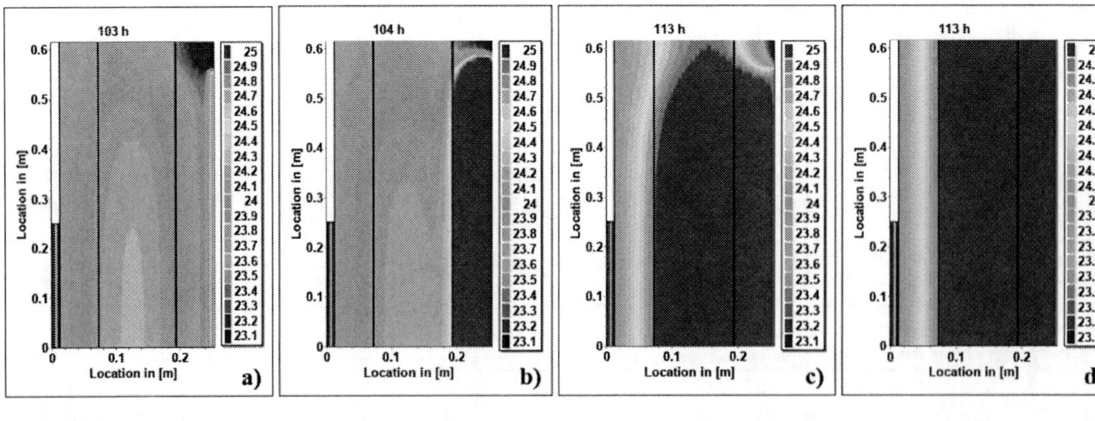

Figure 11: Temperature fields.

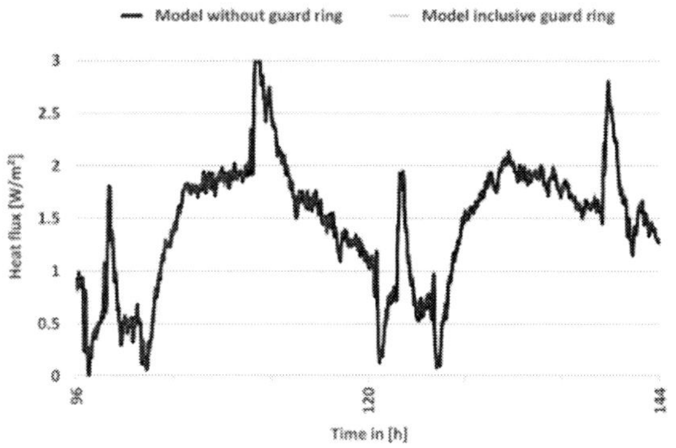

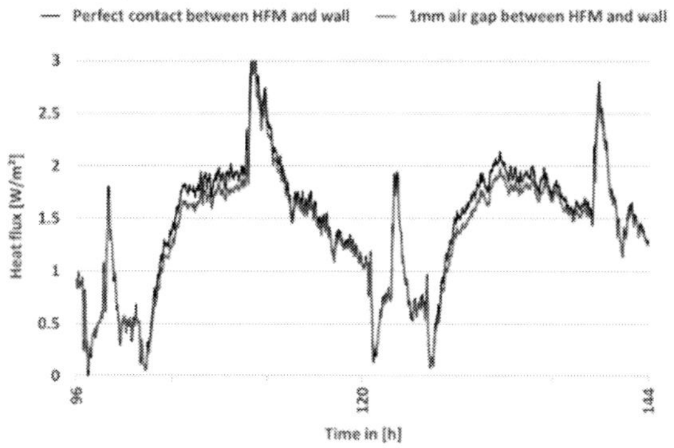

Figure 12: Influence of a guard ring around the *HFM* on numerical results.

Figure 13: Influence of a contact resistance between *HFM* and specimen on numerical results.

For reaching one-dimensional heat flux conditions, a guard ring can be installed around the *HFM.* Within the numerical model, a guard ring with *HFM*-similar properties was simulated around the *HFM*, covering the surface area of our wooden test specimen completely. Differences within the simulated heat flux of initial and modified numerical models are marginal, as it can be seen in Figure 12.

3.4.3 Effects of *HFM* installation issues: If the *HFM* is not attached properly to a specimen's surface, the contact resistance can affect the measurement accuracy. For this reason, in the experimental setup a thin layer of heat conducting paste was used. However, considering the development of in-situ dynamic measurement approaches, imprecise *HFM* installations can more easily occur and, thus, the evaluation of their impact on the final results is worth of investigation. Consequently, the calibrated model was modified by adding a 1 mm thick air gap between the *HFM* and the specimen.
As showed in Figure 13, the main effect is a reduction of heat flux amplitudes and peaks. Periodic thermal transmittance changes substantially, showing a deviation of -8.27 % from the values of the calibrated numerical model while deviations in terms of time shift are much lower (2.27 %).

4. CONCLUSIONS

In this work, a hotbox apparatus equipped with heat flow meter at the Free University of Bozen-Bolzano was modified in order to allow dynamic tests aimed at calculating EN ISO 13786 dynamic parameters (i.e., periodic thermal transmittance, decrement factor and time shift). Exploiting the results of a test on a timber wall, a calibrated numerical model was used to assess the impact of different sources of measurement errors, in the perspective of further developments of methodologies for in-situ evaluations of the dynamic parameters through the data monitored by heat flow meters. In particular, we focused on the impact of different boundary conditions (e.g., convective conditions, contact resistances, homogeneity of heat flow).
We observed that the measurement of the heat flux by the heat flow meter is affected by some noisiness. Further developments will involve CFD analysis of alternative strategies to reduce the impact of turbulences and unsteady air flow conditions adjacent to the sensors first for the use in controlled environment and then in-situ. The impact of some setup elements which may affect the required one-dimensional heat flux, such as the size of the electrical heater inducing the periodic forcing solicitation and the presence of a guard ring around the heat flow meter, was investigated and found negligible. Finally, we assessed the extent to which some installation errors of the heat flow meter can affect the results of the test. Indeed, while heat conducting paste can be used in controlled environment, the same cannot be allowed for in-situ tests. The presence of small air gaps between the sensor and the wall were found relevant, especially concerning the assessment of the periodic thermal transmittance.
Thanks to the findings of this preliminary research, new modifications will be applied to the modified hotbox, which will be used to collect more data exploitable for the development of dynamic testing procedures also to assess the EN ISO 13786 dynamic parameters of the existing building stock.

REFERENCES

Brown, C. W., and D. G. Stephenson. 1993a. A guarded hot box procedure for determining the dynamic response of full-scale wall specimens - Part I. *ASHRAE Transactions,* 99(1): 632–642.

Brown, C. W., and D. G. Stephenson. 1993b. Guarded hot box measurements of the dynamic heat transmission characteristics of seven wall specimens - Part II. *ASHRAE Transactions*, 99(1): 643–660.

European Committee for Standardization (CEN). 1998. *EN 1934: Thermal performance of buildings - Determination of thermal resistance by hot box method using heat flow meter*. Brussels, Belgium.

European Committee for Standardization (CEN). 2007a. *EN ISO 13786: Thermal performance of building components - Dynamic thermal characteristics - Calculation methods*. CEN: Brussels, Belgium.

European Committee for Standardization (CEN) 2007b. *EN ISO 6946: Building components and building elements - Thermal resistance and thermal transmittance - Calculation method*. CEN: Brussels, Belgium.

European Directive 2010/31/EU of the European Parliament and of the Council of 19 May 2010 on the energy performance of buildings, *Official Journal of the European Union*.

Flanders, S. N. 1985. Heat flow sensors on walls- What can we learn? In E. Bales, M. Bomberg, and G. E. Courville (Eds.), *Building applications of heat flux transducers* (pp. 140–159): Philadelphia, U.S..

Flanders, S. N. 1994. Heat flux transducers measure in-situ building thermal performance. *Journal of Building Physics*, 18: 28–52.

Grunewald, J. 1997. *Diffuser und konvektiver Stoff- und Energietransport in kapillarporösen Baustoffen*. PhD Thesis. Dresden, Germany.

Italian Minister Decree 26/06/2015, *National Official Gazette* no. 162, July 15 2015.

Kollmann, F. 1982. *Technologie des Holzes und der Holzwerkstoffe: Band 1: Anatomie und Pathologie, Chemie, Physik, Elastizität und Festigkeit*. Berlin: Springer-Verlag.

Martín, K., I. Flores, C. Escudero, A. Apaolaza and J. M. Sala. 2010. Methodology for the calculation of response factors through experimental tests and validation with simulation. *Energy and Buildings*, 42(4): 461–467.

Martín, K., A. Campos-Celador, C. Escudero, I. Gómez and J. M. Sala. 2012. Analysis of a thermal bridge in a guarded hot box testing facility. *Energy and Buildings*, 50: 139–149.

Nicolai, A. 2008. *Modelling and Numerical Simulation of Salt Transport and Phase Transitions in Unsaturated Porous Building Materials*. PhD Thesis. Printed at Institute of Building Climatology - Dresden University of Technology: Dresden, Germany.

Nicolai, A. 2011. *Towards a Semi-Generic Simulation Framework for Mass and Energy Transport in Porous Materials*. Proceedings of the 9th Nordic Symposium on Building Physics (pp. 559–566).

Pernigotto, G., A. Prada, F. Patuzzi, M. Baratieri and A. Gasparella. 2015. Characterization of the dynamic thermal properties of the opaque elements through experimental and numerical tests. *Energy Procedia*, 78: 3234 – 3239.

Press, W. H., S. A. Teukolski, W. T. Vetterling and B. P. Flannery. 2007. *Numerical Recipes 3rd Edition: The art of scientific computing*. Cambridge University Press: Cambridge, U.K..

Sala, J. M., A. Urresti, K. Martín, I. Flores, and A. Apaolaza 2008. Static and dynamic thermal characterisation of a hollow brick wall: Tests and numerical analysis. *Energy and Buildings*, 40(8): 1513–1520.

Ulgen, K. 2002. Experimental and theoretical investigation of effects of wall's thermophysical properties on time lag and decrement factor. *Energy and Buildings*, 34(3): 273–278.

Cold-Soak Testing Method Based Evaluation of Thermal Performance of a Typical Residential Dwelling

Minh Nguyen[1]*, Fumiaki Baba[2], Anna Sung[3]

[1]Mitsubishi Electric R&D Centre Europe, Energy and Environment Division,
Livingston, West Lothian, United Kingdom
(+44 (0) 1506 446978, m.nguyen@uk.merce.mee.com)

[2]Mitsubishi Electric Corporation, Advanced Technology R&D Centre,
Amagasaki, Hyogo 661-8661, Japan
(+81 (0) 66497 7734, Baba.Fumiaki@ap.MitsuibishiElectric.co.jp)

[3]Mitsubishi Electric Corporation, Advanced Technology R&D Centre,
Amagasaki, Hyogo 661-8661, Japan
(+81 (0) 66497 7734, Sung.Anna@ak.MitsubishiElectric.co.jp)

* Corresponding Author

ABSTRACT

Thermal characteristics of two identical British-style residential houses were evaluated and compared to each other in this paper. The two houses were built in Scotland under the Scottish building regulation 2011. Thermal characteristic evaluation of the houses was accomplished by determining their heat loss coefficient and thermal mass potential. A practical heating measurement method was applied to evaluate heat loss coefficient before a thermal mass storage evaluation was carried out using a cold soak testing method. This cold-soak testing method is based on evaluation approach under the absence of effect of solar radiation that eliminates solar heat gain during the assessment period. Because of this, the uncertainty of analytical result could be reduced and the accuracy of evaluation result would be improved. The evaluation result indicated that heat loss coefficient of the two identical houses was almost the same while their thermal mass storage potential was quite similar with a small difference of 5.5% that is thought to be caused by a different number of heating systems installed in each house as well as small differences in either variability in materials or variability in the batch of material that the houses were made.

1. INTRODUCTION

Understanding thermal characteristics of residential houses can help minimize heating or cooling load and improve efficiency of HVAC systems as well as optimization of their control system for energy savings. In most cases, evaluation of house thermal characteristics is usually to determine the amount of heat that leaks from house fabric and to estimate heat gain from solar radiation. Both heat loss and solar gain can be easily obtained by practical measurement methods that are commonly used in the UK including co-heating test and in-use heat balance test. However the practical measurement methods mentioned here are almost not able to evaluate the potential of house thermal mass storage which is also an important property of house thermal characteristics.

In this paper, the so-called cold-soak testing method is applied to quantify thermal mass storage potential of two identical British-style residential houses. The proposed approach is developed and evaluated based on the test condition of night-time period when the effect of solar radiation is absent. This means that solar heat gain received by the houses during day-time is not included in data analysis. Thus the accuracy of analysis results is thought to be more trusted because it is identified that the largest sources of measurement error in the practical measurement methods is solar heat gain.

2. DESCRIPTION OF EVALUATION FACITIY

The evaluation facility has been set-up in Scotland in the form of two highly instrumented test houses that are mirror-symmetrical and identical detached houses. The test houses were built to comply with Scottish Building Standards (S.B.S) regulations 2011, under the Domestic regulations 2011 section. The houses were also designed to be backward compatible to the S.B.S regulations 1991, 1985 etc. by removing insulation accordingly.

The design U-value of the house elements is shown in Figure 1 in comparison with the standard U-value as assigned in the S.B.S regulations 2011. It can be seen from Figure 1 that the design U-value of some house elements is slightly higher than the standard U-value. This is attributable to the presence of thermal bridges in the design U-value calculation that results in a higher heat transfer through the house elements. The air change rate was also designed in accordance with the British Standard BS EN 12831 with a design value of 1.71 ACH (air change rate per hour).

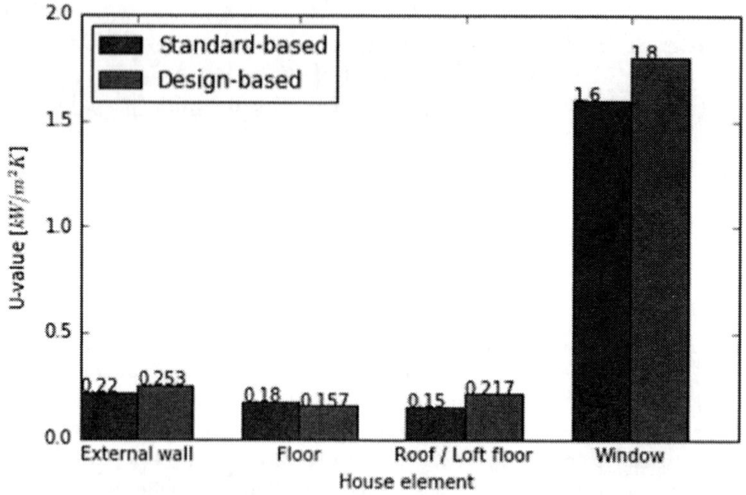

Figure 1: Standard and design U-value of house elements

There are about 1,500 sensors that were installed to measure the detailed indoor and outdoor environment conditions and in-depth monitor house performance as well as the operation of heating systems. The front view of the test houses in the evaluation facility can be seen in Figure 2. Each house consists of two stories with three bedrooms and one combined living and dining room which are main living spaces. Besides the main rooms, the houses also include smaller rooms such as kitchen, hall, bathroom, toilet etc. Number of rooms and floor plans are shown in Figure 3.

Figure 2: Mitsubishi Electric House-type Evaluation Facility

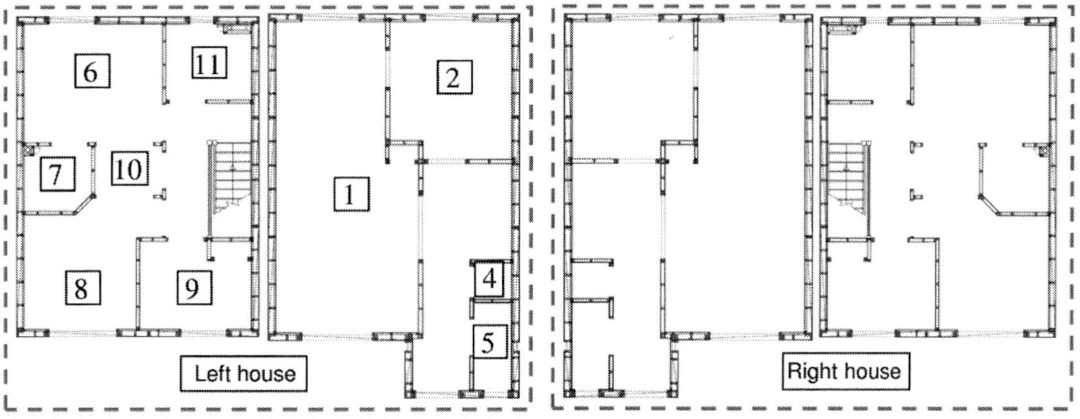

Figure 3: First-floor and ground-floor plans of test houses

Individual room volume is shown in the following table in comparison with whole house volume. Each room is assigned with its reference number including rooms 1-5 in ground floor and rooms 6-11 in first floor. Total area of ground- and first floors are 49.1 m^2 and 45.5m^2 respectively.

Table 1: Individual room volume in comparison with whole house volume

	Description	Volume of room / volume of the house [%]
Room 1	Living and dining room	29.5
Room 2	Kitchen	12.1
Room 3	Hallway and landing	11.0
Room 4	Shower	1.1
Room 5	Toilet	2.6
Room 6	Bedroom 1	12.9
Room 7	En-suite	3.5
Room 8	Bedroom 2	11.6
Room 9	Bedroom 3	7.8
Room 10	Utility room	2.6
Room 11	Bathroom	5.2

3. PROPOSED APPROACH

To evaluate thermal mass storage potential of the two houses by a cold-soak testing method, heat loss coefficient of whole house need to be determined. Assuming that the evaluation period is during night-time, amount of heat supplied from heat pump to heat the houses is equivalent to heat loss from indoor to outdoor. Thus whole-house heat loss coefficient is determined by the following equation:

$$\alpha = Q_{HP}/(T_h - T_a) \tag{1}$$

where,

$$Q_{HP} = \rho \times q \times C_p^w \times (T_{out}^w - T_{in}^w) \tag{2}$$

$$T_h = \sum(T_r \times V_r)/\sum V_r \tag{3}$$

where, T_r is the average temperature of each room in the house which is the mean of representative temperatures measured at many positions in the room and calculated as follows:

$$T_r = \frac{\sum T_i}{n}$$

After heat loss coefficient is obtained from a heating test using heat pumps, cold-soak test is applied to quantify apparent heat capacity of the houses for thermal mass evaluation. Based on this approach, test houses are isolated without heating for several days until trends of indoor temperature variation stabilize following the trends of outdoor temperature variation. Home appliances that are potential to dissipate heat to the houses such as computers, fridge, TV, lighting etc. must be switched off during the cold-soak test. Thermal mass storage of the houses is evaluated by the equation as follows:

$$C_r\left(\frac{dT_r}{dt}\right) = Q_s - Q_L \quad (4)$$

Due to the evaluation period of thermal mass storage that is also during nighttime from 11:00pm to 4:00am, solar heat gain Q_s is negligible while heat loss Q_L can be calculated based on whole-house heat loss coefficient from equation (1). Thus equation (4) becomes:

$$C_r\left(\frac{dT_r}{dt}\right) = -\alpha\,(T_h - T_a) \quad (5)$$

Based on equation (5), heat loss $-Q_L$ is plotted against hourly room temperature drop $\frac{dT_r}{dt}$ to give apparent heat capacity C_r as the gradient of the linear regression. Thus apparent heat capacity C_r can be obtained from the regression plot and representative for thermal mass storage of the house as follows:

$$C_r = -\alpha\,(T_h - T_a) \times \left(\frac{b}{c}\right) / \left(\frac{dT_r}{dt}\right) \quad (6)$$

where, b is a conversion factor from kW to kJ/h and c is a conversion factor from kJ to MJ.

4. EVALUATION RESULTS

4.1. Heat loss coefficient determination

For heating test to determine heat loss coefficient, both the left and right houses were heated by heat pumps. The house indoor were maintained at a stable high temperature while also taking the advantage of stable low temperature of the ambient in order to obtain a good condition of determining heat loss coefficient. Figure 4 shows a steady-state condition of indoor and outdoor temperatures and night-time periods at the winter time when heat loss coefficient was determined.

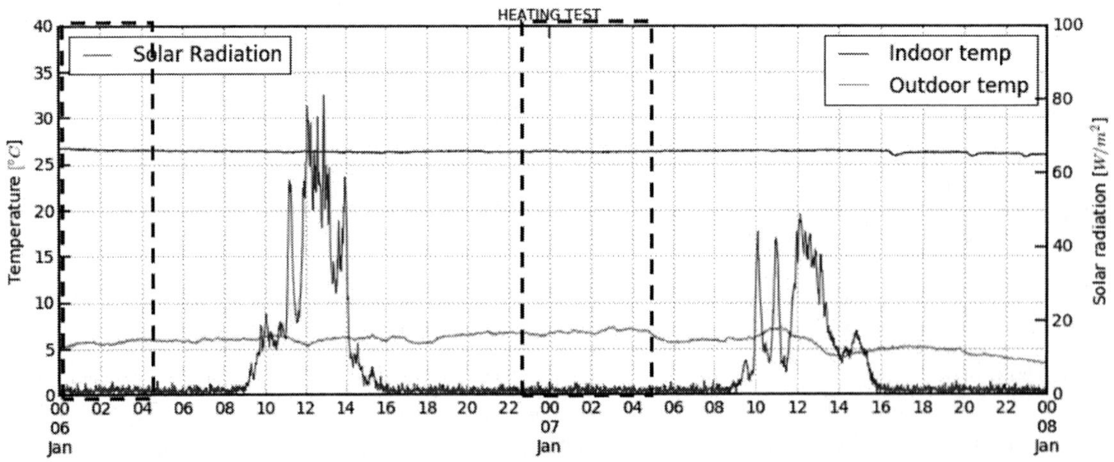

Figure 4: Data analysis period for whole-house heat loss coefficient (left house as a representative)

Due to the absence of solar gain during the night-time and no heat dissipation from home appliances, supplied heat produced by heat pumps is equivalent to heat loss of the whole house. As shown in Figures 5 and 6, heat loss coefficient is the gradient of the linear regression line in the plot that is forced through the origin due to the assumption of no supplied heat when the indoor and outdoor temperature difference is zero.

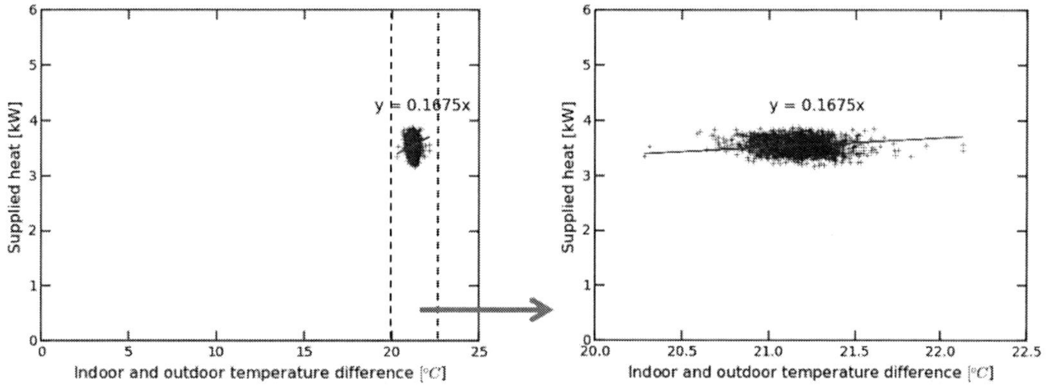

Figure 5: Regression plot for determining whole-house heat loss coefficient of left house

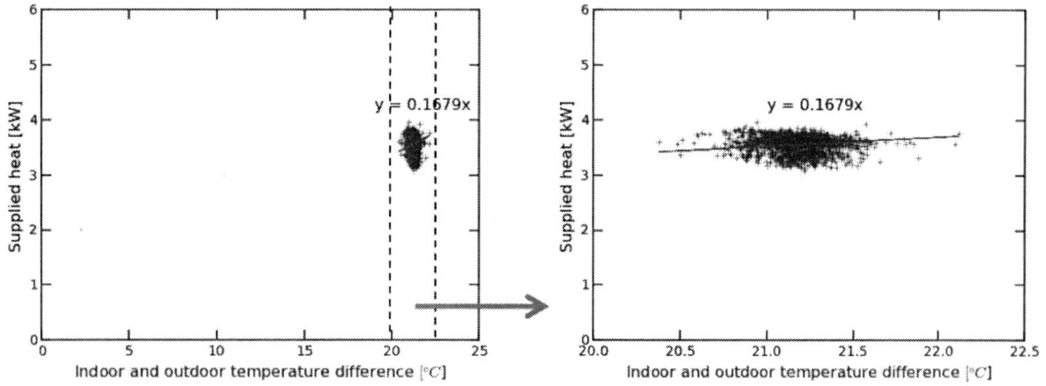

Figure 6: Regression plot for determining whole-house heat loss coefficient of right house

It can be seen from Figures 5 and 6 that heat loss coefficients of the left house ($\alpha = 0.1675$ kW/K) and of the right house ($\alpha = 0.1679$ kW/K) are almost identical. This is due to the same house design and architecture used for construction of the two test houses.

4.2. Thermal mass evaluation by cold-soak testing method

The cold-soak test was carried out for nearly ten days under non-heating condition in summer time from 01/07/2015 to 10/07/2015 when effect of solar radiation is seen to be significantly high. It took about 3 days from beginning of the test before indoor temperature variation reached a steady state together with steady-state variation of outdoor temperature and solar radiation.

The steady-state period for data analysis is highlighted in the dashed line as seen in Figure 7. It is a consecutive 5-day test period from 04/07/2015 to 09/07/2015 when the magnitude of indoor temperature fluctuation was quite similar throughout 5 testing days while the outdoor temperature fluctuation matched well with the variation of solar radiation.

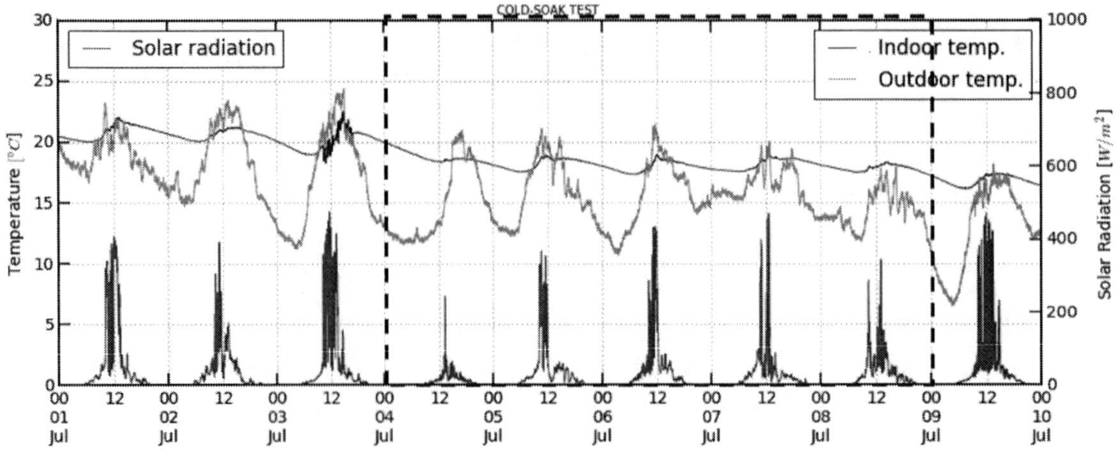

Figure 7: Evaluation period for thermal mass evaluation under cold-soak test (left house as a representative)

The evaluation period is taken with only night-time data from 11:00pm to 4:00am when solar radiation is absent. This means that the houses were being cooled down during the evaluation period due to heat loss from indoor to outdoor that causes indoor temperature to drop. Figure 8 shows the approach to quantify thermal mass storage of the two houses.

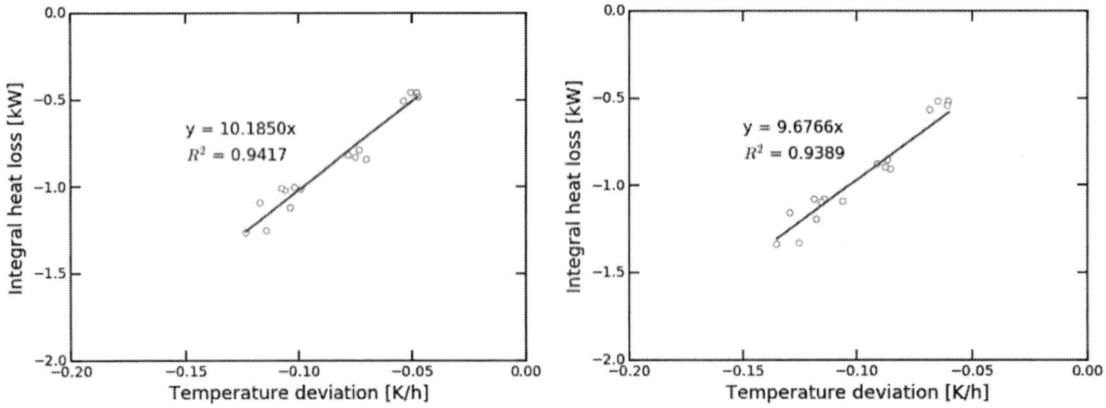

Figure 8: Regression plots for determining apparent heat capacity of left house (left plot) and right house (right plot)

Similar to the approach used for obtaining heat loss coefficient in Figures 5 and 6, if assuming no heat loss from house indoor to outdoor, room temperature has no drop. Thus the linear regression line must be passed through the origin (0, 0). By applying the equations (5) and (6), apparent heat capacity of the left and right houses is derived and summarized in table 2.

Table 2: Apparent heat capacity for thermal mass storage potential evaluation

	House	Apparent heat capacity [MJ/K]
1	Left	36.7
2	Right	34.8

Based on the analysis result of apparent heat capacity as shown in table 2, thermal mass storage is seen to be quite similar between the two houses with a small difference of 5.5%. This would be caused by a larger number of heating systems installed in the left house compared to the right house as well as small differences in either variability in materials or variability in the batch of material that the houses were made.

4. CONCLUSIONS

Thermal characteristics of two identical British-style residential houses in Mitsubishi Electric House-type Evaluation Facility are analysed and compared to each other. Thermal characteristic evaluation was accomplished by determining heat loss coefficient from a steady-state heating test and thermal mass storage potential from a cold-soak test. Due to the evaluation period of heat loss coefficient and thermal mass storage that was night-time when solar radiation was negligible, solar gain was not included in data analysis. Because of this, measurement error of solar radiation was eliminated from analysis results. The uncertainty of analytical result could be reduced and the accuracy of evaluation result would be improved. Comparing the evaluation results between the two houses, the heat loss coefficients were almost the same while thermal mass storage potential were quite similar with only a small difference of 5.5%.

NOMENCLATURE

Symbols

C_r	Apparent heat capacity [MJ/K]
C_p	specific heat capacity [kJ/kg°C]
Q	heat capacity [kW]
t	time [hour]
T	temperature [°C]
U	heat transfer coefficient [W/m^{2}K]
V	volume of room [m^3]
a, b	conversion factors

Greek

α	heat loss coefficient [kW/K]
ρ	water density [kg/m^3]

Subscripts

a	ambient
i	position for room temperature measurement
in	inlet
L	heat loss
h	house
HP	heat pump
out	outlet
S	solar
r	room

Superscripts

w	water

REFERENCES

Allinson, D., (2007), Evaluation of aerial thermography to discriminate loft insulation in residential housing, *Doctoral thesis*, University of Nottingham.

Allinson, D., Stephen P. (2014), Ad-hoc assessment of infiltration, thermal bypasses and thermal bridging, *Technical report*, School of Civil and Building Engineering, Loughborough University.

Anderson, B., (2006), Conventions for U-value calculations, *2006 edition*, BRE Press, Watford.

Butler, D., Dengel, A., (2013), Review of co-heating test methodologies, *Research report of NHBC Foundation*, BRE Environment, BRE.

Jack, R., Loveday, D., Allinson, D., & Porritt, S. M. (2015), Practical measurement of the thermal performance of houses, *Innovation and Research Focus*, (100), 6–7.

Johnston, D., Miles-Shenton, D., Farmer, D., Wingfield, J., (2013), Whole house heat loss test method (Co-heating), *Leeds Beckett University, Leeds*.

Richard J., (2015), Building diagnostics: Practical measurement of the fabric thermal performance of houses, *Doctoral Thesis*, Loughborough University.

ACKNOWLEDGEMENTS

The authors wish to acknowledge the cooperation and support of Mitsubishi Electric Air-Conditioning System Europe Ltd (M-ACE) during the construction of test houses. The authors would also like to acknowledge Mr. Gordon McDonald for his excellent installation of monitoring and measurement systems and his kind help during the test process.

Energy Savings Potential of Phase Change Material Integrated Building Envelope in South Texas

Hui SHEN[1]*, Xiaoyu LIU[1]

[1]Texas A&M University - Kingsville, Department of Civil and Architectural Engineering
Kingsville, Texas, USA, Hui.Shen@tamuk.edu

* Corresponding Author

ABSTRACT

The performance of phase change material (PCM) integrated building envelope in terms of peak load reduction, peak load time delay and thermal demand improvement was investigated for a prototype detached single family house in a hot climate location. Five market available PCM layers, which can be easily inserted into traditional wall construction, were selected and coupled into different walls and different locations within the wall. Annual simulation results from EnergyPlus indicated the significant effect of PCM integrated envelope for the studied case. The PCM layer of melting temperature of 23 C is determined as the best one for south wall implementation. When it is placed to the interior of the insulation component of the wall, heat flux at heating load peak is reduced by 22.3% and heat flux at cooling load peak is reduced by 12.3%. For ceiling application, the best performing one is the PCM layer of melting temperature of 25 C and when it is placed below the insulation component of the ceiling. Reduction in peak heat flux is 30.2% for heating and 25.8% for cooling. The greater reduction in peak heat flux of ceiling than south wall is due to its larger area. For peak heat flux time delay, all PCM cases show similar effect of about 30 minutes for typical heating time and 60 minutes for typical cooling time. The performance of PCM layers in reducing annual thermal demand is not as evident as in heat flux: reduction in heating demand ranges from 7.9% to 54.34% and reduction in cooling demand ranges from 1.2% to 7.2%.

1. INTRODUCTION

The thermal properties of building envelope are critical with regard to building energy consumption and occupants' thermal comfort. As a result, different versions of standards and codes have been developed to guide the design of building envelope. For example, ASHRAE standards (2007) and International Energy Conservation Code (IECC) have published prescriptive criteria of thermal resistance for envelop elements including ceilings, walls, floors and fenestrations according to weather characteristics of different climate zones. Increased thermal resistance can reduce heat transfer between building inside and outside, and consequently decrease energy consumption. However, there is a tradeoff point beyond which additional increase in thermal resistance will instead result in higher energy consumption. To further reduce peak loads of building air-conditioning and save energy, phase change materials (PCMs) have been integrated into building envelope. PCMs are substances change phase at a certain temperature range with capabilities of storing and releasing large amounts of energy. Being an integrated component of envelope, they absorb heat when outdoor temperature rises and solar energy strikes the building until complete phase change occurred. When surrounding temperature is lower than their phase-change temperature, heat will then flow out of PCMs and reverse phase change occurs. Previous researches have shown that PCMs integrated building envelope can mitigate indoor air temperature swings, decrease cooling loads and greatly reduce or shift the building peak loads (Zhu et al., 2009; Gracia and Cabeza, 2015).

Halford and Boehm (2007) simulated reductions in peak cooling load by 11-25% and 19-57% via installing encapsulated PCM within the ceiling or wall insulation when compared to envelopes of the same mass and of the same insulation but no phase change, respectively. Kondo and Ibamoto (2006) experimented the effects of the peak shaving control of air-conditioning systems using PCM ceiling board in an office building located in Tokyo. They reported that the maximum thermal load reduced to 85.2% of that of using rock wool ceiling board after adding microcapsulate PCM. But the integrated thermal load increased by 5.3%. Kissock et al. (1998 and 2006) investigated the thermal performance of phase change wallboard experimentally and numerically. They measured a reduction in indoor peak air temperature up to 10 °C comparing the PCM test cell with the conventional test cell (1.22 m × 1.22 m × 0.61 m). A finite difference numerical mode was developed and modified using their measured data. For the location of Dayton Ohio, the model simulated that the addition of PCM gypsum wallboard in frame walls reduced the peak and annual cooling loads through the wall by 16% and 9% respectively. The performance of PCM frame wall in reducing peak air-conditioning demand in residential buildings was studied by Zhang et al. (2005) with two test cells (1.83 m × 1.83 m × 1.22 m). Their experimental results showed that the retrofit test cell with a 10% PCM concentration reduced the peak heat flux through the wall and space cooling load by about 15% and 8.6% respectively, as compared with the control test cell. They also reported that the west and north walls achieved higher reductions in wall heat flux than the south wall and the load shifting was spread over many hours from about midnight until about 1 pm. Later the test cells were used again to access the integration of a thin PCM layer in a residential building wall for heat transfer reduction and management (2015). The PCM layer was installed at different locations within the wall cavities of south and west facing walls of the test cells to identify the optimal integration locations. They reported the maximum peak heat flux reduction of 51.3% for the south wall and 29.7% for the west wall when the PCM layers were at their respective optimal locations. Also, their test showed the maximum peak heat flux time delay of 6.3 hours and 2.3 hours for the two walls. Through hot box experiments, Cao et al. (2010) concluded that the attenuation effect of PCM layer on the mean air and interior surface temperatures was noticeable. The benefits of peak heat flux reduction and time delay can be reflected as reduced size of air-conditioning equipment, and improved system operation and efficiency.

In order to enable whole building dynamic simulation with PCM presents, several studies have been performed to develop or refine the PCM modeling procedure. EPS-r models the behavior of PCM with its special materials facility (Heim and Clarke, 2004; Heim, 2006). The effect of phase transition is expressed as a latent heat generation term based on effective heat capacity method adding to the energy balance equation. Ibanez et al. (2005) presented a simple methodology for the energetic simulation of buildings with PCM components in TRNSYS (Type 56). The PCM model in EnergyPlus was verified and validated using a similar approach as dictated by ASHRAE Standard 140 (2007) consisting of analytical verification, comparative testing and empirical validation (Tabares-Velasco et al., 2012). This valuable work identified and fixed two bugs in the program and enabled validated and reliable simulations to be delivered by versions beyond 7.1, along with its outstanding strengths in whole building dynamic modeling. Besides, the program was also tested before by an additional set of eight cases, a combination of analytical cases and numerical cases, in Tabares-Velasco and Griffith (2012). Based on these work, researchers should feel confident about using EnergyPlus in studies involving PCM components.

Despite the researches reviewed above, the studies on energy performance of buildings using PCMs are still insufficient. The effectiveness of PCM components in envelope is strongly influenced by several factors including climate, building characteristics, PCM integration location in the wall, PCM phase changing temperatures, application wall orientation, and etc. Texas is a highly populated and energy-intensive state with long and hot summer. Residential buildings in the area typically have light construction and present low energy efficiency. Therefore, a comprehensive study of PCMs integrated residential building envelope is needed. In this paper, five market available PCM layers are coupled into the walls and ceiling of a prototype detached single-family house in San Antonio, Texas. Annual simulations are performed using EnergyPlus to evaluate the improvements of peak load and thermal demand of cases with PCM layers at different locations in the space surfaces over the no PCM case. Peak load time delay and free-floating uncomfortable indoor air temperature mitigation are also investigated for representative cases with optimal PCM layer at optimal location.

2. PCM MODELING IN ENERGYPLUS

The titles of the main sections have to be centered, numbered and in 12-point bold type capital letters. With the exception of the abstract, nomenclature, references, and acknowledgements section headings have to be numbered. Blank lines need to be placed above and below each main section title. To model building envelope with PCM components in EnergyPlus, the conduction finite difference (CondFD) solution algorithm should be employed. The algorithm assumes one-dimensional energy flow through envelope construction and discretizes all elements into thermal nodes depending on the thermal diffusivity of the material and simulation time step (EnergyPlus, 2016). There are two options of implicit finite difference scheme for the ConFD algorithm, and the fully implicit scheme is used in this study. Heat balance of an internal thermal node (such as one of the PCM layer nodes) for the fully implicit scheme is expressed in equation (1) (EnergyPlus, 2016; Tabares-Velasco et al., 2012).

$$c_p \rho \Delta x \frac{T_i^{j+1} - T_i^j}{\Delta t} = k_W \frac{\left(T_{i+1}^{j+1} - T_i^{j+1}\right)}{\Delta x} + k_E \frac{\left(T_{i-1}^{j+1} - T_i^{j+1}\right)}{\Delta x} \quad (1)$$

In equation (1), k_W is thermal conductivity for interface between the ith node and the $(i+1)^{th}$ node; k_E is the thermal conductivity for interface between the ith node and the $(i-1)^{th}$ node. They are calculated as the average of the conductivities of the two materials at the interface. T is temperature; i is the node being modeled; $i+1$ is the adjacent node to the interior; $i-1$ is the adjacent node to the exterior; $j+1$ is the new time step; $j-1$ is the previous time step; Δt is time step; Δx is the finite difference layer thickness; ρ is density of material and c_p is specific heat of material.

For PCMs c_p is varying depending on the state of the material, so equivalent specific heat (equation (2)) obtained from the enthalpy-temperature function is used and updated at each time step in equation (1).

$$c_p^*(T) = \frac{h_i^j - h_i^{j-1}}{T_i^j - T_i^{j-1}} \quad (2)$$

where, h is enthalpy which is also a function of T.
The verification and validation studies indicated that time steps equal to or shorter than three minutes should be used in EnergyPlus for PCM modeling for accuracy concern. So in this study, time step is set to three minutes in the following simulations.

3. STUDIED PCM LAYERS

PCMs integration in building envelope has several different methods and they were all studied in existing literature. For example, imbibing, macro-encapsulation, direct mixing (of PCM with blown-in insulation). A relatively new integration method studied (Jin et al., 2013) is to encapsulate PCM in thin sealed polymer pouches, arrange in sheets laminated with aluminum foil on both sides and perforated around the PCM pouches. Similar products of this type are available from market with different and customable thickness, various densities, melting temperatures and specific heats. In this paper, five PCM layers, which can be easily inserted into traditional wall construction, are investigated. They share some common properties: a moderate thickness of 0.0208 *m*, thermal conductivity of 0.2 *W/m·K*, specific heat of 1970 *J/kg·K*, and density of 235 kg/m^3. Differences among them lie in the enthalpy-temperature function shown in Figure 1 and melting temperature. PCM1 melts around 21 °C as can be seen from Figure 1, or in other words, 21 °C is the peak melting temperature of PCM1. The peak melting temperatures of the other four PCM layers are 23 °C for PCM2, 25 °C for PCM3, 27 °C for PCM4 and 29 °C for PCM5.

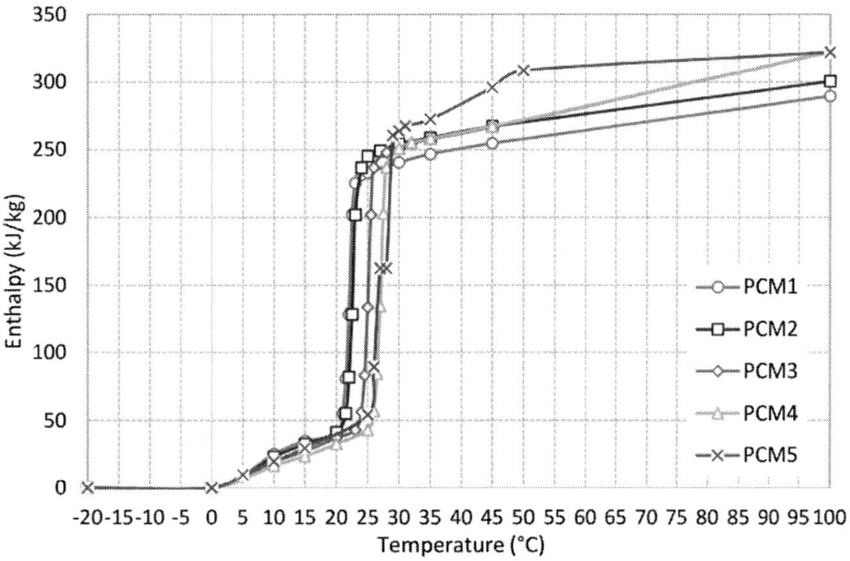

Figure 1: Variation of enthalpy as a function of temperature for the five investigated PCM layers

4. PROTOTYPE SINGLE-FAMILY HOUSE IN SAN ANTONIO TEXAS

The city of San Antonio is selected as a representative of the south Texas area in the study. It features a hot and long summer, comfortably warm or mild winter with a dozen subfreezing nights. Its weather information including monthly total solar energy incident on horizontal surface, monthly average outdoor air temperature and monthly average diurnal temperature range is illustrated in Figure 2. It's worth noticing that the diurnal temperature range is varying around 11 °C with a maximum of 25 °C and a minimum of 1.6 °C throughout the year.

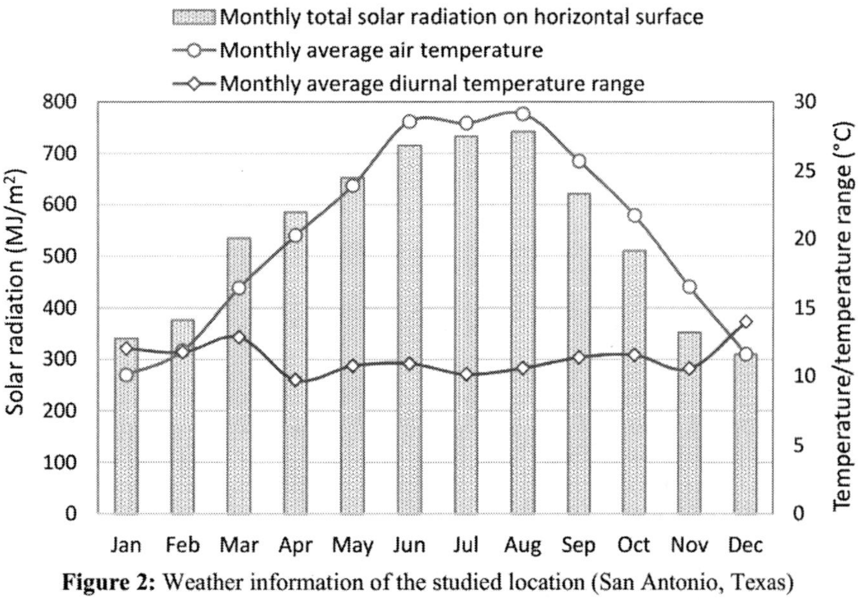

Figure 2: Weather information of the studied location (San Antonio, Texas)

A prototype of detached single-family house for this area is shown in Figure 3. It's from EnergyPlus example file consisting of three zones: the air-conditioned living zone, and the unconditioned attic zone and garage zone. The floor area of the three zones are 185 m^2, 228 m^2, and 42 m^2, respectively. There is one window on each wall, occupying

27% of the north wall, 17% of the south wall, and 26% of the east wall and west wall. Heating and cooling are always available for the living zone throughout the year with temperature setpoint changing with seasons: from Apr 1 to Oct 31, setpoints are 24 °C for cooling and 18 °C for heating; for the rest of the year, setpoints are 26 °C for cooling and 20 °C for heating.

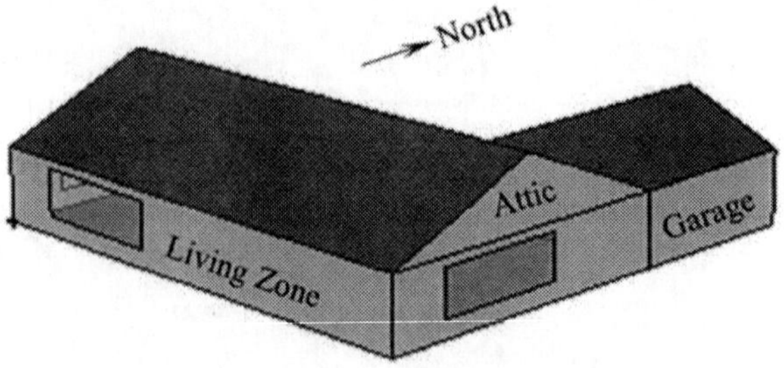

Figure 3: 3D view of the studied single-family house

For this type of residential house in the studied area, IECC (2006) and ASHRAE Standard 90.2 (2007) require thermal resistances of the ceiling (with attic space) and exterior walls to be R-30 and R-15. The Pacific Northwest National Laboratory (PNNL) have developed constructions satisfying this requirement and are adopted in this study. Details of the ceiling, roof and exterior wall constructions are listed in Table 1. Please refer to the residential prototype building models developed by PNNL for detailed construction of each envelope element.

Table 1: Construction details of the studied house

	Exterior wall (living zone)		**Ceiling (living zone)**
Outside layer	Stucco_1in	Outside layer	Ceil_consol_layer: 0.266 m
Layer 2	Bldg_paper_felt	Layer 2	Drywall_1/2in
Layer 3	Sheathing_consol_layer		
Layer 4	OSB_5/8in		**Roof**
Layer 5	Wall_consol_layer: 0.0889 m	Outside layer	Asphalt_shingle
Layer 6	Drywall_1/2in	Layer 2	OSB_1/2in

The 'Wall_consol_layer' of exterior wall and 'Ceil_consol_layer' of ceiling are selected as potential locations to insert the PCM layer and their thicknesses are indicated in Table 1 as well. Depending on thickness and thermal properties, 3 locations are investigated for exterior wall: PCM layer is placed to the exterior, at the middle and to the interior of 'Wall_consol_layer'. Similarly, 5 locations are investigated for ceiling: PCM layer is placed above, at the depth of ¼, ½ and ¾ of the thickness of 'Ceil_consol_layer' within and below the 'Ceil_consol_layer'. Note that complete schedules and internal heat gains for typical single-family houses follow the same settings from PNNL models.

5. RESULTS AND DISCUSSION

5.1 No PCM Integration Case

Load simulation is first performed for the original design of the house without PCM. Annual thermal demands per floor area, peak loads, corresponding heat flux through walls at load peaks and annual peak heat flux through walls for heating time and cooling time are summarized in the first row of Table 2 and Table 3. Cooling is dominating in San Antonio. The annual cooling demand is 63.01 $kWh/m^2 \cdot year$ while the annual heating demand is only 5.32 $kWh/m^2 \cdot year$. Peak cooling load almost doubles peak heating load, and they are 6545.97 W and 3361.69 W respectively. It is also illustrated that the heat transfer through south wall is more intense than that of ceiling. With three minutes as the simulation time step, there are 949.4 hours during the year that the air-conditioning system is in heating mode. The corresponding number of hours for cooling is 5538.8 (also listed in Table 2 and Table 3). Because heating and cooling are always available, it is simulated that there are 105 days with heating needs and 323 days with cooling needs, indicating 63 days of both heating and cooling requirements within one day.

5.2 PCM Integrated Into South Wall

In total, 15 cases are simulated for integrating PCM layer into the south wall of living zone, combining the 5 samples of PCM layer and three integration locations in the wall. Simulation results are listed in Table 2. The influence of PCM layer on heating demand is significant with reductions ranging from 7.9% to 20.4% comparing with the case without PCM. However, the influence on cooling demand is minimal ranging from 1.2% to 2.9%. The third main column shows small decrease in peak thermal loads, about 0.6% ~ 3.7% for heating and 0.9% ~ 6.6% for cooling. It can be seen from the fourth and fifth main columns that heat transfer through the south wall is greatly decreased at both heating time and cooling time. The range of reduction in heat flux through wall at load peaks is from 8.3% to 52.7% during heating time and from 12.2% to 130% during cooling time. Reductions of annual peak heat flux through the south wall show similar trend. It may seem confusing that the PCM layers reduce heat flux through wall significantly but only result in minimal reduction in cooling demand. This can be explained by looking at the last main column of Table 2 and considering the inadequate area of south wall (42.04 m^2). For the studied amount of PCM in south wall, it benefits by decreasing heat transfer through the implementation wall, but also harms by shifting heat transfer to a later time which results in longer time with cooling requirement. Like the south wall integration cases here, the benefits are not superior greatly to penalties, so PCM integration presents great reduction in heat transfer but minimal decrease in annual cooling demand. If the amount of PCM increases, for example by using thicker layers or applying on larger wall area, the reduction in thermal demands could be more notable.

Use annual thermal demands as criterion, the best integration location (highlighted in Table 2) shows itself as inside the 'Wall_consol_layer' for all studied PCM layers. Comparing the PCM layers with each other, PCM2 is the best performing considering thermal demands and typical system efficiency.

Besides of heat flux decrease, another very important effect of PCM is peak time delay. To show this effect clearly, the days of monthly peak heating load and monthly peak cooling load for the case without PCM are identified and peak heat flux times of best-performing case are shown for these days. It is found that, on average the peak heat flux time is about 30 minutes for heating and 60 minutes for cooling later than the corresponding peak heat flux time of no PCM case.

Table 2: Summary of annual simulation results for all studied cases with PCM integrated into south wall

Cases	Thermal demand (kWh/m²·year)		Peak load (W)		Heat flux through south wall at load peaks (w/m²)		Peak heat flux through south wall (W/m²)		Total number of hours with	
	Heating	*Cooling*	*Heating*	*Cooling*	*Heating*	*Cooling*	*Heating*	*Cooling*	*Heating*	*Cooling*
NoPCM	5.32	63.01	3361.69	6545.97	-11.88	11.89	-13.14	17.23	949.4	5538.8
PCM1S1	4.85	62.27	3337.22	6486.86	-10.39	10.38	-11.87	15.48	907.2	5560.3
PCM1S2	4.69	62.06	3238.07	6460.40	-5.62	9.66	-11.04	14.98	897	5554.4
PCM1S3	4.24	61.63	3271.79	6487.63	-7.76	10.44	-11.78	16.02	846.3	5505.1
PCM2S1	4.85	62.23	3338.06	6485.79	-10.49	10.34	-11.94	15.46	905.7	5568.1
PCM2S2	4.69	61.99	3269.17	6459.10	-6.51	9.61	-11.05	14.97	895.3	5567.1
PCM2S3	4.23	61.45	3326.5	6486.8	-9.23	10.42	-12.11	16.02	838.4	5536.3
PCM3S1	4.87	62.18	3340.31	6477.86	-10.72	10.1	-12.11	15.23	904.05	5590.4
PCM3S2	4.74	61.93	3332.65	6440.9	-9.36	9.11	-11.68	14.64	892.6	5591.75
PCM3S3	4.5	61.26	3340.6	6398.6	-10.88	8.91	-12.61	15.82	839.35	5631.15
PCM4S1	4.9	62.16	3340.96	6455.83	-10.78	9.53	-12.15	14.7	904.2	5599.3
PCM4S2	4.83	61.96	3338.07	6365.09	-10.44	7.13	-11.92	12.53	896.25	5605.7
PCM4S3	4.83	61.67	3341.04	6116.69	-10.9	-3.56	-12.62	10.45	876.2	5627.65
PCM5S1	4.89	62.15	3340.76	6433.85	-10.76	8.93	-12.14	14.22	903.25	5603.2
PCM5S2	4.8	61.93	3337.62	6296	-10.4	4.73	-11.9	11.87	894.6	5609.85
PCM5S3	4.76	61.55	3340.47	6125.15	-10.8	-1.71	-12.63	10.02	866.35	5637.25

Note: PCM1S1 indicates the case when PCM1 is integrated into the south wall at the first integration location (to the outside of 'Wall_consol_layer').

5.3 PCM Integrated Into Ceiling

For PCM integration into ceiling, 25 cases are simulated and results are shown in Table 3. Comparing with south wall integration cases, the influences of PCM layer on both heating demand and cooling demand are more evident. Reductions range from 10.9% to 54.3% for heating and 2.6% to 7.2% for cooling comparing with the case without

PCM. This is because the area of ceiling is larger than the area of south wall. Likewise, annual peak loads are smaller by 0.8%~14.9% for heating and 2.3%~21.1% for cooling respectively. If we look at the changes in heat flux through ceiling, reductions are smaller than the south wall cases. Heat flux from ceiling at load peaks are reduced by 5%~45% for heating and 6.3%~76.1% for cooling. For annual peak heat flux through ceiling, more complex trend presents. They are reduced for most of the studied cases, but with one or two cases increased. Similar to the south wall integration cases, the time with heating requirement is shorter for all cases while the time with cooling requirement could be longer or shorter.

The best-performing case of each PCM layer sample could be different depending on the criterion. Based on the annual thermal demand, case PCM3C5 is the best one. With the third PCM layer sample integrated into the fifth location in ceiling, peak heat flux time delay is about 30 minutes for heating and 60 minutes for cooling.

Table 3: Summary of annual simulation results for all studied cases with PCM integrated into ceiling

Cases	**Thermal demand ($kWh/m^2 \cdot year$)**		**Peak load (W)**		**Heat flux through ceiling at load peaks (w/m^2)**		**Peak heat flux through ceiling (W/m^2)**		**Total number of hours with**	
	Heating	*Cooling*	*Heating*	*Cooling*	*Heating*	*Cooling*	*Heating*	*Cooling*	*Heating*	*Cooling*
NoPCM	5.32	63.01	3361.69	6545.97	-5.86	9.11	-7.69	9.97	949.4	5538.8
PCM1C1	4.52	61.4	3334.6	6396.9	-5.55	8.54	-7.23	10.25	845.4	5621.35
PCM1C2	4.22	60.99	3041.07	6119.55	-3.77	7.04	-6.51	9.38	827.65	5643.55
PCM1C3	4.04	60.78	3000.2	6039.2	-3.52	6.51	-6.3	8.89	813.2	5639.35
PCM1C4	3.66	60.52	2944.51	6094.03	-3.26	6.88	-6.54	9.22	797.7	5600.25
PCM1C5	2.43	59.51	2861.41	6340.36	-3.22	8.16	-7.21	10.24	523.8	5351.8
PCM2C1	4.54	61.35	3335.11	6390.93	-5.55	8.51	-7.25	10.23	845.35	5642.1
PCM2C2	4.25	60.93	3053.91	6115.03	-3.85	7	-6.6	9.34	829.7	5661.05
PCM2C3	4.1	60.72	3007.58	6035.84	-3.56	6.48	-6.37	8.86	816.1	5656.8
PCM2C4	3.74	60.4	2967.46	6088.43	-3.39	6.83	-6.63	9.18	791.95	5631.8
PCM2C5	2.58	59.01	2973.71	6336.16	-3.73	8.14	-7.64	10.22	522.75	5444.85
PCM3C1	4.66	61.29	3336.08	6365.01	-5.57	8.37	-7.29	10.14	856.9	5693.05
PCM3C2	4.43	60.88	3263.94	6090.75	-4.05	6.84	-6.76	9.2	841.75	5710.6
PCM3C3	4.32	60.64	3192.26	6007.08	-3.65	6.29	-6.54	8.7	827.35	5718.4
PCM3C4	4.15	60.27	3183.86	6017.52	-3.64	6.41	-6.79	8.86	802.5	5735.1
PCM3C5	3.68	58.48	3219.18	6013.45	-4.09	6.76	-7.84	9.42	684.3	5725.3
PCM4C1	4.74	61.29	3336.22	6324.82	-5.57	8.17	-7.3	10.02	867.35	5743.25
PCM4C2	4.53	60.92	3303.31	6046.39	-3.89	6.56	-6.77	8.98	857.05	5752.15
PCM4C3	4.45	60.74	3247.05	5941.09	-3.93	5.88	-6.54	8.36	852.5	5757.15
PCM4C4	4.35	60.45	3263.21	5797.76	-4.04	5.1	-6.79	7.9	839.2	5770.6
PCM4C5	4.24	58.68	3328.88	5542.65	-4.66	2.29	-7.84	6.69	798.8	5782.45
PCM5C1	4.72	61.27	3336.14	6300.2	-5.57	8.16	-7.3	9.98	865.6	5750.1
PCM5C2	4.49	60.87	3294.55	5977.31	-3.84	6.14	-6.75	8.65	852.15	5753.25
PCM5C3	4.38	60.67	3226.83	5875.95	-3.82	5.43	-6.5	7.96	846.7	5754.6
PCM5C4	4.25	60.36	3227.77	5656.61	-3.85	5.02	-6.71	7.57	832.5	5760.25
PCM5C5	4.08	59.29	3324.8	5166.12	-4.58	2.18	-7.9	6.83	778.6	5790.9

6. CONCLUSIONS

This paper investigated the performance of phase change material (PCM) integrated building envelope in terms of peak load reduction, peak load time delay and thermal demand improvement for a prototype detached single family house located in San Antonio, Texas. The thermal properties of the house envelope satisfy the prescriptive criteria of thermal resistance for ceiling, exterior wall, roof and etc, set by IECC 2016. Five market available PCM layers, which can be easily inserted into traditional wall construction, were selected and coupled into different walls and different locations within the wall. They share some common properties such as thickness, density and specific heat but with have different melting temperatures. Annual simulations were performed in EnergyPlus using 3 minutes as time step. The results indicated the significant effect of PCM integrated envelope for the studied case. The PCM layer of melting temperature of 23 °C is determined as the best one for south wall implementation. When it is placed to the interior of the insulation component of the wall, heat flux at heating load peak is reduced by 22.3% and heat flux at cooling load peak is reduced by 12.3%. For ceiling application, the best performing one is the PCM layer of melting temperature of 25 °C and when it is placed below the insulation component of the ceiling. Reduction in peak heat flux is 30.2% for heating and 25.8% for cooling. The greater reduction in peak heat flux of ceiling than south wall is due to its larger area. For peak heat flux time delay, all PCM cases show similar effect of about 30 minutes for typical heating time and 60 minutes for typical cooling time. The performance of PCM layers in reducing annual thermal demand is not as evident as in heat flux: reduction in heating demand ranges from 7.9% to 54.34% and reduction in cooling demand ranges from 1.2% to 7.2%. It is worth noticing that even though the reduction in thermal demand is small, the decreased load peaks still provide chances for energy saving because: i) air-conditioning system can operate at a later time with lower outdoor air temperature than it would without PCM during peak hours; and ii) lower outdoor air temperature implies better system efficiency (COP). Besides, the load profile is flatter. With a size reduced air-conditioning system, it enables the system to operate at relatively high cop range.

REFERENCES

American Society of Heating, Refrigerating, and Air-Conditioning Engineers, Inc. ANSI/ASHRAE/IESNA Standard 90.2: Energy-Efficient Design of Low-rise Residential Buildings, Atlanta, 2007.

American Society of Heating, Refrigerating, and Air-Conditioning Engineers, Inc., ANSI/ASHRAE Standard 140: Standard Method of Test for the Evaluation of Building Energy Analysis Computer Programs.

IECC: International Energy Conservation Code.

PNNL, https://www.energycodes.gov/development/residential/iecc_models, (retrieved April 20 2016)

Halford C.K., Boehm R.F, 2007. Modeling of phase change material peak load shifting. Energy and Buildings, vol. 39, 298-305.

Kondo T., Ibamoto T., 2006. Research on thermal storage using rock wool phase-change material ceiling board. ASHRAE Transactions, vol. 112, 526-531.

Kissock J.K., Hanning J.M., Whitney T.I., Drake M.L., 1998. Testing and simulation of phase change wallboard for thermal storage in buildings. Proceedings of International Solar energy Conference, Solar Engineering. 45-52.

Kissock J.K., Limas S., 2006. Diurnal load reduction through phase change building components. ASHRAE Transactions, vol. 112, 509-517.

Zhang M., Medina M.A., King J.B., 2005. Development of a thermally enhanced frame wall with phase change materials for on-peak air conditioning demand reduction and energy savings in residential buildings. International Journal of Energy Research, vol. 29, 795-809.

Lee K.O., Medina M.A., Raith E., Sun XQ, 2015. Assessing the integration of a thin phase change material (PCM) layer in a residential building wall for heat transfer reduction and management. Applied Energy, vol. 137, 699-706.

Zhu N., Ma Z., Wang S., 2009. Dynamic characteristics and energy performance of buildings using phase change materials: A review. Energy Conversion and Management, vol. 50, 3169-3181.

Cao S., Gustavsen A., Uvslokk S., Jelle B.P., Gilbert J., Maunuksela J., 2010. The effect of wall-integrated phase change material panels on the indoor air and wall temperature – hot box experiments. Proceedings of Renewable Energy Conference. Trondheim, Norway.

Heim D., Clarke J.A., 2004. Numerical modelling and thermal simulation of PCM-gypsum composites with ESP-r. Energy and Buildings, vol. 36, 795-805.

Heim D., 2006. Phase-change material modelling within whole building dynamic simulation. ASHRAE Transaction, vol. 112, 518-525.

Ibanez M., Lazaro A., Zalba B., Cabeza L.F., 2005. An approach to the simulation of PCMs in building applications using TRNSYS. Applied Thermal Engineering, vol. 25, 1796-1807.

Gracia A. de, Cabeza L.F., 2015. Phase change materials and thermal energy storage for buildings. Energy and Buildings, vol. 103, 414-419.

Tabares-Velasco P.C., Christensen C., Bianchi M., 2012. Verification and validation of EnergyPlus phase change material model for opaque wall assemblies. Building and Environment, vol. 54, 186-196.

Tabares-Velasco P.C., Griffith B., 2012. Diagnostic test cases for verifying surface heat transfer algorithms and boundary conditions in building energy simulation programs. Journal of Building Performance Simulation, 5:5, 329-346, DOI: 10.1080/19401493.2011.595501.

Jin X., Median M.A., Zhang X., 2013. On the importance of the location of PCMs in building walls for enhanced thermal performance. Applied Energy, vol. 106, 72-78.

EnergyPlus, 2016. Engineering Reference. U.S. Department of Energy.

Investigation of Dynamic Thermal Parameters of Various Insulation Filled Bricks Exposed to Periodic Thermal Variations for Energy Efficient Stuffed Bricks Design

Ashok Babu TALANKI PUTTARANGA SETTY[1]*, Saboor SHAIK[2]

[1]National Institute of Technology Karnataka, Mechanical Engineering,
Mangalore, Karnataka, India
(+91 9986548546, tpashok@rediffmail.com , tpashok@gmail.com)

[2] National Institute of Technology Karnataka, Mechanical Engineering,
Mangalore, Karnataka, India
(+91 9880452903, saboor.nitk@gmail.com)

* Corresponding Author

ABSTRACT

The space cooling systems account for nearly 30% of energy consumption in India. The highly insulated external walls are essential prerequisites for low energy consumption of buildings. Filling shell of the brick with insulation material is one of the methods to reduce cooling loads in buildings. This paper aims to investigate the dynamic thermal parameters of various insulation filled bricks. The dynamic thermal parameters highlighted in the study include thermal transmittance, admittance, decrement factor, time lag and areal thermal heat capacity. The cyclic admittance method was employed in computer simulation program to compute dynamic thermal parameters of stuffed bricks. The four brick materials such as mud bricks, burnt bricks, concrete blocks and fly ash bricks were selected. The insulation materials used for filling shells of the brick are foam glass and asbestos fiber. The total forty eight stuffed bricks were investigated. The shell of the brick is filled by the insulation and the each layer of insulation is separated by a brick web. The shell and web thickness of the brick were maintained as per the Indian standards. The six configurations of the stuffed bricks were investigated (1. Solid brick, 2. Bricks with the shell of the brick filled with single layer of insulation, 3. Bricks with the shell of the brick filled with two layers of insulation with each insulation layer separated with a brick web, 4. Bricks with the shell of the brick filled with three layers of insulation with each insulation layer separated by a brick web, 5. Bricks with the shell of the brick filled with four layers of insulation with each insulation layer separated by a brick web and 6. Bricks with the shell of the brick filled with five layers of insulation with each insulation layer separated by a brick web). From the results, it is observed that the stuffed bricks significantly improve time lag values as compared to the conventional solid bricks. The decrement factor decreases, admittance increases and areal thermal heat capacity increases with the increase in the no. of insulation layers in the bricks.

1. INTRODUCTION

According to the U.S. green building council, building sector worldwide consumes about 40% of energy for cooling, heating and day lighting and it is also responsible for 40% of global greenhouse gas emissions. Heating, ventilation and air-conditioning consume about 33% of total energy in both residential and commercial buildings in India. Building walls with elevated thermal insulation and thermal mass can reduce cooling loads in buildings. Bricks are the chief building materials for building construction. The appropriate design of bricks can improve thermal insulation and thermal mass of the buildings. The dynamic thermal parameters are significant to evaluate thermal performance of building enclosures. The reduction of heat wave amplitude as it moves from outside of the wall to the inside of the wall is called decrement factor and time taken by that heat wave to pass through the thickness of the wall from outside to inside is called the time lag (Duffin and Greg, 1984). The building materials with lower decrement factor and higher time lag values maintain comfortable indoor conditions (Knowles, 1983). The researchers also studied the effect of thermal properties and thickness of the wall on time lag and decrement factor using Crank Nicolson method (Asan and Sancaktar, 1998). The studies on optimum location of insulation material were also reported in the literature (Asan, 1998). The Crank-Nicolson method was used to compute decrement factor and the time lag for composite wall (Asan, 2000). The unsteady state thermal characteristics of ten homogeneous building and insulating materials and their composite walls were studied and the optimum position of insulation within the wall based on time lag and decrement factor were presented (Ashok Babu and Saboor, 2014). The cyclic admittance method was used to optimize insulation location within the roof based on decrement factor and time lag (Saboor and Ashok Babu, 2015). Unsteady thermal performance of sustainable laterite rocks exposed to various temperature and relative humidity were reported in the literature (Saboor and Ashok Babu, 2016). The Crank-Nicolson method under periodic boundary condition was used to calculate decrement factor and time lag of twenty six homogeneous building materials (Asan, 2006). The unsteady state thermal characteristics of hollow bricks were investigated and reported in the literature (Saboor and Ashok Babu, 2015). This paper presents dynamic thermal parameters of various stuffed bricks when they are exposed to ambient periodic thermal variations.

2. THERMAL PARAMETERS OF STUFFED BRICKS EXPOSED TO PERIODIC THERMAL VARIATIONS

The cyclic admittance technique is used to compute the dynamic thermal parameter values of the bricks. This technique uses matrices to simplify the temperature and energy cycles for a brick that is subjected to dynamic sinusoidal temperature variations at the outside. The diffusion equation for the temperature distribution in a homogeneous brick subjected to one dimensional heat flow is given by (CIBSE, 2006),

$$\frac{\partial^2 T(X,t)}{\partial X^2} = \frac{\rho Cp}{k}\frac{\partial T(X,t)}{\partial t} \quad (1)$$

The Fourier analysis gives the solution to the diffusion equation as,

$$T(x,t) = [P_1 sinh(\delta x + j\delta x) + P_1 cosh(\delta x + j\delta x)]exp(j2\pi t/P) \quad (2)$$

Where, $\delta = \sqrt{\pi \rho c_p/\lambda P}$.

When the conducting medium finite thickness slab X, temperature and flows at the two surfaces are considered then the above equation can be related as, $T_0 = T_i cosh(y + jy) + q_i\,(sinh(y + jy))/a$ and $q_0 = T_i(sinh(y + jy)) \times a + q_i cosh.$

The cyclic heat fluxes occurring at the outside and inside surfaces of the brick can be written as a function of the cyclic surface temperatures as follows (Davies, 2004):

$$\begin{bmatrix} T_0 \\ q_0 \end{bmatrix} = \begin{bmatrix} cosh(y + jy) & (sinh(y + jy))/a \\ (sinh(y + jy)) \times a & cosh(y + jy) \end{bmatrix} \begin{bmatrix} T_i \\ q_i \end{bmatrix} \quad (3)$$

Where, cyclic thickness$(y) = \sqrt{\pi \rho c_p X^2/\lambda P} = \sqrt{\pi cr/P}$ and thermal effusivity $(a) = \sqrt{j2\pi\lambda\rho c_p/P} = \sqrt{j2\pi c/rP}$. the letter j is the imaginary unit. these two parameters consist of the information of brick thermal properties, thickness of the brick and period of the periodic energy transfer.

Transmission matrix of single layer solid brick can be written as,

$$\begin{bmatrix} P_1 + jP_2 & (P_3 + jP_4)/a \\ (-P_4 + jP_3).a & P_1 + jP_2 \end{bmatrix} \quad (4)$$

Where, constants $P_1 = \cosh(y)\cos(y)$,$P_2 = \sinh(y)\sin(y)$, $P_3 = [\cosh(y)\sin(y) + \sinh(y)\cos(y)]/\sqrt{2}$ and $P_4 = [\cosh(y)\sin(y) - \sinh(y)\cos(y)]/\sqrt{2}$.

Transmission matrix of surface internal and external film resistances can be written as,

$$M_{si} = \begin{bmatrix} 1 & -R_{si} \\ 0 & 1 \end{bmatrix} \; and \; M_{se} = \begin{bmatrix} 1 & -R_{se} \\ 0 & 1 \end{bmatrix} \quad (5)$$

Transmission matrix for solid brick (BC-1) can be written as,

$$\begin{bmatrix} T_i \\ q_i \end{bmatrix} = \begin{bmatrix} 1 & -R_{si} \\ 0 & 1 \end{bmatrix} \begin{bmatrix} b_1 & b_2 \\ b_3 & b_1 \end{bmatrix} \begin{bmatrix} 1 & -R_{se} \\ 0 & 1 \end{bmatrix} \begin{bmatrix} T_e \\ q_e \end{bmatrix} \tag{6}$$

Transmission matrix for Shell of the brick filled by single layer of insulation (BC-2) can be written as,

$$\begin{bmatrix} T_i \\ q_i \end{bmatrix} = \begin{bmatrix} 1 & -R_{si} \\ 0 & 1 \end{bmatrix} \begin{bmatrix} s_1 & s_2 \\ s_3 & s_1 \end{bmatrix} \begin{bmatrix} i_1 & i_2 \\ i_3 & i_1 \end{bmatrix} \begin{bmatrix} s_1 & s_2 \\ s_3 & s_1 \end{bmatrix} \begin{bmatrix} 1 & -R_{se} \\ 0 & 1 \end{bmatrix} \begin{bmatrix} T_e \\ q_e \end{bmatrix} \tag{7}$$

Transmission matrix for Shell of the brick filled by five layers of insulation with each insulation layer separated by a brick web (BC-6) can be written as

$$\begin{bmatrix} T_i \\ q_i \end{bmatrix} = \begin{bmatrix} 1 & -R_{si} \\ 0 & 1 \end{bmatrix} \begin{bmatrix} s_1 & s_2 \\ s_3 & s_1 \end{bmatrix} \begin{bmatrix} i_1 & i_2 \\ i_3 & i_1 \end{bmatrix} \begin{bmatrix} w_1 & w_2 \\ w_3 & w_1 \end{bmatrix} \begin{bmatrix} i_1 & i_2 \\ i_3 & i_1 \end{bmatrix} \begin{bmatrix} w_1 & w_2 \\ w_3 & w_1 \end{bmatrix} \begin{bmatrix} i_1 & i_2 \\ i_3 & i_1 \end{bmatrix} \begin{bmatrix} w_1 & w_2 \\ w_3 & w_1 \end{bmatrix} \begin{bmatrix} i_1 & i_2 \\ i_3 & i_1 \end{bmatrix} \begin{bmatrix} w_1 & w_2 \\ w_3 & w_1 \end{bmatrix} \tag{8}$$

$$\begin{bmatrix} i_1 & i_2 \\ i_3 & i_1 \end{bmatrix} \begin{bmatrix} s_1 & s_2 \\ s_3 & s_1 \end{bmatrix} \begin{bmatrix} 1 & -R_{se} \\ 0 & 1 \end{bmatrix} \begin{bmatrix} T_e \\ q_e \end{bmatrix}$$

Where, b is brick layer, s is brick shell layer, i is brick insulation layer and w is the brick web layer thickness. Transmission matrix can be further reduced as follows,

$$\begin{bmatrix} T_i \\ q_i \end{bmatrix} = \begin{bmatrix} Z_1 & Z_2 \\ Z_3 & Z_4 \end{bmatrix} \begin{bmatrix} T_e \\ q_e \end{bmatrix} \tag{9}$$

The steady thermal transmittance (U) can be computed using the following correlation,

$$U = \frac{1}{\frac{1}{h_{so}} + \left(\frac{X_1}{k_1}\right) + \left(\frac{X_2}{k_2}\right) + \cdots \frac{1}{h_{si}}} \tag{10}$$

The dynamic thermal transmittance (u) is the ration of the dynamic heat flux released on the inner side of the brick to the dynamic temperature excitation on the outer surface of the brick, while indoor temperature is maintained constant. The dynamic thermal transmittance of the bricks can be calculated by the following correlation,

$$u = 1/Z_2 \tag{11}$$

The decrement Factor (f) is the ratio of dynamic thermal transmittance to the steady thermal transmittance and it can be computed using the following equation,

$$f_c = -\frac{1}{UZ_2} \text{ and } f = |f_c| \tag{12}$$

The decrement delay (ϕ) is the delay between outside peak temperature profile and the corresponding peak inside heat flux released. It can be calculated by

$$\phi = \frac{12}{\pi} arctan\left(\frac{Im(f_c)}{Re(f_c)}\right) \tag{13}$$

As far as the thermal admittance (Y) is concerned, it is the amount of energy leaving the internal surface of the brick into the room per unit degree of temperature swing. It can be calculated by,

$$y_i = \left(\frac{q_i}{\theta i}\right)_{Te=0} = -Z_1/Z_2 \text{ and } Y = |y_i| \tag{14}$$

Areal thermal heat capacity (χ) is the amount of energy stored in the brick over the first half period of the heat flow swing per unit area of brick element per unit degree of temperature swing. The same amount of heat is released in the following half period. It can be calculated using the following equation,

$$\chi = \frac{t}{2\pi}\left|\frac{Z_4 - 1}{Z_2}\right| \tag{15}$$

Table 1 shows the thermal properties of brick and insulation materials considered for this study. The thermal properties of the brick and insulation materials are as per the Indian standards (SP: 41, 1987). Four brick materials such as mud brick, burnt brick, concrete block and fly ash bricks were selected and two insulation materials such as foam glass and asbestos fiber were selected as filling materials in the shell of the bricks. Figure 1 shows configurations of the bricks. The various configuration of the bricks are solid brick (BC-1), bricks with the shell of the brick filled with single layer of insulation (BC-2), bricks with the shell of the brick filled with two layers of insulation with each insulation layer separated by a brick web (BC-3), bricks with the shell of the brick filled with three layers of insulation with each insulation layer separated by a brick web (BC-4), bricks with the shell of the brick filled with four layers of insulation with each insulation layer separated by a brick web (BC-5) and bricks with the shell of the brick filled with five layers of insulation with each insulation layer separated by a brick web (BC-6). Table 2 shows the brick configurations with thicknesses. The nominal thickness of brick configurations is considered as 0.14 m and the web thickness is maintained 0.01 m (not less than 0.008 m) as per the Indian standards. The dynamic thermal parameters of bricks include thermal transmittance, thermal admittance, decrement factor, time lag and areal thermal heat capacity. The computer simulation program was developed and used to compute dynamic thermal parameters of various brick configurations. The correlations used for the computer simulation are shown from Equation (1) to Equation (15). The accuracy of the computer simulation results was validated with the CIBSE standard values and its accuracy is confirming to CIBSE standards. The outside and inside surface resistances of the bricks considered for the study are 0.04 m^2 K/W and 0.13 m^2 K/W, respectively as per the CIBSE standards. Table 3, Table 5, Table 7 and Table 9 show the dynamic thermal parameters of mud bricks, burnt bricks, concrete and fly ash bricks filled with foam glass insulation, respectively. Table 4, Table 6, Table 8 and Table 10 show the dynamic thermal parameters of mud bricks, burnt bricks, concrete and fly ash bricks filled with asbestos fiber insulation, respectively.

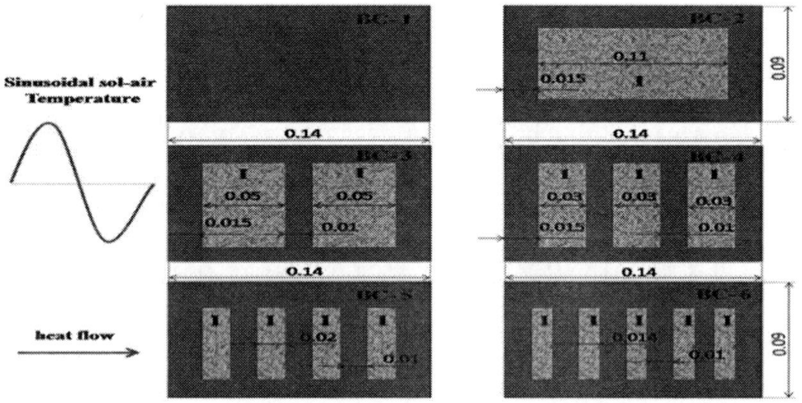

Figure 1: Configurations of bricks

Table 1: Thermal properties of brick and insulation materials

Brick and insulation materials	Code	Thermal conductivity k (W/mK)	Density ρ (kg/m³)	Specific heat capacity Cp (J/kgK)
Mud brick (B)	MB	0.75	1731	880
Burnt brick (B)	BB	0.811	1820	880
Concrete block (B)	CB	1.74	2410	880
Fly ash brick (B)	FAB	0.360	1700	857
Foam glass (I)	FG	0.055	160	750
Asbestos fiber (I)	AF	0.06	640	840

Table 2: Brick configurations and Thickness

S.No.	Configuratio n	Thickness of the brick from outside to inside (m)
1.	BC-1	0.14B
2.	BC-2	0.015B+0.11I+0.015B
3.	BC-3	0.015B+0.05I+0.01B+0.05I+0.015B
4.	BC-4	0.015B+0.03I+0.01B+0.03I+0.01B+0.03I+0.015B
5.	BC-5	0.015B+0.02I+0.01B+0.02I+0.01B+0.02I+0.01B+0.02I+0.015B
6.	BC-6	0.015B+0.014I+0.01B+0.014I+0.01B+0.014I+0.01B+0.014I+0.01B+0.014I+0.015 B

Table 3: Dynamic thermal parameters of mud bricks filled with foam glass insulation

MB with FG Configuration	U (W/m²K)	f	φ (h)	Y (W/m²K)	ω (h)	χ (J/m²K)
BC-1	2.8048	0.75336	3.9118	4.2737	1.4119	60797
BC-2	0.4525	0.9386	2.5657	1.8469	4.0568	29867
BC-3	0.4899	0.7764	4.0927	1.9555	3.7253	33034
BC-4	0.5339	0.7019	5.0626	2.1046	3.4474	36459
BC-5	0.5867	0.6519	5.7125	2.2575	3.2165	39883
BC-6	0.6511	0.6184	6.1351	2.4142	3.0153	43478

Table 4: Dynamic thermal parameters of mud bricks filled with asbestos fiber insulation

MB with AF Configuration	U (W/m²K)	f	φ (h)	Y (W/m²K)	ω (h)	χ (J/m²K)
BC-1	2.8048	0.75336	3.9118	4.2737	1.4119	60797
BC-2	0.4894	0.6513	6.3317	2.4181	3.3967	41956
BC-3	0.5291	0.5658	6.7176	2.3838	3.2561	41452
BC-4	0.5759	0.5377	7.0252	2.4437	3.0710	43079
BC-5	0.6316	0.5270	7.1740	2.5391	2.9136	45498
BC-6	0.6994	0.5255	7.1997	2.6534	2.7718	48446

Table 5: Dynamic thermal parameters of burnt bricks filled with foam glass insulation

BB with FG Configuration	U (W/m²K)	f	φ (h)	Y (W/m²K)	ω (h)	χ (J/m²K)
BC-1	2.9197	0.74693	3.9355	4.4045	1.3641	62491
BC-2	0.4531	0.9362	2.6091	1.9177	4.0603	31051
BC-3	0.4908	0.7644	4.2029	2.0250	3.7233	34235
BC-4	0.5354	0.6865	5.2147	2.1742	3.4395	37688
BC-5	0.5888	0.6349	5.8896	2.3278	3.2043	41151
BC-6	0.6541	0.6006	6.3261	2.4858	2.9999	44809

Table 6: Dynamic thermal parameters of burnt bricks filled with asbestos fiber insulation

BB with AF Configuration	U (W/m²K)	f	φ (h)	Y (W/m²K)	ω (h)	χ (J/m²K)
BC-1	2.9197	0.7469	3.9355	4.4045	1.3641	62491.44
BC-2	0.4901	0.6504	6.3650	2.4806	3.4052	43053
BC-3	0.5303	0.5584	6.7915	2.4429	3.2632	42468
BC-4	0.5775	0.5275	7.1358	2.5027	3.0729	44115

BC-5	0.6340	0.5150	7.3104	2.5999	2.9103	46599
BC-6	0.7028	0.5123	7.3535	2.7168	2.7639	49643

Table 7: Dynamic thermal parameters of concrete blocks filled with foam glass insulation

CB with FG Configuration	U (W/m²K)	f	ϕ (h)	Y (W/m²K)	ω (h)	χ (J/m²K)
BC-1	3.9947	0.73609	3.6727	5.3355	0.98347	70521
BC-2	0.4572	0.9196	2.8710	2.3803	4.0335	38878
BC-3	0.4973	0.6877	4.8589	2.4692	3.6794	41928
BC-4	0.5450	0.5921	6.1191	2.6108	3.3733	45397
BC-5	0.6028	0.5342	6.9372	2.7634	3.1219	49033
BC-6	0.6744	0.4987	7.4480	2.9257	2.9038	53053

Table 8: Dynamic thermal parameters of concrete blocks filled with asbestos fiber insulation

CB with AF Configuration	U (W/m²K)	f	ϕ (h)	Y (W/m²K)	ω (h)	χ (J/m²K)
BC-1	3.9947	0.73609	3.6727	5.3355	0.98347	70521
BC-2	0.4949	0.6435	6.5613	2.8919	3.4254	50411
BC-3	0.5378	0.5123	7.2328	2.8303	3.2819	49238
BC-4	0.5887	0.4653	7.8014	2.8848	3.0674	50929
BC-5	0.6503	0.4441	8.1286	2.9887	2.8795	53766
BC-6	0.7263	0.4359	8.2666	3.1181	2.7090	57366

Table 9: Dynamic thermal parameters of fly ash bricks filled with foam glass insulation

FAB with FG Configuration	U (W/m²K)	f	ϕ (h)	Y (W/m²K)	ω (h)	χ (J/m²K)
BC-1	1.7897	0.64005	5.34	3.6448	1.764	56658
BC-2	0.4438	0.9344	2.6821	1.7703	4.0152	28722
BC-3	0.4764	0.7760	4.1785	1.8725	3.6895	31644
BC-4	0.5141	0.6990	5.1548	2.0099	3.4177	34747
BC-5	0.5583	0.6448	5.8348	2.1480	3.1928	37764
BC-6	0.6109	0.6060	6.3065	2.2865	2.9984	40839

Table 10: Dynamic thermal parameters of fly ash bricks filled with asbestos fiber insulation

FAB with AF Configuration	U (W/m²K)	f	ϕ (h)	Y (W/m²K)	ω (h)	χ (J/m²K)
BC-1	1.7897	0.6400	5.34	3.6448	1.764	56658
BC-2	0.4793	0.6341	6.5313	2.3147	3.3238	40125
BC-3	0.5134	0.5548	6.9026	2.2827	3.1949	39608
BC-4	0.5529	0.5257	7.2138	2.3331	3.0231	40934
BC-5	0.5988	0.5119	7.3887	2.4137	2.8770	42917
BC-6	0.6532	0.5060	7.4591	2.5093	2.7461	45296

3. RESULTS AND DISCUSSIONS

3.1 Thermal transmittance and admittance of insulation filled bricks

Thermal admittance and thermal transmittance have the same units. Thermal admittance is the ability of a brick to admit, absorb and emit heat. Thermal admittance mainly depends upon thermal properties of inner side of the brick.

For thin bricks, thermal transmittance and thermal admittance are the same. Thermal admittance becomes constant for bricks more than 0.2 m thick. Figure 2 shows the thermal transmittance and admittance of bricks filled with foam glass insulation. From the results, it is observed that the thermal transmittance of the stuffed bricks is less than the conventional solid bricks and hence the stuffed bricks have more thermal resistance than the solid bricks. It is also noticed from the results that the thermal admittance of the stuffed bricks increases with the increasing number of insulation layers in the brick. The Bricks with the shell of the brick filled with five layers of insulation with each insulation layer separated by a brick web have more thermal mass than the Bricks with the shell of the brick filled with single layer of insulation. From Figure 2, it is observed that the concrete blocks with the shell of the brick filled with five foam glass layers of insulation with each insulation layer separated by a brick web have the highest admittance value (2.9257 W/m^2K) among all the studied bricks filled with foam glass insulation. It is also clear from the results that the fly ash bricks with the shell of the brick filled with five foam glass layers of insulation with each insulation layer separated by a brick web have the lowest admittance value (2.2865 W/m^2K) among all the studied bricks filled with foam glass insulation. Figure 3 shows the thermal transmittance and admittance of bricks filled with asbestos fiber insulation. From the results, it is noticed that the concrete blocks with the shell of the brick filled with five asbestos fiber layers of insulation with each insulation layer separated by a brick web have the highest admittance value (3.1181 W/m^2K) among all the studied bricks filled with asbestos fiber insulation. It is also clear from the results that the fly ash bricks with the shell of the brick filled with five asbestos layers of insulation with each insulation layer separated by a brick web have the lowest admittance value (2.5093 W/m^2K) among all the studied bricks filled with asbestos fiber insulation.

3.2 Decrement factor and time lag of insulation filled bricks

The dampening of the outside sinusoidal heat wave as it passes through the brick (Decrement factor) should be as low as possible to reduce the outside heat wave effect on indoor comfort conditions. The time taken by heat wave to pass from outer side of the brick to the inner side of the brick (Time lag) should be as high as possible. The decrement factor and time lag are the measures of thermal mass. Figure 4a shows the decrement factor of bricks filled with foam glass insulation. From the results, it is observed that the decrement factor decreases and time lag increases with the increasing number of insulation layers in the brick. It is also observed that the mud bricks with the shell of the brick filled with five foam glass layers of insulation with each insulation layer separated by a brick web have the highest decrement factor value (0.6184) among all the studied bricks filled with foam glass insulation. It is also clear from the results that the concrete blocks with the shell of the brick filled with five foam glass layers of insulation with each insulation layer separated by a brick web have the lowest decrement value (0.4987) among all the studied bricks filled with foam glass insulation. Figure 4b shows the time lag of bricks filled with foam glass insulation. From the results, it is observed that the concrete blocks with the shell of the brick filled with five foam glass layers of insulation with each insulation layer separated by a brick web have the highest time lag value (7.448 h) among all the studied bricks filled with foam glass insulation. It is also clear from the results that the mud bricks with the shell of the brick filled with five foam glass layers of insulation with each insulation layer separated by a brick web have the lowest time lag value (6.1351 h) among all the studied bricks filled with foam glass insulation. Figure 5a shows the decrement factor of bricks filled with asbestos fiber insulation. From the results, it is noticed that the mud bricks with the shell of the brick filled with five asbestos fiber layers of insulation with each insulation layer separated by a brick web have the highest decrement factor value (0.5255) among all the studied bricks filled with asbestos fiber insulation. It is also clear from the results that the concrete blocks with the shell of the brick filled with five asbestos fiber layers of insulation with each insulation layer separated by a brick web have the lowest decrement value (0.4359) among all the studied bricks filled with foam glass insulation. Figure 5b shows the time lag of bricks filled with asbestos insulation. From the results, it is observed that the concrete blocks with the shell of the brick filled with five asbestos fiber layers of insulation with each insulation layer separated by a brick web have the highest time lag value (8.266 h) among all the studied bricks filled with asbestos fiber insulation. It is also clear from the results that the mud bricks with the shell of the brick filled with five asbestos fiber layers of insulation with each insulation layer separated by a brick web have the lowest time lag value (7.199 h) among all the studied bricks filled with asbestos fiber insulation.

3.3 Areal thermal heat capacity of insulation filled bricks

The areal thermal heat capacity of the brick is the amount of heat stored in the brick over the first half period of the heat flow swing per unit area of brick element per unit degree of temperature swing. This gives the information about thermal mass of the brick. Figure 6a shows areal thermal heat capacity of bricks filled with foam glass insulation. From the results, it is noticed that the concrete blocks with the shell of the brick filled with five foam glass layers of insulation with each insulation layer separated by a brick web have the highest areal thermal heat capacity (53053 J/m^{2}K) among all the studied bricks filled with foam glass insulation. It is also clear from the results that the fly ash bricks with the shell of the brick filled with five foam glass layers of insulation with each insulation layer separated by a brick web have the lowest areal thermal heat capacity value (40839 J/m^{2}K) among all the studied bricks filled with foam glass insulation. Figure 6b shows the areal thermal heat capacity of bricks filled with asbestos fiber insulation. From the results, it is noticed that the concrete blocks with the shell of the brick filled with five asbestos fiber layers of insulation with each insulation layer separated by a brick web have the highest areal thermal heat capacity (57366 J/m^{2}K) among all the studied bricks filled with asbestos fiber insulation. It is also observed from the results that the fly ash bricks with the shell of the brick filled with five asbestos fiber layers of insulation with each insulation layer separated by a brick web have the lowest areal thermal heat capacity value (45296 J/m^{2}K) among all the studied bricks filled with asbestos fiber insulation.

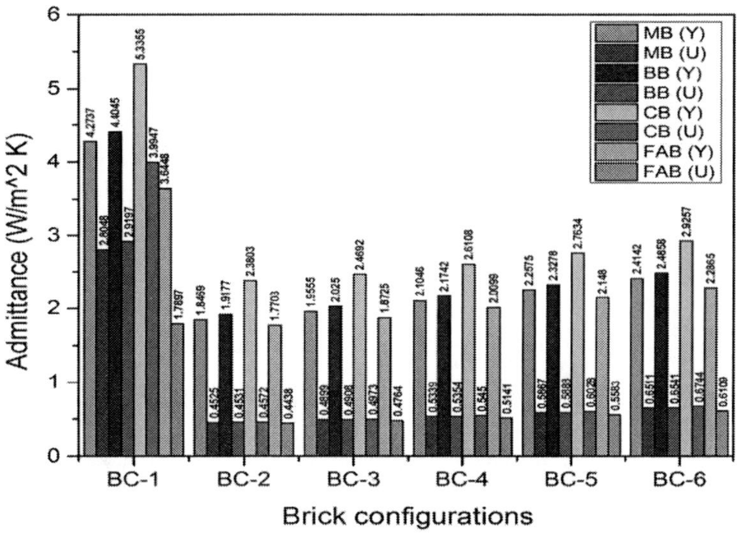

Figure 2: Thermal transmittance and admittance of bricks filled with foam glass insulation

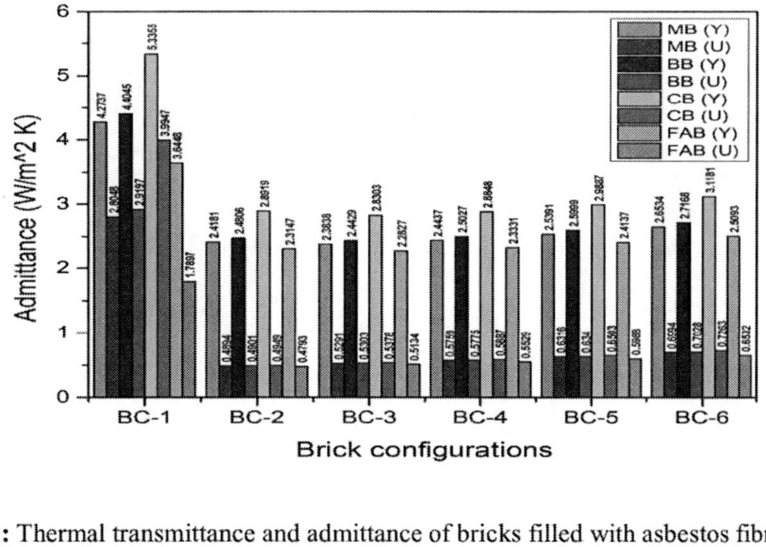

Figure 3: Thermal transmittance and admittance of bricks filled with asbestos fibre insulation

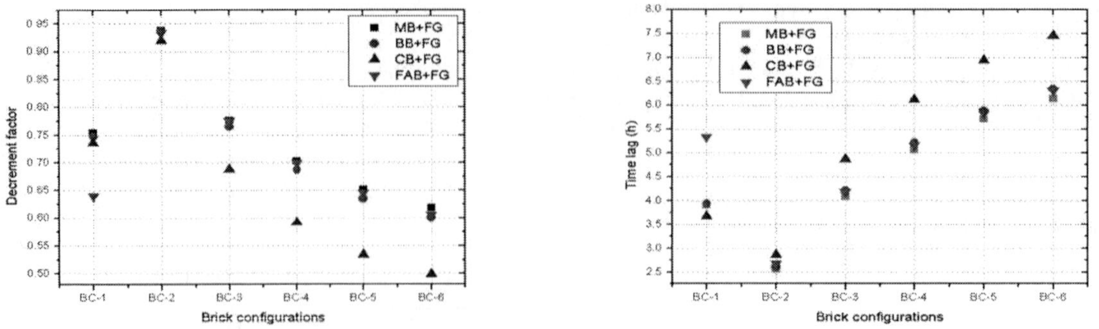

(a) (b)

Figure 4: Decrement factor and time lag of bricks filled with foam glass Insulation

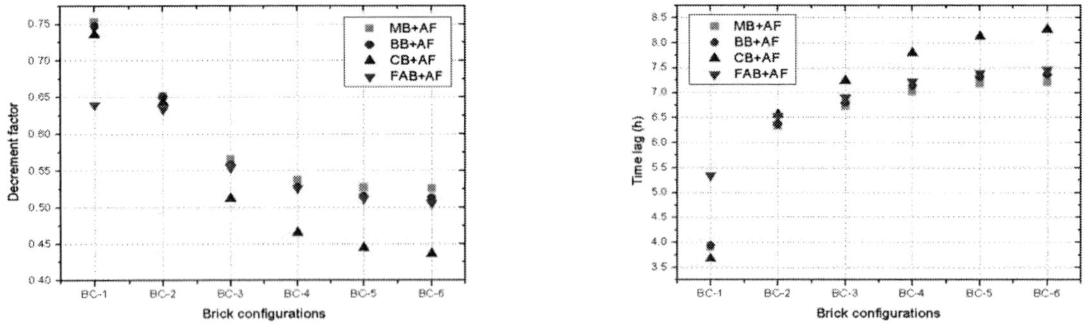

(a) (b)

Figure 5: Decrement factor and time lag of bricks filled with asbestos fibre Insulation

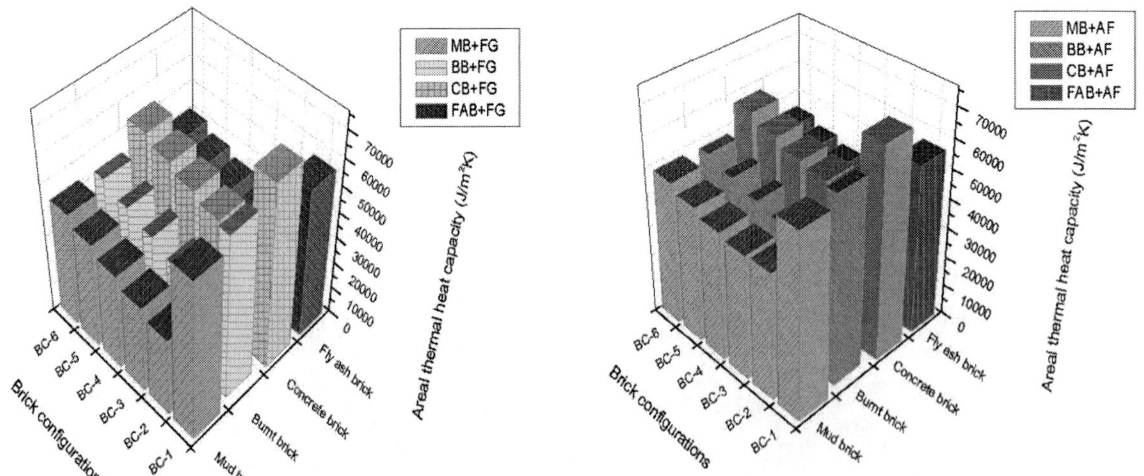

(a) (b)

Figure 6: Areal thermal heat capacity (a) Bricks filled with foam glass insulation and (b) Bricks filled with asbestos fiber insulation

4. CONCLUSIONS

- Thermal admittance, decrement time lag and areal thermal heat capacity of the insulation filled bricks increase with the increase in the number of insulation layers in the bricks due to improved thermal mass for all materials and all combinations. These enhanced dynamic parameters make the stuffed bricks more energy efficient than the ordinary solid bricks.
- The decrement factor of the insulation filled bricks decreases with the increase in the number of insulation layers in the bricks for all configurations. The lower decrement factor values are essential to reduce the effect of outdoor climatic changes on indoor conditions.
- The concrete blocks with the shell of the bricks filled with five layers of asbestos fiber insulation offer the highest admittance (3.11Wm^2K), the lowest decrement factor (0.435), the highest time lag values (8.26 h) and the highest areal thermal heat capacity (57366 J/m^2K) among all insulation filled bricks studied.
- From the comparison of hollow bricks and stuffed bricks, it is observed that the hollow bricks are suitable for reducing heating loads (Winter air conditioning) whereas, the stuffed bricks are suitable for reducing cooling loads (Summer air conditioning) from thermal mass and thermal insulation point of view.
- The results of the paper help in designing energy efficient stuffed bricks for reducing cooling loads in buildings.

REFERENCES

Asan, H., (1998). Effects of wall's insulation thickness and position on time lag and decrement factor. *Energy Build.*, 28,299-305. http://doi.org/10.1016/S0378-7788(98)00030-9

Asan, H., (2000). Investigation of wall's optimum insulation position from maximum time lag and minimum decrement factor point of view. *Energy Build.*,32(2),197-203. http://doi.org/10.1016/S0378-7788(00)00044-X

Asan, H., (2006). Numerical computation of time lags and decrement factors for different building materials. *Build Environ.*, 41(5),615–620. http://doi.org/10.1016/j.buildenv.2005.02.020

Asan, H., Sancaktar, Y.S., (1998) Effects of wall's thermo physical properties on time lag and decrement factor. *Energy Build.*, 28(2),159-166. http://doi.org/10.1016/S0378-7788(98)00007-3

Ashok Babu, T.P., Saboor, S., (2014). Study of Unsteady State Thermal Characteristics of Homogeneous and Composite Walls of Building and Insulating Materials for Passive Cooling. Proceedings of 3rd international high performance buildings conference, Purdue University, U.S.A. http://doi.org/10.13140/2.1.3459.0402

CIBSE, (2006). *CIBSE Environmental Design Guide-A* (8-16). Chapter 3, Chartered Institution of Building services engineers, London.

Davies, M.G., (2004). *Building Heat Transfer* (335-350), Chapter 15, U.K., John-Wiley & sons Ltd.

Duffin, R.J., Greg, K., (1984). A passive wall design to minimize building temperature swings. *Sol. Energy*, 33(3/4), 337-342.

Knowles, T.R., (1983). Proportioning composites for efficient thermal storage walls. *Sol. Energy*, 31(3), 319-326.

Saboor, S., Ashok Babu, T.P., (2015). Optimizing the Position of Insulating Materials in Flat Roof Exposed to Sun Shine to Gain Minimum Heat into Building under Periodic Heat Transfer Conditions. *Environ. Sci. Pollut. Res.*, (Online). http://doi.org/10.1007/s11356-015-5316-7

Saboor, S., Ashok Babu, T.P. (2015). Investigation of Unsteady Thermal Response Characteristics of Various Hollow Bricks Exposed to Sinusoidal Solar Thermal Excitation. Proceedings of 14th International Conference of the International Building Performance Simulation Association IBPSA-2015, Dec 7-9th, IIIT Hyderabad, Telangana, India.

Saboor, S., Ashok Babu, T.P., (2016). Influence of ambient air relative humidity and temperature on thermal properties and unsteady thermal response characteristics of laterite wall houses. *Build. Environ.*, 99, 170-183. http://doi.org/10.1016/j.buildenv.2016.01.030

SP:41 (S&T) (1987). *Handbook on functional Requirement of Buildings other than industrial buildings*. Bureau of Indian Standards, India, 205.

Autonomous Optimization and Control for Central Plants with Energy Storage

Michael J. WENZEL[1]*, Robert D. TURNEY[1], Kirk H. DREES[1]

[1]Johnson Controls Inc., Technology and Advanced Development,
Milwaukee, WI, USA
mike.wenzel@jci.com*, robert.d.turney@jci.com, kirk.h.drees@jci.com

* Corresponding Author

ABSTRACT

An economic model predictive control (MPC) framework is used to determine how to optimize the allocation of energy resources across a central energy facility including chillers, hot water generators, and thermal energy storage; present the results to an operator; and execute the plan. The objective of this MPC framework is to minimize cost in real-time in response to both real-time energy prices and demand charges as well as allow the operator to appropriately interact with the system. Operators must be given the correct intersection points in order to build trust before they are willing to turn the tool over and leave it in fully autonomous mode. Once in autonomous mode, operators need to be able to intervene and impute their knowledge of the facilities they are serving into the system without disengaging optimization.

1. INTRODUCTION

Design and operation of central energy facilities is becoming an increasingly difficult problem. High efficiency products are available; but it is necessary to plan for and control the products properly to realize the potential cost savings (Ma & Wang, 2011)(Yu et al., 2008). Recently, model predictive control for central energy facility was developed (Wenzel et al., 2014). It was established as a planning tool to help engineers choose a type of facility and the equipment types and capacities in that facility. Simulation showed the potential for savings of over 10%. In this paper, an operational tool is described that is capable of distributing the load across the various subplants, selecting the equipment to turn on, determining optimal setpoints, and most importantly communicating this information to the building automation system (BAS) for execution. In this way, the operational tool is capable of realizing this 10 to 15% savings estimated by the planning tool.

In order for any of the savings to be realized, the operational tool must remain in autonomous mode. A system with energy storage that is capable of supplying the campus load for as many as 10 hours has too many variables for any operator to efficiently manage. However, with the proper interface an operator is capable of augmenting the system with any knowledge that may be missing, and allowing the model predictive control algorithm to incorporate that additional knowledge into the algorithm's decisions.

Cascaded control logic is ubiquitous in building control systems. Typically, low level feedback controllers produce the actuator input for a system (e.g., drive speed), whereas supervisory controllers produce setpoints for these low level controllers. The reasons for this hierarchy are numerous: the low level requires feedback to respond to disturbances, the low level operates a much faster sample rate to keep control of the equipment, there is a decoupling of purpose to improve debugging capability when there is an issue, etc.

The operational tool presented in the present paper continues the idea of cascading the control algorithm. It describes how several cascaded sub-algorithms in a supervisory economic model predictive control algorithm provide a natural interface for the operators to interact with the system, thereby increasing the probability the system remains in autonomous mode. Of course, adding interfaces and cascading the control problem cannot improve optimal performance (as each level does not have all the information of the previous level). Cascaded control is still often used to convert a computationally unsolvable (in real-time) mixed-integer, non-linear optimization problem into several smaller, solvable, problems. With the current advances of computing technology and optimization techniques, it is possible to see a future where the hierarchy of control is not necessary and a single omniscient system can calculate all required system inputs for optimal control. There may still be debate as to whether this control would be "optimal" in a system where humans are still ultimately responsible for the facilities operations. Facility owners and operators of large facilities (where the most energy savings can be obtained) seem unmotivated to move to a system that gives them no control or only gives them the control to either "live with what the computer says" or "override the outputs". When posed with this type of system, operators may try to outthink the optimal control because there is no information as to why the controller made the decisions it did. Thus, the system will remain in manual override and no optimization will be performed. A cascaded system, however, decouples the purpose of each individual sub-algorithm, allowing the operator to understand purpose of each output, thus allowing him or her to comprehend the decisions made by the system. A system that the operator understands and has the ability to add knowledge to is more likely to be left in autonomous mode and capture the savings projected during planning.

A single omniscient control system has not only a problem with the user's ability to understand the output, but also the user's ability to change portions of the output. The operator of a large energy facility will likely still have more information about the plants operation than the control system, simply because it not possible to individually program all factors that could affect the optimal control decisions.. Furthermore, a control system that comes with a more general artificial intelligence where the inputs are essentially anything that it can be exposed to will fail in combinations of inputs that it has not yet seen. As long as there is an operator that has been on the job longer than the AI, the operator may have relevant information to improve the control systems information. Thus, it is necessary that contemporary central energy facility control systems provide an easy to understand interface for the operators to both gather information and to augment the system with their own knowledge.

In the current paper, an operational tool is described which is capable of autonomous model predictive control and optimization of a central energy facility. The goal is not to provide a tool that runs the plant in the background with no potential for operator interaction, but to provide a tool the is capable of running a plant while providing for human-in-the-loop interactions where the operator can view future decisions of the control algorithm and easily provide additional information to the system as necessary. It is shown how the cascaded optimization problem provides a natural interface for the operators to make changes to the system. First the optimal control problem is described, the inputs and outputs of the system are enumerated and the optimization objective defined. Then a five layer cascaded model predictive control system is described. Finally, it is shown how this cascaded control problem is made into a tool for online plant operation with all the proper intersection points where the operators can interact with the system. Results from a live deployment of the operational tool are shown, illustrating that the operators are able to control the plant while leaving the system in autonomous mode. Thus, the majority of the potential savings are realized.

The first sub-algorithm is the prediction of the energy loads of the campus; i.e., the inputs to the optimization system. The predictions are made for a week in advance, giving the operator ample time to react to predictions they do not agree with and override the predictions if they feel it necessary. The predictions are inputs to the subplant energy distribution optimization. The energy distribution sub-algorithm determines the optimal distribution of energy across major equipment classes (subplants and storage) for the prediction horizon and sends the current distribution to the equipment selection and setpoint sub-algorithms. The operators are able to use the subplant-level optimization for "advisory" only and enter their own load distribution into the equipment level optimization. This could be done if they feel that they need to be conservative with the charge of the tank, etc. Finally, the equipment level optimization determines the devices to turn on and their setpoints in each subplant and sends those setpoints to the building automation system. These decisions can be overridden, but should be rare as the system takes device availability, accumulated runtime, etc. as inputs. The final tier is the low level control layer. To override this layer the operator has to put the building automation system (BAS) in "hand" mode and write specific outputs actuator inputs: valve positions, drive speeds, etc. This is likely to only be done during testing or maintenance of the system.

2. CASCADED CONTROL

As stated in the introduction the model predictive control and optimization of central energy facilities is broken into five distinct subalgorithms each with a specific purpose. By limiting the scope of any particular sub-algorithm of the control scheme, it is easier for an operator supervising the control to decide if intervention is necessary, and if necessary, how to intervene. The five sub-algorithms: prediction, energy distribution, equipment selection, system setpoints, and equipment control are described below. Much literature can be found for individual many of the sub-algorithms. It is the combination of these building blocks cascaded that makes possible a model predictive control scheme for the operational tool that gives the operator the ability to properly interact with the system. A flow chart of the control algorithm is shown in Figure 1 with the inputs and outputs of each of the five sub-algorithms clearly defined.

2.1 Prediction

Regardless of the method, to perform economic model predictive control, it is necessary to have the loads of each of the energy sinks and the pricing for purchasing energy from each of the energy sources for the length of the horizon. These predicted loads and rates then become the input to the economic model predictive control algorithm.

To perform prediction several factors contributing to the load/rate can be considered. For loads (and rates) it is typical to use the day-type (day of week, occurrence of special events), time of day, and weather conditions. Historical load data along with the independent factors used are collected and used to train a model. Model training can be done offline first to load the initial model. Subsequent model updates should be done automatically and online at regular intervals or continuously to account for changes in the model. However, the updates should be done in a separate process to avoid delaying the process performing the control.

In the present operational tool, finding an estimate of the future load, $\hat{l}_{k+h}$, prediction is done by adding a deterministic part, d_k, and a stochastic part, s_k, (ElBsat & Wenzel, 2016)

$$\hat{l}_{k+h} = d_{k+h}(m_{k+h};\theta) + s_{k+h}(e_k;\phi) \quad \forall h \in horizon = \{0,1,2,\ldots,N_h\}, \tag{1}$$

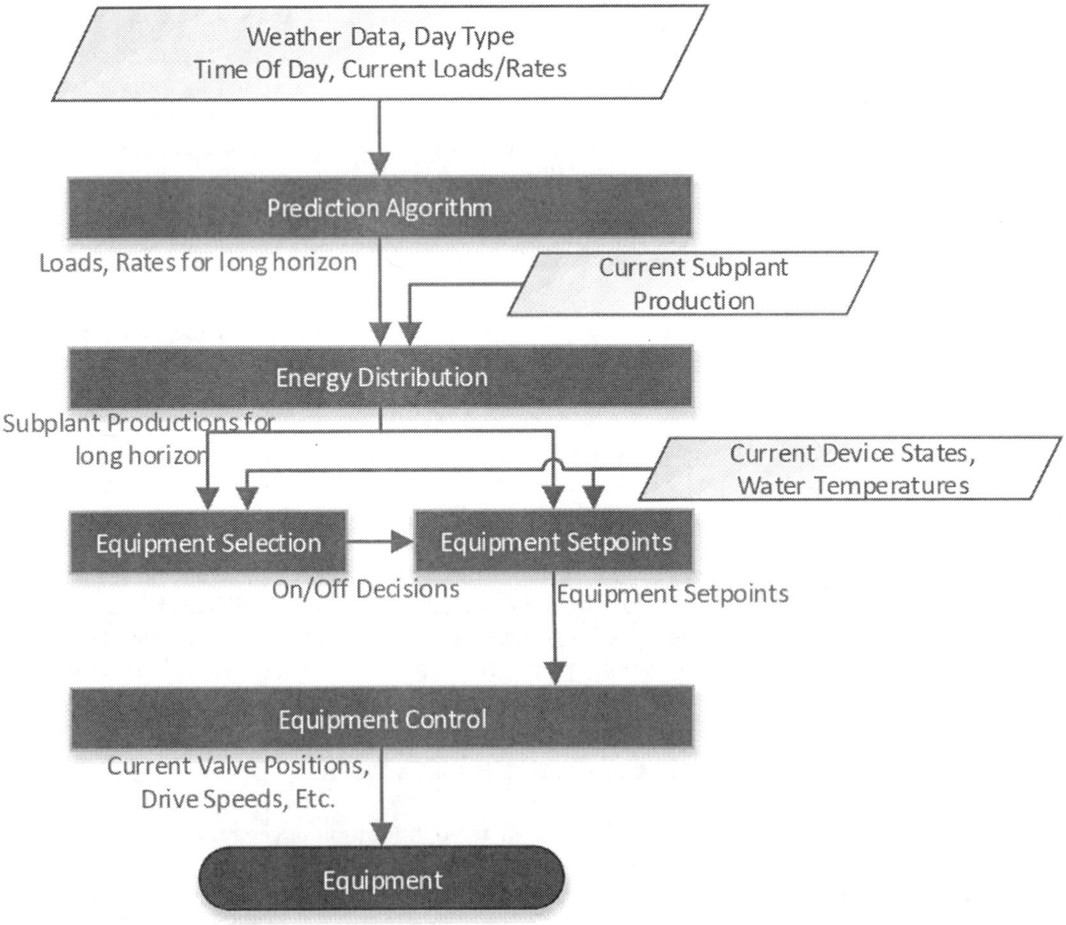

Figure 1: Flow chart of an operational tool using cascaded model predictive control.

where,

$$e_k = l_k - d_k(m_k;\theta). \tag{2}$$

The purpose of the deterministic model is to account for factors (m_k) that have a known relationship to the load. The relationship between the factors, such as weather, day type, and time-of-day, is a steady-state function. The parameters θ can be found by various curve fitting strategies. The purpose of the stochastic model is to update the prediction of the deterministic model base on its current error. Because it is impossible to account for every factor that can affect the load, it is necessary to update the prediction for short term prediction errors. To perform this an auto-regressive model is used. The difference between the predictions by the deterministic model is passed through an auto-regressive predictor and the output is added back to the prediction. Thus allowing for un-modeled disturbances to be accounted for in the overall prediction.

2.2 Energy Distribution

The predicted loads and rates from the prediction sub-algorithm are then directed to the energy distribution sub-algorithm where they are coupled the current subplant production in order to produce the optimal energy distribution over the entire horizon.

At a high-level, a central energy facility serving a large campus can be represented by the manner in which the energy can be distributed across the campus. An energy facility consists of subplants, formed of like equipment, that are capable serving campus by converting purchased energy (electricity, natural gas, etc.) into the energy required by the buildings (chilled water, steam, etc.). Mathematically, subplants describe a mapping from the space of independent decisions of that subplant to the space of the subplants productions and consumptions:

$$f:\Re^n \mapsto \Re^{n_p+n_c} \quad where \; f_{subplant}(\theta) = \begin{bmatrix} p \\ c \end{bmatrix}. \tag{3}$$

An energy distribution diagram is shown in Figure 2. The energy facility pictured consists of five subplants capable converting utilities: electricity, natural gas, and water to chilled water and hot water. The job of the energy distribution sub-algorithm is to optimize the cost of purchasing energy, *J*, subject to constraints including: subplant capacities, equipment models, and energy balance equations at each resource:

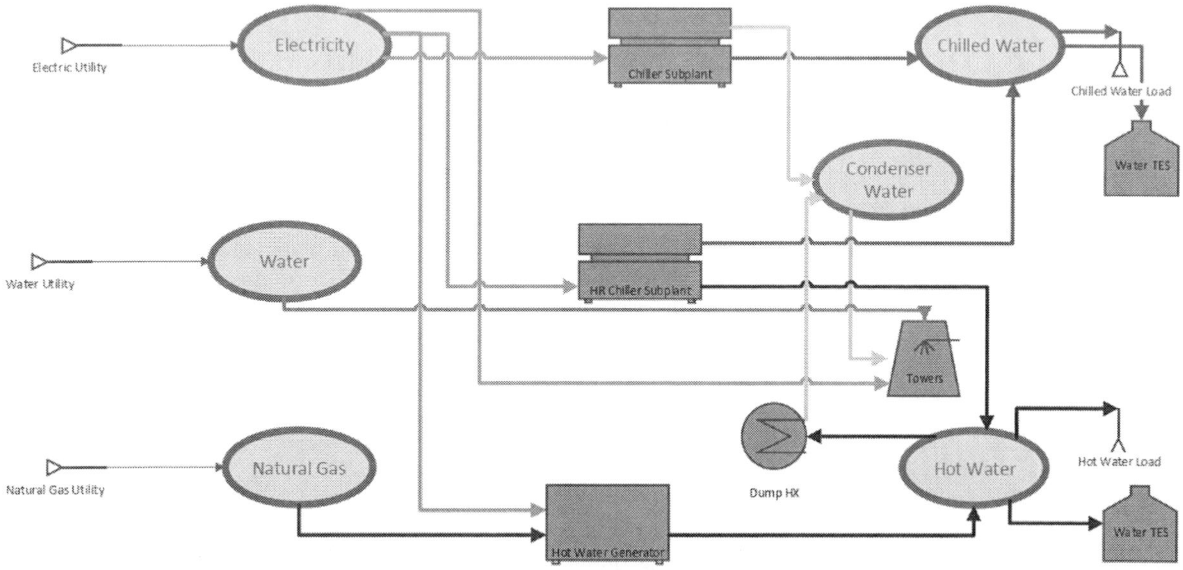

Figure 2: Energy distribution diagram for a plant with five subplants and two energy storage elements.

$$J = \sum_{utility} \sum_{horizon} cost\left(purchase_{resource,\, time}, time\right), \tag{4}$$

subject to,

$$\sum_{sources} purchase_{resource,\, time} + \sum_{subplants} production_{resource,time} - \sum_{subplants} consumption_{resource,time} + \cdots$$
$$\sum_{storages} discharge_{resource,time} - \sum_{sinks} requests_{resource} = 0, \tag{5}$$
$$\forall resources, \forall time \in horizon.$$

The formulation of this in a linear programming framework is described in (Wenzel et al., 2014). In general, the output of the energy distribution sub-algorithm is the amount of resource (typically energy, except in the case of water for the cooling towers) that is disturbed on the connections in energy distribution diagram of Figure 2.

The energy distribution level is responsible for determining the energy distributed at every hour during a prediction horizon for several days. If the prediction horizon is one week, the prediction horizon contains 168 hours. It is this that causes the economic model predictive control to become difficult to solve. For a modest sized energy facility there may be around 25 pieces of equipment that can be turned on/off as well as a similar number of setpoints to control. While a mixed integer, non-linear program with 25 binaries and 25 continuous variables can be solved in a reasonable amount of time, when that it is multiplied by the 168 hours in the horizon there become 4200 binaries and 4200 continuous variables and problem becomes much more difficult. By limiting the problem to only the energy distribution, the problem can be posed as a linear program there by greatly reducing the computational time for the sub-algorithm that has the most decision variables due to the multiplicity of the long horizon.

2.3 Equipment Selection

Given the request for the subplant to produce a certain amount of energy for the entire length of the long horizon, it is possible to decide which equipment should be turned on/off. The purpose of the equipment selection sub-algorithm is to determine a method for mapping the independent decisions of a subplant to the on/off decisions of the equipment inside that subplant. There is not a need for a horizon at the equipment decision level. Its job is to minimize the energy use while producing the load request from the energy distribution level. However, it may be desired to display the expected device decisions for a short horizon. Assuming this is the case, then the equipment selection sub-algorithm can simply be called several times using the future values requested by the energy distribution level.

There are several approaches to decide which devices should be turned on (Deng et al., 2013). A dynamic programming version is to build the optimal decision points from a single device up. This done by taking two devices and generating the optimal performance of the subplant with these two. Combining the optimal performance of the first two devices with the next device and finding the optimal performance with these three devices. The process of adding one more device is continued until all devices have been added. An example of the process for 3 chillers is shown in Figure 3. First chillers 1 and 2 are combined by comparing the electricity used for all each chiller and both chillers (it was assumed that each chiller must run at the same part load ratio). The minimum electricity used is found for each amount of production and the device selection that produces this is stored, Figure 3b. The optimal place to on chiller 2 is at approximately 3150 kW. Now the third chiller is combined by comparing the energy use of the result from combining chillers 1 and 2 with chiller 3 and the sum of the two of them, Figure 3c. Now it is possible to see that it optimal selection strategy is to run chiller 1 up to 3150 kW then run chillers 1 and 3 to 4000 kW, chillers 1 and 2 from 4000 to 5250, and all chillers above that load.

2.4 Equipment/System Setpoints

The purpose of the setpoint sub-algorithm is to output the setpoints that are used in order to obtain the loads specified by the energy distribution algorithm with the equipment determined by the device selection. Again, in this sub-algorithm, there is not a need for a horizon. If a display horizon is required then the setpoint sub-algorithm can be called several times using the future values requested by the energy distribution level. This layer requires a plant model and needs to optimize the setpoints given the constraints representing the optimal energy distribution and equipment selection. Conceptually, this may be one of the simpler sub-algorithms, as it is a call to a nonlinear optimization routine where the cost function, $J(x_{SP})$, is the cost calculated by a steady-state simulation of the plant model given the current setpoints being evaluated. While conceptually simple, expressing the steady-state cost of the energy facility given several setpoints and device selections in a manner that can solved quickly enough for the nonlinear optimizer to converge in short amount of time is difficult. The equipment/system setpoints that produce the lowest cost are computed and set to the equipment control sub-algorithm for final execution of the plan.

2.5 Equipment Control

The equipment control layer exists in the BAS and is typically implemented by PID controllers. Its job is to manipulate the actuation points so that the measured values are at the setpoints calculated by the equipment/system setpoints sub-algorithm. This layer of control is likely to exist at any existing central energy plant.

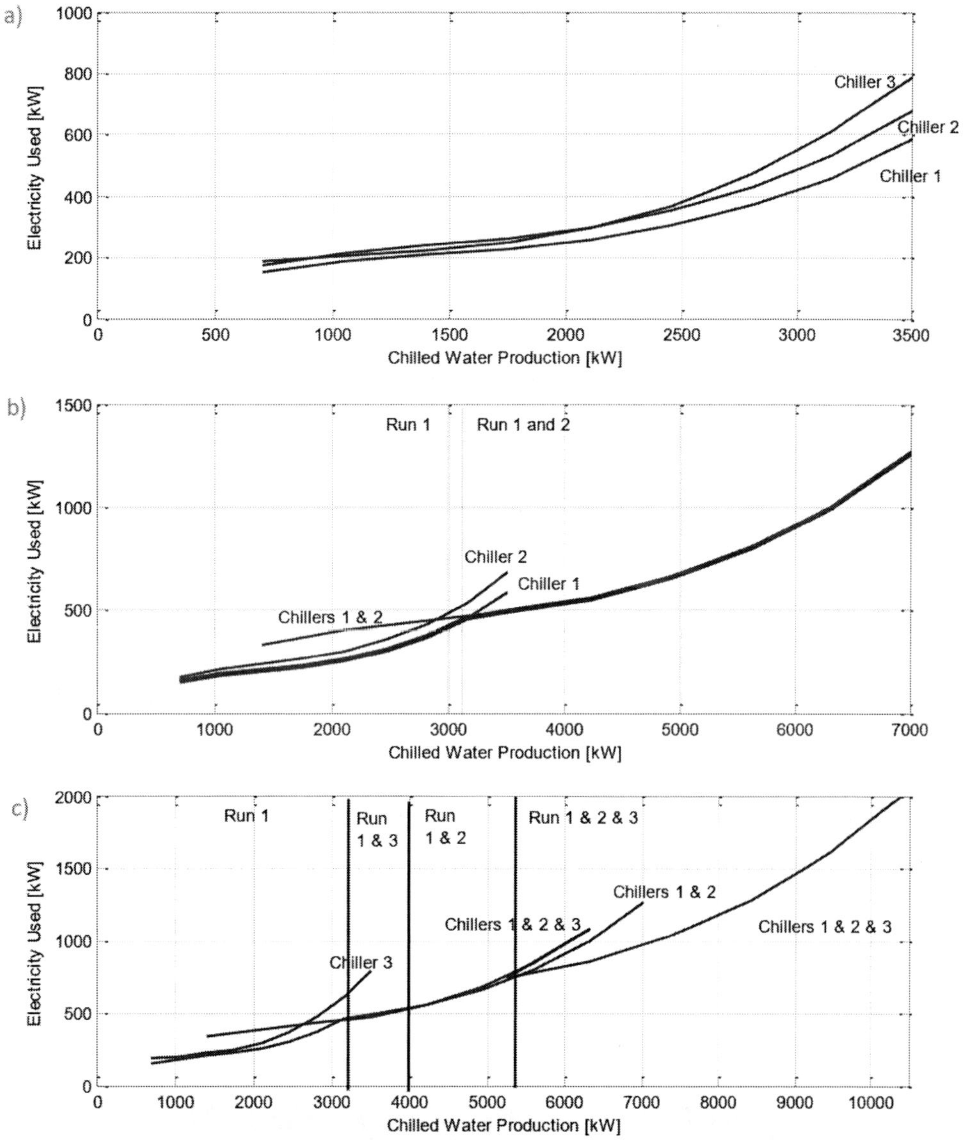

Figure 3: Determining the optimal equipment selection using the principle of optimality. a) The performance of chillers 1, 2, and 3 each with a 3510 kW (1000 ton) capacity. b) locus of possible chiller production and useage using only chillers 1 and 2. c) locus of possible chiller productoion and useage when chiller 3 is combined with the minimal use of chillers 1 and 2

3. OPERATOR INTERACTION

Empowering the operators to impart their knowledge through human-in-the-loop decisions while remaining in autonomous mode is likely the most powerful tool the engineer design the optimization system has in maximizing the time that the system spends in automatic mode. It is likely impossible for the optimization algorithm to take into account explicitly all factors that can influence the optimal control sequence. In the near term, it is equally unlikely that an AI could be placed on the site and learn all the factors that operators already know.

3.1 Predictor Overrides

Periodically, the prediction algorithm will not take into account a factor that has a significant effect on the load. A common example is special events. When this occurs, the operator would have to take the system out of autonomous mode if no interface was given to allow the user to override the prediction. Instead, the current system allows the operator to view the prediction and add an override that will be used during optimization.

For the optimization and control algorithm to be autonomous, the system must produce outputs on a regular interval. It is impossible to allow the system to wait for an operator to override any predictions. Instead, the system must output load and rate predictions for several hours (days) into the future. The operator then has the ability to make a decision about the load and override it if necessary. If the override is far enough in the future it should have little effect on the current decisions and the system will have the time to optimally respond to the new information.

Figure 4 shows a capture of the operational tool. The cooling load has been predicted for the next 7 days of operations. Here an operator decided that the load in the morning of June 23rd was too small, and decided to override the load to be larger. Overridden predictions are shown in yellow in the figure. The knowledge of the increased loads provided by the operator is already being used to optimally distribute energy across the subplants and the storage, shown in Figure 5. By giving the operator the time to decide if an override is required for a future dispatch it is also possible to give the model predictive control algorithm time to blend the new information into the dispatch between the current time and the override without having a large change or disturbance in current operations when the override is added.

3.2 Energy Distribution Overrides

In the case that the operator is not satisfied with the high level load distribution created by the energy distribution sub-algorithm, constraints can be entered in order to force the energy distribution sub-algorithm to allocate a specific load to a specific subplant. Again, optimization can continue as expected subject to the additional constraints,

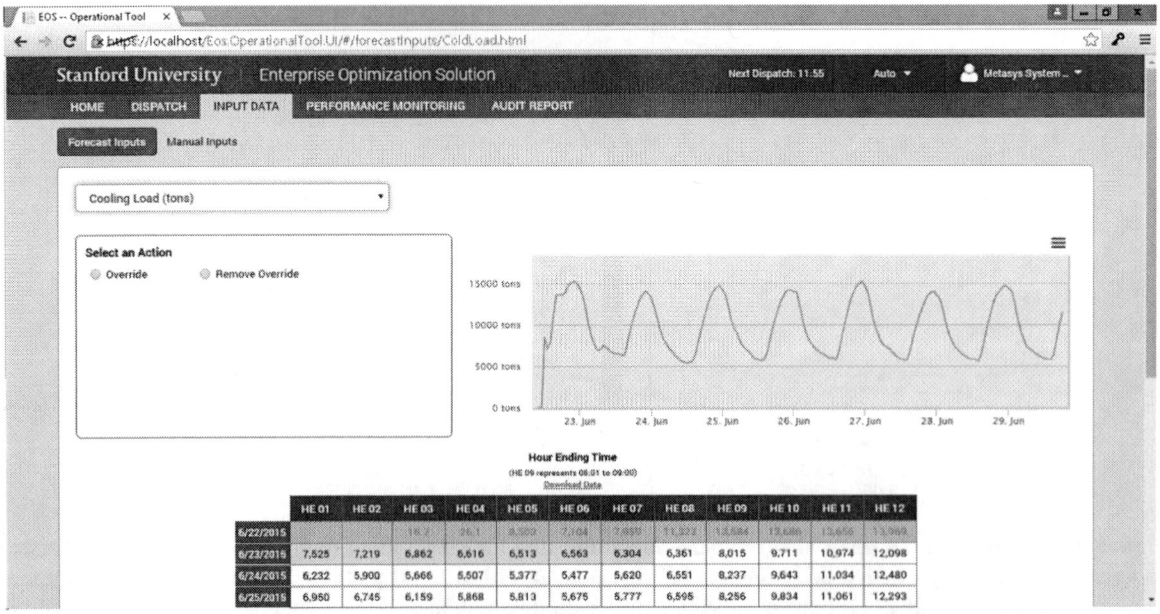

Figure 4: Screen capture of an operational tool while the operator is using the predictor override interface.

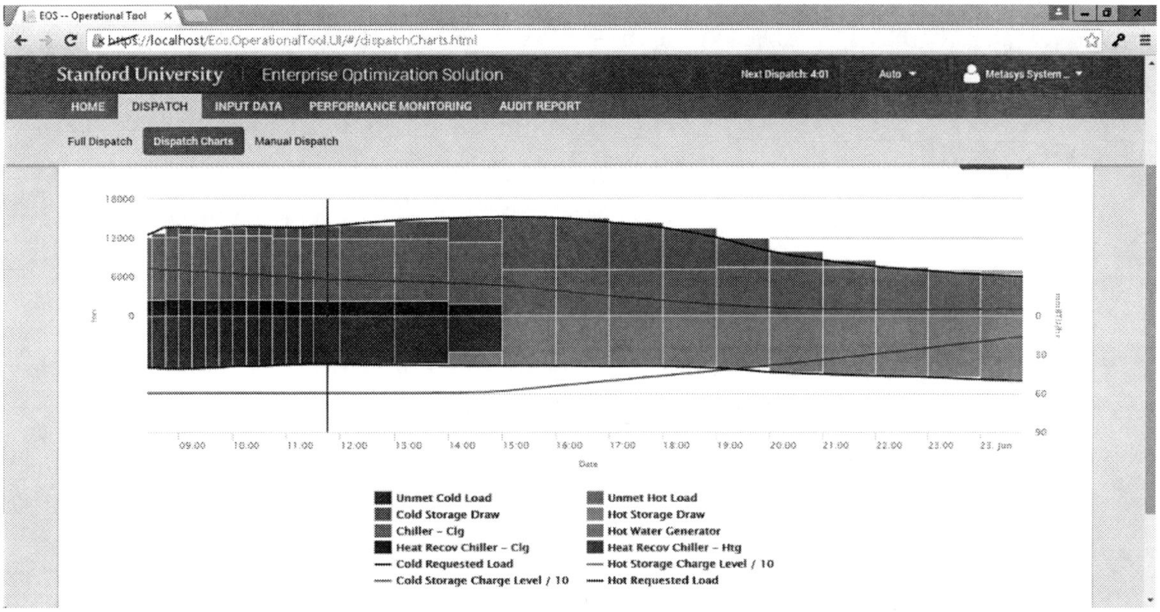

Figure 5: Operational tool dispatch zoomed to show the next 12 hours of energy distributions.

thereby allowing the operator to augment the system with his knowledge rather than simply taking command of control. If the constraint is added with ample time before it occurs, the model predictive control algorithm should have the time to blend the new information into the dispatch without creating a disturbance in current operations.

Additionally, the operators are able to use "manual mode" in which they are capable of specifying the production they want performed by each subplant. In manual mode, equipment selections, setpoints, and equipment control still follow their normal path. However, the dispatch is only updated on the operator's request. The allocation developed by the energy distribution sub-algorithm is used as advisory information.

3.3 Equipment Selection Overrides

In most cases, operators should not have a preference of which equipment runs to produce the desired load. The exception to this rule is when a device needs to be removed from service for maintenance. Again, it is undesirable to force the operator to remove the system from autonomous mode when equipment must be taken out of service. In a large plant there is constantly maintenance to be done. Requiring all equipment available to perform economic MPC would drastically reduce the amount of time the system spends in autonomous mode.

When equipment is not to be run the operator can remove the equipment from service using the user interface. Using this interface the operator tells the operational tool that the equipment cannot be used during a certain time period. Both the energy distribution and equipment selection sub-algorithms use this information to produce a dispatch that is optimal subject to the new constraints. As in the previous examples, if the equipment is scheduled out of service in advance it should not cause much of a disturbance in current operations. In Figure 6 two chillers have been scheduled out of service.

The expected dispatch is available for a horizon so that the operator has time to decide if anything must be overridden. This also gives the operator the ability to plan short term maintenance. If chiller 2 is not scheduled to be turned on for the next two hours, it would be a good time to perform any maintenance that requires a short amount of time. Likewise, if the operator sees that the chiller subplant is not scheduled to run for the next five days then it is possible for all the chillers to be taken out of service and more substantial maintenance performed with no substantive change to the central energy facility's operations cost.

3.4 Setpoint and Equipment Control Overrides

Changes in the setpoints determined by an operational tool and the low level equipment control that realize those setpoints should rarely be overridden. No user interface is provided to override these items in the operational tool. However, these points can be overridden in the building automation system. If a point in the building automation

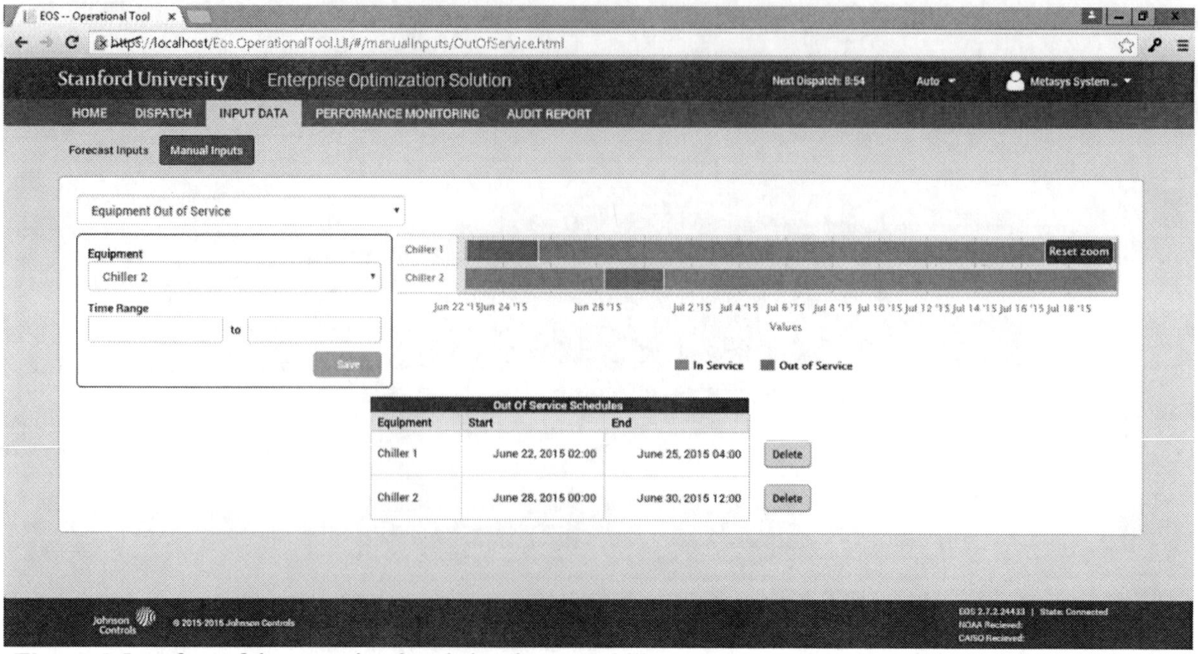

Figure 6: Interface of the operational tool showing two chillers taken out of service. This will cause the model predictive control algorithm to choose the optimal dispatch without using this equipment at the specified times.

system is overridden an operational tool will not be able to write the setpoint and it will be impossible to carry out an unadjusted dispatch. While it is not optimal for any of the setpoints or low level control logic to be overridden and in "hand" mode, it is likely to happen. Instead of forcing the operator to take the operational tool offline when overriding the BAS logic, it is better to respond to the overrides. The operational tool looks to verify that none of the logic it must interact with has commands that are sent at a higher priority than the priority at which the operational tool writes. If there is a higher priority command, the equipment controlled by that logic is "unavailable" to the operational tool and optimization must be performed with the new constraints.

4. REMARKS AND CONCLUSIONS

Allowing the operator the ability to override has meant that the system is in autonomous mode a large majority of the time. The operational tool was installed on a new construction site during the commissioning of the new plant. Figure 7 shows a plot of system mode over a three week period from February 22 until March 15. The system was in autonomous mode for approximately 95% of the time. While it is desired that the amount of time spent in autonomous mode approach 100%, given that the plant was new construction and the operators were still managing chillers that tripped periodically, 95% autonomous mode shows that the operators trust their ability to make the proper control changes by modifying the constraints through the operational tool interface rather than completely removing it from manual mode.

Because of the large amount of time spent in autonomous mode the operational tool is able to save the money that was originally expected during the planning phase. Figure 8 shows the comparison of the central energy facility cost as run on site to the expected cost assuming the control algorithm had perfect knowledge of the loads and electrical rates and was always in automatic mode (i.e., a planning mode style run). The chart shows that the central energy facilities operational cost for this three week period is only 4% greater than the theoretical best the economic model predictive control algorithm could have done (with perfect prediction of loads and rates).

The model predictive control framework used to produce an operational tool for autonomous control of central energy facilities was shown to capture the majority of the 10 to 15% savings estimated by the planning tool. This would not be possible if the system did not remain in autonomous mode. However, by giving the operators the appropriate interfaces to augment their knowledge with the operational tools model predictive control kernel it was possible to keep the time spent in autonomous mode above 90%.

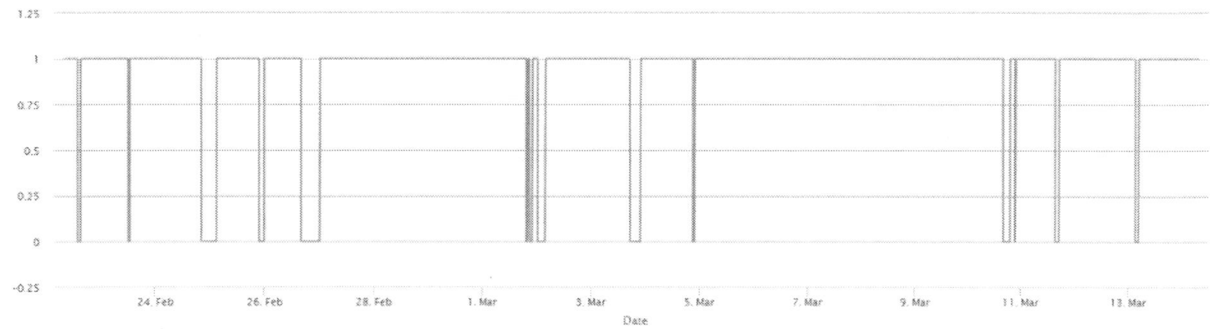

Figure 7: Screen capture of an operational tool while the operator is using the predictor override interface. A value of "1" incidates the economic MPC algorithm is in autonomous mode.

REFERENCES

Deng, K., Yu, S., Chakraborty, A., Lu, Y., Brouwer, J., Mehta, P.G. (2013). Optimal scheduling of chiller plant with thermal energy storage using mixed integer linear programming, *American Control Conference June 2013*, 2958 – 2963.

ElBsat, M.N., Wenzel, M.J. (2016). Load and electricity rates prediction for building wide optimization applications, *Submitted to 4th International High Performance Buildings Conference.*

Ma, Z., Wang, S. (2011). Supervisory and optimal control of central chiller plants using simplified adaptive models and genetic algorithm, *Applied Energy*, *88*(1), 198-211.

Wenzel, M.J., Turney, R.D., Drees, K.H. (2014). Model predictive control for central plant optimization with thermal energy storage, *International High Performance Buildings Conference.* Paper 122.

Yu, F.W.., Chan, K.T. (2008). Optimization of water-cooled chiller system with load-based speed control, *Applied Energy*, *85*(10), 931-950.

ACKNOWLEDGEMENT

Special thanks are due to Joe Stagner, Executive Director of Sustainability and Energy Management Department, Stanford University for his collaboration in defining the problem statement and uses cases that helped drive the interfaces for the operators seen in this document.

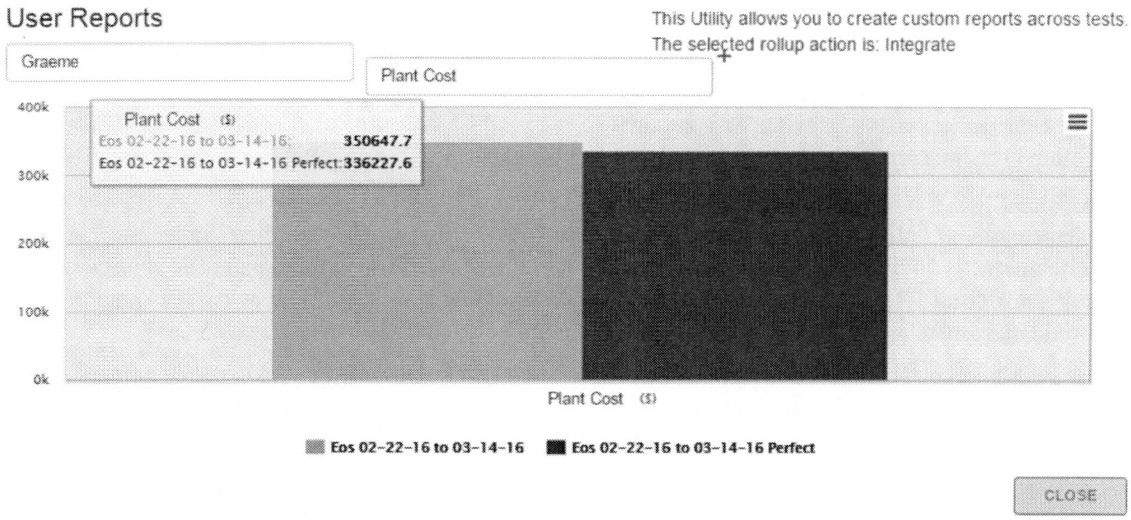

Figure 8: Comparison of energy facility operational cost as run on site (blue) to simulated run where all loads are perfectly predicted (grey).

Towards an Automated ToolChain for MPC in Multi-zone Buildings

Filip JORISSEN[1,2*], Lieve HELSEN[1,2]

[1] KU Leuven, Department of Mechanical Engineering,
Leuven, Belgium
filip.jorissen@gmail.com

[2] EnergyVille,
Genk, Belgium
lieve.helsen@kuleuven.be

* Corresponding Author

ABSTRACT

Due to climate change energy efficiency is becoming increasingly important. This is particularly true for the building system control sector, where a large energy saving potential exists. Model Predictive Control (MPC) is a new control technique in this sector that minimises the building energy use while taking into account comfort and other constraints through the use of a simplified building model. Setting up the required MPC 'controller' model for each building is however challenging and requires a lot of expertise work. In this paper we present a Linear Automated Toolchain for MPC (LAT-MPC) that allows highly automating the process of setting up a controller model and running a linear MPC. Due to this automation new research such as integrated optimal control and design of buildings comes into the picture.

1. INTRODUCTION

Building Heating, Ventilation and Air Conditioning (HVAC) systems use large amounts of energy to provide comfort to its occupants. According to Pérez-Lombard et al. (2008) HVAC represents 20% of the total energy use in the USA. Since the production of the required thermal or electrical energy typically produces CO_2, HVAC operation contributes to climate change. Energy efficiency allows to limit the primary energy consumption and therefore the corresponding CO_2 emissions and operational costs.

Model Predictive Control (MPC) is a control technique that allows significant energy savings to be obtained without jeopardizing comfort or other constraints. MPC uses a simplified building model, the controller model, to predict the (future) impact of possible control actions on the building. Depending on the level of detail of the controller model, disturbances such as occupancy and weather forecasts can be included. Using this information the MPC controller computes the control action that uses the least amount of energy, while satisfying a set of constraints. These constraints are most often 1) technical, e.g. they may limit the mass flow rate of a fan to its nominal value, or 2) related to occupancy comfort, e.g. there may be a constraint on the building zone temperature(s).

Real implementations of MPC have shown energy savings of 14% in a Swiss office building (Sturzenegger et al., 2013), 17% in a Belgian office building in Hasselt (Váňa et al., 2014), more than 30% in a Belgian office building in Brussels (De Coninck & Helsen, 2016), 19% and 27% in two Australian office buildings (West et al., 2014), and 15%-28% in a Czech building (Široký et al., 2011; Prívara et al., 2011; Cigler et al., 2013).

Clearly, there is a large potential for energy savings using MPC, although the potential depends on the building type and the reference compared with (Cigler et al., 2013). However, exploiting this potential is difficult due to the large amount of expertise work incurred by setting up an MPC controller model (Sturzenegger et al., 2014; Cigler et al., 2013). Research should therefore automate this process as much as possible.

To this end De Coninck et al. (2015) presented a toolbox for identifying the parameters of low-order 'grey-box' building models using building measurement data. This approach however requires high-quality measurement data and is not easily scalable to larger problems. Sturzenegger et al. (2012, 2014) present Building Resistance-Capacitance Modelling (BRCM), a Matlab toolbox for setting up building MPC controller models and input data using basic building construction data contained in an EnergyPlus input file.

To further simplify the creation of controller models, tools should exist that allow the use of preconfigured (blocks of) models that can be connected together in an intuitive, drag-and-drop fashion. Modelica (Elmqvist & Mattsson, 1997) is a modelling language that allows the creation of such models. Although the initial scope of Modelica is to perform dynamic simulations, tools such as JModelica (Åkesson et al., 2010) and OpenModelica (Fritzson et al., 2005) also allow optimising such models, which can be used to implement an Optimal Control Problem (OCP) (Bonvini & Wetter, 2015) or MPC (De Coninck & Helsen, 2016). These tools are however general-purpose solvers that do not exploit the linear nature of our MPC optimisation problems, reducing computation speed, while also requiring multiple OCPs to be run in sequence to correctly implement MPC. This approach also does not directly support the coupling of the MPC controller to a detailed building model, e.g. in JModelica the MPC controller *is* the detailed building model, which is a simplification of reality and does not allow to take into account aspects such as model mismatch.

The goal of this paper is to describe a Linear Automated Toolchain for MPC (LAT-MPC) that 1) derives exportable 'white box' MPC controller models in an automated way, similar to BRCM, 2) facilitates the fast simulation of detailed building models implementing an MPC controller *into the model itself*, 3) has an easily configurable MPC interface and 4) an intuitive modelling environment using a single simulation tool, and 5) has components that can be reused in real MPC implementations.

The toolchain consists of the following steps. Firstly, a Modelica model of the building envelope needs to be created, using for instance the open source Modelica library IDEAS (Baetens et al., 2015). Note that other libraries can be used for other applications as long as they result in a model with linear dynamics. Secondly a linear version of the building model is set up, for which a linear HVAC system, constraints and cost function needs to be added. This linear model is exported into state space format using the `Linearize()` command found in Modelica simulation tools. When using IDEAS, the approach presented by Picard et al. (2015) can be used. A third step generates an input file containing the disturbances and time-dependent constraints matching this state space model. Fourthly, a model needs to be set up that maps the optimal control results to real HVAC component set points. In the final step the MPC controller must be configured using basic parameters such as the optimal control horizon duration. The tool-chain contains an MPC controller implementation using efficient code generated by CasADi 2.4 (Andersson, 2013) such that even large MPC problems can be solved efficiently.

2. METHODOLOGY

This section describes how firstly a detailed emulator model and secondly a linear controller model of the building can be set up using Modelica. The controller model is then exported and a corresponding input data file is generated. We further explain how this data is used to construct the MPC controller and how this MPC controller is again integrated into the building model.

Throughout this section the model structures required in the different steps are illustrated using 1) a schematic and 2) an example Modelica implementation.

2.1 Creating the emulator model

The first step of the toolchain is to create an 'emulator' model of the building. Figure 1a illustrates the typical structure of such a model. The most important block is the 'building envelope', which models all thermal dynamics related to conductive, convective and radiative heat transfer of the structural building materials. The building envelope is connected to HVAC components such as radiators, concrete core activation, boilers, heat pumps and a ventilation system. The controller determines the HVAC model control signals such as valve and Variable Air Volume (VAV) positions using measurement data from the building envelope and the HVAC system. Building envelope boundary conditions are typically computed using a weather data file. The weather data are needed to compute the incident solar radiation and sky radiative and dry-bulb temperatures for all surfaces of the building envelope.

The models required for setting up the emulator have been made available in IDEAS (Baetens et al., 2015), an open source Modelica library and extension of the Annex 60 library (Wetter et al., 2015). When setting up the emulator it is important to create some structure in the model such that the HVAC and controller models can easily be removed for the building envelope linearisation step.

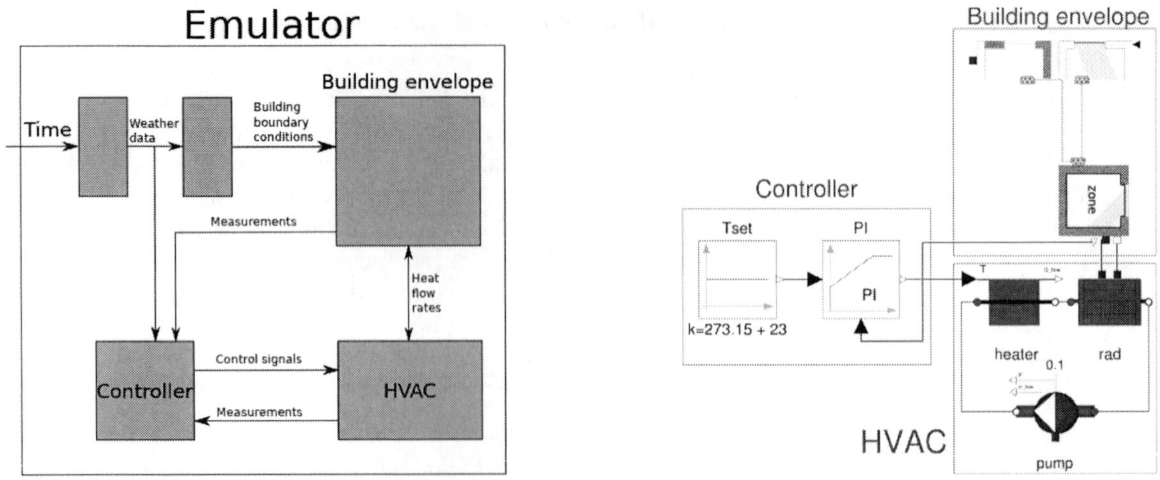

(a) Schematic of a typical building model containing a building envelope, HVAC models and a controller model. HVAC exchanges heat with the building model. The HVAC models are controlled by the controller, which uses measurement data from the building envelope and HVAC models.

(b) Example containing one zone, one window and one wall. Temperature is controlled using a radiator, which is connected to a pump and a heater. The outlet temperature of the heater is controlled using a PI controller with a fixed temperature set point of 23 °C. The building boundary condition computation is not shown since it is integrated into the building envelop models.

Figure 1: Emulator model.

Figure 1b shows an example of a very simple emulator model implementation using IDEAS. It will be used throughout this paper to illustrate the methodology. Key component groups are 1) the building envelope, containing a zone, wall (top left) and window (top right), 2) the HVAC system containing a pump, heater and radiator and 3) the controller: a PI controller that tracks a fixed temperature set point.

2.2 Generating the linear controller model

In this section the emulator model is modified such that it becomes a Linear Time Invariant (LTI) model as shown in Figure 2a. This LTI is then passed to the Dymola function `Linearize()` such that a linear state space model is generated. This model is a 'white box' model since it only uses building construction and HVAC equipment information.

To obtain an LTI model the building envelope model equations need to be linearised, since by default IDEAS contains non-linear convective and radiative heat transfer equations. Furthermore, non-linear building boundary condition computations are considered to be model inputs. A dedicated version of IDEAS has been created that allows to create input variables for all boundary conditions by changing the value of a single parameter. For more details see Picard et al. (2015).

The linear emulator model must also contain a linear representation of the HVAC system. In practice this often means that the optimisation variables must be heat flow rates since the explicit modelling of mass flow rates and temperatures can lead to bilinear equality constraints, which are non-linear and non-convex. All optimisation variables must be inputs of the linear HVAC model such that they are correctly detected when linearising the model. Linear combinations of the optimisation variables may then be used to create constraints and to interact with the building envelope.

Since the used Modelica solver, Dymola, does not support the declarations of inequalities, a work-around is proposed for defining constraints. Whenever a constraint is added, the constraint is reformulated such that it has the structure

$$f(x,u) \geq 0 \tag{1}$$

where $f(\cdot)$ is a linear combination of the states x and inputs u of the LTI model. Note that u contains an element whose value is fixed to 1 such that constant terms in the inequalities are also detected when linearising around x and u. A Modelica output variable must then be created whose value equals $f(x,u)$ and whose name starts with the keyword

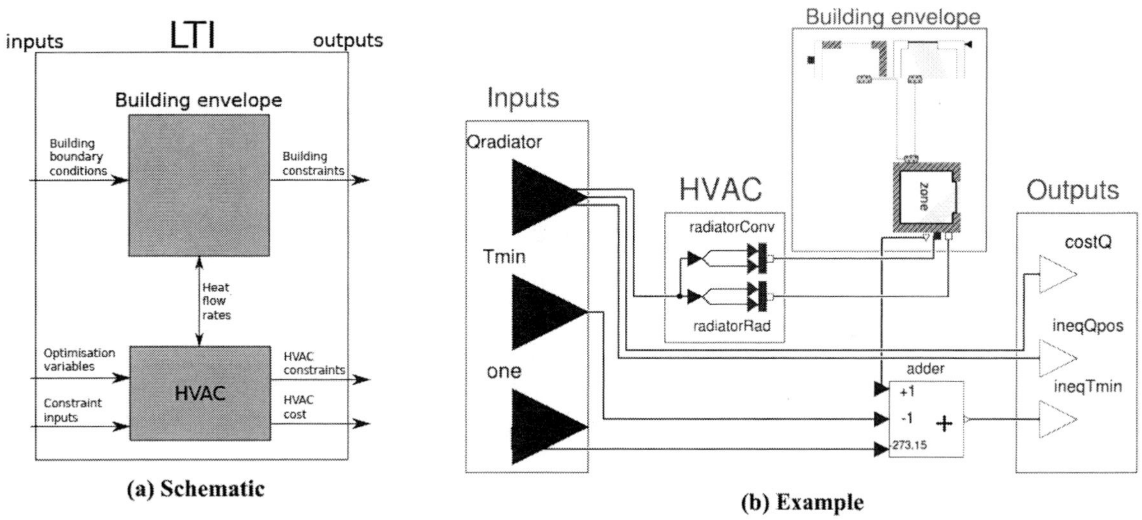

Figure 2: Schematic and example of a linear controller model. The building envelope model is retained albeit in a linearised version, while the HVAC model is simplified into a linear format. Optimisation variables, constraint inputs and building boundary conditions are model inputs. Constraint and cost variables are model outputs.

`ineq` such that the MPC-tool can identify that this output is an inequality whose value should be larger than or equal to zero.

Linear cost functions $g(x,u)$ can be formulated in a similar way, but by creating an output variable whose name starts with a different keyword, `cost`.

Figure 2b presents the result after linearisation starting from the example of Figure 1 and corresponds to the scheme shown in Figure 2a. The model is a simplified, linear version of Figure 1b with a single optimisation variable `Qradiator` that injects heat in the convective and radiative heat port of the zone model, representing the radiator. Furthermore the model contains two inequality outputs: 1) `ineqTmin` that sets a lower bound for the zone operative temperature and 2) `ineqQpos` that constrains `Qradiator` to be larger than zero. To illustrate the use of variable `one` we choose `Tmin` to have as unit degrees Celsius while the zone temperature has the unit Kelvin. The constraint represented by `ineqTmin` is then

$$zone.T - Tmin - 273.15 \geq 0 \tag{2}$$

or equivalently

$$zone.T \geq Tmin + 273.15 \tag{3}$$

where `zone.T` is the zone temperature and `Tmin` is a 'constraint input' as indicated on Figure 2a.

Finally, the output `costQ` results in the cost function J, i.e. $J = Qradiator$.

Now that the linear Modelica model has been set up, it needs to be exported. Dymola contains a built-in function for this purpose, `Linearize()`. This function linearises the model around a working point and returns four matrices and three vectors. The matrices are state space matrices A, B, C, D such that

$$\frac{\mathrm{d}x}{\mathrm{d}t} = A \cdot x + B \cdot u \tag{4}$$

$$y = C \cdot x + D \cdot u \tag{5}$$

where x are the model states, u are the model inputs and y are the model outputs. The three vectors *xNam*, *uNam* and *yNam* are the variable names of respectively x, u and y.

Matrices A and B can be used after discretisation to set up the MPC optimal control problem. Constraints can be integrated by extracting the linear constraint equations from matrices C and D. Constraints are denoted as

$$A_{ineq} \cdot x + B_{ineq} \cdot u \geq 0 \tag{6}$$

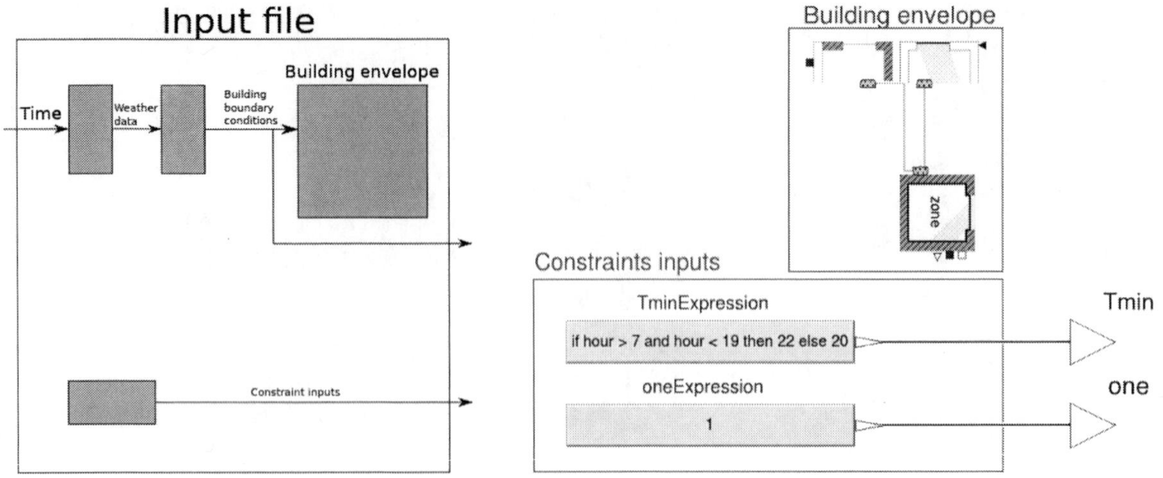

(a) Schematic

(b) Example

Figure 3: Schematic and example of the model required to generate the MPC input file. The model outputs constraint inputs and building boundary conditions, which are not illustrated in the example since these computations are embedded in the building envelope model.

where A_{ineq} and B_{ineq} consist of all rows i of C and D whose name in $yNam[i]$ starts with the predefined keyword `ineq`, since this name indicates that output i should be interpreted as a constraint.

Similarly the cost function J equals

$$J = c_x \cdot x + c_u \cdot u \tag{7}$$

where c_x and c_u are row vectors that equal the *sum* of all rows i of C and D whose name in $yNam_i$ starts with the predefined keyword `cost`.

When applied to the example model, the linearisation results in a model with 7 states, 27 inputs (`Qradiator`, `Tmin`, `one`, 5 boundary conditions for the window and 19 other boundary conditions) and 3 outputs (2 constraints and 1 cost function).

2.3 Generating an input file

The state space model derived in the previous step contains many inputs. Most of these inputs are however known disturbances. Other inputs, like constraint inputs in Figure 2a also need to be defined. In the current step the values of these known disturbances are stored such that they can be used as inputs in the MPC optimisation problem.

To achieve this we create an extended version of the emulator model as illustrated in Figure 3. In this version all boundary conditions are outputted automatically by setting a flag. Additional variables need to be specified manually. In this example the minimum temperature bound `Tmin` is set to 22 °C during the day and 20 °C at night. Note that the variable name `Tmin` must be the same as in the linearisation model in Figure 2b.

This emulator model is then simulated to generate the MPC input data matrix $K[i,j]$ for the desired points in time $t(i)$ and variable names $uNamIn[j]$.

2.4 Post-processing of MPC-results

Since we require that the optimisation problem is linear, typically heat flow rates will be optimised, whereas the building HVAC typically needs valve set points and/or temperature set points to be provided. A mapping therefore needs to be made between the optimal, linear control variables and the HVAC set points. This mapping is problem-specific and is therefore not part of the scope of this paper.

The structure of the emulator model including the MPC-controller and corresponding post-processing step is shown in Figure 4a. Figure 4b shows an example of what this post-processing may look like. In this example the optimal heat

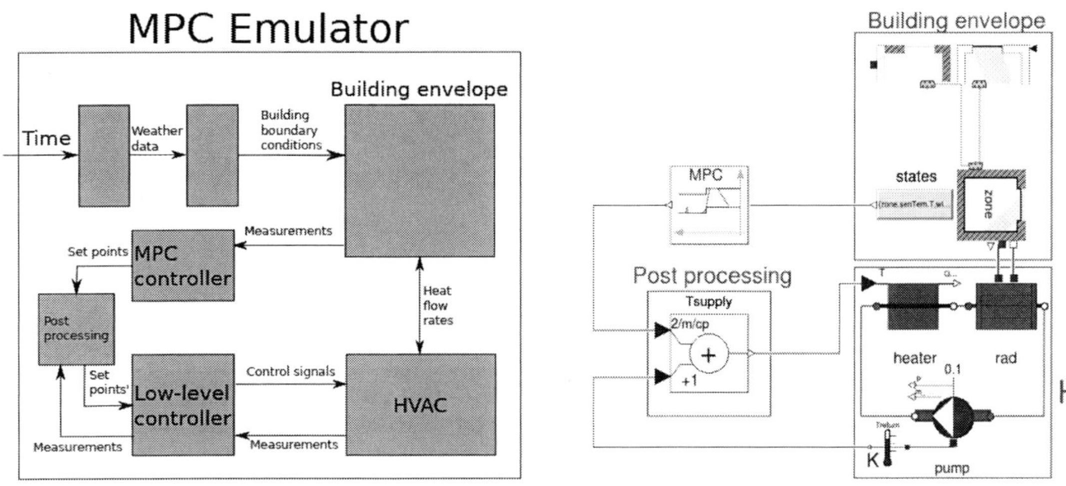

(a) Schematic (b) Example

Figure 4: Emulator model where the MPC controller replaces the existing control strategy. The state values in block `states` are used to update the state of the MPC-controller. The MPC controller computes optimal control signals, which are post-processed to provide meaningful set points for the HVAC controller. The example shows a temperature sensor and an adder that computes that temperature set point. The low-level controller is empty in this example.

flow rate `Qradiator` must be converted to a supply temperature. Since the pump has a constant mass flow rate in this example the supply temperature T_{supply} is computed as

$$T_{supply} = T_{return} + \frac{2\,\dot{Q}_{radiator}}{\dot{m} \cdot c_p} \tag{8}$$

with $\dot{m}$ the mass flow rate and c_p the specific heat capacity of water. T_{return} is the return water temperature, which is measured at the pump in this example. The factor 2 appears because `Qradiator` is injected twice: once in the radiative port and once in the convective port of the zone model.

2.5 Configuring the MPC controller

Figure 4b also shows the Modelica implementation of an MPC controller. This model calls a C++ library that performs the actual optimisation on-line using CasADi (Andersson, 2013) and CPLEX. The Modelica simulation therefore does not need to be stopped and/or restarted to perform the MPC optimisation. The MPC controller can be configured (through parameter settings) to choose between several state update types, different MPC control horizons and control intervals and solver options, whether or not to use warm starting, etc.

The remainder of this section explains how the actual MPC optimal control problem is formulated.

Discrete time dynamics The library uses the matrices from section 2.2 and the input data from section 2.3 to set up an optimisation problem. Matrices A and B are discretised into matrices $\tilde{A}$ and $\tilde{B}$. Using these matrices the states x_i at optimisation intervals $i \in [1, n]$ are computed as

$$x_0 = x0 \tag{9}$$

$$x_{i+1} = \tilde{A} \cdot x_i + \tilde{B} \cdot u_i \qquad \forall i \in [0, n-1] \tag{10}$$

with $x0$ a vector of initial state values.

Due to the large number of states in a typical Modelica building model we use direct single shooting (Kirches, 2011) and eliminate the states such that they are not optimisation variables but instead functions of u_i and the initial states $x0$.

$$x_i = f(u_i, u_{i-1}, \cdots, u_0, x0) \equiv x_i(u, x0) \tag{11}$$

Free and fixed inputs We make a distinction between 'free' and 'fixed' elements of u_i.

Fixed inputs are those elements $u_i[k]$ whose name $uNam[k]$ has an identical copy in $uNamIn[j]$ for some index j, e.g. they are variables such as boundary conditions that may not be optimised. These inputs are denoted using vector $u_{i,fixed}$. Other inputs are free inputs, which are the optimisation variables, denoted using $u_{i,free}$. When a fixed input is used in an equation, e.g. in equation (14), then we use its numerical value by making a linear interpolation between the two values of $K[\cdot,j]$ that match time index i.

Simple and linear bounds The library transforms matrices C and D into A_{ineq}, B_{ineq}, c_x and c_u as earlier described. Matrices A_{ineq} and B_{ineq} are then further split up into 'simple bounds' that constrain a single *free* input variable:

$$\underline{u}_i \leq u_{i,free} \leq \bar{u}_i \tag{12}$$

and 'linear bounds' that constrain a linear combination of multiple states $x_i(u, x0)$ and inputs u_i.

Constraint and cost function format A practical problem is how to pass the constraints to CasADi since CasADi expects a constraint of the format $Q \cdot u + b \geq 0$ with u_{free} the concatenation of all $u_{i,free}$, Q a matrix and b a vector of constants. We solve this by defining a CasADi function $g(u_{free}, u_{fixed}, x0)$ that contains the concatenation of all constraints in the format of equation (13):

$$g(u_{free}, u_{fixed}, x0) = [A_{ineq} \cdot x_i(u, x0) + B_{ineq} \cdot u_i \geq 0] \quad \forall i \tag{13}$$

We then compute

$$b = g(0, u_{fixed}, x0) \tag{14}$$

and

$$Q = J(g) \tag{15}$$

with $J(g)$ the Jacobian of $g(\cdot,\cdot,\cdot)$ with respect to u_{free}. Note that since A, B, C and D are constant matrices, Q is also constant. This means that the Jacobian only needs to be evaluated once, significantly reducing the overhead of the used single shooting approach.

Similarly, the Jacobian of equation (7) is computed to obtain the cost function in the format below:

$$c_{u,free} \cdot u_{free} \tag{16}$$

State estimation method A state estimation method needs to be implemented. For the example considered here we assume that perfect state updates are possible, by reading the state values on-line using block `states` in Figure 4b, which directly reads their values using their names in vector *xNam*.

Optimal control problem formulation Using this information the MPC optimisation problem is compactly formulated as

$$\min \quad c_{u,free} \cdot u_{free} \tag{17}$$
$$\text{s.t.} \quad \underline{u}_i \leq u_{i,free} \leq \bar{u}_i \quad \forall i \tag{18}$$
$$Q \cdot u + b \geq 0 \tag{19}$$

which is solved using the CPLEX QP solver. The optimal control variables $\tilde{u}_{i,free}$ are periodically read by Modelica using C-function calls and used to control the HVAC models.

Results Figure 5 shows the simulation result of the example presented in Figure 4b. The Figure shows overheating due to solar radiation that enters the window, which is the consequence of not having an upper temperature bound in this example. During the morning and at night the radiator needs to supply heating to avoid violating the temperature bound. Even in this idealised case, the lower temperature bound `Tmin` is violated. This is due to the model mismatch introduced by linearising the convective and radiative heat transfer equations and by not taking into account the thermal dynamics of the radiator.

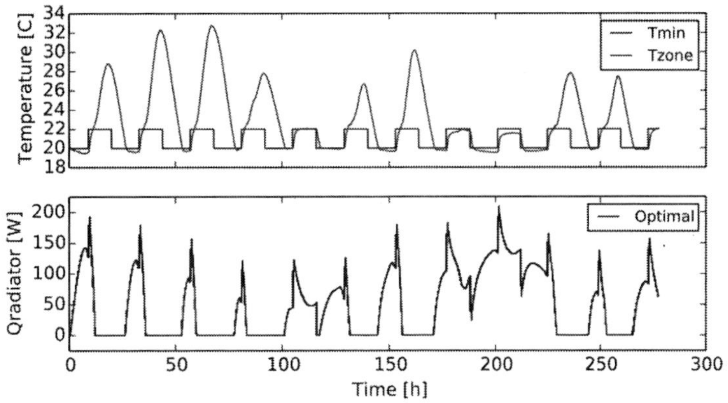

Figure 5: Illustrative optimisation result of the MPC controller

3. CASE STUDY

The presented approach was applied to a more complex emulator model of an office building with 24 zones, two heat recovering air handling units (AHUs) that each have a heating battery, summer bypass, indirect evaporative cooling and an active chiller. The AHUs are connected to each zone through VAVs that control the air mass flow rate. Each zone also contains a concrete core activation circuit that allows free cooling or heating using four heat pumps. The AHU uses heat from solar collectors or a pellet boiler. The VAV heating batteries use heat from the heat pumps. The zone temperatures and zone CO_2 concentrations were constrained using soft constraints.

For this enulator model an MPC controller model was derived using the approach described above. It resulted into a controller model with 1330 states, 177 free inputs, 475 fixed inputs 155 simple bounds and 387 linear bounds per control interval. We currently use 24 control intervals of one hour, which takes approximately 2 seconds to solve using the CPLEX Dual Simplex QP solver. A comparison was made with QpOases (Ferreau et al., 2014), where the problem took multiple hours to solve. When rerunning the MPC every three hours (receding horizon), it takes 3 hours to simulate the numerically optimised (Jorissen et al., 2015) emulator model for one year. The initialisation code that computes $\tilde{A}$, $\tilde{B}$, $c_{u,free}$ and Q takes approximately 15 minutes to evaluate. Preliminary results show electrical energy savings of approximately 30% compared to a conventional rule based control strategy, with similar thermal comfort and indoor air quality.

Picard et al. (2016) present large energy savings when comparing the proposed (white-box) methodology with a grey-box system identification technique from Prívara et al. (2013).

4. CONCLUSIONS

A Linear Automated Toolchain for MPC (LAT-MPC) is demonstrated. It is a stepwise methodology for automating the derivation of MPC controller models for building HVAC applications. The first step is to set up an emulator model of the building. Then a linear version of the emulator is exported into a linear state space model. For the building envelope this step is already automated. Linear HVAC models need to be added manually at this stage. Afterwards an input data file for the state space model inputs is generated. A mapping then needs to be made from the MPC control variables to the physical HVAC component model set points. Finally some MPC formulation and parameter options, such as the horizon length, need to be specified. The methodology was illustrated using a simple Modelica example and was also tested on a complex building model, proving its scalability.

This methodology allows a new research line where the design and control of a building are optimised simultaneously, taking into account their mutual influence. Furthermore these first steps towards an automated derivation of building MPC models contribute to the practical implementations of MPC.

5. FUTURE WORK

Future work should focus on simplifying the step of setting up the emulator model. Automated scripts to convert Building Information Modelling (BIM) files into Modelica models are being developed within the scope of the IEA EBC Annex 60 project (Wimmer et al., 2015).

A library containing linear versions of common HVAC components should be developed, such that HVAC models no longer need to be added manually.

The MPC library can be extended to implement 1) multiple state estimation techniques 2) multiple shooting techniques as opposed to single shooting 3) other, open source solvers.

This MPC framework will be included in integrated optimal design and control of building systems. Implementation in real buildings, both in operation and design phase, allows quantification of real gains.

REFERENCES

Åkesson, J., Årzén, K.-E., Gäfvert, M., Bergdahl, T., & Tummescheit, H. (2010, nov). Modeling and optimization with Optimica and JModelica.org—Languages and tools for solving large-scale dynamic optimization problems. *Computers & Chemical Engineering*, *34*(11), 1737–1749. doi: 10.1016/j.compchemeng.2009.11.011

Andersson, J. (2013). *A General-Purpose Software Framework for Dynamic Optimization* (PhD thesis). Arenberg Doctoral School, KU Leuven, Department of Electrical Engineering (ESAT/SCD) and Optimization in Engineering Center, Kasteelpark Arenberg 10, 3001-Heverlee, Belgium.

Baetens, R., De Coninck, R., Jorissen, F., Picard, D., Helsen, L., & Saelens, D. (2015). OPENIDEAS - An Open Framework for Integrated District Energy Simulations. In *Proceedings of building simulation 2015.* Hyderabad, India.

Bonvini, M., & Wetter, M. (2015). Gradient-Based Optimal Control of Batteries and Hvac in District Energy Systems. In *Proceedings of building simulation 2015.* Hyderabad, India.

Cigler, J., Gyalistras, D., Siroky, J., Tiet, V.-N., & Ferkl, L. (2013). Beyond Theory: the Challenge of Implementing Model Predictive Control in Buildings. *Clima 2013 - 11th REHVA World Congress & 8th International Conference on IAQVEC - "Energy Efficient, Smart and Healthy Buildings"*, 1008–1018.

De Coninck, R., & Helsen, L. (2016). Practical implementation and evaluation of model predictive control for an office building in Brussels. *Energy and Buildings*, *111*, 290–298. doi: 10.1016/j.enbuild.2015.11.014

De Coninck, R., Magnusson, F., Åkesson, J., & Helsen, L. (2015). Toolbox for development and validation of grey-box building models for forecasting and control. *Journal of Building Performance Simulation*(June), 1–16. doi: 10.1080/19401493.2015.1046933

Elmqvist, H., & Mattsson, S. E. (1997). Modelica – the next generation modeling language-an international design effort. In *Proceedings of first world congress of system simulation* (pp. 1–5). Singapore.

Ferreau, H., Kirches, C., Potschka, A., Bock, H., & Diehl, M. (2014). qpOASES: A parametric active-set algorithm for quadratic programming. *Mathematical Programming Computation*, *6*(4), 327–363.

Fritzson, P., Aronsson, P., Lundvall, H., Nyström, K., Pop, A., Saldamli, L., & Broman, D. (2005, December). The OpenModelica Modeling, Simulation, and Software Development Environment. *Simulation News Europe*, *44*(45).

Jorissen, F., Wetter, M., & Helsen, L. (2015). Simulation Speed Analysis and Improvements of Modelica Models for Building Energy Simulation. In *11th international modelica conference* (pp. 59–69). Paris, France. doi: 10.3384/ecp1511859

Kirches, C. (2011). *Fast numerical methods for mixed-integer nonlinear model-predictive control.* Springer.

Pérez-Lombard, L., Ortiz, J., & Pout, C. (2008, jan). A review on buildings energy consumption information. *Energy and Buildings*, *40*(3), 394–398. doi: 10.1016/j.enbuild.2007.03.007

Picard, D., Jorissen, F., & Helsen, L. (2015). Methodology for Obtaining Linear State Space Building Energy Simulation Models. In *11th international modelica conference* (pp. 51–58). Paris, France. doi: 10.3384/ecp1511851

Picard, D., Sourbron, M., Jorissen, F., Váňa, Z., Cigler, J., Ferkl, L., & Helsen, L. (2016). Comparison of model predictive control performance using grey box and white box controller models, submitted. In *2016 Intelligent Building Operations Workshop.* West Lafayette.

Prívara, S., Cigler, J., Váňa, Z., Oldewurtel, F., Sagerschnig, C., & Žáčeková, E. (2013, jan). Building modeling as a crucial part for building predictive control. *Energy and Buildings*, *56*, 8–22. doi: 10.1016/j.enbuild.2012.10.024

Prívara, S., Široký, J., Ferkl, L., & Cigler, J. (2011, feb). Model predictive control of a building heating system: The first experience. *Energy and Buildings*, *43*(2-3), 564–572. doi: 10.1016/j.enbuild.2010.10.022

Široký, J., Oldewurtel, F., Cigler, J., & Prívara, S. (2011). Experimental analysis of model predictive control for an energy efficient building heating system. *Applied Energy*, *88*(9), 3079–3087. doi: 10.1016/j.apenergy.2011.03.009

Sturzenegger, D., Gyalistras, D., Gwerder, M., Sagerschnig, C., Morari, M., & Smith, R. S. (2013). Model Predictive Control of a Swiss office building. In *Clima-rheva world congress*. Prague.

Sturzenegger, D., Gyalistras, D., Morari, M., & Smith, R. S. (2012). Semi-automated modular modeling of buildings for model predictive control. In *Proceedings of the fourth acm workshop on embedded sensing systems for energy-efficiency in buildings - buildsys '12* (pp. 99–106). Toronto. doi: 10.1145/2422531.2422550

Sturzenegger, D., Gyalistras, D., Semeraro, V., Morari, M., & Smith, R. S. (2014). BRCM Matlab Toolbox : Model Generation for Model Predictive Building Control. In *2014 american control conference* (pp. 1063–1069). Portland.

Váňa, Z., Cigler, J., Široký, J., Žáčeková, E., & Ferkl, L. (2014). Model-based energy efficient control applied to an office building. *Journal of Process Control*, *24*(6), 790–797. doi: 10.1016/j.jprocont.2014.01.016

West, S. R., Ward, J. K., & Wall, J. (2014). Trial results from a model predictive control and optimisation system for commercial building HVAC. *Energy and Buildings*, *72*, 271–279. doi: 10.1016/j.enbuild.2013.12.037

Wetter, M., Fuchs, M., Grozman, P., Helsen, L., Jorissen, F., Lauster, M., ... Thorade, M. (2015). IEA EBC Annex 60 Modelica Library – An International Collaboration to Develop a Free Open-Source Model Library for Buildings and Community Energy Systems. In *Proceedings of building simulation 2015*. Hyderabad, India.

Wimmer, R., Cao, J., Remmen, P., Maile, T., Donnell, J. O., Frisch, J., ... Treeck, C. V. (2015). Implementation of Advanced BIM-Based Mapping Rules for Automated Conversions to Modelica. In *Proceedings of building simulation 2015*. Hyderabad, India.

ACKNOWLEDGMENT

We thank Joris Gillis for his help with implementing CasADi and Damien Picard for the fruitful discussions that have contributed to this work. The authors acknowledge the financial support by the Agency for Innovation by Science and Technology in Flanders (IWT) for the PhD work of F. Jorissen (contract number 131012). This work emerged from the Annex 60 project, an international project conducted under the umbrella of the International Energy Agency (IEA) within the Energy in Buildings and Communities (EBC) Programme. Annex 60 will develop and demonstrate new generation computational tools for building and community energy systems based on Modelica, Functional Mockup Interface and BIM standards.

Comparison of Model Predictive Control Performance Using Grey-Box and White-Box Controller Models of a Multi-zone Office Building

Damien PICARD[1*], Maarten SOURBRON[1], Filip JORISSEN[1], Zdeněk VÁŇA[2], Jiří CIGLER[2], Lukás FERKL[3], Lieve HELSEN[1,4]

[1] KU Leuven, Department of Mechanical Engineering,
Leuven, Belgium
damien.picard@kuleuven.be

[2] Feramat Cybernetics s.r.o.,
Prague, Czech Republic

[3] University Centre for Energy Efficient Buildings, Czech Technical University,
Prague, Czech Republic

[4] EnergyVille,
Waterschei, Belgium
* Corresponding Author

ABSTRACT

Model Predictive Control (MPC) is a promising control method to reduce the energy use of buildings. Its commercialization is, however, hampered by the difficulty of obtaining a reliable controller model. This paper compares two approaches to obtain such controller model: (1) a white-box model approach for which a detailed first-principles building model is linearized, and (2) a system identification method using a grey-box model approach. The MPC performance using both model approaches is evaluated on a validated 12 zones model of an existing office building. The results indicate that the MPC performance is very sensitive to the prediction accuracy of the controller model. This paper shows that both approaches can lead to an efficient MPC as long as very accurate identification data sets are available. For the considered simulation case, the white-box MPC resulted in a better thermal comfort and used only 50% of the energy used by the best grey-box MPC.

1. INTRODUCTION

Model based Predictive Control (MPC) is a control technique for dynamic systems that computes optimal control set points in order to minimize a predefined cost. For this, the controller contains a dynamic model of the system that is used in an optimization routine that solves at each control step an optimal control problem (Maciejowski, 2002). Building HVAC systems present an interesting application for MPC, especially when emission systems with a large thermal mass such as Thermally Activated Building Systems (TABS) are applied. In this case, the cost that is minimized is typically a weighted sum of energy use and thermal discomfort.

Although the potential of MPC in building control has been widely demonstrated (Cho & Zaheer-uddin, 2003; Gwerder & Tödtli, 2005; Ihm & Krarti, 2005; Karlsson, 2006; Prívara et al., 2011; Ma et al., 2012; Gyalistras & Gwerder, 2009), the commercial application of MPC in buildings is not widespread. One of the bottlenecks in a reliable application of MPC in real life buildings is deriving an accurate model of that particular building that at the same time is suitable to be applied in the optimal control problem (Morari & Lee, 1999; Dounis & Caraiscos, 2009; Hilliard et al., 2015).

In this paper a comparison is presented between two approaches to derive a building controller model to be used in an MPC for a multi-zone office building. The first approach starts from a detailed first principles dynamic building model which is linearized to obtain a state space controller model (Picard et al., 2015, 2016). In the second approach, measurement data are used to identify a suitable grey-box controller model (Privara et al., 2013). This paper follows the following outline. Several sets of real building data are gathered as validation and identification data. On the one hand, a detailed dynamic building model (emulator) is validated using these measurement data and is linearized into a state space controller model (further referred as *Lin-Mod*). On the other hand, the same sets of measurement data are used to identify and validate a grey-box model (*IE-Mod*). A last controller model is identified using simulation data instead of the measurement data sets (*IS-Mod*). The simulation data is obtained by a full-year simulation of the emulator using a rule-based controller, to imitate real building operation. The three controller models are compared regarding their prediction performance, after which the MPC performance is evaluated using the same MPC framework but the three different controller models.

2. BUILDING EMULATOR MODEL

This section describes the real multi-zone office building, the building emulator model and the validation of the model using measurement data.

2.1 Description of Building and Building Model.

'Hollandsch Huys', located in Hasselt, Belgium, is an office building with approximately 4000 m^2 of conditioned floor area. The building consists of 5 floors: an underground parking, 3 floors with offices and a rooftop apartment used as private office (Fig. 1). Both landscape and single offices are used in the building. The two main façades are south-west and north-east oriented. The building fulfils all passive house criteria except air tightness (n50=0.99 instead of 0.6).

All transparent parts of the façade are equipped with triple glazing. The window surface lies 40 cm deeper than the façade. Each of them is equipped with an external slat shading device whose angle is adjusted automatically to the solar radiation intensity.

The building is equipped with TABS on the first and second floors. The TABS have a double water piping circuit in each concrete floor. Both circuits can be controlled individually and can be used for heating or cooling. On the ground floor, floor heating is installed, which can only be used for heating. TABS and floor heating cover the major part of the heating and cooling load. The ventilation is used mostly for hygienic purposes. The apartment on the third floor is heated using floor heating and TABS and cooled using the TABS. The heating load is mainly provided by a ground coupled heat pump of 158 kW, while cooling is provided with a direct ground coupled heat exchanger of 72 kW. Reversed heat pump operation for cooling is possible, but has never been used. The Air Handling Unit (AHU) uses a 60 kW gas back up boiler to supply heat in the mid-seasons (spring/autumn) when the building itself is still in cooling mode.

The building is modelled using the open-source Modelica Library IDEAS (Baetens et al., 2015). For a detailed description of its governing equations, we refer to Baetens et al. (2015); Picard et al. (2015). For this study, only the building envelope is considered, i.e. the walls, windows, roofs and TABS. An abstraction is made of the HVAC components, i.e. we assume that 1) the HVAC injects thermal power in the water-circuit of the TABS and 2) all zones have perfectly controllable ventilation supply temperatures. The supply temperature of each zone must, however, be the same. Twelve thermal zones are used to model the three floors as shown in Figure 1. The roof apartment is omitted and the basement is modelled as one extra zone without any control input.

2.2 Validation

A Modelica model of the building is created and validated using real measurement data. The following paragraphs describe the building data, the weather data, the tuning of the model and the model validation.

Measurement data. The Modelica model of the Hollandsch Huys building is validated using 5 sets of measurement data. The first data set was obtained from an experiment conducted on the building during the Christmas holidays of 2012. The building was not or only partially occupied during 11 consecutive days and this opportunity was used to excite the HVAC system with several step inputs: two sequences of whole building cooling down (all HVAC off) and heating up (resp. TABS on and TABS+AHU on), and sequences of active cooling down (TABS cooling on) and heating up (TABS on or TABS+AHU on) of specific zones. These last step inputs are generated to evaluate the intra-zone effect of thermal conditioning. During these experiments, extra zone temperature sensors were installed in

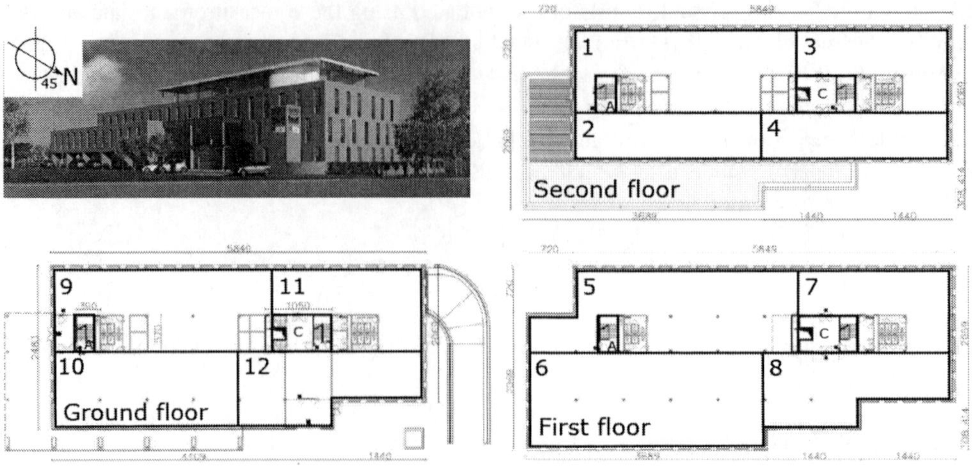

Figure 1: Picture of the *Hollandsch Huys* building and zones layout of the building model.

addition to the Building Management System (BMS) zone temperature sensors, allowing a more accurate validation for each of the 12 zones.

The other four data sets each contain approximately one month of measurement data during the year 2014: Jan 14-Feb 10; May 9-June 2; June 13-July 7; Aug 23-Sept 16. These periods were the only error-free data sets for zones 1, 2 and 4 (Figure 1) and reflect a winter, a mid-season and two summer periods. Due to corruption or failure of many sensors of the BMS, no data during 2014 is available to validate the 9 other zones.

Weather data. The building does not have a local weather station, although during the Christmas experiment the ambient temperature was measured using a dedicated temperature sensor. Therefore, the identification and validation data sets from the BMS are extended with historical data of a weather station in Hasselt obtained from the website `forecast.io`. In order to verify this data, the Christmas ambient temperature from the weather station in Hasselt (F-Hasselt) was compared with those from the airport of Maastricht from both `forecast.io` (F-Maastricht) and `www.wunderground.com` (WG-Maastricht). The comparison shows that the weather data are similar except between the 31st of December and the first of January. The data F-Hasselt is chosen because it is the closest match and because no solar information is available for historical data from `www.wunderground.com`. The direct normal irradiation and the diffuse irradiation on a horizontal surface are computed using the cloud coverage factor from the weather data and the theoretical cloudless solar radiation, which depend on the position of the sun and on the geographical location (*PyEphem*, 2016; Brinsfield et al., 1984). Using this conversion, the ambient temperature, direct normal and diffuse horizontal irradiation are known. The Modelica model then makes a geometrical projection on the building surfaces.

Model parameter tuning and validation results. The data obtained during the Christmas experiment (D-Xmax) are used to fine-tune the Modelica model, while the data sets of the year 2014 (D-Y14) are used as validation data. In order to obtain a good temperature fit on D-Xmax, the following tuning is done: the insulation thickness is increased below the apartment and decreased for all roof surfaces that are not below the roof-apartment. The heat capacity of the TABS concrete is doubled from 840 J/kgK to 1680 J/kgK.

For the validation using D-Y14, the tuning made with D-Xmas is kept unchanged and only the internal gains are reduced by 40% compared to their theoretically estimated values. Figure 2 shows boxplots of the air temperature error for each validated zone. The horizontal line corresponds to the median, the box to the first and third quartiles, the whiskers to the 95% confidence interval and the crosses to the outliers. Figure 2b shows that the errors on the validation data mostly stay below 1 K for the entire data set, which indicates that the Modelica model is a realistic representation of the real building.

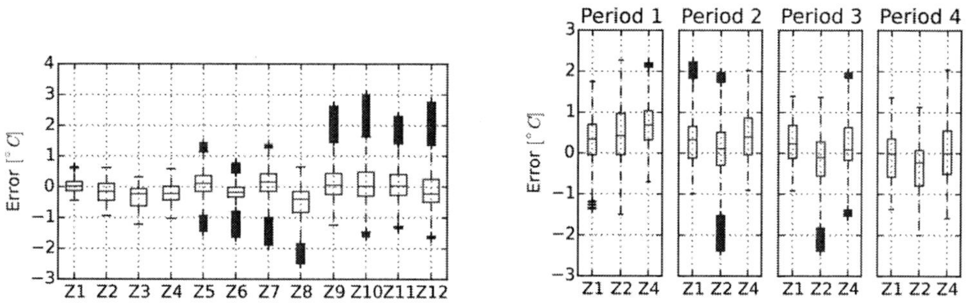

(a) Christmas experiment. **(b) Year 2014.**

Figure 2: Boxplot of the air temperature error for each zone for the tuning data-set (a) and for zones 1, 2 and 4 using the four validation data sets of one month each (b).

3. CONTROLLER MODEL

This section describes the different controller models. Section 3.1 describes the controller model obtained by linearizing the emulator model, Section 3.2 the ones obtained by system identification, and Section 3.3 compares their prediction performance.

3.1 Linearized Controller Model

The heat transfer equations describing the thermal behavior of the building envelope are non-linear. For example, the convection coefficient depends on the wind speed or on the temperature gradient due to buoyancy effect, the radiative heat transfer depends on the temperature to the fourth power and the amount of solar radiation through the windows depends on the solar position, on the shading and on the glass properties. However, the convective and radiative heat transfer equations can be approximated using linear functions and the solar radiation through the windows can be pre-computed since it does not depend on the building states.

In this paper, the building linearization method proposed by Picard et al. (2015) is applied. The authors demonstrated an example where the non-linear building model can be approximated using a linear time-invariant (LTI) State Space Model (SSM) with an error band typically below $\pm 1K$. Using this approach a SSM of 821 states is obtained for the building envelope model of Hollandsch Huys. The SSM inputs are the thermal powers injected in the water circuit of the TABS, the supply ventilation temperature to the zones, the convective and radiative occupancy heat gains, the various ambient conditions, and the solar radiation through each window. Section 3.3 shows that the linearized model is able to predict the zone temperatures of the emulator model with an error mostly smaller than ± 0.1 K.

3.2 Grey-box Controller Model

Grey-box system identification is a technique which pre-defines the model structure based on physical knowledge but which optimizes its parameter values such that the model response fits some measurement data.

In this paper, a method based on the simplified discretization of the continuous model structure is used, as described by Privara et al. (2013) in the *deterministic semi-physical modeling* section of their paper. The method boils down to parameter estimation of a linear SSM. The SSM structure is constructed based on physical knowledge about the building and its discretization is then approximated by its first order Taylor expansion. The model parameters are estimated by solving a quadratic programming problem in which the one step ahead prediction error is minimized and the parameter values are constrained within physically meaningful bounds defined by the user. By keeping the model linear, the identification procedure can be formulated as a convex optimization problem.

In this paper, the model structure corresponds to the so-called RC-model depicted by Fig. 3. The model is composed of thermal resistances (representing thermal conduction, convection and radiation resistances) and capacitances (representing the heat capacity of the building mass). Each zone is represented by 3 nodes (zone temperature $T_{z,i}$, floor concrete temperature $T_{f,i}$, ceiling concrete temperature $T_{c,i}$). The energy delivered by the AHU $\dot{Q}_{AHU,i}$, internal gains

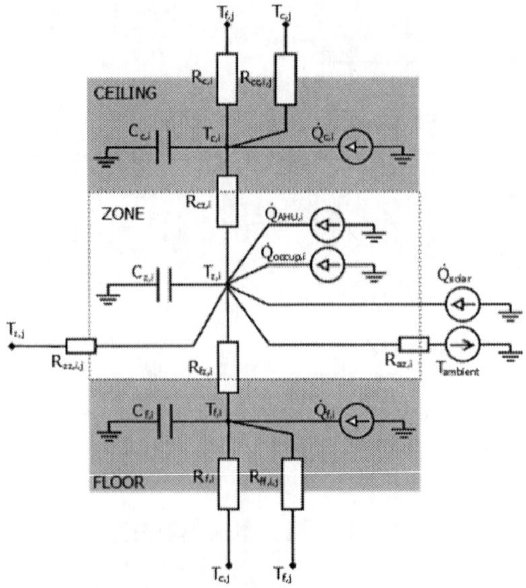

Figure 3: RC network of a zone. Subscripts f,z,c stand for floor, zone, ceiling respectively, and subscripts i and j stand for zone number. T stands for temperature (analogous to voltage) and $\dot{Q}$ stands for heat flux (analogous to electric current). The sources represent the inputs to the model – either temperatures or heat fluxes. Note that subscript j may represent more than one zone – it covers all neighboring zones

from occupants $\dot{Q}_{occup,i}$ and the solar radiation entering the zone through its windows $\dot{Q}_{solar}$ are injected in the zone capacity while the energy delivered to the TABS is injected in the floor and ceiling capacities. Thermal resistances exist between each zone and i) the ambient temperature $T_{ambient}$, ii) all neighboring zones $T_{z,j}$, iii) its ceiling and its floor. Ceilings and floors are coupled both vertically and horizontally to each others.

Two controller models with the same structure are identified: *IS-Mod* is identified using data obtained by simulation using the emulator building with a reference controller from January to July, and *IE-Mod* is identified using D-Xmas. The identification data is sampled every $T_s = 320s$ and the inputs are converted such that all inputs are either expressed in kW or °C. This ensures that the estimation problem is not ill-posed. The global solar radiation on the horizontal plane, available in the identification data set, is transformed into solar radiation per façade taking the solar blind into account. Window area and window properties are considered only in the case of identification from real data. For this model, the coupling between neighboring zones is not considered as it increases the number of parameters to estimate and information contained in the identification data is insufficient to identify all parameters correctly.

3.3 Validation of Controller Models

In this section, the controller models are validated against the reference emulator model (*Ref-Mod*). The operative temperatures (i.e. a weighted temperature composed of the air temperature and the zone surface temperatures) of *Ref-Mod* are compared to the values of the linearized model (*Lin-Mod*), the identified model using simulation data (*IS-Mod*), and the identified model using experiment data (*IE-Mod*). The controller models are excited with the same (or the equivalent) inputs as *Ref-Mod*. *Ref-Mod* is simulated over a full year using a typical meteorological year from Uccle, Belgium (Meteotest, 2009) and typical control inputs. The simulation integrator step for *Ref-Mod* is 30 seconds and the outputs are sampled each 900 seconds. The controller models, however, are transformed to discrete SSM with the same sampling time as the MPC ($T_s = 3600$s) and the inputs are sampled accordingly using zero-hold. For model consistency, *IS-Mod* and *IE-Mod* are transformed such that their inputs and outputs are in Watt and Kelvin instead of kilo-Watt and degrees Celsius. Further, as only the global supply ventilation temperature ($T_{sup,ven}$) can be controlled and not the ventilation thermal power to each zone ($\dot{Q}_{ven,i}$), the controller models are extended to use $T_{sup,ven}$ and the nominal ventilation mass flow rates instead of the $\dot{Q}_{ven,i}$'s. As explained in Section 4.1, this transformation introduces a model mismatch for each zone but it becomes zero when the average is taken over all zones.

Figure 4a and 4b show 3-day winter and summer validation sets for the average operative temperature $T_{op,av}$ of the

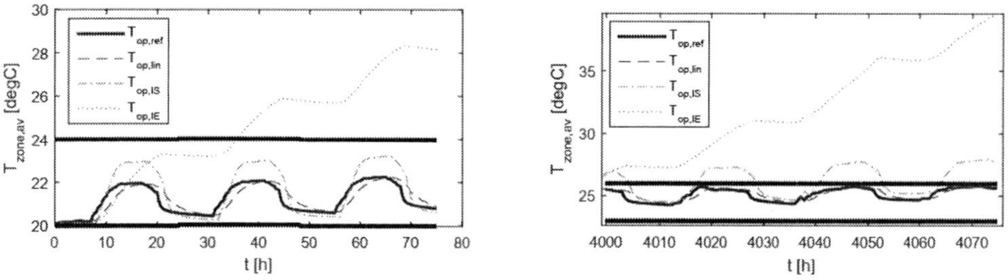

(a) Winter period. **(b) Summer period.**

Figure 4: Average operative temperature of all zones for the reference model ($T_{op,ref}$), the linearized model ($T_{op,lin}$), the identified model using simulation data ($T_{op,IS}$), and the identified model using experiment data ($T_{op,IE}$) for a winter period (a) and for a summer period (b).

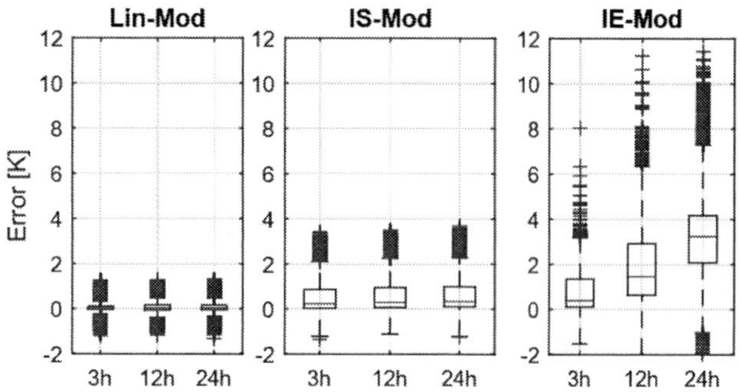

Figure 5: Prediction error for horizons of 3, 12 and 24 hours for the linearized model (*Lin-Mod*), the identified model using simulation data (*IS-Mod*), and the identified model using experiment data (*IE-Mod*).

12 zones. *Lin-Mod* and *IS-Mod* are able to accurately predict $T_{op,av}$. The error made by *Lin-Mod* is mainly due to its zero-hold discretization as decreasing T_s significantly improves the predictions. For the *IS-mod* and *IE-mod*, no operative temperature is available and the temperature of the zone is used instead. This is the main cause of model mismatch for the *IS-mod* as its zone temperatures do coïncide better with the air temperature of the reference model. However, both *IS-mod* and *IE-mod* show a temperature drift causing a serious prediction error for *IE-mod*. The drift probably originates from the system identification procedure which is carried out in the discrete time domain. An estimation error on the parameters can then lead to an integration error, which means the numerical creation of energy within the model. Simulating *IS-Mod* and *IE-Mod* with zero heat inputs and constant temperature input does indeed not result to convergence of the states to the temperature input value.

Figure 5 shows a boxplot of the prediction error of *Lin-Mode*, *IS-Mod* and *IE-Mod* for the prediction horizons of 3, 12 and 24 hours. The prediction errors are obtained by taking 1000 points from the reference simulation, evenly spaced over the whole year. For each point, the controller models are initialized with the corresponding state values of the reference simulation and the temperatures at the end of the prediction horizon are computed. Figure 5 gives the errors between these prediction temperatures and the reference temperatures for all zones together. Figure 5 shows that the *Lin-Mod* predictions are for 95% within ± 0.5 K for all horizons and the average prediction error is zero. The medians for *IS-Mod* and *IE-Mod* are positive which indicates a systematic overestimation of the temperature. However, *IS-Mod* is still able to predict the temperature with an error band of $[0, 1]$ K for 75% of the time. This is not the case for *IE-Mod* which shows an average prediction error of 3 K for a horizon of 24 hours.

4. MPC CONTROLLER

This section describes the MPC formulation and the algorithm and software used to solve the MPC problem.

4.1 MPC cost function, constraints and parameters

In this work, the MPC optimization problem boils down to minimizing the heating and cooling TABS energy ($\mathscr{T}_{H,i}^{(k)}$, $\mathscr{T}_{C,i}^{(k)}$), and the ventilation energy $\mathscr{V}_i^{(k)}$, while the operative zone temperature $T_{op,i}$ stays within the time varying comfort band $[T_{\text{lb}}^{(k)}, T_{\text{ub}}^{(k)}]$ by minimizing the slack variable $\mathscr{S}_c^{(k)}$ (Eq. 1a). The subscript i represents the zone number, H and C refer to heating and cooling and (k) refers to the time step within the MPC horizon of length $N \times T_s$ with T_s the sampling time. The sampling time in this work is 1 hour and the horizon is 1 day. The MPC uses perfect state update every 4 hours. Note that using perfect state updates is not possible in reality as not all states (such as the TABS core-temperature) are measurable and, moreover, sensors are not perfect. In order to limit the influence of the state update, the update only takes place every 4 hours, while the typical sampling time for building measurements is between 5 to 30 minutes.

Eq. 1 formulates the MPC cost function and constraints. The optimization variables are $\mathscr{T}_{H,i}^{(k)}$, $\mathscr{T}_{C,i}^{(k)}$, and the ventilation supply temperature $T_{\text{sup,ven}}$. It should be noted that the current ventilation system imposes nominal ventilation flow for each zone when the building is occupied and zero otherwise. $T_{\text{sup,ven}}$ is furthermore the same for all zones, which means that the $\mathscr{V}_i^{(k)}$ computed by the MPC can not be exactly delivered to each zone. The exact value of $\mathscr{V}_i^{(k)}$'s cannot be formulated in the linear MPC as it is bilinear (multiplication of the time varying ventilation mass flow rate $\dot{m}_{\text{i}}^{(k)}$ with $T_{\text{sup,ven}}$ or $T_{\text{air,i}}$). $\mathscr{V}_i^{(k)}$ is approximated in this MPC by assuming that $\dot{m}_{\text{i}}^{(k)}$ is equal to its nominal value (Eq. 1b) and that the supply temperature equals the average of all air zone temperatures $\bar{T}_{\text{air}}$ when $\dot{m}_1^{(k)}$ is zero (Eq. 1i and 1j). Note that all $\dot{m}_i^{(k)}$ are turned on or off at the same time. This forces the MPC to limit $\mathscr{V}_i^{(k)}$ as much as possible when the ventilation is off as the real ventilation system can then not supply the $\mathscr{V}_i^{(k)}$'s to the zones. Due to this approximation, the MPC makes a prediction error on the amount of ventilation power injected when the ventilation is off. The global cost function and constraints can now be formulated as:

$$J = \min_{T_{\text{sup,ven}}, \mathscr{T}_{H,i}, \mathscr{T}_{C,i}} \sum_{k=0}^{N-1} \left[\sum_{i=1}^{12} \left(\underbrace{\mathscr{T}_{H,i}^{(k)} + \mathscr{T}_{C,i}^{(k)}}_{\text{TABS}} + \underbrace{\mathscr{A}_i^{(k)}}_{\text{Ventilation}} \right) + \underbrace{10^6 \mathscr{S}_c^{(k)}}_{\text{Discomfort}} + \underbrace{10^6 \mathscr{S}_v^{(k)}}_{\text{Soft ven.}} \right] \tag{1a}$$

$$s.t. \quad \mathscr{V}_i^{(k)} = \dot{m}_{\text{nom,i}} c_p \left(T_{\text{sup,ven}} - T_{\text{air},i}^{(k)} \right) \tag{1b}$$

$$\mathscr{A}_i^{(k)} - \mathscr{V}_i^{(k)} > 0 \,, \; \mathscr{A}_i^{(k)} + \mathscr{V}_i^{(k)} > 0 \tag{1c}$$

$$\mathscr{T}_{H,i}^{(k)} > 0 \,, \; \mathscr{T}_{C,i}^{(k)} > 0 \,, \; \mathscr{S}_c^{(k)} > 0 \,, \; \mathscr{S}_v^{(k)} > 0 \tag{1d}$$

$$\sum_{i=1}^{12} \mathscr{T}_{H,i}^{(k)} < \mathscr{T}_{\text{H,nom}} \,, \; \sum_{i=1}^{12} \mathscr{T}_{C,i}^{(k)} < \mathscr{T}_{\text{C,nom}} \,, \; \sum_{i=1}^{12} \mathscr{V}_i^{(k)} < \mathscr{V}_{\text{H,nom}} \,, \; \sum_{i=1}^{12} \mathscr{V}_i^{(k)} > -\mathscr{V}_{\text{C,nom}} \tag{1e}$$

$$\mathscr{T}_{H,i}^{(k)} < \mathscr{T}_{\text{H,i,nom}} \,, \; \mathscr{T}_{C,i}^{(k)} < \mathscr{T}_{\text{C,i,nom}} \tag{1f}$$

$$T_{\text{ub}}^{(k)} - T_{\text{sup,ven}}^{(k)} + \varepsilon + \mathscr{S}_v^{(k)} > 0 \,, \; T_{\text{sup,ven}}^{(k)} - T_{\text{lb}}^{(k)} + \varepsilon + \mathscr{S}_v^{(k)} > 0 \tag{1g}$$

$$T_{\text{ub}}^{(k)} - T_{\text{op},i}^{(k)} + \mathscr{S}_c^{(k)} > 0 \,, \; T_{\text{op},i}^{(k)} - T_{\text{lb}}^{(k)} + \mathscr{S}_c^{(k)} > 0 \tag{1h}$$

$$\frac{\dot{m}_1^{(k)}}{\dot{m}_{\text{nom,1}}} \mathscr{V}_{\text{nom,H,1}} - \dot{m}_{\text{nom,1}} c_p \left(T_{\text{sup}} - \bar{T}_{\text{air}} \right) > 0 \tag{1i}$$

$$-\frac{\dot{m}_1^{(k)}}{\dot{m}_{\text{nom,1}}} \mathscr{V}_{\text{nom,C,1}} + \dot{m}_{\text{nom,1}} c_p \left(T_{\text{sup}} - \bar{T}_{\text{air}} \right) > 0 \tag{1j}$$

$$T_{\text{air}}^{(k)} = f\left(x^0, [T_{\text{sup,ven}}, \mathscr{T}_{H,i}, \mathscr{T}_{C,i}]^{(0)\cdots(k)} \right) \tag{1k}$$

with nominal working condition indicated by subscript *nom*, ventilation mass flow rates $\dot{m}_{\text{i}}^{(k)}$, air heat capacity c_p, maximum TABS powers $\mathscr{T}_{H,max}$, $\mathscr{T}_{C,max}$ and $\bar{T}_{\text{air},i}$ the average of all zone air temperature.

The minimization of $\mathscr{A}_i^{(k)}$ and Eq. 1b and 1c are the linear equivalent to the minimization of $|\mathscr{V}_i^{(k)}|$ with $|\cdot|$ representing the absolute value. Eq. 1e constraints the total thermal powers such that the nominal production power of the heat pump is not exceeded. Eq. 1f limits the power of each TABS circuit, while Eq. 1g confines $T_{\text{sup,ven}}$ within a band ε broader than the comfort band. The relaxation ε is set to 1 K to avoid a too cold or too warm ventilation air flow. The constraint is furthermore relaxed by the slack variable $\mathscr{S}_v^{(k)}$ to improve the robustness of the algorithm. Finally, Eq. 1k represents the dynamics of the controller SSM.

4.2 Software and algorithm

The tool chain used for implementing the MPC controller is described in detail by Jorissen and Helsen (2016). The first step in the tool chain is to convert a detailed emulator model into a linear state space model and to create a file with pre-computed boundary conditions, as described in section 3.1. The tool chain contains an MPC implementation in Modelica that calls a C++ library that performs the actual computations. This library uses the state space and input data to efficiently (Jorissen & Helsen, 2016) formulate an Optimal Control Problem (OCP) using CasADi (Andersson, 2013). This OCP is solved using the CPLEX Dual Simplex QP solver. The Modelica model periodically calls the library to retrieve the optimal control results and to provide data for the state update algorithm within the controller. These optimal control results are used to provide set points for the building HVAC model, after which the simulation can advance in time. After an integer multiple of the MPC time step the states are updated and the OCP is rerun.

5. RESULTS

The MPC formulation is used with the different controller models to control the building climate during a full year simulation. Fig. 6 shows the average of the 12 operative temperatures as calculated by the emulator for the different MPCs together with the lower and upper indicators temperature bounds. Table 1 summarizes the MPC performance factors for the whole building for each MPC, i.e. the heating and cooling energy used by the TABS (TABS-Heat, TABS-Coo) and the ventilation (AHU-Hea, AHU-Coo) and the total discomfort (DC) together with the minimal and maximal deviation from the comfort range.

Figure 6 and Table 1 show that the *Lin-Mod* MPC is able to keep the operative temperatures within the comfort bounds with only a negligible discomfort of 147 Kh/year and with a punctual maximal deviation of 1 K. The discomfort is mainly due to the model mismatch introduced by Eq. 1b which culminates when the air temperatures differ the most between the zones. When $T_{air,i}$ is higher than $\bar{T}_{\text{air}}$ and the ventilation is turned off (i.e. $\dot{m}_1 = 0$), the MPC constraints set $T_{\text{sup,ven}} = \bar{T}_{\text{air}}$. Eq. 1b becomes then negative and the MPC supposes that zone i is cooled by the ventilation while this is in reality not the case. Nevertheless, the discomfort caused by *Lin-Mod* MPC is only 7% and 0.4% of the discomfort caused by *IS-Mod* MPC and *IE-Mod* MPC, respectively, while its energy use is 41% and 8% of the energy used in the *IS-Mod* and ES-Mod approaches, respectively.

While less energy efficient and comfort are guaranteed, *IS-Mod* MPC still achieves good comfort with only 2064 Kh /year of discomfort (172 Kh / zone / year) and its energy use is only 20% of those of *IE-Mod*. Due to the systematic overestimation of the temperatures as shown by Figure 5, *IS-Mod* MPC is not able to work close to the upper temperature comfort bound and therefore it cools more than necessary. Furthermore, the TABS control signals exhibit (not shown) too fast oscillations between heating and cooling actions which lead to an inefficiency energy supply to the building. This bang-bang behavior is caused by the model mismatch and the linear formulation of the objective function. Adding a smoothing term on the inputs in the cost function as proposed by (Cigler et al., 2013) would probably significantly improve the MPC robustness against its prediction errors and so improve its global performance. Figure 5 also shows that errors on the 3 hours ahead prediction are non-negligible which causes regular discomfort when the MPC operates close to the bounds. The maximum deviation of -1.3 K remains, however, acceptable.

Finally, *IE-Mod* MPC shows poor comfort and a total energy use of 476 MWh while the energy use when the building model is controlled by the original rule-based-controller (not explained in the text) stays below 240 MWh. This confirms that due to the lack of 'rich' measurement data, the identified model *IE-Mod* is not reliable enough to be used in an MPC. The collection of 'rich' data (containing enough information for system identification) in real practice is thus a topic for further research, next to testing these approaches in real buildings.

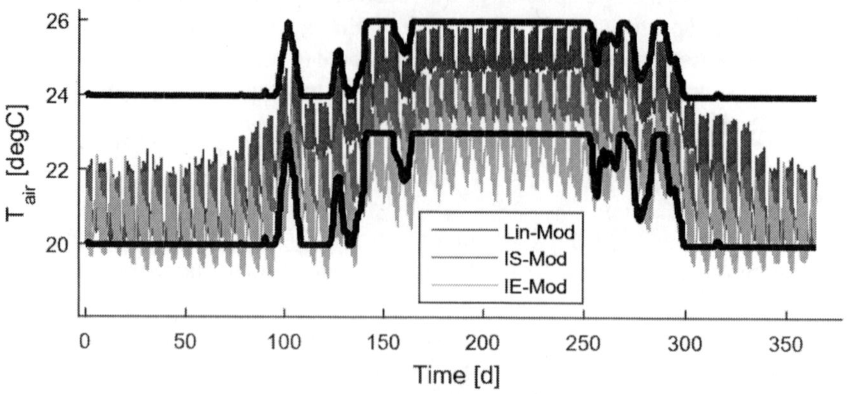

Figure 6: Comparison of the average of all operative zone temperatures for a full year simulation with MPC's using the *Lin-Mod*, the *IS-Mod* and the *IE-Mod* controller models.

Table 1: MPC performance indicators for the three approaches (*Lin-Mod, IS-Mod, IE-Mod*) in MPC.

	TABS-Heat	TABS-Coo	AHU-Hea	AHU-Coo	DC	Min DC	Max DC
	[kWh_{th} / y]	[kWh_{th} / y]	[kWh_{th} / y]	[kWh_{th} / y]	[Kh / y]	[K]	[K]
Lin-Mod	4360	19931	753	15159	147	-0.4	1.0
IS-Mod	26926	57262	589	14387	2064	-1.3	0.0
IE-Mod	205670	265070	2375	2788	37721	-3.6	0.0

6. CONCLUSION

In this paper, we compared two approaches to obtain a controller model for MPC: a system identification method using a grey-box model approach and a white-box model approach for which a detailed building model is linearized. The MPC performance using both models is evaluated on a validated 12 zones emulator model of an existing office building. The results indicate that the MPC performance is very sensitive to the prediction accuracy of the controller model. This paper shows that both approaches can lead to an efficient MPC as long as very accurate identification data sets are available. For the considered simulation case, the white-box MPC resulted in a better thermal comfort and used only 50% of the energy used by the best grey-box MPC. Tests in real buildings, however, are still needed to confirm the strength of the white-box approach in presence of all uncertainties (weather predictions, state measurement and estimations, user behaviour, ...).

REFERENCES

Andersson, J. (2013). *A General-Purpose Software Framework for Dynamic Optimization* (PhD thesis). Arenberg Doctoral School, KU Leuven, Department of Electrical Engineering (ESAT/SCD) and Optimization in Engineering Center, Kasteelpark Arenberg 10, 3001-Heverlee, Belgium.

Baetens, R., De Coninck, R., Jorissen, F., Picard, D., Helsen, L., & Saelens, D. (2015). OpenIDEAS - An open framework for integrated district energy simulations. In *IPSBA building simulation 2015*. Hyderabad.

Brinsfield, R., Yaramanoglu, M., & Wheaton, F. (1984). Ground level solar radiation prediction model including cloud cover effects. *Solar Energy*, *33*(6), 493–499.

Cho, S., & Zaheer-uddin, M. (2003, May). Predictive control of intermittently operated radiant floor heating systems. *Energy Conversion and Management*, *44*(8), 1333–1342.

Cigler, J., Siroky, J., Korda, M., & Jones, C. (2013). On the selection of the most appropriate mpc problem formulation for buildings. In *11th rehva world congress clima 2013*.

Dounis, A., & Caraiscos, C. (2009, August). Advanced control systems engineering for energy and comfort management in a building environmentA review. *Renewable and Sustainable Energy Reviews*, *13*(6-7), 1246–1261.

Gwerder, M., & Tödtli, J. (2005). Predictive Control for Thermal Storage Management in Buildings. In *8th REHVA world congress for building technologies - CLIMA* (pp. 1–6). Lausanne.

Gyalistras, D., & Gwerder, M. (2009). *Use of weather and occupancy forecasts for optimal building climate control (opticontrol), two years progress report.* (Tech. Rep.). Terrestrial Systems Ecology ETH Zurich.

Hilliard, T., Kavgic, M., & Swan, L. (2015). Model predictive control for commercial buildings: trends and opportunities. *Advances in Building Energy Research*, 1–19.

Ihm, P., & Krarti, M. (2005). Optimal Control Strategies for Heated Radiant Floor Systems. *ASHRAE Transactions*, 535–546.

Jorissen, F., & Helsen, L. (2016). Towards and automated tool chain for MPC in multi-zone buildings, submitted. In *2016 intelligent building operations workshop.* West Lafayette.

Karlsson, H. (2006). *Thermal system analysis of embedded building integrated heating* (Unpublished doctoral dissertation). Chalmers University of Technology.

Ma, Y., Kelman, A., Daly, A., & Borrelli, F. (2012, Feb). Predictive control for energy efficient buildings with thermal storage: Modeling, stimulation, and experiments. *IEEE Control Systems*, *32*(1), 44-64. doi: 10.1109/MCS.2011.2172532

Maciejowski, J. M. (2002). *Predictive Control with Constraints.* Harlow, Essex: Pearson Education Limitedl.

Meteotest. (2009). Meteonorm version 6.1 - edition 2009 [Computer software manual].

Morari, M., & Lee, J. H. (1999). Model predictive control : past , present and future. *Computers & Chemical Engineering Engineering*, *23*, 667–682.

Picard, D., Drgoňa, J., Helsen, L., & Kvasnica, M. (2016). Impact of the plant and controller model complexity on model predictive control performance evaluation for buildings, submitted. *Applied Energy.*

Picard, D., Jorissen, F., & Helsen, L. (2015). Methodology for obtaining linear state space building energy simulation models. In *11th international modelica conference* (pp. 51–58).

Privara, S., Cigler, J., Vana, Z., Oldewurtel, F., Sagerschnig, C., & Zacekova, E. (2013, JAN). Building modeling as a crucial part for building predictive control. *Energy & Buildings*, *56*, 8-22. doi: {10.1016/j.enbuild.2012.10.024}

Prívara, S., Široký, J., Ferkl, L., & Cigler, J. (2011, February). Model predictive control of a building heating system: The first experience. *Energy & Buildings*, *43*(2-3), 564–572.

Pyephem. (2016). `https://pypi.python.org/pypi/pyephem/`. (Accessed: 2016-04-28)

ACKNOWLEDGEMENT

The Authors gratefully acknowledge the financial support of the of the Scientific Grant Agency of the Slovak Republic under the grants 1/0403/15. The authors also acknowledge the financial support by the Agency for Innovation by Science and Technology in Flanders (IWT) and WTCB in the frame of the IWT-VIS Traject SMART GEOTHERM focusing on integration of thermal energy storage and thermal inertia in geothermal concepts for smart heating and cooling of (medium) large buildings, as well as the financial support of the Ph.D. work of Filip Jorissen (IWT 131012 Ph.D. Grant), and the EU for funding the research work of Maarten Sourbron.

A Multi-Level MPC Simulation Study in a School Building

Vahid. R. DEHKORDI [1]*, José A. CANDANEDO[2]

[1]CanmetENERGY-Varennes, Natural Resources Canada,
Varennes, Québec, Canada
(Ph.: 450-652-3126, Fax: 450-652-5177, jose.candanedoibarra@canada.ca)

* Corresponding Author

ABSTRACT

This paper discusses a proof of concept application of a multi-level modelling concept for the treatment of predictive control in an institutional building. The multi-level approach consists in tackling the control problems of a building hierarchically in several control layers. By setting aside internal details and focusing only on the problem at hand, it is possible to describe the dynamics of the system under consideration with simple low-order models; this strategy facilitates model calibration and the solution of predictive control problems. The case study building considered is a two-story school with a floor area of 24,000 m^2. A benchmark model of the building, created in EnergyPlus, was used to generate data to identify models at three different levels: building level (one model), wing (7 models) and thermal zone (46 models). The multi-level representation enables an effective treatment of the different time scales involved. An MPC strategy was investigated by optimizing the use of two energy storage devices, one for heating (brick thermal energy storage) and another for cooling (ice bank). The MPC strategy enables reducing cost under a time-of-use tariff.

1. INTRODUCTION

The use of predictive models in control strategies has emerged as a promising path to improve energy performance, load management and thermal comfort in buildings. Nevertheless, significant obstacles remain regarding the practical implementation of the concept. For instance, the issue of model development for control applications is still a challenge. One possibility is to create a detailed model of the entire building with a building simulation tool (e.g., EnergyPlus) in order to predict the dynamic response of the building. Although in principle a building simulation model enables taking into account the effect of small changes in conditions (e.g., opening a window, sudden increase in the number of occupants in a room), in practice the difficulty of calibrating hundreds or thousands of parameters so as to match the real building dynamics *at every scale* (e.g., temperature fluctuations within a room, overall heating load of the building) presents nearly unsurmountable obstacles. Assuming that the calibration of a detailed model for control purposes is possible, the question remains of how to perform this calibration within a reasonable time frame and at a reasonable cost so that it is an attractive option for building operators.

It might be helpful to reflect at this point on the role of a model. All models are imperfect representations of reality that help in making a prediction of a phenomenon with the purpose of finding a practical solution for a specific problem. Models allow comparing different "what if" scenarios, either for design (very long term planning) or control (short term planning) applications. In essence, this kind of cognitive reasoning is employed by human brains in decision-making: an approximate image of reality is used to plan actions (Battaglia *et al.*, 2013). Thus, when using a model it is essential to keep in mind the question the model intends to answer. In a building, these questions might be phrased as "what will be the heating profile over the next 24 hours", "how much will the temperature of this office rise over the next hour?", "should the motorized blinds be closed?", "when should I start pre-cooling this meeting room if the assembly is supposed to start at 2 pm?", etc. A model may perform quite well according to one criterion (e.g., long term energy use) and yet be inadequate for a control application.

The multi-level control (MLC) approach discussed in this paper (Figure 1) focuses on viewing the building in a hierarchy of control levels, so that a simple low-order model is enough to make useful predictions for control purposes (Dehkordi and Candanedo, 2014; Candanedo *et al.*, 2014; Candanedo and Dehkordi, 2013). In this sense, the MLC methodology mimics human organizations. The MLC modelling approach has the following features:

- It is recognized that different time scales may be needed for different applications. For example, the building level may require prediction horizons of 48 hours at time steps of 2 hours; at a room level, the prediction horizon may be 1 hour, with time steps of 10 minutes.
- Different kinds of low-order models ("grey box" or "black box").
- Models do not need to be of the same kind for every control level.
- Models can be easily calibrated with measured data.
- Significantly reduced computational burden in the solution of optimization problems.
- Uncertainty can be incorporated relatively easily.
- The predictions of adjacent layers (e.g., heating load) should "match" approximately, although this correspondence is not necessarily exact.
- The recognition of discrepancies between predictions at different levels requires the use of "negotiation" algorithms.
- The multi-level concept may be extrapolated to include a "building cluster" or "community layer".

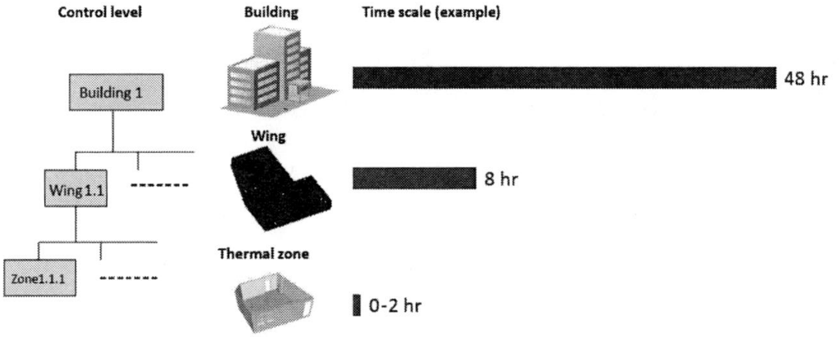

Figure 1: Multi-level control structure

In recent years, distributed MPC (a concept sharing similarities with MLC) has been used to tackle predictive control problems in buildings. In this approach, information is shared between the controllers of adjacent rooms in order to solve the predictive control between zones more effectively (Morosan *et al.*, 2010; Putta *et al.*, 2014). Distributed MPC is associated with a "horizontal" information transfer. The concept of a hierarchical control structure, implying a "vertical" arrangement was applied by Lefort *et al.* (2013) for the predictive control of a house, including the idea of different horizons for two different hierarchical layers: "scheduling control", for a long term horizon; "piloting control", for a short term horizon. A similar two-tier approach (scheduling vs short time scale) was investigated by Touretzky and Baldea (2014).

2. METHODOLOGY

2.1 Case study building

A two-story secondary school, one of DOE Commercial Reference Buildings (Deru *et al.*, 2011), was used as a benchmark model (Figure 2). The school building has a total footprint of 24,000 m^2 (258,000 ft^2) with 46 thermal zones, with a usable floor area of about 19,600 m^2 (211,000 ft^2) and a glazing fraction of 33%. This building was chosen considering its size, the diversity of thermal spaces (classrooms, offices, gymnasia, long hallways). The EnergyPlus detailed model was used to perform "virtual experiments". Output data (indoor operative temperatures) were used together with input signals (exterior temperature, solar gains, internal gains) in order to create simplified low-order models at different control levels.

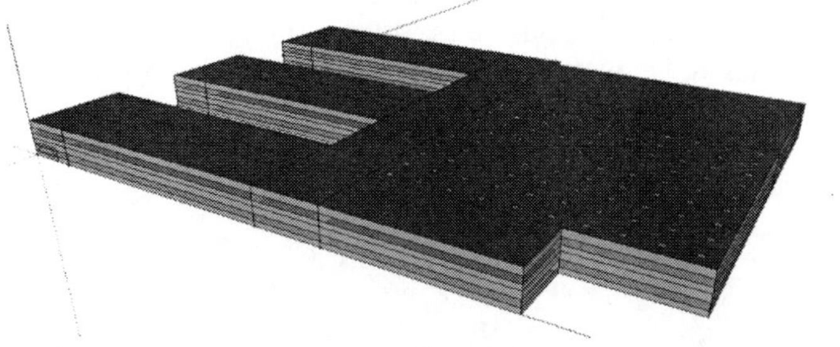

Figure 2: Secondary school building used as a benchmark.

It is assumed that the building is located in Montreal, Canada. The corresponding EnergyPlus weather file (EPW) is used to create weather inputs in MATLAB.

Three control levels were chosen for this case study building: building level (one model), "wing" level (7 models), and thermal zones level (46 models). More details about the system identification procedure are available in a complementary paper presented in this conference (Dehkordi and Candanedo, 2016). The control levels are also associated with different time scales, as shown in Table 1. The "wing models" distribution is shown in Figure 3.

Table 1

Control level	Prediction horizon	Time step	Vector length	Recalculation frequency
Zone level	2 hours	10 min	12 samples	30 min
"Wing" level	8 hours	1 hour	8 samples	4 hours
Whole Building	48 hours	2 hours	24 samples	12 hours

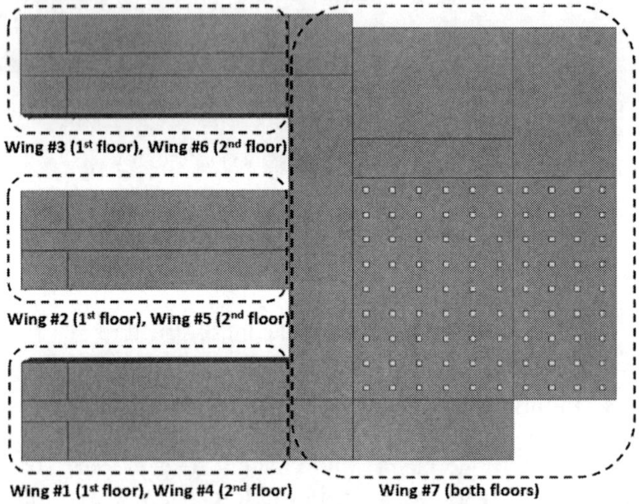

Figure 3: Wing models.

2.2 HVAC and heat distribution system

To illustrate the concept of multi-level control, a simple mechanical system is assumed. Heating and cooling are distributed to the thermal zones by means of a piping system connected to local air handling units. It is assumed that two energy storage devices are available at the building level (i.e., for the entire school): a large ice bank (over 15,000 kWh of storage capacity) for storing cooling energy, and a hot water tank (100 m^3) for storage of heating energy. Cooling is provided by a centralized 3-stage chiller with a maximum cooling capacity of 850 kW (2,900 kBTU/hr). The electricity power use of the chiller and its cooling output are both functions of the "partial load ratio" (PLR) requested by the user. Heating energy is provided, for the purposes of this exercise, with an electric furnace of 1,000 kW (3,410 kBTU/hr).

The heating/cooling load is calculated independently for each of the 46 thermal zones, for the 7 wings and for the entire building by using the models developed through the system identification routine. For the purposes of this study, in which no co-simulation is conducted with a building simulation tool, the loads calculated by the 46 zonal models are used as the benchmarks (i.e., it is assumed that the "real loads" are those calculated by the zonal models). The heating/cooling is distributed to the different wings proportionally according to the loads calculated by each of the wing models (Figure 4).

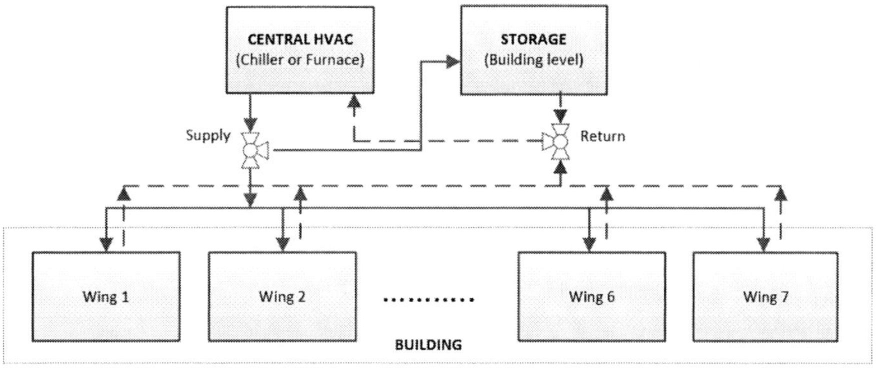

Figure 4: Connection of central plant heating and cooling plant, thermal storage and building "wings".

2.2 Role of different control levels: negotiation rules

The hierarchical division in control layers simplifies the optimization calculation required by predictive control, while at the same time there is a continuous corroboration of the comfort conditions at the zonal level. This facilitates calculations, and provides a mechanism to implement a negotiation between economic optimization and comfort requirements. The simulation can be summarized as follows:

A. Load calculation and optimization carried out at the "building level"

The "building level" model is used along with the weather forecast to run the optimization algorithm (see description in section 2.3). The prediction horizon in this case is 48 hours, and the "states" of the furnace, chiller, ice bank and hot water tank are calculated a 2 hour intervals, thus yielding vectors with 24 elements each.

B. Wings-Zones interaction ("negotiation")

At this step, a negotiation takes place between the heating/cooling required by the zones and the estimate made by the wing models. The 46 zones are used as the calculation benchmark (i.e., the "accurate" temperature calculation). The order of the models depends on factors such as their exposure to neighboring zones, solar radiation, geometry, etc. (Dehkordi and Candanedo, 2016). The zonal models employ a prediction horizon of 2 h at 10 min intervals (also used as the basic simulation time step), while the wing models use a prediction horizon of 8 h at intervals of 1 hour. Therefore, the "time overlap" occurs only over the first two hours predicted by the wing model.

If the difference between the heating/cooling load predicted by a wing model and the aggregated heating/cooling load predicted by its zones is more than a certain threshold (in this case 20 kW), then the wing model prediction is adjusted to match the aggregated load. If the difference is $\leq$ 20 kW, then this "discrepancy" is tolerated, and the zones having a non-zero heating or cooling load share the difference evenly. In essence this will result in some minor deviations from optimal comfort. The trade-off between economics and comfort can be adjusted by playing with the threshold value.

C. Wing models used to correct building level prediction

It is assumed that the "right value" for the heating and cooling loads is provided by the wing models as determined in the previous step. This corrected heating/cooling load calculation is then used to revise the scheduling of the operation of the main equipment.

- In heating mode:
  - If *more* heating is needed, then the electric furnace provides the difference
  - If *less* heating is needed, then (a) first, reduce heating provided by furnace; (b) if necessary, reduce heat taken from the hot water tank
- In cooling mode:
  - If *more* cooling is needed, then increase the cooling provided by the chiller
  - If *less* cooling is needed, then (a) first, reduce the contribution of the chiller; (b) if necessary, reduce the contribution from the ice bank.

This algorithm is summarized in Figure 5.

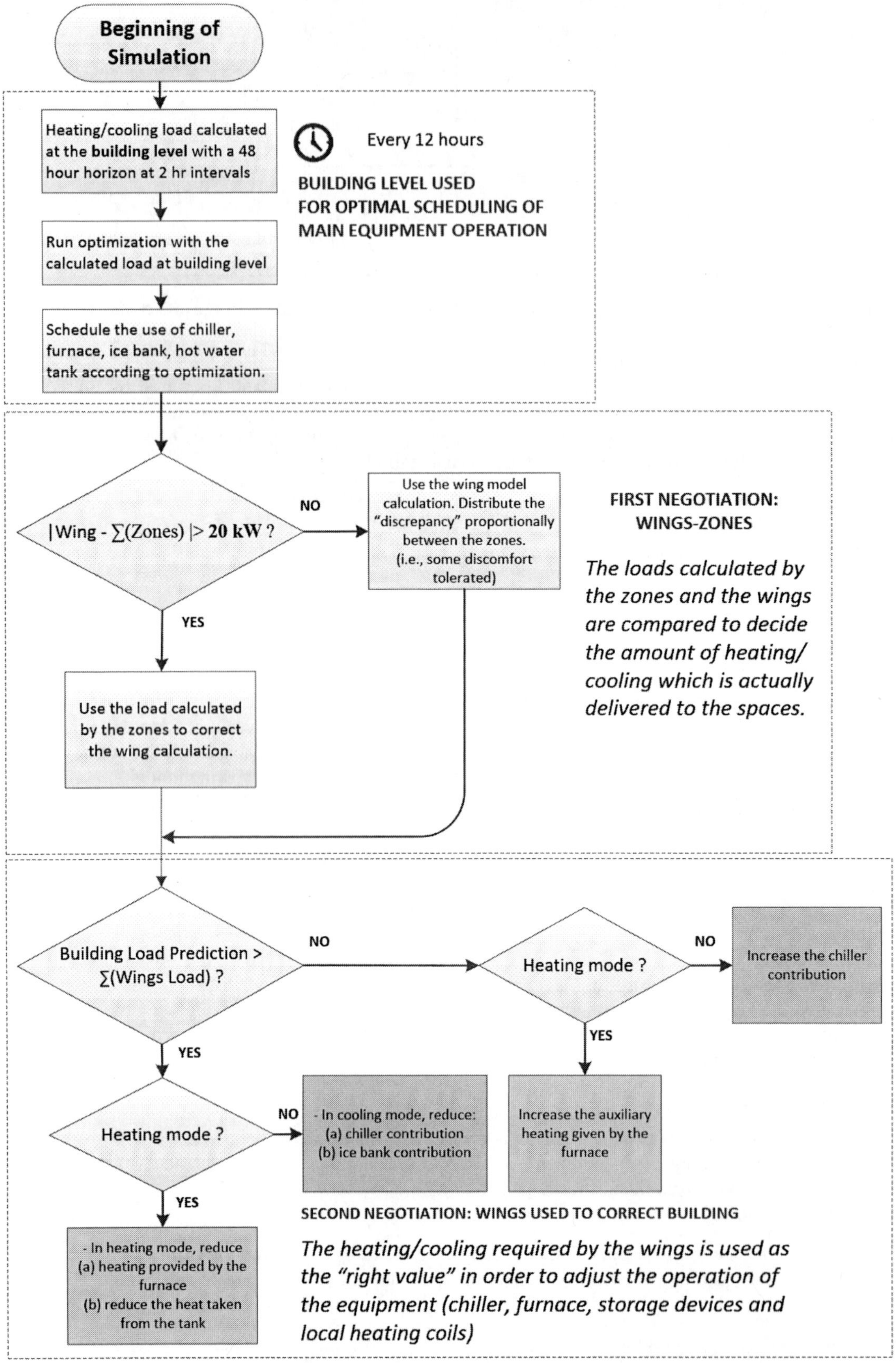

Figure 5: Summary of algorithm used in the simulation.

2.3 Cost function and optimization

A time-of-use rate (TOU) profile, with 24 different hourly rates is used in the calculation of the energy cost in the building by using the "building level" model. The optimization problem to be solved is then:

$$\min \sum_{\langle N \rangle} C(n) \cdot P(n) \cdot \Delta T \tag{1}$$

For the **cooling loop**, both the latent and sensible cooling loads must be provided. The latent load is estimated based on some typical schedules for occupancy rate. At any given time step:

$$\begin{aligned} q_{\text{bldg.c}} &= q_{\text{bldg.sens}} + q_{\text{bldg.lat}}, && \text{at any given time step} \\ \mathbf{q} &= \mathbf{q}_{\text{sens}} + \mathbf{q}_{\text{lat}}, && \text{in vector form} \end{aligned} \tag{2}$$

Considering that the total cost is the dot product of the hourly cost vector and the average electric power of the chiller, the total cost can be calculated as:

$$\text{cost}(\mathbf{P}_c) = \mathbf{c}^T \cdot \mathbf{P}_c, \quad \mathbf{P}_c = f(\mathbf{q}_{ch}) \tag{3}$$

The optimization can then be written as follows:

$$\begin{aligned} &\min_{\mathbf{q}_h, \mathbf{q}_{hwt}} \ \mathbf{c}^T \cdot \mathbf{P}_h, \\ &\text{s.t.} \begin{cases} \mathbf{q}_{ch} + \mathbf{q}_{IB} = -\mathbf{q}_c^* \\ \mathbf{q}_{ch} \le 850\text{kW} \\ 0 \le \mathbf{PLR} \le 1 \\ 0 \le \mathbf{x}_v \le 1 \\ 0 \le \mathbf{x}_{IB} \le 1 \\ 150\dfrac{\text{kg}}{\text{s}} \le \dot{\mathbf{m}}_{ch} \le 450\dfrac{\text{kg}}{\text{s}} \\ -7^\circ\text{C} \le \mathbf{T}_{chwr} \le 15^\circ\text{C} \\ -7^\circ\text{C} \le \mathbf{T}_{chws} \le 10^\circ\text{C} \end{cases} \end{aligned} \tag{4}$$

The optimization constraints make sure that the cooling load is provided by the sum of the ice bank and the chiller, the respect of the operational limits of the chiller (maximum capacity and PLR), the positions of the three-way valves remain between 0 and 1. This also makes sure that the flow rates, and return and supply temperatures remain within realistic limits. The ice banks and chiller models used by the authors in a smaller buildign (Candanedo *et al.*, 2013) were properly scaled for the school building.

Regarding the **heating loop**, the problem is formulated in a similar way, except that in this case an electric furnace linked to a hot water tank is employed. In this case, only sensible heating is considered. The total cost over a period of time is:

$$\text{cost}(\mathbf{P}_h) = \mathbf{c}^T \cdot \mathbf{P}_h \tag{5}$$

Then, the optimization problem can be written as:

$$\min_{q_h, q_{hwt}} \mathbf{c}^T \cdot \mathbf{P}_h ,$$

$$\text{s.t.} \begin{cases} \mathbf{q}_h + \mathbf{q}_{hwt} = \mathbf{q}_L^* \\ 30\frac{kg}{s} \leq \dot{\mathbf{m}}_h \leq 100\frac{kg}{s} \\ \dot{\mathbf{m}}_{hwt} \in \{0,10\}\frac{kg}{s} \\ 19^\circ C \leq \mathbf{T}_{hwt} \leq 40^\circ C \end{cases} \tag{6}$$

At any given moment, the heating load is provided by combining the heat provided by the furnace and the hot water tank. The hot water tank is modeled according to the procedure presented in (Candanedo *et al.*, 2015). Minimum and maximum flow rates are established for the furnace. Limits for the hot water tank temperature are also defined.

3. RESULTS AND DISCUSSION

3.1 Wing/zones interaction

According to the algorithm presented in Figure 5, the heating/cooling load is calculated for each of the wings, and their corresponding zones. Figure 6 shows the results of the load calculation at Wing #2 for a period of 3 hours in January, and the aggregated load of the first two hours for its zones (the prediction horizon at the zone level is 2 h). Note that the calculations for the zone level take place at 10 min intervals, whereas the load prediction at the wing level yields results at hourly intervals. The wing level prediction horizon is 8 h, of which only 3 hours are shown.

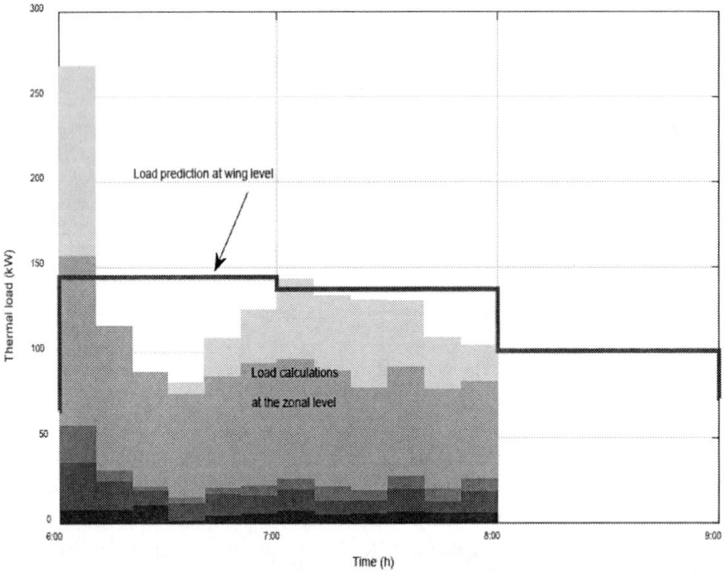

Figure 6: Load prediction for wing #2 and the aggregated load of its zones (before "negotiation")

These calculations for the wing/zone level are then revised according to the "negotiation" rules explained above. The revised calculations are shown in Figure 7. When the difference between the wing load and the aggregated load is larger than the 20 kW threshold, the aggregated load prevails. In contrast, when the difference is smaller than the threshold, the wing calculation is used as the "right value", and the load for the zones is adjusted accordingly.

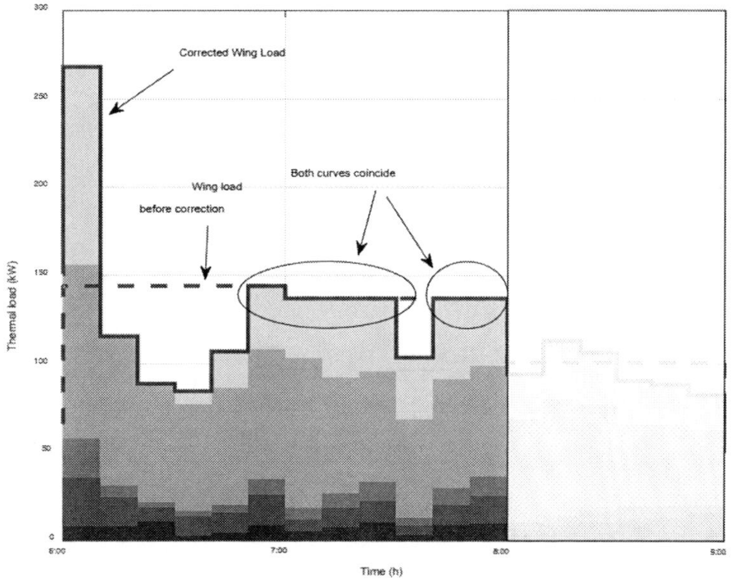

Figure 7. Revised calculations after "negotiation".

3.2 Building/wings interaction

Figure 8 shows the load prediction for the building level and the aggregated load of the wings. To make it consistent with the previous graph, the graph is shown starting at 6:00. There is, however, a non-zero load from 4:00 to 6:00 (preheating). The calculations at the building level are used in the optimization algorithm described in Section 2.3. The building load calculation extends much longer than shown, since a prediction horizon of 48 h is used.

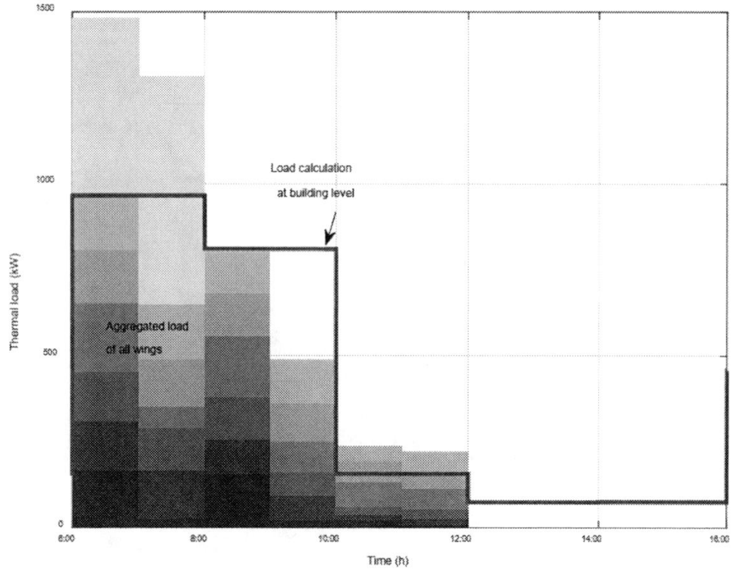

Figure 8: Load calculations at the building level (before negotiation) vs aggregated load of the wings.

As a result of the optimization for the building level, the optimum states of operation for the main mechanical equipment (total heating load, heat delivered to the hot water tank, heat taken from the hot water tank, and heat provided by the furnace) are given at 2-hour intervals (Figure 9).

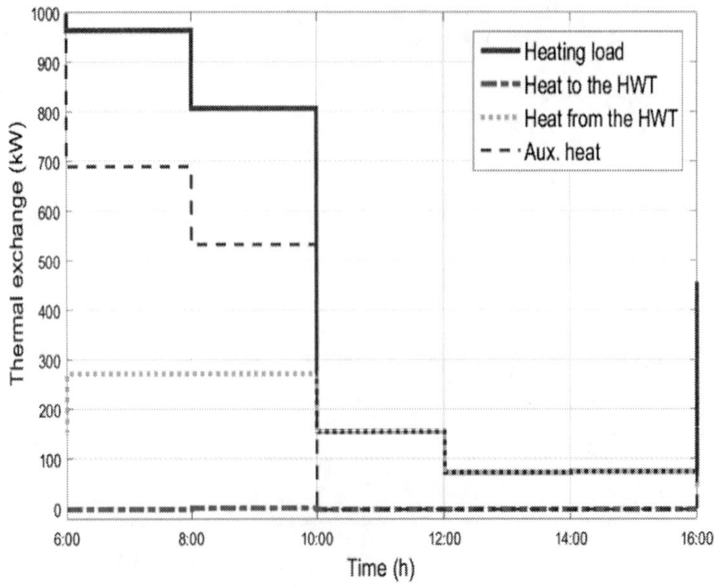

Figure 9: State of the thermal exchanges (heating load, heat delivered to the hot water tank, heat taken from the hot water taken, and auxiliary heating provided by the furnace)

Finally, when running the actual simulation at 10 min time steps, the load delivered to the wings is the sum of the load of the zones, and the *actual* building load is the sum of the heat delivered to the wings (Figure 10).

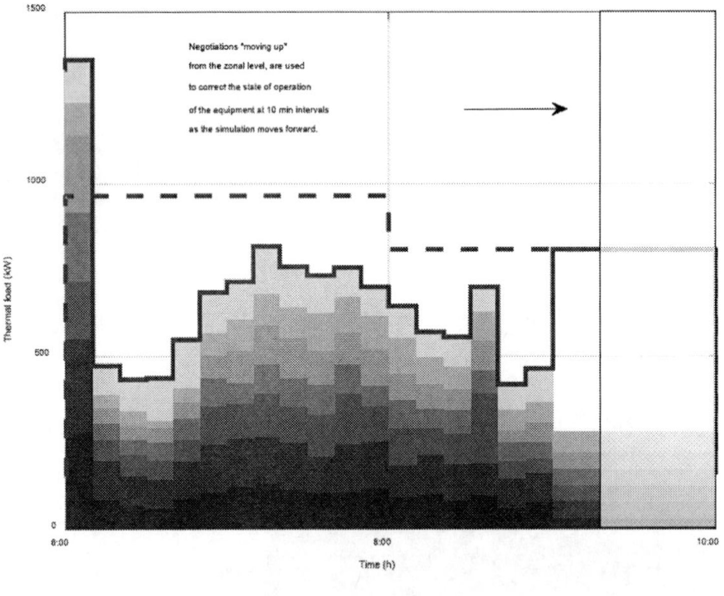

Figure 10: Adjustment of the load according to the zonal/wing negotiation.

6. CONCLUSIONS

This paper has illustrated the concept of multi-level control as an alternative to the more conventional method of modeling the entire building with a high-resolution, detailed, all-encompassing representation. It has been shown that this multi-level approach makes it possible to use a simple model for the planning of an optimization strategy (predictive control) while simultaneously tackling issues at local control levels. Minor discrepancies in the predictions of controllers at different levels are addressed through a "negotiation" algorithm, which implements a trade-off between the long-term planning intended for cost optimization, and the immediate needs of local zones to guarantee thermal comfort. Although, the proposed methodology works well, an issue that adds some complexity is the handling of the time scales; a specifically designed simulation environment could significantly simplify this task.

NOMENCLATURE

The nomenclature should be located at the end of the text using the following format:

C	total cost	(CAN$)
$\dot{m}$	mass flow rate	kg/s
P	electric power	
q	heat flux	(W)
T	temperature	(°C)

Subscript

bldg	building
c	cooling, cooling power from chiller
chwr	return water into chiller
chws	supply water from chiller
h	heating, heat from furnace

REFERENCES

Battaglia, P. W., Hamrick, J. B., Tenenbaum, J. B., 2013. Simulation as an engine of physical scene understanding. Proceedings of the National Academy of Sciences 110(45), pp. 18237-18332.

Candanedo, J. A., Dehkordi, V. R., Multi-level approach for model-based predictive control (MPC) in buildings: a preliminary overview. International Conference for Enhanced Building Operation (ICEBO) 2013, October 8-10 2013 Montreal, Quebec.

Candanedo, J. A., Dehkordi, V. R., Lopez, P., A multi-level architecture to facilitate MPC implementation in commercial buildings: basic principles and case study. eSim 2014, May 7-10 2014 Ottawa.

Candanedo, J. A., Dehkordi, V. R., Stylianou, M., 2013. Model-based predictive control of an ice storage device in a building cooling system. Applied Energy 111, pp. 1032–1045.

Candanedo, J. A., Dehkordi, V. R., Tamasauskas, J., Impact of heat pumps and predictive control on residential energy use, load and grid interaction: a Canadian perspective. 14th International Building Simulation Conference (IBPSA 2015), December 7th-9th 2015 Hyderabad, India.

Dehkordi, V. R., Candanedo, J., Multi-level model predictive control of a building. ASME 2014 Dynamic Systems and Control Conference (DSCC 2014), October 22-24 2014 San Antonio, Texas.

Dehkordi, V. R., Candanedo, J. A., State-space modeling of thermal spaces in a multi-zone building. 4th High Performance Buildings Conference, 2016 Purdue University, West Lafayette, IN.

Deru, M., Field, K., Studer, D., Benne, K., Griffith, B., Torcellini, P., Liu, B., Halverson, M., Winiarski, D., Rosenberg, M., Yazdanian, M., Huang, J., Crawley, D.,2011. U.S. Department of Energy Commercial Reference Building Models of the National Building Stock. NREL - DOE.

Lefort, A., Bourdais, R., Ansanay-Alex, G., Guéguen, H., 2013. Hierarchical control method applied to energy management of a residential house. Energy and Buildings 64, pp. 53-61.

Morosan, P.-D., Bourdais, R., Dumur, D., Buisson, J., 2010. Building temperature regulation using a distributed model predictive control. Energy and Buildings 42, pp. 1445-1452.

Putta, V., Kim, D., Cai, J., Hu, J., Braun, J. E., Distributed model predictive control for building HVAC systems: a case study. Third High Performance Buildings Conference, July 14-17 2014 Purdue University, Indiana.

Touretzky, C. R., Baldea, M., 2014. Integrating scheduling and control for economic MPC of buildings. Journal of Process Control 24, pp. 1292-1300.

ACKNOWLEDGEMENT

Funding for this work was provided by Natural Resources Canada through the Office of Energy Research and Development.

Reducing energy consumption for buildings under system uncertainty through robust MPC with adaptive bound estimator

Hao Huang[1*], Lei Chen[2], and Eric Hu[3]

[1] The University of Adelaide, School of Mechanical Engineering,
Adelaide, Australia
Contact Information (h.huang@adelaide.edu.au)

[2] The University of Adelaide, School of Mechanical Engineering,
Adelaide, Australia
Contact Information (lei.chen@adelaide.edu.au)

[3] The University of Adelaide, School of Mechanical Engineering,
Adelaide, Australia
Contact Information (eric.hu@adelaide.edu.au)

* Corresponding Author

ABSTRACT

Model Predictive Control (MPC) has emerged as an alternative to traditional control method to reduce building energy consumption. With the presence of model uncertainty, such as mismatch between the plant and control-oriented model, the use of MPC may result in thermal comfort violation or energy waste. The influence of model uncertainty becomes even more significant as the size and complexity of the investigated building increase. Robust MPC (RMPC), which requires knowledge on the system uncertainty, has been investigated for enhancing the stability of MPC. However, the implementation possibility of the RMPC is prevented by increased computational burden and conservativeness of controller performance. This paper deals with the latter issue by presenting a novel adaptive RMPC scheme for temperature regulation in commercial buildings. The novelty comes from the development of a comparison model built based on a nonlinear autoregressive model for worst-case analysis. This comparison model enables us to transform a linear, robust MPC problem into an adaptive one with a time-varying uncertainty bound. The proposed method is tested on a simulation model developed from building data collected from a spacious hall at an airport terminal. By conducting simulation using different MPCs, it is found that the proposed RMPC method is able to behave robustly against uncertainty with the least performance loss. This means the maximum energy saving and the least thermal comfort violation.

1. INTRODUCTION

Recently, the application of model predictive control (MPC) as a supervisory controller for optimising building-energy usage has received an increasing attention. The advantage of MPC over the traditional control strategy is that the former one is able to handle systematically and effectively constraints on control inputs and states, by taking advantage of weather forecast to perform disturbance prediction.

The performance of an MPC is largely dependent on how accurately the specific model describes the real building process. While different model types are available for building modelling, most researchers choose resistance-capacitance (RC) models as the control-oriented model for optimisation (Cai & Braun, 2015a; Razmara et al., 2015; Bengea et al., 2014). This is because the RC models can be accurately identified using experimental data while maintaining physical significance. Even with limited training data, the RC models can be accurately identified by using a component-based approach (Cai & Braun, 2015b).

However, generating highly accurate RC models by using experimental data is theoretically not always possible, because of persistent disturbances and simplification of heat transfer process. Because the MPC saves energy by maintaining the indoor temperature close to the upper (in summer) or lower (in winter) comfort limits, an inaccurate prediction of indoor temperature can result in thermal comfort violations or energy waste.

Conventionally, the MPC employs linear, time-invariant (LTI) models to predict future dynamics of the system, despite the fact that controlled systems (thermal zones) are non-linear and non-deterministic. The feedback nature of the MPC allows the controller to reject a small degree of uncertainty. Kim *et al.*(2015) model roof top units in a mid-size commercial using multi-input and multi-output (MIMO) auto-regressive with exogenous input (ARX) model. The simple ARX model could be employed in this case study because the prediction horizon is short and the control input is not strongly correlated to the unmeasured disturbance. However, for systems with large uncertainties and MPC that requires a long prediction horizon, the explicit consideration of the uncertainties becomes especially crucial. This motivates the research on robust MPC (RMPC). The main idea of RMPC is to consider the model–plant mismatch as an uncertain term and address it explicitly in the control algorithm.

The RMPC has been studied to optimise the energy efficiency of the HVAC systems. For example, Kim (2013) designed an RMPC to improve the stability of traditional MPCs under uncertainty conditions. He found that robust MPC outperforms a nominal MPC when uncertainty is dominant and the model mismatch is significant. Maasoumy *et al.* (2014) compared the performance of a closed-loop robust MPC with rule-based control and nominal MPC under different levels of uncertainties. They suggest that robust MPC should only be chosen within a specific range of uncertainties. Gondhalekar *et al.* designed a least-restrictive robust MPC law for indoor temperature regulation. The proposed method eliminates the conservativeness of traditional min–max open-loop prediction MPC while guaranteeing reasonable computational complexity. Past studies show that correctly choosing the uncertainty bound is crucial for the design of RMPC: a too narrow bound would cause thermal comfort violation while a too wide uncertainty bound cause performance lose.

An unattended issue in the previous studies (Kim, 2013; Maasoumy et al., 2014) on RMPC is how to design the uncertain bound properly to reduce conservatism and guarantee robustness. If the uncertainty bound is chosen to be fixed, it will inevitably lead to conservative solutions, because the RMPC has to work conservatively at all times to guarantee that the thermal comfort is satisfied. In this work, we develop a comparison model for worst-case analyse and provide uncertainty bound to the RMPC. The comparison model is built upon a recursive neural-network model (RNN). The idea is to use the recursive nature and nonlinear approximation ability of the RNN model to capture the uncertainty dynamics and nonlinear properties of the buildings. This algorithm allows one to estimate the error bound in an adaptive fashion over the given prediction horizon. Once the uncertainty bounds are obtained, the optimisation problem is solved as a closed-loop min–max RMPC problem, based on nominal model predictions and tightened constraint sets. The proposed method is demonstrated at the terminal building of Adelaide Airport, South Australia. The investigated zone is uncertain due to frequent variation of passenger flow, as well as unknown coupling from both controlled and uncontrolled neighbouring space. The complexity and uncertainty of the investigated building make the case study suitable for testing proposed control methods.

The remainder of this paper will proceed as follows: Section 2 introduces grey-box and RNN modelling methods for building thermal dynamics modelling. The design of RMPC and its adaptive bound estimator are introduced in Section 3. Section 4 discusses the results of using the proposed controllers, including the control performance comparison between different types of MPC. The paper ends with a conclusion and a description of future work.

2. SYSTEM MODELLING

2.1 Case Study Building

The test building is the Terminal-1 (T1) building of Adelaide Airport, South Australia. The perimeter zones of the second floor were selected as the test site for the experiment. The layout of the test area is shown in Figure 1. This area serves as the check-in hall from where most passengers enter or leave the building. A large glass facade is installed to the north to ensure good lighting conditions, while two motorised blinds are installed to reduce the effects of solar gain. The position of the blinds is programmed into BMS and controlled with a fixed schedule. The hall is divided into four adjacent zones. Each zone has an associated CAV (Constant Air Volume) box to condition the space and a sensor to measure the zone temperature. Different from other airport terminals, T1 has no flights during the night, so the HVAC system is scheduled to be switched off during the night. The occupied period for the testing area is 7:00 am to 10:00 pm.

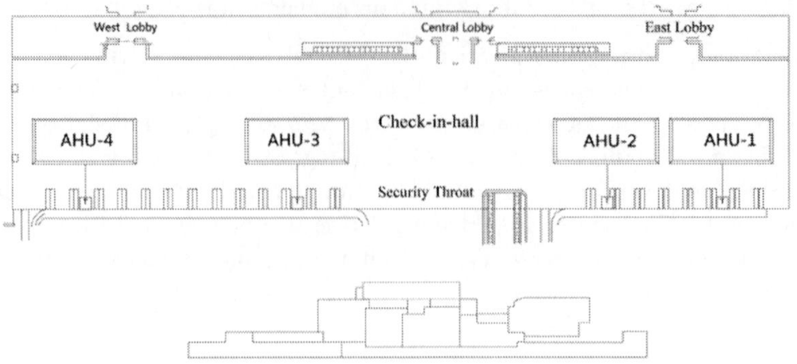

Figure 1: Layout of the check-in hall of Adelaide airport

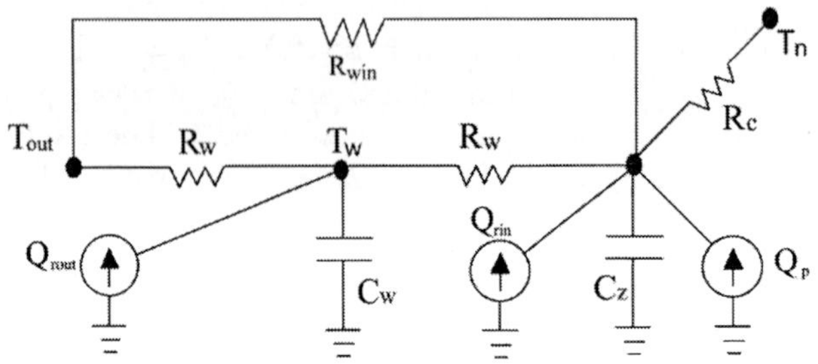

Figure 2: RC modelling of the building thermal model.

2.2 RC Modelling

Theoretically speaking, buildings' thermal dynamic can be best described by time-varying nonlinear partial differential equations, which are not suitable for control optimisation. Therefore, simplified RC models are more often used. Past research has compared different structures of RC models in modelling the thermal dynamics of buildings (Sourbron et al., 2013; Fux et al., 2014; Gouda et al., 2002). For control purposes, this study employs a second order RC model to represent the thermal dynamics of the building. In this work, we focus on minimising the supplied thermal energy Q_u at the AHUs level and will not model the efficiency of the AHU system.

The modelling work is a typical multi-input (thermal energy supplied to individual zones) and multi-output (zone temperatures) problem. The parameters that are of concern are the heat transfer coefficients between the air handling units (AHUs) , adjacent zones, internal walls and uncontrolled spaces. Zones 1 and 2 located at the east end of the building were selected as the experimental area in this study. Zones 3 and 4 are not considered in this study because they have similar dynamics with zones 1 and 2.

The structure of the RC model is depicted in Figure 2. The corresponding energy and mass balance governing equations for a single zone can be written as:

$$C_z \frac{dT_z}{dt} = \dot{m} C_a (T_{sa} - T_z) + \frac{T_{out} - T_z}{R_{win}} + \frac{T_w - T_z}{R_w} + \frac{T_n - T_z}{R_c} + Q_{rin} + Q_p + Q_{inf}, \tag{1}$$

$$C_w \frac{dT_w}{dt} = \frac{T_z - T_w}{R_w} + \frac{T_{out} - T_w}{R_w} + Q_{rout}, \tag{2}$$

where C_z is the overall thermal capacitance of the air and other fast-response elements, C_w is the thermal capacitance of the interior walls and ceiling, C_a is the specific heat of the air, T_z is the temperature of the investigated zone, T_n is the temperature of the neighbouring zone(s), T_{out} is the outdoor air temperature, T_w is the mean surface temperature of the interior walls and ceiling, R_{win} is the thermal resistance of the windows, R_w represents the convective resistances between the building envelope and the outdoor and indoor air, R_c is the convective heat transfer coefficient between adjacent zones, Q_{rin} and Q_{rout} denote the inside and outside surface solar radiation heat flux, respectively, Q_p is the internal heat gain generated by the presence of occupants and their behaviours, and Q_{inf} is the internal heat gain from leakage and door openings.

Observing Equation (1), we found that the only non-linear component in the simplified model is the bilinear term $\dot{m}C_a(T_{sa} - T_z)$. Using feedback linearisation, the control input is redefined as $Q_u = \dot{m}C_a(T_{sa} - T_z)$. The new control input Q_u represents the thermal energy supplied to the individual zones. Equations (1) and (2) are discretised using the Euler forward method to obtain:

$$\begin{pmatrix} T_z(k+1) \\ T_w(k+1) \end{pmatrix} = \begin{pmatrix} 1 - \frac{\Delta t}{C_z R_{win}} - \frac{\Delta t}{C_z R_w} & \frac{\Delta t}{C_z R_w} \\ \frac{\Delta t}{C_w R_w} & 1 - \frac{2\Delta t}{C_w R_w}, \end{pmatrix} \begin{pmatrix} T_z(k) \\ T_w(k) \end{pmatrix} + \begin{pmatrix} \frac{\Delta t}{C_z} \\ 0 \end{pmatrix} \left(Q_u(k)\right) + \begin{pmatrix} \frac{\Delta t}{C_z R_{win}} & \frac{\Delta t}{C_z R_c} & \frac{\Delta t}{C_z} & 0 & \frac{\Delta t}{C_z} & \frac{\Delta t}{C_z} \\ \frac{\Delta t}{C_w R_w} & 0 & 0 & \frac{\Delta t}{C_w} & 0 & 0 \end{pmatrix} \begin{pmatrix} T_{out}(k) \\ T_z(k) \\ Q_{rin}(k) \\ Q_{rout}(k) \\ Q_p(k) \\ Q_{inf}(k) \end{pmatrix}, \tag{3}$$

where k is the time step, and Δt is the sampling time. There are three disturbance inputs in Equation (3). The first input is the incident solar radiation on the inside and outside of the building envelope. The data used for calculating the solar radiation heat flux is global horizontal irradiation I_r (w/m^2). We use $Q_{rin} = \alpha I_r$ and $Q_{rout} = \beta I_r$ for calculating the solar radiation heat flux, where α and β denote the coefficients associated with the area and the absorption rate of the inside and outside building envelop. β also includes the effect of window transmittance.The second input to the system is the internal gain generated by the occupants. The internal gain is proportional to the number of occupants in the hall, so it can be indicated by the carbon dioxide concentration (ppm). Therefore, it is presented by $Q_p = \gamma CO_2$. Here γ denotes the coefficient associated with the number of occupants. The two variables are identified together with Equation (3). The third input Q_{inf} is not measurable so they are regarded as unmeasured uncertainty. Taking into account the uncertainty and measurement noise, Equation (3) can be re-written as state-space form as:

$$x(k+1) = Ax(k) + Bu(k) + Ed(k) + Fe(k), \tag{4}$$

where A, B, E, and F are the corresponding matrices, x is state vector, u is input vector, d denotes measurable disturbance, and e denotes the uncertainty and measurement noise that is assumed to be Gaussian distributed. The unknown parameters in Equation (4) are identified using measured building data using a Trust-Region reflective algorithm (Moré & Sorensen, 1983). The data used for model identification were collected from between the 1st and 20th of January, 2013. The input is the supplied cooling energy Q_u, the meteorological data of the Bureau of Meteorology of Australia provide disturbance forecast (Q_r and T_{out}) to the thermal zones, the output is indoor temperature.

A source of uncertainty in building control comes from the errors in forecasting. The outside ambient temperature forecast uncertainty increases with longer prediction horizon. However, this study does not specifically investigate the effects of weather forecast error on the modelling results. Instead, we use measured outside air temperature instead of the predicted value to perform analyse. All the other uncertainties are lumped into the error terms.

2.3 Recursive Neural Network Modelling

In this section, the thermal dynamics of the same zone will be modelled by an RNN model. Nonlinear autoregressive with exogenous inputs (NARX) model is used to express the RNN structure:

$$y(k) = f_2(y(k-1), ...y(k-n_y), u(k-1), ...u(k-n_u) + e(k), \tag{5}$$

where $y(k) = [y_1(k), ...y_p(k)]^T$, $u(k) = [u_1(k), ...u_m(k)]^T$, and $e(k) = [e_1(k), ...e_m(k)]^T$ are the system output, input and noise, respectively; p and m are the number of outputs and inputs, respectively; n_y and n_u are the maximum lags in the outputs and inputs, respectively; k is the process dead time; and f_2 is a vector-valued non-linear function. Model order selection is an important step of system identification, but it will not be elaborated this paper. For detailed procedures see (Huang et al., 2015a).

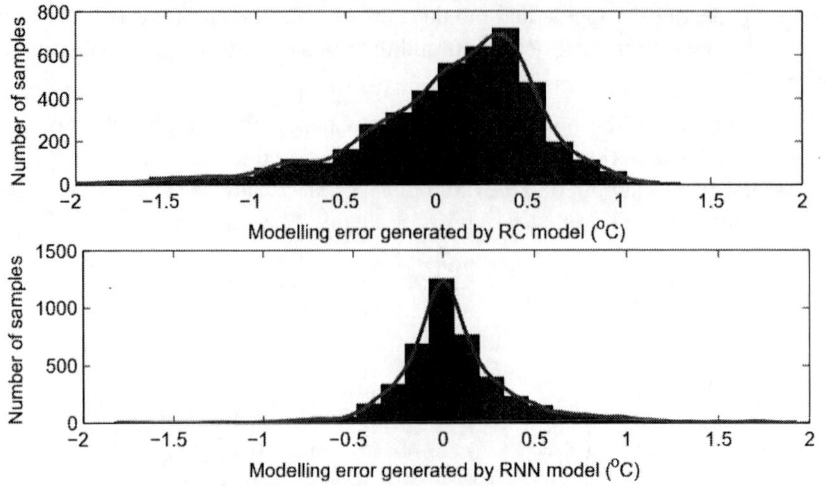

Figure 3: Histogram of residuals (31 days) generated by the two models.

A neural network with three layers of neurons was employed. The network function is expressed with the following equation:

$$\hat{y}(t) = F\sum_{i=1}^{n_h} W_{j,u} f_3(\sum_{i=1}^{n_u} w_{u,i}\phi_i(k) + b_{u,0}) + B_{j,0}, \tag{6}$$

$$f_3(x) = \frac{1}{1+\exp(-x)}, \tag{7}$$

where $W_{j,u}$ and $w_{u,j}$ are weights vector to the hidden layer and output layer respectively. $b_{j,0}$ and $B_{u,0}$ are the bias of the hidden units and the output layer, respectively, $\phi_i(k)$ indicates the vector that contains the regression of the Equation (5) at time step k, f_3 is sigmoid function expressed by Equation (7), and F uses a linear function and $j = 1$ as only one output is considered. The weight vector w and bias vector b at the hidden layer were initialised using the Nguyen-Widrow method to keep the trained model more consistent. Levenberg-Marquardt algorithm was employed to train the neural networks, which minimises mean square error (MSE). For training the RNN, the input-output data cannot be directly fed into the network but should be arranged to follow the structure of the NARX model.

2.4 Model Validation and Comparison

The data used for validation purposes were a completely different set of data, obtained from 21st January 2013 to 24th January 2013. Because the primary purpose of building the models is to achieve predictive control, the multi-step-ahead prediction is needed. However, both of the aforementioned training algorithms only minimise the error over a single step ahead. A more reasonable way to validate our model is to consider multi-step-ahead prediction by taking into consideration the recursive feature of the model (Huang et al., 2015b). The histogram of the residuals generated by the two models is plotted in Figure 3. Overall speaking, compared with the RC model, the residuals generated by the RNN model are smaller and closer to a normal distribution. This is because the recursive nature and nonlinear approximation ability of the RNN model enable it to capture some uncertainty and nonlinear dynamics of the buildings. The complexity and uncertainty of the investigated building make the case study suitable for comparing these two modelling methods.

3. CONTROL DESIGN

3.1 Closed-loop min–max Robust MPC

Conventionally, the MPC employs linear, time-invariant models to predict future dynamics of the system, despite the fact that controlled systems (thermal zones) are non-linear and non-deterministic. This type of MPC is referred to as a deterministic MPC (DMPC). In designing the DMPC, it is assumed that the model can predict the real-world building plant perfectly. In the context of RMPC, which is designed to address model mismatch explicitly, the effects of model mismatch on state estimations should be addressed. Therefore, Equation (4) is rewritten as:

$$x(k+1) = Ax(k) + Bu(k) + Ed(k) + Fw(k), \tag{8}$$

where $w_k \in \mathbb{W}^2$ denotes the model uncertainty. w_k is bounded but its exact value is not known. The basic idea of min–max RMPC is to determine all possible evolutions of the disturbance sequence over the control horizon, and to minimise worst-case cost. Min–max RMPC can be either in the form of open-loop predictions (Lee, 2011) or closed-loop predictions (Lee, 2011; Löfberg, 2003). The open loop approach fails to take into account that feedback is presented in the receding-horizon implementation of the control, therefore leads to conservative solutions and could even make the optimisation problem infeasible (Löfberg, 2003; Lucia et al., 2014).

In this study, we focus on the application of a closed-loop RMPC to building-energy control. Different from the open-loop approach, the closed-loop RMPC considers the feedback over the prediction horizon and incorporates it into the prediction. In particular, at each step, the closed-loop RMPC considers the future value X under different disturbance trajectories. The controller generates a family of control sequences, each one corresponds to a different measured state. The maximum costs can be calculated along with some of the worst-case predictions.

Considering the following min–max optimisation problem:

$$\min_{u} \max_{w} \sum_{j=0}^{N-1} Q(u_{k+j|k}) + R(y_{k+j|k} - r_{k+j|k}) + S\sum_{k=1}^{N}(|\underline{e}_{k+j|k}| + |\overline{e}_{k+j|k}|), \tag{9}$$

subject to:

$$\begin{aligned} x_{k+j+1|k} = Ax_{k+j|k} + Bu_{q,k+j|k} + Ed_{k+j|k} + Fw_{k+j|k}, \quad & \forall j = 0, \dots, N-1 \\ y_{k+j|k} = Cx_{k+j|k}, \quad & \forall j = 1, \dots, N \\ T_{min,k+j|k} - \underline{e}_{k+j|k} \leq y_{k+j|k} \leq T_{max,k+j|k} + \overline{e}_{k+j|k}, \quad & \forall j = 1, \dots, N \\ \underline{e}_{k+j|k} > 0, \overline{e}_{k+j|k} > 0, \quad & \forall j = 1, \dots, N \\ \overline{w}_{k+j|k} \leqslant w_{k+j|k} \leqslant \underline{w}_{k+j|k}, \quad & \forall j = 0, \dots, N-1, \\ U_{max,k+j|k} \leq v_{k+j|k} \leq U_{min,k+j|k}, \quad & \forall j = 0, \dots, N-1, \end{aligned} \tag{10}$$

The constraints that should be met are:

1. $T_{oc} \in [\ 21°C\ ,\ 24°C]$ Thermal comfort during occupied hours.
2. $T_{uo} \in [\ 19.5°C,\ 26°C]$ Thermal comfort during unoccupied hours.
3. $u_q \in [\ -10\text{ kW},\ 12\text{ kW}]$ Maximum cooling energy that can be supplied to each zone.

where the double indices $k+j|k$ denote the prediction value at time $k+j$ made at time k, $U = [u_{k|k}, u_{k+1|k}, \dots, u_{k+N-1|k}]$ is a vector of the control inputs (supplied thermal energy) applied to the model, $\hat{y}_{k+j|k}$ is the predicted output at time k, $T_{max} = [T_{max,k|k}, T_{max,k+1|k}, \dots, T_{max,k+N-1|k}]$ is a vector of the upper comfort temperature band within the horizon, T_{min} is a vector of the lower comfort temperature band, variants $\underline{e}$ and $\overline{e}$ denote the temperature violation from the upper and lower bounds, respectively, N is the prediction horizon, d is the measured disturbance, r is the set-point temperature, U_{min} and U_{max} denote the maximum cooling and heating energy that the system can supply, respectively, Q denotes the penalty on the energy usage, R denotes the penalty on the comfort constraint violation, and S denotes the penalty on the set-point temperature deviation. If S is choosen to be 1, the controller strives to maintain set point temperature during occupied hours; If S is chosen to be 0, the controller maintains the zone temperature closed to the upper comfort limit. $\underline{w}_{k+j|k}$ and $\overline{w}_{k+j|k}$ denote the lower and upper uncertainty bounds, respectively. All possible trajectories are included in bands that depend on $\underline{w}_{k+j|k}$ and $\overline{w}_{k+j|k}$. The values of $[\overline{w}; \underline{w}]$ depends on the future state of the real system, which is not known. A possible way of estimating the error bounds is to obtain them directly from the historical residuals, e.g. determine the numerical range where a certain probability of the modelling errors happens. However, this method still relies on the choice of the probability density. If the uncertainty set is chosen to be too large, the controller becomes very conservative and control performance will be lost. The conservatism will be demonstrated with an example in a later section.

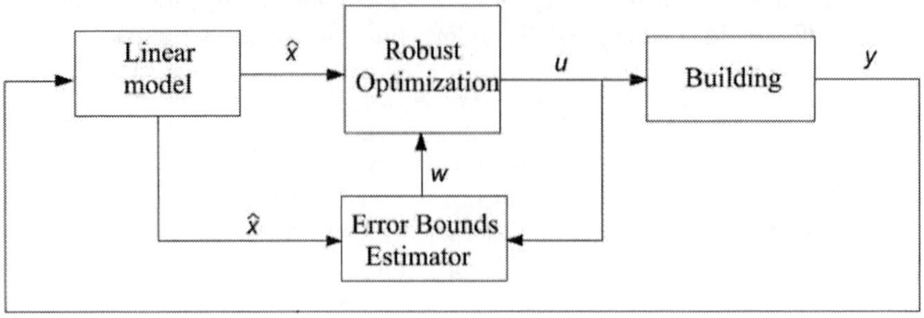

Figure 4: Control structure of the ARMPC scheme.

3.2 RMPC with Adaptive Uncertainty Bound

The above-mentioned bounding method can result in conservative solutions, because it is purely based on the historical data but fails to take into account that the errors are also related to the occurrence of future disturbances. The uncertainty can be predicted using a comparison model (Fukushima & Bitmead, 2005). Illustrated by this idea, we propose an uncertainty bound estimator in this study, constructed based on the RNN model presented in the previous sections. The idea is to use the recurrent nature of the RNN model to capture the uncertainties that exist in the building system. The unknown model mismatch between the RC model and the actual building plant can therefore be approximated by the difference between the RC model and the RNN model. The positive uncertainty can be calculated as

$$\bar{z} = \max_{k \in 1 \ldots n} (\hat{y}_{nn}(k) - \hat{y}_{rc}(k)), \tag{11}$$

where $\hat{y}_{rc}$ and $\hat{y}_{nn}$ are open-loop prediction results generated by RC and RNN models, respectively. $\bar{z}$ denotes the maximum error within n steps, $n \leqslant N$ is the number of steps considered. Figure 4 shows the adaptive RMPC (ARMPC) procedure and the adaptive RMPC algorithm is summarised as follows:

Algorithm: Adaptive Robust MPC

1. Choose the initial bounds using sampling method introduced in the previous section.
2. DMPC computes the open loop input trajectory $U = [u_1, u_2 \ldots u_n]$.
3. The RNN performs n-steps ahead prediction using input vector U and disturbance vector D to obtain a comparison set of output trajectory $Y = [y_1, y_2 \ldots y_{1+N}]$.
4. Calculate the error bounds based on Equations (11).
5. During occupation period, if z is positive (underestimation), then update uncertainty bounds to $[-\overline{w}, \overline{w}]$ by setting $n = 3$. Else the uncertainty bound is set to the minimum range.
6. During transitional period, if z is positive (underestimation), then update uncertainty bounds by setting $n = N$. Else the uncertainty bound is set to the minimum range.
7. Increment k. Go to Step 2.

4. RESULTS AND DISCUSSION

The programs for model training, validation and control optimisation were coded in Matlab, which runs on a PC with Intel Core i7 CPU 2.4 GHz. The optimisation problem are solved using Yalmip (Löfberg, 2004). The optimised control parameter is shown in Table 1.

We present a comparison study on three different MPC approaches, which are DMPC, RMPC and ARMPC, in the presence of model uncertainty. Our goal is to compare the performance of these three controllers in terms of conservatism, stability and computational speed. As illustrated in the previous section, MPC can save energy by keeping the zone temperature close to its upper or lower comfort limit. Because cooling is considered in this study, only underestimation $\hat{T} > T$ will cause thermal comfort violation on the upper comfort bound.

Table 1: Parameters of the RMPC.

RMPC parameters	
Sampling interval	10 min
Prediction horizon, N	30
Control horizon, P	30
Weight for uncertainty term	0.1
Fixed uncertainty bound	$[-0.2, 0.25]$
Number of look-ahead steps	10
Penalty on energy input Q	0.8
Penalty on soft-constraint violation in R	1
Penalty on set-point temperature deviation S	1
Occupancy hour	5:00 am to 9:30 pm

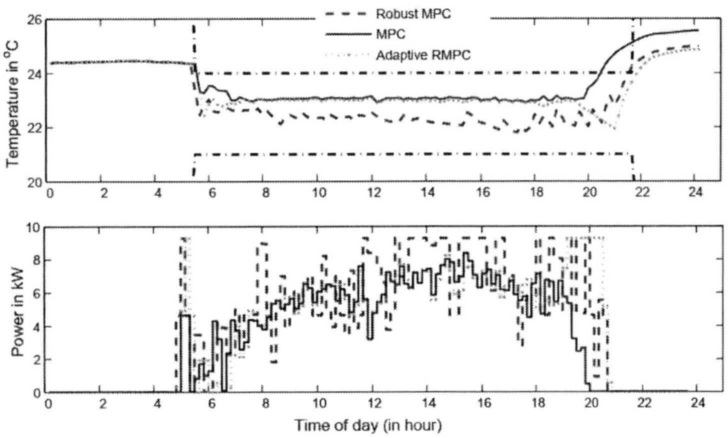

Figure 5: Performance of DMPC, RMPC and ARMPC under model uncertainty.

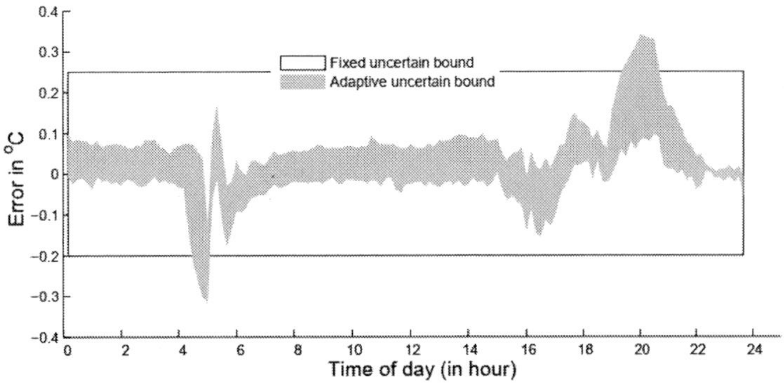

Figure 6: Comparison of fixed bound and the adaptive bound.

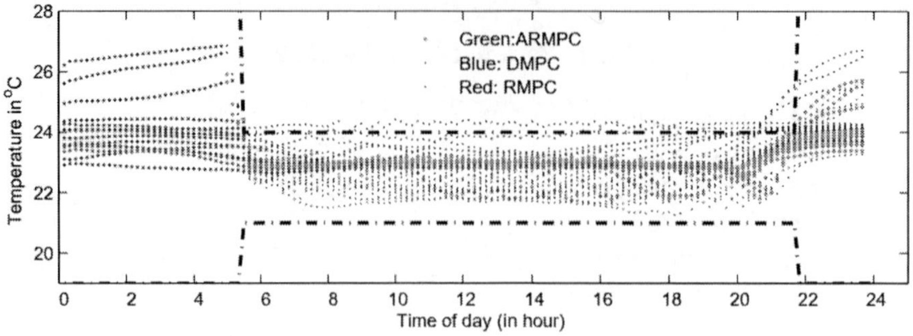

Figure 7: Two weeks' control performance of DMPC, RMPC and ARMPC.

Table 2: Performance comparison between DMPC, RMPC and ARMPC.

	Baseline	DMPC	RMPC	ARMPC
Total energy (kWh)	3580	2440	3052	2656
Number of infeasible days	3	9	0	0
Simulation time per step	n/a	0.4 s	4.4 s	7.1 s

First, we choose a day on which the LTI control model suffers from a considerable amount of uncertainty to conduct the comparison. The uncertainty is illustrated in Figure 5. The uncertainty bound for RMPC is set to be [-0.2, 0.25], based on the simple sampling methods and the histogram of residuals. The set-point value is chosen to be close to the upper comfort temperature (23 °C). Figure 5 compares the performance of RMPC and DMPC when a uncertainty is presented. The red, dashed line indicates the output of DMPC. It can be seen that, although the uncertainty happened during the steady state, it does not impose an unstable influence on the performance of the DMPC. This is because the closed-loop nature of the DMPC makes it robust to some degree of uncertainty. However, when the temperature is moved towards the upper bound, the comfort constraint is violated by the DMPC, which is due to the large error happens before the end of the occupancy. The blue, solid line shows the performance of the RMPC. The thermal comfort is satisfied by the RMPC all the time, as the modelled uncertainty is located within the designated uncertainty bound for most of the time. However, the RMPC consumes more energy (32% with respect to DMPC), by maintaining the temperature below the set-point value during steady states. This is expected because the RMPC performs conservatively by lowering the indoor temperature to make sure that constraints are safely satisfied at all times. The RMPC works in such a way to prevent a potential comfort violation due to the positive uncertainty. The green, dotted line shows the performance of the ARMPC. It can be seen that the ARMPC tracks the set-point temperature well during the steady states, without wasting too much energy. This is because of the use of a smaller uncertainty bound, as shown in Figure 6. This proves the effectiveness of the proposed method, because a smaller, adaptive uncertainty bound indicates a lower degree of conservatism without violating thermal comfort for RMPC. The ARMPC also satisfies the thermal comfort requirement before the end of occupancy, when the uncertainty bound is expanded to allow a larger uncertainty to happen. Similarly, this is because the ARMPC expand the uncertainty bound during the transitional period to ensure a better robustness.

The simulation period was then extended to two weeks, from 4th to 28th Feb, 2015. Figure 7 shows the simulation results using the three above-mentioned control methods. A summary of the performance of each controller for the investigated days is reported in Table 2. It can be seen that the baseline control consumes the most energy and violated the thermal comfort on 3 days. DMPC consumes the least energy which leads to the greatest energy savings. However, the zone temperature regulated by DMPC violates the comfort constraints in 9 days. This is because the DMPC does not take into account the occurrence of positive uncertainty. The RMPC does not violate any comfort constraints, but consumes 25% more energy than does the DMPC. The ARMPC does not violate the thermal constraints on an days and at the same time consumes 12% less energy than the RMPC. This is because the ARMPC controls the zone temperature much closer to the set-point value as compared to the RMPC. In terms of computational speed, DMPC is the most efficient one. Both RMPC and ARMPC have slower computational speed, because the exponential increased complexity with the increase of prediction horizon. The ARMPC is slower than RMPC because estimating the uncertainty bounds using RNN estimator requires extra computational efforts.

5. CONCLUSIONS

In this paper, we have presented an adaptive bounds estimator, which allows the uncertainty bound of RMPC to vary according to the dynamic changes of the system. It is shown that the proposed method results in a smaller uncertainty bound when the actual uncertainty is low, which greatly reduces the conservatism of the traditional RMPC. A comparison between two modelling methods shows that when the uncertainty is significant, the RNN model achieves more accurate long-term prediction result as compared with the linear RC model. More interestingly, it is also found the errors generated by the former model are closer to a normal distribution, which makes statistical analysis possible. We believe this is because the RNN model captures the uncertainty and nonlinearity of the system using its recursive property. Additionally, because of the closed-loop nature of the DMPC, the modelling errors do not always lead to poor control performance. The influence of modelling error on control performance also depends on the types of uncertainty and the time when this might happen. It seems that the DMPC usually performs more robustly during the steady states as compared with the transition period. Conclusively, by combining neural network modelling technology with the classical control method, this paper opens up a new path for solving complicated real-world building energy control problem. As a future study, the proposed control method will be tested experimentally at the investigated building.

REFERENCES

Bengea, S. C., Kelman, A. D., Borrelli, F., Taylor, R., & Narayanan, S. (2014). Implementation of model predictive control for an HVAC system in a mid-size commercial building. *HVAC&R Research*, *20*(1), 121-135.

Cai, J., & Braun, J. E. (2015a). A generalized control heuristic and simplified model predictive control strategy for direct-expansion air-conditioning systems. *Science and Technology for the Built Environment*, *21*(6), 773-788.

Cai, J., & Braun, J. E. (2015b). Gray-box modeling of multistage direct-expansion units to enable control system optimization. *ASHRAE Transactions*, *121*(2), 203-216.

Fukushima, H., & Bitmead, R. (2005). Robust constrained predictive control using comparison model. *Automatica*, *41*(1), 97 - 106.

Fux, S. F., Ashouri, A., Benz, M. J., & Guzzella, L. (2014). EKF based self-adaptive thermal model for a passive house. *Energy and Buildings*, *68, Part C*(0), 811 - 817.

Gouda, M., Danaher, S., & Underwood, C. (2002). Building thermal model reduction using nonlinear constrained optimization. *Building and Environment*, *37*(12), 1255 - 1265.

Huang, H., Chen, L., & Hu, E. (2015a). A neural network-based multi-zone modelling approach for predictive control system design in commercial buildings. *Energy and Buildings*, *97*(0), 86 - 97.

Huang, H., Chen, L., & Hu, E. (2015b). A new model predictive control scheme for energy and cost savings in commercial buildings: An airport terminal building case study. *Building and Environment*, *89*(0), 203 - 216.

Kim, S. H. (2013). An evaluation of robust controls for passive building thermal mass and mechanical thermal energy storage under uncertainty. *Applied Energy*, *111*(0), 602 - 623.

Lee, J. (2011). Model predictive control: Review of the three decades of development. *International Journal of Control, Automation and Systems*, *9*(3), 415-424.

Löfberg, J. (2003). *Minimax approaches to robust model predictive control* (Unpublished doctoral dissertation). Linköping University, Automatic Control, The Institute of Technology.

Löfberg, J. (2004). Yalmip : A toolbox for modeling and optimization in MATLAB. In *Proceedings of the CACSD conference*.

Lucia, S., Andersson, J. A., Brandt, H., Diehl, M., & Engell, S. (2014). Handling uncertainty in economic nonlinear model predictive control: A comparative case study. *Journal of Process Control*, *24*(8), 1247 - 1259.

Maasoumy, M., Razmara, M., Shahbakhti, M., & Vincentelli, A. S. (2014). Handling model uncertainty in model predictive control for energy efficient buildings. *Energy and Buildings*, *77*(0), 377 - 392.

Moré, J. J., & Sorensen, D. C. (1983). Computing a trust region step. *SIAM Journal on Scientific and Statistical Computing*, *4*(3), 553-572.

Razmara, M., Maasoumy, M., Shahbakhti, M., & Robinett, R. (2015). Optimal exergy control of building HVAC system. *Applied Energy*, *156*, 555 - 565.

Sourbron, M., Verhelst, C., & Helsen, L. (2013). Building models for model predictive control of office buildings with concrete core activation. *Journal of Building Performance Simulation*, *6*(3), 175-198.

A Distributed Model Predictive Control Approach for Optimal Coordination of Multiple Thermal Zones in a Large Open Space

Xiaodong HOU[1*], Yingying XIAO[1], Jie CAI[2], Jianghai HU[1], and James E. BRAUN[2]

[1] School of Electrical and Computer Engineering, Purdue University,
West Lafayette, IN 47907, United States
{hou39, xiao106, jianghai}@purdue.edu

[2] School of Mechanical Engineering, Purdue University,
West Lafayette, IN 47907, United States
{cai40, jbraun}@purdue.edu

* Corresponding Author

ABSTRACT

Model Predictive Control (MPC) based approaches have recently seen a significant increase in applications to the supervisory control of building heating, ventilation and air-conditioning (HVAC) systems, thanks to their ability to incorporate weather, occupancy, and utility price information in the optimization of heating/cooling strategy while satisfying the physical constraints of HVAC equipment. Many of the proposed MPC solution approaches are centralized ones that often suffer from high computational cost caused by the large number of decision variables and the overhead in information gathering and distribution. This paper investigates a distributed MPC approach based on a variant of the Alternating Direction Method of Multipliers (ADMM). The proposed method is highly scalable, and facilitates a device level plug-and-play implementation. A case study is carried out on one of the Purdue Living Labs, which is an open office space of multiple thermal zones with individual thermostat controls. In view of significant thermal couplings due to direct air exchange and noticeable load gradient between zones, a multiple thermal zones coordination problem is formulated with the objective of optimally scheduling the different thermostat setpoints for energy minimization and comfort delivery while satisfying actuation constraints. Simulation results demonstrate the effectiveness of the proposed method.

1. INTRODUCTION

The building sector consumes about 40% of the energy used and is responsible for about 40% of the green house gas emissions in the United States (McQuade, 2009). Hence, reducing the energy consumption of buildings is of huge economical and environmental importance. Many commercial buildings consist of multiple thermal zones. Each zone may represent an individual room, a partially enclosed space, or even a part of a room in an open space. The thermal dynamics of different zones in a building are typically coupled, either through shared HVAC equipments, or via heat transfer across walls, door openings or direct air exchange in the case of open space.

The problem of controlling multi-zone buildings with coupled zone thermal dynamics has received increased attention in the building control community. Various intelligent control techniques have been proposed for reducing the overall energy consumption of such buildings while maintaining and improving the occupant comfort. To name a few, genetic algorithms are used in (Mossolly, Ghali, & Ghaddar, 2009) to obtain an optimal control strategy for a multi-zone air conditioning system. Particle Swarm Optimization (PSO) is utilized to optimize the building energy management in (Yang & Wang, 2013).

In particular, a control technique that has been especially popular in the recent years is Model Predictive Control (MPC). Due to its ability of incorporating weather forecasts and occupancy profiles in real time decision making, MPC provides a powerful and practical solution to mamy building control problems (Ma et al., 2012), (Privara, Široký, Ferkl, & Cigler, 2011), (Oldewurtel et al., 2012). In the MPC framework, A receding horizon optimal control problem is formulated and solved at each time step, but only the first control action will be applied to the building system while the procedure is repeated at the next time step with updated information (weather, solar radiation, occupancy gains, etc).

However, centralized MPC approaches often have difficulty in dealing with large-scale building clusters or even a single building consisting of multiple thermal zones due to the large number of decision variables that are distributed across different parts of the system. In addition, repetive engineering costs of designing controllers for different building systems with similar multiple thermal zone structure is formidable. Therefore, a distributed control strategy that utilizes the special structure of multi-zone buildings, while at the same time have good scalability is highly desired.

This paper aims at designing a general Distributed MPC (DMPC) strategy that is capable of dealing with not only coupling in the thermal dynamics, but also coupling in HVAC equipments. The DMPC method takes advantage of a decomposition technique called Proximal Jacobian Alternating Direction Method of Multipliers (PJ-ADMM) and is highly scalable with respect to the size of the building system.

The proposed method is applied to a case study on one of the Purdue Living Labs. A optimal coordination problem of multi-zones with individual thermostat controls is formulated and solved with the proposed controller. Simulation results demonstrates the effectiveness of the proposed distributed controller, and the electricity saving potential of inter-zonal coordination.

The rest of this paper is organized as follows. In Section 2, the case study building and HVAC system models are discussed. The centralized optimal multiple thermal zone coordination problem is formulated in Section 3. The proposed distributed MPC approach is introduced and explained in Section 4. Section 5 presents the simulation results and discusses their implications. Finally, some concluding remarks and future directions are given in Section 6.

2. CASE STUDY MODEL

In this section, a detailed description of the case study model is given. The case study model consists of two parts, namely the room envelope model and the rooftop units (RTUs) equipment model. The room envelope model captures the thermal dynamics of the room space under the influence of sensible cooling/heating provided by HVAC systems. The RTU equipment subsection explains the input-output relation of the RTU model used as well as some realistic simplifications and assumptions.

2.1 Envelope Model

The Purdue Living Lab is a large open space located at the Center for High Performance Buildings, West Lafayette, IN. It has a south-facing double façade. The room serves not only as a test-bed for building thermal, HVAC system research, but also as the office for 16 graduate students. A multi-zone thermal network state-space model of the room is obtained with estimated model parameters from building construction information. The structure of the room is given in Figure 1.

As can be observed from Figure 1, the room space is divided into three zones based on their relative distance to the south facing double façade. The double façade system is equivalent to a "thermal buffer" between room and ambient environment from the heat transfer point of view. In addition, solar radiation, occupancy schedule and even local thermal comfort preference may be significantly different from one zone to another. All these factors could cause a load imbalance between different zones. Indeed, preliminary experiment results show that there is a noticeable 2°C zone temperature gradient across the office space under normal operation conditions.

A general linear discrete-time thermal dynamics of one individual zone in a multiple zone building is given by

$$\begin{aligned} x_i(k+1) &= A_{ii}x_i(k) + \sum_{j\in\mathcal{N}_i} A_{ij}x_j(k) + B_{ii}Q_i(k) + \sum_{j\in\mathcal{M}_i} B_{ij}Q_j(k) + F_iw_i(k), \\ T_i(k) &= C_ix_i(k), \end{aligned} \tag{1}$$

where $k \in \mathbb{N}$, $x_i(k) \in \mathbb{R}^{n_i}$ is the local state variable of zone i, including average zone temperature and a lumped wall/floor temperature. $Q_i(k) \in \mathbb{R}^{m_i}$ is the sensible cooling or heating directly injected into zone i by the HVAC, which is the controllable input. $w_i(k) \in \mathbb{R}^{p_i}$ are the exogenous inputs including solar radiation, ambient temperature, occupancy gains, etc., which are uncontrollable but assumed to be predictable. $T_i(k)$ is the average temperature of zone i, and C_i capture its dependency on $x_i(k)$. $A_{ij} \in \mathbb{R}^{n_i \times n_j}$ are constant matrices denoting the thermal coupling between different zones. In our case, this coupling is caused by the direct air exchange between adjacent zones; in other cases, the heat

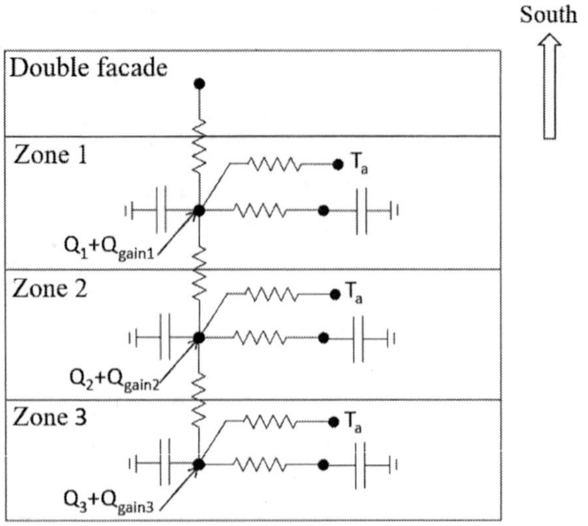

Figure 1: Purdue Living Lab multi-zone thermal network structure

exchange channel may be an open door, a shared corridor or windows between two rooms. $B_{ij} \in \mathbb{R}^{n_i \times m_i}$ are constant matrices that representing the coupling in the HVAC system (zero matrices in this case study). $\mathcal{N}_i$ and $\mathcal{M}_i$ represent the set of neighbouring zones that have influence on x_i through x_j and Q_j, respectively.

The lumped envelope model of the whole thermal system can be obtained by combining the dynamics of individual zones

$$\begin{aligned} x(k+1) &= Ax(k) + BQ(k) + Fw(k), \\ T(k) &= Cx(k), \end{aligned} \tag{2}$$

where $x(k)$, $Q(k)$, $w(k)$ and $T(k)$ are the concatenation of corresponding zone-level local variables. Because of the thermal couplings between zones, A is not block diagonal. The thermal interactions between zones give opportunity for coordination between zones in order to minimize the energy consumption of the HVAC system, more details will be discussed in later sections.

2.2 RTU Equipment Model

In the case study, we assume that each zone in the Purdue Living Lab is served by a dedicated RTU with individual thermostat control. RTUs are packaged direct-expansion type cooling systems that regulate the temperature and circulate air as part of a HVAC system. Such air-conditioning systems have straightforward configurations and are easy to install and maintain. Thus, they are widely used in small to medium commercial buildings. In the US, RTUs are responsible for approximately 60% of the installed cooling capacity in the commercial building sector. The proposed DMPC methodology provides a generic means to coordinate the different RTUs serving a large open space.

In this study, a correlation-based RTU model is developed from manufacturer catalogue data and the capacity of RTUs have been scaled down to meet the requirements of this specific study. However, the characteristics of the model remain unchanged. The input and output relation of the RTU model used is given in

$$[Power, PLR] = RTU(Q_{sen}, T_{amb}, T_{wb}, T_{db}), \tag{3}$$

where inputs of the RTU are: (1) Q_{sen}, sensible cooling load of the building; (2) T_{amb}, ambient temperature; (3) T_{wb}/T_{db}, nominal wet/dry bulb temperatures and are assumed to take constant values of 15.3/28.9°C. The outputs are the instant power consumption *Power* of unit and its part-load ratio *PLR*. The coefficient of performance of the unit, namely, *COP* varies with the ambient temperature T_{amb}, and in general lower ambient temperature provides higher unit efficiency.

Note that at part load conditions, RTUs primarily rely on cycling to match the building loads and the RTU models use degradation curves (coefficients of degradation assumed to equal 0.2) to capture the efficiency decrease caused by

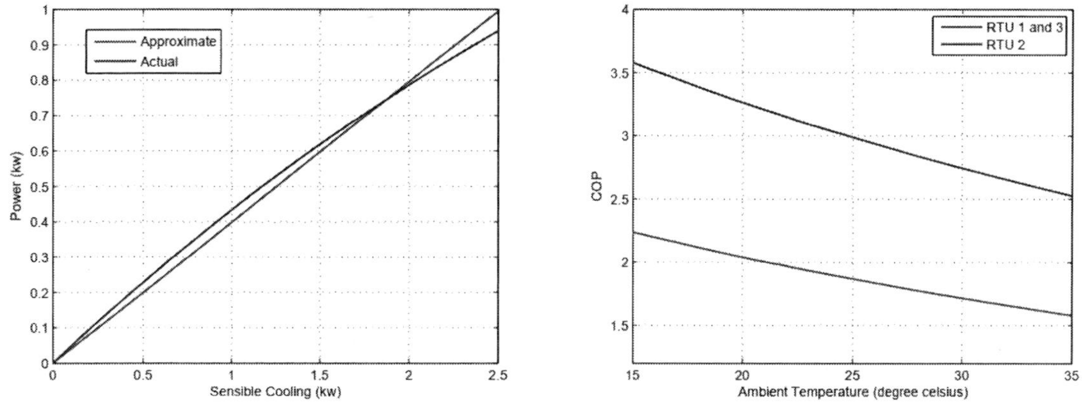

Figure 2: RTU characteristics left: ***Power*** **vs.** Q_{sen} **(**T_{amb} **= 17.9°C) right:** ***COP*** **vs.** T_{amb}

RTU cycling. However, such degradation effect creates a slightly concave mapping from Q_{sen} to the RTU power. To convexify the problem, linear approximations of the power consumption function with respect to Q_{sen} were obtained under various ambient temperatures T_{amb}. Note that this approximation captures the averaged unit efficiency within a range of cooling load and is different than a zero coefficient of degradation. This approximation will not cause noticeable inaccuracy in practice, but can significantly help guarantee the theoretical performance of the controller to be designed. The power consumption vs. sensible cooling relation under T_{amb} = 17.9 °C is plotted in Figure 2 (left) as an example.

Among the three RTUs, RTU 2 (serving zone 2) has the largest capacity (around 3kw), while RTU 1 and 3 are relatively smaller in size (around 2.5kw). In addition, under the same T_{amb}, RTU 2 is the most efficient unit, that is, it has a larger *COP* compared to RTU 1 and 3 under the same ambient temperature. The plot of *COP* vs. T_{amb} is given in Figure 2 (right) to demonstrate the efficiency difference between RTUs under various ambient temperatures.

3. OPTIMAL MULTI-ZONE COORDINATION

As mentioned before, RTUs are prevalent in many open-spaced commercial buildings (restaurant, retail stores, etc). Most of the current RTU operation strategies are based on simple feedback control logic in response to individual thermostat setpoints for the zone temperature. The lack of inter-zonal coordination in the traditional control strategy usually results in a higher total energy consumption of the building.

It is particularly beneficial to seek coordination between different units in a large open space, especially when unit efficient differences are significant. A multiple thermal zone coordination problem is a supervisory control problem, with individual thermostat setpoint temperatures as well as sensible coolings provided by RTUs being the decision variables. It is assumed that the RTU equipment level local (PID) controller is fast enough to respond to the optimal Q_{sen} and thermostat setpoint temperatures. In other words, we are trying to find the optimal thermostat setpoint temperature *trajectories* and the corresponding RTUs operations.

The optimal multiple thermal zone coordination problem will be formulated as a finite horizon optimal control problem with the objective of minimizing the total electricity consumptions in a receding horizon fashion, i.e. via model predictive control (MPC). Besides incorporating exogenous inputs information (weather, occupancy, etc.) into the optimization, another major reason for using MPC is that it is able to take advantage of the thermal storage capability of building itself so that load shifting from peak hours (possibly higher utility rates) to off-peak hours becomes a possibility.

3.1 Centralized MPC Formulation

For notational simplicity, constant variables T_{wb}, T_{db} will be neglected from the RTU model. Also, index of the current time step will be omitted, and we will use $x_i(k)$ to represent the predicted state variables of zone i that are k steps ahead of the current step. If we denote the approximate power consumption function for RTU i as $P_i(Q_i(k), T_{amb}(k))$, the centralized optimal multi-zone coordination problem can be formulated under the MPC framework as

$$\underset{T(k),Q(k)}{\text{minimize}} \quad J = \sum_{k=0}^{N-1} \sum_{i=1}^{3} \mu(k) \cdot P_i(Q_i(k), T_{amb}(k)) \tag{4}$$

$$\text{subject to} \quad x(k+1) = Ax(k) + BQ(k) + Fw(k), \quad T(k) = Cx(k), \tag{5}$$

$$T_{min,i}(k) \leq T_i(k) \leq T_{max,i}(k), \tag{6}$$

$$0 \leq PLR_i(Q_i(k), T_{amb}(k)) \leq 1, \tag{7}$$

$$i = 1, 2, 3, \quad k = 0, 1, \ldots, N-1,$$

where N is the look-ahead horizon (prediction horizon), $\mu(k)$ is the utility price (possibly time-varying). $T_{min,i}$ and $T_{max,i}$ are the lower and upper bounds of the temperature of zone i, respectively, which together specify the "comfort interval" for the occupants sitting in zone i, and can be locally customized. PLR_i is the part-load ratio of RTU i, and it should be less or equal to 1 at all times.

In this study, we assume that the ambient temperature information is predictable through weather forecast. Therefore, at each time instant, a new linear approximation of P_i can be obtained for the current $T_{amb}(k)$. Similarly, unit capacities $Q_{max,i}(k)$ at each time instant could be obtained according to $T_{amb}(k)$. With this simplification, the cost function (4) can be re-written as a single linear function in $Q_i(k)$, and RTU operation constraint (7) can be replaced by

$$0 \leq Q_i(k) \leq Q_{max,i}(k),$$

which consists of two linear constraints on part of the decision variables. Notice that the other two constraints (5) and (6) are also linear constraints on the decision variables $T(k)$ and $Q(k)$. Therefore, the centralized MPC problem (4) for optimal multi-zone coordination under the current formulation is a linear program.

Optimization problem (4) can be more compactly written as

$$\underset{\mathbf{x},\mathbf{Q}}{\text{minimize}} \quad J = \sum_{i=1}^{3} \tilde{f}_i^{\mathsf{T}} \mathbf{Q}_i \tag{8}$$

$$\text{subject to} \quad \mathbf{x} = \mathbf{\Omega} x(0) + \mathbf{\Phi}\mathbf{Q} + \mathbf{\Psi}\mathbf{w}, \tag{9}$$

$$\mathbf{x}_{min,i} \leq \mathbf{x}_i \leq \mathbf{x}_{max,i}, \quad i = 1, \ldots, 3, \tag{10}$$

$$\mathbf{Q}_{min,i} \leq \mathbf{Q}_i \leq \mathbf{Q}_{max,i}, \quad i = 1, \ldots, 3, \tag{11}$$

where $\tilde{f}_i$ is the linear coefficient for the power consumption approximation function of RTU i, and

$$\mathbf{x} = [x^{\mathsf{T}}(1) \cdots x^{\mathsf{T}}(N)]^{\mathsf{T}}, \quad \mathbf{Q} = [Q^{\mathsf{T}}(0) \cdots Q_i^{\mathsf{T}}(N-1)]^{\mathsf{T}}, \quad \mathbf{x_i} = [x_i^{\mathsf{T}}(1) \cdots x_i^{\mathsf{T}}(N)]^{\mathsf{T}}, \quad \mathbf{Q}_i = [Q_i^{\mathsf{T}}(0) \cdots Q_i^{\mathsf{T}}(N-1)]^{\mathsf{T}},$$

$$\mathbf{x}_{min,i} = [x_{min,i}^{\mathsf{T}}(1) \cdots x_{min,i}^{\mathsf{T}}(N)]^{\mathsf{T}}, \quad \mathbf{Q}_{min,i} = [Q_{min,i}^{\mathsf{T}}(0) \cdots Q_{min,i}^{\mathsf{T}}(N-1)]^{\mathsf{T}},$$

$$\mathbf{x}_{max,i} = [x_{max,i}^{\mathsf{T}}(1) \cdots x_{max,i}^{\mathsf{T}}(N)]^{\mathsf{T}}, \quad \mathbf{Q}_{max,i} = [Q_{max,i}^{\mathsf{T}}(0) \cdots Q_{max,i}^{\mathsf{T}}(N-1)]^{\mathsf{T}},$$

$$\mathbf{\Omega} = \begin{bmatrix} A \\ A^2 \\ \vdots \\ A^N \end{bmatrix}, \quad \mathbf{\Phi} = \begin{bmatrix} B & 0 & \cdots & 0 \\ AB & B & \cdots & 0 \\ \vdots & \vdots & \ddots & \vdots \\ A^{N-1}B & A^{N-2}B & \cdots & B \end{bmatrix}, \quad \mathbf{\Psi} = \begin{bmatrix} F & 0 & \cdots & 0 \\ AF & F & \cdots & 0 \\ \vdots & \vdots & \ddots & \vdots \\ A^{N-1}F & A^{N-2}F & \cdots & F \end{bmatrix}, \quad \mathbf{w} = \begin{bmatrix} w(0) \\ w(1) \\ \vdots \\ w(N-1) \end{bmatrix}.$$

Observing optimization problem (8), it should be noted that the cost function (8), as well as the constraints (10) and (11) are all readily separable with respect to each zone and their RTUs. However, the thermal dynamics equality constraint (9) is a coupling constraint between all three zones.

Linear programs in general can be solved very efficiently. However, in practice, as the number of zones and its corresponding variables grow, the communication costs in collecting and diffusing variables between local zones and central computing facility through Building Management System (BMS) could become prohibitive. In particular, if we are dealing with building clusters, in which case a zone may be a single building or a group of buildings (optimal coordination problem becomes coordinating demands between buildings to reduce total demand charge), this problem becomes more challenging.

In addition, we are seeking local control strategies that can be potentially embedded into HVAC devices (RTUs in this case), such that instead of transmitting all local information to a central computer for computation, the optimization problem could be solved distributedly and cooperatively by local agents through simple information exchange with each other. This type of distributed control strategies will have much better scalability compared to centralized ones when dealing with large-scale building systems. One of the possible ways to design such distributed control strategies is through distributed MPC (DMPC), and a novel DMPC algorithm for solving problems in the form of (8) is proposed in Section 4.

4. DISTRIBUTED MPC VIA PROXIMAL JACOBIAN ADMM

In a general DMPC formulation, the centralized problem are solved cooperatively by a group agents through information exchanging and broadcasting. Each agent takes responsibility for a small part of the decision variables (corresponding to one zone in this case) and may not be aware of the global system model. Many existing works on DMPC are based on agent negotiation (Maestre, De La Pena, Camacho, & Alamo, 2011) or distributed optimization. Dual decomposition method was applied to DMPC in (Wakasa, Arakawa, Tanaka, & Akashi, 2008) and (Farokhi, Shames, & Johansson, 2014), while (Farokhi et al., 2014) also used Gauss-Seidel alternating direction method of multipliers (ADMM). However, dual decompostion needs very strong conditions for convergence, and the convergence speed is often slow in practice due to improper tunings of stepsizes for gradient update of the dual problem. As for standard Gauss-Seidel ADMM (Boyd, Parikh, Chu, Peleato, & Eckstein, 2011), it uses a sequential update scheme between agents instead a parallel one, which can be more efficiently implemented. We refer interested readers to (Scattolini, 2009) and (Christofides, Scattolini, de la Peña, & Liu, 2013) for more recent work on DMPC.

In this study, a new DMPC method using Proximal Jacobian ADMM (PJ-ADMM), a variant of the traditional ADMM is proposed. PJ-ADMM is based on primal-dual optimization scheme, and it incorporates the parallel update scheme from dual decomposition while at the same time shares robust convergence behavior as Gauss-Seidel ADMM.

4.1 Distributed Primal-Dual Optimization Scheme

First, we concatenate decision variables $\mathbf{x}$ and $\mathbf{Q}$ into a single variable $\mathbf{z}$, and rewrite optimization problem (8) as

$$\underset{\mathbf{z}}{\text{minimize}} \quad J = \sum_{i=1}^{3} f_i^{\mathsf{T}} \mathbf{z}_i \tag{12}$$

$$\text{subject to} \quad \mathbf{A}\mathbf{z} = \mathbf{b}, \tag{13}$$

$$\mathbf{z}_{min,i} \le \mathbf{z}_i \le \mathbf{z}_{max,i}, \quad i = 1, \dots, 3, \tag{14}$$

where $\mathbf{z} = [\mathbf{x}^{\mathsf{T}}, \mathbf{Q}^{\mathsf{T}}]^{\mathsf{T}}$, and

$$f_i = \begin{bmatrix} 0^{\mathsf{T}} & \tilde{f}_i^{\mathsf{T}} \end{bmatrix}^{\mathsf{T}}, \quad \mathbf{A} = \begin{bmatrix} \mathbf{I} & -\mathbf{\Phi} \end{bmatrix}, \quad \mathbf{b} = \mathbf{\Omega} x(0) + \mathbf{\Psi}\mathbf{w}$$

Notice that, the elements in $\mathbf{z}$ can be re-ordered as $\mathbf{z} = [\mathbf{z}_1^{\mathsf{T}}, \mathbf{z}_2^{\mathsf{T}}, \mathbf{z}_3^{\mathsf{T}}]^{\mathsf{T}}$, where $\mathbf{z}_i = [\mathbf{x}_i^{\mathsf{T}}, \mathbf{Q}_i^{\mathsf{T}}]^{\mathsf{T}}$ is the decision variables related to zone i. In the following development, we want to design a parallel update protocol of $\mathbf{z}_i$, implemented by the computing agent of RTU i. In this regard, constraint (14) can be thought of as the local constraint for zone i and RTU i, while constraint (13) is a shared constraint across zones.

For optimization problem (12), an augmented Lagrangian function is formulated as,

$$\begin{aligned}
\mathcal{L}_\rho(\mathbf{z}, \lambda) &= \sum_{i=1}^{3} f_i^{\mathsf{T}} \mathbf{z}_i + \lambda^{\mathsf{T}}(\mathbf{A}\mathbf{z} - \mathbf{b}) + \frac{\rho}{2}\|\mathbf{A}\mathbf{z} - \mathbf{b}\|^2 \\
&= \sum_{i=1}^{3} f_i^{\mathsf{T}} \mathbf{z}_i + \lambda^{\mathsf{T}}\Big(\sum_{i=1}^{3} \mathbf{A}_i \mathbf{z}_i - \mathbf{b}\Big) + \frac{\rho}{2}\Big\|\sum_{i=1}^{3} \mathbf{A}_i \mathbf{z}_i - \mathbf{b}\Big\|^2 \\
&= \sum_{i=1}^{3} \mathcal{L}_i(\mathbf{z}_i, \lambda) - \lambda^{\mathsf{T}}\mathbf{b} + \varphi(\mathbf{z})
\end{aligned}$$

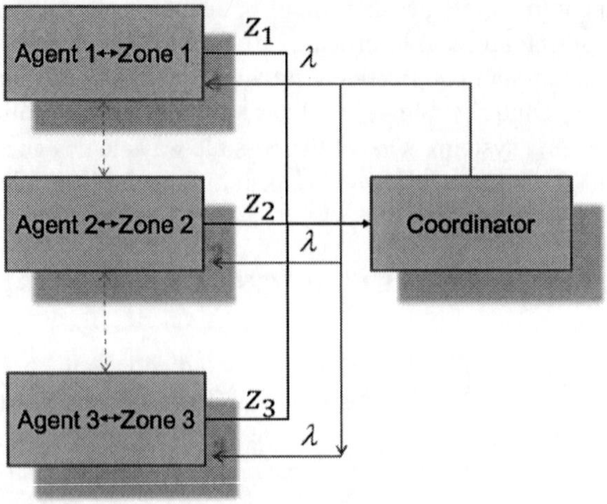

Figure 3: Distributed primal-dual optimization scheme: information structure

where λ is the lagrange multiplier, or dual variable, $\rho > 0$ is a penalty parameter, $\mathcal{L}_i(\mathbf{z}_i, \lambda) = f_i^{\mathsf{T}}\mathbf{z}_i + \lambda^{\mathsf{T}}\mathbf{A}_i\mathbf{z}_i$ and $\varphi(\mathbf{z}) = \frac{\rho}{2}\|\sum_{i=1}^{3}\mathbf{A}_i\mathbf{z}_i - \mathbf{b}\|^2$. Notice that $\mathbf{A}_i$ is the columns of $\mathbf{A}$ that correspond to the elements of $\mathbf{z}_i$. The dual function is obtained by minimizing the Lagrangian function with respect to the primal variable $\mathbf{z}$,

$$\begin{aligned} d(\lambda) &= \inf_{\mathbf{z}} \mathcal{L}_\rho(\mathbf{z}, \lambda) \\ &= \sum_{i=1}^{3}\left(\inf_{\mathbf{z}_i \in \mathcal{Z}_{Ni}} \mathcal{L}_i(\mathbf{z}_i, \lambda)\right) - \lambda^{\mathsf{T}}\mathbf{b} + \varphi(\mathbf{z}), \end{aligned}$$

and the corresponding dual problem is

$$\underset{\lambda}{\text{maximize}}\ d(\lambda)$$

4.2 Algorithm

The proposed PJ-ADMM based DMPC algorithm is an iterative algorithm that contains two steps in each iteration: primal optimization and dual update. The primal optimization step will be carried out by local (zone-level) agent in parallel, and each agent will do a best-response coordinate descent for the Lagrangian function with respect to only part of the decision variables, namely $\mathbf{z}_i$. Dual update will be carried out by a "coordinator", and the objective of dual update is to maximize the dual function $d(\lambda)$. Interestingly, one can also think of the dual variable as an "enforcer" that facilitates the satisfaction of the shared constraint (13).

A diagram illustrating the information structure is given in Figure 3. The overall structure of the proposed algorithm is given in Algorithm 1.

Notice that in Algorithm 1, a proximal term $\frac{\phi_i}{2}\|\mathbf{z}_i - \mathbf{z}_i^{\nu}\|^2$ is added to regularize each agent's subproblem, where $\mathbf{z}_i^{\nu}$ is the value of $\mathbf{z}_i$ from the previous iteration. Intuitively, the larger ϕ_i is, the slower $\mathbf{z}_i$ should be moving away from its previous value (or equivalently, the less selfishly agent i responds to other agents' actions). With this extra term, the global convergence of the algorithm could be established.

Theorem 1. *(Global Convergence Theorem) Suppose the parameters in Algorithm 1 satisfy that* $\rho > 0$, $0 < \gamma < 2$, *and*

$$\phi_i > \frac{1+\gamma}{2-\gamma} \cdot \rho\|\mathbf{A}_i\|^2, \quad i = 1,2,3,$$

then the sequence $\{\mathcal{S}^{\nu} = (\mathbf{z}_1^{\nu}, \mathbf{z}_2^{\nu}, \mathbf{z}_3^{\nu}, \lambda^{\nu})\}$ *generated by Algorithm 1 converges to a solution* $\mathcal{S}^*$ *of problem* (12).

Proof. See (Hou, Xiao, Cai, Hu, & Braun, 2016). Algorithm 1 is a special case of Theorem 1 therein. □

Theorem 1 provides us the theoretical guarantee of algorithmic convergence. However, in practice, one often should expect some fine tuning of parameters for a better convergence speed.

Algorithm 1 Distributed MPC via Proximal Jacobian ADMM

1: Initialize $(\mathbf{z}^0, \lambda^0, \mu^0)$, set $v = 0$;
2: **repeat**
3: Update $\mathbf{z}_i$ (in parallel) according to

$$\begin{cases} \mathbf{z}_1^{v+1} = \arg\min_{\mathbf{z}_{min,1} \leq \mathbf{z}_1 \leq \mathbf{z}_{max,1}} \left(\mathcal{L}_1(\mathbf{z}_1, \lambda^v) + \frac{\phi_1}{2}\|\mathbf{z}_1 - \mathbf{z}_1^v\|^2 + \varphi(\mathbf{z}_1, \mathbf{z}_2^v, \mathbf{z}_3^v)\right); \\ \mathbf{z}_2^{v+1} = \arg\min_{\mathbf{z}_{min,2} \leq \mathbf{z}_2 \leq \mathbf{z}_{max,2}} \left(\mathcal{L}_2(\mathbf{z}_2, \lambda^v) + \frac{\phi_2}{2}\|\mathbf{z}_2 - \mathbf{z}_2^v\|^2 + \varphi(\mathbf{z}_1^v, \mathbf{z}_2, \mathbf{z}_3^v)\right); \\ \mathbf{z}_3^{v+1} = \arg\min_{\mathbf{z}_{min,3} \leq \mathbf{z}_3 \leq \mathbf{z}_{max,3}} \left(\mathcal{L}_3(\mathbf{z}_3, \lambda^v) + \frac{\phi_3}{2}\|\mathbf{z}_3 - \mathbf{z}_3^v\|^2 + \varphi(\mathbf{z}_1^v, \mathbf{z}_2^v, \mathbf{z}_3)\right) \end{cases} \quad (15)$$

4: Update λ according to $\lambda^{v+1} = \lambda^v + \gamma\rho(\mathbf{A}\mathbf{z}^{v+1} - \mathbf{b})$;
5: $v \leftarrow v + 1$;
6: **until** some stopping criterion is satisfied.

4.3 Extention to a More General Setting

This subsection discusses some possible extensions of our problem formulation in (4). All of these extensions could still be handled by the DMPC algorithm proposed in this paper, and may lead to more applications other than the multiple thermal zone coordination problem.

The first possible extension is on the objective function. Although in this case study, the objective function is approximated by a linear function, the distributed control strategies is applicable to general convex HVAC cost functions. Namely, the only requirement on function $P_i(\cdot)$ in (4) is convexity (we do not require strong convexity).

In addition to the shared equality constraint in the form of $\sum_{i=1}^{L} \mathbf{A}_i\mathbf{z}_i = \mathbf{b}$, where L is the number of subproblems, we can also deal with shared inequality constraint in the form of $\sum_{i=1}^{L} \mathbf{D}_i\mathbf{z}_i \leq \mathbf{e}$. Many "shared resource" kind of constraints can be modelled by this form. For instance, in a building where one central air handling unit (AHU) serves multiple zones through VAVs, the total amount of sensible cooling provided to all zones should be smaller or equal to the capacity of the AHU. Another example is demand response management. Assuming $\mathbf{z}_i$ is the load of a single building in a building cluster, we can use this type of shared inequality constraint to enforce an upperbound for the demand of the whole cluster.

In the situation where one also has the shared inequality constraint in the form of $\sum_{i=1}^{L} \mathbf{D}_i\mathbf{z}_i \leq \mathbf{e}$, the Lagrangian function should be modified by adding a corresponding penalty term (with an extra dual variable μ) for the violation of the shared inequality constraint. Details for this formulation can be found in (Hou et al., 2016).

5. SIMULATION RESULTS

In the following simulations, the proposed DMPC with 24-hour prediction horizon is implemented and the sampling time is 30 minutes. The thermal comfort interval for all zones are set to be [21.5, 23.5]°C during occupied hours (9am to 17pm) and [20.5, 24.5]°C during unoccupied hours (17pm to 9am). A seven day warm up period is implemented to obtain the intitail states. Actual weather measurements from May 2015 are used in the simulations and perfect weather prediction is assumed in the DMPC optimization. Electricity price is assumed to be $0.1/kw. Optimization problems are numerically solved using CVX (Grant & Boyd, 2014) and the SDPT3 solver (Toh, Todd, & Tütüncü, 1999).

To evaluate the energy saving potential of our proposed optimal coordination formulation, a baseline control strategy is implemented for comparison. The baseline control is based on simple feedbacks on the zone temperatures. Whenever the temperature in zone i is about to go above the pre-specified comfort interval, RTU i turns on to maintain it at its upperbound. This greedy control strategy is not expected to explore inter-zonal coordination.

Table 1: Electricity costs comparison

Control Strategy	Baseline	Optimal Coordination with DMPC	**Saving**
Electricity Cost ($)	55.95	48.49	13.3%

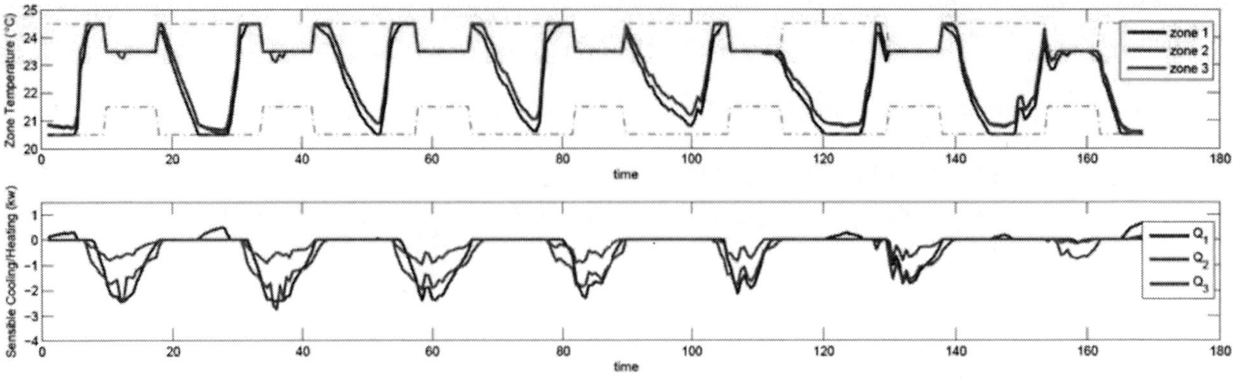

Figure 4: Baseline control

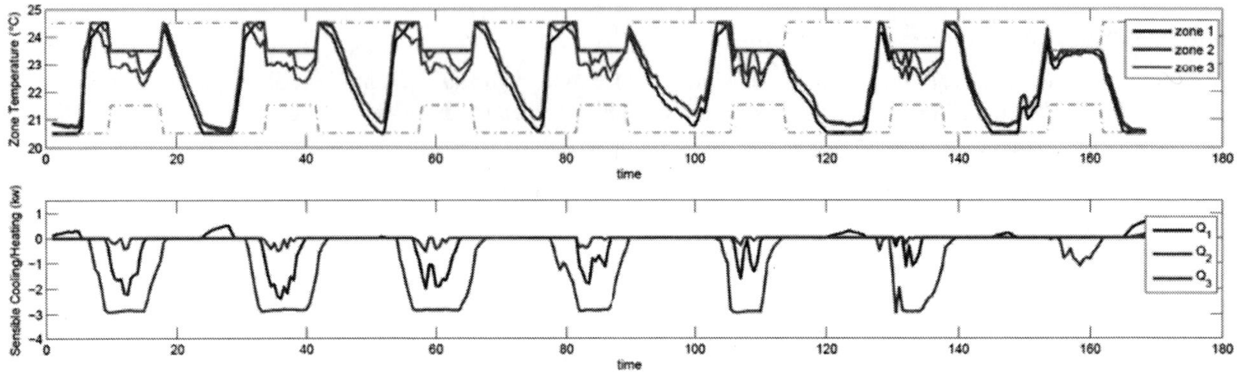

Figure 5: Optimal coordination with proposed DMPC

Simulation results with the proposed DMPC are plotted in Figure 5 and the baseline results are plotted in Figure 4. The total electricity costs of the two controllers are given in Table 1. Comparing these results, we can make the following observations and comments.

- As expected, the double façade adjacent zone 1 has the highest load during most of the days simulated. All three zone temperatures are well maintained in the pre-specified comfort interval.
- In the baseline control, each RTU only cares about the corresponding zone temperature it is "looking after", but does not take the thermal interactions between zones into consideration.
- In the optimal coordination formulation with DMPC control, the most efficient unit, RTU 2, is more heavily utilized during the peak load hours. Through direct air exchange between zones, RTU 2 over cools zone 2 to help out RTU 1 and 3, whose loads are reduced compared to the baseline case. This can also be verified from the fact that the setpoint temperature for zone 2 is noticeably lower than the comfort upperbound.
- It is confirmed that the simulation results with the proposed DMPC is almost identical to the results obtained with a centralized controller.

Some other issues that could be explored but omitted due to space limitation include,

- Time-of-use price. Some preliminary results have already been obtained considering time-of-use price. The proposed coordination formulation with DMPC control is capable of responding to the higher peak hour electricity price by pre-cooling the room at off-peak hours.
- Local comfort delivery. Instead of using a uniform zone temperature constraints for all zones, customized comfort intervals based on local occupants' preferences could easily be included in our formulation.
- Demand response. (Cai, Braun, Kim, & Hu, 2016) proposed a novel multi-agent based demand reduction strategy. A similar method could also be incorporated in our formulation.

6. CONCLUSIONS

This paper investigated a case study for the optimal coordination of multiple thermal zones in a large open space. A general distributed MPC method based on Proxiaml Jacobian ADMM is proposed to solve the optimal control problem. The proposed DMPC algorithm can not only deal with the current case study, but also a family of other building control problems. One possible future direction is a more sophisticated case study that includes a more realistic envelope model and at the same time investigates demand charge reduction.

REFERENCES

Boyd, S., Parikh, N., Chu, E., Peleato, B., & Eckstein, J. (2011). Distributed optimization and statistical learning via the alternating direction method of multipliers. *Foundations and Trends® in Machine Learning*, *3*(1), 1–122.

Cai, J., Braun, J. E., Kim, D., & Hu, J. (2016). A multi-agent control based demand response strategy for multi-zone buildings. In *American control conf.* (Accepted)

Christofides, P. D., Scattolini, R., de la Peña, D. M., & Liu, J. (2013). Distributed model predictive control: A tutorial review and future research directions. *Computers & Chemical Engineering*, *51*, 21–41.

Farokhi, F., Shames, I., & Johansson, K. H. (2014). Distributed mpc via dual decomposition and alternative direction method of multipliers. In *Distributed model predictive control made easy* (pp. 115–131). Springer.

Grant, M., & Boyd, S. (2014, March). *CVX: Matlab software for disciplined convex programming, version 2.1.* `http://cvxr.com/cvx`.

Hou, X., Xiao, Y., Cai, J., Hu, J., & Braun, J. E. (2016). Distributed model predictive control via proximal jacobian alternating direction method of multipliers. In *Proc. 55th ieee int. conf. decision and control.* (Submitted)

Ma, Y., Borrelli, F., Hencey, B., Coffey, B., Bengea, S., & Haves, P. (2012). Model predictive control for the operation of building cooling systems. *Control Systems Technology, IEEE Transactions on*, *20*(3), 796–803.

Maestre, J., De La Pena, D. M., Camacho, E., & Alamo, T. (2011). Distributed model predictive control based on agent negotiation. *Journal of Process Control*, *21*(5), 685–697.

McQuade, J. (2009). A system approach to high performance buildings. *United Technologies Corporation, Tech. Rep.*

Mossolly, M., Ghali, K., & Ghaddar, N. (2009). Optimal control strategy for a multi-zone air conditioning system using a genetic algorithm. *Energy*, *34*(1), 58–66.

Oldewurtel, F., Parisio, A., Jones, C. N., Gyalistras, D., Gwerder, M., Stauch, V., ... Morari, M. (2012). Use of model predictive control and weather forecasts for energy efficient building climate control. *Energy and Buildings*, *45*, 15–27.

Privara, S., Široký, J., Ferkl, L., & Cigler, J. (2011). Model predictive control of a building heating system: The first experience. *Energy and Buildings*, *43*(2), 564–572.

Scattolini, R. (2009). Architectures for distributed and hierarchical model predictive control–a review. *Journal of Process Control*, *19*(5), 723–731.

Toh, K.-C., Todd, M. J., & Tütüncü, R. H. (1999). Sdpt3—a matlab software package for semidefinite programming, version 1.3. *Optimization methods and software*, *11*(1-4), 545–581.

Wakasa, Y., Arakawa, M., Tanaka, K., & Akashi, T. (2008). Decentralized model predictive control via dual decomposition. In *Decision and control, 2008. cdc 2008. 47th ieee conference on* (pp. 381–386).

Yang, R., & Wang, L. (2013). Multi-zone building energy management using intelligent control and optimization. *Sustainable cities and society*, *6*, 16–21.

ACKNOWLEDGMENT

This work was supported by the National Science Foundation under Grant No. 1329875.

Financial Analysis of University Investments in Solar Photovoltaics

William Arnett[1]*, Praneet Arshi[2], William Hutzel[3]*, Brittany Newell[4], John Piller[5]

School of Engineering Technology, Purdue University
West Lafayette, IN, United States of America

[1]arnettw@purdue.edu
[2]parshi@purdue.edu
[3]hutzelw@purdue.edu
[4]bnewell1@purdue.edu
[5]jpiller@purdue.edu

* Corresponding Author

ABSTRACT

The global demand for electrical energy is increasing as a result of population growth and a higher standard of living that is enjoyed by many people. However, the availability of electricity is often limited by fuel supplies and/or infrastructure for generating and distributing power. In addition, the looming threat of greenhouse gas emissions and the collateral damage to the environment has encouraged efforts to diversify methods of electricity production. These factors have led to the increased use of renewable energy, particularly solar and wind, to help meet the demand for energy. The shift towards solar energy has been accelerating due to the decreasing cost of materials and installation.

A project funded by the U.S Department of Energy's Sunshot Initiative is looking at the economics of solar energy for universities. In particular the project is looking at whether university endowments, which seek to make a long term financial return to the university, can view solar electricity as an investment opportunity. This financial incentive, along with the mission of a university to showcase new technologies, could be helpful for deploying more utility-scale solar electricity at university campuses. The university that is the case study for this research enjoys exceptionally low cost for electricity due to its on-campus combined heat power plant. This low cost of electricity generation makes it difficult to justify utility scale solar energy unless creative financing strategies are used.

1. INTRODUCTION

This research is a case study of the financing options and factors for implementing photovoltaics at a university in Indiana. This university offers a unique set of challenges for financing due to the low rate that is paid for the electricity that is used because of the on-campus Combined Heat Power plant (CHP). To justify the cost of solar photovoltaic electricity, the cost of electricity ultimately has to be designed around the total production value, matching the timing of production with the peak rates seen in order to minimize the payback of the investment. (Energy Information Administration, 2015). In an ideal situation, the demand for electricity would peak at the same time that a solar photovoltaic array generates its peak output.

Peak electrical demand and output typically occurs in the afternoon, according to the university's Energy Office. This peak develops when the campus is active and the sun is directly overhead. Following the economic rules of supply and demand, the value of this peak electricity should be higher than electricity used during periods when less energy is being used. This research focuses on 1) understanding how electricity is used on a university campus and 2) evaluating the available financing options to determine which option is most compatible with the university's operating strategy. This research can be a template for other universities that are considering similar renewable energy installations. While each university will have varying criteria, the methodology that is presented can help others understand the financing considerations. It is expected that universities will have different priorities and thus different financing strategies will be used.

2. HISTORICAL PV COST TRENDS

Figure 1 is a graph developed by Lawrence Berkeley National Laboratory in their annual report called "Tracking the Sun" that illustrates the decreasing cost of photovoltaics (PV), while also showing the economy of scale. The independent (horizontal) axis is the year of PV installation and the dependent (vertical) axis is the project cost in $/Watt using 2014 as the base year. The graph clearly shows that as technology progresses the cost of manufacturing photovoltaic panels decreases. The installation cost of PVs has also decreased, which plays a significant role in the financial feasibility of implementing photovoltaics. This decreased cost is also complimented by the economy of scale, which makes the installation of a larger system relatively cheaper on a unit cost basis.

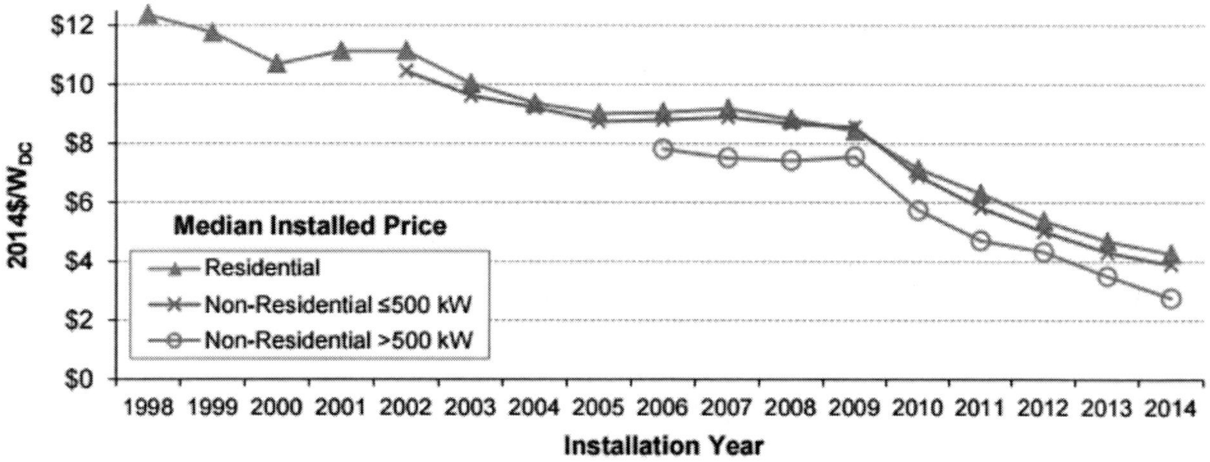

Figure 1: The installed cost of solar PV is trending downward. (Feldman *et al.,* 2014)

The two non-residential trends from Figure 1 are the key aspects to consider. A larger system size, signified by the blue line in Figure 1, costs less than a smaller system size (red line) for any given year. A larger size of installation means a larger total cost but a lower unit electricity cost, which increases the Return on Investment (ROI) for the system. Figure 1 indicates that any system greater than 500 kW is considered utility scale, but larger systems achieve greater economies of scale and thus are easier to finance.

For the case study in question, a 5 MW photovoltaic system is being considered. Local photovoltaic contractors in the state of Indiana estimate that the current installation cost of a 5 MW system would be approximately $1.60-$1.70/W. As a point of comparison, the Sunshot Initiative that is managed by the U.S. Department of Energy's National Renewable Energy Lab estimates that when solar PV achieves an installed price of $1/Watt it will have reached grid parity with other sources of electricity. If the trend of decreasing prices continues according to Figure 1, utility scale PV could be financially competitive with traditional fossil fuels in the next 5 to 10 years. Thus a solar installation that is not quite cost competitive today can still be viewed as a smart investment over its 25 year life span as a hedge against the steadily increasing cost of electricity from traditional fuel sources.

3. ELECTRIC RATES

The basic premise for net metering sets the value of electricity generated by a solar photovoltaic array equal to the retail rate provided by the local electric utility. Although this seems like an obvious assumption, recent actions by electric utilities in the U.S. have cast doubt on this approach. Legislators in Indiana recently considered a bill that would reduce the net metering rates paid for solar electricity because of its intermittent nature and because local solar installations do not contribute to the costs of operating and maintaining the distribution/transmission grid that is provided by the local electric utility.

The university in this case study owns a combined heat and power (CHP) plant, which provides approximately 1/3 of the total campus electricity annually. Due to the university owned CHP, electricity is generated and purchased at a low rate. The university also has an agreement with the local utility so that they can make flexible economic decisions on when to increase production, or purchase more energy. This drives the average price down to \$0.06/kWh, which is what the university uses to determine the payback of energy conservation and improvement measures (Hutzel *et al.*, 2015).

Figure 2 compares the national and state average electrical rates to the cost of electricity at the university with a CHP. The vertical axis is the cost of generation in cents per kWh. The national average (blue) is approximately 13 cents per kWh. The cost of generation in Indiana (red) is about 11 cents per kWh due to an abundance of low cost coal. The university's electrical average cost of generation (green) is less than half of the national average. This low rate, approximately 6 cents per kWh, does not imply that a solar project cannot be justified financially. The low value only implies that the return on investment and payback period will take longer and emphasizes the need for innovative financing to make the project profitable.

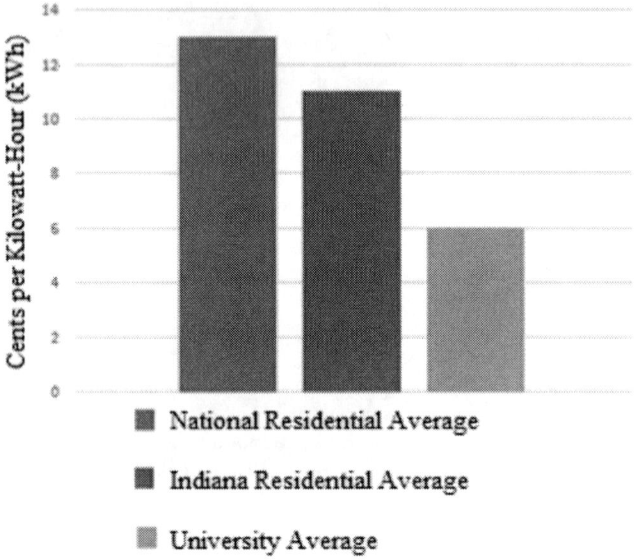

Figure 2: Comparing U.S. electric rates (Landry, 2015)

4. ELECTRICITY CONSUMPTION ANALYSIS

Understanding how photovoltaics integrates in the university's electric consumption is crucial to quantifying the financial impact. This case study assumes that the utility-scale solar energy system is located on or near the campus so that all of the energy is consumed locally by the university. This "inside the fence" option is easier to implement because it does not require the negotiation of a power purchase agreement with the local electric utility. It has the additional advantage of providing a direct hedge against future cost increases for energy. The operators of the power plant also like the "inside the fence" option because two local sources of electricity (CHP and solar) potentially offers more flexibility in terms of scheduling down time for maintenance at the power plant.

This case study considered a 5 MW solar energy system, which would provide up to 3% of the university's annual consumption in 2014. Part of this selection was based on the physical size of the solar installation. 1 MW of solar electricity requires approximately 6 acres of land in the Midwest (Ong *et al., 2013*). Scaling up this size shows that 5 MW of solar would take approximately 30 acres of land. Finding the physical space for anything larger than 30 acres close to the university campus would be difficult if not impossible. The ideal location is one that is close to an existing electric distribution network for easy grid connectivity and also on land that cannot be developed for other purposes, such as the restricted space surrounding the runway of an airport.

The university in this case study acquires electricity in three ways. One method is purchasing electricity at a "base" rate. A base value is a set amount of electricity purchased a year in advance by the university from the local utility company. This value is negotiated between the university's energy office and the local utility. The fact that the university has the ability to generate its own electricity enhances its ability to negotiate low costs with the utility.

Another way of getting electricity is purchasing it at a Real Time Pricing (RTP) rate. RTP is a rate that changes every hour and the price is determined a day in advance. The rate varies over the course of a day and also varies due to weather. The RTP price is generally higher during times of peak demand, like during hot weather. The RTP values are provided to the university power plant operators, who then make the decision to generate or purchase electricity based on whatever option is the cheapest.

Generating electricity through the university's CHP is the third method for meeting campus electricity demands. The university's power plant produces not only electricity, but is also the main supply of heating and cooling for the campus. Some amount of electricity is always being generated as a by-product of producing heating or cooling. Any additional electricity generation is based off of the RTP prices from the electric utility. Each day when the RTP prices are provided, the university energy office determines if it's cheaper to purchase or generate electricity for each hour of the next day.

Figure 3 shows the interplay of the three main electrical sources for 2014, with a solar component added. The horizontal axis represents each day of the 365 day calendar year. The vertical axis shows the sources of electricity for each day expressed as a percentage. Each day of electricity consumption is represented as a combination of base, RTP, or CHP. Solar energy (yellow) for a 5 MW system is added into the graph to show its potential contribution. On any given day, the contribution of RTP, CHP, Base, and Solar adds to 100%.

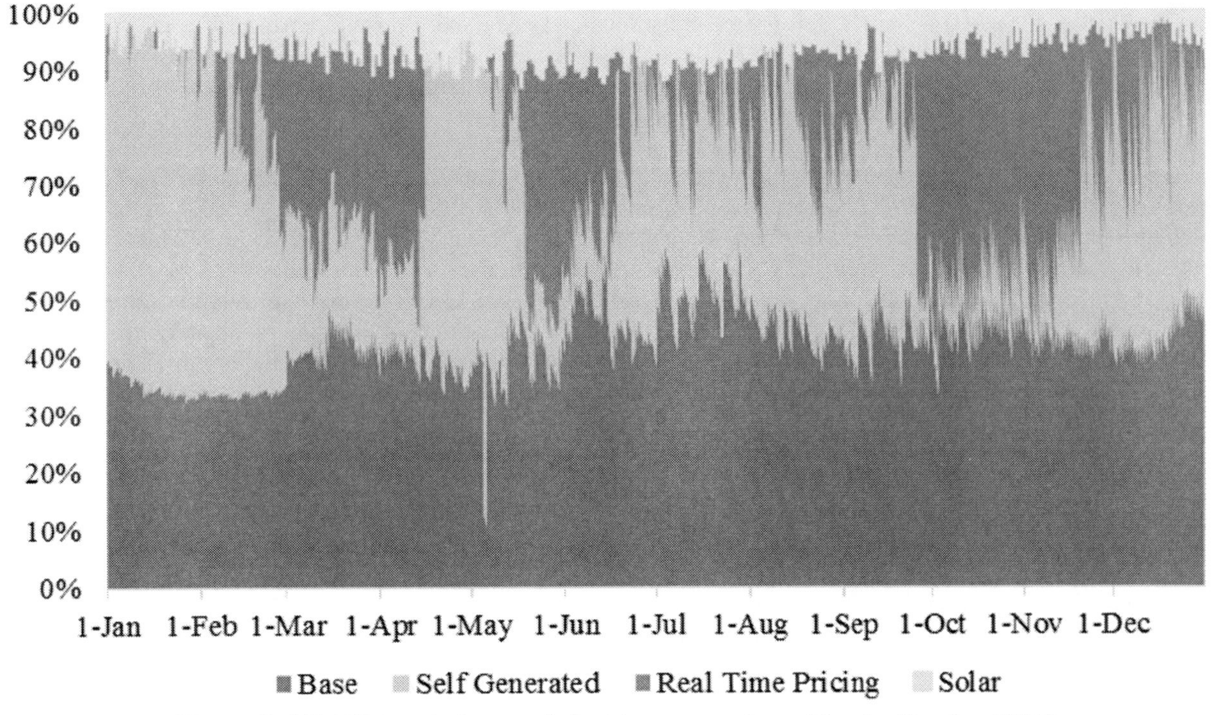

Figure 3: 2014 University's annual electric consumption with solar (Landry, 2015)

Figure 3 illustrates how the university makes decisions about energy generation. The impact of the very cold weather associated with the polar vortex that occurred in January of 2014 can be easily seen. During this time period the RTP price for electricity was quite high. As a result the university opted to self-generate, which is shown by the large amount of CHP (green) power in early January. The weather had moderated by March, so the price signal for RTP dropped by a corresponding amount. As a result the university purchased more power, which is shown by the large amount of RTP (orange) energy in April and May.

Figure 3 also shows the impact of a 5 MW array in the context of the university's energy consumption. This is illustrated by the yellow band at the top of the figure. The band of solar is wider in the summer, because the long summer days allow for the solar array to supply a greater percentage of the electricity consumed. The yellow band is narrower from November through February because of the relatively short days and lack of sunshine.

Figure 3 also clearly shows the intermittent nature of solar energy because the yellow band in not smooth, but instead has peaks and valleys due to daily weather related anomalies. This simulated 5 MW solar array would provide roughly 3% of the university's total electricity requirements. The solar energy predicted is based off scaling a small PV array located on a university building. Energy production values were collected for a year, and then scaled to a 5 MW system size. Degradation was accounted during these calculations to normalize the values (Landry, 2015).

5. FINANCING OPTIONS AND MECHANICS

One of the major criteria in evaluating photovoltaic implementation is project financing. There are three potential methods for the university to invest in a PV system, and four decision factors that accompany each method. Each method provides a unique set of benefits and concerns, which can be weighted differently based on the priorities of the university. The following narrative explains each option in greater detail.

University owned is one of the most basic financing options. Under this scenario, the university would directly purchase the solar array using the investment capital in its endowment. This approach has the advantage of putting the university in full control of the operation of the array, including ownership of the Renewable Energy Credits (RECs) that accumulate to any green source of electricity. All energy produced by the system would be used by the university to reduce electric purchases from the local utility, or be sold to the local utility through a negotiated power purchasing agreement (PPA) depending on grid interconnect location. The main drawback to this approach is the perceived risk of the investment in terms of whether the solar array would live up to its performance expectations over its 25 year life span. There is additional uncertainty as to whether this type of investment could live up to the ROI expectations from the university's investment office, particularly since the installation may not automatically qualify for the 30% solar tax credit that is available to entities with a tax appetite.

Third party owned is another financing option where the solar array is funded by a university alumni or other generous benefactor. The university provides land near a utility interconnect to the third party, with the value of the land being calculated into the power purchasing agreement. Here the third party could take advantage of the federal tax incentives and rapid depreciation for solar projects, but still sell the electricity back to the university at a low rate. By largely eliminating the upfront cost to the university, the third party PPA could be set to a lower value while still maintaining a positive net present value (NPV) that benefits both groups. This opportunity offers a unique appeal since a generous alumni would be less motivated by profit, and more by promoting renewable energies and the image of the university. All things being equal, this option would likely deliver the greatest financial benefit to the university.

Land lease is a financing option where the university leases land to the local utility and then the utility constructs the PV system. This method is attractive because it has minimal risk to the university. The utility is responsible for all expenses and costs due to equipment malfunctions. With this structure the utility would own the system, making them accountable for all operation and maintenance costs. These expenses would be calculated into the land lease arrangement. One negative side effect is that the utility would also own the RECs generated by the solar array, limiting the university's ability to claim responsibility for reducing greenhouse gas emissions. While this financing option offers the lowest financial returns, the revenue generated can be used wisely. Some potential uses for the land leasing revenue include using the funds to establish additional renewable energy systems, purchase other low cost RECs to benefit sustainably and to help keep tuition costs from increasing.

Once the preferred financing options were identified each one was valued based on the operating priorities of the university. Four types of decision factors were presented; financial, risk mitigation, energy security, and sustainability. Table 1 provides a list of each decision factor and values them according to "high", "medium", or "low" values. A "high" value means that the decision factor embodies that attribute to a high degree. For example, a "high" value for financing means that strategy would likely have a high return on investment. On the other hand, "high" risk mitigation means that option carries investing with little or no risk. This type of ranking system helps identify financing options that align closely with the operational goals for the university.

Table 1: Financing options and decision factors

Decision Factors	University Owned	3rd Party (Donor) Owned	Land Leasing – Local Utility
Financial (Payback/ROI)	M	H	L
Risk Mitigation	L	M	H
Energy Security	H	H	L
Sustainability	H	H	M

H=High, M=Medium, L=Low (Levels)

Table 1 demonstrates the rankings for each financing method. With all things weighted equally, the 3rd party option has the best overall ranking because it has three "high" rankings and one "medium" ranking. The findings are not quite that simple, because the decision factors may not be weighted equally. The university considering these methods seems more interested in risk mitigation than financial payback. With this additional consideration, the land leasing option is the most beneficial, 3rd party owned is second, and university owned is last. A detailed list of the decision factors is listed below to better define each term.

5.1 Financial (Payback, ROI)

This decision factor is purely based on financial value generated by the system. The total revenue generated from the system, such as net present value or annual annuity, is what defines this factor.

A land leasing agreement would allow for positive cash flow from year 1, but the lease value will generally have a lower NPV and ROI than the other two options. The donor option would likely have the highest return to the university since financing is actually a form of gift to the university.

5.2 Risk Mitigation

Risk mitigation is a decision factor based off how risky the investment is to the university. This includes which party makes the initial investment and pays all Operations and Maintenance expenses. The ideal, or high value, for this factor would consist of another party, other than the university, being liable for these costs.

For land leasing, the system is owned by the local utility, so the university is not responsible for upfront or maintenance costs. A third party owner, potentially a university alumni, takes on all responsibilities of the system. While this still doesn't affect the university directly, the funding from the donor could potentially be invested in another area, creating a "medium" risk avoidance scenario. The university owned array scenario has the lowest rating since the university would be responsible for any damages and so forth.

5.3 Energy Security

Energy security is defined by the PV system's grid connection. A desired interconnect would be on the university grid, with unrestricted access to the power in times of need.

Depending on the financing method the system could also be tied into the local utility grid. By tying into the university grid, the system would provide a consistent electrical source for the campus, which would create a high energy security value. Any additional electricity not consumed by the university could be sold back to the local utility, through net metering or a PPA.

Both university and 3rd party owned options could potentially be tied into the university's electrical grid, which would offset any RTP or CHP power consumption. These options also have the potential to be tied into the local utility, depending on negotiations and university preference. On the other hand, the land leasing option would be located in the local utilities grid. In addition, the local utility would treat the system as a power generating piece of equipment, and have no PPA with the university.

5.4 Sustainability

The sustainability decision factor relates to the environmental benefits the university could indulge in by owning the Renewable Energy Certificates. For the case study, Renewable Energy Certificates can be beneficial. If the university can own or generate the RECs it will be valued as "high". Since the university owns a combined heat and power plant, there is a basic need/desire for reducing carbon emissions. A REC is proof that an institute is purchasing green energy, since 1 REC = 1 MWh of "green" energy.

As seen in Table 1, both university owned and 3rd party owned financing options have a high sustainability rating. This is because the university would have, at least the option to purchase or generate the RECs and benefit from them. The land leasing option is ranked at a medium value, since the local utility will keep the RECs, but the generated revenue could purchase RECs from another renewable source. The potential RECs for purchase depend on the negotiated land leasing rate, and where the RECs would be purchased.

6. CONCLUSIONS

The results of an analysis of energy consumption patterns and financing options suggest that a 3rd party donor would be the most financially beneficial option. That being said the weight values of each section, such as risk mitigation, is subject to change according to the university's operational priorities. Since the university has not invested in renewable energies of this magnitude, they may value the risk avoidance overall. With that in mind, the land leasing option appears to have the highest probability of implementation.

Future steps for this research include doing an economic analysis on the various financing methods. This will require direct interaction and cooperation from the university's physical facilities, along with external sources such as the local utility and PV contractors. Each method will be reviewed for payback time frame, return on investment, and the net present value.

REFERENCES

Energy Information Administration, "Electric Power Monthly", with Data for April 2015, June 2015, http://www.eia.gov/electricity/monthly/pdf/epm.pdf

Feldman D., G. Barbose, R. Margolis, T. James, S. Weaver, N. Darghouth, R. Fu, C. Davidson, & R. Wiser. (2014). *Photovoltaic System Pricing Trends: Historical, Recent, and Near-Term Projections*. 2014 Edition. Golden, CO: National Renewable Energy Laboratory.

Hutzel, William, Landry, Samuel, Schuster, Daniel, and Lynall, Matthew, (2016), *University investments in solar photovoltaics – the solar endowment*, Proceedings of the 2016 ASHRAE Annual Meeting, Orlando, FL, 8 p.

Landry, S. J. (2015). *Process development for evaluating utility scale solar at a combined heat and power facility* (Order No. 10056389). Available from Dissertations & Theses @ CIC Institutions; ProQuest Dissertations & Theses A&I; ProQuest Dissertations & Theses Global. (1776459167). Retrieved from http://search.proquest.com/docview/1776459167?accountid=13360

Ong S., Campbell C., Denholm P., Margolis R., & Heath G. (2013). *Land-Use Requirements for Solar Power Plants in the United States*. National Renewable Energy Laboratory.

ACKNOWLEDGEMENT

The information, data, and work presented herein was funded in part by the Office of Energy Efficiency and Renewable Energy (EERE), U.S. Department of Energy, under Award Number DE-EE0006910.

The Potential of Liquid-based BIPVT and Ice Storage for High Performance Housing in Canada

Justin TAMASAUSKAS[1]*, Michel POIRIER[1], Radu ZMEUREANU[2], Martin KEGEL[1], Roberto SUNYE[1], Eric MCDONALD[1]

[1]CanmetENERGY, Natural Resources Canada,
Varennes, Quebec, Canada
Tel.:450-652-4822 Email: justin.tamasauskas@canada.ca

[2]Concordia University, Department of Building, Civil and Environmental Engineering,
Montreal, Quebec, Canada

* Corresponding Author

ABSTRACT

This paper presents an innovative combination of liquid-based building integrated photovoltaic and thermal (BIPVT) panels with heat pump and ice thermal storage technologies. The system cools roof-integrated PV panels while simultaneously using the recovered thermal energy to efficiently supply heating and hot water to the building. Performance of the proposed system is examined for high performance housing in Calgary and Montreal. Simulated results show strong energy savings potential, with annual net energy use reductions of 13%-34% in comparison to an air-source heat pump + PV only case. BIPVT technologies increase annual PV electrical production by up to 10%, while the integration of ice storage reduces heating/hot water energy use by approximately 26%.

1. INTRODUCTION

ASHRAE Vision 2020 has defined market viable net-zero energy buildings as a key objective for new construction in North America (ASHRAE, 2008). Designing for this target requires the effective integration of renewable energy generation into the building. However, many buildings have limited roof and façade area to incorporate such technologies, making it difficult to achieve a net zero energy design. Building Integrated Photovoltaic and Thermal (BIPVT) systems could facilitate the greater adoption of net zero energy targets by converting the building envelope into a simultaneous producer of both thermal and electrical energy.

Photovoltaic (PV) electrical efficiencies decrease with higher cell temperatures. BIPVT systems address this issue by cooling PV panels with a heat transfer fluid, and then using recovered heat to supplement building heating and/or hot water systems. Commonly, BIPVT systems in North America use air as a working fluid. While this offers easy integration with building ventilation systems and minimizes potential damage from leaks, air also has a lower thermal capacitance which reduces the amount of thermal energy extracted from a collector. Additionally, there is no simple method of directly storing thermal energy for later use. This becomes an important issue in residential buildings, where there is often a time discrepancy between solar supply and peak building demands.

Liquid based BIPVT (and also non-building integrated PVT) systems offer improved heat transfer characteristics and a simpler integration with existing thermal storage technologies. These systems can generally use two collector types (Zondag, 2008):

i. Glazed Modules. Glazed systems prioritize the recovery of thermal energy by adding an additional glass cover over the PV cells. This reduces thermal losses, allowing glazed configurations to achieve fluid temperatures high enough to directly meet heating or domestic hot water (DHW) loads. However, the additional glass cover and higher fluid temperatures combine to reduce PV electrical efficiencies, and can lead to high stagnation temperatures which damage the PV modules (Dupeyrat *et al.,* 2011).

ii. Unglazed Modules. Unglazed collectors remove the glass covering in an effort to simplify collector design and improve electrical efficiencies. Working fluid temperatures must be kept low in order to minimize thermal losses, especially in cold climates. As such, these systems require the integration of heat pump and thermal storage technologies to store and then deliver recovered thermal energy to the building.

The strong potential of combining low-grade solar thermal energy with heat pump systems has been previously noted (Zondag, 2008). In particular, several authors have proposed the combination of solar collectors, ice storage, and heat pumps as a means of improving energy storage densities and collector thermal efficiencies (Trinkl *et al.* 2009, Tamasauskas *et al.* 2012, Dott & Afjei 2014, Carbonell *et al.* 2016). In these concepts, a cold working fluid is circulated to the solar collectors to reduce losses and improve thermal efficiencies. The low grade thermal energy recovered is then upgraded via a heat pump to meet heating and DHW demands. However, while results are promising, there has been little work to date on the potential of these systems in combination with BIPVT technologies for the Canadian climate, where cold temperature working fluids could have the dual benefit of increasing both the thermal and electrical efficiency.

This paper examines the potential of using liquid based BIPVT panels with ice storage and heat pump technologies to meet the thermal and electrical demands of high performance Canadian homes. Detailed simulation models are developed and used to examine system performance in two Canadian cities. Simulated results are compared to two base cases to quantify the benefits of BIPVT and ice storage technologies. Finally, the performance of key system components is discussed to better understand system operations.

2. DEVELOPMENT OF BASE CASE MODELS

Two base systems were developed to identify the impact of heat pumps and ice storage on annual energy use.

2.1 Housing Model

High performance housing models were developed for Calgary, AB and Montreal, QC. Each climate is summarized below in Table 1 (data from (Pelland *et al.*, 2006) and (CCBFC, 2010)). Heating degree days are based on temperatures below 18°C, while solar potential is based on a surface tilted to the latitude of the location. While both cities experience cold winters, they differ significantly in terms of solar availability, with Calgary amongst the sunniest cities in Canada. It is also important to note that while Calgary has more heating degree days, mean air temperatures from January through March are actually higher than for Montreal. Calgary, however, experiences cooler spring and summer conditions which increases the duration of time spent below 18°C.

Table 1: Selected Climate Characteristics

	Calgary	Montreal
Heating Degree Days	5000	4200
Heat T_{Design} (°C,DB)	-30	-23
Annual Solar (kWh/m^2)	1292	1185

The shell of the home used in this study was based on the Canadian Centre for Housing Technology (CCHT) test home located in Ottawa, Ontario (Swinton *et al.*, 2003). This home is typical of single family Canadian housing, and consists of two above ground floors and a finished basement, with a total heated floor area of 284 m^2. The envelope and systems of the home were modified to represent a high performance design achieving a minimum ERS-86 rating on the EnerGuide Rating Scale (OEE, 2005). Further details on the development of this model can be found in Kegel *et al.* (2012). Table 2 summarizes the key performance characteristics of the building.

Table 2: Housing Characteristics

Roof RSI	8.93 m^2°C/W
Wall RSI	5.65 m^2°C/W
Basement Wall RSI	4.95 m^2°C/W
Basement Slab RSI	2.58 m^2°C/W
Window U-Value	1.01 W/m^2°C
Infiltration	0.75 ACH_{50}
Generation	PV, 40 m^2, 6.0 kW_p

Each home is assumed to be grid-tied, with the local electricity network acting as an unlimited source/sink of electrical energy. PV panels are integrated into the south facing roof of the home.

2.2 Base Mechanical Systems

Two mechanical systems were selected to represent both conventional and more efficient approaches to supplying building heating and DHW. In all cases, the above ground floors are maintained at a minimum 21°C, while the basement is maintained at 18°C. Cooling is met using natural ventilation, with cooling available Apr. 15 to Oct. 15.

Base Case 1

Heating is provided using electric baseboards on each level of the home. DHW loads are met using a conventional DWH tank with electric resistance elements.

Base Case 2

Heating and DHW are met using a cold climate air-water heat pump operating against a warm storage tank (Figure 1). Cold climate heat pumps are specifically designed for colder winter climates like those in Canada, and can operate down to an outdoor air temperature of -25°C while still maintaining a significant portion of heating capacity. In this system, warm water is drawn from the top of the tank to serve radiant flooring on all three levels of the home, while DHW loads are met by passing cold water from the city mains through a vertical coiled heat exchanger internal to the tank. Auxiliary heating (if needed) is provided using electric resistance elements in the floor loops and DHW loop.

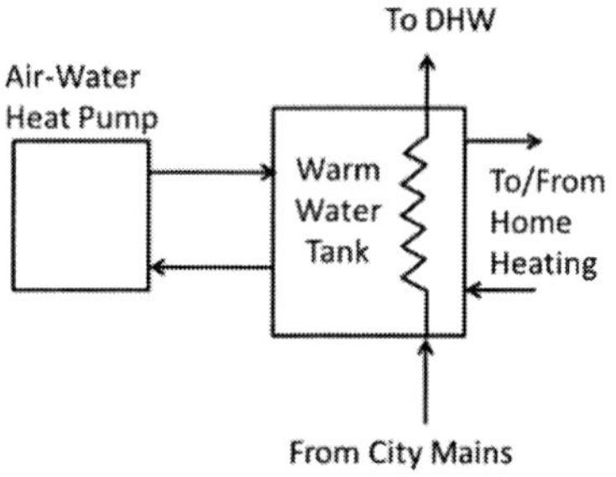

Figure 1: Air-Water Heat Pump Integration

A summary of key system parameters is provided in Table 3. The warm tank volume for Base Case 2 is larger because this tank serves both heating and DHW loops.

Table 3: Base Case Mechanical System Parameters

	Base Case 1	Base Case 2
Heating	Electric baseboards (BB)	Air-Water HP (COP 3.7[1])
DHW	Electric resistance	HP + tankless electric auxiliary (if needed)
Notes	Conventional DHW tank (V=0.23 m^3)	Radiant flooring (Basement, 1st, 2nd) Warm storage tank (V=0.5 m^3)

[1]Rated at 8.3°C outdoor dry bulb, 30°C inlet water temperature

3. PROPOSED SYSTEM CONCEPT AND OPERATIONS

The proposed BIPVT system layout is shown in Figure 2. Priority was given to a simple configuration to reduce system complexity and equipment requirements.

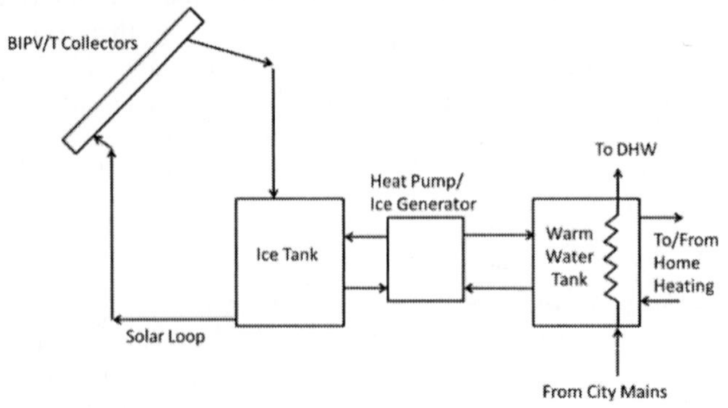

Figure 2: Proposed System Integration

3.1 System Description

The proposed system combines BIPVT panels with a heat pump and ice-based thermal storage. In this study, ice storage is based on the use of ice slurry (a mixture of small diameter ice particles and water) as described in Tamasauskas *et al.*, 2015. BIPVT collectors (both glazed and unglazed are examined) operate in series with the ice tank, improving panel thermal and electrical efficiencies. A water to water heat pump links the ice tank with the warm water storage, upgrading low-temperature fluid obtained from the collectors for use within the building. The warm tank serves as the main heat distribution point for the building. Hot water is drawn from the top of the tank to serve radiant floor heating loops on each level of the home. DHW demands are met by passing cold water from the city mains through a coiled heat exchanger located internal to the warm tank. Auxiliary heaters in the floor loops and DHW loop operate as needed to supplement system operations.

BIPVT panels should be cooled year round to maximize PV electricity generation. During the winter, building loads far outweigh thermal energy gains from the collectors, increasing ice mass in the ice tank and allowing a consistently cold fluid to circulate and cool the BIPVT array. However, this situation is reversed in the summer. Excess thermal gains from the BIPVT array substantially increase fluid temperatures in the ice tank and reduce its available cooling capacity. To counter this, a simple cooling mode is added for the summer months for systems operating with unglazed BIPVT collectors. During the night-time, fluid can circulate to the BIPVT array, where it is cooled via radiative heat exchange with the cold sky. This builds cooling capacity for the following day, allowing PV panels to operate more efficiently.

3.2 System Operations

System operations can be divided into solar, heat pump, and radiant flooring loops.

<u>Solar Loop</u>

Control of the solar loop is based on the combination of three control variables:

i. *BIPVT Cooling*
The solar loop operates only when collector cooling is required. In this study, a predicted initial fluid temperature rise ($\Delta T_{BIPVT,I}$) of 3°C in the winter and 5°C in the summer is required to start the solar loop.

ii. *Ice Tank Temperature*
Ice tank fluid temperatures ($T_{IT,Fluid}$) are limited to avoid exceeding the maximum heat pump evaporator inlet temperature. This limit also ensures that fluid entering the collector will have some cooling effect on the panels. For this study, the maximum temperature was set to 26°C (ClimateMaster, 2010).

iii. *Heat Rejection Potential* (Unglazed collectors, Cooling only)
For the unglazed system only, the night cooling mode is activated during the cooling season (Apr 15 to Oct 15) whenever the BIPVT plate temperature ($T_{BIPVT,PLT}$) falls at least 1°C below the ice tank temperature.

Flow rates in the solar loop are proportional to collector area, and were set to 65 L/h/m^2 (Solimpeks, 2011).

Heat Pump Loop

Heat pump operations are based on two control variables:

i. *Ice Mass*

Heat pump operations are limited by the maximum ice capacity of the ice storage tank. For this study, the maximum ice mass ($M_{IT,Max}$) is set to 60% of the tank fluid mass, based on experimental results derived from a test bench (Tamasauskas *et al.*, 2015).

ii. *Warm Tank Temperature*

The heat pump operates to maintain a warm tank fluid temperature ($T_{WT,Fluid}$) of 45°C at the top of the tank. This value is selected to meet a significant portion of both space heating and DHW demand.

Radiant Floor Loop

Radiant flooring loops in the basement, first, and second floors are used to meet space heating requirements. The flow rate through each loop is varied using a PID controller to maintain an air temperature of 21°C on the two above ground floors, and 18°C in the basement. Supply temperatures were varied using an outdoor air temperature reset.

Loop controls are summarized by mode/season below in Table 4

Table 4: Control Summary for BIPVT Cases

	Mode	Control Summary
Solar Loop	Heating Season	$T_{IT,Fluid}$<26°C, and $\Delta T_{BIPVT,I}$>3°C
	Day, Cooling Season	$T_{IT,Fluid}$<26°C, and $\Delta T_{BIPVT,I}$>5°C
	Night, Cooling Season	$T_{IT,Fluid}-T_{BIPVT,PLT}$>1°C
Heat Pump	All Seasons	$M_{Ice}<M_{Ice,Max}$, $T_{WT,Fluid}$<45°C
Radiant Floor	All Seasons	$T_{RF}=-0.3659T_{Out}+31.59$, 25°C≤$T_{RF}$≤ 40°C

4. SIMULATION METHODOLOGY

TRNSYS (Klein *et al.*, 2010) is used to model the selected systems because of the complexity and range of component models required. A time step of 3.75 minutes is selected to properly represent system control decisions. Systems are simulated with a one month warm up period to reduce the impact of assumed initial conditions.

4.1 Summary of Simulation Components

For all cases, Type 56 was used to develop a detailed model of the building based on parameters outlined in Table 2.

Key HVAC simulation components for each base case are provided in Table 5. Storage tank volumes are based on Kegel *et al.* (2012), Tamasauskas *et al.* (2015) and mechanical space considerations. Heat pump performance is based on manufacturer data (Ecologix, 2016).

Table 5: Key Simulation Components for Base Cases

	Component	TRNSYS Type	Notes
Base Case 1	Hot Water Tank	Type 4	V=0.23 m^3
	Baseboards (BB)	Equation Type	Separate BB for each level
	Heating Control	Type 23	Separate BB PID control for each level
Base Case 2	Heat Pump	Type 941	4 ton, COP_{Rated}=3.70[1]
	Hot Water Tank	Type 534	V=0.5 m^3, 4 nodes
	Heating Control	Type 23	Separate PID control for each RF Loop
	Radiant Flooring	Type 56 Active Layer	Separate loops for all three levels
Common	PV	Type 562	A_{Panel}=1.37 m^2/panel, N_{Panels}=29

[1]Rated at 8.3°C outdoor dry bulb, 30°C inlet water temperature

Table 6 summarizes selected HVAC components for the BIPVT analysis. Heat pump data and tank volumes are derived from an experimental test bench (Tamasauskas *et al.*, 2015). Pump performance is obtained from manufacturer data (Wilo, 2016).

Table 6: HVAC Simulation Components for BIPVT Cases

Component	TRNSYS Type	Notes
Heat Pump	Type 927	4 ton, COP_{Rated}=4.28[1]
Ice Generator	Equation Type	Derived from Guilpart & Fournaison (2005)
Ice Tank	Type 217	V=5 m^3, custom model
Warm Tank	Type 534	V=1.5 m^3, 4 nodes
BIPVT Array	Type 563 (unglazed)/Type 50 (glazed)	A_{Panel}=1.37m^2 /panel, N_{Panel}=29
BIPVT Pump	Type 654	Constant speed, 60W, 65 L/h/m^2 collector area
Heating Control	Type 23	Separate PID control for each RF Loop
Radiant Flooring	Type 56 Active Layer	Separate Loops for all three levels

[1] Rated at 0°C evaporator water temperature, 28°C condenser water temperature

4.2 Simulation of Solar Systems

Table 7 summarizes key PV and BIPVT panel parameters for each case. All solar systems are integrated into the south facing roof of the home. PV type, area, and rated cell efficiency were held constant across all systems to provide a common base of comparison. Listed PV efficiencies are for the cell only, and do not include corrections for the glass covering (these are, however, accounted for in the simulations). Provided information is derived from Solimpeks (2011), Notton *et al.* (2005), and Boubekri *et al.* (2009).

Table 7: PV and BIPVT Properties

PV Type	Mono-Crystalline
PV Cell Efficiency	14.88%
PV Temperature Efficiency Modifier	-0.0044
PV and BIPVT Dimensions (L x W)	1.66 m x 0.828 m
PV and BIPVT Glass Covering(s)	3.2 mm Low-Tempered Glass
	Transmittance-Absorptance:0.855
Slope/Orientation	40°/Due South

5. RESULTS

Results are presented on both a system and component level to better assess the energy performance of the proposed design.

5.1 System Level Performance

Table 8 summarizes annual energy performance in Calgary. A significant decrease in energy use is evident from Base Case 1 (Electric Baseboards) to Base Case 2 (Air-Water HP), which results from the efficiency of the heat pump and its ability to address both space heating and DHW loads. Integrating BIPVT and ice storage also has a clear impact on system energy performance, with both BIPVT cases reducing net annual electricity use by 28%-34% from Base Case 2. The unglazed configuration increased PV generation by 10% with a 13% decrease in heating/DHW in comparison with Base Case 2. The glazed configuration reversed this trend, with strong heating/DHW energy savings of 29%, but a decrease in PV generation of 7%.

Table 8: Annual Energy Use in Calgary

	Base 1: Electric BB	Base 2: Air-Water HP	BIPV/T (Unglazed)	BIPV/T (Glazed)
Heating+DHW (kWh)	13087	6273	5440	4454
Fans+Pumps (kWh)	964	1552	1747	1620
Lighting+Receptacles (kWh)	4476	4476	4476	4476
Total Electricity Use (kWh)	18526	12300	11664	10550
Total PV Generation (kWh)	8053	8053	8883	7513
Net Electricity Draw (kWh)	10473	4247	2780	3037

Table 9 summarizes annual energy performance in Montreal. The integration of an air-water heat pump again shows strong benefits, with a 54% reduction in net energy use in comparison to electric baseboards. The impact of the BIPVT system is less evident in Montreal because of reduced available solar radiation. Unglazed BIPVT collectors improve PV electricity production by 10% but reduce net annual electricity use by only 13% in comparison to Base Case 2. This can be attributed to the poor thermal performance of the unglazed collectors during the cold winter months. The glazed BIPVT system offers a greater net annual electricity savings of 20% in comparison to Base Case 2, but with a reduction in PV production of 5%.

Table 9: Annual Energy Use in Montreal

	Base 1: Electric BB	Base 2: Air-Water HP	BIPV/T (Unglazed)	BIPV/T (Glazed)
Heating+DHW (kWh)	12479	6102	6044	4653
Fans+Pumps (kWh)	964	1538	1720	1593
Lighting+Receptacles (kWh)	4476	4476	4476	4476
Total Electricity Use (kWh)	17918	12116	12241	10722
Total PV Generation (kWh)	7227	7227	7975	6834
Net Electricity Draw (kWh)	10691	4889	4266	3888

Each BIPVT configuration reduces energy use in both cities in comparison to a simple PV+HP system. However, the selection of an optimal BIPVT configuration is highly dependent on several additional factors, including initial costs, and utility rate structures. Given similar performance characteristics, it is likely that the unglazed model would be preferred, as this model represents a simpler and less expensive collector design. The unglazed model also offers the additional benefit of increased PV production, financially benefiting the homeowner in areas where feed-in tariffs are in place.

It is also important to view system performance within the context and operational objectives of the target building. Homeowners simply looking to reduce energy use are likely to achieve acceptable performance and payback periods with the HP + PV-only combination. The true benefit of the proposed BIPVT system is seen for buildings attempting to achieve near/net zero energy, where the building envelope must play an important role in the generation and management of energy. In these situations, the simulated savings of the proposed systems (13%-34%) represent a major step towards meeting these performance objectives in a simple and cost effective manner.

5.2 BIPVT Performance

While Table 8 and Table 9 present the thermal and electrical demands of the house, Table 10 summarizes BIPVT performance in both regions. Integrating BIPVT substantially increases energy output from the solar array, with the unglazed unit in particular offering strong annual electrical, thermal, and heat rejection performance. Interestingly, thermal energy outputs are higher for the unglazed unit than the glazed configuration. This primarily relates to summer collector performance: The unglazed unit integrates a night cooling mode, which allows the system to have greater capacity to cool the BIPVT panels during the day. Conversely, the glazed system does not include this feature as the additional cover would greatly reduce the night cooling effect. This results in less capacity to cool the BIPVT panels during the day and subsequently, less thermal energy inputs into the system. For both configurations, it is clear that operating the BIPVT loop improves system performance, with useful heating/cooling extracted from the panels far greater than pump energy use.

Table 10: BIPVT Annual Energy Performance

	Calgary			Montreal		
	PV	Unglazed BIPV/T	Glazed BIPV/T	PV	Unglazed BIPV/T	Glazed BIPV/T
PV Electrical (kWh)	8053	8883	7513	7227	7975	6834
PV Thermal (kWh)	NA	16381	9269	NA	14548	8581
PV Heat Rejection (kWh)	NA	8254	0	NA	7542	0
PV Total Useful Energy (kWh)	8053	33517	16782	7227	30065	15415
BIPV/T Pump Day (kWh)	NA	106	85	NA	103	85
BIPV/T Pump Night (kWh)	NA	104	NA	NA	114	NA

Figure 3(a) compares the monthly breakdown of energy generation for the PV-only and unglazed BIPVT configurations in Calgary. The unglazed BIPVT module consistently offers a small boost in electricity production throughout the year, primarily resulting from average PV cell temperatures 3.1°C lower than a PV-only system. Thermal energy production is lowest in November and December when the duration and intensity of incident solar radiation is at a minimum.

Figure 3(b) compares the monthly performance of the unglazed and glazed BIPVT modules in Calgary. As explained above, although the glazed unit offers an improvement in thermal gains in the winter months, the glass covering prevents the possibility of using passive night-time cooling, subsequently reducing its capacity to cool the PV cells during the summer months. The added glazing also has a clear impact on electrical generation, which is consistently lower throughout the year because of higher cell temperatures and a reduced amount of incident solar radiation transmitted through the covering.

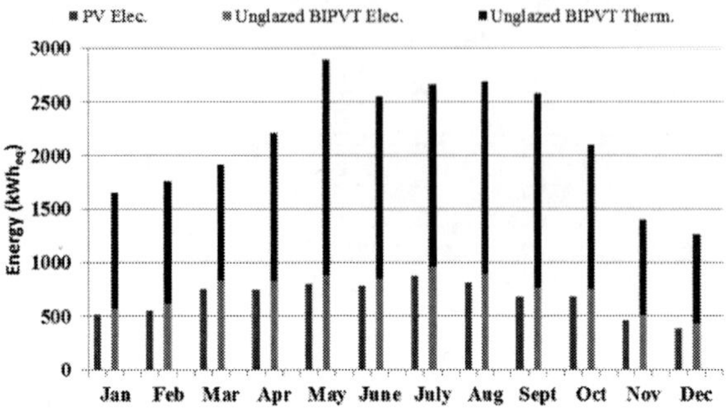

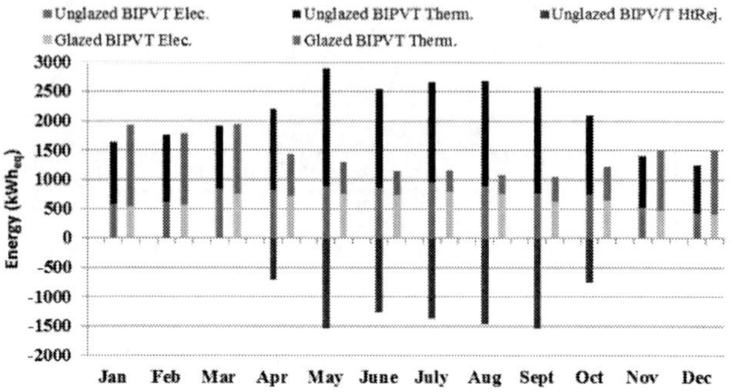

Figure 3: (a) System Energy Use Comparison: Base Case (b) System Energy Use Comparison: Collector Types

5.3 Ice Tank Performance

Figure 4(a) shows the annual change in ice mass for the unglazed system in Calgary. The ice capacity of the tank is used significantly throughout the core heating months when building thermal demands outweigh solar energy inputs. Maximum ice fractions are reached briefly from November through February, suggesting that the system is sized to effectively use ice storage while allowing consistent heat pump operations. In fact, an additional simulation for unglazed BIPVT in Calgary shows that the integration of ice storage yields a 26% reduction in heating/DHW energy use (1954 kWh) in comparison to an identically sized system using only sensible storage.

Figure 4(b) shows the average monthly fluid temperature in the ice tank. Temperatures remain low during the winter months when ice is built up, helping to improve BIPVT electrical and thermal efficiencies. Although temperatures rise significantly in the summer, fluid temperatures remain well below those seen when BIPVT is coupled directly to a warm storage such as a DHW tank (Haurant *et al.*, 2015).

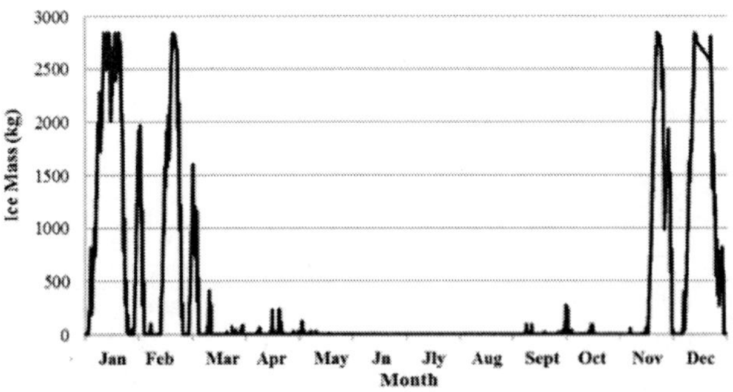

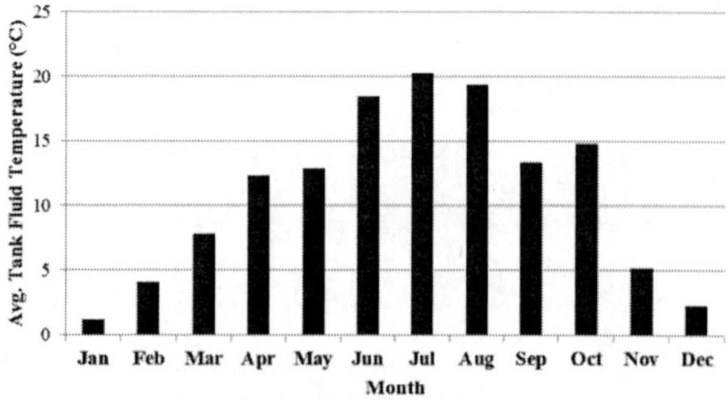

Figure 4: (a) Ice Mass in Ice Tank (b) Monthly Average Fluid Temperatures in Ice Tank

6. CONCLUSIONS

This paper presents an analysis of liquid-based BIPVT combined with heat pump and ice thermal storage technologies. Roof integrated BIPVT panels provide both electricity and low-grade thermal energy to the system. The ice tank serves as a cold temperature storage reservoir, increasing energy storage densities and BIPVT electrical and thermal efficiencies. A heat pump upgrades collected thermal energy as needed to meet space heating and DHW loads.

The proposed system is integrated into high performance housing in Calgary and Montreal. Simulated results demonstrate a clear reduction in net annual energy use, with predicted savings ranging between 13% and 34% depending on the city and collector configuration. Cooling and recovering heat from the PV array improved annual electrical production by up to 10%, while integrating the ice tank increased system storage capacity and reduced heating and DHW energy use by 26% in comparison with a sensible only system design.

NOMENCLATURE

Symbols

Δ	Difference	(-)
BB	Electric baseboards	(-)
COP	Coefficient of performance	(-)
M	Ice mass	(kg)
T	Temperature	(°C)
V	Volume	(m^3)

Subscript

BIPVT	Building integrated photovoltaic and thermal
Fluid	Fluid in tank
I	Initial
Ice	Ice mass
IT	Ice tank
Max	Maximum
Out	Outdoor
P	Peak
PLT	BIPVT plate
RF	Radiant floor
WT	Warm water tank

REFERENCES

ASHRAE. (2008). *ASHRAE Vision 2020: Producing Net Zero Energy Buildings*. Atlanta,GA:ASHRAE <https://www.ashrae.org> [Accessed Oct. 2015]

Boubekri, M., Chaker, A., & A. Cheknane. (2009). Numerical approach for performance study of hybrid PV/Thermal collector. *Revue des Energies Renouvelables*, 12(3), 355-368.

Canadian Commission on Building and Fire Codes (CCBFC), (2010). *National Building Code of Canada 2010 Volume 2*. Ottawa, CA: National Research Council.

Carbonell, D., Phillippen, D., Haller, M.Y., & S. Brunold. (2016). Modeling of an ice storage buried in the ground for solar heating applications: Validations with one year of monitored data from a pilot plant. *Solar Energy*, 125(2), 398-414.

Dott, R., & T. Afjei. (2014). Evaluation of Solar & Heat Pump System Combinations. In *Proceedings of IEA HPC 2014*. IEA HP.

Dupeyrat, P., Menezo, C., Rommel, M., & H.M. Henning. (2011). Efficient single glazed flat plate photovoltaic–thermal hybrid collector for domestic hot water system. *Solar Energy*, 85(7), 1457-1468.

ClimateMaster. (2010). TMW *Water-To-Water Series*. < http://www.climatemaster.com> [Accessed Jan. 2012]

Ecologix. (2016). *Cold Climate Air-Water Heat Pump*. < http://www.ecologix.ca> [Accessed Feb. 2016]

Haurant, P., Menezo, C., Gaillard, L., & P. Dupreyrat. (2015). Dynamic numerical model of a high efficiency PV–T collector integrated into a domestic hot water system. *Solar Energy*, 111(1), 68-81.

Guilpart, J., & L. Fournaison. (2005). The Control of Ice Slurry Systems. In Kauffeld, M., Kawaji, M., & P. Egolf. (Eds.) , *Handbook on Ice Slurries* (273-285). Paris, France: International Institute of Refrigeration.

Kegel, M., Sunye, R., & J. Tamasauskas. (2012). Lifecycle Cost Comparison and Optimisation of Different Heat Pump Systems in the Canadian Climate. In *Proceedings of eSim 2012 Halifax* (492-505). Ottawa, CA: IBPSA Canada.

Klein S. A., et al. (2010). *TRNSYS 17 – A TRaNsient SYstem Simulation program user manual*. Madison, WI: University of Wisconsin-Madison Solar Energy Laboratory.

Notton, G., Cristofari, C., Mattei, M., & P. Poggi. (2005). Modelling of a double-glass photovoltaic module using finite differences. *Applied Thermal Engineering*, 25(17-18), 2854-2877.

Office of Energy Efficiency (OEE). (2005). *EnerGuide for New Houses: Administrative and Technical Procedures*. Ottawa, CA: Natural Resources Canada.

Pelland, S., McKenney D.W., Poissant, Y., Morris, R., Lawrence, K., Campbell, K., & P. Papadopol. (2006). The Development of Photovoltaic Resource Maps for Canada. In *Proceedings of 31st Annual SESCI Conference*. Ottawa,CA: SESCI.

Solimpeks. (2011). *Installation, Operation, and Maintenance: Solar Panels*. <http://www.solimpeks.com>, [Accessed Mar. 2016]

Swinton, M.C., Entchev, E., Szadkowski, F., & R. Marchand. (2003). *Benchmarking twin houses and assessment of the energy performance of two gas combo heating systems*. Ottawa, CA: Canadian Centre for Housing Tech.

Tamasauskas, J., Poirier, M., Zmeureanu, R., & R. Sunye. (2012). Modeling and optimization of a solar assisted heat pump using ice slurry as a latent storage material. *Solar Energy*, 86(11), 3316-3325.

Tamasauskas, J., Poirier, M., Zmeureanu, R., Kegel, M., & R. Sunye. (2015). Development and validation of a solar-assisted heat pump using ice slurry as a latent storage material. *Science and Technology for the Built Environment*, 21(6), 837-846.

Trinkl, C., Zorner, W., & V. Hanby. (2009). Simulation study on a Domestic Solar/Heat Pump Heating System Incorporating Latent and Stratified Thermal Storage. *J. Sol. Energ.-T. ASME*, 131(4), 041008-1 to 041008-8.

Zondag, H.A. (2008). Flat-plate PV-Thermal collectors and systems: A review. *Renewable and Sustainable Energy Reviews*, 12(4), 891-959.

Wilo. (2016). *Wilo Pump Selection Tool*. < http://www.wilo-canada.com/>, [Accessed Mar. 2016]

ACKNOWLEDGEMENT

The authors acknowledge the financial support provided by Natural Resources Canada through the Program of Energy Research and Development (PERD) built environment heat pumps program.

Influence Of Building Management On Optimality Definition In Residential Buildings Retrofitting

Paola PENNA*[1], Alessandro PRADA[2], Francesca CAPPELLETTI[3], Andrea GASPARELLA[4]

[1]Free University of Bozen-Bolzano, Faculty of Science and Technology, Bolzano-Bozen, Italy
Phone +39 0471017617, Fax +39 0471 017009, paola.penna@unibz.it

[2] Università degli Studi di Trento, Trento, Italy
Phone +39 0461 282653, alessandro.prada@unitn.it

[3]University IUAV of Venezia, Dept. of Design and Planning in Complex Environments, Venezia, Italy
Phone +39 041 2571295, Fax +39 041 257 1282, francesca.cappelletti@iuav.it

[4]Free University of Bozen-Bolzano, Faculty of Science and Technology, Bolzano, Italy
Phone +39 0471 017200, Fax +39 0471 017009, andrea.gasparella@unibz.it

* Corresponding Author

ABSTRACT

Energy efficient retrofit strategies should not only enhance the energy performance of the building in a cost effective way, but also improve, or at least, not compromise the indoor thermal comfort during the all year, even when system is off. Even if the indoor thermal comfort is commonly not included among the objectives of the optimization process, a well-known problem of highly insulated buildings is that, if not correctly managed, they can undergo to important overheating issues. On the other hand, it is also true that, under discomfort conditions, occupants tend to react, making adjustments (operating the windows and/or the shading devices), to improve their thermal comfort sensation. In this paper, the influence of the occupants' adaptation on the definition of the optimal solutions for building retrofits has been investigated considering a set of reference building modules, representative of the Italian building stock and located in two typical Italian climates. In a first stage, the optimization of retrofits has been carried out evaluating the nominal performance, without considering the building management operated by the occupants as adaptation measure to uncomfortable conditions. In a second stage, the performance of the optimal solutions has been recalculated including some user-operated management actions and the differences between nominal and adapted performance assessed. This allows evaluating the efficacy of an appropriate management of the building by the occupants, through windows and shading operation, and its impact on nominal estimated performance. Finally, the optimization has been repeated including the adaptive management actions to investigate the possibility that some of the retrofit configurations neglected in the first optimization could reveal better performance than the selected ones.

1. INTRODUCTION

The enhancement of the existing buildings' performance requires the choice among a wide selection of energy efficiency measures. The rational approach, by means of a cost-optimization process, is able to find the configurations representing the best trade-off between energy and economic performance. Although the cost-optimal analysis represents a multi-objective optimization tool, useful to guide the investment in building renovation, the occupants' comfort is not usually included in the analysis. However, since any retrofit strategy should not only enhance the energy performance of the building in a cost effective way, but also improve, or at least, not deteriorate the indoor thermal comfort during the all year, optimization should be extended to comfort aspects. Moreover, as highlighted in recent works (Mlakar and Stryncar 2011, McLeord et al. 2013, Penna et al. 2015), one of the well-known problems in highly insulated buildings, including renovated ones, is that, if not correctly managed, they can undergo overheating issues, hampering the indoor environment livability. Not only this could raise complains about the quality of the renovation, but also induce occupants' reaction, making adjustments (operating the windows and/or the shading devices) in order to improve their conditions, and possibly compromising the building energy

performance. Indeed, occupants are not passive actors but tends to adapt themselves to changing conditions in their environment and "to react in ways which tends to restore their comfort" (Nicol and Humphreys, 2002), which could lead, in case of building renovation, to unexpectedly high energy needs. Some authors have described this kind of outcomes under the name of rebound effect (Herring, 2006). According to Hirst et al. (1985) occupants tend for instance to increase the indoor setpoint in winter to feel more comfortable in new buildings. This was also evidenced by other studies as summarized in a review by Sorrell et al (2009).
In this paper, the influence of the occupants' behavior on the definition and on the actual performance of the optimal retrofit solutions has been investigated for a set of reference buildings, representative of different construction periods and windows orientations, located in two typical Italian climates, Milano and Messina, characteristic of respectively heating and cooling dominated climatic regions. A wide selection of energy efficiency measures has been evaluated, using a genetic algorithm for the multi-objective optimization to select the configurations assessed through a dynamic simulation code. The search of the best combination of measures has been conducted optimizing energy efficiency, global costs and the indoor thermal comfort at once. In the first stage, the optimization has been carried out evaluating only the nominal performance of the building, assuming a standard occupation profile and use of the building, such as in the typical asset rating procedure foreseen for energy labelling. In a second stage, the performance of the optimal solutions has been evaluated again, this time including some user-operated building adaptation management actions, through windows and shading operation in relation to the indoor and outdoor conditions. This allows assessing the impact of these interactions, not only on the comfort conditions, but also on the actual performance of the building. Finally, the optimization has been repeated including the adaptive management actions to investigate the possibility that some of the retrofit configurations neglected in the first optimization could reveal better performance than the selected ones.

2. CASE STUDY

In this study, a set of simplified building modules have been analyzed, originated from a shoebox-like configuration, ideally representative of the top of a multistorey building (penthouse typology, S/V=0.63 and infiltration rate set to 0.13 ACH), with a floor area of 100 m^2, an internal height of 3 m and a window to floor ratio equal to 14.4%. The modules' heating system is a standard boiler (efficiency, η= 89%) coupled with radiators and on-off control system. The set of buildings has been defined from the above model by the variation of:

a) Envelope thermal transmittance, to analyze two of the most diffuse Italian building structure, a typical building structure dated back before the first energy legislation (Italian parliament, 1976) (REF 1) and one between the first and the second one (Italian parliament, 1991) (REF 2);
b) Windows orientation, assuming two options, South and East;
c) Climate, considering two localities representative of the Italian territory, Milano (HDD_{20} = 2404 K d), as a heating dominated climate, and Messina (HDD_{20} = 707 K d), as a cooling dominated one.

Table 1 summarizes the values variations and the characteristics of the buildings' set. More detailed information about the building modules can be found in Penna et al. (2015).

Table 1: Variation of the reference building module

Variation	Values	
Construction period	REF 1 (before 1979) REF 2 (1979-1991)	U_{opaque}=1.03 W m^{-2} K^{-1} $U_{glazing}$=5.7 W m^{-2} K^{-1} U_{opaque}=0.49 W m^{-2} K^{-1} $U_{glazing}$=3.2 W m^{-2} K^{-1}
Windows orientation	Southern Eastern	
Climatic context	Milano Messina	HDD_{20} = 2404 K d - Cfa, Köppen Classification HDD_{20} = 707 K d - Csa, Köppen Classification

2.1 Energy Efficiency Measures (EEMs)

To improve the energy performance of the buildings set, the following Energy Efficiency Measures (EEMs) have been considered:

i) Walls external insulation with a thickness from 1 to 20 cm, in 1 cm steps;
ii) Roof external insulation with a thickness from 1 to 20 cm, in 1 cm steps;
iii) Replacement of glazing systems with the ones reported in Table 2. Also the windows' frames are replaced with an improved aluminum frames with thermal break (U_{frame}=1.2 W m^{-2} K^{-1});

iv) Substitution of the heat generator with modulating or condensing boilers with a climatic control system;
v) Installation of a mechanical ventilation system with heat recovery.

Some of the listed EEMs bring some energy performance improvements without any additional costs. In particular:

- The linear thermal transmittances of thermal bridges are reduced according to different insulation thickness and glazing types;
- The air tightness of the building is assumed to be improved in the case of substitution of the windows by the half of the starting value;
- Since the substitution of the radiators is not planned, the nominal capacity of the emission system does not change, although the boiler is substituted. This means that a climatic control of the radiator supply temperature, enables lowering this temperature under the design value, further increasing the boiler effectiveness.

Table 2: Technical specifications of the Energy Efficiency Measures considered in the analysis

Thermal characteristics of External Insulation: Polystyrene EPS		
Thermal conductivity λ (W m^{-1} K^{-1})	0.04	
Specific heat c (J kg^{-1} K^{-1})	1470	
Density ρ (kg m^{-3})	40	
Thermal characteristics of Glazing system		
	U (W m^{-2} K^{-1})	**SHGC**
DH – Double, high SHGC (4/9/4, krypton, low-e)	1.140	0.608
DL – Double, low SHGC (6/16/6, krypton, low-e)	1.099	0.352
TH – Triple, high SHGC (6/12/6/12/6 krypton, low-e)	0.613	0.575
TL – Triple, low SHGC (6/14/4/14/6 argon, low-e)	0.602	0.343
Nominal Efficiency of the Heating system		
Standard (STD)	89 %	
Modulating (MD)	96 %	
Condensing (CD)	101 %	
Technical characteristics of the Mechanical Ventilation System		
Ventilation Rate (m^3 h^{-1})	150.0	
Power (W)	59.7	

2.2 Building Energy Simulation Models

To analyze the impact of the building management on the definition of optimal solutions, in terms of energy, costs and comfort, two simulations or models have been implemented in TRNSYS (Solar Energy Laboratory, 2012a). The simulation time step of has been set to 10 minutes, in order to catch a detailed behavior of the building. The first one, called Nominal Model (NM), assumes standard utilization profiles, consistently with what is usually assumed in an asset rating for building energy labelling. No specific reactions to the indoor conditions are considered from the occupants. The second one, the Adaptive Model (AM), includes some rough user-operated management actions, such as those related to blinds and windows opening. This model can be assimilated to a tailored rating approach, because it tries to take into account the real operation of the building. In particular, based on the thermal sensations of the occupants, those actions that could reasonably prevent or minimize indoor thermal discomfort, have been considered. First a comfort range has been defined, in terms of a lower and a higher operative temperature bounds (CEN, 2007a). These are constant in the heating season, when the minimum indoor air temperature is maintained by the heating system, and vary with the running mean outside air temperature during summer, since no cooling system is considered. In particular, the comfort bounds consider a normal level of expectation (Category II) and an activity level ranging from 1 to 1.3 met. During the heating season, from 15th October to 15th April in Milan and from 1st December to 31st March in Messina according to the D.P.R. 74/2013 (Italian Parliament, 2013), the lower and upper bounds are 20 °C and 25 °C, respectively. During the rest of the year, they are related to the external running mean temperature, Θ_{rm}, according to the following equations:

$$\Theta_{o,limit,upper} = 0.33\,\Theta_{rm} + 18.8 + 3 \qquad (1a)$$

$$\Theta_{o,limit,lower} = 0.33\,\Theta_{rm} + 18.8 - 3 \qquad (1b)$$

The running mean temperature, is calculated as an exponentially weighted running mean of the daily outdoor mean air temperature, Θ_{ed}, for the seven days immediately before the analyzed one:

$$\Theta_{rm} = (1-\alpha)\ (\Theta_{ed-1} + \alpha\ \Theta_{ed-2} + \alpha^2\ \Theta_{ed-3} + \ldots + \alpha^6\ \Theta_{ed-7}) \quad (2)$$

According to this approach, discomfort occurs when the indoor operative temperature comes out of the comfort range, and specifically, providing cold sensation for values lower than the lower bound, or hot for temperatures exceeding the upper bound.

2.2.1 Nominal Model (NM): As weather data files, the national Test Reference Years (TRY) of Milan and Messina (Comitato Termotecnico Italiano) is used. Trnsys Multi-zone building subroutine, Type 56 (Solar Energy Laboratory, 2012b), with Type 869 (Haller at al. 2011a, Haller et al. 2011b) model the building and the heating system respectively. A thermostat is set to switch on the boiler when the indoor air temperature is lower than 20°C, and switch it off, when it overcomes 22°C. When replacing the boiler, the water supply temperature is assumed to be regulated in relation to the outside one. Internal gains, half radiative and half convective, are modelled according to the Italian technical specification UNI/TS 11300 (UNI, 2008). The air change rate, is set to 0.5 ACH during the occupancy time. If the mechanical ventilation is considered, heat recovery is used to pre-heat the outdoor inlet air in winter, while in summer mechanical ventilation is operated to avoid the indoor overheating. Specifically, if the indoor operative temperature overcomes the upper bound of the comfort range and the outdoor temperature can cool down the indoor one, the mechanical ventilation system turns on, bypassing the heat recovery. When the building is occupied, if the outdoor conditions are worse than inside (too cold or too hot), the mechanical ventilation is operated with a fixed airflow rate of 0.5 ACH and to pre-condition the inlet air.

2.2.2 Adaptive Model (AM): Adaptive Model tries to simulate a more realistic operation of the building by the occupants, to prevent indoor overheating and restore thermal comfort. Thus some control actions to operate the shutters and the windows opening, have been added to the NM. Those control actions are inspired by common-sense reactions to discomfort conditions. Since people tend to operate actively on the building according to their thermal perception, in order to prevent discomfort conditions (Nicol and Humphreys, 2002, Mahdavi 2011), it is reasonable to assume they react differently in summer and winter conditions. Then, a "summer adaptation" period when occupants tends to actively react to discomfort conditions, has been defined according to the EN 15251 (CEN, 2007a). The start of the summer adaptation period is set when outdoor running mean temperature exceed the 10 °C, which corresponds to an indoor comfort operative temperature of 25 °C. Considering the TRYs, this period is 28th April – 14th October for Milan and 3rd April – 30th November for Messina. During this period, the occupants are presumed to be more incline to adapt, operating actively shading devices and increasing the ventilation rate by opening the windows. The applied shading factor is assumed 0.8 and two shadings controls are considered: when the building is not occupied, the shades are closed, while during the occupied time, they are closed only when the beam solar radiation incident on the window exceeds 150 W m^{-2}. The control strategy for windows opening has been set according to thermal sensation of the occupants. During all the year, when the building is occupied and the occupants feel hot (operative temperature is higher than the higher comfort bound), if the outside temperature can improve the indoor comfort (outside temperature is lower than operative temperature), the windows are considered to be opened. The ventilation rate, due to windows' opening, has been modelled according to EN 15242 (CEN, 2008), which considers wind speed, temperature difference between inside and outside and windows opening angle. Two opening angles have been set according to the season. During the winter, the windows are considered partially opened, with an angle of 5°, while during the summer they are considered completely opened, i.e. with an opening angle of 90°. Windows are then closed when the operative temperature decrease below the lower comfort bound.

3. MULTI-OBJECTIVE OPTIMIZATION

The best combinations of retrofit strategies have been defined taking into account different aspects, namely energy savings, global costs and occupants wellbeing. For this purpose, a multi-objective approach has been used to minimize simultaneously those three target functions expressed by the Energy Performance for Heating (EPH), the Net Present Value of the total cost (NPV) and the Weighed Discomfort Time (WDT). Section 3.1 reports a description of these three objective functions. The solutions identified by this approach are the so-called Pareto front, which represents the best trade-off among different competitive targets. Considering that the optimization problem with three objective functions has three dimensions, the result of the optimization is a "Pareto surface". In our case, the solutions laying on the Pareto surface are those with a lower EPH than the initial one, which for a given EPH, minimize the WDT at any given NPV.

3.1 Objective functions

3.1.1 Energy Performance for Heating (EP_H): EP_H represents the primary energy for heating per heated floor area to maintain the set temperature conditions during a year (CEN 2007b). This indicator takes into account natural gas consumed by boiler, and electricity, due to pumps or mechanical ventilation system.

3.1.2 Net Present Value (NPV): NPV is the actualized cash flow, generated by a new construction or a retrofit action considering the building lifespan of 30 years. According to the comparative framework methodology proposed by EU 244/2012 (European Commission, 2012), the following costs have been taken into account:
- the initial Investment Cost (IC) for the retrofits. The costs of different EEMs are reported on Table 3 and are define according to regional price list.
- the annual running costs, composed of the annual Energy Cost (EC) for heating and the Maintenance Cost (MC) for restoring building components. Energy costs and energy prices rising are reported in Table 4;
- the replacement cost (RC), for the periodic substitution of building/system elements;
- the residual value (RV) for the pieces of equipment with longer lifespan according to EN 15459 (CEN, 2007c).

Table 3: Investment costs for the Energy Efficiency Measures

Energy Efficiency Measures	Costs
Insulation of Vertical wall (* thickness (cm))	IC_{VW} = 1.6 x* + 38.53 EUR m^{-2}
Insulation of Horizontal wall (* thickness (cm))	IC_{HW} = 1.88 x* + 8.19 EUR m^{-2}
DH – Double glazing, high SHGC	IC_{DH} = 404.33 EUR m^{-2}
DL – Double glazing, low SHGC	IC_{DL} = 439.06 EUR m^{-2}
TH – Triple glazing, high SHGC	IC_{TH} = 477.65 EUR m^{-2}
TL – Triple glazing, low SHGC	IC_{TL} = 454.49 EUR m^{-2}
Standard (STD)	IC_{STD}=1000 EUR
Modulating (MD)	IC_{MDL}=1500 EUR
Condensing (CD)	IC_{MDL}=2000 EUR
Mechanical Ventilation System	IC_{MV} = 6000 EUR

Table 4: Parameters for the economic analysis

Parameters for the economic analysis			
Fuel Cost [(1)]	0.85 EUR S m^{-3}	Electricity Cost [(1)]	0.25 EUR kWh_{el}^{-1}
Lower Heating Value [(2)]	32.724 MJ S m^{-3}	Annual increase of electricity price [(3)]	1.71 %
Annual increase of fuel price [(3)]	2.8 %		
VAT	10 %	Real Interest Rate	3 %

[(1)] Autorità per l'Energia Elettrica e il Gas, 2011, *Relazione annuale sullo stato dei servizi e sull'attività svolta,* Milan, Italy
[(2)] *Energetico* Ministry of Economic Development, 2011, *Bilancio Nazionale 2010*, Rome.
[(3)] EU Enery Trends to 2030, update 2009. European Union 2010.

3.1.3 Weighted Discomfort Time (WDT): WDT, through degree-hours criterion (CEN, 2007a), indicates how much and for how long the operative temperature lies outside of the comfort range. To calculate WDT, the occupied hours during which the operative temperature exceeds the comfort bounds are weighted by the deviation *wf* from the range (Equations 3 and 4).

$$WDT=\sum wf \cdot \tau \,(K\,h) \tag{3}$$

$$wf= \Theta_o - \Theta_{o,limit} \quad (K) \text{ when } \Theta_o < \Theta_{o,limit,lower} \text{ or } \Theta_o > \Theta_{o,limit,upper} \tag{4}$$

3.2 Genetic Algorithm

The algorithm used to perform the optimization is the Non-dominated Sorting Genetic Algorithm, NSGA (Deb K. *et al.* 2002). The set parameters for the Genetic Algorithm (GA) are a fraction of 0.5 of tournament selection, 0.8 of arithmetic crossover and a mutation rate of 0.1. The initial population is composed by 128 individuals, defined through the Sobol's Method, a quasi-random number generator. This method defines random points uniformly distributed on the problem's space, having the advantages of improving the outcomes of the genetic algorithm by giving a good individuals' collection as initial population (Saltelli et al., 2004).

4. RESULTS AND DISCUSSION

To evaluate the impact of the building management on the definition of the optimal solutions, the analysis has been split in three different steps. A first optimization has been run using the Nominal Model, evaluating the nominal performance of the building without considering any human adaptive reaction to the indoor conditions. Secondly, the performance of the optimal solutions found in the first optimization, has been assessed using the Adaptive Model. By including adaptation management actions, it has been possible to evaluate how building management affects the performance of the selected optimal solution, highlighting the performance gap between asset and tailored rating of refurbished buildings. Finally, a second optimization run, using the Adaptive Model, has allowed investigating the possibility that some of the retrofit configurations neglected in the first optimization could reveal better performance than the selected ones, if considering the adaptation management since the optimization phase.

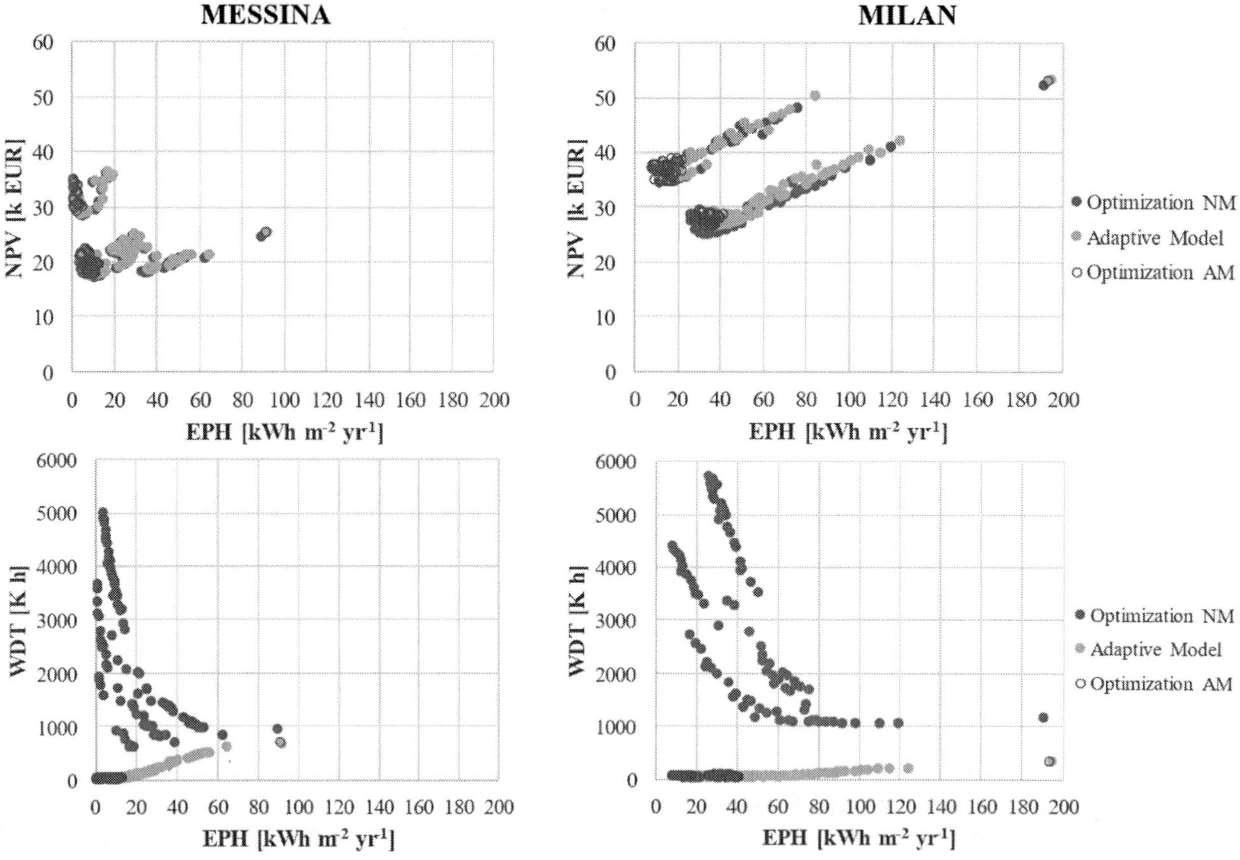

Figure 1: Pareto surfaces solutions of the case studies REF 1, windows east exposed. The graphs on the left refer to Messina, the ones on the right to Milan.

4.1 Comparison of the Pareto front solutions simulated with NM and AM

Figure 1 reports the results of the three different evaluation steps for the case studies REF 1 (building built before 1976) with windows East oriented, located in Messina and in Milan. The first two graphs report the relationship between energy and economic performances (EP_H – NPV), the second two show the relationship between energy and comfort performances (EP_H – WDT). The grey dots are the Pareto optimal solutions identified by the GA using the NM, the yellow ones are the same Pareto solutions simulated with the AM and the orange circles represents the results of the second optimization run with the AM. By comparing the grey and yellow dots is visible how considering the user-operated control actions affect the energy, cost and comfort performance.

One first finding concerns the increase of energy and costs of the Pareto frontier solutions when user adaptive management is included. This trend is clearer in Figure 2, where regression lines show the entity of the differences in energy and economic performance of the two models. In particular, the energy needs of the building increases by about 5% in Messina and 10% in Milan with user interactions. Consequently, the NPV increases by 1.5% in Messina and about 4% in Milan. Similar trends are visible in the other cases and not represented. Generally, two trends are highlighted by the results: the first one related to the building characteristics, such as construction periods and windows orientations, the second one related to the location of the building, in a heating and cooling dominated climatic region. The first trend is related to overheating: the lower are the heating needs, such as for (REF 2) and windows South oriented, the larger is the overheating issue. As a consequence, the more a case is affected by overheating, the more is the differences in energy and costs performance. This is due to the control actions modelled into the AM, and in particular with the windows' opening. When the user feels hot, windows opening increases the ventilation rate. Particularly when this occurs in winter, as it happens more frequently for the most efficient solutions, the indoor air temperature decreases and the heating system starts, increasing the building energy consumptions, and consequently the energy costs. The second trend is related to the location of the building: in Milan the differences in energy performance and costs are larger between the two models, if compared to the ones located in Messina. At the first sight, this could be seem in contrast with the previous statement, in fact, Milan presents higher heating needs. However, since in Milan, during the heating season, the outdoor temperature is considerably lower, when the indoor conditions are characterized by overheating, increasing the ventilation rate cools down more rapidly the indoor air temperature, worsening more consistently the building energy performance.

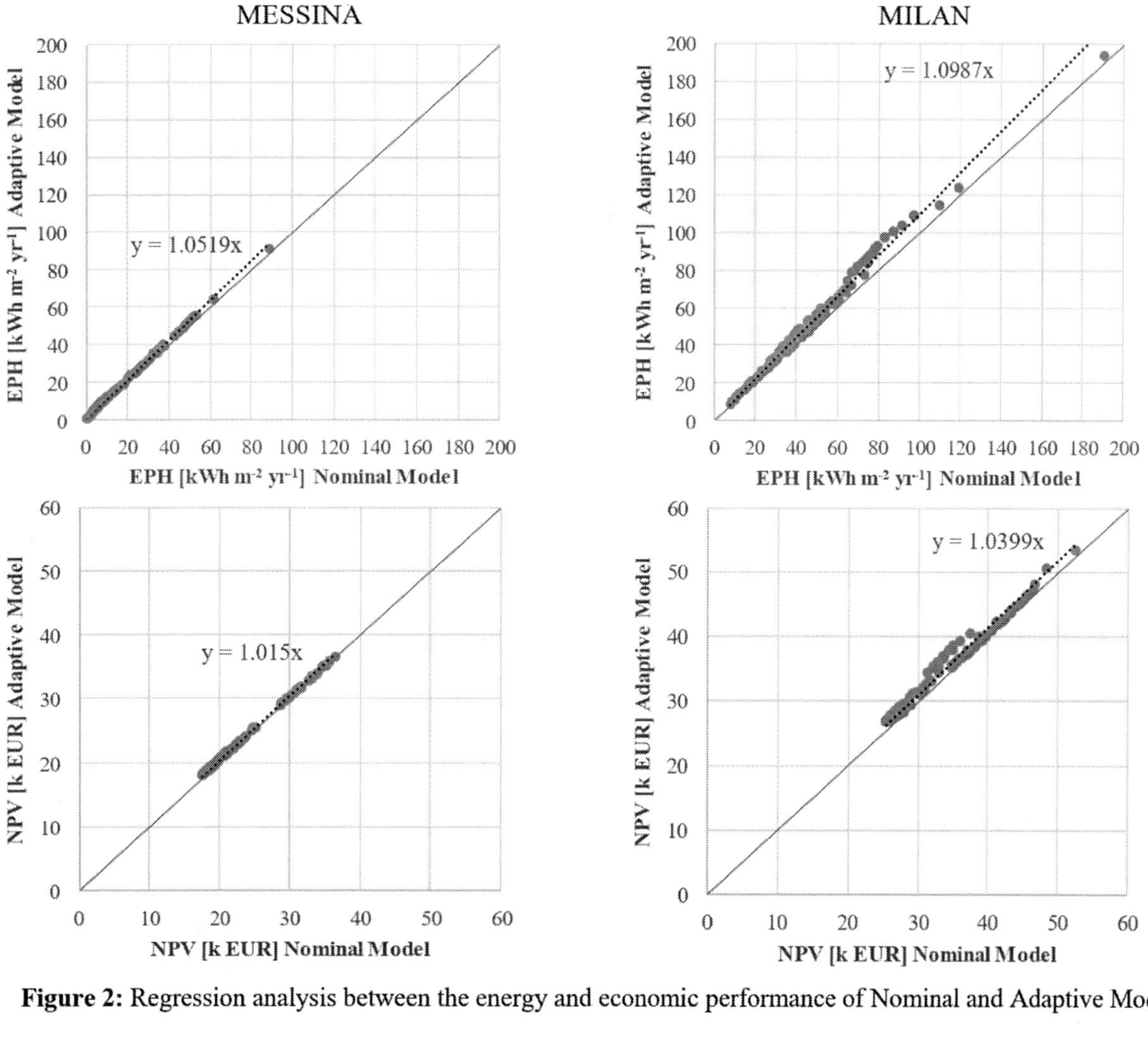

Figure 2: Regression analysis between the energy and economic performance of Nominal and Adaptive Models.

A further aspect underlined by the results is related to the comfort performance. The NM shows how improving the building energy performance leads to deteriorating the indoor thermal comfort. In particular, increasing the external insulation thickness increases the indoor thermal discomfort. In fact, as already highlighted by Penna et al. (2015), in highly insulated buildings a small energy input raises significantly the internal temperature and if the extra heat is not dissipated, the indoor livability can be hampered. According to the NM, the higher is the nominal building energy efficiency improvement, the higher is the indoor discomfort. However, adaptive reactions reveal capable not only to address this problem, but even to improve the situation particularly for the best energy performing configurations. The highest discomfort is for the base case, where no retrofits are considered. For this case, introducing the building management reduces the overheating risk, but not the discomfort caused by low indoor operative temperature. In fact, in poorly insulated building, even if the air temperature reaches the set-point of the heating system, the operative temperature can exit the comfort range because of the low mean radiant temperature.

4.2 Comparison of the Pareto front solutions identified with the two optimizations

A second optimization has been run using the AM to investigate the possibility that some of the retrofit configurations, neglected in the first optimization, could reveal better performance than the selected ones (orange circles in Figure 1). The graphs show that, if the user-operated control actions are considered since the optimization phase, the EEMs with EPH higher than the cost optimal ones are not selected by the algorithm. This means that the overheating problem, related to solutions with high energy efficiency, can be controlled more effectively by using the shading devices and by increasing the ventilation rate to dissipate the excess of heat when different configurations are selected. Solutions that in NM optimization were chosen, because of the better comfort despite of the high EPH, in the AM optimization are dominated and not included on the Pareto frontier. Conversely, solutions excluded when assessed according an asset rating approach are more easily managed and better performing also from an energy point of view.

Table 5: Combination of EEMs for the optimal solutions in terms of cost, energy and comfort evaluated with the Nominal and Adaptive Model for the set of reference buildings.

	REF 1								REF 2							
	EAST				SOUTH				EAST				SOUTH			
	ME (NM)	ME (AM)	MI (NM)	MI (AM)	ME (NM)	ME (AM)	MI (NM)	MI (AM)	ME (NM)	ME (AM)	MI (NM)	MI (AM)	ME (NM)	ME (AM)	MI (NM)	MI (AM)
	COST OPTIMAL (CO)															
Wall	11	13	16	17	12	10	16	14	11	10	10	12	10	10	11	11
Roof	10	10	14	16	10	11	15	14	11	10	11	12	9	9	11	11
Win	DH	DH	DH	DH	0	0	DH	DH	0	0	0	DH	0	0	0	0
Boiler	STD	STD	STD	STD	STD	STD	STD	STD	STD	STD	STD	STD	STD	STD	STD	STD
Vent	STD	STD	STD	STD	STD	STD	STD	STD	STD	STD	STD	STD	STD	STD	STD	STD
EPH	10	11	34	36	23	25	22	29	24	27	56	40	14	16	43	54
NPV	18	18	25	27	15	15	22	24	15	16	24	26	12	12	20	23
WDT	3473	51	4995	85	877	392	4255	62	2123	188	2866	86	1361	193	2189	74
	ENERGY OPTIMAL															
Wall	18	17	18	20	20	15	18	18	12	12	12	12	12	12	12	12
Roof	19	19	20	20	18	20	19	20	12	12	12	12	12	12	12	12
Win	TH	TH	TH	TH	TH	TL	TH	TL	TH	TH	TH	TH	TH	TH	TH	TH
Boiler	CD	STD	CD	MD	CD	MD	CD	CD	CD	STD	MD	CD	MD	MD	CD	CD
Vent	MVS	MVS	MVS	MVS	MVS	MVS	MVS	MVS	MVS	MVS	MVS	MVS	MVS	MVS	MVS	MVS
EPH	0.3	0.3	8	8	0.3	0.4	3	6	1	1	17	20	0	0	9	10
NPV	35	32	37	37	35	34	36	37	33	30	37	38	32	32	35	36
WDT	3675	29	4418	80	3166	21	3598	41	2908	39	3638	81	2307	26	2930	61
	COMFORT OPTIMAL															
Wall	1	19	0	9	1	18	0	13	0	12	0	8	0	12	0	8
Roof	10	20	1	19	16	20	4	20	10	12	3	12	0	12	10	12
Win	TL	TL	DL	TL	0	TL	DL	TL	TL	TL	TL	TL	DL	TL	TL	TL
Boiler	CD	MD	STD	MD	STD	MD	STD	CD	MD	MD	MD	MD	MD	MD	MD	MD
Vent	MVS	STD	STD	MVS	MVS	STD	STD	STD	MVS	MVS	MVS	MVS	MVS	MVS	MVS	MVS
EPH	19	6	119	22	29	2	119	26	28	3	72	29	24	1	53	21
NPV	36	22	41	39	30	21	41	28	34	30	45	39	32	32	41	37
WDT	640	26	1061	50	423	18	1061	37	541	30	951	54	376	23	748	41

Table 5 reports the configuration of the Costs Optimal (CO), Energy Optimal (EO) and Comfort Optimal (CMFO) configuration defined by considering (AM) or not (NM) the user's management actions in the optimization process. Generally, the CO solutions identified for the AM, presents higher insulation thickness than the NM original ones. This allows improving comfort conditions while limiting the drawbacks on the energy performance.
NM and AM EO solutions have almost the same configuration, sometimes differing for the heat generator. What is significantly different is the WDT, which with AM optimization not only improves with respect to NM, but is also better than the CO one.
As for the CMFO solution, for the AM optimization it is always way more insulated than the corresponding one with NM. If the NM selects, as CMFO, solutions with lower insulation thickness, by introducing some control actions to prevent the overheating risk, solutions with higher insulation and, consequently, with higher energy performance are also the most comfortable ones. The energy performance (EPH) is always better than for the CO, sometimes quite close to zero, and the NPV is almost always better than that of the EO, thus smoothing the path to the nearly zero energy building.

5. CONCLUSIONS

In this paper, the influence of the occupants' behavior on the definition and on the actual performance of the optimal retrofit solutions has been investigated for a set of reference buildings. The search for the best combination of measures has been conducted optimizing energy efficiency, global costs and the indoor thermal comfort at once. Firstly, the optimization has been carried out evaluating only the nominal performance of the building, assuming a standard occupation profile and use. This has highlighted how increasing the building energy performance leads to deteriorate the indoor thermal comfort, because of the overheating issue. Then the performance of the optimal solutions has been evaluated again, including some user-operated building adaptation actions to prevent overheating. In particular, including the adaptive management radically enhances the building comfort performance, reducing the overheating risk, but also increases energy consumptions and global costs.
Finally, the optimization has been repeated including the adaptive management actions showing that retrofit configurations neglected in the first optimization reveal better performance than the selected ones.
This paper highlights how high energy performance buildings, if not correctly managed, can presents really poor performance in terms of comfort. On the one hand, underestimating this issue can lead to improper estimation of the actual performance of the building, which is likely to be affected by the occupant's adaptive reactions. On the other hand, overheating and discomfort can be avoided or considerably reduced through an appropriate management of shadings and ventilation rates. Those strategies have to be considered when optimizing the combination of retrofit measures if the actual cost, energy and comfort optimal solutions have to be defined.
Finally, as a future development of this work, the evaluation of the potential of optimized building management strategies, also based on automated systems, to maintain adequate comfort condition while reducing the energy demand, seems worth investigating, either in combination with other energy efficiency measures, or on their own.

REFERENCES

CEN, 2007a. EN 15251, Indoor Environmental Input Parameters for Design and Assessment of Energy Performance of Buildings Addressing Indoor Air Quality, Thermal Environment, Lighting and Acoustics. Bruxelles, Belgium

CEN, 2007b. EN 15217:2007. Energy Performance of Buildings. - Methods for ex- pressing energy performance and for heating certification of buildings. Bruxelles, Belgium

CEN, 2007c. EN 15459:2007, Energy Performance of Buildings—Economic Evaluation procedure for Energy Systems in Buildings. Bruxelles, Belgium

CEN, 2008. EN 15242:2008. Ventilation for buildings - Calculation methods for the determination of air flow rates in buildings including infiltration. Bruxelles, Belgium

Comitato Termotecnico Italiano, www.CTI2000.it

Deb, K., A. Pratap, and S. Agarwal. 2002. A fast and elitist multi- objectives genetic algorithm: NSGA-III. *IEE Transactions on Evolutionary Computation 6* (2):182–97.

European Commission (EC). 2012. Commission delegated regulation (EU) No. 244/2012 of 16, January. Brussels: European Commission

Haller M.Y., Paavilainen J., Konersmann L., Haberl R., Dröscher A., Frank E., 2011. A unified model for the simulation of oil, gas and biomass space heating boilers for energy estimating purposes. Part I: Model development, *J. Build. Perform. Simul. 4* (1): 1–18.

M.Y. Haller, J. Paavilainen, L. Konersmann, R. Haberl, A. Dröscher, E. Frank, 2011. A unified model for the simulation of oil, gas and biomass space heating boilers for energy estimating purposes. Part II: Parameterization and comparison with measurements, *J. Build. Perform. Simul. 4* (1): 19–36

Herring H., 2006. Energy efficiency, a critical view. *Energy 31* (1):10–20.

Hirst E., White D., Goeltz R., 1985. Indoor temperature changes in retrofit homes. *Energy 10*: 861–70.

Italian Parliament. 1976. Law 373 Norme per il contenimento del consumo energetico per usi termici negli edifici.

Italian Parliament. 1991. Law 10, Norme per l'attuazione del piano energetico nazionale in materia di uso nazionale dell'energia, di risparmio energetico e di sviluppo delle fonti rinnovabili di energia.

Italian Parliament, 2013, D.P.R. n.74/2013, Regolamento recante definizione dei criteri generali in materia di esercizio, conduzione, controllo, manutenzione e ispezione degli impianti termici per la climatizzazione invernale ad estiva degli edifici e per la preparazione dell'acqua calda per usi igienici sanitari

Mahdavi A., 2011. People in building performance simulation, in Hensen J.L.M. & Lamberts R., *Building performance simulation for design and operation*, Spon Press, New York, USA, p.56-83

McLeod R.S., Hopfe C.J., and Kwan A., 2013. An investigation into future performance and overheating risks in Passivhaus dwellings. *Building and Environment 70*:189–209.

Mlakar J., and Strancar J., 2011. Overheating in residential passive house: Solution strategies revealed and confirmed through data analysis and simulations. *Building and Environment 43*: 1443–51

Nicol J.F., Humphreys M.A. 2002. Adaptive thermal comfort and sustainable thermal standards for buildings. *Energy and Buildings 34*: 563-72

Penna, P., Prada, A., Cappelletti, F. and A. Gasparella. 2015. Multi-objectives optimization of Energy Efficiency Measures in existing buildings. *Energy and Buildings 95*: 57-69

Saltelli A., Tarantola S., Campolongo F., Ratto M., 2004. Sample generation, in: *Sensitivity Analysis in Practice. A Guide to Assessing Scientific Models*, John Wiley & Sons, Chichester, UK, pp. 193–204

Solar Energy Laboratory, 2012a, TRNSYS 17, TRNSYS. A transient system simulation program, 2012, http://sel.me.wisc.edu/trnsys

Solar Energy Laboratory, 2012b, TRNSYS 17, Multizone Building modeling with Type 56 and TRNBuild, Vol. 5, Madison, WI, USA

Sorrell S., Dimitropoulos J., Sommerville M., 2009. Empirical estimates of the direct rebound effect: a review. *Energy Policy 37*:1356–71.

UNI 2008, UNI/TS 11300-1, Energy performance of buildings – Part 1: Evaluation of energy need for space heating and cooling.

Recovery of Waste Thermal Energy in U.S. Residential Appliances

Stephen L. CASKEY[1]*, Eckhard A. GROLL[1], Eric J. BOWLER[2]

[1]Ray W. Herrick Laboratories, School of Mechanical Engineering, Purdue University
West Lafayette, IN, USA
scaskey@purdue.edu, groll@purdue.edu

[2] Sustainability Group, Whirlpool Corporation,
Benton Harbor, MI, USA
Eric_J_Bowler@whirlpool.com

* Corresponding Author

ABSTRACT

With the United States being the world's second largest consumer of primary energy, research into areas of significant consumption can provide large impacts in terms of the global energy consumption. Buildings account for 41% of US total energy consumption with the residential sector making up a majority. Household appliances account for the second largest site energy consumption at 27%, after the HVAC system for the U.S. residential sector. By quantifying the expected energy available in the waste stream for five major appliances; household refrigerator, clothes dryer and washer, dishwasher, and cooking oven, a potential energy source is presented. A cold water cooling stream is applied to the waste stream of each appliance and an estimated amount of energy can be recovered. The household refrigerator is modeled having an increase in cooling capacity of about 12% and a reduction on compressor power consumption of about 26%. A sample operation of the clothes dryer has the exhaust air stream being cooled down to 30.5°C (86.9°F) or on the other side, is able to heat 19 liter (5 gal) of water up to about 54.5°C (130.1°F). Large volumes of water are available by the clothes washer, but due to typical operation characteristics, low wash and rinse temperatures, the waste stream was not high in temperature. While the dishwasher provided higher heat source temperatures, 40°C (104°F), than the clothes washer, 36°C (97°F), the opposite was true. The volume of waste water drained is very low compared to the clothes washer 11.7 liter (3.1 gal) to 155 liter (41 gal). Thus water temperatures in the storage tank did not reach above 30°C (86°F) even with low storage volumes. The cooking oven can generate very high water temperatures depending how small of a storage tank is connected. Further work in this area is recommended due to the potential of high water temperatures generated from residential waste energy streams not currently being captured, and thus can offset some site-energy usage.

1. INTRODUCTION

The United States has fallen to second place behind China as the largest consumer of energy in the world with less than a quarter of the population. The US consumed 103.2 EJ (97.8) quads of primary energy in 2010 (Buildings Energy Data Book, 2011). Reducing the energy consumption of the US provides an effective mechanism to achieve significant reductions to the world total energy consumption. The U.S. DOE breaks down the energy picture across four sectors: transportation, commercial, residential and industrial. The building sector which combines both the commercial and residential sectors accounts for the largest consumer at 41% of the US total compared to 29% from the transportation sector and 30% from the industrial sector. A further look into the energy consumption of the building sector shows 41 EJ (39 quads) of primary energy provided only 21 EJ (20 quads) of site energy at an energy loss of 49%. A large source of these losses occurs from on-site electricity usage. From the production at the power plant to the transmission and distribution of electricity through the grid, many conversion and resistive losses occur. Exploring methods to reduce the on-site electrical usage will expand the impact of the local energy savings. Due to the residential sector having a slightly larger consumption over the commercial sector, 22% versus 19%, the focus will be on addressing the residential energy picture.

The largest percentage of residential, site energy usage is from the heating, air conditioning, and ventilation (HVAC) system, 54%. The next significant percentage is from the household appliances; water heater, refrigerator, wet cleaning (dishwasher, clothes washer and dryer) and cooking equipment at 27% (Table 2.1.5 Buildings Energy Data Book, 2011). Countless studies have been conducted to reduce building HVAC energy by exploring various improvements on the HVAC equipment to upgrading building materials obtaining reductions in the thermal demand of the home. For household appliances, the research has been focused only on the appliance itself and less on the external factors to the appliance. The use of federal standards propels exploration and adoption of new technologies by appliance manufacturers. One example for household refrigerators is the average, annual electricity use in 1980 of about 1300 kWh drops to about 950 kWh in 1990 and then is under 700 kWh by 1992 due to the introduction of new energy efficiency standards in 1990 and 1993 (Meyers *et al.,* 2003). While the use of federal standards can create reductions in the individual appliance energy usage, they force appliance manufacturers to only consider improvements within the scope of one appliance. New savings exist when one considers the waste streams available from each appliance that is not covered by the standards. The refrigerator rejects heat to air behind or beneath the unit. Clothes washers and dishwashers drain hot or warm water after each wash and rinse cycle. Clothes dryers exhaust hot, moist air to the exterior of the building. Cooking ovens have thermal mass with very high temperatures at the end of a baking cycle. The goal is to identify the amount of heat and the associated temperatures available from these waste streams and to explore the amount of reuse possible through heat recovery.

2. U.S. APPLIANCE CHARACTERISTICS

To develop an understanding of the waste energy profile of household appliances, their standard operation and typical usage characteristics had to be understood. The total number appliances installed in the United States are identified. If available, a distinction is made between the different technologies available driving the appliance. For example, clothes dryers heated by gas combustion or electric resistors. The energy usage per appliance is reported as an annual or per cycle energy consumption. Usage characteristics are applied for appliances with cycle energy consumption to estimate annual energy consumption. Published or manufacturer data provides typical operational parameters of the appliance; volume and temperature of water being drained by the clothes washer and dishwasher, temperature and humidity of exhaust air by the clothes dryer, vapor compression cycle of a household refrigerator, and the metal cavity mass of the cooking oven.

2.1 Household refrigerator

Out of 113.6 million American homes, 113.4 are listed as using a refrigerator with 87 million having only one while 26 million have two or more (Table HC3.1 RECS, 2009). Two configurations are the most common: top-mount freezer accounts for 49% of all refrigerators while the side-by-side accounts for 34%. In 2008 the median volume and annual electricity consumption for the top-mount freezer and side-by-side were 510 liter (18 ft^3), 454 kWh and 710 liter (25ft^3), 580 kWh respectively (Refrigerator Market Profile, 2009). A higher annual consumption of 660 kWh is projected when under certain test conditions specified by the Department of Energy (Table 2.1.16 Buildings Energy Data Book, 2011). The refrigerator size for this larger value is not mentioned and the test standard version is unknown, both of which would impact this estimated annual energy consumption. The current EPA Energy Star program requires a 10% reduction from the 2014 federal minimum standards (ENERGY STAR V.5, 2013). As an example, the maximum annual energy consumption under the current standards; a top-mount freezer with no icemaker and a rated capacity of 510 liter (18 ft^3), can use 403 kWh for the federal standard and 363 kWh for Energy Star. For the other common configuration, a side-by-side with an icemaker and a rated capacity of about 710 liter (25 ft^3), must have a maximum annual consumption of 705 kWh for the federal standard and 634 kWh for Energy Star. Comparing the median values from 2008 to the 2014 federal standard, a reduction of 11% is reported for the top-mount freezer, while for the side-by-side, an increase of almost 10% is found. One potential reason for the allowable increase for the side-by-side is having an icemaker puts the refrigerator into a different category for the federal standard, and thus provides a higher annual consumption to handle ice making.

Experimental data collected by the manufacturer during a standard DOE test run was obtained for a 750 liter (26.5 ft^3) rated volume, side-by-side refrigerator with ice maker. Referencing an older Energy Star version, the refrigerator was certified with a 552 kWh annual energy usage falling under the federal standard of 737 kWh. The data was processed using EES to determine the heat transfer rates and COP of the refrigeration cycle (Klein *et al.*, 2002). Many cycles of the compressor turning on and off are captured during the 48 hour test. The ambient and refrigerated cavity conditions are relatively constant during the testing span and only one cycle is needed to understand the operation of the refrigerator. The ambient and freezer compartment (location of the evaporator) temperatures measured were 32.2°C (90°F) and -16.6°C (2.1°F) respectively. Air flow rates of 85 m^3/hr (50 CFM) and 68 m^3/hr (40 CFM) were assumed for the evaporator and condenser. Air side measurements were used to obtain an estimated heat transfer rate for each heat exchanger. The temperature difference across the evaporator was 3.9°C (7°F) providing about 148 Watts of cooling. For the condenser, the temperature difference was 5.6°C (10°F) requiring 256 Watts of heat rejection to the ambient. The measured power consumption of the entire refrigerator was 103 watts and the rated compressor displacement was 5.56 cm^3 (0.34 in^3). Cooling and heating COPs were calculated to be 1.44 and 2.49 respectively. Using the ambient and freezer compartment temperatures for heat sink and source reservoir temperatures, a Carnot COP provides the ideal cycle efficiency for cooling and heating. Dividing the measured cooling and heating COPs by the Carnot COP generates a 2nd law efficiency temperature. The data shows a cooling second law efficiency of 27% and a heating second law efficiency of 40%. A refrigerator with the performance characteristics from the experimental data, 552 kWh per year, is referenced for the waste heat analysis. The number of units in the U.S is roughly 38.6 million from 34% of all refrigerators are side-by-side. The total annual energy consumption is therefore 0.077 EJ (0.073 quads) for all side-by-side refrigerators.

2.2 Clothes dryer

The number of U.S. households that use a dryer at home is 90.2 million with 80% having electrically heated ones versus 20% using natural gas or propane/LPG (Table HC3.1 RECS, 2009). Only electrically heated clothes dryers will be considered due to the small percentage of combustion heated dryers. 74.4 million households report the dryer is used every time clothes are washed which provides some insight to the correlation between clothes dryer and washer usage. Until the DOE test procedure was adjusted in 2011, the number of drying cycles a year was assumed to be 416 (ENERGY STAR, 2011). With new data on usage characteristics from housing surveys, the number of cycles a year was adjusted to 283, or 32% less (Table HC3.1 RECS, 2009). This number is also lower than the assumed 359 cycles per year by the DOE, (Table 2.1.16 Buildings Energy Data Book, 2011).

The electrical consumption of clothes dryers depends on a number of inputs; some are specified by user settings on the interface of the appliance and the others depend on the moisture content of the clothes. Different drying cycles can be run: permanent press, delicates, or auto-termination using moisture detection. Low, medium or high temperature heat settings can be selected. The moisture content of the clothes being loaded directly correlates with the required heating energy to evaporate and remove all the stored water. The type of clothing, the amount of clothes or load size, and the water extraction efficiency of the washing machine all determine the clothing moisture content. While all these factors directly impact the amount of heating energy delivered by the electric resistor, the motor electricity consumption will be relatively constant regardless of dryer inputs. One source reports an annual electric consumption of 1000 kWh for electric dryers (Table 2.1.16 Buildings Energy Data Book, 2011). Referencing the previously mentioned 359 cycles per year by the Buildings Energy Data Book, the average power consumption is estimated at 2.78 kWh per cycle.

Due to the large number of different inputs available and the associated number of different combinations for the dryer operation, experimental data from a published study presents exhaust air conditions used for analysis. An electric clothes dryer is first monitored and recorded with no modifications to develop a baseline operation before running different failure mode tests as a safety evaluation (Butturini *et al.*, 2004). Two plots identify the baseline operation for measured temperatures at various locations throughout the dryer starting from the ambient air intake to the exhaust air, and relative humidities of the ambient, interior cabinet, and exhaust. The parameters of interest here are the exhaust air temperature and relative humidity leaving the appliance. Curve-fits are generated off the two plots in the study by picking points evenly along the curves of interest and applying a polynomial fit to the points found using a spreadsheet software. A load of wet towels weighing 10.1 kg (22.2 lbs) is loaded into an electric dryer where 4.5 kg (10 lbs) is removed during the drying process. The average exhaust velocity was measured to be about 6.8 m/s (1337 ft/min) and assuming a 10 cm (4 in) diameter exhaust duct, the exhaust volumetric air flow rate is 200 m^3/hr (117 CFM). The heating element drew an average of 22.8 amps while the electric motor drew an average 4.35 amps. The power supply for electric dryers is typically higher voltages, 220 volts, for a lower amp draw. The baseline test lasted 1 hour and with the known power consumption would result in 5.97 kWh consumed over the entire cycle. Clothes dryers are often rated by their moisture extraction rate, MER, a ratio of the drying energy used divided by the wet clothing weight at the start of the drying cycle. For the data shown here, a MER value of 0.5914 kWh/kg is calculated. If 283 cycles per year are run with a power consumption of 5.97 kWh per cycle, an annual power consumption of 1690 kWh is determined. Including the total number of homes with an electric dryer, the total annual energy consumption with this drying profile is 0.44 EJ (0.416 quads).

2.3 Clothes washer

The total number of U.S. households having a clothes washer is 93.2 million homes, slightly more than the number of households with clothes dryers (Table HC3.1 RECS, 2009). While there are two different clothes washers to consider, top loading, vertical axis machines versus front loading, horizontal axis ones, the predominate type is the top loading washer at 81% of all clothes washers used. Many sources were found for the number of wash cycles per year. The smallest value reported was 289 for North American in a worldwide study of washing machine usage which is close to the number of dying cycles assumed (Pakula *et al.*, 2010). The largest value mentioned, 1.21 loads per day or 442 cycles per year, was obtained from a two-month, field demonstration in Bern, Kansas comparing vertical and horizontal washing machines (Tomlinson *et al.*, 1998). One value was found in two reports, 392 cycles per year, from the Texas Water Development Board and the DOE (7.2 Residential Clothes Washer, 2013) (Table 2.1.16 Buildings Energy Data Book, 2011). A final source provided a breakdown of the number homes for a range of loads per week (Table HC3.1 RECS, 2009). Converting the weekly total to a yearly value, 43.9 million households use the washing machine at least 260 to 468 cycles per year versus 84.9 million homes using it at least 104 to 208 cycles per year. To be conservative on estimating the waste energy source, assuming an annual number of 289 wash cycles covers at least 43.9 million households and is larger than the annual number of drying cycles assumed.

The energy consumption per cycle of clothes washers is sometimes reported with the energy required for generating the hot water used during the wash cycle in spite of the energy being spent externally to the appliance. Low average wash temperatures for North American result in low energy consumption levels due to reduced energy required for heating water, 0.43 kWh per cycle (Pakula *et al.*, 2010). Annual energy consumption at this level with 289 cycles is 124 kWh. On the same magnitude, 110 kWh is reported when excluding energy required to heat water used by the appliance (Table 2.1.16 Buildings Energy Data Book, 2011). If the water heating energy is included, a top loader had an average, measured energy consumption of 2.26 kWh per cycle and a front loader had 0.96 kWh per cycle (Tomlinson *et al.*, 1998). Following the same procedure with 289 cycles per year, the annual energy consumption is 653 kWh for top loading washers and 277 kWh for front loading. Considering only top loading washers covering 76 million households, the total annual energy usage when including water heating energy (653 kWh/yr) is 0.18 EJ (0.17 quad) or excluding water heating energy (110 kWh/yr) is 0.03 EJ (0.028 quad).

The waste stream of clothes washers is the elevated temperature water at the drain of the appliance. To accurately determine the amount of energy available, the volume of water drained and the associated temperature is required. The type of washing machine greatly impacts the volume of waste water leaving due to the different methods employed to agitate the clothing. Top loading washers require larger volumes of water to achieve the desirable cleaning performance compared to front loading machines. With a significant number of clothes washers in the U.S. being top loaders, a value of 144 liter (38 gal) per cycle is provided from one study (Pakula *et al.*, 2010). The extensive field demonstration is able to identify average water usage per cycle for top loaders at 157 liter (41.5 gal) covering a range of 68 liter (18 gal) to 227 liter (60 gal), and for front loaders at 98 liter (26 gal) covering a range of 64 liter (17 gal) to 140 liter (37 gal) (Tomlinson *et al.*, 1998). One report lists top loaders at 155 liter (41 gal) per cycle and high-efficient clothes washers between 42 liter (11 gal) and 95 liter (25 gal) per cycle (7.2 Residential Clothes Washer, 2013). A water usage of 155 liter (41 gal) per cycle from top load washers is assumed due to the large percentage, 81%, of all U.S. washing machines used. The remaining moisture content, RMC, of the clothes at the end of a wash cycle reduces the amount of water supplied to the appliance from making it down the drain. The type of clothes, load size, and maximum spin speed of the washer all impact the RMC. Front load washers operate at much higher spin speeds and result in lower RMC values than their top load counterparts. Comparing the total volume of water remaining in the clothes to the total volume of water used during the entire cycle results in a small percentage, 2-5%, of total volume left in the clothes. Thus the impact of the RMC on the energy available in the waste stream will not be accounted for.

The other piece of information needed on the waste stream is the temperature of the water being drained. Two separate processes occur during the entire washing machine cycle, a wash and then a rinse step. 48% of households report selecting a warm water, wash cycle and 46% select a cold water wash (Table HC3.1 RECS, 2009). For rinsing, a large majority, 80% of households, report selecting a cold water rinse. Similar trends are presented in another study with 58% to 67% of all washing machine cycles using a warm water wash, cold water rinse (Tomlinson *et al.*, 1998). While for a hot water wash, and cold water rinse, the same study showed higher percentages of all washing machine cycles, 17% to 25%. To associate the relative terms: cold, warm, and hot water with temperatures, typical housing water supply temperatures and experimental data from the manufacturer were referenced. Cold water typically is around ground temperatures and is assumed to be 12.8°C (55°F). Hot water supply temperature depends on the water heater set-point used by the homeowner and is assumed to be 49°C (120°F). Appliance testing data from the manufacturer identified warm water used during the wash step at a temperature of 36°C (97°F).

2.4 Dishwasher

The market penetration of dishwashers in U.S. households is not as high as the other appliances covered so far. About 59% or 67.4 million households use a dishwasher (Table HC3.1 RECS, 2009). The largest number of dishwashing cycles per year is listed at 365 or one per day (Table 2.1.16 Buildings Energy Data Book, 2011). The current Energy Star rating for dishwashers reduced the average, annual number of cycles from 264 to 215 (ENERGY STAR V.6, 2015). From the housing survey, 104 cycles per year covers 67% of all households using dishwashers while higher usage rates, 208 cycles per year only covers 35% (Table HC3.1 RECS, 2009). For a conservative estimate, 215 cycles per year is selected for dishwasher usage covering 35% or 23.7 million households.

The energy usage of a dishwasher is heavily weighted by the required heating energy to maintain the water temperature at a high temperature of around 50°C (122°F). One study on dishwasher energy usage reports 88% of a dishwasher cycle energy input of 1.05 kWh is used for heating the inlet water, the dishes and cookware, and the physical cabinet of the appliance (Persson, 2007). Another source that does not include the energy required for water heating lists an annual energy consumption of 120 kWh or by including the associated 365 cycles per year, the energy usage per cycle is 0.33 kWh (Table 2.1.16 Buildings Energy Data Book, 2011). When considering Energy Star requirements, a standard size dishwasher cannot use more than 270 kWh per year or 1.26 kWh per cycle with 215 cycles per year (ENERGY STAR V.6, 2015). And for a compact dishwasher (less than 8 place settings) the annual energy usage must be less than 203 kWh or 0.94 kWh per cycle. A standard size, Energy Star dishwasher with 215 cycles a year, is selected by referencing an annual energy consumption of 270 kWh. With 23.7 million households, the total annual energy consumption of a standard, Energy Star dishwasher is 0.023 EJ (0.022 quad).

The amount of water used by the dishwasher per cycle is listed in one study at 10.2 liter (2.7 gal) per cycle (Persson, 2007). The current Energy Star requirements for a standard size limits water usage to 13.2 liter (3.5 gal) per cycle and for a compact dishwasher, limits usage to 11.7 liter (3.1 gal) per cycle (ENERGY STAR V.6, 2015). Minor losses of water volume to the ambient through evaporation and being trapped in the dishwasher on cookware or in the sump for the pump are all neglected. The water usage of a standard, Energy Star dishwasher is used for the volume of waste water available. With a working water temperature around 50°C (122°F) being maintained by the internal electric heater, assuming some but minimal losses, result in an estimated drain temperature of 40°C (104°F).

2.5 Cooking oven

The number of U.S. households that have a stove is 102.3 million where 60% are heated by electricity and the remaining 40% are heating by direct combustion (Table HC3.1 RECS, 2009). The number of cooking cycles a year depends on the usage per day. Only 3.5% of all households are using an oven 3 or more times a day, while 31% are using the oven at least once a day. The largest jump in percentage of households occurs when the oven is only used 2-3 times a week or 104-156 cycles a year. To capture a large number of ovens, 104 cycles a year is assumed to capture 72.1 million households. In the U.S. electric ovens annually consumed an average of 444 kWh (Hager *et al.*, 2013). An annual usage of combustion driven ovens could not be found in the literature. Therefore only electrically driven ovens will be considered. With the 104 cycle and 444 kWh per year assumption, the oven is estimated to use 4.27 kWh per cooking cycle. To determine the mass of metal creating the oven cavity, a cavity size must be determined. Referencing the local, building supply retailers identifies the most common oven size at about 150 liters (5.3 ft^3) for a free-standing range having a self-cleaning feature. The waste heat recovery of the cooking oven will utilize only the energy from an elevated temperature, metal cavity. Self-cleaning ovens offer the consumer a convenience and for waste heat, a bonus with a higher temperature heat source at the end of the cooking process. The rough cabinet dimensions for the oven cavity selected are 71 cm by 66 cm by 33 cm (28 inches by 26 inches by 13 inches). Assuming a wall thickness of 0.635 cm (1/4 inch) and using the calculated surface area of the oven, not including the door, of 1.6 m^2 (17.3 ft^2), the total estimated volume of the metal oven cavity is 0.0102 m^3 (0.36 ft^3). Assuming the oven cavity is made from steel with a density of 7,900 kg/m^3 (493.5 lbs/ft^3); the resultant mass of the cavity is about 80 kg (178 lbs). A cooking temperature of 204°C (400°F) is assumed as the initial temperature of the oven cavity at the end of a cooking cycle.

3. METHODS AND PROCEDURES

3.1 Transient Analysis

Both the household refrigerator and the clothes dryer generate a waste heat steam that is occurring over time. When the refrigerator compressor turns on, it rejects heat either to the ambient air as in a normal application or is rejected to a cooling water loop with a water to refrigerant heat exchanger. For the clothes dryer, the exhaust air is typically sent outside the home, but for heat recovery, a fin and tube heat exchanger is placed in this exhaust stream and a cooling water loop provides an ability to recover the sensible and latent heat leaving the dryer.

To simplify the analysis, a lumped parameter, heat transfer effectiveness of 80% is assumed for all heat exchangers, waste heat and the vapor compression cycle heat exchangers. The maximum heat transfer is determined by cooling or heating the refrigerant to the inlet conditions of the air or water streams. The experimental data from the manufacturer for the refrigerator is used to provide the compressor displacement and the cycle improvements implemented in a standard household refrigerator. Modeling the suction line heat exchanger attached to the capillary tube was captured by further cooling the subcooled liquid leaving the condenser with the leaving refrigerant vapor of the evaporator. The expansion process then was assumed to be isenthalpic. The subcooling leaving the condenser and the superheat leaving the evaporator were both assumed to be 5°C (9°F). Due to the suction line heat exchanger, this means additional subcooling is created for a larger evaporator capacity while sacrificing some of the compressor efficiency with higher suction superheat. Additional properties of the compressor were assumed to roughly model the compression process. A volumetric efficiency of 95%, an isentropic efficiency of 80%, and 20% of the input power into the compressor is lost as waste heat to the ambient. The evaporation and condensing conditions are free to move depending on the heat sink and source temperatures being exposed to the respective heat exchangers. For the clothes dryer, the exhaust air stream dew point temperature is checked with the incoming cooling water temperature to determine if the coil is wet or dry. If the inlet water temperature is below the exhaust air dew point, then the exit air temperature of the waste heat exchanger is at saturated conditions. Otherwise when the cooling water temperature is above the dew point, the humidity ratio is kept constant across the waste heat exchanger.

The strategy for measuring the available heat recovery during the transient process is to explore different volume sizes of water storage tanks and cooling water flow rates. The storage tank supplies the waste heat exchanger with cool water where it picks up heat and returns to the storage tank. Over time the tank temperature should rise in temperature depending on amount of heat being rejected to the water stream. Smaller water volume tanks and flow rates should obtain higher temperature at the end of the transient process. While this higher temperature water has more potential to offset energy usage in another location of the home, less total energy is recovered from the waste stream due to the higher sink temperatures for cooling.

3.2 Bulk Analysis

Due to the infrequent available heat source from the clothes washer, dishwasher, and cooking oven, all the available thermal energy is considered as a bulk source. A fixed volume of cooling water is brought into contact with the fixed volume of waste water or thermal mass as in the case of the cooking oven, to reach thermal equilibrium. Decreasing the volume of cooling water used causes the final temperature to approach the original temperature of the heat source. Maximizing the temperature of cooling water provides a higher availability energy input to another heat demand, and thus reducing the amount of external energy required. A trade-off exists between maximizing the heat recovery from the waste stream of the appliance with larger volumes of cooling water, and maximizing the supplemental heat provided to another process from smaller cooling water volumes providing higher temperatures.

4. MODELING RESULTS

4.1 Transient Heat Generation

Four different tank volumes were run for the refrigerator model covering a typical cycle time of 50 minutes and a cooling water flow rate of 1.9 liter/min (0.5 gal/min). A summary of the water tank temperature at the end of the cycle run for each tank volume is shown in Table 1. The tank volume sizes were selected based off a desirable water temperature.

Table 1: Summary of household refrigerator performance with water cooling and air cooling condenser

Cooling Stream	Tank Volume [liter/gal]		Maximum Tank Temperature [C/F]		Total Evaporator Energy [kJ]	Total Compressor Energy [kJ]	Avg. Cooling COP [-]
Water	19	5	28.58	83	753.6	198.4	2.71
Water	15	4	32.5	90	745.8	205.3	2.62
Water	11	3	39	103	730.8	217.8	2.47
Water	7.6	2	55.9	132	694.2	246.8	2.17
Air	-		-		669.6	267.6	2.20

To see the transient analysis of the tank temperature for the same tank sizes, Figure 1 shows 4 curves with the same conditions shown in Table 1.

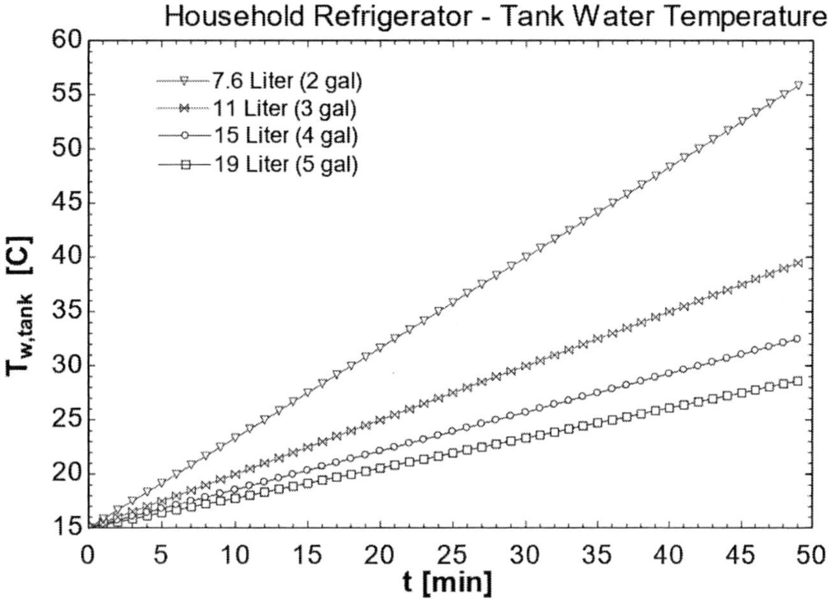

Figure 1: Storage tank temperature for water cooled condenser of household refrigerator

A summary of the clothes dryer simulation runs is shown in Table 2. As the tank size increases the percent of energy recovered increases but the maximum tank temperature decreases. Also the air outlet temperature of the heat recovery process decreases, providing a better possibility of allowing the dryer exhaust to be vented indoors instead of outdoors.

Table 2: Summary of clothes dryer heat recovery

Tank Size [liter/gal]		Start Tank Temperature [C/F]		End Tank Temperature [C/F]		Max Air Outlet Temperature [C/F]		Total Heat Recovered [kJ]	Total Energy Input [kJ]	Percent Energy Recovered
378	100			22.2	72.0	30.5	86.9	11408		51.8%
113	30			33.0	91.4	39.1	102.5	8369		38.0%
76	20	15	59	37.7	99.8	42.9	109.2	6937	22041	31.5%
38	10			45.4	113.7	49.1	120.3	4471		20.3%
19	5			54.5	130.1	56.3	133.4	2668		12.1%

To understand the impact of the storage tank size for the same water flow rate as the refrigerator, 1.9 liter/min (0.5 gal/min), Figure 2 shows the tank temperature as a function of time. As the volume decreases from 113 liter (30 gal) to 19 liter (5 gal), the final tank temperature at the end of the drying cycle goes from 33°C (91.4°F) to 54.5°C (130°F).

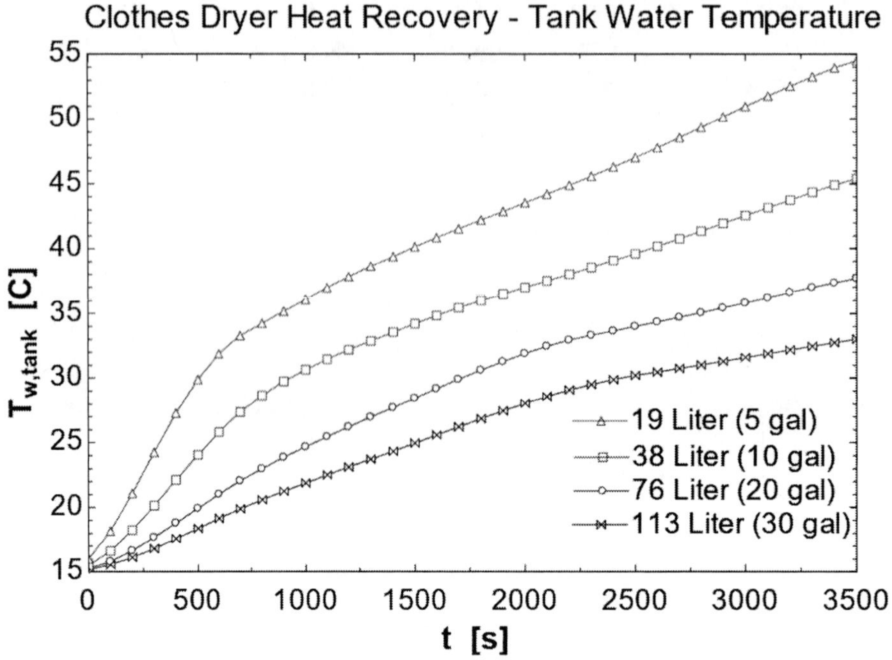

Figure 2: Impact of the storage tank volume size on the water temperature stored in the tank

A secondary benefit of having heat recovery at the exhaust of the dryer is the reduction of the dryer air exhaust temperature and the removal of moisture from the exhaust stream. Figure 3 provides an indication of the exhaust air stream leaving the heat exchanger changes as the storage tank increases in temperature over time.

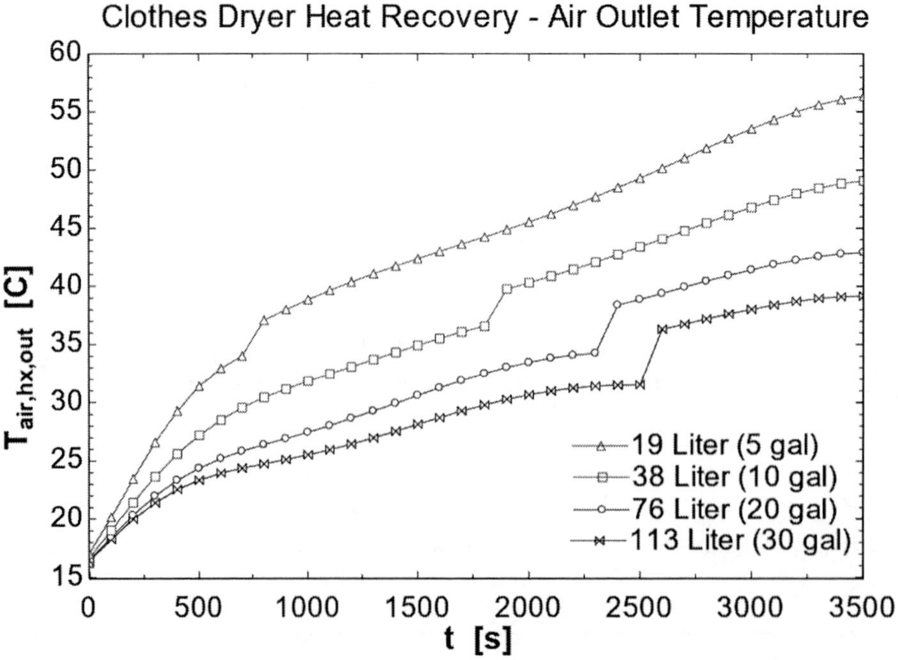

Figure 3: Cooled dryer exhaust air leaving the waste heat exchanger as a function of the storage tank volume

4.2 Bulk Heat Generation

Using the assumed cycle parameters for each appliance providing a bulk energy source, the amount of heat recovered by the cooling water source is calculated for a number of tank volumes. A summary of all the results is listed in Table 3. The assumed cycle energy per cycle presented in section 2 is converted from kWh to kJ to calculate a percent of energy input recovered by the cooling water source.

Table 3: Summary of bulk energy recovery from the clothes washer, dishwasher and cooking oven

Tank Volume [liter/gal]		Clothes Washer [8136 kJ/cycle] Final Tank Temperature [C/F]		Heat Recovered [kJ]	% Recovery [-]	Dishwasher [4536 kJ/cycle] Final Tank Temperature [C/F]		Heat Recovered [kJ]	% Recovery [-]	Cooking Oven [15372 kJ/cycle] Final Tank Temperature [C/F]		Heat Recovered [kJ]	% Recovery [-]
7.6	2	34.9	94.8	700.7	8.6%	30.0	86.1	546.2	12.0%	123.9	255.0	3530	23.0%
15.1	4	33.9	93.1	1339	16.5%	25.4	77.7	799.6	17.6%	90.5	194.9	4923	32.0%
22.7	6	33.0	91.4	1922	23.6%	22.7	72.9	945.8	20.9%	72.3	162.1	5652	36.8%
30.3	8	32.2	89.9	2457	30.2%	21.0	69.8	1041	23.0%	60.9	141.7	6096	39.7%
37.9	10	31.4	88.6	2951	36.3%	19.8	67.6	1108	24.4%	53.2	127.8	6398	41.6%
45.4	12	30.7	87.3	3407	41.9%	18.9	66.0	1158	25.5%	47.6	117.7	6615	43.0%
53.0	14	30.1	86.1	3829	47.1%	18.2	64.7	1196	26.4%	43.4	110.1	6776	44.1%
60.6	16	29.5	85.0	4222	51.9%	17.6	63.7	1226	27.0%	40.0	104.1	6902	44.9%
68.1	18	28.9	84.0	4588	56.4%	17.2	62.9	1251	27.6%	37.4	99.3	7004	45.6%
75.7	20	28.4	83.0	4929	60.6%	16.8	62.2	1271	28.0%	35.2	95.3	7086	46.1%

5. DISCUSSION

There exists a strong trade-off between the tank volume of water used to circulate for heat recovery and the final temperature at the end of the appliance cycle. While smaller tank volumes provide higher water storage temperatures which are useful as a direct input to another appliance cycle, smaller percentages of the energy recovery result due to the higher water temperatures being delivered to the appliance. An improved analysis would be to calculate the availability or exergy of the heat recovered by the water cooled heat exchanger. Most of the input energy for all appliances is pure exergy in the form of electricity since it can be used to perform any task. Once it is converted by the appliance into useful work through refrigeration, cooking or wet cleaning, much of this potential energy is destroyed. The leaving waste stream is now low in availability due to the low temperature and its close to a dead state where no more useful work can be extracted.

An improvement of the work presented here would be to have one fixed volume tank and conduct a large transient analysis where different appliances would come on and off. This approach would explore potential benefits or pitfalls when larger storage tanks are used to combine all the waste heat into one location. Additionally, the ability to offset some of the household domestic hot water would be introduced by adding in a hot water schedule relative to the appliance schedule.

While the experimental data referenced for the clothes dryer was very detailed and provided an easy way to calculate the available energy from the waste stream, the drying conditions seen might not be fully representative of a typical cycle. The clothing load was large and the heat setting was on high. Obtaining exhaust air data for different dryer conditions would properly identify an optimal tank size for circulating the cooling water.

6. CONCLUSIONS

Household appliances account for the second largest site energy consumption, 27%, after the HVAC system for the U.S. residential sector. Many approaches explored in the past to address this area were mainly focused on only the appliance itself. If the scope for improvement also considers factors external to the appliance, there is the potential for greater energy savings. By quantifying the expected energy available in the waste stream for five major appliances; household refrigerator, clothes dryer and washer, dishwasher, and cooking oven, a potential energy source is presented. A cold water cooling stream is applied to the waste stream of each appliance and an estimated amount of energy can be recovered. The household refrigerator is modeled having an increase in cooling capacity of about 12% and a reduction on compressor power consumption of about 26%. A sample operation of the clothes dryer has the exhaust air stream being cooled down to 30.5°C (86.9°F) or on the other side, is able to heat 19 liter (5 gal) of water up to about 54.5°C (130.1°F). A different modeling approach is used for the bulk energy sources. A large volume of water was available by the clothes washer, but due to typical operation characteristics, low wash and rinse temperatures, the waste stream was not high in temperature. The highest temperature of water generated was at 35°C (95°F). While the dishwasher provided higher heat source temperatures, 40°C (104°F), than the clothes washer, 36°C (97°F), the opposite was true in that the volume of waste water drained is very low compared to the clothes washer 11.7 liter (3.1 gal) to 155 liter (41 gal). Thus storage tank water temperatures did not reach above 30°C (86°F) even with low storage volumes. The cooking oven can generate very high water temperature depending how small of a storage tank is connected. There exists a controls risk when preventing the cooling water stream from reaching the boiling point and creating steam. Further work in this area is recommended due to the potential of high water temperatures generated from waste energy streams not currently being captured.

REFERENCES

Buildings Energy Data Book. (2011). Pacific Northwest National Laboratory, Buildings Technologies Program, Energy Efficiency & Renewable Energy, U.S. Department of Energy. Retrieved from http://buildingsdatabook.eren.doe.gov/

Butturini, R., Lee, A., & Dunmore, F. (2004). Detection of Abnormal Operating Conditions in Electric Clothes Dryers. U.S. Consumer Product Safety Commission.

ENERGY STAR Market & Industry Scoping Report Residential Clothes Dryers. (2011). U.S. Environmental Protection Agency. Retrieved from: https://www.energystar.gov/sites/default/files/asset/document/ENERGY_STAR_Scoping_Report_Residential_Clothes_Dryers.pdf

ENERGY STAR Version 5.0 Specification for Residential Refrigerators and Freezers. (2013). U.S. Environmental Protection Agency. Retrieved from: https://www.energystar.gov/products/spec/residential_refrigerators_and_freezers_specification_version_5_0_pd

ENERGY STAR Version 6.0 Specification for Residential Dishwasher. (2015). U.S. Environmental Protection Agency. Retrieved from: https://www.energystar.gov/products/spec/residential_dishwasher_specification_version_6_0_pd

Hager, T.J., & Morawicki, R. (2013). Energy consumption during cooking in the residential sector of developed nations: A review. *Food Policy 40*. 54-63.

Klein, S. A., & Alvarado, F. L. (2002). Engineering equation solver. F-Chart Software, Madison, WI.

Meyers, S., McMahon, J.E., McNeil, M., & Liu, X. (2003). Impacts of US federal energy efficiency standards for residential appliances. *Energy 28,* 755-767.

Pakula, C., & Stamminger, R. (2010). Electricity and water consumption for laundry washing by washing machine worldwide. *Energy Efficiency 3*, 365-382.

Persson, T. (2007). Dishwasher and washing machine heated by a hot water circulation loop. *Applied Thermal Engineering 27*, 120-128.

Refrigerator Market Profile. (2009). Energy Star Program, U.S. Department of Energy. Retrieved from http://apps1.eere.energy.gov/states/pdfs/ref_market_profile.pdf

Residential Energy Consumption Survey (RECS). (2009). U.S. Energy Information Agency. Retrieved from http://www.eia.gov/consumption/residential/data/2009/

Section 7.2 Residential Clothes Washer Incentive Program. (2013). Texas Water Development Board. Retrieved from http://www.twdb.texas.gov/conservation/BMPs/Mun/doc/7.2.pdf

Tomlinson, J. J., Rizy, D. T. (1998). Bern Clothes Washer Study Final Report. Energy Division, Oak Ridge National Laboratory, U.S. Department of Energy.

ACKNOWLEDGEMENT

The authors would like to thank Whirlpool Corporation for their support on the research presented here.

Net Zero Energy House Evaluation

Meher TALEYARKHAN[1], William HUTZEL[2]*

School of Engineering Technology, Purdue University
West Lafayette, Indiana, United States of America
[1]mtaleyar@purdue.edu
[2]hutzelw@purdue.edu

*Corresponding Author

ABSTRACT

A net zero energy home was designed and built by Purdue University for the 2011 Solar Decathlon in Washington, DC. After earning a 2nd place finish in this international competition, the home was permanently located in Lafayette, IN and became a private residence. Since that time, energy monitoring equipment has measured both energy consumption of the home and energy generation by the solar photovoltaic panels. Three full years of data (2013 to 2015) have been collected to show that the home has met its original design goal for net zero energy. Beyond net zero, the data also shows interesting seasonal trends for HVAC, water heating, lighting, and other energy consuming devices. A survey of the current occupants provides further insight into how well features like indoor air quality, comfort, and aesthetics were incorporated into the overall design. The results suggest that a net zero energy home can be accomplished without sacrificing quality of life.

1. INTRODUCTION

A net zero energy home was created and built by a Purdue University student team for the 2011 Solar Decathlon. The goal of this competition was to design and build a home that took into account aesthetics, comfort, practicality, affordability, and energy efficiency. The term net-zero means that the total amount of energy used on an annual basis is equal to the amount of energy produced by a renewable energy source, such as solar power in this scenario. The Purdue entry was called the INhome, short for Indiana Home, because of the stated goal of showcasing advanced technologies in a practical residence that would appeal to the people from the Midwest.

After the Solar Decathlon competition, the INhome was moved to an affordable neighborhood in Lafayette, Indiana. Figure 1 is a street view of the home as it looks today. The south-facing roof in the photograph is covered in solar panels and is part of the INhome's 9.0 kW solar photovoltaic array that is comprised of 36, 240 watt solar panels. Although the INhome is still tied to the local electric grid, it was designed to provide all of its own electricity. The purpose of this paper is to evaluate the INhome's net zero performance after 3 full years of occupancy (2013-2015).

Figure 1: Street view of INhome

The INhome has many passive features, such as high levels of insulation and air sealing, which minimize energy consumption. Figure 1 shows the operable clerestory windows that provide day lighting and also, when open, allow for fresh air flow. The clerestory windows work in tandem with the large double-hung windows at ground level to encourage natural ventilation during appropriate seasons. The front porch and large roof overhangs provide external shading to the walls of the house are further examples of passive features that promote energy efficiency.

The INhome also has several active features that are installed to reduce energy consumption. Several web-based controllers automate the HVAC (heating, ventilating, and air conditioning) and security systems in the INhome. An energy recovery ventilator helps reduce heating and cooling loads. Also, high efficiency appliances were used to minimize electric loads in the kitchen and laundry. Together, the active and passive features were designed to make it possible for the INhome to be net zero.

Figure 2 is a floor plan for the INhome. It is a 92.9 square meter house that consists of a living room, kitchen, bathroom, 2 bedrooms, garage, and a utility room. The kitchen and living room is a combined great room with high vaulted ceilings that makes the space seem larger than it really is. The overall layout is compact, but offers a sustainable design without sacrificing modern comforts and amenities.

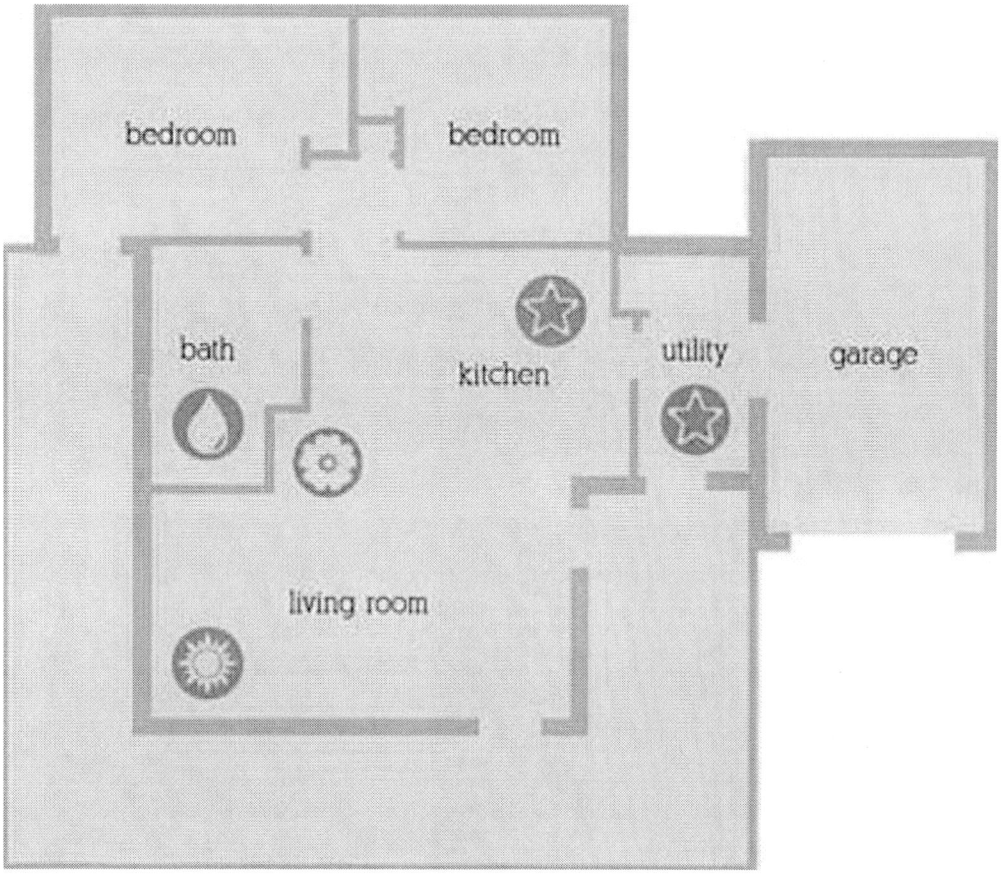

Figure 2: Floor Plan of the INhome

One of the primary energy consuming devices is the electric air source heat pump that provides heating and cooling. This heat pump has two individual compressors that allows for fewer on and off cycles. This not only extends the life of the compressor but also results in a higher efficiency. The heat pump works by using the outdoor air to produce heating and cooling for the INhome, yielding a SEER rating of 19.

The indoor air quality of the INhome is controlled by the energy recovery ventilator (ERV) that removes airborne contaminants and allergens. The ERV provides fresh air inside the home and also pre-conditions the incoming air, reducing the energy used by the HVAC heat pump to condition the air to room temperature.

Four different web-based platforms provide the "smart" features of the INhome. One is an advanced system for monitoring component level electrical consumption of the home. The second system monitors the performance of the solar panels. The third systems provides home security, such as the ability to lock/unlock doors and turn on/off lights remotely using a smart phone app. The final system provides web-based monitoring and control of the HVAC system in the home.

2. EVALUATION

Figure 3 shows the electric energy in the INhome over a three-year period. The data summarizes the electrical consumption consumed by the home (blue) and the solar production of the solar panels (orange) measured in kilowatts per hour (kW-h) for the years 2013, 2014, and 2015. The home is all electric so this summary provides a comprehensive view of total energy consumption. The data was collected using a web-based electricity monitoring system that was attached to the main electric panel and the solar array.

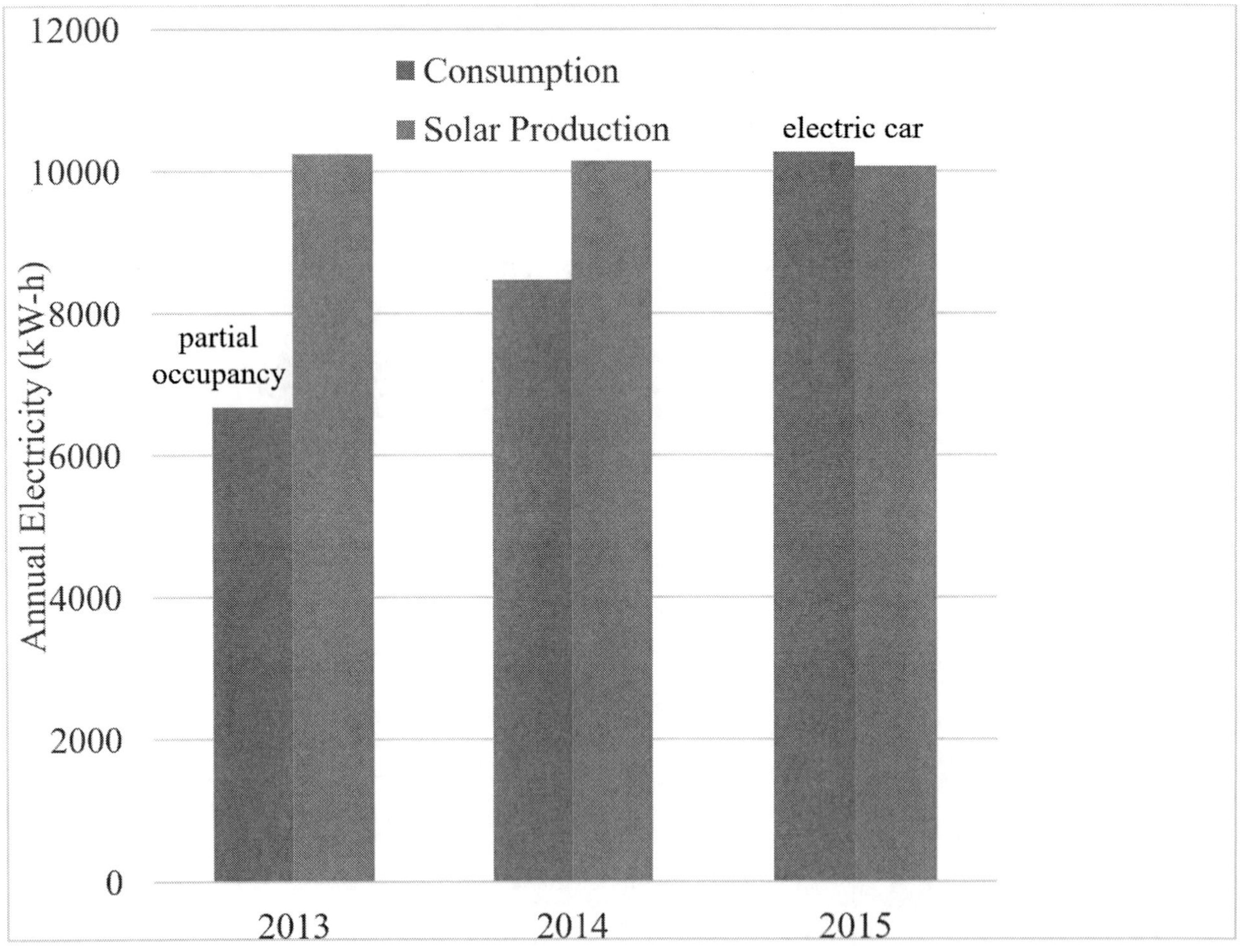

Figure 3: The INhome has been net zero for the past 3 years.

Figure 3 shows that the INhome has been net zero for each of the last three years. The solar production (orange) has been consistent at about 10,000 kWh each year. The production has been surprisingly constant, despite small-scale weather anomalies like the "polar vortex" that covered the INhome's solar panels with snow for much of January, 2014.

The electrical consumption (blue) has increased in each of the past 3 years. As noted on Figure 3, 2013 was not quite a full year of occupancy. The 2013 energy consumption was low because the homeowners were still moving in. 2014 was a full year of occupancy, but the solar panels still generated approximately 15% more energy than the homeowner actually used. For 2015, the electrical consumption and solar production were nearly equal.

The 2015 match between consumption and production was not an accident, but a smart decision by the homeowner to take full advantage of the electricity being generated by the INhome. The homeowners purchased an electric lawnmower and even an electric car to use all of the solar electricity produced. This is a by-product of the Indiana Utility Regulatory Commission's net metering rules, whereby a homeowner is credited for the solar power they produce but there are no additional benefits for an annual surplus. Thus it is wise to use all of the renewable electricity as it is generated.

Figure 4 shows the individual components of the electrical consumption of the INhome for 2015. The data was taken from the main electric panel by a system that monitors the energy use of each circuit in the house. The five broad categories are HVAC, Kitchen, Laundry, Controls, and Lights & Plugs. The HVAC component includes the electric consumption of the heat pump (outdoors), air handler (indoors), and the ERV system. The kitchen component is the refrigerator, oven, dishwasher, and other devices used for cooking. The laundry component is a washer and dryer. The controls_component includes four different types of web-based energy monitoring equipment. The lights & plugs is all the remaining electricity in the home.

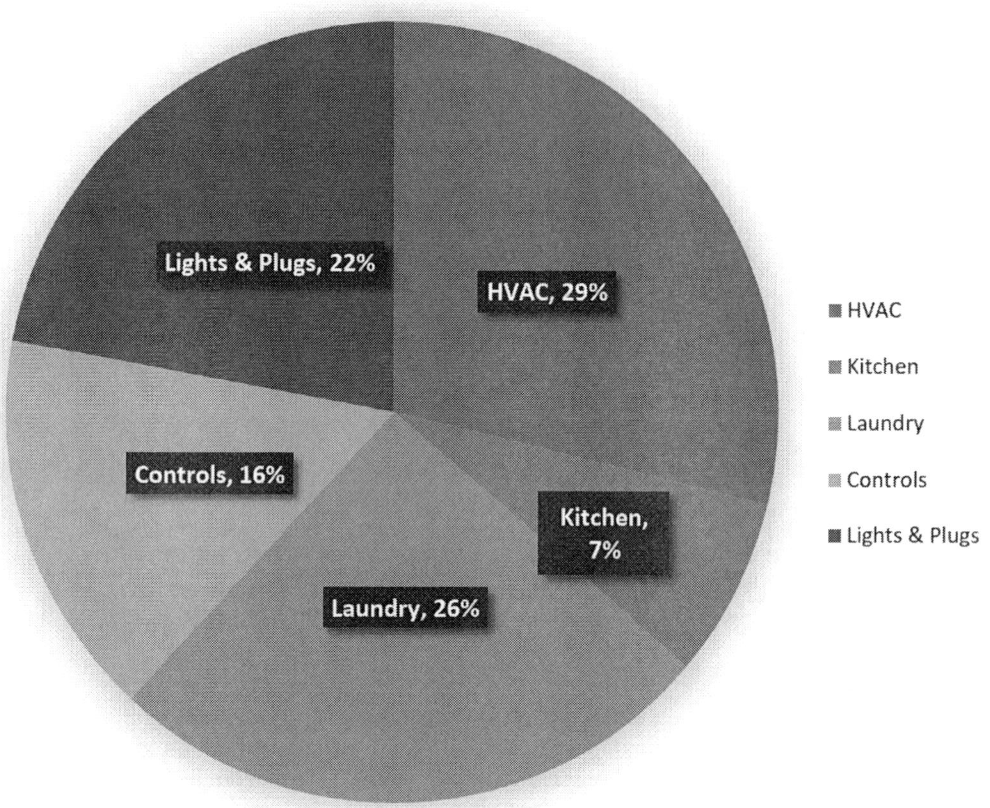

Figure 4: 2015 Electricity Use of the INHome

Figure 4 shows that the HVAC system is the single biggest consumer of energy. That is not surprising due to the sometimes extreme weather conditions of Indiana. The laundry used the 2nd most energy, followed by the lights & plugs. The controls used a surprisingly large amount of energy, but this is there are 4 separate monitoring systems that operate 24 hours a day and 365 days a year. It is also interesting that the kitchen consumes the smallest amount of energy, probably due to the high efficiency appliances used throughout.

Figure 5 is a graph of annual energy consumption and solar panel production for 2015 on a cumulative basis. The vertical axis is a % of annual electricity to provide insight about the interaction between production (blue) and consumption (orange) on a monthly basis. The graph shows that electrical energy is consumed at a steady rate throughout the year. The blue trend line is mostly linear. Most of the electric loads in the INhome are fairly constant. Even the electricity used in space conditioning, heating in the winter and cooling in the summer, is essentially at constant rate.

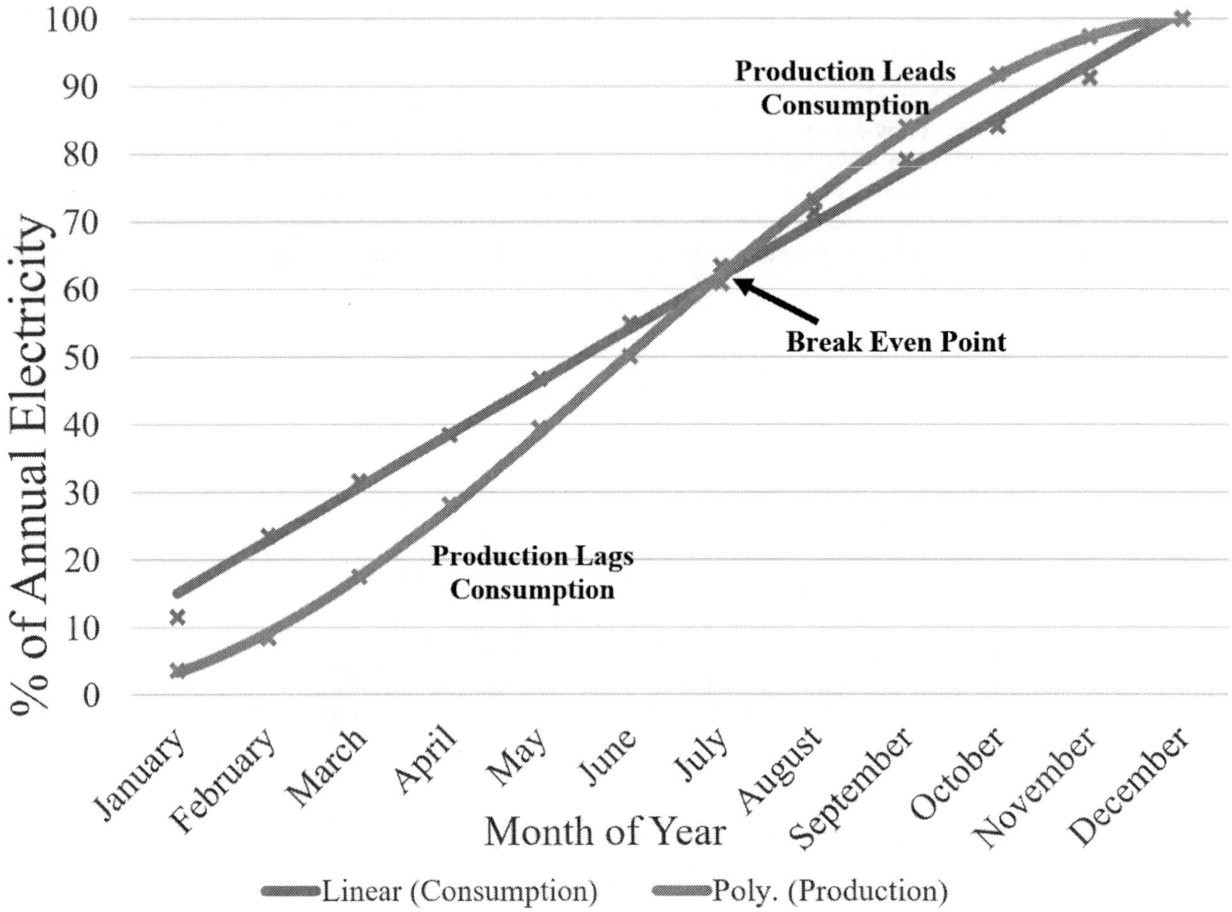

Figure 5: 2015 Annual Energy Consumption vs. Solar Panel Production

Figure 5 shows that energy production by the solar panels (orange) varies over the course of a year. In January when the days are short and snow sometimes covers the panels, solar production lags consumption so some energy is taken from the grid. Solar production starts increasing in March. By the July/August time frame solar production reaches the break-even point where consumption and production are in balance. Solar production leads consumption until December when the days are short and solar energy is at its minimum.

Although it is not clearly visible from Figure 5, a closer inspection of the electricity consumption shows that it does increase slightly for the winter months, January and February. Then as the weather conditions become warmer, the electrical consumption decreases. Electricity use then increases sharply from October onwards, when the weather conditions start to become much colder. The electrical consumption increases primarily due to the heat pump. During the warmer months of summer, the heat pump works best due to using the warm air from outside. Though, during the winter there is very little heat that can be used by the heat pump. Thus, the heat pump then relies on electricity to heat and cool the INhome during the colder months. As a result, more electricity is consumed during these colder months.

Overall, Figure 5 shows that the electrical consumption and solar panel production both offset one another over the course of a year. When solar panel production increases the electrical consumption decreases and vice versa. The overall result is a net-zero energy home.

3. HOMEOWNER SURVEY

An informal survey was given to the current homeowners of the INhome to gage their overall satisfaction after 3 years of occupancy. The 10 question survey about the installation of solar panels, energy efficient appliances, and overall comfort of the home was submitted to the homeowners before a short interview at their home. This non-scientific survey revealed that overall the homeowners very much enjoy living in the home. In regards to the solar photovoltaic panels, the homeowners recognize that these panels produce more power than what the home is using. As a result of this, the only energy cost the homeowners pay is the service fee to the electric utility for being connected to the grid. The homeowners are thrilled by this and have even recommended the installation of solar photovoltaic panels to several others.

The energy efficient appliances were also highly recommended. One of the more unique energy efficient appliances is an induction cooktop that was well received by the homeowners, because it is both aesthetically pleasing and performs very well. The only drawback was that the induction cooktop requires certain types of cookware that only a few companies sell. The air source heat pump for heating and cooling was another appliance that was rated highly by the homeowners because it provides consistent temperature and humidity control inside the home. The ERV system was very well received in terms of reliability, functionality, and practicality for providing adequate fresh air and controlling humidity levels in the INhome. It was consistent in automatically turning on and off when needed, and was described as being nearly flawless in its functionality.

The overall comfort of the INhome was described by the homeowners as a "delight to live in". They cited the overall layout, architecture, and outside wraparound deck as a few of the most enjoyable features. The open-concept layout of the home in particular was a favorite feature, in that it manages to let in large amounts of natural light. This was primarily accomplished by the large south-facing triple-pane and clerestory windows. The triple-pane windows help keep the home from any heat losses and the clerestory windows provide additional fresh air into the home when open.

There was one interesting development in terms of the smart home controls that provided detailed energy monitoring and smart phone apps for door locks and lights. Although they had a high "wow" factor to the new homeowner, these features were not really used on a day to day basis. As years passed, this monitoring equipment became more of a nuisance because it took up space and used un-necessary energy. That is why these smart monitoring components were ultimately removed. The homeowners placed a higher value on simplicity than high tech gadgets.

4. CONCLUSIONS

The INhome achieved one of its main design goals by being consistently net zero from 2013 to 2015. The first years of occupancy showed that the homeowners were not utilizing all the energy produced by the solar panels, resulting in excess energy being delivered to the electric grid. However, electricity consumption has increased as the homeowners made a conscious decision to use more "free" electricity. By 2015, the production and consumption levels were nearly equal. In addition, it was also found that the HVAC appliances used the most energy every year, due in part to the fluctuating weather conditions. The overall functionality and livability of the INhome was described by the homeowner as being flawless and very enjoyable to live in and that the INhome is a perfectly designed home to showcase energy efficiency and solar energy of today.

NOMENCLATURE

INhome	Indiana home
HVAC	heating, ventilating, and air conditioning
ERV	energy recovery ventilator

REFRENCES

(2011). *Team Purdue INhome*. Retrieved from http://www.purdue.edu/inhome/

(2012, July 29). *Solar Decathlon 2011 Purdue University*. Retrieved from http://www.solardecathlon.gov/past/2011/team_purdue.html

Hutzel, W. J., Shaurette, M.E. and Loss, B.W., (2012). Competing in the 2011 Solar Decathlon. *Proceedings of the 2012 ASEE Annual Conference*, American Society for Engineering Education, San Antonio, TX, 12 p.

Wallpe, Jordan and Hutzel, William, (2012), Assessing the Engineering Performance of Purdue University's INhome as an Example of Affordable Net-Zero Energy Housing, *Proceedings of the 2012 Summer Study on Energy Efficiency in Buildings*, American Council for an Energy Efficient Economy, Pacific Grove, CA.

ACKNOWLEDGEMENT

The authors acknowledge the assistance of Marifran and Jim Mattson, the owners of the Indiana Home, for allowing us to monitor energy use and for providing us feedback on the experience of living in a net zero home

A Non-Iterative Balancing Method for HVAC Duct System

Haoran CHEN[1], Wenjian CAI[2]*, Qingguo WANG[3]

[1]EXQUISITUS, Centre for E-City, School of Electrical and Electronic Engineering,
Nanyang Technological University, Nanyang Avenue, Singapore 639798, Singapore
Chen0804@e.ntu.edu.sg

[2]EXQUISITUS, Centre for E-City, School of Electrical and Electronic Engineering,
Nanyang Technological University, Nanyang Avenue, Singapore 639798, Singapore
Ewjcai@ntu.edu.sg

[3]Institute for Intelligent Systems, the University of Johannesburg,
South Africa
wangg@uj.ac.za

* Corresponding Author

ABSTRACT

Building Heating, Ventilation and Air Conditioning (HVAC) system maintain comfortable indoor environment by supplying processed air to each terminal precisely through duct system. Testing, Adjusting and Balancing (TAB) plays critical role in achieving desired air distribution. Traditional TAB method is inaccurate and inefficient due to its trail-and-error natural, which forces people to pay high but expect low. Recently, it has been proposed that non-iterative approach to TAB is promising to improve performance and reduce cost. In this paper, a novel non-iterative balancing method is developed and implemented for TAB engineers to adjust dampers systematically and efficiently. Different from other TAB methods, this method is based on modeling and optimization. The mathematical model for duct system is firstly developed from its components including fan, duct segments and dampers to predict flow rates and pressures in the duct system for any damper positions. To identify the parameters in the model, flow rate measurements are taken for each terminal on real system under different damper positions. With the obtained model, optimal damper positions that gives desired air distribution are calculated by minimizing a specific objective function. To facilitate the adjusting process in real duct system, a sequential tuning instructions are generated which can help engineers to adjust dampers to their proper position using flowmeter as indicators. In this sequential tuning process, each damper only adjusts once to reach balance. Because the pressure and airflow dynamics of the duct system has been modeled, the entire TAB procedure is deterministic and non-iterative. Simulations are performed to validate the effectiveness of this method in Matlab/Simulink environment. Comparison study with existing methods shows that the proposed TAB method significantly shorten the duration of process and reduces balancing error while using easily-accessible equipment like pressure sensor and flowmeter only. It can be expected that the TAB service contractor will apply this method for advanced duct system where accurate air distribution is strictly required.

1. INTRODUCTION

Due to the impending shortage of energy and growing concern on environment, modern heating, ventilation and air conditioning (HVAC) systems in buildings are required to catch up with rising threshold on energy efficiency while pursuing higher indoor air quality (IAQ). Commercial building uses central HVAC system to maintain comfortable indoor environment, which is usually quantified by a combination of temperature, humidity and CO2 concentration level, by supplying a specific amount of process air to each terminal through duct system. Unbalanced duct systems that are unable to deliver desired amount of air to each terminal cause unsatisfactory thermal comfort and bad air circulation pattern. Consequently, occupants may suffer from sick building syndrome. Moreover, unbalanced duct system may give misleading signal to control system, causing the performance and energy efficiency of the system degraded. In fact, it is suggested that balancing should be performed regularly for all constant/variable air volume (CAV/VAV), induction, return air and even toilet and kitchen exhaust systems [5]. Moreover, as the system aging and changing room configurations during the operation, actual flow rates no longer match the demands and re-balancing the system are required. Therefore, testing, adjusting and balancing (TAB) plays critical role in balancing the system to achieve desired air distribution. However, many building owners are reluctant to perform TAB regularly, as the traditional TAB methods are costly and time-consuming, which largely interrupts normal operating. These disadvantages of the conventional TAB methods are due to the lack of solid theoretical analysis. Because of their inefficient trail-and-error nature, the results are largely depending on engineers' skill and experience.

A few researchers proposed new TAB methods to overcome these disadvantages. Federico Pedranzini et al. [3] designed a progressive flow method by utilizing the characteristic of duct resistance and air distribution. This method adjusts dampers progressively from the furthest terminals (largest pressure drop) to the closest terminals (smallest pressure drop). In order to decoupling the interactive effect in the ducts, variable speed drive and a controller for fan speed control at reference terminal are required. It is estimated that about 43% to 33% adjustments can be reduced. However, when the system is rebalanced due to the change of desired air distribution, the whole process must be repeated. Small [4] proposed developing mathematical model and balancing the system based on the obtained model. In order to determine the model parameters, the flow rates of each terminal under two different settings of damper positions are measured. After model identification, the damper positions for balancing are calculated. This method has advantage in rebalancing because the obtained model can be reused. However, due to the inevitable sensor noise and asymmetry of choosing the settings of damper positions, the model accuracy is poor and the sensitivity is biased distributed in the model. Hence the final balancing results is not very accurate.

In this paper, a novel non-iterative balancing method is developed and implemented for TAB engineers to adjust dampers systematically and efficiently. A more accurate model based on Darcy-Weisbach equation is developed to better characterize the duct system. New measuring process is designed to obtain data more accurately while simplify the operation procedure for TAB engineers. Differential and algebraic equation (DAE) software package and maximum a posteriori (MAP) estimation is applied in the model identification algorithm to reduce the uncertainties cause by sensor noise. A sequential tuning procedure using flow meter as indicators for adjusting dampers is implemented. The results are validated by simulation and compared to existing methods. The use of this method is also possible to be extended to balancing other system based on similar physical principles like exhaust duct network.

2. MODEL DEVELOPMENT

A schematic of duct model is shown in Figure 1. To model the air flow distribution behavior in duct system for balancing, the pattern of pressures and flow rates under a given duct system must be analyzed. Duct system consists of many components including duct segments, elbows, dampers and terminal outlets. Therefore, model of duct system must be built up from these components.

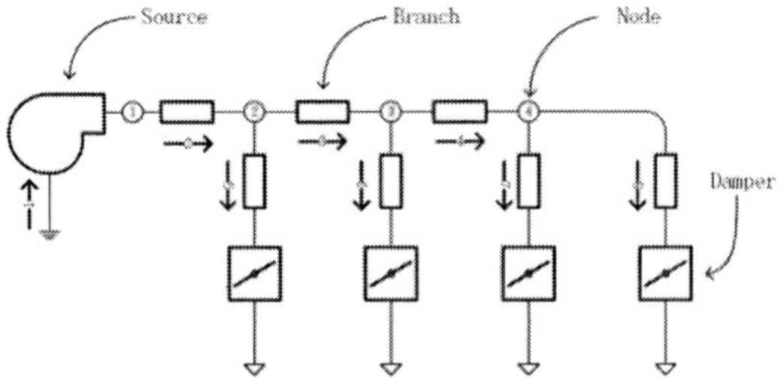

Figure 1: Schematic of duct model

Darcy-Weisbach equation has been used to estimate pressure drop along duct due to friction for fully developed flow in straight conduit for long [2], which is given by:

$$\Delta P = C_f \left(\frac{L}{D}\right)\left(\frac{\rho}{2}V^2\right) \tag{1}$$

where C_f is friction coefficient, L is duct length, D is duct diameter, ρ is air density and V is the average air velocity. The friction coefficient C_f depends on Reynolds Number and roughness of duct wall, the Colebrook's equation estimates C_f implicitly:

$$\frac{1}{\sqrt{C_f}} = -2\log\left(\frac{\varepsilon}{3.7D} + \frac{2.51}{\mathrm{Re}\sqrt{C_f}}\right) \tag{2}$$

where ε is the roughness of duct wall, Re is the Reynolds Number. Since Colebrook's equation is an implicit equation, instead of approximated formulas, DAE toolbox is employed to calculate the values.

For elbows and other fittings in the duct system, the pressure drop is estimated empirically by:

$$\Delta P = \frac{1}{2} C_f \rho V^2 \tag{3}$$

where local friction coefficient C_f is tested in experiments and collected as a look-up table in ASHRAE Duct Fitting Database. Many types of components have been tested including elbows, junctions, reducers and dampers at different positions.

The fan curve which characterize the relationship between pressures and flow rate is approximated by quadratic model:

$$\Delta P = P_{\max}\left(1 - \left(\frac{q}{Q_{\max}}\right)^2\right) \tag{4}$$

The entire duct system model is assembled in Simscape environment, which fully supports physical modeling by DAE approach. In Simscape, the physical domain is defined by the through and across variables. Through variables are subjected to conservation laws that sum up to zero among interconnected components, while across variable shares values for these interconnected components. In this application, flow rates through each branch are defined as through variables and pressures are defined as across variables. By building up the entire duct system in the Simscape environment, simulations can be performed to estimate the pressures and flow rates in the duct system. Different from usual blocks in Simulink, the ports of blocks built in DAE model are non-directional, which means the variables satisfy both sides. The equation (1)-(4) which relate through and cross variables of the two ports are encoded into different blocks including dampers, terminals and fans. Figure 2 demonstrates the Simscape environment to integrate the model of duct system.

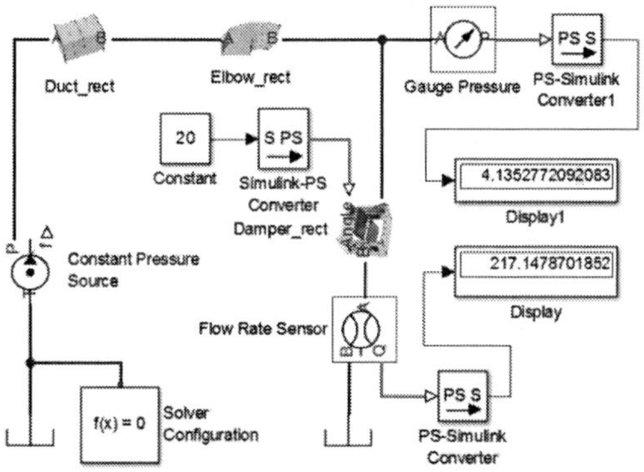

Figure 2: Simulink/Simscape duct system modeling environment

3. MODEL IDENTIFICATION

In order to obtain accurate prediction from this model, the parameters in the model must be identified by measuring data. Measurements are taken in $N+1$ steps for a N terminal duct system. In each step, the flow rates of each terminal are measured and recorded. In the first step, all dampers are fully open. For the following N steps, each damper is closed one time. Therefore, the overall data are symmetrically sampled to avoid biased calibration results. The purpose of keeping exactly one damper closed in each measurement are three-folds. First, cover the largest range of testing by setting to fully closed and opened position. Second, simplify the analysis because the fully closed damper completely stops the airflow. Third, avoid any potential damage to the system due to the high plenum pressure when too many dampers are closed. The detail procedure is shown in Figure 3.

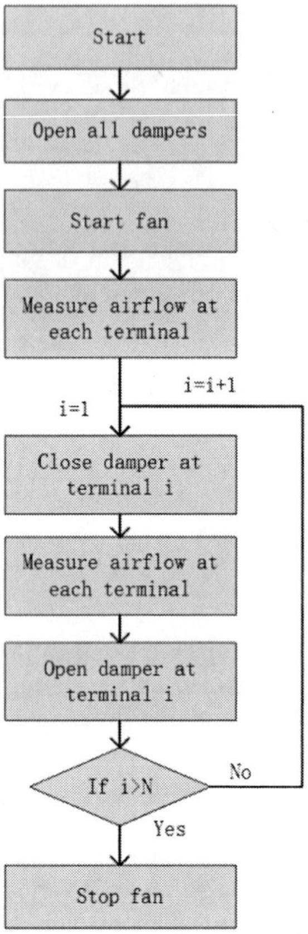

Figure 3: Measuring procedure

To identify the model parameters, Maximum a posteriori (MAP) estimation is used. Define the undetermined parameters $\boldsymbol{\beta}$ and the damper positions $\boldsymbol{\theta}$, then the simulation process can be represented as a function $\mathbf{X}=\left[\mathbf{q}^T \ \ \mathbf{P}^T\right]^T = f(\boldsymbol{\beta},\boldsymbol{\theta})$. To derive the formula, the following assumptions are applied:

I) Measurement $\mathbf{Z}$ is conditional independent to parameters $\boldsymbol{\beta}$ given states $\mathbf{X}$ and damper position $\boldsymbol{\theta}$.

II) Measurement $\mathbf{Z}$ follows multi-variable normal distribution with mean $\mathbf{CX}$ and covariance $\boldsymbol{\Sigma}$, where $\mathbf{C}$ is the observation matrix.

III) The probability of states $\mathbf{X}$ given system parameters $\boldsymbol{\beta}$ is a single peak probability mass function:

$$P(\mathbf{X}|\boldsymbol{\beta})=\begin{cases}1 & ,\mathbf{X}=f(\boldsymbol{\beta},\boldsymbol{\theta})\\ 0 & ,\mathbf{X}\neq f(\boldsymbol{\beta},\boldsymbol{\theta})\end{cases} \tag{5}$$

IV) The prior probability of parameters $\boldsymbol{\beta}$ follows logarithmic normal distribution with given mean $\boldsymbol{\beta}_0$ and covariance $\boldsymbol{\Omega}$.

By applying these assumptions, the posteriori probability distribution can be estimated by:

$$P(\mathbf{X}|\boldsymbol{\beta})=\begin{cases}1 & ,\mathbf{X}=f(\boldsymbol{\beta},\boldsymbol{\theta})\\ 0 & ,\mathbf{X}\neq f(\boldsymbol{\beta},\boldsymbol{\theta})\end{cases} P(\boldsymbol{\beta}|\mathbf{Z})\propto P(\mathbf{Z}|\mathbf{X},\boldsymbol{\beta})P(\mathbf{X}|\boldsymbol{\beta})P(\boldsymbol{\beta})=P(\mathbf{Z}|f(\boldsymbol{\beta},\boldsymbol{\theta}))P(\boldsymbol{\beta})$$
$$\propto \exp\left(-\frac{1}{2}(\mathbf{Z}-\mathbf{C}f(\boldsymbol{\beta},\boldsymbol{\theta}))^{\mathrm{T}}\boldsymbol{\Sigma}^{-1}(\mathbf{Z}-\mathbf{C}f(\boldsymbol{\beta},\boldsymbol{\theta}))-\frac{1}{2}(\ln\boldsymbol{\beta}-\ln\boldsymbol{\beta}_0)^{\mathrm{T}}\boldsymbol{\Omega}^{-1}(\ln\boldsymbol{\beta}-\ln\boldsymbol{\beta}_0)\right) \tag{6}$$

To find the maximum a posteriori, it is equivalent to minimize the following objective function:

$$\hat{\boldsymbol{\beta}}=\underset{\boldsymbol{\beta}}{\operatorname{argmin}}\frac{1}{2}(\mathbf{Z}-\mathbf{C}f(\boldsymbol{\beta},\boldsymbol{\theta}))^{\mathrm{T}}\boldsymbol{\Sigma}^{-1}(\mathbf{Z}-\mathbf{C}f(\boldsymbol{\beta},\boldsymbol{\theta}))+\frac{1}{2}(\ln\boldsymbol{\beta}-\ln\boldsymbol{\beta}_0)^{\mathrm{T}}\boldsymbol{\Omega}^{-1}(\ln\boldsymbol{\beta}-\ln\boldsymbol{\beta}_0) \tag{7}$$

Since the model is nonlinear and implicit, it is difficult to derive the explicit form for estimated parameters in general. Therefore, global optimization algorithm may apply to estimate the parameters. In this work, particle swarm optimization is used to estimate the best-fit model parameters.

4. DAMPER ADJUSTMENT

For a duct system with N terminals, only N-1 dampers are need to be adjusted and one damper is fully open. This is because the total flow rate supply is controlled by fan speed so the degree of freedom to balance the air distribution is only N-1. To balance the duct system, the fully open damper must be located first. Define the damper position $\boldsymbol{\theta}$ to be 0 for fully open and 90 for fully closed. When all dampers fully open, the pressures and flow rates of the system is:

$$\mathbf{X}_0=f(\hat{\boldsymbol{\beta}},\boldsymbol{\theta}=0) \tag{8}$$

In this case, the flow rates on each terminal can be obtained by defining the terminal matrix $\mathbf{C}_T$ such that the flow rate vector for terminals is $\mathbf{q}_0=\mathbf{C}_T\mathbf{X}_0$. Suppose the desired flow rates are denoted by vector $\mathbf{q}_d$, and the normalized desired flow rates vector $\hat{\mathbf{q}}_d$ is defined as:

$$\hat{\mathbf{q}}_d=\frac{\mathbf{q}_d}{\|\mathbf{q}_d\|_1}=\frac{\mathbf{q}_d}{\sum_i q_{d,i}} \tag{9}$$

Similarly, the fully open flow rates on each terminal $\mathbf{q}_0$ can also be normalized as $\hat{\mathbf{q}}_0=\mathbf{C}_T\mathbf{X}_0/\|\mathbf{C}_T\mathbf{X}_0\|_1$. By comparing their difference, the largest gap between $\hat{\mathbf{q}}_0$ and $\hat{\mathbf{q}}_d$ indicates the location of fully open dampers:

$$k=\underset{i}{\operatorname{argmax}}\,\hat{q}_{d,i}-\hat{q}_{0,i} \tag{10}$$

Therefore, damper k is the fully open damper throughout the balancing process. For other dampers, the optimal damper positions are calculated by:

$$\begin{aligned}&\underset{\boldsymbol{\theta}}{\text{Minimize}} && \frac{1}{2}\left(\hat{\mathbf{T}}-\mathbf{C}_T f(\hat{\boldsymbol{\beta}},\boldsymbol{\theta})/\|\mathbf{C}_T f(\hat{\boldsymbol{\beta}},\boldsymbol{\theta})\|_1\right)^T\left(\hat{\mathbf{T}}-\mathbf{C}_T f(\hat{\boldsymbol{\beta}},\boldsymbol{\theta})/\|\mathbf{C}_T f(\hat{\boldsymbol{\beta}},\boldsymbol{\theta})\|_1\right)\\ &\text{subject to} && 0<\theta_i<90 \text{ for } i=1,\cdots,k-1,k+1,\cdots,\mathrm{N}\\ &&& \theta_k=0\end{aligned} \tag{11}$$

Again, the particle swarm optimization is applied to solve this nonlinear optimization problem to obtain the optimized damper position $\boldsymbol{\theta}^*$. Once the damper position for all dampers are determined, the engineers are able to adjust dampers in real duct system accordingly. Considering the imprecise damper position indicator in most of dampers, we proposed a sequential tuning procedure with the help of flowmeter as follow:

I) Initialize the settings of damper positions as fully open, and $i=1$.
II) Set the damper *i* as its optimized position θ_i^* and compute the corresponding target airflow rate q_i^*.
III) Adjust the damper position such that the airflow rate in this terminal is equal to q_i^*.
IV) Inherit the current settings of damper position and set $i=i+1$.
V) Repeat step II to IV as long as $i \le N_t$.
VI) After the last damper is adjusted, the desired airflow distribution is achieved.

During this process, each damper is only adjusted once. Therefore, no iterative adjusting process is needed. Noticed that during the adjusting process, the target airflow rate q_i^* is not equal to the desired airflow rate $q_{d,i}$ because of the coupling effect in the duct system. Once the last damper is adjusted properly, the airflow rate in all terminals will reach the desired airflow rate $q_{d,i}$ at the same time.

5. VERIFICATION

The ASHRAE has created a duct network in Example 8 of the ASHRAE handbook 2001 Chapter 34 [1] to illustrate the duct calculation procedures. For convenience, the schematic of Example 8 is represented in Figure 4. The duct system consists of two parts: the induced ventilation system under negative pressure before fan consists of three inlet terminals; the supply air system under positive pressure after fan consists of six outlet terminals. Desired air flowrates through all nine terminals are listed in Table 1.

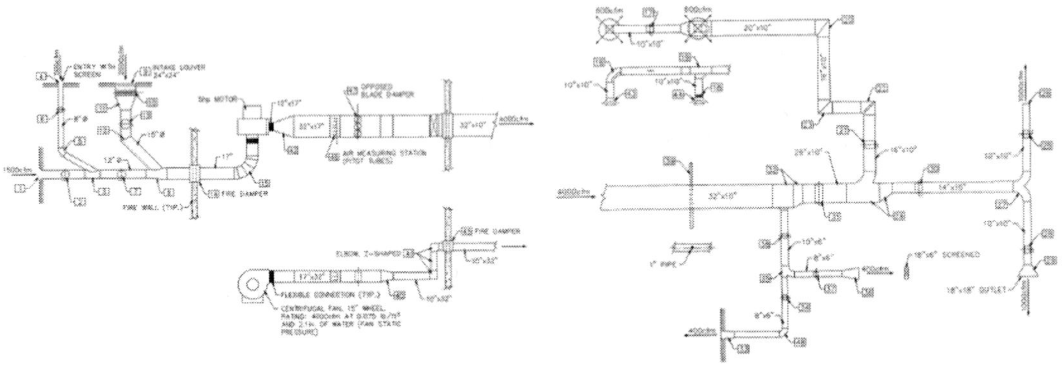

Figure 4: Schematic for Example 8

Table 1: Designed airflow for terminals in Example 8

Induced terminal	1	2	3
Designed flowrate	700 L/s	250 L/s	950 L/s
Supply terminal	4	5	6
Designed flowrate	275 L/s	275 L/s	475 L/s
Supply terminal	7	8	9
Designed flowrate	475 L/s	200 L/s	200 L/s

However, this duct system is unable to achieve desired air flow rate in each terminal without any adjustment of dampers. This fact is revealed by the mismatch of nodal pressure when calculating pressure drop by using designed airflow, which is demonstrated in Figure 5 (blue lines). By modeling this duct system by the proposed duct DAE approach in Simscape, shown in Figure 6 (i and ii), the actual airflow distribution can be calculated and the pressure drop pattern is obtained in the same time, shown in Figure 5 (green lines). In Figure 5, different nodes are distinct from each other by the locations in x-axis. The y-axis is the corresponding nodal pressure. Each line segment represents a part of duct. Air flows through the duct from high pressure to low pressure. Some nodes that have two lines merging together or one line separating apart are the T-junctions.

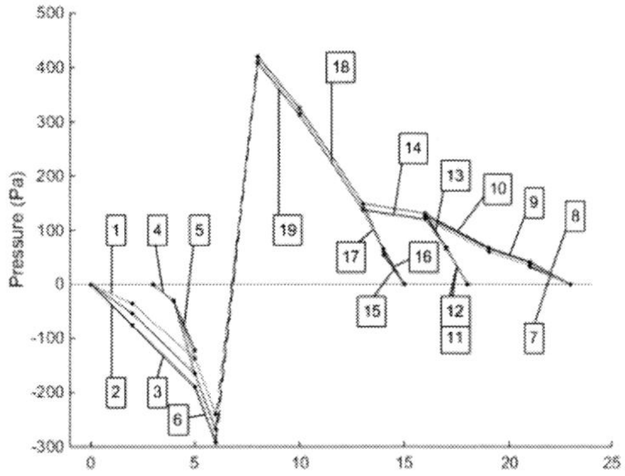

Figure 5: Nodal pressure before and after balancing in Example 8

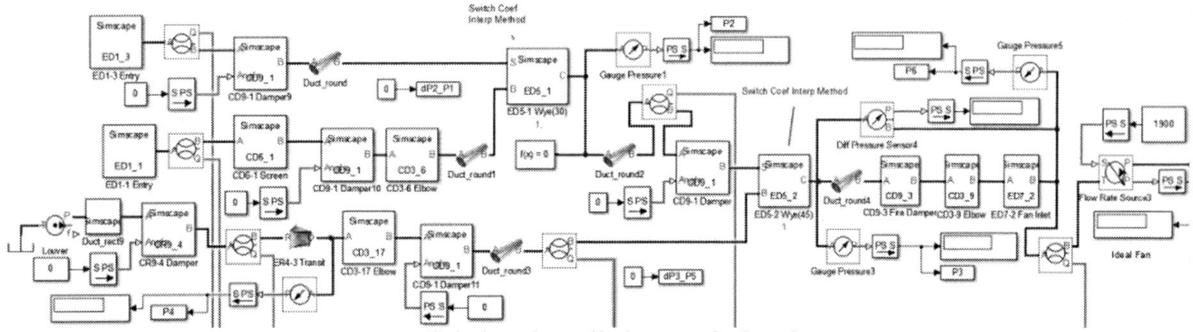

(i) induced ventilation part before fan

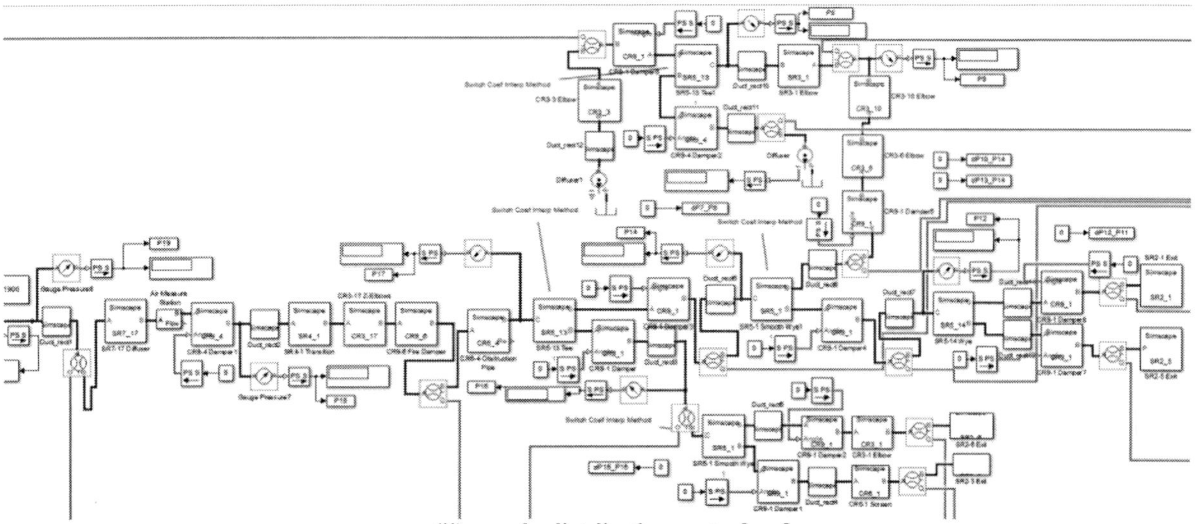

(ii) supply distribution part after fan

Figure 6: DAE model for Example 8 in Simscape

By applying the balancing procedure described in previous sections, the balanced duct system is achieved, shown in Figure 5 (red lines). The detailed flowrates, nodal pressures and damper positions for balancing is listed in Table 2. To evaluate the level of imbalance, we calculate the maximum absolute percentage error (MAPE) of each terminal. The MAPE is calculated by $\max_i |q_i - q_{d,i}| / |q_{d,i}|$, where q_i is actual airflow for terminal i and $q_{d,i}$ is the corresponding designed airflow. It can be seen that using balancing method does achieve satisfactory results.

Table 2: Airflow and pressures details before and after balance for Example 8

Branch	Designed Case		Actual Case		Balanced Case		
	Q (L/s)	dP (Pa)	Q (L/s)	dP (Pa)	Q (L/s)	dP (Pa)	Damper
1	700	35.0129	690.1346	35.1671	702.7297	51.7309	5.7139°
2	250	57.0221	212.0255	35.1671	245.7971	51.7309	0.3983°
3	950	111.1274[a]	902.1602	101.9938	948.5268	110.7122	0.0005°
4	950	29.5259	997.8398	29.9911	951.4732	30.6775	3.5831°
5	950	92.5145[b]	997.8398	107.1698	951.4732	131.7657	14.0113°
6	1900	102.4568	1900.0	102.4568	1900.0	102.4568	N.A.
7	275	32.2125	335.6966	34.8530	275.4957	40.8245	15.6997°
8	275	40.7355	218.1797	34.8530	274.6282	40.8245	0.1520°
9	550	24.5789	553.8764	24.9200	550.1239	24.5898	N.A.
10	550	64.5241*	553.8764	61.4334	550.1239	64.5230	0.0444°
11	475	66.3677*	472.5109	65.6802	474.5057	68.3618	2.4564°
12	475	67.9720*	466.6739	65.6802	474.4890	68.3618	0.6127°
13	950	55.5232*	939.1848	55.5261	948.9946	61.5754	3.4866°
14	1500	17.4490	1493.1	17.2650	1499.1	17.4417	N.A.
15	200	55.5232*	216.5671	63.0497	200.4184	67.9792	0.0007°
16	200	66.9949	190.3718	63.0497	200.4630	67.9792	12.5280°
17	400	58.0070[c]	406.9388	75.4217	400.8814	79.3997	1.0215°
18	1900	176.2132*	1900.0	176.2132	1900.0	176.2132	N.A.
19	1900	95.7082*	1900.0	95.7082	1900.0	95.7082	10.6706°
MAPE			22.07%		1.68%		

[a]Due to different value of loss coefficient, value in Example 8 is 103.
[b]Due to different value of loss coefficient, value in Example 8 is 103
[c]Due to the change of loss coefficient as the database updates, value in Example 8 is 79
*Due to rounding error, the values are different from Example 8

The flow rates for each branch is plotted in the bar chart, shown in Figure 7. It can be seen clearly that the calculated damper position successfully balanced the duct system (yellow bars) to the desired flowrate (blue bars). Therefore, conclusion can be drawn that the proposed methods can calculate the suitable damper position for HVAC duct balancing.

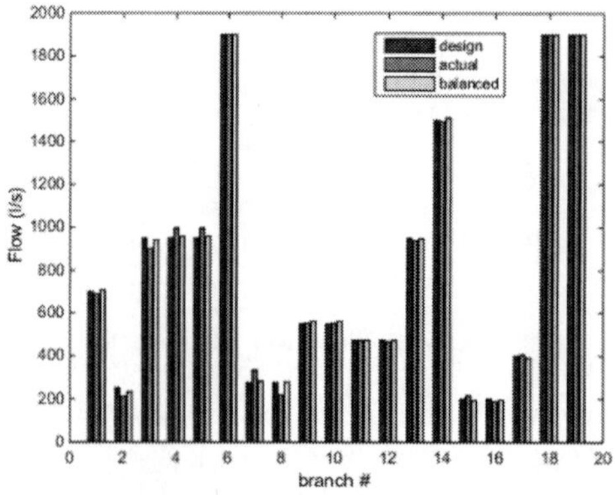

Figure 7: Airflow distribution before and after balancing in Example 8

6. CONCLUSIONS

We present a new non-iterative balancing method for HVAC duct system. The method contains three steps: 1) developing mathematical model for the duct system; 2) measuring pressures and flow rates to identify model parameters; 3) calculating optimal damper positions and adjusting dampers sequentially. This method distinguishes from other non-iterative TAB methods by its DAE model. By using Simscape modeling environment, developing duct system modeling becomes easier, and implementing optimization algorithms to calculate damper position for balancing becomes more convenient. This method has four main advantages: 1) the process is non-iterative and efficient; 2) the balancing result is highly accurate; 3) the procedure is easy to perform for engineers; and 4) the equipment requirement is only capture hood. The performance of this new TAB method is verified by the ASHRAE's example. The MAPE towards desired of the duct system reduces from 22.07% to 1.68%. This method saves the cost and effort of balancing duct systems. Hence this method has strong potential in providing balancing services for building HVAC system. In the future, the feasibility of this method in large scale HVAC system with hundreds of terminals must be studied. Besides, the obtained DAE model by this method can be further exploited. We will seek more applications on control and fault detection and diagnosis.

NOMENCLATURE

ΔP	Pressure drop	(Pa)
C_f	friction coefficient	(Dimensionless)
L	duct length	(m)
D	duct diameter	(m)
ρ	air density	(kg/m^3)
V	average air velocity	(m^3)
ε	wall roughness	(mm)
Re	Reynolds number	(Dimensionless)
$\boldsymbol{\beta}$	Model parameters	(-)
$\boldsymbol{\theta}$	Damper positions	(-)
$\mathbf{Z}$	Flowrate measurements	(m^3/s)
$\mathbf{C}$	Observation matrix	(-)
$\mathbf{X}$	State vector	(-)
$\boldsymbol{\Sigma}$	Measurement covariance	(-)
$\boldsymbol{\Omega}$	Parameter covariance	(-)
$\boldsymbol{\beta}_0$	Prior value of parameter	(-)
$\mathbf{C}_T$	Observation matrix	(-)

Subscript

0	Fully open state
d	Desired valued

REFERENCES

ASHRAE Committee, (2001) *ASHRAE HANDBOOK: Fundamentals*: Atlanta: ASHRAE.

ASHRAE Committee, (2013) *ASHRAE HANDBOOK: Fundamentals*, Atlanta: ASHRAE.

Pedranzini, F., Colombo, LPM., Joppolo, CM., (2013) A non-iterative method for Testing, Adjusting and Balancing (TAB) air ducts systems: Theory, practical procedure and validation. *Energ Buildings*, 65, 322-330

Small, M., (2002) *Non-iterative Techniques for Balancing and Air Distribution System*. Blacksburg, Virginia: Virginia Polytechnic Institute and State University.

Sugarman, S. C., (2006) *Testing and balancing HVAC air and water systems*, Lilburn, GA: Fairmont Press, Inc.

ACKNOWLEDGEMENT

This work is funded by National Research Foundation of Singapore (NRF-CRP8-2001-03).

Assisted Point Mapping to Enable Cost-effective Deployment of Intelligent Building Applications

Francesco LEONARDI[1]*, Hayden M. REEVE[2], Timothy C. WAGNER[2], Ziyou XIONG[2], June W. PARK[3]

[1]United Technologies Research Center,
Berkeley, California, USA,
LeonarF@utrc.utc.com

[2]United Technologies Research Center,
East-Hartford, Connecticut, USA,
ReeveHM@utrc.utc.com, WagnerTC@utrc.utc.com

[3]Center for Building Performance and Diagnostics, Carnegie Mellon University,
Pittsburgh, Pennsylvania, USA,
juneyoup@andrew.cmu.edu

* Corresponding Author

ABSTRACT

Over the past years intelligent building applications that promise to dramatically reduce energy consumption, improve occupant comfort and streamline maintenance have been proposed. However their adoption has met a steep barrier in the unexpectedly high cost of mapping data from building automation systems into these applications' data models. In fact the industry does not have a common convention on how to name points. Generally, names are correlated with the semantic of the variables they represent, but typically engineers have the freedom to set up variable names according to their preferences. In the last few years the research community has devoted increasing attention in automating the process of mapping data points from existing BAS. In order to meet market requirements, an "ideal" algorithm must be accurate, able to infer complex relationships between data points, easy to use, and scalable. However, none of the published work meets all the market requirements. Most need an expert user, some are not "easy to use" and none can automatically infer complex relationships.

This paper presents a novel algorithm to tackle the mapping problem. This work contributes to the state-of-the-art by providing an algorithm that can be successfully applied to various datasets with minor or no modification to the algorithm (i.e. no expert in the loop). This is also the first to automatically infer complex relationships with only point names as input (i.e. ease of use).

The algorithm was tested against a large and diversified dataset comprising points from five buildings, two vendors, three distributors and more than 20K points. The algorithm correctly mapped over 90% of the points required by a test application and successfully identified over 90% of VAVs and 80% of AHUs.

1. INTRODUCTION

There is an increasing recognition within both industry and academia (Project Haystack; Bhattacharya, Ploennigs, & Culler, 2015) that semantic metadata associated to points can reduce the maintenance cost of BAS and make software applications more portable across BAS, buildings and vendors. Semantic metadata captures knowledge about points, equipment and buildings. They describe information such as: point types (e.g. is a point a sensor, set-point, control parameter, trend, alarm, etc.), the physical quantity measured/controlled (e.g. temperature, velocity, pressure, etc.), relationship with the physical world (e.g. is the sensor is measuring air, water, etc.), relationship within the system/equipment where the point belongs to (e.g. return air, return water, etc.), relationship with the system/equipment (e.g. to which equipment a point belongs to), relationship between equipment (e.g. hierarchy of an HVAC system, relationship between the physical world and the equipment/system: e.g.: the location of the sensor (which room in a building), list of zones served by an AHU and physical relationship between points (e.g. distance between sensor/actuators).

The process of manually annotating points with semantic metadata is costly and error prone. Typical BASs have thousands of points and would need several times more metadata. A trained field engineer would have to spend many hours to fully annotate a BAS and this is cost prohibitive in many cases. This is the major driving force in the research for techniques and algorithms to assist in the generation of metadata from data points.

Even though most of the knowledge required to assign the correct metadata is not explicitly represented in existing BAS, several algorithms have recently been developed to infer it by analyzing data point values and properties (such as point names).

In order to meet market requirements, an "ideal" algorithm to infer semantic metadata would have five properties: 1) high accuracy and especially low false positive rate (i.e. small number of erroneously mapped points); 2) ability to infer complex relationships between data points (e.g. grouping points by equipment, classifying equipment, establishing relationship between equipment); 3) ease of use: it should leverage only readily available knowledge and data about the system (i.e. not requiring the installation of additional sensors); 4) minimal need of having a domain expert using it; and 5) scalability: the algorithm infers semantic metadata across a variety of BAS, buildings, and vendors with satisfactory results at most requiring minimum human involvement in adjusting settings and parameters. Any approach requiring extensive human effort faces a steep market resistance.

1.1 State of the Art

The last few years witnessed an increasing interest from academia in automatically inferring semantic metadata from points. The algorithm described in (Schumann *et al.*, 2014) and in (Schumann *et al.*, 2015) learns how to map point names to a prebuilt dictionary from a set of example point names presented by an operator. This algorithm assumes the existence (and the knowledge) of a naming convention of point names. The algorithm attempts to match each point with a definition in the dictionary. The richer the dictionary and its definitions the easier it is to find the right match.

An alternative technique is presented in (Bhattacharya *et al.*, 2014). It iteratively learns the naming schema of point names and associates this schema to a common semantic model like Haystack (Project Haystack). Similar to the previous approach the algorithm generates questions (or examples) that a user needs to answer. A characteristic of this solution is that the operator has to provide the "translation" between the naming schema and the semantic model. Therefore this approach is well suited for buildings with limited naming schemas. The problem of "transfer learning" i.e. the ability to learn a model in one or few building and to reuse the same model across several buildings to classify/label points has been addressed only recently. A promising approach is presented in (Hong *et al.*, 2015) where the algorithm combines both sensor readings and point names to learn a classifier that can be applied to other buildings. While results are encouraging in labelling point types there is no attempt in inferring relationships between points. Similar conclusions can be drawn from (Balajiy *et al.*, 2015) where an active learning algorithm can achieve very high accuracy in establishing point types. In the same paper it was also noted that for many practical uses, identifying point types is not sufficient. Invaluable information such as identifying equipment types and grouping points per equipment can lead to a drastic simplification in the deployment of building applications. The algorithm illustrated in this paper addresses this last problem.

2. Algorithm Description

The primary intent of this inference algorithm is to reduce the effort in deploying intelligent applications on top of existing building automation systems by assisting in the laborious task of mapping embedded controller data to application data.

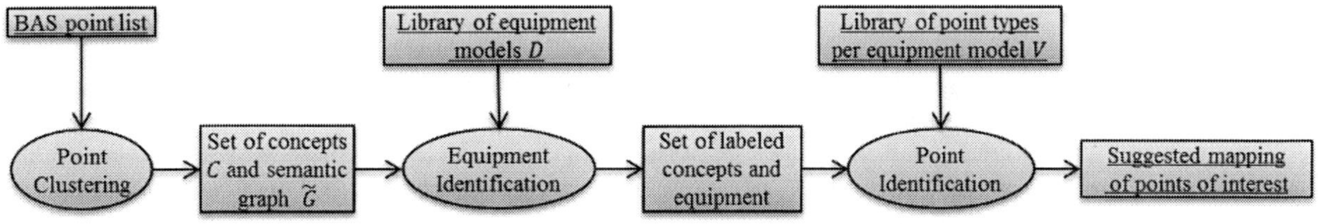

Figure 1: Block diagram of the inference algorithm.

The algorithm digests a list of point names from a building automation system as input and generates suggestions of the likely meaning of points as output. The overall algorithm leverages the fact that point names are usually correlated with the semantic of the variables they represent. The inference algorithm can be logically divided into three steps: Point Clustering, Equipment Identification, and Point Identification (Figure 1). The algorithm takes in a list of point names from a building automation system, a library D of equipment models, and a library V of point types per equipment model. Models are created offline by domain experts or by learning from samples in a training set (as accomplished in this paper). Customized libraries that take into consideration particular characteristics of vendors, distributors and regions can be adopted, trading portability for enhanced accuracy of the algorithm.

2.1 Point Clustering

In this unsupervised step the algorithm first transforms each point name into a list of its constitutive parts (i.e. tokens). Tokens in point names are usually delimited by special characters (e.g.: punctuation, '/', '\', '-') or other tokens. Numbers are also considered as tokens. After this initial step, points are represented in a directed graph $G = \{N, E\}$ where each node $n \in N$ is a token. A directed edge $e_{ij} \in E$ between two nodes (n_i, n_j) indicates that the nodes belong to at least one point and n_i strictly precedes n_j in those point names. The set of edges originating in node n_i is denoted by E_i^- and $\|E_i^-\|$ represents the cardinality of this set. Set $F \subseteq N$ contains leaf nodes: i.e. nodes representing the last tokens in point names. Each path starting from a source node $n_s \in N$ (i.e. a node without edges pointing to it) and ending on a $n_f \in F$ uniquely identifies a point name.

We introduce the function $T(e)$ that returns the label of the token of the sink node of edge e. We define the equivalence function of two edges as:

$$\epsilon(e_{ij}, e_{nm}) = \begin{cases} 1 \; if \; T(e_{ij}) = T(e_{nm}) \\ 0 \; otherwise \end{cases} \tag{1}$$

The similarity metric of two nodes is defined as:

$$S(n_i, n_j) = \begin{cases} 1 \; if \; \dfrac{0.5}{max(\|E_i^-\|, \|E_j^-\|)} \displaystyle\sum_{e \in E_i^-, f \in E_j^-} \epsilon(e, f) \geq \vartheta \\ 0 \; if \; \|E_i^-\| \leq 2 \; or \; \|E_i^-\| \leq 2 \\ 0 \; otherwise \end{cases} \tag{2}$$

where $\vartheta \in [0,1]$ is a parameter. When $S(n_i, n_j) = 1$ the two nodes are considered similar and placed in the same cluster. Similar nodes are likely to belong to the same equipment type (e.g. VAVs in a building are likely to have more or less the same point names encoded in similar naming conventions) or other logical groups (e.g. list of equipment, subcomponent of equipment etc.).

The algorithm proceeds by transforming clusters into concepts. A concept is an abstraction of a cluster: it contains the list of tokens that are common across nodes in the cluster. Nodes in a cluster are also instances of the concept of the cluster. Concepts are organized in the semantic graph $\tilde{G}(C, R)$, where C is the set of concepts and R is the set of relationships between concepts. For the purpose of this paper, only the inclusion relationship is relevant. This type of relationship is inferred from C and G. A concept c_i is included by c_j if there is a directed path connecting at least one instance of c_j to one instance of c_i.

2.2 Equipment Identification

The second step of the inference algorithm attempts at automatically labeling (i.e., assigning semantic meaning to the concepts in C. In other words, the algorithm establishes whether a concept represents HVAC equipment and the label of the equipment (e.g. VAV, AHU, etc.). This is accomplished in two separate stages. Stage one considers each individual concept; stage two leverages global knowledge of concepts present in the semantic graph $\tilde{G}$.

2.2.1 Stage one: stage one starts by extracting features from the information present in the concepts (i.e. tokens) and presented in a format that is readily consumed by off-the-shelf machine-learning algorithms. Features are computed by comparing concepts with equipment models D. Each equipment model $M_k \in D$ consists of a set of model-points: i.e. points that are typically present in that particular equipment type. Each model-point $m \in M_k$ is identified by a label expressed in a formal semantic (e.g. Haystack tags, custom tags, etc.) and carries a set of possible alternative token representations that are usually found in the field for that point. To increase classification robustness, likely variations of tokens (e.g., by removing vowels) are also included. The feature for concept j and equipment model i is computed according to:

$$f_{ij} = \left(\beta \sum_{m \in M_i} \max_{p \in c_j} \{G(m,p)\} \right)^2 \tag{3}$$

Where p is a point in concept j, β is a parameter ($\beta = 3$ in this paper) and

$$G(m,p) = \max_{a \in A(m)} \left\{ \frac{\sum_l \sum_n d_j(t_l^a, t_n^p)}{\max(\|p\|, \|a\|)} \right\} \tag{4}$$

Where d_j is the Jaro–Winkler distance, a is an alternative token representation for model point m, $A(m)$ returns the alternatives for the m-th model point, t_l^a is the l–th token in the alternative a, and t_n^p is the n-th token of concept point p. $\|p\|$ and $\|a\|$ indicate the number of tokens in concept-point p and alternative a respectively. In general, any off-the-shelf classifier can be used to assign a label to a concept. In particular the results in this paper are obtained with a linear Support Vector Machine (SVM) classifier.

2.2.2 Stage two: the output of stage one can be further refined by leveraging global knowledge of the relationships between concepts present in the semantic graph $\tilde{G}$. In particular new types of concepts (e.g. collections) can be only labeled at this stage. For example a concept including only concepts of one type can now be labeled as a collection. Similarly some classes of erroneous classifications generated at stage one can now be corrected. Two examples: a VAV is unlikely to include an AHU and a concept including both types of equipment is likely to be neither of them.

2.3 Point Identification

The final step in the algorithm consists in identifying points in labeled concepts. In order to achieve this, the algorithm finds for each point in a labeled concept the best matching point in the corresponding equipment-model in V. Library V is derived from D and it can also include additional model-points that, while less common, might still be useful for some intelligent applications. The label of each point p in a labeled concept c_i is determined by:

$$L\left(M_i^v, \underset{m \in M_i^p}{\operatorname{argmax}} \{G(m,p)\} \right) \tag{5}$$

Where M_i^v represents the model in V of c_i and $L(M_i^v, n)$ returns the label of the m-th model-point in M_i^v .

3. Results

This section demonstrates the performance of the algorithm on a large dataset comprising data points from 9 buildings with BASs from multiple vendors and dealers.

Table 1: Dataset composition and cluster results.

Building ID	Building Vendor	#BAS points	#C	#Points in $\tilde{G}$	#Un-clustered points
1	JCI	1160	26	126 (10%)	0
2	Siemens	4308	36	145 (3.3%)	121 (2.8%)
3	JCI	3545	50	329 (9.3%)	1
4	ALC	5015	93	432 (8.6%)	200 (4.0)
5	Siemens	1808	29	101 (5.6%)	82 (4.5%)
6	Siemens	5981	49	213(3.6%)	119(2.0%)
7	Siemens	8247	38	205 (2.5%)	321 (3.9%)
8	ALC	3448	44	211 (6.1%)	53 (1.5%)
9	ALC	3428	33	143 (4.2%)	85 (2.5%)

Table 2: Equipment identification results.

ID	#VAV			#AHU		
	GT	FP	FN	GT	FP	FN
1	107	0	0	8	0	1
2	51	1	1	-	-	-
3	148	0	1	6	0	0
4	101	0	0	7	2	0
5	181	0	16	6	0	0
6	82	0	5	-	-	-
7	421	3	0	15	3	0
8	52	3	0	2	0	0
9	100	0	0	1	0	1

3.1 Point Clustering

Table 1 summarizes the point clustering results. The first three columns report the building ID, vendor name, and the number of points in the building. Column 4 lists the number of concepts obtained by the algorithm. Column five represents the number of points in the semantic graph $\tilde{G}$. The last column summarizes the number of points that the algorithm is not able to assign to any cluster. The number of points per building ranges from about 1000 to 8000. The dataset is divided in a training set (building 1 to 4) and a test set (building 5 to 9). The first set was used to create library models and tune parameters. The second set was used to assess the algorithm performance. The test set consists of data points collected from 5 buildings with 2 BMS vendors and 3 dealers.
Depending on the considered building, the number of clusters identified by the algorithm ranges from 26 to 93. More importantly the number of points in the semantic graph $\tilde{G}$ is on the order of 3 to 7% of the total number of points and the number of un-clustered points is less than 5% of the total number of points. This means that by assigning a semantic definition to each concept and to each concept-point it is possible to annotate most of the dataset. In other words clustering by itself can drastically reduce the effort required to label a dataset.

3.2 Equipment Identification

Equipment identification results are shown in Table 2. The first columns represents the building ID, columns 2 and 5 report the ground truth (GT) of the number of VAV and AHU per building. Columns 3 and 4 summarize the number of false positives (FP) and false negatives (FN) of VAV classified by the algorithm. Columns 6 and 7 list similar results for the AHUs. The algorithm accurately recognizes more than 90% of the VAVs and more than 80% of the AHUs present in a building. It is worth mentioning that the classifier was unable to detect the only AHU of building 9. This happened because of a current limitation of the clustering algorithm that requires the presence of at least two similar groups of points/equipment in order to create a cluster. In this case, having only one AHU, the clustering algorithm does not create a cluster for the AHU points and therefore the classification algorithm is unaware of it. In general it appears that correctly classifying AHUs is more difficult than for VAVs. A possible explanation is that the training set has fewer examples of AHUs and this makes it more challenging to generalize a robust classifier.

Table 3: Point Classification VAV

ID	Sns Dmpr Position			Sns Flow Supply Air			Air Flow Supply SP			Sns Zone Temp		
	GT	FP	FN	GT	FP	FN	GT	FP	FN	GT	FP	FN
1	107	0	0	107	0	0	-	-	-	107	0	0
2	51	0	1	51	0	1	51	0	1	51	1	1
3	148	0	1	148	0	1	148	0	1	144	-	5
4	101	0	0	101	0	0	101	0	0	4	0	0
5	-	-	-	181	0	16	-	-	-	181	0	16
6	82	0	5	82	0	5	82	0	5	82	0	5
7	421	3	0	421	3	0	421	3	0	421	3	0
8	52	3	0	52	3	0	52	3	0	21	0	0
9	-	-	-	100	0	0	100	0	0	98	0	0

Table 4: Point Classification AHU

ID	Sns Rtrn Air Temp			Sns Mixed Air Temp			Outside Air Dmpr			Sns Rtrn Air Hum			Sns Outside Air Temp		
	GT	FP	FN	GT	FP	FN	GT	FP	FN	GT	FP	FN	GT	FP	FN
1	8	0	0	8	0	0	8	0	0	8	0	0	8	0	0
2	-	-	-	-	-	-	-	-	-	-	-	-	-	-	-
3	2	-	-	6	0	0	6	0	0	2	4	0	-	-	-
4	7	2	0	7	2	0	7	0	0	7	-	-	-	-	-
5	6	0	0	-	-	-	-	-	-	-	-	-	6	0	0
6	-	-	-	-	-	-	-	-	-	-	-	-	-	-	-
7	15	0	0	15	0	0	-	-	-	15	0	0	-	-	-
8	2	0	0	2	0	0	-	-	-	1	0	1	-	1	-
9	1	0	1	1	0	1	1	0	1	1	0	1	-	-	-

3.3 Point Identification

Point identification results for VAVs and AHUs are reported in Table 3 and 4 respectively. The first row of both tables indicates some of the supported point types expressed with custom semantic tags. For each point type three columns summarize the number of occurrences of the point (GT), the number of false positives (FP) and false negatives (FN) identified by the algorithm. In most cases, the algorithm classifies points with more than 90% accuracy. Similar to the equipment identification, building 9 represents an exception. Since a cluster was not created, it is not possible to detect any of the AHU points for building 9. Additional work is required to enable the algorithm to properly identify clusters and points for buildings with single pieces of equipment.

6. PRELIMINARY COST MODEL

In order to truly understand the impact of any tool for semantic inference of BAS points, a cost model for mapping data points from a BAS into an application must be introduced. Assuming that an operator requires a constant amount of time to assign each semantic tag, a simplified cost model can be written as a function of the number of tags (τ) that need to be assigned:

$$\varphi(\tau) = \|B\|\tau_B + \sum_{b\in B}\sum_{e\in E} I(e,b)\left(\tau_e + \mathrm{P(e)}\tau_{\mathrm{p}}\right) \tag{6}$$

Where B represents a set of buildings, τ_B is the number of tags that are specific to a building (examples of building tags are name, location, area, type, climate zone, etc.), E is the set of equipment types that are of interest of the application, function $I(e,b)$ returns the number of instances of an equipment type e in a building b, τ_e is the number of tag per equipment type (examples of equipment tags are type, location, relationship with zones and other equipment, etc.), function $\mathrm{P(e)}$ returns the number of points in equipment type e, and τ_p is the number of tag per point (examples of point tags are: point definition and reference to containing equipment).

The first term of the cost model ($\|B\|\tau_B$) is usually small compared to the second one. In fact in a typical campus there are many more pieces of equipment and data points than buildings and $\tau_B \approx \tau_e$. It is understandable that the research presented so far in the open literature attempts to minimize the number of point tags that need to be manually assigned. In particular most activities are directed towards automatically inferring point definition tags. A few papers present results on establishing the location of sensors and relationships of equipment. The algorithm described in these pages is the first introduction of a viable mechanism to automatically infer both point definition tags, references to containing equipment and equipment tags.

7. CONCLUSIONS

This paper presented a novel algorithm for inferring semantics of BAS data points. This algorithm is the first to automatically infer complex relationships with only point names as input. It can also successfully operate on data point from various buildings with minor or no modification to the algorithm. Results obtained from a large and diversified dataset comprising points from five buildings, two vendors, three distributors and more than 20K points indicate that the algorithm can correctly map over 90% of the points required by a test application and successfully identified over 90% of VAVs and 80% of AHUs.

While knowing the equipment type and the associated points is important in many applications, there are cases where other relationships are relevant too. In fact, there have been attempts to infer the hierarchy of VAVs and AHUs (Pritoni *et al*, 2015) and spatial location of sensors in (Ortiz *et al.*, 2013) and in (Koc *et al.*, 2014). Future work should definitely revisit and incorporate these approaches. Future studies might also use additional knowledge about points (e.g. values, measurement units, etc.) to increase the robustness and the accuracy of the output as suggested by (Balajiy, 2015) and (Hong, 2015).

NOMENCLATURE

BMS	Building Management System
BAS	Building Automation System
VAV	Variable Air Valve
AHU	Air Handling Unit
HVAC	Heating Ventilation Air Conditioning
GT	Ground Truth
FP	False Positive
FN	False Negative
Sns Rtrn Air Temp	Data point reporting the return air temperature
Sns Mixed Air Temp	Data point reporting the mixed air temperature
Outside Air Dmpr	Data point reporting the position of a damper for the outside air
Sns Rtrn Air Hum	Data point reporting the return air humidity
Sns Outside Air Temp	Data point reporting the outside air temperature
Sns Dmpr Position	Data point reporting the position of a damper
Sns Flow Supply Air	Data point reporting the air supply flow
Air Flow Supply SP	Data point reporting the value of the air supply flow set-point
Sns Zone Temp	Data point reporting the temperature of a zone

REFERENCES

Balajiy, B., Vermay, C., Narayanaswamyy, B., & Agarwal, Y. (2015). Zodiac: Organizing Large Deployment of Sensors to Create Reusable Applications for Buildings. *Proceedings of the 2nd ACM Conference on Embedded Systems for Energy-Efficient Buildings.*

Bhattacharya, A., Culler, D. E., Ortiz, J., Hong, D., & Whitehouse, K. (2014). *Enabling portable building applications through automated metadata transformation.* University of California: University of California.

Bhattacharya, A., Ploennigs, J., & Culler, D. (2015). Short Paper: Analyzing Metadata Schemas for Buildings: The Good, the Bad, and the Ugly. *Proceedings of the 2nd ACM International Conference on Embedded Systems for Energy-Efficient Built Environments* (pp. 33-34). ACM.

Hong, D., Wang, H., Ortiz, J., & Whitehouse, K. (2015). The Building Adapter: Towards Quickly Applying Building Analytics at Scale. *Proceedings of the 2nd ACM Conference on Embedded Systems for Energy-Efficient Buildings.*

Koc, M., Akinci, B., & Berges, M. (2014). Comparison of linear correlation and a statistical dependency measure for inferring spatial relation of temperature sensors in buildings. *Proceeding BuildSys '14 Proceedings of the 1st ACM Conference on Embedded Systems for Energy-Efficient Buildings.*

Ortiz, J., Dezhi, H., Whitehouse, K., & Culler, D. (2013). Towards automatic spatial verification of sensor placement in buildings. *Proceedings of the 5th ACM Workshop on Embedded Systems For Energy-Efficient Buildings.*

Pritoni, M., Bhattacharya, A. A., Culler, D., & Modera, M. (2015). Short Paper: A Method for Discovering Functional Relationships Between Air Handling Units and Variable-Air-Volume Boxes From Sensor Data. *Proceedings of the 2nd ACM International Conference on Embedded Systems for Energy-Efficient Built Environments* (pp. 133-136). ACM.

Project Haystack. (n.d.). Retrieved 04 26, 2016, from Project Haystack: http://project-haystack.org/

Schumann, A., & Lécué, F. (2015). Minimizing User Involvement for Accurate Ontology Matching Problems. *Twenty-Ninth AAAI Conference on Artificial Intelligence.*

Schumann, A., Ploennigs, J., & Gorman, B. (2014). Towards automating the deployment of energy saving approaches in buildings. *Proceedings of the 1st ACM Conference on Embedded Systems for Energy-Efficient Buildings.* ACM.

ACKNOWLEDGEMENT

This material is based upon work supported by the Consortium for Building Energy Innovation, sponsored by the U.S. Department of Energy under Award Number DE-EE0004261. The authors would like also to acknowledge the generous contribution of Intellimation (http://www.intellimation.net/) for providing some of the datasets used in this study.

DISCLAIMER

This work was prepared as an account of work sponsored by an agency of the United States Government. Neither the United States Government nor any agency thereof, nor any of their employees, makes any warranty, express or implied, or assumes any legal liability or responsibility for the accuracy, completeness, or usefulness of any information, apparatus, product, or process disclosed, or represents that its use would not infringe privately owned rights. Reference herein to any specific commercial product, process, or service by trade name, trademark, manufacturer, or otherwise does not necessarily constitute or imply its endorsement, recommendation, or favoring by the United States Government or any agency thereof. The views and opinions of authors expressed herein do not necessarily state or reflect those of the United States Government or any agency thereof.

Automated Fault Diagnostics for AHU-VAV Systems: A Bayesian Network Approach

Adam REGNIER[1]*, Jin WEN[2]

[1]Drexel University,
Department of Civil, Architectural, and Environmental Engineering,
Philadelphia, PA, USA
ar626@drexel.edu

[2] Drexel University,
Department of Civil, Architectural, and Environmental Engineering,
Philadelphia, PA, USA
jinwen@drexel.edu

* Corresponding Author

ABSTRACT

Although it is widely accepted that 20%-30% of total HVAC energy in commercial buildings is wasted due to faulty or inefficient operation, there presently exist no widely adopted solutions to identify and remedy this waste. This lack of adoption is due primarily to the high upfront costs associated with the manual process of customizing commissioning solutions for each individual building. In order to reduce these costs, diagnostic technologies must automate the process of installing and customizing solutions for each implementation. A novel approach to addressing this problem for air handling units (AHUs) and variable air volume (VAV) boxes is presented here. This strategy utilizes a Bayesian network to identify and understand the system operation, identify faults that are wasting energy or impacting occupant comfort, and then generate performance baselines against which future operation will be compared. When this algorithm is first connected to a new building, an adaptive diagnostic Bayesian network identifies the components and configuration of the AHUs and VAVs, and then generates probabilistic outputs indicating the root causes of potential faults and inefficiencies. Once these issues have been rectified to the satisfaction of the building operator, the algorithm then begins to accumulate training data with detailed information about the system operation (while simultaneously continuing to monitor for additional faults). As this training data is accumulated, the diagnostic confidence of the Bayesian network is continuously improved. Additionally, this use of an operational baseline allows for accurate detection and diagnosis of faults causing gradual performance degradation in addition to faults that abruptly occur.

1. INTRODUCTION

It is well known that the energy costs of operating commercial buildings are high. Commercial buildings represent nearly one-fifth (18.9%) of all U.S. energy consumption, of which 43% is estimated to be spent on HVAC for space conditioning (DOE, 2010). In office buildings specifically, Pérez-Lombard et al. (2008) provided estimates that HVAC energy comprises 48%-55% of the total energy used and that HVAC consumption across all sectors accounts for approximately one fifth of all energy use in developed countries. While HVAC will inevitably comprise a substantial portion of our energy consumption, a portion of this energy is actually being squandered as a result of faulty operation of these systems.

The percentage of total energy wasted as a result of undiagnosed faults is conservatively estimated in the range of 15%-30% of total HVAC energy. Studies have estimated 20-30% energy savings can be achieved by retro-commissioning (Annex 34, 2006) and the most comprehensive longitudinal case study available found a 44% reduction in the use of electricity and a 78% reduction in natural gas consumption across the 10 year study period

(Annex 47, 2010). These data clearly demonstrate that cost-effective and accurate automated fault detection and diagnosis (AFDD) can significantly mitigate the energy wasted in buildings. This study focuses exclusively on AFDD for AHU-VAV systems, but many of the concepts investigated here are applicable to a wider range of systems. In the United States, AHU-VAV systems serve over 30% of commercial floor space (EIA, 2003; EIA, 2013), and they are a known source of many common faults.

This study investigates the use of a Bayesian network for fault isolation in an automated-commissioning (ACx) environment. ACx combines and automates both retro-commissioning (RCx), and continuous-commissioning (CCx) practices. As their names reflect, RCx refers to the identification of existing faults in a building, while CCx refers to continuous monitoring of a building to identify new faults as they occur in real time. These two practices are noted separately because CCx has the added benefit of being able to automatically customize baseline thresholds for each building once RCx has identified any existing faults.

Bayesian networks have been successfully employed for decision support and diagnostics applications in many fields and have recently been included in a number of publications in the HVAC AFDD field for RCx applications. Specifically, the utility of Bayesian networks for AHU-VAV systems were most recently demonstrated for VAV-boxes in Xiao et al. (2014). This work followed up on the paper using the same methodology for AHU fault detection and chiller fault detection (Zhao et al., 2013). This paper utilizes a similar Bayesian network approach, but expands upon the existing approaches to suggest a method for including automated threshold customization to improve diagnostic accuracy in a CCx setting. Additionally, this paper suggests that monitoring multiple levels of severity in the observed evidence in order to improve overall diagnostic accuracy, as well as the inclusion of multiple levels of severity in the fault node in order to provide more accurate and actionable information to a building operator. Beyond the modifications to the network structure, this paper also presents a novel approach for the network parameter estimation. In all previous studies, the network was trained and tested utilizing a single building, but it was found that the networks utilized in these studies suffered when tested on other buildings. During the course of this study, a formal method for identifying generalizable parameters for the Bayesian network conditional probability tables (CPTs) was developed and tested in order to avoid the problem of over-fitting.

2. METHODS

The first step in this study was to generate the overall AHU Bayesian network structure. This was performed based upon review of the literature and the first-principles applicable to an AHU. The Bayesian network is structured such that *fault nodes* represent the different pieces of equipment that can fail, while *evidence nodes* represent the symptoms, or the effects of faults that are observable/measurable in typical building automation system (BAS) data. These fault nodes are then connected via *arcs* to the evidence nodes with which there exists a causal relationship.

2.1 Fault Nodes

Each fault node corresponds to an individual component within an AHU that is being monitored for faults. These components primarily consist of dampers, valves, fans, and sensors. Each node can take on a variety of different values, one of which is always "fault free". For example, the fault node for an outdoor air (OA) damper can take the following basic values:

- Fault-free (operating normally)
- Stuck too far open
- Stuck too far closed

In order to improve diagnostic accuracy, the "too far open" and "too far closed" were then split to include *moderate* and *extreme* levels of "faultiness". This is due to the fact that the severity of most faults can have different impacts on the observable evidence as the effects propagate to downstream components. Without this modification, the network would either omit useful diagnostic information, or erroneously identify corroborative evidence as contrary evidence.

2.2 Evidence Nodes

Each evidence node corresponds to a diagnostic *rule* that is based on the operational mode of the system. These rules are typically comprised of a relationship between two or more measured BAS values that can be predicted based upon the operational mode of the system. For example, if the system is operating at 100% outdoor air, it is expected that the mixed air temperature and the outdoor air temperature are approximately equal. Originally these nodes were constructed with three possible values: *higher than expected*, *within normal operational range*, and *lower than expected*. During initial testing it was noted that, while discretizing the evidence nodes was required for computational tractability, the simplification in this manner was perhaps overly-simplified. For example, a temperature deviation of 3°F is much "weaker" evidence than a temperature deviation of 10°F, yet this differentiation was not reflected in the Bayesian network conditional probability tables. As a result, the evidence was binned more finely to include this additional information, so a new node may have values such as *within normal operational range, moderately higher than expected,* and *significantly higher than expected.*

Illustrated in **Figure 1** is a section of the Bayesian network that is utilized to diagnose faults related to the outdoor air damper (OA damper) and mixed air temperature (MAT) sensor. Faults with either of these items can cause significant energy waste, negatively impact indoor air quality, and can also affect occupant comfort during economizing operation. Additionally, faults with these two components are known to go undetected for long periods of time, and have similar fault "evidence", making it potentially difficult to rapidly identify the root cause of the fault. The term "fault evidence" is used interchangeably with the term "fault symptom" and refers to the behavior of a fault that can be observed in the BAS data.

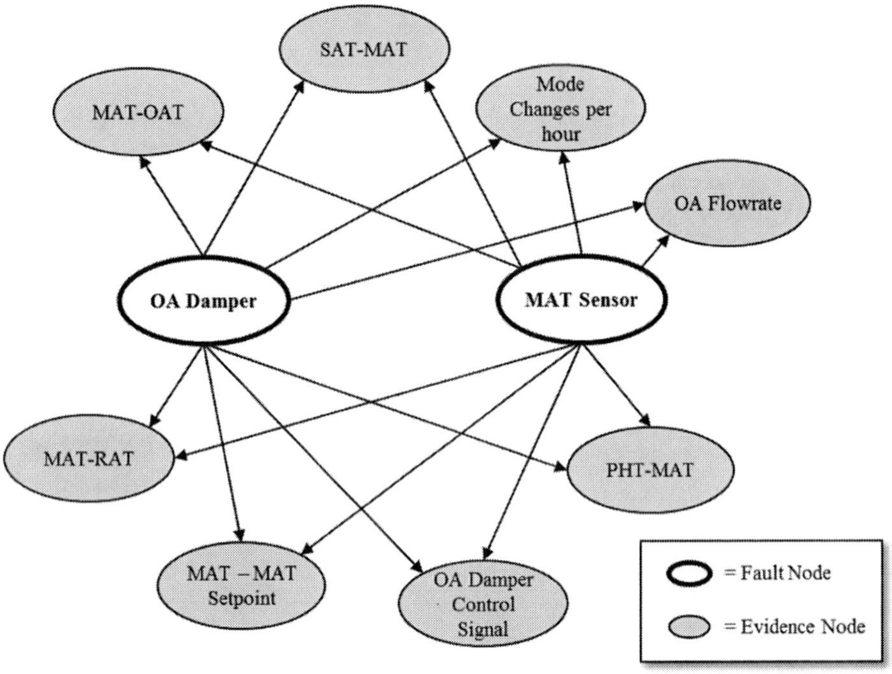

Figure 1: Bayesian network excerpt diagram (economizer section)

Figure 1 utilizes the following abbreviations: OAT for *outdoor air temperature*, MAT for *mixed air temperature*, SAT for *supply air temperature*, RAT for *return air temperature*, PHT for *preheat (between-coil) temperature.*

2.3 Bayesian Network Parameter Estimation

The models used in this study are considered graded, leaky, causal, noisy-max models. (Depending on which reference is utilized, there is some variation in nomenclature, but they will be referred to simply as noisy-max models throughout this document.) The "noisy" aspect takes into account that the fault may not always cause the expected effect in the evidence. The "leaky" aspect refers to the fact that there is some probability that a faulty evidence state will be observed without any underlying fault. The causal aspect refers to the fact that all of the arcs connecting the nodes must be utilized only for direct cause-effect relationships. And finally, the graded aspect refers to the fact that the noisy-max model can only be used for graded variables, since it is a max function. Good explanations of the practical applications of noisy-max models can be found in Zagorecki & Druzdel (2013). Use of the noisy-max model provides two key advantages in the use of generating Bayesian networks: (1) a reduction in the number of probability states required to be specified to generate the conditional probability tables, and (2) the ability to be reduced for more efficient computation (Díez & Galán, 2003). This use of the canonical model significantly simplifies the process of selecting the CPT parameters and standardizing these parameters for multiple buildings. Previously, use of Bayesian networks for HVAC diagnostics required customized for a specific building, but in order for this technique to be viable for commercial applications it must be applicable to all AHU-VAV systems.

The first step was to significantly reduce the number of possible options that the network designer would have to choose from for each combination of fault and evidence. For a typical evidence node, if random probability values were selected for analysis, there could be nearly two million different potential combinations to evaluate. Evaluating the efficacy of these values is very time consuming for each potential combination, so performing a traditional optimization is not feasible. As a result, the probabilities are binned, and the values are based upon the known properties of AHU-VAV systems. To perform this simplification, a set of bins were defined as follows:

- The evidence state is the definition of the fault state in question
- The evidence state is likely to be observed, given the fault state in question
- The evidence state has a "split-likelihood", given the fault state in question. For example, in an instability fault, the evidence is equally likely to be too high as it is to be too low.
- The evidence state is unlikely to be observed, given the fault state in question – but there exists some causal link between the fault and evidences states in question.
- There is no causal link between the fault and evidence states in question.

Since these definitions are well understood for most fault/evidence combinations, assigning each arc to one of these combinations is possible based on first-principles. There may be some advantage to adding some other options beyond the five described above, and that can be investigated in the future. Once these states are assigned, the next step was to provide a standard value for each possible definition. The default values selected are included in **Table 1**, below.

Table 1: Bayesian network parameters

Abbrev	Description	Prob. Value
	3 Evidence States (noisy-max)	
D	Evidence is definition of fault, or fault-free state	1
L_f	Likely, extreme severity	0.5
L_m	Likely, moderate severity	0.5
c_2	Likely remainder	= 1 - L_f - L_m
S_f	Split likelihood, extreme severity	0.25
S_m	Split likelihood, moderate severity	0.25
c_3	Split likelihood remainder	= 1 - S_f - S_m
U_f	Unlikely, but possible, extreme	0.05
U_m	Unlikely, but possible, moderate severity	0.05
c_1	Unlikely, but possible, remainder severity	= 1 - U_f - U_m
N	Not related at all	0
	2 Evidence States (noisy-or)	
L_2	Likely (2 states)	= L_f + L_m
S_2	Split likelihood (2 states)	= S_f + S_m
U_2	Unlikely (2 states)	= U_f + U_m

In this table it can be observed that only six parameters (highlighted in black) are eligible to be adjusted in order to perform refinement of the method. Since the analysis of the network across many different fault scenarios and buildings is computationally- and time-intensive, limiting the number of parameters allows for a future improvement via sensitivity analyses and informal optimization approaches. Since no tuning of these parameters was performed using real building data – only fundamental knowledge of general AHU-VAV variable interactions – the model is not over-fit for any specific building, but generally constructed to apply to all buildings. Any future improvements or refinements should be performed utilizing testing across multiple buildings to ensure that this remains the case.

2.5 Field Testing

Field testing of this approach was performed at three occupied commercial buildings in the Philadelphia area with AHU-VAV systems. In each of these buildings a variety of faults were artificially introduced to mimic commonly observed faulty conditions. For different fault scenarios, this was performed either via control system override, physical manipulation of components, or by manually applying a voltage control signal to a system component. The method utilized for each fault experiment was dictated by the ability to implement each fault in a manner that would accurately mimic not only the physical effects of the fault, but also accurately mimic what the BAS data would be if the fault had been naturally occurring.

Building 1

The first building in which fault experiments were conducted was Building 101 at the Philadelphia Naval Yard. It is a three-story 55,000 ft^2 commercial building located on the Navy Yard campus at the southern edge of Philadelphia at the confluence of the Delaware and Schuylkill rivers. The Building 101 HVAC system consists of three AHUs serving 26 VAV boxes. The heating is supplied by a natural gas boiler and the cooling is provided by three direct expansion units. The HVAC system is controlled by a BACnet system, with a typical AHU sensor arrangement. This building has a standard AHU-VAV configuration, except the AHUs do not have mixing boxes – instead having a small outdoor air damper with no return or mixed air dampers. The outdoor air flowrate is minimal, so the building ventilation is primarily due to infiltration through the envelope.

Building 2

Building 2 is the Swope Music Building, a three-story, 90,000 ft^2 school building at West Chester University. Located approximately 23 miles west of Philadelphia, this building was constructed in 2007 and received a silver LEED certification at that time. This building contains a 375-seat theater, a 125-seat recital hall, a music library, and numerous rehearsal spaces, classrooms, and offices. The HVAC system is served by 7 AHUs, with space-specific AHUs for the Theater Lobby, the Theater, the Instrument Rehearsal Room, the Recital Hall, and the Stage. Two other AHUs serve the 90 VAV-boxes that condition the spaces throughout the rest of the building. In addition to the typical AHU-VAV configuration, this building requires careful humidity control so there are additional humidity sensors and controls utilized with this system.

Building 3

Building 3 is Stratton Hall at Drexel University, a three-story 50,000 ft^2 building constructed in 1955, with a complete renovation of the HVAC system and improvements to the building envelope performed in 2013. This building has a typical AHU-VAV system, with separate AHUs serving the

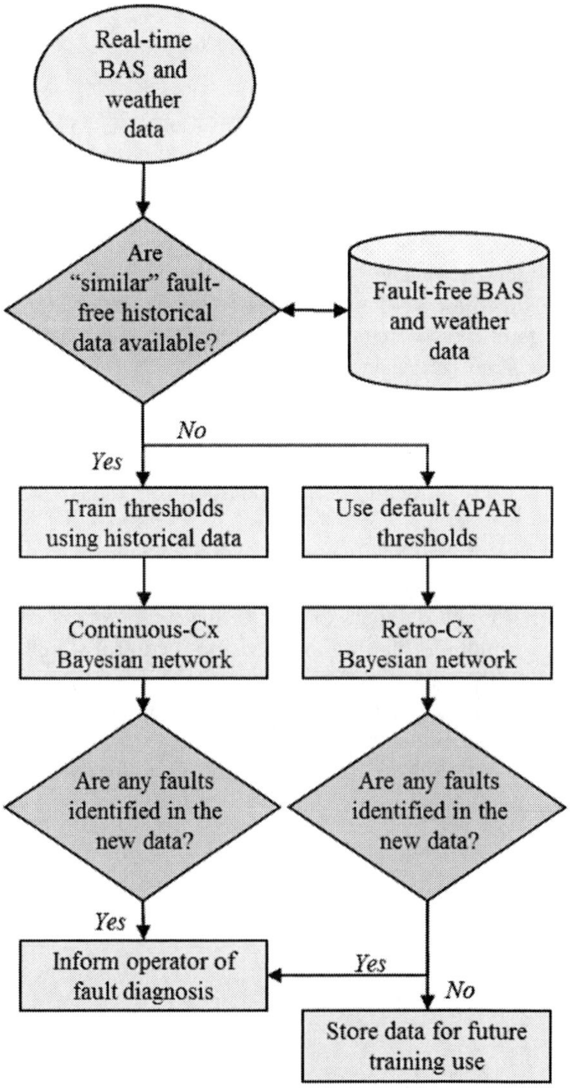

Figure 2: RCx versus CCx scenarios

second and third floors of the building. These AHUs are served by a district chilled water plant and a heat exchanger connected to Philadelphia's steam heating infrastructure. The first floor of this building utilizes small window units and direct expansion units for cooling and radiant heat, that are not connected to the BAS, so only the second and third floors were included in this study.

Fault Experiments
To perform initial testing of this method, 15 fault experiments were selected from each building for analysis. Each building was studied in two ways: simulating the RCx scenario, and simulating the CCx scenario. Due to space constraints, a comprehensive listing of all faults studied is not included, but the faults were selected to obtain a cross-section of different fault "types" and severities. The faults selected included fan faults, coil/valve faults, damper faults, and sensor faults (temperature and pressure sensors) at each building. These faults were implemented across a variety of severities, with some faults causing noticeable impact to the system operation and other faults causing negligible impact to overall system operation.

In a commercial deployment setting, the RCx scenario would be the first to be tested, and then once the data is identified as a suitable baseline, this information would be used to train the CCx network, as depicted in **Figure 2**. In this figure, the term "similar fault free historical data" refers to data from the same mode of operation as is being analyzed that has already been found to be free of significant faults using the RCx network.

3. RESULTS

To analyze the results, the day during which each fault was implemented was analyzed, and the results of the Bayesian network were summarized over the duration of each fault experiment. If a fault was identified with an average probability exceeding 60% during the time in which it was implemented, it was considered to be *accurately diagnosed*. If a different fault was diagnosed with higher than 60% confidence, and this incorrect diagnosis was in-part due to the effects of the fault that was artificially introduced, it was categorized as *mis-diagnosed*. If a different fault was identified during the course of the fault experiments that was not due to the fault experiment, it was categorized as a *false-alarm*. If no fault was detected with greater than 60% confidence, the fault was categorized as *not detected*. The faults that were not detected were also sub-classified as either *impactful* faults of faults with *no impact*. A fault is only categorized as having no impact if:

- It is not in the control-loop of the system, so it does not directly impact any operation
- If an experienced building operator was reviewing at the building data, with knowledge that the fault exists, it would not be apparent from the BAS data.

The first bullet above is objectively determined, but the second bullet requires a subjective assessment of whether the fault has any system impact. The faults that were found to have no impact were sensors with very small biases (2-3°F), and dampers or valves that were "stuck" at their normal operational range for a given operational mode. For example, if an outdoor air damper control signal is requesting the damper to be 45% open, but the fault experiment has the value fixed at 50% open, the difference in the data would be very difficult to discern and the energy, cost, and comfort impacts would be negligible. However, when the system shifts to a different mode and requests that the damper is closed, or moved to 100% open, then the fault would become immediately apparent to a building operator, and would have measurable impact on the system operation.

A summary of the results of the faults utilizing the RCx approach (baseline threshold values) is included in **Table 2**, and a summary of results utilizing the CCx approach (training the Bayesian network thresholds based on known fault-free operation) is included in **Table 3**.

Table 2: Retro-commissioning summary

Retro-Commissioning Results						
Building	Total Faults	Accurately Diagnosed	Misdiagnosed	False Alarms	Not Detected	
					Impactful	No Impact
Building 1	15	10	1	1	0	4
Building 2	15	9	2	2	0	4
Building 3	15	12	0	0	0	3

Table 3: Continuous-commissioning summary

Continuous-Commissioning Results						
Building	Total Faults	Accurately Diagnosed	Misdiagnosed	False Alarms	Not Detected	
					Impactful	No Impact
Building 1	15	11	0	0	0	4
Building 2	15	11	0	0	0	4
Building 3	15	13	0	0	0	2

From these tables, it can be firstly observed that no *impactful* faults were missed in either network, and that when using the continuous commissioning network, all faults that impacted energy, comfort or indoor air quality were accurately detected and diagnosed. To illustrate the concept of a fault that is not impactful, the operational data for the fault in which the Building 2 return fan was stuck at a fixed speed is shown in **Figure 3** and the diagnostic result is included in Figure 4. As observed in **Figure 3**, the fan speed was fixed at the normal operational range, resulting in negligible impact to the system, and limited evidence to use for diagnosis.

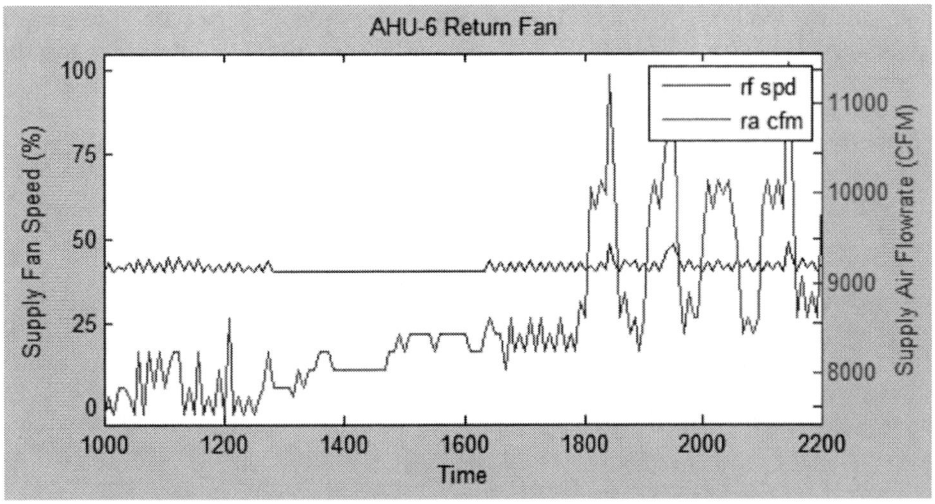

Figure 3: Return fan stuck at fixed speed (minimal impact)

In the above figure, "rf spd" refers to the return fan speed, and "ra cfm" refers to return air flowrate.

As demonstrated in **Figure 4**, below, the most likely diagnosis was correctly found to be a fan stuck at fixed speed ("frozen"), but due to the limited impact the diagnosis was not strong enough to meet the 60% threshold used to evaluate the efficacy of the fault diagnostic algorithm.

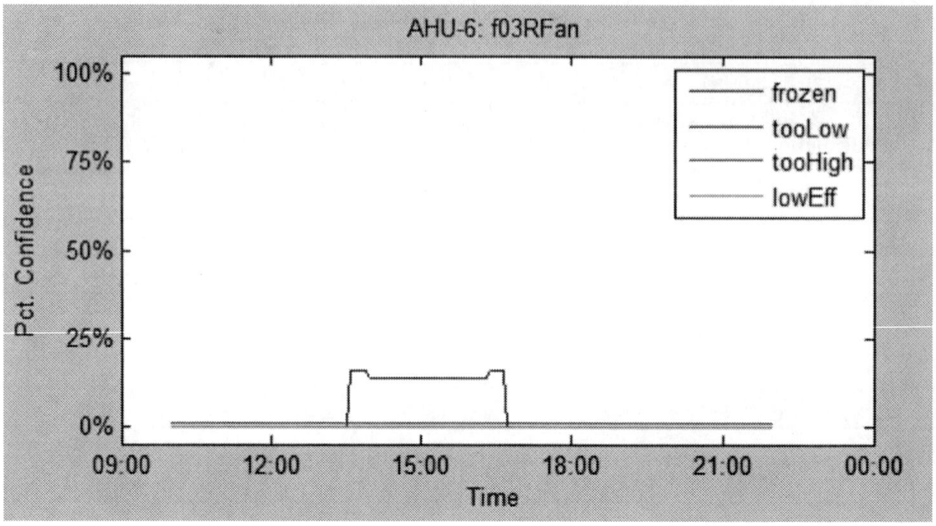

Figure 4: Return fan diagnostic result, 2015-12-17 (Swope AHU-6)

In the retro-commissioning network, there were three false alarms and three misdiagnosed faults. In reviewing the results, it was found that the misdiagnoses corresponded directly with the false alarms. The false alarm and misdiagnosis in Building 1 were due to the fact that there is no return air (RA) or mixed air (MA) damper. As a result, when economizing is expected, little to no outdoor air is actually being introduced into the system. While this was not technically a fault (since no piece of equipment was malfunctioning), it is behavior that differs from typical AHU operation, so it is useful information for a building operator or service provider. However, when previous operational data was used to train the thresholds, these faults were not falsely flagged.

Similarly, the misdiagnoses and false alarms identified in Building 2 were due to the non-traditional humidity-based control that was utilized in this building. In one instance a fault was flagged due to simultaneous operation of the heating and cooling coils. While this was done intentionally at this site for the purpose of dehumidification, this is indicative of faulty operation in many buildings and is a reasonable operational condition of which to alert a building operator or service provider. The other false alarm, and associated misdiagnosis was due to the temperature differential created by the steam humidification. Again, once these items were incorporated in the training data as "normal" operation, the faults were properly diagnosed.

4. DISCUSSION

This demonstration of the AHU-VAV AFDD tools provides evidence that the use of a Bayesian network approach can be an effective diagnostic tool and can be utilized across multiple AHUs to diagnose all different categories of faults in both RCx and CCx applications. The use of a "passive" diagnostic algorithm (not manipulating system controls) will always be limited by the impact of the fault on system operation, but the method utilized by this AFDD tool allows for accurate detection and diagnosis of faults across a vast majority of AHU-VAV configurations and applications.

The results presented here are the initial results from the Bayesian network investigation, prior to any tuning or refinement being performed. The network parameters for this initial study were selected based upon expert knowledge and first principles. An effort is underway to refine the parameters using the three existing "training" buildings, and tested on additional buildings to see if the results are improved. These initial results are very promising, but further testing and validation of this method is required in order to properly quantify the strengths and weaknesses. This can only be accomplished through extensive additional testing across multiple buildings.

5. CONCLUSIONS

The primary conclusions from this study include:

- The use of a Bayesian network is an effective method for combining multiple rules in order to identify the root cause (source) of abnormal/faulty/sub-optimal AHU operation.
- By generalizing the parameter selection based on first-principles of a typical AHU system, it is possible to create a diagnostic network that is effective across multiple buildings with very limited tuning.
- Allowing the Bayesian network to automatically customize thresholds for individual buildings fault-free baselines can improve accuracy over a general Bayesian network.
- There appears to be some advantage to including different levels of severity for fault and evidence nodes in terms of diagnostic accuracy. Further work needs to be done to quantify this improvement.

REFERENCES

ANNEX 25, (1996) *Building Optimization and Fault Diagnosis Source Book, ANNEX 25 Real Time Simulation of HVAC Systems for Building Optimization, Fault Detection and Diagnosis*, International Energy Agency, Finland.

Díez, F. J. & Galán, S.F. (2003). Efficient computation for the noisy max. *International Journal of Intelligent Systems, 18*(2), 165-177.

Katipamula, S., & Brambley, M. R. (2005a). Methods for fault detection, diagnostics, and prognostics for building systems - A review, Part I. *HVAC and R Research, 11*. 3-25.

Katipamula, S., & Brambley, M. R. (2005b). Methods for fault detection, diagnostics, and prognostics for building systems - A review, Part II. *HVAC and R Research, 11*. 169-187.

Pérez-Lombard, L., J. Ortiz, et al. (2008). A review on buildings energy consumption information. *Energy and Buildings 40*(3). 394-398.

Xiao, F., Zhao, Y., Wen, J., & Wang, S. (2014). Bayesian network based FDD strategy for variable air volume terminals. *Automation in Construction, 41*, 106-118.

Zhao, Y., Xiao, F., & Wang, S. (2013). An intelligent chiller fault detection and diagnosis methodology using Bayesian belief network. *Energy and Buildings, 57*, 278-288.

Zagorecki, A. & Druzdzel, M. J. (2013). Knowledge Engineering for Bayesian Networks: How Common Are Noisy-MAX Distributions in Practice?. *IEEE Transactions on Systems, Man, and Cybernetics: Systems, 43*(1), 186-195.

Current Based HVAC Systems Air Filter Diagnostics and Monitoring

Fadi ALSALEEM[1]*, Michael MUNROE[2], Mostafa RAFAIE[1]

[1]Wichita State University, Department of Mechanical Engineering,
Wichita, Kansas, USA
Contact Information (316-518-5825, mxhhasan@wichita.edu)

[2]Emerson Climate Technologies, Residential Solution
Sidney, Ohio, USA
Contact Information (937-494-7874, Michael.Munroe@Emerson.com)

* Corresponding Author

ABSTRACT

This paper addresses the use of continuous indoor motor current to detect filter blockage in HVAC system. A commonly known phenomenon exists in the loading of a typical indoor motor blower that results in a power consumption decrease and hence less current draw for PSC motor, and power and currant draw increase for constant torque motor. Testing using Akaike information criterion (AIC), classification and regression tree (CART) models, and using both fixed radius basis function and linear basis function was described and performed against field of installed systems to determine if candidate data filter was sufficient, or to motivate use of Mann-Kendall to determine trend existence, strength, and transition in existence or strength. The bases, commonly used in practice, were found to have cumulative effectiveness against only 50.4% of installed systems, and were strongly differentiated in performance against motor type. The Mann-Kendall approach was found to have performance of ~88% of evaluated systems. This approach calculates the confidence trending level corresponds to the nonparametric correlation coefficient for the indoor current daily averages. Trend levels will be accumulated over time and will be used to declare filter blockage once they suggest a strong trend in the direction of filter blockage.

1. INTRODUCTION

Most heating, ventilation, and air-conditioning (HVAC) systems in commercial buildings and residential homes are equipped with air filters to improve indoor air quality. Filter is intended to be used to collect dirt, debris, and dust and prevent their accumulation anywhere else. To prevent excessive dirt accumulation and possible HVAC performance degradation and malfunction, air filters need to be replaced once in periodically. The problem is that there is wide variation, even in the nearby houses, in the rate of accumulation of debris in the air filter. This can be driven by pets, by construction (infiltration rate), or by exposure. Current manufacturer recommendation a replacement of air filter every 3 months per inch of filter. While this approach minimizes the worst case, it also allows the inference with uniform rate of deposition, that one average day of use consumes less than an average of about 1.1% of filter capacity. In this work, however, we developed a framework to warn the need for filter replacement based on scientific data rather using the 3 months rule.

The impact of filter quality and its current conditions on HVAC system performance has received great deal of attention in the literature. For example, Yang et al. (2007) studied this impact using different filter types (six different level of filtration) on three different HVAC systems. Two series of tests were performed while the filters are in healthy and dirty states. The study showed that a dirty filter could cause significant reduction in HVAC cooling capacity, increase in the pressure drop across the filter, and increase in the indoor fan power usage. Another study (Stephens et al., 2010) was performed on a residential home in Austin, Texas. While the HVAC system energy use was monitored, different filters MERV (MERV < 4, MERV 8. and MERV 11) were tested in two different HVAC systems. The study shows that energy consumption increases with dirty filter but did not differ much with high-efficiency filters compared to low-efficiency filters.

2. DIRTY FILTER DETECTION ALGORITHMS

2.1 Hardware and Experiment Set-up

A sample of 225 HVAC systems that are installed in real homes were used in this investigation. Of these systems, 118 were charged with R410A refrigerant and the rest were charged with R-22 refrigerant. The tonnage size and SEER information for this sample size is given in Figure 1. The hardware that is used to collect data from these systems is shown in Figure 2 and is described in extensively in (Alsaleem et al., 2016). Few key points are to be mentioned about the hardware. (1) It consists of an indoor kit that is equipped with voltage and current sensors as well as temperature sensors to track air supply, air return, indoor liquid line, and suction line temperatures. The outdoor kits measure the aggregate outdoor current and voltage. (2) the data for each system run is transmitted in chunks, each of fifteen minutes maximum length and at sampling rate of one sample per five seconds (0.2 HZ),(3) a cloud structure is built to host algorithms to receive and processes the data from the indoor and outdoor kits.

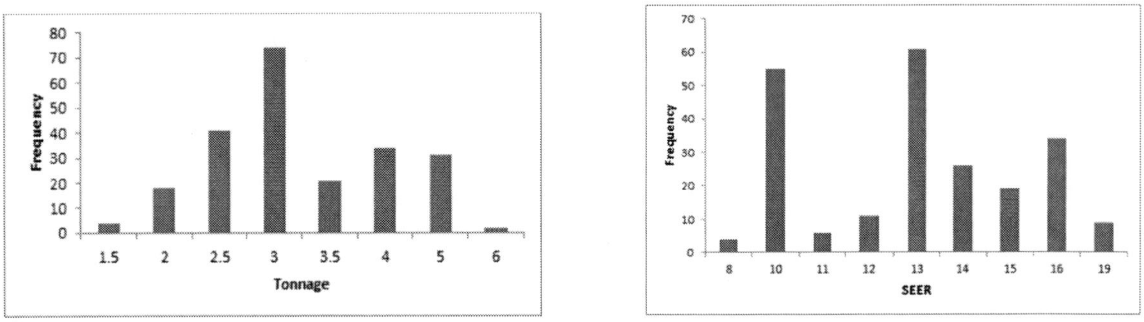

(a) (b)

Figure 1: The tonnage size (a) and SEER ratting (b) for the 225 HVAC systems used in this study

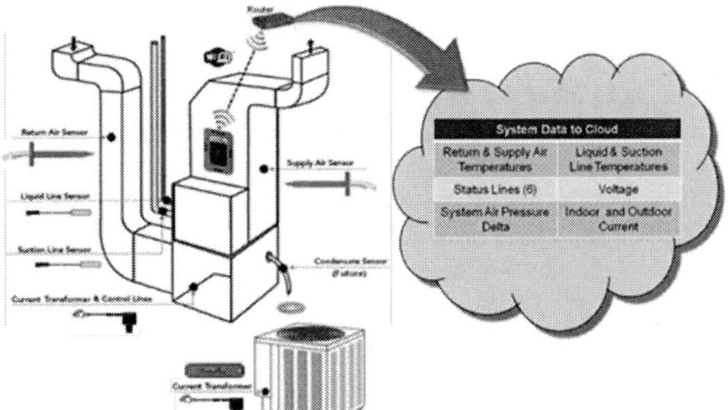

Figure 2: Hardware shows the indoor and outdoor kits to gather and send sensors data to the cloud

2.2 Dirty Filter Detection Basic Idea

A commonly known phenomenon exists in the loading of a typical indoor motor blower that results in a power consumption decrease for PSC motor and power increase for constant torque motor. A dirty filter-detection algorithm benefits from this simple fact to capture filter blockage over time by monitoring the daily averaged indoor motor power. Figure 3(a) illustrates the normalized daily average of the indoor current of one of the filed monitored HVAC system correlated to the expected airflow percentage drop as the filter developed dirt over time. The figure demonstrates that filter degradation over time can be detected only by using current measurement. This chart also shows the time when the homeowner changed the filter, by the sudden jump by the end of September. Figure 3(b) visually confirms the predicted poor filter performance before replacement.

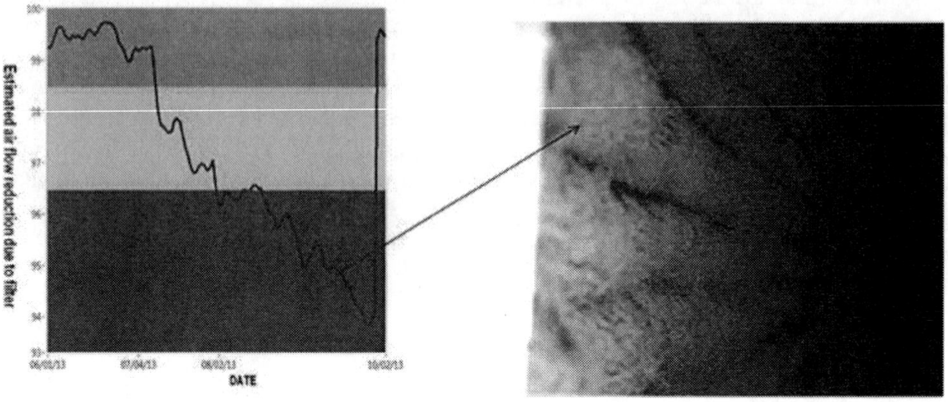

(a) (b)

Figure 3: Indoor current correlation to filter blockage, (a) data of the normalized indoor current over time correlated to the airflow reduction, (b) the dirty filter before replacement.

In order to reduce the effect of special cause variation on disposition some level of signal conditioning precedes analysis. The major axes of variation in indoor current measurement are going to include run-to-run, solar hour-of-day, week-of-year, by year, conditioning-mode (AC/HP, number stages, etc.), by indoor temperatures (wet and dry bulb), by equipment make/model/manufacturer, by installed configuration, by installer and installation, by home size/insulation/exposure, by geographic location, by maintenance history, by operation history, and by current homeowner configuration and operation. The following examples are around air filter health, so the variables most strongly related to that are considered.

Common types of variation that confound with air filter indication include variation of load by hour of day, by indoor temperatures, by operation (count of stages) and by changes to configuration including occlusion of returns or changes to registers by the homeowner. The first is accounted for by averaging across a full day, including both daytime and nighttime use. The second is accounted for using normalizations like using historic data to approximate the relationship between indoor current and indoor temperature as a polynomial. The third can be accounted for by tracking control signal. The fourth has the challenge that there is no leading variable, and that it is a discontinuous jump. On either side of the discontinuity, the polynomial relationship between power and time is consistent, and acts as if one added a scaled unit step at that point in time. When there is a fundamental change, such as a replacement of a dirty filter with a clean one, the coefficients of the polynomial, after accounting for the jump, change in a sustained manner.

2.3 Linear and Redial Fitting Approaches

Two simple but effective bases for "appropriate function" to model filter behavior over time included fixed variance radial clustering (a cousin of k-means), and purely linear least squares fit. The assumption of the constant diameter radial model is that jumps outside the "radius" were not indication of filter behavior. The background, derivation, and proper use AIC or corrected AIC as used herein is more comprehensively described in (Akaike, 1992).

The "Information Criteria" measures of fit are derived from Kullback-Leibler divergence and via approximations for log-likelihood, and account not only for error as R^2, but model complexity and sample size. One condition of use to compare two or more models is that the same data must be used for all model evaluations. A lower value of criteria is indicative of higher likelihood. The AIC was used to compare the performance of radial clustering and linear fitting.

The expressions for AIC used in this work was:

$$AIC = n \cdot ln\left(\frac{RSS}{n}\right) + 2 \cdot k \quad (1)$$

Where "RSS" is the residual sum of squares (or sum of squared error), "n" is the number of observations, and "k" is the number of parameters in the model. Each split was considered a piecewise Heaviside function, so it has only the parameter of the location of the split plus the sum of parameters of the fit-model within each split. A three term linear expression would then be described as:

$$Amps(t) = (linear1 + H(t_1) * linear2) \cdot (1 - H(t_2)) + H_2 \cdot linear3 \quad (2)$$

This expression has two parameters for each linear expression and the two parameters for the two Heaviside functions for a total "k" value of "8". Number of samples was large enough that correction was not required in this case.

A Classification and Regression Tree (CART) model was used to evaluate performance of these for ability of the basis to account for jumps by testing non-split then split transformations for AIC. In another words, as sudden jump in indoor current are expected to happen, the CART model will be used to remove the effect of jumps in the accuracy. The premise of the CART is that the leaf-tip is constant, and splitting can be single axis at a time to improve either GINI coefficient or accuracy. While the CART premise is incompatible with detection of slope, the greedy splitting is optimal for detection of univariate jumps. In analogy to nontransitive dice, while this CART approach is inadequate for directly filtering the data, it is sufficient for evaluation of the filtering algorithms. While a full discussion of CART models is substantially beyond the scope of this text, there are excellent references including (Breiman et al 1984).

If the CART imposed splits improved the AIC by a factor less than 2.198, then the basis was deemed "capable" otherwise, it was deemed "incapable". This means that if the CART imposed splits did not make the old model contain any less than 25% of the weight, then the old model was considered "capable". This balanced, non-greedy approach to the setting of the weight threshold was made in consideration of the technical confidence of high levels of noise in the overall system and using the game-theoretic idea of "equal division of the contested value" (Talwakar 2014).

The log of the likelihood of a model given a difference in AIC compared to a model with minimum AIC is negative one half the difference; the weight for addition of models is the likelihood of the individual divided by the sum of likelihoods in the ensemble and the constant of proportionality divides out. A delta AIC of 5.882, as can be seen in Table 1, indicates a relative (Akaike) weight below 5%. Table 1 indicates a few common AIC thresholds and weight, and then indicates common weights and their thresholds, for 2-component weighted mixture of models.

Δ AIC	Weight
0	50.0%
0.5	43.78%
1	37.75%
2	26.89%
3	18.24%

Weight	Δ AIC
33%	1.416
25%	2.198
10%	4.395
5%	5.889
1%	9.191

Table 1: Useful list of values relating difference in AIC with component weight in additive model when there are two components under consideration.

It was found, using the above criteria, that the linear was capable for 39.0% of 225 HVAC systems and not capable for 61.0% of them. Similarly, the radial approach was capable for 27.5% of systems but was incapable for 72.5% of them, the details of these calculation are beyond the scope of this paper and will be discussed in a separate work. When combined they were capable for 50.4% of all systems. These very simple bases, while easy in practice to execute, are substantially incapable of handling the naturally occurring variation. This motivates the use of the Mann-Kendall test.

For the remainder of this section, we first review the Mann-Kendall test, and then discuss the trending analysis use in warning of dirty filter.

2.4 Mann-Kendall Trending Test

The purpose of the Mann-Kendall test is to assess for a given data if there is a monotonic trend over time. It has been used in many applications, such as water quality monitoring (McLeod et al., 1990) and more recently for weather trending and prediction (Soltani et al., 2013). Its advantages over linear regression approach include it does not require a normal distribution for the data, it is independent of the data magnitude, and it can deal well with missing or irregularly spaced data. The trending test compares each data point to all subsequent values. If the data from newer values is greater than earlier data, then the Mann-Kendall statistics value S is increased by 1.0; otherwise, it decreased by 1.0. In mathematical terms, this could be described as:

$$S = Sign\left(\sum_{i=1}^{m-1} \sum_{j=i+1}^{m} \begin{cases} 1\ if\ (T_j - T_i) > 0 \\ 0\ if\ (T_j - T_i) = 0 \\ -1\ if\ (T_j - T_i) < 0 \end{cases} \right) \tag{3}$$

where T_i, T_j are the daily average values for a given measurement at a given day where j > i. A very high positive S value indicates an increasing trend, whereas a very low negative value indicates a decreasing trend. The sign function, introduced by the authors, adjusts the sign of the S value to always indicate a positive value if the measurement trend direction support the fault.

To account for sample size, the magnitude of S was adjusted by the following scale (assuming sample data are not similar) as follow (HydroGeoLogic, 2004):

$$S_N = \frac{S \mp 1}{g(S)^{1/2}} \tag{4}$$

where, $g(S) = \frac{1}{18}[n(n-1)(2n+5)]$, and n is the sample size.

Mann-Kendall use for dirty filter detection

Figure 4 shows a flowchart for the algorithm of performing the Mann-Kendall to detect dirty filter. The process starts when the filter detection algorithm receives a new daily average for the indoor motor current data. Next, the algorithm adds the data to the existing dataset of previous data and applies the Mann-Kendall analysis described above. Based on the Mann-Kendall analysis, the algorithm calculates the trend confidence level, in the range of -1 to 1. The trend confidence level corresponds to the nonparametric correlation coefficient (S_N) described above.

The algorithm scales the estimated value based on duration in time of the new data set and then adds it to a cumulative sum. The scaling is based on the time duration represented by the current data sample being added to the data set versus the time duration represented by the existing data set of previous data. For example, a confidence level corresponding to one week will be appropriately scaled or weighted when added to, for example, previous data corresponding to six weeks. Next, the algorithm determines whether the absolute value of the trend confidence sum is greater than a predetermined threshold. In this implementation, the predetermined threshold used is 2.0. This value was chosen based on correlation analysis between the empirically confirmed amounts of dirt for multiple filled and dirty filters to the absolute trend sum. When the absolute value of the trend confidence sum is greater than 2.0, the algorithm generates a dirty filter alert, resets the trend confidence cumulative to 0.0, and loops back to starts the trend analysis again. On the other hand, if the absolute value of the trend confidence sum is less than 2.0, the algorithm applies a new filter detection algorithm. In this case, the algorithm determines whether the current data is within three sigma (standard deviation) of the average data over time. When the current data is not within three sigma of the average data, and has stayed outside of three sigma of the average data for a predetermined duration of two days, the algorithm determines that the filter has been replaced and generates a new filter acknowledgment and resets the trend confidence sum to 0.0.

With reference to Figure 4, a graphical representation of using the Mann-Kendall algorithm, depicted in the flowchart of Figure 5, in warning of dirty filter is illustrated for a sample unit data. Figure 5(a) shows a plot of daily averages of indoor blower fan current over time in which each square represents one full day's average value. Similar to Figure 5(a), the plot indicates that the filter has been replaced with a new filter when the current spiked from 6.05 Amp to 6.50 Amp, seven days before the last recorded day average. Figure 5(b), shows a plot of the trend confidence levels, calculated using the Mann-Kendall analysis for the data of Figure 5(a), over time. As depicted in the figure, the trend confidence level is generally a negative number, indicating a downward trend, until the last trend confidence level, when the trend confidence level moves to a positive level, corresponding to the filter being replaced with a new filter.

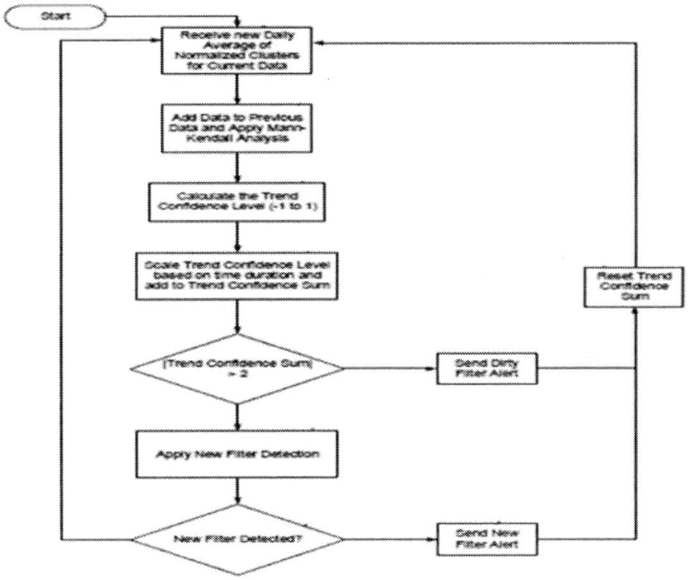

Figure 4: Flowchart for using the Mann-Kendall of warning dirty filter

At Figure 5(c), a graphical representation of the trend confidence sum over time is shown. As shown in the figure, the time interval for which each trend confidence sum is made generally corresponds to the width of each step in the graph. However, the trend confidence sum can be updated more frequently, or less frequently, as desired. The figure shows that the trend confidence sum increases in negative magnitude until it reaches a point at which the trend confidence absolute sum is greater than the predetermined threshold, which in this case is 2.0. With the trend confidence absolute sum greater then 2.0, a dirty filter alert is generated, and the trend confidence sum is reset to *0.0*. The trend confidence sum again begins to increase in negative magnitude, until it reaches the point when the homeowner finally responded to the alert and changed the filter. The algorithm detected this filter change event and the trend confidence sum is reset to *0.0*. The logic to detect the need for a new filter is built to search for a sudden change in the trend with a direction opposite to the normal direction of filter dirt development. Specifically, a moving average of the standard deviation for the indoor current is calculated. If at any given time the current measured value jumped from the previous day value a threshold of at least three times the current averaged standard deviation, then this is reason for a potential reset. If the values for at least three new days in a row still deviate by that threshold, then a trend reset is triggered. The three-day condition is to eliminate a false reset trigger due to measurement noise.

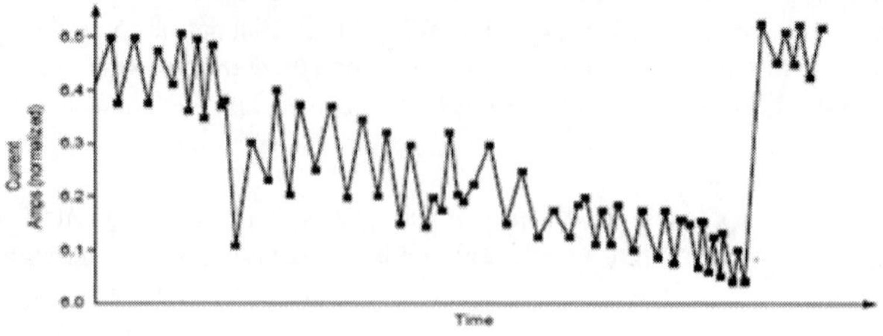

(a) Daily averaged indoor current

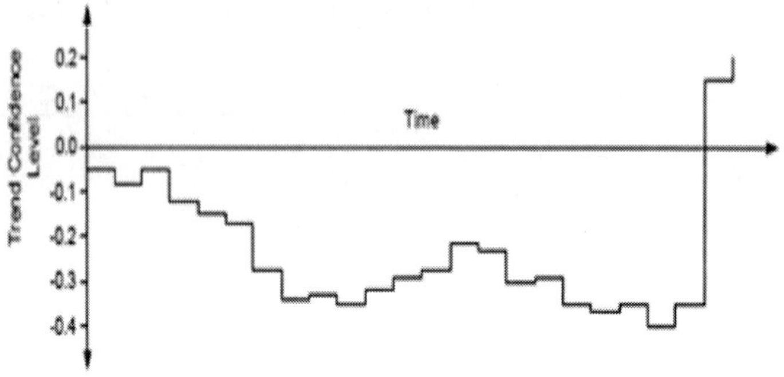

(b) Trend confidence level

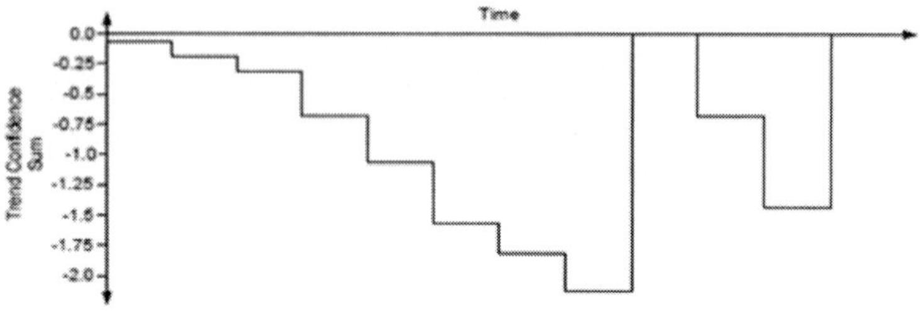

(c)Trend confidence sum over time

Figure 5: A graphical representation of using the Mann-Kendall based algorithm, depicted in the flowchart of Figure 4, in warning of dirty filter for a sample unit, (a) is the indoor current daily average values over time, (b) is the corresponding trend confidence level using the Mann-Kendall test, (c) is the trend sum that is used to compare with a predetermine threshold for potential filter alert.

3. RESULTS AND DISCUSSION

Figure 6 show results for the application of the Mann-Kindle algorithm over the daily averaged indoor current of a HVAC system for more than a year. The system has a PSC motor type. In the figure, the averaged current is in white, the weekly based trend measure is in green, the weekly trend sum accumulation is in red, and the actual filter replacement signal is in blue. The figure demonstrates the idea of using trends confidence level sum to trigger the dirty filter alert. As shown in the figure, the filter alert is triggered if the trend sum exceeds a threshold value of -2 or its mutiples. These events are marked by purple arrows. In average, the filter alert for this system is triggered for at least 1 time per two months. The trend sum accumulation is only rested to zero if the algorism detect a new filter replacement.

The charts in Figure 7 show the filter alerts status generated in a month for the 225 systems. The filter alerts status were obtained from homeowners responses to an online survey, immediately sent after receiving the filter alert. The survey responses reveal that 88% of the filter alerts were valid. System with electrostatics filter is among the highest population for invalid filter alert. Due to the unique dynamics for dirt accumulation and its effect on indoor current, a different algorithm might be needed to tackle electrostatics filter.

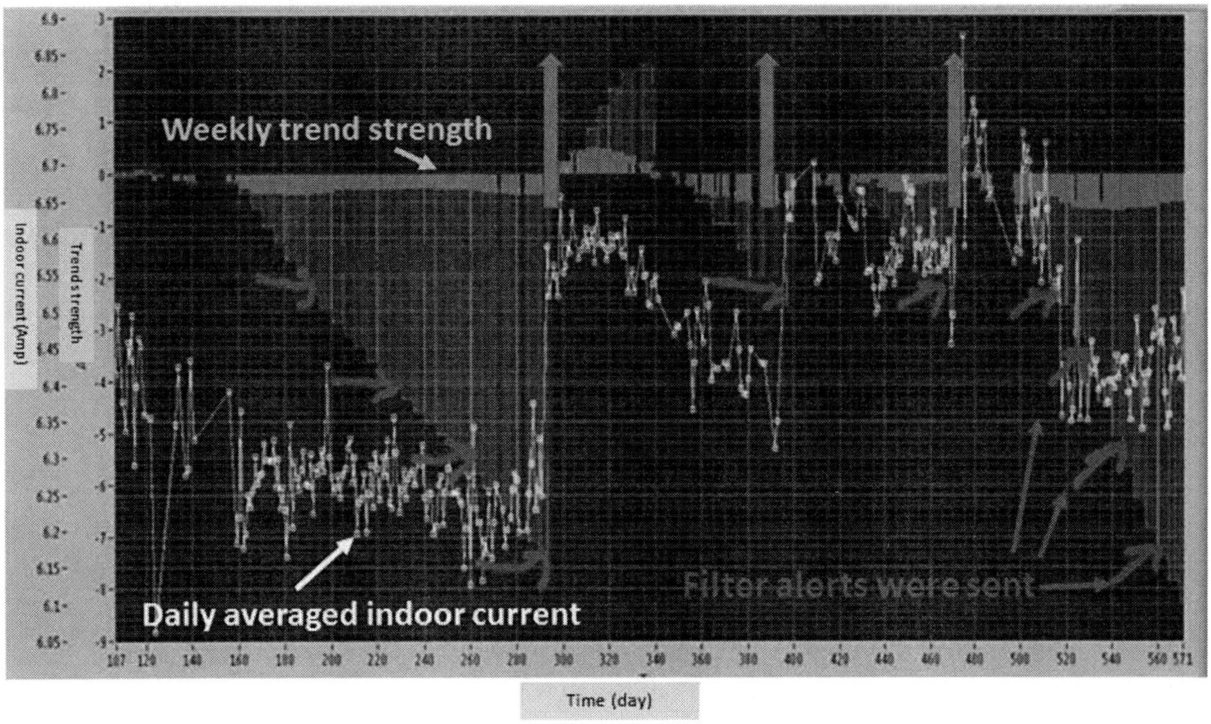

Figure 6: Results for the application of the Mann-Kindle algorithm over the daily averaged indoor current of a HVAC system for more than a year

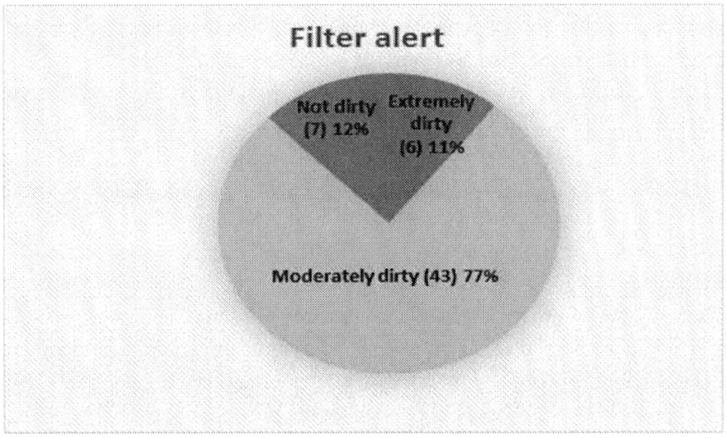

Figure 7: Trending filter alert statistics

4. CONCLUSIONS

This paper addressed the use of continuous indoor motor current to detect filter blockage in HVAC system. Different filter-detection algorithms were designed to capture filter blockage by monitoring the daily averaged indoor current over time. 225 HVAC field units were used in this investigation. An indoor kit that is equipped with current sensor was used to transmit these units' data for each system run in chunks, each of fifteen minutes maximum length and at sampling rate of one sample per five seconds (0.2 HZ). For the candidate models within the duration of the analysis, the Mann-Kendall approach was found to be sufficient for alerting dirty filter. The accuracy of this approach can be further enhanced by combining the trends of other indoor measurements such as the split and evaporator suction temperatures.

REFERENCES

Akaike, H. (1992). Information Theory and an Extension of the Maximum Likelihood Principle. *Springer Series in Statistics*, Perspectives in Statistics. Springer-Verlag: New York, pp. 610-624.

Alsaleem, F., Abiprojo, R., Arensmeier, J., &Hemmelgarn, G. (2014). HVAC System Cloud Based Diagnostics Model, *International Refrigeration and Air Conditioning Conference*, Paper No. 2496, West Lafayette, IN.

Alsaleem, F., &Quedan, A., (2016). Low Refrigerant Algorithm Detection for Cooling Systems Relying on Trending and Data Analysis, to be presented in the 4th *International High Performance Buildings Conference at Purdue*, West Lafayette, IN.

Alsaleem, F., (2015) Current based air filter diagnostics and monitoring, *USA pending patent application* (14/709658)

ASHRAE Standard, (1987). 41.2, Standard methods for laboratory air-flow measurement. *American Society of Heating, Refrigerating and Air-Conditioning Engineering*, Atlanta.

ASHRAE Standard, (1986). 41.1Standard method for temperature measurement. *Trans. of the ASHRAE*.

ASHRA E Handbook, (1996). HVAC systems and equipment. *Trans. of the ASHRAE*

Breiman, L., Friedman, J., Olshen, R., and Stone C., (1984). Classification and Regression Trees. Belmont, California: Wadsworth.

Deru, M., Field, K., Studer, D., Benne, K., Griffith, B., Torcellini, P., and Yazdanian, M. (2011). US Department of Energy *commercial reference building models of the national building stock*

Hipel, K.W. & McLeod, A.I. (1994). Time Series Modelling of Water Resources and Environmental Systems, Amsterdam: *Elsevier*. ISBN:0-444-89270-2.

Hirsch, R.M., Slack, J.R. & Smith, R.A. (1982). Techniques for Trend Assessment for Monthly Water Quality Data, *Water Resources Research*, vol.18, P.107-121.

HydroGeoLogic, Inc. (2004). Annual Groundwater Monitoring Report – *Former Fort Ord*, California.

Montgomery, J. F., Green, S. I., Rogak, S. N., & Bartlett, K. (2012). Predicting the energy use and operation cost of HVAC air filters. *Energy and Buildings*, vol. 47, 643-650.

Nassif, N. (2012). The impact of air filter pressure drop on the performance of typical air-conditioning systems. *In Building Simulation* (vol. 5, No. 4, pp. 345-350). Tsinghua Press.

Priotomo, R., and Alsaleem F., (2015) HVAC system air filter diagnostics and monitoring, *USA pending patent application* (14/712049).

Soltani, M., Rousta I., & Modir T. (2013). Using Mann-Kendall and Time Series Techniques for Statistical Analysis of Long-Term Precipitation in Gorgan Weather Station, *World Applied Sciences Journal*, vol. 28, no. 7, pp. 902-908.

Stephens, B., Novoselac, A., & Siegel, J. A. (2010). The effects of filtration on pressure drop and energy consumption in residential HVAC systems (RP-1299). Hvac&R Research, vol. 16(3), 273-294.

Talwalkar, P. (2014). The joy of game theory. *CreateSpace Independent Publishing Platform*, pp. 32-39.

Yang, L., Braun, J. E., & Groll, E. A. (2007). The impact of evaporator fouling and filtration on the performance of packaged air conditioners. *International Journal of Refrigeration*, vol.30 (3), pp.506-514.

A Probabilistic Framework to Diagnose Faults of Air Handling Units

Debashis DEY[1], Bing DONG[1], Zhaoxuan LI[1]*

[1]University of Texas at San Antonio, Department or Mechanical Engineering,
San Antonio, Texas, United States of America
Zhaoxuan.li@utsa.edu

* Corresponding Author

ABSTRACT

Air handling unit (AHU) is one of the most extensively used equipment in large commercial buildings. This device is typically customized and lacks quality system integration, which can result in, hardwire failures and controller errors. Air handling unit Performance Assessment Rules (APAR) is a fault detection tool that uses a set of expert rules derived from mass and energy balances to detect faults in air handling units. Although APAR has many advantages over other methods, for example, no training data required and easy to implement commercially, most of the time it is unable to provide the diagnosis of the faults. There is no established way to have the correct diagnosis for rule based fault detection system. In this study, we developed a new way to detect and diagnose faults in AHU through combining APAR rules and Bayesian Belief Network. BBN is used as a decision support tool for rule-based expert system. BBN is highly capable to prioritize faults when multiple rules are satisfied simultaneously. The proposed model tested with real time measured data of a campus building at University of Texas at San Antonio (UTSA).

1. INTRODUCTION

Heating, ventilation and air conditioning (HVAC) equipment commonly fails to satisfy performance expectations from design due to problems caused by inadequate maintenance, improper installation, or equipment failure. These problems of faults include mechanical failures such as dampers or actuators, stuck broken or leaking valves; control problems related to frozen or biased sensors, poor feedback loop tuning of incorrect sequencing logic; fouled heat exchangers; design errors; or inappropriate operation intervention. By identifying and diagnosing faults to be repaired, FDD techniques can benefit building owners by reducing energy consumption improving operation and maintenance and increasing effective control over environmental conditions in occupied spaces. Approximately 50% of a commercial's building energy consumption is associated with HVAC systems [1] and more specifically Air handling unit's (AHU) energy consumption counts for around 40% of industrial sites total energy consumption [2] because of its inefficient operation. Overall it is estimated that HVAC energy consumption accounts for 10-20% of total energy consumption in developed countries with AHU associated energy use associated with the majority of this [3]. The energy potential saving of FDD is estimated at 10-40% of HVAC energy consumption depending on the age and condition of the equipment, maintenance practices, climate and building use [4]. In this study we tried to detect and diagnose faults of AHU through a probabilistic way by combining expert rules and Bayesian Belief Network (BBN). GENIE and SMILE decision support toolbox published by University of Pittisburg [5] are used to develop Bayesian Belief network.

2. LITERATURE REVIEW

There were many researchers conducted on FDD of AHU and VAV in the last decades. Most of them are focused on AHUs. Among data driven based models, Principle Component Analysis (PCA) is most commonly used to detect and diagnose sensor faults like fixed bias, drifting bias or complete failure. Wang [6], Wu [7], Du [8] and many others developed PCA based FDD method for Air Handling Unit (AHU), Variable Air Volume (VAV) box and centrifugal chiller. Three kinds of diagnostic methods were used in those studies, which are Q-contribution plot,

Joint Angle Analysis (JAA) and Fisher Discriminant Analysis (FDA). However, it is found that JAA is more efficient than Q-contribution plots in diagnosing complex faults. The studies focused on both simulation and field data.

The use of Artificial Neural Network (ANN) as FDD strategy was developed [9-10]. Study used ANN to detect and diagnose faults in AHU can be proposed as two stages ANN, one is to detect the abnormality in the system and other is to isolate the subsystem. Residuals of different parameters are given as input and the trained network was used to diagnose root cause of the faults. The Neural Network trained with normal operating data to predict the parameters. The deviation of actual data from predicted value considered as a sign of fault. Similarly, Liang [11] used support vector regression to detect and diagnose sensor faults of AHU.

In summary, many of the methods usually provide good fault detection results. However, in certain cases fault may cause similar symptoms and propagate to other components .So, it becomes very hard to provide proper diagnosis. For example, a fault in mixed air temperature can influence parameters like cooling coil and damper signal etc. Therefore, it is more reasonable to give probabilities of faults at given symptoms in FDD analysis. Symptoms can be considered as abnormal parameter change or abnormal residuals, expert rules etc. Here, in this study we tried to use expert rules (APAR) as symptoms of faults. APAR is an established set of rules which are able to trigger an alarm if the system is running abnormally means if any fault happens. FDD of AHU can be very efficient and effective if the FDD strategy can work in a similar way as that used by FDD experts. In this study, a Bayesian Belief network (BBN) simulates diagnostic thinking of experts, where APAR rules are used as symptoms of fault.

3. METHODOLOGY

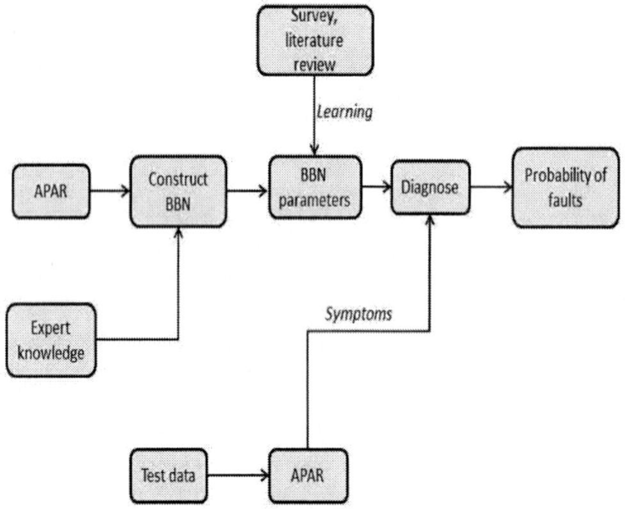

Figure 1: Flowchart of the proposed FDD model.

The flowchart of the proposed BBN based FDD strategy is illustrated in Figure 1.The steps of the proposed strategy are described below:

Step 1: First using the APAR rules (as symptoms) and expert knowledge (cause-effect) Bayesian Belief Network (BBN) is constructed.

Step 2: Then from literature reviews, survey results and maintenance records (AHU1 of AET campus building, University of Texas at San Antonio) parameters of BBN are identified. There are two types of parameters. Prior probability of faults and conditional probabilities.

Step 3: In the third step, we placed a reference outdoor air temperature sensor. We were very careful to place the two outdoor air temperatures close to each other. Also checked if the sensors are showing real outdoor air temperature that is entering into AHU as intake air. If there is a difference between the two outdoor air temperature sensors outdoor air temperature sensor fault is diagnosed. In addition, if supply air temperature set point is unreasonably low or high set point fault will be diagnosed.

Step 4: Testing data are passed through rule-based analysis. In this process, at first based on the control signals of dampers and cooling/heating coils, the mode of operation is selected. Then the appropriate rule sets are applied.
Step 5: The results of rule-based analysis are passed through BBN for diagnosis. The states of the rules (True/False) are the inputs of BBN. If any rule or multiple rules are satisfied Bayesian Network will provide a list of faults with probability of occurrence.

4. SYSTEM DESCRIPTION

The expert rules were developed for application to single duct variable volume or constant volume air handlers with hydraulic heating and cooling coils and economizer capabilities and logic. The rules focus on temperature control in an AHU. Hence, the system description will be restricted to components and control strategies directly related to this purpose.The control loops of AHU include damper control, temperature control and pressure control. The fresh air through the fresh air damper is mixed with the return air with recirculation air damper. Typical but advanced control strategies are implemented to provide adequate outdoor air ventilation and suitable supply air temperature and indoor pressure, and to minimize energy use. PID controllers are employed to control the supply air temperature, supply static pressure, outdoor air flow rate and return fan speed. Optimal control strategies are used to reset the set points of the local PID control loops of supply air temperature and supply static pressure.
AHUs have four primary modes of operation during occupied periods for maintaining the supply air temperature at set point. Sequencing logic determines the mode of operation of an AHU at any particular time. The four modes of operation are: 1) Mode 1: Heating, 2) Mode 2: Cooling with outdoor air, 3) Mode 3: Mechanical cooling with 100% outdoor air, 4) Mode 4: Mechanical cooling with minim outdoor air, and 5) Mode 5: Simultaneous heating and cooling.

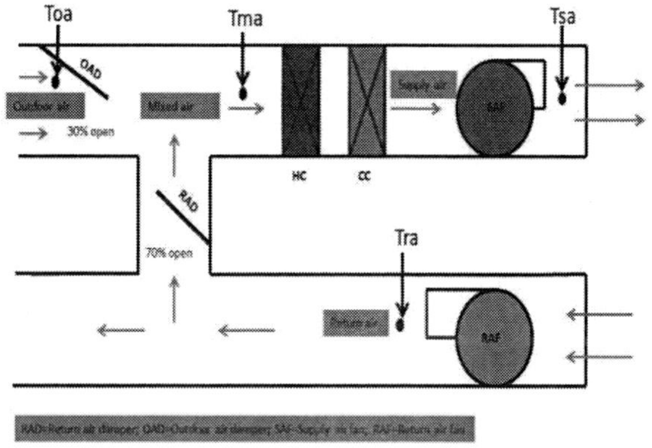

Figure 2: AHU Operation on Mode-4.

Our scope of the study only limited to mode-4, which is mechanical cooling with minimum outdoor air. In mode-4 cooling coil valve is modulated to satisfy the temperature set point. Heating coil valve remains closed. Mixing box dampers are open and set to meet the minimum outdoor air requirements which is 30% in our study. Figure 2 shows an AHU operating on Mode-4.

5. AHU PERFORMANCE ASSESSMENT RULES (APAR)

APAR is a set of 28 rules that uses control signals and occupancy information to identify the mode of operation of the AHU, thereby identifying a subset of the rules that specify temperature relationships that are applicable for that mode. The two main mode classifications are occupied and unoccupied. The four modes of operation are applied for occupied periods. Because the direct digital control output to the actuators of the heating and cooling coil valves and the mixing box dampers are known, the mode of operation can be ascertained. Table 1 lists all the expert rules associated with Mode-4 which is cooling with minimum outdoor air. Also some rules are applied to all modes of operation. Rule based detection is computationally simple and can be easily implemented in commercial buildings. According to a study by NIST (2001), number of false positives and negatives are low in rule-based detection. In addition, different rules can be considered symptoms of different faults. For example, 'True' state of Rule-18 is a symptom of mixing box related faults and sensor related faults. 'True' state of Rule-15 is a symptom of controller logic error.

Table 1: Rules associated with mode-4

Cooling with minimum outdoor air temperature	**15**	$T_{oa} < T_{co} - E_t$
	16	$T_{sa} > T_{ma} + \Delta T_{sf} - E_t$
	17	$T_{sa} > T_{ma} - \Delta T_{sf} + E_t$
	18	$\lvert T_{ra} - T_{oa} \rvert \geq \Delta T_{min}$ where $\lvert Q_{oa/sa} - T_{(oa/sa)min} \rvert \geq E_f$
	19	$\lvert U_{cc} - 1 \rvert \leq E_c$ and $T_{sa} - T_{sa},s > E_t$
	20	$\lvert U_{cc} - 1 \rvert \leq E_c$
	25	$\lvert T_{sa} - T_{sa},s \rvert > E_t$
	26	$T_{ma} < \min(T_{oa}, T_{ra}) - E_t$
	27	$T_{ma} > \max(T_{oa}, T_{ra}) + E_t$
	28	Max. num of transitions > MTmax

6. BAYESIAN BELIEF NETWORK

Bayesian Belief Network (BBN) is a probabilistic graphical model that represents a set of random variables and their conditional dependencies via a directed acyclic graph. It has been successfully applied in the domain of knowledge discovery and probabilistic inference since it was introduced by Pearl in early 80's [12-13].It is very powerful to represent and to diagnose complex systems with uncertain incomplete and even conflicting information. there is very little application in HVAC systems. Najafi [14] and Wall [15] introduced BBN to detect and diagnose faults of AHU. Both works required a full set of faulty dataset to learn the fault patterns. However, faulty datasets are very hard to get in real building applications. Therefore In this study we used expert knowledge based Bayesian Belief Network (BBN), which does not require any training datasets.

There are two basic components of BBN Structure and parameters. The structure of the network should capture qualitative relationships between variables. In particular, two nodes should be connected directly if one affects or causes the other, with indicating the direction of the effect. Cause and effect can be termed as parent and child node respectively. An arc points from a parent to child node. In this study node presents faults and fault symptoms or expert rules. The arc indicates direct probabilistic dependencies among nodes. The node without any input arc is called root node. A node may have several states. For example, a fault node has two states Faulty & Not Faulty. Each state is an event. When an event occurs, it is evidence for diagnosing faults. Each root node has a prior probability corresponding to its each state. A conditional probability table is used to specify all parameters or probabilities of a child node, considering all possible combinations of its own states and its parent node states. The number of parameters needed in a conditional probability table exponentially grows with the number of its parent nodes. It is very difficult to obtain all the conditional probabilities in the FDD applications.

Once the structure of a BBN is defined, the posterior probability can be obtained by Bayesian inference. The Bayesian theory plays a very important role in Bayesian inference. The equation of the Bayesian theorem can be written as:

$$P(A|B) = \frac{P(AB)}{P(B)} = \frac{P(A)P(B|A)}{P(B)} \tag{1}$$

Where P(AB) is the joint probability of the event A and B. For any given event A the marginal probability can be calculated by:

$$P(A) = \sum_{i=1}^{n} P(B_i)P(A|B_i) \tag{2}$$

The Bayesian theorem can be obtained based on the conditional probability and marginal probability.

$$P(B_i|A) = \frac{P(AB_i)}{P(A)} = \frac{P(B_i)P(A|B_i)}{\sum_{i=1}^{n} P(B_i)P(A|B_i)} \tag{3}$$

Items on right hand side called prior probabilities and on left hand, side is called posterior probabilities. The Bayesian network provides the posterior probabilities from the prior probabilities.

7. LEAKY NOISY-OR GATES

Considerable reduction in effort can be obtained by using noisy-OR gate [12-13]. The noisy-OR influence structure applies when there are several possible logical causes, xi, i=1, 2, 3, 4…n of a binary effect variable y. Where (a) each of which has a probability Pi, is being sufficient to produce the effect in the absence of all other causes and (b) the probability of each cause sufficient is independent of the presence of other causes. Pearl showed that the probability of y, given a subset of x of the xi, which is present:

$$p(y|X) = 1 - \prod_{i,Xi \in X} (1 - Pi) \tag{4}$$

The complete conditional distribution for n binary predecessors would require the specification of 2n parameters, but if the Noisy-OR assumption applies, we need to specify only n for each Pi. Like any model, Bayes network is never complete and so there will often be possible causes of an effect that are not explicitly modelled. To allow for this in a Noisy-OR it is useful to access a base rate probability, Po for all other causes i.e. the event that the effect will occur apparently spontaneously in the absence of any of the cause is modelled explicitly. Base rate probability is often called leak probability. If the expert assesses the probability that each predecessor is sufficient to cause the effect variable in the absence of any other, explicit cause the number needs readjustment to obtain the formulae for the probability of y. .

$$p(y|X) = 1 - (1 - Po) \prod_{i,Xi \in X} \frac{1 - Pi}{1 - Po} \tag{5}$$

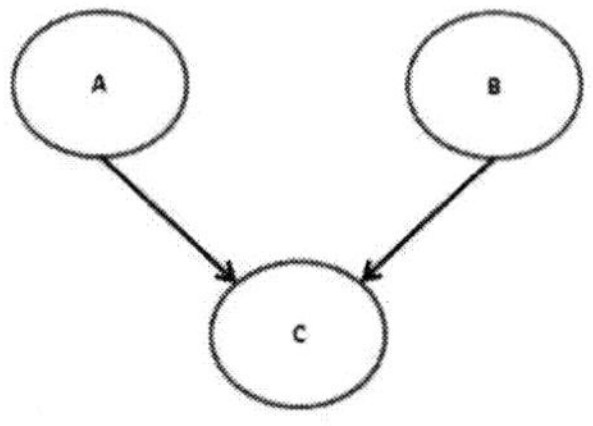

Figure 3: Example of Leaky Noisy OR Gate.

Figure 3 shows a potential example of the leaky Noisy-OR gate. If each node has, two states named Yes/No, also, A & B both independently can cause C with a probability then we can assume this node as noisy-OR gate. C has two states True/False. If Leak probability is given then we can calculate the Conditional Probability Table (CPT) from equation (5). In our study, we assumed the Leak probability of the rules is .01, which can be increased based on different system design and efficiency.

8. TYPICAL FAULTS AND PRIOR PROBABILITIES

International Energy Agency (IEA) published some survey results listing most important faults of HVAC system based on frequency of occurring, environmental effect, and recovery cost and energy impact [16]. The survey included fabricators, designers, constructors and maintenance stuffs. After that through Annex-34 and Annex-40 projects IEA provided a comprehensive list of frequently occurring natural faults [17-18].In 2003, National Institute of Standards and Technology (NIST) also published a project result where they used a rule based detection tool to detect and diagnose faults of 13 AHUs monitored from eight different sites [19]. Most of the field studies are based on artificially created faults in AHU. Therefore, the studies are not in our scope of study.

Based on all the literature reviews, a list of most frequently occurring natural faults can be found. We considered five major components/type of faults, which are heating coil, cooling coil, mixing box, controller logic error, and sensor related faults. Table 2 shows the frequency of occurrence of faults. In addition, we collected one-year maintenance records of "HVAC trouble calls" of Applied Engineering and Technology (AET) building at University of Texas at San Antonio (UTSA) to find the most frequent faults for whole AET. Table 2 also shows the calculated prior probabilities, which are simply the proportion of each fault.

Table 2: Probabilities of faults

Faults	Counts	Probability
HC-Leak	5	0.108695652
HC-Stuck	2	0.043478261
HW Temperature Low	4	0.086956522
CC-Leak	3	0.065217391
CC-Stuck	3	0.065217391
CC-Fouled	3	0.065217391
CW Temperature High	4	0.086956522
Logic Error	1	0.02173913
Controller Signal Error	1	0.02173913
T_{sa} Biased	4	0.086956522
T_{ma} Biased	4	0.086956522
T_{ra} Biased	4	0.086956522
RAD Stuck	5	0.1
OAD Stuck	3	0.074
Total	46	1

9. STRUCTURE OF BBN

The structure of BBN is a graphical illustration of expert's diagnostic thinking which can illustrate qualitatively the relationships among fault and fault symptoms. To train BBN we can either use probabilities developed by experts or machine learning methods using full datasets including both normal and faulty data. The second approach is impractical because full datasets for AHU is quite hard to achieve (both normal and faulty data).Therefore first approach is adopted in this study. In developing the BBN for FDD of AHU, the diagnostic thinking process of FDD experts are reflected and simulated. FDD experts usually diagnose faults based on the fault symptoms. Therefore, BBN should consist two types of node at least, one is fault node and another is fault symptom node. Here fault nodes represent different typical faults of AHU represented in table-2. Symptom nodes are the event (True/False) of the nine rules. Figure 4 shows the constructed BBN with nine rules and twelve fault nodes. Rules and fault nodes are connected with each other by cause and effect relationship.

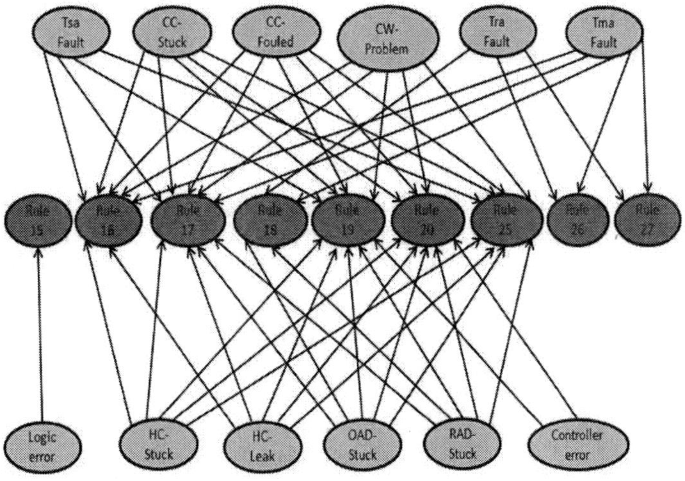

Figure 4: Proposed BBN.

10. CONDITIONAL PROBABILITIES

Parameter of a BBN represents the quantitative dependencies among faults and symptoms using Probabilities. A conditional probability table is used to define the probabilities of all states of child node given the state of its parent nodes. In these study nodes with only one parent node is considered general nodes. Nodes with more than one parent nodes are considered to be Leaky Noisy-OR nodes. For the Leaky Noisy-OR nodes, the conditional probabilities and leak probabilities are needed. Therefore, the parameters of DBN are the prior probabilities for all the root nodes, conditional probability table for fault category nodes and conditional probability for general nodes as well as Leaky Noisy-OR nodes. Table 3 and 4 provide the conditional probability of different rules given the faults.

Table 3: Conditional Probability for Rule 15-17.

Faults/Rule	Rule 15	Rule 16	Rule 17
Tsa biased	N/A	0.167	0.142
Tma biased	N/A	0.167	N/A
Tra biased	N/A	N/A	0.142
HC-Stuck	N/A	0.167	0.142
HC-Leak	N/A	0.167	0.142
CC-Stuck	N/A	0.167	0.142
CC-Leak	N/A	N/A	0.285
CC-Fouled	N/A	0.167	N/A
Logic error	0.99	N/A	N/A

Table 4: Conditional Probability for Rule 18-20.

Faults/Rule	Rule 15	Rule 16	Rule 17
Tma biased	0.285	N/A	N/A
Tra Biased	0.285	N/A	N/A
Tsa biased	N/A	.153	N/A
HC-Stuck	N/A	0.076	0.11
HC-Leak	N/A	0.153	0.22
CC-Stuck	N/A	0 .076	0.11
CC-Leak	N/A	0.076	0.11
CC-Fouled	N/A	0.076	0 .11

CW temp. too high	N/A	0.076	0.11
OAD Stuck	0.142	0.076	0.11
RAD Stuck	0.285	0.153	0.11
Controller error	N/A	0.076	0.11

For example, the probability of Rule-15 is 'True' when sensor fault happened is expressed as P(Rule-15|sensor fault)). The entire conditional probability is obtained from different FDD project reports based on APAR rule. At first, the different faults found against each rules are listed not found against a rule then a minimum amount of fault is considered to be happened against that rule. Later every column is normalized to one to get the conditional probabilities developed by experts and machine learning using full datasets including both normal and faulty data. The second approach is impractical because full datasets for AHU is quite hard to achieve (both normal and faulty data).Therefore first approach is adopted in this study. Again, Figure 4 shows the constructed BBN with nine rules and twelve fault nodes. Rules and fault nodes are connected with each other by cause and effect relationship.

11. DATA COLLECTION

Real time data are collected from one of the campus buildings of University of Texas at San Antonio (Main Campus) named Applied Engineering and Technology Building (AET).One month's data are collected from June 2014 to July 2014 and further analyzed. The results will be discussed in the next section.

12. RESULTS

As the first step testing data from AET are conducted through rule based analysis. From the control signals of cooling coil, heating coil and dampers the mode of operation is ascertained which is mode-4. After that the appropriate rule set are applied. The results of rule based analysis can be seen through figure-5.Y-axis value1 represents the true state of the rule. From Figure 5(a) we can see rule-16 was true four different times. That means supply air temperature was greater than mixed air temperature in the cooling period. The next figures (b) and (c) show that rules 25 and 26 were also true for multiple time periods. That means for some period of time supply air set point did not met set point and mixed air temperature was showing out of range temperature. Figure (d) shows that minimum outdoor air requirement was not met at the later period of the day.

According to Figure 5 (a), (b), (c), (d) rule-16, 18, 25 and 26 were true for some period. Sometimes rules were simultaneously true. This information was given to BBN as inputs and BBN prioritized the faults, which can be seen through Figure 6 (a) and (b). Every time step is of 5 minutes interval. Figure 6(a) shows the BBN diagnosis from 4:00-5:00pm period. During this one-hour period, Tma had the highest probability (.66) to be biased. The next highest probable faults are Tra biased (.30), cooling coil stuck (.19) and Tsa biased (.18) respectively. Between 18:00pm-4:00am. Rules 18, 25, 26 were true. However, most of the time rule-18 was true. The diagnosis results can be seen through Figure 6(b). Also, between 21:10pm-21:50pm rule-18, 25, 26 were true. During the period, Tma biased (.75) had the highest probability of occurrence. In other periods RAD-Stuck (.25), Tra biased (.25) and Tma biased (.25) had the highest probability. The proposed model is mainly effective when multiple rules are true simultaneously. When a single fault affects other parameters severely then multiple rules can be simultaneously true. In addition, when two or more faults simultaneously exist that can also cause multiple rules to be true. Further study include to analyze the multiple faults probability by expanding the Bayesian formulae.

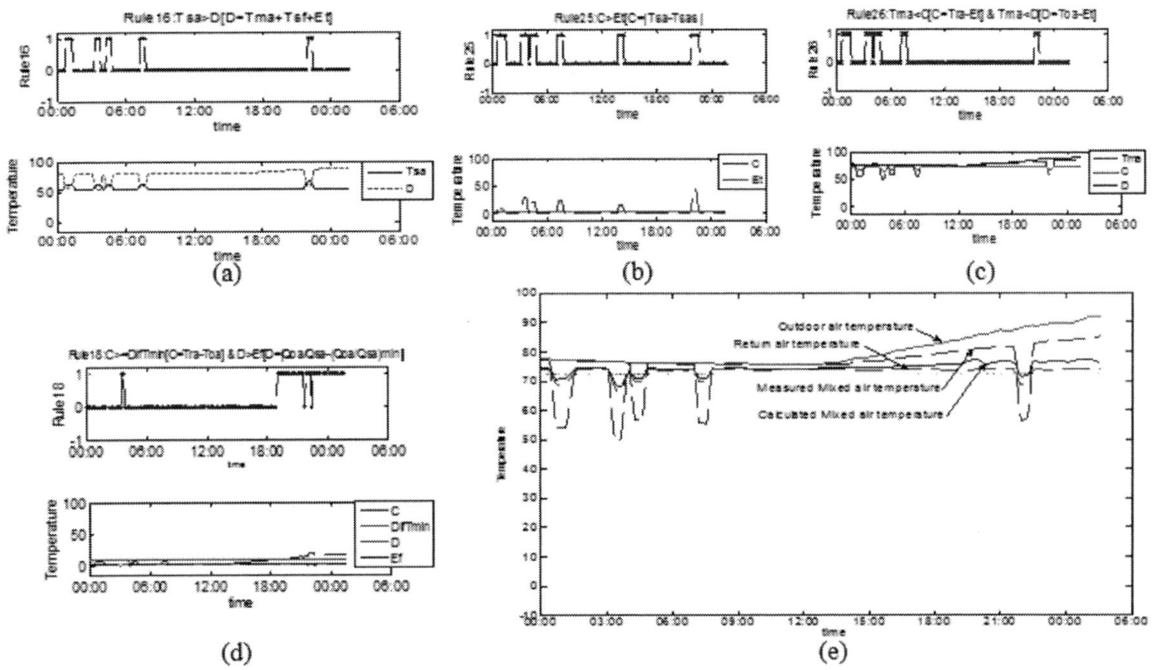

Figure 5 : Rule based Detection (15th June).

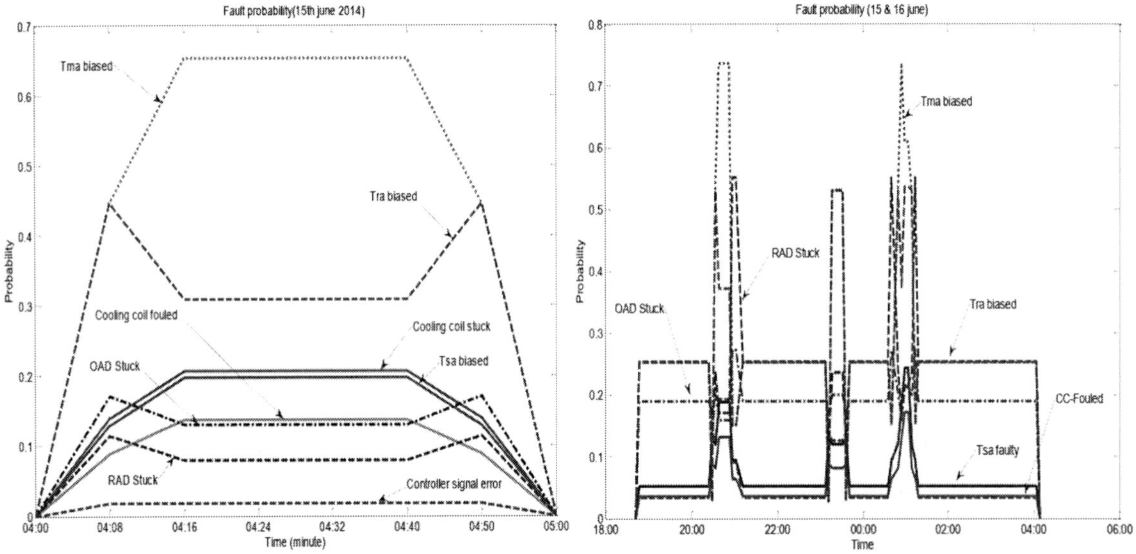

Figure 6 : BBN Results.

13. VALIDATION

Figure-5 (e) shows a plot of mixed, outdoor and return air temperature. As AHU is running on mode-4, the damper signal is set for minimum outdoor air. Based on the damper control signal mixed air temperature should be very close to return air temperature. However actual Tma is more towards Toa. For that reason most of the time rule-18 was true. Also sometimes, other rules were simultaneously true. So based on the priors and conditional probability BBN calculates that many of the times mixed air and return air temperature sensor biased were the highest probable faults. To validate our result we arranged onsite investigation Form the investigation, it is confirmed that sometimes-mixed air temperature is biased toward outdoor air temperature. Figure-7 shows two case scenario when multiple rules satisfied. In a simulation study [28] it was seen that when rule 19, 20 and 25 are simultaneously true then HC-Leakage was the actual fault. In addition, when rule 18, 19 and 25 were simultaneously true RAD-Stuck was the actual fault. We tried to validate our model with these two Combinations .The result shows that HC-leakage was the diagnosed faults for the first combination with .55 probability. For the second combination, RAD-Stuck was the diagnosed fault with .7 probability.

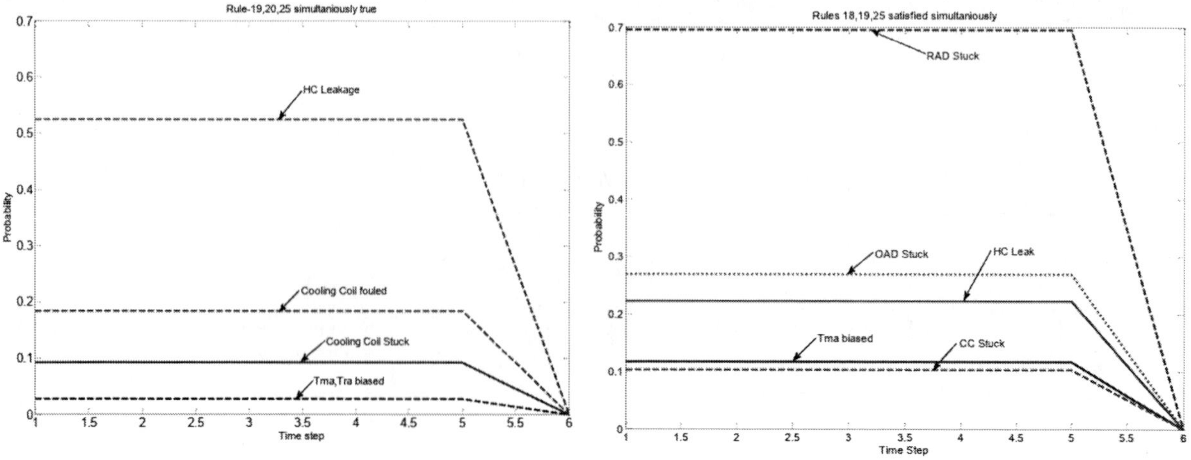

Figure 7 : Validation of the Model.

14. CONCLUSION

Expert rule based fault detection is proved to be very robust and computationally simple for commercial implementation. Bayesian Belief Network can be a very effective tool to diagnose the best possible root cause behind the satisfaction of a rule or multiple rules. In addition, it is proved effective in the current study. In future, the study can be expanded to different modes of operation. Future scopes of study can be,

- The FDD tool can be implemented to all modes of operation.
- More survey and projects can be performed to improve the parameters of BBN.

NOMENCLATURE

Tco	Changeover temperature
TRA	Return air temperature
TOA	Outdoor air temperature
TMA	Mixed air temperature
TSA	Supply air temperature
Et	Threshold parameter accounting for errors in temperature measurements
Ef	Threshold parameters accounting for errors related to airflows
Tsf	Temp. rise with supply fan
Trf	Temperature rise across return fan
TSA, S	Supply air set point
UCC	Cooling coil control signal
UHC	Heating coil control signal
Q	Outdoor air fraction

MTMAX	Maximum allowed mode transition per hour
OEC	Threshold parameter for cooling coil valve control signal

REFERENCES

[1] Fu, X. (2009). Progress and methodologies of lifecycle commissioning of HVAC systems to enhance building sustainability. *Renewable and sustainable energy reviews*, 13(5), 1144–1149.
[2] HVAC SWG – Spin I. (2007).Accessed at: http://www.seai.ie/Your_Business/Large_Energy_Users/Special_Initiatives/Special_Working_Groups/HVAC_SWG_07/.
[3] Pérez-Lombard, L. (2008). A review on buildings energy consumption information. *Energy and Buildings*, 40(3), 394–398.
[4] Kang, C., & Golay, M. (1999). A Bayesian belief network-based advisory system for operational availability focused diagnosis of complex nuclear power systems. *Expert Syst. Appl.*, 21–32.
[5] GeNIe and Smile toolbox, Decision systems Laboratory, University of Pittsburgh, weblink: https://dslpitt.org/genie/.
[6] Wang, H., Qin, J., &Wang, J. (2012). Online model-based fault detection and diagnosis strategy for VAV air handling units. *Energy and Buildings*, 55, 252-263.
[7] Wu, S., & Sun, J.Q. (2011). Cross-level fault detection and diagnosis of building HVAC systems. *Building and Environment*, 46(8), 1558–1566.
[8] Du, Z.M., & Xin, X.Q. (2006). Fault detection and diagnosis based on improved PCA with JAA method in VAV systems. *Building and Environments* 42(9), 3221-3232.
[9] Du, Z.M., (2014). Fault detection and diagnosis for buildings and HVAC systems using combined neural networks and subtractive clustering analysis. *Buildings and Environment*, 73, 1-11.
[10] Zhu, Y., Jin, X, & Du, Z.M. (2012). Fault Diagnosis in sensors for Air handling unit based on artificial neural network pre-processed by wavelet and fractal. *Energy and buildings*, 44, 7-16.
[11] Liang, J., & Du, R. (2007). Model-based fault detection and diagnosis of HVAC systems using support vector machine method. *International Journal of Refrigeration*, 30(6), 1104–1114.
[12] Pearl, J. (1985). Bayesian Networks: a Model of Self-Activated Memory for Evidential Reasoning. Computer Science Department, University of California (1985).
[13] Pearl, J. (1986). Fusion, propagation, and structuring in belief networks. *Artif. Intell*, 29, 241–288.
[14] Najafi, M., & Auslander, D.M. (2012). Application of machine learning in the fault diagnostics of air handling units. *Appl. Energy*, 96, 347–358.
[15] Wall, J., & Guo, Y. (2011). A dynamic machine learning-based technique for automated fault detection in HVAC systems. *ASHRAE Trans.*, 117, 449–456.
[16] IEA Annex 25. (1996). Building Optimization and Fault Detection Source Book
[17] Vaezi-Nejad, H. (2004). Using Building Control System for Commissioning (Annex-40). Energy Systems Laboratory (http://esl.tamu.edu), Texas A&M University.
[18] Annex-34. (1997). Computer Aided evaluations of HVAC system Performance.
[19] Schein, J. (2010). Results from Field Testing of Air Handling Unit and Variable Air Volume Box Fault Detection Tools. National Institute of Standards and Technology.

Improving the Performance of PCA-Based Chiller Sensor Fault Detection by Sensitivity Analysis for the Training Data Set

Yunpeng HU*, Jiani LIU, Li ZHOU, Yang LIU, Qingling QIU

Wuhan Business University, Department of Building Environment and Energy Engineering, 430056, Wuhan, Hubei, PR China
* Corresponding Author : +86 27 8429 3835, YunpengHu@hust.edu.cn

ABSTRACT

An improved approach of fault detection for chiller sensors is presented based on the sensitivity analysis for the original data set used to train the Principal Component Analysis (PCA) model. Sensor faults are inevitable due to the aging, environment, location and so on. Meanwhile, because of the wide range of operational conditions, the fault of a certain sensor is very difficult to be directly detected by its own historical data. PCA is a multivariate data-based statistical analysis method and it is very useful for the sensor fault detection in HVAC&R. The undetectable zone of a certain sensor by Q-statistic is derived from the definition of Q-statistic which is usually employed as a boundary to detect the sensor fault situation. Due to the similar style between Q-statistic and Hawkins' T_H^2, the undetectable zone by Hawkins' T_H^2 is also obtained. Undetectable zone is a predictive index to indicate the detectability of different sensors by different statistics. Since undetectable zone is the character of the original training data set, it can indicate the quality for the selected training data. One field data set is employed to validate the presented approach. Results show that the undetectable zone of a certain sensor by Q-statistic is quite different from that by Hawkins' T_H^2. Therefore, the undetectable zone can be used to improving the performance of PCA-based chiller sensor fault detection by choosing different fault detection statistics with less undetectable zone for different sensor.

1. INTRODUCTION

Due to long term operation and severely working environment, sensor faults are inevitable in HVAC&R. There are much disadvantage because of sensor fault, including ineffective control, unsafe operation, unreasonable energy consumption and so on (Lee and Yik, 2010; Yoon et al., 2011). For energy saving and conservation, researches on sensor fault detection, diagnosis and erroneous sensor data reconstruction (FDDR) for HVAC&R system have been paid more attention to in the last decade.

Usually, the model-based methods and the data-driven methods are the two typical classes of FDDR methods. Any faulty sensor cannot be easily identified just only from the historical data of its own. Thus, various multi-dimensional data-based methods have been introduced to the FDDR of HVAC&R system in the recent years, such as fuzzy inference systems(Koçyigit, 2015), data fusion(Sun et al., 2010), neural network(Du et al., 2014; Lee et al., 2004), support vector machine(Han et al., 2011), principal component analysis(Li et al., 2016), fisher discriminant analysis(Du et al., 2007), Bayesian network(Zhao et al., 2015), etc.

Recent years, principal component analysis (PCA) (Härdle and Simar, 2007; Jackson, 1991), a multivariate statistical analysis method, was presented in the sensor FDDR, including the whole system(Wang et al., 2010), AHU(Li and Wen, 2014; Xiao et al., 2009), VAV(Du et al., 2009), chiller(Chen and Lan, 2009; Hu et al., 2016; Xu et al., 2008) and so on. By the different assignment of sensors or the combination with other algorithms, PCA-based approaches were successfully applied into sensor FDDR for chiller.

Many researchers were dedicated in applying novel data-driven methods into the sensor FDDR of HVAC&R system. For any data-driven method, the analysis results highly depend on the character of the training data set. Because the fielded data relies on sampling interval, sampling location, measurement principles, and so on, the quality of training data is much worse. However, rare work was reported on how to predict the quality of the training data so as to enhance the FDDR results in detail. In this paper, the undetectable zone for each sensor assigned in PCA model is presented to evaluate the detectability for chiller sensor fault. It is derived from the training data and the definition of statistics employed as fault detection boundary. The undetectable zone for each sensor can be used to evaluate the fault detection ability and reliability with clear physical or thermodynamic meanings. A fielded data set of a real screw chiller was employed to validate the detectability of different statistics in detail.

2. PCA-BASED SENSOR FDDR

2.1 PCA-based sensor FDDR

In PCA method, the original data matrix $X^0 \in R^{m \times n}$ usually consists of m samples (rows) and n process variables (columns) obtained from the field measurements. The training data $X^0 \in \Re^{m \times n}$, which is consisted of the original measured data, is transferred to a normalized matrix $X = \{\vec{x}_1^T, \cdots, \vec{x}_m^T\}^T$ with zero mean and unit variance due to engineering units and orders of magnitude. After the eigenvalue decomposition of the covariance matrix, $R \approx X^T X/(n-1)$, any normalized samples $\vec{x}$ can be expressed as

$$\vec{x} = \hat{\vec{x}} + \tilde{\vec{x}} \tag{1}$$

where $\hat{\vec{x}}$, the estimation of $\vec{x}$, is the projection vector of $\vec{x}$ onto the PC subspace, and $\tilde{\vec{x}}$, the residual of $\vec{x}$, is the projection vectors of $\vec{x}$ onto the Residual subspace.

A common FDDR strategy for sensor fault based on PCA is illustrated as Figure 1. Its detailed structure can be referred in reference (Hu et al., 2012). It needs to emphasize that the original operational data used to train PCA model is included with many outliers inevitably due to measurement errors, hardware failure and so on. The aim of PCA modeling is to establish a fault boundary to detect whether there is a faulty sensor in system or not.

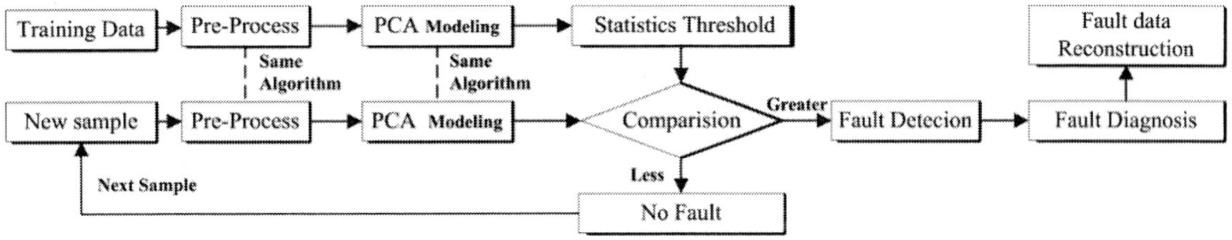

Figure 1: A common FDDR strategy for sensor fault based on PCA

Several statistics can be employed as the boundary to detect sensor fault, such as Q, Hawkins' T_H^2 and so on. When the value of statistics for the tested sample is greater than the boundary, the fault can be detected successfully. Therefore, the below equations mean the sensor fault cannot be detected.

$$Q \leq Q_\alpha \tag{2}$$

$$T_H^2 \leq \chi^2_{n-k;\alpha} \tag{3}$$

Where, Q_α is the threshold of Q statistic and $\chi^2_{n-k;\alpha}$ is the threshold of Hawkins' T_H^2.

2.2 Undetectable Zone by Q-statistic

If there is a faulty sensor, the measurement data of this sensor make the value of some statistic is greater than the threshold. Assuming the certain faulty sensor is the ith sensor, there must be a pair of limited upside and downside of ith sensor measurement data, which can just satisfy the Equation (1) or Equation (2). When the measurement data of the ith sensor is outside the pair of limitations, the fault can be detected. From the threshold and the other sensors measurement data, we derived the calculation of the limitations for the *i*th sensor not to be detected. Obviously, the pair of limitations is a predictive zone to demonstrate the fault detectability for the *i*th sensor.

Assuming the *j*th sensor is the faulty one, x_j, the *j*th entry of $\vec{x}$, is the erroneous measurement value. e_i, the *i*th entry of $\tilde{\vec{x}}$, can be rewritten as

$$e_i = y_{ij}^{RS} x_j + \sum_{k=1}^{j-1} y_{ik}^{RS} x_k + \sum_{k=j+1}^{n} y_{ik}^{RS} x_k = y_{ij}^{RS} x_j + Y_{i,:}^{RS} \Xi_j \vec{x} \tag{4}$$

Where, y_{ij}^{RS} is the *j*th entry of the *i*th row of Y^{RS}. $\mathbf{Y^{RS}}$ is the projection matrix of RS. Ξ_j is used to indicate the direction of the erroneous senor and be written as

$$\Xi_j = \begin{pmatrix} 1 & & 0 \\ & 0 & \\ 0 & & 1 \end{pmatrix}_{1,\cdots,j-1 \quad j \quad j+1,\cdots,n} \tag{5}$$

Therefore, *Q*-statistic can be derived as

$$Q = \left(\sum_{i=1}^{n} (y_{ij}^{RS})^2 \right) x_j^2 + \left(\sum_{i=1}^{n} 2 y_{ij}^{RS} (Y_{i,:}^{RS} \Xi_j \vec{x}) \right) x_j + \sum_{i=1}^{n} (Y_{i,:}^{RS} \Xi_j \vec{x})^2 \tag{6}$$

Due to the *j*th sensor is faulty one, its *Q*-statistic will satisfy the following equation (7)

$$\left(\sum_{i=1}^{n} (y_{ij}^{RS})^2 \right) x_j^2 + \left(\sum_{i=1}^{n} 2 y_{ij}^{RS} (Y_{i,:}^{RS} \Xi_j \vec{x}) \right) x_j + \sum_{i=1}^{n} (Y_{i,:}^{RS} \Xi_j \vec{x})^2 - Q_\alpha > 0 \tag{7}$$

From the style of Equation, it's a one-variable quadratic inequalities with the form of $ax^2 + bx + c > 0$. Where,

$$\begin{aligned} a &= \sum_{i=1}^{n} (y_{ij}^{RS})^2, \\ b &= \sum_{i=1}^{n} 2 y_{ij}^{RS} (Y_{i,:}^{RS} \Xi_j \vec{x}), \\ c &= \left(\sum_{i=1}^{n} (Y_{i,:}^{RS} \Xi_j \vec{x})^2 \right) - Q_\alpha \end{aligned} \tag{8}$$

The pair of solutions of Equation (7), $x_{j,min}$ and $x_{j,max}$, are the limitations for the normalized sensor fault boundary. A zone, $[x_{j,\min}^0, x_{j,\max}^0]$, can be obtained by de-normalizing. If the original measured data x_j^0 is outside of $[x_{j,\min}^0, x_{j,\max}^0]$, we can easily find the faulty sensor. Therefore, the area, $[x_{j,\min}^0, x_{j,\max}^0]$, can be defined as the undetectable zone of the *i*th sensor by *Q*-statistic. The undetectable zone can be employed as the index to evaluate the sensor fault detectability.

2.3 Undetectable Zone by T_H^2

Similar with the definition of *Q*-statistic, the Hawkins' T_H^2 can be rewritten as

$$T_H^2 = \left(\sum_{i=1}^{n} (y_{ij}^{\mathrm{T2H}})^2 \right) x_j^2 + \left(\sum_{i=1}^{n} 2 y_{ij}^{\mathrm{T2H}} (Y_{i,:}^{\mathrm{T2H}} \Xi_j \vec{x}) \right) x_j + \sum_{i=1}^{n} (Y_{i,:}^{\mathrm{T2H}} \Xi_j \vec{x})^2 \tag{9}$$

Where, $Y^{T2H} = \Lambda_{k+1,m}^{-1/2} \tilde{P}^T$ is the projection matrix of Hawkins' T_H^2. Therefore, the undetectable zone of the *i*th sensor by Hawkins' T_H^2 can be the solutions with the style of $ax^2 + bx + c > 0$, where

$$
\begin{aligned}
a &= \sum_{i=1}^{n} (y_{ij}^{\mathrm{T2H}})^2 \\
b &= \sum_{i=1}^{n} 2 y_{ij}^{\mathrm{T2H}} (Y_{i,:}^{\mathrm{T2H}} \Xi_j \vec{x}) \\
c &= \left(\sum_{i=1}^{n} (Y_{i,:}^{\mathrm{T2H}} \Xi_j \vec{x})^2 \right) - \chi_{n-k;\alpha}^2
\end{aligned}
\tag{10}
$$

The undetectable zone by Hawkins' T_H^2 is quite different to that by Q-statistic. Therefore, the compared results can indicate the different sensitivity for these two different statistics.

2.4 PCA model for a water-cooled chiller

From the consideration of the energy balance principle, there are eight important sensors in the water-cooled chiller and its control logic. The PCA model of a typical water-cooled chiller is

$$
X = \begin{bmatrix} T_{chw}^{i} & T_{chw}^{o} & M_{chw} & T_{cw}^{i} & T_{cw}^{o} & M_{cw} & W & V \end{bmatrix} \tag{11}
$$

Where, T_{chw}^{i} and T_{chw}^{o} are the temperature sensors of inlet node and outlet node of evaporator, respectively. T_{cw}^{i} and T_{cw}^{o} are the condenser-water system inlet node temperature and outlet node temperature, respectively. M_{chw} is the water flowrate of chilled-water system and M_{cw} is the water flowrate of condenser-water system, respectively. W is the electrical power. V is the position of the slide valve to indicate the mass flowrate of the refrigerant into the screw compressor.

3. VALIDATION

3.1 Cases study

A fielded data (Hu et al., 2012; Hu et al., 2016) of a water-cooled screw chiller were used to validate the sensitivity of Q and Hawkins' T_H^2 for different sensors. The undetectable zones of different sensors by a same training data set were investigated. The results of sensor fault detection were used to validate the predictive ability of undetectable zone for the faulty sensor.

CASE I: T_{chw}^{i} with -1.5℃ bias fault

Undetectable zone of T_{chw}^{i} for the training data set is illustrated in the Figure 2.

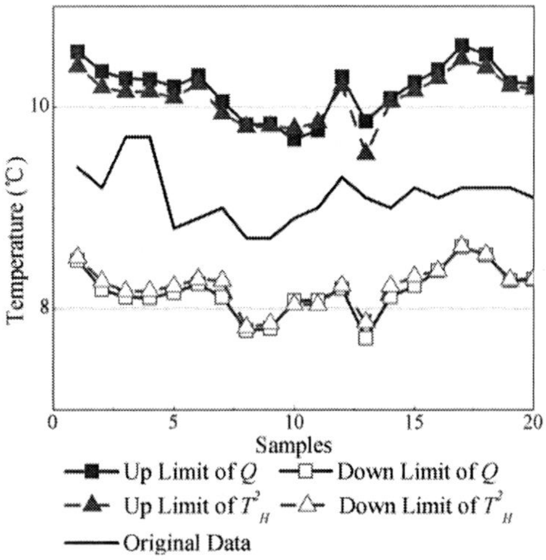

Figure 2: Undetectable zone of T_{chw}^{i} for the training data set

The up limits of undetectable zone by Q is almost equal to that by Hawkins' T_H^2, as well as the down limits of two statistics. Obviously, the fault detectability of T_{chw}^i by Q is equal to the ability by Hawkins' T_H^2. The undetectable zone of T_{chw}^i by Q is ±1.39 ℃, while that by Hawkins' T_H^2 is ±0.67℃.There are just the former 20 samples shown in the horizontal axis in order to make the figure clear.

A bias fault with -1.5℃ was introduced into T_{chw}^i to test the predictive ability. The fault detection results by Q and T_H^2 are illustrated in Figure 3 (a) and (b). All the Q-statistics values of tested samples are greater than the Q_α and the detection efficiency of sensor fault by Q is 100 %. Meanwhile, the -1.5℃ bias fault of T_{chw}^i is completely detected by Hawkins' T_H^2. Therefore, the detectability indicated by the undetectable zone of T_{chw}^i is according to the fault detection results of tested samples of T_{chw}^i. The undetectable zone successfully predicted the fault detection results.

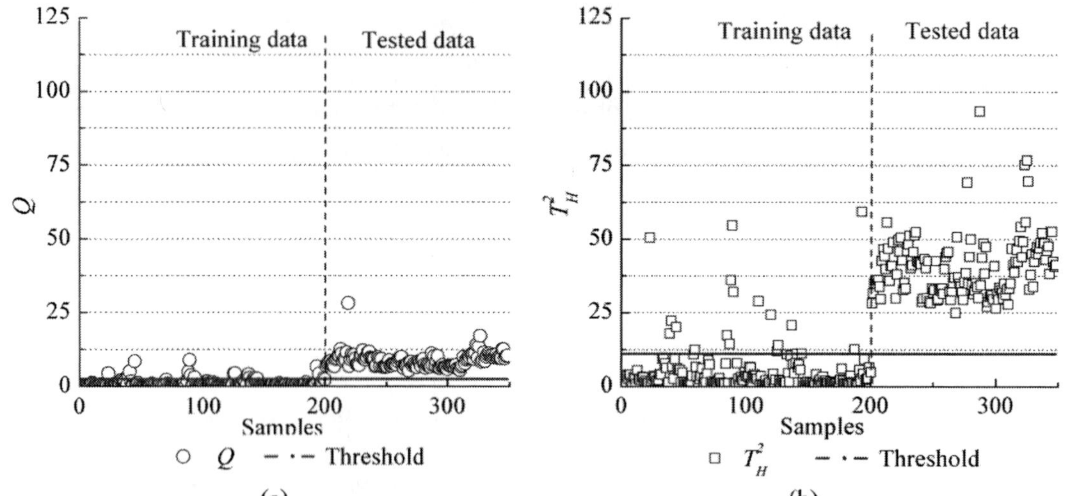

Figure 3: Fault detection for T_{chw}^i with -1.5℃ bias fault: (a) by Q (b) by T_H^2

CASE II: T_{cw}^o with -2.0℃ bias fault

Undetectable zone of T_{cw}^o for the training data set is illustrated in the Figure 4.

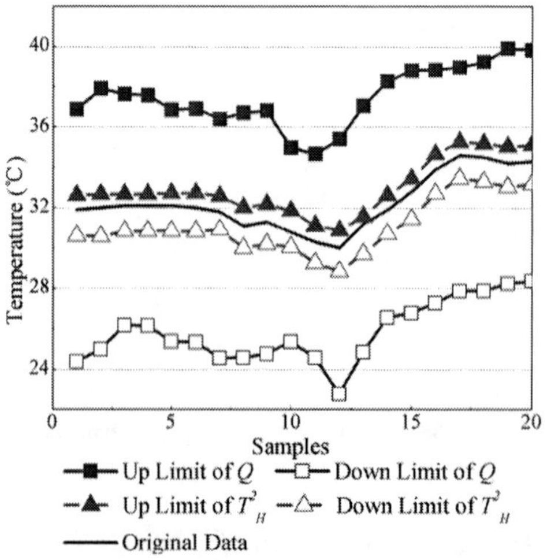

Figure 4: Undetectable zone of T_{cw}^o for the training data set

Unlike to the results of T^{i}_{chw}, the up and down limits of undetectable zone by Q is quite different to that by Hawkins' T^2_H. The undetectable zone by Q is much greater than that by Hawkins' T^2_H. The undetectable zone of T^{i}_{chw} by Q is over ±6.0 ℃, while that by Hawkins' T^2_H is less than ±1℃. Consequently, the detectability of T^{o}_{cw} fault by Hawkins' T^2_H is much better than the ability by Q.

The fault detection results by Q and T^2_H are illustrated in Figure 5 and Figure 6, respectively, when a 2.0℃ bias fault was introduced into T^{o}_{cw}. There are only 21% of Q-statistics values in the tested samples are greater than the Q_α. It means that the fault detection by Q did not working. Meanwhile, the 2.0℃ bias fault of T^{o}_{cw} is completely detected by Hawkins' T^2_H. Due the introduced fault level, 2.0℃, is less than the undetectable zone by Q, ±6.0 ℃, the detectability by Q is completely worse than that by Hawkins' T^2_H. Therefore, it is better that the Hawkins' T^2_H is employed to detect the T^{o}_{cw} fault.

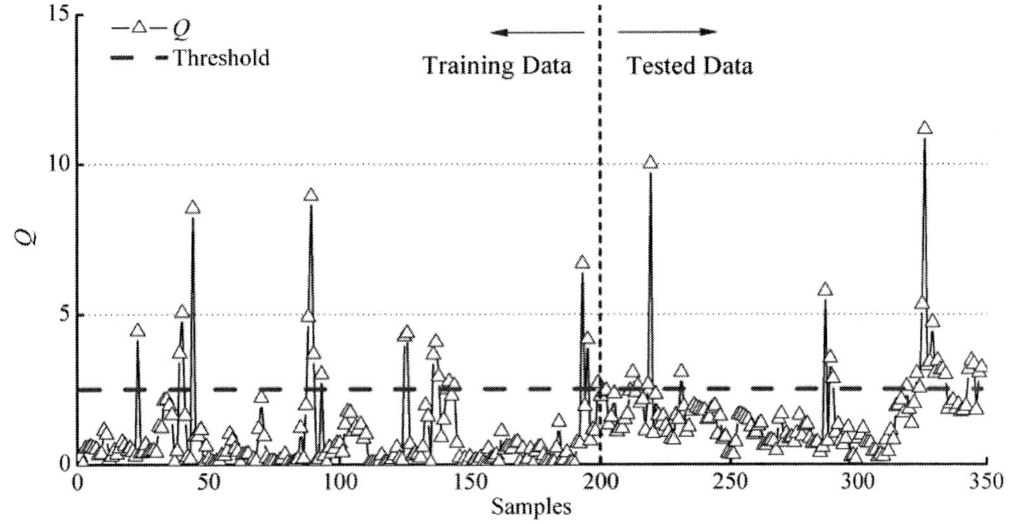

Figure 5: Fault detection for T^{o}_{cw} with 2.0℃ bias fault by Q

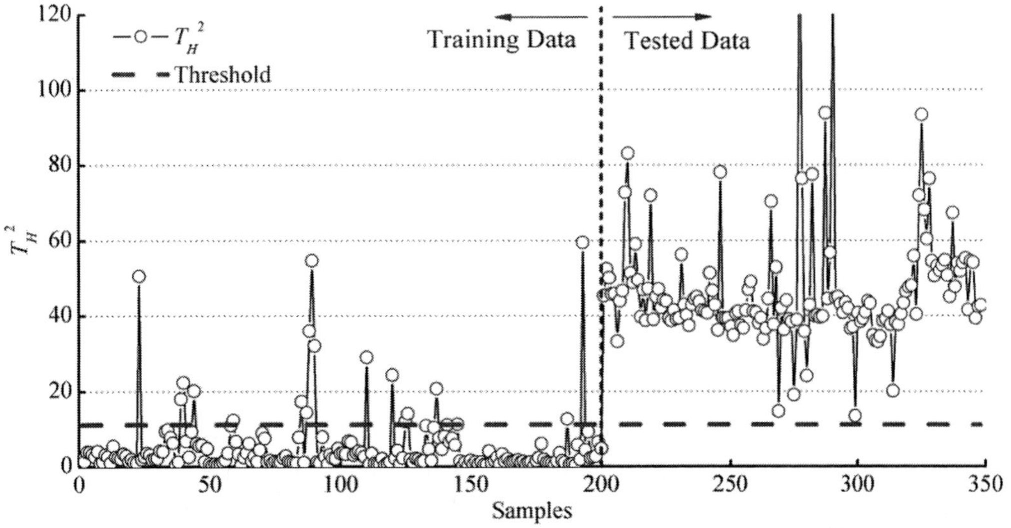

Figure 6: Fault detection for T^{o}_{cw} with 2.0℃ bias fault by T^2_H

CASE III: M_{cw} with +10% bias fault

The predictive results, undetectable zone, of M_{cw} for the training data set is illustrated in the Figure 7. Meanwhile the fault detection results of M_{cw} by Q and T_H^2 are illustrated in Figure 8. The undetectable zone by Q is nearly equal to that by Hawkins' T_H^2, so the fault detection results of M_{cw} by Q are in accordance with that by T_H^2.

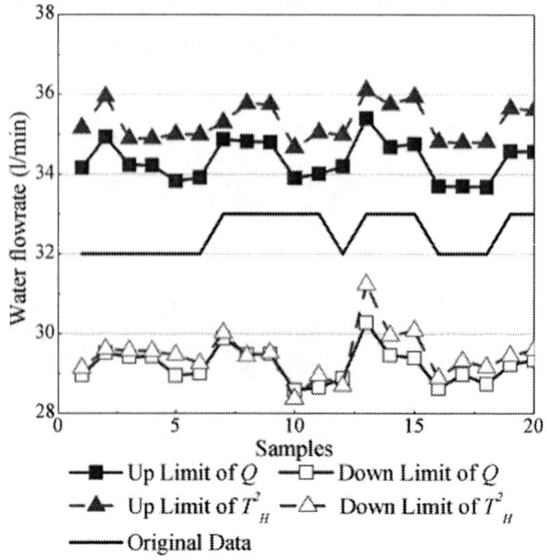

Figure 7: Undetectable zone of M_{cw} for the training data set

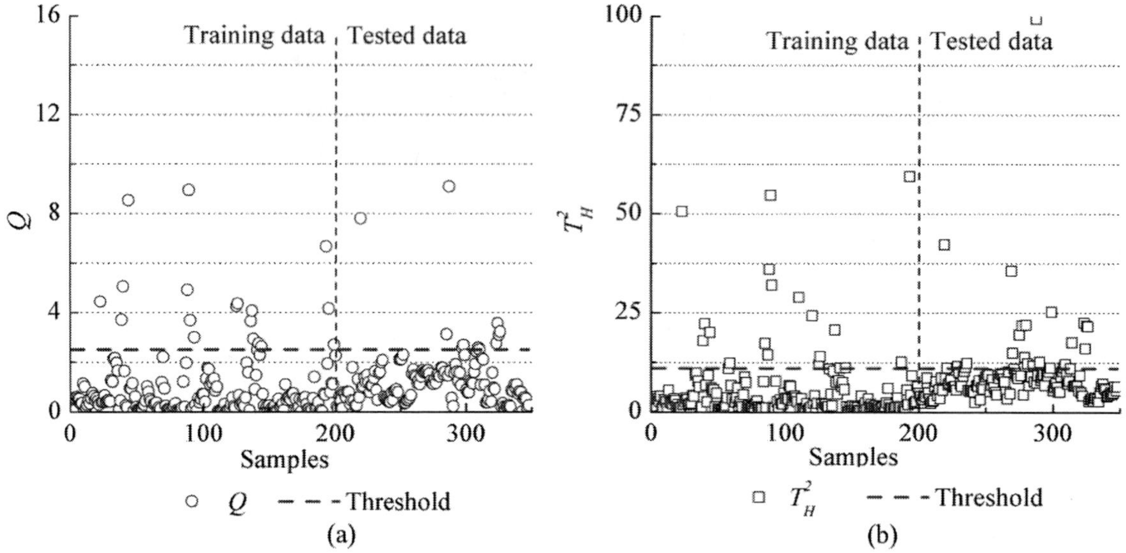

Figure 8: Fault detection for M_{cw} with +10% bias fault: (a) by Q and (b) by T_H^2

3.2 Summary

The undetectable zones for all sensors in the PCA model by Q and by T_H^2 are summarized in the Table 1. The detection abilities for different sensor by Q and by T_H^2 are quite different. At the step for choosing the optimal statistics to obtain the fault boundary, the undetectable zone can directly predict the detection ability for the sensor by a certain statistics, such as Q or T_H^2. Therefore, the performance of PCA-based sensor Fault detection can be improved by choosing the statistics with higher fault detection efficiency by the sensitivity analysis for the certain training data set.

Table 1: Summary for all sensors' Undetectable zones by Q and by T_H^2

Sensor	Unit	Undetectable zone by Q	Undetectable zone by T_H^2
T_{chw}^o	℃	2.78 (±1.39)	1.74 (±0.67)
T_{chw}^i	℃	2.47 (±1.24)	1.68 (±0.84)
M_{chw}	l/min	6.28 (±3.14)	6.35 (±3.18)
T_{cw}^o	℃	11.77 (±5.89)	1.66 (±0.84)
T_{cw}^i	℃	12.77 (±6.39)	1.61 (±0.81)
M_{cw}	l/min	7.18 (±3.59)	5.30 (±2.65)
W	kW	137.84 (±68.92)	68.07 (±34.04)
M_{ref}	%	41.49 (±20.75)	20.33 (±10.17)

There is an important point illustrated by the case study and the summary. The solution of undetectable zone is derived from the matrix calculation process of different statistics. The results of undetectable zone only rely on the original training data. Therefore, the undetectable zone demonstrates the original feature of the training data.

4. CONCLUSIONS

Sensor fault detection, diagnosis and erroneous data reconstruction is the fundamental work for the thermodynamic fault isolation, the optimal control, the safety operation and so on. In this paper, an evaluation index, undetectable zone, is presented to predict the detectability of sensor fault so as to improve the performance of sensor fault detection. Different calculation algorithm is derived to obtain the undetectable zone by different statistics.

Undetectable zone can be employed as an index to predict and to evaluate the detectability of sensor fault by some statistics for a certain training data set. It can be used to choose the optimal statistics of fault detection for each sensor. From the evaluation of detectability for each sensor by different statistics, the online sensor fault detection can be more flexible by choosing the most sensitive statistics. Therefore, the detection efficiency can be promoted by the prediction of undetectable zone.

NOMENCLATURE

The nomenclature should be located at the end of the text using the following format:

T	temperature	(℃)
M	water flow rate	(l/min)
W	chiller electrical-power input	(kW)
V	position of the slide valve	(-)
$\boldsymbol{X}$	original matrix	(-)
$\boldsymbol{X}^0$	normalized original matrix	(-)
$\boldsymbol{R}$	covariance matrix	(-)
U	eigen vector matrix	(-)
VE	variance explained	(-)
CV	cumulative contribution of variance	(-)
FDD	fault detection and diagnosis	(-)
FDDR	fault detection, diagnosis and reconstruction	(-)
HVAC&R	heating, ventilating, air-conditioning and refrigeration	(-)

PC	Principal Component	(-)
PCA	Principal Component Analysis	(-)
SPCA	Principal Component Analysis with a statistical data-cleaning	(-)
P	PC subspace projection matrix	(-)
$\tilde{P}$	Residual subspace projection matrix	(-)
Q_α	threshold of the Q-statistic	(-)
$\vec{x}$	a sample	(-)
$\hat{\vec{x}}$	estimate of a sample	(-)
$\tilde{\vec{x}}$	residual of a sample	(-)
$\vec{x}_{rc}$	reconstruction of a sample	(-)
x_j	the jth entry of $\vec{x}$	(-)
e_i	the ith entry of $\vec{e}$	(-)

Greek letters

μ	mean
σ	standard deviation
$\lambda_1, \cdots, \lambda_n$	eigenvalues

Superscript

i	inlet node
o	outlet node

Subscript

chw	chilled-water system
cw	condenser-water system

REFERENCES

Du, Z. M., Jin, X. Q. & Wu, L. Z. (2007). PCA-FDA-Based Fault Diagnosis for Sensors in VAV Systems. *HVAC&R Research, 13*(2), 349-367.

Du, Z. M., Jin, X. Q. & Yang, X. B. (2009). A robot fault diagnostic tool for flow rate sensors in air dampers and VAV terminals. *Energy and Buildings, 41*(3), 279-286.

Du, Z., Fan, B., Chi, J. & Jin, X. (2014). Sensor fault detection and its efficiency analysis in air handling unit using the combined neural networks. *Energy and Buildings, 72*(0), 157-166.

Han, H., Gu, B., Kang, J. & Li, Z. R. (2011). Study on a hybrid SVM model for chiller FDD applications. *Applied Thermal Engineering, 31*(4), 582-592.

Härdle, W. & Simar, L. (2007). *Applied Multivariate Statistical Analysis* (Second ed.). New York: Springer Berlin Heidelberg.

Hu, Y., Chen, H., Xie, J., Yang, X. & Zhou, C. (2012). Chiller sensor fault detection using a self-Adaptive Principal Component Analysis method. *Energy and Buildings, 54*, 252-258.

Jackson, J. E. (1991). *A User's Guide To Principal Components* (First ed.). New York: John Wiley & Sons, Inc.

Kocyigit, N. (2015). Fault and sensor error diagnostic strategies for a vapor compression refrigeration system by using fuzzy inference systems and artificial neural network. *International Journal of Refrigeration, 50*(0), 69-79.

Lee, S. H. & Yik, F. W. H. (2010). A study on the energy penalty of various air-side system faults in buildings. *Energy and Buildings, 42*(1), 2-10.

Lee, W., House, J. M. & Kyong, N. (2004). Subsystem level fault diagnosis of a building's air-handling unit using general regression neural networks. *Applied Energy, 77*(2), 153-170.

Li, G., Hu, Y., Chen, H., Shen, L., Li, H., Hu, M., Liu, J. & Sun, K. (2016). An improved fault detection method for incipient centrifugal chiller faults using the PCA-R-SVDD algorithm. *Energy and Buildings, 116*, 104-113.

Li, S. & Wen, J. (2014). A model-based fault detection and diagnostic methodology based on PCA method and wavelet transform. *Energy and Buildings, 68, Part A*(0), 63-71.

Sun, Y. J., Wang, S. W. & Huang, G. S. (2010). Online sensor fault diagnosis for robust chiller sequencing control.

International Journal of Thermal Sciences, 49(3), 589-602.
Wang, S. W., Zhou, Q. & Xiao, F. (2010). A system-level fault detection and diagnosis strategy for HVAC systems involving sensor faults. *Energy and Buildings, 42*(4), 477-490.
Xiao, F., Wang, S. W., Xu, X. H. & Ge, G. M. (2009). An isolation enhanced PCA method with expert-based multivariate decoupling for sensor FDD in air-conditioning systems. *Applied Thermal Engineering, 29*(4), 712-722.
Yoon, S. H., Payne, W. V. & Domanski, P. A. (2011). Residential heat pump heating performance with single faults imposed. *Applied Thermal Engineering, 31*(5), 765-771.
Zhao, Y., Wen, J. & Wang, S. (2015). Diagnostic Bayesian networks for diagnosing air handling units faults - Part II: Faults in coils and sensors. *Applied Thermal Engineering, 90*, 145-157.

ACKNOWLEDGEMENT

The research work presented in this paper is supported by National Natural Science Foundation of China (Project 51576074), and is funded by Wuhan Science and Technology Bureau of China (Project 2015061705011607), and is funded by Beijing Key Lab of Heating, Gas Supply, Ventilating and Air Conditioning Engineering (Project NR2016K02), China, and is supported by state key laboratory of compressor technology, China.

Low Refrigerant Algorithm Detection for Cooling Systems Relying on Trending and Data Analysis

Fadi ALSALEEM[1]*, Amro QUEDAN[2]

[1]Wichita State University, Department of Mechanical Engineering,
Wichita, Kansas, USA
Contact Information (316-719-0450, fadi.alsaleem@wichita.edu)

[2]Wichita State University, Department of Electrical Engineering,
Wichita, Kansas, USA
Contact Information (940-595-4627, amquedan@wichita.edu)

* Corresponding Author

ABSTRACT

A hybrid algorithm of an enhanced version of Mann-Kendall trending and data analysis is proposed to solve the limitations of current technology in detecting and diagnosing cooling system refrigerant faults in general and refrigerant leakage specifically. A data abstraction mechanism is applied over feed of temperatures and power measurement to calculate and store only the significant information for further analysis. Next, an enhanced version of Mann-Kendall trending is applied periodically over the stored data to calculate the trend strength (upward or downward) for each measurement. Finally, a harmonic mean is utilized to balance the trends contribution and evaluate the result against a threshold value for potential faults. Such an algorithm is expected to have an important positive impact, because it is designed to accurately detect low refrigerant at an early stage. This should help in the following ways: (a) to reduce the impact of refrigerant emissions on climate, and (b) to potentially reduce the U.S. energy use by more than 0.1–.02 quad per year. This algorithm is a robust first step towards leveraging the latest technology advancements, especially in computer science and mathematics, in order to vertically advance the field of cooling systems.

1. INTRODUCTION

Refrigerant leakage is one of the most common problems with cooling systems, causing, for example, supermarket refrigeration systems to lose up to 25% of their nominal refrigerant per year (David Cowan et al., 2010) and 60% of existing cooling systems and one-third of new cooling system installations (Rossi, 2004), (Mowris, 2004), and (Piotr et al., 2014) to be inefficient mostly due to low charge (David Cowan et al., 2010). Refrigerant leakage is one of the dangers responsible for climate change, creating a double adverse effect due to the increase in direct emissions as well as indirect CO_2 emissions. In addition to climate impact, refrigerant leakage creates a negative snowball financial impact (David Cowan et al., 2010), as shown in Figure 1. In the early stages, leakage affects energy cost, whereby the system runs longer to satisfy the cooling load and the repair cost is fairly constant. The repair and energy costs rise exponentially as the system persists in losing refrigerant. At a certain low refrigerant level, catastrophic system failures may occur. These failures, such as compressor failure due to high discharge temperature or food damage in the case of a reach-in cooler at a supermarket, dramatically increase the financial burden of refrigerant leakage.

There are two major components for the low refrigerant detection model; the processes to initiate the need for refrigerant evaluation (processes) and the method /algorithm to detect and diagnose low refrigerant (method). Next, we review the major limitations of these two components.

1.1 Refrigerant-Leakage Detection Process

For the past decades, most refrigerant-leakage detection processes were limited to on-site, expensive, and time-consuming diagnosis, with less attention on cost and flexibility. The on-site diagnosis may occur during a regular maintenance check-up or after an emergency call. Both scenarios have major drawbacks. Although regular maintenance schedules tend to be proactive and usually detect refrigerant leakage in the early stages, the cost of regular check-up is high and its reoccurrence is inconvenient, especially when the cooling system itself is fine. Moreover, repetitive refrigerant checking using a pressure measurement, during these checkups, results in lowering the refrigerant level. Emergency visits are usually initiated by users, and by that time, leakage is very severe and has already impacted the environment and the energy bill.

1.2 Refrigerant-Leakage Detection Method

Different methods were proposed in the literature to detect refrigerant leakage (Katipamula and Brambley, 2005a). Some of these methods rely on the underlying physics of HVAC systems (Quantitative methods). Others are rule-based methods that were developed based on prior knowledge of the system (Qualitative methods). In addition, if the system operation data are available, during normal operation and under fault condition, a learned relationship (black or gray box models) is used. Each of these methods suffers from a certain type of weakness. For example, the quantitative-based methods are complex and computationally intensive and may require input measurement that may not be readily available. The qualitative methods, on the other hand, are specific to a system type and rely on rules and thresholds values that will be difficult to be used for other systems. Finally, the history learned-based models require a huge amount of training data and may fail in dealing with data that are beyond that training data.

Most refrigerant leakage-detection methods require intrusive sensing of the refrigerant's pressure. One exception is the work of Li and Braun who they invented the application of virtual sensor concept in the HVAC domain to replace the intrusive expensive pressure measurements by the cheap temperature sensors (Li and Braun, 2006). This concept was further enhanced and evaluated in recent research work (Li and Braun, 2009) and (Kim, W.H and Braun, 2013). Different methods were suggested to estimate the temperature to pressure correlation parameters. However, most of these methods (i) require technical data that are provided by manufacturers and (ii) the temperature sensors are required to be mounted at a specific location to provide accurate estimation (Li and Braun, 2009) and (Kim, W.H and Braun, 2013).

Due to the above shortcoming, most of the current leakage detection methods are not accurate or can't be widely applied over majority of HVAC systems (Yuill, 2014). To assess current cooling system fault diagnostic methods and algorithms, including refrigerant leakage detection, recently a team of researchers at Purdue University conducted a rigorous study and developed an evaluation protocol. Applied over thousands of faulty cooling systems data, the protocol concluded that current generic cooling system fault diagnostics algorithms and hand-held FDD tools are ineffective. For example, their study shows that the protocol of California's Title 24 for HVAC fault diagnostics for refrigeration and air conditioning (RAC) has produced more than 50% false positive faults and has missed 32% of the real faults (Yuill and Braun, 2013), (Yuill et al., 2014), and (Yuill, 2014).

Form the aforementioned literature review, we first note that while there are a lot of work on improving the methods of detecting and diagnosing refrigerant-leakage fault, less attention has been given to advance the processes for applying these methods. It is noted that also most of the refrigerant-leakage detection methods are not accurate or not general or practical enough to cover wide range of systems. In this work, we are exploring a new processes of refrigerant-leakage ,online -monitoring, benefiting from recent advances in computing technology that can facilitate detecting refrigerant leakage in the early stages, similar to a regular check-up but without the need to check the system on site. Also, the new processes demonstrably reduces the need for expensive emergency calls. Utilizing this new processes, a hybrid algorithm of an enhanced version of Mann-Kendall trending and data analysis is proposed to solve the limitations of current methods in accurately detecting and diagnosing refrigerant leakage.

1.3 Paper Scope and Organization

In this work we highlight the use of online monitoring and cloud computing platform described in (Alsaleem, et al., 2014) in providing continuous refrigerant leakage detection capabilities for residential vapor compression systems. Specifically, a data abstraction mechanism is applied over a feed of temperatures and power measurement to calculate and store only the significant information for further analysis. Next, a sample size adjusted version of Mann-Kendall trending is applied over the stored data to calculate the trend strength (upward or downward) for each measurement. Finally, a harmonic mean is utilized to aggregate the rate trends contribution and this result is evaluated against a threshold value to determine whether to signal for potential faults. The organization of this paper is as follows. In section 2, we introduce the new refrigerant detection model, the processes and the method. In section 3, we present initial result for applying the new model over hundreds of real filed HVAC systems. In section 4, we summarize the paper and give some conclusions.

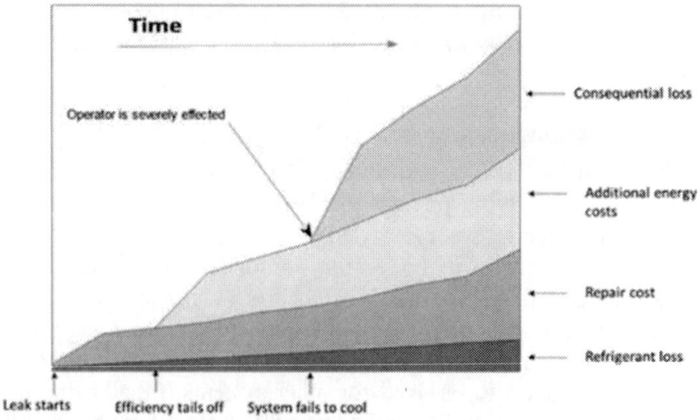

Figure 1: Financial impact of refrigerant leak over time, (David Cowan et al., 2010)

2. THE REFRIGERANT LEAKAGE DETECTING MODEL

The complete hardware for this new diagnostics model is described in (Alsaleem et al., 2014) and is shown in Figure 2. Few key points are to be mentioned here. (1) The hardware consists of an indoor kit that is equipped with voltage and current sensors as well as temperature sensors to track air supply, air return, indoor liquid line, and suction line temperatures. The outdoor kits measure the aggregate outdoor current and voltage. (2) the data for each system run is transmitted in chunks, each of maximum of fifteen minutes and at sampling rate of one sample per five seconds (0.2 HZ),(3) a cloud structure is built to host the low charge algorithms to receive and processes the data from the indoor and outdoor kits.

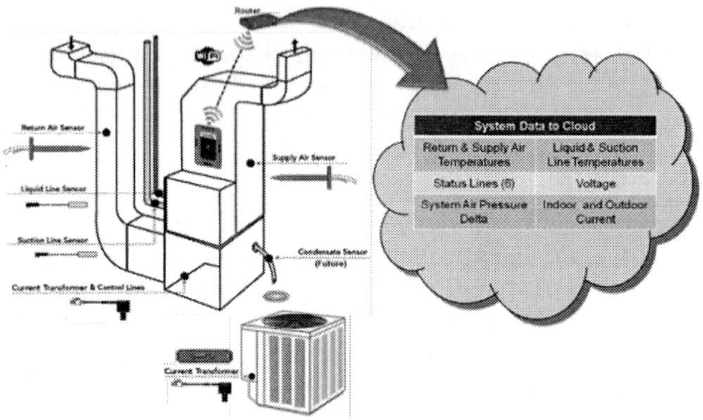

Figure 2: Hardware shows the indoor and outdoor kits to gather and send sensors data to the cloud

1.2 Mann-Kendall Trending Test

The purpose of the Mann-Kendall test is to assess for a given data if there is a monotonic trend over time. It has been used in many applications, such as water quality monitoring (McLeod et al., 1990) and more recently for weather trending and prediction (Soltani et al., 2013). Its advantages over linear regression analyses includes: it does not require a normal distribution for the data, it is independent of the data magnitude, and it can deal well with missing or irregularly spaced data. The trending test compares each data point to all subsequent values. If the data from newer values is greater than earlier data, then the Mann-Kendall statistics value S is increased by 1; otherwise, it decreased by 1. In mathematical terms, this could be described as:

$$S = Sign\left(\sum_{i=1}^{m-1} \sum_{j=i+1}^{m} \begin{cases} 1 \; if \; (T_j - T_i) > 0 \\ 0 \; if \; (T_j - T_i) = 0 \\ -1 \; if \; (T_j - T_i) < 0 \end{cases} \right) \tag{1}$$

where T_i, T_j are the daily average values for a given measurement at a given day where $j > i$. A very high positive S value indicates an increasing trend, whereas a very low negative value indicates a decreasing trend. The sign function, introduced by the authors, adjusts the sign of the S value to always indicate a positive value if the measurement trend direction support the fault. To account for sample size, the magnitude of S was adjusted by the following scale (assuming sample data are not similar) as follow (HydroGeoLogic, 2004):

$$S_N = \frac{S \mp 1}{g(S)^{1/2}} \tag{2}$$

where, $g(S) = \frac{1}{18}[n(n-1)(2n+5)]$, and n is the sample size.

2.2 Mann-Kendall Trend Calculation

Here we discuss the use of trending analysis to track and warn of capacity degradation due to refrigerant leakage over time. As a HVAC system loses it is capacity, it is expected that some measurements (such as split temperature: return air minus supply air, and averaged total power) to trend down and others (such as suction temperature) to trend up, as shown in Figure 3. These trend directions uniquely characterize capacity drop and are different for other faults such as evaporator fouling, in which the split temperature is expected to trend up. In such a situation, it is rational to design an algorithm that watches the trend of each measurement over time and use the trends conurbation that aligns with a capacity drop behavior to generate a confident decision about the status of system capacity.

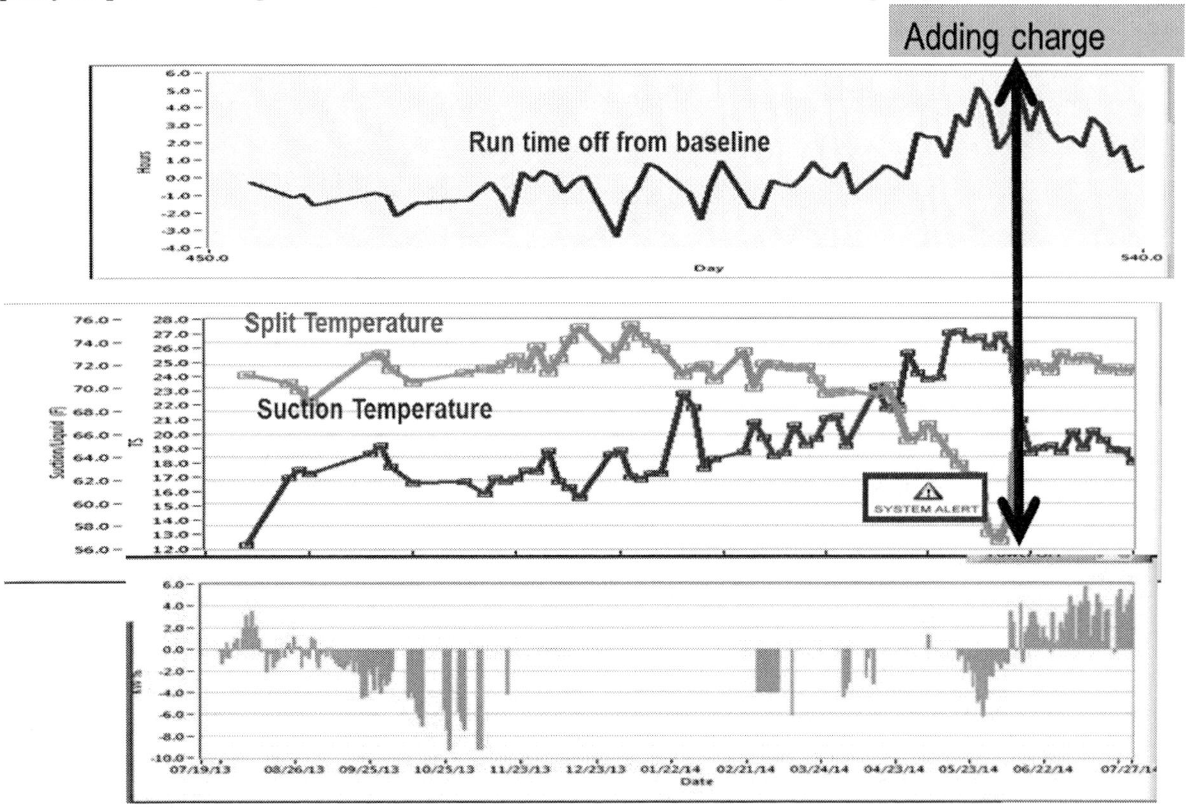

Figure 3: HVAC capacity drop, due to refrigerant leakage, impact on split temperature, suction temperature, run time, and power. First upper plot shows the run time in hour's deviation from baseline data. Middle plots are; red for suction and green for split temperatures in F°, and bottom plot is the daily average power deviation from baseline. The figure shows that as system loses refrigerant over time, run time deviation from baseline and the suction line temperatures increase while the split temperature and power deviation from baseline decrease.

Figure 4 shows a schematic for the proposed algorithm. Here, the average temperatures, power, and total run time deviation from a bassline of each day are accumulated and stored in a database. The Mann-Kendall is then applied weekly to calculate the trend strength/sum of each component, the Mann-Kendall application detail over one measurement is described in (Alsaleem et al., 2016). It is worth to mention that (2) is modified to include the normalized difference between the recent measurement and the average of this measurement for all monitored systems as follows:

$$S_{NA} = \frac{S \mp 1}{VAR(S)^{\frac{1}{2}}} + sign \times \frac{T_m - \overline{T}.}{\overline{T}} \tag{3}$$

where $\overline{T}$ is the average of a measurement under consideration for all systems, T_m is the current measurement value for a given system. The sign is positive for a measurement that a decreasing trend supports a fault and is negative for a measurement that an increasing trend supports the fault.

Equation 3 represents one of the major contribution of this paper, that is to come up with an adaptive approach to customize a measurement threshold to trigger a certain fault based on knowledge from the big data of monitored systems (second term) and a specific signature from the monitored system itself (first term). This solution is in contrast with most of the current HVAC systems fault diagnostics methods that were built around comparing a HVAC system measurements to fixed thresholds values. These thresholds were either chosen wide enough to cover system variations in the field (threshold 2 in Figure 5) at the expense of missing real faults or chosen very tight range at the expense of producing false alerts (system 1 using threshold 1).

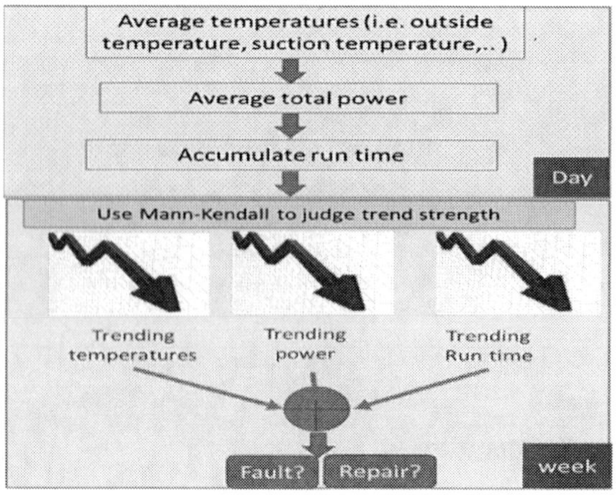

Figure 4: Algorithm schematic for detecting low charge using trending analysis

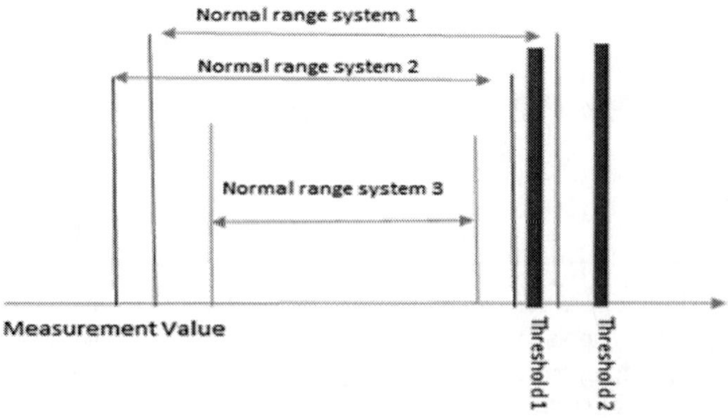

Figure 5: Fault threshold variation in typical HVAC systems

The importance of adding the normalized difference term to (3) can be understood easily considering the following scenarios; a strong decrease trend for the split temperature might exist for a cooling system, as a sign of continuous loss of refrigerant, while its current value is around the average split temperature of all monitored cooling systems. This should have less impact on triggering a low-refrigerant fault, compared to another system with the same trend strength but with a lower split temperature. The contribution of the calculated trends sum is averaged using a harmonics mean approach that is explained in the next section.

2.3 Mann-Kendall Trends Accumulation

Multiple trends algorithms are needed to capture the measurements behavior effected by capacity degradation. The combined effect of these trends is averaged using the F-measure, borrowed from the information retrieve science (Rijsbergen, 1979). The F-measure, given in (4) for two inputs values *X, Y,* calculates the harmonic mean of two values. The harmonic mean differs from the traditional arithmetic mean by keeping an equal influence of the two values, especially when their difference is high. Using the capacity degradation problem as an example, if the trend for split temperature is very strong decreasing (0.9 confidence level out of 1.0) but suction temperature has poor increasing up trend (0.1 confidence out of 1.0). The arithmetic mean of these trends is 0.5. However, the harmonic mean for the same trends is 0.18. As the system suction temperature doesn't strongly support capacity degradation, the 0.18 measure is more physically realistic measure to infer low capacity fault.

$$F = \frac{2XY}{X+Y} \tag{4}$$

A slight modification to *(4)* is done by introducing β:

$$F_\beta = \frac{(\beta^2+1)XY}{\beta^2 X + Y} \tag{5}$$

β, is a parameter to control the balance between the two inputs variable *X* and *Y*. for example, when β>1 then *F* becomes more influenced by *Y* value. In the capacity drop calculation, it is a good practice to weight the split temperature higher compared to suction temperature. This is true as split temperature is the one observed directly by homeowner and it is a stronger indicator for capacity drop. Equation *5* can be generalized so it could be used for *n* variable inputs:

$$F = \frac{n}{\frac{1}{x_1}+\frac{1}{x_2}\cdots+\frac{1}{x_n}} = \frac{n}{\sum_{i=1}^{n}\frac{1}{x_i}} = \frac{n.\prod_{j=1}^{n}x_j}{\sum_{i=1}^{n}\frac{\prod_{j=1}^{n}x_j}{x_i}} \tag{6}$$

The use of the multiple trends with the harmonic mean for capacity degradation detection is as follows: at the end of each week, if the harmonic mean of the confidence level of the measurement trends sum is greater than a specific threshold value, then a low-refrigerant fault is declared.

3. RESULT AND DISCUSSION

A sample of 225 HVAC systems installed in real homes were used in this investigation. Of these systems, 118 were charged with R410A refrigerant and the rest were charged with R-22 refrigerant. The tonnage size and SEER information for this sample size is given in Figure 6. The Pie charts in Figure 7 captures the low refrigerant alerts status statistics in a two years period. The pie chart to the left, divides the alerts to: valid not sent alerts, invalid alerts, and sent alerts. The chart shows 0.0% invalid alerts. The pie chart to the right, classifies the valid sent alerts to two groups; fixed and not fixed alerts. Alerts are considered to be valid only if a fix was performed after sending the alert or, in the case of no fix was reported, if the capacity drop was confirmed by noticeable increase of power consumption for at least a week after declaring the fault at slimier environment conditions, refer to Figure 8 as an example. The chart in Figure 7 shows that 21.0% of the valid alerts were not sent. Alert is not severe enough and from past experience was not easy for on-site technician to confirm, or homeowner chose not to be notified about efficiency alerts are just few reasons for not sending this valid alert. An alert is considered to be a fixed alert only if the owner of the HVAC system received the alert and responded by calling a service technician, the technician showed up and confirmed the diagnostics, the homeowner agreed to pay for the fix, and finally the technician did the right fix!. If any of these conditions is not met, such as the homeowner chose not to acknowledge or to react on the alert, technician failed to confirm the diagnostics, technician confirmed the diagnostics but homeowner chose not to perform the repair due to cost issue, or technician confirmed the diagnostics, homeowner agreed on the fix cost and technician performed the fix, but it was not complete, then the alert considered to be a not fixed valid alert. In this study, only 44% of the valid alerts were fixed. This raise the difficult question on how to grantee the right fix to be performed for systems with valid alerts. We believe *understanding and accounting for the complex relationship between technician and homeowner, while designing the HVAC FDD algorithms is a key in answering this question.*

Work in progress is to establish a way to compare results obtained by this algorithm with results from other traditional generic HVAC FDD algorithms. Initial screening shows good signs that this new method of alerting refrigerant leakage is more accurate compared to the on-site checkup performed by an average experienced technician. The evidence came from multiple HVAC installs, where the new algorithm has trigged the low capacity drop due to refrigerant leakage and it took multiple on-site service visits to confirm these leakage using temperature and pressure measurements.

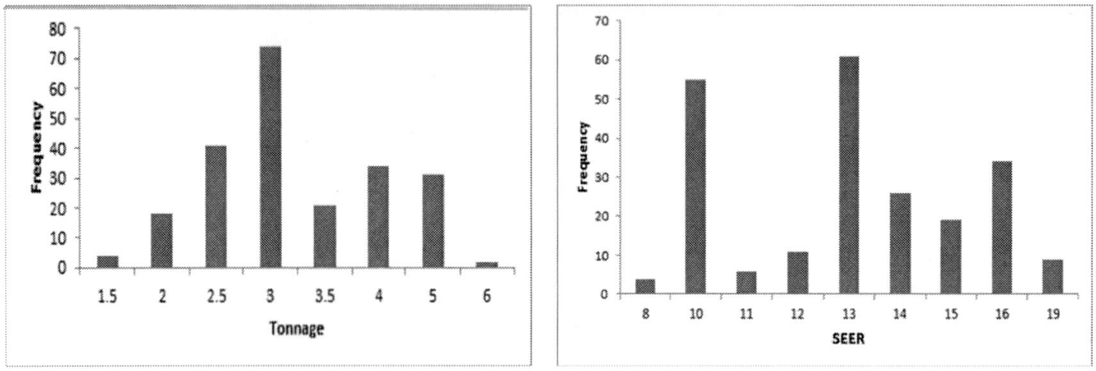

(a) (b)

Figure 6: The tonnage size (a) and SEER ratting (b) for the 225 HVAC systems used in this study

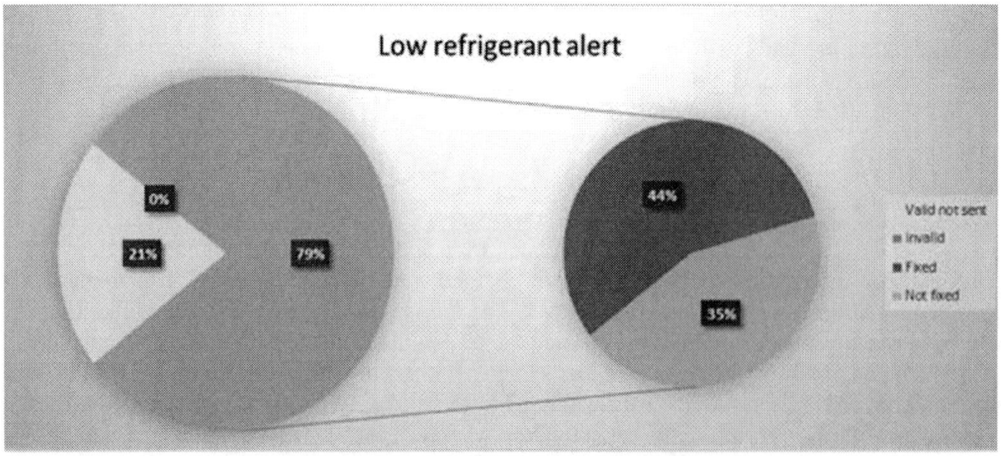

Figure 7: Trending low refrigerant alerts statistics

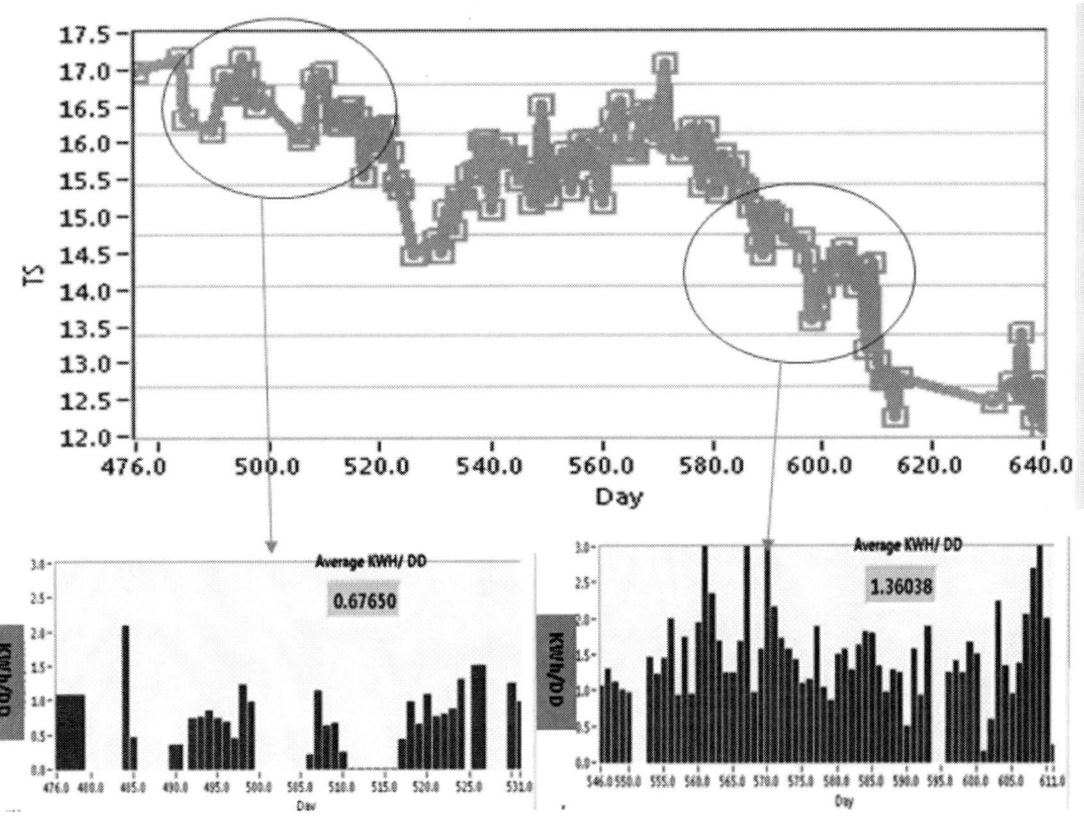

Figure 8: An example of using power consumption to confirm the generated low refrigerant alert for a system that an on-site diagnostics/fix was not performed. Top plot is the trend plot of split temperature, and bottom plots are the normalized KWh for the system before and after the alert was sent. In these plots, the KWh is normalized to the degree per day temperature index

4. CONCLUSIONS

The presented work attempts to bridge the research gap between a cost-effective, practical implementation, and an accurate and effective diagnostic model for HVAC systems by combining multidisciplinary fields of science: data analytics, cloud computing, and mechanical engineering. As an example, the use of modified trending analysis in addressing the challenging problems of refrigerant-leakage detection was presented. The application of this research may not only impact the fault detection and diagnosis within HVAC and other energy transporting system, but can also more generally inform innovation in other domains with fault phenomenology indicated by trends over time including medical, automotive, and economic policy fields.

Finally, the presented work touched on some of the HVAC systems FDD algorithms filed implementation issues. These issues such as, the complex relationship between a home owner and a service technician and the multiple possible scenarios and interactions that could result from sending even a valid alert, are hardly discussed in other research work. Future work is planned to explore further this complex relationship, benefiting from the cloud-based system flexibility to communicate, store, and analyze data, that can potentially feed and enhance the overall field of HVAC system FDD.

REFERENCES

Afram, A., & Janabi-Sharifi F., (2014). Theory and Applications of HVAC Control Systems – A Review of Model Predictive Control (MPC), *Building and Environment*, vol. 72, P. 343–355.

Akaike, H. (1992). Information Theory and an Extension of the Maximum Likelihood Principle. Pages 610-624 In: Kotz, S., and N.L. Johnson, editors. Breakthroughs in Statistics Volume 1. Foundations and Basic Theory. *Springer Series in Statistics, Perspectives in Statistics.* Springer-Verlag: New York.

Alsaleem, F., Abiprojo, R., Arensmeier, J., & Hemmelgarn, G. (2014). HVAC System Cloud Based Diagnostics Model, *International Refrigeration and Air Conditioning Conference*, Paper No. 2496, West Lafayette, IN.

Alsaleem, F & Rafaie, M. (2016). Current Based HVAC Systems Air Filter Diagnostics and Monitoring, to be presented in the 4th International High Performance Buildings Conference at Purdue, West Lafayette, IN.

Braun, J. (2003). Automated fault detection and diagnostics for vapor compression cooling equipment. *ASME Trans. Journal of Solar Energy Engineering*. 125, pp. 266-74.

Braun, J.E., Yuill, D.P., & Cheung, H. (2012). Test methods for evaluating fault detection and diagnostics systems. NIST Grant #104867 Final Report. *National Institute of Standards and Technology*, Gaithersburg, MD.

Breiman, Leo, Friedman, J., Olshen, R., & Stone, C. (1984). *Classification and Regression Trees*. Belmont, California: Wadsworth.

Breuker, M., & Braun, J.E. (1998). Common faults and their impacts for rooftop air conditioners. *HVAC&R Research,* vol.4 (3):p.303–18.

Cheung, H. & Braun, J.E. (2013a). Simulation of Fault Impacts for Vapor Compression Systems by Inverse Modeling Part I: Component Modeling and Validation, *HVAC&R Research*, vol. 19, no. 7: p. 892-906.

Cheung, H. & Braun, J.E. (2013b). Simulation of Fault Impacts for Vapor Compression Systems by Inverse Modeling Part II: System Modeling and Validation, *HVAC&R Research*, vol. 19, no. 7: p. 907-921.

Cowan, D., Gartshore, J., Chaer, I., Francis, C., & Maidment, G. (2010). REAL Zero: Reducing Refrigerant Emissions and Leakage: Feedback from the IOR Project, *Institute of Refrigeration*.

Halm-Owoo, A., & Suen, K. (2002). Applications of fault detection and diagnostic techniques for refrigeration and air conditioning: A review of basic principles. *Journal of Process Mechanical Engineering*. 216, (3), pp. 121-132.

Hipel, K.W. & McLeod, A.I. (1994). Time Series Modelling of Water Resources and Environmental Systems, Amsterdam: *Elsevier*. ISBN:0-444-89270-2.

Hirsch, R.M., Slack, J.R. & Smith, R.A. (1982). Techniques for Trend Assessment for Monthly Water Quality Data, *Water Resources Research*, vol.18, P.107-121.

HydroGeoLogic, Inc. (2004). Annual Groundwater Monitoring Report – *Former Fort Ord*, California.

Rossi, T. (2004). Unitary Air Conditioner Field Performance. *International Refrigeration and Air* Conditioning Conference at Purdue, Paper No. R146, West Lafayette, IN.

Katipamula, S., & Brambley, M. (2005). Methods for Fault Detection, Diagnostics, and Prognostics for Building Systems—A Review, Part I. *International Journal of HVAC&R Research*. 11, (1),: pp. 3-25.

Katipamula, S., & Brambley, M. (2005). Methods for Fault Detection, Diagnostics, and Prognostics for Building Systems—A Review, Part II. *International Journal of HVAC&R Research*. 11, (1),: pp. 169.

Kim, M., Payne, W.V., Domanski, P.A., Yoon, S.H., & Hermes, C.J.L. (2009). Performance of a Residential Heat Pump Operating in the Cooling Mode with Single Faults Imposed. *Applied Engineering*, vol.29, P.770–8.

McLeod, A., Hipel, K., & Bodo, B., (1990). Trend Analysis Methodology for Water Quality time Series, *Environmetrics*, vol.2, 169-200.

Mowris, R., Blankenship, A., & Jones, E. (2004). Field Measurements of Air Conditioners with and without TXVs. ACEEE 2004 summer study on Energy Efficiency in Buildings, *American Council for an Energy Efficient Economy*, Washington, DC. http://www.aceee.org.

Piotr, A., Domanski, I., Henderson, W., & Payne, V. (2014). Sensitivity Analysis of Installation Faults on Heat Pump Performance, *NIST Technical Note 1848*, http://dx.doi.org/10.6028/NIST.TN.1848.

Rijsbergen, C. (1979). Information Retrieval, *Butterworth-Heinemann Newton*, MA, USA.

Roth, K., Westphalen, D., & Brodrick, J. (2006). Residential Central AC Fault Detection & Diagnostics. *ASHRAE Journal*. 48, (5), pp.96-97.

Shuhua, H. (2007). "Akaike Information Criterion", Publications, 2007. [Online]. Available: http://www4.ncsu.edu/~shu3/Presentation/AIC.pdf. [Accessed: 27- Mar- 2016].

Soltani, M., Rousta I., & Modir T. (2013). Using Mann-Kendall and Time Series Techniques for Statistical Analysis of Long-Term Precipitation in Gorgan Weather Station, *World Applied Sciences Journal*, vol. 28, no. 7, pp. 902-908.

Talwalkar, P. (2014). The joy of game theory. *CreateSpace Independent Publishing Platform*, 2014, pp. 32-39.

Wemhoff A., & Frank, M. (2010). Predictions of Energy Savings in HVAC Systems by Lumped Models. *Energy Build*; 42:1807e14.

Yuill, D. (2014). Development of methodologies for evaluating performance of fault detection and diagnostics protocols applied to unitary air-conditioning equipment, *PhD dissertation,* Purdue University.

Yuill, D. & Braun, J. (2013). Evaluating the performance of fault detection and diagnostics protocols applied to air-cooled unitary air-conditioning equipment, - *HVAC&R Research*.

Yuill, D., Cheung, H., & Braun, J. (2014). Evaluating Fault Detection and Diagnostics Tools with Simulations of Multiple Vapor Compression Systems, *International Refrigeration and Air Conditioning Conference*, Paper No. 1543, West Lafayette, IN.

Daylight Glare Evaluation when the Sun is Within the Field of View Through Window Shades

Iason KONSTANTZOS[1,2]*, Athanasios TZEMPELIKOS[1,2], Nicole MURCHISON[3], Robert PROCTOR[3]

[1]Lyles School of Civil Engineering, Purdue University
West Lafayette, Indiana 47907 USA, (ikonsta@purdue.edu)

[2]Center for High Performance Buildings, Ray W. Herrick Laboratories, Purdue University
West Lafayette, Indiana 47907 USA

[3]Department of Psychological Sciences, Purdue University
West Lafayette, Indiana 47907, USA

* Corresponding Author

ABSTRACT

Shading fabrics have the ability to reduce daylight glare and provide privacy when needed. Recent studies have shown that glare indices such as DGP and its simplified version can be used to predict daylight glare through shades when the sun is not within the field of view. However, there are no comprehensive studies on glare sensation with the sun visible through the fabric – a situation that happens in office buildings – and therefore the applicability of glare indices for such conditions is uncertain. Shades with very low openness factors transmit only a small amount of direct sunlight due to their weave density; nevertheless, existing glare metrics may show intolerable conditions for these cases, while specific studies with human subjects are nearly non-existent.

This paper presents an experimental study on daylight glare evaluation for the case of shading fabrics with the sun within the field of view. 41 human subjects (n=41) were tested while performing specific office activities, with 14 shade products of different openness factors and visible transmittance values (direct and total light transmission characteristics). The measured variables and survey results were used to: (i) associate discomfort glare with measured and modeled parameters for these cases (ii) evaluate the robustness of existing glare indices for these cases (iii) examine recently suggested alternate (direct and total vertical illuminance) criteria for glare assessment through fabrics, extract discomfort thresholds and suggest a new related index and (iv) propose corrections in the DGP equation when the sun is visible through the shades, which could be generalized for other systems following a similar approach. Combining illuminance-based metrics and existing glare indices can result in a more realistic glare evaluation covering all cases with and without the sun through shading fabrics. The new results can be inversely used as thresholds for selecting optical properties of shades to ensure glare protection, as well as for the development of glare-based shading controls.

1. INTRODUCTION

Visual comfort is one of the main concerns in a human-centered design of interior spaces and is mostly associated with effective control of daylight glare. Several indices developed to quantify glare, given its subjective nature and the several factors involved with its evaluation (Clear, 2013, Kent *et al.*, 2015). Examples include DGI (Hopkinson, 1972), originally proposed to describe glare from a large source as a window, and UGR (CIE, 1992), originally developed to describe glare from artificial sources. Vertical illuminance on the eye has been found to be the most significant factor in some studies (Van Den Wymelenberg and Inanici, 2014, Karlsen *et al.*, 2015), outperforming more complex metrics, while other studies support luminance-based metrics. The Daylight Glare Probability or DGP (Wienold and Christoffersen, 2006) is considered a realistic metric to assess daylight discomfort glare, as it simultaneously considers the overall brightness of the visual field as well as the impact of glare sources and contrast (Eq. 1), extracted from experiments with human subjects.

$$DGP = 5.87\times10^{-5}E_v + 9.18\times10^{-2}\cdot\log_{10}\left(1+\sum_{i=1}^{n}\frac{L_{s,i}^2\omega_{s,i}}{E_v^{1.87}p_i^2}\right)+0.16 \quad (1)$$

$$DGPs = 6.22\times10^{-5}E_v + 0.184 \quad (2)$$

, where Ev is the total vertical illuminance, Ls, ωs and, P are the luminance, the solid angle and the position index for each identified glare source respectively and Lb the background average luminance.

Shading fabrics are widely used in office spaces to improve visual and thermal comfort, control solar gains and also induce privacy when necessary. They are available in a variety of different colors, materials and weave densities and they can be manually or automatically controlled. The main optical properties that characterize shading fabrics are the openness factor (OF) and the visible transmittance (Tv); the first is an indicator of the weave density and the direct light transmission, whereas the latter indicates the portion of the visible light transmitted through the fabric, characterizing also the fabric's color (Konstantzos *et al.*, 2015a). Literature focusing on the impact of roller shades on visual comfort is scarce and mostly consists of simulation studies; moreover, studies investigating the glare under the presence of sunlight, with or without shades, are certainly needed (Rodriguez *et al.*, 2015). Wienold (2009) used roller shades among other shading systems to present the simplified and enhanced simplified approaches of DGP and investigate the applicability of DGPs, the simplified version of DGP (Eq. 2) as developed by Wienold (2007), for low-openness fabrics, while Van den Wymelenberg and Inanici (2014) noted that direct sunlight on the work plane can increase task area luminance and result in misleading use of glare sources when following the standard DGP calculation approach. Jakubiec and Reinhart (2015) concluded that more than one metric is required for quantifying glare if direct sunlight is present.

Konstantzos *et al.* (2015) presented a comprehensive analysis of DGP for roller shades, and concluded that for all instances when the sun is not visible by the occupant, DGPs can be used to approximate daylight glare, including cases with sunlight on various surfaces in the space, for any fabric openness and control type. However, it is still not clear whether the full DGP index is applicable for the cases when the sun is visible through shading fabrics (Konstantzos *et al.* 2015b). Due to the extreme values of the solar corona's luminance, the luminance term of DGP is inflated, predicting discomfort levels that are rather incompatible with everyday practice, especially for the cases of low openness fabrics.

To account for that inconsistency, Chan *et al.* (2015) suggested an alternative dual visual discomfort criterion for roller shades based on direct and total vertical illuminance on the eye. The reasoning behind the dual criterion is that (i) a threshold for direct eye illuminance could be used to capture the effect of sunlight (or contrast), potentially substituting the luminance terms, and (ii) the total vertical eye illuminance would still be used for the overall brightness term, excluding the effect of sunlight. The proposed threshold values were 2760 lux for the total vertical illuminance (equal to DGPs=0.35) and 1000 lux for the direct vertical illuminance, as a modification of IES Standard LM-83-12 (2012). In this way, the fabric openness factor is directly associated with direct illuminance and the fabric visible transmittance is directly associated with the total vertical illuminance –therefore guidelines for selecting shade optical properties based on glare protection may be developed. Chan *et al.* (2015) used this approach and identified the appropriate ranges of fabric properties in order to mitigate glare for different orientations, locations, glazing properties and distances from the window. For instance, it was found that, to entirely eliminate glare when seated close to the window, a fabric of maximum OF = 2% and maximum Tv = 5% should be used on south-facing facades.

Fewer studies have included human subjects assessing glare through roller shades (Konis, 2013; Hirning *et al.*, 2014; Sadeghi *et al.*, 2016). However, there are no studies with human subjects, either evaluating glare using different shade fabrics, focused on the actual impact of their properties to the sensation of glare with the sun within the visual field, or exploring the applicability of known metrics in such cases.

This paper analyzes daylight glare through shading fabrics with the sun within the field of view (through the shades). Fourteen shading fabrics with different light transmission characteristics were evaluated by 41 human subjects. The measured and survey results were used to associate discomfort glare with measured and modeled parameters, test the usability of existing glare indices, examine the efficiency of alternate illuminance-based criteria, and propose corrections in the DGP index for the cases when the sun is visible through the shades. These results can be used for overall glare assessment through roller shades, as well as thresholds for selecting optical properties of shades to ensure glare protection.

2. METHODOLOGY

2.1 Experimental setting and Instrumentation

The experiments were conducted in the Architectural Engineering Laboratories of Purdue University in West Lafayette, Indiana. Two identical, side-by-side office spaces with reconfigurable south-facing facades, specifically designed for quantifying the impact of facade design options and related controls on indoor environmental conditions and energy use, were modified in order to host six isolated workstations for testing different fabrics, all facing the exterior façade from a distance of 1.30 m. The placement of the workstations was decided in order to capture the "worst case" scenario for daylight glare in office spaces (view direction), while also providing an adequate time frame of view of the sun through the fabric (distance from window) for each measuring day. All façade sections were equipped with a SB70XL-clear high performance glazing unit (60% window-to-wall ratio) with a selective low-emissivity coating (visible transmittance: $\tau_v = 65\%$ at normal incidence). The partitions were approximately 1.70 × 2.40 m, separated by fully opaque dividers having the same color as the walls of the facility. Fig. 1 shows a typical experimental setting for each office space, separated in three workstations (6 total).

Figure 1: Experimental layout of one of the two identical offices used for the study with three partitioned workstations, each equipped with a different shade.

Several LI-COR calibrated photometers were used to measure illuminance levels during the experiments, for data acquisition and validation purposes. Illuminance sensors have a cosine correction for incidence angles up to 80°, a response time of 0.01 ms and an absolute error of 3%. Exterior sensors (mounted on the roof and south wall) measured the exterior horizontal and vertical illuminance. An illuminance sensor mounted on the interior of the glass measured the transmitted vertical illuminance through the window. The total vertical illuminance on the eye of the subjects was measured with vertical photometers mounted next to the subjects' head (visible in Fig. 1), as well with a handheld Konica T-10s illuminance meter and a calibrated Canon T2i HDR camera. Direct and diffuse portions of incident solar radiation on the façade were also measured with a SPN-1 solar pyranometer, mounted vertically on the exterior south wall. These readings were used to calculate the direct portion of vertical illuminance on the eye, after also correcting for the angular properties of the glass and each fabric (see section 2.3). All sensors are connected to a data acquisition and control system (HP Agilent and Labview), accessible through remote access in order to run experiments without interfering with indoor lighting conditions.

2.2 Shading Fabrics

The purpose of this experiment was to evaluate glare sensation with the sun visible through roller shades (fabrics), therefore the selection of fabric types and properties was critical in order to be able to produce results that cover a wide range of products/optical properties, to generalize the study findings. Roller shades consist of different fabric materials with varying degrees of openness and transmission characteristics, both affecting direct and diffuse light transmission, which in turn have an impact on daylight provision, visual comfort and energy use.

Table 1: Fabric codes and respective optical properties

Code	A	B	C	D	E	H	I	J	O	P	Q	R	S	T
OF (%)	2.6	0.7	1.6	3.7	2.3	3.9	7	6.7	1.65	4.36	1.15	0.85	1.87	0.95
Tv (%)	2.8	6.4	13.7	4.1	6.6	15.9	7.5	13	7.63	8.57	1.43	12.29	2.18	6.62

A careful selection of 14 different fabrics was made for the tests. Their properties covered a wide range of OF and Tv values, and shades of white, black and grey. The selected combinations capture the entire space of interested with no specific pattern/relationship between the properties, given realistic limitations. 12 of the fabrics were in the lower range of openness factor (0.7% to 4.3%), while two of them were selected to be very open (OF~7%). The reasoning behind this selection was to confirm that fabrics of high openness would always lead to conditions of glare, and essentially focus on the lower end of the spectrum to closely observe patterns and thresholds. The selected fabrics were measured in detail in terms of their basic properties (openness factor and visible transmittance) using an integrated sphere. They were codenamed using letters for reasons of procedural flexibility. Their basic properties are listed in Table 1.

2.3 Angular fabric transmission properties and direct vertical illuminance on the eye

The angular optical properties of the glazing system were calculated by WINDOW 7.0 software (LBNL, 2009). Shades also have angular light transmission characteristics, which can be modeled either using detailed BSDF data or the semi-empirical model originally proposed by Kotey *et al.* (2009). The latter is further discussed in detail and validated using integrated sphere measurements and full-scale experiments (Tzempelikos and Chan, 2016). This model, which proved to be accurate and reliable for several types of standard (PVC-coated and vinyl) fabrics, calculates the beam-beam and beam-total visible transmittance angular variation as a function of the incidence angle and the normal incidence OF and Tv properties, provided by manufacturers. In summary, the angular beam-beam shade transmittance (τ_{bb}) is calculated from:

$$\tau_{bb-norm} = \frac{\tau_{bb}(\theta)}{\tau_{bb}(\theta = 0)} = \cos^{b}\left(\frac{\theta}{\theta_{cutoff}} \cdot \frac{\pi}{2}\right) \tag{3}$$

, where θ is the solar incidence angle, $\tau_{bb}(0)$ is the beam-beam transmittance at normal incidence, assumed equal to the OF of the fabric (provided by manufacturers), and b and $\theta_{cut\text{-}off}$ are parameters that depend on $\tau_{bb}(0)$, as explained in Kotey *et al.* (2009). The angular beam-total transmittance (τ_{bt}) is calculated from:

$$\tau_{bt}(\theta) = \tau_{bt}(0) \cdot (\cos\theta)^{d}, \{\theta < \theta_{cutoff}\} \tag{4}$$

, where $\tau_{bt}(0)$ is the beam-total transmittance at normal incidence (total visible transmittance provided by manufacturers) and d is a parameter that depends on openness factor and total visible transmittance. The cut-off angle should not be applied to light-colored fabrics, to account for direct light scattering at higher angles, while small corrections might be needed for dark-colored fabrics (Tzempelikos and Chan, 2016). The beam-diffuse transmittance, necessary for accurate modeling of light transfer through shades, is then equal to $\tau_{bt} - \tau_{bb}$ for each angle. Finally, integrating τ_{bt} over the hemisphere yields the diffuse-diffuse shade transmittance (τ_{dd}), which cannot be measured or calculated otherwise.

In contrast with total vertical illuminance on the eye, which is directly measured, there is no standard way to measure direct vertical illuminance on the eye of the observer (through the glazing and shading) without interfering with the experiment. Instead, the measured transmitted illuminance through the glazing was separated into direct and diffuse parts using the direct/diffuse ratio obtained by the SPN1 pyranometer, while the incidence angle θ was computed for each respective measured data point. The shade angular transmission model, described above, was then used to calculate the direct and diffuse illuminance through each fabric at each measurement time. In this way, we achieved a reliable estimation of the direct portion of vertical illuminance on the eye with any of the tested fabrics, for the selected position and view direction of the observers. This measurement is needed to evaluate vertical illuminance thresholds and alternate glare criteria when the sun is within the field of view.

2.4 HDR Imaging

For the calculation of the luminance-based metrics, the overall luminance mapping of the visual field of the subjects was measured once for each case of subject and partition (355 images in total). A Canon Rebel T2i camera was used, equipped with a Sigma 4.5 Fisheye lens. Due to the main objective of this study, all measurements were taken with the sun being included within the visual field, leading to severe overexposure problems. To overcome the latter, a Wratten ND 3.0 neutral density filter was used, mounted between the fisheye lens and the CCD sensor of the camera. To compensate for the dark conditions created by the filer, a strategy of slower exposures (9 in total) was selected for creating the HDR images. A Konica LS110 luminance sensor, a calibrated LMK Canon 550 HDR measurement system and a Color checker color chart were used for obtaining the camera's response curve. A script was created in order to automatically perform all the stages of HDR imaging, from creating the images from the pictures based on the extracted response function, cropping and resizing appropriately and then running Evalglare (Wienold, 2012) to calculate the metrics of interest such as DGP. As for each measuring point the camera had be to set up at the exact same point with the occupant's head, each case showed slight differences in terms of absolute camera position (due to height differences of the subjects, minor differences in the distance from the screen each subject was choosing etc.). For that reason, and to avoid inconsistencies between observations due to assuming a fixed uniform task area for all DGP calculations (as the task area including part of the window or not, etc.), a fixed glare identification threshold approach was followed instead in Evalglare, using the threshold of 2000 cd/m^2. As validating the readings would require an instrument with luminance measuring range exceeding the order of 10^7

cd/m², a validation in terms of vertical illuminance (Fig. 2) was performed for all of the images. A MSE of 197 lux was calculated and considered satisfactory given the severe conditions (extreme solar corona luminance being partly diffused through the fabric).

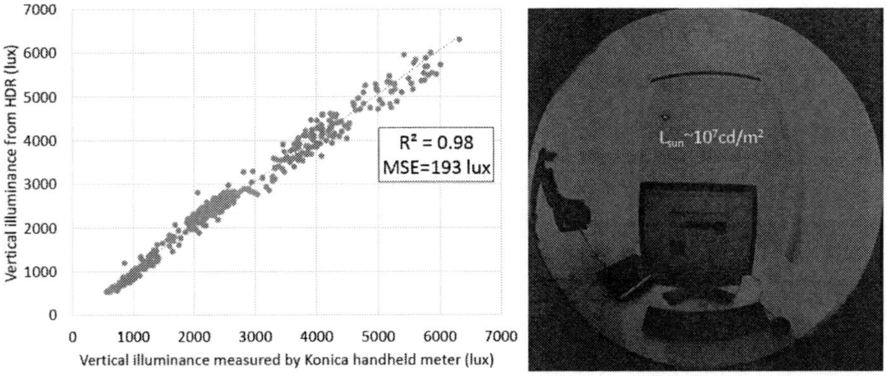

Figure 2: Validation of HDR imaging in terms of vertical illuminance (left); HDR image indicating the luminance of the sun after application of filter and calibration (right)

2.5 Experimental procedure, tests, questionnaires and surveys

The experiments were conducted during sunny days from December 2015-March 2016. Winter conditions were selected to utilize low sun angles, so that the sun is visible through the fabric during the entire test periods, in order to evaluate glare sensation under the worst case scenario situations. An IRB approval (#1410015323) was obtained in order to recruit human subjects to participate in the study. In total, 41 different subjects participated in the experiment, 25 male and 16 female, all graduate students, while care was taken to achieve the maximum possible diversity in terms of ethnicity. On each test day, the experiments lasted 2-4 hours depending on the number of test subjects and variability in the sky conditions (stable clear sky conditions were necessary for these tests). The test subjects evaluated 6-14 fabrics during each measurement day. The shades were randomly deployed in each workstation every day, although care was taken to ensure a complete range of fabric properties to be present in each measuring day, diminishing the possibility of bias as much as possible.

Each subject was initially assigned to a workstation/partition, in which they would spent 15 minutes. The duration should satisfy the need for proper adaptation to the conditions while also providing adequate time to perform specific tasks. The objective was to simulate regular office activities, including free and time-sensitive tasks. For that reason, the 15 minutes were split in to 3 main parts, including free web browsing, a character count test and a reading comprehension task, in which subjects had to complete a short questionnaire (Fig. 3). As the main reason for including the specific tasks was to force the subjects to be focused on their screens (performing computer-related activities while the sun is within the field of view through the fabric), the task performance of the subjects is outside of the scope of this study. At the end of the 15 min, the subjects were asked to proceed to the second part of the short questionnaire (Questions 4-8), commenting about their visual comfort sensation. At the same time, illuminance and DGP measurements were acquired for every workstation.

- Question 4 was a 7-point scale satisfaction with the visual conditions (from very unsatisfied to very satisfied)
- Question 5 was a 4-point glare vote (imperceptible to intolerable) about their overall perception of glare (including any possible source of it, the sun, the fabric, reflections on the desk or within the room, etc.). Keeping the same scale as in the original DGP study (Wienold and Christoffersen, 2006) allows a systematic comparison with previous results.
- Question 6 asked about the level of distraction because of the presence of the solar disc within the visual field. (4-point scale from imperceptible to intolerable)
- Question 7 requested the subjects to indicate the sources of visual discomfort, if any, on a photo of their workstation, and
- Question 8 asked about whether the subjects felt they were affected by the heat from the sun during their stay in the partition (7-point scale from not affected at all to very affected). The contents are out of the scope of this study, but may be useful in future for thermal comfort evaluation near roller shades.

In total, 425 data points were recorded, with each point being a subject evaluating a fabric. After a careful analysis, 355 observations were considered reliable and were used for the main part of the glare evaluation results. The rest were not used, mainly due to (i) subjects that appeared to be entirely insensitive to any change of conditions (ii) data with Fabrics I and J in Table 1 that were used just to confirm that high OF or Tv will always result in uncomfortable conditions.

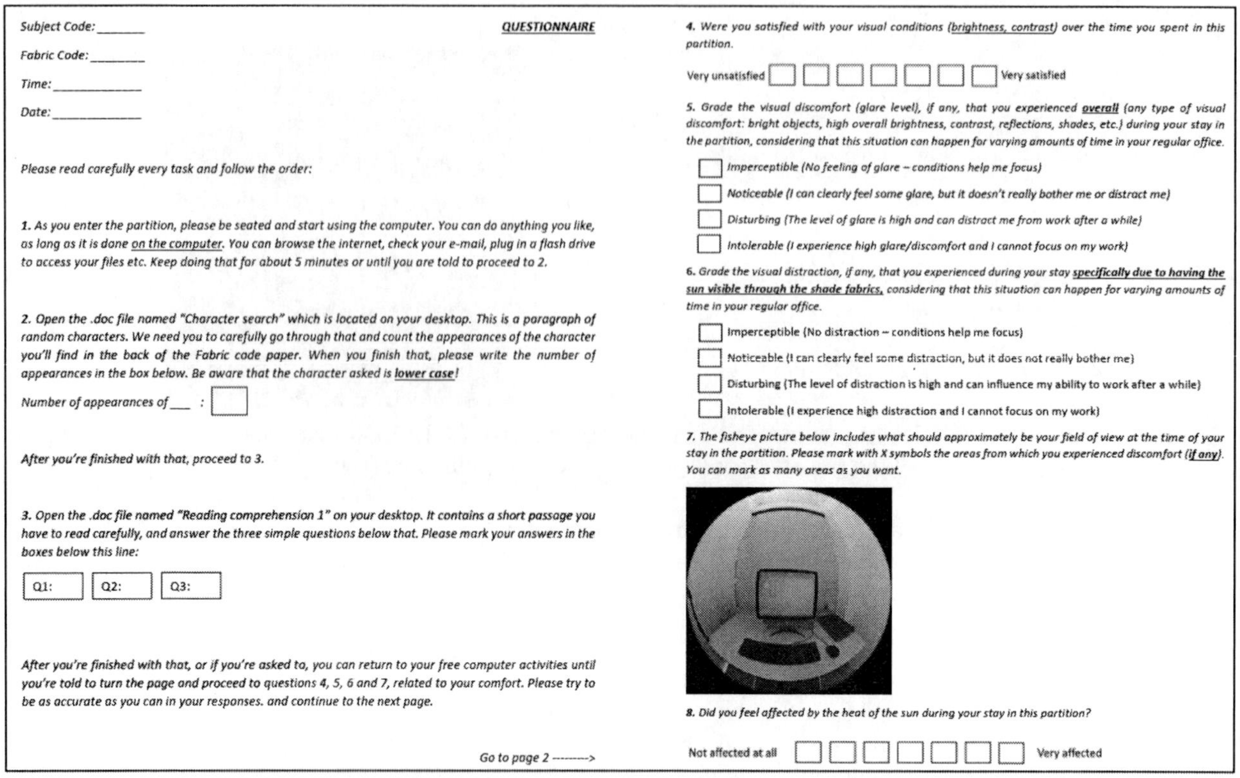

Subject Code:______

Fabric Code:_______

Time:___________

Date:___________

QUESTIONNAIRE

Please read carefully every task and follow the order:

1. As you enter the partition, please be seated and start using the computer. You can do anything you like, as long as it is done on the computer. You can browse the internet, check your e-mail, plug in a flash drive to access your files etc. Keep doing that for about 5 minutes or until you are told to proceed to 2.

2. Open the .doc file named "Character search" which is located on your desktop. This is a paragraph of random characters. We need you to carefully go through that and count the appearances of the character you'll find in the back of the Fabric code paper. When you finish that, please write the number of appearances in the box below. Be aware that the character asked is lower case!

Number of appearances of___ :

After you're finished with that, proceed to 3.

3. Open the .doc file named "Reading comprehension 1" on your desktop. It contains a short passage you have to read carefully, and answer the three simple questions below that. Please mark your answers in the boxes below this line:

Q1: Q2: Q3:

After you're finished with that, or if you're asked to, you can return to your free computer activities until you're told to turn the page and proceed to questions 4, 5, 6 and 7, related to your comfort. Please try to be as accurate as you can in your responses. and continue to the next page.

Go to page 2 -------->

4. Were you satisfied with your visual conditions (brightness, contrast) over the time you spent in this partition.

Very unsatisfied ☐ ☐ ☐ ☐ ☐ ☐ ☐ Very satisfied

5. Grade the visual discomfort (glare level), if any, that you experienced overall (any type of visual discomfort: bright objects, high overall brightness, contrast, reflections, shades, etc.) during your stay in the partition, considering that this situation can happen for varying amounts of time in your regular office.

☐ Imperceptible (No feeling of glare – conditions help me focus)

☐ Noticeable (I can clearly feel some glare, but it doesn't really bother me or distract me)

☐ Disturbing (The level of glare is high and can distract me from work after a while)

☐ Intolerable (I experience high glare/discomfort and I cannot focus on my work)

6. Grade the visual distraction, if any, that you experienced during your stay specifically due to having the sun visible through the shade fabrics, considering that this situation can happen for varying amounts of time in your regular office.

☐ Imperceptible (No distraction – conditions help me focus)

☐ Noticeable (I can clearly feel some distraction, but it does not really bother me)

☐ Disturbing (The level of distraction is high and can influence my ability to work after a while)

☐ Intolerable (I experience high distraction and I cannot focus on my work)

7. The fisheye picture below includes what should approximately be your field of view at the time of your stay in the partition. Please mark with X symbols the areas from which you experienced discomfort (if any). You can mark as many areas as you want.

8. Did you feel affected by the heat of the sun during your stay in this partition?

Not affected at all ☐ ☐ ☐ ☐ ☐ ☐ ☐ Very affected

Figure 3: Questionnaire used in the study and task-related information

3. RESULTS

3.1 General impact of fabric properties on glare

As expected, the impact of fabric properties in glare sensation is apparent. A general overview of the effect of fabrics on glare through the responses of Questions 5 is shown in Figure 4, with the fabrics appearing in order of increasing openness factor. The yellow dots indicate the averaged responses, while the boxplots effectively illustrate the distribution of the results.

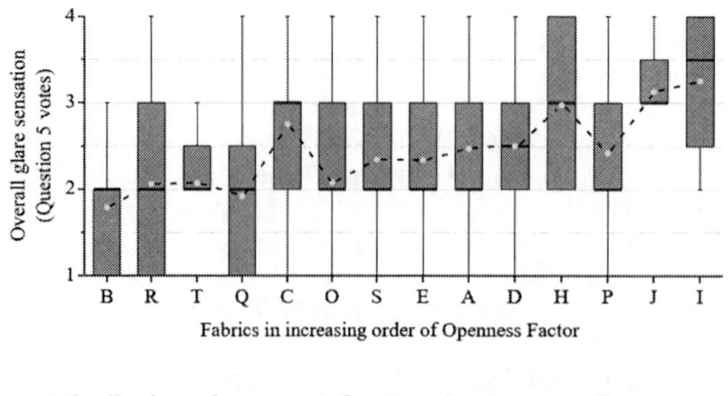

Figure 4: Distribution of responses for Question 5 – overall glare sensation.

The two fabrics of high openness factor, I and J, are the only ones with averaged responses that always lie within the discomfort zone for glare sensation, while their distribution is clearly distinguished from the rest of the fabrics. In addition, it has to be mentioned that, while all other 12 fabrics produce a relatively continuous range of direct illuminance, from 119 to 2228 lux, the latter two produce values that lie on an entirely different range (from 2940 to 3558 lux). This inconsistency in terms of ranges, combined with the clear discomfort responses obtained for the two fabrics led to their elimination when it came to the calculation of thresholds or other quantification attempts presented in the next sections.

3.2 Vertical illuminance, DGP and respective comfort ranges

As described in chapter 2.5, there were three questions including classic comfort votes, one from a positive aspect (visual satisfaction) and two from a negative aspect (glare and visual distraction). The authors consider the four - point glare scale as defined by Wienold and Christoffersen (2006) to be an effective way to assess discomfort. This, combined with the fact that one of the metrics of interest was DGP, led to a 4-point range extraction for the three main metrics investigated (total vertical eye illuminance, direct vertical eye illuminance and DGP) –and their combinations, as shown later. As the objective was to associate the overall sensation of glare with measurable metrics, Question 5 was considered to be most suitable.

Figure 5 shows the association of the three measured metrics with visual discomfort according to the responses of Question 5. Note from the results that the selected fabrics indeed resulted in a wide variation for all metrics, which was important for the analysis. More specifically, direct vertical illuminance ranged between 120-2228 lux; total vertical illuminance varied between 588-5940 lux; and DGP ranged between 0.26-0.62. The standard way to extract thresholds based on the four-point scale requires the mean, standard deviation and upper and lower bound confidence intervals need to be calculated for each vote and for each metric. Although there are clear differences between the different votes, the distribution of the data for each vote did not always approach normality at the desired level. Therefore it was preferred to follow a dichotomous approach (grouping the votes into two groups – comfort for votes 1 and 2 and discomfort for votes 3 and 4), which gives two data groups of 211 and 144 points respectively, with distributions that approach normality in the case of all three metrics. The results are shown in Fig. 6. Table 2 shows the descriptive statistics for these dichotomous votes, including the 95% upper bound confidence interval for the comfort group and the 95% lower bound confidence interval for the discomfort group, which indicate the corresponding thresholds for the border of discomfort.

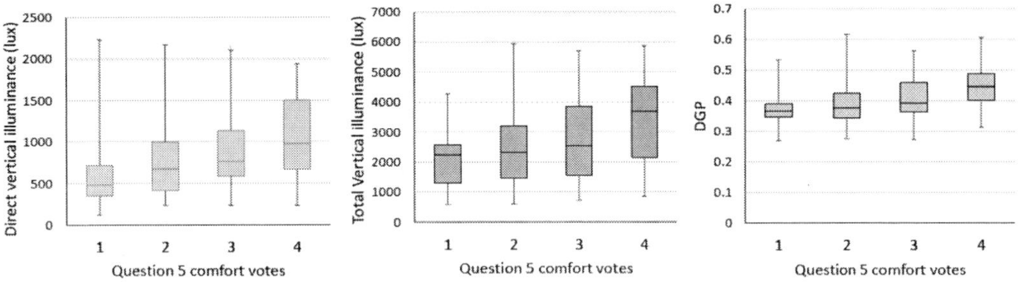

Figure 5: Boxplots associating the responses of Question 5 with direct vertical illuminance, total vertical illuminance and DGP respectively for four-point responses

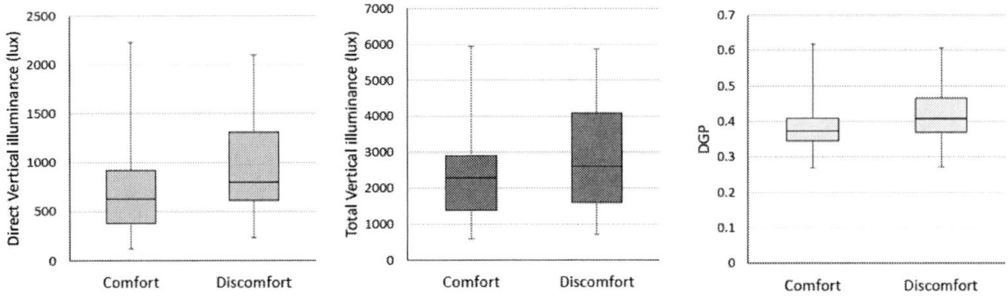

Figure 6: Boxplots associating the responses of Question 5 with direct vertical illuminance, total vertical illuminance and DGP respectively using dichotomous approach

The results show that the vertical illuminance discomfort threshold for the cases of roller shades and the sun present is around 870 lux. This result provides a first validated insight about the acceptable ranges of direct vertical illuminance, as recent studies attempt to use a discomfort threshold of around 1000 lux (Chan *et al.*, 2015, Jakubiec and Reinhart, 2015) based on recommendations for direct illuminance on the work plane found in with IES LM-83-12 (IESNA, 2012). The threshold cannot be generalized however for other types of shading or daylighting devices without conducting similar studies with these systems.

Table 2: Descriptive statistics and thresholds extraction for dichotomous approach

	Comfort			**Discomfort**		
	Evdir	Evtot	DGP	Evdir	Evtot	DGP
Mean	741.95	2410.39	0.39	947.28	2896.14	0.42
St.Deviation	470.42	1195.53	0.07	490.36	1420.34	0.07
n	211	211	211	144	144	144
CI (UP/LB)	63.47	161.31	0.01	80.09	231.99	0.01
Threshold	805	2572	0.40	**867**	2664	0.41

For the other two metrics, the results show a noticeable agreement with some previous studies; the discomfort threshold for total vertical illuminance lies in the order of the discomfort threshold for DGPs (~2800 lux), while the discomfort threshold for DGP was found to be at the level stated by Wienold's original DGP study (~0.4). For reference, there are three (and recent) published studies that propose total vertical illuminance discomfort thresholds: Karlsen *et al.* (2015) suggested 1700 lux; Van Den Wymelenberg and Inanici (2014) proposed 1250 lux and Konis (2014) suggested 1600 lux. However, these studies used different shading systems and different discomfort scales, the presence of the sun was not consistent and the vertical illuminance ranges were not similar.
Note that due to the high deviation, as expected, none of the three presented metrics is considered an appropriate comfort/discomfort predictor by itself, for the studied case of roller shades with the sun present. Therefore, the work presented in the next sections aims to bridge this gap.

3.3 Correlation of discomfort sensation with existing illuminance- and luminance-based metrics

The results presented in 3.2 indicate the extracted thresholds from this study's dataset. However, to effectively evaluate the extent to which a metric can capture the fluctuation of discomfort sensation, the method of ordered grouping of the data points is used. According to that, the data set is grouped in n groups of *m* points per group, and then the correlation of the ordered averages of the groups and the percentage of discomfort per group is evaluated. There is some ambiguity in related literature about the correct approach for creating groups in that sense. Wienold (2006) splits the total 349 data points into 12 groups of 29 points per group. Hirning (2014) states that if the number of groups exceeds the number of data points per group, the system will be underdetermined, leading to lower correlation results, while for number of data points per group exceeding the number of groups, over-determination will occur. This topic was discussed by Karlsen *et al.* (2015) who presented both grouping approaches and found differences in the results. The authors of the present study consider that an increased number of data points per group would improve the validity of the results as long as for each group a relatively low deviation around the mean would be observed for all metrics of interest. Within that scope, the 355 total data points were split into 12 groups in total, divided to 11 groups of 30 points and 1 group of 25 points. The deviation was only slightly increased in the boundary points (1st and 12th) with that being a result of the continuity of the data set and the fabrics selection (discrete properties).
A script was created in order to sort all data points (for each metric of interest) from lowest to highest, along with the respective comfort votes, transform the four votes to a binary approach of comfort and discomfort, split the groups according to the approach described above, calculate the averages and standard deviation for each group, along with the percentage of discomfort, and then produce the respective coefficient of determination for each case. Several metrics were evaluated, including vertical illuminance (total and direct), average luminance of the visual field, DGP and DGI with different thresholds of glare sources identification. The results are shown in Table 3.

Table 3: Evaluation of the fit of some existing illuminance- and luminance-based metrics

	No glare sources			**Threshold: 2000 cd/m^2**		
Metric	Evdir	Evtot or DGPs	Lave	DGP	DGI	UGR
R^2	0.43	0.49	0.36	0.65	0.79	0.82

As expected, metrics that were not able to describe the influence of the peak luminance of the solar disc did not manage to behave satisfactory. Similar poor results were observed for metrics that could describe the influence of the sun but not the overall brightness (such as the vertical illuminance). UGR, which was found to perform well in the study of Hirning *et al.* (2014) and DGI, which was not found to be an adequate metric in related studies (Van Den Wymelenberg and Inanici, 2014) are both performing better than DGP. Table 3 shows the coefficient of determination results for the evaluated metrics.

3.4 Modification of the DGP equation for cases with the sun visible through roller shades

DGP is considered a generalizable glare index, as it simultaneously takes into account the overall brightness of the scene, expressed with the vertical illuminance term, as well as the individual glare sources using the luminance term, and can be applied to roller shades (Konstantzos *et al.*, 2015b). The overall brightness is important when it comes to cases of high vertical illuminance conditions with limited glare sources (such a fully open window inflating the task area luminance). However, the results of Fig. 7 show a relatively poor fit of the existing DGP index for the studied cases. For that reason, and assuming that this inconsistency might be a consequence of the specific cases met in this study (glare through fabrics with the sun visible), it was investigated whether the same form of equation could describe the current data set with a modification of its four coefficients. The number of data points was equivalent to the one in the original DGP study (355 compared to 349), therefore such an investigation would show whether indices should be fixed or if a 'cluster-like' approach should be followed, having different sets of coefficients for fundamentally different kinds of environmental conditions.

An optimization algorithm was created, reading the detailed output of Evalglare, using a 2000cd/m^2 identification threshold and applying the genetic algorithm approach, with objective to maximize the coefficient of determination for the ordered groups of the modified DGP and the corresponding percentages of discomfort. This investigation showed that the four DGP coefficients can be indeed modified in order to describe our dataset better than any of the metrics evaluated in Table 3. The resulting new equation with the modified coefficients is shown in Eq. 5.

$$DGP_{mod} = 8.40\cdot 10^{-5}\cdot Ev + 11.97\cdot 10^{-2}\cdot \log(1+\sum \frac{L_{s,i}^2\cdot \omega}{Ev^{2.12}\cdot P^2}) + 0.16 \qquad (5)$$

Figure 7 (left) shows the correlation between DGP for each group and respective percentage of discomfort for the original equation and for the modified one, with obvious improvements. In addition, Figure 7 (left) shows the extracted discomfort threshold based on the techniques used in section 3.2, equal to 0.44, slightly higher than the discomfort threshold assumed for the original DGP (0.40). This structure allows utilizing the same fundamental index for cases with and without direct sunlight on the occupant, as a dual function with different coefficients. The authors believe that the general form of the DGP equation is reasonable and adequate, and can be adjusted to account for different cases. Similar approaches may be followed for other shading or daylighting systems and further studies with human subjects are needed for that purpose.

3.5 Formulation of a new illuminance-based metric for predicting discomfort glare cases with the sun visible through roller shades

The final part of this study attempts to assess the efficiency and applicability of a new metric for discomfort glare evaluation, for the cases studied here, based only on vertical illuminance. While DGP (especially in its modified aspect presented above) can adequately describe discomfort with roller shades, it requires both extensive field measurements and complicated procedures (calibrated cameras with filters, automation, processing etc.), or, in the case of simulations, heavy computational load for accurate luminance mapping. Although recent computational efforts made it possible for fast calculations of luminance and DGP (Konstantzos *et al.*, 2015, Chan *et al.*, 2015, Xiong *et al.*, 2016), illuminance-based metrics would allow faster and simpler calculations and can be directly associated to shade optical properties for development of design guidelines.

As shown in section 3.3 and in Konstantzos *et al.* (2015), total vertical illuminance or DGPs are not applicable in cases like the focus of this study. However, a combination of a metric that solely describes the effect of the sun (direct vertical on eye illuminance) and another that captures the overall sensation of brightness (total vertical on eye illuminance) was hypothesized to adequately capture cases including the sun in the visual field but not directly looking at it (as that case would have to be assessed as disability glare). So the algorithm presented for the modified DGP equation was used to associate the direct and total parts of vertical illuminance with their corresponding comfort votes of question 5, in order to find an equation that would predict discomfort glare based on these two metrics. The two independent variables were chosen to be (i) the direct part of vertical illuminance, to capture the visible sun's impact (capturing also the position of the sun, as function of the incidence angle) and (ii) a fraction of the total and the direct part of vertical illuminance in order to capture this interdependence between the color of a fabric, overall brightness, and the apparent intensity of the visible sun. Other combinations of direct and total vertical illuminance were also tried without satisfactory results. The best fit was found for the following equation:

$$Glare(Ev) = 0.13 \cdot Evdir^{0.27} + 0.04 \cdot \left(\frac{Evtot}{Evdir} \right)^{0.84} - 0.48 \tag{6}$$

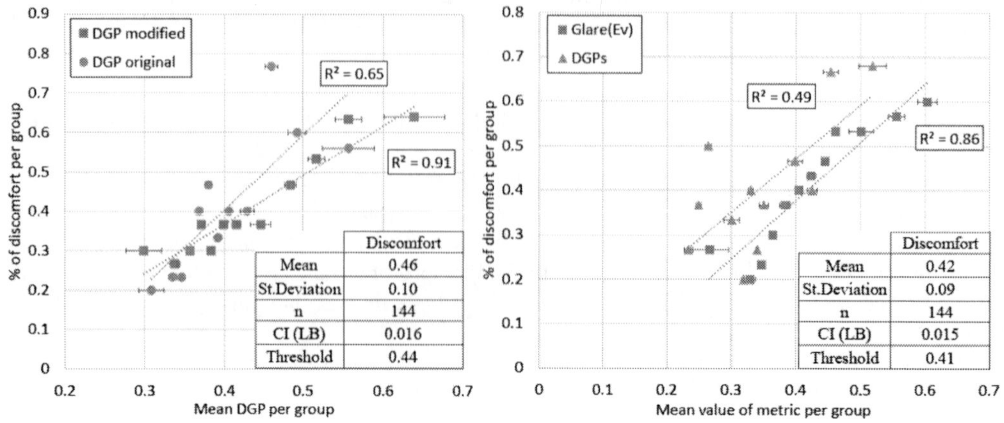

Figure 7: Improvement of fit for the modified DGP compared to original (left) and for the new illuminance-based metric compared to DGPs (right)

Figure 7 (right) shows the satisfactory fit of equation 6 ($R^2 = 0.86$) when regressed with the fluctuation of discomfort in ordered groups, as well as the improvement compared to the standard metric -DGPs or total vertical illuminance - for the entire usable data set of this study.

One of the indications of the effective behavior of a metric is the normality observed in the respective comfort and discomfort groups, a fact that allows the direct extraction of a discomfort threshold equal to 0.41 (Fig. 7 right). Although the fit is poorer compared to the modified DGP equation, an illuminance metric on the basis of Eq. 6 would simplify annual simulations, eliminating the need for a detailed luminance mapping of the interior, compensating the slight compromise in terms of fit with increased convenience of use. At the same time, it can be much more effective than the only other existing illuminance based glare metric (Evtot or DGPs) for cases with direct sunlight through fabrics. Therefore, the authors want to clarify that an index in the form of equation 6 is not proposed as a successor to any of the luminance-based glare metrics, but only as one (the only one for the cases studied here) alternative to vertical illuminance or DGPs, to be used for simpler calculation for cases of direct sunlight through fabrics and for practical applications. Considering that the fabric OF relates to direct vertical illuminance and the fabric Tv relates to total vertical illuminance, the discomfort glare thresholds can be directly used to provide design guidelines for selecting fabric properties, as suggested by Chan *et al.* (2015).

4. CONCLUSION

This paper provides new insights on daylight glare evaluation for the case of shading fabrics with the sun within the field of view. 41 human subjects were tested while performing office activities near a south facing façade equipped with 14 different shade products of different openness factors and visible transmittance values. The experimental data was used to extract discomfort glare thresholds for direct vertical illuminance (870 lux), total vertical illuminance (2800 lux) and DGP (0.4). As none of these metrics was proven to be entirely adequate to be a discomfort predictor for the studied case, a modified version of DGP was proposed, using optimized coefficients for this specific conditions. The new modified index allows utilization of the same fundamental index as a dual function with different coefficients for the case with and without direct sunlight on the occupant through shades. To further explore other opportunities, the correlation of discomfort sensation with an illuminance based metric was investigated, aimed to predict discomfort using only the direct and total portions of vertical illuminance on the eye. An index on the basis of equation 6 shows a very improved fit, compared to the only available illuminance based glare predictor (DGPs). Such an index can simplify annual simulations, eliminating the need for a detailed luminance mapping of the interior, and can directly associated with fabric optical properties for development of design guidelines. Note that the results presented in this paper are only applicable to roller shades. Combining illuminance-based metrics and existing glare indices can result in a more realistic glare evaluation covering all cases with and without the sun through shading fabrics, and potentially through other systems.

5. ACKNOWLEDGEMENTS

This work is supported by Alcoa Foundation and Lutron Electronics Co Inc.

6. REFERENCES

Chan Y.-C., Tzempelikos A., Konstantzos I. (2015). A systematic method for selecting roller shade properties for glare protection. *Energy and Buildings 92*, 81-94.

CIE (1992) Discomfort Glare in the Interior Lighting, Comission Internationale de l'Eclarage (CIE), Technical committee TC-3.13, Division 4, Interior Environment and Lighting Design, Vienna, Austria.

Clear R.D. (2013). Discomfort glare: what do we actually know? *Lighting Research and Technology 45*, 141-158.

Hirning M., Isoardi G., Cowling I. (2014). Discomfort glare in open plan green buildings. *Energy and Buildings 70*, 427-440.

Hopkinson, R.G. (1972). Glare from daylighting in buildings. *Applied Ergonomics 3*(4), 206-214.

IESNA: IES Standard LM-83-12 (2012). Approved method: IES Spatial Daylight Autonomy (sDA) and Annual Sunlight Exposure (ASE), New York.

Jakubiec J.A., Reinhart C.F. (2015). A concept for predicting occupants' long term visual comfort within daylit spaces. *Leukos 2724*, 1-18.

Karlsen L., Heiselberg P., Bryn I., Johra H. (2015). Verification of simple illuminance based measures for indication of discomfort glare from windows. *Building and Environment 92*, 615-626.

Kent M.G., Altomonte S., Tregenza P.R., Wilson R. (2015). Temporal variables and personal factors in glare sensation, *Lighting Research and Technology*, 0: 1-22. 10.1177/1477153515578310

Konis K. (2013). Evaluating daylighting effectiveness and occupant visual comfort in a side-lit open-plan office building in San Francisco, California. *Building and Environment 59*, 662-677.

Konis, K. (2014). Predicting visual comfort in side-lit open-plan core zones: results of a field study pairing high dynamic range images with subjective responses, *Energy and Buildings* 77, 67-79.

Konstantzos I., Chan Y-C., Seibold J., Tzempelikos A., Proctor R.W., Protzman B. (2015a). View Clarity Index: a new metric to evaluate clarity of view through window shades. *Building and Environment 90,* 206-214.

Konstantzos I., Tzempelikos A., Chan Y-C. (2015b). Experimental and simulation analysis of daylight glare probability in offices with dynamic window shades. *Building and Environment 87*, 244-254.

Kotey N.A., Wright J., Collins, M.R. (2009). Determining off-normal solar optical properties of roller blinds. *ASHRAE Transactions 117* (1), 10 pages.

LBNL (2013). WINDOW 7 simulation manual. Lawrence Berkeley National Laboratory. http://windows.lbl.gov/software/window/7/

Rodriguez RG, Yamı´n Garreto´ JA, Pattini AE (2015). Glare and cognitive performance in screen work in the presence of sunlight. *Lighting Research and Technology* 0:1-18. 10.1177/1477153515577851

Sadeghi, S.A., Karava, P., Konstantzos, I., Tzempelikos, A. (2015). Occupant interactions with shading and lighting systems using different control interfaces: A pilot study. *Building and Environment 97*, 177-195.

Tzempelikos A., Chan Y-C., (2016). Estimating detailed optical properties of window roller shades from basic available data and modeling implications on daylighting and visual comfort. Energy and Buildings, in press.

Van Den Wymelenberg K., Inanici M. (2014). A critical investigation of common lighting design metrics for predicting human visual comfort in offices with daylight. *Leukos 10*(3), 145-164.

Wienold J., Christoffersen J. (2006). Evaluation methods and development of a new glare prediction model for daylight environments with the use of CCD cameras. *Energy and Buildings 38*(7), 743-757.

Wienold J. (2007). Dynamic simulation of blind control strategies for visual comfort and energy balance analysis. *Proc. of IBPSA 2007 Conference*, Beijing, 1197-1204.

Wienold J. (2009). Dynamic daylight glare evaluation. *Proceedings of IBPSA 2009 conference*, Glascow, Scotland, 944-951.

Wienold J. (2012). EvalGlare Version 1.0. Fraunhofer Institute for Solar Energy Systems, Freiburg.

Xiong J., Tzempelikos A. (2016). Model-based shading and lighting controls considering visual comfort and energy use. Solar Energy, in press.

Solar Irradiance Modelling and Uncertainty on Building Hourly Profiles of Heating and Cooling Energy Needs

Giovanni PERNIGOTTO[1]*, Alessandro PRADA[2], Paolo BAGGIO[3], Andrea GASPARELLA[4], Ardeshir MAHDAVI[5]

[1]Free University of Bozen-Bolzano, Faculty of Science and Technology,
Bolzano, Italy
Tel: +39 0471 017632, Fax: +39 0471 017009, email: giovanni.pernigotto@unibz.it

[2]University of Trento, Department of Civil Environmental and Mechanical Engineering,
Trento, Italy
Tel: +39 0461 282653, Fax: +39 0461 282672, email: alessandro.prada@unitn.it

[3]University of Trento, Department of Civil Environmental and Mechanical Engineering,
Trento, Italy
Tel: +39 0461 282630, Fax: +39 0461 282672, email: paolo.baggio@unitn.it

[4]Free University of Bozen-Bolzano, Faculty of Science and Technology,
Bolzano, Italy
Tel: +39 0471 017201, Fax: +39 0471 017009, email: andrea.gasparella@unibz.it

[5]Vienna University of Technology, Department of Building Physics and Building Ecology,
Vienna, Austria
Tel: +43 (1)58801-27030, email: ardeshir.mahdavi@tuwien.ac.at

* Corresponding Author

ABSTRACT

Building energy simulations require a detailed characterization of the boundary conditions to solve the air heat balance problem. For dry and dew bulb temperatures and wind speed, measured hourly profiles can be easily included in weather files. On the contrary, beam and diffuse solar irradiance values incident on the building envelope are not directly available. This requires the adoption of solar irradiance models, which are often based on statistical correlations derived from empirical data. Since the samples of solar irradiance measurements used for the models development have been collected mostly in North America and European localities, no model can provide an adequate worldwide representativeness or be precisely defined as the best one. In this research we investigate the impact of the choice of solar irradiation models on simulated hourly energy needs in five European climates (Berlin, Vienna, Trento, Rome and Messina). The full combination of 22 horizontal diffuse irradiance models and 12 irradiance models for tilted surfaces has been considered for the development of hourly solar irradiation profiles, used as input in building energy simulation (i.e., TRNSYS) for a set of 72 simplified reference buildings. The results show that the variability of the estimation of solar irradiation leads to different levels of uncertainty in hourly energy model predictions, also depending on the building characteristics.

1. INTRODUCTION

Building Energy Simulation (*BES*) can be exploited for different purposes, ranging from energy to thermal and visual comfort analyses. Especially for multi-objective studies integrating and optimizing concurrent goals, the solar irradiation incident on the building envelope requires an accurate estimation. However, only global solar irradiation on a horizontal plane is usually recorded in most of meteorological stations and just in few cases horizontal beam and diffuse components are measured separately. Moreover, the solar irradiation incident on tilted planes is monitored rarely and only for some orientations of particular interest, such as for south-oriented surfaces. For all these reasons, empirical and mathematical models have been developed and included in *BES* codes. The variety of models proposed in the literature can be grouped into those aiming to distinguish the horizontal solar irradiation into beam and diffuse components (*horizontal diffuse irradiance models*) and those used to calculate the irradiation on tilted surfaces (*irradiance models for tilted surfaces*).

Every model has been developed trying to minimize the differences seen by contrasting with experimental data collected in specific locations. For this reason, constant and accurate reliability cannot be taken for granted by changing location and climate. In the literature, this issue is well-known and some research groups assessed the capabilities of some models by comparison with experimental data of locations different with respect to those used in their definitions in order to make some considerations about the best model for a given locality (Dervishi and Mahdavi, 2012). The uncertainty of *BES* output can be affected in different ways: some building characteristics can emphasize the inaccuracies of the solar irradiation models and some climatic conditions can be more sensitive to the selected model. Consequences concerns imprecisions of building energy labelling, unsuitable sizing of the energy system controls and incorrect optimizations of the retrofit measures (Prada *et al.*, 2015).

Further expanding previous analyses, in this research work we investigated the effect of the choice of solar irradiation models on the uncertainty of the predicted hourly energy performance. 22 *horizontal diffuse irradiance models* were coupled with 12 *irradiance models for tilted surfaces* in order to develop hourly profiles of solar irradiance for 5 European localities (Berlin, Vienna, Trento, Rome and Messina). The 264 alternatives were used as input in TRNSYS for the simulation of the hourly energy performances of a set of 72 simplified residential buildings, built varying parametrically insulation and thermal inertia of opaque components, windows surface and orientation and solar heat gain coefficient of glazing. The distributions of hourly heating and cooling energy needs along the year and for the different configurations in the sample were studied in order to identify the building features enhancing the uncertainty due to the solar irradiance modelling on short-term outputs.

2. METHODS

The procedure followed in the current research work is based on two phases. In the first one, the hourly solar irradiation profiles were elaborated in order to get all inputs necessary to run *BES*: for any couple of models, the irradiation was evaluated for each vertical surface and orientation. Then, for each building in the sample, the hourly distributions of heating and cooling energy needs were calculated and analyzed.

2.1 Solar irradiance models

In this research, we selected 22 *horizontal diffuse irradiance models* and 12 *irradiance models for tilted surfaces*. As regards the first group, the analysis included models presented in the literature as milestones, such as those by Orgill and Hollands (1977), Erbs *et al.* (1982), Muneer *et al.* (1984), Spencer (1982), Skartveit and Olseth (1987), the three models by Reindl *et al.* (1990a) and that by Boland *et al.* (2008). Other models, based on the correlations of the previous researches, were considered as well. These further models often implement some modifications to those listed above in order to adjust the correlations to specific climates and sky conditions (Hawlader, 1984; Maxwell, 1987; the three models by Perez *et al.*, 1992; the two models by Chendo and Maduekwe, 1994; Chandrasekaran and Kumar, 1994; Lam and Li, 1996; De Miguel *et al.*, 2001; Oliveira *et al.*, 2002; Karatasou *et al.*, 2003; Soares *et al.*, 2004).

As regards the 12 *irradiance models for tilted surfaces*, both isotropic and anisotropic models were considered: the models by Liu and Jordan (1960), Temps and Coulson (1977), Burgler (1977), Klucher (1978), Hay and Davies (1980), Skartveit and Olseth (1986), Reindl *et al.* (1990b), Ma and Iqbal (1983), Gueymard (1986), Perez *et al.* (1990) and the two models by Muneer (2006).

2.2 Building configurations and climates

BES were repeated with any couple of solar models on a set of simplified buildings, with the aim of understanding the effects of solar irradiation estimation on the energy performance of buildings. The set of buildings was not defined to represent the actual European building stocks but to catch a sufficiently wide range of sensitivities to the external environment solicitations (Pernigotto *et al.*, 2014).

Each simplified building consists of a square thermal zone with a floor area of 100 m^2, an internal height of 3 m and the façades oriented towards the main cardinal directions. The floor is modelled as a crawl space and the thermal bridges are neglected. For each case, all windows are positioned on the same façade and the transparent area is 80 % of the whole window's area. Both sides of the vertical walls and the internal side of the roof have a solar absorptance of 0.3 while the external side of the roof and the internal side of the floor have 0.6.
All opaque components have the same composition, characterized by a two-layer structure with insulation on the external side and a massive layer whose thermal resistance is around 0.8 m^2 K W^{-1}. The insulating layer is polystyrene (thermal conductivity: 0.04 W m^{-1} K^{-1}; specific heat capacity: 1470 J kg^{-1} K^{-1}; density 40 kg m^{-3}) while the massive layer can be timber (thermal conductivity: 0.13 W m^{-1} K^{-1}; specific heat capacity: 1880 J kg^{-1} K^{-1}; density 399 kg m^{-3}; thickness: 0.10 m) or concrete (thermal conductivity: 0.37 W m^{-1} K^{-1}; specific heat capacity: 840 J kg^{-1} K^{-1}; density 1190 kg m^{-3}; thickness: 0.30 m). Windows are composed by a double-pane glazing with thermal transmittance of 1.1 W m^{-2} K^{-1} and a timber frame, whose transmittance is equal to 1.2 W m^{-2} K^{-1}.
Internal gains are assumed constant and equal to 4 W m^{-2}, half radiative and half convective, according to EN ISO 13790 (CEN, 2008) suggestions for residential dwellings. A constant ventilation rate of 0.3 air changes per hour are imposed in accordance with the Italian technical specification UNI/TS 11300-1:2014 (UNI, 2014). An ideal system provides all the power needed to maintain the zone internal air temperature between the heating and the cooling setpoints of 20 °C and 26 °C. In order to simplify the analysis of the results, heating and cooling seasons were delimited by 1st October and 31st March, independently of the climate.
Each building configuration differ for insulation level (5 cm or 15 cm of polystyrene, i.e., with thermal transmittances of the vertical walls of, respectively, 0.45 W m^{-2} K^{-1} or 0.21 W m^{-2} K^{-1}), thermal inertia of the opaque components (area specific internal heat capacity equal to 75 kJ m^{-2} K^{-1} for the timber structure and to 300 kJ m^{-2} K^{-1} for the concrete), size of windows (14.56 or 29.12 m^2), orientation (east, south or west) and typology (*SHGCs* equal to 0.35, 0.49 or 0.61). Combining all the possible alternatives, the set of 72 simplified buildings was defined.
Five European locations were considered in this study: Berlin, Germany (Köppen classification: Cfb; heating degree-days with 18 °C as base temperature HDD_{18} = 3156 K d; cooling degree-days with 18 °C as base temperature CDD_{18} = 170 K d), Vienna, Austria (Köppen classification: Dfb; HDD_{18} = 3158 K d; CDD_{18} = 223 K d), Trento, Italy (Köppen classification: Cfa; HDD_{18} = 2610 K d; CDD_{18} = 391 K d), Rome, Italy (Köppen classification: Cfa; HDD_{18} = 1444 K d; CDD_{18} = 649 K d) and Messina, Italy (Köppen classification: Cfa; HDD_{18} = 758 K d; CDD_{18} = 1085 K d). The meteorological data of test reference year IWEC, developed by ASHRAE, were used as data source for all localities except Trento, for which the typical year presented in Pernigotto *et al.* (2014) was employed.

2.3 Analysis of hourly irradiation profiles and energy needs

The 264 series of incident hourly solar irradiation were analyzed in every climate and cardinal vertical orientations. For each hour during the daytime, the median of the 264 estimations was calculated and used as reference to determine the number of models within an acceptable range of error. Differently from the criteria adopted in our previous research (Prada *et al.*, 2014a; Prada *et al.*, 2014b; Pernigotto *et al.*, 2015), we chose the same reference threshold (i.e., 10 % deviation from the median) for the assessment of the variability of both solar irradiance and energy needs. Indeed, while for solar irradiance a 20 % error represents the difference often found between experimental data and models (Dervishi and Mahdavi, 2012), 10 % is the uncertainty expected from *BES* results according to the current state of the art of building simulation. The fraction of solar irradiation models within 10 % deviation from the median were calculated along all daytime hours belonging to both heating and cooling seasons. Then, their distribution functions were represented against the normalized daytime in order to allow for an easier comparison and, thus, to identify climates and orientations more sensitive to the model choice.
As regards the hourly energy needs, the implemented procedure is similar: for each climate, building and hour, 264 values were calculated and from them the hourly median was identified. When larger than a minimum of 0.1 kWh, the median was used as reference otherwise the series was neglected. As mentioned before, a threshold of 10 % deviation from the median was chosen to categorize each hour into four performance classes. An hour belongs to class "A" if more than 75 % of models ensure a deviation of heating or cooling demand within 10 % from the series median (i.e., more than 198 models), "B" if the percentage is between 50 % and 75 % (i.e., between 132 and 198), "C" between 25 % and 50 % (i.e., between 66 and 132) and "D", if less than 25 % models are able to satisfy the 10 % deviation target (i.e., less than 66 models). The time-distributions of the four classes were analyzed, looking for correlations between climate and building characteristics in the propagation of the uncertainty due to the solar irradiation modelling. In particular, we focused on the distribution of hours belonging to class "A". The cumulative distribution functions were calculated and normalized with respect to the actual length of heating and cooling seasons (i.e., number of hours with heating and cooling load within a season) for the different building

configurations and climates, simplifying the comparison among the different cases. Finally, the normalized frequencies of class "A" during the entire actual heating or cooling seasons were calculated for each building.

3. RESULTS AND DISCUSSION

3.1 Hourly solar irradiation profiles

As explained in the methods, the 264 hourly irradiation profiles were calculated, as well as the median for each daytime hour. Comparing each result with the median, the fractions of models within 10 % of deviation were calculated and represented in Figure 1, distinguishing "summer" and "winter season", respectively delimited by 1st October and 31st March. The normalization of the daytime hours allowed an easier comparison between the five localities, which are characterized by different total of daytime hours per season because of their latitudes. Indeed, during the so-called "summer season" we have 2882 daytime hours for Berlin, 2736 h for Vienna, 2537 h for Trento, 2671 h for Rome and 2656 h for Messina. During the "winter season", instead, there are 1763 daytime hours for Berlin, 1872 h for Vienna, 1725 h for Trento, 2000 h for Rome and 2005 h for Messina. Trento has the minimum number of daytime hours because of the orographic characteristics of the locality: thus, due to the mountains surrounding the city in the Adige Valley, actual times of dawn and dusk occur later and earlier.
Especially for the "summer season", it can be observed that south vertical orientation is the one showing the best agreement among the models: except for Trento, at least around 30 % of model are within 10 % deviation from the median. We registered at least 50 % of models respecting the chosen accuracy threshold for 92.3 % of hours in Berlin, 92.7 % in Vienna, 81.1 % in Trento, 86.5 % in Rome and 80 % in Messina. East and west vertical orientations have very similar trends in all localities excluding Trento and, in particular, for Berlin and Messina. For these two orientations, we found at least 50 % of solar models in good agreement for almost 49 % of summer daytime in Berlin, 46.2 % (east) and 52.2 % (west) in Vienna, 49.3 % (east) and 57.1 % (west) in Trento, 47.1 % (east) and 42.7 % (west) in Rome and almost 45 % in Messina. North vertical orientation is the most critical in all localities during the summer season: only 41.3 % of daytime for Berlin, 39.8 % for Vienna, 39.5 % for Trento and 35 % for both Rome and Messina are characterized by at least 50 % of solar irradiation models giving estimations within 10 % deviation from the median. The peculiar trends in Trento can be explained because of the orography issues, affecting the accuracy of the estimations on east and north-oriented vertical planes.
For the "winter season", the south-oriented vertical plane is remarkably better than the other cardinal orientations only for Messina. The performance of solar models for the south is very close to other vertical planes in the other localities and even worse for Berlin. The minimum of 50 % of models within 10 % distance from the median in case of South vertical orientation is reached for 53.1 % of winter daytime in Berlin, 65.9 % in Vienna, 69.4 % in Trento, 59.8 % in Rome and 63 % in Messina. East and west orientations are almost overlapped in every climate except Trento: around 73 % of winter daytime has at least 50 % of solar models in good agreement in Berlin, 69.3 % (east) and 67.5 % (west) in Vienna, 51 % (east) and 62.7 % (west) in Trento, 52.1 % (east) and 49 % (west) in Rome and around 45 % in Messina. While in Berlin and Vienna, the estimation of solar irradiation on north-oriented vertical planes is as good as for east and west orientations, in the other climates it is the most critical. The target of 50 % models within 10 % deviation from the median is ensured for 71.4 % of daytime in Berlin and 63.2 % in Vienna, but only for 41.3 % in Trento, 39.4 % in Rome and 33.1 % in Messina. Regarding Trento, also during the winter season the orographic effect on the models' accuracy can be observed: west and south vertical orientations on a hand and east and north on the other hand have similar trends but very different levels of uncertainty.
As a whole, we can recapitulate that the best agreement among hourly estimates of solar irradiation is found during the summer season for south-oriented vertical walls while the most uncertain estimations are generally found for north-oriented vertical planes during the winter season. We can conclude, consequently, that when the beam irradiation is prevailing, the uncertainty due to the choice of solar model is smaller while the opposite is true for irradiation mostly diffuse. Models' outputs for east and west vertical planes are generally characterized by similar performances except in case of regional orographic obstacles, as it is in Trento. In that case, large discrepancies can be found among solar irradiation values predicted by the different models.

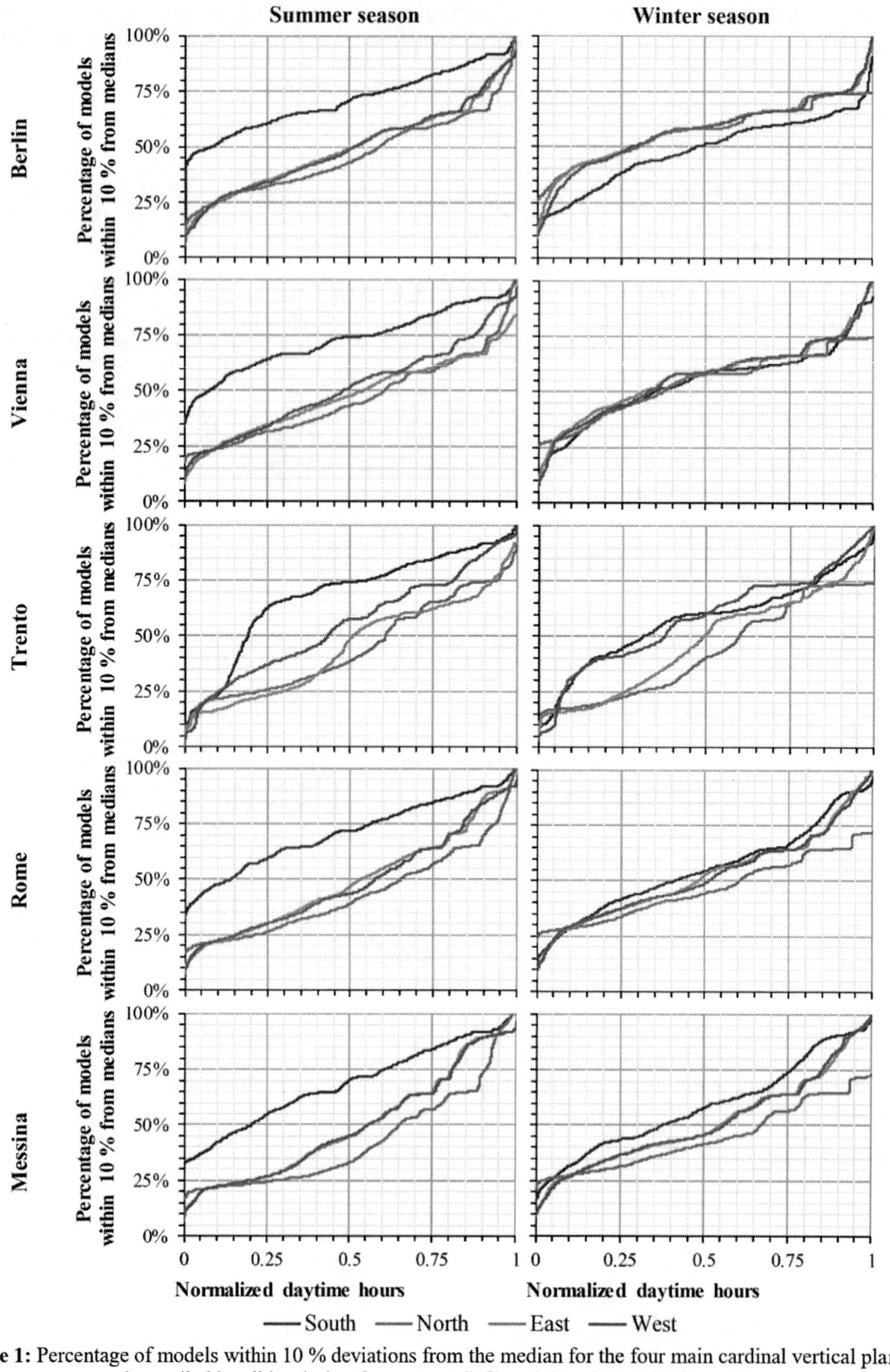

Figure 1: Percentage of models within 10 % deviations from the median for the four main cardinal vertical planes in the studied localities during the summer (left) and the winter (right) seasons.

3.2 Hourly energy needs profiles

The variability of solar irradiation predictions clearly affects the variability of hourly energy need predictions. For example, Figure 2 shows a comparison between the level of agreement found in hourly solar irradiation predictions for east and west-orientation vertical walls in Rome during the whole year and the variability of heating and cooling hourly energy needs for two building cases with a well-insulated timber structure and large windows with *SHGC* equal to 0.608 but different orientations. The daytime hours are divided into four groups depending on the amount of models whose solar irradiation estimations are within a 10 % deviation from the median: more than 75 % (very light grey), between 50 % and 75 % (light grey), between 25 % and 50 % (dark grey) or less than 25 % (very dark grey). Similarly, the graphs reporting the energy needs classes distinguish the time belonging to class "A" (very light grey), class "B" (light grey), class "C" (dark grey) and class "D" (very dark grey). As it can be seen, inaccuracies in solar irradiation calculations are specular moving from east to west vertical planes and the same trends can be recognized comparing the cooling need of the building with windows on the east façade to that with windows on the west façade. Indeed, while small impact can be observed for the heating needs, for the cooling ones the largest uncertainty is registered in the afternoon in case of east orientation and in the morning in case of west orientation, when the solar irradiation entering into the thermal zone is mostly diffuse.

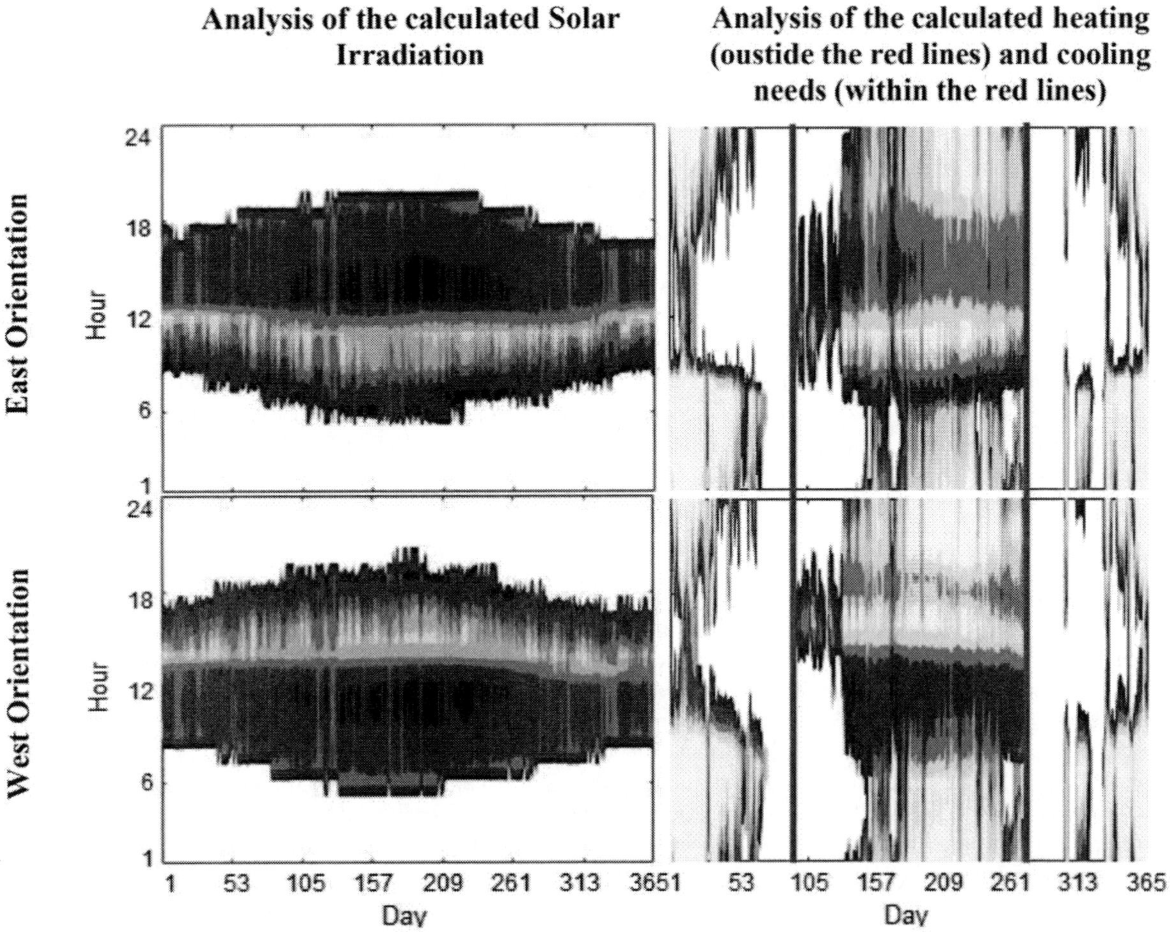

Figure 2: On the left, the density of solar models with normalized deviations lower than 10 % with respect to the medians are reported for Rome east and west orientations. On the right, the variability of the energy need predictions. Irradiation and energy values are categorized into the four performance based on the number of models respecting the 10 % deviation: more than 75 % (very light grey), between 50 % and 75 % (light grey), between 25 % and 50 % (dark grey) or less than 25 % (very dark grey).

3.3 Hourly cooling needs

In the five climates, the set of 72 buildings present different length of the actual cooling season. In Berlin, the fraction of summer hours with positive cooling load is only 10.7 %, as average, with a standard deviation of 9.4 %,

ranging from a minimum of 0.2 % to a maximum of 34 %. As regard the other climates, in Vienna the average percentage of hours with cooling load is 28.9 % ± 11.2 % (ranging from 11.1 % to 55.5 %), in Trento 44.8 % ± 15.6 % (from 16.5 % to 86.7 %), in Rome 62.8 % ± 9.1 % (from 43.8 % to 81.5 %) and in Messina 76.5 % ± 6 % (from 64.5 % to 90.2 %). Lowering the latitude, the cooling needs increase as well as their occurrence during the season. The building configurations with the minimum number of cooling hours are different from location to location but common characteristics can be observed. As expected, a low value of *SHGC* (i.e., 0.351), small windows (i.e., 14.56 m^2) and poorly-insulated opaque component (i.e., 5 cm of polystyrene) help in reducing the frequency of the cooling loads. On the contrary, high *SHGC* (i.e., 0.608), large windows (i.e., 29.12 m^2) and well-insulated opaque component (i.e., 15 cm of polystyrene) lead to higher frequencies. Massive concrete walls are always present in case of maximum number of hours with cooling load while, for the configurations with minimum occurrence, concrete structures are found in colder climates and timber structures in the Mediterranean ones (i.e., Rome and Messina).
Figure 3 reports the cumulative distribution functions of class "A" for the cooling needs of all buildings in the five climates with respect to the normalized time with positive cooling needs. Of all hours with load during the summer season, in Berlin only 11.3 % ± 9.1 %, as average, belongs to the best performing class, ranging from cases with no occurrences in class "A" to a highest frequency of 38.9 %. In Vienna the average is 36.4 % ± 7.9 % (from 20.1 % to 53.5 %) and, in Trento, 27.9 % ± 13.6 % (from 5.3 % to 50.9 %). As regard the Mediterranean localities, Rome has an average of 53.4 % ± 8.6 % (from 35.6 % to 74.5 %) and Messina 68.6 % ± 10.7 % (from 45.2 % to 91.5 %). It can be observed that climates with larger cooling needs have higher frequency of class "A". However, looking at the building cases with the lowest frequencies in class "A", all of them have large glazing with high *SHGC* and timber structures, except for Messina. On the contrary, the occurrence of class "A" hours on the total hours with cooling load is maximized when *SHGC* is low and windows are small; for those cases, it can be seen that in northern localities the structure is in concrete while in the Italian ones it is in timber.
In the right part of Figure 3, the seasonal frequencies of class "A" are reported for each building, distinguished by orientation of the windows. For the simulation in Berlin, the configurations with highest frequency of class "A" have often south-oriented windows, while those with lowest frequencies are often east-oriented, coherently with the findings on solar irradiation estimations. Similarly, in Trento the worst cases have east-oriented windows while the best ones have windows exposed towards south or west. For Rome and Messina, instead, many cases with windows in the southern façade are characterized by lower accuracy on the cooling need estimation even if south is the orientation with the best agreement among the solar models' outputs. It can be concluded that for those climates where specific discrepancies have been identified among the results given by the different solar irradiation models (e.g., Trento) or where the cooling load is not particularly high (e.g., Berlin), a direct correlation between the accuracy of the solar inputs and that of the hourly cooling needs can be identified while for hotter climates (e.g., Rome and Messina) the interaction is more complex and the uncertainty propagation is more altered by the combined effect of the different building features.

3.4 Hourly heating needs

As regards the heating needs, many trends are opposite of those observed in the previous paragraph. As well-known, the actual heating season is longer in Berlin and shorter in Messina. Specifically, in Berlin the fraction of time of the winter season with heating load is 97.2 % ± 3.4 %, as average, and ranges from 84.8 % to 100 %, in Vienna we found an average of 90.6 % ± 6.4 % (from 72.2 % to 99.1 %), in Trento 79.3 % ± 16.2 % (from 38.8 % to 100 %), in Rome 57.8 % ± 7.8 % (from 7.8 % to 82.6 %) and in Messina 27.3 % ± 18.5 % (from 0 % to 63.7 %). The building configurations with the minimum number of heating hours are characterized by a high value of *SHGC*, large windows towards south and well-insulated opaque components, with concrete massive layers in the Italian localities and timber in Berlin and Vienna. Except for Berlin, all cases maximizing the hours with heating needs have low *SHGC*, small windows and poorly-insulated concrete opaque component.
In Figure 4 the cumulative distribution functions of class "A" are represented for the heating needs of all buildings in the five climates. Also this time they were compared to the normalized time with positive load. The majority of hours of the winter season with loads belongs to class "A" for Berlin (92.6 % ± 6.8 %, from 70.5 % to 99.9 %), Vienna (90.5 % ± 7.2 %, from 62.9 % to 98.9 %), Trento (77 % ± 12 %, from 30.2 % to 95.9 %) and Rome (66.5 % ± 21.1 %, from 9.7 % to 94.6 %) and for Messina their percentage is close to half of occurrences as average (46.8 % ± 23 %, from 0 % to 79.2 %). For all localities the same buildings are characterized by either the minimum frequency of class "A" or the maximum one. Respectively, the first is characterized by large south-oriented windows with high *SHGC* and well-insulated concrete walls while the latter has small east-oriented windows and poorly-insulated concrete walls. About the seasonal frequencies of class "A" for heating needs distinguished by orientation of the windows (right side of Figure 4), we can see similar behaviors in all locations except Trento and the best performances are registered for east and west orientation. In Trento, the locality with the most marked

differences of accuracy in the solar irradiation estimation for the different orientations, cases with east-oriented windows appear to be the most robust to the choice of solar model and those with south or west-oriented windowed façades appear more influenced.

4. CONCLUSIONS

In this work, we investigated how the choice of solar irradiation models can affect the reliability of *BES* for the calculation of hourly energy results. Hourly series from different models have been analyzed considering a tolerance level of 10 % of deviation from the hourly median. The fraction of solar models combinations leading to an acceptable deviation of hourly heating and cooling needs was estimated for a set of 72 simplified buildings and five European climates (Berlin, Vienna, Trento, Rome and Messina). We found that:

- Regarding the elaboration of solar irradiation, the best agreement is encountered when the beam irradiation is prevailing (e.g., during the summer season and for south-oriented vertical walls) while uncertainty is increased for irradiation mostly diffuse (e.g., for north-oriented vertical planes during the winter season). Large discrepancies can be detected in case of regional orographic obstacles (e.g., Trento).
- The hourly cooling needs simulated for buildings with small windows and low *SHGC* are less sensitive to the choice of the solar irradiance models, especially in hotter climates. Moreover, the windows' orientation can emphasize or reduce the effects of solar irradiation uncertainty, in particular in those climates where large discrepancies have been identified among the results given by the different solar irradiation models (e.g., Trento) or where the cooling load is not particularly high (e.g., Berlin).
- The hourly heating needs of buildings with large south-oriented windows, high *SHGC* and well-insulated concrete walls are more sensitive to the choice of the solar irradiance models, while those cases with small east-oriented windows and poorly-insulated concrete walls are more robust. Regarding the window's orientation, the best accuracy is registered for east and west in all localities except for in Trento, once again influenced by the local orography.

REFERENCES

Boland, J., Ridley, B., & Brown, B. (2008). Models of diffuse solar radiation. *Renewable Energy*, *33*, 575–584.

Bugler, J.W. (1977). The determination of hourly insolation on an inclined plane using a diffuse irradiance model based on hourly measured global horizontal insolation. *Solar Energy*, *19*(5), 477–491.

Chandrasekaran, J., & Kumar, S. (1994). Hourly diffuse fraction correlation at a tropical location. *Solar Energy*, *53*(6), 505-510.

Chendo, M. A. C., & Maduekwe, A. L. (1994). Hourly global and diffuse radiation of Lagos, Nigeria - correlation with some atmospheric parameters. *Solar Energy*, *52*(3): 247–251.

de Miguel, A., Bilbao, J., Aguiar, R., Kambezidis, H.D., & Negro, E. (2001). Diffuse solar irradiation model evaluation in the North Mediterranean Belt area. *Solar Energy*, *70*(2), 143–153.

Dervishi, S., & Mahdavi, A. (2012). Computing diffuse fraction of global horizontal solar radiation: A model comparison. *Solar Energy*, *86*, 1796–1802.

Ente Nazionale Italiano di Normazione (UNI). (2014). *UNI/TS 11300-1:2014 – Energy performance of buildings Part 1: Evaluation of energy need for space heating and cooling*, Milan, Italy: UNI.

Erbs, D. G., Klein, S. A., & Duffie, J. A. (1982). Estimation of the diffuse radiation fraction for hourly, daily and monthly-average global radiation. *Solar Energy*, *28*(4), 293–302.

European Committee for Standardization (CEN). (2008). *EN ISO 13790:2008 – Energy performance of buildings - Calculation of energy use for space heating and cooling*, Brussels, Belgium: CEN.

Gueymard, C.A. (1986). An anisotropic solar irradiance model for tilted surfaces and its comparison with selected engineering algorithms. *Solar Energy*, *38*(5), 367 – 386.

Hay, J.E., & Davies, J.A. (1980). Calculation of the solar radiation incident on an inclined surface. *Proc. First Canadian Solar Radiation Data Workshop*, 59 – 65.

Hawlader, M.N.A. (1984). Solar diffuse, global and extraterrestrial solar radiation for Singapore. *International Journal of Ambient Energy*, *5*(1), 31 – 37.

Karatasou, S., Santamouris, M., & Geros, V. (2003). Analysis of experimental data on diffuse solar radiation in Athens, Greece, for building applications. *International Journal of Sustainable Energy*, *23*(1), 37–41.

Klucher, T. M. (1979). Evaluation of models to predict insolation on tilted surfaces. *Solar Energy*, *23*(2), 111–114.

Lam, J.C., & Li, D.H.W. (1996). Correlation between global solar radiation and its direct and diffuse components. *Building and Environment*, *31*(6), 527–535.

Liu, B.Y.H., & Jordan, R.C. (1960). The interrelationship and characteristic distribution of direct, diffuse and total solar radiation. *Solar Energy*, *4*(3), 1 – 19.

Ma, C.C.Y., & Iqbal, M. (1983). Statistical comparison of models for estimating solar radiation on inclined surfaces. *Solar Energy*, *31*(3): 313–317.

Maxwell, E.L. (1987). A quasi-physical model for converting hourly global horizontal to direct normal insolation. *Technical Report of Solar Energy Research Institute, SERI/TR-215-3087.*

Muneer, T., Hawas, M., & Sahili, K. (1984). Correlation between hourly diffuse and global radiation for New Delhi. *Energy Conversion and Management*, *24*(4), 265–267.

Muneer, T., & Younes, S. (2006). The all-sky meteorological radiation model: proposed improvements. *Applied Energy*, *83*(5), 436–450.

Oliveira, A., Escobedo, J.F., Machado, J.A., & Soares, J. (2002). Correlation models of diffuse solar-radiation applied to the city of São Paulo, Brazil. *Applied Energy*, *71*(1), 59–73.

Orgill, J.F., & Hollands, K.G.T. (1977). Correlation equation for hourly diffuse radiation on a horizontal surface. *Solar Energy*, *19*(4), 357 – 359.

Perez, R.R., Ineichen, P., & Maxwell, E.L. (1992). Dynamic global-to-direct irradiance conversion models. *ASHRAE Transactions*, *98*(1), 354-369.

Perez, R.R., Ineichen, P., & Seals, R. (1990). Modeling daylight availability and irradiance components from direct and global irradiance. *Solar Energy*, *44*(5), 271–289.

Pernigotto, G., Prada, A., Cóstola, D., Gasparella, A., & Hensen, J.L.M. (2014). Multi-year and reference year weather data for building energy labelling in north Italy climates. *Energy and Buildings*, *72*, 62-72.

Pernigotto, G., Prada, A., Baggio, P., Gasparella, A., & Mahdavi, A. (2015). Impact of solar irradiation models on simulated hourly energy performance of buildings, *BS 2015 - 14th International Conference of the International Building Performance Simulation Association*, Hyderabad, India.

Prada, A., Pernigotto, G., Gasparella, A., & Mahdavi, A. (2014a). Combined effects of diffuse fraction and tilted surface radiation models, *ECPPM 2014 - 10th European Conference on Product & Process Modelling*, Vienna, Austria.

Prada, A., Pernigotto, G., Baggio, P., Gasparella, A., & Mahdavi, A. (2014b). Effect of Solar Radiation Model on the Predicted Energy Performance of Buildings, *III International High Performance Buildings Conference at Purdue*, West Lafayette, Indiana, U.S.

Prada, A., Pernigotto, G., Cappelletti, F., & Gasparella, A. (2015). Impact of solar irradiation models on building refurbishment measures from multi-objective optimization, *BS 2015 - 14th International Conference of the International Building Performance Simulation Association*, Hyderabad, India.

Reindl, D., Beckman, A., & Duffie, J. A. (1990a). Diffuse fraction correlations. *Solar Energy*, *45*(1), 1–7.

Reindl, D. Beckman, A. & Duffie, J. A. (1990b). Evaluation of hourly tilted surface radiation models. *Solar Energy*, *45*(1), 9–17.

Skartveit, A., & Olseth, J.A. (1986). Modelling slope irradiance at high latitudes. *Solar Energy*, *36*(4), 333 – 344.

Skartveit, A., & Olseth, J.A. (1987). A model for the diffuse fraction of hourly global radiation. *Solar Energy*, *38*(4), 271–274.

Soares, J., Oliveira, A., Boznar, M., Mlakar, P., Escobedo, J.F., & Machado, J.A. (2004). Modeling hourly diffuse solar radiation in the city of São Paulo using a neural-network technique. *Applied Energy*, *79*(2), 201–214.

Spencer, J. (1982). A comparison of methods for estimating hourly diffuse solar radiation from global solar radiation. *Solar Energy*, *29*(1), 19–32.

Temps, R.C., & Coulson, K.L. (1977). Solar radiation incident upon slopes of different orientations. *Solar Energy*, *19*(2), 179 – 184.

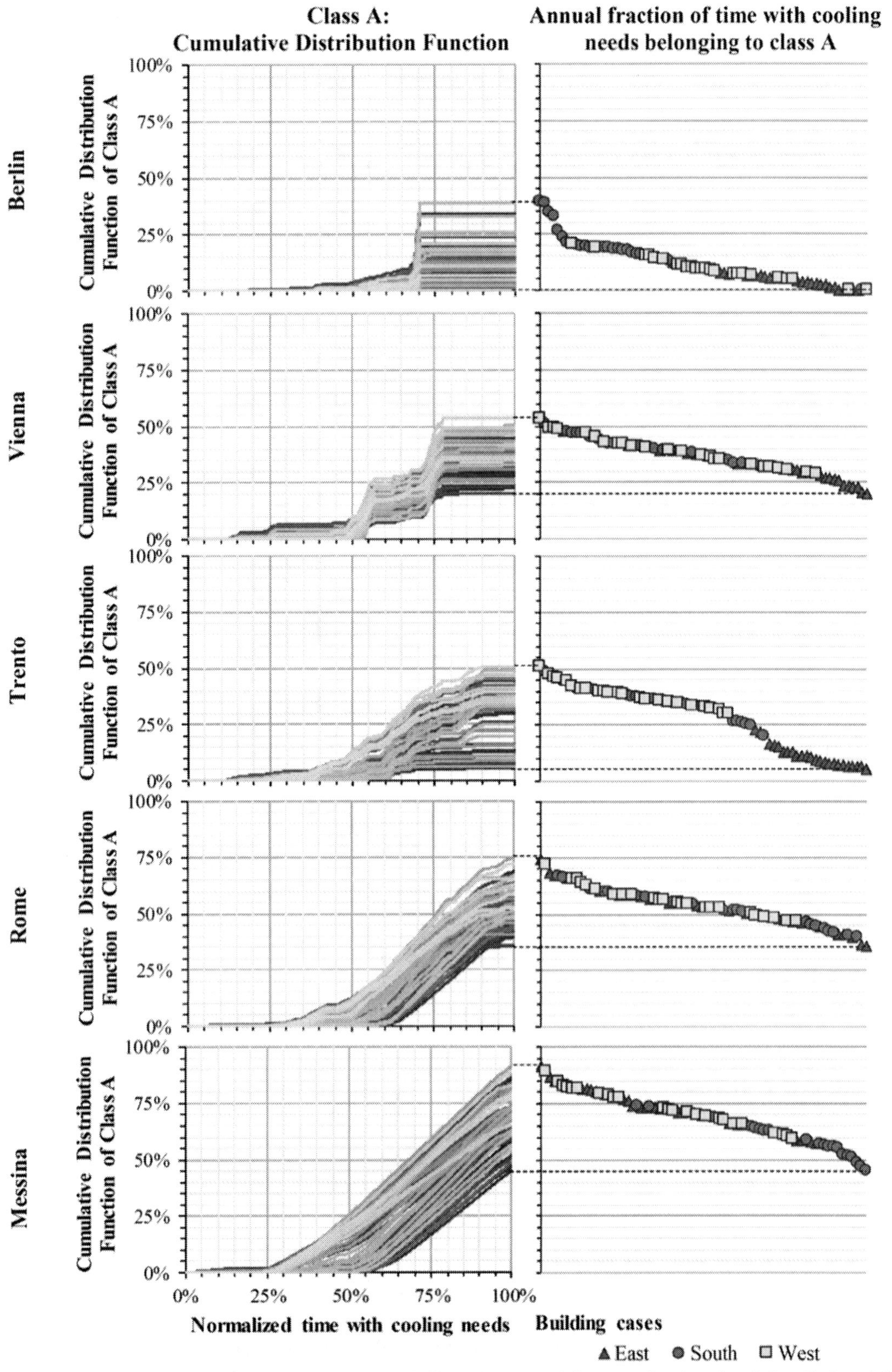

Figure 3: Cumulative distribution functions (on the left) and seasonal frequencies (on the right, distinguished by window orientation) of hours with cooling needs belonging to class "A" with respect to the normalized hours with cooling load during the summer season, for each building configuration and climate.

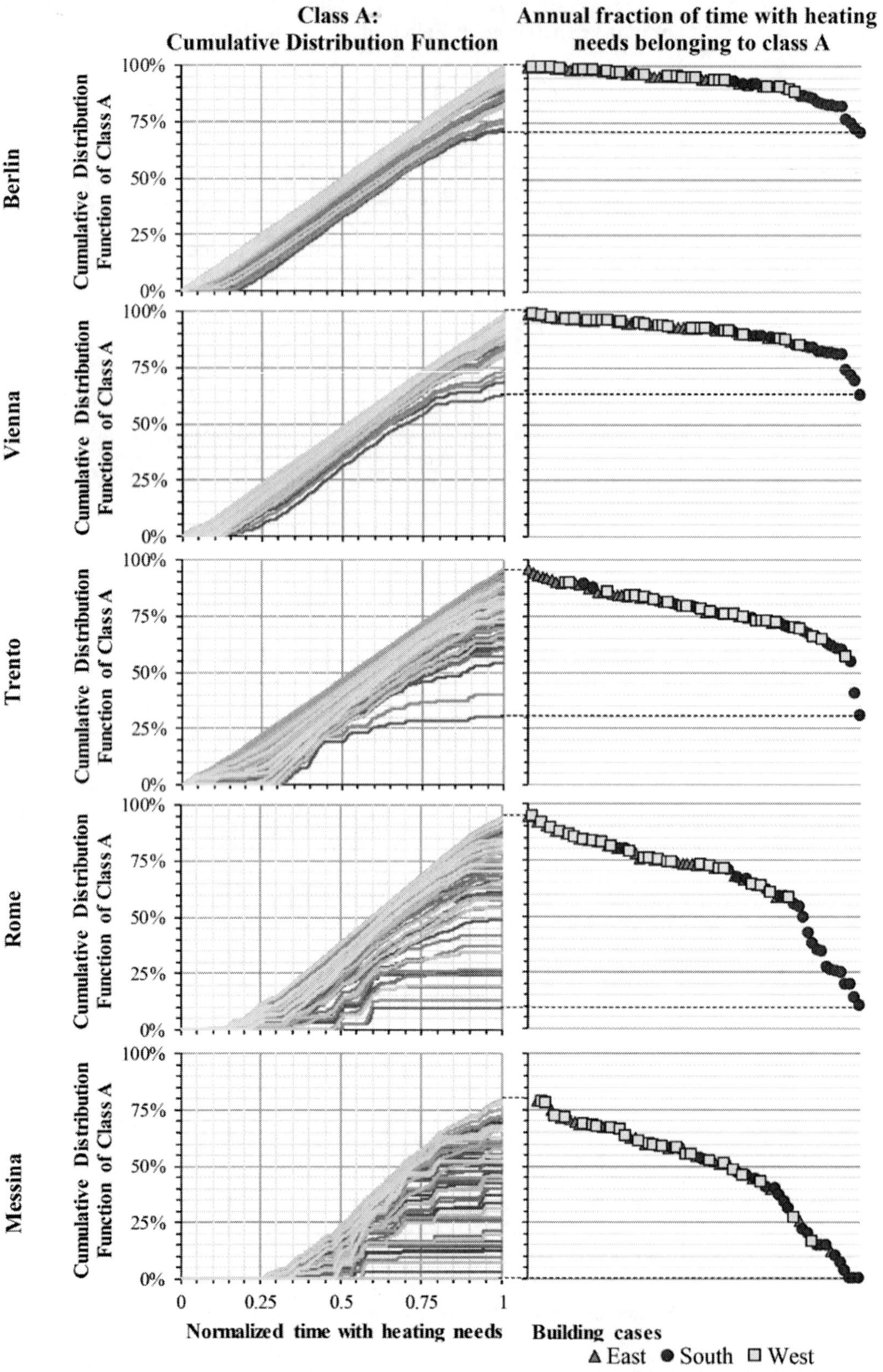

Figure 4: Cumulative distribution functions (on the left) and seasonal frequencies (on the right, distinguished by window orientation) of hours with heating needs belonging to class "A" with respect to the normalized hours with heating load during the winter season, for each building configuration and climate.

On The Representation Of The Thermal And Visual Behavior Of A Roller Shade Material: Comparison Between Different Simulation Models

AnnaMaria ATZERI[1]*, Francesca CAPPELLETTI[2], Andrea GASPARELLA[1], Athanasios TZEMPELIKOS[3]

[1]Free University of Bolzano/Bozen, Faculty of Science and Technology,
Bolzano,Italy
annamaria.atzeri@unibz.it
andrea.gasparella@unibz.it

[2] University IUAV of Venezia, Dpt. of Design and Planning in Complex Environments
Venezia, Italy
francesca.cappelletti@iuav.it

[3]Purdue University School of Civil Engineering and Ray W. Herrick Labs
West Lafayette, Indiana (USA)
ttzempel@purdue.edu

* Corresponding Author

ABSTRACT

Shading systems, if efficiently operated, can improve the internal environmental quality, namely both thermal and visual comfort, and reduce the energy consumption due to cooling needs. Roller shades represent one of the most commonly used types of shading systems, in particular in the tertiary sector. Not only they can be easily installed and maintained, but also they often represent the only design choice when existing buildings are considered. Although roller shades are characterized by a beam-beam and by a beam-diffuse transmittance, both changing according to the incidence angle, as the transmitted solar radiation decreases with the increase of it, they are typically modelled assuming equal reflectance and emissivity for both sides and perfect diffuser behavior, with transmittance and reflectance independent from the solar radiation incidence angle. Neglecting the daily variability of these properties can lead to underestimate their impact on the occupants' comfort conditions.
In this paper, different models for representing the roller shades behavior, embedded in two widely diffuse simulation codes have been compared with a set of measured data, recorded at the Bowen laboratories of the Purdue University (Indiana – USA), combining thermal (Energy Plus) and lighting simulation (Energy Plus or DIVA for Rhino). The thermal properties of the building materials and the internal gains have been calibrated for the thermal simulation, in order to evaluate better the models' capability of predicting the roller shades behavior. Then, starting from the simplest daylighting model, which assumes the roller shades as perfect diffusers, more complex characterizations have been considered and validated through the comparison with the measured data.

1. INTRODUCTION

Different studies showed that a proper operation of shading devices can reduce the energy consumption due to cooling needs and, at the same time, improve internal environmental conditions, related to both thermal and visual comfort. As underlined in Kirimtat et al. (2015), shading devices can be operated to control direct sunlight, limiting solar gains just in the cooling period. Through shading devices, it is also possible to manage the daylight distribution making it as much homogenous as possible, all over the year. At the same time, considering that thermal and visual requisites can be contrasting, the solar shading devices effects on the global building performance have to be analyzed since the early design stages, in order to define the best trade off and optimize the control strategy.
As underlined in (ANSI/ASHRAE Standard 55-2013), fenestrations with shading devices have a degree of thermal and optical complexity by far greater than that of unshaded fenestrations. Currently, different types of shading devices exist: overhangs, roller shades, venetian blinds, etc. Among them, roller shades, regardless of their internal or external installation, represent one of the shading devices most commonly used in buildings, in particular in the tertiary sector. Not only they can be easily installed and maintained, but also they often represent the only design choice when existing buildings are considered. Kirimtat et al. (2015) pointed out that roller shades constitute the

third most common type of shading devices studied in literature. Although accurate simulation approaches about their visible transmission, absorption and reflection would be highly necessary to really maximize the positive effects that they could provide, common agreement about the best way of simulating them, both from a thermal and visual point of view, still lacks.
In most of the studies roller shades are modelled as perfect diffusers (Kapsis et al. 2010; Bessoudo et al. 2010; Tzempelikos et al. 2010; Shen & Tzempelikos 2011 and 2012; Tzempelikos & Shen 2013; Yao 2013 and 2014; Ye et al. 2015), even if this can lead to inaccurate evaluations of the illuminance in certain positions (Kapsis et al. 2010; Tzempelikos & Shen 2013). Often the WindowMaterial:Shade EnergyPlus model has been used, but in some cases integrated thermal and lighting simulation codes have been expressly written. Appelfeld et al. (2012) and Hoffmann et al. (2015) simulated complex fenestration systems through Radiance to generate a bi-directional scattering distribution function (BSDF), while Chan et al. (2015) applied the Kotey's model (Kotey, Wright, & Collins, 2009), in order to correct shades' solar-optical properties and to evaluate their variation with the incidence angle.
In this paper, different models for representing the roller shades behavior, embedded in two widely diffuse simulation codes have been compared with a set of measured data, recorded at the Bowen laboratories of the Purdue University (Indiana – USA), combining thermal (Energy Plus) and lighting simulation (Energy Plus or DIVA for Rhino). The thermal properties of the building materials and the internal gains have been calibrated for the thermal simulation, in order to evaluate better the models' capability of predicting the roller shades behavior. Then the WindowMaterial:Shade and WindowMaterial:Shade:EquivalentLayer models of EnergyPlus, and the *trans* and *transdata* Radiance models embedded in DIVA, have been considered.

2. SIMULATING ROLLER SHADES MATERIAL – STATE OF THE ART

As specified in Kotey et al. (2009), roller shades consist of strands of yarn that may be woven loosely, leaving open areas, or woven tightly, with no open areas. Their specific composition operates in such a way that the incident direct solar radiation is split in two components: a portion directly transmitted through the openings, and a scattered portion that, regardless from the fact that it can be transmitted or reflected, is considered as purely diffuse. Nevertheless, the major part of the regulations or simulation codes dealing with the calculation of complex fenestration system adopting roller shades does not take into account this aspect.
The ANSI/ASHRAE Standard 55 (2013) proposes a simplified approach for calculating the solar shading system effects on thermal exchange introducing the indoor solar attenuation coefficient (IAC). It represents the fraction of heat flow, direct and diffuse, that enters the room, considering the effects of the shades. In other words, it represents the ratio of the solar heat gain coefficients (SHGC) of the glazing system, considering and not considering the shades. IAC values have been determined using the ASHWAT models (Wright J.L., 2008), and can be used in order to obtain the shaded SHGC. As concerns roller shades, the Standard assumes the IAC value is independent of incident angle of irradiation, even if it underlines that generally shades are able to both transmit and diffuse solar radiation, and, for this reason, more complex models would be needed.
As regards Europe, the reference standards are EN 13363-1:2007 (CEN, 2007) and the EN 13363-2:2005 (CEN, 2005). Both assume only the beam solar radiation coming from the sun, neglecting the sky vault's component, and a totally diffuse transmission through the shading systems (Zinzi, Agnoli, & Fasano, 2014), and can be applied to all types of solar protection devices parallel to the glazing (louvre, venetian or roller blinds), wherever installed (externally, inside glazings or internally). EN 13363-1:2007 (CEN, 2007) describes a simplified method to estimate the total solar energy transmittance of a solar protection device combined with glazing, suggested for solar energy transmittance of the glazing between 0.15 and 0.85, solar transmittance and solar reflectance of shades are within 0 and 0.5, and 0.1 and 0.8 respectively. The simplified method does not take into account either the angular dependence or the differences of spectral distribution. EN 13363:2-2005 (CEN, 2005)specifies a detailed method, based on the spectral transmission data of solar protection devices and glazings. Again, the angular dependence of transmittance or reflectance of the materials is neglected.
As concerns the simulation codes, EnergyPlus offers three different models for roller shades. The WindowMaterial:Shade model (named in the following as model A) assumes that the transmission, absorption and reflection are independent of the incidence angle. Moreover, reflectance and emissivity properties are assumed to be the same on both sides of the shades. In contrast, the WindowMaterial:Shade:EquivalentLayer model (model B, in the following) is able to consider that roller blinds can have also a beam-beam transmittance. This is assumed to be the same for both sides of the shade and equal to the openness area fraction (OF). Beam-diffuse transmittance and reflectance, and emissivity can be different for front and back side of the shade. The off-normal solar and optical property calculation of shades is based on a set of correlations developed from measurement of samples of commercially produced roller blind material with openness fraction less than 0.14 (Kotey et al., 2009). The model is

not intended for materials with unusually high values of openness and should be limited to a maximum openness fraction of 0.20. Unfortunately, at the moment Energy Plus implementation of model B does not provide daylight calculation, due to the software limitations, although the method itself has been proven to work for daylighting simulation (Chen et al. 2015). Finally, the WindowMaterial:ComplexShade can be used for modeling shades whose properties are represented by a BSDFs file, which contains the optical properties of the Complex Fenestration layers. Generally, the optical properties are provided as a two-dimensional matrix describing the basis and four two-dimensional matrices of system bidirectional optical properties. All these objects are commonly exported directly from LBNL WINDOW program.

Also Radiance provides different options for simulating roller shades, each one characterized by a different level of complexity (Chan et al., 2015). The simplest one, called *trans* (model C, in the following), is able to trace direct source rays through a semi-specular surface in order to determine the diffuse and specular transmitted components (Larson & Shakespeare, 1998). However, the *trans* model does not account for the incidence angle.

Radiance provides another model able to adjust the direct or the diffuse part of the transmitted component according to the incidence angle: the *transdata* or *transfunction* model (model D), validated for a translucent glass with diffuse characteristics (Reinhart & Andersen, 2006). According to Apian-Bennewitz (2013) the *transdata* or *transfunction* model represents the more precise model to simulate non-redirecting and forward scattering materials when BSDFs data are not available. Even if BSDFs data should be available, the geometrical radiosity method used in WINDOW does not show good agreement between simulated and measured data (Chan et al., 2015), while the *genBSDF* Radiance function, with which these data could be generated, performs well for micro-perforated shading system, but it has not been validated yet for open-weave shading fabrics.

In this paper, models_A, B, C and D have been considered and compared in order to investigate their reliability and accuracy in the evaluation of radiation and daylighting aspects in combined thermal and lighting simulation. Model A and B are considered for thermal simulation, while model A, C and D for daylighting. The Energy Plus and Radiance models based on BSDFs such as the WindowMaterial:ComplexShade have not been considered since the data needed to build the model were not available for the roller shades used in the Bowen Laboratories (Chan et al., 2015).

3. METHODS

3.1 Experimental setup

The measured data used for the models calibration and validation have been recorded throughout five days in June 2015, from the 2nd to the 8th, with a one minute measurement time-step. During this period, the roller shades have been maintained continuously closed in one of the two laboratories (named in the following as LAB1) and continuously open in the other (named in the following as LAB2). Both the heating, cooling and lighting systems have been kept switched off. Only the acquisition system, composed of a laptop and a data-logger for each room, has been maintained switched on. Regarding the external environment the global and diffuse vertical solar radiation, the global vertical illuminance, the global horizontal solar radiation and illuminance and the air temperature have been measured. The transmitted illuminance and solar radiation just after the window's pane, the air and surface temperature, the work-plane illuminance and solar radiation and the vertical illuminance have been measured in different points in the rooms (Figure 1).

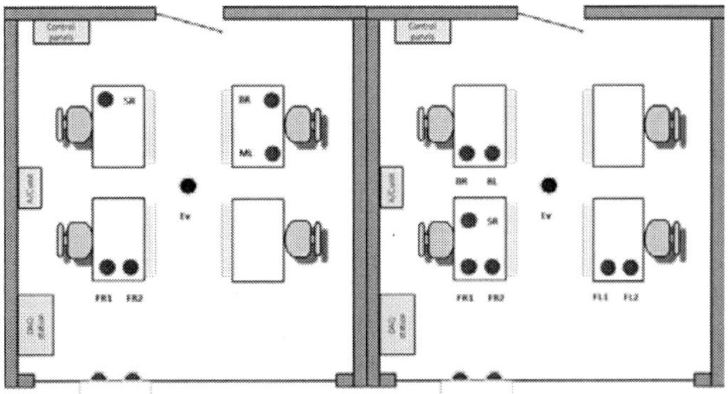

Figure 1: sensors' location in LAB1 (right) and LAB2 (left): transmitted or work-plane illuminance = red dots; vertical illuminance = black dots; transmitted or work-plane solar radiation = blue dots

3.2 Thermal model calibration and validation

Thermal simulation has been conducted through an EnergyPlus model. The external air temperature and the global horizontal solar radiation recorded during the measurements period have been used as input. A first simulation has been run for the LAB2 with open shades, starting from the nominal properties for opaque and transparent components and internal gains, as from technical documentation (ASHRAE HOF 2005) or data sheets by the manufacturers. The first day (June 2nd) has been used as a simulation warm up period, considering the internal air temperature as setpoint with and ideal HVAC. In order to improve the accuracy of the thermal model, a calibration process has been carried out by means of jEplus+EA, minimizing the differences between the internal air temperature measured and simulated in LAB2, considering the Coefficient of Variation of the Root Mean Square Error CV(RMSE) and the Normal Mean Bias Error NMBE. Only 3rd, 5th and 7th of June have been considered for calibration, while June 4th, 6th and 8th have been used for validation. Four parameters have been calibrated, as reported in Table 1, considering a variation range of approximately +/-20% across the nominal value, with the exception of the internal gains.

Table 1: Calibration input parameters

Input parameters	Material	Nominal value	Range value (*levels)	Unit measure
Specific heat	Concrete	840	672-1000	$J\ kg^{-1}\ K^{-1}$
	Acoustic Tile	590	472-708	
	Gypsum	1090	872-300	
Infiltration airchanges		0.1	0.05-0.20	ACH
Thermal conductivity	Concrete	0.53	0.42-0.64	$W\ m^{-1}\ K^{-1}$
	Acoustic Tile	0.06	0.048-0.072	
	Gypsum	0.16	0.13-0.19	
Internal gains (electric equipment)		36	40;50;60*	W

3.3 Roller shades modeling

The properties of the roller shades to be provided to the different models have been derived from measurements with an integrated sphere device, as specified in Chan et al. (2015), and reported in Table 2.

Model A (Table 3), as already underlined, treats the fabric material as a perfect diffuser, and it does not allow distinguishing between the internal or external shade's sides. In this case the solar and visible transmittance have been set equal to the total value, direct plus diffuse, in order to consider all the radiation passing through the material regardless from the way in which it has been transmitted.

On the contrary, model B allows separating between the direct and diffuse components of both the transmitted and reflected part of the solar radiation, even if this is not possible yet for the visible spectrum. The model is not available to consider visible radiation, so it can be used only for the thermal simulation.

The parameters for model C and model D have been calculated using two different approaches. In one case (C_1 and D_1), according to Kotey et al. (2009), the specular component of the reflected solar radiation has been considered null. However, imposing this parameter to zero it is impossible to consider the color characteristics of the shades consistently, so we tried to include (Model C_2 and D_2) in this value aspects related to locally specular reflection, which, randomly distributed because of the disposition of the yarns, accordingly to the balance equations as described in Jacobs (2014).

Moreover, model D allows changing the direct-direct component of the transmitted radiation according to the incidence angle (Reinhart & Andersen, 2006). In this work 17 off-normal values have used, obtained from the relations suggested in Kotey et al. (2009), for the specific correction factor (cf):

$$cf = \frac{\frac{\tau_{b-b}}{180^\circ}}{\tau_{b-tot}} \tag{1}$$

where $\tau_{b\text{-}b}$ represents the direct part of the total radiation transmitted through the material and $\tau_{b\text{-}tot}$ the total amount (direct plus diffuse) for each incidence angle.

Table 2: Roller shade solar and optical properties (SS-EC02-4 dark color silver screen). Values in %

	Sol	Opt	Sol	Opt	Sol	Opt	Sol	Opt	Sol	Opt	
	Normal		15°		30°		45°		60°		Side
Beam-Total Transmission	5.0	5.0	5.1	4.9	5.1	5.0	4.3	4.1	2.9	2.7	
Beam-Beam Transmission	4.2	4.2	3.8	3.1	3.1	3.0	1.7	1.7	0.6	0.6	
Beam-Diffuse Transmission	0.8	0.8	1.3	1.0	2.0	2.0	2.6	2.4	2.3	2.1	
Beam-Total Reflectance	74.5	72.3	74.0	71.7	74.9	72.2	74.8	72.4	75.9	74.2	ext
	28.3	28.3									int

Table 3: Model A simulation values

Quantity	Value
Solar Transmittance	0.050
Solar Reflectance	0.745
Visible Transmittance	0.050
Visible Reflectance	0.723

Table 4: Model B simulation values

Quantity	Value
Shade Beam-Beam Solar Transmittance	0.042
Front Side Shade Beam-Diffuse Solar Transmittance	0.008
Back Side Shade Beam-Diffuse Solar Transmittance	0.008
Front Side Shade Beam-Diffuse Solar Reflectance	0.723
Back Side Shade Beam-Diffuse Solar Reflectance	0.283

4. RESULTS AND DISCUSSION

Three main aspects of the modelling performance have been assessed, namely temperatures as regards thermal simulation (Fig. 2), work-plane (Fig. 3) and vertical eye illuminance (Fig. 4) as for the lighting. Table 5 illustrates the models and simulation codes validated against the experimental data.
As regards thermal simulation, it is quite clear that:

1. Calibration improves the performance of the model (Fig. 2b).
2. When adding shades (Fig. 2c) the performance keeps quite good, even if the Model B seems to overestimate the temperature profile.
3. As for the entering radiation behind the glazing surface without shades (Fig. 2d), both models A and B seems to provide a good response, considering that their performance is just analyzing the glazings.

Regarding the work-plane illuminance, for space reasons, only the results obtained for the FR1 sensor located 1.5 meter from the window has been reported in the figures. The charts reported in the following underline some aspects:

Table 5: Measured quantities versus simulation output comparison

Validation aspect	Shade's model	Model	Software
Internal air temperature and transmitted solar radiation (Fig. 2)	WindowMaterial:Shade	A	Energy Plus
	WindowMaterial:Shade:EquivalentLayer	B	Energy Plus
Work-plane illuminance (Fig. 3)	WindowMaterial:Shade	A	Energy Plus
	Trans	C_1 C_2	DIVA
	Transdata	D_1 D_2	DIVA
Vertical eye illuminance (Fig. 4)	*Trans*	C_1 C_2	DIVA
	Transdata	D_1 D_2	DIVA

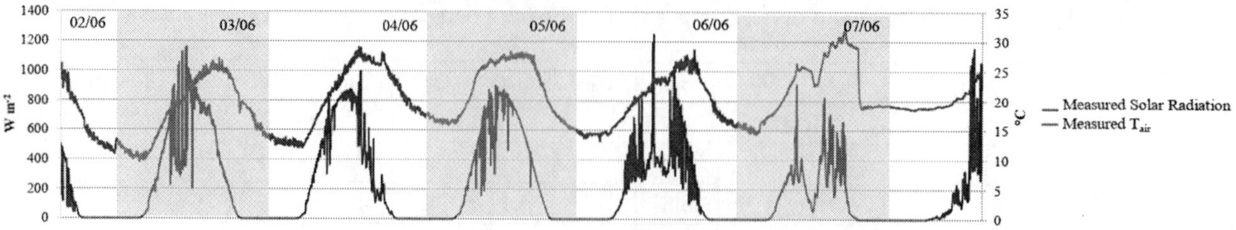

a: Measured global horizontal solar radiation and external air temperature

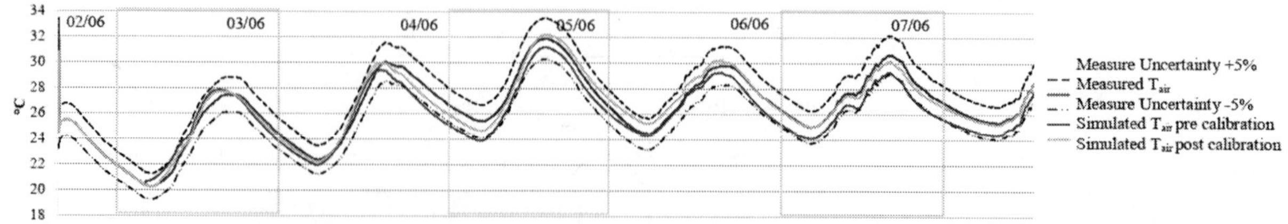

b: LAB2 measured and simulated, pre and post calibration, internal air temperature

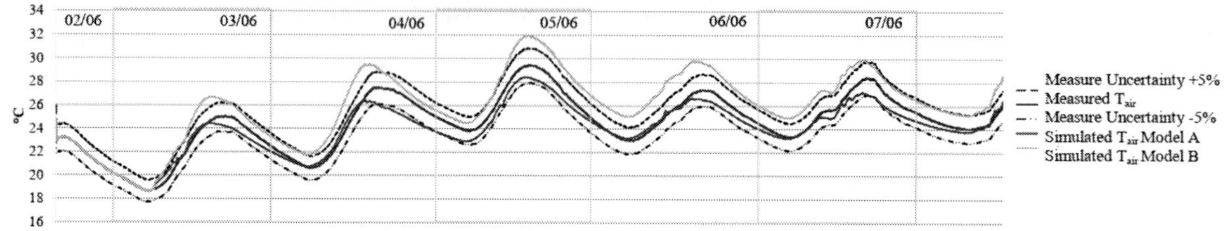

c: LAB1 measured and simulated internal air temperature Model A and Model B

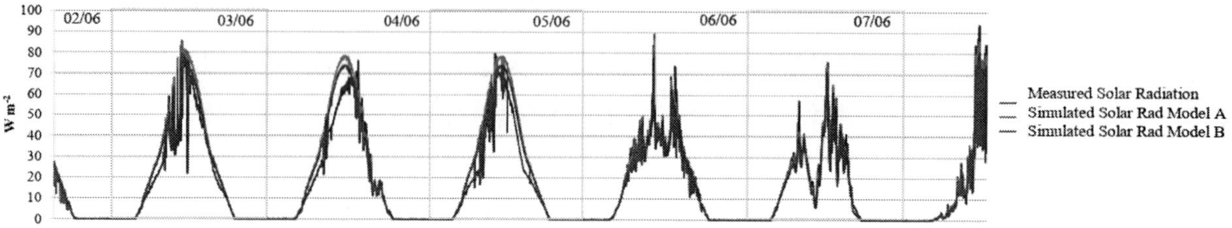

d: LAB2 simulated and measured transmitted solar radiation Model A and Model B

Figure 2: Thermal simulation validation

1. A poor agreement between simulated and measured work-plane illuminance values using Model A (Fig. 3a), Model C_1 (Fig. 3b) or Model D_1 (Fig. 3c) ($\rho_{b\text{-}b}=0$) when the direct-direct component prevails. In clear sky conditions, this fact can determine:
   a. an overestimation of the available natural light
   b. a wrong evaluation of glare occurrence
   c. a possible underestimation of artificial light consumptions if a lighting system dimmed according to the natural light availability is considered.
2. A good agreement between simulated and measured work-plane illuminance values using Model C_2 (fig. 3d) and Model D_2 (fig. 3e) ($\rho_{b\text{-}b}\neq0$) when the direct-direct component prevails.
3. A good agreement between simulated and measured work-plane illuminance values when the direct-diffuse component prevails.
4. A good agreement between simulated and measured vertical eye illuminance using Model D_2 (fig. 4d).
5. Also for the other positions in the room, the same trends can be underlined. As expected, increasing the distance between the sensor and the window, as happens considering the ML and the BR sensors located respectively 2.5 and 3.7 meters from the daylight source, the discrepancy between measured and simulated trend decreases. Actually, these points, especially the BR sensor, are located far enough not to be affected by the direct-direct component of the visible radiation, which is less efficiently calculated by the shade's models applied.

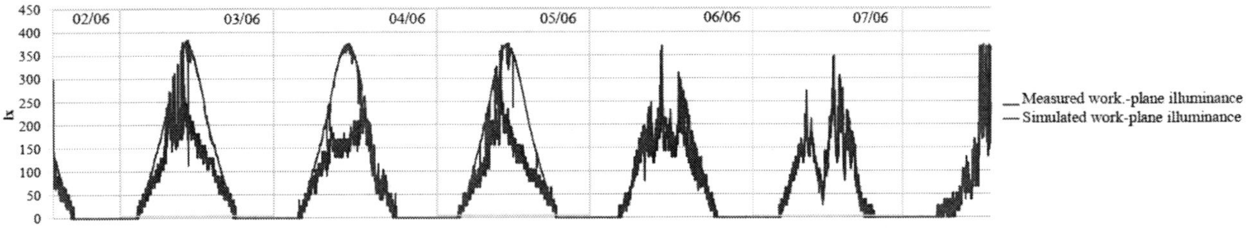

a: LAB1 measured and simulated work-plane illuminance for sensor FR1 – Model A

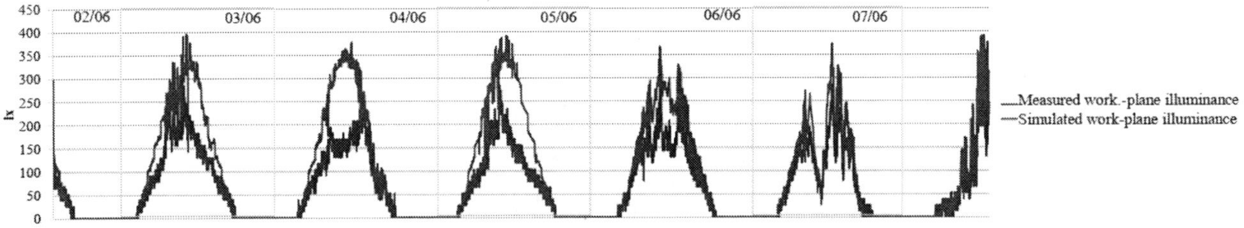

b: LAB1 measured and simulated work-plane illuminance for sensor FR1 – Model C_1

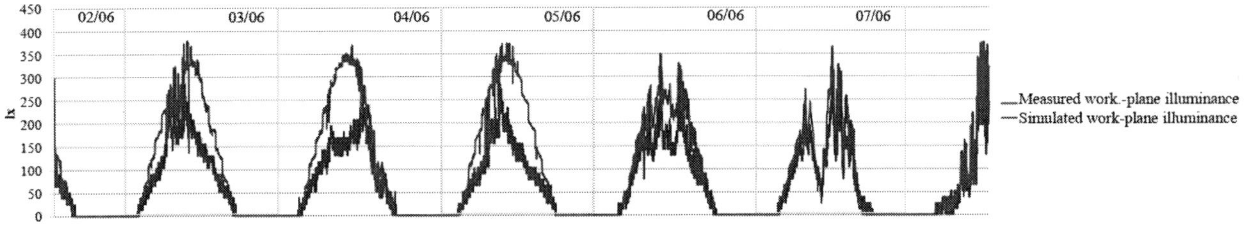

c: LAB1 measured and simulated work-plane illuminance for sensor FR1 – Model D_1

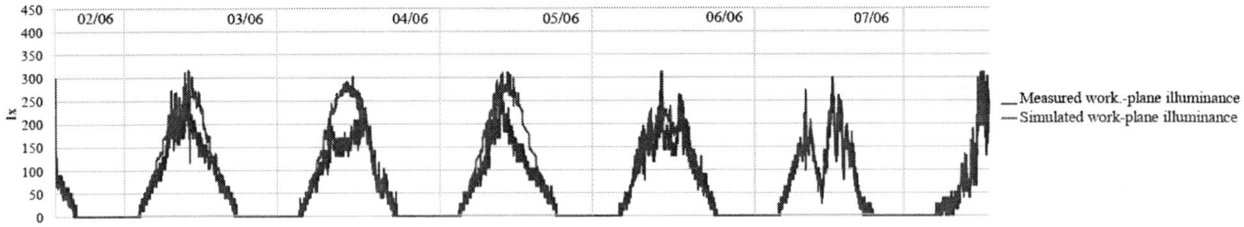

d: LAB1 measured and simulated work-plane illuminance for sensor FR1 – Model C_2

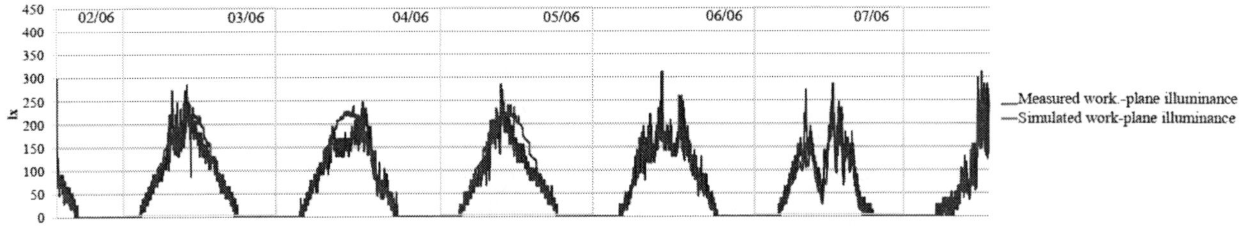

e: LAB1 measured and simulated work-plane illuminance for sensor FR1 – Model D_2

Figure 3: Lighting simulation validation – work-plane illuminance

Summarizing, it is possible to affirm that when a Lambertian diffuser material is used for simulating roller shades influence on visible light, the illuminance in the confined space is always overestimated when the direct-direct component prevails. At the same time, if overcast sky conditions or points enough far from the window are considered, even a simplified model which does not distinguish between the direct-direct and the direct-diffuse component of the solar radiation can provide reliable results.

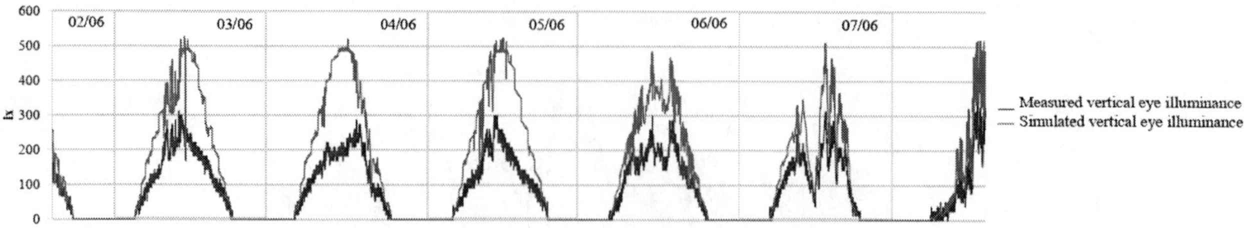

a: LAB1 measured and simulated vertical eye illuminance for sensor Ev – Model C_1

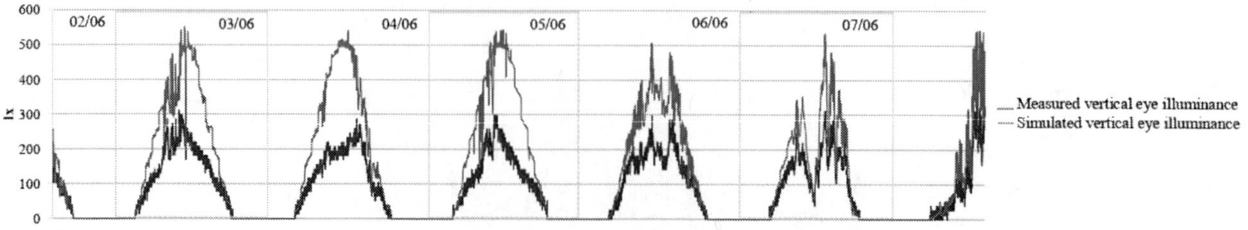

b: LAB1 measured and simulated vertical eye illuminance for sensor Ev – Model D_1

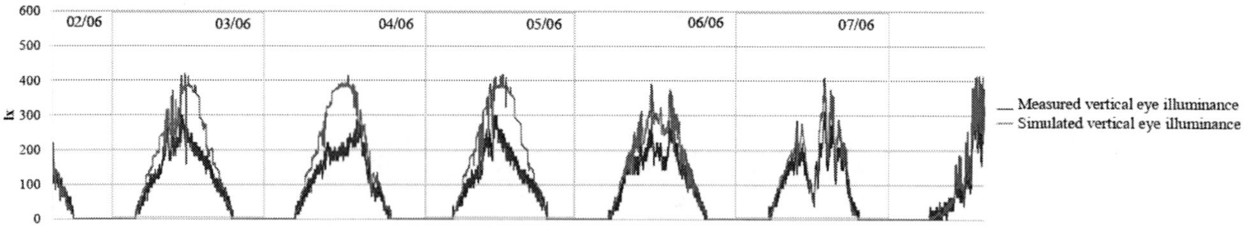

c: LAB1 measured and simulated vertical eye illuminance for sensor Ev - Model C_2

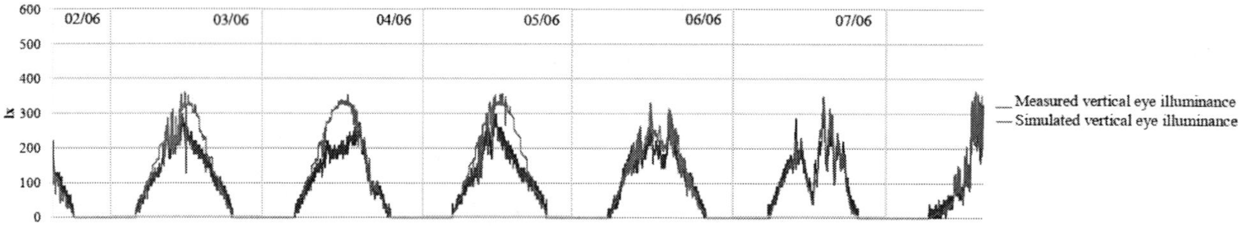

d: LAB1 measured and simulated vertical eye illuminance for sensor Ev – Model D_2

Figure 4: Lighting simulation validation – Vertical eye illuminance

6. CONCLUSION

Roller shades can have several positive effects on the building's global performance, reducing the energy consumption and improving the comfort conditions. However, in order to maximize their contribution it is necessary to consider realistically their effects since the design phase. At the moment, an agreement does not exist about which should be considered the best modelling approach.

In this paper, different models for representing the roller shades behavior, embedded in two widely diffuse simulation codes, have been compared with a set of measured data, recorded at the Bowen laboratories at Purdue University (Indiana – USA), coupling thermal (Energy Plus) and lighting simulation (Energy Plus or DIVA for Rhino). The thermal properties of the building materials and the internal gains have been calibrated for the thermal simulation, in order to evaluate better the models' capability of predicting the roller shades behavior and their contribution to the thermal balance. Then, starting from the simplest daylighting model (Model A), which assumes the roller shades as perfect diffusers, more complex characterizations have been considered and validated through the comparison with the measured data.

The results coming from the thermal simulation underlines that adding the shades' model the simulation performance keeps quite good, even if the Model B seems to overestimate the temperature profile, underestimating consequently the shades ability to control solar gains with respect to Model A. This inaccuracy could be particularly critic when the aim of the analysis consists in assessing the internal thermal comfort conditions and/or calculating the heating or cooling consumptions.
On the contrary, Chan et al. (2015) demonstrated that, thanks to the calculation approach suggested in Kotey et al. (2009) and used in the Model B just for thermal simulation, it is possible to model accurately the visual performance of a roller shading system. Further analysis is then necessary in order to understand the model overall potential.
The analysis related to the optical behavior of the roller shades, has underlined, once again, the strong correlation between the solar radiation incidence angle and the amount of visible light able to pass through the fabric material. From this point of view, the *trans* or *transdata* model (C and D) showed a good agreement between simulated and measured trends, but only considering the specular reflected component different from zero. This aspect appears even clearer considering the vertical eye illuminance trends. In this case, the Model D_2 shows the best behavior, while all the other models tend to overestimate the daylight contribution. Assuming the hypothesis that the specular reflection component is different from zero, apparently diverges from what has been established through the measurement campaign described in Kotey et al. (2009), and suggests that more measured data coming from different roller shades and from different periods of time, both in terms of duration and season of the year, are needed.
Once identified weaknesses and strengths of the different lighting models and validating them against experimental data, a further step in the work is evaluating their influence in assessing comfort and energy aspects, with the aim of understanding to which extent a more sophisticated model can improve the design decisions.

REFERENCES

ANSI/ASHRAE Standard 55. 2013 ASHRAE Handbook -Fundamentals - Chapter 15 Fenestration (2013). http://doi.org/10.1163/ej.9789004155947.i-937.23

Apian-Bennewitz, P. (2013). Review of simulating four classes of window materials for daylighting with non-standard BSDF using the simulation program Radiance, 1–24. Retrieved from http://arxiv.org/abs/1307.4214

Appelfeld, D., McNeil, A., & Svendsen, S. (2012). An hourly based performance comparison of an integrated micro-structural perforated shading screen with standard shading systems. *Energy and Buildings*, *50*, 166–176. http://doi.org/10.1016/j.enbuild.2012.03.038

Bessoudo, M., Tzempelikos, A., Athienitis, A. K., & Zmeureanu, R. (2010). Indoor thermal environmental conditions near glazed facades with shading devices – Part I: Experiments and building thermal model. *Building and Environment*, *45*(11), 2506–2516. http://doi.org/10.1016/j.buildenv.2010.05.013

Chan, Y.-C., Tzempelikos, A., & Konstantzos, I. (2015). A systematic method for selecting roller shade properties for glare protection. *Energy and Buildings*, *92*, 81–94. http://doi.org/10.1016/j.enbuild.2015.01.057

European Committee for Standardization (CEN). EN 13363-2: 2005, Solar protection devices combined with glazing. Calculation of total solar energy transmittance and light transmittance. Part 2: Deteiled calculation method (2005). Brussels, Belgium.

European Committee for Standardization (CEN). EN 13363-1:2007, Solar protection devices combined with glazing. Calculation of total solar energy transmittance and light transmittance. Part1: simplified method (2007).

Hoffmann, S., Lee, E. S., McNeil, A., Fernandes, L., Vidanovic, D., & Thanachareonkit, A. (2015). Balancing daylight, glare, and energy-efficiency goals: An evaluation of exterior coplanar shading systems using complex fenestration modeling tools. *Energy and Buildings*, *112*, 279–298. http://doi.org/10.1016/j.enbuild.2015.12.009

Jacobs, A. (2014). *RADIANCE Cookbook*. Retrieved from http://www.jaloxa.eu/resources/radiance/documentation/docs/radiance_cookbook.pdf

Kapsis, K., Tzempelikos, A., Athienitis, A. K., & Zmeureanu, R. G. (2010). Daylighting performance evaluation of a bottom-up motorized roller shade. *Solar Energy*, *84*(12), 2120–2131. http://doi.org/10.1016/j.solener.2010.09.004

Kirimtat, A., Koyunbaba, B. K., Chatzikonstantinou, I., & Sariyildiz, S. (2015). Review of simulation modeling for shading devices in buildings. *Renewable and Sustainable Energy Reviews*, *53*, 23–49. http://doi.org/10.1016/j.rser.2015.08.020

Kotey, N. A., Wright, J. L., & Collins, M. R. (2009). Determing off-normal solar optical properties of roller blinds.

ASHRAE Transactions, *115*(1), 10.
Larson, G. W., & Shakespeare, R. A. (1998). *Rendering with Radiance - The Art and Science of Lighting Visualization*. Retrieved from http://radsite.lbl.gov/radiance/book/
Reinhart, C. F., & Andersen, M. (2006). Development and validation of a Radiance model for a translucent panel. *Energy and Buildings*, *38*(7), 890–904. http://doi.org/10.1016/j.enbuild.2006.03.006
Shen, H., & Tzempelikos, A. (2011). Daylighting and energy analysis of private offices with automated interior roller shades. *Solar Energy*, *86*(2), 681–704. http://doi.org/10.1016/j.solener.2011.11.016
Shen, H., & Tzempelikos, A. (2012). Sensitivity analysis on daylighting and energy performance of perimeter offices with automated shading. *Building and Environment*, *59*, 303–314. http://doi.org/10.1016/j.buildenv.2012.08.028
Tzempelikos, A., Bessoudo, M., Athienitis, A. K., & Zmeureanu, R. (2010). Indoor thermal environmental conditions near glazed facades with shading devices – Part II: Thermal comfort simulation and impact of glazing and shading properties. *Building and Environment*, *45*(11), 2517–2525. http://doi.org/10.1016/j.buildenv.2010.05.014
Tzempelikos, A., & Shen, H. (2013). Comparative control strategies for roller shades with respect to daylighting and energy performance. *Building and Environment*, *67*, 179–192. http://doi.org/10.1016/j.buildenv.2013.05.016
Wright J.L. (2008). Calculating Centre-Glass Performance Indices of Glazing Systems with Shading Devices. *ASHRAE Transactions*, *114 (2)*, 199–209. http://doi.org/10.1017/CBO9781107415324.004
Yao, J. (2013). An investigation into the impact of movable solar shades on energy, indoor thermal and visual comfort improvements. *Building and Environment*, *71*, 24–32. http://doi.org/10.1016/j.buildenv.2013.09.011
Yao, J. (2014). Determining the energy performance of manually controlled solar shades: A stochastic model based co-simulation analysis. *Applied Energy*, *127*, 64–80. http://doi.org/10.1016/j.apenergy.2014.04.046
Ye, Y., Xu, P., Mao, J., & Ji, Y. (2015). Experimental study on the effectiveness of internal shading devices. *Energy and Buildings*, *111*, 154–163. http://doi.org/10.1016/j.enbuild.2015.11.040
Zinzi, M., Agnoli, S., & Fasano, G. (2014). *Comparazione tra standard e strumenti di calcolo per le prestazioni solari e luminose di componenti vetrati con accoppiamento di schermature solari.*

The Effect of the Configuration of the Absorber on the Performance of Flat Plate Thermal Collector

Moyu YAN[1], Ming QU[1]*, Steve PENG[2]

[1]Purdue University, Department of Civil Engineering,
West Lafayette, IN, United States of America
E-mail: yan170@purdue.edu, mqu@purdue.edu

[2] California State University, College of Business and Economics,
East Bay, CA, United States of America
E-mail:steve@ziretech.com

* Corresponding Author

ABSTRACT

In this study, a numerical thermal analysis for a new designed flat plate thermal collector was conducted through modeling. The new flat plate thermal collector has ellipse shaped tubes inside a wavy shaped absorber, which is made of stainless steel. For the comparison, the conventional flat plate thermal collector with circular copper tubes served as a base case was also modeled. Hottel-Whillier equations were utilized to formulate thermal networks for both models developed in Engineering Equation Solver. According to simulation results, for a given solar radiance, the thermal efficiency of both solar collectors decreases along the increase of the operating temperature of the fluid. Given a same operating temperature of fluid, higher solar radiances lead to higher efficiency of both thermal collectors. Comparably, the new flat plate thermal collector generally obtains a 10% higher efficiency than the conventional one at different solar radiance level and average operation temperature of towing to its new design of the absorber. This concludes that the new flat plate solar collectors would be a better option for typical solar domestic hot water system.

1. INTRODUCTION

Conventional flat plate thermal collector is widely utilized in domestic hot water system as it provides reliable and satisfactory performance with a low price. However, the efficiency of flat plate thermal collector can be promoted by several methods and were validated by experimental data and simulation results from Mansour and Khamis's work (2013). Preceding studies explored the underlying pattern of the design of flat plate thermal collector in terms of configuration of absorber, tube, and bond and its associated impacts. The design of irregular shaped absorber is a tradeoff as in Dović *et al.* (2012) and Varol *et al.* (2008)'s research, heat transfer between fluid and absorber was stimulated as corrugated and wavy shaped absorber provided larger contact area, but more thermal loss from absorber to ambient air was also observed due to the fact that the shape of absorber enlarged convective and radiative heat transfer between absorber and cover. Still, a conclusion can be drawn from Karim *et al.* (2006) and Karsli and Suleyman (2007)'s research that a well-designed corrugated absorber confirmed outstanding performance than conventional flat plate thermal collector. Besides the circular tube in conventional thermal collector, the possibility of constructing various shaped tube such as ellipse tube and rectangular tube were also investigated by Demianiuk *et al.* (2012) and Shelke *et al.* (2015).

This study presents a numerical thermal analysis of a new flat plate thermal collector whose absorber is configured as a wavy shape with many ellipse tubes inside of the absorber. To evaluate the performance of this new thermal collector, a conventional thermal collector configured with welded tubes beneath a flat absorber serves as a base case. Apart from the configuration of the absorber and the tube in two thermal collectors, the housing of collector, fluid condition, and weather condition are the same. Assuming a quasi-steady state condition, thermal network was developed in Engineering Equation Solver (EES) to solve simultaneous thermal balance equations and to determine thermal resistances between or within each component in collector as a function of temperature. The performance of the new flat plate thermal collector is evaluated by its instantaneous energy efficiency at different operating temperature of fluid for some given solar radiance. The study serves as a potential renovation on flat plate thermal collector for domestic hot water system.

2. CONFIGURATION OF THE NEW THERMAL COLLECTOR

The objective of this study is to evaluate the performance of the new designed flat plate thermal collector. Contrast to the conventional flat plate thermal collector, which has a flat absorber with many circular tubes beneath the absorber through welded bonds, the absorber in new thermal collector is configured as a wavy shape with many ellipse tubes inside of the absorber. Figure 1 depicts the cross section of one tube in new thermal collector alone its thermal network. The top segment is the tempered glass cover. Below the cover is the absorber. The gap between the cover and the absorber is filled with air. The ellipse shaped tubes are conjected with the next tubes via the flat part of absorber. The absorber is placed on the insulation. The bottom part of the collector is the back sheet. The vertical edge of collector is insulated by the same insulation underneath absorber. Solar radiation transmits through the glass cover and is received by the absorber. Due to the structure of the collector, the received solar energy is lost from the absorber to ambient air through the glass cover, the back sheet, and the edge. The remaining solar energy converts into thermal energy by transferring to the fluid in tube.

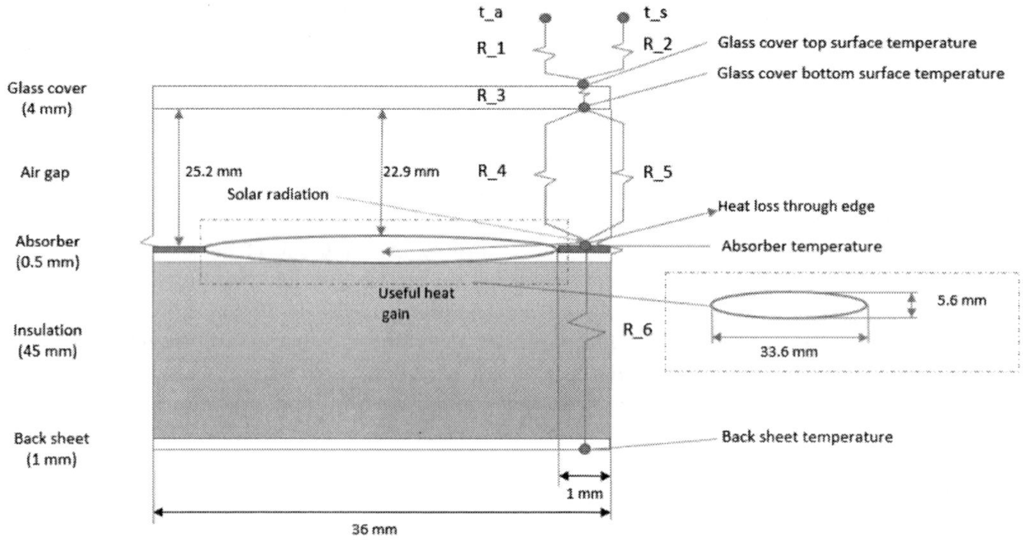

Figure 1: Cross section of new thermal collector

* t_a: ambient air temperature; t_s: sky temperature; R_1: convective thermal resistance between collector cover glass and air; R_2: the radiative thermal resistance between collector cover glass and sky; R_3: the conductive thermal resistance within cover glass; R_4: the convective thermal resistance between absorber and cover glass; R_5: the radiative thermal resistance between absorber and cover glass; R_6: the conductive thermal resistance within absorber, insulation and back sheet.

Typical conventional flat plate thermal collector is selected to compare the performance with the new collector. Figure 2 is the configuration and thermal network of the conventional flat plate thermal collector serves as base case. Apart from the configuration and material of absorber and tube, the housing, consisted of cover glass, insulation, back sheet, and edge insulation, are constructed the same as new collector.

3. NUMERICAL MODEL

Thermal analysis of solar thermal collector is a complicated process, but one simplified numerical model developed by Hottel and Whillier gives useful results. As the useful heat collected by fluid in Hottel-Whillier's model is defined by the difference between received solar radiance by absorber and the overall heat loss from absorber to ambient air, the performance of a thermal collector is influenced by the configuration of absorber, tube, and its housing. The useful heat gain can be expressed as a function of inlet temperature of fluid, property and geometry of thermal collector and ambient weather condition shown in equation (1). The property and geometry of thermal collector are factored in the area of flat plate collector Ac, heat removal factor F_R, and overall heat transfer coefficient U_L. Outlet fluid temperature of fluid can be inferred by a heat balance equation shown in equation (2). Efficiency of collector in equation (3) is defined as a ratio of useful heat gain to solar radiation on the plane of collector.

$$Q_u = A_c \times F_R \times [S - U_L \times (T_i - T_a)] \tag{1}$$

$$Q_u = \dot{m} \times C_p \times (T_o - T_i) \tag{2}$$

$$\eta = \frac{Q_u}{I \times A_c} \tag{3}$$

,where: Q_u is the useful heat gain, W; Ac is the area of thermal collector, m^2; FR is the heat removal factor; S is the solar radiance received by absorber W/m^2; I is solar radiation on the plane of collector, W/m^2;U_L is the overall heat transfer coefficient, W/ (C*m^2); T_i is inlet temperature of fluid, C; To is the outlet temperature of fluid, C; T_a is the ambient air dry bulb temperature, C; $\dot{m}$ is the mass flow rate of fluid, kg/s; C_p is the specific heat of fluid, J/(kg*C). To compare the new and conventional thermal collector, the first impact from the new configured absorber is the thermal resistance between the absorber and the cover. Assuming that the shape of insulation will not change under the weight of absorber and the tube. Due to the wavy shape of the new absorber, the distance of air gap varies while in conventional collector stays a uniform. The changing air gap results the changes of the radiative thermal resistance between cover and absorber. In addition, the wavy shape of the absorber disturbs natural convection above the absorber. The second impact is the convective heat transfer between the fluid and the tube. In new thermal collector, the convective heat transfer occurs between the fluid and the ellipse tube while in conventional thermal collector, the convective heat transfer occurs between fluid and circular tube. The third impact is the conductive thermal resistance of bond. In new thermal collector, heat flux is from the absorber to the fluid via the tube while in the conventional thermal collector, heat flux is from the absorber to the fluid via the bond and the tube. The bond is eliminated in the new thermal collector to increase heat flux to the fluid.

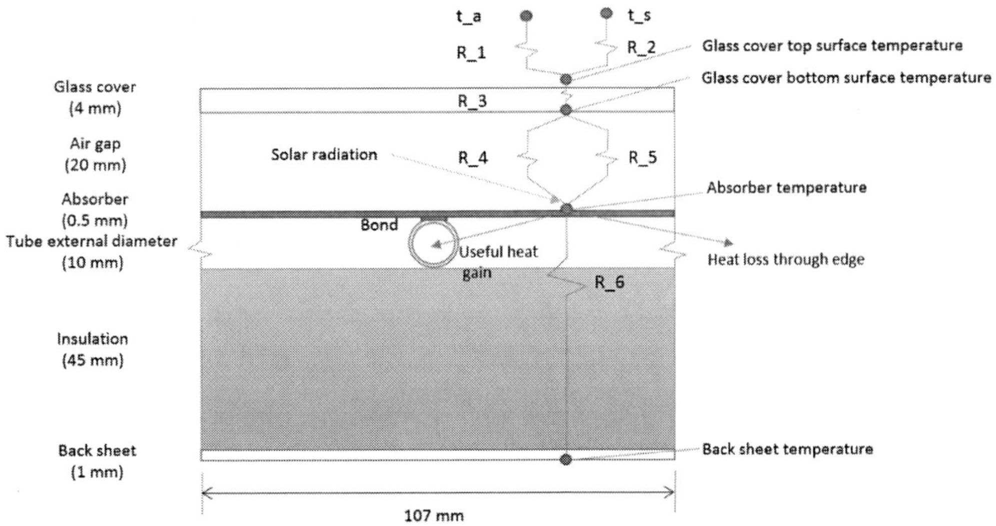

Figure 2: Cross section of conventional thermal collector

3.1 Overall Heat Transfer Resistance

The overall heat transfer coefficient is an equivalent heat transfer coefficient from the absorber to ambient air via the top cover, back sheet, and the edge insulation. The definition of overall heat transfer resistance is given in equation (4). The following section gives more details of the radiative and the convective thermal resistance between the absorber and the cover, the convective thermal resistance between the fluid and the tube.

$$R_L = \frac{1}{U_L} = \frac{1}{\frac{1}{\frac{1}{\frac{1}{Rc\_ac}+\frac{1}{Rr\_cs}}+\frac{1}{\frac{1}{Rc\_sc}+\frac{1}{Rr\_sc}}}+\frac{1}{R\_ins+R\_bs}+\frac{1}{R\_edge}} \quad (4)$$

, where R is the overall thermal resistance, C*m²/W; Rc_ac is the convective thermal resistance between cover and ambient air, C*m²/W; Rr_cs is the radiative thermal resistance between cover and sky, C*m²/W; Rc_sc is the convective thermal resistance between absorber and cover, C*m²/W; Rr_sc is the radiative thermal resistance between absorber and cover, C*m²/W; R_ins is the conductive thermal resistance in insulation, C*m²/W; R_bs is the conductive thermal resistance in back sheet, C*m²/W; R_edge is the conductive thermal resistance in edge insulation, C*m²/W.

3.1.1 Convective and radiative heat transfer between absorber and cover: Abdelkader et al. investigated the natural convection between a wavy surface and a flat surface in a horizontal enclosure. They described a wavy surface as expressed in equation (5) (Slimani *et al*, 2007).

$$F(x) = A \times [1+\cos(2\times\pi\times x)]C_P \quad (5)$$

The average Nusselt number for the convective heat transfer between a wavy surface and a flat surface was obtained under different Rayleigh number and aspect ratio AR in Slimani *et al* (2007)'s work by simulation, where aspect ratio is defined as AR = (2 × b) / a and b is the amplitude of cosine function and a is the wave length.

To determine the convective thermal resistance between the absorber and the cover, a tube shown in Figure 3 is taken as an example for illustration. The absorber can be divided into three separate parts due to its geometry: one ellipse surface and two flat surfaces. R_c_1 and R_c_3 in Fig.3 are the convective thermal resistance between the flat part of absorber and the cover which can be treated as the convective thermal resistance between two parallel flat surfaces (Duffie *et al*,1980). R_c_2 in Figure 3 is the convective thermal resistance between an ellipse surface and a flat surface. R_c_2 can be obtained from Slimani *et al*, (2007)'s work by approximating the ellipse tube as a cosine wave. Therefore, the convective thermal resistance between absorber and cover R_c for one tube equals:

$$Rc\_sc = \frac{1}{\frac{1}{R\_c\_1}+\frac{1}{R\_c\_2}+\frac{1}{R\_c\_3}} \quad (6)$$

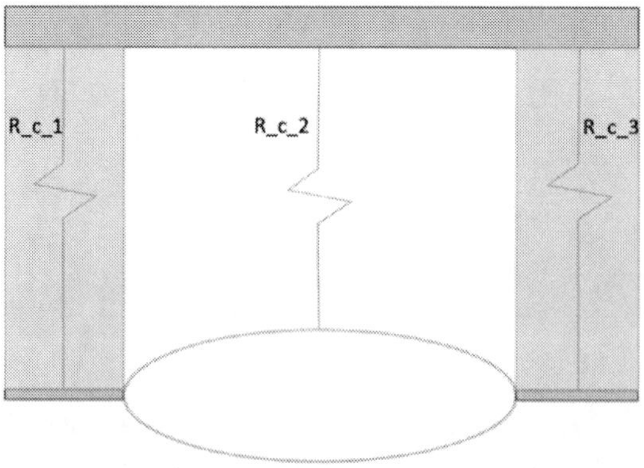

Figure 3: Convective heat transfer between absorber and cover

Following the same approach, the radiative thermal resistance between absorber and cover can be determined. The radiation between two surfaces with any shapes is given in Duffie *et al* (1980)'work as the effect of shape is factored in view factor. The view factor between the ellipse shaped tube and the flat cover approximates 0.48. The radiative thermal resistance between the flat part of absorber and the cover can be treated as the radiative thermal resistance between two flat surfaces.

3.1.2 Convective thermal resistance between fluid and tube: The flow in the conventional thermal collector can be treated as the flow in a circular tube. While the flow in the ellipse tubes in the new thermal collector can be approximated as the flow in noncircular tubes by taking hydraulic diameter as the characteristic length (Bergman *et al*, 2011).

$$D_h = \frac{4 \times A_C}{p} \quad (7)$$

, where D_h is the hydraulic diameter, m; A_C is the area of cross section of ellipse tube, m^2; p is the perimeter of cross section of ellipse tube, m.

3.2 Heat Removal Factor

Classical fin analysis method is implemented to associate collector's heat removal factor with the configuration of absorber. The flat part of absorber that separates tubes can be treated as a fin with a thickness of δ shown in Figure 4. The length of fin shown in Figure 3 equals (W-D)/2. The distance between tubes is W and the width of ellipse tube is D/2.

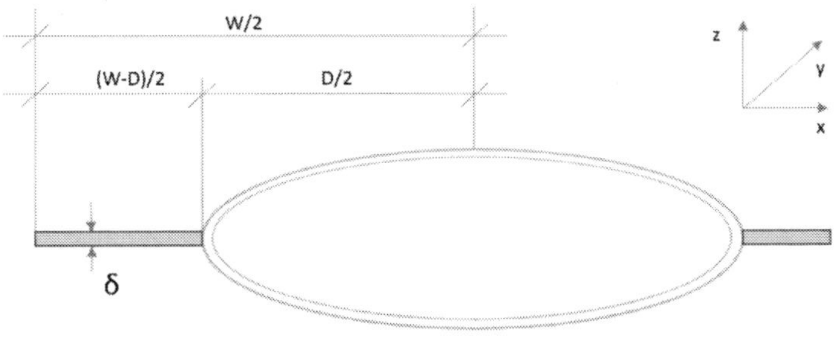

Figure 4: tube and sheet dimensions

The temperature variation alone z and y axis was negligible due to the good conductance of absorber (Duffie and Beckman, 1980). The ellipse tube was assumed at a uniform temperature Tt. Applying the energy balance with boundary conditions and introducing new variables m, the heat transfers from the fin to the fluid from both sides per unit length in flow direction is shown in equation (8).

$$q'_{fin} = (W - D) \times F \times [S - U_L \times (T_t - T_a)] \quad (8)$$

, where

$$F = \frac{\tanh[m \times (W-D)/2]}{m \times (W-D)/2} \quad (9)$$

$$m = \sqrt{U_L / k \times \delta} \quad (10)$$

, where k is the thermal conductivity of absorber, W/m*°C.

The useful heat collected by the fluid also includes the heat from the ellipse tube shown in equation (11). Here one assumption was made: the overall heat transfer coefficient for the ellipse tube part and the flat part of absorber were the same.

$$q'_{tube} = D \times [S - U_L \times (T_t - T_a)] \tag{11}$$

Therefore, the useful heat gain per unit length in flow direction equals:

$$q'_u = [(W-D) \times F + D] \times [S - U_L \times (T_t - T_a)] \tag{12}$$

Finally, the useful heat must be transferred to the fluid. In the new thermal collector, the effect of the bond is eliminated. As a result, the thermal resistance results from the convective thermal resistance between the fluid and the tube.

$$q'_u = \frac{T_t - T_f}{\dfrac{1}{h_{fi} \times \pi \times D_i}} \tag{13}$$

, where h_{fi} is the convective heat transfer coefficient between fluid and tube, W/m^2-°C, D_i is the tube's inner diameter, m.

By solving Tt in equation (13) and substituting it to equation (12), equation (14) is obtained.

$$q'_u = W \times F' \times [S - U_L \times (T_f - T_a)] \tag{14}$$

, where F' is the collector efficiency factor:

$$F' = \frac{1/U_L}{W \times [\dfrac{1}{U_L \times [D + (W-D) \times F]} + \dfrac{1}{\pi \times h_{fi} \times D_i}]} \tag{15}$$

Obtaining F', heat removal factor can be expressed as:

$$F_R = \frac{\dot{m} \times C_p}{A_c \times U_L} \times [1 - \exp(-\frac{A_c \times U_L \times F'}{\dot{m} \times C_p})] \tag{16}$$

While for conventional flat plate thermal collector, the collector efficiency factor F' equals:

$$F'_c = \frac{1/U_L}{W \times [\dfrac{1}{U_L \times [D + (W-D) \times F]} + \dfrac{1}{C_b} + \dfrac{1}{\pi \times h_{fi} \times D_i}]} \tag{17}$$

, where C_b is the conductance of bond on a per unit length basis known the bond thermal conductivity k_b, W/m-C; average bond thickness γ, m; and width of bond b, m.

$$C_b = \frac{k_b \times b}{\gamma} \tag{18}$$

4. CALCULATION OF NUMERICAL MODEL

To solve this numerical model, thermal network model is formulated in Engineering Equation Solver, developed by the University of Wisconsin, Madison. Its ability to solve simultaneous equations and the richness of built-in functions to calculate physical and thermal property of working fluid make it widely utilized to deal with thermodynamic and heat transfer problems.

As the space between two tubes in conventional thermal collector is three times the space between two tubes in new thermal collector, a sample area that contains one tube for conventional thermal collector and three tubes for new thermal collector was selected for numerical simulation to make two cases comparable. The fluid and weather information used in the model were summarized in Table.1 and collector information was summarized in Table 2.

Table 1: Fluid and weather information

Fluid information		Weather information	
Matter of fluid	Water	Wind velocity (m/s)	3
Specific heat of fluid (J/(g °C)	4.2	Dry bulb temperature (°C)	15
Density of fluid (kg/m^3)	1000	Wet bulb temperature (°C)	10
Inlet fluid temperature (°C)	From 20°C to 90°C	Solar radiance (W/m^2)	From 600 to 1000

Table 2: Collector information

Parameter	Conventional collector	New collector
Width of simulated model (m)	0.107	0.107
Length of simulated model (m)	0.54	0.54
Tube inner diameter (m)	0.008	-
Tube outside diameter (m)	0.1	-
Major axis length of tube (m)	--	0.0168
Minor axis length of tube (m)	--	0.0028
Length of flat part of absorber (m)	--	0.002
Gap between cover and oval surface (m)	--	0.02
Gap between cover and flat surface (m)	--	0.023
Thickness of absorber (m)	0.0005	0.0005
Thickness of insulation (m)	0.045	0.045
Thickness of back sheet (m)	0.001	0.001
Thickness of cover glass (m)	0.004	0.004
Conductivity of insulation (W/m °C)	0.04	0.04
Conductivity of back sheet (W/m °C)	205	205
Conductivity of cover glass (W/m °C)	1.7	1.7
Conductivity of absorber (W/m °C)	205	26.5
Cover emittance	0.88	0.88
Cover transmittance	0.8	0.8
Absorber emittance	0.05	0.05
Absorber absorbance	0.95	0.95
Mass flow rate for one tube (kg/s)	0.001106	0.001106
Tilted angle (degree)	0	0

5. RESULTS AND DISCUSSION

In this section, analysis of performance of the new thermal collector is performed at different inlet temperature of fluid and solar radiance comparing with the conventional flat plate thermal collector. Solar radiance and inlet temperature of fluid are two key inputs and the efficiency of thermal collector is a critical output. Solar radiance was varied from 1100 W/m^2 to 600 W/m^2 with an interval of 100 W/m^2. The inlet temperature of the fluid was varied from 20°C to 90°C with an interval of 5°C to prevent the phase change of fluid in tubes.

5.1 Thermal Efficiency and Solar Radiation

The thermal efficiency of the new and the conventional thermal collectors are shown in Figure 5 and Figure 6 separately. In Figure 5 and Figure 6, horizontal axis represents the average operating temperature of fluid above ambient air temperature. The vertical axis stands for the thermal efficiency, which is the ratio of useful heat gain to the solar radiance on collector's plane, as expressed in equation (3). For the two thermal collectors placed horizontally, solar radiation significantly affects energy efficiency. As shown in Figure 5 and Figure 6, the models predict that for a given operating temperature of fluid, thermal efficiency of the both collectors increase with the increase of solar radiance. This can be translated that larger solar radiance contributes to higher absorbed solar energy but does not enlarge overall heat transfer coefficient of collector notably. For a given solar radiance, the energy efficiency decreases as the average operating temperature of fluid increases. This can be explained as the average operating temperature of fluid increases, the mean temperature of absorber and overall heat transfer coefficient also increase. Therefore, resulting more heat loss and a lower energy efficiency.

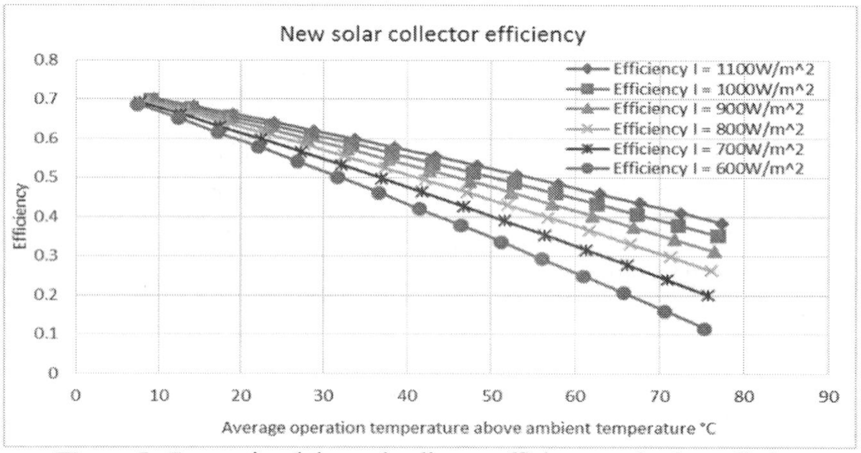

Figure 5: Conventional thermal collector efficiency and solar radiation

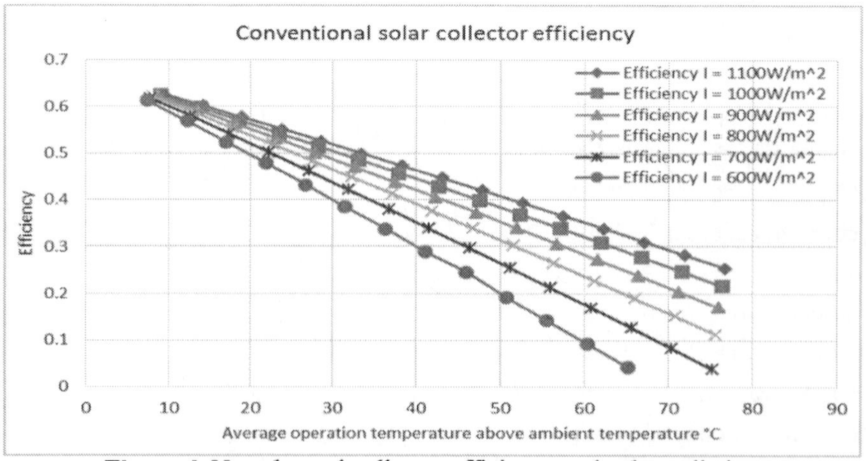

Figure 6: New thermal collector efficiency and solar radiation

For comparison, Table 3 lists the coefficients to present the linear relationship between efficiency and average operating temperature of fluid above ambient air temperature. Additionally, one chart illustrates energy efficiency at a given solar radiance equals 800 W/m^2 is shown in Figure 7. Simulation results confirm a better performance of new thermal collector than the conventional one. As it can be noticed that for different solar radiance level, both thermal collectors offer similar trend as the operating temperature of fluid increases but the new thermal collector achieves an average 10% higher thermal efficiency.

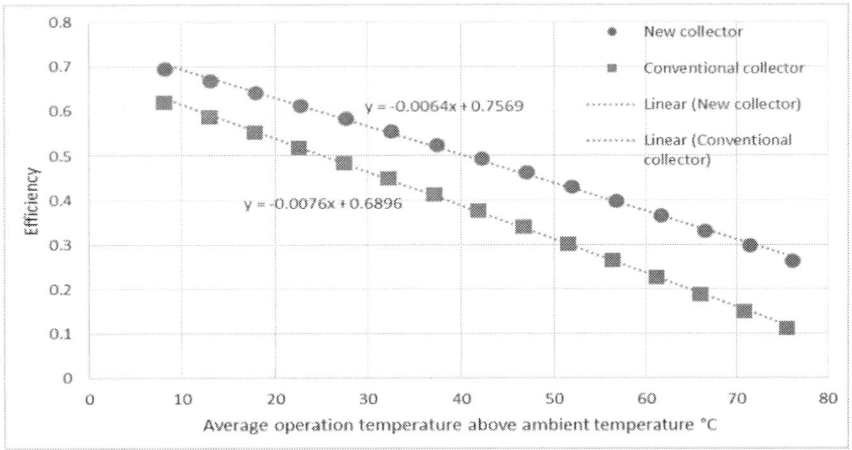

Figure 7: New and conventional thermal collector efficiency under solar radiation I = 800 W/m^2

Table 3: Energy efficiency of two thermal collectors

	New collector		Conventional collector	
Solar radiance (W/ m^2)	a	b	a	b
1100	-0.0047	0.7521	-0.0055	0.6832
1000	-0.0051	0.7534	-0.0061	0.685
900	-0.0057	0.7549	-0.0067	0.6871
800	-0.0064	0.7569	-0.0076	0.6896
700	-0.0072	0.7593	-0.0086	0.6926
600	-0.0084	0.7625	-0.0099	0.6936

* Efficiency = a × average operation temperature above ambient temperature + b

Table 4 is prepared to illustrate the difference of energy efficiency between the two collectors when inlet temperature of fluid is 40°C at a given solar radiance equals 1000 W/m^2. At the same working condition, the thermal efficiency of the new thermal collector is about 10% higher than the efficiency of the conventional collector. As the useful heat gain and thermal efficiency are defined, the better performance of new thermal collector can be translated from three aspects: (1) the configuration of the new thermal collector contributes a higher heat removal factor while a lower overall heat loss coefficient comparing with the conventional thermal collector. (2) The temperature of the absorber in new collector is lower than the temperature of absorber in conventional collector which results less heat loss from absorber to ambient air. (3) The temperature of the glass cover in new thermal collector is closer to ambient air temperature. While the higher temperature of the glass cover in conventional collector leads to a higher heat loss comparing with the novel collector.

To make the simulation results comparable, two collectors worked at the same flow rate. Therefore, the heat transfer coefficient inside tubes of the new thermal collector was slightly higher than that of the conventional collector due to the slow-moving fluid in new thermal collector (0.00278 m/s for new thermal collector while 0.022 m/s for conventional thermal collector).

Table 4: Performance of two thermal collectors

	New	Conventional
Heat transfer coefficient inside tubes (W/m °C)	333	294
Fluid mean temperature (°C)	43.2	42.9
Absorber mean temperature (°C)	47.1	52.7
Cover glass temperature (°C)	15.9	18.7
Overall heat loss coefficient (W/m^2 °C)	4.96	6.32
Heat removal factor	0.94	0.85
Instantaneous efficiency	0.5624	0.4579
Useful heat gain (W/m^2)	562.4	457.9

6. CONCLUSION

In this study, a numerical thermal model for a new designed flat plate thermal collector with a wavy shaped absorber and ellipse tubes insides the absorber is presented. The main objective of this study is to evaluate the thermal performance of the new thermal collector by modeling. For evaluation, thermal efficiency of the new thermal collector is compared with a conventional flat plate thermal collector with similar dimensions and same working conditions. The simulation results disclose that due to the configuration of the new thermal collector, it provides an approximate 10% higher thermal efficiency than the conventional thermal collector under different operating temperatures of fluid for given solar radiances from 600 W/m^2 to 1100 W/m^2. As indicated by the modeling results, new thermal collector can be served as a better option than convention collector for a typical domestic hot water system.

REFERENCES

Varol, Y., & Oztop, H. F. (2008). A comparative numerical study on natural convection in inclined wavy and flat-plate solar collectors. *Building and Environment*, *43*(9), 1535-1544.

Dović, D., & Andrassy, M. (2012). Numerically assisted analysis of flat and corrugated plate solar collectors' thermal performances. *Solar energy*, *86*(9), 2416-2431.

Shelke, V. G., & Patil, C. V. Analyze the Effect of Variations in Shape of Tubes for Flat Plate Solar Water Heater.

Demianiuk, A., & Sorko, S. A. (2012). Analysis of Flow and Thermal Phenomena in Evacuated Tube Collectors. *acta mechanica et automatica*,*6*(4), 5-10.

Karim, M. A., & Hawlader, M. N. A. (2006). Performance evaluation of a v-groove solar air collector for drying applications. *Applied Thermal Engineering*, *26*(1), 121-130.

Karsli, S. (2007). Performance analysis of new-design solar air collectors for drying applications. *Renewable Energy*, *32*(10), 1645-1660.

Abdelkader, S., Mebrouk, R., Abdellah, B., & Khadidja, B. (2007). Natural convection in a horizontal wavy enclosure. *Journal of Applied Sciences*, *7*, 334-341.

Duffie, J. A., & Beckman, W. A. (1980). *Solar engineering of thermal processes* (Vol. 3). New York etc.: Wiley.

Bergman, T. L., Incropera, F. P., & Lavine, A. S. (2011). *Fundamentals of heat and mass transfer*. John Wiley & Sons.

Mansour, M. K. (2013). Thermal analysis of novel minichannel-based solar flat-plate collector. Energy, 60, 333-343.

A Multi-objective Approach to Optimal Battery Storage in The Presence of Demand Charges

Reza Kamyar[1*], Matthew M. Peet[2]

[1] Arizona State University, School for Engineering of Matter, Transport & Energy, Tempe, Arizona, USA
rkamyar@asu.edu

[2] Arizona State University, School for Engineering of Matter, Transport & Energy, Tempe, Arizona, USA
mpeet@asu.edu

* Corresponding Author

ABSTRACT

We propose a multi-objective optimization algorithm for optimal energy storage by residential customers using Li-Ion batteries. Our goal is to quantify the benefits of optimal energy storage to solar customers whose electricity bills consist of both Time of Use charges ($/kWh, with different rates for on-peak and off-peak hours) and demand charges ($/kW, proportional to the peak rate of consumption in a month). We first define our energy storage optimization problem as minimization of the monthly electricity bill subject to certain constraints on the energy level and the charging/discharging rate of the battery, while accounting for battery's degradation due to cycling and depth of discharge. We solve this problem by constructing a sequence of parameterized multi-objective dynamic programs whose sets of non-dominated solutions are guaranteed to contain an optimal solution to our energy storage problem. Unlike the standard formulation of our energy storage problem, each of the parameterized optimization problems satisfy the principle of optimality - hence can be solved using standard dynamic programming algorithms. Our numerical case studies on a wide range of load profiles show that in the presence of demand charges, optimal energy storage using the existing residential batteries can reduce the monthly electricity bill by up to 52% relative to the case where no energy storage is used.

1.INTRODUCTION

One concern of electric utilities is that rapid integration of renewables such as distributed solar generation may change customers' consumption behavior in ways that current generating units cannot accommodate for these changes. The demand peaks are of particular concern to regulated public utilities as these utilities are required to maintain reserve generating capacity as determined by the demand peak. According to a recent report by Arizona Public Services (APS (2014)), the demand peak in Arizona is projected to increase by 40% over a period of next 15 years (See Fig. 1). In order to cover the cost of increasing the generating capacity and influence the consumption behavior of rational customers, Arizona's utilities Salt River Project (SRP) and APS have recently incorporated *demand charges* in some of their pricing plans. Demand charges are proportional to the maximum rate at which the electricity is consumed by the customer during the on-peak hours within a month. Thus, in this new form of pricing, the amount paid by the customer within a month is

$$J(q) := 30\,p_{\text{on}} \int_{T_{\text{on}}} q(t)dt + 30\,p_{\text{off}} \int_{T_{\text{off}}} q(t)dt + p_d \sup_{t \in T_{\text{on}}} q(t), \tag{1}$$

where $q(t)$ is the power supplied by the utility company. The first two terms in (1) are Time of Use (ToU) charges, where p_{on} ($/kWh) and p_{off} ($/kWh) are the on-peak and off-peak prices. The third term is the monthly demand charge, where p_d ($/kW) is called the demand price. In (1), T_{on} and T_{off} denote the on-peak and off-peak periods respectively.

One approach for rational customers to reduce their consumption peaks (as a response to the integration of demand charges) is energy storage (Dunn and Tarascon (1997)). Several papers have studied optimal use of battery storage for residential customers (Bozchalui et al. (2012); Boaro et al. (2013); Huang and Liu (2013); Wei et al. (2015)). Majority of the existing papers use Model Predictive Control (Giorgio et al. (2012)), dynamic programming-based algorithms (Boaro et al. (2013); Huang and Liu (2013)) or neural networks (Wei et al. (2015)) to find optimal storage programs under real-time or ToU pricing. Li et al. (2011) have coupled the problem of optimal scheduling of battery and other controlled appliances with a social welfare optimization problem at the utility level to find optimal battery schedules and optimal real-time prices. Although optimal battery scheduling is a well-developed research area, to the best of our knowledge, there has been no studies on optimal scheduling of residential-sized batteries in the presence of demand charges.

From the mathematical standpoint, the presence of demand charges poses a significant challenge in solving the underlying optimization problem. In particular, because of the existence of the term sup $q(t)$ (a *non-separable* term with respect to time) in the electricity bill defined in (1), the problem min $J(q)$ violates the principle of optimality (Bellman and Dreyfus (1962)). The principle of optimality provides sufficient conditions for optimal control algorithms such as dynamic programming and Hamiltonian-based algorithms to converge to an optimal solution. While works such as Oudalov et al. (2007) and Rowe et al. (2014) have included some forms of demand charges in their storage optimization problems, they have not addressed the problem of inseparability of the objective function and violation of the principle of optimality.

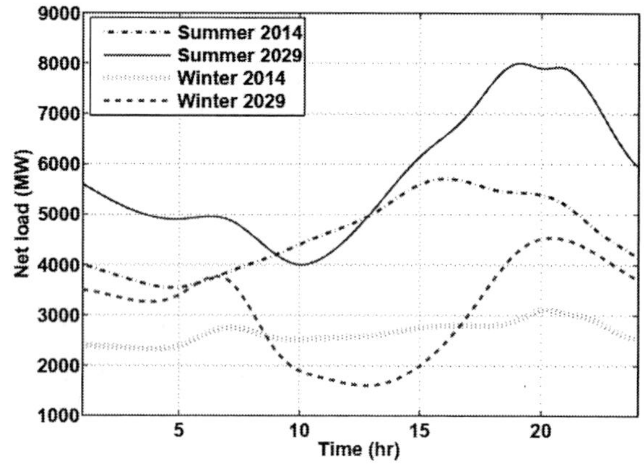

Figure 1: Net retail load for typical summer and winter days in Arizona for years 2014 and 2029 (projected)

In this paper, we establish a provably convergent algorithm for optimal battery scheduling for residential customers, under a combined ToU and demand pricing plan. We achieve this in two steps: First, we replace sup $q(t)$ in the objective function with an l_p approximation of $q(t)$ for some large p. Then, we construct a sequence of multi-objective optimization problems (similar to Li and Haimes (1991, 1987)) whose sets of non-dominated solutions contain an optimal solution to the residential battery optimization as p tends to infinity. The objective functions of these multi-objective problems are separable with respect to time - hence can be minimized using the existing dynamic programming algorithms. In a number of numerical case studies, we applied our algorithm to the residential battery optimization problem for a wide range of customer's load, while considering the effects of solar generation and battery's degradation over 5 years. We used one of the existing residential-sized battery in the market as the source of energy storage. Our numerical analysis shows that optimal energy storage can reduce the monthly electricity bill by up to 52% using SRP's pricing plan (SRP (2015)). Moreover, we observed that under a combined ToU and demand pricing, optimal energy storage is most effective (in terms of saving over 5 years) for those customers who have the highest monthly peak load within our range of analysis.

2.OPTIMAL ENERGY STORAGE PROBLEM

In this section, we first define a model describing the amount of energy stored in the battery and the battery degradation due to cycling and depth of discharge. We then use this model to define our residential battery storage optimization problem under a mixed ToU and demand pricing plan.

2.1 A Model for Battery Storage and Degradation

To model the amount of energy stored in the battery, we use the difference equation

$$e(d,k+1) = e(d,k) + \eta u(d,k)\Delta t, \tag{2}$$

where $e(d,k)$ denotes the amount of energy stored in the battery at time $t = k\,\Delta t$ of day d, $u(d,k)$ denotes the charging/discharging (+/−) rate at time $t = k\,\Delta t$ of day d, and η denotes the charging efficiency. We denote the initial capacity of the battery by C_I, meaning that the initial energy level $e(0,0)$ is bounded by C_I. Moreover, we denote the maximum charging and discharging rates by $\bar{u}$ and $\underline{u}$, meaning that $u(k) \in [\underline{u}, \bar{u}]$ for all k.

We model the capacity of the battery at day d and time-step k as

$$c_b(d,k) = (1 - \beta(d,k))C_I, \tag{3}$$

where $\beta(d,k)$ denotes the battery's degradation rate. We model the degradation rate as the sum of cycling degradation and depth of discharge degradation:

$$\beta(d,k) = \underbrace{\alpha\, n(d,k)}_{\text{cycling degradation}} + \underbrace{\sum_{k=0}^{T} \gamma(d,k)}_{\text{depth of discharge degradation}} . \tag{4}$$

In (4), $n(d,k)$ is the total number of cycles completed till time-step k of day d. We calculate the number of cycles recursively as

$$n(d,k+1) = n(d,k) + \frac{\eta\,\Delta t\,|u(d,k)|}{2\,c_b(d,k)}. \tag{5}$$

In (4), $\gamma(d,k)$ is the depth of discharge degradation at time-step k of day d. We use a model similar to the Li-Ion battery model by Yoshida et al. (2016) to calculate $\gamma(d,k)$ as

$$\gamma(d,k) = \begin{cases} X\,\zeta(d,k) + Y & \zeta(d,k) \geq 0.7 \\ 0 & \text{otherwise} \end{cases}, \tag{6}$$

where $\zeta(d,k)$ is the depth of discharge at time-step k and is calculated as

$$\zeta(d,k) = \frac{c_b(d,k) - e(d,k)}{c_b(d,k)}.$$

In this paper, we consider one of the existing residential-sized batteries in the market as the battery used by the residential customer. As declared by the manufacturer, this battery degrades to 80% of its initial capacity $C_I = 10$ kWh after completing 5000 round-trip cycles. To capture the 20% drop in the capacity, we set the parameters α, X and Y in (4) and (6) to 1×10^{-4}, 5.8×10^{-6} and -4×10^{-6} respectively.

2.2 Problem Statement: Residential Battery Storage Optimization

We define our battery storage optimization problem as minimizing the monthly electricity bill subject to constraints on the battery's energy-level and charging/discharging rates. To define the electricity charges within a month, we divide each day into two off-peak periods from 12 AM to t_{on} and t_{off} to 12 AM. Let us denote the electricity price during the of-peak periods by p_{off} (\$/kWh). The on-peak period starts at t_{on} and ends at t_{off}. We denote the electricity price during this period by p_{on} (\$/kWh). In addition to the on-peak and off-peak charges, we assume that customers are charged proportional to their maximum rate of consumption during the on-peak hours in a month. We denote the proportionality ratio by p_d (\$/kW) and call it the *demand price*. Given the prices $p_{\text{off}}, p_{\text{on}}$ and p_d, the ToU charge in a month is calculated as

$$J_t(q, p_{\text{off}}, p_{\text{on}}) = \sum_{d=1}^{N=30} \left(p_{\text{off}} \sum_{k=0}^{t_{\text{on}}-1} q(d,k)\Delta t + p_{\text{on}} \sum_{k=t_{\text{on}}}^{t_{\text{off}}-1} q(d,k)\Delta t + p_{\text{off}} \sum_{k=t_{\text{off}}}^{24/\Delta t} q(d,k)\Delta t \right),$$

and the demand charge is calculated as

$$J_d(q, p_d) = p_d \sup_{\substack{k \in \{t_{\text{on}}, \cdots, t_{\text{off}}-1\} \\ d \in \{1, \cdots, 30\}}} q(d,k), \tag{7}$$

where $q(d,k)$ is the power supplied by the grid at time-step k of day d.

We now define the problem of optimal residential battery storage for a month as

$$\min_{\substack{u(d,k),\beta(d,k)\in\mathbb{R}\\ e(d,k),n(d,k)\in\mathbb{R}}} \quad J_t(q,p_{\text{off}},p_{\text{on}}) + J_d(q,p_d) \quad \text{subject to}$$

$$\begin{aligned}
& q(d,k) = q_a(d,k) - q_s(d,k) + u(d,k) && \text{for } d \in \Phi_d \text{ and } k \in \Phi_k\\
& e(d,k+1) = e(d,k) + \eta\, u(d,k)\Delta t && \text{for } d \in \Phi_d \text{ and } k \in \Phi_k\\
& e(d,0) = e\left(d-1, \frac{24}{\Delta t}\right) + \eta\, u\left(d-1, \frac{24}{\Delta t}\right)\Delta t && \text{for } d \in \Phi_d\\
& 0 \le e(d,k) \le (1-\beta(d,k))C_I && \text{for } d \in \Phi_d \text{ and } k \in \Phi_k\\
& \beta(d,k) = \alpha\, n(d,k) + \sum_{k=0}^{T} \gamma(d,k) && \text{for } d \in \Phi_d \text{ and } k \in \Phi_k\\
& n(d,k+1) = n(d,k) + \frac{\eta\,\Delta t\,|u(d,k)|}{2\,c_b(d,k)} && \text{for } d \in \Phi_d \text{ and } k \in \Phi_k\\
& \underline{u} \le u(d,k) \le \bar{u} && \text{for } d \in \Phi_d \text{ and } k \in \Phi_k\\
& e(0,0) = e_0,\ n(0,0) = 0, && \qquad (8)
\end{aligned}$$

where we define $\Phi_d := \{1, \cdots, N\}$ and $\Phi_k := \{0, \cdots, 24/\Delta t\}$. The second line of Problem (8) constrains the sum of the power, u, given to the battery and the power, q_a, consumed by the appliances to be equal to the total power supplied by the grid and the solar Photovoltaics. Lines 3 and 4 enforce the change in the battery's energy level to be proportional to the power given to or take from the battery. Line 5 constrains the energy level of the battery to stay below the battery's capacity. Lines 6 and 7 model the battery's degradation. Finally, Line 8 constrains the charging/discharging rate to stay within a pre-specified bound determined by the manufacturer.

3.SOLUTION METHODOLOGY: MULTI-OBJECTIVE DYNAMIC PROGRAMMING

The residential battery optimization defined in (8) is an instance of the general optimal control problem

$$\begin{aligned}
J^\star(z) := \min_{u_k\in\mathbb{R}^m, x_k\in\mathbb{R}^n} \quad & J(x_0, u_0, \cdots, x_{N-1}, u_{N-1}, x_N) && \\
\text{subject to} \quad & x_{k+1} = f(x_k, u_k), x_0 = z && \text{for } k = 0, \cdots, N\\
& x_k \in X && \text{for } k = 1, \cdots, N\\
& u_k \in U && \text{for } k = 0, \cdots, N-1, \qquad (9)
\end{aligned}$$

where $z \in \mathbb{R}^n$ is a given initial state, and $f : \mathbb{R}^n \times \mathbb{R}^m \to \mathbb{R}^n$, $X \subset \mathbb{R}^n$ and $U \subset \mathbb{R}^m$ are known. It can be shown that if the objective function, J, of Problem (9) is *time-separable*, i.e., there exist maps $\psi_k : \mathbb{R}^n \times \mathbb{R}^m \times \mathbb{R} \to \mathbb{R}$ for $k = 0, \cdots, N-1$ and $\psi_N : \mathbb{R}^n \to \mathbb{R}$ such that

$$J(x_0, u_0, \cdots, x_{N-1}, u_{N-1}, x_N) = \psi_0\Bigg(x_0, u_0, \psi_1\bigg(x_1, u_1, \psi_2\Big(\cdots, \psi_{N-1}\big(x_{N-1}, u_{N-1}, \psi_N(x_N)\big)\Big)\bigg)\Bigg), \qquad (10)$$

for all $u_k \in \mathbb{R}^m$, where $x_k = f(x_{k-1}, u_{k-1})$ for $k = 1, \cdots, N$, then one can apply dynamic programming to Problem (9) to find $J^\star(z)$ for any $z \in X$. Dynamic programming uses the principle of optimality (Bellman and Dreyfus (1962)) to reduce Problem (9) to a finite sequence of optimization problems indexed by the time-steps $k = 0, \cdots, N$ (also known as Bellman's recursive formula):

$$\begin{aligned}
& V_k(z) = \min_{v\in U}\big(g(z,v) + V_{k+1}(f(z,v))\big) \text{ for } z \in X \text{ and } k \in \Phi_k \backslash N\\
& V_N(z) = h(z) \qquad \text{for all } z \in X. \qquad (11)
\end{aligned}$$

It can be shown that $J(z) = V_0(z)$. Moreover, at each time-step k and state z, the minimizer v in (11) is an optimal control. Unfortunately, the objective function of Problem (8) is not time-separable. Specifically, the supremum function in J_d as defined in (7), can not be represented in the Form (10). Moreover, it can be shown that the objective function of Problem (8) violates the principle of optimality. As a result, dynamic programming will not necessarily converge to an optimal solution. In this section, we resolve this issue in two steps. First, we approximate the supremum function in J_d using the l_p-norm of $q(d,k)$, i.e.,

$$\sup_{\substack{k\in\{t_{\text{on}},\cdots,t_{\text{off}}-1\}\\ d\in\{1,\cdots,30\}}} q(d,k) \approx \sqrt[p]{\sum_{d=1}^{30}\sum_{k=t_{\text{on}}}^{t_{\text{off}}-1} q(d,k)^p}$$

for some large $p \in \mathbb{N}$. We then use Theorem 1 to define a multi-objective optimization problem whose set of *non-dominated* solutions contains an optimal solution to Problem (9) when $p \to \infty$. Before presenting the theorem, we define the set of non-dominated solutions as follows.

Definition 1 *Consider the multi-objective optimization problem*

$$\begin{aligned}
\min_{u_k\in\mathbb{R}^m, x_k\in\mathbb{R}^n} \quad & \left[D_1(x_0,u_0,\cdots,u_{N-1},x_{N-1},x_N),\cdots,D_M(x_0,u_0,\cdots,u_{N-1},x_{N-1},x_N)\right] \\
\text{subject to} \quad & x_{k+1} = f(x_k,u_k), x_0 = z \qquad \text{for } k = 0,\cdots,N \\
& x_k \in X \qquad \text{for } k = 1,\cdots,N \\
& u_k \in U \qquad \text{for } k = 0,\cdots,N-1,
\end{aligned}$$

for some $z \in \mathbb{R}^n, X \subset \mathbb{R}^n, U \subset \mathbb{R}^m$, where the maps $D_i : \mathbb{R}^n \times \mathbb{R}^m \times \cdots \times \mathbb{R}^n \times \mathbb{R}^m \times \mathbb{R}^n \to \mathbb{R}$ are known. A solution $(x_0,u_0^\star,\cdots,u_{N-1}^\star,x_{N-1}^\star,x_N^\star)$ to the multi-objective problem is called non-dominated if there exists no other feasible $(x_0,u_0,\cdots,u_{N-1},x_{N-1},x_N)$ such that $D_i(x_0,u_0,\cdots,u_{N-1},x_{N-1},x_N) \le D_i(x_0,u_0^\star,\cdots,u_{N-1}^\star,x_{N-1}^\star,x_N^\star)$ for $i = 1,\cdots,M$, with strict inequality for at least one i.

We now present the main theorem.

Theorem 1 *Suppose there exist maps $f : \mathbb{R}^n \times \mathbb{R}^m$,*

$$G : \underbrace{\mathbb{R}\times\cdots\times\mathbb{R}}_{M\text{-times}} \to \mathbb{R} \quad \text{and} \quad D_i : \underbrace{(\mathbb{R}^n\times\mathbb{R}^m)\times\cdots\times(\mathbb{R}^n\times\mathbb{R}^m)}_{(N-1)\text{-times}} \times\mathbb{R}^n \to \mathbb{R}$$

for $i = 1,\cdots,M$ such that:

I. ***(Time-separability of** D_i) For each $i \in \{1,\cdots,M\}$, there exist maps $\psi_{i,j} : \mathbb{R}^n \times \mathbb{R}^m \times \mathbb{R} \to \mathbb{R}$ for $j = 1,\cdots,N-1$ and $\psi_{i,N} : \mathbb{R}^n \to \mathbb{R}$ such that*

$$D_i(x_0,u_0,\cdots,x_{N-1},u_{N-1},x_N) = \psi_{i,0}\Bigg(x_0,u_0,\psi_{i,1}\Bigg(x_1,u_1,\psi_{i,2}\Big(\cdots,\psi_{i,N-1}\big(x_{N-1},u_{N-1},\psi_{i,N}(x_N)\big)\Big)\Bigg)\Bigg) \tag{12}$$

for all $u_k \in \mathbb{R}^m$, where $x_k = f(x_{k-1},u_{k-1})$, $k = 1,\cdots,N$;

II. ***(Backward monotonicity** D_i) For each $i \in \{1,\cdots,M\}$ and any $j \in \{1,\cdots,N\}$, if $\psi_{i,j}$ in (12) satisfies*

$$\begin{aligned}
&\psi_{i,j}\Bigg(x_j,u_j,\psi_{i,j+1}\Bigg(x_{j+1},u_{j+1},\psi_{i,j+2}\Big(\cdots,\psi_{i,N-1}\big(x_{N-1},u_{N-1},\psi_{i,N}(x_N)\big)\Big)\Bigg)\Bigg) \\
&\qquad > \psi_{i,j}\Bigg(x_j,v_j,\psi_{i,j+1}\Bigg(y_{j+1},v_{j+1},\psi_{i,j+2}\Big(\cdots,\psi_{i,N-1}\big(y_{N-1},v_{N-1},\psi_{i,N}(y_N)\big)\Big)\Bigg)\Bigg)
\end{aligned}$$

for all $x_j \in \mathbb{R}^n$ and for some $(u_j,\cdots,u_{N-1})$ and $(v_j,\cdots,v_{N-1})$, where $x_k = f(x_{k-1},u_{k-1})$ for $k = j+1,\cdots,N$ and $y_k = f(y_{k-1},v_{k-1})$ for $k = j+1,\cdots,N$ with $y_j = x_j$, then

$$\begin{aligned}
&\psi_{i,j-1}\Bigg(x_{j-1},u_{j-1},\psi_{i,j}\Bigg(x_j,u_j,\psi_{i,j+1}\Bigg(x_{j+1},u_{j+1},\psi_{i,j+2}\Big(\cdots,\psi_{i,N-1}\big(x_{N-1},u_{N-1},\psi_{i,N}(x_N)\big)\Big)\Bigg)\Bigg)\Bigg) \\
&\qquad > \psi_{i,j-1}\Bigg(x_{j-1},u_{j-1},\Bigg(x_j,v_j,\psi_{i,j+1}\Bigg(y_{j+1},v_{j+1},\psi_{i,j+2}\Big(\cdots,\psi_{i,N-1}\big(y_{N-1},v_{N-1},\psi_{i,N}(y_N)\big)\Big)\Bigg)\Bigg)\Bigg)
\end{aligned}$$

for all $x_{j-1} \in \mathbb{R}^n$, $u_{j-1} \in \mathbb{R}^m$;

*III. (**Monotonicity of** J) G, D_i and the objective function J in Problem (9) satisfy*

$$J(x_0, u_0, \cdots, x_{N-1}, u_{N-1}, x_N) = G\Big(D_1\big(x_0, u_0, \cdots, x_{N-1}, u_{N-1}, X_N\big), \cdots, D_M\big(x_0, u_0, \cdots, x_{N-1}, u_{N-1}, X_N\big)\Big), \tag{13}$$

where $\dfrac{\partial G(D_1, \cdots, D_M)}{\partial D_i} > 0$ *for* $i = 1, \cdots, M$.

Then, an optimal solution to Problem (9) is a non-dominated solution of the multi-objective problem

$$\begin{array}{lll} \min\limits_{u_k \in \mathbb{R}^m, x_k \in \mathbb{R}^n} & [D_1(x_0, u_0, \cdots, u_{N-1}, x_{N-1}, x_N), \cdots, D_M(x_0, u_0, \cdots, u_{N-1}, x_{N-1}, x_N)] & \\ \text{subject to} & x_{k+1} = f(x_k, u_k), x_0 = z & \text{for } k = 0, \cdots, N \\ & x_k \in X & \text{for } k = 1, \cdots, N \\ & u_k \in U & \text{for } k = 0, \cdots, N-1. \end{array}$$

See Li and Haimes (1987) for a proof. To apply Theorem 1 to our residential battery optimization problem, defined in (8), let us start by defining the maps G and D_i in theorem 1 as

$$D_1\big(q(1,0), \cdots, q(30, 24/\Delta t)\big) := \sum_{d=1}^{30} \left(p_{\text{off}} \sum_{k=0}^{t_{\text{on}}-1} q(d,k)\Delta t + p_{\text{on}} \sum_{k=t_{\text{on}}}^{t_{\text{off}}-1} q(d,k)\Delta t + p_{\text{off}} \sum_{k=t_{\text{off}}}^{24/\Delta t} q(d,k)\Delta t \right), \tag{14}$$

$$D_2\big(q(1, t_{\text{on}}), \cdots, q(30, t_{\text{off}} - 1)\big) := p_d^p \sum_{d=1}^{30} \sum_{k=t_{\text{on}}}^{t_{\text{off}}-1} q(d,k)^p \tag{15}$$

and

$$G(D_1, D_2) := D_1 + \sqrt[p]{D_2}, \tag{16}$$

where recall that $\sqrt[p]{D_2}$ is the l_p-norm approximation of the monthly demand charge, J_d, as defined in (7). Moreover, let us define $x := [q, e, n]$ and $v := [u, \beta]$, where recall that q is the power supplied by the utility company, e is the energy stored in the battery, n is the number of cycles, u is the power given to or taken from the battery and β is the battery's degradation rate. Then, Condition I of Theorem 1 holds if we define

$$\psi_{1,N}(x) := p_{\text{off}}\, \Delta t\, x_1,$$

$$\psi_{1,j}(x, u, \psi_{1,j+1}(y, v)) := \begin{cases} p_{\text{on}}\, \Delta t\, x_1 + \psi_{1,j+1}(y, v, \psi_{1,j+2}) & \text{if } j \in \Gamma \\ p_{\text{off}}\, \Delta t\, x_1 + \psi_{1,j+1}(y, v, \psi_{1,j+2}) & \text{if } j \in \{0, \cdots, N\} \backslash \Gamma \end{cases} \tag{17}$$

and

$$\psi_{2,j}(x, u, \psi_{1,j+1}(y, v)) := \begin{cases} p_d^p x_1^p + \psi_{2,j+1}(y, v, \psi_{2,j+2}) & \text{if } j \in \Gamma \\ 0 & \text{if } j \in \{0, \cdots, N\} \backslash \Gamma, \end{cases} \tag{18}$$

where $N = 30 \times 24/\Delta t$ and Γ is defined as $\Gamma := \bigcup_{i=1}^{30} \{i \cdot t_{\text{on}}, \cdots, i\,(t_{\text{off}} - 1)\}$. Condition II follows immediately by using $\psi_{i,j}$ as defined in (17) and (18). Finally, D_1, D_2 and G defined in (14), (15) and (16) satisfy Condition III:

$$\frac{\partial G(D_1, D_2)}{\partial D_1} = 1 > 0, \quad \frac{\partial G(D_1, D_2)}{\partial D_2} = \frac{1}{p} D_2^{\frac{1-p}{p}} > 0$$

if we choose p to be even and

$$\sup\big\{|q(1, \cdots, t_{\text{on}})|, \cdots, |q(30, t_{\text{off}} - 1)|\big\} = \sup\big\{q(1, \cdots, t_{\text{on}}), \cdots, q(30, t_{\text{off}} - 1)\big\} \geq 0.$$

This corresponds to the typical case where the consumption peak during the on-peak period is positive and is greater than or equal to the maximum power sent back to the grid during the on-peak period. Since all the conditions of Theorem 1 hold, the multi-objective optimization problem

$$\min_{\substack{u(d,k),\beta(d,k)\in\mathbb{R}\\ e(d,k),n(d,k)\in\mathbb{R}}} \left[D_1\left(q(1,0),\cdots,q\left(30,\frac{24}{\Delta t}\right)\right), D_2\left(q(1,0),\cdots,q\left(30,\frac{24}{\Delta t}\right)\right)\right]$$

$$\begin{aligned}
&\text{subject to}\\
&q(d,k) = q_a(d,k) - q_s(d,k) + u(d,k) && \text{for } d\in\Phi_d \text{ and } k\in\Phi_k\\
&e(d,k+1) = e(d,k) + \eta\, u(d,k)\Delta t && \text{for } d\in\Phi_d \text{ and } k\in\Phi_k\\
&e(d,0) = e\left(d-1,\frac{24}{\Delta t}\right) + \eta\, u\left(d-1,\frac{24}{\Delta t}\right)\Delta t && \text{for } d\in\Phi_d\\
&0 \le e(d,k) \le (1-\beta(d,k))C_I && \text{for } d\in\Phi_d \text{ and } k\in\Phi_k\\
&\beta(d,k) = \alpha\, n(d,k) + \sum_{k=0}^{T}\gamma(d,k) && \text{for } d\in\Phi_d \text{ and } k\in\Phi_k\\
&n(d,k+1) = n(d,k) + \frac{\eta\,\Delta t\,|u(d,k)|}{2\,c_b(d,k)} && \text{for } d\in\Phi_d \text{ and } k\in\Phi_k\\
&\underline{u} \le u(d,k) \le \bar{u} && \text{for } d\in\Phi_d \text{ and } k\in\Phi_k\\
&e(0,0) = e_0,\ n(0,0) = 0,
\end{aligned} \tag{19}$$

yields an optimal solution to our residential battery optimization problem defined in (8). Various algorithms (e.g., *ε-constraint* by Miettinen (2012) and *Envelope* by Li and Haimes (1987)) can be applied to Problem (19) to compute the set of non-dominated solutions. In this paper, we choose to use the *linear scalarization* approach in Boyd and Vandenberghe (2004). In this approach, the set of non-dominated solutions to Problem (19) is the parameterized set of solutions to

$$\min\ D_1\big(q(1,0),\cdots,q(30,24/\Delta t)\big) + \lambda\, D_2\big(q(1,0),\cdots,q(30,24/\Delta t)\big) \tag{20}$$

subject to the constraints of Problem (19), where λ is a positive parameter. Fortunately, the objective function of Problem (20) is time-separable - implying that applying dynamic programming for each $\lambda > 0$ yields a non-dominated solution to the multi-objective Problem (19).

4.NUMERICAL CASE STUDIES

In this section, we apply our multi-objective approach, described in Section 3, to our residential battery optimization problem, defined in Section 2, using a wide range of load profiles and two pricing plans. The results are optimal charging/discharging rates as a functions of time, minimum electricity bills, and the amounts which the customers save from optimal energy storage in each scenario.

We consider three load profiles q_{a1}, q_{a2} and q_{a3} (corresponding to small, medium and large houses) with 5.66 kW, 8.09 kW and 12.14 kW as their maximum values over the summer and 3.89 kW, 5.55 kW and 8.32 kW as their maximum values over the winter. These load profiles were synthesized based on the measured load of a group of Salt River Project's (SRP) retail customers in Arizona. As for the electricity pricing, we use SRP's E-22 plan and a modified version of E-27 P plan(SRP (2015)). The first plan only includes ToU charges, whereas the second plan charges include both ToU and demand charges. The on-peak period for E-22 plan consist of those hours from 4 PM to 7 PM. The on-peak period of plan E-27 P is from 1 PM to 8 PM for the months May to October. From November to April, the on-peak period of plan E-27 P consist of those hours from 5 AM to 9 AM and from 5 PM to 9 PM. The electricity prices for all the seasons are presented in Table 1. To synthesize a solar generation profile for each day, we created a random generator which uses a time-varying mean and standard deviation of solar generation for a typical summer day and a time-varying mean and standard deviation for a typical winter day. The means and standard deviations for summer and winter were calculated based on the solar radiation data measured at Equestrian Manor station in Scottsdale, AZ for the entire months July and January. In our scenarios, we consider Tesla's Powerwall battery as the source of energy storage. The two existing models of Powerwall battery possess C_I =7 kWh and C_I =10 kWh as storage capacities, $\bar{u}$ = 3.3 kW as the maximum charging/discharging rate and η = 0.92 as the charging/discharging efficiency.

Table 1: SRP's pricing plans (SRP (2015)): E-22 (ToU) and modified E-27 P (ToU & demand). Summer: May, June, Sep. & Oct. Summer peak: July & Aug. Winter: Nov. through April.

Pricing plan	Off-peak price ($/kWh)	On-peak price ($/kWh)	Demand price ($/kW)
E-22 (Summer)	0.0840	0.3033	0
E-22 (Summer peak)	0.0864	0.3588	0
E-22 (Winter)	0.0758	0.1215	0
E-27 P (Summer)	0.0371	0.0486	14.63
E-27 P (Summer peak)	0.0423	0.0633	17.82
E-27 P (Winter)	0.0390	0.0430	5.68

4.1 Case Study I: Analysis of The Benefits of Optimal Energy Storage Over a Month

We applied our multi-objective method to the battery optimization problem defined in (8), except that we reduced the 30-day period ($N = 30$) in Problem (8) to 3 Summer peak days in order to simplify the presentation of results. Accordingly, we scaled the demand price in Table 1 by a factor of $\frac{1}{10}$, i.e., $p_d = \frac{1}{10} \times 17.82$ \$/kW. In Table 2, we have presented the resulting ToU and demand charges for 3 customers with loads q_{a1}, q_{a2} and q_{a3}, and for pricing plans E-22 and modified E-27 P. Assuming that the three-day simulation is repeated ten times, we calculated the monthly bill by multiplying the sum of 3-day charges by ten. For comparison, we have also included the same results for the case where no battery storage is used. For all the cases associated with modified E-27 P pricing, we solved Problem (20) for 500 values of the weighting parameter λ in (20), using time-step $\Delta t = 0.5$ hour and l_{20}-norm approximation for the supremum function in (7). The resulting non-dominated solutions corresponding to the load q_{a2} are shown in Fig. 2. A Matlab implementation of our method calculated the 500 non-dominated solutions in 14.11 minutes using a Core i7 machine with 16 GB of RAM. From Table 2 we observe that in the presence of demand charges (as in modified E-27 P pricing), the customer with the largest house (heavy load) achieves the greatest saving from optimal energy storage, whereas in the absence of demand charges (E-22 pricing), savings from optimal storage are independent of load. In Fig. 3, we have shown the resulting optimal charging/discharging rates and optimal energy storage corresponding to the load q_{a2} for plans E-27 P and E-22 respectively. Fig. 3 (Left) shows that in the presence of demand charges, optimal strategy for battery storage involves precise peak shaving (power going into the battery decreases in response to an increase in appliances load) during the on-peak hours. In the absence of demand charges, optimal strategy is straightforward: Apply maximum discharging rate during the on-peak hours (see Fig. 3 (Right)).

4.2 Case Study II: Long-term Analysis of The Benefits of Optimal Energy Storage

To quantify the long-term benefits of optimal energy storage for residential customers, and assessing battery's degradation under optimal operation, we applied our method to Problem (8) for the duration of 5 years. To reduce the computation, we used the appliances load of 3 representative days for each month, assuming that the same load is repeated 10 times during the month. We accounted for the battery's degradation during the entire month by multiplying the degradation rate β (defined in (4)) by a factor of 10. The resulting total bills (for 5 years) corresponding to using q_{a1}, q_{a2} and q_{a3} as the load profiles and modified E-27 P as the pricing plan are shown in Table 3. For comparison, in Table 3, we have also included the bills corresponding to no battery storage. Moreover, we have included the cost of operating the battery for 5 years as $(C_I - C_F)/C_I \times \$3750$ considering that \$3750 is the market price of Tesla's Powerwall battery with $C_I = 10$ kWh of capacity. C_F denotes

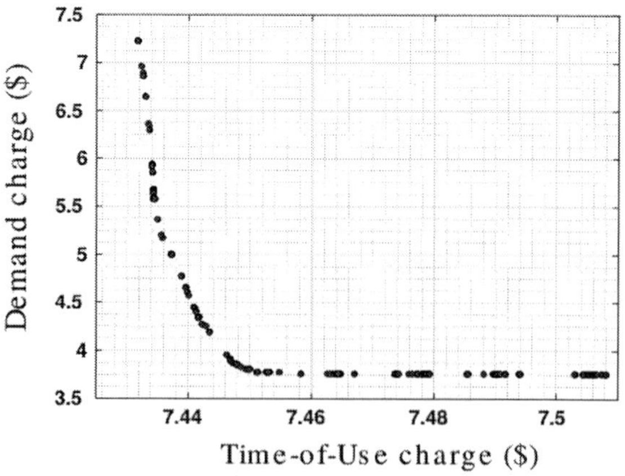

Figure 2: Non-dominated solutions for the multi-objective battery storage optimization problem using medium load profile q_{a2}

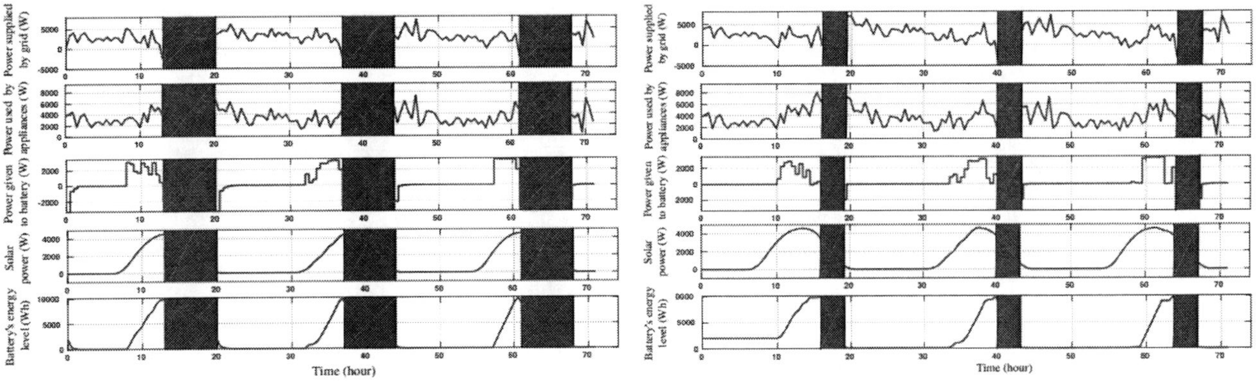

Figure 3: Optimal charging/discharging rates and energy storage during three Summer peak days for q_{2a} appliances load and E-27 P (Left), E-22 (Right) pricing plans

Table 2: ToU & demand charges for 3 days using Summer peak plans E-27 P (top) and E-22 (bottom), and light, medium & heavy loads: q_{a1}, q_{a2} & q_{a3}. (Left: with battery - Right: no battery)

Load profile	ToU charge($)	Demand charge($)	Monthly bill ($)	Monthly saving ($)
q_{a1} (light)	3.30 - 4.03	1.37 - 5.86	46.70 - 98.95	52.25
q_{a2} (medium)	7.45 - 8.20	3.78 - 9.49	112.30 - 176.91	64.61
q_{a3} (heavy)	14.39 - 15.15	9.95 - 16.38	243.43 - 315.04	71.61

Load profile	ToU charge($)	Demand charge($)	Monthly bill ($)	Monthly saving ($)
q_{a1} (light)	2.38 - 11.26	0	23.88 - 112.64	88.76
q_{a2} (medium)	13.59 - 22.47	0	135.94 - 224.71	88.77
q_{a3} (heavy)	32.27 - 41.14	0	322.71 - 411.47	88.76

the remaining capacity after 5 years. We have also calculated the total saving from the battery (see Table 3) as $B_b - (C_b + B_n)$, where B_b denotes the electricity bill when optimal energy storage is used, B_n denotes the electricity bill when no energy storage is used, and C_b denotes the cost of operating the battery. From the table we observe that similar to our analysis in Section 4.1, the customer with the highest load (q_{a3}) achieves the greatest benefit from optimal energy storage. Specifically, the customer with load q_{a3} gains 21% and 32% greater benefits relative to the customers with q_{a2} and q_{a1} respectively.

5.CONCLUSIONS

We addressed the problem of optimal energy storage for residential customers who are charged for both total energy consumed and peak rate of consumption (demand) during a month. Because of the presence of demand charges, this problem violates the principle of optimality - a sufficient condition for Hamiltonian-based algorithms and dynamic programming to converge to an optimal solution. We approached this problem by defining a sequence of multi-objective dynamic programs, indexed by p, whose sets of non-dominated solutions are guaranteed to contain an optimal solution to the original problem as $p \to \infty$. Since the principle of optimality holds for each multi-objective problem in the sequence, standard policy iteration and linear scalarization can be used to find sub-optimal non-dominated solutions for each p. Our numerical case studies over a wide range of customers' load profiles show that optimal energy storage using the existing residential batteries and solar generation can reduce the monthly electricity bill by up to 52%, in the presence of demand charges. Moreover, our long-term numerical analysis over the load range, 5.6 kW to 12.1 kW, shows that in the presence of demand charges and battery degradation, a customer with twice the peak load gains 32% higher benefit from optimal energy storage over 5 years.

Table 3: Total bill (for 5 years), final capacity of battery, cost to battery & saving from storage using light, medium and heavy load: q_{a1}, q_{a2} & q_{a3}. (Left: with battery - Right: no battery)

Load profile	Total bill ($)	Final Capacity (kWh)	Cost to Battery ($)	Total saving ($)
q_{a1} (light)	1599.13 - 2352.73	9186.1	305.21	448.39
q_{a2} (medium)	3493.93 - 4318.25	9098.3	338.13	486.19
q_{a3} (heavy)	6730.27 - 7657.36	9103.2	336.37	590.72

REFERENCES

(2014). Integrated resource plan: Executive summary. *Arizona Public Service.*

(November 2015). Standard electric price plans. *Salt River Project Agricultural Improvement and Power District Corporate Pricing.*

Bellman, R. E. and Dreyfus, S. E. (1962). *Applied Dynamic Programming.* Princeton University Press.

Boaro, M., Fuselli, D., De Angelis, F., and Piazza, F. (2013). Adaptive dynamic programming algorithm for renewable energy scheduling and battery management. *Cognitive Computation*, 5(2):264–277.

Boyd, S. and Vandenberghe, L. (2004). *Convex optimization.* Cambridge university press.

Bozchalui, M. C., Hashmi, S. A., Hassen, H., Cañizares, et al. (2012). Optimal operation of residential energy hubs in smart grids. *IEEE Transactions on Smart Grid*, 3(4):1755–1766.

Dunn, Bruce, H. K. and Tarascon, J.-M. (1997). Electrical energy storage for the grid: a battery of choices. *Science*, 334(6058):928–935.

Giorgio, A. D., Pimpinella, L., and Liberati, F. (2012). A model predictive control approach to the load shifting problem in a household equipped with an energy storage unit. In *Mediterranean Conference on Control & Automation*, pages 1491–1498.

Huang, T. and Liu, D. (2013). A self-learning scheme for residential energy system control and management. *Neural Computing and Applications*, 22(2):259–269.

Li, D. and Haimes, Y. Y. (1991). Extension of dynamic programming to nonseparable dynamic optimization problems. *Computers & Mathematics with Applications*, 21(11):51–56.

Li, D.-J. and Haimes, Y. Y. (1987). The envelope approach for multiobjeetive optimization problems. *IEEE Transactions on Systems, Man and Cybernetics*, 17(6):1026–1038.

Li, N., Chen, L., and Low, S. (2011). Optimal demand response based on utility maximization in power networks. *IEEE Power & Engineering Society*, pages 1–8.

Miettinen, K. (2012). *Nonlinear multiobjective optimization*, volume 12. Springer Science & Business Media.

Oudalov, A., Cherkaoui, R., and Beguin, A. (2007). Sizing and optimal operation of battery energy storage system for peak shaving application. pages 621–625.

Rowe, M., Yunusov, T., Haben, S., Singleton, C., Holderbaum, W., and Potter, B. (2014). A peak reduction scheduling algorithm for storage devices on the low voltage network. 5(4):2115–2124.

Wei, Q., Liu, D., and Shi, G. (2015). A novel dual iterative-learning method for optimal battery management in smart residential environments. *IEEE Transactions on Industrial Electronics*, 62(4):2509–2518.

Yoshida, A., Sato, T., Amano, Y., and Ito, K. (2016). Impact of electric battery degradation on cost-and energy-saving characteristics of a residential photovoltaic system. *Energy and Buildings, In press.*

ACKNOWLEDGEMENT

This material is based upon work supported by the National Science Foundation under Grant #CMMI-1301851 and by Salt River Project under the FY 2016 ASU Cooperative Agreement.

Assessing the Potential for Reduction in Peak Residential Electrical Loads Using a Heat Pump and Thermal Storage

Christopher BALDWIN*, Cynthia A. CRUICKSHANK

Carleton University, Department of Mechanical and Aerospace Engineering,
Ottawa, Ontario, Canada
christopher.baldwin@carleton.ca
* Corresponding Author

ABSTRACT

Demand for space cooling in Canada has significantly increased in the past 20 years, and in conjunction with space heating loads in the winter are placing larger peak loads on the electrical grid. As a result, utilities must increase their generating and transmission capacity to meet the peak annual demand, with much of the capacity going unused for large portions of the year. Additionally, base loads are typically met using cleaner technologies including hydro and nuclear, while the variable peak loads are more commonly met using fossil fuel generation, increasing the greenhouse gas emissions per kilowatt-hour of electrical generation. To reduce this peak load, demand side management strategies are becoming more common, with one potential method for reducing the peak load produced by residential buildings is the pairing a heat pump with thermal storage. This paper outlines the first stage of a multi-stage research project to develop a comprehensive system and control strategy for a residential heat pump with sensible hot and cold thermal storage tanks. It outlines the steps that were taken to optimize the control strategy, with a focus on reducing consumption during peak periods while remaining cost and greenhouse gas emission neutral on an annual basis. It was found that using small scale sensible storage and a standard geothermal heat pump, a reduction in the percent of energy used during peak periods is realized, however the annual consumption, electrical costs, and greenhouse gas emissions increase. This was primarily the result of a significant decrease in heat pump performance as the result of lower source and higher load temperatures into the heat pump.

1. INTRODUCTION

As the demand in Canada for space cooling continues to increase, an ever growing peak load is being placed on the utility grid (Natural Resources Canada - Office of Energy Efficiency, 2013). As electrical utilities must size generating and transmission capacity to the peak load the grid experiences, a much greater emphasis is now being placed not only on reducing electrical consumption, but the time of day electricity is used (Ontario Ministry of Energy, 2013). A number of strategies are being employed by utility providers to reduce the peak load, including time of use billing which charges a premium for energy used during peak times and providing incentives to turn off high consuming devices (most notably air conditioners) during peak periods (Independent Electrical Service Operator, 2016). In addition to incentives and policy changes being implemented, the development of new systems and control strategies to shift energy consumption to off peak periods are being developed. One proposed method is using a liquid-to-liquid heat pump paired with thermal storage system(s). Before the wide scale implementation of heat pumps with thermal

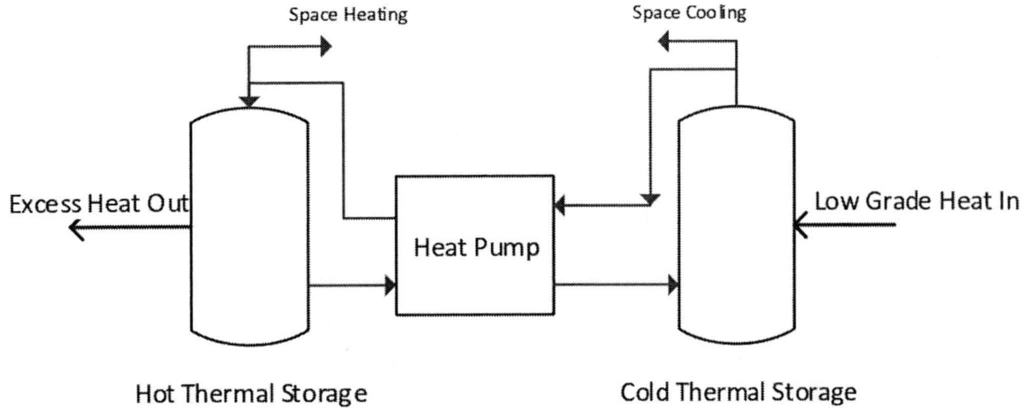

Figure 1: High level schematic of proposed system

storage systems is possible, extensive study must be conducted on their potential for reducing peak loading, utility cost savings and effect on overall energy consumption.

This paper examines the potential for peak load reduction using an integrated system consisting of a small, 6 $kW_{thermal}$ heat pump with a water tank based hot thermal storage and a cold thermal storage tank with a glycol/water mixture as the storage medium. A schematic of the proposed system is shown in Figure 1. A model of the heat pump and cold thermal storage tanks has been previously developed and calibrated to an experimental test system (Baldwin & Cruickshank, 2016). This previously validated heat pump model was then integrated into a house model developed in TRNSYS and , with the composite model used to determine the impact of different control strategies of the heat pump and thermal storage systems on total energy consumption, electrical consumption during peak periods, total electrical costs and total greenhouse gas emissions. This paper represent the first phase in a long term project, and examines whether a standard geothermal heat pump and sensible thermal storage will provide the necessary storage capacity and performance to efficiently offset a significant portion of peak energy consumption while reducing the annual costs and greenhouse gas emissions from space heating and cooling.

2. BACKGROUND AND THE ONTARIO ENERGY LANDSCAPE

This study was conducted for a house located in Ottawa, Ontario, Canada, and as a result, the electrical landscape in Ontario is of vital importance to the work. Electricity in Canada is handled and managed at a provincial level, with each province having distinct electrical grid and generating capacity. Ontario's electrical generation is very diverse, with a significant portion of the annual electrical generation coming from 4 different sources (nuclear, hydro, wind and natural gas) (Independent Electrical Service Operator, 2016). Consequently, the hour by hour generating mix is unique each hour of the year, and as such, the greenhouse gas intensity per kilowatt-hour changes based on the province wide electrical demand and the availability of variable renewable sources (wind and solar). As the peak load is predominantly met using natural gas power plants, meaning that energy used during peak periods has a much higher greenhouse gas intensity on average compared to periods of low demand.

To promote the reduction in electrical consumption, Ontario introduced time of use billing, in which a premium is paid for electricity used during peak periods, while electricity used during off-peak periods is considerably cheaper. For the ease of the customer, peak, mid-peak and off-peak periods are pre-defined based on historical consumption patterns, allowing consumers to tailor their consumption to these time periods. During the summer months, which last from May 1st to October 31st, peak periods occur Monday to Friday, 11am to 5pm while mid-peak occur from 7am-11am and 5pm-7pm. During the winter months, from November 1st to April 30th, peak periods occur from 7am-11am and 5pm-7pm, while mid-peak occurs from 11am-5pm. In both the summer and winter periods, off-peak periods occur from 7pm to 7am, and 24hr on weekends. As of May 1st, 2016, peak rates in Ontario are 18 cents, mid-peak rates are 13.2 cents, and off-peak rates are 8.7 cents (Ontario Energy Board, 2015).

As Ontario has clearly defined peak and off-peak periods, for the remainder of this work, peak and off-peak periods will be as defined by the electrical providers in Ontario. Using data from the Independent Electrical Services Operator of Ontario, the amount of electricity generation from each sources on an hourly basis was calculated. From these values, and using the 50th percentile of greenhouse gas emissions by source as stated by the 2011 United Nations Intergovernmental Panel on Climate Change (Edenhofer, et al., 2011), the average greenhouse gas emissions per kilowatt-hour of electrical generation for peak, mid-peak and off-peak periods in both the summer and winter periods were calculated and are shown in Table 1.

Table 1: Greenhouse gas intensity in Ontario by usage period-2015

Period	Greenhouse Gas Intensity (g/kWh)		
	Off-Peak	Mid-Peak	Peak
Summer	54.9	70.5	76.0
Winter	51.8	59.9	60.2

The cost of electricity provides significant incentive to shift electrical consumption from peak to off-peak periods. For example, for each kilowatt-hour shifted, the consumer saves 9.3 cents (this represents a reduction greater than 50% in cost). In addition, each kilowatt-hour shifted provides a reduction in greenhouse gas emissions, that is, 21.1 g of CO_2 in the summer and 8.2 g of CO_2 in the winter. These reductions benefit both the consumers and utility provides.

3. HOUSE MODEL

Before the proposed system could be modelled to determine the annual performance, a house model had to be developed in TRNSYS (TRNSYS: A Transient Simulation Program, 2015) to provide building loads for the heating and cooling system. The house has modelled to represent a newly built house located in Ottawa, Ontario, Canada. The modelled house is a two-story, single detached house with a full basement. Each level has a floor area of 110 m^2 with a total volume of 835 m^3. The house was broken into 4 air zones, being the basement, main floor, 2nd floor and the attic, with air exchange between the basement, main and 2nd floor. The heating was an air based system, where 20% of distribution air goes to the basement, 35% to the main floor and 45% the 2nd floor, allowing the increased heating and cooling demand in the 2nd floor to be met as a result of the heat loss through the ceiling into the attic. A single thermostat has been placed in the main floor with a heating set-point of 20°C and a cooling set-point of 23°C. The remaining specifications for building insulation and windows are provided in Table 2.

Table 2: House construction specifications

Type	Parameter	Units	Value
Thermal Resistance	Above Grade Walls	m^2K/W	4.5
	Attic	m^2K/W	11.5
	Below Grade Walls	m^2K/W	2.7
	Under Slab	m^2K/W	1.9
Windows	U-Value	W/m^2K	1.27
	Solar Heat Gain		0.624
House	Air Leakage	ACH	0.05
	Occupancy	Number of People	4

3.1 Baseline Energy Consumption

Once the house model was developed within TRNSYS, a baseline energy consumption for heating and cooling had to be determined. A fluid heater and fluid chiller, both with a coefficient of performance (COP) of 1 and with no losses where integrated into the model to represent an ideal heating system. The model was then run for 1 year (8760 hours), with the electrical consumption being recorded every 2 minutes. Based on these simulation results, the house being modelled was found to have an annual space heating load of 16,215 kWh and an annual cooling load of 4686 kWh. Of more importance than the total energy consumption, was the time at which the energy was used, the cost to the consumer and the total greenhouse gas emissions as a result of the space heating and cooling. The energy consumption was broken down by peak, mid-peak and off-peak as defined by Ontario's time of use billing, while the greenhouse gas emissions were calculated using the average for each time period for winter and summer periods. A summary of these results are presented in Table 3.

Table 3: Baseline energy consumption by time, heating costs and greenhouse gas emissions

Energy Type	Summer (kWh)			Winter (kWh)			Cost (CAD$)	Greenhouse Gas Emissions (kg)
	Peak	Mid-Peak	Off-Peak	Peak	Mid-Peak	Off-Peak		
Heating	17	87	2091	1975	752	11,293	1580	321
Cooling	2399	857	1430	0	0	0	664	871
Total	2416	944	3521	1975	752	11,293	2444	1192

Based on these results, a number of conclusions can be noted. The baseline energy consumption for space heating is predominantly off peak, with over 80% of the winter space heating load occurring during off-peak periods. This was expected, as during the overnight period, no solar gains are present to reduce the space heating load and the outdoor temperature is at its lowest. This results in the bulk of the heating required during the overnight, off-peak period. If a night-time temperature set-back was introduced, the energy consumption would shift towards a larger percentage during peak periods, particularly if the morning heating occurs at or after 7am. Although a large percentage of the heating load occurs during off peak periods, almost 2000 kWh of heating occurs during peak periods, accounting for almost 23% of the total space heating costs, of which most occurs early in the morning. This peak load could be easily shifted to off-peak periods through the use of thermal storage overnight.

When looking at the space cooling demand, a much larger percentage of the load occurs during peak periods, with 51% of the cooling load (and 65% of cost) occurring during peak periods, when peak periods only account for 18% of the summer period. As such, space cooling shows the greatest potential to see a meaningful benefit from shifting electricity consumption from peak to off-peak periods using heat pumps and thermal storage.

3.2 Baseline Energy Consumption Using a Heat Pump

The previous section indicated the total space heating and cooling loads, and the time of day they occurred using a generic auxiliary heater and chiller with a COP of 1. This provided useful data to recognize energy trends and a true baseline, however to make a true comparison of the proposed system that integrates both a heat pump and thermal storage systems, a baseline energy consumption using the same heat pump that will be integrated with the thermal storage had to be determined. Using the same house model and air distribution system, a 6 $kW_{thermal}$ heat pump was coupled with the heating and cooling coils. This work focused on the heat pump and thermal storage systems and not the potential heat source or sink for the system, and as such to remove any discrepancies or errors that could be introduced by these, a constant heat source was provided on the source side of 15°C and 450 L/hr during the heating season and a heat rejection of 25°C and 450 L/hr. The same thermostat set-points were used as for the baseline energy consumption and the electrical consumption for space heat and space cooling was independently reordered and is presented in Table 4.

Table 4: Baseline energy consumption using a heat pump for space heating and cooling

Energy	Summer (kWh)			Winter (kWh)			Cost	Greenhouse Gas
Type	Peak	Mid-Peak	Off-Peak	Peak	Mid-Peak	Off-Peak	(CAD$)	Emissions (kg)
Heating	4	22	510	485	219	2282	351	190
Cooling	536	196	322	0	0	0	149	72
Total	540	218	832	485	219	2282	500	262

These results held the trends of the first baseline energy consumption, however, the introduction of the heat pump saw the energy consumption reduced by 78% for space heating and by 77% for space cooling. The values obtained from this base model incorporating the heat pump with no thermal storage will be used as the baseline for all future simulations incorporating thermal storage.

4. INTEGRATING THERMAL STORAGE – TEMPERATURE BASED CONTROLS

After determining the baseline energy consumption for the modelled house with and without a heat pump, the complete systems including the heat pump, hot thermal storage and cold thermal storage was integrated with the house model. The specifications for each of the thermal storage systems are provided within Table 5.

Table 5: Baseline energy consumption using a heat pump for space heating and cooling (DOW Chemical Compnay)

Tank Property	Hot Storage	Cold Storage
Fluid	Water	Water/Glycol
Density (kg/m^3)	1000	1046.7
Thermal Conductivity (W/m·K)	0.594	0.378
Specific Heat (kJ/kg·K)	4.19	3.32
Viscosity (cP)	1	5
Thermal Expansion (1/K)	0.00026	0.000495
Capacity (L)	400	270

Once the thermal storage tanks were integrated into the model, a control strategy had to be developed to control the complete systems. The thermostat within the house model would still control when heating or cooling was provided to the house, however the control of heated water or chilled water/glycol solution to the air handler and the charging of the two thermal storages using the heat pump must be controlled independently. To accomplish this, four independent controllers were implemented.

The first controlled the flow of hot water through the air handling unit, providing space heat to the building. As the tank temperature at any given point is variable, the flow rate through the air handler must vary to produce a constant delivery temperature, and therefore, a variable speed pump, controlled using a PID controller to control the flow to meet a delivery output temperature of 30°C. A similar control strategy was implemented for space cooling, with a PID controller providing a control signal to a variable speed pump to meet the output temperature of 12°C. These set-points were used as the default values and the optimal value determined. A schematic of the air handler and heat pump control system is shown in Figure 2.

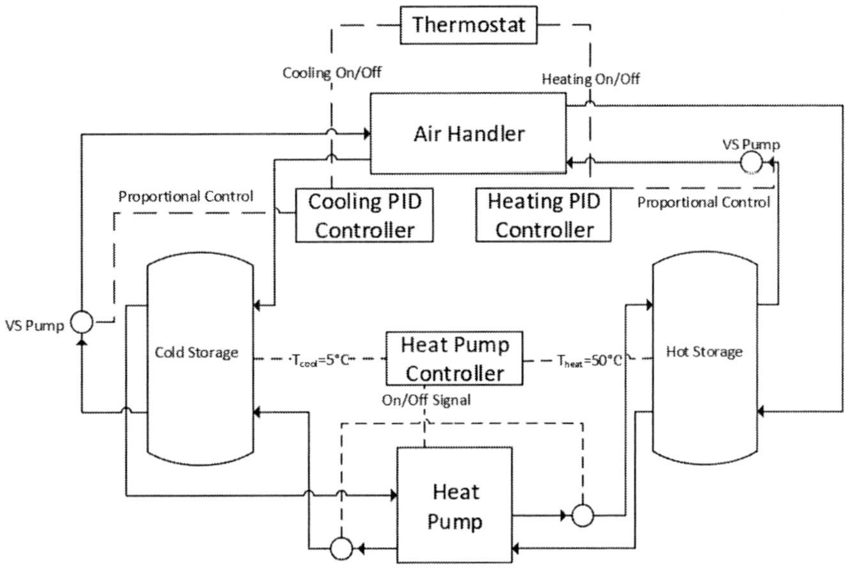

Figure 2: Control schematic for the integrated heat pump and thermal storage system

To control the heat pump, thermostats were placed within both tanks, with the thermostat in the hot tank placed at a height of 80% of the tank, while the thermostat is placed at a height of 20% of the tank. The hot tank was initially set to 55°C and the cold tank set to 5°C. For the heat pump to turn on, the hot tank must drop below the set-point, and be during the heating season or the cold tank must rise above the set-point and be during the cooling season. When the heat pump turns on, two single speed pumps (one on the load side, one on the source side) are simultaneously switched on at a default flow rate of 360 L/hr.

To ensure the hot tank does not overheat, or the cold tank temperature drops too low, a supplementary control system was implemented. This allowed the heat pump to draw heat from a supplementary source (e.g., from a ground loop) when the cold tank dropped below the set-point, or to reject heat when the hot tank was above the set-point (e.g., rejected outdoors). In these cases, when the cold tank is bypassed, a return temperature to the heat pump is set at 15°C and when the hot tank is bypassed, a constant return temperature of 20°C is set.

4.1 Preliminary Results

The integrated system has a number of variables that are being simultaneously controlled for both the hot and cold settings. Values for each of the variables were chosen to determine preliminary results, which will provide a baseline for optimization of the system. These variables, and the initially selected values are shown in Table 6. In addition to the initial value, a high and low value are provided to show the range of interest for each variable that will be simulated to determine the impact of each variable on the overall system.

Table 6: Control variables within the integrated system

Variable	Units	Initial	High	Low
Hot Tank Set-Point	°C	50	60	40
Cold Tank Set-Point	°C	5	10	0
Hot Flow Rate	L/hr	360	480	240
Cold Flow Rate	L/hr	360	480	240
Cooling – Delivery Set-Point	°C	12	16	8
Heating – Delivery Set-Point	°C	30	35	25

As the heating and cooling systems are integrated, and the heat pump can simultaneously provide energy for space heating and space cooling, the electrical energy consumption cannot be differentiated between heating and cooling. As a result, the electrical consumption can only be reported as the total electrical energy required for space heating and cooling in the winter, summer and as an annual total. Using the initial values, the total electrical consumption, time of electrical consumption, total electricity costs and the greenhouse gas emissions are listed in Table 7.

Table 7: Results using the initial values for control variables

Variable	Summer	Winter	Annual
Electrical Consumption (kWh)	3265	4954	8219
Peak Consumption (kWh)	841	864	1705
Mid-Peak Consumption (kWh)	549	746	1295
Off-Peak Consumption (kWh)	1876	3344	5221
Electrical Costs ($CAD)	$379	$532	$911
Greenhouse Gas Emissions (kg)	206	270	476
Percent Consumption During Peak Periods (%)	25.7%	17.4%	20.7%

The initial results provided mixed results, with the amount of electricity being used for space heating and cooling seeing an electrical consumption increase of 105% in the summer, 66% in the winter and 80% over the whole year. As a result, greenhouse gas emissions and electrical costs have increased 82%. On the promising side however, the percentage of electrical consumption that occurs during the peak period in the summer period was reduced from 34% to 26%, however a large percentage of this energy was shifted to mid-peak as opposed to off-peak, as the percent of consumption during mid-peak almost doubled, from 9% in the base case to 16% with thermal storage.

From the baseline simulation with the complete system, an advantage can be seen in terms of shifting energy consumption, however significant optimization is required before the system out-performs the system without thermal storage. The first step is to determine which of the 6 control variable has the greatest impact on the annual performance of the system. To assess this, 12 additional simulations were conducted, changing one of the variables each time, first to the low value, followed by the high value, with the percent increase or decrease in total energy consumption, cost, greenhouse gas emissions and percent consumption during peak periods are shown in Figure 3.

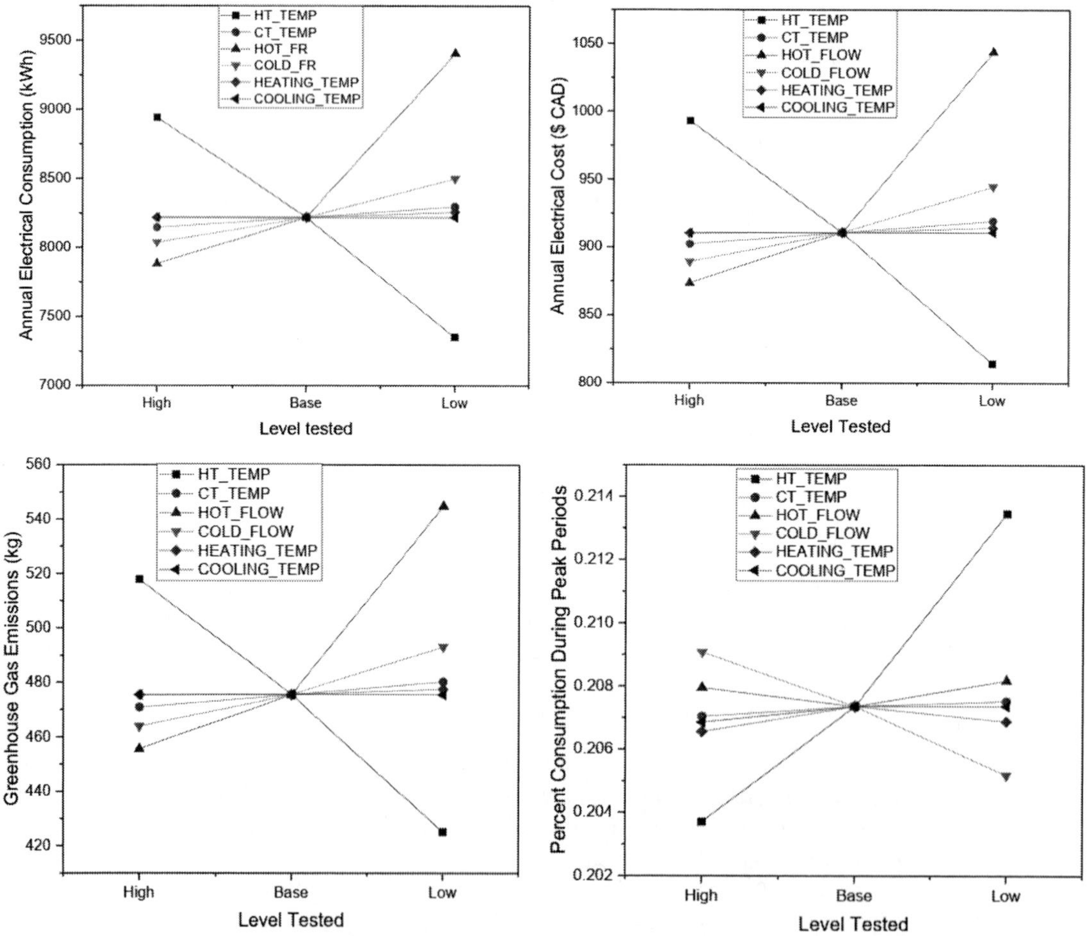

Figure 3: Comparison of high and low values of each of the 6 control variables

Based on the results of varying the 6 control variables, the parameter that provided the lowest annual energy consumption was selected (in every case this also provided the lowest cost and the least amount of greenhouse gas emissions). A hot tank temperature of 40°C, a cold tank temperature of 12°C, a hot side flow rate of 480 L/hr, a cold side flow rate of 480 L/hr, a heating distribution temperature of 30°C and a cooling distribution temperature of 8°C. The simulation was run using this control scheme and the results for all of the results of interest are shown in Table 8, and compared to the results using just the heat pump.

Table 8: Results using the optimal set-point for each control variable

Variable	Heat Pump with Thermal Storage	Heat Pump Only	Difference with Thermal Storage
Electrical Consumption (kWh)	6656	4576	+45%
Peak Consumption (kWh)	1411	1025	+38%
Mid-Peak Consumption (kWh)	871	437	+99%
Off-Peak Consumption (kWh)	4373	3114	+40%
Electrical Costs ($CAD)	$732	$500	+40%
Greenhouse Gas Emissions (kg)	382	262	+45%
Percent Consumption During Peak Periods (%)	21.2%	22.4%	-5%

These results showed that even when the optimal temperature based control strategy is implemented, a significant increase occurs in energy consumption on an annual basis and when each rate period is considered. Additionally, as a result of the increased consumption through each rate period, the annual electrical cost and the total greenhouse gas emissions also increased. This increase in consumption was primarily the result of decreased performance of the heat pump when charging the thermal storage systems as compared to directly heating and cooling the space. This was as result of the increased load temperatures returning from the hot thermal storage tank during the heating season and the much lower source temperatures returning from the cold thermal storage tank. Although the total electricity consumption during peak periods increased, the percentage of total consumption that occurred during peak periods decreased, showing that there is potential for this system in reducing peak loads, however further optimization of the control strategy is required.

5. TIME BASED CONTROLS

The results presented in the previous section clearly demonstrated that a simple temperature based control strategy significantly increased energy consumption and had a negligible impact on shifting energy consumption to off-peak periods. As such, a control strategy that takes into account the time of day is required when peak and off-peak periods exist. As such, a controller with a variable set-point for both the hot tank and cold tank was implemented, with the set-points for peak and off-peak periods for the two thermal storage systems indicated in Table 9.

Table 9: Thermal storage set-points for time based control

	Off-Peak Set-Point (°C)	Peak and Mid-Peak Set-Point (°C)
Hot Thermal Storage	50	30
Cold Thermal Storage	5	15

The simulation was run with this new control strategy, with the remaining control parameters left unchanged from the final simulation using a temperature based controlled strategy. Results utilizing the new control strategy on an annual basis are shown in Table 10, with a comparison to heating and cooling with the heat pump.

Table 10: Base results with a time based control strategy

Variable	Heat Pump with Thermal Storage	Heat Pump Only	Difference with Thermal Storage
Electrical Consumption (kWh)	6324	4576	+38%
Peak Consumption (kWh)	1112	1025	+8%
Mid-Peak Consumption (kWh)	506	437	+16%
Off-Peak Consumption (kWh)	4706	3114	+51%
Electrical Costs ($CAD)	$657	$500	+31%
Greenhouse Gas Emissions (kg)	356	262	+36%
Percent Consumption During Peak Periods (%)	17.6%	22.4%	-22%

From these results, it can be seen that switching to a time based control strategy shows significant benefit when compared to using a temperature based control strategy, however still uses more energy in each of the time periods when compared to space conditioning with just the heat pump. To ensure the selected temperature settings are optimal a range of values from 0°C to 10°C and 40°C to 60°C were simulated as the set points during the peak periods, with the results for total consumption and percent during peak periods shown in Figure 4.

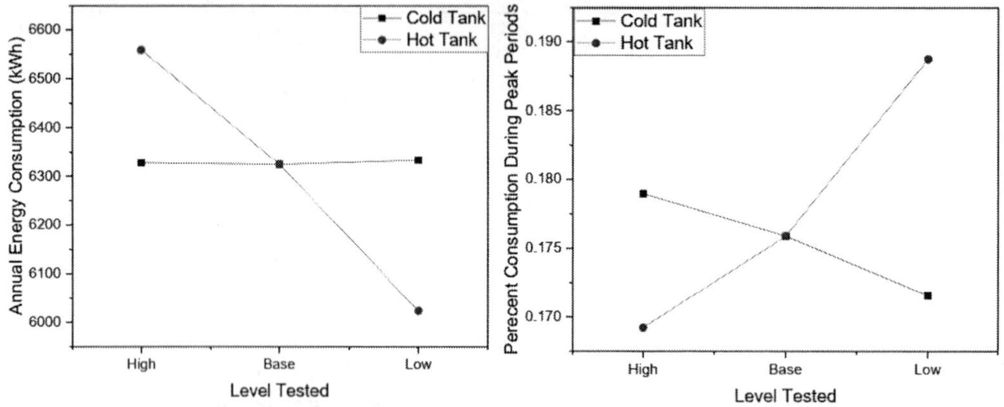

Figure 4: Comparison of high and low values for the set-point in the hot and cold tank

Based on these results, the set-point of the cold tank has minimal impact on the total energy consumption, with a less than a 0.25% change in the total consumption, however, the set-point did account for a 2.5% decrease in the percentage of energy consumption during peak periods. The temperature set-point on the hot tank had a much larger impact on the overall energy consumption, with decreasing the hot tank set-point to 40°C reducing total energy consumption by almost 5%, however the percent consumed during peak periods increased by almost 7%, while setting the hot tank to 60°C had the opposite effect, increasing the total energy consumption by 4%, while decreasing the percent consumption during peak periods by 4%. Although this looks beneficial in terms of reducing the percent consumed during peak periods, most of the change actually came from increasing the total consumption as opposed to actually reducing the consumption during the peak periods with the actual consumption during peak periods decreasing only 3 kWh from the base case. As such, the simulation was rerun using a hot tank set-point of 40°C and a cold tank set-point of 0°C and the results of this simulation are presented in Table 11, and compared to when only the heat pump was used for space conditioning.

Table 11: Results using the optimal control set-points for both the hot and cold thermal storage

Variable	Heat Pump and Thermal Storage using Time Based Controls	Heat Pump Only	Difference with Thermal Storage
Electrical Consumption (kWh)	6034	4576	+32%
Peak Consumption (kWh)	1112	1025	+8%
Mid-Peak Consumption (kWh)	511	437	+17%
Off-Peak Consumption (kWh)	4412	3114	+42%
Electrical Costs ($CAD)	$633.63	$500.45	+27%
Greenhouse Gas Emissions (kg)	341	262	+30%
Percent Consumption During Peak Periods (%)	18.4%	22.4%	-18%

From these results, it can be seen that a time based control not only reduces the overall annual consumption compared to temperature based controls, but considerably reduces the peak power consumption and the percentage of power used during peak periods. This shows that the control strategy has potential to reduce the peak load consumption, however the annual electrical consumption, costs and greenhouse gas emissions are still much greater than when using just a heat pump. It is important to note that the consumption during peak periods is now close in both scenarios and the percentage of total consumption during peak periods has decreased considerably.

6. DISCUSSION

Through the optimized control of the proposed system, mixed results were observed through this preliminary study of the potential for shifting peak consumption from off-peak to peak periods. The total electrical consumption of the system with the optimized control strategy has increased by 32% when compared to directly using the heat pump for space conditioning. The only parameter which decreased when going from the heat pump only to the heat pump and thermal storage was the percent of energy consumption during peak periods, which decreased by 18%, however this is somewhat misleading as it was as much the result of increasing the off-peak consumption as it was in reducing the peak consumption. These results indicate that there is potential for reducing peak consumption through thermal storage while remaining cost neutral, however other factors other than just the control strategy must be considered.

This study looked only at optimizing the control strategy with the stated components, and did not in any way look at optimizing the parameters of the components or the size and capacity of the system. It is anticipated that these factors could have a significant impact on the overall system performance. Of particular interest is the heat pump. Moving forward, the heat pump's thermal output should be increased as the current model just meets the heating demand in the winter, but must be on almost continuously during the coldest periods of the year. As a result, there is no time for the heat pump to charge the hot thermal storage overnight, and consequently almost no heating load is shifted from peak to off-peak periods when the heating loads are greatest.

The other factor that should be considered when selecting the heat pump is the performance at the load and source temperatures provided from the thermal storage systems. As this project focused on using a standard ground source heat pump, the source temperature from the cold storage is much lower than the design temperatures. This significantly decreases the performance of the system as a whole as the COP of the heat pump drops significantly. The same effect to a lesser extent is also observed with the hot thermal storage, however storage temperature are closer to the design temperatures for the load side of the heat pump. When using just the heat pump, the overall COP of the heat pump for space heating and cooling for the year is 4.5, while the overall COP when integrating thermal storage is 3.3. This amounted to a 25% reduction in heat pump performance. This phenomenon was discussed by (Dincer & Rosen, 2011), where they state that a decrease from 0°C to -10°C for a cold storage causes a reduction in COP of 70%, and a decrease in cooling capacity of 56%, which would account for the 25% annual reduction seen in this study. To achieve the goal of being cost and greenhouse gas emission neutral while shifting all or a substantial proportion of energy use to off peak periods, a heat pump that is designed to perform at lower temperatures must be employed.

In addition to changes required to the heat pump to improve performance of the system, the cold thermal storage is significantly undersized, and currently is unable to meet the cooling demand during peak and mid-peak periods during the day. As one of the objectives to this study was to determine whether small scale, sensible thermal storage systems can be utilized, it was determined that a thermal storage of this size is unable to provide the required load shifting, and alternative methods for cold storage needs to be explored, allowing for a greater thermal storage density. The most promising method for this is the future study of implementing ice storage into the system, capitalizing on the phase change from water to ice as the primary cold thermal storage mechanism. The introduction of ice storage would see a further decrease in the source temperature entering the heat pump, and as such, the implementation of ice storage would require a heat pump designed for low temperatures.

7. CONCLUSIONS

Based on the findings of this study, a number of conclusions can be drawn from the results. The use of a standard heat pump designed for a geothermal heating and small scale, sensible thermal storage, as tested in this study, increased energy consumption during each rate period on an annual basis, but did decrease the percentage of electrical consumption during peak periods. This proposed system, with the optimization of the control strategy implemented, did not achieve the primary goal of shifting a substantial amount of electrical consumption to off peak periods while staying cost and greenhouse gas neutral. Although this system was unable to reach the desired outcome, this study showed that there is potential for pairing of a heat pump and thermal storage systems. Once the complete design is optimized, including heat pump selection and thermal storage capacity, it is anticipated that the system can reduce peak loads within single detached housing, with a decrease in annual electrical costs and greenhouse gas emissions.

This was the first phase of a multi-phase project. Based on these results, further study will be conducted on determining the required performance from a heat pump to make the systems viable. Based on those results, a new heat pump will be specified and experimentally evaluated. Concurrently, work on increasing the storage density of the cold storage will be explored, with the end goal of developing a small scale, residential sized ice storage system that can provide adequate cooling potential to meet the maximum daily cooling load.

REFERENCES

Baldwin, C., & Cruickshank, C. A. (2016). Using TRNSYS Types 4, 60 and 534 to model residential cold thermal storage units using water and water/glycol solutions. *eSim 2016*, (pp. 512-522). Hamilton, Ontario.

Dincer, I., & Rosen, M. (2011). *Thermal Energy Storage; Systems and Applications, 2nd Edition.* West Sussex, United Kingdom: John Wiley and Sons, Ltd.

DOW Chemical Compnay. (n.d.). *Guide to Glycols.* Midland, Michigan: Dow Chemical Company.

Edenhofer, O., Pichs-Madruga, R., Sokona, Y., Seyboth, K., Matschoss, P., Kadner, S., . . . Stechow, C. v. (2011). *IPCC Special Report on Renewable Energy Sources and Climate Change Mitigation.* Cambridge, United Kingdom: Cambridge University Press.

Independent Electrical Service Operator. (2016, January 1). *Generator Output by Fuel Type Hourly Report.* Retrieved from http://reports.ieso.ca/public/GenOutputbyFuelHourly/

Independent Electrical Service Operator. (2016). *Peaksaver PLUS.* Retrieved from https://www.saveonenergy.ca/Consumer/Programs/Peaksaver-Plus.aspx

Natural Resources Canada - Office of Energy Efficiency. (2013). *Energy Use Data Handbook.* Ottawa, Canada.

Ontario Energy Board. (2015, November 02). *Electrcity Proces.* Retrieved March 2016, from http://www.ontarioenergyboard.ca/oeb/Consumers/Electricity/Electricity%20Prices

Ontario Ministry of Energy. (2013). *Achieving Balance - Ontario's Long Term Energy Plan.* Toronto.

(2015). *TRNSYS: A Transient Simulation Program.* Madison, Wisconsin: University of Wisconsin Solar Energy Laboratory.

ACKNOWLEDGEMENT

The authors would like to acknowledge the financial support of the Natural Sciences and Engineering Research Council (NSERC) of Canada in completing this research project.

Simulation Based Design and Testing of a Supervisory Controller for Reducing Peak Demand in Buildings

James NUTARO[1]*, Ozgur OZMEN[2], Jibonanada SANYAL[3], David FUGATE[4], Teja KURUGANTI[5]

[1]Oak Ridge National Laboratory
Oak Ridge, Tennessee, USA
nutarojj@ornl.gov
*Corresponding author

[2]Oak Ridge National Laboratory
Oak Ridge, Tennessee, USA
ozmeno@ornl.gov

[3]Oak Ridge National Laboratory
Oak Ridge, Tennessee, USA
sanyalj@ornl.gov

[4]Oak Ridge National Laboratory
Oak Ridge, Tennessee, USA
fugatedl@ornl.gov

[5]Oak Ridge National Laboratory
Oak Ridge, Tennessee, USA
kurugantipv@ornl.gov

ABSTRACT

We describe a supervisory control strategy for limiting peak power demand by small and medium commercial buildings while still meeting the business needs of the occupants. The objective of the supervisory control is to operate no more than N loads at all times while satisfying equipment limits on acceptable operation, and to exceed N only when these constraints cannot otherwise be satisfied. This is accomplished with a scheduling algorithm that prioritizes the operation of equipment according to data provided with each request to operate. We demonstrate this algorithm using a simulation model of a gymnasium with four roof top air conditioning units. Notably, this simulation is performing using the control software as it will be installed in the building by having the simulation software expose an interface identical to that used for interacting with the air conditioning equipment. The performance of the proposed algorithm relative to uncoordinated operation based on a typical on/off thermostat is also explored via simulation. These simulations suggest that the scheduling algorithm can reduce peak demand under a modest load and that it degrades gracefully under heavy loads while maintaining the temperature of the gymnasium within the desired range.

1. INTRODUCTION

Reducing the peak power demand in a building can reduce electricity expenses for the building owner and contribute to the efficiency and reliability of the electrical power grid. For the building owner, reducing peak power demand can reduce expenses by eliminating peak power charges from electricity bills. For the power system operator, reducing peak power demand leads to a more predictable load profile. Reducing peak power demand favors scheduling the operation of cost-efficient but inflexible power generating resources such as coal and nuclear plants.

This can reduce the need for more flexible, but less efficient and more expensive, generating resources used to meet unanticipated, short-term demand.

This article describes a supervisory control strategy for limiting peak power demand by small and medium commercial building while still meeting the business needs of the occupants. This control strategy has two features that make it relevant to new and existing buildings. First, it is designed to operate with building equipment, such as air condition and refrigeration systems, as they are presently installed in most small and medium commercial buildings. Because of this, the supervisory control could be realized as a software-only retrofit to existing building management systems. Second, the proposed control acts as a supervisory management layer over existing control systems, rather than replacing them outright. The primary idea of this approach is that the controls for individual building equipment request energy resources and the supervisory control examines these requests and decides which to allow while satisfying a limit on peak power demand.

Our approach to developing the control software departs from the common approach of separate modeling and software development tasks. Instead, we build and test the control algorithm from the start in a form that is suitable for deployment by using a simulated environment in all stages of development. Our approach is similar to the model continuity techniques described by Hu and Zeigler (2004), which were introduced to minimize the effort and costs associated with shifting from algorithm development in simulations to the writing of software that will actually implement the control.

Our simulation environment is structured like those that have been used in the aerospace industry to develop high reliability software (see Brown et. al. (2005) and Shankar et. al. (2014)). The modeled environment is built to expose the interfaces that will appear in the fielded hardware or operating system. In our case, the model is built to replace the operating system clock services (i.e., sleep, gettimeofday, etc.) and a Modbus protocol library for talking to HVAC equipment. An important advantage of this approach is that it can enable faster than real time execution of the software, thereby permitting more comprehensive testing than could be achieved in a hardware laboratory. This aspect of our development method will become increasingly important as building control software grows in complexity with a corresponding increase in the likelihood of software defects and the economic impact of software faults in fielded control systems.

2. CONTROL ALGORITHM

The proposed control strategy borrows key concepts from prior theoretical work on scheduling electrical loads to limit power consumption while meeting criteria for the satisfactory performance of the loads (see, in particular, Vedova and Facchinetti (2013) and Nghiem et. al. (2011)). Our novel contributions are (i) the introduction of a priority scheme that ensures graceful degradation of the control process during unusually hot or cold weather when the desired peak power demand cannot be satisfied while maintaining an acceptable temperature; and (ii) the ability to effectively manage the HVAC load without requiring a predictive model of the building.

The essential ingredients of ours and prior control strategies are assumptions concerning the behavior of the electrical loads and the information that is available regarding their operation. In our approach, we make the following assumptions:

1. Each electrical load has its own control strategy that determines when the load should be idle and when it should be active according to its application. Information concerning this strategy is not available to the supervisory control. In particular, the supervisory control does not know when the loads will desire to operate and what the desired duration of that operation will be.
2. The electrical load supplies a priority with its request to operate. A priority of zero indicates that the load does not need to operate, and there is a maximum priority M that indicates the load must begin operating immediately. Other priority values indicate the urgency for operation within these extremes, with higher priority indicated a greater need to operate.
3. There is some number N of electrical loads that can operate simultaneously while still reducing peak demand charges.

The supervisor maintains a list of equipment sorted in decreasing order of priority, where priority values may be any integer between zero and M inclusive. Upon receiving a change in priority from any piece of equipment, the order of the list is updated and the scheduling algorithm is run. The scheduling algorithm has three steps. First, it scans the list of equipment from front to back, requesting that each equipment item shutdown if it satisfies one of two conditions:

(1) if the equipment item has a priority equal to zero; or
(2) the equipment item has a priority less than M and its position in the list is greater than N.

These requests can be disregarded by the equipment if necessary to avoid violating some operating constraint; for example, an HVAC unit may ignore this request to avoid rapid cycling of its compressor. Therefore, the second step is to scan the list again and count the number R of equipment items that are currently running. In the third step, equipment is activated according to its priority and position in the list. The list is scanned again from front to back and equipment is instructed to operate if it satisfies one of two conditions:

(1) if $R < N$ and the item occupies a position in the list less than $N - R$ or;
(2) the item has a priority equal to M.

Under extreme conditions, this control strategy will favor satisfactory operation of the equipment over reducing peak demand. This occurs when more then N units must operate because too many have reached the maximum priority. Otherwise, the system will respect the N unit limit and keep peak demand beneath N times the maximum power consumption of any equipment item.

To demonstrate that there exists a set of conditions under which the control can satisfy its limits, let us make the following assumptions. Each item of equipment can switch between on and off with a frequency of at most $1/T$. Furthermore, there is h such that the priority $p(t)$ of the equipment evolves according to

$$p(t+h) = \begin{cases} p(t)+1 & \text{if the equipment is off} \\ p(t)-1 & \text{if the equipment is on and } p(t) > 0 \\ 0 & \text{otherwise} \end{cases}$$

and the initial priority is $p(0) = 0$. We can normalize time by the smallest allowed switching interval and rewrite this expression as

$$p(t^* + h/T) = \begin{cases} p(t^*)+1 & \text{if the equipment is off} \\ p(t^*)-1 & \text{if the equipment is on and } p(t^*) > 0 \\ 0 & \text{otherwise} \end{cases}$$

where $t^* = t/T$.

To satisfy the constraint that the priorities remain less than or equal to M, it is necessary for the sum of the priorities over all equipment items to be bounded for all t^*. Let there be E equipment items in total. Clearly, $\sum p(0) = 0$ and $\sum p(h/T) = E$. Because $N < E$, it will be the case at each subsequent step in time that $E - N$ equipment items are off and increase their priority by one and N equipment items are on and decrease their priority or have it remain at zero. Therefore, at each step in time we add $E - N$ to the sum of priorities and subtract N, for a total change in the sum of $E - 2N$. The total priority at time step kh/T will be

$$\sum p(kh/T) = kE - 2(k-1)N$$

We prove this by induction on k. For $k = 1$ we already know that $\sum p(h/T) = E$ which agrees with the above expression. Suppose it is true for kh/T and consider $(k+1)h/T$. Adding $E - 2N$ to $\sum p(kh/T)$ gives

$$kE - 2(k-1)N + E - 2N = (k+1)E - 2kN + 2N - 2N = (k+1)E - 2kN = \sum p((k+1)h/T)$$

as expected.

To bound the sum of the priorities, it is necessary for there to be some k such that for all subsequent steps

$$2(k-1)N = kE$$

Rearranging this expression for N we have

$$N = \frac{kE}{2(k-1)} = \frac{E}{2 - 1/k}$$

and as $k \to \infty$ we get $N = E/2$. Hence, we can expect the control to bound the priority of each equipment item when $N \geq E/2$. In this case, the sum of the priorities is

$$\sum p(kh/T) = kE - 2(k-1)E/2 = kE - \frac{2kE}{2} + \frac{2E}{2} = E$$

and so if $M \geq E$ and $N \geq E/2$ the control scheme can satisfy its constraints.

This derivation demonstrates that there are plausible conditions under which the proposed control limits peak demand. We will show in the next section that this derivation can also serve as a practical guide for selecting N when installing the control in a building.

3. SIMULATION BASED DEVELOPMENT AND TESTING

The new control algorithm was implemented in software that reuses large parts of a prior, deployed control system (see Nutaro et. al. (2014)). Briefly, this existing system is installed in the basketball gymnasium of the Family Life Center at the Central Baptist Church in Fountain City, Tennessee (see Figure 1). This gymnasium has four 10 ton HVAC roof top package units each having two-stage gas heat and two-stage conventional refrigerant cooling. The prior, deployed control software is hosted on a Linux PC with connections via Modbus to four Tempstat 6 thermostats from Temco Controls. The parts of this software that we have reused to implement the new control enables direct switching of the thermostat relay contacts, reading the temperature sensors, and reading the temperature set point. This portion of the software uses the libmodbus library (see libmodbus.org) to communicate with the thermostats.

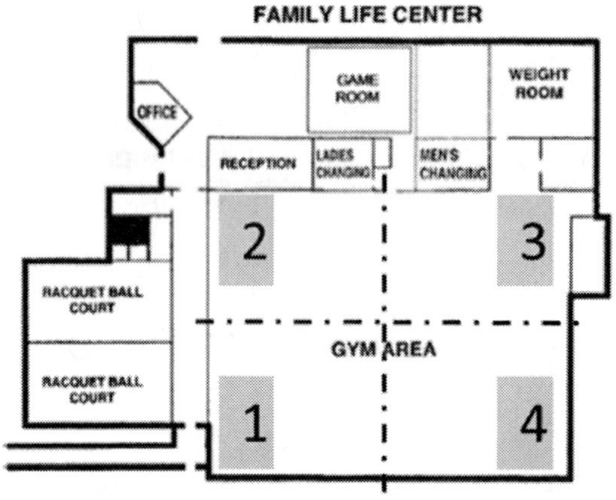

Figure 1: Floor plan for the gymnasium. Dotted lines in the gym show the HVAC zone boundaries.

To facilitate testing prior to deployment, we implemented a simulated building that is accessed by the new control software via simulated Modbus library calls and operating systems calls. The simulated Modbus library and system calls are implemented in a software library that is linked with the control software for testing. For deployment, the control software would be linked with the actual Modbus library and operating system functions.

In testing, a call to one of the simulated Modbus or operating systems functions causes the building model to be advanced in time by the amount of wall clock time that has passed since the prior, simulated call. Then the effect of the call is determined by referencing the appropriate element of the building model. For instance, if the call was to send a Modbus message that closes a relay, the corresponding HVAC unit is activated in the model; if the call was to read a temperature sensor, then the corresponding model variable value will be returned. Calls to sleep are unique in that they advance the simulation clock as described above plus an amount of time equal to the requested sleep period. This approach is illustrated in Figure 2 for a code segment that periodically samples the temperature sensor.

Testing in this way allows the simulated software to operate more quickly that real time, greatly facilitating testing by compressing days of simulated software operation into several minutes of real time.

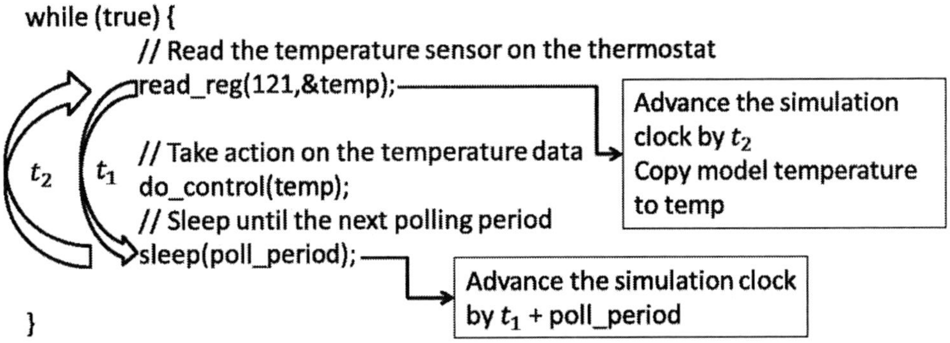

Figure 2: Redirecting calls to read_reg and sleep to interact with a simulated building

To demonstrate the operation of the proposed control and its software implementation, we constructed a model of a four zone building that is based on our prior experience with the Family Life Center (see Nutaro et. al. (2014)), and we consider only summer months when electrical power is used to cool the building. The model HVAC equipment has two cooling stages, which gives its three modes of operation – off, stage 1 on, and both stages on. An HVAC unit will not change its mode of operation more often than once every ten minutes. The HVAC equipment has 10 priority levels, and the priority p of the air conditioner is based on its thermostat temperature T [C] and temperatures set point S [C] as follows:

$$p = \begin{cases} \operatorname{ceil}\left(\frac{T-S}{0.1}\right) & \text{if } 0 < T - S < 1 \\ 0 & T \leq S \\ 10 & T - S \geq 1 \end{cases}$$

The first stage is activated if $T \leq S + 1$ and the second stage is activated if $T > S + 1$.

The model equations that describe temperature change are listed below. The R_{ik} terms model heat flow between the ith and kth zones or the outside air; the T_k terms are the zone air temperatures; T_a is the outside air temperature with daily low T_{low} and daily high T_{high}; the Q_{sk} is solar heating, occupants, and other sundry sources with maximum daily heating of Q_k in zone k; the h_k is cooling power provided by the HVAC unit in zone k; u_k is the operating condition of the HVAC unit which can be 0 (off), 1 (stage 1 on), or 2 (stages 1 and 2 on); and C_k is the zone's thermal capacitance. The time t is in seconds.

$$T_1 = \frac{1}{C_1}\left(\frac{T_2 - T_1}{R_{12}} + \frac{T_4 - T_1}{R_{14}} + \frac{T_a - T_1}{R_{a1}} + Q_{s1} + u_1 h_1\right)$$

$$T_2 = \frac{1}{C_2}\left(\frac{T_1 - T_2}{R_{12}} + \frac{T_2 - T_3}{R_{23}} + \frac{T_a - T_2}{R_{a2}} + Q_{s2} + u_2 h_2\right)$$

$$T_3 = \frac{1}{C_3}\left(\frac{T_2 - T_3}{R_{23}} + \frac{T_4 - T_3}{R_{34}} + \frac{T_a - T_3}{R_{a3}} + Q_{s3} + u_3 h_3\right)$$

$$T_4 = \frac{1}{C_4}\left(\frac{T_1 - T_4}{R_{14}} + \frac{T_3 - T_4}{R_{34}} + \frac{T_a - T_4}{R_{a4}} + Q_{s4} + u_4 h_4\right)$$

$$T_a = T_{low} + (T_{high} - T_{low})\frac{1 + \cos\left(2\pi \frac{t}{86400}\right)}{2}$$

$$Q_{sk} = Q_k \frac{1 + \cos\left(2\pi \frac{t}{86400}\right)}{2}$$

We simulate the operation of this building with several choices of parameters selected at random from the uniform distributions listed in Table 1; only the set point is fixed across all runs at 21.111 C (70 F). Though the model is based roughly on the Family Life Center such that it yields similar high and low temperatures for a summer day and roughly approximates the rates of temperature change seen in the gymnasium, no attempt has been made to validate the model against experimental data. Nonetheless, the randomized parameter sets represent a plausible range of values for this building and serve to demonstrate the robustness of the control across a range of similar buildings and circumstances.

Table 1: Ranges for parameter values

Parameter	Value range	Units
C_k	252,628.5 ± 10%	Joules / Degrees Centigrade
R_{ik}	1000 ± 10%	Watts / Degrees Centigrade
R_{ai}	10 ± 10%	Watts / Degrees Centigrade
Q_k	1,710 ± 10%	Watts
h_k	17,500 ± 10%	Watts
S	21.111 (70)	C (F)

To evaluate the performance of the control algorithm, we conducted one thousand simulations with this model using a randomly selected set of parameters in each simulation. This was repeated with $N = 1, \ldots, 4$. The case with $N = 4$ is equivalent to uncoordinated control of the loads, as would occur in a building with common thermostat equipment. The simulations each spanned 10 days and we recorded the maximum number of simultaneously operating units (peak load) and the maximum temperature above the thermostat set point (temperature error) that was observed over those ten days.

The distribution of peak load and temperature error for each choice of N is shown in Table 2. In these results it is clear that the control reduces the expected peak load from 3 to approximately 2 with only a small change in the maximum and average errors. In all cases these metrics are within 2/10 of a degree Centigrade. Referring to the prior theoretical consideration of the minimum value for N that can be reliably achieved, the simulations show that $N = 2$ is maintained across 91% of the scenarios considered. On the other hand, this simulation data also suggests an optimistic strategy in which $N = 1$ is selected and the natural degradation of the control is relied upon to maintain an acceptable temperature error. This strategy produces an expected peak load only slightly worse than the $N = 2$ case, while occasionally beating that case by running just a single unit.

Table 2: Simulation results

N	Distribution of peak load across runs				Expected peak load	Temperature error [C]		
	1	2	3	4		Min.	Avg.	Max.
1	5.7%	79%	14.5%	0.8%	2.104	0.4069	0.7601	1.343
2	0%	91.4%	7.9%	0.7%	2.093	0.0590	0.6784	1.332
3	0%	1.8%	97.1%	1.1%	2.993	0.0587	0.6114	1.321
4	0%	2.5%	82.3%	15.2%	3.127	0.0587	0.6276	1.344

4. CONCLUSIONS

The simulation results show that theoretical limits on performance calculated for an idealized model are satisfied in the majority of variations of a model derived from a real building. This is encouraging, but not conclusive. We intend to install the control software in the Family Life Center and observe its performance through the summer of 2016. This experiment will be greatly facilitated by having developed and tested the control software in a simulated environment that looks, from the point of view of the control application, indistinguishable from the real building. Indeed, the simulation results reported in the previous section have also provided us with the equivalent of 4,000 days of operational testing for the primary control logic, and this should substantially reduce the amount of experimental time lost to software errors. Though we have demonstrated the proposed control in the context of limiting peak demand caused by HVAC equipment, it can be readily extended to coordinate refrigeration, ventilation, lighting and other energy intensive systems by defining suitable priority schemes. These types of simple extensions are possible chiefly because of the limited information that is required by the control, and this feature most strongly distinguishes the proposed system from prior work on load scheduling in small and medium commercial buildings.

REFERENCES

Vedova, M. L. D., Facchinetti, T. (2013). Real-time scheduling for peak load reduction in a large set of HVAC loads. In *Proceedings of the third international conference on smart grids, green communications, and IT energy-aware technologies* (161-166). Lisbon, Portugal: IARIA

Nghiem, T., Behl, M, Pappas, G.J., & Mangharam, R. (2011). Green scheduling: scheduling of control systems for peak power reduction. In *Proceedings of the 2011 international green computing conference and workshop* (1-8). Orlando, Florida, USA: IEEE

Nutaro, J., Kuruganti, T., Fugate, D., & Starke, M. (2014). An inexpensive retrofit technology for reducing peak power demand in small and medium commercial buildings. In *Proceedings of the 3rd international high performance buildings conference at Purdue* (1-7). West Lafayette, Indiana, USA: Purdue University.

Hu, X., Zeigler, B.P. (2004). Model Continuity to Support Software Development for Distributed Robotic Systems: A Team Formation Example. *Journal of Intelligent and Robotic Systems*, 39(1), 71-87.

Brown, J., Lam, D., Chang, L., Burk, T., & Wette, M. (2005). The role of the Flight Software Development System simulator throughout the Cassini mission. In *Proceedings of the AIAA Guidance, Navigation, and Control Conference and Exhibit* (1-13). San Francisco, California: AIAA

Shankar, S., Desai, K., Dutta, S., Chetwani, R., Ravindra, M., & Bharadwaj, K.M. (2014). Mission critical software test philosophy: a SILS based approach in Indian Mars Orbiter mission. In *Proceedings of the 2014 International Conference on Contemporary Computing and Informatics* (414-419). Mysore, India: IEEE

ACKNOWLEDGEMENT

This manuscript has been authored by UT-Battelle, LLC under Contract No. DE-AC05-00OR22725 with the U.S. Department of Energy. The United States Government retains and the publisher, by accepting the article for publication, acknowledges that the United States Government retains a non-exclusive, paid-up, irrevocable, world-wide license to publish or reproduce the published form of this manuscript, or allow others to do so, for United States Government purposes. The Department of Energy will provide public access to these results of federally sponsored research in accordance with the DOE Public Access Plan (http://energy.gov/downloads/doe-public-access-plan).

Load modulation strategies of residential heat pumps for demand-response programs with different thermal storage options

Emeline GEORGES[1]*, Vincent LEMORT[1]

[1]Energy systems, Thermodynamics Laboratory, University of Liège,
Liège, Belgium

* Corresponding Author: emeline.georges@ulg.ac.be

ABSTRACT

This research work presents a methodology to assess the potential of load modulation strategies of HVAC systems for demand response programs at different scales. Two different demand response programs are considered: feed-in tariff signals for renewable production load matching and signals from the distribution grid operator for operational planning and congestion management at the distribution level. The resulting control problems are solved using optimal control formulations. First, the strategies are applied to four typical Belgian houses to match on-site PV production. Different thermal storage options are considered: on the one hand, thermal storage in the building envelope and in water tanks for domestic hot water, and, on the other hand, additional water tanks for space heating needs either in a parallel four-pipe or in a parallel two-pipe configuration. According to the type of house and the modulation strategy considered, a ranking of the most suitable storage options is proposed. Secondly, the method is extended to the scale of a distribution feeder with 50% PV penetration rate. Results show that with 20% heat pumps and suitable storage options, the residual load reduction reaches 28% to 73.4%. Upward load modulation in response to grid operator signals shows better performance for short modulation intervals, but can lead to up to 13% additional overconsumption if the chosen thermal storage option is not adapted to the house insulation level.

1. INTRODUCTION

The objectives of reduction of greenhouse gas emissions to mitigate the impact of human activities on global climate have highlighted the need for increased energy efficiency policies and energy transition towards renewable energy sources. The massive penetration of decentralized renewable energy sources and electrification of systems put additional stresses on the electricity grid and require changes in consumption profiles. As the building sector accounts for approximately 40% of the total energy demand in Europe, and with the increasing electrification of systems providing space heating (SH) and domestic hot water (DHW), such as heat pumps, there exists a large potential to control, enhance and adapt the related consumption profiles. In this context, there is a rising interest for demand response programs, which consist in the modification of consumption profiles in response to different incentives such as forward price signals, feed-in tariffs or direct signals from the grid operator. Several studies have focused on the potential of using the building thermal mass for load shifting. Results have shown that overheating the room temperature during off-peak hours can considerably reduce (by up to 20% according toWolisz et al., 2013) the heating needs for the following peak hours. However, this leads to significant overconsumption (Reynders et al., 2013) and doesn't reduce peak demand amplitude but shifts in time (Masy et al., 2015). The use of DHW water tanks for demand response programs has also been investigated, for example, to reduce curtailing losses from PV panels in NZEBs neighborhood (De Coninck et al., 2014). The use of an extra water tank for space heating needs is less widespread in practice so far, but has been the interest of the several studies. Studies of Miara et al. (2014) and Palacio et al. (2014) focused on control strategies, and showed a possible reduction of up to 1.5% per every 10%

TES penetration of the overall power system's operation costs. Floss et al. (2015) studied the impact on the system performance of several hydraulic integrations (see Section 3.2) of SH thermal storages in heat pump driven heating systems. Hedegaard et al. (2012) compared the use of water tanks for SH to thermal storage in building structures for wind power integration. Finally, Patteeuw et al. (2016) propose an integrated approach to compare load shifting strategies with heat pumps to improve electricity generation efficiency at the system level. The present study compares two optimal control formulations for load shifting with residential heat pumps following a decentralized approach defined in Section 2. Three different storage options, presented in Section 3, are compared for four typical Belgian freestanding houses. A ranking of the most suitable option is presented in Section 4.1. In Section 4.2, the method is extended to a distribution feeder with 63 houses in order to reduce curtailment losses of PV production.

2. LOAD MODULATION STRATEGIES

This section proposes two optimal control formulations to trigger demand flexibility of residential buildings at a given time period and over set of consecutive periods. The load modulation amplitude is defined from a baseline profile that minimizes the electricity cost for the flexible consumer. The first formulation aims at activating an upward load modulation over a time interval. The second formulation aims at promoting load matching with local electricity production source over a given time interval by mean of a reduced electricity surplus selling tariff, as proposed by Georges et al. (2014).

2.1 Baseline consumption

The investigated system is represented by a discrete state-space formulation. The state transition is summarized by

$$x_{t+1} = f(x_t, u_t, w_t) \quad (1)$$

where x is the state space variable vector, u is the vector of decision variables, i.e. the modular electric power, and w is a vector of disturbances. The total electricity consumption of the consumer at time t, P_t^{cons}, is composed of modular components, $u_{i,t}$, i.e., the consumption of systems that can be adjusted by the optimal load management scheme, and the exogenous consumption, Γ_t, i.e. the share of the electricity consumption that is not modulable,

$$P_t^{cons} = \sum_i u_{i,t} + \Gamma_t \quad (2)$$

A baseline consumption is obtained by minimizing the electricity costs for the flexible consumer for a given retail electricity price profile, π_{ret},

$$\min \sum_{t=1}^{H} \left(P_t^{cons} \pi_{ret,t} \right) \quad (3)$$

with respect to the decision variables, $u_{i,t}$, on the optimization horizon H, and subject to the following constraints

$$x_t^{\min} \leq x_t \leq x_t^{\max} \quad (4)$$

$$u_{i,t}^{\min} \leq u_{i,t} \leq u_{i,t}^{\max} \quad (5)$$

$$u_{i,t} + u_{j,t} \leq \max(u_{i,t}, u_{j,t}) \quad (6)$$

where Equation (4) specifies the upper and lower limits for power modulation of the modular components and Equation (6) ensures that two related decision variables are not activated simultaneously, as further explained in Section 3.3. The resulting baseline consumption profile is denoted $\hat{P}^{cons}$.

2.2 Upward modulation (ADR #1)

From the baseline consumption, a first mechanism can be implemented to trigger load flexibility in the desired time interval. It consists in carrying out an upward modulation of the electrical consumption at a given time period

τ over a number n of periods. The modulation profile, P_t^{mod}, is defined from the baseline consumption by the decision variable δ_t

$$P_t^{\text{mod}} = \hat{P}_t^{cons} + \delta_t \qquad \text{for } t \in [\tau, \tau + n - 1] \tag{7}$$

The objective function is written as

$$\max(I_{\text{mod}}) \tag{8}$$

where

$$\delta_t \geq I_{\text{mod}} \qquad \text{for } t \in [\tau, \tau + n - 1] \tag{9}$$

and subject to Constraints (4) to (6). After the n periods of upward modulation, a new consumption profile that minimizes the consumer's costs is determined as defined by Equations (3) to (6).

2.3 Load matching with buy-back ratio (ADR #2)

This optimal control formulation corresponds to a metering program that promotes load matching between production and consumption. The net electricity flow, P_t^{net}, i.e. the instantaneous difference between the power consumed and produced on site is determined. If positive, the billing tariff is the retail tariff, π_{ret}. If negative, the excess production is bought back at a buy-back tariff, π_{bb}. One can therefore define a buy-back ratio $\alpha_{bb} = \pi_{bb} / \pi_{ret}$. From a given time period τ and over n periods, the buy-back ratio is set to a value of zero. The objective is to minimize the electricity cost for the end-user for a given net metering buy-back ratio, which can be expressed as

$$\min \sum_{t=\tau}^{H} \left(\max(P_t^{net}, 0) + \min(P_t^{net}, 0)\alpha_{bb,t} \right) \tag{10}$$

and subject to Constraints (4) to (6).

3. THERMAL MODELS

3.1. Buildings and Heat Pumps

Two types of building geometries, representative of the Belgian residential building stock, are considered in the study: a single-story building representative of freestanding houses built between 1971 and 1990 and a two-story building representative of freestanding houses built after 1991. For each geometry, two insulation levels are considered. Buildings characteristics are taken from the study of Gendebien et al. (2013) and summarized in Table 1. The building thermal behavior is modeled by an equivalent thermal network consisting of thermal resistances, R, in K/W, and lumped thermal capacitances, C in J/K. For the purpose of this study, two different single zone 5R3C models are used. They are illustrated in Figure 1. The first model structure (left), is used to model well-insulated buildings. The second structure (right), is a modification of the ISO13790 model (ISO, 2007) by the introduction of two additional lumped thermal capacitances to take into account the non-homogeneous air distribution in the zone and the thermal capacity of the internal walls. It is well-suited to model standard or poorly insulated buildings. Each building is equipped with an inverter air-to-water heat pump, whose characteristics are given in Table 1. The heat pump model is based on "ConsoClim" (Bolher et al., 1999) which expresses the heat pump coefficient of performance (COP) as a function of the ambient temperature, T_a, the water supply temperature, $T_{w,su}$ and the part-load ratio, PLR, through polynomial laws

$$COP = f(T_a, T_{w,su}, PLR) \tag{11}$$

Regression coefficients are identified based on performance curve from manufacturer data. The emitters installed in the building are conventional radiators. The emitted power is modeled with an empirical emission law function of the difference between the average water temperature in the emitter and the room temperature. The water supply temperature is adapted to the house insulation level and adjusted throughout the year following a heating curve.

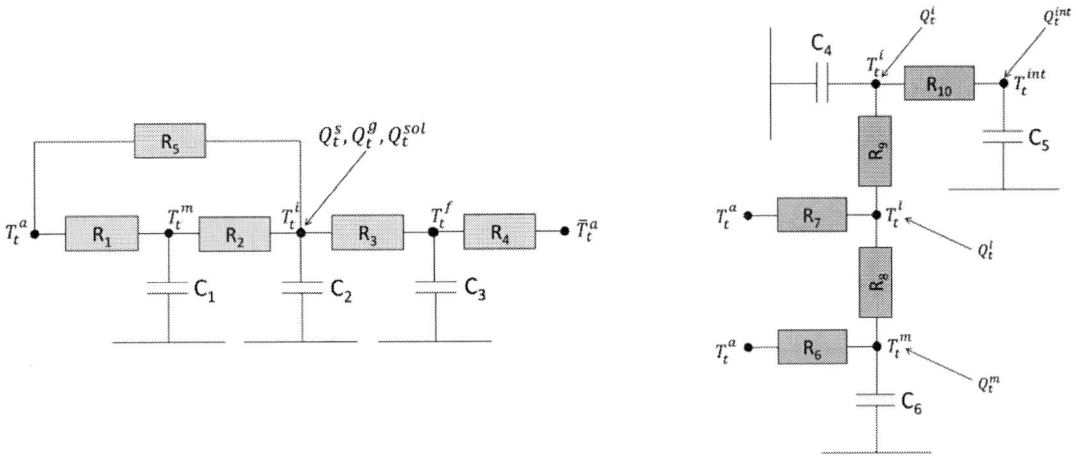

Figure 1: Grey-box model structures.

Table 1: Buildings and heat pumps characteristics

Year of construction	N# of floors	Heated volume [m³]	Ground floor area [m²]	Average U value [W/m²K]	Design heat demand @ - 10°C [kW]	Nominal HP temperatures A/W [°C]	Auxiliary heater [kW]
2007-2014 (A)	2	457	75	0.31	6.5	7/45	3
1991-2006 (B)	2	457	75	0.46	8.0	7/45	3
1971-1990 (C)	1	423	148.5	1.24	17.0	7/65	5
1971-1990 (D) Retrofit	1	423	148.5	0.77	12.0	7/45	5

3.2 Thermal storage

Each house is equipped with a domestic hot water (DHW) tank. Additionally, three thermal storage options are investigated for space heating (SH):

- storage in the building envelope,
- storage in a water tank following the parallel four-pipe configuration (Figure 2 – right),
- storage in a water tank following the parallel two-pipe configuration (Figure 3).

As illustrated in Figure 2 (left), both storage tanks for SH and DHW are installed in parallel, and the heat pump can only supply one of them at a time. Priority is always given to the DHW tank. Both DHW and SH tanks are modeled by one-node models with homogeneous water temperature.

Table 2: Water storage tank volumes

House type	DHW tank Volume [m³]	SH tank Volume [m³]	SH Tank configuration
A1	0.25	/	/
A2		0.35	Two-pipe
A3		0.42	Four-pipe
B1		/	/
B2		0.40	Two-pipe
B3		0.52	Four-pipe
C1			/
C2		0.60	Two-pipe
C3		0.90	Four-pipe
D1		/	/
D2		0.50	Two-pipe
D3		0.75	Four-pipe

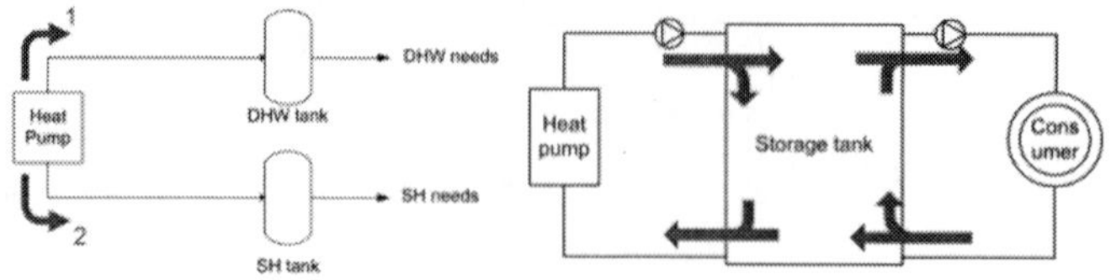

Figure 2: Left: Hydraulic scheme for DHW and SH water storages. Right: Parallel four-pipe configuration for SH water storage.

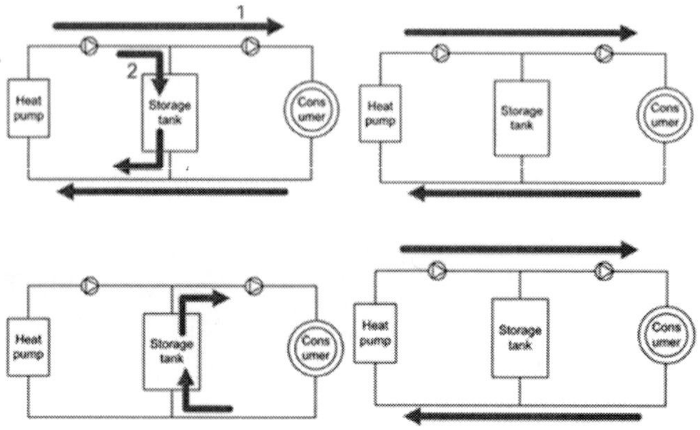

Figure 3: Parallel two-pipe configuration for SH water tank. Top left: the heat pump is used to supply the house directly if needed or to store heat in the storage tank. Top right: the heat pump is used to meet the house heating needs only. Bottom left: the storage tank, if charged enough, is used to supply the house. Bottom right: the storage tank charge level is too low to supply the house, and the heat pump is used to supply the house directly.

3.3 Constraints

Constraints (4) to (6) translate as follows, depending on the thermal storage option for SH:

- for thermal storage in building envelope, the building zone temperature, should remain within a dead band of ΔT equals 1K around the set point profile during occupancy hours and between the minimum and maximum daily set point for unoccupied hours. An intermittent heating strategy is chosen. The set point is set to 18°C during the night and unoccupied periods and to 21°C during occupied hours of the day.
- for the four-pipe SH tank configuration: the water temperature, T_{SH}, is always maintained above the house heating curve temperature, T_{hc}:

$$T_{hc} \leq T_{SH} \leq T_{hc} + \Delta T_{4-pipe} \tag{12}$$

  where ΔT_{4-pipe} is set to 15K during grid-constrained periods, i.e. $[\tau, \tau+n-1]$, and to 10K otherwise.
- for the two-pipe SH tank configuration: the water temperature has to be higher than the house heating curve temperature when the tank is used to supply the house, which corresponds to y_{SH} equals one:

$$T_{hc} y_{SH} \leq T_{SH} \leq T_{hc} + \Delta T_{2-pipe} \tag{13}$$

  where ΔT_{2-pipe} is equal to 10K. The tank can only be supplied during grid-constrained periods.
- for both four-pipe and two-pipe configurations, the indoor temperature strictly follows the set point.
- the heat delivered to the house should not exceed the full load capacity of the heat pump and auxiliary heater combined or the limits imposed by the temperature in the tank and by the emission system.
- for DHW, the water tank temperature should remain within an imposed dead band:

$$50°C \leq T_{DHW} \leq 57°C \tag{14}$$

- the heat pump can only supply either the house, or the SH tank or the DHW tank at a time, which introduces Boolean variables and turns the optimization problem into a mixed-integer linear program.

4. RESULTS AND DISCUSSION

In this section, the two load modulation strategies ADR#1 and ADR#2 are applied to mitigate voltage fluctuations entailed by photovoltaic (PV) production in distribution networks. The objective consists in minimizing the residual load, i.e., the difference between the electricity production and consumption on a time interval $[\tau, \tau+n-1]$. The strategies are first applied to each typical building A to D to compare the different storage options, and then extended to simulate the response of a set of buildings in a distribution feeder.

4.1 Influence of thermal storage options

Each association of house and thermal storage summarized in Table 2 is simulated for two typical winter days and one day in the spring. The activation takes place at time period 49 and the modulation lasts over 10 consecutive periods. The incentive signal for ADR #2 is illustrated in Figure 4 (left). Figure 5 illustrates the temperature and power consumption profiles obtained for simulations B1 to B3 for load strategy ADR#2 for a sunny winter day (Figure 4 - right). During the investigated time interval, the occupation profiles are such that there are no occupants in the house, and the allowed indoor temperature range is 17°C to 22°C. For all three storage associations, both DHW production and SH contribute to demand flexibility. During unoccupied periods, using thermal envelope for demand flexibility allows the best match with PV production. The power consumption modulation from the baseline reaches up to 3kW. Both the DHW tank and the zone can be preheated to meet comfort requirements imposed during occupied periods later in the day. With the parallel four-pipe configuration, most of the flexibility is provided by the SH tank. Its upper set point is raised by 5K to increase flexibility. Load matching is achieved during two concentrated time frames due to the limited storage capacity of both SH and DHW tanks and the limited heating demand of the house during unoccupied hours. The residual load, initially of 4.6 kWh for the baseline profile, is reduced by 47%. For the two-pipe configuration, load matching is ensured either by DHW production, or by direct space heating or by heat storage in the SH tank. The latter is mostly heated during the last periods of the interval. This is due to the fact that the house can only be supplied either by the SH tank or directly by the heat pump and also to the control strategy that anticipates the use of the SH tank later in the day. The residual load is reduced by 43%.

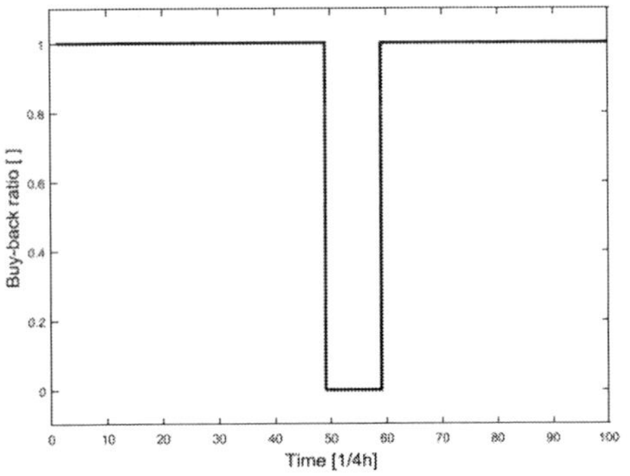

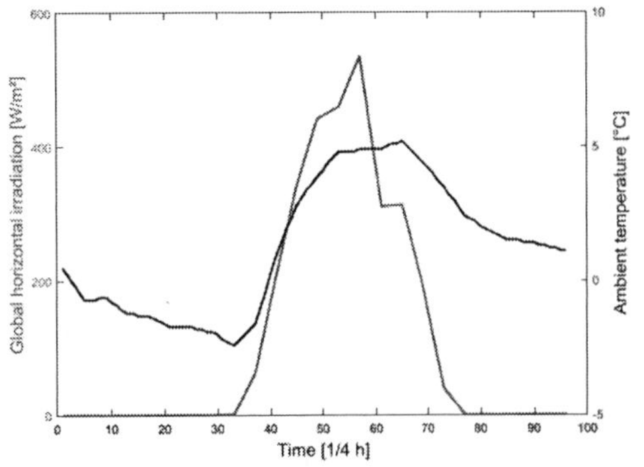

Figure 4: Global horizontal irradiation and ambient temperature for February 20th, 2012.

Figure 6 generalizes the findings of Figure 5 to houses A to D for the two control strategies. Results are presented in terms of average residual load reduction and additional overconsumption. The overconsumption is defined as the difference between the integrated daily electricity consumption of the baseline and the flexible profile. The additional overconsumption is the share of overconsumption that does not occur during the activation time interval. The determination of the most suitable storage option for each house depends on the criteria to satisfy. If the only objective is to reduce the residual load, heat storage in building envelope is the best option with both ADR#1 and ADR#2 for buildings A to D, due to the large storage capacity available. Residual load reduction reaches 51% to 85%. ADR#1 outperforms ADR#2 for poorly insulated houses such as C and D, whereas ADR#2 is more efficient for well-insulated houses such as A and B for which the good insulation of the envelope allows an efficiency use of the stored heat. For ADR#1, the parallel two-pipe configuration allows an additional reduction of the residual load by 4% to 16% compared to the four-pipe configuration. This additional flexibility increases with decreasing house insulation levels. Indeed, for poorly insulated houses, the SH tank in the four-pipe configuration is often close to its highest temperature level at the beginning of the activation interval. For ADR#2, the difference in performance between the two-pipe and four-pipe vary strongly with the type of house. Indeed, for houses A and B, both storage options present similar ability to reduce the residual load. For houses C and D, contrariwise, trends are opposite. In the case of house C, the indoor temperature fluctuations induced by the supply of the SH tank reduce its potential to be used in a parallel two-pipe configuration, compared to better insulated houses such as D. If a condition regarding the overconsumption outside the modulation interval is added to the residual load reduction, then using thermal envelope as storage option becomes detrimental for house C for both ADR#1 and #2. The two-pipe configuration for houses C and D with ADR#1 also leads to significant overconsumption, as the energy stored in the tank may not be stored at a sufficient high temperature to supply the house afterwards. The four-pipe configuration becomes a suitable option for houses C and D.

4.2 Application to a semi-urban feeder

The methodology is applied to a typical semi-urban feeder in Belgium. The feeder has an average of 63 load points corresponding to 63 households, mostly composed of freestanding buildings (Gonzalez et al., 2012). Baetens (2015) has outlined the sensitivity of such feeders to high penetration of distributed generation sources and heat pumps. A penetration rate of PVs superior to 50% entails over-voltages, which could be reduced by load compensation. The four typical houses presented above are distributed in the feeder according to their share in the overall Belgian residential buildings stock. The number of inhabitants in each house is drawn from a normal distribution of average three and a standard deviation of two with a maximum of five occupants. The exogenous consumption profiles associated to lighting and appliances are obtained from article (Gendebien et al., 2015), as well as the domestic hot water draw-off events. It is assumed that the optimal controller knows the average demand profiles for all houses. Indoor temperature set points schedules are intermittent temperature profiles generated based on normal distribution laws for morning, midday and evening start-up times. All profiles have a weekly average indoor set point above 18°C. Occupancy profiles are derived from the latter. Simulations are run for both 10 and 20% of houses equipped with heat pumps. Based on the results from Section 4.1, the best suited storage option that satisfies both minimum residual load and minimum overconsumption is chosen for each of them. Additionally, 50% of the buildings are equipped with South, East or West-oriented PV panels with thirty-five degrees tilt angle and 18% average efficiency. The installed surface area is drawn from a discrete uniform distribution over an interval ranging from ten to forty square meters. Houses with PVs and heat pumps are distributed randomly over the 63 houses and their occurrence may not coincide. Load management strategies ADR#1 and ADR#2 are compared in Figure 7 for average exogenous consumption and domestic hot water profiles. For ADR#1, the input signal is to perform an upward modulation. For ADR#2, the same buy-back ratio as in Figure 4(left) is used, as well as the mean PV production per buildings. For a modulation interval located from period 49 to 58, ADR#1 and ADR#2 allows to reduce the peak residual power of period 57 from 63.1 kW with no heat pumps down 20.2kW and 30.1 kW with 20% heat pumps, respectively. The integrated residual load over the interval is reduced by 73.4% and 51.4% with ADR#1 and ADR#2 respectively, compared to no load modulation. The sensitivity in performance of ADR#2 to the PV production input used in Equation (10) is illustrated in Figure 8. If the modulation interval is extended from period 37 to period 71, and the input PV production provided to ADR#2 is the maximum PV production per house rather than the average production per house, then ADR#2 shows the best performance and provides an additional 23% reduction of the integrated residual load compared to ADR#1.

5. CONCLUSION

This paper proposes two optimal control formulations to compare load shifting strategies for residential heat pumps. Three storage options combining a DHW tank and either the thermal envelope of the building or a SH tank with two hydraulic configurations are investigated for four typical Belgian houses characterized by different insulation levels.

The method can be used to relieve network congestions, such as those caused by excess PV production. Results show that, for both control strategies, the use of the building envelope as storage option allows a reduction of the residual load by 51 to 75% for well-insulated houses. As the house insulation level decreases, different behaviors are observed according to the control strategy. To maximize the upward modulation while limiting overconsumption, the parallel four-pipe scheme is the most suitable, whereas the best load shaping is obtained with a parallel two-pipe or a parallel four-pipe configuration. At the scale of a distribution feeder with 63 houses, if 50% of the houses are equipped with PV panels, a penetration rate of 20% heat pumps with appropriate thermal storage allows to reduce the residual load by 28 to 73.4% depending on the control strategy and the modulation interval length. An upward modulation strategy shows good result for short interval lengths with relatively constant power production, whereas the load shaping strategy increases the residual load reduction by an extra 21% on longer intervals.

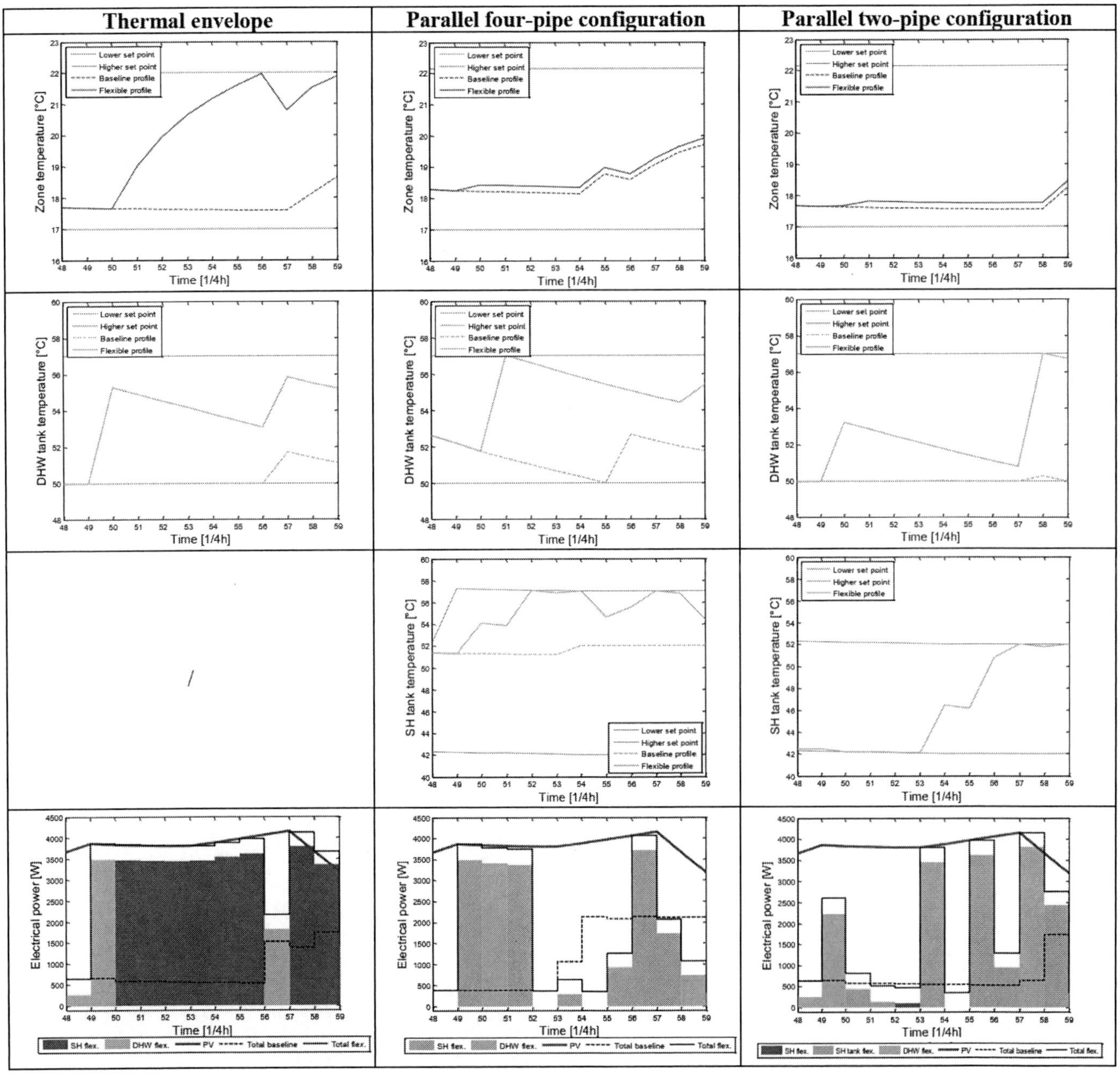

Figure 5: Illustration of load modulation formulation ADR#2 for house B and three thermal storage options for a winder day (Day 51 – February 20th).

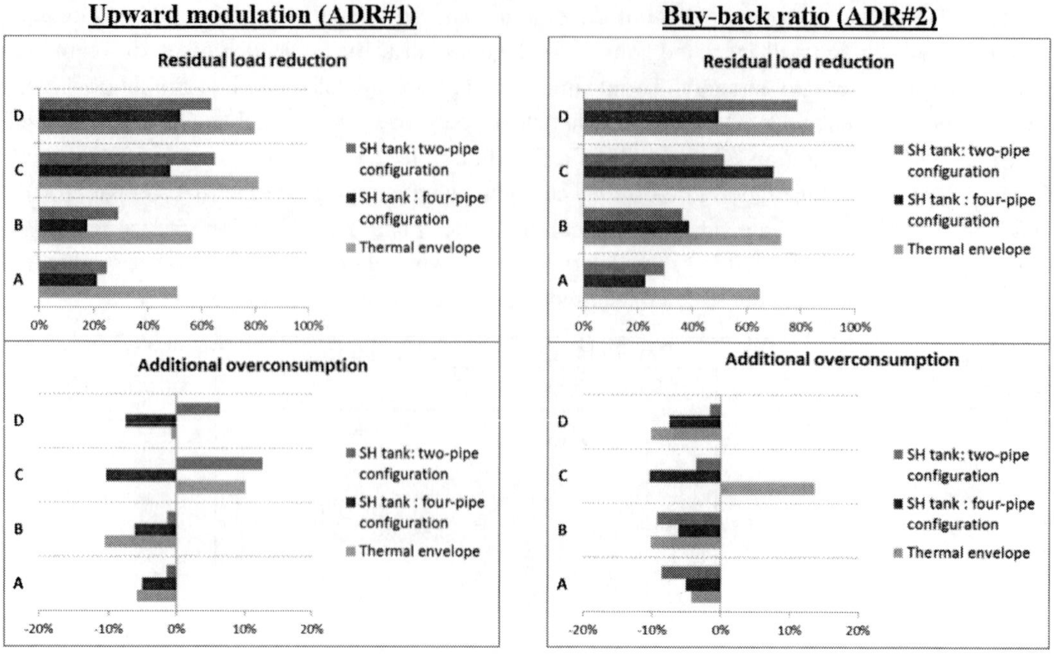

Figure 6: Comparison of ADR#1 and ADR#2 for houses A to D and three storage options. Average values over two winter days an one day in the spring.

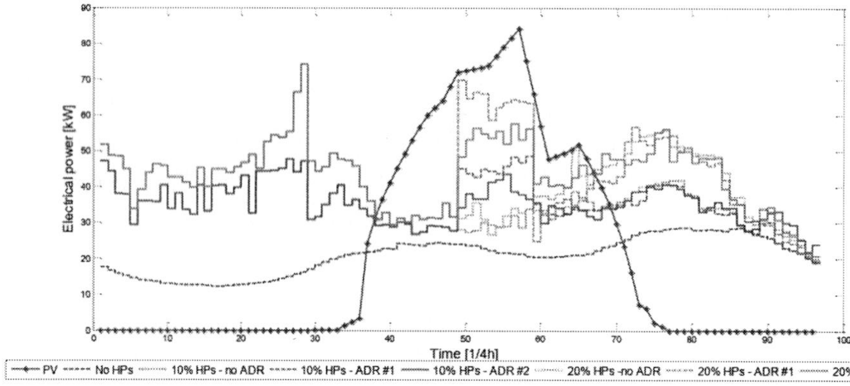

Figure 7: Load modulation strategies with 50 % PVs and 10 and 20% penetration of heat pumps (HPs) for time period 49 to 58.

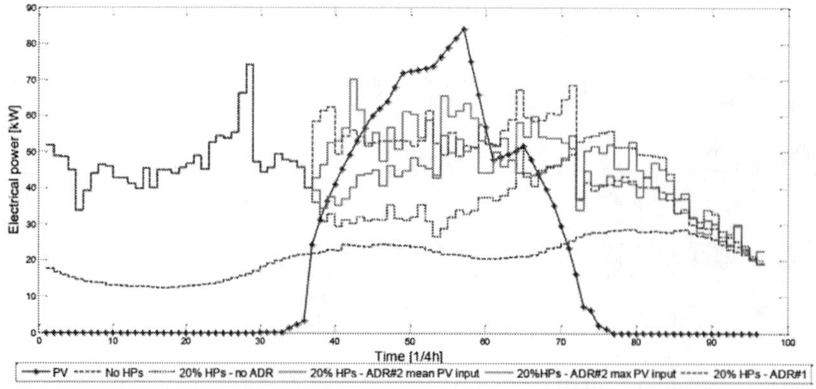

Figure 8: Impact of PV input for ADR#2 and of the modulation length on residual load reduction.

NOMENCLATURE

Q_t^g	Internal heat gains	**Superscripts**	
		g	gain
Q_t^{sol}	Solar gains	sol	solar
T	Temperature	s	space heating
Q	Thermal power	i	indoor
		m	massive
Subscripts		l	light
t	at time t	int	internal

REFERENCES

Baetens, R. (2015). On Externalities of Heat Pump-based Low-Energy Dwellings at the Low-Voltage Distribution Grid. PhD dissertation, KUL, Belgium.

Bolher A., Casari R., Fleury E., Marchio D.. (1999). Méthode de calcul des consommations d'énergie des bâtiments climatisés : ConsoClim. Technical report, Ecole des Mines (Paris).

De Coninck, R., Baetens, R., Saelens, D., Woyte, A., & Helsen, L. (2014). Rule-based demand-side management of domestic hot water production with heat pumps in zero energy neighbourhoods. *Journal of Building Performance Simulation*, 7(4), 271-288.

Floss, A., & Hofmann, S. (2015). Optimized integration of storage tanks in heat pump systems and adapted control strategies. *Energy and Buildings*, 100, 10-15.

Gendebien, S., Georges, E., Bertagnolio, S., & Lemort, V. (2015). Methodology to characterize a residential building stock using a bottom-up approach: a case study applied to Belgium. *International Journal of Sustainable Energy Planning and Management*, *4*, 71-88.

Georges, E., Braun, J. E., Groll, E., Horton, W. T., & Lemort, V. (2014). Impact of net metering programs on optimal load management in US residential housing–a case study. *Proceedings of the 9th International Conference on System Simulation in Buildings (SSB2014),* Liège, Belgium.

Gonzalez, C., Geuns, J., Weckx, S., Wijnhoven, T., Vingerhoets, P., De Rybel, T., & Driesen, J. (2012). LV distribution network feeders in Belgium and power quality issues due to increasing PV penetration levels. *2012 3rd IEEE PES International Conference and Exhibition* on *In Innovative Smart Grid Technologies (ISGT Europe),* (pp. 1-8). IEEE.

Hedegaard, K., Mathiesen, B. V., Lund, H., & Heiselberg, P. (2012). Wind power integration using individual heat pumps–analysis of different heat storage options. *Energy*, 47(1), 284-293.

International Standard Organization. (2007). ISO13790:2007: energy performance of buildings – calculation of energy use for space heating and cooling, Geneva, Switzerland.

Masy, G., Georges, E., Verhelst, C., Lemort, V., & André, P. (2015). Smart grid energy flexible buildings through the use of heat pumps and building thermal mass as energy storage in the Belgian context. *Science and Technology for the Built Environment*, 21(6), 800-811.

Miara, M., Günther, D., Leitner, Z. L., & Wapler, J. (2014). Simulation of an Air-to-Water Heat Pump System to Evaluate the Impact of Demand-Side-Management Measures on Efficiency and Load-Shifting Potential. *Energy Technology*, 2(1), 90-99.

Palacio, S. N., Valentine, K. F., Wong, M., & Zhang, K. M. (2014). Reducing power system costs with thermal energy storage. *Applied Energy*, 129, 228-237.

Patteeuw, D., Henze, G. P., & Helsen, L. (2016). Comparison of load shifting incentives for low-energy buildings with heat pumps to attain grid flexibility benefits. *Applied Energy*, 167, 80-92.

Reynders, G., Nuytten, T., & Saelens, D. (2013). Potential of structural thermal mass for demand-side management in dwellings. *Building and Environment*, 64, 187-199.

Wolisz, H., Harb, H., Matthes, P., Streblow, R., & Müller, D. (2013). Dynamic simulation of thermal capacity and charging/discharging performance for sensible heat storage in building wall mass. In *Proceedings of Building Simulation Conference*.

Control-Oriented Modelling of Thermal Zones in a House: A Multi-Level Approach

Jennifer DATE[1#], José A. CANDANEDO[2*], Andreas K. ATHIENITIS[1%]

[1]Deparment of Building, Civil and Environmental Engineering, Concordia University,
Montréal, Québec, Canada
[#]+1 514-848-2424x7070, j_date@encs.concordia.ca
[%]+1 514-848-2424x8791, aathieni@encs.concordia.ca

[2]Natural Resources Canada – CanmetENERGY,
Varennes, Québec, Canada
+1 450 625-3126, jose.candanedoibarra@canada.ca

* Corresponding Author

ABSTRACT

This paper investigates the use of a multi-level simplified linear thermal modelling approach based on the electrical analogy for the development of control strategies in conventional detached residential homes equipped with convective electric heating systems. These models are developed from parameter identification of results obtained through comparison with whole building measured data. Although detailed building simulation models can be used directly for testing control strategies, this approach can be quite computationally intense and time consuming thus simplified models become advantageous. The aim of this work is to present a methodology to allow a user to switch back and forth between thermal models representing different control levels according to the modelling objectives. Different control levels include, but are not limited to, community simulation studies, whole building studies, or zone-level studies. Zone-level models take into account inter-zonal heat transfer. From these simple models, useful information can be extracted without performing any simulation, and this is also explored in this paper.

1. INTRODUCTION

This paper presents a multi-level approach to the problem of modelling different thermal zones in a house for control applications. This problem has been treated before by modelling the whole house with a single, all-inclusive RC thermal circuit which may have different levels of resolution. The core feature of the proposed methodology allows the user to switch back and forth between models representing different control levels according to the modelling objectives.

For the development of specific control algorithms for each zone, the house can be treated as a collection of interconnected zonal models, as opposed to a single, large model. This modelling approach has the advantage of maintaining a simple structure for each zone, while also taking into account the heat transfer between zones; at this control level, issues such as occupancy, thermal comfort or setpoint profiles can be examined in more detail. On the other hand, if the user is interested in a quick estimate of global variables (e.g., overall thermal load over the next 24h) then different zones or even the entire house may be combined into a single low-order model. In summary, this multi-level approach allows the user to "zoom in and out" so that models at each control level remain manageable, easy to calibrate and easy to physically interpret.

Suitable simplified multi-zone thermal models enable a rapid assessment of control strategies targeting energy reduction, or occupant thermal comfort (Bacher and Madsen, 2011; Candanedo *et al.*, 2010; Lin *et al.*, 2012; Wang and Xu, 2006) and advanced control strategies could greatly benefit from adequate, simple models (Sturzenegger *et al.*, 2016). One common approach for simplified building modelling is grey-box RC thermal networks (Inderfurth *et al.*, 2015), where system identification techniques are used to determine effective resistance and capacitance values for the model. Model predictions should be meaningful for energy and power results for the whole building level or at the zone level. Zone level detail allows for even greater potential for advanced controls. Besides energy conservation measures, there is interest in ways to reduce peak power demand (due to space heating or cooling) at critical times (Fournier and Leduc, 2014; Leduc *et al.*, 2011). Simplified thermal models of buildings also offer advantages for district modelling (Baetens *et al.*, 2015; Lauster *et al.*, 2014). Simplified models allow for rapid simulation of complex and/or large systems with acceptable accuracy and benefit from quicker calibration procedures (Kummert *et al.*, 2006).

This paper uses data from an existing unoccupied test house, representative of a typical family home in Québec, Canada, as a case study. Four zones are considered: basement, main floor, upper floor and attached garage. All interior doors were kept closed during experimentation and data collection, so as to minimize horizontal zonal heat transfer. For the most detailed analysis, these zones are modelled with four detailed interconnected zone models; alternative methods of connecting zones are investigated in a previous paper (Date *et al.* 2016). A global low-order whole house model is used to calculate the thermal load of the house. Results of thermal load predictions are compared and resolution of the global whole house model is investigated.

2. METHODOLOGY

The methodology employed for the identification, inspection and validation of simplified multi-zone models consists of the following steps:

- Experiments were conducted at unoccupied test homes to get data for comparison with models.
- A global low-order house model was developed and used to calculate the thermal load of the building.
- A detailed zone-level model was developed to represent the real building and used as a benchmark.
- A Simplified zone-level model was created and the connections between zones are studied.
- Unknown values of parameters of the building models were identified through system identification.
- The simplified thermal model predictions were compared with measured experimental data and the zone-level detailed model predictions.

2.1 Building Thermal Modelling Assumptions

Thermal models based on the physics of the system (typically in the form of resistance-capacitance (RC) models) are useful for control studies in buildings. Values of parameters are identified through an optimization technique, and should be interpreted as "effective" values rather than "exact" physical parameters (Candanedo *et al.*, 2013). Model details could be added or taken away depending on the needs of the user. Some important assumptions used to construct simplified thermal RC networks include:

- The temperature of each surface or cross section is uniform.
- The air in each zone is well mixed.
- Radiative and convective heat transfers are combined and constant.
- Air is a non-participating medium with respect to radiation.
- Conduction between each window and window frame is neglected

An optimization algorithm is used to determine unknown parameters, therefore fewer equations is helpful. Several methods are available to reduce the complexity of a model: merging thermal zones, reducing the discretization of the walls, and merging several walls to combined surfaces.

2.2 Benchmark Model (Detailed Model)

A benchmark detailed model (DM) has been developed consisting of 4 zones, with separated walls, windows, doors, resulting in 32 capacitances. A zone represents a level/storey of the house, shown in Figure 3. Figure 1 shows one zone (floor) of the detailed model (DM) for floor-level modelling. A zone model such as that in Figure 1 represents, for example, the main floor (Zone 2) which is connected to a model depicting the upper floor (Zone 1) and another model representing the basement level (Zone 3) via inter-zonal convection and conduction terms, thus creating a very detailed multi-zone model of the building. Several models can be connected to create a multi-zone model via "$T_{adjacent}$" terms. The thermal mass of the envelope is modeled as a single layer (i.e. one capacitance). This model can be seen as the benchmark model or used as the "real building" in MPC simulation-based studies.

Models of similar detail/structure can represent either the whole building or just a section of the building (floor, room, etc). In this study, models are connected via inter-zonal convection and conduction terms to create a multi-zone model at the floor-level detail with a total of four zones. This section outlines two modelling levels investigated in this paper: (i) the floor-level and (ii) the whole building-level. This approach can be expanded in either direction of detail, to zooming out to the community-level, or with further detail at the individual room-level.

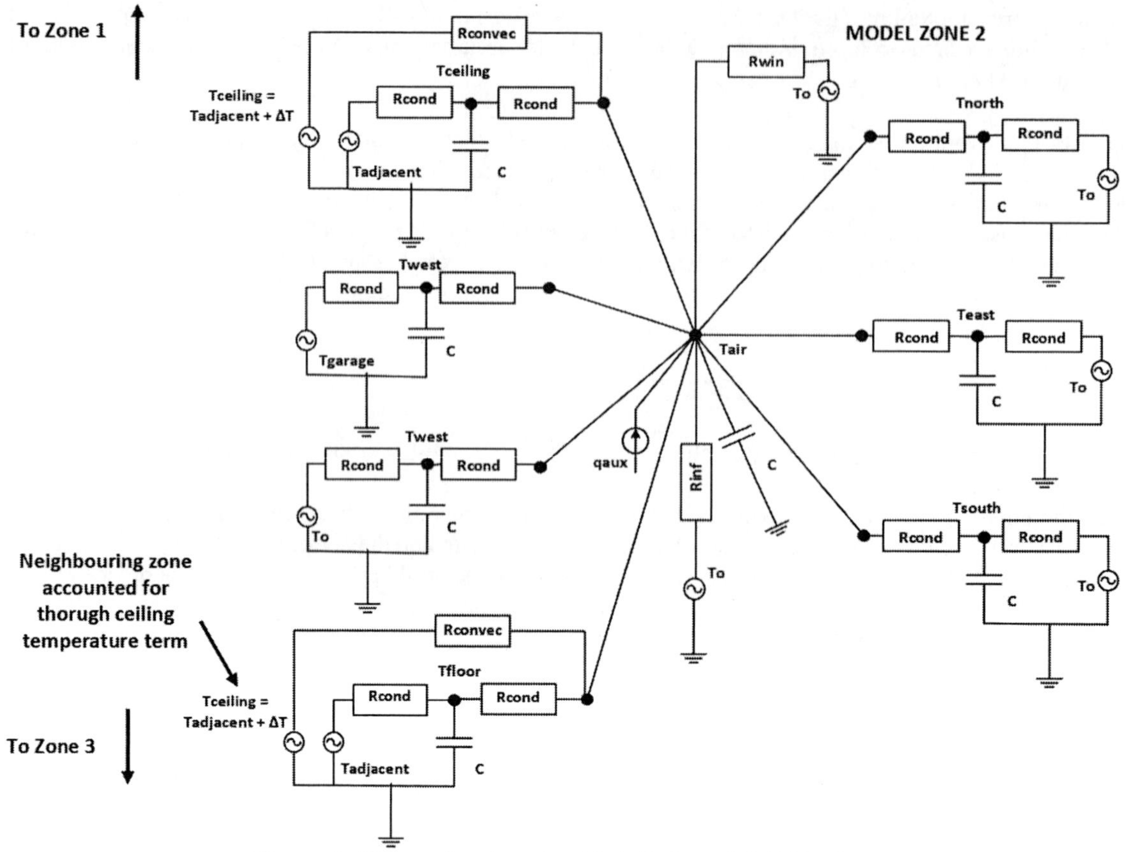

Figure 1: Thermal network of detailed model (DM) for one zone

Figure 3 depicts the system inputs and outputs for the floor-level model. Each zone (floor) is modelled separately and these individual zone models are then connected to one another via inter-zonal convective and conductive terms to create a multi-zone model of the whole building. The system outputs of one zone (i.e. zone air temperature) effectively become system inputs to another adjacent zone. For this study, solar radiation as an input was neglected since the experimental setup effectively blocked the influence of solar radiation (Date *et al.,* 2016).

3. MULTI-LEVEL THERMAL MODELLING APPROACH

3.1 Simplified Floor-Level Model

The simple floor-level model combines surfaces into effective areas, creating 14 capacitances for the whole house model. The thermal mass layer (gypsum board, concrete foundation etc) of the envelope is modeled as a single capacitance. Figure 2 shows an example for the main floor (Zone 2). In a previous study (Date *et al.,* 2016) it was found that a simple floor-level model should include a thermal mass term for the structure between zones (i.e. ceiling/floor material) and the convective and conductive terms should be separated, if one is interested in accuracy at the floor level (rather than building level).

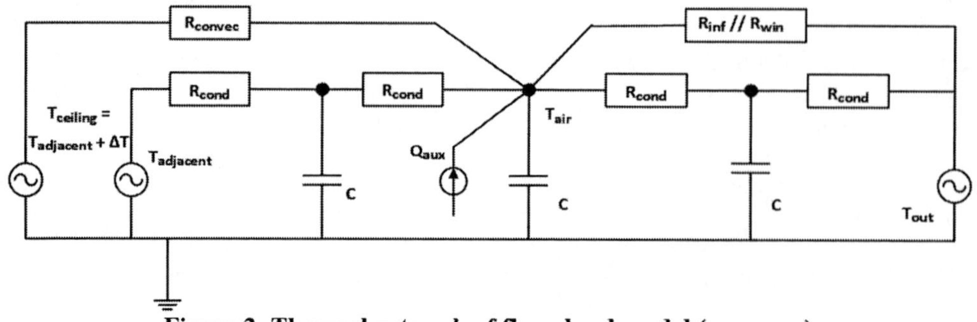

Figure 2: Thermal network of floor-level model (one zone)

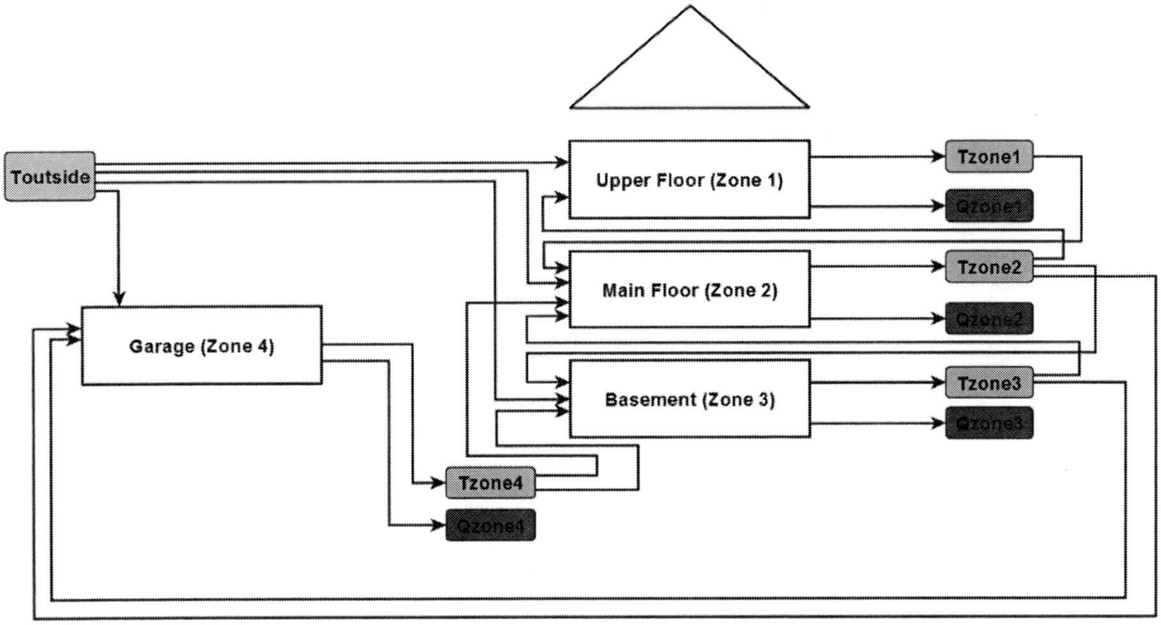

Figure 3: Inputs and outputs of all zones

In all models, experimentally determined data of the zone air temperature stratification was used for the calculation of one-way vertical inter-zonal convective heat transfer (heat transfer driven by a temperature difference). In this case, the models uses a T_{ceiling} term instead of T_{adjacent} temperature. Therefore, in this study, ΔT represents the average temperature difference between the center of the room (height = 1220mm) and the temperature measured near the ceiling (height = 2440mm) obtained from experimental measurements. This is used for temperature difference convective energy flow from vertically adjacent connected zones (i.e. Zone 1 to Zone 2, or Zone 2 to Zone 3). This approach is an attempt at a simplified method for the convective heat transfer between floors in a multi-story building. Further work on simplified inter-zonal modelling for controls includes taking into consideration the techniques and correlations developed by Riffat (1989).

Figure 4 shows the overall thermal network schematic of a whole building of the simple floor-level model structure. It shows the individual zone models (rectangles) and how each zone model is connected to adjacent zone models by convection and/or conduction terms. In all models, experimentally determined data of the zone air temperature stratification was used for the calculation of vertical inter-zonal convective heat transfer (heat transfer driven by a temperature difference), depicted by the ΔT source terms. Parameters are identified using the simplex algorithm.

3.2 Whole Building-Level Model

At the whole building-level, a first order (i.e. one capacitance) model is developed and model parameters are identified using the optimization algorithm. Figure 5 depicts the system inputs and outputs for the whole building thermal model. Figure 6 shows the thermal circuit at the whole building-level, where the building is modeled as one equivalent zone with one effective capacitance and one effective resistance, thus creating a 1R1C model.

As the setpoint of each zone within the house may not be the same, an effective whole house setpoint temperature can be defined as:

$$T_{eff,setpoint} = \sum_{j=1}^{n} \left(\frac{T_{setpoint_j} \cdot A_j}{A_{total}} \right) \tag{1}$$

Using equation (1) to estimate the effective setpoint for the building (when individual zones are controlled differently), this model structure can be useful to obtain quick estimates of whole building loads and operation or for district/community scale simulation studies.

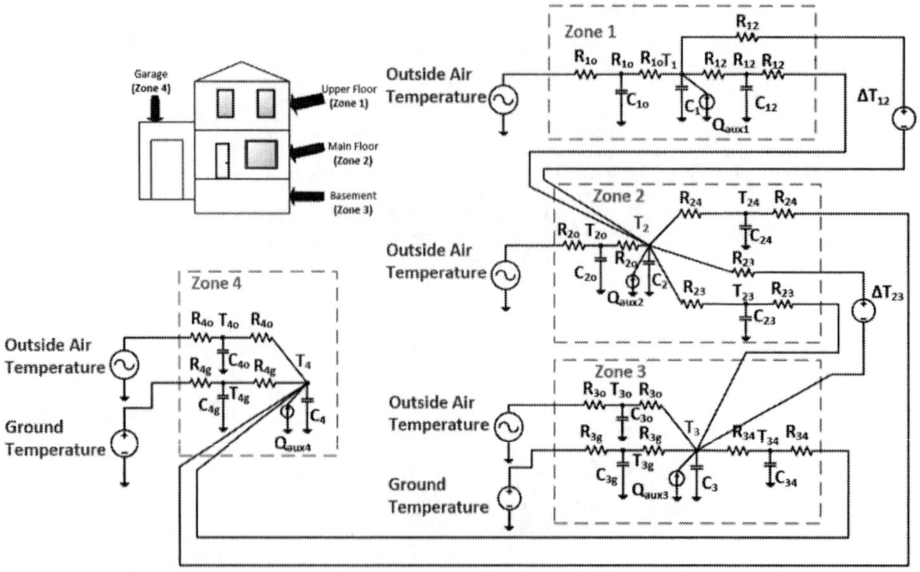

Figure 4: Multi-zone thermal network whole building schematic

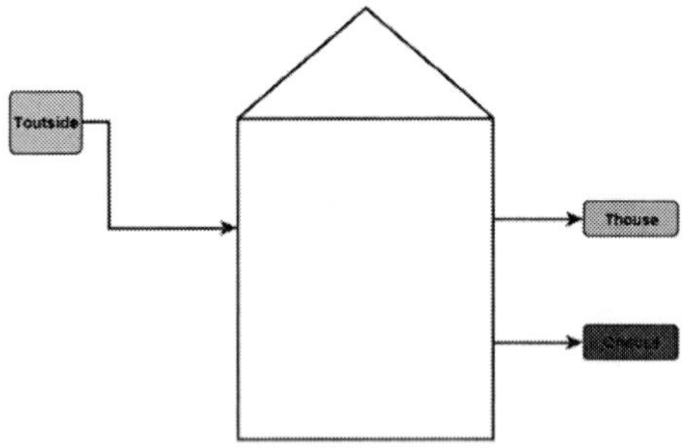

Figure 5: Inputs and outputs of whole house model

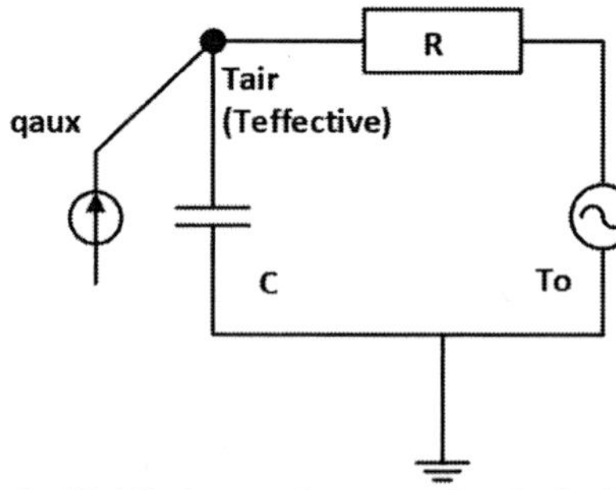

Figure 6: 1R1C thermal network of whole house

4. CALIBRATION OF MODELS

An optimization routine is used to find the parameter values that minimize an objective function. In this case, the objective function chosen was the coefficient of variation of the root-mean-square error (CVRMSE) between measured power and the prediction at 15 minute intervals, similar to (Lavigne *et al.*, 2014). The training data length was 5.4 days of data. Nelder-Mead Simplex was used for this study using the Python programming language. The Simplex algorithm is used here; other algorithms can replace it depending on the user's preference. Since the individual results of each zone and whole building power use are of importance, the CVRSME of each individual zone was minimized (for the floor-level models), and then whole building results were investigated. The objective function (CVRMSE) used is shown in:

$$J(y,\hat{y}) = \frac{1}{\bar{y}}\sqrt{\frac{1}{n}\sum_{i=1}^{n}(\hat{y}_i - y_i)^2} \tag{2}$$

where y is the experimentally measured data (thermal power) and $\hat{y}$ represents the model predictions. The building was modeled using the fully explicit finite difference method to solve the energy balance equations. Initial values of model parameters are based off of the known building material properties and estimates for infiltration, inter-zonal convective transfer and air capacitance multipliers.

4.1 Calibration of Benchmark Model and Floor-Level Model

The parameter values of the benchmark (detailed) floor-level model and simplified floor-level model were identified and the power use predictions were compared. The focus was on accuracy of power prediction rather than room temperature prediction, as most building simulation models are calibrated against power or energy consumption of the whole building. Results after calibration using 5.4 days of model training data are found in Table 1.

Table 1: Calibration results for detailed (benchmark) and simple floor-level models

	Detailed Model	Floor Level Model
CVRMSE – Zone 1 (Top Floor)	23%	26%
CVRMSE – Zone 2 (Main Floor)	25%	23%
CVRMSE – Zone 3 (Basement)	19%	26%
CVRMSE – Zone 4 (Garage)	7%	6%
CVRMSE – Whole Building	13%	15%

4.2 Calibration of Building-Level Model

Table 2 shows the values for the parameters R and C of the 1R1C thermal model of the whole house (Figure 6). Initial and second guesses were made based on geometry and material properties of the building and then an optimization routine was conducted to identify R and C values which results in the lowest CVRMSE, when model predictions are compared to experimentally measured data of 5.4 days. For the initial guess of the 1R1C model, R and C were determined by adding the resistances according to the series and parallel circuit laws and the capacitances were simply added together. In the second guess, a capacitance multiplier of 15 was chosen.

Table 2: 1R1C whole building thermal network model parameter values before and after identification

	Initial Guess	Second Guess	After Identification
R (Kelvin/kW)	7.1	7.1	6.4
C (MJ/Kelvin)	49.1	57.7	17.2
Effective air capacitance multiplier	1	15	N/A
CVRMSE	51%	55%	20%
Time Constant ($\tau = RC$)	97 Hours	114 Hours	31 Hours

One interesting result from the identification process of the 1R1C model is the value of the C parameter, as it is much smaller (almost 3 times smaller) than what the expected "effective" C value should be based on the geometry and material properties of the building. These results suggest that new, revised methods are needed to estimate effective capacitances of buildings. Figure 7 shows the predictions of the floor-level model and experimental heating power data. For visual clarity, only one day of data and predictions is shown here. The predictions of each zone's power demand contribution are shown, and the simple whole house 1R1C power profile is overlaid (dashed line).

The concept of the multi-level thermal modelling approach for different control applications can be summarized as follows. The top benchmark model is the most detailed version of the building thermal model. From there, one can choose to look at optimal control at the zone level by using the simplified zone-level models or at global whole building operation by using the building level model. It is a simple procedure to interchange between the different modelling levels depending on the needs or interests of the model user. The characteristics and results of the three thermal modelling levels are summarized in Table 3.

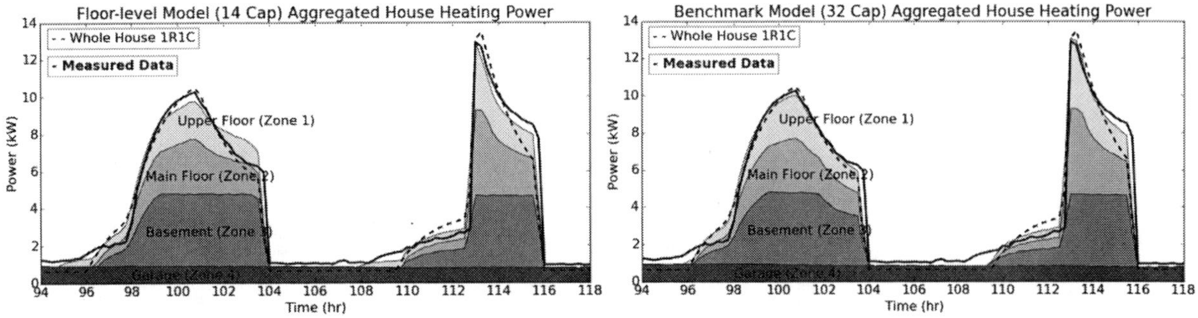

Figure 7: Model verification results of left: simplified floor-level model, and right: benchmark detailed model

Table 3: Summary of three thermal modelling levels

	Benchmark Model	Zone-Level Model	Building-Level Model
Model Order (number of capacitances)	32	14	1
Number of Resistances	69	26	1
Building CVRMSE	13%	15%	20%

5. WHOLE-BUILDING MODEL: IS 1R1C ENOUGH?

From the preceding discussion, it seems apparent that a 1R1C model provides an accurate representation of the heating load of the whole house over a period of several days (and when calibrated with training data of several days). Several questions arise: Is there a need to add more capacitances (e.g. 2 or 3) to the model? Do these models predict the heating load accurately in the short term (say, the next 3 hours)? How does the model training period affect the accuracy of the model? What kind of information can be obtained about the house dynamics from a quick inspection of the models (i.e. without simulation)?

5.1 Alternative Whole House Models

Three model structures for the whole house load calculation are shown below in Figure 8, Figure 9 and Figure 10. The first model on the left corresponds to the 1R1C presented in the previous section. As more capacitances are introduced, the model is more capable to capture the shorter term dynamic behavior of the system outputs based on inputs. CVRMSE values shown correspond to calibration using 5.4 days of training data, and evaluated over a prediction horizon of the same length.

5.2 Effect of (i) Training Data Period and (ii) Length of Prediction Horizon

The three whole building models were trained with datasets of different lengths: 5.4 days (the entire measurement period), 1 day, 6 hours and 3 hours. Different R and C parameters are found depending on the training data period. The predictions of the models were then evaluated in terms of their CVRSME values over different prediction horizons; results are shown in Figure 11. Each bar graph in Figure 11 corresponds to the different training data lengths and the models are then evaluated for different prediction horizon lengths shown in the four tables.

One notable result seen in Figure 11 is that with just 3 hours of training data, two of the simple whole building models (3R2C and 5R3C) can predict the total 5.4 days of operation with acceptable accuracy, while the simplest 1R1C performs poorly. In general, long training data periods result in a model that is satisfactory for long time scales, but not necessarily good for short term predictions (although this is largely dependent on the particular calibration period chosen). In a control application of a real building, simple models could prove useful when short lengths of training data for calibration are available, particularly if the models could be "checked" and systematically re-calibrated at specific times (e.g. once per week or once per day) with the buildings sensor data.

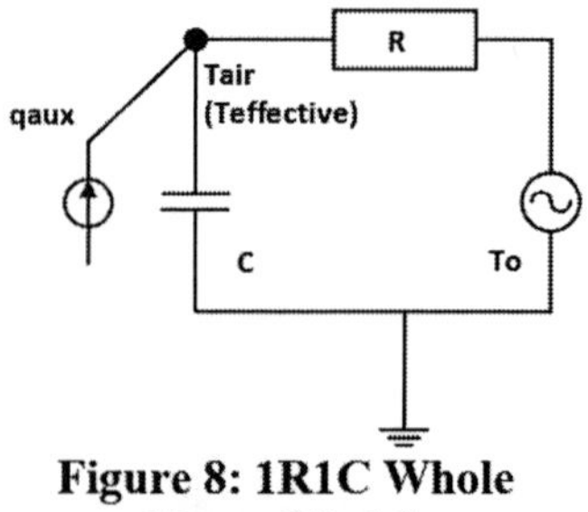

Figure 8: 1R1C Whole House Model, CVRMSE = 21%

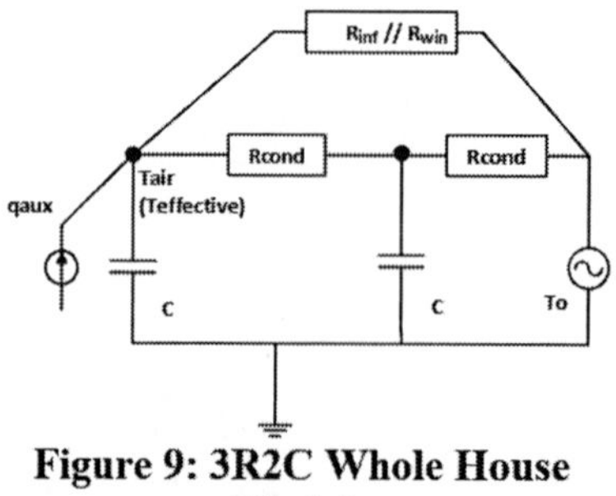

Figure 9: 3R2C Whole House Model, CVRMSE = 20%

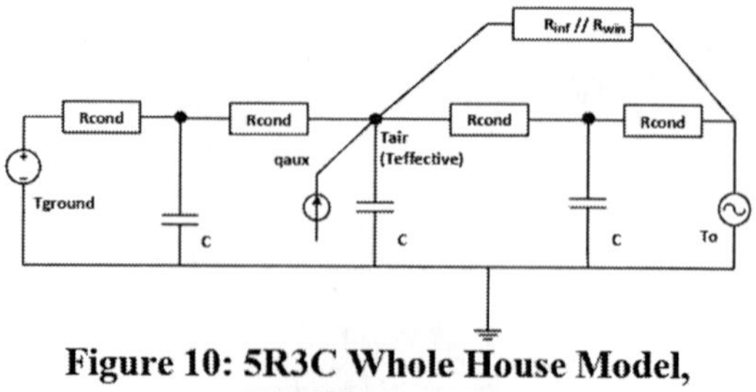

Figure 10: 5R3C Whole House Model, CVRMSE = 19%

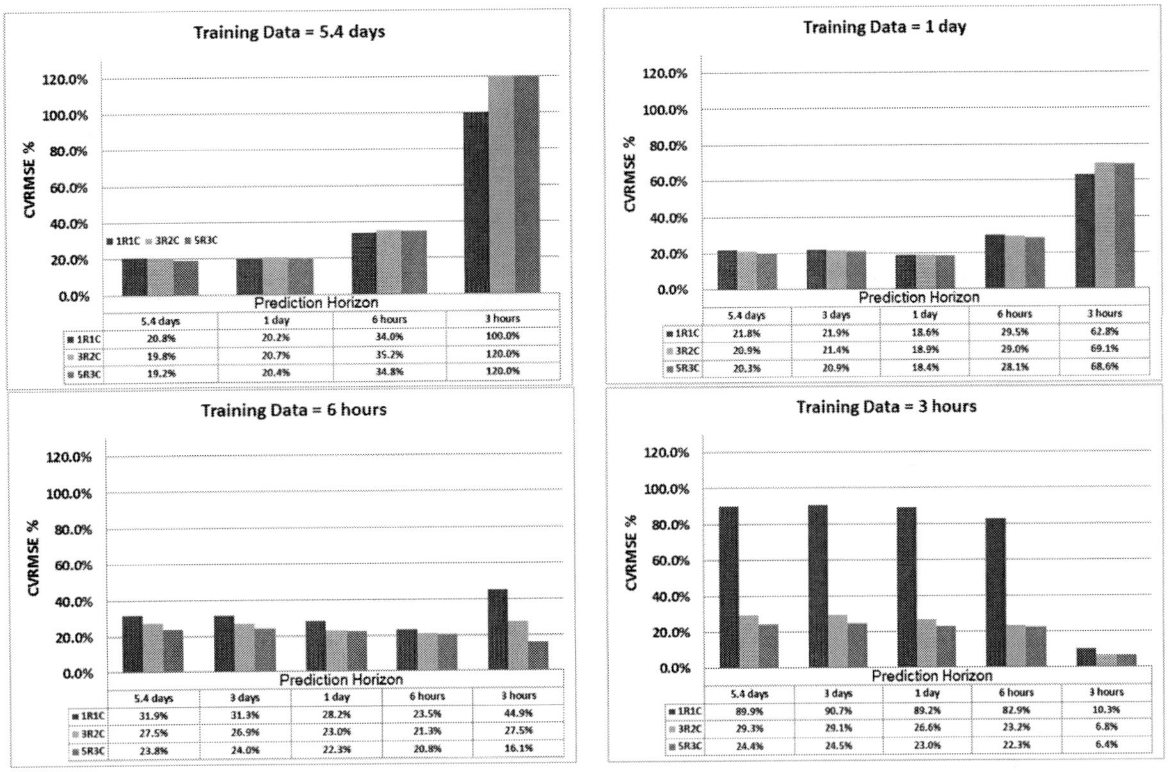

Figure 11: Model Error (CVRMSE) for different prediction horizons and training data lengths

5.3 Frequency Domain Analysis

Frequency domain analysis of these systems could prove useful in evaluating a model's accuracy in a comparative manner, almost by inspection, without the need for a simulation. After solving for the state-space representation of the three models (see Candanedo *et al.* (2013) for procedure), the frequency response of the indoor air temperature (T_{in}) output to the electric heating input (q_{HVAC}) for the three whole-house models were plotted and are shown in Figure 12. The first graph on the left shows the results of the 1R1C model for different training period lengths, the middle graphs shows results for the 3R2C model and the third graph corresponds to the 5R3C model. The magnitude on the left axis refers to °C/Watt.

The following example illustrates how frequency domain analysis enables a quick and simple assessment of a model. The 1R1C model based on 3-hour long training data predicts that, while keeping other inputs at zero, a continuous heating input of 1000 W would result in a steady room air temperature of roughly 16°C (Figure 12). Of course, it is not realistic that such little amount of heating would result in such a high air temperature; this result indicates that 1R1C model trained with only 3 hours of data is far from accurate to represent steady-state or low-frequency effects phenomena. This important result is found without performing any simulation.

In the 1R1C models calibrated with different training periods (Figure 12) there are significant differences in the steady state values predicted (e.g., 16°C/kW vs 8°C/kW); nevertheless, the lines nearly overlap for training data periods of 6 hours and longer. For the 3R2C and 5R3C models, there is significant variation in the phase angle results (corresponding to different time delays) between different training data lengths.

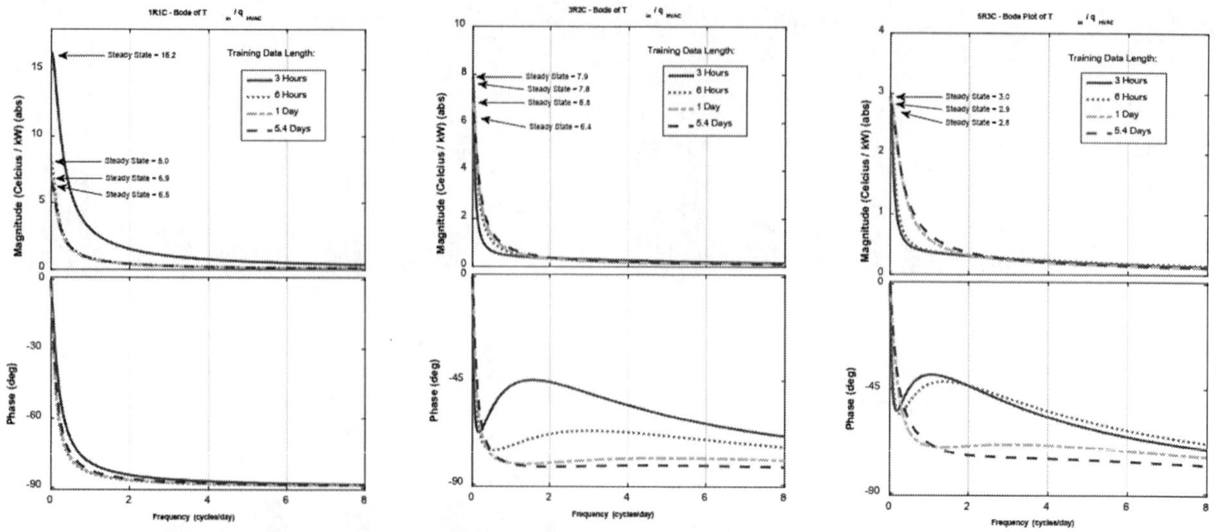

Figure 12: Frequency response - indoor air temperature to heating power

6. CONCLUSIONS

This paper outlined a methodology for multi-level control-oriented modelling for buildings with several zones. This multi-level approach allows the user to "zoom in and out" so that models at each control level remain manageable, easy to calibrate and easy to physically interpret. A global low-order model (1R1C) is developed and used to rapidly calculate the thermal load of the building, while a very detailed benchmark floor-level model is developed and can be used for verification and MPC-based simulation studies. For the development of specific control algorithms for each zone, an adequate simplified zone-level model must be identified. It was found that if zone-level accuracy is of importance, one must incorporate into the model the thermal mass of the structure between zones.

Three building-level models were then evaluated to investigate the effect of incorporating additional capacitance terms. Using these three building-level models, the effects of different lengths of training data periods on the accuracy of different prediction horizons were explored. It was found that a 1R1C whole house model can perform well for either longer horizon or short ones, but not simultaneously for both. Frequency analysis was used to quickly evaluate the whole building models without the need to perform a simulation. Interesting differences emerged in the phase angle predicted by the different models. Work remains to be done on how to improve the guidelines for the initial guess of the grey-box model parameters.

NOMENCLATURE

R	thermal resistance	(K/watt)
C	thermal capacitance	(Joule/K)
T	temperature	(K/°C)
A	floor area	(m^2)
J	objective function	(-)
τ	building time constant	(hour)
CVRMSE	coefficient of variation of the root-mean-square error	(%)
MPC	model predictive control	(-)

Subscript

j	zone
i	time interval

REFERENCES

Bacher, P., and Madsen, H. (2011). Identifying suitable models for the heat dynamics of buildings. *Energy and Buildings*, 1511–1522. http://doi.org/10.1016/j.enbuild.2011.02.005

Baetens, R., De Connick, R., Jorissen, F., Picard, D., Helsen, L., and Saelens, D. (2015). OPENIDEAS - An Open Framework for Integrated District Energy Simulations. *14th International Building Simulation Conference.*

Candanedo, J. A., Allard, A., and Athienitis, A. K. (2010). Predictive Control of Radiant Floor Heating and Transmitted Irradiance in a Room with High Solar Gains. *ASHRAE Transactions.*

Candanedo, J. A., Dehkordi, V. R., and Lopez, P. (2013). A control-oriented simplified building modelling strategy. In *13th Conference of International Building Performance Simulation Association* (pp. 3682–3689). Chambery, France.

Date, J., Candanedo, J., Athienitis, A., and Fournier, M. (2016). Simplified Multi-Zone Thermal Modelling of a House for Demand Reduction & Control Applications. *CLIMA Conference 2016*, 1–10.

Date, J., Fournier, M., Chen, Y., and Athienitis, A. K. (2016). Impact of Thermal Model Resolution on Peak Heating Demand Calculation under Different Setpoint Profiles. *ASHRAE Transactions*, *122*, 1–11.

Fournier, M., and Leduc, M.-A. (2014). Study of Electrical Heating Setpoint Modulation Strategies for Residential Demand Response. In *Proceedings of eSim Conference 2014*. Ottawa, Ontario.

Inderfurth, A., Nytsch-geusen, C., and Ribas Tugores, C. (2015). Parameter Identification for Low-Order Building Models Using Optimization Strategies. *14th Building Simulation Conference.*

Kummert, M., André, P., and Argiriou, A. (2006). Comparing control strategies using experimental and simulation results: Methodology and application to heating control of passive solar buildings. *HVAC and R Research*, 553–575.

Lauster, M., Brüntjen, M., Leppmann, H., Fuchs, M., Teichmann, J., Streblow, R., and Müller, D. (2014). Improving a Low Order Building Model for Urban Scale Applications. *Fifth German-Austrian IBPSA Conference RWTH Aachen University IMPROVING*, 511–518. Retrieved from http://www.ibpsa.org/proceedings/bausimPapers/2014/p1192_final.pdf

Lavigne, K., Daoud, A., Sansregret, S., and Leduc, M.-A. (2014). Demand Response Strategies in a Small All-Electric Commercial Building in Quebec. In *Proceedings of eSim Conference 2014*. Ottawa, Ontario.

Leduc, M., Daoud, A., and LeBel, C. (2011). Developing winter residential demand response strategies for electric space heating. In *Proceedings of 12th Conference of IBPSA* (pp. 1111–1118). Sydney, NSW, Australia.

Lin, Y., Middelkoop, T., and Barooah, P. (2012). Issues in identification of control-oriented thermal models of zones in multi-zone buildings. *51st IEEE Conference on Decision and Control*, 6932–6937. http://doi.org/10.1109/CDC.2012.6425958

Riffat, S. (1989). Measurement of heat and mass transfer between the lower and upper floors of a house. *International Journal of Energy Research*, *13*(January 1988), 231–241. Retrieved from http://onlinelibrary.wiley.com/doi/10.1002/er.4440130212/abstract

Sturzenegger, D., Gyalistras, D., Morari, M., and Smith, R. S. (2016). Model Predictive Climate Control of a Swiss Office Building. *Implementation, Results, and Cost–Benefit Analysis*, *24*(1), 1–12. http://doi.org/10.1109/TCST.2015.2415411 Y3 - 07.01.2016 U6 - http://ieeexplore.ieee.org/ielx7/87/7361777/07087366.pdf?tp=&arnumber=7087366&isnumber=7361777 M4 - Citavi

Wang, S., and Xu, X. (2006). Simplified building model for transient thermal performance estimation using GA-based parameter identification. *International Journal of Thermal Sciences*, 419–432. http://doi.org/http://dx.doi.org/10.1016/j.ijthermalsci.2005.06.009

ACKNOWLEDGEMENT

Support provided through a NSERC/Hydro-Québec Industrial Research Chair in Optimized Building Operation and Energy Efficiency: Towards High Performance Buildings and by CanmetENERGY is gratefully acknowledged.

Optimization of electricity make/buy decisions for Purdue's Wade Power plant

Sugirdhalakshmi RAMARAJ[1]*, James E. BRAUN[2], W. Travis HORTON[3]

[1*]Lyles School of Civil Engineering
Purdue University, West Lafayette, IN, USA
E-mail: sramaraj@purdue.edu

[2]Herrick Professor of Engineering and Director of CHPB, School of Mechanical Engineering
Purdue University, West Lafayette, IN, USA
E-mail: jbraun@purdue.edu

[3]Associate Professor of Civil Engineering with Courtesy Appointment in Mechanical Engineering
Purdue University, West Lafayette, IN
E-mail: wthorton@purdue.edu

* Corresponding Author

ABSTRACT

The Wade power plant at Purdue University produces chilled water, steam and electricity using CCHP (Combined Cooling, Heating and Power) systems to meet the campus cooling, heating and electricity demands. Steam generated from utility boilers is not only distributed through a steam tunnel system for campus heating but also used for power generation, chilled water production and in-plant auxiliary usage. Chilled water is generated using both steam driven chillers and electric chillers and is delivered through a closed water circulation loop to campus to meet the time-varying cooling demand. The electricity generated using two steam turbine driven generators provide 30-50% of the electricity required to meet campus needs. The remainder of the electricity is purchased from the local electric utility and includes a real-time pricing component that varies with time. Plant primary energy use and costs depend on decisions regarding generation and/or purchase of electricity in response to time varying prices, loads and environmental conditions. The operational staff tries to make decisions that minimize operating costs while meeting campus electricity, heating, and cooling demands but only have limited analysis tools for making those decisions in real time. This paper presents an approach for optimizing the operation of the CCHP system using a multimodal genetic algorithm based on load forecasts and fuel pricing along with a model of the plant. The tool is used along with the assumption of perfect forecasts to evaluate the benefits of optimal control for the Purdue CCHP plant as a function of different (possibly future) utility rate incentives.

1. INTRODUCTION

Combined Cooling, Heat and Power (CCHP), also known as trigeneration, has great potential to minimize primary energy consumption in distributed energy generation systems due to its ability to recover low-grade thermal energy resulting in higher energy efficiencies, reduced emission rates and lower operating costs. Another benefit of trigeneration is that, it offers a higher level of energy security and control by removing the reliance on centralized power grids. CCHP systems are composed of many different components with diverse dynamic response characteristics at various time-scales, and wide-ranging operational strategies to meet both electrical and thermal demands. It is very complex to effectively design optimal control strategies because of varying operational conditions such as the costs of fuel and electricity, building loads, and component design and operational limitations. Maximizing the benefits of CCHP systems involves several different aspects: 1) thermodynamics (maximum energy efficiency, minimum fuel consumption, minimum irreversibility), 2) economics (minimum operational cost), and 3) environmental (e.g., emissions reduction). A proper optimization scheme is required to control and operate CCHP systems. A number of papers have been published on optimizing the performance of cogeneration and trigeneration systems. Linear programming was employed to solve long term cost optimization of a Combined Heat and Power (CHP) or cogeneration system (Gustafsson and Karlsson, 1990 and Lahdelma and Hakonen, 2002) and a trigeneration system (Rong and Lahdelma, 2004). Chandan *et al.* (2012) adopted reduced order thermo-economic models of CCHP components integrated with non-linear programming for a CCHP plant at the University of California, Irvine. An evolutionary genetic algorithm has been used to maximize primary energy savings, annual total cost savings, and carbon dioxide emission (CDE) reductions of a CCHP system for a hotel building in Beijing, China (Wang *et al.*, 2010). This approach was also utilized to minimize the total cost rate of the CHP plant using an exergo-economic approach (Ahmadi and Dincer, 2010). An energy dispatch algorithm employs network flow models using linear programming to minimize the operational cost, primary energy consumption and CDE (Cho *et al.*, 2009).

Detailed models for CCHP system components are complex and as such they are typically not employed in plant-wide optimization activities due to the large computation time, non-linear characteristics of components, thermodynamic constraints, mixed-integer variables to represent dispatch states (on/off) of the components, and other associated challenges (Chandan *et al.*, 2012). In this paper, we developed a detailed thermodynamic model including the non-linear characteristics of components, and an optimization procedure for minimizing the total operational cost of the Wade power plant (CCHP system) at Purdue University. A deterministic network energy flow model is developed based on the capacity and operation of the CCHP system and an energy dispatch algorithm is employed for the nonlinear problem using a multimodal genetic algorithm. The energy dispatch algorithm provides control signals to determine the components to be operated (outermost supervisory control) based on the energy (thermal and electric) demand, and in turn the algorithm activates the inner layer of components (pumps, fans, cooling tower and other auxiliaries). Results from the simulation are presented in the paper to demonstrate the economic advantages resulting from optimal CCHP operation. Furthermore, a sensitivity analysis is presented to show how the optimal operation strategy varies with the changes in prices of electricity and natural gas.

2. CCHP MATHEMATICAL MODEL

2.1 System Description

The Wade power plant at Purdue University produces electricity, steam, and chilled water through a CCHP (Combined Cooling, Heating, and Power) system to meet the campus electricity, heating and cooling demands. The steam generated from the utility boilers is used for campus heating, power generation, chilled water production and in-plant auxiliary component usage. The steam, which is distributed through a steam tunnel system, one line at 125 psig and the other line at 15 psig, is the sole heating source for 13.5 million gross square feet of Purdue campus buildings. Chilled water can be generated using steam driven chillers and/or electric chillers, and is delivered through a closed water circulation loop to campus to meet the time-varying cooling demand. The electricity generated using steam turbine generators provides 30-50% of the electricity required to meet campus needs while the remainder of electricity is purchased from the local electric utility, which includes a real-time pricing component. Purdue's environmental footprint has significantly reduced as the result of trigeneration. Figure 1 shows a simple schematic of the Wade power plant providing heating, cooling and electricity to Purdue campus. Apart from major components, there are other auxiliary equipment such as boiler fans, feed water pumps, chilled water pumps, system pumps and cooling tower fans and pumps that are activated depending on the major components to which they are linked.

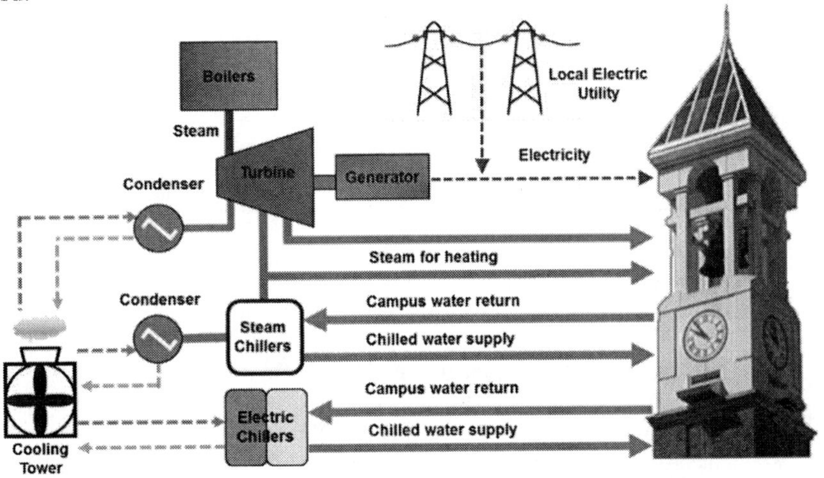

Figure 1: Wade power plant operation schematic

2.2 Energy Flow in the CCHP System

The CCHP system contains separate components for heating, cooling and electricity production. Control of the CCHP system is realized through a hierarchical paradigm. The outermost supervisory control layer determines which components should be operating (on/off states) depending on the fuel cost and electric, cooling, and heating energy demand, together with the energy flow and efficiency constraints of each component. Depending on the results of this outer layer, the inner layer of component controllers activate other auxiliary equipment associated with the major components in the CCHP system. The thermal and electrical demand of campus is met by the combination of all components in the plant. The CCHP system contains components that are operated using steam, electricity, or both as the input.

Superheated steam is generated at 600 psig from four boilers (three water-tube natural gas boilers and one circulating fluidized bed coal boiler) and sent through a common 600 psig steam line. There are two steam driven turbine generators (a 30 MW extraction/condensing turbine and a 10 MW extraction/backpressure turbine) that utilize 600 psig steam to generate electricity. In the extraction/condensing turbine, some portion of the steam is extracted at 125 psig to meet campus steam demand and to run steam chillers and other auxiliaries while the remaining steam is condensed. In the extraction/backpressure turbine, one portion of steam is extracted at 125 psig while the other is exhausted at 15 psig. The extracted steam at 125 psig and 15 psig are used for campus heating and other in-plant auxiliaries. There are four feedwater pumps, three chilled water pumps, a coal boiler fan and a fan for one natural gas boiler. These pumps and fans are driven by 600psig steam and they extract steam at 125 psig except for the fan of natural gas boiler that extracts steam at 15 psig. The extracted 125 psig steam is utilized by three centrifugal chillers, fans of natural gas and coal boiler which are steam-turbine driven. The steam from the fans are extracted at 15psig while the steam from chillers are condensed. The steam is sent to Purdue campus from 125 psig and 15 psig steam line to meet the campus heating demand. The condensate from campus and the various plant components is collected and sent to the boilers again using the feedwater pumps. There are pressure reducing valves in both the 600 psig steam line and the 125 psig steam line used to reduce steam pressure to meet the heating requirements when the demand is high so the other components are bypassed. Figure 2 shows the steam flow across various components in the CCHP system.

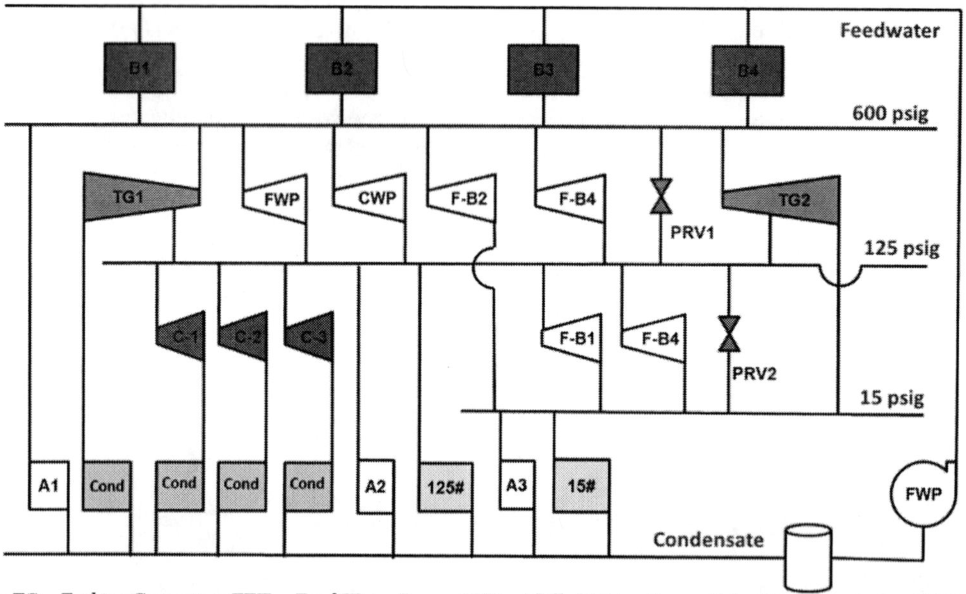

B – Boiler; TG – Turbine Generator; FWP – Feed Water Pump, CWP – Chilled Water Pump; F-B – Fan of the boiler; PRV – Pressure Reducing Valve; C- Chillers; Cond – Condenser; A – Auxiliaries; 125# - 125 psig campus steam line; 15# - 15 psig campus steam line.

Figure 2: Wade power plant steam-driven components and steam flows

The electricity generated using the two turbine generators and the reminder purchased from local electric utility is used to meet the campus electricity demand and to operate the other components within the power plant: four electric chillers at Wade power plant and six electric chillers at North West Chiller plant, boiler feedwater pumps, chilled water pumps, condensate transfer pumps, system pumps, boiler fans, cooling tower pumps and fans. The operation of pumps and fans depends on the states of main components to meet campus energy demands.

2.3 Network Flow Model

Based on the energy flows of steam and electricity described in the previous section, a deterministic network flow model has been developed that connects the supply to the demand. Network optimization can effectively deal with both continuous and discrete variables that are encountered throughout the system. The network flow model helps in visualizing the electric and thermal energy flows through the CCHP equipment and helps in interpreting the results. The network flow model for the CCHP system is shown in Figure 3. The nodes in this network represent sources of energy and energy demand points. It can be seen that the demand drives the activation of individual components throughout the network. Mass and energy conservation has been applied to develop the energy dispatch algorithm with the network model.

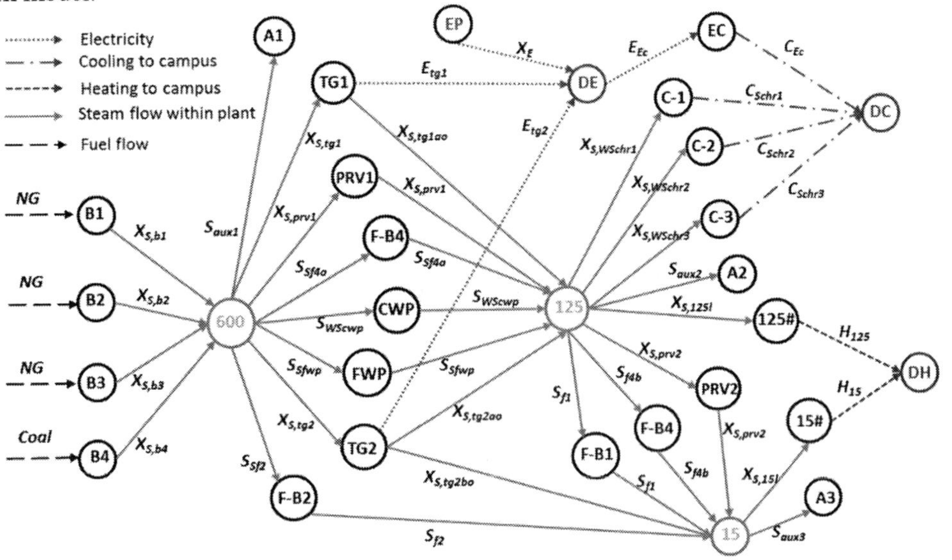

Node B: Boilers
Node 600, 125, 15: 600, 125, 15 psig steam line
Node: A: Auxiliaries
Node TG: Turbine generators
Node F-B: Steam driven fan of the boiler
Node FWP: Steam driven feed water pumps
Node CWP: Steam driven chilled water pumps
Node PRV: Pressure reducing valve
Node EC: Electric chillers
Node C: Steam chillers
Node 125#, 15#: Steam line to campus
Node EP: Electricity purchased
Node DE, DC, DH: Electricity, Cooling, Heating demand

$X_{s,b}$: Steam from boilers
$X_{s,tg}$: Steam to turbine generators
$X_{s,tgo}$: Steam from turbine generators
$X_{s,prv}$: Steam from /to PRV
$X_{s,WSchr}$: Steam to steam chillers
$X_{s,125l}$: 125 psig steam to campus
$X_{s,15l}$: 15 psig steam to campus
X_E: Electricity purchased

E_{tg}: Electricity generated from turbine generators
E_{Ec}: Electricity to electric chillers
C_{Ec}: Cooling capacity from electric chillers
C_{Schr}: Cooling capacity from steam chillers
H: Heating capacity from 125/15 psig steam line
S_{sf}: Steam from/to steam driven F-B
S_{Sfwp}: Steam from/to steam driven FWP
S_{Scwp}: Steam from/to steam driven CWP
S_{aux}: Steam to auxiliaries
NG: Natural Gas

Figure 3: Network energy flow model

2.4 Objective Function and Constraints

Plant primary energy use and costs depend on decisions regarding generation and/or purchasing of electricity in response to time varying factors so as to keep the operating cost at a minimum while also meeting the campus electricity, heating and cooling demands subject to time-varying prices, loads, and environmental conditions. Given the electrical and thermal (heating and cooling) load behavior of campus, the tariff structure for grid-supplied electricity, the price of primary fuel (e.g., natural gas & coal), the operating strategy and characteristics of the CCHP system, and an assumed set of installed CCHP system capacities (e.g., installed capacity of boilers, chillers and generators), classic make-or-buy decision characteristics in response to time varying factors can be analyzed. The main objective is to find the lowest possible operational cost of the CCHP system in terms of the cost of natural gas, coal and electricity. An hourly interval is assumed and no dynamics are considered in the plant modeling. With this in mind, the static (steady-state) optimization problem for the each hour's operating cost is described as:

$$\text{Minimize} \qquad \mathrm{Cost}(x) = \sum_{i=1}^{3} c_{NG} * x_{S,bi} * \frac{\Delta h_{bi}}{\eta_{bi}} + c_C * x_{S,b4} * \frac{\Delta h_{b4}}{\eta_{b4} * HHV_C} + c_E * x_{E,pur} \qquad (1)$$

Here the efficiency of each boiler, η_b, is a quadratic function of boiler steam load, which makes the objective function non-linear. The boiler characteristic curves were determined from power plant operational data.

The energy flow model of Figure 3 is useful in formulating steady-state constraints. Equality constraints are applied to mass and energy balances on each of the nodes that involve decision variables for steam supply and the impacts of those decisions on the supply of energy to meet the campus energy demands. Mass balances applied to the 600 psig, 125 psig and 15 psig steam line nodes in Figure 3 are enforced by driving equations (2), (3) and (4), respectively, to zero in the problem solution.

$$h(1) = x_{S,b1} + x_{S,b2} + x_{S,b3} + x_{S,b4} - x_{S,tg1} - x_{S,tg2} - s_{Sfwp} - s_{WScwp} - s_{Sf2} - s_{Sf4a} - x_{S,prv1} - s_{aux1} \tag{2}$$

$$h(2) = x_{S,tg1ao} + x_{S,tg2ao} + s_{Sfwp} + s_{WScwp} + s_{Sf4a} + x_{S,prv1} - x_{S,WSchr1} - x_{S,WSchr2} - x_{S,WSchr3} - s_{Sf1} - s_{Sf4b} - x_{S,prv2} - x_{S,125l} - s_{aux2} \tag{3}$$

$$h(3) = x_{S,tg2bo} + s_{Sf1} + s_{Sf2} + s_{Sf4b} + x_{S,prv2} - x_{S,15l} - s_{aux3} \tag{4}$$

Mass balance equality constraints for the steam feeding the boiler and turbine generator nodes in Figure 3 are given by equations (5), (6) and (7) respectively, assuming no losses.

$$h(4) = x_{W,Sfwp} + x_{W,Efwp} - x_{S,b1} - x_{S,b2} - x_{S,b3} - x_{S,b4} \tag{5}$$

$$h(5) = x_{S,tg1} - x_{S,tg1ao} - x_{S,tg1bo} \tag{6}$$

$$h(6) = x_{S,tg2} - x_{S,tg2ao} - x_{S,tg2bo} \tag{7}$$

Similarly energy balance constraints across 600 psig, 125 psig and 15 psig steam line have been applied in the set of equations (8)-(10) to determine the input conditions to each steam line.

$$\begin{aligned} h(7) = {} & x_{S,b1} * h_{b1} + x_{S,b2} * h_{b2} + x_{S,b3} * h_{b3} + x_{S,b4} * h_{b4} \\ & - \left(x_{S,b1} + x_{S,b2} + x_{S,b3} + x_{S,b4}\right) * h_{600} \end{aligned} \tag{8}$$

$$\begin{aligned} h(8) = {} & x_{S,tg1ao} * h_{tg1ao} + x_{S,tg2ao} * h_{tg2ao} + s_{Sfwp} * h_{Sfwp} + s_{WScwp} * h_{WScwp} + s_{Sf4a} * h_{Sf4a} \\ & + x_{S,prv1} * h_{prv1} - \left(x_{S,tg1ao} + x_{S,tg2ao} + s_{Sfwp} + s_{WScwp} + s_{Sf4a} + x_{S,prv1}\right) * h_{125} \end{aligned} \tag{9}$$

$$\begin{aligned} h(9) = {} & x_{S,tg2bo} * h_{tg2bo} + s_{Sf1} * h_{Sf1} + s_{Sf2} * h_{Sf2} + s_{Sf4b} * h_{Sf4b} + x_{S,prv2} * h_{prv2} \\ & - \left(x_{S,tg2bo} + s_{Sf1} + s_{Sf2} + s_{Sf4b} + x_{S,prv2}\right) * h_{15} \end{aligned} \tag{10}$$

Energy balances on the heating, cooling and electricity demand nodes in Figure 3 are also treated as constraints so that the differences in supply and demand for campus heating, cooling and electricity are driven to zero using equations (11)-(14). The electricity component includes the purchase of electricity depending on the cost. The electricity demand D_e represents both campus demand and in-plant demand of electric components.

$$h(10) = D_{h,125l} - H_{125} \tag{11}$$

$$h(11) = D_{h,15l} - H_{15} \tag{12}$$

$$h(12) = D_c - C_{Schr1} - C_{Schr2} - C_{Schr3} - C_{Ec} \tag{13}$$

$$h(13) = D_e - x_{E,pur} - E_{tg1} - E_{tg2} \tag{14}$$

where,

$$H_{125} = \Delta h_{125l} * x_{S,125l}; \; H_{15} = \Delta h_{15l} * x_{S,15l} \tag{15}$$

$$C_{Schr} = \sum_{i=1}^{3} cop_{Schr} * x_{S,WSchr} * \Delta h_{WSchr}; \; C_{Ec} = cop_{Echr} * x_{E,WEchr} + cop_{Echr} * x_{E,NWEchr} \tag{16}$$

$$E_{tg1} = \eta_{tg,1} * \Delta h_{tg1a} * x_{S,tg1} + \eta_{tg,1} * \Delta h_{tg1b} * x_{S,tg1bo}; \; E_{tg2} = \eta_{tg,2} * \Delta h_{tg2a} * x_{S,tg2} + \eta_{tg,2} * \Delta h_{tg2b} * x_{S,tg2bo} \tag{17}$$

Additional models for the boiler, turbine generator, chiller and other auxiliary nodes of Figure 3 are employed to convert steam flows to energy inputs (e.g., boiler natural gas or coal requirements) or outputs (e.g., turbine generator electricity outputs) and are not treated as constraints in the formulation. The A, F-B, CWP, FWP and PRV nodes in Figure 3 represent steam loads that are either fixed or dependent on the other decision variables and are also represented with additional models and not included in the constraints.

Additional inequality constraints deal with peak capacity limitations of the components. The two turbine generators have a peak capacity of 30MW and 10MW. The steam chillers have a total capacity of 12,500 tons and the electric chillers have a total capacity of 25,700 tons. Also, the amount of steam required to operate the steam turbine driven pumps and fans, electricity for fans, pumps and other auxiliaries, water flowing through the pumps are a function of variables of the components to which they are connected. Table 1 gives the list of decision variables and parameters used in this energy dispatch algorithm. The CCHP model is complex and involves 22 design variables (5 that directly impact the objective function and 17 that are embedded in constraints) with 13 equality constraints and 14 inequality constraints.

3. OPTIMIZATION

3.1 Data required and assumptions for the model

Data required for the CCHP cost optimization and performances are listed as follows:

- Hourly load (demand) data of Purdue campus for electricity, heating, and cooling
    - End-use loads vary by application type, building size, location, season, work week, and hour
- Utility electricity prices
- Price of on-site fuel (e.g., natural gas, coal)
- Range of "effective" operation of CCHP components for a given installed capacity

Some additional assumptions associated with the model are:

- Neglecting effect of other auxiliary support equipment on costs
- Neglecting the costs associated with maintenance and ownership of the equipment

Table 1: Energy demands (heating, cooling electrical), decision variables, and other parameters

$x_{S,b}$	Steam output from each boiler (kg/s)	c_{NG}	Fuel cost of Natural Gas ($/DT)	D_h	Heating Demand (kW)
$x_{E,pur}$	Electricity purchased from utility (kW)	c_C	Fuel cost of Coal ($/ST)	D_c	Cooling Demand (kW)
$x_{S,tg}$	Steam input to turbine generator (kg/s)	c_E	Cost of electricity purchased ($/kWh)	D_e	Electricity Demand (kW)
$x_{S,tgao}$	Steam output from stage 1 of turbine generator (kg/s)	Δh_b	Heat energy to produce steam in boiler (kJ/kg)	H_{125}	Heating provided by 125 psig steam (kW)
$x_{S,tgbo}$	Steam output from stage 2 of turbine generator (kg/s)	η_b	Boiler efficiency (-)	H_{15}	Heat energy provided by 15 psig steam (kW)
$x_{S,prv}$	Steam input to pressure reducing valve (kg/s)	HHV_C	Higher heating value of coal(kJ/kg)	C_{Schr}	Cooling capacity from steam chillers (kW)
$x_{S,WSchr}$	Steam input to Wade steam chillers (kg/s)	s_{Sfwp}	Steam input to steam-driven feedwater pump (kg/s)	C_{Ec}	Cooling capacity from electric chillers (kW)
$x_{S,125l}$	125psig Steam output to campus (kg/s)	s_{Scwp}	Steam input to steam-driven chilled water pump (kg/s)	E_{tg}	Electricity generated by turbine generator (kW)
$x_{S,15l}$	15psig Steam output to campus (kg/s)	s_{Sf}	Steam input to steam-driven fan (kg/s)	cop_{Schr}	Coefficient of performance of steam chillers (-)
$x_{W,Sfwp}$	Water input to steam-driven feedwater pump (kg/s)	s_{aux}	Steam input to auxiliaries (kg/s)	cop_{Echr}	Coefficient of performance of electric chillers (-)
$x_{W,E,fwp}$	Water input to electric feedwater pump (kg/s)	h	Enthalpy (kJ/kg)	Δh_{125l}	Heat energy provided by 125 psig steam (kJ/kg)
$x_{E,WEchr}$	Electricity to Wade electric chillers (kW)	Δh_{tg}	Enthalpy change across turbine generator (kJ/kg)	Δh_{15l}	Heat energy provided by 15 psig steam (kJ/kg)
$x_{E,NWEchr}$	Electricity to North West electric chillers (kW)	η_{tg}	Turbine generator efficiency (-)	Δh_{WSchr}	Enthalpy change across turbine of steam chillers (kJ/kg)

3.2 Optimization framework

This CCHP optimization problem has a non-linear objective function with linear and nonlinear equality and inequality constraints and strong coupling to the three energy demand components (electricity, heating and cooling). Some of the design variables are continuous while others are discrete. Because of the multimodal and discontinuous nature of this problem, genetic algorithm (GA) was chosen as the solution methodology. Implementation of the genetic algorithm along with the energy dispatch algorithm is illustrated in Figure 4.

The CCHP model has 22 decision variables and 27 constraints. Lower and upper bounds on the decision variables are given as inputs and are readily handled by the genetic algorithm. A relative tolerance for the solution resolution was set as 0.025 for all the decision variables. From that, the number of bits is calculated and rounded off. The number of bits for encoding each design variable was considered in parametric studies to determine a near-optimal optimum value. The resolution, chromosome length, population size, and mutation rate are determined for uniform crossover and tournament selection. The GA uses fitness function values that reflect all objectives and constraints.

There is no need for derivatives or gradients. A fitness function that uses a penalty function to enforce the constraints was developed and is given by:

$$\phi(x) = f(x) + r_p * c_j\{\max[0, g_j(x)]\} + r_p * c_j\{abs[0, h_j(x)]\} \quad (18)$$

where $f(x)$ represents the objective function, $g_j(x)$ represents the inequality constraints, and $h_j(x)$ represents the equality constraints. Parametric studies were performed to determine a good penalty multiplier, r_p, that ensures satisfaction of the constraints. A value of r_p =1000 was employed with a scaling factor c_j =1.

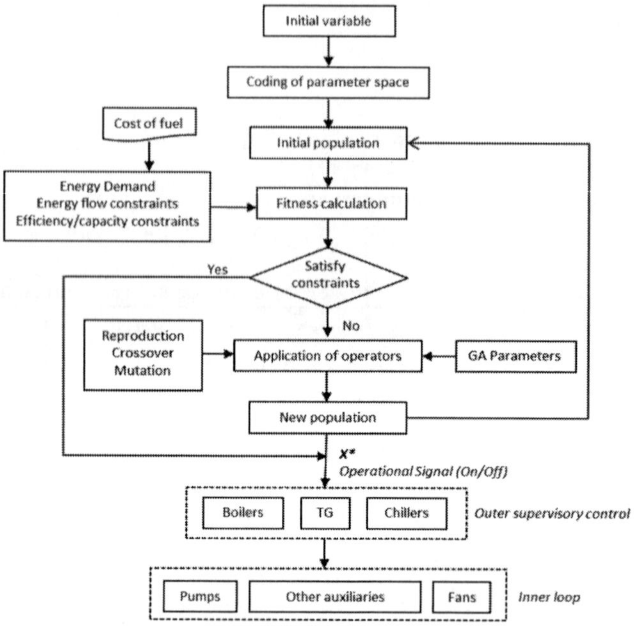

Figure 4: Flow chart of the optimal energy dispatch algorithm solved by GA

4. RESULTS AND DISCUSSION

4.1 Integrated hourly model

An example optimization was performed for a 24-hour period with known cooling, heating, and electricity demand for the campus, and based on actual real time prices (RTP) for purchasing electricity. Figure 5(a) represents thermal and electrical demand and Figure 5(b) shows the real-time price of electricity for a particular day in April. The price of natural gas was 3.00 ($/DT) and the cost of coal was 70.80 ($/ST), which includes the cost of limestone, ash handling and so on. These details were given as inputs to the 24-hour model to compare the optimal performance with the actual operation of the plant.

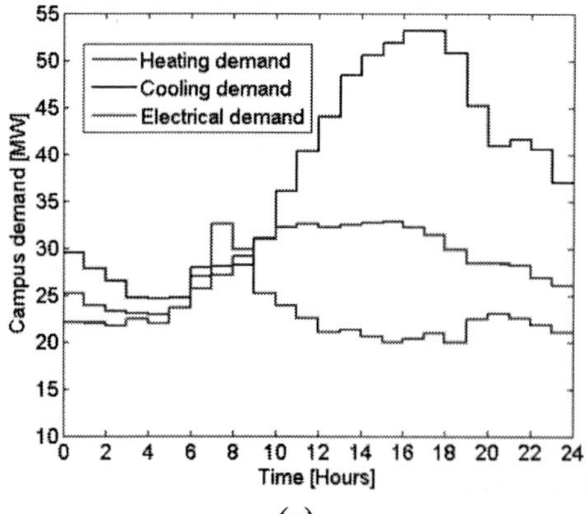

(a)

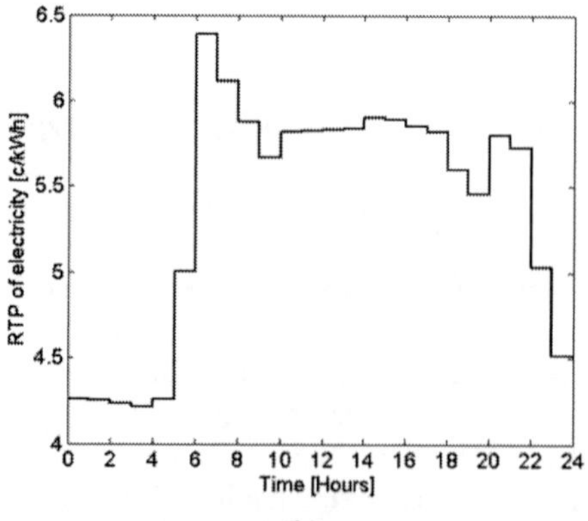

(b)

Figure 5(a): Energy demand of Purdue campus; **(b):** Real time electricity price.

Figure 6(a) shows comparisons of electricity produced and purchased for the actual operation (simulated performance using the actual operating strategies) with the optimum results. It can be seen that the optimum results predict that more electricity should have been generated compared to the actual operation to meet the total electrical demand. However, the optimum still leads to purchasing of some electricity during peak hours of the day. Figure 6 (b) shows comparisons of cooling capacity produced by steam chillers and electric chillers. In the actual operation, all the campus cooling demand was satisfied with the electric chillers. The optimum results predict the usage of both electric and steam chillers to meet the campus cooling load especially when the demand is high. Figure 6 (c) shows the total amount of steam produced in the boilers for the actual operation and optimal results. It can be seen that more steam is produced in optimized operation with the additional steam used for both operation of turbine generators to generate more electricity and steam chillers. In the actual operation, only two natural gas boilers were used, whereas the optimized results include the usage of a coal boiler along with the two natural gas boilers. The selection between coal and natural gas boilers depends on fuel cost, boiler efficiencies and their operating conditions. For the same cost of coal, natural gas and RTP of electricity and the time varying campus electrical and thermal load, the estimated cost of operation for the actual control strategy was $54,006 for 24 hours while the cost for optimal operation was estimated to be $52,190 for 24 hours. This is a cost savings of $1816 for a 24 hour period. Table 2 gives a summary of comparisons between actual operation and optimized results for this 24 hour period.

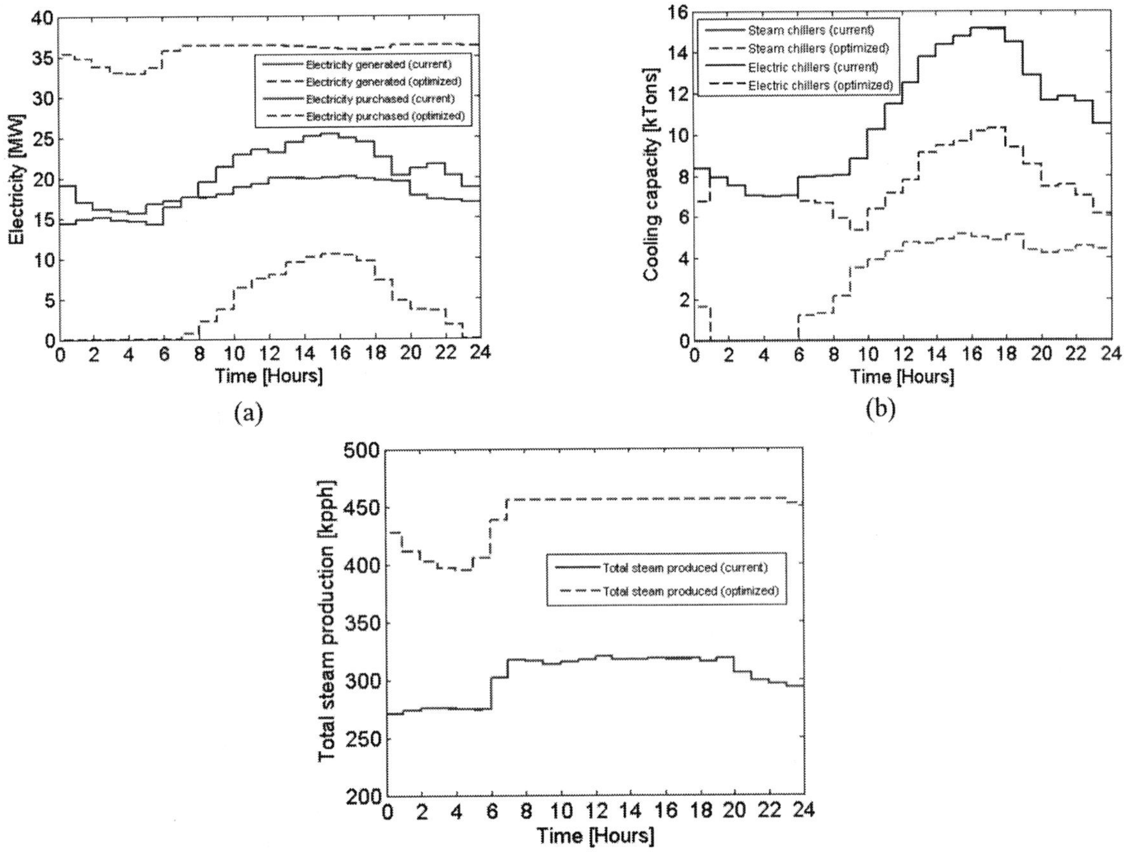

(c)

Figure 6(a): Amount of electricity generated and purchased; **(b)** Cooling capacity of steam and electric chillers (c) Total steam produced in boilers for 24 hour time period [C_{coal} = 70.80 ($/ST); C_{NG} = 3.00 ($/DT)]

4.2 Fuel price sensitivity analysis

The sensitivity of the predicted results to natural gas and electricity prices was studied and typical results are presented in this section. The campus energy demand for a particular hour of the day in summer season was used for the sensitivity analysis, where the heating demand, D_h was 97MMBtu (28,310 kW), cooling demand, D_c was 25,385Tons (8,9354 kW) and the electrical demand, D_e was 29,406 kW. Since there is only one coal boiler and the coal is purchased ahead of time, the cost of coal was set as 70.80 ($/ST) for the analysis.

For the studying the effect of electricity prices, the cost of natural gas was set as 3.00 ($/DT) and the cost of electricity was varied from 0 to 10 (¢/kWh). Figure 7 (a) shows comparisons of electricity generated and purchased for the varying cost of electricity. It can be seen that a higher quantity of electricity is purchased at lower costs of electricity. As the price of electricity increases above 3.90 ¢/kWh, there is a reduction in the purchase of electricity and an increase in the generation of electricity from turbine generator 1 (TG-1). Above 4.20 ¢/kWh of electricity cost, more electricity is generated than is purchased. However some amount of electricity is purchased during the day to meet the electrical demand of campus and all electricity cannot be generated because of limited availability of steam for TG-1 due to a low campus heating demand. Figure 7 (b) shows comparisons of cooling capacity produced by steam chillers and electric chillers for over the range of electricity rates. The control switches from maximizing electric chiller operation at low rates to steam chillers when rates are above 4.20 ¢/kWh in order to meet the campus cooling demand. However, Wade electric chillers are operated on this summer day due to a high cooling demand. From figure 7 (c), we can see that as the price of electricity increases, more steam is produced to meet the steam demand of the turbine generators and steam chillers. At lower electricity prices some amount of steam is still produced to meet the campus heating demand. Boilers 1, 2 and 3 are natural gas boilers while boiler 4 is a coal boiler. The boilers are brought online depending upon their efficiency at different steam loads and operating conditions.

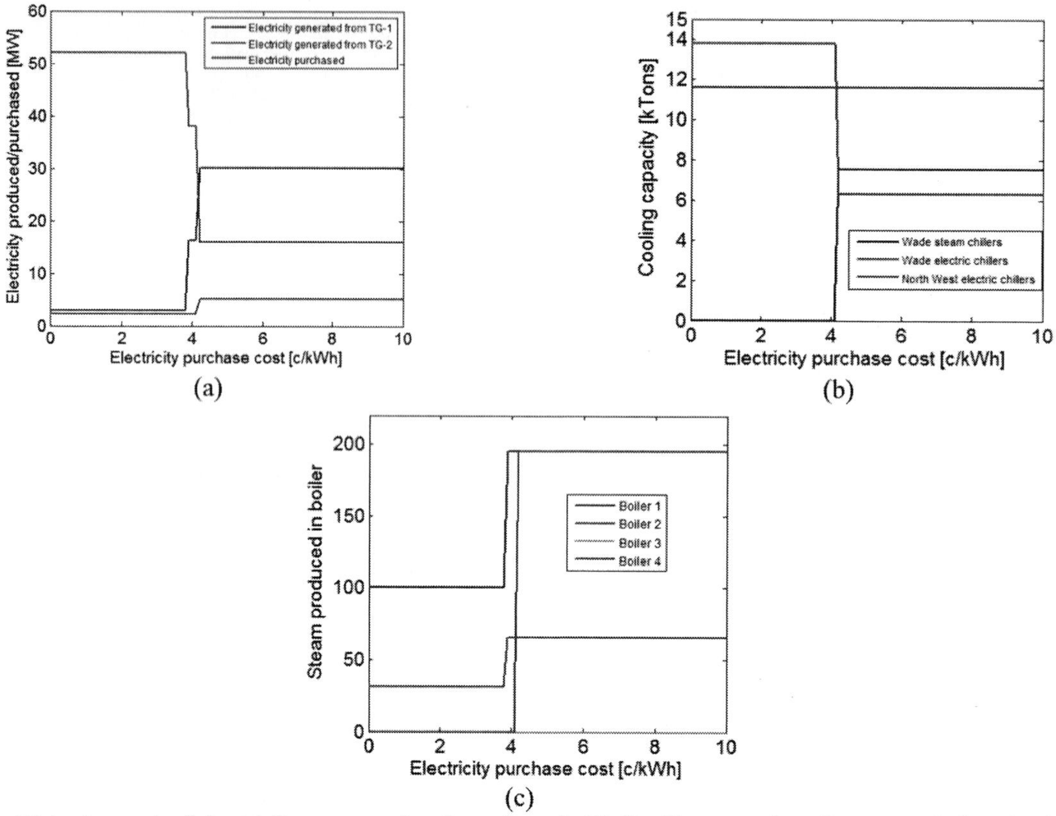

Figure 7(a): Amount of electricity generated and purchased; **(b)** Cooling capacity of steam and electric chillers (c) Total steam produced in boilers with the change in electricity purchase cost [D_h = 97MMBtu; D_c = 25385Tons; D_e = 29,406 kW]

For the second case study, the cost of electricity was set as 4.50 ¢/kWh and the cost of natural gas was varied from 0 to 10 ($/DT). Figure 8 (a) shows comparisons of electricity generated and purchased for the varying cost of natural gas. It can be seen that more electricity is generated at the lower natural gas prices, but above 3.60 $/DT there is an increase in the purchase of electricity and decrease in the generation of electricity. Even at the high prices some amount of electricity is generated that goes along with the production of steam to meet campus heating demand. Figure 8 (b) shows comparisons of cooling capacity produced by steam chillers and electric chillers for the increasing electricity cost. The steam chillers are operated when the price of natural gas is low, but operation is switched to electric chillers above 3.60 $/DT to meet the campus cooling demand. From Figure 8 (c), we can see that as the price of natural gas increases, there is a decrease in the total amount of steam produced. However, some amount of steam must be produced to meet the campus heating demand. The heating, cooling and electrical demand drives the operation of equipment depending on the fuel cost. Considering the interactions between steam-driven and electricity-driven components and fuel costs is essential in making operational decisions.

Table 2: Comparison between current operation and optimized results

For 24 Hours	Current Operation	Optimized Results
Total steam produced [kpph]	7254	10611
Total electricity generated [MW]	422	854
Total electricity purchased [MW]	492	98
Total cooling from steam chillers [kTons]	0	74
Total cooling from electric chillers [kTons]	258	184
Total cost of operation [$]	54006	52190
Total cost savings [$]		**1816 [3.36%]**

5. CONCLUSION

A non-linear genetic algorithm (GA) was applied to a detailed network energy flow model of a large CCHP system in order to determine the benefits of optimizing operation. The optimal energy dispatch algorithm provides operational signals associated with resource allocation that results in minimum operating cost for the CCHP system. Example optimized results were compared with costs for the current operating strategies to illustrate the potential for cost savings. Savings are on the other of $2000 per day. Furthermore, the control decisions and savings potential are sensitive to the fuel costs indicating that the benefits of implementing this type of control optimization could be very significant. Future work will focus on evaluating the yearly savings and developing practical implementation approaches.

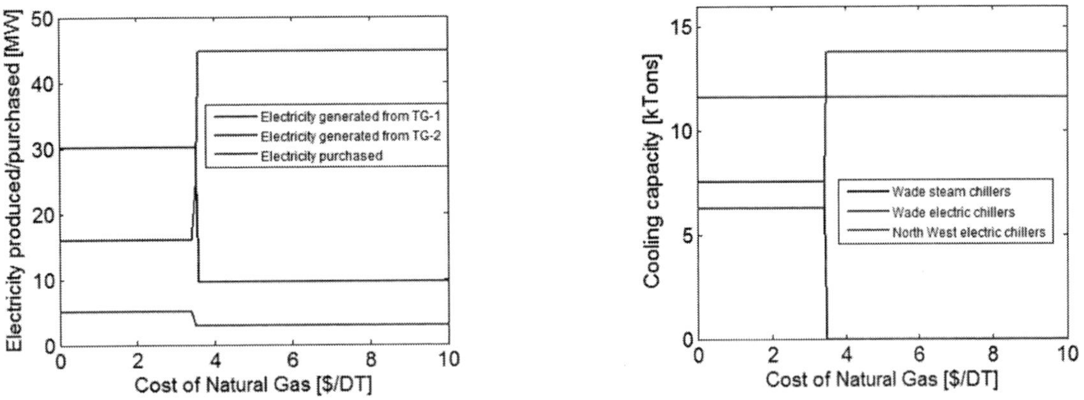

(a) (b)

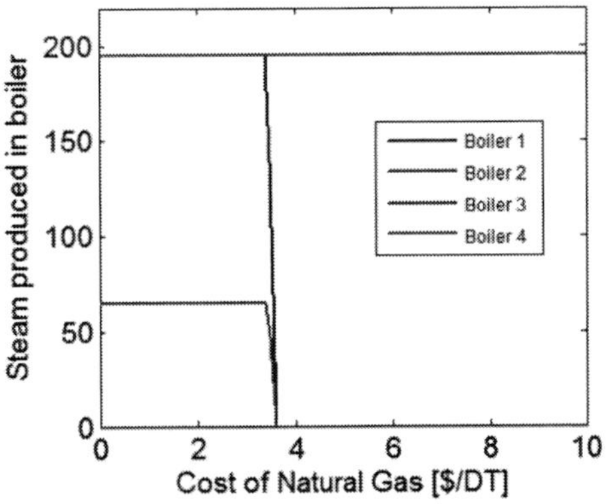

(c)

Figure 8(a): Amount of electricity generated and purchased; **(b)** Cooling capacity of steam and electric chillers (c) Total steam produced in boilers with the change in cost of natural gas

REFERENCES

Ahmadi, P. & Dincer, I. (2010). Exergoenvironmental analysis and optimization of a cogeneration plant system using Multimodal Genetic Algorithm (MGA). *Energy, 35*(12), 5161–5172.

Chandan, V., Do, A.T., Jin, B., Jabbari, F., Brouwer, J., Akrotirianakis, I., Chakraborty, A. & Alleyne, A. (2012). Modeling and optimization of a combined cooling, heating and power plant system. *Proceedings of the American Control Conference.* Montréal, Canada.

Cho, H., Mago, P.J., Luck, R. & Chamra, L.M. (2009). Evaluation of CCHP systems performance based on operational cost, primary energy consumption, and carbon dioxide emission by utilizing an optimal operation scheme. *Applied Energy, 86*(12), 2540–2549.

Gustafsson, S. & Karlsson, B. (1991). Linear programming optimization in CHP networks. *Heat Recovery Systems & CHP, 11*(4), 231-238.

Lahdelma, R. & Hakonen, H. (2003). An efficient linear programming algorithm for combined heat and power production. *European Journal of Operational Research, 148* (1), 141–151.

Goldberg, D. E. (1989). Genetic Algorithms in Search, Optimization and Machine Learning. Addison-Wesley Professional.

Rong, A. & Lahdelma, R. (2005). An efficient linear programming model and optimization algorithm for trigeneration. *Applied Energy, 82*(1), 40–63.

Wang, J., Jing, Y. & Zhang, C. (2010). Optimization of capacity and operation for CCHP system by genetic algorithm. *Applied Energy, 87*(4), 1325–1335.

ACKNOWLEDGEMENT

The authors would like to acknowledge the collaboration of Purdue Physical Facilities on this project.

Characteristics of Energy Consumption by building type of a U.S. Navy Installation in Hot and Humid Climate

Xiaoyu LIU[1], Ahmed AL BARWANI[1], Hui SHEN[1]*

[1]Texas A&M University - Kingsville, Department of Civil and Architectural Engineering
Kingsville, Texas, USA, Hui.Shen@tamuk.edu

* Corresponding Author

ABSTRACT

In fiscal year 2014, the facility energy consumption of the Department of Defense (DOD) amounted to 214,164 billion British thermal units (Btus), which was about 1.2 percent of the total U.S. commercial sector's energy consumption (Department of Defense 2015). DOD spent $4.2 billion on facility energy, which included $4 billion to power, heat, and cool buildings. Several research works have been conducted to save energy and study the energy consumption of military installations in a general view. However, few research works discussed the characteristics of energy consumption by building type in a high temperature and high relative humidity climate.

This paper will investigate the characteristics of energy consumption by building type under the specific hot and humid climate at a selected U.S. Navy installation. The building types comprise of office building, residential building, supermarket and data center. Utility bills of electricity and natural gas for each building will be analyzed to show the actual energy consumption status. Weakness of current building operation in the selected military installation will be discussed based on field test data. Aiming to the weakness discussed, energy conservation measures will be proposed and the estimated energy savings will be presented.

1. INTRODUCTION

Facility energy comprised approximately 19 percent of the total Federal energy consumption in fiscal year 2014 which was 1.2 percent of the total U.S. commercial sector's energy consumption. DoD spent $4.2 billion on facility energy, including $4 billion to power, heat and cool buildings and $0.2 billion to supply fuel to the fleet of NTVs. Facility energy represented 23 percent of the Department's total energy expenditures. In FY 2014, DoD consumed approximately 187,00 billion Btus of energy in its goal-subject buildings (Department of Defense 2015).

The majority of DoD investments are in the military departments' operations and maintenance accounts, to be used for sustainment and recapitalization projects. Such projects typically involve retrofits to incorporate improved lighting, high-efficiency heating, ventilation, and air conditioning (HVAC) systems, double-pane windows, energy management control systems, and new roofs.

LBNL analyzed the energy consumption by building type and end use at 12 U.S. army installations nation-wide by End-use Disaggregation Algorithm (EDA) (Konopacki and Akbari 1996). NREL investigated an assessment and planning process to examine military installations for net zero energy potential for military installations. The work summarized the problems in developing the military installations to net zero energy sites (Booth et al. 2010; Callahan et al. 2011). PNNL investigated the energy reduction goals for federal buildings by analyzing the past performance of buildings, future energy savings potential from retrofits and alternative approaches (Judd et al. 2014).

DoD will give greater focus to more capital-intensive projects that yield greater life-cycle savings, because the low- and no-cost energy efficiency and conservation initiatives continue to diminish and the decreasing budget. This paper will study the energy consumption characteristics of several building types under the specific hot and humid climate at a selected U.S. Navy installation. The building types comprise of office building, residential building, supermarket and data center. Based on the study, low- and no-cost energy saving measures still can be easily acquired.

2. BUILDING DESCRIPTION

The four representative buildings investigated in this study are residential, office, supermarket and data center. Their basic information including building type, schedules and HVAC systems are summarized in Table 1.

Table 1: Basic information of the five investigated buildings at NASK

Building Type	HVAC System	Occupancy Schedule	HVAC Schedule
Residential	Air handling unit, fan coil unit, chiller, boiler	24/7	24/7
Office	Air handling unit, VAV box, chiller, boiler	8 AM to 5 PM	24/7
Data Center	Air handling unit, VAV box, chiller, boiler	7:35 AM to 11:45 PM (M-TR) 7:35 AM to 5 PM (F)	24/7
Supermarket	Air handling unit, split system heat pump, heat pump	7 AM to 6 PM	24/7

2.1 Residential building

The residential building is a 3-floor unaccompanied housing for NASK military. It has around 53 rooms serviced by four handling units (AHUs) and fan coil units (FCUs) for both heating and cooling purpose. The chilled water system consists of one high efficiency air cooled scroll chiller with a total capacity of 60 tons. The chilled water loop is a traditional primary only configuration with a two way valves installed on the return pipe of each Air Handling Unit (AHU). One primary chilled water loop pump is installed and dedicated to the air-cooled chiller at a constant speed.

The heating system consists of one gas fired water heating boiler with a total capacity of 1.2 MMBtu/hr. The hot water is directly sent to heating coils and fan coils. The building has one hot water loop. One primary hot water pump with no VFD installed for this boiler.

This building has 4 AHUs serving the conditioned spaces. The total installed fan power is 7 hp with the total designed airflow rate of 7600 CFM. They are all single duct fresh air only units. Two of them are cooling only units. The building has a total of 63 fan coil units which are five-pipe type. The cold air supplied by AHUs mix with return air at fan coil units and then can be further cooled down or heated up before entering the room.

To save energy, two energy recovery units (ERVs) are installed to exchange heat between outdoor air and exhaust air from the building. However, they are not in use now because of technician reason.

2.2 Office building

The office building is a three-floor building. There are currently about 20 people working in the building. The chilled water system consists of two high efficiency air cooled scroll chiller with a capacity of 110 tons each. The chilled water system configuration is primary system only. The chilled water system and hot water system share the same supplying water pipe. Two chilled water pumps are constant speed pumps. The heating system consists of one gas fired water heating boiler with a total capacity of 3103 MBtu/hr.

There are three AHUs in this building, each serves one floor. Each AHU has fresh air duct, return air duct and supply air duct and is a unit which cannot provide heating and cooling simultaneously. They all have bypass pipe with three-way valve and electronic controlled actuator.

2.3 Data center building

The data center building has mixed usage of office area and data center. There are two air handling units in its mechanical room: one for the data center and the other one for the rest area of the building. The control system used in this building is Johnson Control. The chilled water system consists of two high efficiency air cooled scroll chillers with a capacity of 40 tons each. The chilled water loop is a traditional primary only configuration. Two chilled water pumps with VFDs are installed in parallel. The heating system consists of two gas fired water heating units with a total capacity of 748 MBtu/hr. The hot water pump is a constant speed pump. This building has 2 AHUs serving the conditioned spaces. One of them is cooling only unit and the other one is heating and cooling unit. The fans of both AHUs are equipped with VFD. The building has in total 11 VAV boxes, 6 of which are fan powered VAV boxes. They are all equipped with reheat hot water coil.

2.4 Supermarket

The last investigated building is a medium size supermarket. It has large glazed façade facing east and south. The whole building is divided into the main shopping area, office area (with several offices) and a storage area. The main cooling system in this building is comprised of two split system heat pump units. The split system heat pump unit has one air handler unit and one outdoor condenser unit. The cooling capacity of the two split system heat pump units are 15 Tons and 20 Tons, respectively. Supply air duct has two split branches as observed in the mechanical room, one branch of which has gas heating unit for reheat purpose. Each AHU is equipped with an electrical heater installed before supply air duct split point.

3. UTILITY DATA ANALYSIS

Figures 1 to 4 present one-year period electricity energy consumption and natural gas consumption for four buildings. The metered data at 15-minute interval is processed and plotted as monthly electricity consumption and natural gas consumption. Monthly average outdoor air temperatures were obtained from underground weather website for local installation and plotted in the same figures as well.

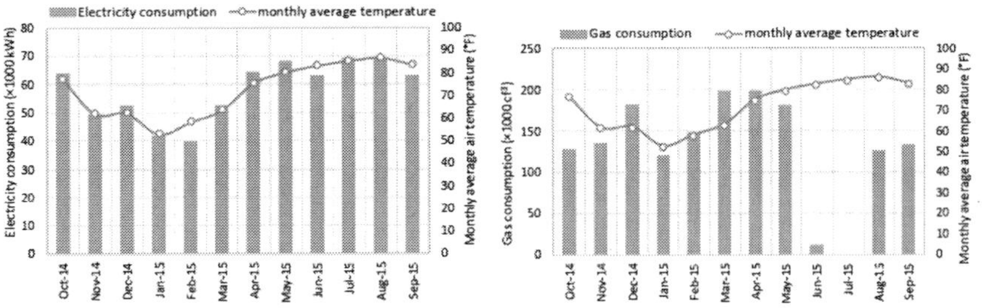

(a) (b)

Figure 1: Monthly total electricity consumption and natural gas consumption of residential building and monthly average local outdoor air temperature, Oct 2014 – Sep 2015

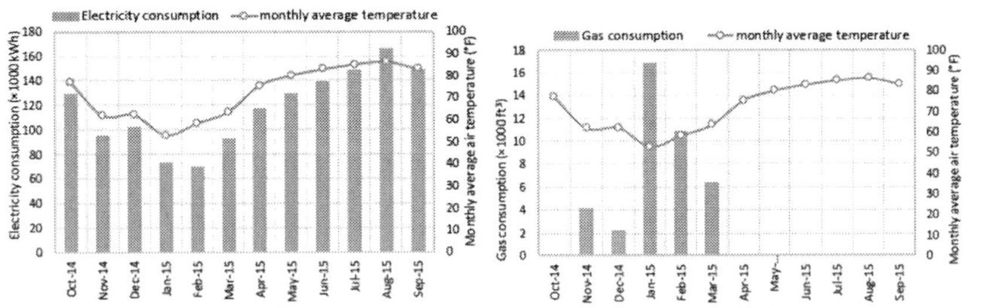

(a) (b)

Figure 2: Monthly total electricity consumption and natural gas consumption of office building and monthly average local outdoor air temperature, Oct 2014 – Sep 2015

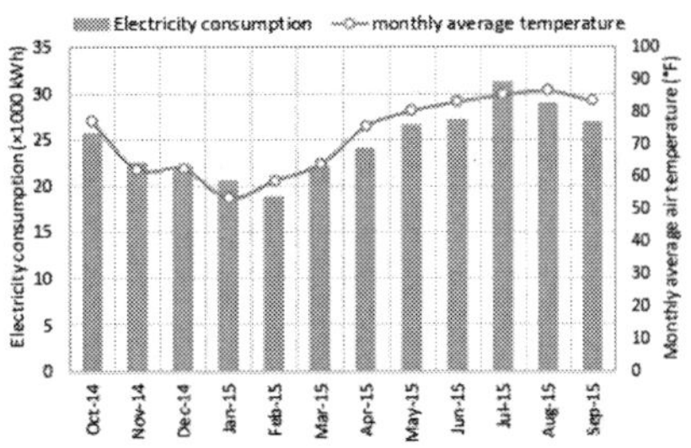

Figure 3: Monthly total electricity consumption and natural gas consumption of data center building and monthly average local outdoor air temperature, Oct 2014 – Sep 2015

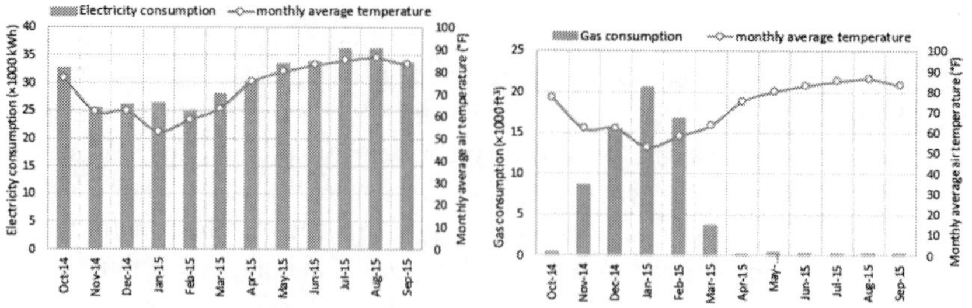

Figure 4: Monthly total electricity consumption and natural gas consumption of supermarket building and monthly average local outdoor air temperature, Oct 2014 – Sep 2015

Based on the figures 1(a) to 4(a), the monthly electricity consumption varies with the change of local outdoor air temperature. The change curve of monthly electricity consumption matches the change curve of outdoor air temperature. Different building type has different change magnitudes. The office building has the largest difference between maximum and minimum of electricity consumption, which is 95 thousand kWh. The residential building has the second largest difference and supermarket building has the smallest difference. The minimum electricity consumption can be regarded as the base load. Generally, base load keeps constant and is not affected by outdoor air. So, the difference indicates the energy consumption for cooling load. Larger difference represents larger cooling load. It also indicates the potential energy savings on reducing energy consumption for cooling load.

Figures 1(b) to 4(b) show the variation of natural gas consumption with the change of local outdoor air temperature. Unlike the electricity consumption, gas was consumed only in limited time period (winter). For the office building and data center building, natural gas was consumed from November to January and the peak natural gas consumption occurred in January. For residential building, natural gas was demanded whole year except July and the peak consumption occurred in June.

For commercial buildings, heating, cooling and hot water account for almost 70% of the total building energy (Book 2010). The ratio of the maximum electricity consumption to the minimum presents the total energy savings potential. Smaller ratio means that the base load is comparably large and this part can be further reduced. The ratio of four buildings are 1.75 for the residential building), 2.357 for the office building, 1.78 for the data center building and 1.48 for the supermarket building. The office building has the largest ratio. It means its base load can be reduced in a smaller magnitude compared with the other three buildings. The next section of this paper will discuss the ways to reduce energy consumption from the viewpoints of reducing energy consumption for base load and cooling/heating load.

4. ENERGY CONSERVATION MEASURES (ECMs)

Currently, major energy efficiency improvement investments in this installation focus on replacing old heating/cooling units with new units. Replacing old HVAC equipment with higher efficiency equipment will save energy. However, it is costly and may not bring the expected energy savings if the control strategies of HVAC system are not carefully addressed. This section discusses the common control strategies, necessary capital investment and the energy savings potential due to the implementation.

4.1 Energy conservation measures for saving total energy consumption

(a) Install central control system

The control system of the four buildings are not integrated and most of them do not work properly. Table 2 summarizes the control systems of each building and their problems.

Table 2: Control system status

Building type	Control system	Problems
Residential	Trane IQ3xcite	Access to the control system is lost. Boiler and chiller are not connected to the system.

Office	Dos-based Honeywell system	Access to the control system is lost. System cannot work properly.
Supermarket	No control system	Units are controlled by thermostats.
Data center	Johnson Control System	Control platform only permit limited functions without the ability to optimize control strategies and record trend data.

According to the problems and current status of control systems in the four buildings, following measures are proposed. A functional control system will be the platform to conduct optimal energy conservation measures.

- Replace old malfunctioning control system with new control system.
- Integrate the control system across the whole campus buildings.

4.2 Energy conservation measures for saving base energy consumption

(a) Optimize operation schedule

Reduce the operation time of HVAC equipment is very important to avoid unnecessary energy consumption and will extend the service life through reducing operation time of equipment. The data center building requires all units to be on for 24/7 for safety concerns. And the supermarket building adopts split heat pump system. So, considering the functions and the types of HVAC systems of each building, only residential building and office building have the potential to optimize the operation schedule.

Table 3: Operation schedule optimization

Buildings	Equipment	Original	Proposed
Residential	Chiller	On for 24/7	Shut down chillers when outdoor air temperature is below 50 °F.
	Boiler	On for 24/7	Shut down boilers when outdoor air temperature is below 50 °F.
	Pump	On for 24/7	Pump will be off as the chiller is off.
Office	Chiller	Manually turn off and changeover	Chillers will changeover every month and will be shut down when outdoor air temperature drops below 50 F or when the boiler is on.

4.3 Energy conservation measures for saving cooling/heating energy consumption

Optimize the operation strategies of HVAC system in buildings will greatly save energy consumption and avoid high capital cost in the meanwhile. Therefore, field test was conducted from December to April in all the four buildings to get their current system operations and energy performance. HOBO data loggers were adopted and deployed to measure major operation parameters including supply air temperature, return air temperature and outdoor air temperature of AHUs, chilled water supply/return temperature (CHWS/RT) and hot water supply/return temperature (HWS/RT). Three weeks test data was collected for each test point. This section discusses the proper energy conservation measures for each type of buildings based on their specific HVAC systems and field test data.

Figures 5 to 7 show the chilled water temperature difference between CHWS and CHWR and hot water temperature difference between HWS and HWR. For all buildings, the chilled water temperature difference is very small. This means the cooling load was very low during the testing period and the pump should not fully run. However, the chilled water pumps were always in operation according to current operation schedule. Hot water temperature difference of the residential building and data center building varies with outdoor air temperature. The smallest difference was around 4 °F indicating that the HW pump was fully running in partial load situation. Installing VFDs on hot water pumps of these two buildings should increase the temperature difference to reduce energy consumption of pumps.

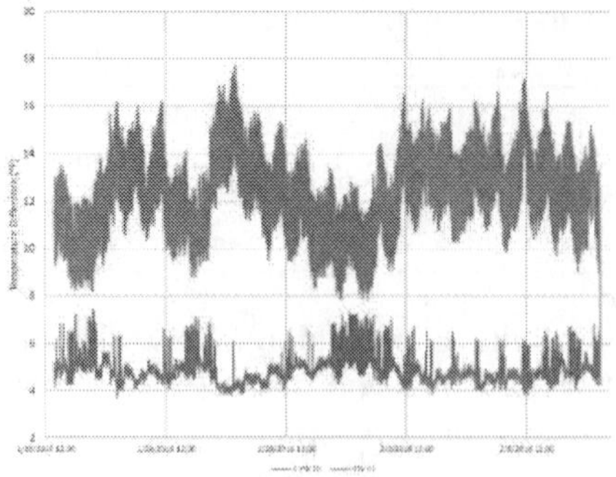

Figure 5: CHW and HW Temperature difference of residential building

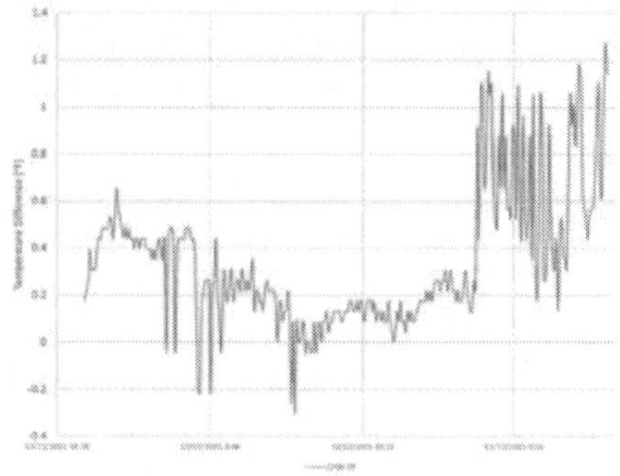

Figure 6: CHW Temperature difference of office building

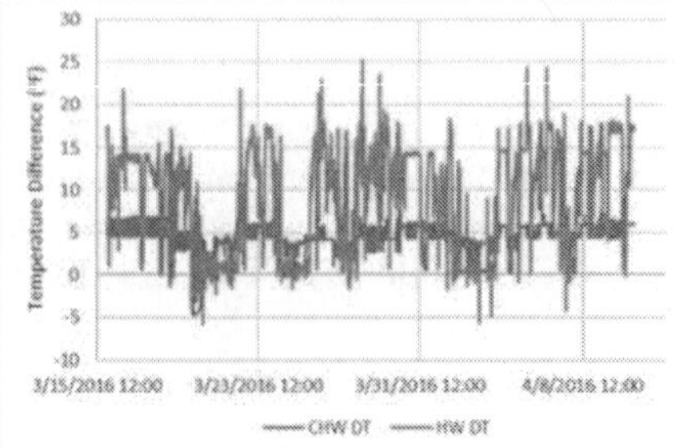

Figure 7: CHW and HW Temperature difference of data center building

According to the walk-through and field test data, the Tables 4 to 7 summarize the optimized control strategies in four types of buildings.

Table 4: Existing and proposed control strategies for residential building

Equipment	Existing	Proposed
Chiller	Constant chilled water supply (CHWS) temperature setpoint	Reset CHWS temperature based on OAT
Chilled water pump	Constant speed	Install VFD and reset differential pressure (DP) setpoint based on VFD speed
Boiler	Constant hot water supply (HWS) temperature setpoint	Reset HWS temperature based on OAT
Hot water pump	Constant speed	Install VFD and reset differential pressure (DP) setpoint based on VFD speed
AHU	Constant supply air temperature (SAT) Setpoint	Reset SAT setpoint based on OAT
	Constant fan speed	Install VFD and reset static pressure (SP) setpoint based on average room temperature setpoint
Fan coil units (FCUs)	Reheat coil is controlled by the thermostat	Reset the reheat temperature setpoint based on room temperature setpoint and VFD speed

Table 5: Existing and proposed control strategies for office building

Equipment	Existing	Proposed
Chiller	Constant chilled water supply (CHWS) temperature setpoint	Reset CHWS temperature based on OAT
Chilled water pump	Constant speed	Install VFD and reset differential pressure (DP) setpoint based on VFD speed
Boiler	Constant hot water supply (HWS) temperature setpoint	Reset HWS temperature based on OAT
AHU	Constant supply air temperature (SAT) Setpoint	Reset SAT setpoint based on OAT

	Constant fan speed	Install VFD and reset static pressure (SP) setpoint based on average room temperature setpoint
Variable air volume terminal box (VAVs)	Reheat coil is controlled by the thermostat	Reset the reheat temperature setpoint based on room temperature setpoint and VFD speed

Table 6: Existing and proposed control strategies for supermarket building

Equipment	Existing	Proposed
AHU	Constant supply air temperature (SAT) Setpoint	Reset SAT setpoint based on OAT
	Constant fan speed	Install VFD and reset static pressure (SP) setpoint based on average room temperature setpoint

Table 7: Existing and proposed control strategies for data center building

Equipment	Existing	Proposed
Chiller	Constant chilled water supply (CHWS) temperature setpoint	Reset CHWS temperature based on OAT
Chilled water pump	Pump speed is controlled to maintain constant differential pressure setpoint	Reset differential pressure (DP) setpoint based on VFD speed
Boiler	Constant hot water supply (HWS) temperature setpoint	Reset HWS temperature based on OAT
Hot water pump	Constant speed	
AHU	Constant supply air temperature (SAT) Setpoint	Reset SAT setpoint based on OAT
	Fan speed is controlled to maintain constant static pressure setpoint	Reset static pressure (SP) setpoint based on average room temperature setpoint

5. ENERGY SAVINGS ESTIMATION

To estimate the potential energy savings upon the implementation of ECMs, bin method is applied. Tables 8 to 11 presents the energy savings estimation based on different ECMs.

Table 8: Energy savings based on operation schedule optimization

Buildings	Equipment	Proposed	Energy Savings (kWh)
Residential	Chiller	Shut down chillers when outdoor air temperature is below 50 °F.	24,317
	Boiler	Shut down boilers when outdoor air temperature is below 50 °F.	1,086,628
Office	Chiller	Chillers will changeover every month and will be shut down with chilled water pumps when outdoor air temperature drops below 50 °F or when the boiler is on.	23,379

Table 8 shows the significant energy savings via optimizing operation schedule. The total energy savings can be about 1.13 million kWh. The office building has the largest ratio as discussed above, so it has the smallest potential to save energy in operation schedule optimization.

As shown in Tables 9 to 12, without big capital investment, in total about 1.7 million kWh can be saved by optimizing control strategies.

Table 9: Energy savings based on control strategies optimization - Residential Building

Equipment	ECMs	Energy Savings (kWh)
Chiller	Reset CHWS temperature based on OAT	32,836

Chilled water pump	Install VFD and reset differential pressure (DP) setpoint based on VFD speed	13,153
Boiler	Reset HWS temperature based on OAT	15,109
Hot water pump	Install VFD and reset differential pressure (DP) setpoint based on VFD speed	5,833
AHU	Reset SAT setpoint based on OAT	8,350
	Install VFD and reset static pressure (SP) setpoint based on average room temperature setpoint	27,696
Fan coil units (FCUs)	Reset the reheat temperature setpoint based on room temperature setpoint and VFD speed	98,389

Table 10: Energy savings based on control strategies optimization - office building

Equipment	Proposed	Energy Savings (kWh)
Chiller	Reset CHWS temperature based on OAT	77,770
Chilled water pump	Install VFD and reset differential pressure (DP) setpoint based on VFD speed	43,844
Boiler	Reset HWS temperature based on OAT	13,944
		5,584
AHU	Reset SAT setpoint based on OAT	10,417
	Install VFD and reset static pressure (SP) setpoint based on average room temperature setpoint	169,606
Variable air volume terminal box (VAVs)	Reset the reheat temperature setpoint based on room temperature setpoint and VFD speed	347,878

Table 11: Energy savings based on control strategies optimization - supermarket building

Equipment	Proposed	Energy Savings (kWh)
AHU	Reset SAT setpoint based on OAT	6,526
	Install VFD and reset static pressure (SP) setpoint based on average room temperature setpoint	69,648
	Reheat Savings	278,908

Table 12: Energy savings based on control strategies optimization - data center building

Equipment	Proposed	Energy Savings (kWh)
Chiller	Reset CHWS temperature based on OAT	52,050
Chilled water pump	Reset differential pressure (DP) setpoint based on VFD speed	9,810
Boiler	Reset HWS temperature based on OAT	15,109
Hot water pump		5,833
AHU	Reset SAT setpoint based on OAT	7,081
	Reset static pressure (SP) setpoint based on average room temperature setpoint	78,715
	Reheat savings	294,495

6. CONCLUSIONS

This paper investigates the characteristics of energy consumption by building type under the specific hot and humid climate at a selected U.S. Navy installation. Utility data of four types of buildings is analyzed and shows two main ways to conserve energy: reduce base energy consumption and heating/cooling energy consumption. Based on walk-through and field data, energy conservation measures are proposed for each type of buildings and potential energy savings are estimated by bin method. There is a great improvement of saving energy consumption of buildings in an easy and quick way. This will be a good addition to the energy conservation program conducted by DoD.

REFERENCES

Book, Buildings Energy Data. 2010. *US Department of Energy, 2011.*

Booth, Samuel, John Barnett, Kari Burman, Josh Hambrick, and Robert Westby. 2010. *Net Zero Energy Military Installations: A Guide to Assessment and Planning.*

Callahan, M., K. Anderson, S. Booth, J. Katz, and T. Tetreault. 2011. "Lessons Learned from Net Zero Energy Assessments and Renewable Energy Projects at Military Installations." National Renewable Energy Laboratory (NREL), Golden, CO.

Department of Defense. 2015. "Annual Energy Management Report, Fiscal Year 2014." http://www.acq.osd.mil/eie/Downloads/Reports/Tab%20B%20-%20FY%202014%20AEMR_FINAL.pdf.

Judd, K. S., J. A. Dirks, D. M. Anderson, B. E. Ford, and D. B. Belzer. 2014. "Analysis of Federal Agency Facility Energy Reduction Potential and Goal-Setting Approaches for 2025."

Konopacki, Steven J., and Hashem Akbari. 1996. "Estimates of Energy Consumption by Building Type and End Use at US Army Installations," August. http://www.escholarship.org/uc/item/4bd685cx.

Airflow Based Model to Estimate Commercial Building HVAC Energy Use: Analysis to determine principal factors for different climate zones

Stefan GUNNSTEINSSON, Rebecca KAHN, Michael GEVELBER*

Mechanical Engineering, Boston University,
110 Cummington Mall, Boston, MA 02215

* Corresponding Author (gevelber@bu.edu)

ABSTRACT

This paper presents an airflow based modeling method to estimate HVAC energy consumption in large commercial office buildings. The model was developed by analyzing operational data from building automation system, relating load profiles and efficiency of key HVAC equipment, based on economizer control policies that determine outside and return airflow rates. The model predicts annual energy use for buildings HVAC loads based on hourly climate data (temperature and relative humidity), and building airflow requirements. The model determines HVAC energy use in terms of building airflow rates for systems utilizing central air handler units with economizers, and identifies the key consumption drivers. Some parts of the model (economizer performance and fan energy use) were based on data collected from the building automation system, relating major component energy use in terms of airflow rates and control laws. Data was obtained for three commercial scale office-lab buildings at Boston University. Results are given in terms of major contributors to HVAC energy use and cost in terms of heating, cooling, and fan motor power for 4 different US cities. A comparison to the building HVAC model used to disaggregate CBECs data is presented.

1. INTRODUCTION

In medium to large commercial office buildings heating, ventilation and air conditioning (HVAC) systems typically accounts for a large portion of the building energy usage of which typically a third goes to driving the ventilation fans. Historically in the United States this category of buildings have been designed with high minimum air-flow rate set-points that are greater than that needed for ventilation. Among CBECS ((EIA, 2012)) building stock for offices over 100,000 sq.ft. there are 18,541 buildings (0.3% of total number of buildings) with total building area of 5.33 billion sq.ft. (7.4% of total building stock sq.ft.) which means improving efficiency of few buildings can have a relatively large impact in terms of energy savings. It is important to reveal how much actual energy is invested in this class of buildings and identify the key consumption drivers.

Pushing air through buildings round-the-clock turns out to be a major contributor and it is not only related to the fan energy required to transport the air but also the required energy to condition (heat and/or cool). An essential part of studying this phenomena is to be able to, for a particular building or project, predict the energy usage. For example when considering airflow rate reductions the prediction model has to be able to assess the energy requirements of each cubic feet of air that needs to be transported through the building. Consequently a model centered around airflow rates has the potential to be more simple and more suitable than a model based on the whole building energy balances. Additionally it makes it possible to quantify where the major cost drivers for HVAC systems in commercial buildings come from.

From an engineering point of view what determines HVAC energy use is 1) airflow rate 2) equipment efficiency 3) how the air is being used (conditioning). Many of the detailed prediction methods available in literature focus on building energy balances to realize the heating and cooling needs while assuming ventilation requirements proposed by standards. This makes them not suitable to estimate savings based on actual airflows rates. Many of the statistical regression models have the potential to base their equipment sub-models fully or partly on measured operating data but typically do not. These consideration were the main motivation to develop an airflow-driven model that could predict energy consumption of HVAC systems. The model methodology is restricted to modern buildings with AHU, central heating/cooling and can accommodate an economizer. This fits with some medium sized and most large sized commercial buildings. The method should give a conservative estimate of the energy use since it ignores energy losses from air infiltration and external heat gain (people, lights, equipment, solar, etc.). For this reason the level of accuracy should be highest in larger buildings where energy consumed for conditioning air flow is much higher than the buildings losses and external loads.

Zhao and Magoules (Zhao & Magoules, 2012) present an extensive review of different models and distinguish between engineering methods, statistical methods and artificial intelligence methods. The methods in the first group have in common that they are based on physical principles to calculate thermal dynamics and resulting energy usage for the whole building and many have the option to do a breakdown to sub-level components. Sometimes the detailed simulation are only used as benchmarks or complimented by model calibration (see (Yik, Burnett, & Prescott, 2001) (Pan, Huang, Wu, & Chen, 2008) and (Fumo, Mago, & Luck, 2010)). Using these methods typically requires an extensive knowledge of the building being studied, both structurally but also the type and rated performance of installed equipment.

Statistical regression models are based on correlating the output (energy consumption, cooling requirement, etc.) with a set of selected influencing variables (see (Chirarattananon & Taveekun, 2004) and (Catalina, Virgone, & Blanco, 2008)). Some of the variables are constant for a particular building but also historical data is essential which includes at least the relevant climate information. A balance between model complexity (number of variables) and model accuracy has to be selected based on level of usage. The most extensive statistical regression surveys related to commercial building energy usage in the U.S. is the Commercial Buildings Energy Consumption Survey (CBECS). CBECS is a national survey that collects information of commercial buildings, including energy-related building characteristics and energy use. The first survey was done in 1979 and the tenth and most recent in 2013 (for calendar year 2012). As part of the 2003 survey ((EIA, 2012)) an end-use energy consumption (EUC) regression models were developed ((KEMA, 2012)). Eight submodels, each covering different part of the building, are either analyzed independently or added to get estimates of the total building energy consumption. The model data are made available to the public through tables, reports and data files.

The third group includes high level learning algorithms (such as artificial neural networks (ANN) or support vector machines (SVM)) which based on historical data try to "learn" the dynamical behavior of the desired output such as (Neto & Fiorelli, 2008).

Among CBECS 2012 building stock 35% of building sized over 100,000 sq.ft. have central HVAC systems in place. Central systems typically use chilled water as cooling medium and has three major components: air-handling unit(s) (AHU), chilled water plant (chiller) and a boiler plant. The AHU conditions and supplies air to the conditioned building space and can be equipped with or without an airside economizer. By measuring climate conditions and applying a control policy an economizer allows the ventilation system to reduce or eliminate needed cooling when weather is mild and cold. Figure 1 shows the main components of an economizer based AHU where return air mixes with outside air before it is conditioned using heating and/or cooling coils. The hot and cold water flowing inside the coils comes from the chillers and boilers, respectively. End-use estimates from CBECS 2003 indicate that 38% of total energy use in large (> 100k) office buildings is related to HVAC which is equivalent to 33.2 mBtu per each sq.ft. of building space.

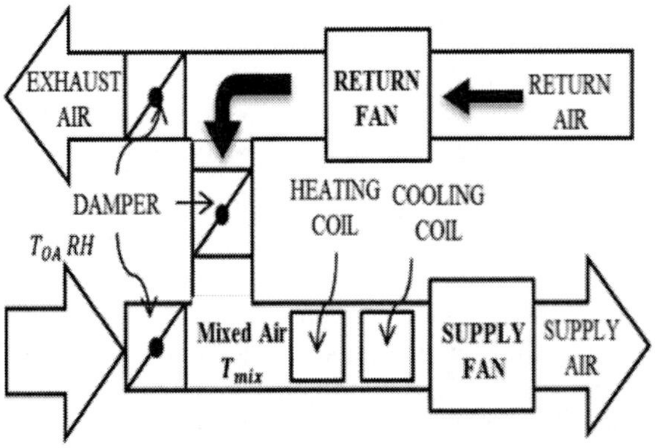

Figure 1: AHU with airside economizers main compoenents

To model the energy consumption and related costs we determine how a buildings total annual HVAC energy use is driven by: a) outside air condition (both temperature and relative humidity) b) the amount of air used c) the economizer control policy (determining the mix of return air in a building with outside air) d) the energy as a function of cubic-feet-per-minute (CFM) of airflow models for fans, chillers and boilers. Based on the climate data and the building CFM demand the economizer control policy determines the amount of outside air (OA) needed. Knowing the mixing temperature set-point allows for the estimation of heating and cooling requirements for the two airstreams. Component models to transform cooling, heating and airflow demand to actual energy consumption are obtained by correlating physical information with measured data from the building automation system (BAS). Data collection from airflow experiments were done on three commercial scale office-lab buildings at Boston University to verify the model.

2. HVAC BUILDING ENERGY MODEL

Based on climate data the economizer control policy sets the heating/cooling requirements as well as the necessary airflow outside air ratio. These results are inputted into the three demand component models for 1) fans 2) cooling plant 3) boiler plant. Each of these models were developed based on data obtained from detailed BAS logging and provide predictions of daily energy use. This ensures that the models are firmly based on operational data and not just equipment specifications. This is important in buildings since part load conditions are common for many of the different equipment and the performance is normally not well documented from the equipment manufacturers. The energy component models were developed for electricity use by AHU and chillers, and natural gas use for boilers. The fan component model receives special attention since it turns out that half of the total yearly HVAC cost is related to running the fans. This stems from the fact that the fans are always on but heating and cooling is only necessary for parts of the year.

2.1 Climate Data and Airflow Energy Content

As mentioned the AHU conditions and supplies air to the building by either using air from outside or from within the building through the return system. What determines the ratio of outside air to return air and level of conditioning is the outside temperature. The energy content of air has a sensible and a latent part. The sensible part (h_s) is only based on temperature and defines the required amount (BTU/hr) of heating or cooling of air:

$$h_s = 1.08Q\Delta T \tag{1}$$

where Q is the air volume flow in cubic feet per minute (CFM) and ΔT is the temperature difference in degrees Fahrenheit. The latent part represents the required energy to condensate the added moisture defined by the relative humidity. The relationship between latent energy content, temperature and relative humidity can be found from psychometric charts or empirical models.

Figure 2 shows how significant the amount of the latent heat energy required is by normalizing the amount of energy required to condensate moisture above a 50% threshold over the sensible heat required to cool from the climate temperature to the HVAC set point (usually $55°F$). For example at 60% humidity and $90°F$ the latent energy is two times that of the sensible heat at that temperature difference.

Figure 3 shows local climate data for Boston in 2010 on a yearly basis. Three switching temperatures, explained further next section, are shown in the figure as horizontal red lines which correspond to the temperatures that the economizer switches and varies the ratios of outside and return air to minimize to chiller energy usage.

It is apparent that relative humidity has a large range of variation over all seasons which indicates that substantial latent heat energy does not necessarily have to be related to the hottest summer days. These variations make it hard to estimate the total daily energy content without having hourly measurements of both variables. As an example of this nonlinearity climate data for a single day (July 21) is presented in Figure 4. It shows how the relative humidity is high overnight and drops as soon as the temperature starts to rise in the morning. If, for this particular day, we were to use daily average temperature and relative humidity to estimate the energy content in the airstream we would overestimate it by 40%. In contrast the CBECS 2003 end-use estimates ((KEMA, 2012)) for cooling accounts for the latent heat by using a multiplication factor (1.70) which is directly proportional to the total cooling energy end-use. These three observations motivated the decision to base the model on hourly climate data and accurate latent heat energy calculations.

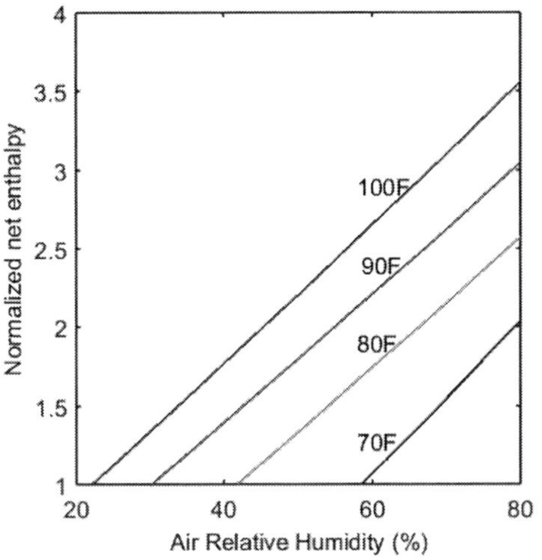

Figure 2: Ratio of relative latent heat and sensible heat change to reach HVAC setpoint (55°F and 100%RH) normalized by sensible heat change as a function of relative humidity.

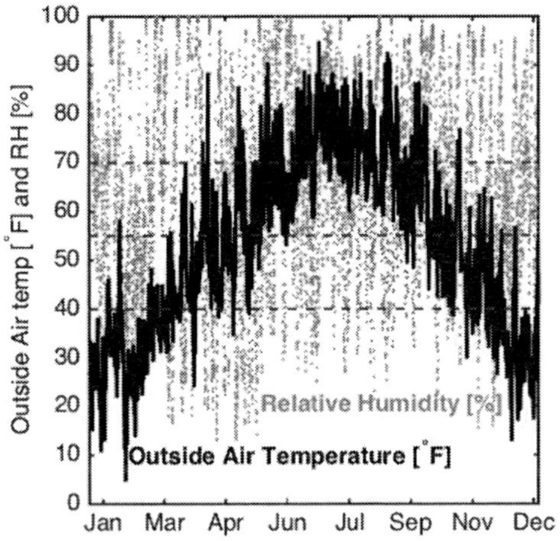

Figure 3: Temperature and relative humidity in Boston for year 2010

2.2 Airside Economizer Model

The economizer model 1) determines, based on outside air temperature, the ratio of outside air versus return air to meet conditioning and ventilation requirements 2) calculates the mixed air temperature which determines the level of conditioning needed. The mixed air temperature set-point, as implemented in the building control policy, is 55°F in the buildings studied (also called the activation temperature) and ensures that the internal building relative humidity objectives are achieved throughout the year. Figure 5 shows how the economizer control policy is implemented as a function of outside air temperature. When outside air temperature is less than the mixed air set-point (55°F) heating is required and we want to use as much of return air as possible so the outside air ratio is set at minimum (here the minimum outside air policy is 50% opening). The outside air temperature break-point where no more heating is required (saturation limit) is based on two factors: 1) the ratio of return air to supply air (based on measured BAS data, this was close to 50% on a yearly basis for the 3 buildings studied, see histogram in Figure 6), and 2) the return air temperature which is observed to be 70°F on a yearly basis.

The model is equipped having return air temperatures as an hourly based dataset which greatly benefits the analysis in buildings where it varies through the year. For outside air ratio of 0.5 and return air temperature of 70°F the break-point

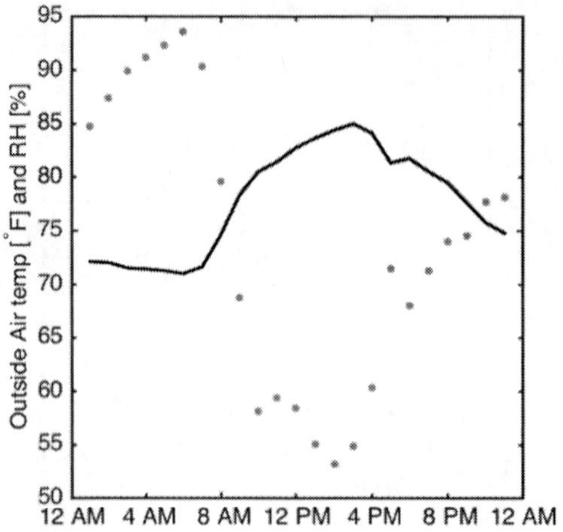

Figure 4: Temperature and relative humidity in Boston on July 21st, 2010

is at 40°F outside air. Between 40°F and 55°F the outside air ratio is increased linearly to accommodate the available free-cooling and is 100% at 55°F all the way to 70°F. When outside air temperatures reach the return air temperature set-point the level of outside air ratio is set to the minimum again. For reference, Figure 3 shows how the switching temperatures 40°F, 55°F and 70°F (red lines) map onto Boston climate data where one can realize for which periods each operating zone is dominant.

Between climate zones the activation temperature should be selected according to the economizer control mode 1) temperature based control 2) enthalpy based control and the variation on outside air temperature and humidity. There is a activation set-point that maximizes energy savings related to the economizer efficiency but currently and in the past it has been selected rather arbitrary. Zhou et. al. (Zhou, Wei, Turner, & Claridge, 2006) discuss common misconceptions related to airside economizer control and explain how to select the "best" activation temperature. As discussed the model being presented here is especially targeted to estimate the benefit of different control strategies and activation temperatures so they can be applied to the building being optimized.

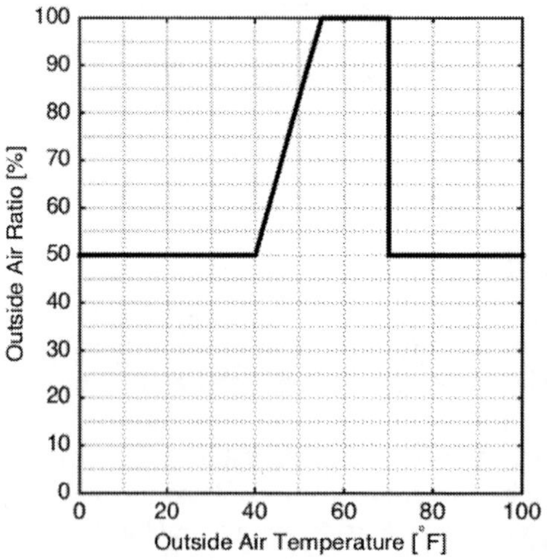

Figure 5: Outside air damper switching curve

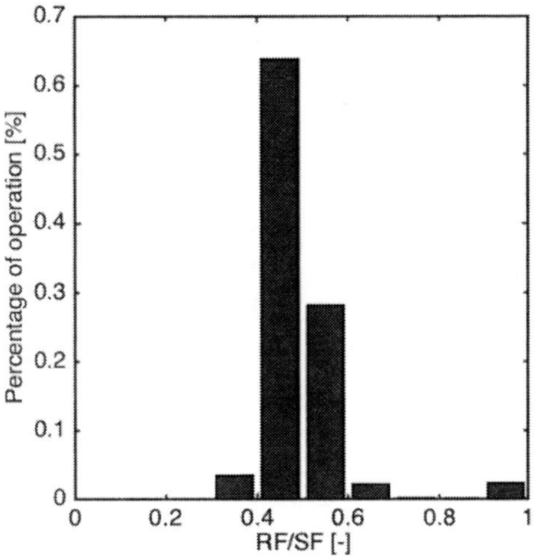

Figure 6: Frequency of return over supply flow ratio for AHU4 in PHO for Mar-Sep 2014

2.3 Energy Demand Model for HVAC Components

Detailed energy models for each of the major HVAC components were developed based on data obtained from detailed BAS logging throughout a full year. This ensures that the model is firmly based on actual operational data and few assumptions. Data based analysis is particularly important since many pieces of equipment operate differently at part load, as well as the overall HVAC system operates differently due to different design practices and control implementation.

Demand models were developed for electricity use by air handlers and chillers, and natural gas use for boilers. All energy demand models were developed as a function of the buildings actual supply airflow rate and are designed to handle hourly input data for the most accurate estimate. This is especially important in buildings were un-occupied mode has been implemented and large changes in airflow are apparent. For others assuming an average constant airflow rate throughout the year will give sufficient results.

Fan Model: Fan energy use depends on 1) fan air power (product of airflow and pressure head across fan 2) fan efficiency 3) motor and drive efficiencies. All of these are non-constant for building HVAC systems where pressures and flows can be varied. This means that the "system curve" is constantly changing and this detailed mapping of the fan performance is seldom available from the fan manufacturer.

Fan energy models were developed as a function of air flow rate (specified in terms of cubic feet per minute (cfm)) for each particular building. They are based on actual fan hourly electricity use data, operating at full and part load conditions throughout the year. The complete models were checked against actual electrical and natural gas bill and sub-metering, and show to estimate HVAC energy use within 10% of actual demand for an entire year. Both supply and return fans are included in the analysis. To develop the relation of motor power consumption (in kilowatts electricity) as a function of airflow rate (in cfm), data form the Building Automation System (BAS) was recorded, and verified by manually comparing data from the fan's variable frequency drives.

These results were also confirmed for airflow reduction by explicitly reducing VAV minimum set-points for a large part of the rooms in the buildings. For example, Figure 7 shows the overall correlation between fan power and supply flow rate for one of three buildings studied. The Photonics Building has identical AHU's which leads to a less complex correlation than for the School of Management Building which has many different sized AHU's. Variations about the correlation functions (shown as black lines), are due to variations in the supply fan pressure set point, which varies due to changes in net duct resistance as VAV boxes open and close.

As a comparison the CBECS submodel (KEMA, 2012) for ventilation requirements uses the regulatory required outside air volume for the building size and assumes 100% outside air for labs and 30% for offices. Using CBECS estimates the amount of return air is assumed to be 35% (0.7 times ratio of static pressure drop for return air over static pressure drop for supply air) compared to the 50% seen from the actual operational data.

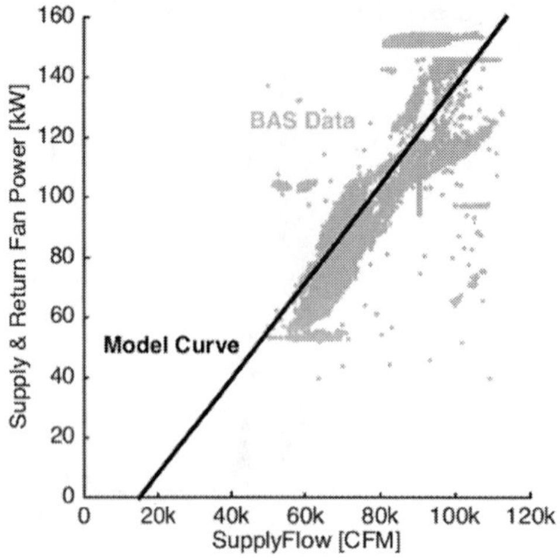

Figure 7: Fan Model Curve (for supply and return fans) for Photonics Center. Data from all 4 identical AHU's is used for the fit yields 1.29 KW/1000 CFM.

Heating and Cooling Model: Cooling and heating (apart from re-heat) is modulated at the AHU level where control valves determine the chilled or hot water flow through two separate heat exchanger coils (see Figure 1). The total supply airflow flows over the coils which makes it a suitable dependent variable. The model estimates the heating and cooling need based on the climate enthalpy and and control policy presented in section **??**. Reheat is assumed to be needed all year to increase the temperature from the mixing set-point at the AHU level to the supply temperature at the VAV-boxes ($55°F$ to $65°F$). Boiler and chiller efficiencies were selected based on specification from manufactures that were adjusted based on actual air reduction experiments. With more experimental data it would have been possible to map the operational profile of both the boiler-plant and chiller-plant for an even more realistic estimate.

3. ANALYSIS OF ANNUAL HVAC ENERGY CONSUMPTION IN 3 BUILDINGS

The 3 Boston University buildings used in this study were chosen due to their different sizes, different HVAC architectures, and different uses. Two of the buildings are large buildings - the School of Management (SMG) building, which is primarily offices (325,000 sq.ft., built in 1996, 1057 rooms, 8 AHU's, 495 VAV-boxes and several dining/cooking areas) and the Photonics Center (PHO), which is a mixed office, classroom, and laboratory space (243,000 sq.ft. built in 1997, 667 rooms, 4 AHU's, 443 VAV-boxes, and 33 fume hoods). The smaller building, Engineering Manufacturing Building (EMB), consists of 2 floors, 50,000 sq.ft. with 2 AHU's, 50 VAV-boxes, and mixed office, research lab, and manufacturing space, constructed in 1997.

First the supply and return airflow trends were analyzed. If a complete set of hourly BAS data is available (without any gaps or missing data) that can be used directly but if not, the airflow characteristics of the building has to be identified. Figure 8 shows a histogram of the total supply flow from one of the buildings revealing constant flow throughout the year (and clearly does not have an un-occupied mode).

Based on the model, we predict the daily energy used to both heat and cool the air for an entire year. The results are presented for PHO in Figure 9. The outside air heating (in red) reflects the economizer dispatch when the maximum amount of return air is used, while the reheat portion (in purple) reflects the use throughout the year to bring the mixed air temperature up to the $65°F$ at the VAV boxes. Cooling energy (in blue) reflects where chillers are needed to provide a $55°F$ mixed air temperature, which is determined by the outside air and relative humidity plots. The total heating load throughout the year per CFM of airflow is 104 kBtu/CFM, while the total demand for cooling is 57.5 kBTU/CFM (averaged over the year). The figure shows how these loads are distributed over the calendar year where cooling loads are depicted on the positive y-axis in terms of But per hour per CFM and heating and reheat are summarized and depicted on the negative half of the y-axis in terms of BTU per hour per CFM . These BTU loads are then converted to therms of natural gas assuming a boiler efficiency of 85%, and the stated efficiency of the chillers (.7kW/Ton). Power

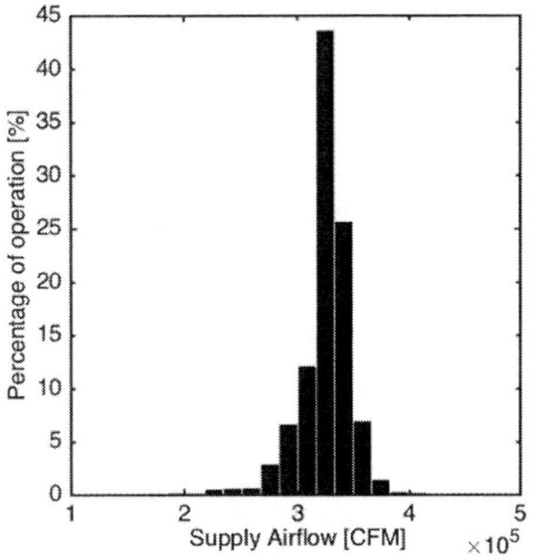

Figure 8: Histogram of Total Supply Airflow (4 AHU's) for Photonics Building (2014).

consumed by AHU's per CFM delivered for each of the three buildings comes from actual fan model curves taking into account supply and return fans and accounting for un-occupied mode where appropriate.

A prediction of heating, cooling and fan energy use for all three buildings is summarized in Table 1. There heating and cooling demands per CFM are identical between the three buildings since they orginate only from the same climate data and economizer control policy. The fan demand per CFM is related to the actual operational fan curves and therefore are buildings specific. PHO and SMG have similarly efficient fans while the smaller capacity equipment in EMB is less efficient. The HVAC energy model is based on current commercial utility rates in Boston ($0.15/kWh and $1.16/Therm), reflecting the reduction of natural gas prices over the past several years. This analysis indicates that these buildings currently operate at an HVAC energy cost of $4.44 - 4.64 per CFM/year, which corresponds to an average of $3.00/sq.ft. of buildings space per annum. Overall it is striking, that the cost of running the fans is for all buildings is at or just above 50% of the total HVAC operating cost.

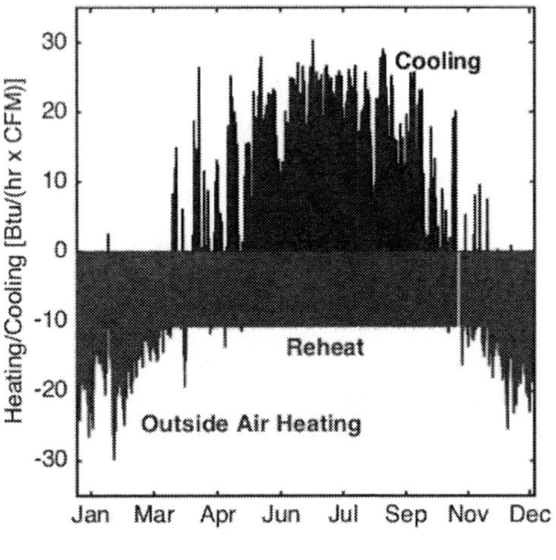

Figure 9: Building HVAC daily heating and cooling energy use per CFM of airflow, model prediction (Boston, 2010.)

Table 1: Model analysis of HVAC Energy consumption per CFM for heating, cooling and fans for the three buildings studied.

Building	Heating		Cooling		Fans		Total
	Therm/CFM	$/CFM	kWh/CFM	$/CFM	kWh/CFM	$/CFM	$/CFM
PHO					11.1	$1.70	$4.44
SMG	1.26	$1.50	7.48	$1.24	11.1	$1.70	$4.44
EMB					12.8	$1.90	$4.64

4. DISCUSSION

Available hourly climate data from three other climate zones (other than Boston) allows for a model comparison of HVAC energy and cost requirements. Keeping all other factors constant (i.e. estimating the consumption of the "same" building in all zones) leads to an enlightening comparison. Table 2 shows the heating, cooling and fan energy estimates of three different metropolitan areas along with Boston. These are New York City, Washington DC and Los Angeles. Local costs of energy (per Therm and per kWh) have been used to populate the table.

Table 2: HVAC energy consumption and cost estimates for heating, cooling and fans for different cities.

	Heating		Cooling		Fan		Totals
CITY	therm/cfm	$/cfm	kWh/cfr	$/cfm	kWh/cfr	$/cfm	$/cfm
Boston	1.26	$1.50	7.48	$1.24	11.1	$1.85	$4.59
NYC	1.15	$0.93	8.92	$1.84	11.1	$2.29	$5.06
DC	1.15	$1.30	11.43	$1.42	11.1	$1.38	$4.10
LA	1.05	$0.93	13.93	$1.92	11.1	$1.53	$4.39

As expected cooling costs are highest in LA and heating cost are highest in Boston and each contribute to a high total HVAC cost per CFM which is slightly higher than the other two cities. Table 3 shows the percentage breakdown for heating, cooling and fans per city. What is striking is the high percentage of the cost that goes into driving the supply and return fans. The percentage cost is lowest for DC at 43% and highest for NY at 55%. This shows that the amount of air that is required to be pushed through the building is the real cost driver for HVAC systems.

Table 3: Percentage cost breakdown for heating, cooling and fans by city.

CITY	Heating %	Cooling %	Fan %
Boston	33%	27%	40%
NYC	18%	36%	45%
DC	32%	35%	34%
LA	21%	44%	35%

We have also compared our results to those of the CBECS end-use energy estimates presented in (KEMA, 2012), which is the basis for breaking out the HVAC energy & cost breakouts presented in CBECS data tables. Our model should be considered to be lower bound estimate since it does not consider air infiltration, heat gain because of people and equipment and heat losses (through windows, open doors, etc.). Looking at the actual building operation in terms of fan energy in the three buildings studied and comparing with the CBECS ventilation estimates shows that on a per CFM basis, both are consistent in terms of the energy required for both the supply and return fan electricity consumption). However, they two models differ significantly in terms of the assumed air flow. In terms of both buildings studied, the CBECS model uses only 55% of the actual air used in the SMG building, and 32% of the Photonics laboratory building. Thus, the CBECS HVAC cost estimates will similarly be understated by 45 to 68% relative to the actual HVAC costs and energy attributable in the study buildings. This large difference is due to the fact that the CBECS model is based on ASHRAE 62.1 airflow design specifications, but many if not most buildings were designed and built with much higher minimum airflow rates. For example, several studies have found that many VAV systems are implemented where the minimum VAV airflow setting is set at 30 to 50% of the maximum flow rate (Arens, Zhang, & Holt, 2015) (Gallagher, Gunnsteinsson, Morse, & Gevelber, 2015).

Historically in the United States this category of buildings have been designed with high minimum air-flow rate set-points that are greater than that needed for ventilation. If supply airflow rates can be reduced for a building not only is less energy required to transport the air but also to condition (heat and/or cool) it. An essential part of studying this phenomena is to be able to, for a particular building or project, predict the energy savings. When considering airflow rate reductions the prediction model has to be able to assess the energy requirements of each cubic feet of air that needs to be transported through the building which is one of the key component of this model presented.

5. SUMMARY

A model to estimate HVAC energy consumption in large commercial office buildings has been developed based primarily on analysis of building airflow rates. The model is based on analysis of three commercial buildings in terms of operational data from building automation system, mapping of load profiles and efficiency of key HVAC equipment, investigating building airflow control policies and deriving buildings loads using hourly climate data. The model reveals how much actual energy is consumed by the HVAC systems in these buildings and to identify the key consumption drivers. In particular, the model analysis reveals that roughly 40% of HVAC energy cost is associated with running supply and return fans. As such, these results indicate that minimizing airflow rates in buildings while meeting required outside air requirements provides a significant opportunity to improve building energy efficiency.

REFERENCES

Arens, E., Zhang, H., & Holt, T. (2015). Effects of diffuser airflow minima on occupant comfort, air mixing, and building energy use. *Science and Technology for the Built Environment RP-15115.*

Catalina, T., Virgone, J., & Blanco, E. (2008). Development and validation of regression models to predict monthly heating demand for residential buildings. *Energy and Buildings, 40*(10), 1825–1832.

Chirarattananon, S., & Taveekun, J. (2004). An ottv-based energy estimation model for commercial building in thailand. *Energy and Buildings, 32*, 680–689.

EIA. (2012). *Commercial buildings energy consumption survey 2003 (cbecs).* Washington, DC: Energy Information Administration (EIA).

Fumo, N., Mago, P., & Luck, R. (2010). Methodology to estimate building energy consumption using energyplus benchmark models. *Energy and Buildings, 42*, 2331–2337.

Gallagher, P., Gunnsteinsson, S., Morse, A., & Gevelber, M. (2015). Airflow system identification tool for optimizing commercial building vav settings for improved energy efficiency. *ASHRAE Winter Conference, CH-15-C036.*

KEMA. (2012). *Energy end-use consumption estimates for commercial buildings.* Washington, DC: Energy Information Administration (EIA).

Neto, A. H., & Fiorelli, F. A. S. (2008). Comparison between detailed model simulation and artificial neural network for forecasting building energy consumption. *Energy and Buildings, 40*(12), 2169–2176.

Pan, Y., Huang, Z., Wu, G., & Chen, C. (2008). *The application of building energy simulation and calibration in two high-rise commercial buildings in shanghai.*

Yik, F. W. H., Burnett, J., & Prescott, I. (2001). Predicting air-conditioning energy consumption of a group of buildings using different heat rejection methods. *Energy and Buildings, 33*, 151–166.

Zhao, H., & Magoules, F. (2012). A review on the prediction of building energy consumption. *Renewable and Sustainable Energy Rev. 16*, 3586–3592.

Zhou, J., Wei, G., Turner, W. D., & Claridge, D. E. (2006). Airside economizer - comparing different control strategies and common misconceptions. *SimBuild 2006.*

ACKNOWLEDGEMENTS

We appreciate the financial support of NSF STTR grant (IIP-1346436) as well as the contributions of A. Morse and R. Yazbeck.

Building Energy Efficiency Assessment of Renewable and Cogeneration Energy Efficiency Technologies for the Canadian High Arctic

Eric MCDONALD*, Martin KEGEL, Justin TAMASAUSKAS

CanmetENERGY, Natural Resources Canada,
Varennes, Québec, Canada
Tel: 450-652-0352 Email: eric.mcdonald@canada.ca

* Corresponding Author

ABSTRACT

This paper presents an energy analysis of the Polar Continental Shelf Program (PCSP) facility located in Resolute, NU. The study focuses on the evaluation of innovative energy efficiency strategies including on-site cogeneration and renewable technologies such as wind and solar. Developed energy models were validated using data collected from installed metering systems recording the electricity, heating fuel and hot water consumption of six buildings that make up the PCSP Resolute facility. A preliminary energy audit identified short term energy efficiency measures that could be implemented to reduce greenhouse gas (GHG) emissions by 15% and reduce utility costs by 12.5% annually. Proposed innovative strategies are designed to build upon the suggested short term energy efficiency measures to achieve significant utility cost savings. From simulations it is estimated that almost all electricity can be produced by on-site generation and the heat recovered from the generators can be used to meet close to 50% of the space heating loads, resulting in close to $600,000 (~50%) savings in annual utility costs from the proposed short term measures. The feasibility of sea water heat pump system is also evaluated to efficiently meet a portion of the building heating loads, while increasing the demand on the combined heat and power system. The use of a sea water heat pump would help save an additional $40,000 (~7.0%) in utility costs annually. In comparison, the use of wind and solar photovoltaic renewable energy technologies can achieve more savings. Sized to meet the base electrical load of the facility, the use of these technologies are estimated to save $175,000 annually.

1. INTRODUCTION

Arctic communities, challenged by the harsh climate and a lack of local energy resources, are often confronted with finding more sustainable solutions for power and energy. Due to their isolated nature, reductions in energy or fuel use can have important implications for operating costs, security, and energy independence. While high performance buildings have received significant attention in more populated areas, there has been less work done on the opportunities and challenges for these buildings in the Canadian North.

The Polar Continental Shelf Program (PCSP) facility in Resolute, NU, (74°70'N, 94°83'W), is one of 17 building groups identified under the Low Carbon Initiative (Low Carbon NRCan 2016) to reduce greenhouse gas (GHG) emissions by 17% from the 2005 levels. Through facility expansions and additional operational requirements, the PCSP facility has seen an increase of close to 500% in electricity use, fuel use and ultimately greenhouse gas (GHG) emissions in comparison to 2005/2006 levels. Combined with a doubling of electricity rates, annual utility costs have risen by 750% over this time period (Table 1). With future expansions planned to add more storage facilities as well as an on-site incinerator, energy consumption, utility costs and GHG emissions will continue to rise.

Table 1: Electricity and fuel consumption with associated utility costs for the PCSP facility

Fiscal Year	Purchased Electricity (kWh)	Heating Fuel (L)	Electricity Cost ($,CAD)	Heating Fuel Cost ($,CAD)	Total Utility Cost ($,CAD)	GHG Emissions (ton CO_2 eq.)
2005/2006	152,095	46,591	$98,007	$40,357	$138,364	243.5
2013/2014	839,516	250,361	$792,832	$300,121	$1,092,953	1324.4
2014/2015	867,031	271,321	$857,557	$338,341	$1,195,898	1403.7

In 1986, the PCSP experienced the first of 3 significant expansions. At this time, the PCSP facility consisted of three buildings, which are still in use today (XR-01, XR-02 and XR-03). In 2011 a second large expansion was undertaken, which nearly doubled the existing infrastructure to provide additional accommodation and living space (XR-12), and a research prep laboratory (XR-04). A third significant expansion was completed in 2013, resulting in two additional buildings (XR-15 accommodations and XR-14 an operations building) to meet the specific needs of the Canadian Armed Forces Arctic Training Centre (Figure 1).

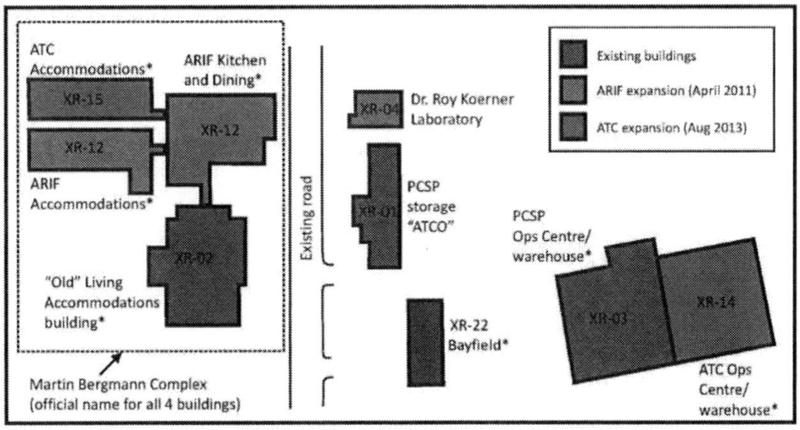

Figure 1: Layout of PCSP facility in Resolute, NU

2. METHODOLOGY

An energy audit conducted in March 2014 identified several easily implementable energy efficiency measures which could reduce GHG emissions by 15% and utility costs by 12.5%, annually. These measures focused primarily on increasing insulation levels in the utilidors, replacing inefficient light fixtures and installing new exterior doors, and are either currently being undertaken and/or planned for future implementation. However, it was evident that a more invasive energy efficiency strategy would be required to achieve significant impacts. In particular, high local electricity rates provided an important opportunity to focus on reducing facility purchased electricity. In order to properly assess opportunities for new energy efficiency measures, a three step methodology was employed:

1. Monitoring of on-site electricity, fuel, and hot water use.
2. Development and calibration of detailed energy models using monitored data.
3. Assessment of selected technologies using calibrated models with the assumption that the more easily implementable energy efficiency measures have already been implemented.

3. BUILDING ENERGY MODEL DEVELOPMENT

An extensive effort was undertaken to develop detailed building energy models of the PCSP facility in order to evaluate the various energy saving measures. TRNSYS v. 17 was selected due to its ability to accommodate non-standard HVAC systems, including renewable energy systems, in building energy models. While this required a detailed component-based approach, it was necessary to have an energy model capable of assessing both standard and innovative efficiency measures in the analysis.

Details and operating characteristics of the buildings were determined during site visits conducted from March 2014 to September 2015. Space temperatures, lighting and occupancy schedules were recorded during the site visits to gain a general idea on how the buildings were being operated. Notes were taken for equipment with significant power draws, and operating schedules were established based on observations and discussions with on-site personnel. Heating equipment details were gathered on-site, and manufacturer rated performance data was used in the energy models. Construction details were inspected in each building and compared with available architectural drawings. Key building characteristics are summarized in Table 2 and Table 3. XR-02, XR-03 and XR-04 are heated via indirect diesel furnaces and air handling units, with some electric unit heaters in the utilidor and entrances.

Currently, there is no central control system in XR-02 and XR-03, so all burners are controlled by local thermostats located throughout the building. XR-12, XR-14 and XR-15 are heated through hot water distribution systems, with diesel boilers in XR-12 being the only building controlled via a central system

Table 2: Summary of key building characteristics for XR-02, XR-03 and XR-04

	XR-02	XR-03	XR-04
Heating System	3 Diesel Fired Furnaces 84.0 % SSE 3 Diesel Fired AHU 80.0 % SSE	4 Diesel Fired Furnaces 86.0 % SSE 2 Diesel Fired AHU 81.0 % SSE	1 Diesel Fired Furnace 81.4 % SSE
DHW	2 Diesel Fired Conventional Tank	1 Electric Conventional Tank	1 Diesel Fired Conventional Tank
Lighting	Compact Fluorescents T12 Light Fixtures	Compact Fluorescents T12 Light Fixtures	Compact Fluorescents T8 Light Fixtures
Ventilation	No HRV	No HRV	No HRV
Heated Floor Area	1,427 m^2	2,693 m^2	191 m^2
Roof RSI	6.4 $(m^2 \cdot °C)/W$	5.3 $(m^2 \cdot °C)/W$	8.8 $(m^2 \cdot °C)/W$
Wall RSI	5.8 $(m^2 \cdot °C)/W$	4.3 $(m^2 \cdot °C)/W$	5.0 $(m^2 \cdot °C)/W$
Utilidor Wall/Floor RSI	3.3/3.1 $(m^2 \cdot °C)/W$	-	3.3/3.1 $(m^2 \cdot °C)/W$
Exposed Floor RSI	4.9 $(m^2 \cdot °C)/W$	3.4 $(m^2 \cdot °C)/W$	7.4 $(m^2 \cdot °C)/W$
Infiltration	0.30 $L/s/m^2$	0.25 $L/s/m^2$	0.25 $L/s/m^2$
SSE: Steady State (Thermal) Efficiency AHU: Air Handling Unit			

Table 3: Summary of key building characteristics for the XR-12, XR-14 and XR-15

	XR-12	XR-14	XR-15
Heating System	2 Two Stage Diesel Fired Boilers 85.6 % SSE	2 Single Stage Diesel Fired Boilers 85.0 % SSE	2 Single Stage Diesel Fired Boilers 83.2 % SSE
DHW	2 Indirect DHW Tanks Served From Boilers	1 Diesel Fired Storage Tank	2 Indirect DHW Tanks Served From Boilers
Lighting	Compact Fluorescents T8 Light Fixtures	Compact Fluorescents T8 Light Fixtures	Compact Fluorescents T8 Light Fixtures
Ventilation	Glycol HRV	Glycol HRV with solar Wall	HRV
Heated Floor Area	2,197 m^2	2,190 m^2	1,232 m^2
Roof RSI	8.8 $(m^2 \cdot °C)/W$	8.4 $(m^2 \cdot °C)/W$	8.0 $(m^2 \cdot °C)/W$
Wall RSI	5.0 $(m^2 \cdot °C)/W$	5.3 $(m^2 \cdot °C)/W$	5.1 $(m^2 \cdot °C)/W$
Utilidor Wall/Floor RSI	3.3/3.1 $(m^2 \cdot °C)/W$	-	3.0/3.8 $(m^2 \cdot °C)/W$
Exposed Floor/Slab RSI	7.4 $(m^2 \cdot °C)/W$	4.7$(m^2 \cdot °C)/W$	5.2 $(m^2 \cdot °C)/W$
Infiltration	0.20 $L/s/m^2$	0.25 $L/s/m^2$	0.20 $L/s/m^2$
SSE: Steady State (Thermal) Efficiency			

4. ENERGY MODEL VALIDATION

Available utility bills were not able to provide an accurate description of each building's electricity and heating fuel consumption, as the electricity bills were given for multiple buildings and often adjusted for incorrectly billed consumption. For heating fuel, only the bulk fuel delivered to the entire facility was recorded. To gain better refinement and an improved understanding of the key energy flows of the PCSP facility, metering equipment was installed in each building to measure the:

- Electricity consumption
- Heating and DHW fuel consumption
- Hot water consumption
- Space temperatures

The electricity, heating fuel and DHW fuel consumption of each building has been recorded since September 2015 via the installed monitoring system and used to validate the simulation results for the complete facility. Figure 2 shows the predictions from the energy model for the electricity and heating fuel consumption for the complete facility. The predictions compare well with the monitored data, with simulations following the same trend in both electricity consumption and fuel consumption over the seven month monitoring period. Electricity consumption is typically within 1% of the measured electricity consumption, with larger differences in January and February attributed to assumed facility occupancy. Heating fuel consumption varies from close to 40,000 L in the colder winter months and to approximately 10,000 L in the warmer summer months for the complete facility. The predicted heating fuel consumption also closely follows the trend of the monitored data, with an average percent difference of 10% from the monitored data, primarily due to assumed infiltration rates.

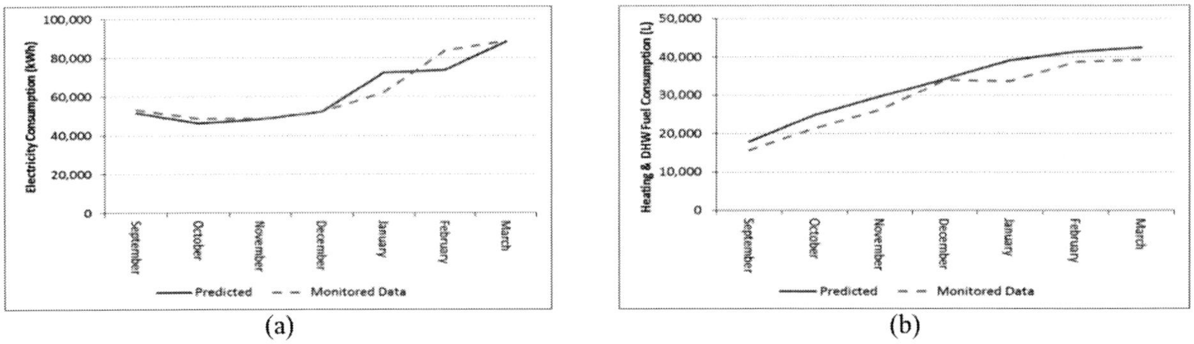

Figure 2: Comparison of simulated and recorded (a) electricity and (b) fuel consumption for the 2015/2016 FY

Using the calibrated energy models, the baseline energy consumption of the facility was estimated with the proposed shorter term energy efficiency measures. The facility energy end use distribution is shown in Figure 3, with 82% of the total energy consumption attributed to diesel fuel for space and hot water heating. Although electricity consumption represents only 18% of the total energy end use of the facility, the high electricity rates yields annual costs that are almost double the fuel costs (Table 4). Thus, efficiency measures targeting a reduction in electricity can have a substantial utility cost savings.

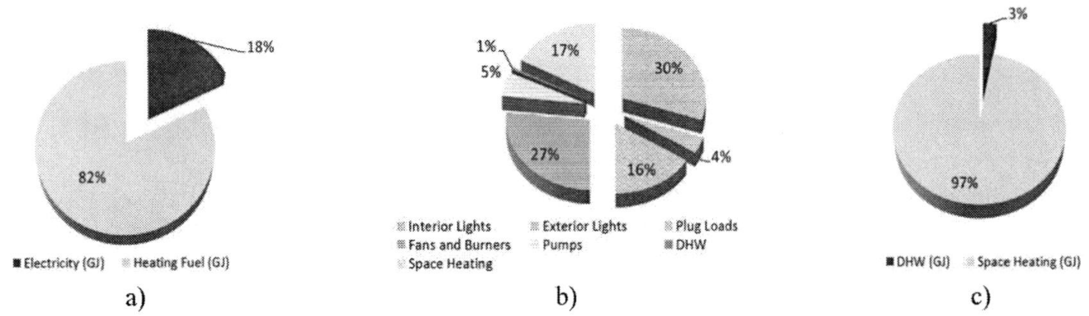

Figure 3: Distribution of a) Energy for the PCSP facility b) Electricity consumption by end-use c) Fuel use

Table 4: Baseline electricity and heating fuel consumption for the PCSP facility

Building	Purchased Electricity (kWh)	Heating Fuel Consumption (L)	Electricity Cost ($, CAD)	Heating Fuel Cost ($, CAD)	Utility Cost ($, CAD)
PCSP Facility	751,075	316,765	$743,773	$395,178	$1,138,951

5. ENERGY EFFICIENCY MEASURES AND ECONOMIC ANALYSIS

A three step process was employed to analyze various energy efficiency measures, as shown in Figure 4. Since electricity accounts for 65% of total utility costs, several on-site electricity generation scenarios were first examined as a means of achieving cost savings. Technologies examined include renewable wind and photovoltaic (PV) systems, and the use of on-site diesel generators. Next, the impact of upgrading the on-site generators to include cogeneration (combined heat and power, CHP) capabilities was examined to reduce on-site fuel use. Finally, to increase the amount of thermal energy available from the CHP units, the use of energy efficient space heating equipment was also investigated.

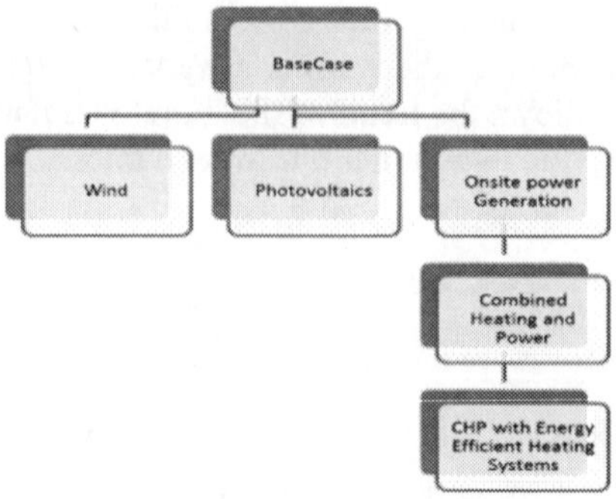

Figure 4: Framework for the energy efficiency measures

5.1 Integration of Renewables (Wind and PV)

Small scale wind turbines and PV systems can be used to reduce the electricity consumption and GHG emissions of each building examined. PV panels were added to the south-facing roof of each building at a tilt angle of 75°. Each PV system was limited to a single row of 240 W_P panels in order to avoid potential shading effects and the need to export excess generation to the Resolute, NU electricity grid. It is also important to note the community experiences 24 hour darkness from October through February, limiting the generating capabilities of the PV panels. Small scale wind turbine systems were sized to meet the minimum base electrical load of each building in multiple 2.5 kW rated systems. Table 5 summarizes the proposed renewable energy system for each building.

Table 5: Nominal system size for wind and PV for each building

Building	Nominal System Size (kW)	
	Wind	PV
XR-02	5	11
XR-03/XR-14	10	16
XR-04	5	-
XR-012/XR-15	15	19

5.2 On-site Power Generation

Electricity for the site is currently supplied via the local power plant, which uses five diesel generators to supply the complete electricity needs of PCSP and the neighboring community of Resolute Bay. A significant portion of the electricity demand of the facility could be met on-site by upgrading the currently installed standby generators to prime power units. This upgrade would effectively reduce the cost of electricity from $0.97/kWh from the local utility to an average of $0.37/kWh by producing the electricity on-site, while maintaining close to the same production of GHG emissions. The size of back-up generators available at the PCSP facility is shown in Table 6.

Table 6: Available on-site back-up generators

Building	Standby Power (kW)	Estimated Prime Power (kW)
XR-02	50	40
XR-03/XR-14	80	64
XR-04	42	34
XR-012/XR-15	100	80

5.3 On-site Combined Heating and Power (CHP)

Modifying the existing back-up generators to prime power units would require constant cooling of the engine jacket to avoid unit over-heating under constant use. With minimal additional investment, this waste heat could be recovered and used in the building, allowing the prime power units to become effective producers of both thermal and electrical energy. These modifications would improve system economics by significantly increasing the overall efficiency of the generators.

5.4 On-site Combined Heating and Power (CHP) with Efficient Heating Technologies

For cogeneration systems to be effective there must be a balance between the electrical loads on the generator, and the thermal loads served by the heat recovery system. In this paper, the potential of adding (1) cold climate air-source heat pumps, and (2) a sea-water heat pump system were examined. In both cases, by adding efficient electrically-driven heating systems, the electrical load on the generator is increased, resulting in more available heat recovery for use within the buildings.

Variable refrigerant flow (VRF) cold climate air-source heat pumps (CC ASHP) were first considered as an efficient heating option. These heat pumps are rated down to -25°C while maintaining 75-80% of their rated capacity (Mitsubishi electric, 2015). From preliminary calculations, it was determined that a heating COP of 1.35 would be required to break even in comparison to utility costs for the existing fuel fired equipment. For the CC ASHP system examined, achieving this COP would require ambient air temperatures above -15°C. However, in the cold Resolute climate, this limits beneficial use from May to October, when available heat recovery from the cogeneration units are capable of meeting the entire facility space heating load. As such, the benefits of this system were limited, and it was not examined further.

An additional option examined was the use of a sea-water heat pump. In this case, the heat pump would use the Arctic Ocean as a constant temperature thermal source for the heat pump system. With the PCSP facility located two kilometers from the Arctic Ocean, this represents a viable and effective option for the facility. At constant seawater temperature of 0°C (U.S. National Oceanographic Data Center 2014) a SWHP system would operate with a COP of 3.5, well above the estimated minimum required efficiency of 1.35.

5.5 Economic Analysis

A cost analysis is provided for the PCSP facility to prioritize efficiency measures and establish a budget. The economic analysis is based online sources and RSMeans with adjustments made based on experience of installing systems in the north (RSMeans 2013). Table 7 summarizes the estimated capital cost per kW of installed power for each efficiency measure. Simple payback periods were given by dividing the costs of the efficiency improvement strategy by the anticipated utility savings. A contingency of 50% was applied because of the remoteness of the facility and variation in design possibilities.

Table 7: Estimates capital costs per kW of installed power

Power Generation System	Costs ($, CAD)	Source
Wind Power	$15,000/kW per installed power	WINEUR 2006
Solar PV	$7.68/kW per installed power	Feldman *et al.*, 2014
On-site Diesel Generator*	$1,240/kW	Darrow *et al.*, 2015
Combined Heating and Power*	$1,665/kW	Darrow *et al.*, 2015
*For a 100 kW of installed generator power		

6. RESULTS

This section presents an analysis of the energy, GHG emissions, and economic performance of each energy efficiency measure outlined above. For all GHG calculations, values of 763 g CO_2 eq./kWh for electricity and 2,735 g CO_2 eq./L for on-site diesel were used. Table 8 outlines the baseline energy performance of the facility after the short term efficiency measures.

Table 8: Baseline electricity and heating fuel consumption for the PCSP facility

Building	Purchased Electricity (kWh)	Heating Fuel Consumption (L)	Electricity Cost ($, CAD)	Heating Fuel Cost ($, CAD)	Utility Cost ($, CAD)	GHG Emission (tons CO_2 eq.)
XR-02	101,194	53,77	$100,499	$66,872	$167,371	224
XR-03/14	269,168	120,170	$268,256	$150,212	$418,468	534
XR-04	78,730	9,481	$77,295	$11,823	$89,118	86
XR-12/15	301,983	133,337	$297,723	$166,271	$463,994	595
All Buildings	751,075	316,765	$743,773	$395,178	$1,138,951	1,439

6.1 Scenario # 1: Renewable Technologies (Wind and PV)

Wind turbine and photovoltaic systems were examined to determine the potential of renewable energy generation at the facility. Figure 5 compares the monthly electricity production of a 2.5 kW-rated wind turbine with a 2.4 kW_P PV array. Wind turbines offer greater annual electricity generation as these systems are able to operate year-round.

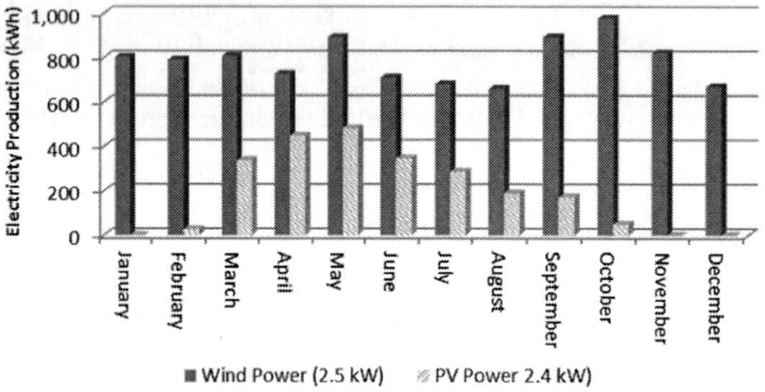

Figure 5: Monthly comparison of electricity production between a 2.5 kW wind turbine and a 2.4 kW PV array

Table 9 compares the annual performance of the wind turbine and PV systems on a facility-wide basis. Wind turbines sized to meet the base electrical loads offset 132,000 kWh of purchased electricity, with an estimated payback of 4 years. PV yields an offset of 46,000 kWh, with a simple payback period of 12 years. Future work will evaluate the life cycle cost of these systems taking into account maintenance requirements, which can be more costly for wind turbines in comparison to photo-voltaic arrays.

Table 9: Electricity, fuel and utility cost savings for small scale wind turbines and PV modules

Building	Purchased Electricity Savings (kWh)	Utility Cost Savings ($, CAD)	GHG Emission Reductions (tons CO_2 eq.)	Estimated Cost ($, CAD)	Simple Payback Period (Years)
Wind Turbines	131,567	$127,370	100	$525,000	4.1
Solar Modules	45,448	$43,998	35	$529,920	12.0

6.2 Scenario # 2: On-site Electricity Production from Diesel Generator

Table 10 summarizes the impact of implementing on-site electricity production from a diesel generator in each building. Adding this on-site generation capability would save nearly $470,000 in annual utility costs, with a payback period of less than one year. These strong cost savings are possible because producing electricity on-site would cost approximately one third of the price of purchased electricity from the local utility. The PCSP facility would produce approximately 750,000 kWh on-site from the upgraded backup generators while consuming close to 210,000 L of fuel. As the associated electricity GHG emission factor is calculated from diesel power generation, minimal reductions are achieved through on-site electricity generation using diesel generators.

Table 10: Electricity, fuel and utility cost savings for on-site electricity generation

Building	Purchased Electricity Savings (kWh)	Generator Fuel Consumption (L)	Utility Cost Savings ($, CAD)	GHG Emission Reductions (tons CO_2 eq.)	Estimated Cost ($, CAD)	Simple Payback Period (Years)
XR-02	101,194	29,002	$64,334	-2	$62,017	<1
XR-03/14	266,037	73,388	$166,419	2	$99,228	<1
XR-04	78,370	22,747	$48,188	-2	$49,930	1.0
XR-12/15	301,983	81,987	$190,458	7	$124,035	<1
All Buildings	747,584	207,124	$469,399	5	$335,210	<1

Currently PCSP receives a fuel delivery approximately every three days during the colder winter months. Converting the generators to prime power units will increase on-site fuel use, with an average daily increase of 40% estimated during the winter. As such, new fuel tanks would likely be required to facilitate this increased fuel use. Alternatively, if the backup generators are kept to maintain emergency power, new generators could be installed to supply the complete electricity consumption of the facility, with a total estimated cost of $1,900,000 and a simple payback period of approximately 4 years.

6.3 Scenario # 3: Combined Heating and Power (CHP)

Table 11 summarizes the energy performance of the proposed cogeneration option. Results are shown based on the implementation of on-site electricity production from diesel generators, as outlined in Scenario #2 above. Incremental savings demonstrate the strong fuel and utility costs savings that this option provides. For the larger buildings (XR-12/15 and XR03/14) there are significant benefits to upgrading the generators to CHP units, with simple payback periods of less than one year. For XR-12 some of the utility cost savings are obtained from being able to operate the boilers on low fire because of the hydronic return loop preheat. For smaller buildings such as XR-02 and XR-04, savings are also achieved. However the simple payback periods are closer to 4 and 5 years, respectively due to the associated cost of installing a hydronic heating system. On a site-wide basis, adding cogeneration capabilities would decrease fuel use by approximately 83,000 L, while reducing GHG emissions by 225 tons CO_2 eq. Economic performance is also strong, with annual utility cost savings of $109,000.

Table 11: Incremental electricity, fuel and utility cost savings for upgrading to on-site CHP from scenario #2

Building	Purchased Electricity Savings (kWh)	Generator Fuel Consumption (L)	Heating Fuel Savings (L)	Utility Cost Savings ($, CAD)	GHG Emission Reductions (tons CO_2 eq.)	Estimated Cost ($, CAD)	Simple Payback Period (Years)
XR-02	0	0	6,796	$8,474	19	$34,035	4.0
XR-03/14	0	0	42,277	$58,863	116	$39,103	<1
XR-04	0	0	5,276	$6,579	14	$32,346	4.9
XR-12/15	0	0	28,736	$35,834	79	$42,481	1.2
All Buildings	0	0	83,085	$109,750	228	$147,965	1.3

6.4 Scenario # 4: Combined Heating and Power with Efficient Heating Technologies

Figure 6 compares available heat recovery and heating loads by building. Results show an imbalance between the two sets of values, suggesting that each building could benefit from the use of an efficient electrically-driven space heating system during the winter months. In particular, heat pumps represent an attractive option because of their high efficiencies and ability to facilitate the integration of renewable energy sources into the building.

Sea water heat pumps (SWHP) represent an interesting option to efficiently meet building space heating loads. Table 12 summarizes the electricity, fuel and utility costs for each building with a SWHP. Estimated costs for each building include the heat pump and associated (heating demand weighted) distribution network costs for the system. For XR-03, new electrical demands with the SWHP exceeded the generating capacity of the previously examined CHP unit. As such, a new 125 kW CHP unit is also included in the costs for this building.

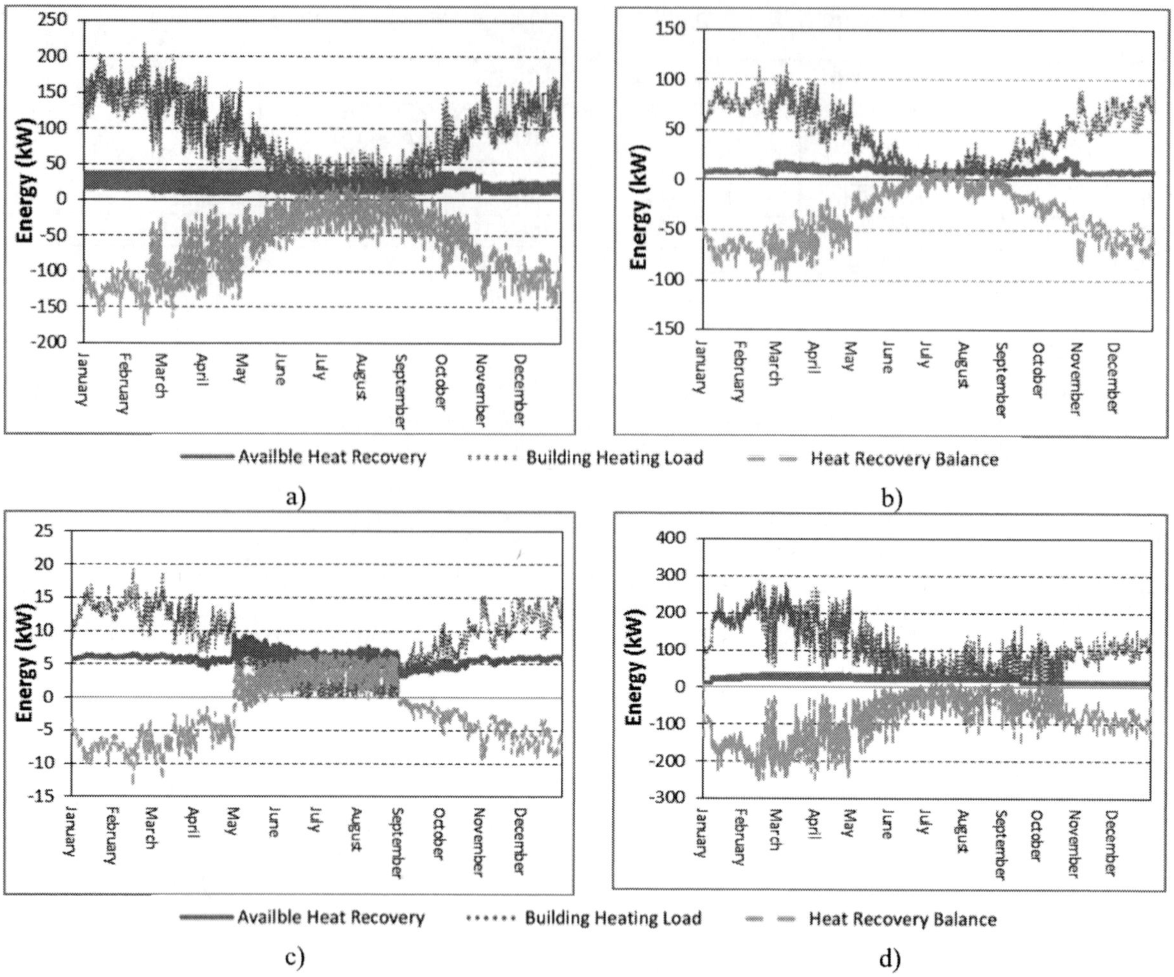

Figure 6: Required heat recovery and availability heat recovery for a) XR-02 b) XR-03/14 c) XR-04 d) XR-12/15

Table 12: Incremental electricity, fuel and utility cost savings for adding SWHP to scenario # 3

Building	Purchased Electricity Savings (kWh)	Increase in Generator Fuel Consumption (L)	Heating Fuel Savings (L)	Utility Cost Savings ($, CAD)	GHG Emission Reductions (tons CO_2 eq.)	Estimated Cost ($, CAD)	Simple Payback Period (Years)
XR-02	-59	18,875	29,484	$13,123	29	$336,808	>20
XR-03/14*	3131	40,393	48,918	$14,809	27	$898,365	>20
XR-12/15	0	11,798	22,484	$13,325	29	$615,308	>20
All Buildings	3072	71,066	100,836	$41,636	85	$1,850,482	>20
*XR-03/14 require a new 125 kW generator to be able to provide savings with the SWHP Scenario							

Adding a SWHP system increases facility fuel consumption by 71,000L. However, the facility space heating fuel consumption is reduced by approximately 100,000 L, resulting in an annual utility cost savings of $42,000. High infrastructure costs associated with the required piping network limit payback periods to over 20 years, which are likely too long for this particular project. However, for communities where the infrastructure is closer to the Arctic Ocean, a SWHP system could represent a viable energy efficiency measure.

7. FUTURE WORK

Future work will explore other innovative efficiency measures to further reduce the heating fuel consumption of the PCSP facility. These efficiency measures will include efficient heating technologies such as ground source heat pumps using advanced refrigerants like CO_2 incorporating the injection of solar thermal energy to reduce borefield sizes. Using solar thermal panels for domestic hot water, building integrated photovoltaic-thermal systems for ventilation preheat, and a facility wide district heating loop incorporating heat recovery from a future incinerator will also be explored. Future work will provide PCSP with a life cycle cost analysis including shipping costs, maintenance costs, additional operational costs, and projected utility cost escalation rates.

8. CONCLUSIONS

A series of long term energy efficiency measures have been assessed for the PCSP facility in Resolute using developed energy models. Proposed measures have been made based on a three step methodology that included (i) extensive monitoring of on-site electricity, fuel, and water use, (ii) development and calibration of detailed energy models using monitored data, and (iii) assessment of selected technologies using calibrated models.

Longer term measures require more planning and initial investment, but yield deep energy, GHG, and utility cost reductions. From monitoring of the energy flows of the facility, the electricity consumption represents only 18% of the total energy end use for the facility. Because of the high local electricity rates, reducing facility electricity consumption became the main target for these more complex measures, as it represents 65% of the total utility costs. To reduce the utility costs related to electricity consumption, renewable technologies such as wind and PV systems, and on-site power generation were first assessed to offset purchased electricity. The second step investigated efficiency strategies to reduce the total diesel consumption by upgrading the on-site power generation to a CHP system using the recovered heat to meet a portion of each buildings heating load. Thirdly, to increase the amount of energy available from the CHP systems, the use of energy efficient space heating equipment was investigated to shift the heating load from diesel fuel to electricity

Based on an extensive analysis of building energy performance, energy efficiency measures targeting electricity consumption were found to yield utility cost savings between $44,000 and $470,000, with simple payback periods ranging from under 1 year to 12 years. On-site power generation produced the greatest amount of annual utility cost savings with a simple payback periods less than 1 year. The second analysis phase targeted reductions in heating fuel consumption by recovering heat from on-site generators. These measures can further decrease utility costs by $109,000, with a simple payback period of 1.3 years The last analysis phase examined efficient heating technologies such as sea water heat pumps, which can be used to boost the heat recovery. Implementing these technologies can provide large reductions in heating fuel of up to 100,000 L and utility cost savings of close to $42,000. However, the high initial cost of the system results in a payback period greater than 20 years.

REFERENCES

Darrow, K., Tidball, R., Wang, J., & Hampson, A. (2015). *Catalog of CHP technologies*, U.S. Environmental Protection Agency Combined Heat and Power Partnership.

Feldman, D., Barbose, G., Margolis, R., James, T., Weaver, S., Darghouth, N., Wiser, R. (2014). Photovoltaic system pricing trends., (No. NREL/PR-6A20-62558).

Klein, S.A. et al. (2010). TRNSYS 17: A Transient System Simulation Program, Solar Energy Laboratory, University of Wisconsin, Madison, USA, http://sel.me.wisc.edu/trnsys.

Low Carbon NRCan (2016). NRCan website, <http://www.nrcan.gc.ca/energy/efficiency/buildings/eefb/leadership/4133> [Accessed Mar. 2016]

Mitsubishi Electric (2015). City Multi H2i™ Y-Series Hyper-Heating Inverter. <http://meus1.mylinkdrive.com/files/02_H2i_-Hyper-heat_Inverter_Unit_Section.pdf>. [Accessed Feb. 2016]

NREL (2016). Distributed Generation Renewable Energy Estimate of Costs. <http://www.nrel.gov/analysis/tech_lcoe_re_cost_est.html> [Accessed Feb. 2016]

RSMeans (2013). Mechanical Cost Data, RSMeans, Kingston, USA.

U.S. National Oceanographic Data Center: Global Temperature–Salinity Profile Programme. (2014). U.S. Department of Commerce, National Oceanic and Atmospheric Administration, National Oceanographic Data Center, Silver Spring, Maryland, 20910, <http://www.nodc.noaa.gov/GTSPP/>.

WINEUR (2006). Wind Energy Integration in the Urban Environment, Techno Economic Report, January 2006.

HVAC Solutions for Small- and Medium-sized Commercial Building Retrofit Opportunities

Bonggil JEON[1]*, W. Travis HORTON[1]

[1]Purdue University, School of Civil Engineering,
West Lafayette, IN, USA
bongway@purdue.edu, wthorton@purdue.edu

* Corresponding Author

ABSTRACT

According to the Commercial Building Energy Consumption Survey 2003 (CBECS 2003) conducted by the U.S. Energy Information Administration, over 70% of existing commercial buildings across the United States are more than twenty years old, with many of these buildings soon in need of renovation. Also, CBECS 2003 reports that existing small- and medium-sized commercial buildings (smaller than 200,000 ft^2 or 18,580.6 m^2) consume about 75% of the energy used in this sector, which means there is a great potential for energy savings with integrated technologies and building retrofit solutions, such as HVAC (Heating, Ventilating, and Air Conditioning) and envelope integration, and window and lighting integration. The primary focus of this study is to compare the annual performance of different types of HVAC equipment in existing small- and medium-sized commercial buildings, and to identify appropriate HVAC systems that could be retrofit into different commercial building types in a cost effective manner. Prototypical building types and characteristics for baseline models are proposed based on the CBECS 2003 microdata; and annual energy simulation results from EnergyPlus are utilized to analyze the different HVAC retrofit technology options.

1. INTRODUCTION

Existing commercial building retrofits have received considerable attention from the U. S. government and the building industry (IBE, 2013). According to the Commercial Building Energy Consumption Survey (CBECS) 2003 (EIA, 2006) conducted by the U.S. Energy Information Administration (EIA), over 70% of existing commercial buildings across the United States are more than twenty years old, with many of these buildings soon in need of renovation. These old buildings consume about 20% of the primary energy in the United States, which means there is a great potential for energy savings with integrated technologies and building retrofit solutions, such as HVAC and envelope integration, and window and lighting integration. Figure 1 shows the ranking of the highest energy consumption for all commercial buildings by Principal Building Activity (PBA). The first seven principal buildings account for 76% of the building energy usage of commercial buildings (EIA, 2008). Since 99% of commercial buildings are smaller than 200,000 ft^2 (18,580.6 m^2), a lot of attention has been paid to retrofitting small- and medium-sized commercial buildings (SMSCB). Furthermore, CBECS 2003 shows that existing small- and medium-sized commercial buildings, within the 1,001 – 200,000 ft^2 (93 – 18,580.6 m^2) range, consume about 75% of the energy used in all commercial buildings.

In retrofit projects of small- and medium-sized commercial buildings, decisions are typically made without a detailed evaluation of the design alternatives due to cost and schedule constraints (IBE, 2011). Also, due to the lack of proper design and application guidelines some small- and medium-sized commercial buildings that have incorporated new technologies in the retrofit process, fail to achieve energy savings targets. This is because inadequate understanding of their operation and poor compatibility between retrofit components are not properly integrated through a holistic design process. Therefore, it is important to understand the features of new technologies and the interaction between potential retrofit components. This study will demonstrate how to develop prototypical building types and characteristics for baseline models based on the CBECS 2003 database and how to analyze annual energy simulation results from EnergyPlus employing the different HVAC retrofit technology options. The objective of this study is to evaluate and select HVAC retrofit solutions for target buildings providing 50% HVAC energy savings with a payback of less than 4 years with available incentive programs.

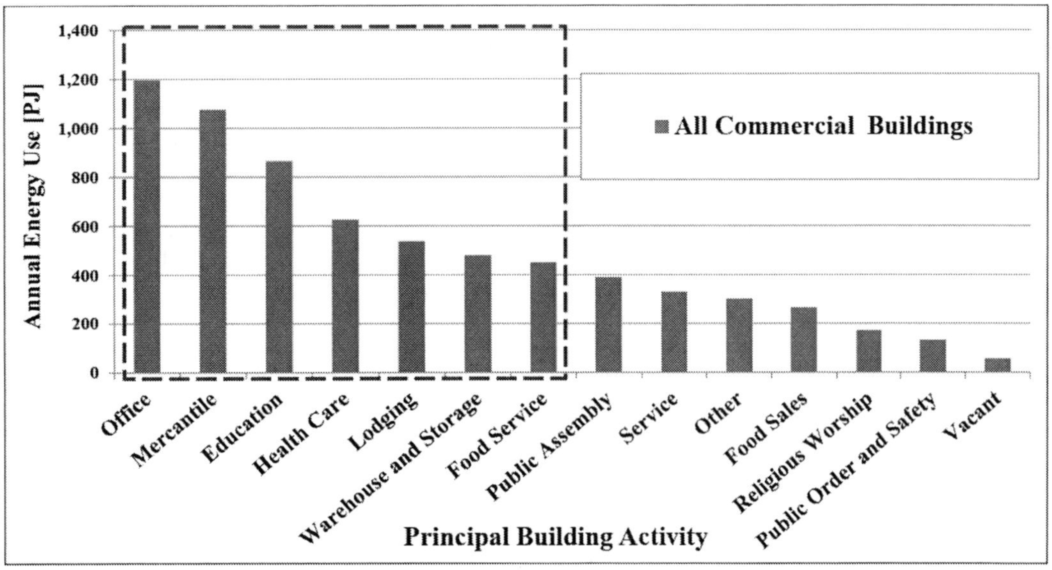

Figure 1: Annual energy use in all commercial buildings (EIA, 2008)

2. TARGET BUILDINGS AND HVAC SYSTEM CHARACTERISTICS

2.1 Target Building Types

Target retrofit building types were identified based on results of the CBECS (Commercial Buildings Energy Consumption Survey) 2003 database (EIA, 2006). For the purposes of this current study, three building types including mercantile, office, and education were selected from among the Principal Building Activities (PBA). Since small- and medium-sized commercial buildings are defined as those buildings with less than 200,000 ft^2 (18,580.6 m^2), a total of five target buildings were selected from within these three buildings types, namely: 1) stand-alone retail, 2) strip mall, 3) small office, 4) medium office, and 5) primary school. These five target buildings are included in both DOE (Department of Energy)'s Commercial Reference Buildings (NREL, 2011) and Commercial Prototype Building Models (PNNL, 2011) and these three building types consumed 46.2% of the energy usage of SMSCB which corresponds to 4,866 trillion Btu (5,134 PJ).

Major fuel consumption for Small- to Medium-Sized Commercial Buildings (SMSCB) within each Principal Building Activity (PBA) was extracted from CBECS 2003. Table 1 details the major fuel consumption for each PBA with different categories of building floorspace. Also, the results show that energy consumption of SMSCB covered about 75% of the energy consumption of all commercial buildings which corresponds to 6,523 trillion Btu (6,882 PJ). Therefore, it is of great importance to investigate SMSCB retrofit opportunities. In Figure 1, the rank order of PBA was defined by the most energy consumption of all commercial buildings; however, it may be different when it comes to SMSCB. Table 1 shows that the first three highest energy consuming building types of SMSCB are mercantile, office, and education.

In this study, it is important to choose the building vintage prior to modeling. Fortunately, in the CBECS 2003 database, it is possible to extract both annual energy consumption and number of building types by year of construction. The results show that the highest annual energy consumption was presented between 1970 and 1999 for all commercial buildings as well as SMSCB. In addition, more than half of buildings were constructed between

1970 and 1999. Accordingly, these two results led to the approach of using DOE's post-1980 reference building models to represent baseline energy consumption.

Table 1: Major Fuel consumption with Principal Building Activity (PBA)

Principal Building Activity	Sum of Major Fuel Consumption [PJ] Building Floorspace [Square Feet] (93-18,580.6 m^2)							All Buildings Total [PJ]	SMSCB Ratio [%]
	1,001 - 5,000	5,001 - 10,000	10,001 - 25,000	25,001 - 50,000	50,001 - 100,000	100,001 - 200,000	SMSCB Total		
Education	31	36	72	149	226	213	725	865	83.8
Food Sales	105	47	45	28	35	4	264	264	100.0
Food Service	217	119	61	33	21	0	450	451	100.0
Health Care	11	23	21	22	67	85	229	626	36.5
Inpatient	0	0	0	4	37	71	112	501	22.4
Outpatient	11	23	21	18	30	13	116	126	92.7
Lodging	8	22	45	91	83	104	353	538	65.7
Mercantile	76	104	219	108	181	191	880	1,078	81.6
Retail (Other Than Mall)	62	55	89	36	19	53	314	336	93.2
Enclosed and Strip Malls	15	49	129	72	163	138	566	741	76.4
Office	112	68	175	142	118	152	767	1,196	64.1
Public Assembly	18	44	62	40	56	108	329	390	84.2
Public Order and Safety	11	12	12	11	18	34	98	133	73.5
Religious Worship	20	37	59	40	6	9	171	172	99.3
Service	77	50	92	38	21	7	286	329	86.9
Warehouse and Storage	24	25	50	52	63	85	300	481	62.3
Other	8	5	33	22	51	118	237	302	78.7
Vacant	4	1	2	7	18	13	45	57	78.8
Total	723	594	948	782	963	1,123	5,134	6,882	74.6

2.2 Representative Climate Zones

To select relevant cities that accurately represent national climatic and energy usage variation, U.S. Census Regions & Divisions as well as U.S. Climate Zones for CBECS 2003 were used. A total of six different locations were selected based on the annual energy consumption of the proposed buildings. The first five locations represent those places where the selected building types exhibit the highest energy consumption. However, West-Zone 4 was also included for sake of diversity even though the annual energy consumption in that climate zone did not rank within the top five. These six combinations of regions and climate zones consumed 32.4% of the total energy (1,576 trillion Btu or 1,663 PJ) of SMSCB. Figure 2 shows the annual energy consumption for three proposed building types by regions and climate zones.

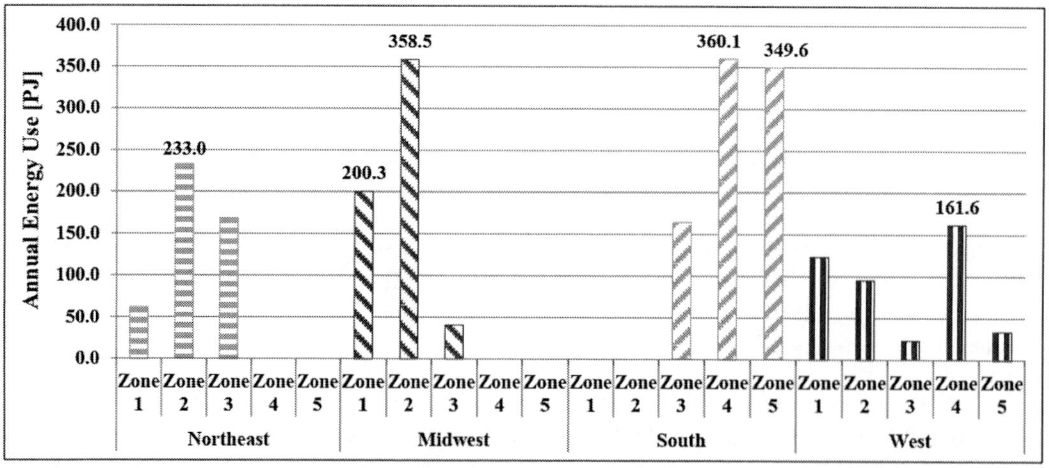

Figure 2: Annual energy consumption for three proposed building types by regions and climate zone

Table 2 shows the rank order of highest energy consumption by region & climate zone, and their representative cities. These representative cities were determined based on population according to the 2010 U.S. Census. Cities with the highest population in each region and climate zone except Midwest-Zone 1 and 2 were selected for use as representative weather data input to the building simulation. Milwaukee and Chicago are the cities with the highest population in Midwest-Zone 1 and Zone 2 respectively. However, Chicago was excluded because it is located close to Zone 1. And also Milwaukee was rejected because it is located near Chicago. Therefore, it is more reliable to use other representative cities, such as Minneapolis, MN and Indianapolis, IN instead. Figure 3 shows the final representative cities in a combination map of regions and climate zones.

Table 2: Ranking of highest energy consumption by regions & climate zones and representative cities

Rank	Regions & Climate	Representative Cities	ASHRAE Climate Zone
1	South, Zone 4	Charlotte, NC	3A
2	Midwest, Zone 2	Indianapolis, IN	5A
3	South, Zone 5	Houston, TX	2A
4	Northeast, Zone 2	Boston, MA	5A
5	Midwest, Zone 1	Minneapolis, MN	6A
6	West, Zone 4	Los Angeles, CA	3B

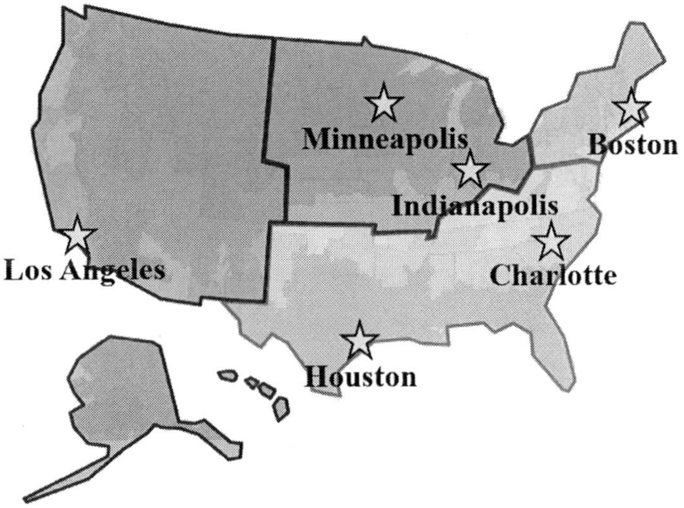

Figure 3: Representative cities in a combination map of regions and climate zones

2.3 Baseline HVAC System Characteristics

According to the CBECS 2003 database, there are seven different heating systems and eight different cooling systems in the building questionnaire. Based on an analysis of the database, baseline heating and cooling systems were determined by building type and climate zone. Although most of the heating and cooling configurations were reasonable in terms of their physical configuration, there were two building types and climate zone where the system configurations were not reasonable based only on the CBECS database. For example, for offices in South-Zone 4, heat pump heating systems are the dominant source for heating equipment and packaged A/C units for cooling equipment. Since heat pumps can provide both heating and cooling, this specific scenario was modified to a heat pump system providing both heating and cooling. Another irrelevant HVAC system configuration was found in Northeast-Zone 2. In nature, Packaged A/C units with a furnace are more appropriate HVAC systems than Individual Room A/C with furnace. Figure 4 shows the dominant heating system for office buildings in each region and climate zone. The results show that the majority of small- and medium-sized office buildings in Northeast-Zone

2 and Midwest-Zone 1 and 2 employ furnaces for major heating system. On the other hand, a heat pump is the main heating system in South-Zone 4 and a packaged heating unit is the primary heating system for South-Zone 5 and West-Zone 4.

In the CBECS 2003 questionnaire, once the responders selected the heat pump system for their heating and cooling equipment, they also needed to answer which heat pump system and what type the heat pump is. In this study, two building types, office and education, in South-Zone 4 have heat a pump system. Detailed heat pump systems and types of heat pumps for those building types are indicated in Table 3.

Since detailed HVAC systems were determined from previous data analysis, the major heat source for each HVAC system may also be filtered from the CBECS 2003 database. Identifying the major heating source is important because source energy prices may have an impact on economic analysis. Natural gas is the most dominant heating source for cold locations. On the other hand, office and education located in South-Zone 4 and 5 where cooling is more important than heating employ electricity for heating source. Also, the major heating source for the proposed building types is presented in Table 3.

For selection of the main air distribution system, a detailed analysis of the raw data was conducted. Air distribution systems are one of the important factors for HVAC system energy consumption. In this study, air distribution systems for the proposed building type were identified using several criteria from the database: region and climate zone; heating and cooling equipment; and buildings within 1,001 – 200,000 ft^2 (93 – 18,580.6 m^2). A detailed raw data analysis from CBECS 2003 reveals that CAV (Constant Air Volume) systems are dominant for most building types in each region and climate zone regardless of whether the data is sorted by the total number of buildings or total floorspace conditioned using CAV systems. However, air distribution systems for mercantile and education in West-Zone 4 is different. For mercantile building types in West-Zone 4, the CAV system is dominant when sorting the data by total building floorspace; however, VAV (Variable Air Volume) systems are dominant when sorted by the total the number of buildings using them. In addition, for education building types in West-Zone 4, VAV is the main air distribution system in both the number of buildings and floorspace.

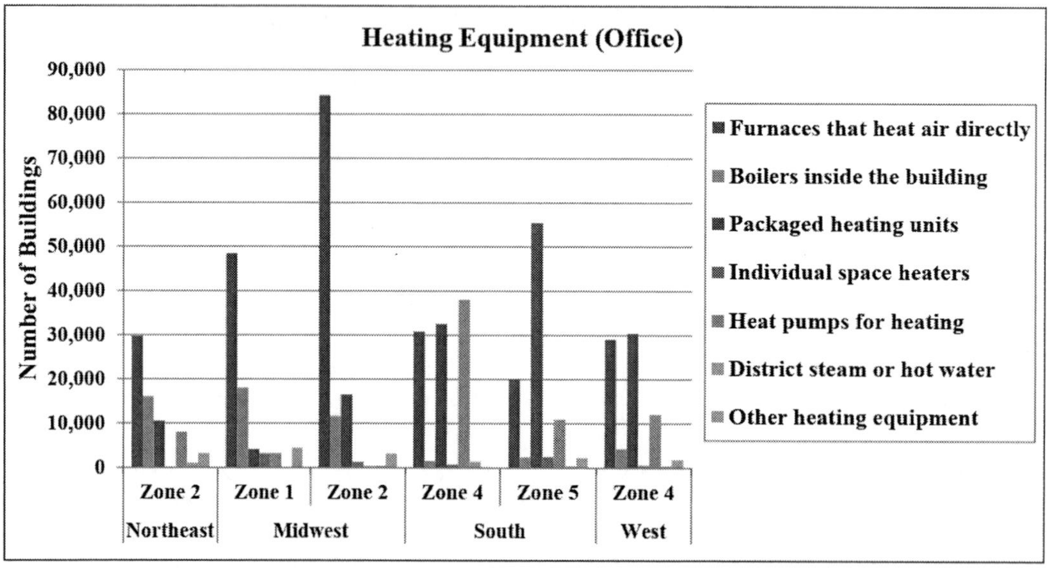

Figure 4: Dominant heating system of office buildings in each region and climate zone

2.4 Baseline SMSCB Models

After selecting the appropriate building types together with their baseline HVAC systems and representative cities, the next step is to develop baseline models for existing SMSCB in EnergyPlus. Even though DOE's post-1980 reference building models (NREL, 2011) are used for baseline models of prototypical SMSCB, they must still be modified because these models have outdated occupancy and operating schedules which are consistent with previous ASHRAE Standard 90.1-1989 (ASHRAE, 1989). Obviously, post-1980 reference building models do not reflect existing commercial buildings in the present era. Thus, in this study baseline prototypical SMSCB models were adjusted with several assumptions; including, 1) the building physical parameters and construction characteristics are the same as post-1980 models, and 2) lighting power density, plug load, and occupancy levels are modified to reflect today's current lifestyle. Table 3 shows a simulation matrix of baseline building models.

Table 3: Simulation matrix of baseline building models

U. S. Census Regions and Divisions			South	Midwest	South	Northeast	Midwest	West
U. S. Climate Zones for 2003 CBECS			Zone 4	Zone 2	Zone 5	Zone 2	Zone 1	Zone 4
Representative City (ASHRAE Climate Zone)			Charlotte, NC (3A)	Indianapolis, IN (5A)	Houston, TX (2A)	Boston, MA (5A)	Minneapolis, MN (6A)	Los Angeles, CA (3B)
Office	Small Office	Heating	Heat Pump (Elect.) - Air Source - Packaged Unit	Furnace (Natural Gas)	Packaged Heating Unit (Electricity)	Furnace (Natural Gas)	Furnace (Natural Gas)	Packaged Heating Unit (Electricity)
		Cooling	Heat Pump - Air Source - Packaged Unit	Packaged A/C Unit	Residential-type Central A/C	Packaged A/C Unit	Packaged A/C Unit	Packaged A/C Unit
	Medium Office	Heating	Heat Pump (Elect.) - Air Source - Packaged Unit	Furnace (Natural Gas)	Packaged Heating Unit (Electricity)	Furnace (Natural Gas)	Furnace (Natural Gas)	Packaged Heating Unit (Electricity)
		Cooling	Heat Pump - Air Source - Packaged Unit	Packaged A/C Unit	Residential-type Central A/C	Packaged A/C Unit	Packaged A/C Unit	Packaged A/C Unit
Mercantile	Stand-Alone Retail	Heating	Furnace (Natural Gas)	Furnace (Natural Gas)	Furnace (Natural Gas)	Furnace (Natural Gas)	Furnace (Natural Gas)	Furnace (Electricity, VAV)
		Cooling	Packaged A/C unit	Packaged A/C unit	Packaged A/C unit	Packaged A/C unit	Residential-type Central A/C	Packaged A/C unit
	Strip Mall	Heating	Furnace (Natural Gas)	Furnace (Natural Gas)	Furnace (Natural Gas)	Furnace (Natural Gas)	Furnace (Natural Gas)	Furnace (Electricity)
		Cooling	Packaged A/C unit	Packaged A/C unit	Packaged A/C unit	Packaged A/C unit	Residential-type Central A/C	Packaged A/C unit
Education	Primary School	Heating	Heat Pump (Elect.) - Air Source - Individual Room HP	Boiler (Natural Gas)	Packaged Heating Unit (Electricity)	Boiler (Natural Gas)	Boiler (Natural Gas)	Furnace (Natural Gas, VAV)
		Cooling	Heat Pump - Air Source - Individual Room HP	Packaged A/C Unit	Packaged A/C Unit	Packaged A/C Unit	Packaged A/C Unit	Packaged A/C Unit

3. HVAC RETROFIT MEASURES AND PACKAGES

Potential HVAC retrofit measures for SMSCB are collected and identified from previous research. Afterwards, each of the HVAC retrofit technology options for each application were ranked for each of the different climate zones. The ranking scores range from 1 to 5, corresponding to not effective to very effective, respectively. Four team members ranked the table of retrofit options and an average score was computed. The final averaged scores for climate zone 1 and 2 are shown in Table 4.

Based on the effectiveness scores and the baseline HVAC system configurations, the most effective technologies were selected to form different permutations or packages for each specific type of building. Figure 5 shows the technology permutations for office buildings in Midwest-Zone 1 (Minneapolis).

The technology packages are configured in consideration of compatibility issues. At each level, technologies that are compatible with each other are placed in the same block, whereas in compatible technologies are in separate blocks. For example, the three cooling & heating technologies, #5, #21 and #22, are incompatible.

Table 4: List of HVAC retrofit technologies

#	HVAC Technology Option Description	Category	Likely Energy Effectiveness	
			Climate Zone 1	Climate Zone 2
1	Switch from CAV to VAV	Air distribution	4	4
2	Customized AC system for hot-dry climates	Cooling	2	2.25
3	High efficiency vapor compression cycle in AC units	Cooling	2.25	2.75
4	High efficiency evaporator fan in AC units	Air distribution	2.75	3
5	High eficiency RTU system	Cooling and heating	4.25	4.25
6	Energy storage shift A/C operation periods	Cooling and heating	3.25	3.5
7	Chilled beams in temperate climates	Air distribution	2.5	3
8	Right-size A/C units and heaters for retrofit	Design	4.25	4.25
9	Use desiccant-enhanced evaporative cooling in both dry and humid climates	Cooling	1.75	2.25
10	Liquid desiccant AC in humid climates	Cooling	2.25	2.75
11	Maisotsenko cycle cooling (e.g., Coolerado) in hot-dry, cold- and mixed-dry climat	Cooling and heating	2.75	3
12	Use spray-cooled evaporators for packaged A/C	Cooling	1.75	2.75
13	Used sprayed mesh to improve A/C efficiency	Cooling	1.5	2.25
14	Best available duct sealing in exist building retrofit	Air distribution	3.5	3.5
15	Low pressure drop air filters in ducted systems	Air distribution	3.5	3.5
16	High efficiency hot water circulation pumps	Heating	3.75	3.75
17	Add stack economizer to boilers	Heating	3.75	3.75
18	Use most efficient furnaces and boilers (i.e., max-tech)	Heating	4	4
19	Replace large-capacity boilers with cascaded-multiple boilers	Heating	4	4
20	Use integrated heat pumps (heating, cooling and hot water)	Cooling and heating	3	3.5
21	Use ground source heat pumps (EER 42.1 by ClimateMaster)	Cooling and heating	3.75	3.75
22	Use dual source (air and ground) heat pumps	Cooling and heating	4	4
23	Use gas engine driven heat pumps	Cooling and heating	2.25	2.25
24	Use dedicated outdoor air systems with energy recovery	Outside air and ventilation	4	3.5
25	Demand controlled ventilation based on CO2	Operation, Control & Diagnostics	3.75	3.25
26	Apply desiccant wheels in hot humid climates	Outside air and ventilation	2	2.25
27	Operational fault diagnostics for key HVAC system components (outside air, RTU	Operation, Control & Diagnostics	3.5	3.5
28	Steam-clean A/C coils regularly for improved performance	Operation, Control & Diagnostics	3.25	3.25
29	Increased subdivision of building spaces for improved zone thermal control	Design	4	4
30	Model-predictive control for small to medium sized buildings	Operation, Control & Diagnostics	4	3.5
31	Night purge for ruduced day-time cooling load and peak demand in shoulder seaso	Operation, Control & Diagnostics	3	3
32	Operable windows link to HVAC control (shut off zone conditioning when window	Operation, Control & Diagnostics	3.25	3.25
33	Variable refrigerant flow HVAC system	Cooling and heating	3.5	3.5
34	High efficiency Multi-split AC units	Cooling and heating	3.25	3.75
35	Predictive thermostats	Operation, Control & Diagnostics	3.25	3.25
36	Automated whole-building diagnostics (AWBD)	Operation, Control & Diagnostics	3.25	3.25
37	Packaged HVAC controls for existing buildings (economizer, multi-speed fan and	Operation, Control & Diagnostics	3.5	3.5
38	Wireless temperature sensors	Operation, Control & Diagnostics	2.5	2.5
39	Develop and deploy low cost wireless sensors.	Operation, Control & Diagnostics	2.5	2.5
40	Develop and deploy a controller pack to RTU accomponied by a gas furnace for d	Operation, Control & Diagnostics	4	4
41	robust optimum start, ventilation lockout during warmup, occupancy sensor standby	Operation, Control & Diagnostics	3.75	3.25
42	Upgrade medium or small sized office buildings (<100,000SF) from standalone the	Operation, Control & Diagnostics	3.75	3.75
43	Move to standardize PSC motors in commercial blowers	Air distribution	3.25	3.25
44	MTAB gas furnaces, comm	Heating	3.5	3.75
45	Regenerative Air source heat pump (The project team, comprised of S-RAM Dyn	Cooling and heating	3.5	3.5
46	Develop and deploy the ThermoLift natural-gas-driven HP. "For heating, the Ther	Cooling and heating	2.75	2.75
47	Develop add-on humidity control packages for retrofit applications	Outside air and ventilation	3.25	3.25

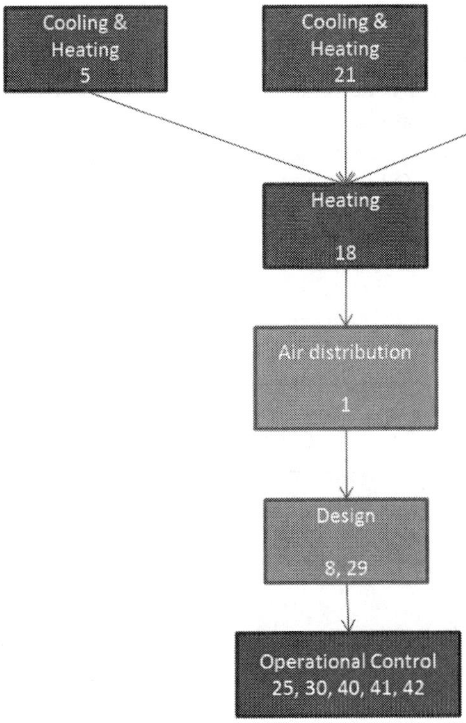

Figure 5: HVAC retrofit technology permutations for "Midwest-Zone 1: Small Office" (Minneapolis)

4. EVALUATION OF THE TECHNOLOGY PACKAGES

To evaluate the energy consumption of the various HVAC retrofit packages, EnergyPlus was selected as the simulation platform. For cost estimation, data in RSMeans (RSMeans, 2013a; RSMeans, 2013b; RSMeans, 2013c) and DOE's P-Tool (NREL, 2012) were used. Also, availability and magnitude of incentives were considered for simple payback analysis. The evaluation involves the following steps.

Step 1: Develop Pre-Retrofit Building Energy Models and Validate Against CBECS 2003 Data

In this step, pre-retrofit baseline models were derived from DOE's Commercial Reference Building Models ("post-1980"). The building envelope information was kept unchanged from the original commercial reference models. The HVAC system configuration was modified according to information shown in Table 3. HVAC equipment performance complies with ASHRAE Standard 90.1-1989 (ASHRAE, 1989). The interior lighting gain was changed to comply with ASHRAE Standard 90.1-2001 (ASHRAE, 2001), assuming that the building has gone through lighting retrofits. Plug load, occupancy level, and schedules were changed to reflect current lifestyle complying with ASHRAE Standard 62.1-2013 (ASHRAE, 2013a) and 90.1-2013 (ASHRAE, 2013b). With the baseline pre-retrofit model, the annual energy consumption was compared with CBECS data to make sure that the baseline energy usage is close to the statistics from CBECS data.

Step 2. Develop Standard Retrofit Energy Models and Evaluate Its Energy Impact

The "standard retrofit" is defined as replacing the HVAC equipment in the pre-retrofit baseline with new equipment of the same style but with higher efficiency. The performance of the new equipment complies with ASHRAE Standard 90.1-2013. Compared with the pre-retrofit scenario, the standard retrofit would demonstrate some energy savings due to the higher efficiency equipment.

Step 3. Develop Integrated Retrofit Energy Models and Evaluate Energy Impact

In this step, the HVAC system configuration and performance from the baseline energy model is modified according to the measures listed in Table 4. The different measures were added until a packaged solution is completed. Then annual simulation is run to obtain the annual energy consumption for each package.

Step 4. Cost Estimation for Standard Retrofit and Integrated Retrofit Packages, Simple Payback Period Calculation, Iterate with Step 3 If Necessary

In this step, the cost to implement the standard retrofit and the other proposed integrated retrofit packages are estimated based on data provided in the RSMeans data book and cost information provided in DOE's P-Tool. Simple payback period is calculated based on the energy savings potential of the integrated retrofit package over the standard retrofit and their cost difference. The cost used for simple payback considers equipment, labor cost, and incentives. Engineer judgement was used to drop certain ineffective measures from the packaged to form a new package.

Table 5: Retrofit package evaluation results

<table>
<tr><td colspan="2">U. S. Census Regions and Divisions</td><td colspan="2">South</td><td colspan="2">Midwest</td><td colspan="2">South</td><td colspan="2">Northeast</td><td colspan="2">Midwest</td><td colspan="2">West</td></tr>
<tr><td colspan="2">U. S. Climate Zones for 2003 CBECS</td><td colspan="2">Zone 4</td><td colspan="2">Zone 2</td><td colspan="2">Zone 5</td><td colspan="2">Zone 2</td><td colspan="2">Zone 1</td><td colspan="2">Zone 4</td></tr>
<tr><td colspan="2">Representative City (ASHRAE Climate Zone)</td><td colspan="2">Charlotte, NC (3A)</td><td colspan="2">Indianapolis, IN (5A)</td><td colspan="2">Houston, TX (2A)</td><td colspan="2">Boston, MA (5A)</td><td colspan="2">Minneapolis, MN (6A)</td><td colspan="2">Los Angeles, CA (3B)</td></tr>
<tr><td rowspan="4">Office</td><td rowspan="2">Small Office</td><td colspan="2">PTHP (high effic.)
DOAS with ERV</td><td colspan="2">Fixed duct leakage
RTU (high effic.)
Gas Heater (high effic.)</td><td colspan="2">VRF multi-split HP
DOAS with ERV</td><td colspan="2">VRF multi-split HP
DOAS</td><td colspan="2">Fixed duct leakage
Economizer
RTU (high effic.)
Gas Heater (high effic.)</td><td colspan="2">Fixed duct leakage
RTU
Economizer</td></tr>
<tr><td>51%</td><td>19</td><td>34%</td><td>6.5</td><td>50%</td><td>29.2</td><td>62%</td><td>71</td><td>31%</td><td>17.1</td><td>52%</td><td>2.9</td></tr>
<tr><td rowspan="2">Medium Office</td><td colspan="2">VRF multi-split HP
ERV unit
DCV</td><td colspan="2">VRF multi-split HP
ERV unit
DCV</td><td colspan="2">VRF multi-split HP
DOAS
DCV</td><td colspan="2">Fixed duct leakage
VAV RTU (high effic.)
DCV</td><td colspan="2">Fixed duct leakage
Heat pump (ground source)</td><td colspan="2">VRF multi-split HP
ERV unit
DCV</td></tr>
<tr><td>40%</td><td>34.7</td><td>58%</td><td>2.3</td><td>58%</td><td>0</td><td>61%</td><td>6.1</td><td>64%</td><td>125</td><td>49%</td><td>0</td></tr>
<tr><td rowspan="4">Mercantile</td><td rowspan="2">Stand-Alone Retail</td><td colspan="2">AC unit (DX)
Gas heat (core)
Multi-split HP (others)</td><td colspan="2">AC unit (DX)
Gas heat (core)
Multi-split HP (others)</td><td colspan="2">AC unit (DX)
Gas heat (core)
Multi-split HP (others)</td><td colspan="2">AC unit (DX)
Gas heat (core)
Multi-split HP (others)</td><td colspan="2">AC unit (DX)
Gas heat (core)
Multi-split HP (others)</td><td colspan="2">AC unit (DX)
Gas heat (core)
Multi-split HP (others)</td></tr>
<tr><td>48%</td><td>0</td><td>51%</td><td>0</td><td>50%</td><td>0.2</td><td>53%</td><td>0</td><td>48%</td><td>1</td><td>50%</td><td>0</td></tr>
<tr><td rowspan="2">Strip Mall</td><td colspan="2">Heat pump
VFD supply fan
2 stage cooling coil</td><td colspan="2">Heat pump
VFD supply fan
2 stage cooling coil</td><td colspan="2">Heat pump
VFD supply fan
2 stage cooling coil</td><td colspan="2">Heat pump
VFD supply fan
2 stage cooling coil</td><td colspan="2">VRF multi-split HP</td><td colspan="2">Heat pump
VFD supply fan
2 stage cooling coil</td></tr>
<tr><td>40%</td><td>0</td><td>30%</td><td>0.3</td><td>48%</td><td>0</td><td>30%</td><td>1.5</td><td>25%</td><td>0.3</td><td>52%</td><td>8</td></tr>
<tr><td rowspan="2">Education</td><td rowspan="2">Primary School</td><td colspan="2">Heat pump (Water source)
DOAS with ERV
Boiler (gas fired)</td><td colspan="2">VAV RTU (classroom)
Enthalpy wheel
CAV RTU (others)
DOAS with ERV</td><td colspan="2">VAV RTU (classroom)
Enthalpy wheel
CAV RTU (others)
DOAS with ERV</td><td colspan="2">VAV RTU (classroom)
Enthalpy wheel
CAV RTU (others)
DOAS with ERV</td><td colspan="2">VAV RTU (classroom)
Enthalpy wheel
CAV RTU (others)
DOAS with ERV</td><td colspan="2">RTU
Economizer
VFD supply fan</td></tr>
<tr><td>31%</td><td>0</td><td>47%</td><td>2.9</td><td>54%</td><td>2.6</td><td>51%</td><td>2.4</td><td>47%</td><td>3.1</td><td>40%</td><td>2</td></tr>
</table>

HVAC energy savings (left value) — Simple payback (year) with incentives (right value)

(medium grey)	Meet both energy savings and payback target
(light grey)	Meet only one of the energy savings or paback target
(dark grey)	Neither targets are met

5. CONCLUSIONS

To assess HVAC retrofit solutions for SMSCB, prototypical building types and their features were developed. Mercantile (stand-alone retail and strip mall), office (small and medium office), and education (primary school) were selected as the baseline models for this study from among the principal building activities based on CBECS 2003 analysis. In addition, an energy model of these building types were upgraded with current lifestyle parameters to represent real existing energy buildings models. The modeling tool used was EnergyPlus.

Nine HVAC solutions were identified to meet the stated objectives, based on 5 building types (small office, medium office, stand-alone retail, strip mall, primary school) in 6 region & climate zone combinations (Table 5). For each of the 30 combinations the baseline standard HVAC retrofit and packaged retrofit solutions were evaluated for their energy savings potentials. Also, the first cost and simple payback of each retrofit were computed based on the incremental cost and annual HVAC energy cost savings of the packaged retrofit solutions over the standard retrofit. Standard retrofits are defined as replacing the old HVAC equipment with new equipment that meets current code efficiency requirements without changing the HVAC system configuration. For each building type and region & climate zone combination, 3-7 retrofit packages were evaluated.

The results show that, for the building types and climate zones analyzed, many of the proposed packaged retrofit solutions can achieve 50% or greater HVAC energy savings. However, most of the packaged retrofit solutions did not achieve the payback target of 4 years or less. Therefore, existing incentives were applied based on the selected locations. These incentives are an important component to reduce the simple payback below the maximum acceptable to most commercial building owners and operators.

REFERENCES

ASHRAE. (1989). *Energy Efficient Design of New Buildings Except Low-Rise Residential Buildings. ANSI/ASHRAE/IESNA Standard 90.1-1989.* Atlanta, GA: American Society of Heating, Refrigerating and Air-Conditioning Engineers.

ASHRAE. (2001). *Energy Standard for Buildings Except Low-Rise Residential Buildings.* ANSI/ASHRAE/IESNA Standard 90.1-2001. Atlanta, GA: American Society of Heating, Refrigerating and Air-Conditioning Engineers.

ASHRAE. (2013). *Energy Standard for Buildings Except Low-Rise Residential Buildings.* ANSI/ASHRAE/IES Standard 90.1-2013. Atlanta, GA: American Society of Heating, Refrigerating and Air-Conditioning Engineers.

ASHRAE. (2013). *Ventilation for Acceptable Indoor Air Quality.* ANSI/ASHRAE Standard 62.1-2013. Atlanta, GA: American Society of Heating, Refrigerating and Air-Conditioning Engineers.

EIA. (2006). *2003 Commercial Buildings Energy Consumption Survey.* Energy Information Administration (EIA).

EIA. (2008). *Overview of Commercial Buildings, 2003.* Energy Information Administration (EIA).

IBE. (2011). *2011 Energy Efficiency Indicator: GLOBAL Results.* Institute for Building Efficiency

IBE. (2013). *Energy efficiency indicator - 2013 U.S. Results.* Institute for Building Efficiency

NREL. (2011). *U.S. Department of Energy Commercial Reference Building Models of the National Building Stock.* (NREL/TP-5500-46861). Golden, CO: National Renewable Energy Laboratory.

NREL. (2012). *A Tool to Prioritize Energy Efficiency Investments.* (NREL/TP-6A20-54799). Golden, CO: National Renewable Energy Laboratory.

PNNL. (2011). *Achieving the 30% Goal: Energy and Cost Savings Analysis of ASHRAE Standard 90.1-2010.* (PNNL-20405). Richland, WA: Pacific Northwest National Laboratory.

RSMeans. (2013). *RSMeans Square Foot Cost Data 2014.* Rockland, MA: RSMeans Publishers.

RSMeans. (2013). *RSMeans Green Building Cost Data 2014.* Rockland, MA: RSMeans Publishers.

RSMeans. (2013). *RSMeans Mechanical Cost Data 2014.* Rockland, MA: RSMeans Publishers.

ACKNOWLEDGEMENT

The authors greatly appreciate the help from Tiejun Wu and Russell Taylor from United Technologies Research Center who provided retrofit measures and cost information and Janghyun Kim from Purdue University who shared his skill to develop the baseline model. This material is based on work supported by the Consortium for Building Energy Innovation (CBEI) sponsored by the U.S. Department of Energy.

The Value of Price Risk Reduction in Energy Efficiency Investments in Buildings

Pekka TUOMINEN[1]*, Tuomas SEPPÄNEN[2]

[1]VTT Technical Research Centre of Finland,
Espoo, Finland
Contact Information (+358407345580, Pekka.Tuominen@vtt.fi)

[2]OP Financial Group,
Helsinki, Finland
Contact Information (+358505749884, Tuomas.Seppanen@op.fi)

* Corresponding Author

ABSTRACT

This paper presents a method for calculating the value of price risk reduction to a consumer that can be achieved with investments in energy efficiency. The examples presented deal with buildings and electricity for heating, but the principles can be applied to other contexts. The value of price risk reduction is largely overlooked so far in literature concerning the costs and benefits of energy efficiency in buildings. The topic is discussed to some length but in the literature reviewed for this paper no methodology for calculating the value was presented. Here we suggest such a method.

The problem of valuating price risk reduction is approached using a variation of the Black–Scholes model by considering a hypothetical financial instrument that a consumer would purchase to insure herself against unexpected price hikes. A rational consumer is prepared for a certain amount of growth in energy prices. If, however, the energy price rises more than expected, it might be very undesirable for the consumer. The consumer could, at least in theory, prepare for the unexpected rise by buying a cap contract on energy prices which would provide compensation for that unexpected rise.

Case examples calculated for typical single family houses in Finland are presented. The calculations concentrate on heating energy only, because there it is relatively easy to find examples of investments that reduce energy consumption, making it well-suited for demonstrating the issues at hand. Moreover, they concentrate on electricity as an energy carrier, as volatility data are needed and such data are available from the relatively well functioning electricity market in the Nordic countries.

The results show that the price risk entailed in household energy consumption can be reduced by a meaningful amount with energy efficiency investments, and that the monetary value of this reduction can be calculated using a variation of the Black–Scholes method. It is argued that this often overlooked benefit of energy efficiency investments merits more consideration in future studies.

1. INTRODUCTION

This paper presents a work in progress for developing a method for calculating the value of price risk reduction that can be achieved with investments in energy efficiency by an energy consumer. The main aim of this paper is to show how such a calculation could be done, while the optimal methodology is still being developed. The value of price risk reduction is largely overlooked so far in literature concerning the costs and benefits of energy efficiency in buildings. The topic is discussed to some length but in the literature reviewed for this paper no methodology for calculating the value was presented. Here we suggest such a method with some preliminary results from case examples.

The calculation examples presented in this paper deal with buildings and electricity for heating, but the principles can as well be applied to other contexts. The calculations presented concentrate on heating energy only because there it is relatively easy to find examples of investments that reduce energy consumption, making it well-suited for demonstrating the issues at hand. Moreover, they concentrate on electricity as an energy carrier, as volatility data is needed and there is a relatively well functioning electricity market in the Nordic countries.

In the literature there is a considerable amount discussion about an energy efficiency gap, meaning underinvestment in energy efficiency due to reasons that are not entirely clear. By this it is meant that there are apparently profitable investment opportunities that are nevertheless not taken by the various decision makers. Possible reasons offered range from information problems to liquidity constraints to misplaced incentives among others. Extensive reviews of the topic have been published by Gillingham et al. (2009), Brown (2004) and Klemick and Wolverton (2004) among others. Moreover, various authors (e.g. Abadie et al., 2013; Jackson, 2010) include the consumers' aversion of risky returns as one possible explanation of the energy efficiency gap.

From the economic point of view investment decisions concerning energy efficiency are evaluated according to the same principles as any other investments: the investment should be made when benefits are greater than costs. In this context savings in energy costs are the equivalent of a cash flow, to which the investment cost is compared. Or, as Gillingham et al. (2009) put it, higher initial capital costs are traded for lower but uncertain future energy operating costs. The uncertainty of operating costs is due to the price of energy which, being in future, is of course unknown.

In practice, according to Jackson (2010), those making the investment decisions do tend to take into account the risk of changing costs when making the decision but, rather than quantifying the risk, they opt to demand a short payback period to safeguard a quick profit. Jackson sees this leading into high-risk but likely profitable energy efficiency investments being overlooked and contributing to the energy efficiency gap.

To tackle the uncertainty of various cost components in energy efficiency investments, Abadie et al. (2013) offer a real options approach to produce a trigger investment cost to help decision making. Taking a similar approach, Jackson (2010) suggests a method for calculating confidence levels for various outcomes based on Monte Carlo simulations of relevant parameters. Vine et al. (2000), on the other hand, interestingly list a number of risk reduction opportunities related to energy efficiency investments and recognize the insurance value of these, but they do not include among them the topic of this paper, price risk reduction.

Thompson (1997) has recognized the reduction of price risk provided by energy efficiency investments and suggests tweaking discount rates in the investment calculation to take it into account. While he gives a sound argument for doing so, he does not provide a method for deciding what the discount rate should be, i.e. what is the value of the reduced risk.

This paper suggests a novel approach to provide a monetary value to the reduction of price risk with energy efficiency investments. This will allow decision makers to compare the value of risk reduction to other cash flows of the investment. Presently the value of reduced price risk, while widely recognized in the literature, is commonly excluded from investment calculations.

1.1 Rationale for Price Risk Reduction

Energy efficiency in buildings can be seen as an investment aiming to decrease uncertainty in future heating costs. From a financial point of view an energy efficient building protects the consumer against rising energy costs. Naturally this does nothing to remove any of the risks that are inherently included in any construction investment, for instance those relating to the interest for the capital or the risk that the investment fails to achieve its goals. Nevertheless, a reduction in the consumer's price risk is an added and commonly overlooked benefit typical to energy efficiency improvements specifically.

People who buy heating energy face the risk of rising energy prices. The standard approach to risky outcomes in economics is the von Neumann-Morgenstern utility function (see e.g. Mas-Colell et al., 1995). The idea is that there are multiple possible outcomes x_i, and for each x_i there is a value $v(x_i)$ that the consumer assigns to them. The sum of all the values v weighted by their likelihood is the value the consumer assigns to the risky undertaking. Let the probability of each x_i be π_i. Then the von Neumann-Morgenstern function for utility u can be stated as follows:

$$u = \sum_{i=1}^{n} \pi_i v(x_i) \tag{1}$$

If energy prices do indeed continue to rise, the payback periods for energy efficiency investments are shortened. After payback, the consumer who invested in an energy efficient building would be accumulating a net profit that would be higher the higher the energy prices. The total cost for heating depending on the initial investment and energy prices can be treated as x_i in Equation (1) and their respective likelihoods as π_i.

The value function v of a typical consumer, shown in Figure 1, is thought to have a concave shape whereby higher levels of consumption provide a diminishing marginal utility. Such consumers will exhibit preference to average outcomes compared to random outcomes. This can be clarified with an example.

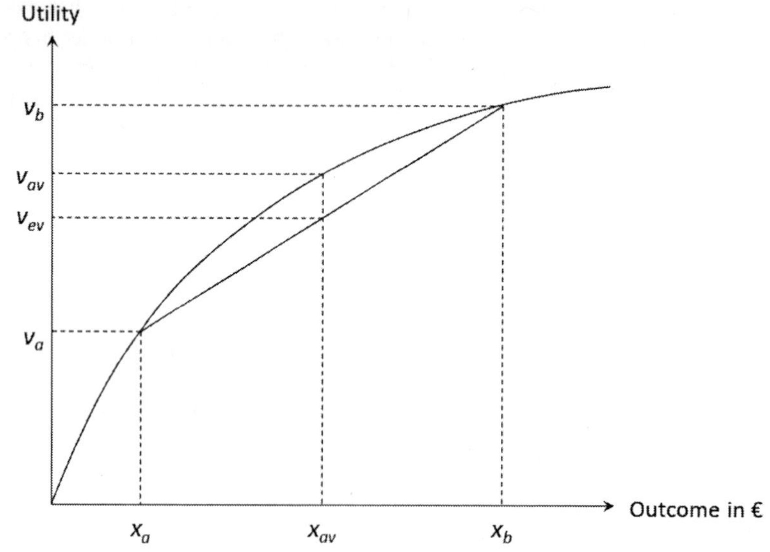

Figure 1: The utility function of a risk-averse consumer.

If the outcomes x_a and x_b in Figure 1 have a 50 % likelihood each and they give the utilities v_a and v_b respectively, then the consumer faces an expected value of utility of v_{ev} which is the average of v_a and v_b. If, however, the consumer can instead choose the average outcome x_{av} with a likelihood of 100 %, then the consumer would receive the utility v_{av}. Since this holds higher utility than v_{ev}, it is preferable compared to the random outcome.

Consumers with such preferences are called risk-averse. This means that the ability to lower the risk holds value to them. Thus the fact that an energy efficient building acts as an insurance against shifts in energy prices provides a premium to risk-averse consumers that does not show in simple investment calculations that concentrate on direct cost savings alone. To include this value of price risk reduction, a wider scope is needed in investment appraisal and a well-founded method to study the said value.

1.2 Review of Price Risk Valuation Methods

In financial economics price risk or market risk is understood as the risk of financial losses caused by movements in market prices (Bank for International Settlements, 2003). The standard approach to price risk management in the energy markets is through the derivatives market where futures and options allow energy buyers and sellers to protect themselves from adverse price fluctuations (James, 2012). The market also allows defining a price for risk reduction by other means, as one has the alternative approach of insuring oneself against price risk using financial products. Any other approach, say an investment to energy efficiency, to reducing price risk makes economic sense only if it costs the same or less than an equivalent financial product in the derivatives market.

The standard approach to derivatives pricing in modern financial theory is the Black–Scholes model. The model is based on the assumption that the prices of traded assets follow a Brownian motion with constant drift and volatility (Profeta et al., 2010). The applications of the Black–Scholes methodology in energy markets concentrates on options trading where the greatest financial interests are at stake (James, 2012). For the purposes of this study an approach that can be used to evaluate the price of price risk reduction in the electricity spot market is needed.

Brennan and Schwartz (1985) approach the problem of valuing uncertain cash flows of an investment project by trying to find a self-financing portfolio of traded securities whose cash flows replicate those which are to be valued. This approach is unsuitable for our case because its output is heavily dependent on market movements and doesn't produce a stable, simple and practical way of evaluating the value of the investment.

Woo, et al. (2001) offer a practical approach from a seller's perspective where the seller's risk premium is calculated by asking the following question: "What is the size of the per MWh risk premium that would allow a positive profit over some pre-determined time period with probability P?" While this is a very practical approach from a company's perspective, it doesn't give any concrete estimate for the value of price risk reduction and hence cannot be used here.

Deng and Oren (2006) suggest that due to the unique physical and operational characteristics of electricity production and transmission processes, classical derivative pricing methods based on Geometric Brownian motion, such as the one used in this paper, may not work very well. Instead, they have suggested two alternative approaches. 'Fundamental approach' relies on simulation of system and market operation to arrive at market prices while 'Technical approach' attempts to model directly the stochastic behavior of market prices from historical data and statistical analysis. Nevertheless, as our goal here is to present how such a calculation could be done in principle, we have chosen to use the most common method in financial industry utilizing geometric Brownian motion suggested by Black. The methods discussed by Deng and Oren (2006) may offer a more well-founded approach in the case of electricity and, thus, offer a clear avenue for further study.

2. METHODS

2.1 Calculating the Value of Price Risk Reduction

The problem of valuating price risk reduction is approached by considering a hypothetical financial instrument that a consumer would purchase to insure herself against unexpected price hikes. A rational consumer is prepared for a certain amount of growth in energy prices. If, however, the energy price rises more than expected, it might be very undesirable for the consumer. The consumer could, at least in theory, prepare for the unexpected rise by buying a cap contract on energy prices which would provide compensation for the unexpected rise. A cap contract bought from the financial markets means that the consumer is guaranteed never having to pay net prices over the agreed cap level. If market prices of energy are higher for a period of time, the consumer only pays a price equal to the cap, the rest being covered by the seller of the cap contract.

The value of a cap contract for the amount of saved energy consumption by an efficiency investment is the value of price risk reduction for the consumer. This is the theoretical fair economical value of the price risk reduction, meaning that in well-functioning financial markets it represents the price of buying capped price security for the same amount of energy consumption that the said energy efficiency investment has allowed to avoid altogether. However, such a cap contract should be viewed as hypothetical, as the practical options for consumers to protect themselves against unexpectedly high energy price rises depend on the market availability of products.

To calculate the value of price risk reduction we use the Black (1976) model for pricing of commodity contacts which is the standard approach used in the financial industry (Zvi et al., 2008). The underlying asset is the Spot price for Nordpool energy price and historical data allows the calculation of volatility in the market. Risk-free interest rate is taken from Euribor yield curve. The expected future prices are estimated based on reasonable energy price inflation expectation, 2.5% for the examples presented here. The strike price for the cap contract depends on the consumer's personal preference of tolerable energy pricing. For the calculation example we use 40% higher than expected as the tolerance level.

The Black model assumes there are no taxes, margins or transaction costs. This assumption can be used for the calculation of the theoretical fair economical value, but should be borne in mind when interpreting the results as real markets of course entail such costs. Moreover, the Black model assumes that the volatility is constant and that the underlying asset price follows a lognormal distribution. These assumptions may not apply exactly for energy prices, but are nevertheless market practice and selected as a practical approach for the purposes of this paper (Eydeland and Wolyniec, 2003). With these caveats in mind, the cap price can be stated as follows:

$$C = NP(0,T)\big(FN(d_1) - KN(d_2)\big)$$
$$d_1 = \frac{\ln\left(\frac{F}{K}\right) + \frac{\delta^2 t}{2}}{\delta\sqrt{t}} \qquad (2)$$
$$d_2 = d_1 - \delta\sqrt{t}$$

Here C is the price of the cap contract, t is the start time of the contract (years from present), T is the end time of the contract, N is the energy consumption between t and T in kWh, P(0, T) is the value of a T maturity zero-coupon bond at present time, F is the expected price of electricity at time T, K is the maximal tolerated price of electricity at time T (the cap value), N(x) is the cumulative normal distribution function and δ is the annualized electricity price volatility.

To calculate P(0, T) we use the current Euribor yield curve. To calculate δ we use the historical Nord Pool prices for Finland. To make the volatility figure realistic for our purposes, we have used average daily prices of working days and calculated the annualized volatility by multiplying the standard deviation of daily logarithmic returns, meaning logarithm of the quotient of the average prices of two consecutive working days, where Monday is considered as the next day after Friday, with the square root of yearly working days.

2.2 Electricity Price

As volatility is central to our reasoning, the consumer price used for the electricity consumed is based on daily prices in Nord Pool, the Nordic electricity exchange, rather than on long-term fixed-price contracts. Even though the latter contract type is more commonplace, consumers in Finland do have the possibility to buy exchange-priced electricity through a number of power companies.

The price data used for our calculations spans five years from 2007 to 2012 (Nord Pool, 2014). During that period, the average daily spot price in Nord Pool for Finland was 4.34 euro cents per kWh. To this we add, from the same period, the average electricity tax 1.39 c/kWh (Energy Authority, 2014), average VAT 0.97 c/kWh (Nord Pool, 2014), typical power company's price marginal 0.25 c/kWh (e.g. Energiapolar, 2013) and average transmission costs 3.16 c/kWh (Nord Pool, 2014), totaling 10.11 c/kWh to be paid by the consumer. This price is used as the starting level for the price of electricity.

2.3 Case Buildings and Energy Consumptions

The case buildings are variations on a typical Finnish single-family house, which is assumed to be electrically heated with direct electric radiators. This mode of heating remains the most common one in newly constructed single-family houses in Finland (Vilhola and Heljo, 2012). The buildings are assumed to have four dwellers, a typical size for a Finnish family, and to have a net area of 147 m^2 which is likely to be very close to the current average for new buildings. Sizes of new houses have been growing and were reported to be on average 144 m^2 in 2010 (Tiihonen, 2011). The buildings' energetic properties were assumed to follow the 2010 government building regulations (Ministry of the Environment, 2010) and they are assumed to be located in southern Finland.

The different case buildings are as follows:

- Business as usual (BAU), with direct electric radiators as the only source of heat,
- BAU + fireplace, with a fireplace supplementing the radiators,
- BAU + heat pump, with an air source heat pump supplementing the radiators and
- BAU + solar collectors, with roof-mounted solar heat collectors supplementing the radiators.

The building energy and financial calculations were done with the heating energy calculation tool of the Finnish government energy efficiency promotion corporation Motiva (2013). The calculation tool estimates that such a building would consume 4000 kWh/a for domestic water heating and 12230 kWh/a for space heating, totaling 16230 kWh/a. For financing, a 3 % interest rate is assumed, which is typical for a Finnish housing loan (Bank of Finland, 2014). Project lifespan used for financing calculations is set at 20 years.

More details about the various heating systems are given in Table 1. Investment costs as well as energy consumption figures are those given as typical in the Motiva tool. In addition to electricity costs, the firewood is estimated to cost 200 €/a. No other running costs are assumed during the 20 year calculation period or, stated differently, it can be assumed that other running costs of the different alternatives are of similar scale and thus do not affect the calculations. In all the cases the total delivered energy stays the same but the extra investments done in cases other than BAU allow a reduction in the amount of purchased electricity and thus, as the argument goes, also a reduction in the price risk the consumer faces in the electricity markets.

Table 1: Key figures concerning the heating systems in the different cases.

	BAU	BAU + fireplace	BAU + heat pump	BAU + solar
Investment cost (€)	4000	9500	6000	10 000
Electricity consumption (kWh/a)	16387	13984	13488	14370
Other heat sources (kWh/a)	0	2403	2899	2017

3. RESULTS

The results of the investment calculation are shown in Table 2 both in the conventional form, excluding the value of risk reduction, and expanded with the said value. Variable costs are in this case cost of electricity purchasing to the consumer, consisting of the electricity spot price, electricity tax, VAT, power company's price marginal and transmission costs. Capital costs are annualized investment costs from Table 1.

Table 2: Annualized values of selected costs and benefits for the different cases. Benefits appear as a negative cost.

	BAU	BAU + fireplace	BAU + heat pump	BAU + solar
Variable cost	1655	1612	1362	1451
Capital cost	269	639	403	605
Subtotal: cost excl. risk reduction	1924	2251	1765	2056
Value of risk reduction	0	-140	-168	-117
Total: cost incl. risk reduction	1924	2111	1594	1939

The results in Table 2 show that the value of price risk reduction for the various investment alternatives appears to be around 10 % of the total annualized costs: 10 % for BAU + fireplace, 12 % for BAU + heat pump and 8 % for BAU + solar.

4. DISCUSSION

This paper presents a method for calculating the value of price risk reduction to a consumer that can be achieved with investments in energy efficiency. It is based on the Black–Scholes model of pricing. Calculation examples are given for a case building with three investment alternatives for conserving purchased energy and thus reducing the risk of adverse effects from future price changes.

The calculation example demonstrates that a working valuation model for quantifying the value of price risk reduction is possible to construct given the information available from the markets. Based on the calculation the value of price risk reduction is estimated to be around 10 % of the total annualized costs for typical energy efficiency investments in typical single family homes in Finland.

In the case studied the value of price risk reduction is large enough to have an effect on the decision whether to invest. Especially in cases where investments appear marginally profitable, even a benefit of about 10 % can have a pivotal role. Moreover, there are circumstances in which the value of price risk reduction can be higher, such as more volatile markets or investments with smart control of the energy consumption to compensate for fluctuating energy prices.

There is also a methodological issue that means that this value estimate can be somewhat conservative: the Black–Scholes model may underestimate the value of price risk reduction in energy markets because it is based on the use of interest rates from the financial markets as reference, and energy commodities often possess more heavy-tailed price distributions in comparison. Also, as was discussed in the introduction, it is recognized that the Black–Scholes model may harbor other shortcomings for analyses dealing with electricity prices. Thus other potential methodological approaches should be further explored, including the possible tweaking of the Black–Scholes equation. To have a fuller accounting of the benefits of energy efficiency investments, the issue of price risk reduction merits further research.

REFERENCES

Abadie, L.M., Chamorro, J.M. & González-Eguino, M. (2013). Valuing uncertain cash flows from investments that enhance energy efficiency. *Journal of Environmental Management*, 116, 113–124.

Bank for International Settlements (2003). *A glossary of terms used in payments and settlement systems*. Basel, Switzerland: Bank for International Settlements.

Bank of Finland (2014). *Monetary Financial Institutions Annual Review 2013*. Helsinki, Finland: Bank of Finland.

Black, F. (1976). The pricing of commodity contracts. *Journal of Financial Economics*, 3, 167–179.

Brennan, M. & Schwartz, E. (1985). Evaluating natural resource investments. *The Journal of Business*, 2, 135–157.

Brown, M. (2004). Obstacles to Energy Efficiency. In C. Cleveland (Ed.), *Encyclopedia of Energy* (465–475), San Diego, CA: Elsevier.

Deng, S. & Oren, S. (2006). Electricity derivatives and risk management. *Energy*, 31, 940–953.

Energiapolar (2013). *Polar Spot hinnat ja ehdot.* Rovaniemi, Finland: Energiapolar.

Energy Authority (2014). *Sähkön hintatilastot.* Retrieved from: http://www.energiavirasto.fi/sahkon-hintatilastot

Eydeland, A. & Wolyniec, K. (2003). *Energy and Power Risk Management: New Developments in Modelling, Pricing and Hedging*. Hoboken, NJ: John Wiley & Sons.

Gillingham, K., Newell, R.G. & Palmer, K. (2009). Energy Efficiency Economics and Policy. *Annual Review of Resource Economics*, 1, 597–620.

Jackson, J. (2010). Promoting energy efficiency investments with risk management decision tools. *Energy Policy*, 38, 3865–3873.

James, T. (2012). *Energy Markets: Price Risk Management and Trading.* Singapore: John Wiley & Sons.

Klemick, H. & Wolverton, A. (2004). Energy-efficiency gap. In Shogren, J. (Ed.), *Encyclopedia of Energy, Natural Resource, and Environmental Economics*, San Diego, CA: Elsevier.

Mas-Colell, A., Whinston, M. & Green, J. (1995). *Microeconomic Theory*. New York, NY: Oxford University Press.

Ministry of the Environment (2010). *D3 Energy management in buildings, Regulations and guidelines 2010.* Helsinki, Finland: Ministry of the Environment.

Motiva (2013). *Lämmitystapojen vertailulaskuri*. Retrieved from: http://lammitysvertailu.eneuvonta.fi

Nord Pool (2014). *Elspot prices*. Retrieved from: http://www.nordpoolspot.com/Market-data1/Elspot/Area-Prices/

Profeta, C., Roynette, B. & Yor, M. (2010). *Option Prices as Probabilities: A New Look at Generalized Black-Scholes Formulae*. Berlin, Germany: Springer.

Thompson, P. (1997). Evaluating energy efficiency investments: accounting for risk in the discounting process. *Energy Policy*, 25, 989–996.

Tiihonen, A. (2011). *Asumisväljyys lisääntyy hitaasti*, Helsinki, Finland: Statistics Finland, Helsinki.

Vine, E., Mills, E. & Chen, A. (2000). Energy-efficiency and renewable energy options for risk management and insurance loss reduction. *Energy*, 25, 131–147.

Vilhola, J. & Heljo, J. (2012). *Lämmitystapojen kehitys 2002-2012.* Tampere, Finland: Tampere University of Technology.

Woo, C-K., Horowitz, I. & Hoang, K. (2001). Cross hedging and forward-contract pricing of electricity. *Energy Economics*, 23, 1–15.

Zvi, B., Kane, A., Marcus, A. (2008). *Investments*. New York, NY: McGraw-Hill/Irwin.

ACKNOWLEDGEMENT

This research has been supported by the project Assessing the intangibles: the socioeconomic benefits of improving energy efficiency (IN-BEE) funded by the Intelligent Energy Europe program of the European Commission.

System Identification for Cascaded Model Predictive Control of Building Region Temperature Applications

Matthew J. ELLIS[1*], Michael J. WENZEL[1], Robert D. TURNEY[1]

[1] Johnson Controls Inc.; Technology and Advanced Development, Building Efficiency, Milwaukee, WI, United States of America
matthew.j.ellis@jci.com*, mike.wenzel@jci.com, robert.d.turney@jci.com

* Corresponding Author

ABSTRACT

A system identification procedure is presented for identifying the models used in a cascaded model predictive control (MPC) architecture for airside demand response. In the cascaded MPC approach, the application considered is temperature control of a region, which is a collection of zones served by the same air handling unit. The outer loop MPC leverages a dynamic model to predict the region temperature and the corresponding power consumption required to produce the cooling duty for the region. Subsequently, the outer loop MPC determines a power profile that optimizes the energy costs by appropriately shifting the cooling load while maintaining occupant comfort constraints. The power profile is sent to the inner loop MPC, which is also formulated with a dynamic model. The inner loop MPC adjusts the region temperature setpoint to force the actual power consumption to track the desired power profile. In the developed system identification approach, two grey-box models are developed that capture the relevant dynamics on the time-scale of interest for the outer and inner loop MPCs. The grey-box models are parameterized, and the resulting model parameters are fit to input-output data for a particular region application so that the resulting model accurately predicts the temperature and power consumption of the region. State and disturbance estimation, which is required by the MPCs, is performed via a Kalman filter with a steady-state Kalman gain.

1. INTRODUCTION

Model predictive control (MPC) is a promising technology for energy cost optimization of buildings because it provides a natural framework for optimally controlling such systems to minimize the energy cost while meeting system constraints (Ma *et al.*, 2012; Mendoza-Serrano & Chmielewski, 2012; Touretzky & Baldea, 2014; Patel *et al.*, 2016, in press). For a review of MPC for heating, ventilating, and air conditioning (HVAC) control systems, refer to (Afram & Janabi-Sharifi, 2014) and for a general reference on MPC, refer to (Rawlings & Mayne, 2009). In particular, economic MPC is an MPC framework where an economically motivated cost function may be specified in the MPC problem (Rawlings *et al.*, 2012; Ellis *et al.*, 2014, in press), which is desirable in airside demand response since the ultimate goal is to reduce the energy costs.

In airside demand response (i.e., using building mass as passive thermal energy storage), the cooling load is shifted through pre-cooling in an optimal way such that energy cost is minimized while meeting the temperature comfort constraints. In a supervisory control layer (i.e., the control layer above the building automation system and zone level control layers), the variable that is modulated in order to minimize cost is the temperature setpoint and the building parameters that are measured are the zone temperatures and the buildings power usage. A model is needed to implement an economic MPC to predict how changes in the temperature setpoints affect the region temperature and building power.

In our previous work (Turney & Wenzel, 2012), a cascaded MPC framework capable of performing airside demand response (i.e., load shifting) to respond to time-varying energy prices was developed. Specifically, the cascaded MPC provides temperature setpoints to a region of a building to minimize the energy cost where a building region is defined as a collection of HVAC zones served by the same air handling unit (Turney & Wenzel, 2012). The outer loop MPC computes a power profile to minimize the energy cost while ensuring that the region temperature is maintained within its comfort constraints. The inner loop MPC receives the power profile from the outer loop MPC and manipulates the region temperature setpoint to ensure that the region power consumption tracks the power profile. The region temperature setpoint are subsequently sent to the zone-level temperature controllers.

Since both MPCs require a predictive model, a modeling framework and system identification (SI) methodology must be developed that is capable of accurately predicting the energy usage and region temperature for a diverse range of building regions. The model is arguably the most important part of MPC because it is used to predict the future system state and output trajectory to determine the optimal input trajectory. Without an accurate model, the computed control actions will be suboptimal and may even lead to an unstable closed-loop system. In this work, a system identification procedure for identifying the models in the outer and inner loop MPCs is developed. The approach leverages two grey-box models developed to capture the relevant dynamics on the time-scale of interest for the outer and inner loop MPCs. The grey-box models are parameterized, and the resulting model parameters are fit to input-output data for a particular region application. State and disturbance estimation, which is required by the MPCs, is performed via a Kalman filter with a steady-state Kalman gain estimated through the developed SI procedure.

In the developed SI procedure, the model parameters and Kalman gains of each grey-box model are updated in a sequential fashion. The significant disturbances affecting the region temperature (e.g., outside temperature and occupancy) may typically be considered as slowly varying disturbances with respect to the control time-scale. To prevent steady-state offset in the identified model caused by the slowly time-varying disturbance, a high-pass filter is applied to the input-output data to filter out the effect of the disturbance. The model parameters are subsequently computed from the filtered input-output data without the Kalman filter applied. The Kalman gain is also adjusted as the model parameters are updated to ensure stability of the resulting state estimator and for optimal estimation. After the model parameters are computed, the steady-state Kalman gain matrix is parameterized, and the parameters are updated using the prediction error method with the unfiltered input-output data and the updated model parameters. The Kalman gain update methodology is advantageous because it avoids the need to estimate the noise statistics. Owing to the fact that there is only a limited number of model parameters, the overall system identification procedure proves to be a practical methodology because it does not require several weeks of input-output data to identify the system parameters.

2. PRELIMINARIES

An overview of the SI concepts and the cascaded MPC design for HVAC region temperature control is presented.

2.1 Notation

The set of integers is denoted by $\mathbb{I}$ and the set of positive integers is denoted by $\mathbb{I}_{\geq 0}$. Boldface letters are used to represent a sequence with cardinality $N \in \mathbb{I}_{\geq 0}$ (i.e., $\mathbf{x} := \{x(0), \ \ldots, x(N-1)\}$). The notation $\hat{x}(i|k)$ is used to represent the predicted value of x at time step i using information (i.e., measurements) obtained at time step k where $i \geq k$. The notation $\cdot^*$ (e.g., x^*) denotes an optimal quantity with respect to some optimization problem.

2.2 Background on System Identification

The class of system models used in both the outer and inner loop MPCs is given by the following discrete-time linear time-invariant system:

$$x(k+1) = Ax(k) + Bu(k) \tag{1a}$$
$$y(k) = Cx(k) + Du(k) \tag{1b}$$

where $k \in \mathbb{I}_{\geq 0}$ is the time index, $x(k) \in \mathbb{R}^n$ is the state vector, $y(k) \in \mathbb{R}^p$ is the measured output, and $u(k) \in \mathbb{R}^m$ is the input vector. The measured output of the system is assumed to be synchronously measured. For simplicity, the measured outputs of the HVAC region temperature control application considered here are assumed to be the region temperature and the electrical power consumption. In practice, flow measurements would be used as a proxy for power consumption, and the region temperature may be determined by aggregating the zone temperatures:

$$T_R = \sum_{i=1}^{n_z} w_i T_{z,i} \tag{2}$$

where T_R is the region temperature, $T_{z,i}$ is the ith zone temperature, $w_i \in [0, 1]$ is a weighting factor such that $\sum_{i=1}^{n_z} w_i = 1$, and n_z is the number of zones in the region.

Two issues need to be addressed to use the system model in a predictive controller. First, a methodology to obtain the model parameters must be developed. Second, a state estimator capable of estimating the state given current and past input and output values must be designed. Using the state estimate, the model (1) may be initialized to make predictions of the system evolution for a given input trajectory within the MPC problem. System identification (SI) is the process used to determine the model and its parameters. SI often relies on manipulating the system inputs in an open-loop fashion, and observing the outputs of the system to subsequently regress model parameters based on the input-output data. Usually, it is also desirable to determine the stochastic characteristics of the disturbances, so that a state estimator, e.g., Kalman filter, may be designed.

Mathematically, SI maps input-output data, denoted by Z, to a model $\mathcal{M}^*$ where the set of all possible models, $\mathcal{M}(\theta)$, is parameterized by the vector $\theta \in D_{\mathcal{M}} \subset \mathbb{R}^d$ and $D_{\mathcal{M}}$ is the domain of $\mathcal{M}$. For the class of linear models described by (1), an approach is to parameterize the system matrices and use a linear state estimator where the gain of the state estimator may also be viewed as parameters that must be determined. The resulting parameterized model, $\mathcal{M}(\theta)$, is given by:

$$\hat{x}(k+1|k) = A(\theta)\hat{x}(k|k-1) + B(\theta)u(k) + K(\theta)\left(y(k) - \hat{y}(k|k-1)\right), \tag{3a}$$

$$\hat{y}(k|k-1) = C(\theta)\hat{x}(k|k-1) + D(\theta)u(k), \tag{3b}$$

$$\hat{x}(0|0) = x_0(\theta) \tag{3c}$$

where $\theta \in D_{\mathcal{M}}$, $K(\theta)$ is the parameterized estimation (Kalman) gain, $x_0(\theta)$ is the parameterized initial state, and $\hat{x}(k|k-1)$ is a state estimated for the kth time step using a measurement obtained at the $(k-1)$th time step. While the state estimator gain may be time-varying in general, a steady-state gain, (i.e., steady-state Kalman gain) is used.

To reduce the space of possible models and model parameters, it is often possible to leverage system knowledge to develop a general modeling framework (e.g., through first principles) that is capable of describing a class or family of similar applications. Then, the resulting modeling framework may be parameterized and the model parameters determined for each specific application through SI. This modeling approach is referred to as grey-box modeling. In a black-box modeling approach on the other hand, each element of the system matrices of (3) are determined, which typically results in substantially more model parameters that have to be identified. Because the amount of input-output data required to properly identify model parameters generally scales with the number of parameters, a grey-box modeling approach is adopted for the HVAC region temperature control application.

With the parameterization of system matrices (A, B, C, and D), the estimator gain (K), and the initial state (x_0) and through the adoption of a grey-box modeling approach, the SI problem involves determining the parameter vector, θ, that minimizes prediction error or some function of prediction error using forward prediction procedure with the model ($\mathcal{M}(\theta)$) of (3). Specifically, the parameter vector is given by:

$$\theta^* = \arg\min_{\theta} J(Z, \theta) \tag{4}$$

that minimizes the objective function:

$$J(Z, \theta) = \sum_{k} \ell(y(k) - \hat{y}(k|k-1)) \tag{5}$$

where ℓ is a non-negative cost function of the prediction error or residual and $\hat{y}(k|k-1)$ is the predicted output using the model (3). To summarize, the goal in offline or batch SI is to identify a model, described by the parameter vector, θ^*, that minimizes the cost function (5), over all the available input-output data Z. This SI approach is referred to as the prediction error method (PEM) (Ljung, 1999).

The most common cost function used in (5) is the square of the residuals:

$$\ell(r(k)) = \ell(y(k) - \hat{y}(k|k-1)) = \|y(k) - \hat{y}(k|k-1)\|_2^2 \tag{6}$$

where $r(k) \in \mathbb{R}^p$ is the residual (i.e., the prediction error) at time k. Using the cost function (6), the objective function (5) represents the standard sum of squares. When the residuals are independently and identically distributed random variables from a normal distribution, the cost function of (6) is optimal from a statistical point-of-view (e.g., (Nocedal & Wright, 2006)).

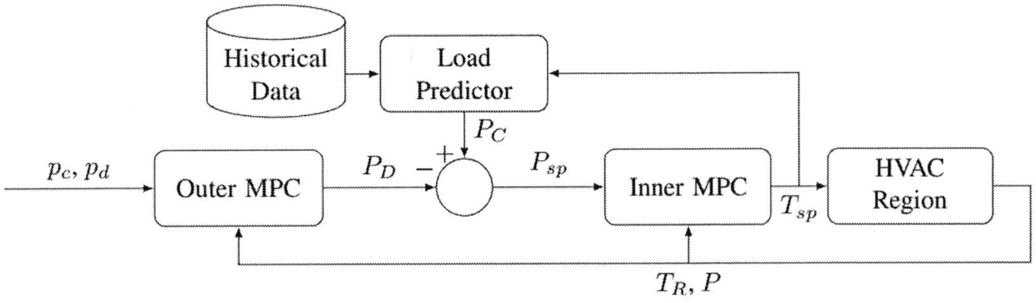

Figure 1: A block diagram of the cascaded MPC scheme developed for airside demand response.

2.3 Cascaded MPC for Airside Demand Response

A brief overview of the cascaded MPC architecture is given to provide the reader with some background on the cascaded MPC approach for airside demand response. A block diagram of the approach is shown in Fig. 1. The outer loop MPC, using the electricity price and demand charge (p_c and p_d, respectively) and the measurements of the region temperature and power, is responsible for computing a power profile that minimizes the energy costs over a prediction horizon. Specifically, the outer loop MPC computes the amount of power the HVAC system should defer, P_D, relative to the predicted power required to maintain the region at a constant temperature. Using this convention, the region temperature is expected to increase if $P_D > 0$, expected to decrease if $P_D < 0$, and expected to stay constant if $P_D = 0$. Leveraging historical data, the prediction power required to maintain a constant region temperature, P_C may be predicted. The power required to maintain a constant region temperature is a function of the difference between the ambient air temperature, T_{OA}, and the region temperature, T_R, ($P_C(T_{OA} - T_R, t)$ where t is the continuous time). A detailed description of this load calculation is beyond the scope of this paper; see, for example, for one such load prediction methodology (ElBsat & Wenzel, 2016, in press). From P_C and P_D, a desired power profile, $P_{sp} = P_C - P_D$, may be computed, which is sent to the inner loop MPC. The inner loop MPC, implemented with a smaller sampling period, manipulates the region temperature setpoint, T_{sp}, to force the region power consumption profile track the desired power profile. The dynamic models used in each MPC are described in the next section where the SI methodology is described.

The outer loop MPC is an economic MPC formulated with an objective function that represents the predicted electrical costs over the prediction horizon:

$$\min_{\mathbf{x},\mathbf{u}} \quad \sum_{i=1}^{N} p_{c,i} y_2(i) + p_d w_{d,k} \max_i y_2(i) \tag{7a}$$

$$\text{s.t.} \quad x(i+1) = Ax(i) + Bu(i) \tag{7b}$$

$$y(i) = Cx(i) + Du(i) \tag{7c}$$

$$H_x x(i) + H_u u(i) \leq g \tag{7d}$$

$$x(0) = \hat{x}(k|k-1) \tag{7e}$$

where $N \in \mathbb{I}_{\geq 0}$ is the prediction horizon, the input being computed in the problem is the deferred power profile, $p_{c,i}$ is the electricity usage price, which may be time-varying, and p_d is the demand charge. A time-varying weight, $w_{d,k}$, may be added to account for a potential discrepancy between the prediction horizon of the MPC and the time horizon over which the demand charge is imposed. The polyhedral constraint (7d) describes the temperature comfort constraints, bound constraints on the amount of power that may be deferred, rate of constraints, etc.

With slight abuse of notation, the formulation of the inner loop MPC is presented using similar notation as the outer loop MPC. Nevertheless, the MPC parameters, model, and constraints used in each formulation are different. The inner loop MPC is a tracking MPC with a standard quadratic cost function:

$$\min_{\mathbf{x},\mathbf{u}} \quad \sum_{i=1}^{N} q_y(y_2(i) - P_{sp}(i))^2 + \sum_{i=0}^{N-1} r_{\Delta u}(u(i) - u(i-1))^2 \tag{8a}$$

$$\text{s.t.} \quad x(i+1) = Ax(i) + Bu(i) \tag{8b}$$

$$y(i) = Cx(i) + Du(i) \tag{8c}$$

$$H_x x(i) + H_u u(i) \leq g \tag{8d}$$

$$x(0) = \hat{x}(k|k-1) \tag{8e}$$

where the input that is determined is the region temperature setpoint. The temperature setpoints are then provided to the HVAC zone-level temperature controllers to ensure that the zones of the region track the region temperature. The cost function used in the inner loop MPC penalizes the deviation of the predicted power from its corresponding desired power profile and penalizes the rate of change of the input.

The two MPCs operate over different time-scales in the sense that the outer loop MPC operates with a larger sampling period and longer prediction horizon to account for the time-scale needed for airside demand response (i.e., it requires a relatively large horizon to properly shift the airside cooling loads). The inner loop MPC operates with a smaller sampling period to ensure that the region power tracks the profile computed by the outer loop MPC. Both MPC schemes are implemented in a receding horizon fashion. At each time step k, a state estimate is received from the state estimator and used to initialize the dynamic model embedded in the optimization problem. The MPC problem is solved to obtain the input trajectory over the prediction horizon, which is denoted by $\mathbf{u}^* = \{u^*(0|k), \ldots, u^*(N-1|k)\}$. The input, $u^*(0|k)$, is applied to the system and remaining inputs of the trajectory are discarded. At the next time step, the problem is resolved using an updated state estimate.

3. SYSTEM IDENTIFICATION FOR MODELS IN CASCADED MPC FOR AIRSIDE DEMAND RESPONSE

The SI procedure adapted for the models used in the cascaded MPC is described in this section. The input-output data was generated with a higher-order region model. Also, substantial measurement noise was added to the output measurements to help demonstrate the robustness of the approach.

3.1 Offset from Unknown Disturbances

When identifying the inner and outer loop MPC models, it is important that the models capture the transients induced by the controlled input. However, the system steady-state is strongly dependent on external forcing factors, which may be considered as disturbances. The heat load generated by the region occupants, the heat from electrical power dissipation, and the heat exchange with the ambient are all driving forces that influence the region temperature and power consumption. Since these factors evolve on a slow time-scale and are difficult to directly measure (with exception to the ambient air temperature), they are considered to be slowly time-varying disturbances. However, these heat sources need to be captured in the model since they affect the steady-state power usage for a given temperature.

To address slowly varying disturbances, one may include an integrating disturbance model in the system model (1). This is equivalent to differencing the data (Ljung, 1999). However, differencing the data may amplify high frequency noise especially near the Nyquist frequency, and the SI algorithm will artificially focus on the high frequency modes. An alternative method is to use a high-pass filter, which filters out the slowly varying disturbance while having a constant gain in the passband. This approach is more desirable than differencing the data, because the filter described by differencing the data has a gain that increases with frequency through the entire spectrum. The high-pass filter selected to filter out the slowly varying disturbance is a high-pass fourth order Bessel filter given by:

$$H(s) = \frac{105s^4}{105s^4 + 105\omega_c s^3 + 45\omega_c s^2 + 10\omega_c^3 s + 10\omega_c^4} \tag{9}$$

where ω_c is the desired cutoff frequency.

To help explain the effectiveness of applying the high-pass filter, consider any linear system. Applying the filter to the input and output only affects the noise and the excited frequencies of the system. If the output for a given input and disturbance is given by:

$$y = L\{u, d\} \tag{10}$$

where L is the linear system mapping the input and disturbance to the output, then the filtered output after applying the filter to the input and disturbance is given by:

$$H(y) = L\{H(u), H(d)\} \; . \tag{11}$$

Thus, the filter, H, may be designed such that $H(u)$ excites the necessary modes of the system, but $H(d)$ is near zero if the disturbance is sufficiently slowly varying. The result is that the disturbance does not affect the results.

Two important considerations should be accounted for when designing the filter and selecting the input trajectory/signal that will be used to drive the system during the identification experiment. First, the input should have a frequency spectrum outside the stopband of H. In other words, $H(u)$ must be nonzero and excite the modes of the system that are to be identified. Additionally, it is important to note that it may be difficult to identify any modes that are at the same frequency as the disturbance. As discussed above, the significant disturbances in the airside demand response models are slowly moving. Therefore, it is possible to identify the system dynamics in the presence of the disturbances by using the high-pass filter (9) prior to using the prediction error method for system identification.

3.2 SI Procedure for Models in the Cascaded MPC

The first step in the SI procedure involves exciting the system by varying the system inputs and collecting the measured outputs. A pseudo-random binary signal (PRBS) is used to generate the setpoint trajectory. Other methods for generating the excitation signal have also demonstrated to be effective, but are not presented due to space limitations. In the region temperature control application, it is desirable to make small changes to the setpoint to prevent discomfort of the region occupants during the identification experiment. However, the magnitude of the region temperature setpoint variation required to properly identify the model parameters depends on the signal-to-noise ratio. After collecting input-output data collected from the identification experiment, the input-output data is divided into two sets: a training data set and a validation data set.

The high-pass filter (9) is applied to the input-output training data to obtain the filtered input-output data (i.e., $y_f = H(y)$ and $u_f = H(u)$). Since the high-pass filter removes the slowly varying disturbance, the resulting system matrices may be identified with the filtered data. The system matrices A, B, C, and D of the grey-box model are parameterized. To initialize the dynamic model within the prediction error method, the initial states may either be estimated or may be pre-specified. The former approach is often a simpler method when prefiltering is used, but adds additional parameters that need to be identified. In the latter approach, care must be taken if prefiltering is performed because the state corresponds to the state of the system with the filtered input and output data. This can cause the states to lose physical meaning making it difficult to supply good initial values. Owing to this reason, the initial states are parameterized and estimated.

For the inner and outer loop MPCs, a state estimator gain will be required so that the state estimate may be adjusted based on measurement errors. The estimator gain can be parameterized and estimated along with the parameters of the system matrix using the prediction error method, which results in an estimate of the optimal steady-state Kalman gain. Since applying a prefilter to the input and output data changes the noise properties, the Kalman gain computed for the filtered system is not generally the same as the optimal one for the unfiltered system. Instead of estimating the system matrices and the Kalman gain simultaneously, the system matrices are computed first. Subsequently, the optimal Kalman gain with the system matrices of the full model, which may include integrating disturbance models, is computed.

Let θ_1 denote all the parameters including the parameters to describe the system matrices and the initial state. With the filtered input-output data, the optimal parameter that minimizes the function of (5) is determined. With the optimal parameter value, θ_1^*, the resulting model is obtained:

$$\begin{aligned} x_f(k+1) &= A(\theta_1^*)x_f(k) + B(\theta_1^*)u_f(k) \\ y_f(k) &= C(\theta_1^*)x_f(k) + D(\theta_1^*)u_f(k) \\ x_f(0) &= x_{0,f}(\theta_1^*) \end{aligned} \tag{12}$$

where x_f is the state of the system resulting with the filtered inputs and outputs, $x_{0,f}$ is the parameterized initial state corresponding to the filtered input and output data. After validating the resulting system matrices $A(\theta_1^*)$, $B(\theta_1^*)$, $C(\theta_1^*)$, and $D(\theta_1^*)$ and augmenting them with appropriate integrating disturbance models, they may be subsequently used in the predictive model of the MPC (i.e., the constraints of (7b)-(7c) and (8b)-(8c) for the outer and inner loop MPCs, respectively).

Augmenting the system matrices with the integrating disturbance models, the optimal Kalman gain and initial condition are parameterized, which is denoted by θ_2. Given the parameters, θ_1^* and augmented system matrices, $\tilde{A}(\theta_1^*)$, $\tilde{B}(\theta_1^*)$, $\tilde{C}(\theta_1^*)$, and $\tilde{D}(\theta_1^*)$, the problem of (4) is solved to compute an estimate of the Kalman gain and initial state for the following system:

$$\begin{aligned} \hat{x}(k+1|k) &= \tilde{A}(\theta_1^*)(\theta_1^*)\hat{x}(k|k-1) + \tilde{B}(\theta_1^*)u(k) + K(\theta_2^*)(\hat{y}(k|k-1) - y(k)) \\ \hat{y}(k|k-1) &= \tilde{C}(\theta_1^*)\hat{x}(k|k-1) + \tilde{D}(\theta_1^*)u(k) \\ \hat{x}(0) &= x_0(\theta_2^*) \end{aligned} \tag{13}$$

where $K(\theta_2^*)$ is the Kalman gain estimated through the SI procedure.

The SI procedure applied to identify both the inner and outer loop MPC models is summarized by the following algorithm:

1. Generate input-output data by varying the region temperature setpoint.
2. Split data set into two sets: one data set for training and one data set for validating.
3. With the training data, filter the input-output data to reduce (ideally, remove) the effect of the slowly varying disturbance by applying the high-pass filter (9) to the input-output data.
4. With the filtered input-output data, apply the prediction error method with the grey-box of (12) to compute θ_1^* to identify the system matrices: $A(\theta_1^*)$, $B(\theta_1^*)$, $C(\theta_1^*)$, $D(\theta_1^*)$, and the initial state $x_{0,f}(\theta_1^*)$.
5. With the (unfiltered) input-output data, apply the prediction error method with the grey-box of (13) to compute θ_2^* to identify $L(\theta_2^*)$ and $x_0(\theta_2^*)$.
6. Validate the identified model.

There are a couple of key advantages of the adopted SI procedure for the models of the cascaded MPC for airside demand response. First, the procedure does not require directly estimating the stochastic characteristics of the noise disturbances, which may be difficult to obtain in practice. Second, the grey-box modeling approach leads to substantially less parameters that need to be identified, which subsequently reduces the amount of training data that needs to be collected.

3.3 Inner Model System Identification Results

As previously mentioned, the model used in the inner loop MPC is a grey-box model. Without including the integrating disturbance models, which are used in the MPC to achieve offset-free control (Rawlings & Mayne, 2009), and presenting the model in terms of time normalized states, the model used in the inner loop MPC is given by:

$$T_R(k+1) = (1-\theta_{1,1})T_R(k) + \theta_{1,1}\theta_{1,2}I(k) + \theta_{1,1}T_{sp}(k) \tag{14a}$$
$$I(k+1) = -T_R(k) + I(k) + T_{sp}(k) \tag{14b}$$
$$P(k+1) = P(k) + \dot{P}(k) \tag{14c}$$
$$\dot{P}(k+1) = \theta_{1,3}T_R(k) - \theta_{1,3}\theta_{1,2}I(k) - \theta_{1,4}\theta_{1,5}P(k) + (1-\theta_{1,4}-\theta_{1,5})\dot{P}(k) - \theta_{1,3}T_{sp}(k) \tag{14d}$$

where T_R is the region temperature, T_{sp} is the region temperature setpoint, P is the power consumed to provide the cooling duty to the region, and $\dot{P}$ is the time-derivative of P, and I is the integral of the error between the temperature setpoint and the region temperature. Physically, the state I models the dynamics of the zone controllers of the region (e.g., integral term of a proportional-integral controller). Recall, the input is the temperature setpoint, while the measured outputs are the region temperature and power ($x^T = [T_R,\ I,\ P,\ \dot{P}]$, $u = T_{sp}$, and $y^T = [T_R,\ P]$).

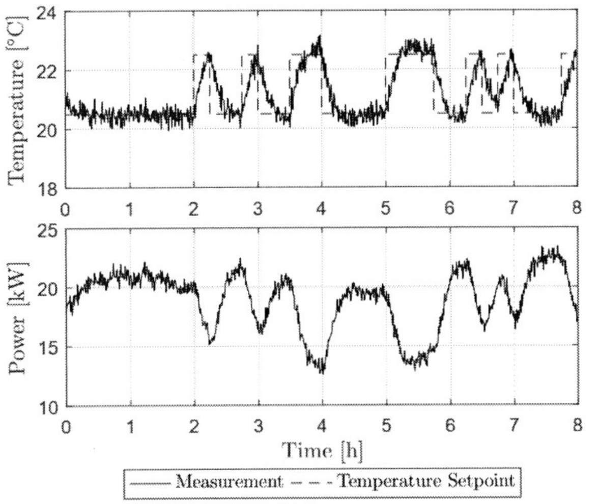

Figure 2: The training input-output data used for the SI procedure to identify the models.

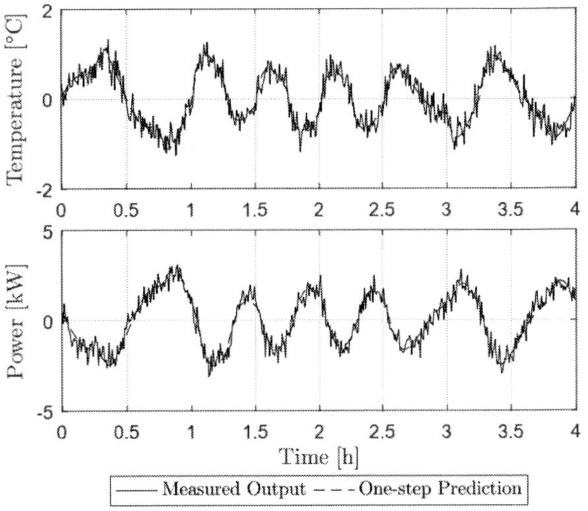

Figure 3: The filtered validation input-output data used to identify the inner loop model system matrices.

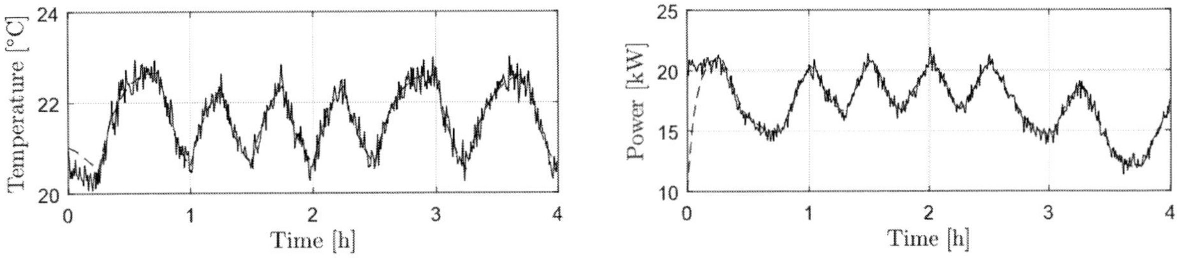

Figure 4: The validation output data (solid trajectory) and the one-step ahead predicted output (dashed trajectory) using the estimated Kalman gain.

Fig. 2 gives the region temperature and power profile for the PRBS setpoint profile used for model training. The measured outputs were collected with a thirty second sampling time. For the inner loop model, it was found through extensive simulation studies that a day of training data and a day of validation data was more than sufficient to properly identify the model parameters. In practice, it is expected that more training data needs to be collected. Nevertheless, there is an advantage of the proposed grey-box modeling approach with respect to the amount of training data needed for the SI procedure because the grey-box model (14) features a limited number of parameters.

To remove the slowly drifting offset, the input-output data is filtered by a high-pass Bessel filter described by the transfer function of (9). Identification is performed to estimate the five elements of the parameter vector θ_1 in (14). The initial states at the beginning of the estimation experiment may also be estimated. If the system is initially at steady-state, the states may be set to zero because the high-pass filter will eliminate the steady-state contribution. Fig. 3 shows the two filtered validation input-output data along with the one-step predicted filter outputs using the identified parameters and the filtered input. A noticeable slowly varying offset is observed in the training and validation power profiles (Fig. 2 and Fig. 4). The effect of the slowly moving disturbance is removed by the filter (Fig. 3).

The system is identified a second time using the prediction error method to estimate the Kalman gain. The unfiltered input-output data is used to complete this part to preserve the noise properties. Care must be taken to insure that the state estimator poles, eigenvalues of $A - KC$, are inside the unit circle. The System Identification Toolbox in MATLAB allows one to ensure stability by setting the z-stability threshold less than one, which is applied here. The Kalman gain is identified for the inner loop MPC model using the training data.

To validate the identified system matrices and Kalman gain, the validation data is used. In particular, convergence of the state estimator to the actual system state is verified. The one-step predicted values resulting from the model of the form (3) and the parameters estimated throughout this section are shown in Fig. 4. From Fig. 4, initially the one-step predicted output do not agree with the measured output owing to the effect of the initial guess on the state. After the

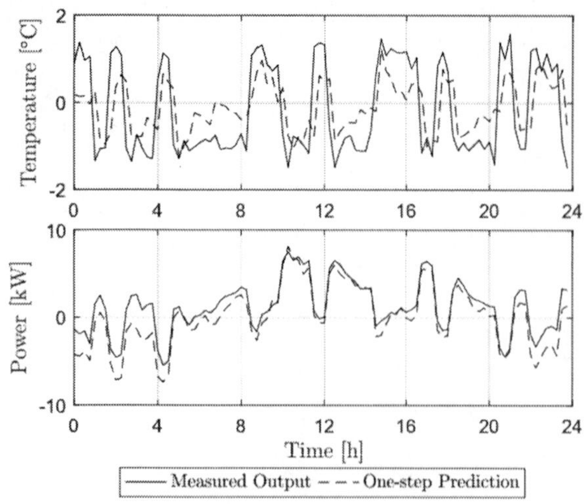

Figure 5: The filtered validation input-output data used to identify the outer loop system matrices.

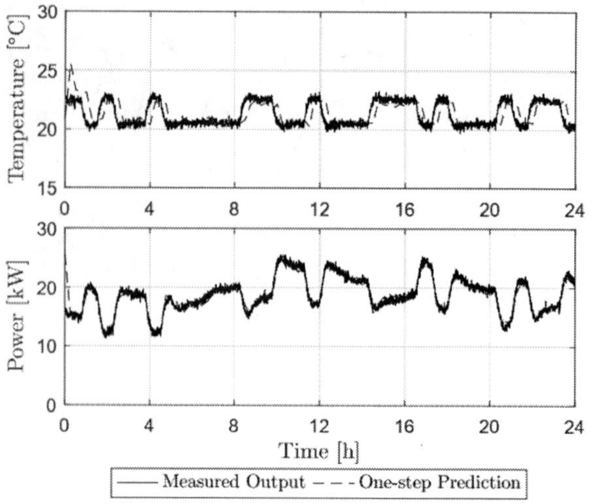

Figure 6: The filtered validation input-output and the one-step predicted output using the estimated Kalman gain.

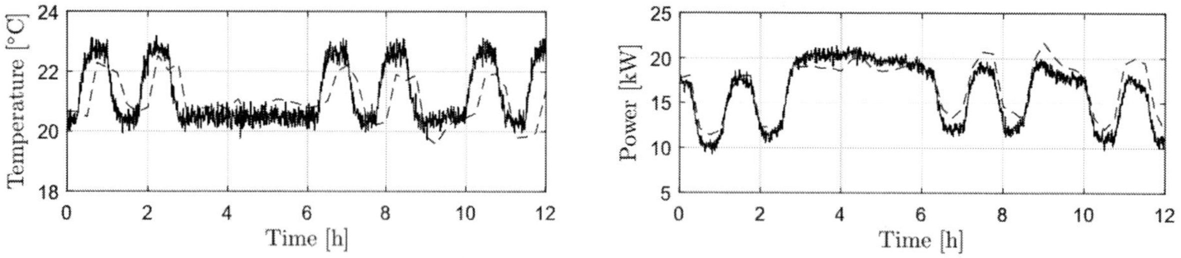

Figure 7: The validation output data (solid trajectory) compared to the predicted output (dashed trajectory) over a twelve hour horizon using the identified outer loop model.

state estimator converges, the state estimate is maintained in a neighborhood of the actual state (these results are not shown). Also, the one-step predicted outputs agree with the measured output (Fig. 4).

3.4 Outer Model System Identification Results

The SI procedure of Section 3.2 is repeated for the outer loop MPC model. The objective of the outer loop MPC model is to predict how the temperature will change based on how much power is deferred with respect to the power to maintain a constant temperature. The parameterized model (not including integrating disturbance models) is:

$$T_s(k+1) = (1 - \theta_{1,1} - \theta_{1,2})\,T_s(k) + \theta_{1,2}T_z(k) + \theta_{1,1}T_d(k) \tag{15a}$$

$$T_z(k+1) = \theta_{1,3}T_s + (1 - \theta_{1,3})\,T_z(k) + \theta_{1,4}P_D(k) \tag{15b}$$

$$P(k) = P_C(k) - P_D(k) + P_{D2}(k) \tag{15c}$$

where T_d is the temperature of the deep mass, T_s is the temperature of the shallow mass, and T_R is the region temperature. The powers in the system model are: the power required to keep the region temperature constant, P_C (recall, P_C is predicted from the temperature difference between the region temperature and the ambient temperature and the time), the deferred power usage, P_D, and a disturbance, P_{D2}, that is not included in P_C.

While the training data is synchronously sampled with a sampling period of thirty seconds (this sampling period is needed for the inner loop model identification), the data is resampled over a longer sampling period since the outer loop MPC uses a longer sampling period. For the outer loop model, two days of training data (with the larger sampling period) is needed to properly identify the model parameters. Again, a high-pass filter is designed and applied to the input-output data. Subsequently, the predictive error method is applied using the filtered input-output data to identify the system matrices. The filtered validation data collected over one day along with the one-step predicted output values are shown in Fig. 5. With the system matrices that are identified using the filtered data, the Kalman gain is determined. The validation data is again used to compare the one-step predicted outputs with the measured outputs (Fig. 6). Although the one-step predicted filtered outputs do not follow the measured filtered outputs (Fig. 5), the one-step predicted outputs in physical coordinates follow closely with the measured outputs (Fig. 6).

Since the outer loop MPC is responsible determining a power profile that optimizes the energy costs, it is important that the outer loop model accurately predicts the actual system response. Thus, the predicted outputs with the identified model is compared to the actual response for a given input trajectory over a twelve hour horizon, which is shown in Fig. 7. Overall, the predicted power trajectory follows closely with the measurements. On the other hand, there is an apparent lag in the predicted temperature profile. This is subject to further investigation.

4. CONCLUSIONS

This paper presented a method for system identification (SI) of region models towards the goal of leveraging the models in a cascaded MPC for airside demand response. The SI procedure included a three step process. First, training data is obtained by modulating the input of the system. The input-output data is then high-pass filtered to remove low frequency disturbances caused by weather, people, etc. The filtered data is then used to identify the system matrices of a linear state-space model (A, B, C, D). Once the system matrices are found, the Kalman gain is estimated by performing system identification on the unfiltered input-output data. Through simulations studies, the SI approach demonstrated that it is possible to identify the model parameters for the inner and outer loop of the cascaded MPC with input-output data. Owing to the adoption of a grey-box modeling framework that featured only a few parameters, a limited amount of training data was needed to identify the model parameters.

REFERENCES

Afram, A., & Janabi-Sharifi, F. (2014). Theory and applications of HVAC control systems – A review of model predictive control (MPC). *Building and Environment*, *72*, 343–355.

ElBsat, M. N., & Wenzel, M. J. (2016, in press). Load and electricity rates prediction for building wide optimization application. In *Proceedings of the International High Performance Buildings Conference.* West Lafayette, IN.

Ellis, M., Durand, H., & Christofides, P. D. (2014). A tutorial review of economic model predictive control methods. *Journal of Process Control*, *24*, 1156–1178.

Ellis, M., Liu, J., & Christofides, P. D. (in press). *Economic Model Predictive Control: Theory, Formulations, and Chemical Process Applications*. London, England: Springer-Verlag.

Ljung, L. (1999). *System Identification: Theory for the User* (2nd ed.). Upper Saddle River, NJ: Prentice Hall.

Ma, J., Qin, S. J., Salsbury, T., & Xu, P. (2012). Demand reduction in building energy systems based on economic model predictive control. *Chemical Engineering Science*, *67*, 92–100.

Mendoza-Serrano, D. I., & Chmielewski, D. J. (2012). HVAC control using infinite-horizon economic MPC. In *Proceedings of the 51st IEEE Conference on Decision and Control* (pp. 6963–6968). Maui, HI.

Nocedal, J., & Wright, S. (2006). *Numerical Optimization* (2nd ed.). New York, NY: Springer Science & Business Media.

Patel, N. R., Risbeck, M. J., Rawlings, J. B., Wenzel, M. J., & Turney, R. D. (2016, in press). Distributed economic model predictive control for large-scale building temperature regulation. In *Proceedings of the American Control Conference.* Boston, MA.

Rawlings, J. B., Angeli, D., & Bates, C. N. (2012). Fundamentals of economic model predictive control. In *Proceedings of the 51st IEEE Conference on Decision and Control* (pp. 3851–3861). Maui, HI.

Rawlings, J. B., & Mayne, D. Q. (2009). *Model Predictive Control: Theory and Design.* Madison, WI: Nob Hill Publishing.

Touretzky, C. R., & Baldea, M. (2014). Integrating scheduling and control for economic MPC of buildings with energy storage. *Journal of Process Control*, *24*, 1292–1300.

Turney, R. D., & Wenzel, M. J. (2012). *Model predictive control (MPC) for demand response* (Tech. Rep.). Milwaukee, WI: Johnson Controls Internal White Paper.

Experiment Design and Training Data Quality of Inverse Model for Short-term Building Energy Forecasting

Liang ZHANG[1]*, Jin WEN[1], Can CUI[2], Xiwang LI[3], and Teresa WU[2]

[1]Drexel University, Department of Civil, Architectural and Environmental Engineering,
Philadelphia, Pennsylvania, USA
E-mail: lz356@drexel.edu
E-mail: jw325@drexel.edu

[2]Arizona State University, School of Computing, Informatics, Decision Systems Engineering,
Phoenix, Arizona, USA
E-mail: ccan1@asu.edu
E-mail: teresa.wu@asu.edu

[3]Harvard University, Graduate School of Design,
Cambridge, Massachusetts, USA
E-mail: xiwang_li@gsd.harvard.edu

*Corresponding author

ABSTRACT

For data-driven building energy forecasting modeling, the quality of training data strongly affects a model's accuracy and cost-effectiveness. In order to obtain high-quality training data within a short time period, experiment design, active learning, or excitation is becoming increasingly important, especially for nonlinear systems such as building energy systems. Experiment design and system excitation have been widely studied and applied in fields such as robotics and automobile industry for their model development. But these methods have hardly been applied for building energy modeling. This paper presents an overall discussion on the topic of applying system excitation for developing building energy forecasting models. For gray-box and white-box models, a model's physical representations and theories can be applied to guide their training data collections. However, for black-box (pure-data-driven) models, the training data's quality is sensitive to the model structure, leading to a fact that there is no universal theory for data training. The focus of black-box modeling has traditionally been on how to represent a data set well. The impact of how such a data set represents the real system and how the quality of a training data set affect the performances of black-box models have not been well studied. In this paper, the system excitation method, which is used in system identification area, is used to excite zone temperature set-points to generate training data. These training data from system excitation are then used to train a variety of black-box building energy forecasting models. The models' performances (accuracy and extendibility) are compared among different model structures. For the same model structure, its performances are also compared between when it is trained using typical building operational data and when it is trained using exited training data. Results show that the black-box models trained by normal operation data achieve better performance than that trained by excited training data but have worse model extendibility; Training data obtained from excitation will help to improve performances of system identification models.

1. INTRODUCTION

Buildings are responsible for 40% of global energy use and contribute towards 30% of the total CO_2 emissions (Ahmad et al., 2016). It is significant to develop better building control and operation strategies to improve building energy efficiency and save energy.

Short-term whole building energy forecasting modeling is a core component of building energy control, operation strategy determining and building-to-grid integration. Inverse modeling (data-driven) approach outperforms forward modeling (physics-based) approach in terms of real-time controllability, simplicity, and engineering practicability. However, the performances of data-driven models are strongly affected by the training data. Because typical buildings often are operated with limited set points (one or two zone set points, for example) and operation modes, operational data collected from typical building operations are not rich enough which result in the poor extendibility often observed in data-driven models (Li et al., 2016).

In recent years, there are many studies applying varieties of inverse modeling strategy to short-term building energy forecasting. Some of them also focus on improving the accuracy as well as simplifying models to make them suitable for on-line control and optimization. Dong et al. (2005) presented support vector machines (SVM), a neural network algorithm, to forecast building energy consumption in the tropical region, which examines the feasibility and applicability of SVM in building load forecasting area. Yang et al. (2005) evaluated the performance of adaptive artificial neural network (ANN) models that are capable of adapting themselves to unexpected pattern changes in the incoming data, and therefore can be used for the real-time on-line building energy prediction. Newsham and Birt developed an ARIMAX model to forecast the power demand of the building in which a measure of building occupancy was a significant independent variable and increased the model accuracy (Newsham and Birt, 2010).

Although plenty of existing studies applied different approaches to model and forecast building energy performance, there is still a lack of systematic analysis about the relationship between data quality and model accuracy (Li and Wen, 2014). Inverse modeling approach is based on data. So data quality directly influences model quality. What's more, the complexity of building energy modeling, with many inputs and nonlinearity require sufficient data to reflect the building dynamics. Good experiment design yields data with rich information, which is the basis of a successful modeling process. Operational data often lacks rich information and may have undesirable properties which make it hard to build or train models. As a result, active input data design or experiment design should be used to ensure sufficient data quality and quantity to reflect the model structure and characteristics.

In this paper, a literature review of experiment design (including system excitation) and training data quality in short-term building energy forecasting models is presented. Difference of experiment design between system identification and black-box models will be illustrated. Finally, the impact of different training data on different building modeling structures will be studied.

2. LITERATURE REVIEW OF EXPERIMENT DESIGN IN BUILDING ENERGY MODELS

There are many review papers introducing building energy forecasting models (Li and Wen, 2014; Zhao and Magoulès, 2012; Swan and Ugursal, 2009), but few summarized the application of experiment design (including system identification) and the evaluation of data quality in this field. In this section, a general literature review will be given to summarize the existing researches about experiment design and data quality of inverse model for short-term building energy forecasting. Gray-box models (including those frequency-based models from system identification) have relatively fixed model structure and known properties of linearity or nonlinearity, only with parameters to be determined. However, black-box models are more complicated: sometimes most of model information, including properties of linearity and nonlinearity, model structure and its parameters, is based on training data. The experiment design theory and process are quite different between system identification and black-box model. In this section, they will be first discussed respectively and then be compared with each other.

2.1 System Identification

In system identification, the requirement that the data should be informative means that the inputs should contains sufficiently many distinct frequencies. The Fisher information matrix is usually used to identify the model parameters, and guide to design experiment for training data collecting (Ljung, 1998). Using the condition number of the Fisher information matrix can determine if the data are sufficiently rich in frequency for identification of a given model, and also can determine the maximum model orders. Cai et al. (2016) generated an optimal training data set for varying zone temperature set-points that maximizes the accuracy of parameter estimates for an intended building model structure. Optimal zone air temperature set-points were determined to maximize the Fisher information matrix. The required size of the training data could be dramatically reduced to achieve a certain accuracy level, which leads to more cost-effective experiments.

To be more specific, deducted from the theory of maximizing Fisher Information Matrix, we should spend the input power at frequencies where the model frequency response is sensitive to parameter variations. Put more leisurely, if a parameter is of special interest, then vary it and check where the frequency response moves, and put the input power there. In many cases this may give sufficient guidance for good input design.

However, when considering cost-effectiveness, frequency alone is not enough to ensure a good experiment design. The crest factor gives an idea of the compactness of the signal. Signals with an impulsive behavior (having a large crest factor) inject much less power into the system than signals having the same peak value and a small crest factor. The crest factor Cr is defined as:

$$Cr^2 = \frac{max_t u^2(t)}{lim_{N\to\infty}\frac{1}{N}\sum_{t=1}^{N} u^2(t)} \quad (1)$$

$max_t u^2(t)$ is the peak value of signal u, and $lim_{N\to\infty}\frac{1}{N}\sum_{t=1}^{N} u^2(t)$ is the average value of signal u. A good signal waveform is consequently one that has a small crest factor. The theoretic lower bound of Cr clearly is 1, which is achieved by binary, symmetric signal: $u(t) = \pm\bar{u}$. This gives a theoretical advantage for binary signals. Braun et al. (2002) demonstrate the application of minimum crest factor signals to accomplish meaningful "plant-friendly" identification and control in the process industries.

The basic issue for design of input signal is clear: achieve a desired input spectrum for a signal with as small crest factor as possible. Unfortunately, these properties are somewhat in conflict: If it is easy to manipulate a signal's spectrum, it tends to have a high crest factor and vice versa. To resolve the conflict between informative frequency and crest factor, Multi-sines signal and Pseudo-Random Binary signals are most widely used.

Multi-sines signal is defined as:

$$u(t) = \sum_{k=1}^{d} a_k \cos(\omega_k t + \Phi_k) \quad (2)$$

d is number of sinusoid; a_k is amplitude; ω_k is frequency; and Φ_k is phase. With d, a_k and ω_k we can thus place the signal power very precisely to desired frequencies. The crest factor is also controllable through choosing the phase Φ_k so that the cosines are "as much out of phase" as possible. Li and Wen (2014) applied Multi-sines signal to excite a system identification model used for short-term building energy forecasting. To guarantee those signals cover the frequency domain around the building, common operation range, Multi-sines signal was used to generate the exciting signal. He decided the parameters of the signal to put power to desired frequency and at the same time minimize the crest factor. Similarly, Privara et al. (2011) presented a comprehensive study of modeling of a large multi-zone office building by physical modeling and RC modeling. In RC modeling part, PRBS signal, Multi-sines signal and MPRS are proposed in black-box subspace identification. Signals are designed to cover required frequency spectrum and meet the demand of switching time.

Pseudo-Random Binary Signal (PRBS) is a periodic, deterministic signal with white-noise-like properties. Since the signal is a deterministic, it can be selected in accordance with the interesting part of the frequency scale of variation. It is generated by the difference equation:

$$u(t) = rem(A(q)u(t), 2) = rem(a_1 u(t-1) + \cdots + a_n u(t-n), 2) \quad (3)$$

Here $rem(x, 2)$ is the remainder as x is divided by 2. Like white random binary noise, PRBS has an optimal crest factor. PRBS has advantages over random binary noise: its covariance matrix will have a very special pattern. It can be analytically inverted, which will facilitate certain computations; PRBS has its second order properties secured when evaluated over whole periods. However, the following caution should be mentioned: A binary input will not allow validation against nonlinearities. There is consequently no way to detect that such a non-linearity is present from a binary input. In light of these advantages, it is widely used in experiment design in building models. Ma et al. (2012) used pseudorandom binary sequence to excite Autoregressive exogenous (ARX) models to build a model with inputs of zone temperature set-points of the cooling system and outputs of actual zone temperature and power measurements. A pseudorandom binary sequence (PRBS) is generated as the excitation input to randomly adjust cooling set-point between 21℃ and 25℃. Bacher and Madsen (2011) suggested a procedure for identification of suitable models for the heat dynamics of a building. The controlled heat input is a PRBS, which has white noise properties and no correlation with the other inputs. It is designed to excite the heat dynamics at several ranges of frequencies in which the time constants of the building is expected to be, such that the information embedded in data is optimized for estimation of the heat dynamic properties of the building. Lara et al. (2015) used PRBS signals as the excitation input signal, a linear time invariant (LTI) model to predict return water temperature of a central chiller plant was identified using grey-box technique. Zubiaga et al. (2015) presented the application of a grey box model to evaluate the thermal performance of a reference social housing building, focusing on its potential to evaluate the thermal performance of building passive elements. The controlled heat input (on and off status) was designed as a combination of a Randomly Ordered Logarithmic distributed Binary Sequence and a Pseudorandom Binary Sequence.

The PRBS input, however, is not always well suited for nonlinear problems. Since the PRBS consists of only two levels, the resulting data may not provide sufficient information to identify nonlinear behavior. Additionally, a PRBS signal of too large a magnitude may bias the estimation of the linear kernel. Multi-level pseudo-random sequences (m-level PRS, or MPRS), in contrast, allow the user to highlight nonlinear system behavior while manipulating the harmonic content of the signal to enable unbiased estimation of the linear dynamics in the presence of nonlinearities (Schoukens et al., 1993). Privara et al. (2011) applied MPRS to excite zone temperatures of a large multi-zone office building.

To sum up, Multi-sines, PRBS and MPRS are widely used in the process of experiment design in building energy modeling: Multi-sines can easily place the signal power very precisely to desired frequencies but with poor crest factor and complex design process; PRBS has an ideal crest factor but with poor nonlinearity application and required period length; MPRS has better nonlinearity application but with complex design process.

Input design should be decided in consideration of model structure and characteristics. If there is a high demand for frequency, Multi-sines signal has its advantage over PRBS and MPRS. If cost-effectiveness matters for the input acquisition, PRBS should be considered first. If the model has severe nonlinearity, Multi-sines and MPRS should be used. The comparison among Multi-sines, PRBS and MPRS in terms of frequency, crest factor, nonlinearity, period length and design complexity is shown in Table 1.

Table 1: Comparison among Multi-sines, PRBS and MPRS

	Multi-sines	**PRBS**	**MPRS**
Frequency	Good	Poor	Poor
Crest Factor	Poor	Perfect	Good
Nonlinearity	Medium	Poor	Good
Period Length	Not required	Required	Not Required
Design Complexity	High	Low	High

The sampling intervals are also important in experiment design: very fast sampling leads to numerical problems, model fits in high-frequency bands, and poor returns for extra work; slower sampling leads to data sets that are subsets of the maximal one, and hence is less informative. As the sampling interval increases over the natural time constants of the system, the variance increases drastically. Optimal choices of T for a fixed number of samples will lie in the range of the time constants of the system. These are, however, not exactly known, and overestimating them may lead to very bad results. Ljung (1998) summarized that a sampling frequency that is about ten times the bandwidth of the system should be a good choice in most cases. However, if the model should be used for control purpose, certain other aspects will enter. The sampling interval for which we build the model should be the same as for the control application. For example, in the building energy forecasting models for building-to-grid integration like demand response, the interval for control is about 15 minutes to 1 hour (Li and Wen, 2014 and Ma et al., 2012), which is the same as the sampling intervals. In the application of zone temperature prediction models, the interval for control is about 5 minutes (2011), and the sampling intervals should be the same.

"What inputs to excite" is another important topic. Modelers should choose the outputs and the inputs so that the predicted output becomes sensitive with respect to parameters that are important for the application in question. The input should have validation power to test possible properties of interest in the system. In most of the papers related to building energy forecasting, zone temperature set-points are usually used as the input to be manipulated. Besides, occupancy (Richardson et al., 2008 and Menezes et al., 2012) and ventilation flow rate are also manipulated in some studies to train building energy models. Other operational inputs like supply air temperature and pressure set points, and chilled water temperature set point have been hardly discussed, which could be the potential topic in the future study.

2.2 Black-Box models

Experiment design in system identification models can be improved by using theories and index like Fisher information matrix and crest factor. In black-box modeling, training data decide most of model information, including properties of linearity and nonlinearity, model structure and its parameters. Also, the diversity of inner structure of black-box models make it hard to have a universal criterion or index for experiment design.

In terms of the topic of "what to excite", it is termed as Feature Selection Process (FSP) in black-box modeling. Canonical Correspondence Analysis (CCA) (Härdle and Simar, 2012), Principal Component Analysis (PCA) (Abdi and Williams, 2010), and Entropy-based Feature Selection are used to find out the most important input variables for the modelling techniques. Jurado et al. (2015) proposed a hybrid methodology that combines feature selection based on entropies with soft computing and machine learning approaches, i.e. Fuzzy Inductive Reasoning, Random Forest and Neural Networks. They are also compared with a traditional statistical technique ARIMA (AutoRegressive Integrated Moving Average). Che et al. (2014) proposed a kernel-based SVR combination model by using a novel

individual model selection algorithm to get the optimal kernel function of STLF problem in the field of short-term load forecasting. Kusiak et al. (2010) proposed a correlation coefficient matrix and the boosting tree algorithm for variable selection when predicting daily steam load of buildings by a neural network ensemble with five Multi-Layer Perceptrons (MLPs) methods. Fan et al. (2014) used the recursive feature elimination technique, an embedded variable selection method, to select the optimal inputs to the base prediction models developed separately using eight popular predictive algorithms when developing ensemble models for predicting next-day energy consumption and peak power demand, with the aim of improving the prediction accuracy. Li et al. (2010) used an improved PCA, called Kernel Principal Component Analysis (KPCA), before training SVMs to predict building cooling load.
In terms of the topic of "whether training is sufficient", usually what modelers do is to run a benchmark model on the training data sets of different sizes, and evaluate the model performance with some metric, for example, the NRMSE, and find the best size. Because different problems may have different design space and different properties, it is difficult to set a universal threshold for determining the training size. Usually researchers will just assume their selected training data is adequate for prevention of under-training, and they will also add cross validation for prevention of over-training. There is also a rule of thumb about the number of training data in machine learning: roughly 10 times as many examples as there are degrees of freedom in the model. Thinking in terms of dimensionality and number of examples is a convenient shortcut.
Data mining techniques are used in improving training data quality in building energy forecasting. Fan et al. (2014) proposed a data mining based outlier detection method to identify abnormal building operating data, which improve the quality of training data.
The topic of pre-training is widely discussed in many researches, including the field of energy forecasting. Erhan et al. (2010) empirically show the influence of pre-training with respect to architecture depth, model capacity, and number of training examples. The observations so far in this paper confirm that starting the supervised optimization from pre-trained weights rather than from randomly initialized weights consistently yields better performance. Pasa et al. (2015) proposed a novel approach to pre-training sequential neural networks that exploits a simpler, first-order Hidden Markov Model to generate an approximate distribution of the original dataset. The learned distribution is used to generate a smoothed dataset that is used for pre-training to drive the connection weights in a better region of the parameter space, where subsequent fine-tuning on the original dataset can be more effective.
To the best of our knowledge, methods to improve the quality of modeling and training data in black-box modeling are limited to feature selection process, data-mining techniques and pre-training,. The focus in black-box modeling has been on how to find the best model (structure and parameters) to represent an existing data set. There is a lack of investigation on how different training data sets would affect a black-box model's performance, especially when the training data is to represent a nonlinear system. Considering that training data sets obtained from a system identification process, i.e., an excitation process, describe a nonlinear system in a more completed sense, we are interested in here the impacts of using such training data sets on a black-box model's performances, when compared with traditional training data set obtained from typical building operation.

3. MATERIAL AND METHODS

3.1 Research Method

The main goal of this research is, in the field of building energy forecasting modeling, to understand the impacts of using training data obtained from well-designed excitation tests which satisfy the maximum Fisher information matrix as well as the minimum crest factor, on the performance of a black-box model. In this study, the training data obtained by exiting building zone temperature set points using Multi-Sine signals are compared with normal operation data without excitation. The black-box models included in this study are Kriging, Support Vector Regression, Radial Basis Function, Multivariate Adaptive Regression Splines, Artificial Neural Network and Polynomial Regression. A system identification model (SID) using frequency is also included in the comparison study.
In lieu of a real building, a virtual building testbed described in Li wand Wen 2014, which is modeled in EnergyPlus, is used to generate all of the training data sets. The first floor of this reference medium size office building has five zones, and the total floor area is 510 m^2. The window-to-wall ratio of this building's facades is approximately 21.2%, and the windows are equally distributed. The U-factor of these single pane windows is3.4 W/m^2K and the solar heat gain factor is 0.36. The solar absorptivity, transmissivity and reflectivity are 0.06, 0.697 and 0.243, respectively. The location of this building selected for this study is Philadelphia, PA, USA. This building is partitioned into five different air conditioning zones, and an unconditioned attic zone. The five conditioned zone are four perimeter zones and one core zone. The roof insulation has an R-value of 15. The roof is covered in an

asphalt membrane, with a solar absorptivity value of 0.9. The exterior wall has the following construction (from outside layer to inside layer): 2.54 cm of stucco; 20.32 cm concrete masonry units; R-6 continuous insulation; 1.27 cm gypsum wallboard.
When comparing the models, the performance of accuracy is measured using Normalized Root Mean Square Error (NRMSE), where

$$\mathrm{NRMSE} = \sqrt{\frac{\sum_{i=1}^{N}(y_i - \hat{y}_i)^2}{N}} / (y_{max} - y_{min}), \tag{4}$$

and y is the true value of the building energy consumption, $\hat{y}$ is the forecast value and N is the number of data sets. Training error and prediction error will be calculated to show the accuracy of each model. Training error means the error in model building process, illustrating how well the model can be trained to reflect the real building data. Prediction error shows how powerful is the model to predict the future cooling energy by using the given inputs.
Two training datasets, i.e., normal operational data and excited operational data are generated. The normal operational data are generated by allowing the virtual building to be operated like a typical commercial building with very limited zone temperature set point variation. Weather inputs from July 1st to 7th are used to generated this seven day normal operation training data set. The exited training data set is described in Sec. 3.2. Seven days of training data are generated using the virtual testbed for both cases.
Two three-day testing datasets, i.e., similar weather testing and extended testing, are used to examine the model performances. In the similar weather testing, a testing period in which the weather conditions are similar to those during the training period is selected. In the extended testing, a testing period in which the weather conditions are very different from those in the training period is selected to examine how accurate a model is. To be more specific, in this TMY3 weather data in Philadelphia, average temperature of seven-day training data (July 1st to 7th) is 24.7℃ and direct solar radiation is 220.5W/m^2. The average outdoor air temperature of similar weather testing is 25.9℃ and direct solar radiation is 210.3W/m^2 (July13th to 15th). The extended weather testing is under weather condition of 28.0℃ and 277.9 W/m^2 (August 16th to 18th), which is hotter than the situation in training data.
Each model is firstly trained with normal operational data and then with the exited operational data, marked as Model I and II, respectively. Each model is then tested using the two testing datasets. The NRMSEs of each model for each testing period, as well as during the two training periods, are compared to examine the model accuracy and extendibility.

3.2 Excited Training Data Algorithms and Acquisition

The inputs used in all of the models in this study are zone temperature, ventilation rate, occupancy schedule ratio. Model disturbances include: outdoor air temperature, direct solar radiation and diffuse solar radiation. Output of the model is zone heating/cooling energy.
Zone temperature set-points are excited when generating the exited training data. The excitation method is a typical Multi-Sine signal generation process, which is to place the signal power very precisely to desired frequencies first, and then minimizing the crest factor. The excitation signal generation procedure is shown in Figure 1.

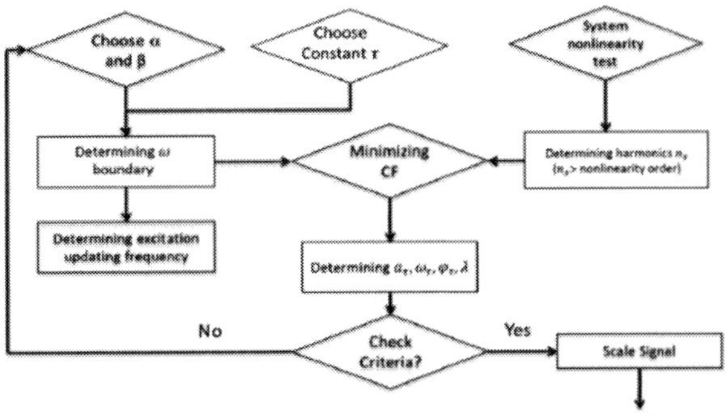

Figure 1: Excitation signal generation procedure (Li and Wen, 2014)

Matlab is used in this study to execute the algorithms to generate excited zone-temperature temperature set-points. A virtual test bed BCVTB connects Matlab to EnergyPlus is used to manipulate the zone temperature set-points in the virtual building testbed in EnergyPlus. According to the theory mentioned in Section 2.1, the sampling interval for which we build the model should be the same as for the control application. In demand response application of building-to-grid operation, the interval of 30 minutes is normally used, thus making the sampling interval of the model to be 30 minutes.

3.3 SID and Black-Box Model Description

Six data-driven models are chosen to test the adaptability of well-trained data from SID model in black-box modeling. They are Kriging, Support Vector Regression, Radial Basis Function, Multivariate Adaptive Regression Splines, Artificial Neural Network and Polynomial Regression.

Kriging (also known as Gaussian process regression) is an interpolation method that assumes the simulation output may be modeled by a Gaussian process. It gives the best linear unbiased prediction of simulation output not yet observed. It generates the prediction in the form of a combination of a global model with local random noise.

Support Vector Regression (SVR) is analogous to support vector classification, which attempts to maximize the distance between two classes of data by selecting two hyperplanes to optimally separate the training data.

Radial Basis Function (RBF) is used to develop interpolation on scattered multivariate data. A RBF is a linear combination of a real-valued radially symmetric function, $\emptyset(x)$, based on distance from the origin.

Multivariate Adaptive Regression Splines (MARS) is a form of regression analysis introduced by Friedman (1991). A set of basis functions, defined as constant, hinge function, or the product of two or more hinge functions, are combined in the weighted sum form, to be the approximation of the response function. A MARS model is built with generalized cross validation regularization in a forward/backward iterative process.

Artificial Neural Network (ANN) is a computational model inspired by an animal's central nervous system. It is apt at solving problems with complicated structures. Due to its promising results in numerous fields, ANN has been extensively applied in stochastic simulation modeling. An ANN model typically consists of three separate layers: the input layer, the hidden layer(s), and the output layer. The neurons across different layers are interconnected to transmit and deduce information. In supervised learning, the output unit is trained to simulate the underlying structure of the input signals and response. The trained structure is depicted by several parameters, the weights on each connection, the biases, the number of hidden layers, the transfer functions, and the number of hidden nodes in each hidden layer.

Polynomial Regression (PR) is a variation of linear regression in which an nth order polynomial is modeled to formulate the relationship between the independent variable x and the dependent variable y. PR models have been applied to various engineering domains such as mechanical, medical and industrial

The system identification model in this paper is derived from the research of Li and Wen 2014. Their study proposed a novel methodology to develop building energy estimation models for on-line building control and optimization using a system identification approach. Frequency domain spectral density analysis is implemented in this on-line modeling approach to capture the dynamics of building energy system and forecast the energy consumption with more than 90% accuracy and less than 2 min computational speed.

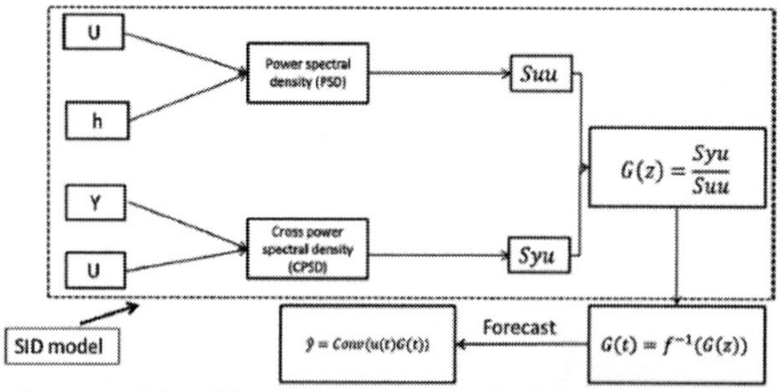

Figure 2: Model structure of system identification method for building energy forecasting (Li and Wen, 2014)

4. Results and Discussion

As mentioned in Section 3, seven models (one SID and six black-box models) and two sets of experiment will be conducted. Two sets of experiments include experiment using normal operation data (non-excited set-points) and well-designed excitation signal in each of the seven models. NRMSE will be calculated in both training period and testing period and it will be used as the criterion of training and prediction accuracy in the following discussion. Table 2 shows the experiment structure as well as the results of the simulation.

Table 2: NRMSE of three testing of two models by seven algorithms

	NRMSE													
Algorithms	**SID[3]**		**Kriging**		**SVR**		**RBF**		**MARS**		**ANN**		**PR**	
Model	I[1]	II[2]	I	II	I	II	I	II	I	II	I	II	I	II
Testing using training data	0.203	0.147	0.000	0.000	0.163	0.180	0.000	0.000	0.039	0.066	0.036	0.083	0.071	0.099
Testing under similar weather	0.153	0.090	0.048	0.098	0.246	0.268	0.231	0.275	0.076	0.171	0.066	0.162	0.080	0.102
Testing under extended weather	0.160	0.089	0.074	0.090	0.250	0.259	0.189	0.199	0.119	0.175	0.092	0.159	0.085	0.093

1 Model I stands for models that trained with normal operation data
2 Model II stands for model that trained with excited operation data
3 SID stands for system identification method by frequency domain spectral density analysis

To compare among seven algorithms, four scenarios are considered: model trained by normal operation data with the best testing result under similar weather testing is Kriging; model trained by normal operation data with the best testing result under extended weather is also Kriging; model trained by excited operation data with the best test result under similar weather testing is SID by frequency domain spectral density analysis; and model trained by excited operation data with the best testing result under extended weather is also SID.

To analyze the simulation results of Model I: by using normal operation training data to train models, most of the black-box models have higher prediction accuracy than SID model (17% better). It performs well when forecasting condition under weather similar to training data, which is better than that under weather much hotter than training data (7% better).

To analyze data of Model II: by using zone-temperate-excited training data to train models, the SID model has higher accuracy than most of the black-box models (48% better). Except RBF model, the rest models perform well under both similar weather and extended weather.

To compare the data in Model I and Model II, we found that training a SID model by using excited training data can raise the accuracy of forecasting (43% better). However, for black-box models, training them by normal operation data can achieve better model accuracy than training them by excited inputs and output (32% worse). As is shown in Figure 3, take ANN as an example, model that trained by normal operation data has much better performance than that trained by excited training data. The fluctuation makes the model trained by excited data unable to capture frequent fluctuation of cooling energy caused by excited zone temperature set-point.

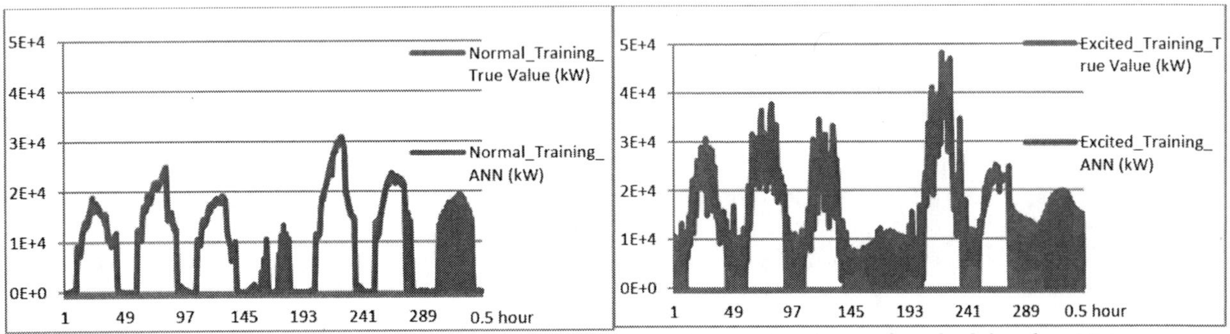

Figure 3. Result of training models excited by normal operation data and excited data in ANN

In the validation period, we could see that model trained by normal operation data predict more accurately than that trained by excited training data, especially when predicting the fluctuation happening in the second day of validation. The second model performs poorly to capture the dynamics. Also, the second model cannot capture the energy consumption pattern during the equipment starting up period.

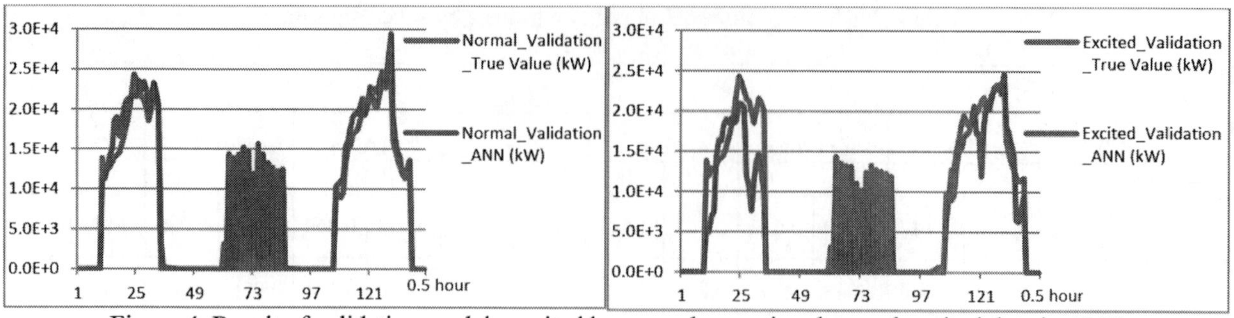

Figure 4. Result of validating models excited by normal operation data and excited data in ANN

To explain this, the experiment design and the excitation theory are used to excite the model to reveal its parameters which are important to the model. However, black-box models do not have parameters with significant meaning. Most black-box models learn only from training data. Excitation will make it hard to capture the normal pattern of the input-output relation, especially in the situation when handling fluctuation of energy and starting up of equipment. Too frequent fluctuation in the excitation input will hinder the black-box model to capture the pattern of equipment starting up period in the real operation. Also frequent on-and-off situation is also hard be detected and well-described in the model trained by exited data.

We also find that in SID models trained by excited data, the difference between NRMSE of similar weather and extended weather is only 0.001. However, this value turns to be 0.024 and 0.018 in black-box models when trained by normal operation data and excited data respectively, which demonstrates the fact that SID models have better extendibility than black-box models. Though the excited training data increase the extendibility of black-box model by 25% but decrease the model accuracy by 32%.

5. CONCLUSIONS

A comparison simulation study using virtual building testbed is conducted to study whether the training data obtained from theory of excitation in SID models can improve the performance of other types of black-box models in the building energy forecasting area. Results show that experiment design, or excitation, is essential to SID modeling: it can increase model accuracy of building energy models. However, for other tested data-driven energy forecasting models, including Kriging, Support Vector Regression, Radial Basis Function, Multivariate Adaptive Regression Splines, Artificial Neural Network and Polynomial Regression, training data obtained from the excitation scheme used in this study can increase the model extendibility. But the overall model accuracy is not improved. In other words, compared with excited operation data generated using the theory of system identification, normal operation data are more suitable for model training in black-box modeling if not considering extendibility.

Future work could further study and compare excitation from different theories, or design specific excitation algorithm that is suitable for black-box models, to increase black-box model accuracy as well as extendibility.

REFERENCES

Abdi, H., & Williams, L. J. (2010). Principal component analysis. Wiley Interdisciplinary Reviews: Computational Statistics, 2(4), 433-459.

Ahmad, M. W., Mourshed, M., Mundow, D., Sisinni, M., & Rezgui, Y. (2016). Building energy metering and environmental monitoring–A state-of-the-art review and directions for future research. Energy and Buildings, 120, 85-102.

Bacher, P., & Madsen, H. (2011). Identifying suitable models for the heat dynamics of buildings. Energy and Buildings, 43(7), 1511-1522.

Braun, M. W., Ortiz-Mojica, R., & Rivera, D. E. (2002). Application of minimum crest factor multisinusoidal signals for "plant-friendly" identification of nonlinear process systems. Control Engineering Practice, 10(3), 301-313.

Cai, J., Kim, D., Braun, J. E., & Hu, J. (2016). Optimizing Zone Temperature Setpoint Excitation to Minimize Training Data for Data-driven Dynamic Building Models. In American control conference, submitted, Boston, July (pp. 6-8).

Che, J., & Wang, J. (2014). Short-term load forecasting using a kernel-based support vector regression combination model. Applied Energy, 132, 602-609.
Dong, B., Cao, C., & Lee, S. E. (2005). Applying support vector machines to predict building energy consumption in tropical region. Energy and Buildings, 37(5), 545-553.
Erhan, D., Bengio, Y., Courville, A., Manzagol, P. A., Vincent, P., & Bengio, S. (2010). Why does unsupervised pre-training help deep learning?. The Journal of Machine Learning Research, 11, 625-660.
Fan, C., Xiao, F., & Wang, S. (2014). Development of prediction models for next-day building energy consumption and peak power demand using data mining techniques. Applied Energy, 127, 1-10.
Härdle, W. K., & Simar, L. (2012). Canonical correlation analysis. In Applied Multivariate Statistical Analysis (pp. 385-395). Springer Berlin Heidelberg.
Jurado, S., Nebot, À., Mugica, F., & Avellana, N. (2015). Hybrid methodologies for electricity load forecasting: Entropy-based feature selection with machine learning and soft computing techniques. Energy, 86, 276-291.
Kusiak, A., Li, M., & Zhang, Z. (2010). A data-driven approach for steam load prediction in buildings. Applied Energy, 87(3), 925-933.
Lara, B. G. V., Molina, L. M. C., & Yanes, J. P. M. (2015). Modeling and identification of the cooling dynamics of a tropical island hotel. Energy and Buildings, 92, 19-28.
Li, X., & Wen, J. (2014). Building energy consumption on-line forecasting using physics based system identification. Energy and Buildings, 82, 1-12.
Li, X., & Wen, J. (2014). Review of building energy modeling for control and operation. Renewable and Sustainable Energy Reviews, 37, 517-537.
Li, X., Wen, J., & Bai, E. W. (2016). Developing a whole building cooling energy forecasting model for on-line operation optimization using proactive system identification. Applied Energy, 164, 69-88.
Ljung, L. (1998). System identification (pp. 163-173). Birkhäuser Boston.
Ma, J., Qin, J., Salsbury, T., & Xu, P. (2012). Demand reduction in building energy systems based on economic model predictive control. Chemical Engineering Science, 67(1), 92-100.
Menezes, A. C., Cripps, A., Bouchlaghem, D., & Buswell, R. (2012). Predicted vs. actual energy performance of non-domestic buildings: Using post-occupancy evaluation data to reduce the performance gap. Applied Energy, 97, 355-364.
Newsham, G. R., & Birt, B. J. (2010, November). Building-level occupancy data to improve ARIMA-based electricity use forecasts. In Proceedings of the 2nd ACM workshop on embedded sensing systems for energy-efficiency in building (pp. 13-18). ACM.
Pasa, L., Testolin, A., & Sperduti, A. (2015). Neural Networks for Sequential Data: a Pre-training Approach based on Hidden Markov Models. Neurocomputing, 169, 323-333.
Prívara, S., VáHa, Z., Gyalistras, D., Cigler, J., Sagerschnig, C., Morari, M., & Ferkl, L. (2011, September). Modeling and identification of a large multi-zone office building. In Control Applications (CCA), 2011 IEEE International Conference on (pp. 55-60). IEEE.
Richardson, I., Thomson, M., & Infield, D. (2008). A high-resolution domestic building occupancy model for energy demand simulations. Energy and buildings, 40(8), 1560-1566.
Schoukens, J., Guillaume, P., & Pintelon, R. (1993). Perturbation signals for system identification. Prentice Hall, New York.
Terés-Zubiaga, J., Escudero, C., García-Gafaro, C., & Sala, J. M. (2015). Methodology for evaluating the energy renovation effects on the thermal performance of social housing buildings: Monitoring study and grey box model development. Energy and Buildings, 102, 390-405.
Xuemei, L., Lixing, D., Jinhu, L., Gang, X., & Jibin, L. (2010, January). A novel hybrid approach of KPCA and SVM for building cooling load prediction. In Knowledge Discovery and Data Mining, 2010. WKDD'10. Third International Conference on. IEEE.
Yang, J., Rivard, H., & Zmeureanu, R. (2005). On-line building energy prediction using adaptive artificial neural networks. Energy and buildings, 37(12), 1250-1259.

ACKNOWLEDGEMENT

This research was supported by project: CPS: Synergy: Collaborative Research: GOALI: SMARTER - Smart Manager for Adaptive and Real-Time decisions in building clustERs. It was also supported by U.S. Department of Energy's Building Grid Integration Research and Development Innovators Program: VOLTTRON Compatible Short-term Transactive Load Control Strategy.

Load and Electricity Rates Prediction for Building Wide Optimization Applications

Mohammad N. ELBSAT[1]*, Michael J. WENZEL[1]

[1]Technology and Advanced Development, Building Efficiency, Johnson Controls, Inc.
Milwaukee, WI, United States of America
(mohammad.elbsat@jci.com*, mike.wenzel@jci.com)

ABSTRACT

A method for predicting building loads, such as hot water load, and utility rates, such as electricity rates, over a horizon into the future is presented. Building loads and utility rates prediction find application in several energy consumption fields, including, but not limited to, optimal control of central energy plants, peak load shaving, peak load shifting, etc.... In this work, the predicted load is characterized by deterministic and stochastic terms. The deterministic term takes into consideration the factors affecting a building load and a utility rate, such as outside air enthalpy. On the other hand, the stochastic term allows for the prediction of future errors using current and past prediction errors. The prediction method developed employs various techniques including clustering analysis, linear regression modeling, auto-regressive modeling, and curve-fitting. The approach taken has been applied to historical cold water, hot water, and electric load and electricity rates data. Simulation results show the feasibility of the method developed.

1. INTRODUCTION

The reduction of energy consumption, use of renewable energy, and preservation of natural resources are becoming increasingly important. Several applications in the energy efficiency field aim at minimizing energy consumption and/or cost. To achieve this, these applications employ optimization techniques that require future prediction of the performance and various loads of a facility, campus, building, or an energy plant, such as hot water, cold water, and/or electric load. The prediction horizon may be as short as few hours to ten days into the future, depending on the application at hand. Wenzel *et al.* (2014) implement a method for optimizing a central plant using model predictive control. That optimization approach makes use of campus load predictions over a horizon for determining the dispatch schedule of the plant. Furthermore, the minimization of utility cost requires an accurate prediction of utility rates, such as electricity rates, over a given horizon. Therefore, a method for predicting building loads and utility rates over a given horizon into the future has been developed.

Building load and utility rates prediction has been addressed in the literature in various types of applications. Seem and Braun (1991) implemented an adaptive method for real-time forecasting of the electrical demand of a building. Seem's method is based on the cerebellar model articulation controller. Nielsen and Madsen (2006) took a grey-box approach in the development of a model linking the heat consumption to climate and calendar information. Catalina and Iordache (2013) developed a method for the fast prediction of heating energy demand using multiple regression models with the building global heat loss coefficient, the south equivalent surface, and the difference between the indoor set point temperature and the sol-air temperature as predictor variables. Most recently, Platon *et al.* (2015) addressed the hourly prediction of a building's electricity consumption using case-based reasoning, artificial neural networks and principal component analysis. In this paper, the building loads and utility rates prediction method developed makes use of clustering analysis, curve-fitting, linear regression modeling, and auto-regressive modeling.

As indicated in Seem and Braun (1991) and other publications in the literature, building loads and utility rates are influenced by several factors such as weather, day of week, time of day, operation schedules, etc.... It is imperative for any real-time load prediction method to account for as many of these factors, if not all. In this work, the method developed takes into consideration the several factors contributing to building loads and utility rates. The load predicted consists of a deterministic term and a stochastic term. The deterministic term is calculated using linear regression models, whose coefficients are determined offline. These models rely on the typical load value for a given time of day and day-type (days with similar load profiles) and weather forecast. The latter is obtained from the National Oceanic and Atmospheric Administration (NOAA) through their National Digital Forecast Database

(NDFD) service. The stochastic term is determined using an Auto-Regressive (AR) model, whose coefficients are also determined offline. The AR model calculates future prediction errors based on the current prediction error. The stochastic term of the predicted load gives the method developed its adaptive property, and thus increases the accuracy of the prediction by updating the forecast using current measurements of the load. The coefficients of the models used in determining both, the deterministic and stochastic terms, are calculated offline using historical weather, building loads, and utility rates data. The model coefficients are updated in real-time adaptively using most recent training data.

For a given set of training data, the method developed generates a set of regression models for each day-type. Day-types are determined by a day-typing algorithm which specifies days with similar load profiles based on cluster analysis techniques. Outside air enthalpy and a typical load profile constitute the predictors variables in each set of regression models. Each day-type is characterized by a different typical load profile which is generated using an optimal data fitting technique. The AR model coefficients are determined using the residuals obtained from different sets of regression models. Given the determined models, the current load measurement, and weather forecast, the future load values are calculated by selecting the appropriate regression model and summing the deterministic and stochastic terms.

This paper is divided as follows. Section 2 provides a high-level description of the load prediction method. Section 3 introduces the techniques used to generate the pertinent linear regression models and the auto-regressive model offline. Section 4 introduces the approach taken to generate the predicted load online at any given time instant using the models generated offline. Simulation results are presented in Section 5. The paper is then concluded in Section 6.

2. High Level Description of the Load Prediction Method

The load prediction method developed generates an hourly load forecast at any time instant over a predetermined horizon into the future. The load forecast generated at time instant n consists of a deterministic term, $\hat{L}_{det}(n)$, and a stochastic term, $\hat{L}_{sto}(n)$, as shown in Equation (1). The deterministic term accounts for the factors affecting the load such as time of day, day of week, weather variables, and operation schedules. Weather variables may include Outside Air Temperature (OAT), Outside Air Humidity (OAH), Outside Air Enthalpy (OAE), Cooling/Heating Degree Days (CDD), and/or Cooling/Heating Energy Days. In this paper, the weather variables used for describing the prediction method are OAT and OAH. Operation schedules may refer to calendar events such as federal holidays, manufacturing schedules of an industrial facility, in-session dates of an Academic institution, etc....The deterministic term is generated using linear regression models, whose coefficients are determined offline. It is dependent on the typical load value for a given time of day and a given day of week. It is also dependent on weather variables such as OAT and OAH. Forecasts of OAH and OAT are available from NOAA through their NDFD service.

The stochastic term acounts for any factors the deterministic term is not able to capture or errors in load measurements. It is generated using an AR model, whose coefficients are initially determined offline and adaptively updated online. At any current time instant and given that current load measurements are available, a current load error or prediction error (residual) is calculated and future prediction error values are iteratively determined using the AR model. The residual forecast constitute the stochastic term of the load forecast. The stochastic term gives the load prediction method its adaptive property. An auto-regressive model, which is shown in more detail in the following section, determines future values of a time-series using its present and past values. Thus, a trend depicted by the past values of the time series, which in this case is the prediction errors or residuals, will affect what the predicted future value is. In the next section, the offline determination of the linear regression models coefficients and the AR model coefficients is presented.

$$load_{forecast}(n) = \hat{L}_{det}(n) + \hat{L}_{sto}(n) \tag{1}$$

3. Offline Determination of Model Coefficients

As mentioned earlier, several factors contribute to the value of a load over a predetermined horizon into the future. These factors are time of day, day of week, weather variables such as OAT and OAH, holidays, and schedules, which are specific to a given facility, building, or campus (for a university, for example, these are in-session and out-of-session schedules). The load prediction method developed captures these factors using historical load and weather data and clustering and modeling methods. The historical data is used offline as training data for determining day-types and the corresponding model coefficients necessary for generating a load forecast online. The offline determination of model coefficients consists of several steps. A high-level flow of the latter is shown in Figure 1.

The training data is first divided into n_{groups} groups based on the given schedules. Then, the data from each group is passed through the day-typing algorithm. The latter determines which days of the week have similar load profiles. Then, the training data of each group is divided into $n_{cluster}$ sub-groups based on the outcome of day-typing. Thus, there will be $n_{group} \times n_{cluster}$ sets of training data, to each of which a set of models and coefficients corresponds. For each set, in order to capture the effect of the time of day, an optimization technique is used to find the optimal typical load profile. The effect of weather variables is captured through linear regression techniques, where a model representing the deterministic term of the load for each set as a function of the weather variables and a typical load profile is determined. In this work, Outside Air Enthalpy (OAE) is used to capture the effect of both OAT and OAH, simultaneously.

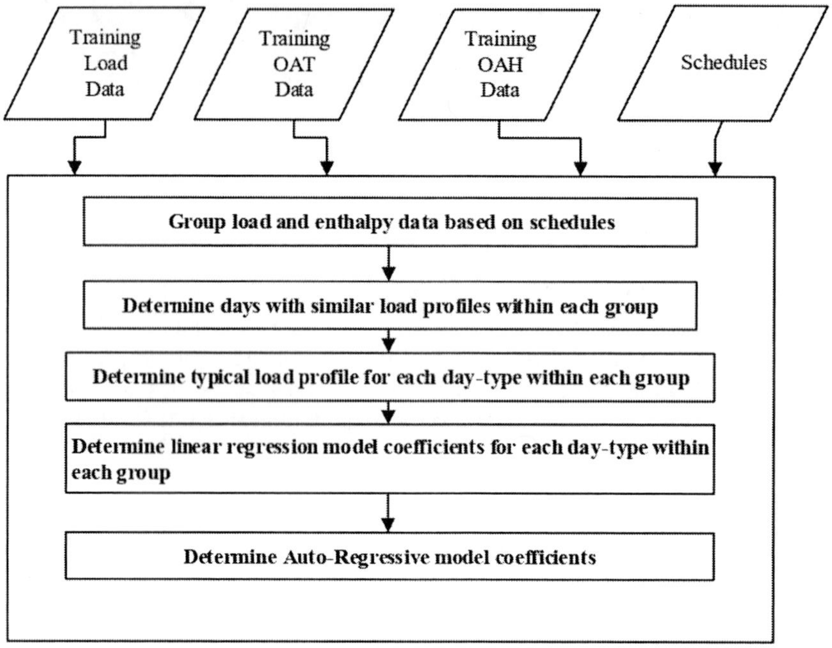

Figure 1: High-level flow of offline determination of model coefficients

After determining all the necessary coefficients pertaining to each set of data, the residuals determined by subtracting the predicted or estimated load from the actual load of all sets are used to train the AR model and determine its coefficients. Thus, for any given problem, there will be $n_{group} \times n_{cluster}$ day-type fits (typical load profiles), $n_{group} \times n_{cluster}$ regression models, and an AR model. These fits and models are, then, used online when generating load forecasts over a predetermined horizon in the future. The next subsections provide in more detail a description of the aforementioned steps of the offline determination of the model coefficients.

3.1 Day-Typing

Day-typing is a pattern recognition algorithm aimed at determining days of the week with similar load profiles. Days having similar load profiles are referred to as one day-type. Building electric loads, for example, tend to have two day-types: weekday and weekend. Figure 2 shows an example of the electric load over several weeks for the Stanford University campus in 2011. As can be observed, the load over the weekend significantly differs from that over the week. Even though it is obvious graphically that there are two day-types, it is not usually the case. This is why a pattern recognition algorithm was developed by Seem (2005), which statistically determines days with similar load profiles. Day-typing captures the effect of day of the week on a load forecast, and thus, it is essential for generating load forecasts. A high-level description of the pattern recognition algorithm developed by Seem (2005) is shown in Figure 3. Given a time series of load data, daily features are generated. The two features considered by Seem are the average daily load and the peak daily load over a one hour period. These features are then transformed in order to alleviate the effect of seasonal load changes. After transforming the feature vectors, seven clusters are created, where each cluster corresponds to each day of the week. Univariate and multivariate outlier identification procedures are then applied to each cluster in order to remove any abnormal or unusual feature vectors in each cluster.

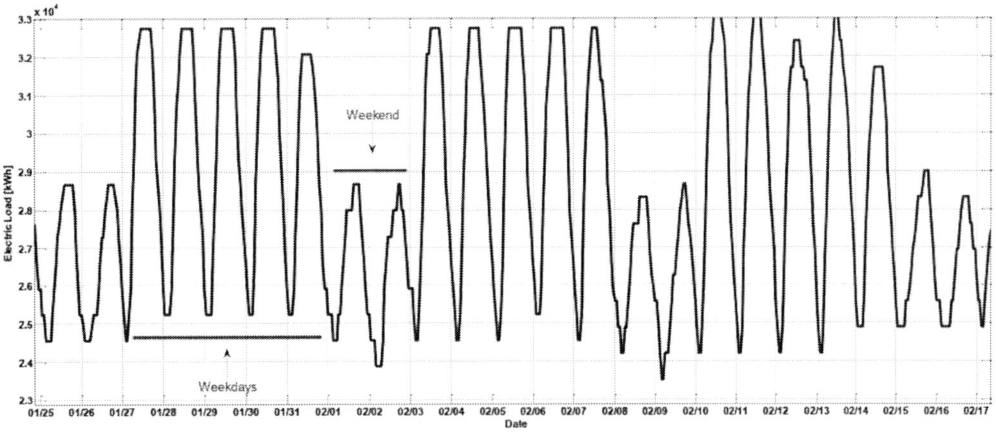

Figure 2: Example of Day-Types

Finally, days of the weeks with similar load profiles are determined using clustering analysis. For more details on the pattern recognition algorithm, refer to Seem (2005).

In this work, the algorithm developed by Seem is used to generate the day-types of a given load. The day-types are used to divide the different groups of training data (historical load and OAE data) into sub-groups, where each sub-group corresponds to a particular day-type. The load data in each sub-group is then passed through a data-fitting method, which finds the optimal load fit for a particular day-type. More details on determining the day-type load fit is presented next.

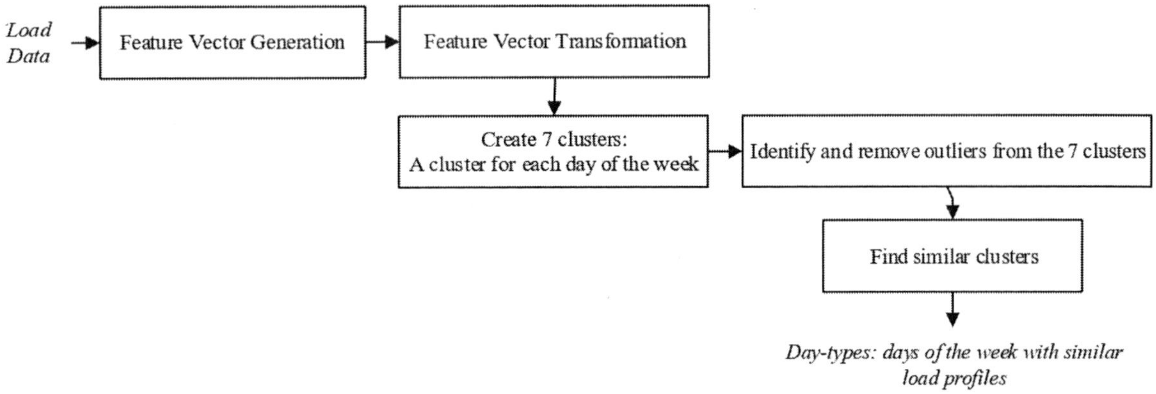

Figure 3: Major steps in pattern recognition algorithm for determining days of the week with similar load profiles. Seem (2005)

3.2 Data Fitting

After dividing the data into groups based on schedules and then into sub-groups based on day-typing results, the result is $n_{group} \times n_{cluster}$ sets of training data, where the days in each set have similar load profiles. The task now is to capture the effect of the time of day factor. Figure 4 shows an example of a sub-group of cold water load training data. The sub-group contains cold water load data of days with similar load profiles. As can be observed, the load profile has a bell curve shape, where the highest consumption takes place in the early to late afternoon. The objective is to find an optimal load data fit for each sub-group corresponding to a particular day-type that will best represent the load profile in that sub-group. This will be called a day-type fit from here on. A day-type fit is dependent on the time of day, therefore there is a need to find a mechanism that allows for calculating the day-type fit at any time of day. In order to achieve an optimal day-type fit for a sub-group and to have the capability of calculating the fit for any time of day, a spline is used to represent the day-type fit, where the spline knots are optimized. A spline is a piecewise-defined function of polynomials. A spline is defined by the knots where these polynomials connect. Predetermining the location of these knots (i.e. the times of day at which these knots are considered), an optimal value of these knots is found using an optimization technique. The location of the knots is fixed and the locations chosen for load prediction are the abscissae of the bold blue circles shown in Figure 4. These locations were chosen as to best capture the general shape of the load profile, especially where there is a change in the curvature. The objective function to be minimized is defined as follows:

$$J = \min_{\mathbf{y}_s} \left(\sum_{i=1}^{m} (\mathbf{y}_i - \hat{\mathbf{y}})^T (\mathbf{y}_i - \hat{\mathbf{y}}) \right) \tag{2}$$

where $\mathbf{y}_s$ is the vector of spline knots, $\mathbf{y}_i$ is the vector of actual load values for the ith full day in a sub-group, m is the total number of full days in a sub-group, and $\hat{\mathbf{y}}$ is the estimated day-type fit of a sub-group.

In this work, the MATLAB 2013b optimization function *fminunc* with the Quasi-Newton optimization algorithm and the function *spline* were used. Figure 4 shows the optimal value $\mathbf{y}_s$ obtained for the example sub-group shown. Thus, a unique day-type fit corresponds to each sub-group. The day-type fit is represented by a set of cubic-splines, which are used to calculate the load value, $L_{fit}(t)$, for a given time of day, t. In the next section, the linear regression models, which make use of the day-type fit discussed in this section, are presented.

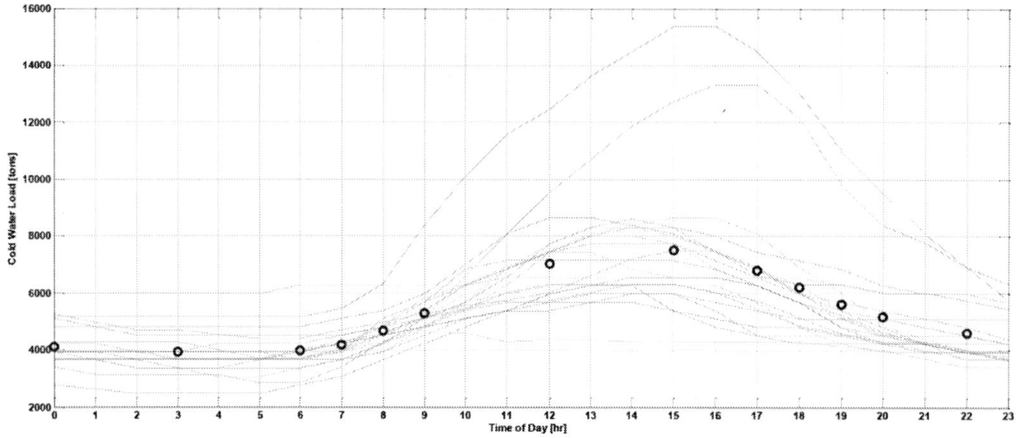

Figure 4: Daily cold water loads of the same day-type

3.3 Linear Regression Models

The remaining factor to be accounted for in calculating the deterministic term $\hat{L}_{det}(t)$ of the load value is the weather. Loads are dependents on OAT and OAH. The effect of OAT and OAH can be accounted for simultaneously by considering OAE. As shown in Figure 5, for example, hot water load is highly correlated to OAE. The same can be said about other loads such as cold water and electric loads. Linear regression modeling (Weisberg, 1985) is implemented to calculate the load value as a function of enthalpy. It is worth noting that instead of using OAE as a predictor variable, degree days such as CDD and HDD may also be used and may improve the performance of load prediction as observed from Figure 5.

For each of the $n_{group} \times n_{cluster}$ sub-groups of training load data, a unique day-type fit exists. The deterministic term of the load value is dependent on both the day-type fit and OAE. In addition, OAE is also correlated to the day-type fit. Therefore, two regression models are necessary for calculating an estimated value of the load. The first regression model has OAE as the response or dependent variable and the day-type fit of a given sub-group as the predictor or independent variable as shown in Equation (3). The objective of this regression is to orthogonalize the predictor variables $\mathbf{OAE}$ and $\hat{\mathbf{L}}fit$. Orthogonalization of predictor variables eliminates the correlation between them.

$$\mathbf{OAE}_k = \beta_0^k + \beta_1^k \hat{\mathbf{L}}fit_k + \mathbf{w} \tag{3}$$

where β_0^k and β_1^k are the model coefficients of the kth sub-group, $\mathbf{OAE}_k$ is the vector of enthalpy observations in the kth sub-group , $\hat{\mathbf{L}}fit_k$ is the vector of day-type fit observations corresponding to those in $\mathbf{OAE}_k$, $\mathbf{w}$ is the model error, and $k = 1 \cdots n_{groups} \times n_{clusters}$. Least Squares Regression is used to estimate the value of the model coefficients. The model coefficient estimates and are used to calculate the estimated enthalpy as shown in Equation (4).

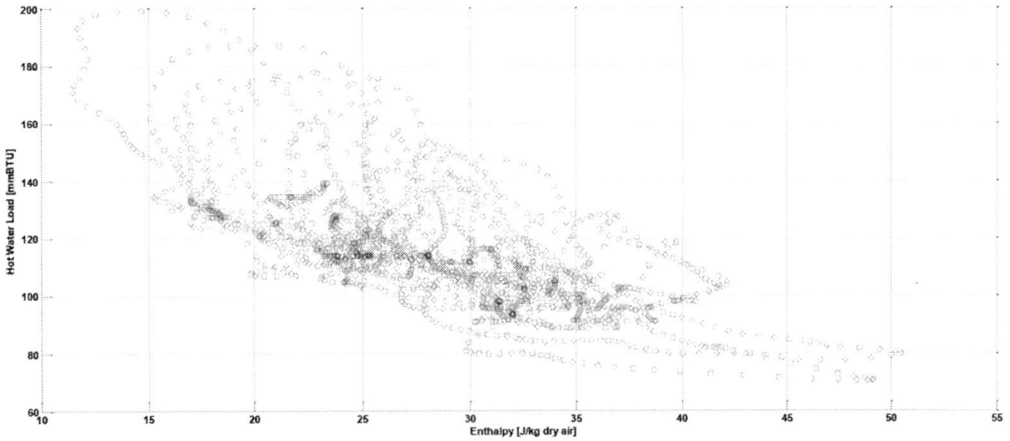

Figure 5: Example of the correlation between load and outside air enthalpy

$$\hat{\mathbf{OAE}}_k = \hat{\beta}_0^k + \hat{\beta}_1^k \hat{\mathbf{L}}fit_k \tag{4}$$

The actual load data $\mathbf{Load}_k$ is then regressed on $\hat{\mathbf{L}}fit_k$ and $\mathbf{OAE}_k - \hat{\mathbf{OAE}}_k$ as follows:

$$\mathbf{Load}_k = \alpha_0^k + \alpha_1^k \hat{\mathbf{L}}fit_k + \alpha_2^k (\mathbf{OAE}_k - \hat{\mathbf{OAE}}_k) + \mathbf{e} \tag{5}$$

where α_0^k , α_1^k , and α_2^k are the model coefficients of the kth sub-group, $\mathbf{OAE}_k - \hat{\mathbf{OAE}}_k$ is the vector of enthalpy residuals, $\mathbf{e}$ is the model error, and $k = 1 \cdots n_{groups} \times n_{clusters}$. Least Squares Regression is used to estimate the value of the model coefficients. The model coefficient estimates $\hat{\alpha}_0^k$, $\hat{\alpha}_1^k$, and $\hat{\alpha}_2^k$ are used to calculate the estimated deterministic term of the load $\hat{\mathbf{L}}_{det_k}$ as shown in Equation (6).

$$\hat{\mathbf{L}}_{det_k} = \hat{\alpha}_0^k + \hat{\alpha}_1^k \hat{\mathbf{L}}fit_k + \hat{\alpha}_2^k (\mathbf{OAE}_k - \hat{\mathbf{OAE}}_k) \tag{6}$$

Each sub-group, then, is characterized by a set of coefficients $\hat{\beta}_0^k$, $\hat{\beta}_1^k$, $\hat{\alpha}_0^k$, $\hat{\alpha}_1^k$, and $\hat{\alpha}_2^k$ and a spline used for calculating the optimal day-type fit corresponding to a sub-group. These coefficients and spline are then used online when generating the load forecast as will be shown later.
As mentioned earlier, the load forecast consists of two terms, a deterministic term and a stochastic term. In the following section, the offline determination of the AR model coefficients, which calculates the stochastic term of the load forecast, is presented.

3.4 Auto-Regressive Models

An auto-regressive model is a model representing a time series and has the following form (Haykin, 2002):

$$r(n)+a_1 r(n-1)+\cdots+a_2 r(n-p)=v(n) \tag{7}$$

where $r(n)$ is the value of the time series at instant n , $a_1,a_2,\ldots,a_p$ are the AR parameters, and $v(n)$ is white noise.
An AR model is basically a regression model of the current value of a time series on its past values. An AR model can be used to predict the value of a time series over a horizon of length h into the future iteratively as follows:

$$\hat{r}(n+j)=-\sum_{i=1}^{p}\hat{a}_i r(n+j-i) \qquad \forall j=1,\ldots,h \tag{8}$$

where $\hat{a}_i$ are estimates of the AR parameters a_i which are determined offline using training data. For more information on AR models refer to Chapter 1 in Haykin (2002).

For the purpose of load prediction, an AR model is used to calculate the stochastic term of the load forecast over a horizon into the future. The time series r , in this case, is the load prediction error or residuals obtained from all the sub-groups and sorted in chronological order. The unsorted AR model training residual data is shown in Equation (9). $r_{unsorted}$ is sorted in chronological order before it is used as training data for estimating the AR parameters. Since an AR model is characterized by its order p and its parameters, an appropriate model order needs to be chosen along with determining estimates of its parameters. In order to determine the order of the model, Akaike's Information Criterion (AIC) may be used along the MATLAB function *arx*, which requires having the System Identification Toolbox. The order of the AR model determines how many past values of the time series are needed. Note that for the Stanford project, the AR model is of order 1 and a least-square estimate of the AR parameter is used. The latter was solution to the fact that the *arx* function cannot be compiled when generating an exe or a kernel.

$$r_{unsorted}=\begin{bmatrix} \mathbf{Load}_1-\hat{\mathbf{L}}_{det_1} \\ \vdots \\ \mathbf{Load}_{n_{groups}\times n_{clusters}}-\hat{\mathbf{L}}_{det_{n_{groups}\times n_{clusters}}} \end{bmatrix} \tag{9}$$

Thus, for a given problem, the offline determination of model coefficients generates, using load training data, OAT and OAH training data, and schedules, a set of day-types, splines, estimates of linear model regression coefficients, and estimates of AR model parameters. These are then used online when generating load forecasts. The next section explains how a load forecast is generated online using the different models developed offline.

4. Online Load Prediction

Online load prediction requires a current load measurement, the current and forecast OAT and OAH, the current date and time, the length of the horizon in hours, the model coefficients, splines, and day-types generated offline, a schedule, and a vector of residuals . The vector of residuals is a persistent vector as it is also an output of the online load predictor. The size of the residual vector is dependent on the order of the AR model generated offline and contains the past residual values necessary for calculating the stochastic term of the load forecast.

For a given problem, day-types, day-type fits (splines), estimates of regressive model coefficients, and estimates of AR model parameters are generated offline. These are then used online to generate a load forecast over a predetermined horizon. Figure 6 shows a flow of the major steps in generating a load forecast online at a given instant in time. Given the current and forecast OAT and OAH obtained from NOAA, the current and forecast enthalpy vector $\mathbf{OAE}_{forecast} = [OAE(1) \;\; \ldots \;\; OAE(h)]$is calculated, where h is the length of the horizon in hours. Each data point is always associated with its corresponding date and time stamp. The day-types and schedules are used to determine which spline and coefficients (These will be referred to from here on as models) are to be used for each data point. Then, the day-type fit value for each data point is calculated using the corresponding splines. This is followed by the calculation of the OAE estimates, $\mathbf{O\hat{A}E}_{forecast} = \left[O\hat{A}E(1) \;\; \ldots \;\; O\hat{A}E(h) \right]$ as shown in Equation (10).

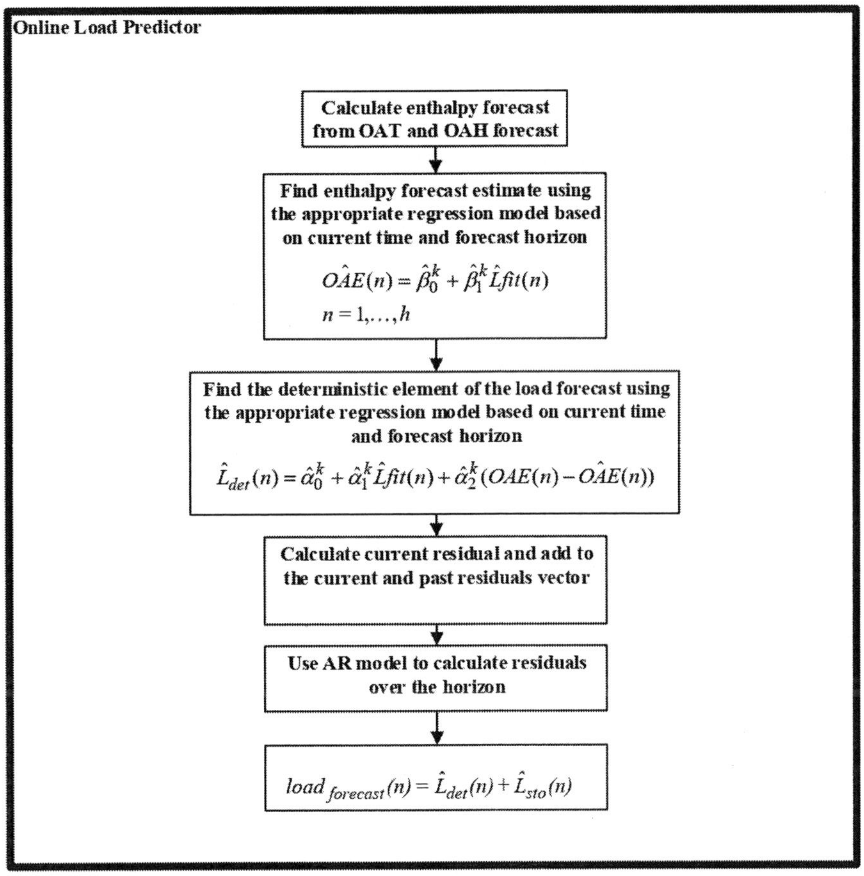

Figure 6: Major steps in online load prediction

$$O\hat{A}E(n) = \hat{\beta}_0^k + \hat{\beta}_1^k \hat{L}fit(n) \qquad n = 1,\ldots,h \tag{10}$$

Then, the deterministic term of the load forecast $\hat{\mathbf{L}}_{det} = \left[\hat{L}_{det}(1) \;\; \ldots \;\; \hat{L}_{det}(h) \right]$ is calculated as shown in Equation (11).

$$\hat{L}_{det}(n) = \hat{\alpha}_0^k + \hat{\alpha}_1^k \hat{L}fit(n) + \hat{\alpha}_2^k (OAE(n) - O\hat{A}E(n)) \qquad n = 1,\ldots,h \tag{11}$$

The current residual is then calculated using the current load measurement as shown in Equation (12) and the stochastic term vector is generated using the AR model shown in Equation (8). Finally, the load forecast is the sum of the deterministic and stochastic terms.

$$r(1) = currenLoad - \hat{L}_{det}(1) \tag{12}$$

5. SIMULATIONS

In this section, simulations of the load prediction method are shown. Figures 7, 8, 9, and 10 show the forecasts for the hot water load, cold water load, electric load, and electricity rates respectively. The models generated offline uses training data corresponding to the month of March. The simulations shown are at the end of day April 1st and the forecast horizon length is 240 hrs. The simulations show that the load prediction method is able to generate load forecasts that are close to the actual loads.

CONCLUSION

A load prediction method capable of generating accurate load forecasts is presented. The method developed is applicable to cold water load, hot water load, electric load, and electricity rates. It takes into consideration several factors affecting a load. These factors consist of schedules, time of day, day of week, and weather variables. The method generates a load forecasts consisting of deterministic and stochastic terms. It makes use of day-typing, regression modeling, and auto-regressive modeling techniques. The method developed is characterized by an adaptive property due to the stochastic term in the generated load forecast. Further improvements to the method is the use of cooling degree days and/or heating degree days instead of outside air enthalpy.

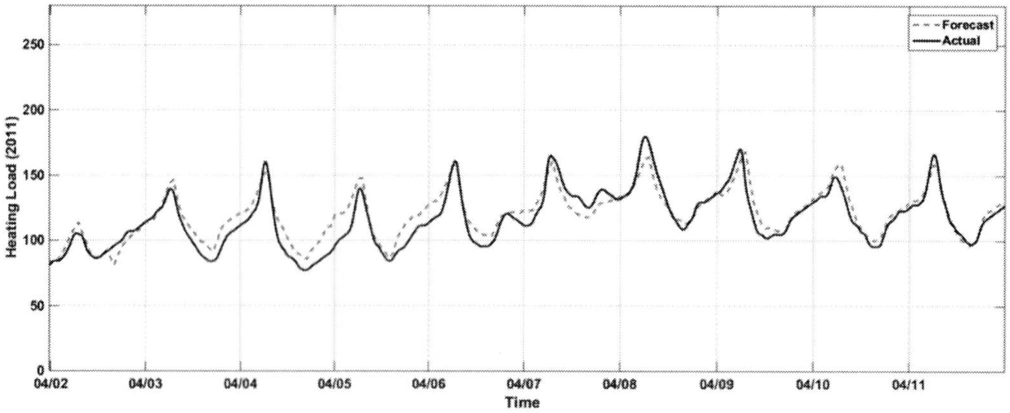

Figure 7: Example of a 10 day-forecast of hot water load

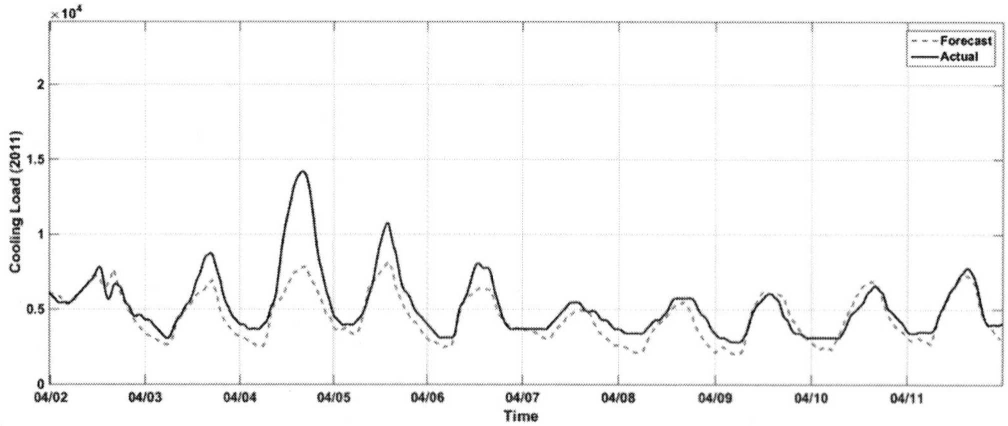

Figure 8: Example of a 10 day-forecast of cold water load

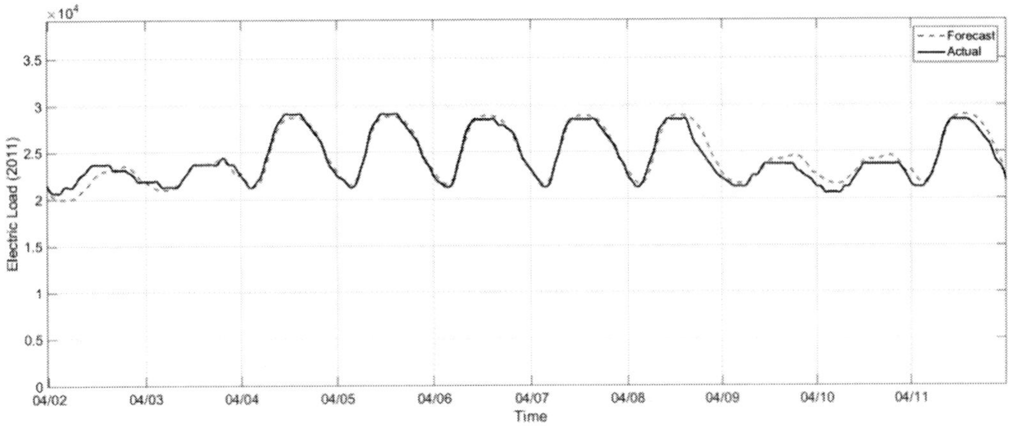

Figure 9: Example of a 10 day-forecast of electrical load

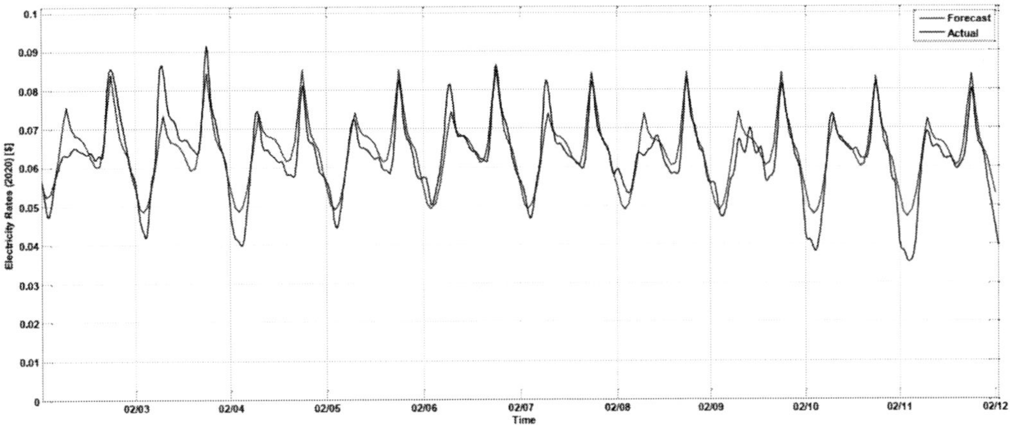

Figure 10: Example of a 10-day forecast of electricity rates

REFERENCES

Catalina, T. & Iordache, V. (2013). Multiple regression model for fast prediction of the heating energy demand. Energy and Buildings. 57. 302 – 312.

Haykin, S. (2002). *Adaptive Filter Theory, 4th Edition* .New Jersey, Prentice Hall.

Nielsen, H. A. & Madsen, H. (2006). Modelling the heat consumption in district heating systems using a grey-box approach. *Energy and Buildings*, 38(1). 63-71

Platon, R., Dehkordi, V. R., & Martel, J. (2015). Hourly prediction of a building's electricity consumption using case-based reasoning, artificial neural networks and principal component analysis. *Energy and Buildings*, *92*. 10 – 15.

Seem, J. E. (2005). Pattern recognition algorithm for determining days of the week with similar energy consumption profiles. *Energy and Buildings*, *37*, 127-139.

Seem, J. E. & Braun, J. E. (1991). Adaptive methods for real-time forecasting of building electrical demand. *ASHRAE Transactions*, *97*(1).

Weisberg, S. (1985). *Applied Linear Regression, 2nd Edition*. New York, NY, John Wiley & Sons, Inc.

Wenzel, M. J., Turney, R. D., & Drees, K. H. (2014). Model predictive control for central plant optimization with thermal energy storage. *Proceedings of the 3rd International High Performance Buildings Conference,* West Lafayette, IN, Paper 122.

Zhao, H. & Magoules, F. (2012). A review on the prediction of building energy consumption. 16(6). 3586 – 3592.

Evaluation of an Extremum Seeking Control Based Optimization and Sequencing Strategy for Multiple-chiller Chilled-water Plant

Zhongfan Zhao[1]*, Yaoyu Li[2], Baojie Mu[3], Timothy I. Salsbury[4], and John M. House[5]

[1]The University of Texas at Dallas, Department of Electrical Engineering
Richardson, Texas 75080, USA
E-mail: zhongfan.zhao@utdallas.edu

[2]The University of Texas at Dallas, Department of Mechanical Engineering
Richardson, TX 75080, USA
E-mail: yaoyu.li@utdallas.edu

[3]Whirlpool, Inc.
Benton Harbor, MI 49022, USA
E-mail: sdmubaojie@gmail.com

[4] Johnson Controls, Inc., Building Efficiency Research Group
Milwaukee, WI 53201, USA
E-mail: Timothy.I.Salsbury@jci.com

[5] Johnson Controls, Inc., Building Efficiency Research Group
Milwaukee, WI 53201, USA
E-mail: John.M.House@jci.com

ABSTRACT

The ventilation and air conditioning systems of large commercial buildings are typically driven by chilled-water plants with multiple chillers. Real-time optimization and sequencing of such plants is thus critical for building efficiency (AHRI 2011). Due to the complexity and cost for calibrating chiller plant in field operation, model-free control has become an attractive solution. In an earlier study, Mu *et al.* (2015) develop an Extremum Seeking Control (ESC) based real-time optimization and sequencing strategy for a chiller-water plant with parallel chillers. Two schemes were proposed. For Scheme A, a chiller is turned on based on the measurement of chilled water valve position and is turned off when a chiller compressor is running at its nominal minimum speed. For Scheme B, a chiller is turned on and off based on the measurement of operating cooling load. Mu (2015) conducts a comprehensive simulation study on Scheme A. In this study, a comprehensive study is conducted for Scheme B under various scenarios and testing conditions, and the simulation results are compared with those from Scheme A.

1. INTRODUCTION

Currently, the ventilation and air conditioning systems of most commercial and institutional buildings are driven by chilled water plants with multiple chillers in parallel or series configuration. In order to meet the building cooling load while achieving desirable energy efficiency, the individual chillers, water pumps of the evaporating and condensing loops, and cooling tower fans are controlled with multiple control loops and sequenced based on process measurements. Considering the high power consumption of chilled-water plants, optimal control and sequencing is thus critical for energy efficiency of commercial buildings. There have been extensive literatures for optimization and sequencing of multiple-chiller plants, including rule-based control (Yu *et al.* 2008, Benton *et al.* 2002, Ahn and Mitchell 2001, Bhattacharjee 2007, Hydeman *et al.* 2002, and Qureshi and Zubair 2006), and model based control (Deng *et al.* 2014, Zhu *et al.* 2013) strategies.

Due the high dimension, nonlinear, and time-varying nature of chilled-water system plant dynamics, it is quite involved and expensive to obtain the plant models as demanded by the model based control and optimization schemes. Therefore, model-free control and optimization approaches would be cost effective for practice. Artificial intelligence and data-driven modeling approaches have investigated (Chow *et al.* 2002, Chang *et al.* 2005, Ma and Wang 2011, Ardakani *et al.* 2008), e.g. genetic algorithm (GA), neural networks and particle swarm optimization (PSO). However, the GA and PSO approaches have inherent limitation in that they need large number of historic data or training samples for offline computation and their solutions and convergence are random.

Extremum Seeking Control (ESC) (Ariyur and Krstic 2003) has emerged as an attractive model-free real-time optimization strategy for HVAC systems (Li et al. 2010, Li et al. 2013, Dong et al. 2014, Hu et al. 2015). Recently, based on a Penalty-function based Extremum Seeking Control (ESC), Mu et al. (2015) propose a model-free real-time optimization and sequencing strategy for chilled-water plant with multiple chillers in parallel, as shown in Fig. 1, where the variable to be optimized is the total power from the chiller compressors, cooling tower fans, condenser and evaporator loop water pumps, while the manipulated inputs include the tower fan airflow, condenser water flows and evaporator leaving chilled-water temperature setpoint.

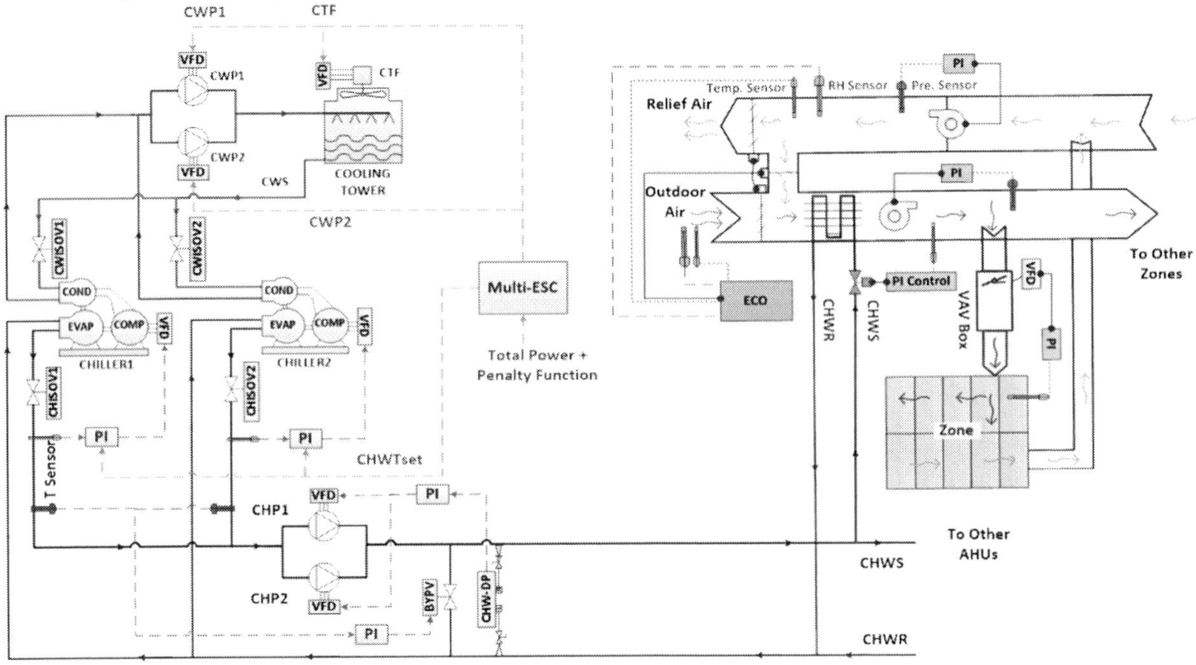

Figure 1: Schematic of a chilled-water plant with two chillers in parallel.

In particular, two schemes (Scheme A and Scheme B) are proposed for chiller sequencing. For Scheme A, a chiller is turned on based on the measurement of chilled water valve position and is turned off when a chiller compressor is running at its nominal minimum speed. For Scheme B, a chiller is turned on and off based on the measurement of operating cooling load. For Scheme A, Mu (2015) performed comprehensive case studies, and the simulation results have well justified the performance of the proposed framework under various ambient, load and equipment conditions. However, for Scheme B, only one scenario is simulated for idea proof purpose. Although the two chiller sequencing schemes have shared physical ground in terms of chiller plant operation, the load based Scheme B will be more appropriate for field implementation.

The objective of this study is to conduct comprehensive study on the performance of Scheme B based optimization and sequencing strategy for various ambient and load conditions. The remainder of this paper is structured as follows. Section 2 presents the details of Scheme A and Scheme B, and the simulation results are given in Section 3. The paper is concluded in Section 4, along with the plan for future work.

2. ESC BASED OPTIMAL CONTROL AND SEQUENCING FOR CHILLED-WATER PLANT WITH PARALLEL CHILLERS

The schematic of a chilled-water plant with *N* chillers in parallel is shown in Figure 2. Each chiller has its own compressor (COMP) regulating the evaporator leaving chilled water temperature (CHWT), a condenser water pump (CWP) adjusting the condenser water flow rate (CWF), and a chilled water pump (CHP) adjusting the supplying chilled water flow rate (CHF) to the cooling coils of the air-handling unit (AHU). As a simplified treatment, the condensing water from all chillers is assumed to share one cooling tower, and the cooling-tower air (CTA) flow is adjusted by a cooling tower fan (CTF).

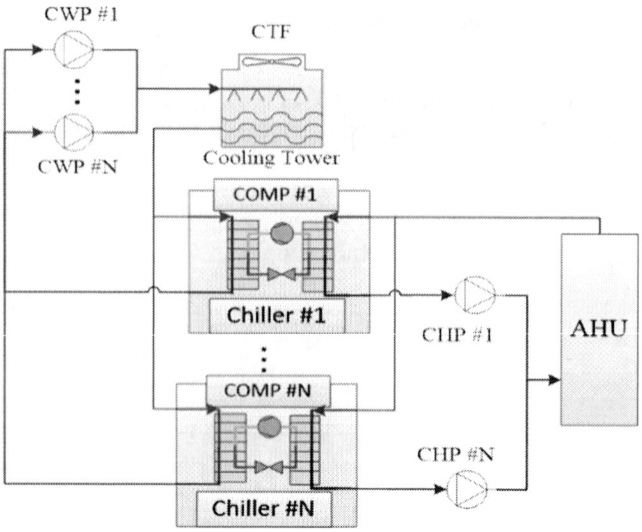

Figure 2: A chilled-water plant with N chillers in parallel.

Chiller plant optimization aims to minimize the total power consumption, provided that regulation of the AHU supply air temperature (SAT) is satisfactory and the suction superheat for the compressors of all operating chillers. The associated ESC aims to find the optimum input u such that

$$u^*(t) = \arg\min_u P_t(t) = \arg\min_u \left\{ P_{CTA}(t) + \sum_{i=1}^{n} \delta_i \left[P_{COMP,i}(t) + P_{CWF,i}(t) + P_{CHF,i}(t) \right] \right\} \quad (1)$$

with $u \triangleq [\dot{m}_{CTA}, \dot{m}_{CWF,1}, \dot{m}_{CHF,1}, T_{set,1}, \cdots, \dot{m}_{CWF,n}, \dot{m}_{CHF,n}, T_{set,n}]$, and the following constraints imposed:

$$\dot{m}_{CTA,\min} \le \dot{m}_{CTA} \le \dot{m}_{CTA,\max} \quad (2)$$

$$\dot{m}_{CWF,i,\min} \le \dot{m}_{CWF,i} \le \dot{m}_{CWF,i,\max} \quad (3)$$

$$\dot{m}_{CHF,i,\min} \le \dot{m}_{CHF,i} \le \dot{m}_{CHF,i,\max} \quad (4)$$

$$\Omega_{COMP,i,min} \le \Omega_{COMP,i} \le \Omega_{COMP,i,max} \quad (5)$$

$$T_{SH,i,\min} \le T_{SH,i} \le T_{SH,i,max} \quad (6)$$

with binary variables δ_i indicating ON or OFF operation for chiller i with value of 1 or 0. The controls inputs for ESC based chiller sequencing for a chilled-water plant with multiple chillers in parallel are selected as: cooling tower air flow rate, condenser water flow rate and leaving chilled water temperature of each chiller. As a simplified treatment, all chillers are assumed to have the same CHWT setpoint. To handle the constraints of plant inputs, the objective function is augmented with a penalty term in addition to the total power consumption, i.e. , $L_p(t) = P_t(t) + y_p(t)$ with respect to the regulation of $\dot{m}_{CTA}, \dot{m}_{CWF}$, and T_{set}, where T_{set} is the CHWT setpoint regulated by the compressor speed. $y_p(t)$ is the penalty function defined as:

$$\begin{aligned} y_p(t) = \sum_{i=1}^{n} \phi_i(\cdot) = & \frac{1}{2}\gamma_{CTA}(\dot{m}_{CTA} - \dot{m}_{CTA,sat})^2 + \frac{1}{2}\gamma_{CWF,i}(\dot{m}_{CWF,i} - \dot{m}_{CWF,i,sat})^2 + \frac{1}{2}\gamma_{COMP,i}(\Omega_{COMP,i} - \Omega_{COMP,i,sat})^2 \\ & + \frac{1}{2}\gamma_{CHF,i}(\dot{m}_{CHF,i} - \dot{m}_{CHF,i,sat})^2 + \frac{1}{2}\gamma_{SH,i}(T_{SH,i} - T_{SH,i,sat})^2 \end{aligned} \quad (7)$$

Subscript '*sat*' refers to the output of the limiter. In Eq. (9), parameters $\gamma_{CWF,i}$, γ_{COMP}, $\gamma_{CHF,i}$, $\gamma_{SH,i}$ and γ_{CTA} are weights of penalty terms for the saturation of $\dot{m}_{CWF,i}$, $\Omega_{COMP,i}$, $\dot{m}_{CHF,i}$, $T_{SH,i}$ and $\dot{m}_{CTA}$, respectively.

2.1 Scheme B for Chiller Sequencing

i) Turn on/off a chiller

The $(n+1)^{\text{-th}}$ chiller is turned on (i.e. $\delta_{n+1}=1$) if $\dot{Q}(t) > \sum_{i=1}^{n+1} \dot{Q}_{i,\min}$ for $t \in [t_0, t_0+\tau]$, where the chiller load is calculated by $\dot{Q} = c_p \dot{m}(T_r - T_s)$ based on the temperature measurements of the supply water and return water, and the supply water mass flow rates for all operating chillers. $\dot{Q}_{i,\min}$ is the nominal minimum chiller capacity for Chiller i based on manufacturer's specification or historical calibration data. Similarly, the $(n+1)^{\text{-th}}$ chiller is turned off (i.e. $\delta_{n+1}=0$) if $\dot{Q}(t) \le \sum_{i=1}^{n+1} \dot{Q}_{i,\min}$, for $t \in [t_0, t_0+\tau]$.

ii) Chiller startup and shutdown.

To turn on the $i^{\text{-th}}$ chiller, the chiller compressor speed is first increased from zero to the nominal minimum speed $\Omega_{COMP,i,\min}$ following a specific (startup) ramp. Then, the compressor speed is used to regulate the evaporator leaving chilled water temperature (CHWT) at its setpoint. Meanwhile, chilled water pump (CHP) is started from zero flow to the nominal minimum flow during the startup period following a specific ramp, and then the chilled water valve is used to regulate the AHU supply air temperature (SAT) setpoint. The condenser water flow rate (CWP) is started from zero flow to the pre-set value during the startup period following a specific ramp. To turn off a chiller, the chiller compressor speed is first disconnected from the CHWT control loop, and then decreased from the current speed to 0 following a specific (shutdown) ramp. Meanwhile, the CHP is slowed from its current speed to zero flow during shutdown. The CWP's are turned off to 0 from its previous values during shutdown.

iii) Enable/Disable ESC

Assume n chillers are in operation. During the startup period of the $(n+1)^{\text{-th}}$ chiller, the ESC for *CH*1 through *CHn* and the cooling tower is disabled, and corresponding ESC control inputs are frozen to their current values, while the condenser water flow rate of the $(n+1)^{\text{-th}}$ chiller will become a new control input for the ESC that includes CH1 through CH(n+1) and the cooling tower. After the startup process, the ESC with increased input dimension is started. Assume $(n+1)$ chillers are on, and the $(n+1)^{\text{-th}}$ chiller is to be shut down. The ESC for *CH*1 through *CHn* and cooling tower is disabled by freezing the control inputs to their current values, while the control inputs of condenser water flow rate (CWF) and the CHWT setpoint for the $(n+1)^{\text{-th}}$ chiller are frozen and disabled. After shutdown of the $(n+1)^{\text{-th}}$ chiller, the ESC for *CH*1 through *CHn* and cooling tower resumes its action, and the initial values of control inputs are the previously frozen values, while the control inputs for the $(n+1)^{\text{-th}}$ remain disabled until conditions in 1) are satisfied.

iv) Bumpless transfer between chiller startup/shutdown and ESC.

To achieve bumpless transfer between the chiller startup/ shutdown operation and the ESC, the conditional integration (CI) scheme (Peng *et al.* 1993) is adopted, see Mu et al. (2015) for details . The CI scheme switches the input of the integrator to zero, and disables the modulation and demodulation signals when ESC is disabled during the chiller startup/shutdown operation. Similarly, the CI scheme switches the input of the integrator to the output of LP filter, and enables the modulation and demodulation signals when ESC is engaged.

3. SIMULATION RESULTS

In order to give a comprehensive evaluation of Scheme B for ESC based chiller sequencing control, simulations are performed for the scenarios defined in Table 1. Three ambient conditions are considered: i) 27°C and 60 % RH ("Mild"), ii) 37°C and 30 %RH ("Dry Hot"), and iii) 37°C and 80 %RH ("Humid Hot"). For each of these ambient conditions, simulation analysis are made in Scenario#1, 2, 5, 6. Scenario#3 is simulated with dynamic ambient and load profiles. The simulations are carried out with the Modelica based dynamic simulation model for a variable primary flow (VPF) chilled-water plant developed by Mu (2015), which has two parallel chillers, one cooling tower, an AHU including a VAV box and a single zone.

Table 1: Simulation scenarios for each ambient condition

Scenario #	Description
1	Two-chiller ESC with no sequencing under fixed ambient conditions

2	Chiller sequencing under variable load and fixed ambient conditions
3	Chiller sequencing with realistic ambient and load profile
4	Penalty Function based ESC Chiller Sequencing
5	ESC for Efficiency Recovery: Chiller A properly charged and Chiller B with a low charge
6	ESC for Efficiency Recovery: Chiller A nominal + Chiller B with heat exchanger fouling

3.1 Simulation Case#1

The multivariate ESC is tested with two chillers in parallel with the fixed ambient conditions and fixed load. The internal heat gain is chosen as 350 kW. Table 2 gives the ESC design parameters. Fig. 3 shows the ESC inputs and out trajectories for the three ambient conditions. For all cases, the cooling tower air flow rate and chilled water leaving temperature start at 23kg/s and 8°C. The initial value of condenser water flow rate for both chillers are 25kg/s. The power consumption of the chilled-water plant are 217.4kW, 302.4kW and 335.9kW respectively. They decrease 12.4%, 6.4% and 6.1%. The superheat, supply-air temperature and zone temperature are well regulated about their respective setpoint of 5°C, 13°C and 25°C.

Table 2: ESC controller parameters for two chiller without sequencing

Parameter	**ESC1**	**ESC2**	**ESC3**	**ESC4**
Dither Freq.(rad/s)	0.005	0.0038	0.0013	0.0023
Dither Amp.	3 kg/s	3 kg/s	0.5℃	3 kg/s
Demodulation Amp.	2/3 kg/s	2/3 kg/s	4 ℃	2/3 kg/s
Demodulation Phase (rad)	0.15	0.1	0.18	0.05
ω_{BPL} (rad/s)	0.0045	0.0034	0.0012	0.0021
ω_{BPH} (rad/s)	0.0055	0.0042	0.0014	0.0025
ω_{LP} (rad/s)	0.005/3	0.0038	0.0013	0.0023
Integrator Gain	0.0006	0.0006	0.000011	0.0006
Input range	[5, 60]kg/s	[5, 30]kg/s	[1, 11]℃	[5, 30]kg/s

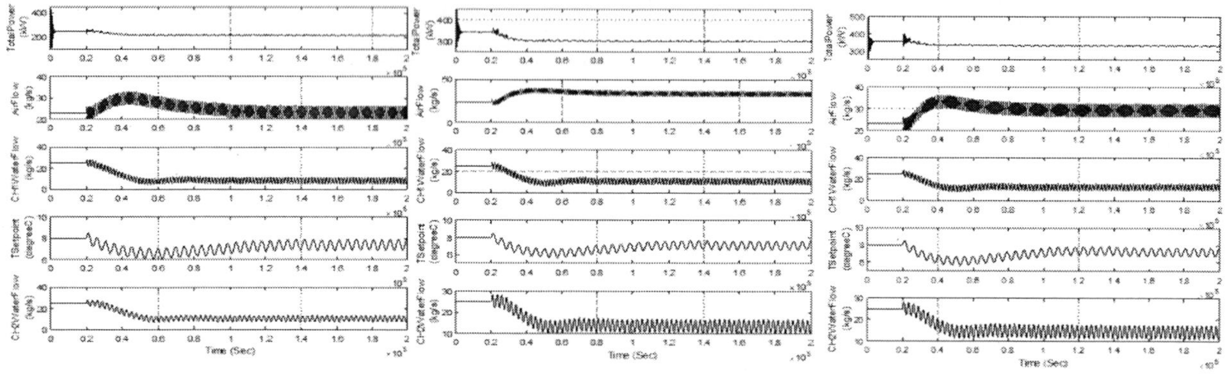

Figure 3: Scenario 1 under mild, dry-hot and humid-hot ambient conditions (from left to right)

3.2 Simulation Case#2

This case is based on two chillers working in parallel with the fixed ambient conditions and variable internal heat load. Scheme B-based optimization and sequencing strategy is applied in this scenario. The ambient conditions are the same with case#1. In the "Mild" condition, the internal heat gain experience a ramp change at 108000 sec (30 hr) from 180 kW to 300 kW with period of 3600 sec, and a 3600-second ramp change at 219600 sec (61 hr) from 300 kW to 180 kW, respectively. In the "Humid Hot" condition, the internal heat gain experience a ramp change at 108000 sec (30 hr) from 150 kW to 250 kW with period of 3600 sec, and a 3600-second ramp change at 219600 sec (61 hr) from 250 kW back to 150 kW, respectively. And in the "Dry Hot" condition, the internal heat gain experience a ramp change at 108000 sec (30 hr) from 180 kW to 300 kW with period of 3600 sec, and a 3600-second ramp change at 219600 sec (61 hr) from 300 kW back to 180 kW, respectively. The ESC controller parameters are summarized in Table 3.

Table 3: ESC controller parameters for two chiller with sequencing

Parameter	ESC1	ESC2	ESC3	ESC4
Dither Freq.(rad/s)	0.005	0.0038	0.002	0.0023
Dither Amp.	3 kg/s	3 kg/s	0.5°C	3 kg/s
Demodulation Amp.	2/3 kg/s	2/3 kg/s	4 °C	2/3 kg/s
Demodulation Phase (rad)	0.15	0.1	0.18	0.05
ω_{BPL} (rad/s)	0.0045	0.0034	0.0018	0.0021
ω_{BPH} (rad/s)	0.0055	0.0042	0.0022	0.0025
ω_{LP} (rad/s)	0.005/3	0.0038	0.002	0.0023
Integrator Gain	0.00072	0.00048	0.000033	0.000288
Input range	[5, 60]kg/s	[0.5, 30]kg/s	[1, 11] °C	[0.5, 30]kg/s

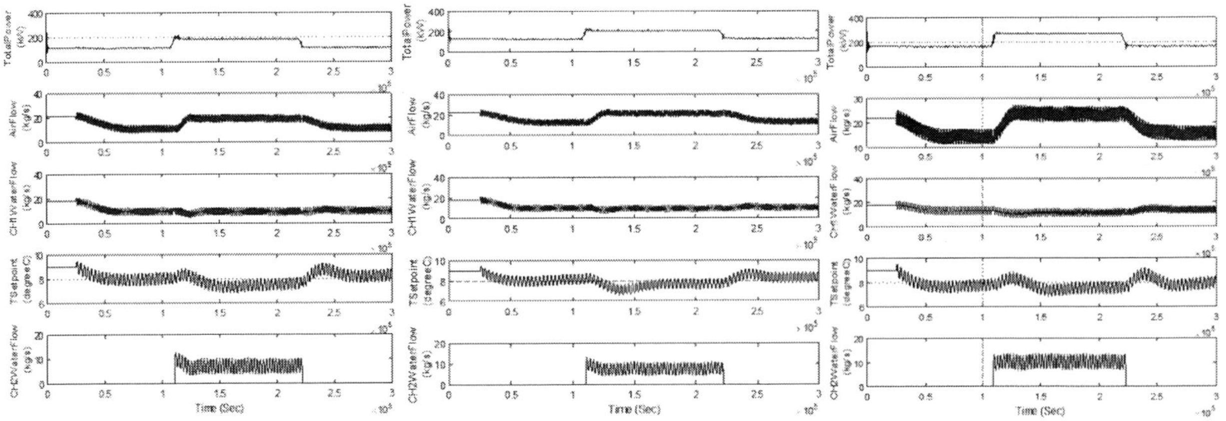

Figure 4: Scenario 2 under mild, dry-hot and humid-hot ambient conditions (from left to right).

In this scenario, the four-input ESC has its input and output trajectories shown in Fig. 4, which also represent the simulation results under three different ambient conditions ('Mild', 'Dry hot' and 'Humid hot') respectively. Under these three ambient conditions, the initial cooling tower air flow rate, chilled water leaving temperature setpoint and condenser water flow rate are 22 kg/s, 9°C and 18 kg/s. The three-input ESC for chiller 1 is turned on at 20000 sec. The ESC search results in an average steady-state total power of 115.1 kW, 124.5 kW and 164.5 kW under the load of 180 kW, 180 kW and 150 kW respectively. The optimal cooling tower air flow rate are 11.5 kg/s, 12.4 kg/s and 14.5 kg/s, chiller 1 condenser water flow rate are 9.8 kg/s, 10.3 kg/s and 12.9 kg/s, and leaving chilled-water temperature setpoint are 8.1°C, 8.0°C and 7.7°C.

With the load increasing, chiller 2 is engaged at around 110000 sec for all three ambient conditions. The four-input ESC reach an average steady-state total power of 184.5 kW, 204.6 kW and 268.1 kW under the load of 300 kW, 300 kW and 250 kW respectively. The optimal cooling tower air flow rate are 19.8 kg/s, 21.2 kg/s and 23.7 kg/s, chiller 1 condenser water flow rate are 9.4 kg/s, 9.7 kg/s and 12.0 kg/s, chiller 2 condenser water flow rate are 7.2 kg/s, 7.8 kg/s and 9.8 kg/s, and leaving chilled-water temperature setpoint are 7.5°C, 7.6°C and 7.4°C. With the load decreasing back to the previous value, chiller 2 is turned off at about 220000 sec. The three-input ESC's average steady-state total power, optimal cooling tower air flow rate, chiller 1 condenser water flow rate and leaving chilled-water temperature setpoint are all back to the previous values.

The gradient trajectories are plotted in second subplot of Fig. 7, 8, 9 to examine the convergence of each ESC channel. The oscillation of gradients of each channel about 0 at steady state indicates that the ESC has converged to an interior optimum. Similar with Case#1, other subplots show that the superheat, supply-air temperature and zone temperature are well regulated about their respective setpoint of 5°C, 13°C and 25°C.

3.3 Simulation Case#3

Case#3 relies on a 32-hr ambient temperature and RH profile developed from the TMY2 data for the Dallas-Fort Worth area (National Solar Radiation Data Base) and a variable load file is also used. Scheme B-based sequencing control is also applied in this scenario. Table 4 shows the ESC controller parameters in this case. The initial condition is set as: cooling tower air flow rate of 22 kg/s, chiller 1 and chiller 2 condenser water flow rate of 18 kg/s, and leaving chilled-water temperature setpoint of 9°C. As shown in Fig. 5, the ESC is turned on at t = 7 hr. In response to the variation of ambient condition and internal cooling load, chiller 2 is turned on at t=20.3 hr.

Table 4: ESC controller parameters for two chiller with realistic ambient and load profile

Parameter	ESC1	ESC2	ESC3	ESC4
Dither Freq.(rad/s)	0.005	0.0038	0.002	0.0023
Dither Amp.	3 kg/s	3 kg/s	0.5°C	3 kg/s
Demodulation Amp.	2/3 kg/s	2/3 kg/s	4°C	2/3 kg/s
Demodulation Phase (rad)	0.15	0.1	0.18	0.05
ω_{BPL} (rad/s)	0.0045	0.0034	0.0018	0.0021
ω_{BPH} (rad/s)	0.0055	0.0042	0.0022	0.0025
ω_{LP} (rad/s)	0.005/3	0.0038	0.002	0.0023
Integrator Gain	0.00096	0.00048	0.000033	0.000288
Input range	[5, 60]kg/s	[0.5, 35]kg/s	[1, 11] °C	[0.5, 35]kg/s

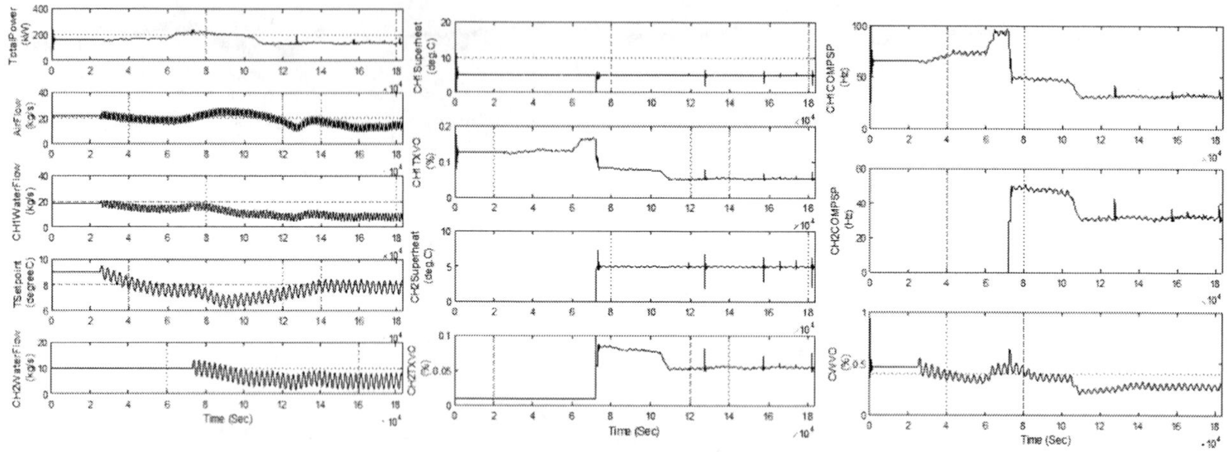

Figure 5: Scenario 3 under realistic ambient condition

3.4 Simulation Case#4

In order to study the input saturation impact on ESC-based chiller sequencing control, the cooling tower air flow was limited to 25 kg/s. The ambient conditions is 37°C and 30 %RH. The internal cooling load follows a 28000s positive ramp at t = 80000s from 250 kW to 400 kW, and a 1-hr negative ramp at t =219600s from 400 kW to 150 kW, respectively. The ESC controller parameters are the same as Case#3. The simulation results are shown in Fig. 6. The ESC feedback is the sum of the total power and penalty terms when input saturation occurs. The weight of penalty term for cooling tower air flow channel is $\gamma_{CTA} = 1.1$.

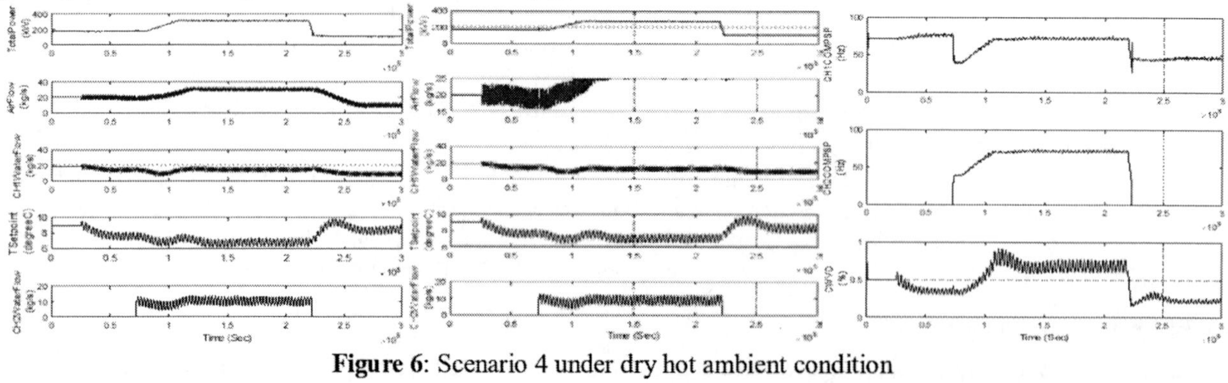

Figure 6: Scenario 4 under dry hot ambient condition

On average, ESC results in cooling tower air flow rate of 19.3 kg/s, chiller 1 condenser water flow rate of 14.1 kg/s, leaving chilled-water temperature setpoint of 6.9°C under the 250 kW load. With the load increased to 400 kW, chiller 2 is engaged and the cooling tower air flow rate is saturated at its upper limit of 25 kg/s while the chiller 1 condenser water flow settles to 14.4 kg/s, the leaving chilled-water temperature setpoint converges to 6.7°C, the chiller 2 condenser water flow rate converges to 10.2 kg/s, respectively. When the load goes down to 150 kW after 31 hour, the ESC converges to the cooling tower air flow rate of 10.5 kg/s, the chiller 1 condenser water flow rate of 8.3 kg/s, and leaving chilled-water temperature setpoint of 8.6°C. The standard ESC without a penalty function get its saturated input stuck at the saturate value regardless of the external condition changing.

3.5 Simulation Case#5

In practice, a major cause for suboptimal operation of HVAC systems is equipment degradation. For chiller plants, equipment degradations may include heat exchanger fouling, low refrigerant charge, cooling tower fill deposits, and compressor wear. In this case, low refrigerant situation is discussed. Chiller 2 properly charged and Chiller 1 with a 40% refrigerant charge. And the ambient conditions and load are the same as Case#1. The design parameters of this four-input ESC are also the same as Case#1.

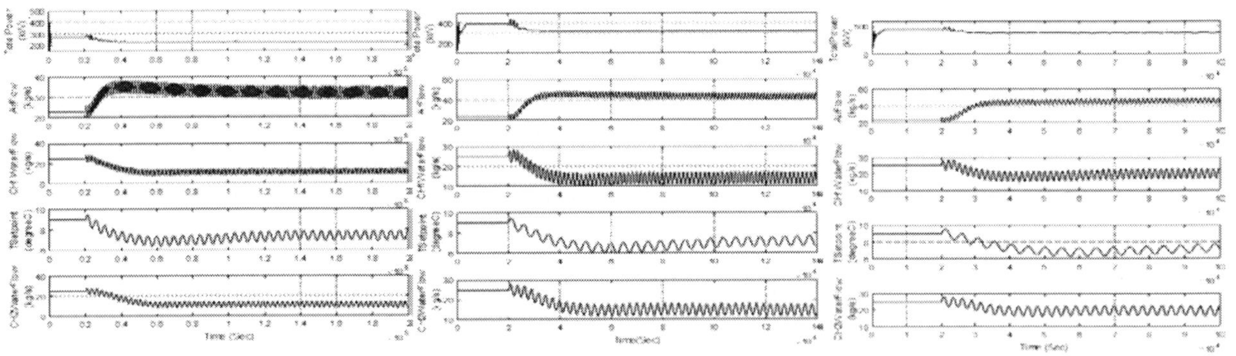

Figure 7: Scenario 5 under dry hot, mild and humid-hot ambient conditions (from left to right).

The average steady-state total power under three different ambient conditions ('Mild', 'Dry hot' and 'Humid hot') are 228.2 kW, 318.5 kW and 381.5 kW respectively shown in Fig. 7. They decrease 16.9%, 18.8% and 14.7%.respectively. Under all three ambient condition, the cooling tower air flow rate, condenser water flow rate for both chillers and chilled water leaving temperature setpoint start at 23kg/s, 25kg/s and 9°C. The four-input ESC converges to the cooling tower air flow rate of 32.5 kg/s, 43.2 kg/s and 45.3 kg/s respectively, the chiller 1 condenser water flow rate of 12.2 kg/s, 13.6 kg/s and 19.7 kg/s respectively, chiller 2 condenser water flow rate of 11.0 kg/s, 14.9 kg/s and 19.7 kg/s respectively and leaving chilled-water temperature setpoint of 7.5°C, 7.3°C and 6.9°C respectively. The superheat, supply-air temperature and zone temperature are well regulated about their respective setpoint of 5°C, 13°C and 25°C. Since the ambient condition and load are the same as Case#1(both chillers are fully charged), the overall power consumption increase with a low refrigerant-charged chiller.

3.6 Simulation Case#6

In this case, another equipment degradation situation for chiller plants is considered. Chiller 2 with nominal operation and Chiller 1 with heat exchanger (HX) fouling. The overall effect of HX fouling is modeled by a fouling factor R_f (Ardakani *et al.* 2008, Mu 2015), Chiller 1 evaporator is set as $R_f = 0.0001$ and the chiller 1 condenser is set as $R_f = 0.00025$. The simulation runs with the same ambient condition, load and ESC parameters as Case#1. The ESC search result in an average steady-state total power under three different ambient conditions ('Mild', 'Dry hot' and 'Humid hot') are 220.6 kW, 310 kW and 345.4 kW respectively, i.e. resulting in decrease of 13.8%, 11.9% and 7.6%, respectively. For all three ambient conditions, the cooling tower air flow rate and chilled water leaving temperature start at 23kg/s and 8°C. The initial value of condenser water flow rate for both chillers are 25kg/s. The four-input ESC converges to the cooling tower air flow rate of 23.5 kg/s, 33.5 kg/s and 30.8 kg/s respectively, the chiller 1 condenser water flow rate of 8.4 kg/s, 11.2 kg/s and 12.27 kg/s respectively, chiller 2 condenser water flow rate of 10.6 kg/s, 13.6 kg/s and 15.5 kg/s respectively and leaving chilled-water temperature setpoint of 7.4°C, 7.1°C and 7.2°C respectively. Moreover, the superheat, supply-air temperature and zone temperature are well regulated about their respective setpoint of 5°C, 13°C and 25°C. Compared with no system degrades Case#1, the total power consumption increase in this case, whileas ESC still can track the optimal input setting for the maximum achievable efficiency when the system degrades with HX fouling under various ambient conditions.

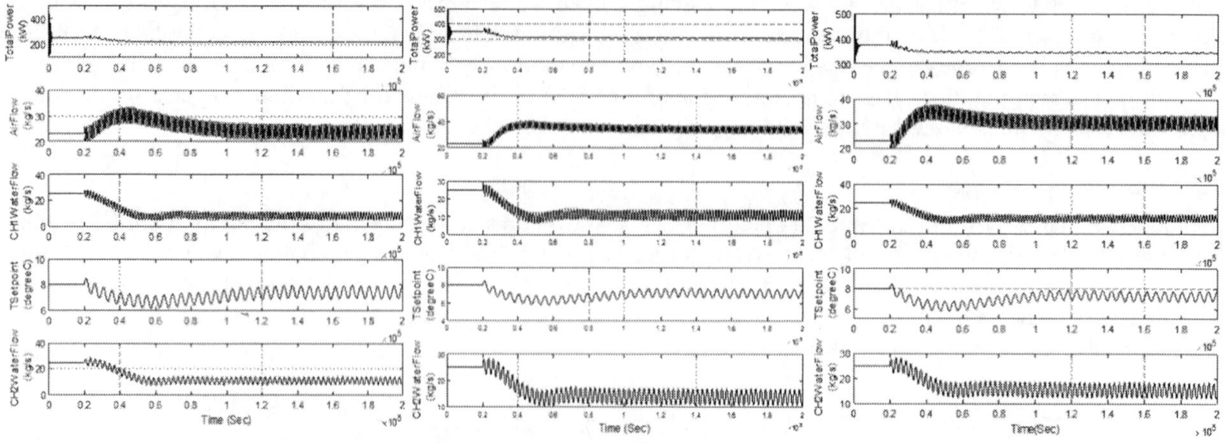

Figure 8: Scenario 6 under mild, dry-hot and humid-hot ambient conditions (from left to right).

3.7 Simulation results comparison of Scheme A and Scheme B

Scheme A and Scheme B are compared with Case #3 of Mu (2015) under the "Mild" condition, for which the internal heat gain experienced a 30000-sec ramp change at 200,000s from 250kW to 500kW, and a ramp change at 350,000 sec from 500kW to 200kW also with a 30000-sec ramp, respectively. The simulation results are shown in Fig. 9. Both Scheme A and Scheme B can activate and deactivate the additional chiller according to the criterion respectively. Since the load varies, the whole simulation process is divided into three periods. In the period 1 (0 to 2×10^5 s) and period 3 (3.8×10^5 to 5.76×10^5 s), only chiller 1 works and in period 2 (2.3×10^5 to 3.5×10^5 s), both chillers are engaged due to the heavy load. The ESC channels comparison of Scheme A and Scheme B is listed in Table 5. During period 2 and period 3, Scheme B achieves better performance, since it has less power consumption. However, during period 1, Scheme A has better efficiency. Moreover, the chiller 2 is turned on 6000s earlier and turned off 13400s earlier when Scheme B is applied.

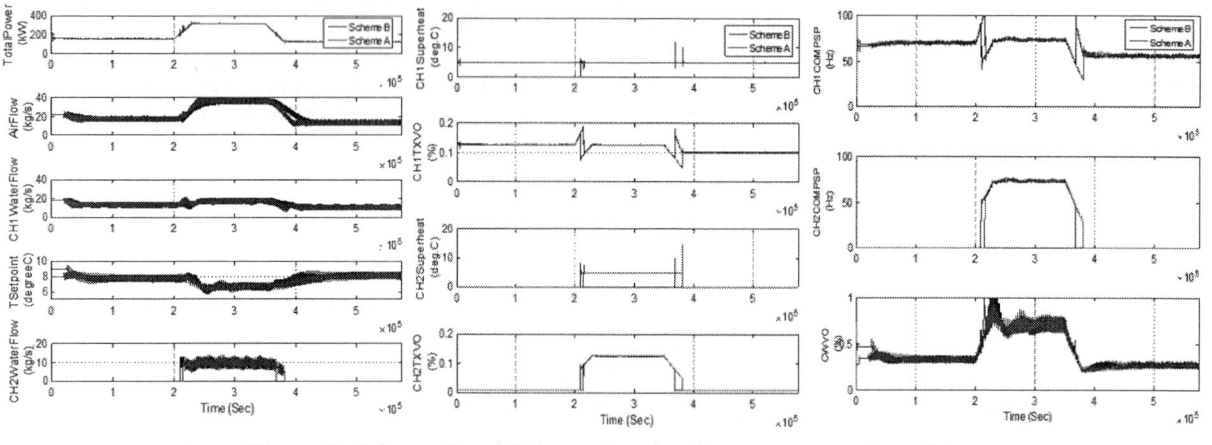

Figure 9: Scheme B and Scheme A under the same external condition

Table 5: ESC channels comparison of Scheme A and Scheme B

	Scheme A			Scheme B		
	Period 1	Period 2	Period 3	Period 1	Period 2	Period 3
Total power (kW)	156.9	316.2	126.3	158.9	315.3	125.7
Air flow (kg/s)	17.2	36.7	13.5	17.6	36.4	13.3
Chiller 1 water flow (kg/s)	13.2	16.7	10.5	13.4	17.4	10.9
Tsetpoint (°C)	7.6	6.6	7.9	7.7	6.7	8.1
Chiller 2 water flow (kg/s)	0	10.1	0	0	9.2	0

4. CONCLUSIONS AND FUTURE WORK

A multivariate ESC based chiller sequencing control algorithm is studied in this paper. It maximizes the chilled water plant energy efficiency in real time under different external conditions. Based on a dynamic chilled water plant simulation model with two chillers in parallel, several scenarios are discussed in order to evaluate the performance of the proposed ESC algorithm. The simulation results show that significant energy saving is obtained and the effectiveness of the proposed Scheme B based chiller sequencing is also validated in various conditions. By comparing the performance of Scheme A and B under the same external condition, the overall power consumption is quite similar. More detailed study is under way.

REFERENCES

AHRI Standard 550/590 (I-P), 2011 Standard for Performance Rating of Water-Chilling and Heat Pump Water-Heating Packages using the Vapor Compressor Cycle, Air-Conditioning, Heating & Refrigeration Institute, 2011.

Ardakani, A.J., Ardakani, F.F., and Hosseinian, S.H. (2008), "A Novel Approach for Optimal Chiller Loading Using Particle Swarm Optimization. Energy and Buildings", 40 (12), pp. 2177-2187.

Mu, B., Li, Y.and Seem, J.E. (2014) "Experimental Study on Extremum Seeking Control for Efficient Operation of Airside Economizer", 2014 Purdue Compressor Engr., Refri. & Air Cond., and High Perf. Bldgs. Conf., 10 pages.

Mu, B., Li, Y., Salsbury, T.I., House, J.M. (2015) "Extremum Seeking Based Control Strategy for a Chilled-Water Plant with Parallel Chillers," ASME Dynamic Systems and Control Conf., Columbus, OH, Paper #9949, 10 pages.

Mu, B, (2015) "Self-optimizing Control for Building Ventilation and Air Conditioning Systems", Ph.D. Dissertation, Department of Electrical Engineering, University of Texas at Dallas, December 2015.

National Solar Radiation Data Base, 1961-1990: Typical Meteorological Year 2, http://rredc.nrel.gov/solar/old_data/nsrdb/1961-1990/tmy2/.

Yu, F.W., Chan, K.T., 2008, "Optimization of water-cooled chiller system with load-based speed control," *Applied Energy*, **85** (10), pp. 931-950.

Benton, D.J., Hydeman, M., Bowman, C.F., Miller, P., 2002, "An Improved Cooling Tower Algorithm for The CoolTools™ Simulation Model," *ASHRAE Transactions*, **108**(1), Paper No. AC-02-9-4.

Hydeman, M., Zhou, G., 2007, "Optimizing chilled water plant control," *ASHRAE Journal*, pp. 45-54.

Ahn, B.C., Mitchell, J.W., 2001, "Optimal control development for chilled water plants using a quadratic representation", *Energy and Buildings*, **33**, pp. 371-378

Bhattacharjee K. 2007, "Reducing energy cost in an industrial chilled water plant", *Energy Engr.*, **107**(4): 42–51.

Hydeman, M., Webb, N., Sreedharan, P., Blanc, S., 2002, "Development and testing of a reformulated regression based electric chiller model," *ASHRAE Transactions* 108(2), pp. 1118 – 27.

Qureshi B., Zubair S. 2006, "Prediction of evaporation losses in wet cooling towers," *Heat Tsfr. Engr.*, **27**(9):86–92.

Deng, K., et al., 2014, "Model predictive control of central chiller plant with thermal energy storage via dynamic programming and mixed- integer linear programming," *IEEE Trans. Auto. Sci. & Eng.* pp. 1-14.

Zhu, J., Yang, Q., Lu, J., 2013, "Model predictive control of chilled water temperature for centrialized HVAC systems", *2013 IEEE Electrical Power & Energy Conference*, 5 pages.

Chow, T.T., Zhang, G.Q., Lin, Z., Song, C.L., 2002, "Global optimization of absorption chiller system by genetic algorithm and neural network," *Energy and Buildings*, 34 (1), pp. 103-109.

Chang, Y.-C., Lin, J.-K., and Chuang, M.-H., 2005, "Optimal chiller loading by genetic algorithm for reducing energy consumption," *Energy and Buildings*, 37 (2), pp. 147-155.

Ma, Z., and Wang, S. 2011, "Supervisory and optimal control of central chiller plants using simplified adaptive models and genetic algorithm," *Applied Energy*, 88 (1), pp. 198-211.

Li, P., Li, Y., and Seem, J.E., 2010, "Efficient Operation of Air-Side Economizer Using Extremum Seeking Control," *ASME J. Dyn. Sys., Meas. & Control*, 132(3): 031009 (10 pages).

Li, X., Li, Y., and Seem, J.E., 2013, "Dynamic Modeling and Self-Optimizing Operation of Chilled Water Systems Using Extremum Seeking Control", *Energy and Buildings*, **58**, pp. 172-182.

Dong, L., Li, Y., Mu, B., Xiao, Y., 2014, "Self-optimizing control of air-source heat pump with multivariable extremum seeking", *Applied Thermal Engineering*, **84**(5), pp. 180-195.

Hu, B., Li, Y., Cao, F. and Xing, Z. "Extremum Seeking Control of COP Optimization for Air-Source Transcritical CO2 Heat Pump Water Heater System," *Applied Energy*, Vol. 147, pp. 361-372, June 2015.

Peng, Y., Vrancic, D., Hanus, R., 1993. "Anti-windup, bumpless and conditioned transfer techniques for PID controllers", *IEEE Trans. Control Sys. Tech.*, **16**(4), pp. 48-57.

A Study of the Effect of Zone Design Parameters on Frequency Domain Transfer Functions for Radiant and Convective Systems

Ali Saberi Derakhtenjani[1]*, Andreas K. Athienitis[2]

[1]Concordia University, building civil and environmental engineering,
Montreal, Quebec, Canada
ali.saberi.mech@gmail.com

[2]Concordia University, building civil and environmental engineering,
Montreal, Quebec, Canada,
aathieni@encs.concordia.ca

* Corresponding Author

ABSTRACT

This paper presents a fundamental parametric study on the effect of a number of room design parameters for radiant and convective heating sources as well as solar gains. This study is performed using frequency domain modeling approach by means of which important room transfer functions are obtained and studied. A room is considered with different types of heating (convective and radiative heating sources) and different levels of thermal mass on the floor. The effect of thermal mass and floor covering on the room thermal response considering different types of heating is investigated. The magnitude of the transfer functions between room air temperature and the convective heating source is a determining factor in the room air temperature fluctuations that affect thermal comfort. Also, in the case of radiant heating, the transfer function between room air temperature and radiant heat source can be used to determine the room air temperature swings due to the floor radiant heating source. The sensitivity of the magnitude of the transfer functions versus different values of convective and radiative heat transfer coefficients is studied and compared. This study will guide future model predictive control (MPC) research by means of frequency domain techniques to optimize key design variable such as thermal mass thickness and properties for floor heating and convective systems. It will contribute to linking building design with MPC.

1. INTRODUCTION

Excessive energy consumption in buildings happens due to a number of reasons including non-optimal building design and material selections considering the building type and its energy supply systems. Energy models developed through various approaches and techniques have been used as the tools to analyze building energy consumption and to optimize their operation. Certain modeling techniques provide valuable information during early stages of design and guide us in choosing the optimum material for the building from the thermal performance point of view.

Frequency domain techniques have been shown to be a useful tool for the design analysis on a relative basis (Athienitis 1994). By means of the frequency domain modeling techniques important building transfer functions can be obtained and studied and for this usually no simulation is required. Also, it has been shown to be a practical tool for building simulation and deriving simplified grey-box building models (Saberi Derakhtejani et al. 2015) (Wang and Xu 2006) (Xu and Wang 2007). Frequency domain techniques are especially useful for periodic analysis of phenomena inside a building. The weather-related inputs affecting energy consumption inside a building such as solar gains, exterior temperature and heating/cooling sources are cyclic phenomena and can be modeled by means of frequency domain techniques assuming periodic conditions in the calculations. These techniques can be efficiently applied to model various building-integrated energy systems (Chen et al. 2013a) (Chen et al. 2013b) and to apply control strategies (Athienitis et al. 1990) (Chen et al. 2014).

Using frequency domain techniques, the transient heat conduction inside the walls can be accurately modeled with no discretization for the thermal mass and an exact solution for 1-D heat conduction inside building envelope or wall is obtained (Athienitis and Santamouris 2002). In the case of convection and radiation heat transfer, the respective heat transfer coefficients are usually linearized in order to have a linear system of equations that can be presented by means of a linear thermal network (Athienitis and O'Brien 2014). Since convection and radiation are inherently non-linear phenomena, linearization produces some errors which are relatively significant in the case of convection between room surfaces and room air compared to the longwave radiant exchanges between surfaces. In this case, a sensitivity analysis on the magnitude of the important room transfer functions considering different values for convective and radiative heat transfer coefficients needs to be done.

2. METHODOLOGY

2.1 Frequency Domain Model

This section presents the methodology to develop frequency domain models for a room. A thermal network for a room is considered which represents the heat transfer between room surfaces and indoor and outdoor temperatures. Walls are modeled using the Norton-equivalent representation. This representation gives an exact solution for the 1-D heat conduction inside the wall (Athienitis et al. 1990). The level of detail in the thermal network depends on how radiation and convection heat transfer are modeled in the room interior zone. Combined radiative-convective heat transfer coefficients are often assumed which can be represented by a star network as shown in Figure 1.a. This network is used when there are small differences between the surfaces and room air temperature and when the heating source is mainly convective. However, in zones with high level of solar gains or in rooms with radiant heating systems, using combined heat transfer coefficients can produce significant error in the results. In those cases, convective and radiative heat transfer needs to be modeled separately (Figure 1.b). The convective and radiative coefficients are calculated as:

$$\text{convective } U_{\text{i1}} = A_{\text{i}} h_{\text{c,i}} \text{ , radiative } U_{\text{ij}} = A_{\text{i}} F_{ij}^{*} (4\sigma T_{m}^{3}) \quad (1)$$

Where:

A_{i} = area of surface i, $h_{\text{c,i}}$ = convective heat transfer coefficent of surface i,

F_{ij}^{*} = radiant exchange factor between surfaces i and j, T_{m} = estimated mean temperature

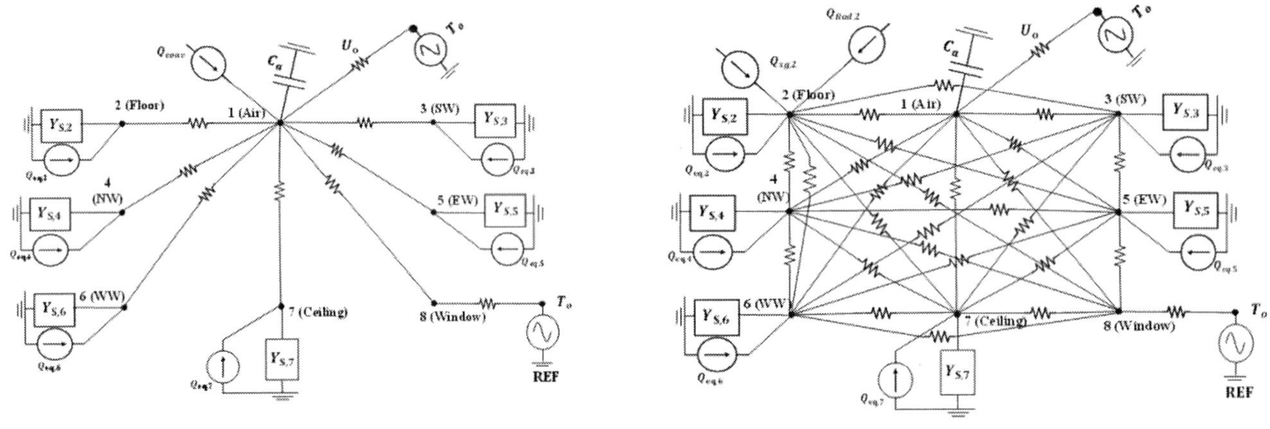

Figure 1. Thermal network for frequency domain considering a) combined heat transfer coefficients (left), b) separate modeling of convective and radiative exchanges (right).

In the thermal network above, all the walls and elements with thermal mass are modeled using the two-port Norton-equivalent subnetwork which consists of the wall self-admittance, Y_s , and an equivalent source, $Q_{eq}=Y_t \times T_{ext}$, which represents the effect of an external temperature on the surface temperature. This representation gives an exact solution for one-dimensional heat conduction inside the wall without spatial discretization. The wall self-admittance (Y_s) and transfer admittance (Y_t) are calculated using the following equations (Athienitis 1994):

$$Y_{\text{s,i}} = \frac{U_{\text{i}} + A_{\text{i}} k_{\text{i}} \gamma_{\text{i}} \tanh(\gamma_{\text{i}} l_{\text{i}})}{\frac{U_{\text{i}}}{A_{\text{i}} k_{\text{i}} \gamma_{\text{i}}} \tanh(\gamma_{\text{i}} l_{\text{i}}) + 1} \quad , \quad Y_{\text{t,i}} = \frac{-A_{\text{i}}}{\frac{A_{\text{i}}}{U_{\text{i}}} \cosh(\gamma_{\text{i}} l_{\text{i}}) + \frac{\sinh(\gamma_{\text{i}} l_{\text{i}})}{k_{\text{i}} \gamma_{\text{i}}}} \quad (2)$$

2.2 Building Transfer Functions

In this section derivation of the building transfer functions for the detailed frequency domain model is presented for which the thermal network is shown in Figure 1.b. Node (1) represents the air inside the zone. Energy balance at node (1) gives:

$$C_a \frac{dT_1}{dt} + U_o(T_1 - T_o) + \sum_{J1} U_{1,J1}(T_1 - T_{J1}) = Q_{aux} \tag{3}$$

In which $U_{1,J1}$ are the convective conductances between the air node and other surfaces (J1=2, 3,…,8), Ca is the thermal capacitance of the room air, T is temperature and Q_{aux} is the auxiliary source at the air node (which can be equal to zero). Using Laplace transform, equation (3) will be:

$$sC_a T_1(s) + U_o T_1(s) + \sum_{J1} U_{1,J1}(T_1(s) - T_{J1}(s)) = \underbrace{Q_{aux}(s) + \underbrace{U_o T_o(s)}_{Q_{eq}}}_{Q_1} \tag{4}$$

where s is the Laplace variable. Now, considering the floor (node (2)), the energy balance equation yields:

$$Y_{s,2}(T_2(s) - \underbrace{T_{ref}(s)}_{=0\,°C}) + \sum_{J2} U_{2,J2}(T_2(s) - T_{J2}(s)) = \underbrace{\underbrace{Q_{sg,2}}_{\text{solar gain}} + Q_{eq,2} + \underbrace{Q_{rad,2}}_{\text{radiant source}}}_{Q_2} \tag{5}$$

where J2=1,3,4,…,8 and $U_{2,J2}$ are the radiative conductances between floor and other surfaces. The energy balance for all the nodes can be written in the form below:

(6)

$$\underbrace{\begin{bmatrix} sC_a+\sum_j U_{1,j}+U_{inf} & -U_{1,2} & -U_{1,3} & -U_{1,4} & -U_{1,5} & -U_{1,6} & -U_{1,7} & -U_{1,8} \\ -U_{1,2} & Y_{s,2}+\sum_j U_{2,j} & -U_{2,3} & -U_{2,4} & -U_{2,5} & -U_{2,6} & -U_{2,7} & -U_{2,8} \\ -U_{1,3} & -U_{2,3} & Y_{s,3}+\sum_j U_{3,j} & -U_{3,4} & -U_{3,5} & -U_{3,6} & -U_{3,7} & -U_{3,8} \\ -U_{1,4} & -U_{2,4} & -U_{3,4} & Y_{s,4}+\sum_j U_{4,j} & -U_{4,5} & -U_{4,6} & -U_{4,7} & -U_{4,8} \\ -U_{1,5} & -U_{2,5} & -U_{3,5} & -U_{4,5} & Y_{s,5}+\sum_j U_{5,j} & -U_{5,6} & -U_{5,7} & -U_{5,8} \\ -U_{1,6} & -U_{2,6} & -U_{3,6} & -U_{4,6} & -U_{5,6} & Y_{s,6}+\sum_j U_{6,j} & -U_{6,7} & -U_{6,8} \\ -U_{1,7} & -U_{2,7} & -U_{3,7} & -U_{4,7} & -U_{5,7} & -U_{6,7} & Y_{s,7}+\sum_j U_{7,j} & -U_{7,8} \\ -U_{1,8} & -U_{2,8} & -U_{3,8} & -U_{4,8} & -U_{5,8} & -U_{6,8} & -U_{7,8} & \sum_j U_{8,j} \end{bmatrix}}_{\mathbf{Y}} \underbrace{\begin{bmatrix} T_1 \\ T_2 \\ T_3 \\ T_4 \\ T_5 \\ T_6 \\ T_7 \\ T_8 \end{bmatrix}}_{\mathbf{T}} = \underbrace{\begin{bmatrix} Q_1 \\ Q_2 \\ Q_3 \\ Q_4 \\ Q_5 \\ Q_6 \\ Q_7 \\ Q_8 \end{bmatrix}}_{\mathbf{Q}}$$

or

$$\mathbf{Y}_{8\times8}\mathbf{T}_{8\times1} = \mathbf{Q}_{8\times1}$$

Where **Y** is the admittance matrix, **T** is the temperature vector and **Q** is the source vector. Elements of equation (6) are in term of the Laplace variable s and can be calculated at different frequencies ($s=j\omega_n$, $j=\sqrt{(-1)}$, $\omega_n=2\pi n/p$, $p=24hrs$, n=harmonic number).

An important parameter is the impedance matrix $\mathbf{Z}=\mathbf{Y}^{-1}$ which contains building transfer functions at different frequencies:

$$Z_{i,j} = \frac{T_i}{Q_j} \quad \rightarrow \quad Z_{1,1} = \frac{T_1}{Q_1}, \; Z_{1,2} = \frac{T_1}{Q_2}, \; \ldots$$

The impedance transfer function represents change in temperature of node i resulting from the effect of the source Q_j at node j. Thus, considering equation (6), temperature of node i for each frequency will be calculated as:

$$T_i = \sum_{j=1}^{8} Z_{i,j} Q_j \tag{7}$$

Studying important building transfer function provides much insight into building thermal dynamics and we can evaluate design alternatives on a relative basis without detailed simulation. In the next section application of this methodology for a thermal zone is demonstrated.

3. MODEL APPLICATION FOR A THERMAL ZONE

3.1 Description of the Thermal Zone

The thermal zone considered here in this study is a typical office room for which the dimensions are shown in Figure 2. with the walls considered to be made of gypsum board with RSI 4 insulation for each wall. The south facing façade has a 2×3m window area (two thirds of the facade area). The floor is made of concrete and represents the main thermal mass in the zone. A similar test room has been built (with different types of auxiliary systems) at Concordia University for experiments at the solar simulator /environmental chamber laboratory of Concordia University.

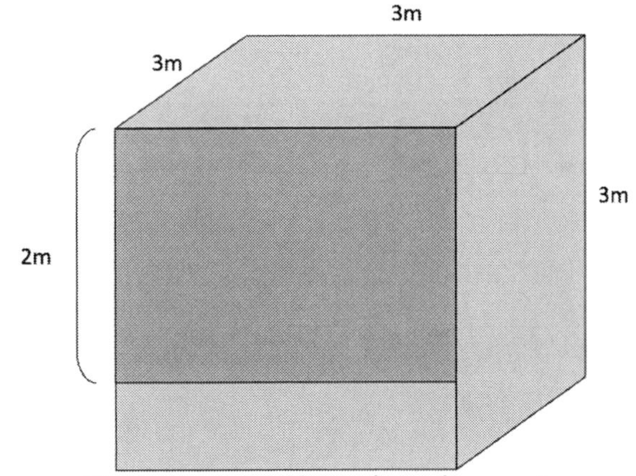

Figure 2. Room (thermal zone) geometry

A detailed frequency domain model shown in Figure 1.b is developed for the room. It is shown how important transfer functions in each case can guide us in the evaluating the dynamic behavior of the thermal zone. As an example, one of the important transfer functions is $Z_{1,1}$ which is between the air temperature and the source at the air node. In the zones with convective heating/cooling systems this transfer function determines the impact of the convective auxiliary load on the temperature of the air in the room.

3.2 Effect of Different Floor Coverings

Figure 3 shows the magnitudes of the room transfer function $Z_{1,1}$ at different frequencies (harmonics) for two cases: 1) without any floor covering that is only concrete floor and 2) when there is a carpet covering on the floor.

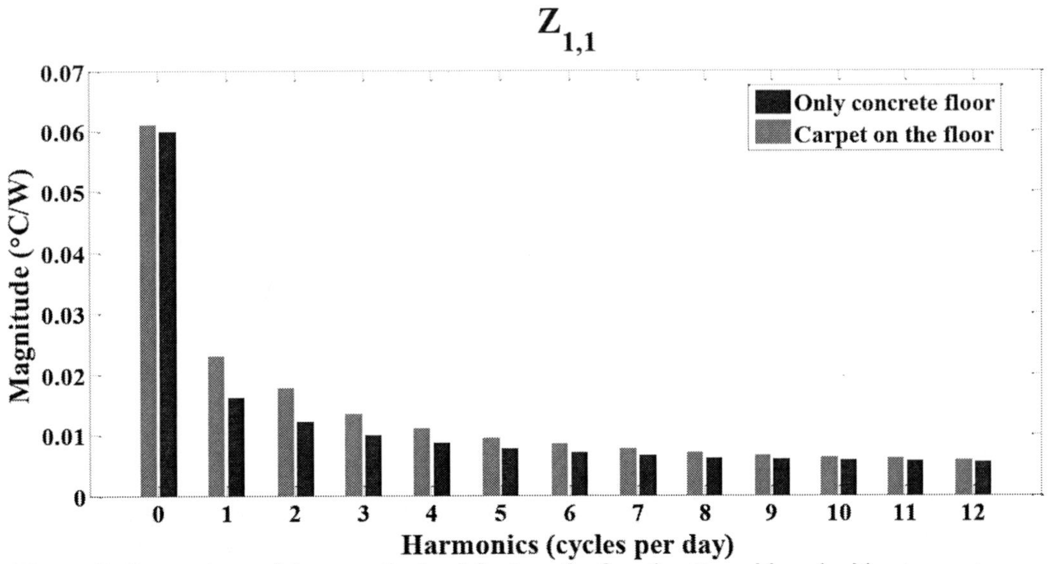

Figure 3. Comparison of the magnitude of the transfer function $Z_{1,1}$ with and without carpet cover

As it can be observed, the magnitude of the transfer function remains at a higher value in the case of carpet compared to the original case with no carpet on the floor. This difference is especially significant for the harmonic range of 1-8 cycles per day. The higher magnitude values means that for a unique input in the air node, there will be larger room air temperature fluctuations when there is a carpet covering on the floor compared to the case of the only concrete floor. This is an important observation. If a sinusoidal excitation (an auxiliary heating source) shown in Figure 4 is inserted at the air node under a constant outdoor air temperature (-10 °C) the following temperature profiles shown in Figure 5 are obtained for the two cases:

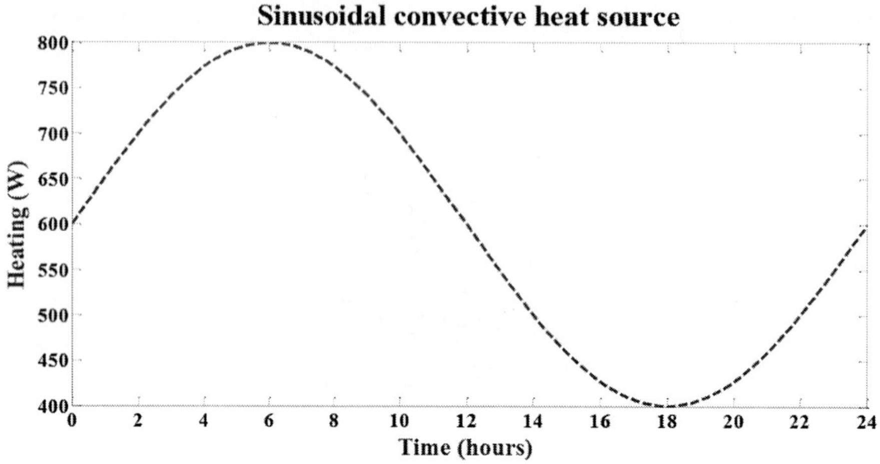

Figure 4. Auxiliary heat source profile

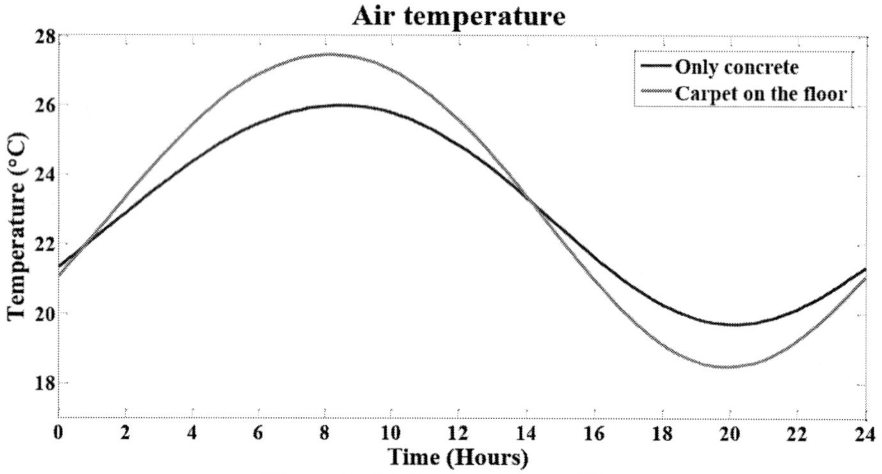

Figure 5. Air temperature profiles

As expected, larger temperature fluctuations are observed in the presence of a carpet on the floor and without carpet a more smooth temperature profile with smaller fluctuations is observed. Smaller fluctuations result in better thermal comfort in the zone. This result shows that a floor covering can have a considerable impact on the thermal performance of the zone and the occupant comfort.

3.3 Effect of Increasing Thermal Mass

Figure 6 shows the magnitude of the transfer function $Z_{1,1}$ considering different thicknesses for the floor concrete. We can see that the main difference between magnitudes in different cases is for the first and second harmonics and there is not a significant difference between magnitudes for the short-term dynamics, high frequency signals. Also, we can see that in the case of the 10cm thick concrete the lowest magnitude (for the first and second harmonics) is observed for the transfer function $Z_{1,1}$ compared to the other cases. Thus, the analysis of this chart gives us an idea for choosing an optimum concrete thickness considering different types of auxiliary systems; convective system in this case.

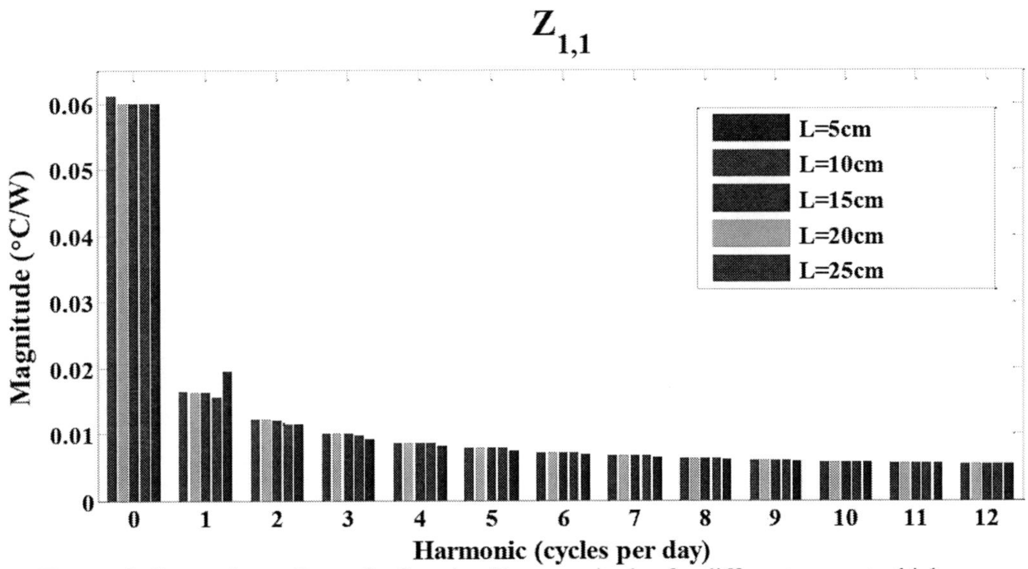

Figure 6. Comparison of transfer function $Z_{1,1}$ magnitudes for different concrete thicknesses

Figure 7 shows magnitudes of the transfer function $Z_{1,2}$ for a frequency range. We can observe that similar to $Z_{1,1}$, the significant difference is between the magnitude in the case of 5cm thick concrete with the other thicknesses for the first and second harmonics. There are no significant differences between other cases.

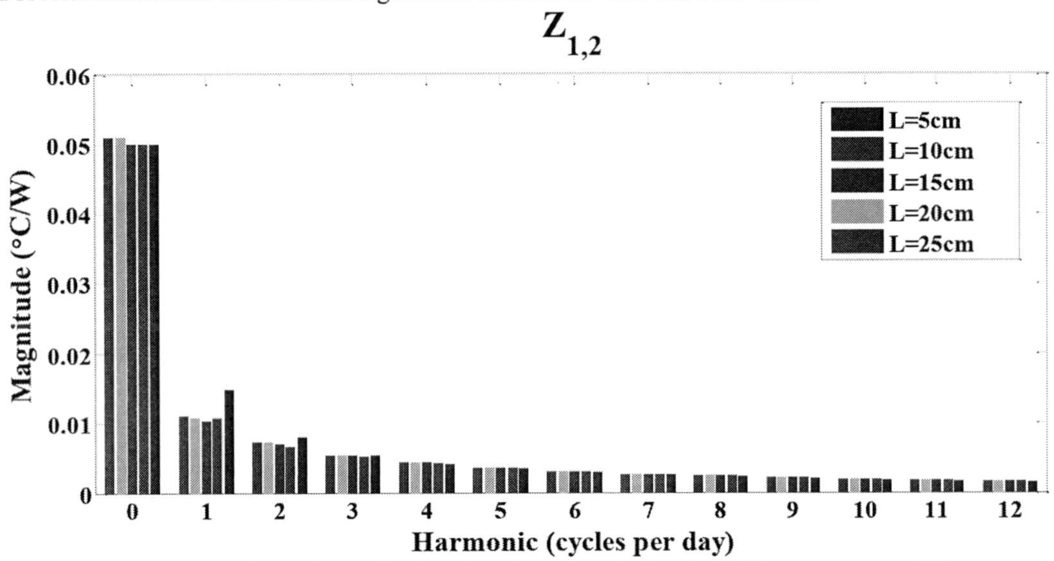

Figure 7. Comparison of transfer function $Z_{1,2}$ magnitudes for different concrete thicknesses

3.4 Sensitivity Analysis of Floor Convective Heat Transfer Coefficient

Convection heat transfer is an inherently non-linear phenomenon and the respective heat transfer coefficient in a room varies significantly with temperature difference between room surfaces and the inside air. As an example for the floor, the value of the convective heat transfer coefficient can be as low as $1 W/m^2K$ for the cold floor and hot air while it can be as high as $3.5 W/m^2K$ for the hot floor and cold air. Therefore, in a room with radiant heating/cooling systems, h_c for the floor can vary a lot during a period of time. A sensitivity analysis on the magnitude of the important room transfer functions can demonstrate the thermal response of the room under different conditions.

Figure 8 shows magnitudes of the transfer function $Z_{1,1}$ over a frequency range for different values of floor convective heat transfer coefficient (h_{cf}). From the graph it can be observed that there is a significant difference in steady-state magnitude of the transfer function (frequency of zero) between all the cases. Also, there are noticeable differences in the magnitudes for the first, second and third harmonics which represents that the smaller the h_{cf}, the larger the temperature fluctuations will be. Therefore, we can see that the impact of the floor convective heat transfer coefficient is quite significant.

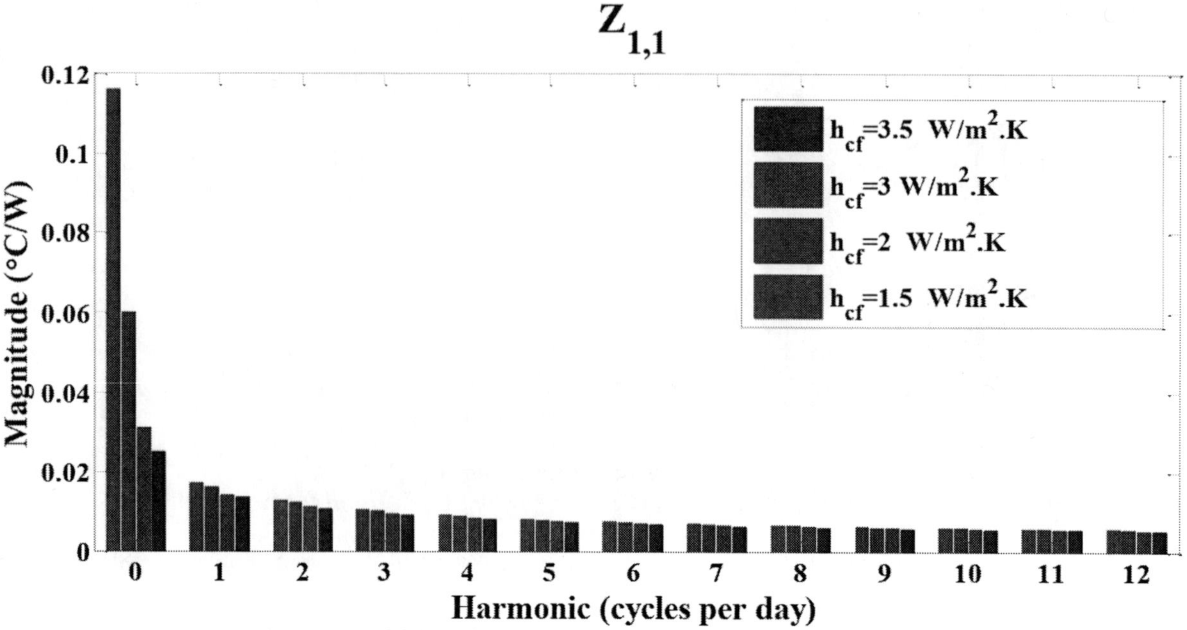

Figure 8. Sensitivity of transfer function $Z_{1,1}$ to different values of h_{cf}

a similar trend in difference between magnitudes for the transfer function $Z_{1,2}$ is observed as well (Figure 9).

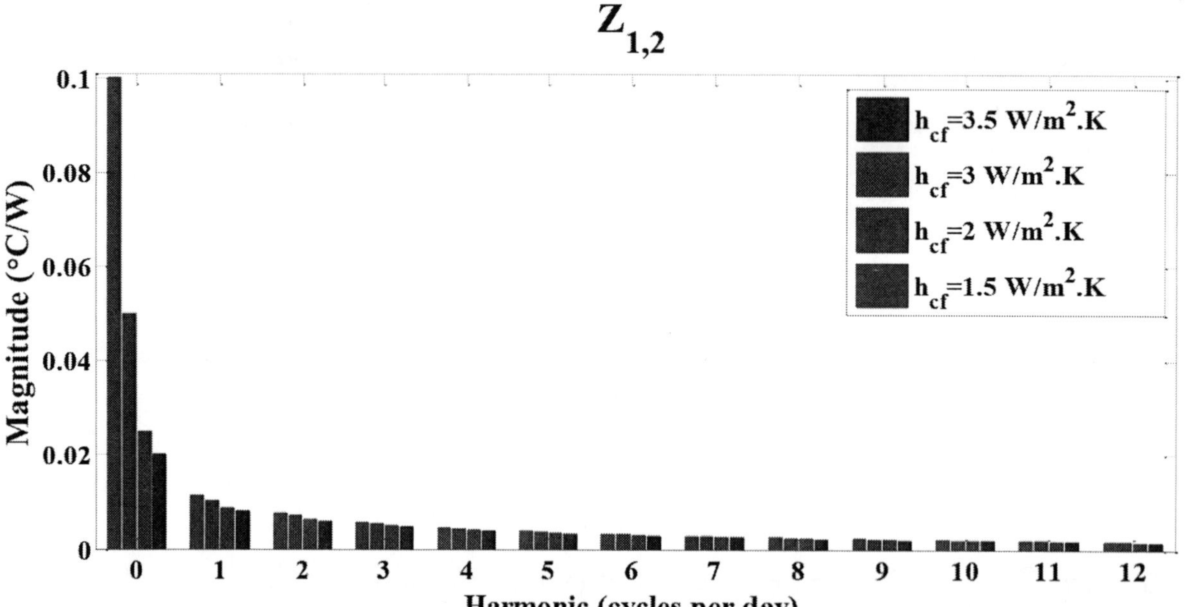

Figure 9. Sensitivity of transfer function $Z_{1,2}$ to different values of h_{cf}

3.5 Analysis of a room set point profile

Figure 10 shows a typical room air temperature setpoint profile with the temperature rising in the morning and a temperature setback in the evening. Using discrete Fourier transform we can see the magnitudes of this profile for different frequencies (amplitude spectrum). It is observed that the highest magnitudes of this setpoint profile are for the first, second and third harmonics.

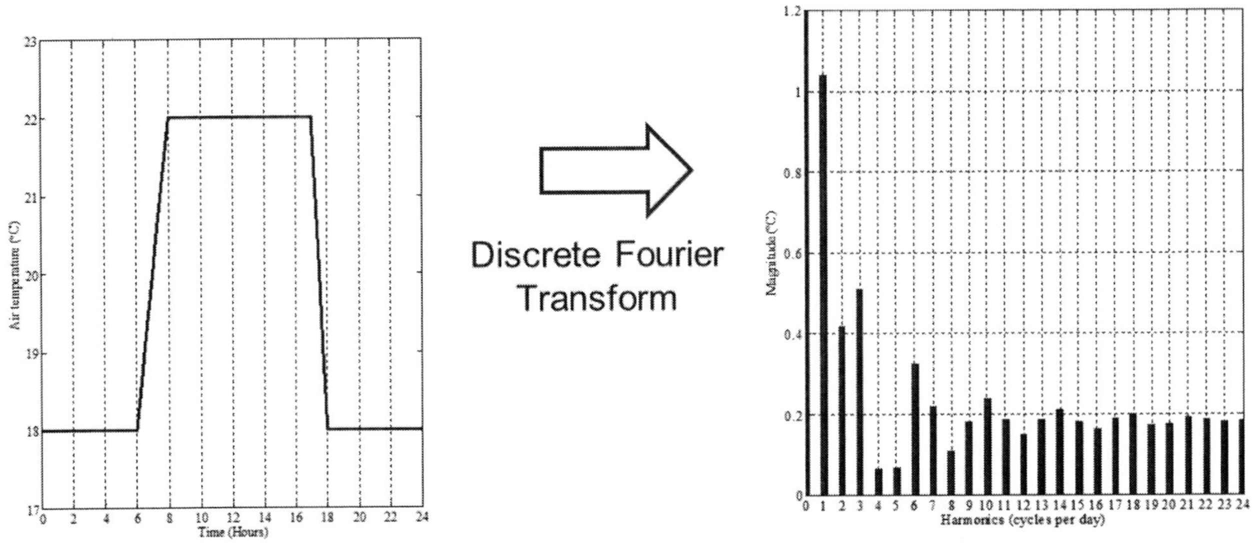

Figure 10. Room air temperature setpoint profile (left) and its amplitude spectrum (right)

Also, the lowest magnitudes for the shown frequency range belongs to the fourth and fifth harmonics. Due to its specific shape, the setpoint profile has noticeable magnitudes in the frequency range of 10-24 cycles per day. Considering the room transfer function equation, $Z \approx dT/dQ$, it be can realized that smaller magnitudes for the transfer function in a certain frequency range lead to less fluctuations in the room temperature for the change in heat gain due to setpoint changes. Therefore, again we can observe that how analysis of the room transfer functions can guide us in evaluating different design alternatives and their impact on the room thermal performance.
With a further look in the energy balance equation of the floor (node 2) we have:

$$T_1 = \sum_{j=2}^{8} Z_{1,j} Q_j = Z_{1,1} Q_1 + Z_{1,2} Q_2 + \ldots + Z_{1,8} Q_8 \,, \quad Q_2 = Q_{\text{floor}} = Q_{\text{solar}} + Q_{\text{radiant}}$$

then:

$$Q_{\text{radiant}} = \frac{T_1 - Z_{1,1}(Q_1) - Z_{1,2}(Q_{\text{solar}}) - \sum_{j=3}^{8} Z_{1,j} Q_j}{Z_{1,2}} \tag{8}$$

In equation (8), the air temperature (T_1) can be considered as a profile (setpoint) and then a radiant floor heating profile will be found to maximize the use of the floor solar gain with respect to the desired air temperature in the room.
From equation (8), it can be see that the transfer function $Z_{1,2}$ is a significant parameter when looking for the optimum heating or air temperature profiles. This reconfirms the significance of the room design parameters on the thermal performance of the room.

4. CONCLUSION

This paper presented a frequency domain analysis of the effect of the different room design parameters on the room thermal response by means of the building transfer functions. It was shown how a certain floor covering (carpet) can affect the room thermal response and increase fluctuations in the room air temperature. Also, considering different types of auxiliary systems in a room, an optimum thickness for the floor thermal mass can be found by looking at the certain transfer functions depending on the type of the auxiliary heating or cooling system used.
It was observed that floor convective heat transfer coefficient value has a very significant effect on the steady-state magnitude of the transfer functions $Z_{1,1}$ and $Z_{1,2}$. In a room with a relatively hot floor compared to the inside air (floor with radiant heating system) where large values for the floor convective heat transfer coefficient are observed, the air temperature fluctuations are smaller and the more uniform air temperature will result in the better thermal comfort.
Then, by means of the discrete Fourier transform (DFS), an analysis has been made on the magnitudes of a typical room air setpoint profile at different frequencies. It was explained that when the room materials are chosen so that the

room transfer functions have smaller magnitudes over a certain frequency range (over which air setpoint profile has high magnitudes), this would lead to less fluctuations in room temperature due to a change in the room heat gain. Frequency domain modeling techniques has been proven to be useful in characterizing building thermal performance. The information obtained in this study will be used in the future model-based control studies and model predictive control (MPC).

NOMENCLATURE

U_i	U-value used in calculation of wall i's self and transfer admittances
A_i	surface area of wall i
$Y_{s,i}$	self admittance for wall i
$Y_{t,i}$	transfer admittance for wall i
k_i	thermal conductivity for wall i
l_i	thickness for wall i
γ_i	angular frequency for wall i, equal to $(s/\alpha)^{1/2}$
C_a	thermal capacitance of the internal air

Subscript

s	self
t	transfer
i/j	node i/j

REFERENCES

Athienitis, A. K. and Santamouris, M. (2002). *Thermal analysis and design of passive solar buildings*. London, UK, James & James.

Athienitis, A.K. (1994). *Building thermal analysis*. Boston,MA, USA, MathSoft Inc.

Athienitis, A.K. and O'Brien, W., Eds. (2014). *Modelling, design and optimization of net-zero energy buildings* Solar heating and cooling, Berlin Ernst, Wilhelm & Sohn.

Athienitis, A.K., Stylianou, M. and Shou, J. (1990). A methodology for building thermal dynamics studies and control applications. *ASHRAE Transactions* 96(2).

Chen, Y., Athienitis , A. K. and Galal, Kh. (2013a). Frequency domain and finite difference modeling of ventilated concrete slabs and comparison with field measurements: Part 1, modeling methodology. *International journal of heat and mass transfer* 66: 948-956.

Chen, Y., Athienitis , A. K. and Galal, Kh. (2013b). Frequency domain and finite difference modeling of ventilated concrete slabs and comparison with field measurements: Part 2. Application. *International Journal of Heat and Mass Transfer* 66: 957-966.

Chen, Y., Athienitis , A. K. and Galal, Kh. (2014). A charging control strategy for active building-integrated thermal energy storage systems using frequency domain modeling. *energy and Buildings* 84: 651-661.

Saberi Derakhtejani, Ali, Candanedo, José, Chen, Yuxiang, Dehkordi, Vahid and Athienitis, A.K. (2015). Modeling approaches for the characterization of building thermal dynamics and model-based control: A case study. *ASHRAE Science and Technology for the Built Environment* 21(6): 824-836.

Wang, Sh. and Xu, X. (2006). Parameter estimation of internal thermal mass of building dynamic models using genetic algorithm. *Energy Conversion & Management* 47: 1921-1941.

Xu, Xinhua and Wang, Shengwei (2007). Optimal simplified thermal models of building envelope based on frequency domain regression using genetic algorithm. *Energy and Buildings* 39(5): 525-536.

ACKNOWLEDGEMENT

The financial and technical support of this work is provided through NSERC/Hydro-Quebec Industrial Research Chair and Concordia University. Also, technical support provided by CanmetENEGY-Varennes, Natural Resources Canada is acknowledged with thanks.

Model Selection for Predicting the Return Time from Night Setback

John E. SEEM[1], John M. HOUSE[2*], Carlos F. ALCALA[3]

[1] High Altitude Trading, Inc.,
Jackson Hole, WY, U.S.A.
John.Seem@gmail.com

[2] Johnson Controls, Inc.,
St-Leonard, QC, CANADA
John.M.House@jci.com

[3] Johnson Controls, Inc.,
Milwaukee, WI, U.S.A.
Carlos.Alcala@jci.com

* Corresponding Author

ABSTRACT

This paper reports a comparison of models for estimating the return time from a night setback condition. Fifty-seven models are compared using simulation data that include the influence of climate, building mass, controller tuning, room orientation, and the unoccupied control strategy on return time. Two-parameter models are recommended for estimating the return time for both heating and cooling. The models use the room air temperature and an EWMA of the normalized heating demand (for the heating model) or cooling demand (for the cooling model) as predictor variables. The outdoor air temperature, a common input for predicting return time, is not used in the recommended models, but influences the return time through its effect on the heating and cooling demands.

1. INTRODUCTION

Night setback is a common strategy used to reduce energy use in buildings. It involves decreasing the heating setpoint and increasing the cooling setpoint throughout a building during unoccupied periods. When the building temperature is within the range defined on the lower end by the heating setpoint and on the upper end by the cooling setpoint, heating and cooling are not needed. Thus widening this range enables energy savings. To ensure occupant comfort and maximize energy savings, the building temperature must be returned to the heating or cooling setpoint at occupancy, but not before. The time required to warm up or cool down a building from a night setback condition is referred to as the *return time* and algorithms for predicting return time are commonly referred to as *optimal start algorithms*.

Optimal start algorithms use a model for predicting return time. The model often has parameters that are learned over time. Numerous studies in the literature describe comparisons of candidate models for predicting return time e.g., (Birtles & John, 1985; Levermore, 2000; John & Salvidge, 1986; Seem et al., 1989; Fraisse et al., 1999; Yang et al., 2003; Vrecko et al., 2009). The models commonly employ the temperature of a representative room in the building and outdoor air temperature as predictor variables. Predicting the return time from night setback can be particularly difficult when the temperature of the representative room reaches either the unoccupied heating or cooling setpoint. In the former case, the heating system will be turned on to provide heating to that room and other rooms served by that system. If the control of the heating system results in the room temperature being maintained at a fixed value (e.g., at the unoccupied setpoint), the room temperature will remain essentially unchanged despite the fact that the load on the room could be changing significantly. Furthermore, since the heating system is already being used to maintain the room temperature at the current condition, the capacity available to raise the temperature to the occupied setpoint (when optimal start begins) will vary depending on the load. In this scenario, the utility of the room temperature as a predictor of return time is reduced because significantly different return times could result for the same room temperature.

The objective of this study is to compare candidate models for predicting return time from night setback using simulation data reflecting the influence of climate, building mass, controller tuning, room orientation, and the unoccupied control strategy on the return time of a room. The ultimate goal of the work is to select return time models for heating (i.e., a model for predicting the return time when heating is required) and cooling. This study expands on the previous body of work on optimal start algorithms by considering a more comprehensive set of models, including traditional model forms that use room and outdoor air temperature as predictor variables for return time, and new model forms that include a measure of the heating demand or cooling demand as a predictor variable.

The paper is organized as follows. The next section describes the simulation study carried out to generate data for comparing candidate models. Next, the models are compared through an analysis of the fit of each model to each data set, and the preferred heating and cooling models are selected. Finally, conclusions and future work are presented.

2. SIMULATION STUDY

To compare models for predicting the return time from night setback, data are needed that relate return time to room air temperature, outdoor air temperature, and room heating or cooling demand at the beginning of the optimal start period, and the room air temperature at the end of the optimal start period. (The *optimal start period* is the period of time between the beginning and end of the Warmup or Cooldown state. These states are introduced in the next paragraph.) Computer simulations were used to generate the data. The simulation platform consists of a Modelica building model implemented in Dymola, and the optimal start algorithm implemented as a Matlab function in Simulink. The Dymola and Simulink models run synchronously and exchange data through the middleware TISC (Kossel et al., 2006).

Figure 1 illustrates the simulation platform and the data exchange between the simulation tools. At each time step the optimal start algorithm determines the state of operation (Unoccupied, Warmup, Cooldown or Normal) appropriate for the current time and conditions and passes a binary variable S_{occ} to Dymola indicating whether the building model should use occupied or unoccupied setpoints for room temperature control during the next time step. The building model executes using the appropriate room setpoints and then passes data back to the optimal start algorithm that are used to determine S_{occ} for the next time step. Included in these data are the occupied ($T_{c,set}$) and unoccupied ($T_{c,set,unocc}$) cooling setpoints, the occupied ($T_{h,set}$) and unoccupied ($T_{h,set,unocc}$) heating setpoints, the room (T_{room}) and outdoor (T_{out}) temperatures, and exponentially weighted moving averages (EWMAs) of the normalized room cooling ($\bar{u}_c$) and heating ($\bar{u}_h$) demands. The data also include the current time (t), the time until the next occupied period (t_{next}), and the occupancy status (O). The following sections provide more detailed descriptions of the building model and optimal start algorithm. Important factors influencing return time that are considered in this study are then described.

2.1 Description of the Building Model

The building model in Figure 1 consists of either a single-room or multi-room model interfaced to a control model, a weather model, an occupancy model, and models needed for data exchange (not shown). At each time step a setpoint model provides the room heating and cooling setpoints to the control model based on the value of S_{occ}. For the multi-room model, a single room was selected from the five controlled rooms to be used in the optimal start algorithm. Thus, when the variable S_{occ} changes, indicating the setpoints for the representative room should be changed from unoccupied

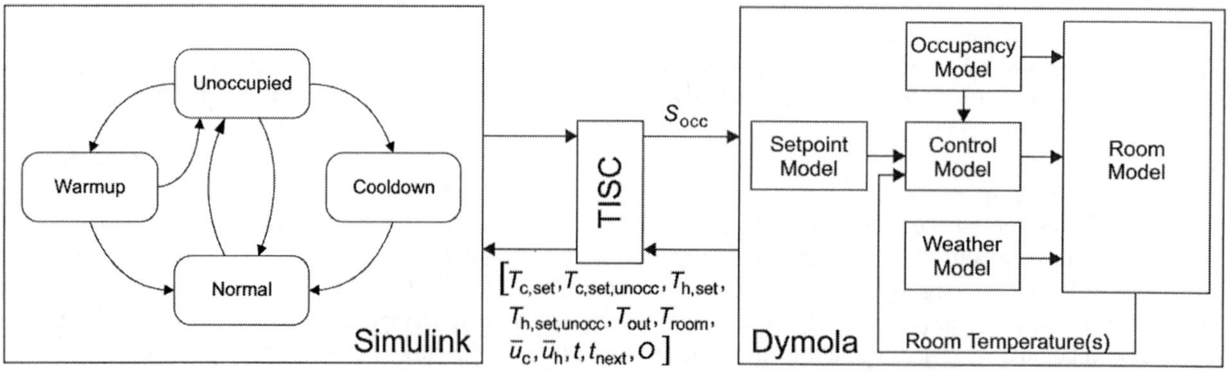

Figure 1: Simulation platform for optimal start data generation

to occupied values (or from occupied to unoccupied), all the room setpoints are changed simultaneously.

The single-room model was adapted from the OneRoom_HVAC model distributed with the HumanComfort Library (Michaelsen & Eiden, 2009) and is modeled as a perimeter room on an intermediate floor of a multi-floor building. The multi-room model was adapted from the VAVReheat model distributed with the Buildings Library (Wetter et al., 2014) and consists of a core room coupled to four perimeter rooms. Collectively, the five rooms are modeled as an intermediate floor of a multi-floor building. Some of the heat transfer mechanisms in the room models are (Wetter et al., 2011; Michaelsen & Eiden, 2009): 1) transient heat conduction through multi-layered walls; 2) heat conduction and solar radiation transmitted through windows; 3) convection at interior and exterior surfaces of walls and windows; 4) solar radiation distribution within the room; and 5) infrared radiation exchange between surfaces in the room.

The room models treat the air as well mixed. Individual rooms have a heat port that connects to the air volume of the room. In the simulations, the room temperature is controlled by adding a sensible convective heat flow to the air volume using this heat port. The air temperature of the individual room is input to a feedback controller (contained in the control model) to determine the sensible heat flow needed to maintain the room temperature within the limits defined by the current temperature setpoints. This sensible heat flow is the room heating (for positive values) or cooling (for negative values) demand.

2.2 Description of the Optimal Start Algorithm

Figure 2 shows a detailed state transition diagram for the optimal start algorithm. As stated previously, there are four states: Unoccupied, Warmup, Cooldown, and Normal. During the Unoccupied state, unoccupied heating and cooling setpoints are used to control the room temperature. At each time step, the algorithm determines whether it is necessary to initiate either heating (Warmup) or cooling (Cooldown) using a model for predicting the return time for the current conditions. The model used for predicting the return time $\hat{\tau}_{set}$ for heating is given by

$$\hat{\tau}_{set} = w_{h,1}\left(T_{h,set} - T_{room}\right) + w_{h,2}\bar{u}_h + w_{h,3}T_{out} \tag{1}$$

where $w_{h,i}$ are the heating model parameters and other variables are defined in the description of Figure 1. Equation 1 is used if the current value of the room temperature T_{room} is less than the occupied heating setpoint $T_{h,set}$. The model used for predicting the return time for cooling is given by

$$\hat{\tau}_{set} = w_{c,1}\left(T_{room} - T_{c,set}\right) + w_{c,2}\bar{u}_c + w_{c,3}T_{out} \tag{2}$$

where $w_{c,i}$ are the cooling model parameters. Equation 2 is used if the current value of T_{room} is greater than the occupied cooling setpoint $T_{c,set}$. The predicted return time $\hat{\tau}_{set}$ calculated from Equation 1 or 2 is compared with the time remaining until the next occupied period t_{next} and operation transitions to the appropriate state (Warmup or Cooldown) if the time remaining until the next occupied period is less than or equal to $\hat{\tau}_{set}$. If the conditions for applying Equations 1 and 2 are not satisfied, or if $t_{next} > \hat{\tau}_{set}$, operation continues in the Unoccupied state unless the occupancy schedule forces a transition to the Normal state. Prior to a transition to either the Warmup or Cooldown state, variables needed to update the parameters of the appropriate model must be stored.

The nomenclature $\hat{\tau}_{set}$ in Equations 1 and 2 emphasizes that the initial prediction of the return time uses the occupied heating setpoint as the target temperature during Warmup and the occupied cooling setpoint as the target temperature during Cooldown. The EWMA of the normalized heating demand $\bar{u}_h$ is determined by calculating the EWMA of the sensible heating divided by the maximum possible sensible heating (determined from sizing simulations). The EWMA of the normalized cooling demand $\bar{u}_c$ is determined in a similar way. One-minute sampled values and a smoothing constant of 0.05 are used to calculate the EWMAs.

During the Warmup state, the HVAC system is heating the room to raise the room temperature to the occupied heating setpoint temperature. If the room temperature rises above the target temperature (in this case the occupied heating setpoint minus an offset ε), control transitions to the Normal state. If the HVAC system has inadequate heating capacity, the room temperature may not reach the target temperature during the Warmup state. In this situation control transitions back to the Unoccupied state at the end of scheduled occupancy for that day. Prior to transitioning from the Warmup state, recursive least squares is used to update the heating model parameters provided the room temperature at the start of Warmup was less than the target temperature.

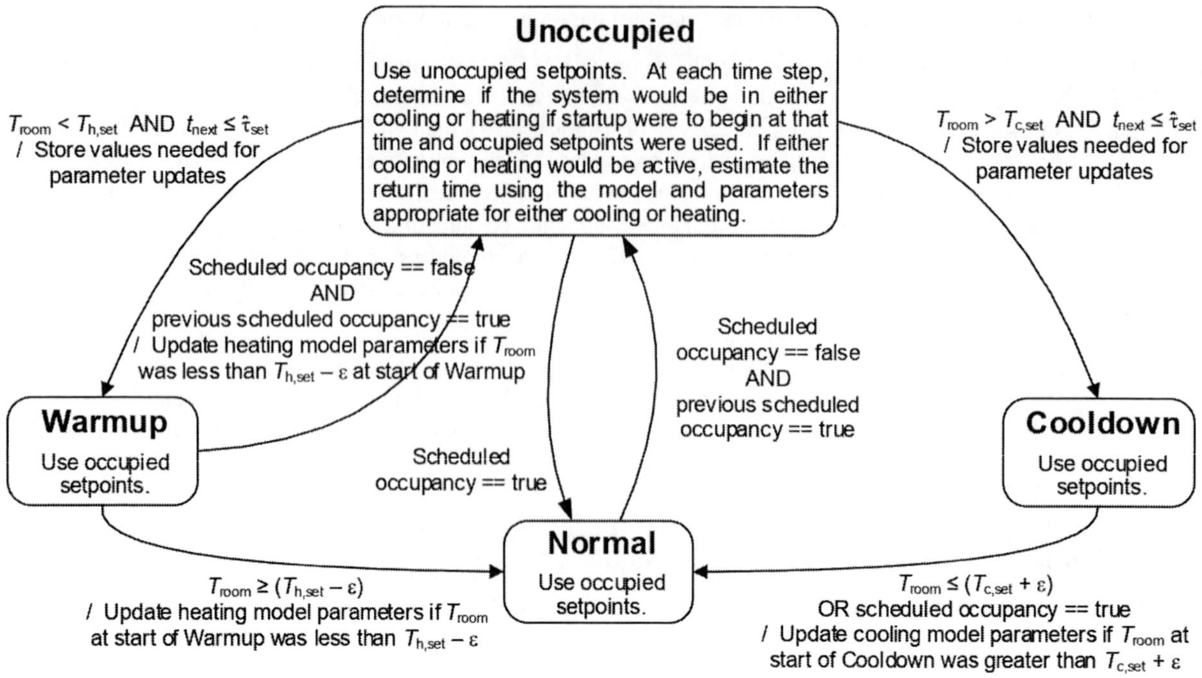

Figure 2: State transition diagram for optimal start algorithm

The offset parameter ε is important for situations when the temperature asymptotically approaches the setpoint. Without the offset, the return time can include a period of time (perhaps significant in length) when the room temperature is nearly equal to the setpoint, but approaching it very slowly. If the model parameters are updated with this data point, the model will learn that it needs to start Warmup earlier (perhaps significantly earlier) for these conditions. The resultant impact on comfort will be minimal, but the impact on energy use could be significant.

During the Cooldown state, the HVAC system is cooling the room to lower the room temperature to the occupied cooling setpoint temperature. If the room temperature drops below the target temperature (in this case the occupied cooling setpoint plus an offset ε), or scheduled occupancy begins, control transitions to the Normal state. Prior to transitioning from the Cooldown state, recursive least squares is used to update the cooling model parameters provided the room temperature at the start of Cooldown was greater than the target temperature.

During the Normal state, the occupied heating and cooling setpoints are used to control the HVAC system. Control transitions to the Unoccupied state when the occupancy schedule changes to unoccupied.

At each time step, the optimal start algorithm uses internal variables and the inputs shown in Figure 1 to determine the state of operation and, based on the state, outputs the binary variable S_{occ} indicating whether occupied or unoccupied heating and cooling setpoints are to be used to control the room temperature. In addition, for each day that requires operation in either the Warmup or Cooldown state, the algorithm updates the appropriate model parameters using the actual return time and predictor variable values for that day. In particular, the room setpoint in Equations 1 and 2 is replaced by the actual room temperature at the end of the Warmup or Cooldown state.

2.3 Factors Affecting Return Time

The optimal start algorithm is intended for use in any building. As such, it is important to consider the main factors that are anticipated to affect the return time, and to assess the performance of the prediction models on a data set representative of these factors. The factors affecting return time considered in this study are climate, building thermal mass, room orientation (e.g., east facing, north facing), controller tuning, and unoccupied control strategy.

To study the influence of climate, simulations were performed using TMY3 weather data (Wilcox & Marion, 2008) for six U.S. cities representing a range of climatic conditions. The six cities considered are Baltimore, Boulder, Chicago, Miami, Phoenix, and Seattle.

The influence of the building mass on return time was studied by considering buildings with heavy and light constructions. For the simulations utilizing the room model from the Buildings Library, the heavy construction buildings have walls and floors that are the same as the Department of Energy (DoE) new construction reference large office building (Deru et al., 2010) for a particular location. The external wall layers of the heavy construction buildings consist of gypsum board, insulation, concrete, and stucco. The light construction buildings have walls and floors that are the same as the DoE reference medium office building for a particular location. The external wall layers of the light construction buildings consist of gypsum board, insulation and wood siding. The floors, ceilings and internal walls for the heavy and light construction buildings are identical.

The simulations utilizing the room model from the HumanComfort Library considered only a heavy construction building. The external wall layer consists of concrete, insulation and a facade plate. The internal walls are hollow cinder blocks and the floor is concrete.

Room orientation can affect return time due to the potential for significant direct solar heat gain through east facing windows in the morning. For cooling, this will lengthen the return time compared to a day that is partly cloudy or cloudy, whereas for heating, it will reduce the return time. East and north facing rooms are included in the simulations.

Controller tuning affects how aggressively the room temperature approaches the setpoint and therefore can influence return time. Cases with "well-tuned" proportional-integral (PI) controller parameters and other cases with controller parameters that result in "sluggish" and "very sluggish" controller behavior were simulated. The control loops of interest are used to regulate the sensible heating and cooling provided to the rooms. In the first step of the tuning process, relay auto-tuning was used to determine the ultimate period and ultimate gain for the rooms. From these values, the Tyreus-Luyben tuning relations (Seborg et al., 2004) were used to calculate the controller gains and integral times representative of "well-tuned" controllers. Sluggish and very sluggish controller behavior are obtained by dividing the well-tuned controller gain by five and ten, respectively.

The strategy used to control the room air temperature during unoccupied periods can impact the estimate of the return time because it affects the room temperature, a common input variable to return time models. Two different strategies are used in this study to control the heating and cooling during unoccupied periods, namely PI control and on-off control. Consider a situation in which the nighttime heating load on a room is sufficient to cause the room temperature to fall below the unoccupied heating setpoint. Assuming there is sufficient heating capacity, PI control will maintain the room air temperature at the unoccupied heating setpoint, whereas on-off control will cause the heating to cycle on and off and the room air temperature will rise and fall accordingly. During this period when the heating is cycling, the room air temperature could vary by several degrees. Because the model for predicting the return time commonly includes the room air temperature, the predicted return time could vary significantly over a 5-10 minute time period despite the fact that the load on the room has likely not changed in any significant way during this time.

Equipment sizing also affects return time. Since the HVAC equipment was not modeled in this study, sizing was performed by determining the maximum sensible convective heating and cooling inputs required to maintain the room temperature between its occupied setpoints for a one-year period (i.e., the building was assumed to be continuously occupied). The maximum sensible heating and cooling inputs for a given climate, building mass, and orientation are scaled by 120% and 140% to simulate equipment oversizing.

3. SUMMARY OF SIMULATION RESULTS

The results described in this section are based on data from 140 simulations. The simulations are one year in length and data were exchanged at one-minute intervals between the optimal start algorithm and building model. The multi-room model was occupied from 7 a.m. until 8 p.m. on weekdays and unoccupied on weekends. The single-room model was occupied from 6 a.m. until 6 p.m. on weekdays and unoccupied on weekends. Holidays were not simulated. All simulations include cooling data (days with return times of greater than 1 minute in the Cooldown state), while 104 of the simulations also had heating data (days with return times greater than 1 minute in the Warmup state). Each simulation took 3-6 hours to complete.

Histograms of the actual return times for heating and cooling are shown in Figure 3. Although difficult to see in Figure 3, the maximum return time for heating (888 minutes) is more than an order of magnitude larger than that for cooling (68 minutes). This is because the Cooldown state occurs early in the morning when the cooling load is relatively small compared to the peak load for which the equipment is sized, while the Warmup state coincides with the period when the peak heating loads are experienced. It is evident that the histograms are heavily weighted toward shorter return times. For heating, 90% of the data have a return time less than 111 minutes, and for cooling, 90% of the data have a return time less than 34 minutes. The longest return times for both heating and cooling occur on Mondays after the building has been unoccupied and in night setback operation for the weekend.

Plots of the return time versus candidate predictor variables for one simulation are shown in Figure 4 for both heating and cooling. The simulation case used the multi-room model, Chicago weather and wall construction, heavy mass, east orientation, very sluggish tuning, PI control during the unoccupied period and 20% oversizing. The plot label ΔT corresponds to $\Delta T_h = T_f - T_i$ for heating, where T_f and T_i are the final and initial room temperatures during Warmup. For cooling, $\Delta T = \Delta T_c = T_i - T_f$ where T_i and T_f are the initial and final room temperatures during Cooldown. The plot label $\bar{u}$ corresponds to $\bar{u}_h$ for heating and $\bar{u}_c$ for cooling. The definitions of ΔT and $\bar{u}$ also apply to tables that follow. Although this is only one simulation out of more than 100, Figure 4 is illustrative when considering the results in the next section.

4. COMPARISON OF MODELS

Table 1 lists the 57 models compared in this study. In Table 1, a_i, b_i and c_i are the model parameters. The statistical indices used for the model comparison and the results of that comparison are presented next.

4.1 Statistical Indices for Model Comparison

Results of the model comparison are reported in terms of two statistical indices. The root mean square of the prediction errors compares the actual and predicted return times for a single simulation and a single model and is given by

$$\text{RMSPE}_{m,s} = \sqrt{\frac{\sum_{i=1}^{n}(\tau_i - \hat{\tau}_{i,-i})^2}{n}} \tag{3}$$

where τ_i is the actual return time for observation i, $\hat{\tau}_{i,-i}$ is the predicted return time for observation i, n is the number of return times for a simulation, and subscripts s and m correspond to a particular simulation case and a particular model. $\hat{\tau}_{i,-i}$ is determined by removing observation i from the data set, performing a regression with the remaining observations, and using the parameter estimates to predict the return time for observation i. This procedure is called cross-validation with one sample left out. To summarize the performance of a model, the average RMSPE for each model over all simulations is determined from

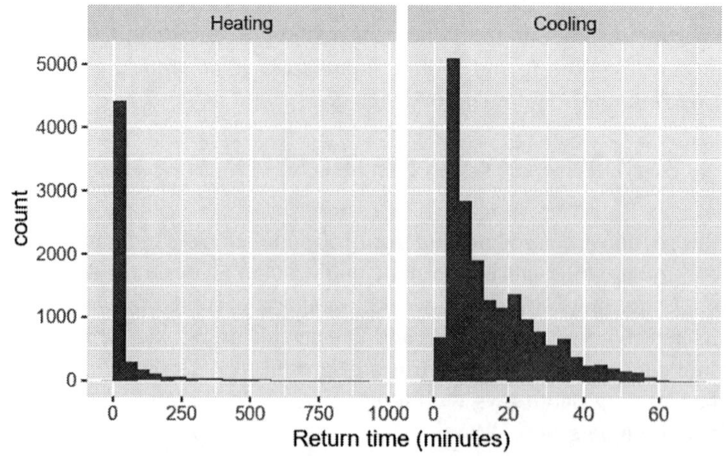

Figure 3: Histogram of actual return times for heating and cooling

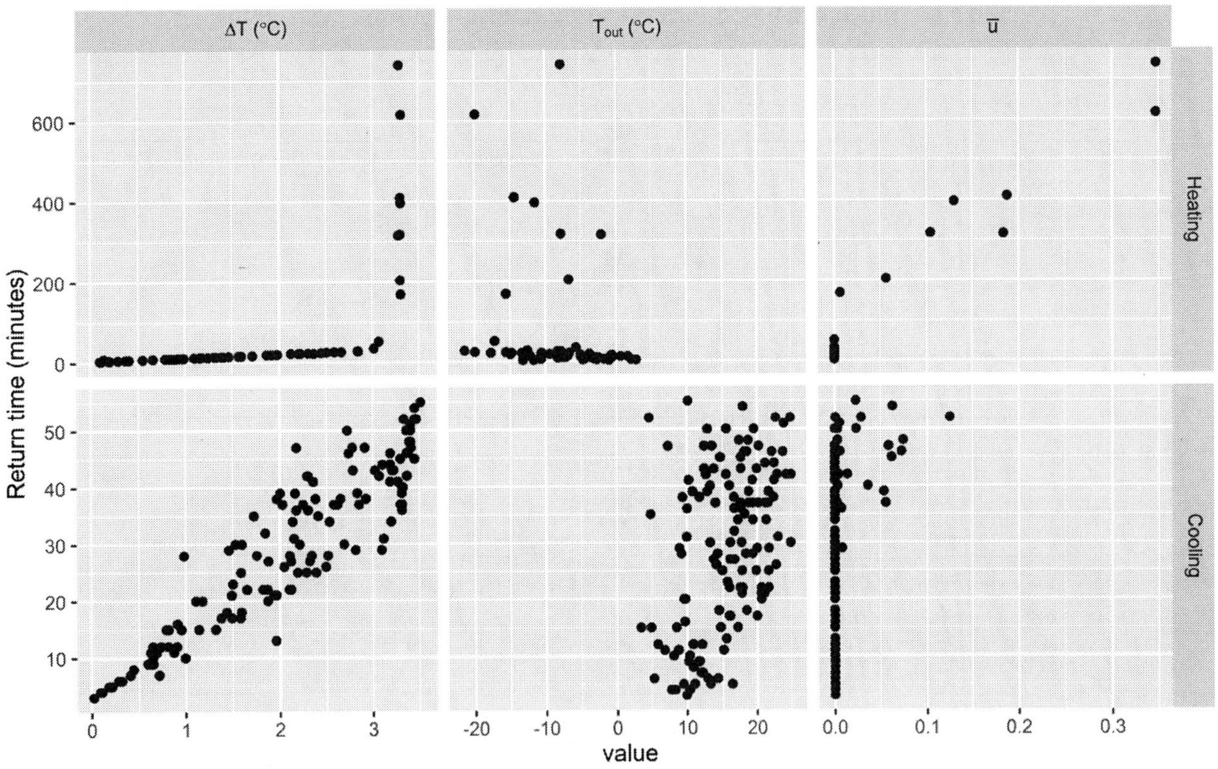

Figure 4: Actual return time versus candidate predictor variables for a single simulation

$$\overline{\text{RMSPE}}_m = \frac{\sum_{s=1}^{S} \text{RMSPE}_{m,s}}{S} \tag{4}$$

where S is the total number of simulations. For each simulation, the difference between the root mean square of the prediction errors for model m and the minimum root mean square of the prediction errors for all models is given by

$$\delta_{m,s} = \text{RMSPE}_{m,s} - \min_{m=1 \text{ to } 57}(\text{RMSPE}_{m,s}) \tag{5}$$

Thus, $\delta_{m,s}$ represents the performance of a particular model for a particular simulation relative to the best performing model for that simulation. For each model, the worst case relative performance across all simulations is given by

$$\delta_{m,\max} = \max_{s=1 \text{ to } S}(\delta_{m,s}) \tag{6}$$

The statistic $\delta_{m,\max}$ can help identify models that have poor performance on one or more simulations despite having good performance on average. All other criteria being equal, preference in the model selection will be given to the model with the smallest value of $\delta_{m,\max}$.

4.2 RESULTS

The top heating models based on their performance over all simulations are compared in Table 2. Each model in Table 2 has the lowest value of $\overline{\text{RMSPE}}_m$ among a subset of models from Table 1. For example, among the models with three parameters having a functional form $\hat{\tau} = f(\Delta T_\text{h}, \bar{u}_\text{h}, T_\text{out})$, model 51 has the lowest $\overline{\text{RMSPE}}_m$.

Table 1: Models compared for estimating return time from night setback

No.	Formula	No.	Formula
1	$\hat{\tau} = a_1(\Delta T)$	30	$\hat{\tau} = a_3(\Delta T)^3 + b_1\bar{u} + b_2\bar{u}^2 + b_3\bar{u}^3$
2	$\hat{\tau} = a_2(\Delta T)^2$	31	$\hat{\tau} = a_2(\Delta T)^2 + a_3(\Delta T)^3 + b_1\bar{u} + b_2\bar{u}^2$
3	$\hat{\tau} = a_3(\Delta T)^3$	32	$\hat{\tau} = a_0 + a_2(\Delta T)^2 + a_3(\Delta T)^3 + b_1\bar{u} + b_2\bar{u}^2$
4	$\hat{\tau} = a_0 + a_1(\Delta T)$	33	$\hat{\tau} = a_1(\Delta T) + c_1 T_{out}$
5	$\hat{\tau} = a_0 + a_2(\Delta T)^2$	34	$\hat{\tau} = a_2(\Delta T)^2 + c_1 T_{out}$
6	$\hat{\tau} = a_0 + a_3(\Delta T)^3$	35	$\hat{\tau} = a_3(\Delta T)^3 + c_1 T_{out}$
7	$\hat{\tau} = a_1(\Delta T) + a_3(\Delta T)^3$	36	$\hat{\tau} = a_0 + a_1(\Delta T) + c_1 T_{out}$
8	$\hat{\tau} = a_0 + a_1(\Delta T) + a_2(\Delta T)^2$	37	$\hat{\tau} = a_0 + a_2(\Delta T)^2 + c_1 T_{out}$
9	$\hat{\tau} = a_0 + a_1(\Delta T) + a_3(\Delta T)^3$	38	$\hat{\tau} = a_0 + a_3(\Delta T)^3 + c_1 T_{out}$
10	$\hat{\tau} = a_0 + a_1(\Delta T) + a_2(\Delta T)^2 + a_3(\Delta T)^3$	39	$\hat{\tau} = a_1(\Delta T) + a_2(\Delta T)^2 + c_1 T_{out}$
11	$\hat{\tau} = a_1(\Delta T) + b_1\bar{u}$	40	$\hat{\tau} = a_2(\Delta T)^2 + c_1 T_{out} + c_2 T_{out}^2$
12	$\hat{\tau} = a_2(\Delta T)^2 + b_1\bar{u}$	41	$\hat{\tau} = a_3(\Delta T)^3 + c_1 T_{out} + c_2 T_{out}^2$
13	$\hat{\tau} = a_3(\Delta T)^3 + b_1\bar{u}$	42	$\hat{\tau} = a_0 + a_1(1-w)(\Delta T) + c_1 w T_{out}$
14	$\hat{\tau} = a_3(\Delta T)^3 + b_2\bar{u}^2$	43	$\log\hat{\tau} = a_0 + a_1(\Delta T) + c_1 T_{out}$
15	$\hat{\tau} = a_3(\Delta T)^3 + b_3\bar{u}^3$	44	$\hat{\tau} = a_0 + a_1(\Delta T) + a_2(\Delta T)^2 + c_1 T_{out}$
16	$\hat{\tau} = a_0 + a_1(\Delta T) + b_1\bar{u}$	45	$\hat{\tau} = a_0 + a_2(\Delta T)^2 + c_1 T_{out} + c_2 T_{out}^2$
17	$\hat{\tau} = a_0 + a_2(\Delta T)^2 + b_1\bar{u}$	46	$\hat{\tau} = a_0 + a_3(\Delta T)^3 + c_1 T_{out} + c_2 T_{out}^2$
18	$\hat{\tau} = a_0 + a_3(\Delta T)^3 + b_1\bar{u}$	47	$\hat{\tau} = a_2(\Delta T)^2 + a_3(\Delta T)^3 + c_1 T_{out} + c_2 T_{out}^2$
19	$\hat{\tau} = a_1(\Delta T) + a_2(\Delta T)^2 + b_1\bar{u}$	48	$\hat{\tau} = a_0 + a_2(\Delta T)^2 + a_3(\Delta T)^3 + c_1 T_{out} + c_2 T_{out}^2$
20	$\hat{\tau} = a_1(\Delta T) + a_3(\Delta T)^3 + b_1\bar{u}$	49	$\hat{\tau} = a_1(\Delta T) + b_1\bar{u} + c_1 T_{out}$
21	$\hat{\tau} = a_2(\Delta T)^2 + b_1\bar{u} + b_2\bar{u}^2$	50	$\hat{\tau} = a_2(\Delta T)^2 + b_1\bar{u} + c_1 T_{out}$
22	$\hat{\tau} = a_2(\Delta T)^2 + b_1\bar{u} + b_3\bar{u}^3$	51	$\hat{\tau} = a_3(\Delta T)^3 + b_1\bar{u} + c_1 T_{out}$
23	$\hat{\tau} = a_3(\Delta T)^3 + b_1\bar{u} + b_2\bar{u}^2$	52	$\hat{\tau} = a_0 + a_1(\Delta T) + b_1\bar{u} + c_1 T_{out}$
24	$\hat{\tau} = a_3(\Delta T)^3 + b_1\bar{u} + b_3\bar{u}^3$	53	$\hat{\tau} = a_0 + a_2(\Delta T)^2 + b_1\bar{u} + c_1 T_{out}$
25	$\hat{\tau} = a_0 + a_1(\Delta T) + a_2(\Delta T)^2 + b_1\bar{u}$	54	$\hat{\tau} = a_0 + a_3(\Delta T)^3 + b_1\bar{u} + c_1 T_{out}$
26	$\hat{\tau} = a_0 + a_2(\Delta T)^2 + b_1\bar{u} + b_2\bar{u}^2$	55	$\hat{\tau} = a_3(\Delta T)^3 + b_1\bar{u} + b_2\bar{u}^2 + c_1 T_{out}$
27	$\hat{\tau} = a_0 + a_3(\Delta T)^3 + b_1\bar{u} + b_2\bar{u}^2$	56	$\hat{\tau} = a_1(\Delta T) + a_2(\Delta T)^2 + b_1\bar{u} + c_1 T_{out}$
28	$\hat{\tau} = a_1(\Delta T) + a_2(\Delta T)^2 + b_1\bar{u} + b_2\bar{u}^2$	57	$\hat{\tau} = a_1(\Delta T) + a_3(\Delta T)^3 + b_1\bar{u} + c_1 T_{out}$
29	$\hat{\tau} = a_1(\Delta T) + a_3(\Delta T)^3 + b_1\bar{u} + b_2\bar{u}^2$		

Among the models with one or two parameters, model 13 is clearly superior based on the statistical measures in Table 2. Performance exceeding that of model 13 can be achieved (e.g., by models 24, 51, 27, 55, and 32), but the improvement comes at a cost of one or more additional parameters and does not appear to be sufficient to justify a more complex model. It is interesting to note that models of the form $\hat{\tau} = f(\Delta T_h, T_{out})$ perform poorly compared with the best performing models of the form $\hat{\tau} = f(\Delta T_h, \bar{u}_h)$. For instance, model 13 is far superior to model 35. Furthermore, the top six models in Table 2 use $\bar{u}_h$ as an input, but only two of the six use T_{out}. Although it is only one simulation, it is apparent from Figure 4 that the return time for the heating data are better correlated to $\bar{u}_h$ than T_{out}. Based on these results, model 13 is recommended for predicting the return time from night setback when heating is needed.

The top cooling models based on their performance over all simulations are compared in Table 3. Note the prediction errors quantified by $\overline{\text{RMSPE}_m}$ are significantly smaller than those for heating in Table 2. This is expected because the return times for cooling are significantly shorter than those for heating.

From Table 3 it can be seen that all the models perform well, with the exception of model 32. For model 32 there is a single simulation for which the RMSPE is 862.5 minutes. This results in the relatively large value of $\overline{\text{RMSPE}_m}$, and the extreme value of $\delta_{m,\max}$. Among the cooling models with one or two parameters, the performances of models 4, 11, and 33 are nearly the same. Model 11 has a slightly larger value of $\overline{\text{RMSPE}_m}$ than the other two models, but a smaller value of $\delta_{m,\max}$ and is recommended over the other two models for this reason. Better performance can be achieved with models having more parameters, but the additional complexity is not justified based on these results. Thus, model 11 is recommended for use for predicting the return time from night setback when cooling is needed.

Table 2: Top performing heating models for different numbers of parameters and inputs (Results in minutes)

Model No.	No. Par.	Formula	$\overline{\text{RMSPE}}_m$	$\delta_{m,\max}$
3	1	$\hat{\tau} = a_3(\Delta T_h)^3$	66.43	141.15
7	2	$\hat{\tau} = a_1(\Delta T_h) + a_3(\Delta T_h)^3$	65.05	142.28
13	2	$\hat{\tau} = a_3(\Delta T_h)^3 + b_1\bar{u}_h$	36.06	80.26
35	2	$\hat{\tau} = a_3(\Delta T_h)^3 + c_1 T_{out}$	65.79	142.40
9	3	$\hat{\tau} = a_0 + a_1(\Delta T_h) + a_3(\Delta T_h)^3$	64.24	142.62
24	3	$\hat{\tau} = a_3(\Delta T_h)^3 + b_1\bar{u}_h + b_3\bar{u}_h^3$	35.29	80.26
38	3	$\hat{\tau} = a_0 + a_3(\Delta T_h)^3 + c_1 T_{out}$	65.53	142.81
51	3	$\hat{\tau} = a_3(\Delta T_h)^3 + b_1\bar{u}_h + c_1 T_{out}$	35.37	75.56
10	4	$\hat{\tau} = a_0 + a_1(\Delta T_h) + a_2(\Delta T_h)^2 + a_3(\Delta T_h)^3$	63.62	146.29
27	4	$\hat{\tau} = a_0 + a_3(\Delta T_h)^3 + b_1\bar{u}_h + b_2\bar{u}_h^2$	34.89	79.63
47	4	$\hat{\tau} = a_2(\Delta T_h)^2 + a_3(\Delta T_h)^3 + c_1 T_{out} + c_2 T_{out}^2$	65.64	145.00
55	4	$\hat{\tau} = a_3(\Delta T_h)^3 + b_1\bar{u}_h + b_2\bar{u}_h^2 + c_1 T_{out}$	34.79	75.56
32	5	$\hat{\tau} = a_0 + a_2(\Delta T_h)^2 + a_3(\Delta T_h)^3 + b_1\bar{u}_h + b_2\bar{u}_h^2$	35.81	61.19
48	5	$\hat{\tau} = a_0 + a_2(\Delta T_h)^2 + a_3(\Delta T_h)^3 + c_1 T_{out} + c_2 T_{out}^2$	65.16	146.48

Table 3: Top performing cooling models for different numbers of parameters and inputs (Results in minutes)

Model No.	No. Par.	Formula	$\overline{\text{RMSPE}}_m$	$\delta_{m,\max}$
1	1	$\hat{\tau} = a_1(\Delta T_c)$	3.44	6.25
4	2	$\hat{\tau} = a_0 + a_1(\Delta T_c)$	2.77	4.98
11	2	$\hat{\tau} = a_1(\Delta T_c) + b_1\bar{u}_c$	2.90	2.85
33	2	$\hat{\tau} = a_1(\Delta T_c) + c_1 T_{out}$	2.88	4.21
9	3	$\hat{\tau} = a_0 + a_1(\Delta T_c) + a_3(\Delta T_c)^3$	2.60	5.04
16	3	$\hat{\tau} = a_0 + a_1(\Delta T_c) + b_1\bar{u}_c$	2.21	1.75
36	3	$\hat{\tau} = a_0 + a_1(\Delta T_c) + c_1 T_{out}$	2.60	4.23
49	3	$\hat{\tau} = a_1(\Delta T_c) + b_1\bar{u}_c + c_1 T_{out}$	2.46	2.15
10	4	$\hat{\tau} = a_0 + a_1(\Delta T_c) + a_2(\Delta T_c)^2 + a_3(\Delta T_c)^3$	2.60	5.12
25	4	$\hat{\tau} = a_0 + a_1(\Delta T_c) + a_2(\Delta T_c)^2 + b_1\bar{u}_c$	2.09	1.74
44	4	$\hat{\tau} = a_0 + a_1(\Delta T_c) + a_2(\Delta T_c)^2 + c_1 T_{out}$	2.44	4.16
52	4	$\hat{\tau} = a_0 + a_1(\Delta T_c) + b_1\bar{u}_c + c_1 T_{out}$	2.02	1.33
32	5	$\hat{\tau} = a_0 + a_2(\Delta T_c)^2 + a_3(\Delta T_c)^3 + b_1\bar{u}_c + b_2\bar{u}_c^2$	8.43	861.90
48	5	$\hat{\tau} = a_0 + a_2(\Delta T_c)^2 + a_3(\Delta T_c)^3 + c_1 T_{out} + c_2 T_{out}^2$	2.58	4.08

To summarize, model 13 is recommended for predicting the return time from night setback for heating and is given by

$$\hat{\tau}_h = w_{h,1}(\Delta T_h)^3 + w_{h,2}\bar{u}_h \tag{7}$$

Model 11 is recommended for predicting the return time from night setback for cooling and is given by

$$\hat{\tau}_c = w_{c,1}(\Delta T_c) + w_{c,2}\bar{u}_c \tag{8}$$

Although the models recommended for predicting the return time for heating and cooling do not use the outdoor air temperature as a predictor variable, the outdoor temperature affects the normalized heating and cooling demand and, therefore, is indirectly included. By including an EWMA of the normalized heating and cooling demand in the models, the recent history of the heating and cooling in the room, including intermittent heating and cooling required to keep the room temperature within the bounds of the unoccupied setpoints, can be captured.

5. CONCLUSIONS AND FUTURE WORK

The performance of 57 models for predicting return time from night setback was compared using simulation data reflecting the influence of climate, building mass, controller tuning, room orientation, and the unoccupied control strategy. The return times for heating were significantly longer than those for cooling and separate models are recommended for the two cases. In each case, a two-parameter model is recommended that uses the room air temperature and an EWMA of the normalized heating or cooling demand as predictor variables. The outdoor air temperature, a common input for predicting return time, is not used in the recommended models, but is indirectly included through its influence on the heating and cooling demands.

Future work will entail testing the preferred models using the same simulation test bed described here and will consider the energy and comfort tradeoffs associated with practical implementation issues such as limiting the maximum allowable return time and changing the unoccupied heating and cooling setpoints.

REFERENCES

Birtles, A. B., & John, R. W. (1985). A new optimum start control algorithm. *Building Services Engineering Research and Technology*, *6*(3), 117-122.

Deru, M., Field, K., Studer, D., Benne, K., Griffith, B., Torcellini, P., ... Crawley, D. (2010). *U.S. Department of Energy commercial reference building models of the national building stock* (Tech. Rep.). Washington, DC: U.S. Department of Energy, Energy Efficiency and Renewable Energy, Office of Building Technologies.

Fraisse, G., Virgone, J., & Yezou, R. (1999). A numerical comparison of different methods for optimizing heating-restart time in intermittently occupied buildings. *Applied Energy*, *62*, 125-140.

John, R. W., & Salvidge, A. C. (1986). The BRE low energy office - five years on. *Building Services Engineering Research and Technology*, *7*(4), 121-128.

Kossel, R., Tegethoff, W., Bodmann, M., & Lemke, N. (2006). Simulation of complex systems using Modelica and tool coupling. In *Proc. of the 5th International Modelica Conference, Vienna, Austria* (p. 485-490).

Levermore, G. J. (2000). *Building energy management systems: Applications to low-energy HVAC and natural ventilation control*. London: E & FN Spon.

Michaelsen, B., & Eiden, J. (2009). HumanComfort Modelica-library: Thermal comfort in building and mobile applications. In *Proc. of the 7th International Modelica Conference, Como, Italy* (p. 403-412).

Seborg, D., Edgar, T. F., & Mellichamp, D. A. (2004). *Process dynamics and control: Second edition*. Hoboken, NJ: John Wiley & Sons, Inc.

Seem, J. E., Armstrong, P. R., & Hancock, C. E. (1989). Algorithms for predicting recovery time from night setback. *ASHRAE Transactions*, *95*, 439-446.

Vrecko, D., Vodopivec, N., & Strmcnik, S. (2009). An algorithm for calculating the optimal reference temperature in buildings. *Energy and Buildings*, *41*, 182-189.

Wetter, M., Zuo, W., Nouidui, T., & Pang, X. (2014). Modelica Buildings library. *Journal of Building Performance Simulation*, *7*(4), 253-270.

Wetter, M., Zuo, W., & Nouidui, T. S. (2011). Modeling of heat transfer in rooms in the Modelica "Buildings" library. In *Proc. of the 12th Conference of International Building Performance Simulation Association, Sydney, Australia* (p. 1096-1103).

Wilcox, S., & Marion, W. (2008). *Users manual for TMY3 data sets* (Tech. Rep. No. NREL/TP- 581-43156). Golden, CO: National Renewable Energy Laboratory.

Yang, I., Yeo, M., & Kim, K. (2003). Application of artificial neural network to predict the optimal start time for heating system in building. *Energy Conversion and Management*, *44*, 2791-2809.

ACKNOWLEDGMENT

The authors would like to thank Tim Salsbury for his contribution to the discussion of the model selection procedure.

Model Predictive Control of a Radiant Floor Cooling System in an Office Space

Jaewan Joe[1,3], Panagiota Karava[1,3]*, Xiaodong Hou[2,3], Jianghai Hu[2,3]

[1]School of Civil Engineering, Purdue University, West Lafayette, Indiana, USA 47907
{jjoe, pkarava}@purdue.edu
[2]School of Electrical and Computer Engineering, Purdue University, West Lafayette, Indiana, USA 47907
{hou39, jianghai}@purdue.edu
[3]Center for High Performance Buildings, Ray W. Herrick Laboratories, West Lafayette, Indiana, USA 47907
* Corresponding Author

ABSTRACT

This study presents an optimal control formulation for the operation of a radiant floor system in an open plan office space with an air-cooled chiller as a source. A simulation case study with different control schemes is used to evaluate the potential of the model predictive control for the radiant floor as well as the optimal control coordination of a radiant and air comfort delivery system. The comparison with a reference case of proportional control shows a saving potential for the radiant floor of around 10 to 15.8 %. This results from maintaining the operative temperature at the upper bound and precooling or load shifting. Optimal control coordination of radiant floor and air system yields additional saving of around 2 %. The proposed intuitive formulation of linear programming can be implemented to other control problems with a linear building model and known COP with respect to weather prediction. The formulation is applicable to other complex systems with two or more control systems such as open-plan spaces with several control units or multiple zones (or buildings) with centralized plant.

1. INTRODUCTION

Radiant floor heating and cooling has been investigated for a long time and its superior performance in terms of energy savings and improved comfort, have been revealed in many studies (Fabrizio, 2012; Kim and Olesen, 2015a; Kim and Olesen, 2015b; Nall, 2013a; Nall, 2013b; Nall, 2013c; Olesen, 2008; Rhee and Kim, 2015). Specific advantages are: (a) The system is operated with moderate temperature so the efficiency of the plant is higher. (b) The room air temperature can be maintained at lower and higher setpoint for the heating and cooling case respectively due to the radiative heat exchange with the large floor surface, thereby less energy is consumed while maintaining equivalent comfort. (c) The large slab surface area yields uniform heat transfer to the room so the thermal comfort is improved.

Many previous studies of the radiant floor system focused on the temperature regulation during the heating season. Conventional feedback control has been implemented to the system (Ahn and Song, 2010; Ahn, 2011; Batista *et al*, 2013; Cho and Zaheer-uddin, 1999; Rhee *et al*, 2011) controlling the valve to maintain the room air temperature (Ahn and Song, 2010; Batista *et al*, 2013; Cho and Zaheer-uddin, 1999), PMV (Rhee *et al*, 2011) and heat flux directly (Athienitis, 1997). However, for all cases, the room air or operative temperature fluctuated more than 2 °C. Controlling the supply water temperature based on outdoor air temperature to prevent the overheating has been suggested (Ahn, 2011) and it has been applied to a cooling case as well (Lin *et al*, 2006), followed by cooling and heating simulation studies (Arteconi et al, 2014; Gwerder, 2008; Gwerder, 2009; Lehmann, 2011; Olesen, 2002; Park *et al*, 2014; Schmelas *et al*, 2015). In some cases, temperature fluctuation was reduced with supply water temperature control (Park *et al*, 2014, Song *et al*, 2008, Seo *et al*, 2014) and with the help of a Dedicated Outdoor Air System (DOAS) in the cooling case (Song *et al*, 2008, Seo *et al*, 2014). However, this improvement was based on a simulation study with a forward modeling approach (Park *et al*, 2014, Song *et al*, 2008), while the test-cell experiment in the same study showed large fluctuation (Song *et al*, 2008). Also, in this case, the cooling load offset by the radiant floor system was reduced due to the cooling rate from the DOAS (Song *et al*, 2008, Seo *et al*, 2014).

Using advanced control methods such as MPC has shown good potential for the radiant floor system due to its ability of incorporating exogenous inputs and predicting thermal dynamics. Although this has been the focus in many studies, its potential has not been fully explored due to the following reasons: (a) The prediction model is not good enough to be used for the MPC and the reason, in some cases, is the use of an unsuitable model structure (Feng *et al*, 2015). (b) The model is not validated with experiment data. Some studies used a forward approach (Nghiem *et al*, 2013) and in others a building simulation program such as EnergyPlus and TRNSYS was used to generate the data for the estimation (Bernal *et al*, 2013; Lehmann *et al*, 2013; Nghiem *et al*, 2012; Oldewurtel *et al*, 2013; Sourbron *et al*, 2013). (c) Overheating for the room temperature was still shown when ANN-based prediction (Lee *et al*, 2002), semi-physical modeling (Váňa *et al*, 2014), and transfer function model (Candanedo *et al*, 2010; Candanedo *et al*, 2011a; Candanedo *et al*, 2011b) were applied.

Another feature of the radiant floor system, besides the advantages discussed so far, is the possibility for load shifting. Many studies focused on the energy saving potential from pre-cooling with air system (Park *et al*, 2014; Turner *et al*, 2015; Braun *et al*, 2001; Braun, 2003; Lee and Braun, 2006a; Lee and Braun, 2006b; Gayeski *et al*, 2011; Favre and Peuportier, 2014), which in some cases included a thermal storage system (Henze *et al*, 2004). Large capacity favors pre-cooling (Favre and Peuportier, 2014), and in the case of radiant floor the potential of the system can be maximized without thermal storage on the plant side. Also, pre-cooling strategies have clearly shown the energy saving potential even with rule-based controls that have been realized with building energy simulation tools (Park *et al*, 2014, Turner *et al*, 2015), well-estimated control-oriented models (Braun *et al*, 2001; Braun, 2003; Lee and Braun, 2006*a*; Lee and Braun, 2006b), and field tests (Braun, 2003). Other recent studies demonstrated the potential of the MPC approach using optimization methods such as pattern search (Gayeski *et al*, 2011) and dynamic programming (Favre and Peuportier, 2014; Henze *et al*, 2004).

The objective of this study is to develop a new optimal control formulation for a radiant floor system, considering a high performance building at Purdue campus as a case-study. The problem is formulated into a linear programming using the capacity of the radiant floor and the HVAC plant. The performance analysis is based on simulations with different control schemes.

2. METHODOLOGY

2.1 Building model

An open plan office space (9.9m by 10.5m) that can host up to 20 occupants is considered as test-bed for this study. Its main features are a radiant floor slab and a south facing double façade system (Figure 1). The building model is Linear Time Invariant (LTI) with 6 states (Figure 2). Details of the model can be found in Joe and Karava (2016). The number of occupants is assumed to be 10 between 08:00 am and 18:00 pm. Occupant and equipment heat gain is 75 W and 100W per person. The minimum outdoor ventilation rate is the summation of the 5 cfm and 0.06 cfm per person and area (ft^2). Initial temperature of all states is assumed to be 24 °C. The operative temperature, which is a linear combination of the air and Mean Radiant Temperature (MRT) is used to control the space.

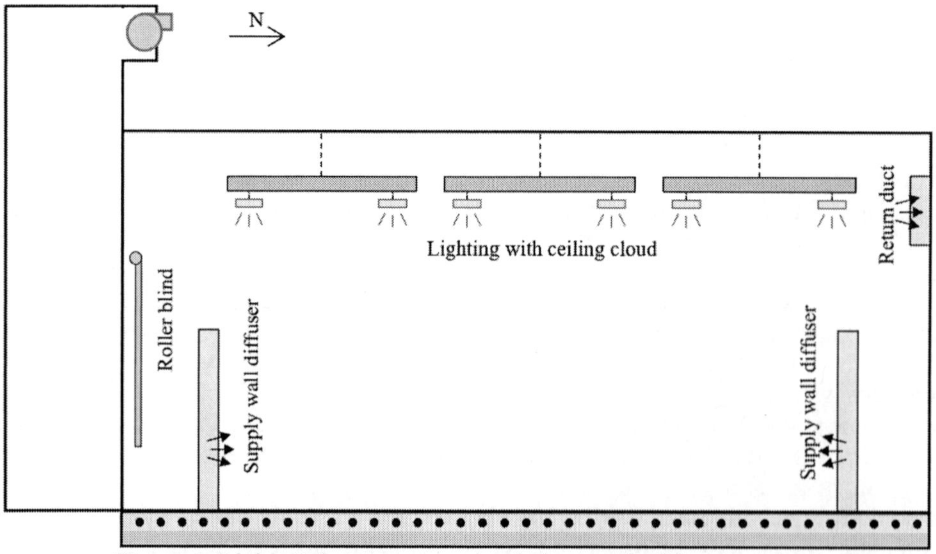

Figure 1: Living Laboratory as a simulation test-bed (section view)

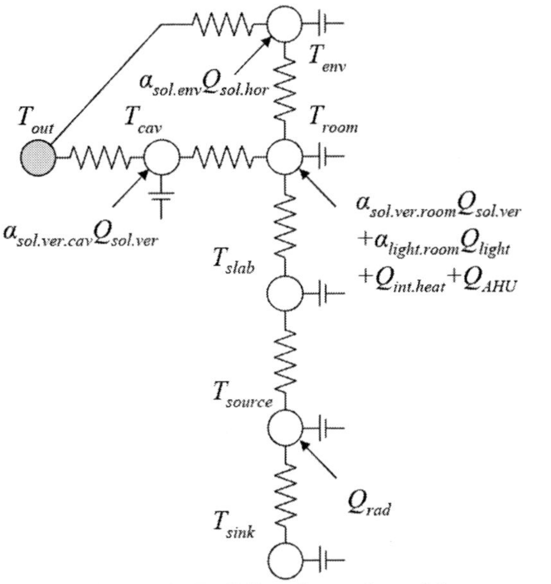

Figure 2: Building thermal model

2.2 HVAC system

An air-cooled chiller is considered as a source for the air and radiant floor system. Performance data are adopted from the EnergyPlus engineering reference and the Energy Input Ratio (EIR) method is applied based on the catalogue data of an actual product, Trane CGAM20, with capacity and Coefficient of Performance (COP) of 68.9 kW and 2.67 COP in nominal condition (EnergyPlus, 2015). For the cases considered in this paper, the chiller capacity is scaled down to 20 %. The electricity consumption of the chiller is a multiplication of three curves that represent the capacity, COP, and Part Load Ratio (PLR). The COP according to different outdoor air temperature and PLR is plotted in Figure 3. Lower outdoor air temperature results in a higher COP for a given PLR. The COP is sharply decreased in lower PLR but slightly decreased in higher PLR, which represents a less than 4 % decrease. In this study, the PLR is neglected in the optimization formulation and the electricity consumption of the chiller is a function of the COP, which can be predicted from the outdoor air temperature, and heat flux from the chiller to the radiant floor. The radiant floor system is assumed to be controlled in different capacity with respect to the concrete temperature, e.g., the capacity is larger for higher concrete temperature. Therefore, the capacity on the radiant floor is a function of the state in the model which is unknown. So the minimum concrete core temperature ($T_{source,LB}$) is implemented in linear programming formulation, which is assumed to be 14 °C in this study. The effectiveness of the radiant floor is found with the NTU method and experimental data from the actual office space. The maximum capacity is calculated as follows:

$$u_{floor.max} = \dot{m}_{water} Cp_{water} (T_{source.LB} - T_{supply.water}) \varepsilon_{effectiveness} \quad (1)$$

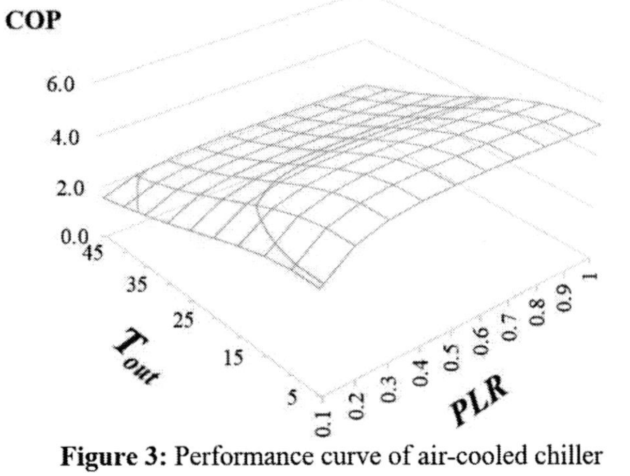

Figure 3: Performance curve of air-cooled chiller

2.3 Optimal control problem formulation

Nonlinear optimization algorithms such as Fmincon, Pattern search, PSO (Particle Swarm Optimization), and Generic Algorithm are easy to be implemented by assigning the control input at each time step as an individual optimization variable but require a lot of computational time and only provide local optimal solutions in the best case scenario. An implementable optimization form in actual controllers is realized with the consecutive state-space equation over the time horizon. When the state-space model is formed, one step ahead temperature is a function of current temperature, and the exogenous (w) and control input (u) as shown in equation 2. Then the temperature trajectory (**X**) is a linear function of a column vector ($\mathbf{\Omega_x}$) consisting of A_d matrix, lower triangle matrix ($\mathbf{\Omega_w}$ and $\mathbf{\Omega_u}$) consisting of A_d, $B_{d,w}$, and $B_{d,u}$ matrix, initial state vector ($\mathbf{X_0}$), and exogenous (**w**) and control input vector (**u**) as shown in equation 3 (capital letter represents the vector). Finally, the input and output trajectories are in an explicit linear relation, which is a suitable form to be implemented in the optimization algorithm to be proposed.

$$x[k+1] = A_d x[k] + B_{d,w} w[k] + B_{d,u} u[k] \quad (2)$$

$$\underbrace{\begin{bmatrix} x[1] \\ x[2] \\ \vdots \\ x[n] \end{bmatrix}}_{\mathbf{X}} = \underbrace{\begin{bmatrix} A_d \\ A_d^2 \\ \vdots \\ A_d^n \end{bmatrix}}_{\mathbf{\Omega_x}} \underbrace{x[0]}_{\mathbf{X_0}} + \underbrace{\begin{bmatrix} B_{d,w} & 0 & \cdots & 0 \\ A_d B_{d,w} & B_{d,w} & \cdots & 0 \\ \vdots & \vdots & \ddots & \vdots \\ A_d^{n-1} B_{d,w} & A_d^{n-2} B_{d,w} & \cdots & B_{d,w} \end{bmatrix}}_{\mathbf{\Omega_w}} \underbrace{\begin{bmatrix} w[0] \\ w[1] \\ \vdots \\ w[n-1] \end{bmatrix}}_{\mathbf{w}} + \underbrace{\begin{bmatrix} B_{d,u} & 0 & \cdots & 0 \\ A_d B_{d,u} & B_{d,u} & \cdots & 0 \\ \vdots & \vdots & \ddots & \vdots \\ A_d^{n-1} B_{d,u} & A_d^{n-2} B_{d,u} & \cdots & B_{d,u} \end{bmatrix}}_{\mathbf{\Omega_u}} \underbrace{\begin{bmatrix} u[0] \\ u[1] \\ \vdots \\ u[n-1] \end{bmatrix}}_{\mathbf{u}} \quad (3)$$

The control input, i.e. the heat flux rate from the plant to the building through the HVAC system is typically considered as a decision variable in previous formulations. However, with the formulation proposed in this study, we can utilize the state which is the temperature of each node together with the control input as decision variables. All dynamics are set through the equality constraint, and bounds for the control input and conditioned zone's temperature are set as bounds on the input and inequality constraint.

2.4 Case study

Four different cases are considered in the simulation study. Case 1 is reference (baseline) case in which the radiant floor system operates with a proportional integral (PI) control. The air system only gives ventilation to the zone by regulating the supply air temperature to be the same with the room air temperature. The coefficients of the PI controller are tuned to maintain the room operative temperature inside the bound.

Case 2 represents a model predictive control strategy of the radiant floor system without considering the COP and with the air system providing only ventilation. In this way, the operative temperature is maintained at the upper bound and the starting time of the radiant floor system is precisely controlled. The objective function is the summation of heat flux from the chiller to the radiant floor system over the prediction horizon. The equality constraint is for the dynamics and the inequality constraints are for the operative and concrete core temperature. Bounds are given to the decision variables considering the capacity of the chiller and radiant floor. The optimization formulation is shown in equation 4. **u**, **X**, and **T** represent the input trajectory, state trajectory, and temperature trajectory vectors. $\mathbf{\Omega}$ represents the lower triangle matrix from equation 3. $\mathbf{C_{op}}$ represents the vector consisting of C matrix calculating the operative temperature in state space equation.

$$\min \begin{bmatrix} -\mathbf{I} & \mathbf{0} \end{bmatrix} \begin{bmatrix} \mathbf{u} \\ \mathbf{X} \end{bmatrix}$$

$$where \begin{cases} \begin{bmatrix} -\mathbf{\Omega_u} & \mathbf{I} \end{bmatrix} \begin{bmatrix} \mathbf{u} \\ \mathbf{X} \end{bmatrix} = \begin{bmatrix} \mathbf{\Omega_X X_0} + \mathbf{\Omega_w w} \end{bmatrix} \\ \begin{bmatrix} \mathbf{0} & \mathbf{C_{op}} \\ \mathbf{0} & -\mathbf{C_{op}} \\ \mathbf{0} & -\mathbf{C_{so}} \end{bmatrix} \begin{bmatrix} \mathbf{u} \\ \mathbf{X} \end{bmatrix} \le \begin{bmatrix} \mathbf{T_{op.UB}} \\ -\mathbf{T_{op.LB}} \\ -\mathbf{T_{so.LB}} \end{bmatrix} \\ \begin{bmatrix} -\min(\mathbf{u_{chiller.max}}, \mathbf{u_{floor.max}}) \\ -\infty \end{bmatrix} \le \begin{bmatrix} \mathbf{u} \\ \mathbf{T_{so}} \end{bmatrix} \le \begin{bmatrix} \mathbf{0} \\ \infty \end{bmatrix} \end{cases} \quad (4)$$

Case 3 represents the optimal control of the radiant floor system with MPC considering the COP. The air system provides ventilation only. The radiant floor system can take advantage of the higher COP with lower outdoor air temperature during the night time, which is typically referred to as pre-cooling or load shifting. The objective function consists of the Energy Efficient Ratio (EER) (1/ COP) and heat flux from the chiller to the radiant floor (equation 5). The same constraints with case 2 are adopted.

$$\min\begin{bmatrix}-\frac{1}{\mathbf{COP}} & \mathbf{0}\end{bmatrix}\begin{bmatrix}\mathbf{u}\\ \mathbf{X}\end{bmatrix}$$

$$where\begin{cases}\begin{bmatrix}-\mathbf{\Omega}_u & \mathbf{I}\end{bmatrix}\begin{bmatrix}\mathbf{u}\\ \mathbf{X}\end{bmatrix}=\begin{bmatrix}\mathbf{\Omega}_X\mathbf{X}_0+\mathbf{\Omega}_w\mathbf{w}\end{bmatrix}\\ \begin{bmatrix}\mathbf{0} & \mathbf{C}_{op}\\ \mathbf{0} & -\mathbf{C}_{op}\\ \mathbf{0} & -\mathbf{C}_{so}\end{bmatrix}\begin{bmatrix}\mathbf{u}\\ \mathbf{X}\end{bmatrix}\le\begin{bmatrix}\mathbf{T}_{op.UB}\\ -\mathbf{T}_{op.LB}\\ -\mathbf{T}_{so.LB}\end{bmatrix}\\ \begin{bmatrix}-\min(\mathbf{u}_{chiller.max},\mathbf{u}_{floor.max})\\ -\infty\end{bmatrix}\le\begin{bmatrix}\mathbf{u}\\ \mathbf{T}_{so}\end{bmatrix}\le\begin{bmatrix}\mathbf{0}\\ \infty\end{bmatrix}\end{cases} \quad (5)$$

Case 4 is a coordinated optimal control of the radiant floor ($\mathbf{u}_1$) and air system ($\mathbf{u}_2$). In this case, the air system provides cooling to the space along with the radiant floor while it is only used for ventilation in case 1, 2, and 3. The optimization formulation is similar to that of case 3 but one more control input is added ($\mathbf{u}_2$), and thereby, an additional inequality constraint is incorporated since the two systems share the same source of energy from the air-cooled chiller (equation 6). Two different inputs, with distinct dynamics, are simultaneously controlled to maintain the operative temperature inside the bound while minimizing the energy consumption.

$$\min\begin{bmatrix}-\frac{1}{\mathbf{COP}} & -\frac{1}{\mathbf{COP}} & \mathbf{0}\end{bmatrix}\begin{bmatrix}\mathbf{u}_1\\ \mathbf{u}_2\\ \mathbf{X}\end{bmatrix}$$

$$where\begin{cases}\begin{bmatrix}-\mathbf{\Omega}_{u1} & -\mathbf{\Omega}_{u2} & \mathbf{I}\end{bmatrix}\begin{bmatrix}\mathbf{u}_1\\ \mathbf{u}_2\\ \mathbf{X}\end{bmatrix}=\begin{bmatrix}\mathbf{\Omega}_X\mathbf{X}_0+\mathbf{\Omega}_w\mathbf{w}\end{bmatrix}\\ \begin{bmatrix}-\mathbf{I} & -\mathbf{I} & \mathbf{0}\\ \mathbf{0} & \mathbf{0} & \mathbf{C}_{op}\\ \mathbf{0} & \mathbf{0} & -\mathbf{C}_{op}\\ \mathbf{0} & \mathbf{0} & -\mathbf{C}_{so}\end{bmatrix}\begin{bmatrix}\mathbf{u}_1\\ \mathbf{u}_2\\ \mathbf{X}\end{bmatrix}\le\begin{bmatrix}\mathbf{q}_{chiller.max}-\mathbf{q}_{vent}\\ \mathbf{T}_{op.UB}\\ -\mathbf{T}_{op.LB}\\ -\mathbf{T}_{so.LB}\end{bmatrix}\\ \begin{bmatrix}-\mathbf{q}_{floor.max}\\ -\infty\\ -\infty\end{bmatrix}\le\begin{bmatrix}\mathbf{u}_1\\ \mathbf{u}_2\\ \mathbf{X}\end{bmatrix}\le\begin{bmatrix}\mathbf{0}\\ \mathbf{0}\\ \infty\end{bmatrix}\end{cases} \quad (6)$$

3 RESULTS ANALYSIS

The prediction horizon for the MPC is 2 days and the optimal control input of the first day is implemented in the simulation. The last state from day one is used as the input to the second day's optimization as initial state. Actual measurements of outdoor air temperature and solar radiation data from the summer of 2015 were used and the deterministic control formulation assumes perfect weather forecast. After a warm up period of 3 days, MPC simulation runs for 8 days. Lower and upper bounds of the operative temperature are 24 and 26 °C, respectively.

For all four cases, temperature profiles and electricity energy consumption of the air-cooled chiller are compared with the results shown in Figure 4. The first graph of each case in Figure 4 represents the temperatures including air, operative, slab, and source, and the second graph shows the capacity and control input. Transportation energy such as pump and fan energy is not considered. In case 1, the operative temperature is inside of the bound as the PI controller was tuned, and the maximum control input is available for every iteration whereas it is not possible in other cases of predictive control. The operative temperature for case 2 stays in the upper bound due to predictive control resulting in a 10% energy saving compared to case 1. In case 3, the potential of the system is maximized by considering the COP which provides an additional saving of 5.8%. The pre-cooling and load shifting is clearly seen by comparing case 2 and case 3. The control input is ON when the EER is high so the temperature during the initial occupied period is lower and it is increased with time. Additional 2% energy saving, compared to case 3, can be achieved if the temperature is close to the upper bound. This is the motivation for case 4 which also includes an air system. The daily (top) and total (bottom) energy consumption are shown in Figure 5.

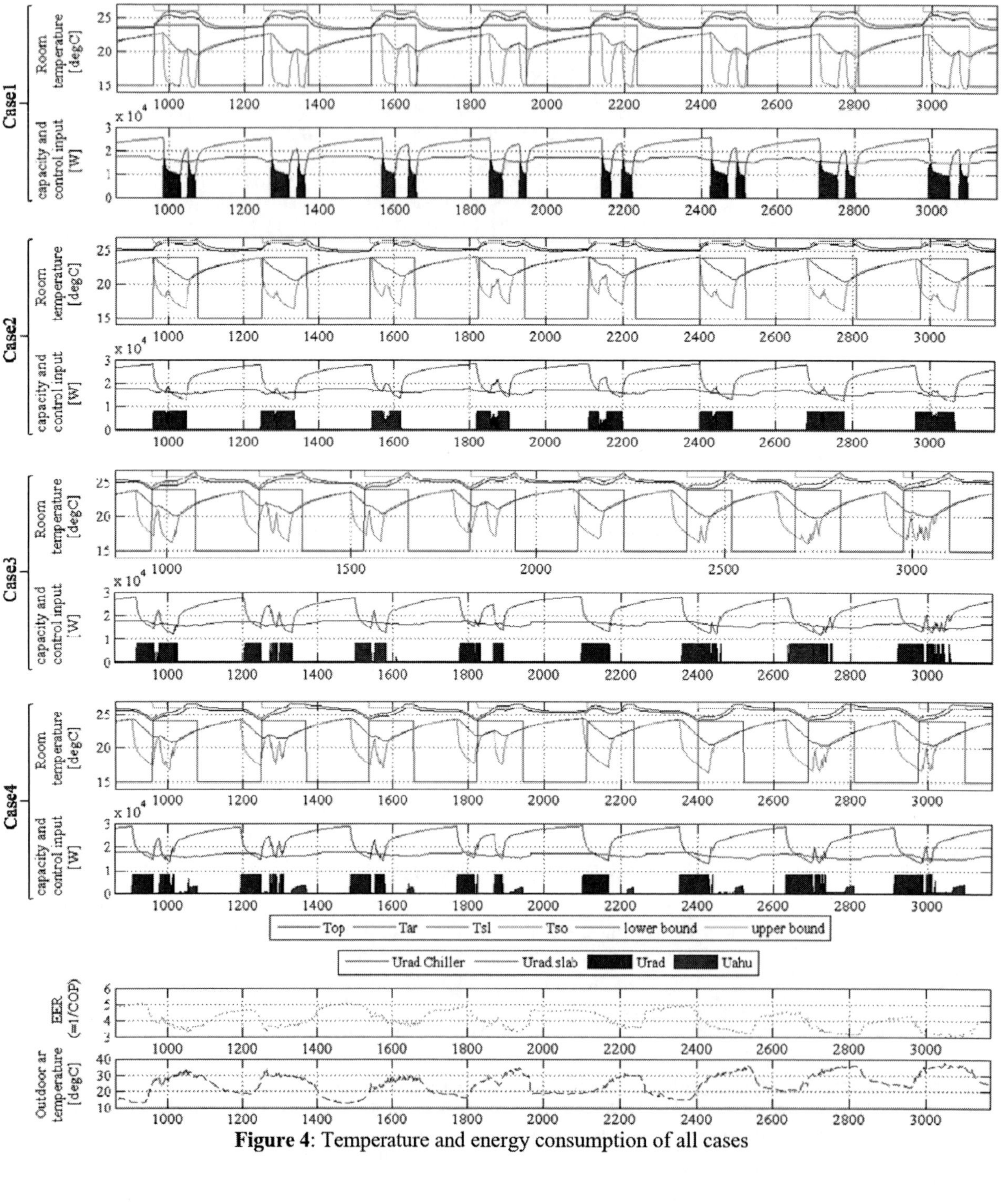

Figure 4: Temperature and energy consumption of all cases

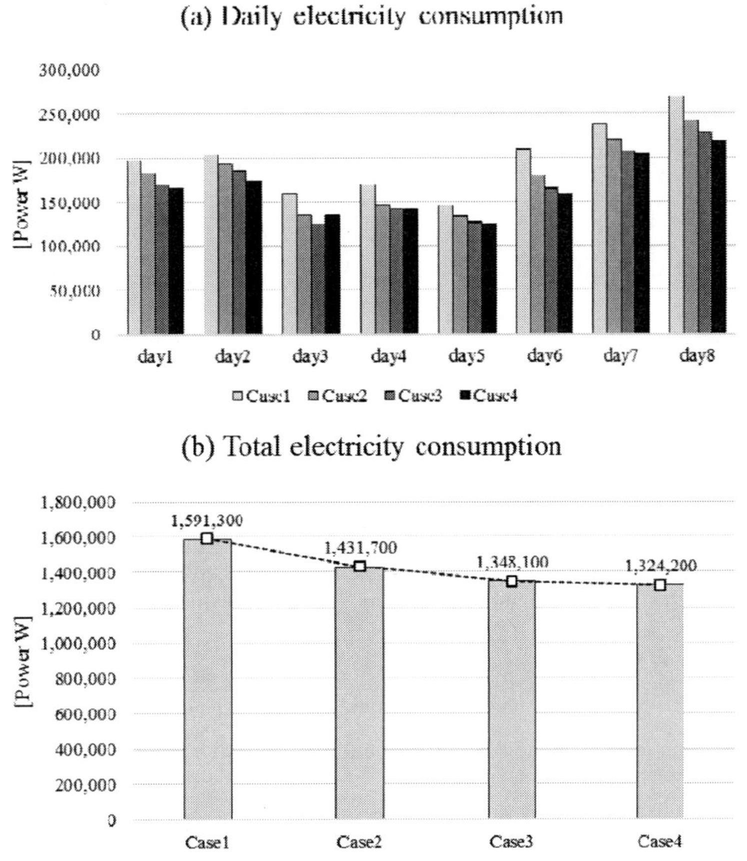

Figure 5: Daily (top) and total (bottom) energy consumption of each case

4. CONCLUSIONS

In this paper, the energy saving potential of a radiant floor cooling system and the coordinated operation of radiant floor and air system in an office building was presented based on a comparative analysis with a simulation study. The optimal control problem was formulated into a linear programming with the decision variables of control input and states considering all constraints of building dynamic and capacity of the plant and radiant floor.

The main findings can be summarized as follows:

- 10 % of energy saving is achieved with MPC compared to the conventional PI control. Additional saving of 5.8 % is achieved from pre-cooling and load shifting.
- Optimal coordination control of radiant floor and air system yields an energy saving of around 2% compared to the optimal control of radiant floor system.

Coordination control is formulated as a centralized optimization problem which requires a central processing unit such as server computer in Building Energy Management System (BEMS). Future work will consider distributed approaches towards plug-and-play building systems with embedded intelligence.

NOMENCLATURE

A	state matrix	(–)
B	input matrix	(–)
Cp	capacitance	(J/kg°C)
k	time step	(–)
m	mass	(kg)
u	control input	(W)

w	disturbance input	(W)
x	state (temperature)	(°C)
ε	effectiveness	(–)
Ω	stacked matrix in state space formulation	(–)

Subscript

d	discrete vector
op	operative temperature
LB	lower bound
UB	upper bound
so	source

REFERENCES

Ahn, Byung-Cheon. (2011). Chaptor7: Radiant floor heating system. Marco Aurelio Dos Santos Bernardes. Developments in Heat Transfer, InTech.

Ahn, Byung-Cheon, and Jae-Yeob Song. (2010). Control characteristics and heating performance analysis of automatic thermostatic valves for radiant slab heating system in residential apartments. *Energy*, 35(4), 1615–1624.

Arteconi, A., Costola, D., Hoes, P., & Hensen, J. L. M. (2014). Analysis of control strategies for thermally activated building systems under demand side management mechanisms. *Energy and Buildings*, 80, 384–393.

Athienitis, A. K. (1997). Investigation of thermal performance of a passive solar building with floor radiant heating. *Solar Energy*, 61(5), 337–345.

Batista, Gilberto, Pedro Dinis Gasper, Pedro D. Silva. (2013). Control, regulation and command system of hydronic radiant floors heating by wireless and energy harvesting sensors and actuators. *Key Engineering Materials*, 543, 389–392.

Bernal, Willy, Madhur Behl, Truong X. Nghiem, Rahul Mangharam. (2013). MLE+: A Tool for Integrated Design and Deployment of Energy Efficient Building Controls. *4th ACM Workshop on Embedded Sensing Systems for Energy-Efficiency in Buildings*, (BuildSys '12), Toronto, Canada.

Braun, James E. and Kyoung-Ho Lee. (2006b). Assessment of Demand Limiting Using Building Mass in Small Commercial Buildings. *ASHRAE Transactions*, 112(1).

Braun, J. (2003). Load Control Using Building Thermal Mass. *Transactions of the ASME*, 125(August), 292.

Braun, J., Montgomery, K., & Chaturvedi, N. (2001). Evaluating the performance of building thermal mass control strategies. *HVAC&R Research*, 7(4), 403–428.

Candanedo, Jose A. and Andreas K. Athienitis. (2011b). Predictive control of radiant floor heating and solar-source heat pump operation in a solar house. *HVAC&R Research*, 17(3).

Candanedo, Jose A., Amelie Allard, Andreas K. Athienitis. (2010). Predictive Control of Radiant Floor Heating and Transmitted Irradiance in a Room with High Solar Gains. *ASHRAE Transactions* 117, December.

Candanedo, Jose A., Amelie Allard, Andreas K. Athienitis. (2011a). Solar-Assisted Radiant Floor Heating in a Net-zero Energy residential Building. *ASHRAE transactions*, Jan.

Cho, S.-H., and Zaheer-uddin, M. (1999). An experimental study of multiple parameter switching control for radiant floor heating systems. *Energy*, 24(5), 433–444.

EnergyPlus. (2015). EnergyPlus engineering reference: the reference to EnergyPlus calculations. Lawrence Berkeley National Laboratory.

Fabrizio, Enrico, Stefano P. Corgnati, Francesco Causone, Marco Filippi. (2012). Numerical comparison between energy and comfort performance of radiant heating and cooling systems versus air systems. *HVAC&R Research*, 18(4).

Favre, B., and Peuportier, B. (2014). Application of dynamic programming to study load shifting in buildings. *Energy and Buildings*, 82, 57–64.

Feng, J., Chuang, F., Borrelli, F., & Bauman, F. (2015). Model predictive control of radiant slab systems with evaporative cooling sources. *Energy and Buildings*, 87, 199–210.

Gayeski, N. T., P., R., Armstrong, L., K., Norford. (2011). Predictive Pre-cooling of Thermo- Active Building Systems with Low- Lift Chillers. Part I : *Control Algorithm. ASHRAE*, (1999), 1–15.

Gwerder, M., Lehmann, B., Tödtli, J., Dorer, V., & Renggli, F. (2008). Control of thermally-activated building systems (TABS). *Applied Energy*, 85(7), 565–581.

Gwerder, M., Tödtli, J., Lehmann, B., Dorer, V., Güntensperger, W., & Renggli, F. (2009). Control of thermally activated building systems (TABS) in intermittent operation with pulse width modulation. *Applied Energy*, 86(9), 1606–1616.

Henze, G. P. (2013). Model predictive control for buildings: a quantum leap? *Journal of Building Performance Simulation*, 6(3), 157–158.

Henze, G. P., Felsmann, C., & Knabe, G. (2004). Evaluation of optimal control for active and passive building thermal storage. *International Journal of Thermal Sciences*, 43(2), 173–183.

Joe, Jaewan, and Panagiota Karava. (2016). Agent-Based System Identification for Control-Oriented Building Models. Under Review.

Kim, Kwang woo and Bjarne Olesen. (2015a). Radiant Heating and Cooling Systems: Part 1. ASHRAE Journal, Feb.

Kim, Kwang woo and Bjarne Olesen. (2015b). Radiant Heating and Cooling Systems: Part2. ASHRAE Journal, Mar.

Lee, Kyoung-Ho and James E. Braun. (2006a). An Experimental Evaluation of Demand Limiting Using Building Thermal Mass in Small Commercial Buildings. *ASHRAE Transactions*, 112(1).

Lee, Jin-Young, Myoung-Souk Yeo, Kwang-Woo Kim. (2002). Predictive control of the radiant floor heating system in apartment buildings. *Journal of Asian Architecture and Building Engineering*, 1(1), 105–112.

Lehmann, B., Dorer, V., Gwerder, M., Renggli, F., & Tödtli, J. (2011). Thermally activated building systems (TABS): Energy efficiency as a function of control strategy, hydronic circuit topology and (cold) generation system. *Applied Energy*, 88(1), 180–191.

Lehmann, B., Gyalistras, D., Gwerder, M., Wirth, K., & Carl, S. (2013). Intermediate complexity model for Model Predictive Control of Integrated Room Automation. *Energy and Buildings*, 58, 250–262.

Lim, J.-H., Jo, J.-H., Kim, Y.-Y., Yeo, M.-S., & Kim, K.-W. (2006). Application of the control methods for radiant floor cooling system in residential buildings. *Building and Environment*, 41(1), 60–73.

Nall, D. H. (2013b). Thermally active floors: Part2. ASHRAE Journal, 55(2).

Nall, D. H. (2013c). Thermally active floors: Part3. ASHRAE Journal, 55(3).

Nghiem, Truong X., George J. Pappas, and Rahul Mangharam. (2013). Event-based Green scheduling of radiant systems in buildings, *Proceedings of the American Control Conference*: 455–460, June 17–19.

Nghiem, Truong X., Madhur Behl, George J. Pappas, and Rahul Mangharam. (2012). Green Scheduling for Radiant Systems in Buildings. *Proceedings of the IEEE Conference on Decision and Control*: 7577–82.

Oldewurtel, F., Parisio, A., Jones, C. N., Gyalistras, D., Gwerder, M., Stauch, V. Morari, M. (2012). Use of model predictive control and weather forecasts for energy efficient building climate control. *Energy and Buildings*, 45, 15–27.

Olesen, Bjame W., Klaus Sommer, Bjorn Duchting. (2002). Control of slab heating and cooling systems studied by dynamic computer simulations. *ASHRAE Transactions*, 108

Olesen, Bjarne. (2008). Radiant Floor Cooling Systems. ASHRAE Journal, Sep.

Park, S. H., Chung, W. J., Yeo, M. S., & Kim, K. W. (2014). Evaluation of the thermal performance of a Thermally Activated Building System (TABS) according to the thermal load in a residential building. *Energy and Buildings*, 73, 69–82.

Rhee, K. N., Yeo, M. S., & Kim, K. W. (2011). Evaluation of the control performance of hydronic radiant heating systems based on the emulation using hardware-in-the-loop simulation. *Building and Environment*, 46(10), 2012–2022.

Rhee, Kyu-Nam. and Kwang Woo. Kim. (2015). A 50 year review of basic and applied research in radiant heating and cooling systems for the built environment. *Building and Environment*, 91, 640–650.

Schmelas, M., Feldmann, T., & Bollin, E. (2015). Adaptive predictive control of thermo-active building systems (TABS) based on a multiple regression algorithm. *Energy and Buildings*, 103, 14–28.

Seo, J.-M., Song, D., & Lee, K. H. (2014). Possibility of coupling outdoor air cooling and radiant floor cooling under hot and humid climate conditions. *Energy and Buildings*, 81, 219–226.

Song, D., Kim, T., Song, S., Hwang, S., & Leigh, S.-B. (2008). Performance evaluation of a radiant floor cooling system integrated with dehumidified ventilation. *Applied Thermal Engineering*, 28(11-12), 1299–1311.

Sourbron, M., Verhelst, C., & Helsen, L. (2013). Building models for model predictive control of office buildings with concrete core activation. *Journal of Building Performance Simulation*, 6(3), 175–198.

Turner, W. J. N., Walker, I. S., & Roux, J. (2015). Peak load reductions: Electric load shifting with mechanical pre-cooling of residential buildings with low thermal mass. *Energy*, 82, 1057–1067.

Váňa, Z., Cigler, J., Široký, J., Žáčeková, E., & Ferkl, L. (2014). Model-based energy efficient control applied to an office building. *Journal of Process Control*, 24(6), 790–797.

ACKNOWLEDGEMENT

This work was supported by the National Science Foundation under Grant number 11329875. Any opinions, findings, and conclusions or recommendations expressed in this material are those of the authors and do not necessarily reflect the views of the National Science Foundation.

From energy signature to cluster analysis: an integrated approach

Lorenza PISTORE[1]*, Giovanni PERNIGOTTO[2], Francesca CAPPELLETTI[3], Piercarlo ROMAGNONI[4], Andrea GASPARELLA[5]

[1]Free University of Bozen-Bolzano, Faculty of Science and Technology, Bolzano, Italy
Phone +39 0471 017700, Fax +39 0471 017009, lorenza.pistore@natec.unibz.it

[2]Free University of Bozen-Bolzano, Faculty of Science and Technology, Bolzano, Italy
Phone +39 0471 017632, Fax +39 0471 017009, giovanni.pernigotto@unibz.it

[3]University IUAV of Venezia, Dept. of Design and Planning in Complex Environments, Venezia, Italy
Phone +39 041 2571295, Fax +39 041 257 1282, francesca.cappelletti@iuav.it

[4]University IUAV of Venezia, Dept. of Design and Planning in Complex Environments, Venezia, Italy
Phone +39 041 2571293, Fax +39 041 257 1282, piercarlo.romagnoni@iuav.it

[5]Free University of Bozen-Bolzano, Faculty of Science and Technology, Bolzano, Italy
Phone +39 0471017200, Fax +39 0471 017009, andrea.gasparella@unibz.it

* Corresponding Author

ABSTRACT

Energy audits of existing buildings are especially important for public buildings and, in particular, for schools, where a more efficient use of energy implies unquestionable benefits to public budgets. Schools audit can drive the public administrator to better address retrofit investments, facilitating the choices of energy efficiency measures in the renovation or operation phases. However, the energy audit of existing buildings can be onerous when the number of buildings is large and requires extensive monitoring campaigns, field surveys and energy performance calculation. A simplified method for building energy diagnosis is the "Energy Signature" (*ES*) method described in the annex B of standard EN 15603:2008. According to this approach, heating and cooling energy uses of a given building are correlated to climatic data over a suitable period. Plotting for several time periods the average heating or cooling power versus the average external temperature provides useful information on building energy performance and allows a fast detection of malfunctions or changes in the building operation/features, as well as the verification of the efficacy of any retrofit intervention. Although this method is preferably adopted in the case of constant internal temperature (e.g., fixed temperature setpoint) and when the external temperature is the most influential parameter (e.g., for buildings with stable and relatively low internal and passive solar gains), it can be applied also recording energy use for heating or cooling and accumulated temperature difference between indoor and outdoor, at average regular intervals (e.g., one hour or, for manual monitoring, a week). The *ES* is the best fitting linear regression between these two quantities and, consequently, can be characterized by means of intercept and slope.

In this paper, the building energy signature parameters have been used to analyze a large set of school buildings and to define the characteristics most influential on the energy needs. In particular, the weekly energy consumptions for heating of a set of 42 school buildings located in the province of Treviso, North East of Italy, have been considered. A cluster analysis based on multiple regression has then been used to identify the buildings' subsets homogeneous as for the features affecting the signature parameters.

1. INTRODUCTION

The Energy Efficiency Directive 2012/27/EU (European Parliament, 2012) has accelerated the spread of energy efficiency strategies by identifying existing buildings as a strategic sector for the reduction of energy consumptions. In particular, the public sector has been set within the European energy policy as the leading one for the promotion of energy efficiency measures. Public authorities represent the first example for the citizens, giving guidance and good

practices to highlight the importance of energy efficiency, changing the behavior and individual choices. According to the directive, in fact, every year starting from 2014, central Government buildings are required to be renovated to meet at least the national minimum energy performance requirements set in application of the Energy Performance of Buildings Directive (European Parliament, 2010). A 3% annual renovation is the target currently prescribed. Achieving this objective requires the management of retrofitting for the whole building stock that means developing strategies suitable for the different building typologies (Geyer *et al.*, 2016). Excluding the climatic conditions, buildings' energy demand, in fact, is strictly related to the geometry shape, the envelope thermal quality, the system efficiency and building management (i.e., opening hours and windows, shading, lighting and system operation). Energy audit of existing buildings can be onerous when the number of buildings is large and it requires extensive monitoring campaigns, field surveys and energy performance calculation. For these reasons, one of the priorities, when a big stock has to be analyzed, is the classification starting from a few kind of data. In the literature, three approaches for energy audit can be distinguished: a case by case approach based on detailed on-site monitoring, the energy signature approach applied to one or more buildings and a clustering method when the number of buildings is high and there is the need of grouping buildings with similar features. In order to compare different retrofit scenarios for a primary school in Moita, Portugal, Brás *et al.* (2015) performed a standard energy audit with data collection by on-site monitoring and interviews. Dall'O and Sarto (2013) analyzed 49 Italian schools to develop three sets of retrofit measures (i.e., standard, cost-effective and high performance) to apply to the whole sample of buildings. As preliminary step, an energy audit campaign was carried out and actual energy consumption for space heating, occupants' behavior and technical characteristics of the buildings were collected. Marinosci *et al.* (2015) used the energy signature as simplified energy audit in the refurbishment of the historical building of the School of Engineering and Architecture of Bologna, Italy. As a whole, energy signature is one of the most utilized tool to investigate and assess controls and management in buildings (Lindelöf *et al.*, 2015; Belussi *et al.*, 2015; Belussi and Danza, 2012; Hitchina and Knightba, 2016), but also to detect building thermal performance information (Danov *et al.*, 2013; Nordström *et al.*, 2012). However, when the sample of buildings is large and a few representative cases have to be found, cluster analysis is the tool mostly applied. For example, Santamouris *et al.* (2007) used clustering techniques to define energy classes based on heating energy consumption of a large sample of schools in Greece. Gaitani *et al.* (2010) used principal component and cluster analysis to group school buildings with similar characteristics and to find the typical school for each level of energy class. Arambula *et al.* (2015) applied cluster analysis in order to group schools with similar characteristics and find representative architectural types and a small number of parameters for an effective description of the energy consumption for heating and hot water production.
In this paper, energy signature and cluster analysis approaches have been combined in order to group a set of 42 school buildings located in the North of Italy. Schools energy signature parameters, i.e., the slope and the zero of linear regression function, have been used as dependent variables to group schools in clusters. The cluster analysis has been used to identify the buildings' homogeneous subsets as for the features affecting the signature parameters.

2. METHODS

2.1 Energy Signatures Implementation

A building energy audit can be performed by the "Energy signature" method described in the annex B of the EN 15603:2008 (CEN, 2008). According to this approach, heating and cooling energy uses of a given building are correlated to climatic data over a suitable period. Plotting for several time periods the average heating or cooling power versus the average external temperature provides useful information on the building energy performance and allows a fast detection of malfunctions or changes in the building operation/features. Generally, the indoor air temperature is considered constant, and assumed to be equal to the setpoint and, for this reason, the external air temperature results the most influential parameter. Energy signature is preferably applied for buildings with stable internal gains and relatively low passive solar gains. Its application requires that energy uses for heating or cooling, as well as average external temperatures or, when possible, accumulated temperature differences are recorded or obtained at regular intervals. These intervals can be as small as one hour, but a week is often used, since this time discretization is long enough to neglect non-linear short-term behaviors due to the building thermal inertia. For this reason, in this work, weekly intervals have been adopted. The indoor air temperature has been fixed at 20 °C during the occupancy time. The average weekly power per unit volume, obtained by dividing the energy use during one week per unit volume by the amount of opening hours per week (Equation 1), has been plotted versus the weekly heating degree-hours during the opening hours (Equation 2).
Energy signatures are characterized by two main parameters: the slope of the regression function and the intersection with the x-axis, hereafter called *zero of the function*. The slope represents the energy performance of the building: the

steeper the slope, the larger is the heating power needed. The zero of the function represents the minimum number of $HDH_{20,occ}$ for which the system has to be turned on: the higher is the intercept, the better is the passive building thermal performance.

$$\Phi = \frac{\sum_{i=1}^{7} EP_{h_i} / V}{\tau} \tag{1}$$

$$HDH_{20,occ} = \sum_{i=1}^{n} (20 - T_{ext})_i \quad with \quad n{=}opening\ hours\ of\ a\ week \tag{2}$$

2.2 Multiple Regression and Clustering

According to the methodology already developed and described in Arambula *et al.* (2015), the first step regards the selection of the most correlated parameters to perform the clustering. The slope and the zero of the function of each building can be correlated to some of them, such as those describing the geometry and the thermal properties of the envelope and the heating system capacity. The influence of each quantity on the slope and the zero is different and the highly correlated parameters and variables can be used to characterize effectively the sample of buildings and to develop the clustering. A list of 12 candidate descriptive quantities has been used: the area of the vertical walls exposed to the external environment, of roof, of floor and of floor in thermal contact with the ground, respectively, A_{vw}, A_r, A_f and $A_{f\text{-}g}$ [m^2]; the total area of opaque and transparent envelope, i.e., $A_{env,o}$ and $A_{env,gl}$ [m^2]; the ratios between windows' area and vertical walls' area A_{win}/A_{vw}, between windows and total floor areas A_{win}/A_f and between transparent and opaque envelope $A_{env,gl}/A_{env,op}$; the average thermal transmittance of the envelope U [W m^{-2} K^{-1}]; the shape factor of the school, expressed in terms of ratio S/V [m^{-1}] between the dissipating surface S and the conditioned volume V and the capacity of the heating system H [kW].

A multiple linear regression has been adopted to find the sets of the candidate quantities which better define homogenous groups, to perform the following clustering. Indeed, the selected quantities can be employed both to identify the groups and to develop linear predictive models for their elements. One regression has been performed for slope and another one for the zero of the function. For each one of the 12 descriptive quantities, the highest value in the whole dataset has been identified and used to normalize the characteristics of each building. The predictors have been grouped in 4083 possible combinations starting from groups with 2 to groups with 12 predictors. For each regression, the adjusted index of determination R^2_{adj} has been calculated and monitored, as well as F-tests and the p-values to check the model's statistical significance and variance inflation factors VIF for the analysis of multi-collinearity issues. Only models with significant p-value with respect to a significance level of 10 % and, preferably, without multi-collinearity issues (i.e., $VIF < 10$) have been considered for the definition of the quantities for the clustering. The combinations of predictors with the highest R^2_{adj} have been selected as set of coordinates to define the "position" of each element in the sample of schools. The next two steps involve the clustering and its validation. K-means approach is one of the most popular techniques in clustering and data mining and it is based on a simple partitional algorithm that tries to find K non-overlapping clusters (Lloyd, 1957; Wu, 2012). By this method, once defined the desired number of clusters K, an equivalent number of centroids is selected and data points are assigned to the closest centroid according to the squared Euclidean distances calculated from the closer centroid. After the definition of the clusters, it is possible to validate them by checking if the combination of predictors with the highest R^2_{adj} with respect to the whole dataset is the best for the cluster as well. If it is not, the combination of predictors with the highest R^2_{adj} is found and used as new coordinate system. If the cluster has an improved but still low R^2_{adj}, and if enough buildings are present to consider sub-clustering (i.e., with at least 25 elements), it is possible to run the algorithm again, but using, this time, the set of parameters with the highest R^2_{adj} for the cluster to split.

Since the whole dataset for the clustering includes 42 elements, and since the objective is to find a matrix of clusters (a slope-zero matrix), we imposed $K^I = K^{II} = 2$ for both clustering and sub-clustering. As it is commonly done in K-means approaches, the initial virtual centroids are randomly generated within the domain of the dataset. After the creation of the initial clusters, the centroids $C_{1,k}$ are calculated and the K-means approach is iterated. The heuristic procedure continues until the determination of the ith combination of centroids $C_{i,k}$ minimizing the global squared Euclidean distances. Since this method is sensitive to the choice of the initial centroids $C_{0,k}$, each clustering has been repeated several times with different random centroid and the solution giving the best improvement with respect to the statistical models has been selected. Once all possible clusters have been defined, the results have been analyzed and the models optimized. The number of predictors, their combination and new regression models have been re-calculated with the elements of each of the cluster in order to find possible improvements to the adjusted index of determination. Also in this case, *F*-values, *p*-values and *VIF* have been analyzed to determine if the models can only describe the data in the clusters or be used also for further extrapolations and predictions. As a final step, the schools closest to the centroids are determined.

3. RESULTS

3.1 Energy Signatures

As specified in the method, the first step has been to elaborate the schools' energy signatures. The results of this approach are shown in Figure 1 where the regression functions of each school energy signature have been plot on a two-axis graph. The slope of the curves represents the normalized heating power per heating degree hour and as it can be seen in Figure 1, almost all the functions have a slope included between 0.005 and 0.075 W m^{-3} K^{-1} h^{-1}. Only school TV032_01 presents an abnormal slope of 0.286, thus revealing a great uncertainty in the reliability of the available data. Concerning the zero of the function, almost all of the curves presents a positive value included between 21.1 and 277 K h, thus revealing a great variability in the schools' thermal inertia towards the outdoor climatic variation. An extreme case is represented by school CN028_09, with an almost null zero of the function value.

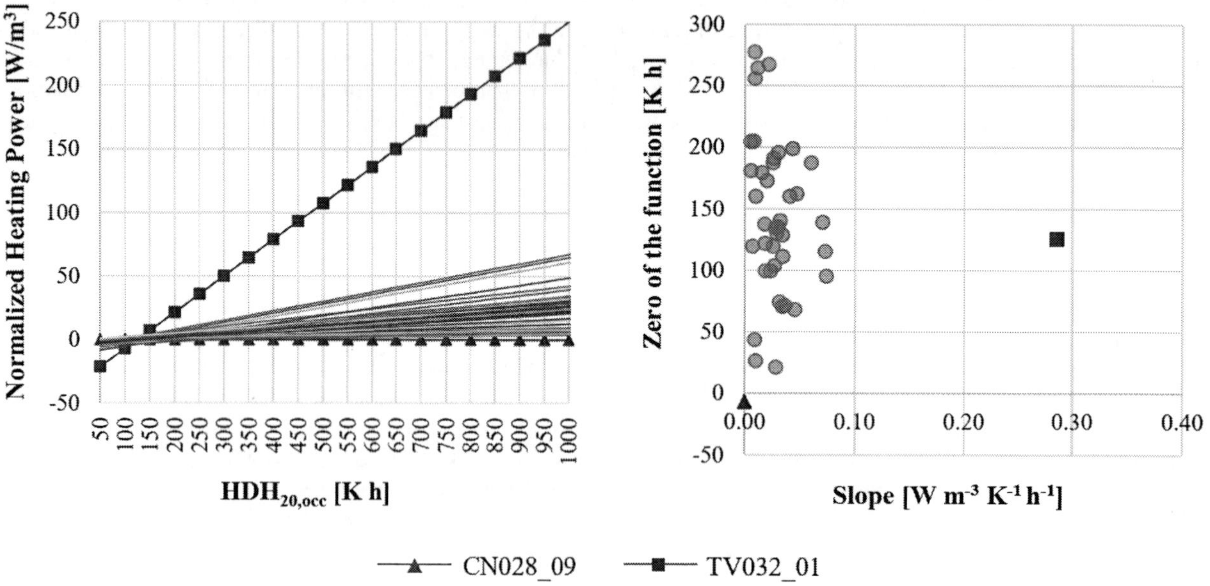

Figure 1: Schools energy signatures regression functions (left) and zero and slope of functions (right)

3.2 Multiple Linear Regression and Clustering

The energy signatures of the set of schools have been used to implement a multiple linear regression taking into account the 12 predictors used to describe the entire stock of buildings as independent variables and the zero and the slopes in turn as dependent variables of a linear function. The linear models have been elaborated starting from the smallest groups (i.e., 2 predictors) and evaluated in terms of R^2_{adj}, *F*-value and *p*-value, taking into account the top-ten configurations output for each set of variables. Results show that the most descriptive configurations have been obtained with a maximum of 6 variables, since the R^2_{adj} tend to increase only from 2 to 6 predictors.

The top-ten configurations among the 924 combinations obtained with 6 predictors, each one identified by an ID number, have been selected for both zero and slopes of the functions (Tables 1 and 2). As it can be seen, the statistical indexes are similar for all of them.

As regards the involved predictors, it can be noticed that floor area, opaque and transparent envelope area and ratio between windows and floor area are the most common variables for the zero of the function's top-10 combinations, whereas for the slope's combinations the most recurring ones are the ratio between vertical walls and windows, the average between the opaque and the transparent envelope, the average thermal transmittance of the envelope, and the compactness ratio *S/V*. In addition to these common predictors, some others descriptive variables differentiate each model.

All the top ten combinations have been used to perform the first clustering and the results are shown in Table 3 and 4. In order to choose the best ID to perform the further steps, a preliminary analysis has been managed according to the statistics and to the number of elements included in each resulted cluster. Looking at their R^2_{adj}, *F*-value, *p*-value and their numeric quantity, the configurations chosen for the further steps are ID805 and ID825 respectively for the zero of the function and the slope.

3.2.1 Zero of the function clustering. The results for the zero of the function clustering are shown in Table 5. ID805 have been grouped in two clusters (i.e., CL_{z1} and CL_{z2}) of 15 and 27 elements. The multiple regression of each cluster has been performed in order to find the best descriptors combination. After this regression, ID819 is the best combination for CL_{z1}, having a R^2_{adj} equal to 0.85, and ID755 the best for CL_{z2}, even with a still very low R^2_{adj}. So, a second clustering has been implemented for CL_{z2}. Combination ID755 has been used and $CL_{z2.1}$ and $CL_{z2.2}$, respectively made of 14 and 13 elements, have been formed. The results have been further optimized with a second regression, performed by ID427 for $CL_{z2.1}$ and ID148 for $CL_{z2.2}$. In this way, three final clusters have been obtained: CL_{z1} with a R^2_{adj} of 0.85, and $CL_{z2.1}$ and $CL_{z2.2}$ with a R^2_{adj} respectively of 0.82 and 0.94, with a considerable improvement of the initial R^2_{adj} values.

Table 1: Zero of the function. Top-10 combinations selected for the first clustering and their descriptors.

	Zero of the function									
ID	**805**	**286**	**861**	**346**	**356**	**614**	**871**	**851**	**872**	**362**
Predictors										
A_{vw}		x		x	x					X
A_r						x				
A_f	x	x	x	x	x	x	x	x	x	X
$A_{f\text{-}g}$	x	x								
$A_{env,op}$			x	x	x	x	x	x	x	X
$A_{env,gl}$	x	x	x	x	x	x	x	x	x	X
A_{win}/A_{vw}	x		x			x	x	x	x	X
A_{win}/A_f	x	x	x		x	x	x	x	x	X
$A_{env,gl}/A_{env,op}$									x	
U				x				x		
S/V	x	x	x	x	x					
H							x			
R^2_{adj}	0.16	0.15	0.15	0.15	0.15	0.14	0.14	0.14	0.14	0.14
F value	2.27	2.24	2.23	2.17	2.16	2.16	2.15	2.15	2.11	2.11
p-value	0.06	0.06	0.06	0.07	0.07	0.07	0.07	0.07	0.08	0.08

Table 2: Slopes. Top-10 combinations selected for the first clustering and their descriptors.

	Slopes									
ID	**902**	**557**	**662**	**761**	**661**	**822**	**627**	**881**	**825**	**762**
Predictors										
A_{vw}										
A_r		x	x		x		x			
A_f										
$A_{f\text{-}g}$		x		x		x			x	X
$A_{env,op}$				x			x	x		X
$A_{env,gl}$	x		x		x					
A_{win}/A_{vw}	x	x	x	x	x	x	x	x	x	X
A_{win}/A_f				x		x				X
$A_{env,gl}/A_{env,op}$	x	x	x			x	x	x	x	X
U	x	x	x	x	x	x	x	x	x	X
S/V	x	x	x		x	x	x	x	x	
H	x			x	x			x	x	
R^2_{adj}	0.17	0.16	0.16	0.15	0.15	0.15	0.15	0.15	0.15	0.15
F value	2.38	2.34	2.28	2.23	2.22	2.21	2.20	2.20	2.19	2.18
p-value	0.05	0.05	0.06	0.06	0.06	0.07	0.07	0.07	0.07	0.07

3.2.2 Slope clustering. The same process has been implemented with the slope combinations (Table 6). For the first clustering, combination ID825 has been used and grouped in 2 clusters (i.e., CL_{s1} and CL_{s2}) of 18 and 14 elements. After the regression for the selection of better predictors, ID442 and ID818 are found optimizing, respectively, CL_{s1} with R^2_{adj} equal to 0.41 and CL_{s2} with a R^2_{adj} of 0.73. While the latter can be considered satisfying, the first cluster has been divided again into 2 clusters, using ID442 which is the best ID that has optimized the results. $CL_{s1.1}$ and $CL_{s1.2}$ are composed of respectively 14 and 13 buildings. The results have been further optimized with a second regression, performed with ID222 for $CL_{s1.1}$ and ID769 for $CL_{s1.2}$. In this way, three final clusters have been obtained: CL_{s2} with a R^2_{adj} of 0.73, and $CL_{s1.1}$ and $CL_{s1.2}$ with R^2_{adj} respectively of 0.60 and 0.99, with a good increasing of the CL_{s1} and CL_{s2} R^2_{adj} values output from the first clustering. In Table 7, the 6 predictors of the best IDs configurations can be observed.

Table 3: Results of the 1st clustering for each ID configuration (zero of the function as dependent variable).

		Zero of the function									
ID		**805**	**286**	**861**	**346**	**356**	**614**	**871**	**851**	**872**	**362**
CL_{z1}	R^2_{adj}	0.516	0.322	0.121	0.558	0.100	0.225	-0.282	-0.011	-0.006	0.004
	F value	3.486	1.870	1.734	3.107	1.558	1.339	0.707	0.941	0.966	1.020
	p-value	0.053	0.254	0.153	0.146	0.203	0.579	0.686	0.483	0.466	0.434
	N	15	12	33	11	31	8	9	34	34	34
CL_{z2}	R^2_{adj}	-0.086	0.102	0.325	0.066	0.286	0.006	0.012	0.597	0.836	0.255
	F value	0.657	1.550	1.641	1.355	1.666	1.034	1.067	2.730	6.937	1.399
	p-value	0.684	0.207	0.426	0.273	0.323	0.425	0.407	0.433	0.283	0.570
	N	27	30	9	31	11	34	33	8	8	8

Table 4: Results of the 1st clustering for each ID configuration (slope as dependent variable).

		Slopes									
ID		**902**	**557**	**662**	**761**	**661**	**822**	**627**	**881**	**825**	**762**
CL_{s1}	R^2_{adj}	0.109	0.200	0.091	0.249	1.00	0.020	0.117	0.141	0.226	0.522
	F value	1.388	1.960	1.469	2.772	-	1.078	1.732	1.929	2.315	2.640
	p-value	0.290	0.129	0.234	0.032	-	0.413	0.152	0.111	0.071	0.228
	N	20	24	29	33	1	24	34	35	28	10
CL_{s1}	R^2_{adj}	0.489	0.566	-0.060	0.896	0.148	0.014	0.763	-	0.629	0.149
	F value	4.349	4.701	0.886	12.484	2.155	1.040	4.753	-	4.676	1.906
	p-value	0.010	0.013	0.556	0.076	0.072	0.450	0.337	-	0.031	0.119
	N	22	18	13	9	41	18	8	7	14	32

Table 5: Results of the clustering and optimization basing on the zero of the function as dependent variable.

Zero of the function	**1st clustering**	**1st regression**	**755- subclustering**	**2nd regression**
ID	**805**	**819**	**755**	**427**
	CL_{z1}	**CL_{z1}**	**$CL_{z2.1}$**	**$CL_{z2.1}$**
R^2_{adj}	0.516	**0.852**	0.285	**0.823**
F value	3.486	14.479	1.862	11.101
p-value	0.053	0.001	0.217	0.003
N	15	15	14	14
ID		**755**		**148**
	CL_{z2}	**CL_{z2}**	**$CL_{z2.2}$**	**$CL_{z2.2}$**
R^2_{adj}	-0.086	0.033	0.906	**0.945**
F value	0.657	1.148	20.211	35.493
p-value	0.684	0.372	0.001	0.000
N	27	27	13	13

Table 6: Results of the clustering and optimization basing on the slope as dependent variable.

Slope	1st clustering	1st regression	442- subclustering	2nd regression
ID	**825**	**442**	**442**	**222**
	CL_{s1}	**CL_{s1}**	**$CL_{s1.1}$**	**$CL_{s1.1}$**
R^2_{adj}	0.226	0.302	0.563	**0.606**
F value	2.315	2.947	4.865	5.614
p-value	0.071	0.030	0.010	0.006
N	28	28	19	19
ID		**818**		**769**
	CL_{s2}	**CL_{s2}**	**$CL_{s1.2}$**	**$CL_{s1.2}$**
R^2_{adj}	0.629	**0.731**	0.850	**0.997**
F value	4.676	6.882	8.548	389.260
p-value	0.031	0.011	0.108	0.003
N	14	14	9	9

Table 7: The 6 predictors resulted from each best ID configuration.

	Zero of the function			Slope		
ID	**819**	**427**	**148**	**818**	**222**	**769**
Predictors						
A_{vw}		x	x		x	
A_r			x	x	x	x
A_f	x	x		x		x
$A_{f\text{-}g}$	x					
$A_{env,op}$				x	x	x
$A_{env,gl}$	x	x	x	x	x	
A_{win}/A_{vw}		x	x	x	x	
A_{win}/A_f	x	x			x	
$A_{env,gl}/A_{env,op}$	x			x		x
U			x			x
S/V		x				
H	x		x	x		x

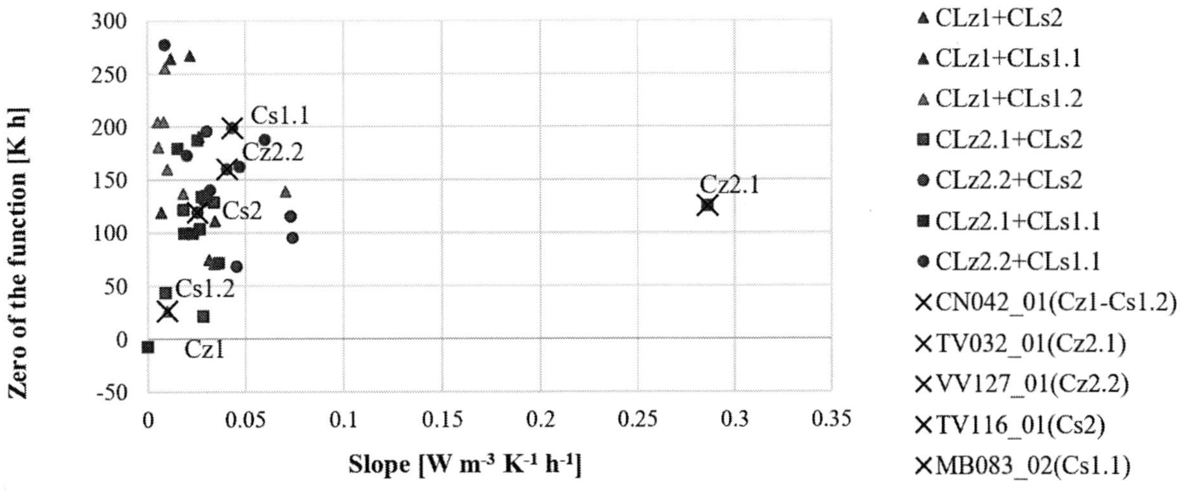

Figure 2: Schools energy signatures slope and zero of function, grouped by clusters

3.2.3 Comprehensive results. In Figure 2, the entire sample of schools grouped by the cluster analysis is represented. Each school position is given according to its belonging to a cluster for both the zero of the function and the slope of its energy signature. The clusters combinations, expressed in the legend, are elaborated using a matrix in which the zero clusters are crossed with the slope ones. The matrix consists of 3 rows and 3 columns, that combined together give a total of 9 possible combinations.

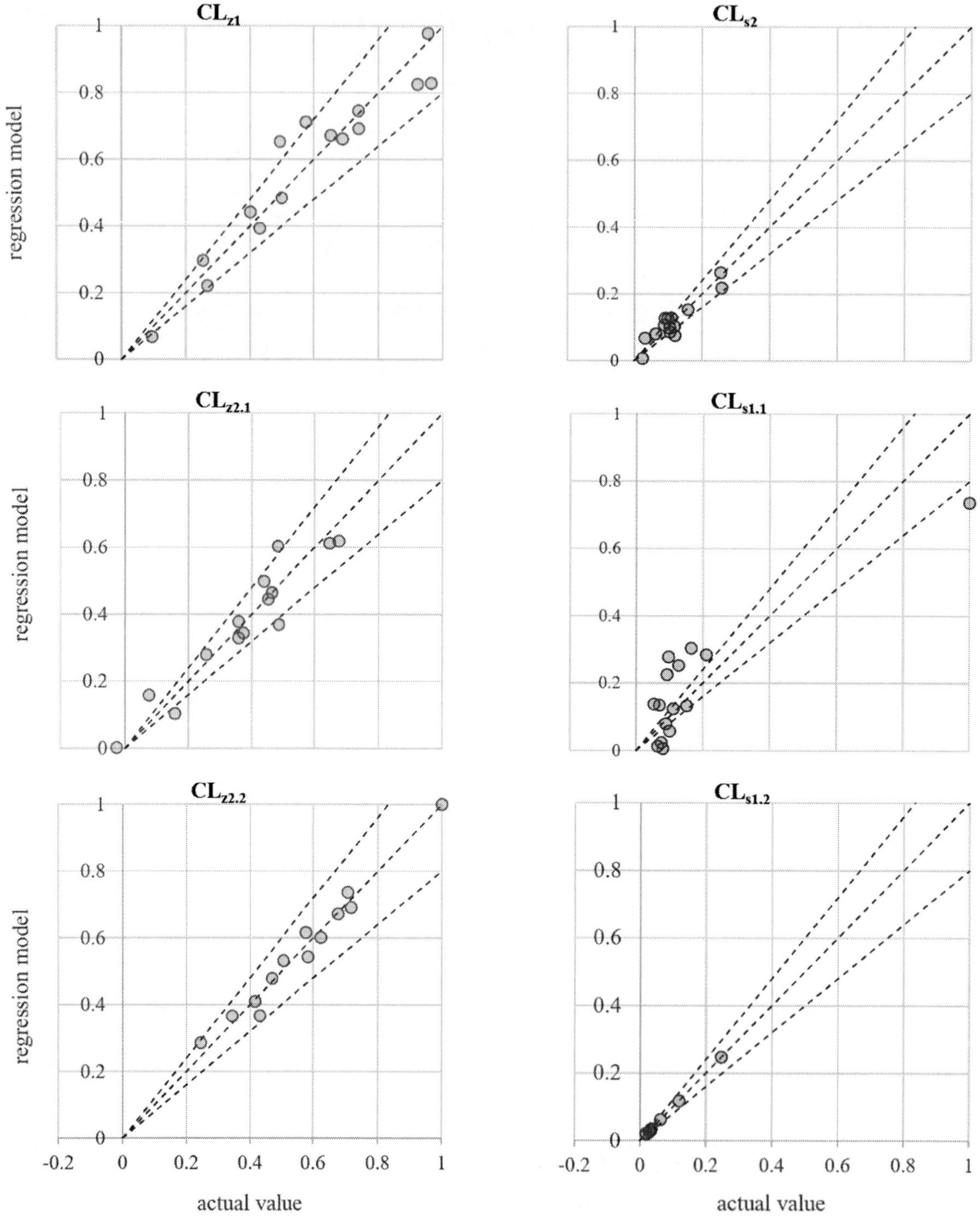

Figure 3: Actual vs. Estimated zero of the function of schools in clusters C_{z1}, $C_{z2.1}$, and $C_{z2.2}$ in green color, and clusters C_{s2}, $CL_{s1.1}$ and $CL_{s1.2}$ in blue color. The dashed lines indicate a deviation of ±20 %.

In the legend, only 7 of these are shown, since no school is included in the $CLs_{1.2}$+$CLz_{2.1}$ and in the $CLs_{1.2}$+$CLz_{2.2}$ cases. Each X indicates the school nearest to the centroid of each cluster, with the exception of school CN042_01 which refers to two clusters, one for the zero and one for the slope.
Finally, Figure3 shows the comparison of the linear regression models to the actual data, for schools included in each cluster, the normalized values of the outputs (i.e., the zero of the function in the left side, the slopes in the right side). In the x-axis there are the measured data, in the y-axis the results of the regression.
As it can be observed, the models fit well for almost all the clusters, since almost all the elements in the stock are within the error band of ±20%. The only exception is $CL_{s1.1}$, in which some points exceed the deviation trend line. This confirms the results of Table 5 and Table 6, where it can be seen that ID222 model regression obtained the lowest value of R^2_{adj}.

4. CONCLUSIONS

In this work the problem of energy auditing a large stock of buildings is discussed. The sample study includes 42 schools in the Province of Treviso, in the North-East of Italy. In order to achieve the purpose, firstly the schools' energy consumptions have been analyzed with the energy signature method, then the K-means cluster analysis has been used for their classification. In addition to this, the multiple linear regression approach has been implemented to validate and optimize the cluster analysis results. The main steps and final goals of the method have been the following ones:

- Starting from relatively few data from a large building stock, using the energy signature method it has been possible to analyze and compare buildings energy performance without onerous and long-term monitoring campaigns.
- Starting from a set of 12 descriptive variables (i.e., predictors), correlated with the normalized zero of the functions and the normalized slopes of the outcome energy signatures, through a first multiple linear regression these predictors have been reduced from 12 to 6.
- After the identification of the best group of predictors, a cluster analysis has been performed and validated through some statistical indices, the adjusted index of determination, *p*-values and *F*-values. This step has been iterated until the clustering output was no more meaningful nor improvable.
- Data in the clusters have been studied and optimized with regressions. For the zero of the function, CL_{z1} has a R^2_{adj} of 0.85, and $CL_{z2.1}$ and $CL_{z2.2}$ with a R^2_{adj} respectively of 0.82 and 0.94, with a considerable improvement from the CL_{z1} and CL_{s2} first R^2_{adj} values. For the slopes, it was obtained a CL_{s2} with a R^2_{adj} of 0.73, $CL_{s1.1}$ and $CL_{s1.2}$ with a R^2_{adj} respectively of 0.60 and 0.99, with also in this case a good increasing of the CL_{z1} and CL_{s2} R^2_{adj} values output from the first clustering.
- For each best ID configuration, it has been possible to select the 6 predictors that describe the set of school included in the clusters, which can be, in additional phase, the areas of intervention to act on.

Further outlooks of this research project are to identify the representative schools for each cluster. Then, taking into account the set of variables most influencing each stock of schools, to compile a list of possible interventions and classify the schools according to a priority of refurbishment. In addition to this, applying a cost-optimal approach, the most convenient solutions can be pointed out. Finally, conducting this approach also in the further years, there is the possibility to check the potentialities of the method described in this paper to highlight the effects of performed energy efficiency measures.

NOMENCLATURE

Symbol		*Subscripts*	
A	area (m²)	*0*	initial
CL	referred to cluster	*20*	referred to an indoor set temperature of 20°C
C	referred to centroid		
EP	Primary Energy	*adj*	adjusted
F	F-test statistic (-)	*env,gl*	referred to transparent envelope
H	capacity of the heating system (kW)	*env,o*	referred to opaque envelope
h	hour (h)	*ext*	external
HDH	heating degree hours (K h)	*f*	referred to floor
K	number of partitions for *K-means* algorithm	*f-g*	floor in thermal contact with the ground

N	number of elements	h	referred to heating
R^2	index of determination (-)	occ	occupancy time
S	dissipating surface (m²)	r	referred to the roof
T	temperature (°C)	s	referred to the slope
τ	time (h)	vw	vertical walls exposed to the external environment
U	thermal transmittance (W m^{-2} K^{-1})		
V	conditioned volume (m³)	win	referred to the windows
VIF	variance inflation factor (-)	z	referred to the zero of the function

REFERENCES

Arambula Lara, R. Pernigotto, G., Cappelletti, F., Romagnoni, P., Gasparella, A. (2015). Energy audit of schools by means of cluster analysis, *Energy Build. 95*, 160-171.

Belussi, L., Danza, L., Meroni, I., Salamone, F. (2015). Energy performance assessment with empirical methods: application of energy signature, *Opto−Electronics Review 23*(1), 83–87.

Belussi L. Danza L., (2012). Method for the prediction of malfunctions of consumption analysis: Holistic and multidisciplinary approach of Energybuildings through real energy Signature, *Energy Build. 55*, 715–720.

Brás A., Rocha A., Faustino P., (2105). Integrated approach for school buildings rehabilitation in a Portuguese city andanalysis of suitable third party financing solutions in EU, *Journal of Building Engineering 3*, 79–93.

CEN, (2008). EN 15603:2008 Energy performance of buildings. Overall energy use and definition of energy ratings, Brussels, Belgium.

Dall'O G. and Sarto L., (2013). Potential and limits to improve energy efficiency in space heating in existing school buildings in northern Italy, *Energy and Buildings 67*, 298–308.

Danov S., Carbonell J., Cipriano J., Martí-Herrero J., (2013). Approaches to evaluate building energy data considering dynamic and solar gain effects performance from daily consumption, *Energy Build. 57*, 110–118.

European Parliament, (2012), Directive 2012/27/EU of the European Parliament and of the Council of 25 October 2012 on energy efficiency, in Official Journal of the European Union, L315/1.

European Parliament, (2010), Directive 2010/31/EU of the European Parliament and of the Council of 19 May 2010 on the energy performance of buildings (recast), in Official Journal of the European Union, L153/13.

N. Gaitani, C. Lehmann, M. Santamouris, G. Mihalakakou, P. Patargias, (2010). Using principal component and cluster analysis in the heating evaluation of the school building sector, *Appl. Energy 87*(6), 2079–2086.

Geyer P., Schlüter A., Cisar S., (2016). Application of clustering for the development of retrofit strategies for large building stocks, *Advanced Engineering Informatics*. In press

Hitchina R. and Knightba I., (2016). Daily energy consumption signatures and control charts for air-conditioned buildings, *Energy Build. 112*, 101–109.

Lindelöf D., Afshari H., Alisafaee M., Biswas J., Miroslav J., Mocellin X., Viaene J., (2015). Field tests of an adaptive, model-predictive heating controller for residential buildings. *Energy Build. 99*, 292–302.

Lloyd S.P., 1957, Least squares quantization in PCM. Bell Telephone Labs Memorandum, Murray Hill, NJ. Reprinted in: *IEEE Trans. Information Theory IT-28* (1982), *vol. 2*, 129–137.

Marinosci C., Morini G.L., Semprini G., Garai M., (2015). Preliminary energy audit of the historical building of the School of Engineering and Architecture of Bologna, *Energy Procedia 81*, 64 – 73.

Nordström G., Johnsson H., Lidelöw S., (2012). Using the Energy Signature Method to Estimate the Effective U-Value of Buildings, in Sustainability in Energy and Buildings, *Proceedings of the 4th International Conference in Sustainability in Energy and Buildings (SEB'12), Volume 22 of the series Smart Innovation, Systems and Technologies, Chapter 4*, pp 35-44.

Santamouris M., Mihalakakou G., Patargias P., Gaitani N., Sfakianaki K., Papaglastra M., Pavlou C., Doukas P., Primikiri E., Geros V., Assi-makopoulos M. N., Mitoula R., Zerefos S., (2007). Using intelligent clustering techniques toclassify the energy performance of school buildings, *Energy Build. 39*, 45–51.

Wu Junjie,(2012). Advances in K-means clustering. A data mining thinking, Springer Thesis, Springer.

ACKNOWLEDGEMENT

The authors would like to thank the Province of Treviso (Provincia di Treviso) for making the schools database available for this research.

A Bayesian Approach for Learning and Predicting Personal Thermal Preference

Seungjae Lee[1, 2], Ilias Bilionis[3], Panagiota Karava[1, 2*], Athanasios Tzempelikos[1, 2]

[1]School of Civil Engineering, Purdue University, West Lafayette, Indiana, USA

[2]Center for High Performance Buildings, Ray W. Herrick Laboratories, Purdue University, West Lafayette, Indiana, USA

[3]School of Mechanical Engineering, Purdue University, West Lafayette, Indiana, USA

* Corresponding Author

ABSTRACT

Typical thermal control systems automated based on the use of "widely acceptable" thermal comfort metrics cannot achieve high levels of occupant satisfaction and productivity since individual occupants prefer different thermal conditions. The objective of this study is to develop environmental control systems that provide personalized indoor environments by learning their occupants and being self-tuned. Towards this goal, this paper presents a new methodology, based on Bayesian formalism, to learn and predict individual occupant's thermal preference without developing different models for each occupant. We develop a generalized thermal preference model in which our key assumption, "Different people prefer different thermal conditions" is explicitly encoded. The concept of clustering people based on a hidden variable which represents each individual's thermal preference characteristic is introduced. Also, we exploited equations in the Predicted Mean Vote (PMV) model as physical knowledge in order to facilitate modeling combined effects of various factors on thermal preference. Parameters in the equations are re-estimated based on the field data. The results show evidence of the existence of multi-clusters in people with respect to thermal preference.

1. INTRODUCTION

Heating and cooling systems in buildings have been automated based on the use of "widely acceptable" thermal comfort metrics and simple heuristic rules. However, field studies show that individual occupants prefer different thermal conditions (Brager & de Dear, 1998; de Dear & Brager, 1998; Fanger, 1967, 1970). As a result, typical thermal control systems cannot achieve high levels of occupant satisfaction and productivity. Moreover, because of the conservative control settings designed for "widely acceptable" conditions, there is high probability of energy waste (Hwang et al., 2009; Nicol & Humphreys, 2009). Studies have recognized that the aforementioned problems can be resolved by incorporating building occupants in sensing and control frameworks, and tuning systems based on individual preferences to achieve personalized indoor environments (Auffenberg et al., 2015; Feldmeier & Paradiso, 2010; Murakami et al., 2007). Learning occupant's preference is an essential part of this innovative concept.

Several studies have been conducted to develop methods for learning individual occupant's thermal preference (Daum et al., 2011; Erickson & Cerpa, 2012; Ghahramani et al., 2014; Ghahramani et al., 2015; Guillemin & Molteni, 2002; Jazizadeh et al., 2014). Although the studies have demonstrated the feasibility of learning occupants' thermal preference and have shown that occupants' thermal satisfaction has been improved by controlling HVAC systems based on it, there are two significant problems associated with the implementation of these methods in real buildings. First, learning requires a long-term data collection, since the rate of data collection is limited by the fact that occupants should not be exposed to potentially uncomfortable conditions for long time. Second, in order to avoid overfitting, the limited data availability imposes model structures that may be too simple to describe the human preference. Therefore, the majority of existing models consider only the effect of air temperature and ignore other parameters, e.g., local airflow, mean radiant temperature that could be important. Although (Humphreys & Nicol, 2002) reported that a prediction made with the air temperature was not inferior to using a complex model in typical buildings, if the other environmental factors either do not highly co-vary with air temperature or not remain constant, the simple model may not provide reliable predictions. For example, in perimeter zones, occupants are affected by both solar radiation and long-wave radiation which may vary significantly according to the sky condition and the position of the sun (La Gennusa et al., 2007). Also, in cases that the HVAC system controls not only air temperature but also other parameters in order to create comfortable thermal environments (e.g., radiant heating/cooling systems or local systems exploiting the effect of increased air velocity), the simple air-temperature based model cannot predict occupant's thermal preference.

In order to overcome these limitations, a new method has been designed for learning and predicting individual occupant's thermal preference based on a Bayesian approach.

2. METHODOLOGY

2.1 Model Structure

Collecting enough data to train a model from an individual occupant is challenging especially if the model structure is complex and many parameters in the model have to be estimated. Therefore, we designed a method to develop a generalized thermal preference model based on a large dataset collected from various people and use the model for learning and predicting an individual occupant's preference by explicitly encoding our key assumption: "Different people prefer different thermal condition" in the model instead of developing different models for each occupant.
In order to develop the model, first, a graph representing connections between occupant's thermal preference and related factors (i.e. environmental, human, contextual, hidden features, and occupant's behaviors) has been developed based on our knowledge and beliefs. Fig. 1 shows a simplified version of the graph. Grey and white nodes correspond to observed and hidden random variables, respectively. The arrows of the graph represent conditional probabilistic relationships between the nodes. An important point is that individual occupant's thermal preference characteristic, which is an unobserved variable, is included in the graph denoting our aforementioned key assumption. This study assumes that occupant's thermal preference is mainly governed by an overall energy balance condition and the individual occupant's thermal preference characteristic (bold elements in Fig. 1) for simplicity.

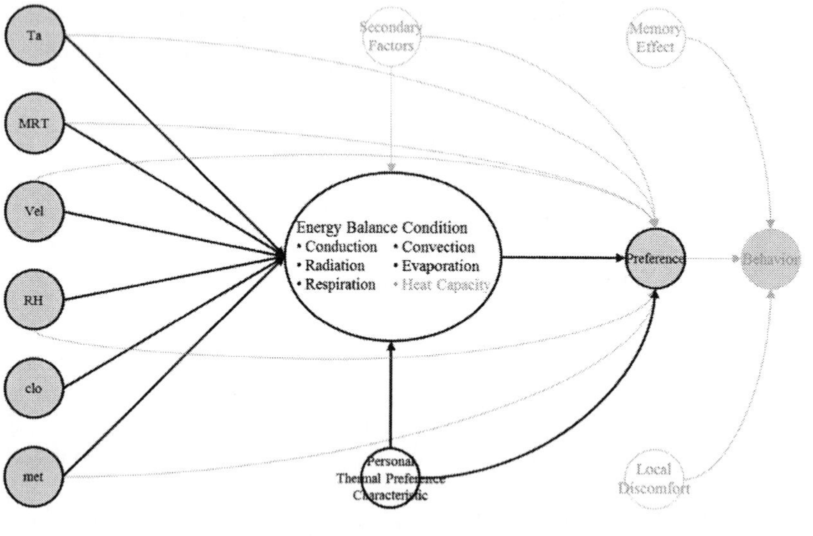

Figure 1: A graph representing our knowledge and beliefs regarding occupants' thermal preference

The graph corresponds to a decomposition of the joint probability density of all the random variables (observed and hidden) (Eq. 1):

$$P(y,E,z,\mathbf{x}) = P(y \mid E,z)P(E \mid z,\mathbf{x})P(z)P(\mathbf{x}) \tag{1}$$

where y denotes occupant's thermal preference, E denotes energy balance condition, z denotes thermal preference characteristic of an individual occupant and $\mathbf{x}$ denotes input variables (i.e. air temperature, MRT, air velocity, relative humidity, clo, met). In this study, y, i.e. occupant's thermal preference, is discretized to take only three values: "want warmer", "no change", "want cooler". Also, by assuming that there is an unknown number K of possible clusters of people who have similar z, i.e. thermal preference characteristic, z is discretized to take K values. In order to quantify the energy balance condition, imbalance between the actual heat flow from the body and the heat flow required for optimum comfort is used.

Eq. 1 represents that the two sub-parts for $P(y \mid E, z)$ and $P(E \mid z, \mathbf{x})$ are required in the model. $P(E \mid z, \mathbf{x})$ predicts the energy balance condition and since the energy transfer processes depend on the combined effects of the input variables, modeling this condition with classical regression methods is challenging and requires estimating many parameters for which we don't have prior knowledge and a large dataset. This problem can be resolved by introducing physical knowledge and results of previous studies in the model and estimating fewer parameters with our prior knowledge. In this study, equations in PMV model developed by Fanger to predict occupant's thermal sensation are used (Fanger, 1970). Although previous studies have shown that occupant's thermal sensation and preference are different (Van Hoof, 2008), and the PMV model with its parameters estimated based on the laboratory study by Fanger cannot explain occupant's thermal preference in real buildings (Barlow & Fiala, 2007; Corgnati et al., 2007; Wagner et al., 2007), it has significant advantages as it calculates the energy balance condition by combining the effects of major environmental factors based on physical mechanisms examined in previous research. Therefore, the equations are introduced in our model as a physical knowledge. However, 15 major parameters (Table 1) are re-estimated since they were originally estimated to predict thermal sensation based on data collected from chamber experiments. Moreover, since six of the major parameters ($\xi_{1:6}$, in Table 1) may relate to the thermal preference characteristic z, the model is designed to allow each cluster to have its own values for these six parameters. $P(y \mid E, z)$ maps the energy balance condition to occupant's thermal preference, and since y is ordered discrete variable, ordered logit function is used for the mapping (McCullagh, 1980). The model allows each cluster to have its own values for parameters in ordered logit function for the same reasons described above.

The aforementioned parts are assembled into one model and the parameters are estimated using a large dataset along with the hidden cluster value of each occupant in the dataset (representing individual occupant's thermal preference characteristic) with a fully Bayesian approach. The rationale behind our modeling choice is related to its inherent advantages: it allows encoding and testing our prior knowledge and beliefs about the relationships of the various variables; it can easily account for hidden (unobserved) variables; and it can seamlessly combine data from heterogeneous sources (Jaynes, 2003). Eq. 2 and Eq. 3 show the posterior density and likelihood of the modeling problem respectively. In order to develop the model in fully Bayesian approach, the parameters and the hidden cluster values should be sampled from the posterior distribution. However, since the posterior distribution is intractable analytically, MCMC (Markov Chain Monte Carlo) is used for the estimation. MCMC provides samples from the posterior distribution for each parameter, and it allows quantifying the uncertainty of the model and its prediction. In this study, we used Python package PyMC and select adaptive Metropolis-hasting algorithm for MCMC sampling (Fonnesbeck et al., 2015).

Detailed information for the parameters and prior distribution for each parameter in the Bayesian estimation is presented in Table 1 and Fig. 2 shows the overall model structure.

$$P(z_{1:D}, \mathbf{w}, \boldsymbol{\theta}, \xi \mid \mathbf{x}_{1:D}, y_{1:D}) = \frac{P(y_{1:D} \mid \mathbf{x}_{1:D}, z_{1:D}, \mathbf{w}, \boldsymbol{\theta}, \xi) P(z_{1:D}) P(\mathbf{w}, \boldsymbol{\theta}) P(\xi)}{P(\mathbf{x}_{1:D}, y_{1:D})} \quad (2)$$

$$
\begin{aligned}
&P(y_{1:D} \mid \mathbf{x}_{1:D}, z_{1:D}, \mathbf{w}, \boldsymbol{\theta}, \xi) = \prod_{d=1}^{D} P(y_d \mid \mathbf{x}_d, z_d, \mathbf{w}, \boldsymbol{\theta}, \xi), \\
&P(y_d \mid \mathbf{x}_d, z_d, \mathbf{w}, \boldsymbol{\theta}, \xi) = \prod_{i=1}^{n_d} P(y_i \mid \mathbf{x}_i, z_d, \mathbf{w}, \boldsymbol{\theta}, \xi), \\
&P(y_i \le j \mid \mathbf{x}_i, z_d, \mathbf{w}, \boldsymbol{\theta}, \xi) = \phi_{z_d,j} = \frac{1}{1+\exp(\mathbf{w}_{z_d} E(\mathbf{x}_i, \xi_{z_d}) - \theta_{z_d,j})}, \quad j = 0,\ldots,2 \\
&P(y_i = j \mid \mathbf{x}_i, z_d, \mathbf{w}, \boldsymbol{\theta}, \xi) = \begin{cases} \phi_{z_d,j} - \phi_{z_d,j-1}, & j = 1,2 \\ \phi_{z_d,j}, & j = 0 \end{cases}
\end{aligned}
\tag{3}
$$

D : The number of people
d : Occupant ID
n_d : The number of data from occupant d
K : The number of clusters
$\mathbf{x}$: Input data
y : Output data (0: want warmer, 1: no change, 2: want cooler)
z : Hidden cluster value (1,...,K)
E : Energy balance condition calculated by PMV equations with input $\mathbf{x}$ and parameters ξ
$\mathbf{w}$: Set of parameters in ordered logit function ($K \times 2$ matrix)
$\boldsymbol{\theta}$: Set of parameters in ordered logit function ($K \times 3$ matrix)
ξ : Set of parameters in PMV equations ($K \times 15$ matrix)

Table 1: Estimated parameters and prior distributions in Bayesian estimation

Parameter		Related Element	Prior Distribution		
$\mathbf{w}, \boldsymbol{\theta}$	Cluster Specific	Ordered logistic function	Normal	Mean: 0	Std.: 100
ξ_1	Cluster Specific	Balance between the actual heat flow from the body and the heat flow required for optimum comfort	Exp.	λ: 3.300	
ξ_2			Normal	Mean: -0.036	Std.: 0.013
ξ_3			Exp.	λ: 35.71	
ξ_4	Cluster Specific	Optimal skin temperature making human comfortable	Normal	Mean: 308.7	Std.: 0.658
ξ_5			Normal	Mean: 0.028	Std.: 0.002
ξ_6	Cluster Specific	Heat emission through sweating	Normal	Mean: 0.42	Std.: 0.099
ξ_7	Shared	Forced convective heat transfer	Normal	Mean: 12.1	Std.: 1.316
ξ_8		Natural convective heat transfer	Normal	Mean: 2.38	Std.: 0.263
ξ_9	Shared	Evaporative heat transfer from the skin	Exp.	λ: 1.0	
ξ_{10}			Normal	Mean: 5733	Std.: 131.6
ξ_{11}			Normal	Mean: 6.99	Std.: 1.656
ξ_{12}	Shared	Clothing area factor estimation based on clo	Exp.	λ: 2.0	
ξ_{13}			Exp.	λ: 0.775	
ξ_{14}			Exp.	λ: 1.818	
ξ_{15}			Exp.	λ: 1.550	

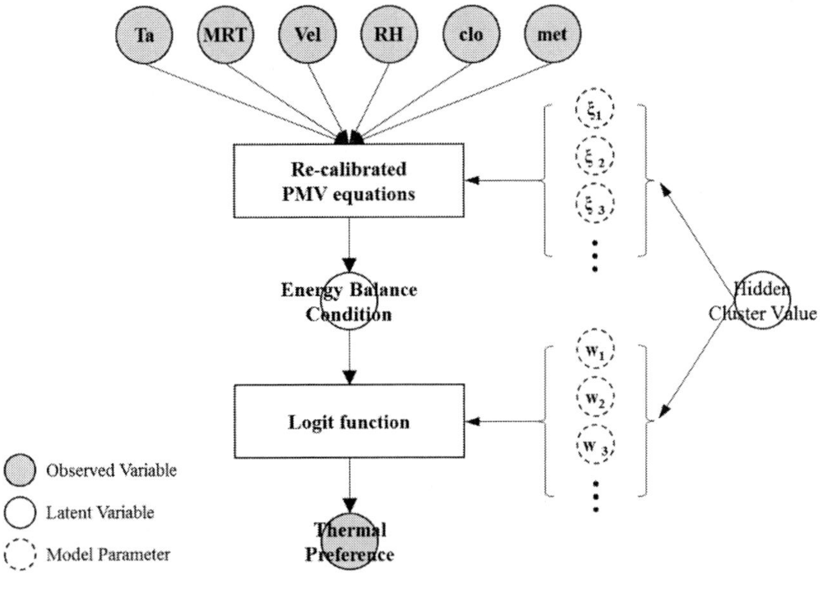

Figure 2: Overall model structure

2.2 Individual Occupant's Thermal Preference

Exactly the same process is used to learn the thermal preference of a new individual occupant. When observations have been collected for the occupant, they are added into the large dataset, and the model is updated based on the new dataset. During this process, $P(z_{new} = k \mid \mathbf{x}_{1:D}, y_{1:D}, \mathbf{x}_{new}, y_{new})$, $k = 1,...,K$, which denotes the probability of the occupant being in cluster k is estimated. Then, occupant's thermal preference under certain condition can be predicted by a mixture of sub-models for each cluster with the set of probabilities following Eq. 4.

$$
\begin{aligned}
&p(y_{\text{inference}} = j|\mathbf{x}_{\text{inference}}, \mathbf{x}_{1:D}, y_{1:D}, \mathbf{x}_{new}, y_{new}, \mathbf{w}, \boldsymbol{\theta}, \xi) \\
&= \sum_{k=1}^{K} p(y_{\text{inference}} = j|\mathbf{x}_{\text{inference}}, \mathbf{x}_{1:D}, y_{1:D}, \mathbf{x}_{new}, y_{new}, \mathbf{w}, \boldsymbol{\theta}, \xi, z_{new} = k) P(z_{new} = k \mid \mathbf{x}_{1:D}, y_{1:D}, \mathbf{x}_{new}, y_{new}), \quad j = 0,\dots,2
\end{aligned} \tag{4}
$$

The overall process for learning and predicting individual occupant's thermal preference is shown in Fig. 3. First, a large dataset collected from various people with which our prior knowledge, beliefs, and hypotheses can be tested is prepared. Then, a cluster model for general people dealing with observed data and unobserved hidden variables are developed. With the dataset and the model, a new individual occupant's thermal preference can be learned and prediction problem can be solved.

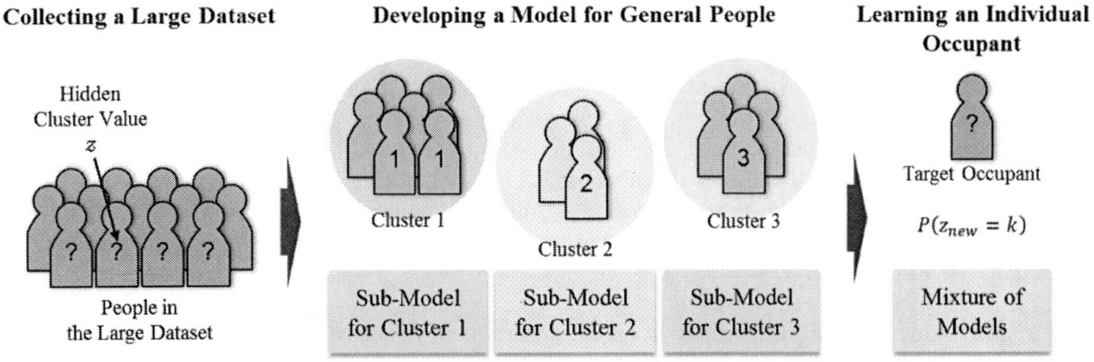

Figure 3: Overall process for learning and predicting individual occupant's preference

2.3 Data

A large dataset regarding occupant's thermal comfort was collected by ASHRAE through RP-884 (de Dear et al., 1997). This dataset consists of physical data (e.g., indoor and outdoor air temperature, MRT, air velocity, etc.), human attributes (e.g., age, gender, etc.), survey data (e.g., thermal preference, sensation, acceptability). In this study, a subset of this dataset with data collected from HVAC conditioned office buildings in North America is used to evaluate the feasibility of our method. The dataset includes 3,070 observations collected from 1,248 occupants and supports the rationale for re-calibrating the parameters in PMV equations instead of using the original PMV model. Fig. 4 shows the thermal preference distribution with respect to the original PMV values. As shown in the figure, some people do not prefer a warmer condition even though the PMV value is quite low. Moreover, a non-negligible percentage of the occupants prefer cooler conditions with PMV of less than -1. In other words, the PMV model with its original parameters cannot predict or explain occupant's thermal preference in real buildings.

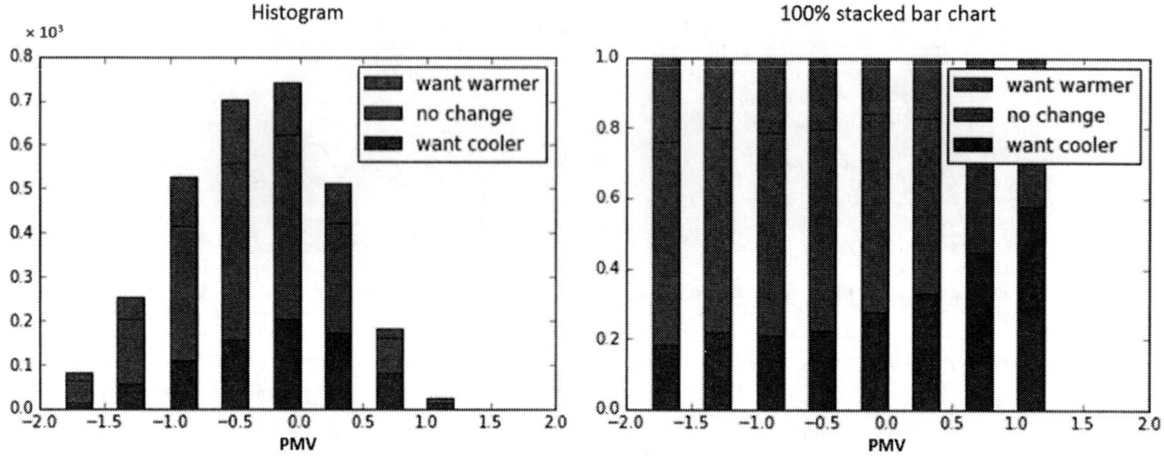

Figure 4: Thermal preference distribution with respect to PMV (ASHRAE RP-884)

3. RESULTS

3.1 Single Cluster Model

Since the thermal preference characteristic is a hidden variable, the number of clusters for the characteristic is also unknown. Therefore, the optimal number of clusters K for this general model should be identified with the large dataset before using this method to learn and predict individual occupant's preference. In order to determine the optimal number of clusters, models which have different number of clusters should be developed and evaluated. First, a model having a single cluster was developed, i.e. modeling without clustering. Fig. 5 shows estimated parameters (histograms) and Fanger's original values (red vertical line) in the equations. As shown in the figure, most of estimated parameters are different from Fanger's values.

Fig. 6 (a) shows the probability distribution of an occupant being in each preference class (i.e., "want warmer", "no change", "want cooler") with respect to the air temperature variation, where MRT is equal to the air temperature, air velocity is 0.1m/s, relative humidity is 50%, met is 1.2, clo is 0.65. In this figure, since each sample from MCMC is equally probable to be a model, the expected probabilities and the associated 95% credible intervals for each set of inputs are calculated with the sample sets, and plotted as solid lines and shaded areas respectively. Since the number of data in both lower and higher temperature ranges are insufficient, the width of the shaded area in these ranges are wider than that of the middle range.

Fig. 6 (b) and (c) show the probability distribution for different air velocity (0.2 m/s) and clo (1.0) respectively. In accordance with general knowledge about thermal comfort, the shape of the probability curves changes. For example, curves move to the right with increased air velocity, i.e. people prefer warmer condition, and curves move to the left with higher clo, i.e. people prefer cooler condition.

However, although the calculated probability distributions follow reasonable trends, since the model is developed with a dataset collected from various people and there are many contradicting observations, the shape of distributions is quite smooth, and the maximum probability of an occupant preferring no change is only around 0.6. As a result, although this model could be used for explaining general thermal preference of people, it cannot be used for learning and predicting individual occupant's thermal preference.

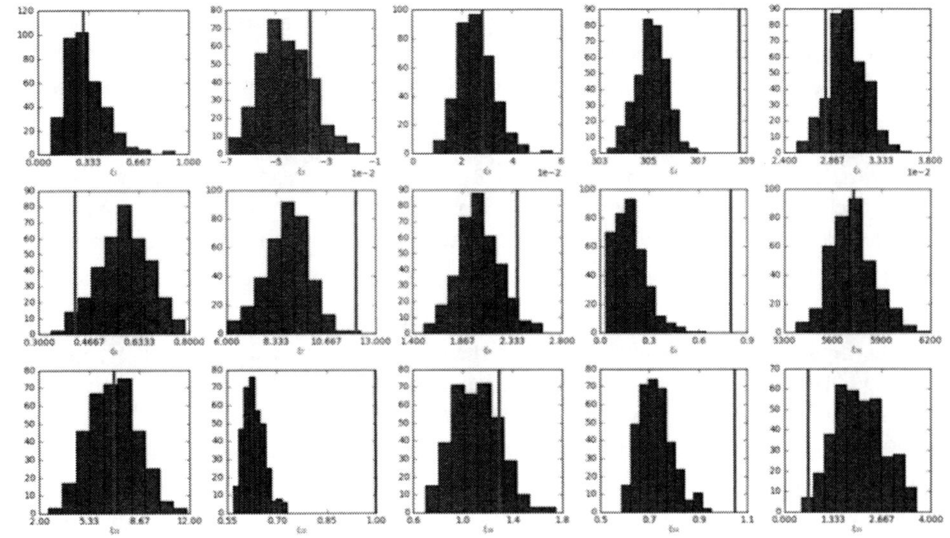

Figure 5: Estimated parameters (histograms) and Fanger's original values (red vertical line)

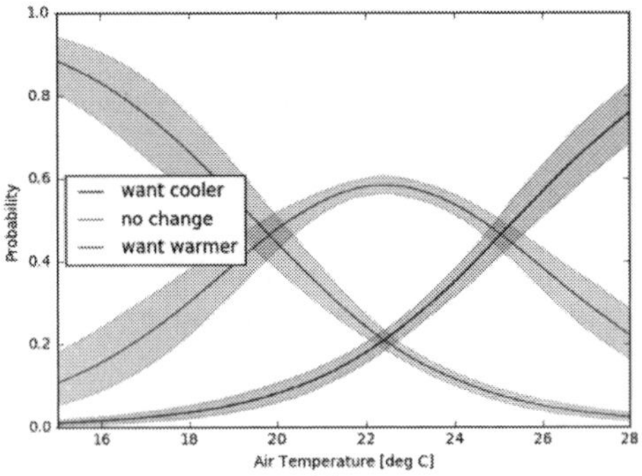

(a)

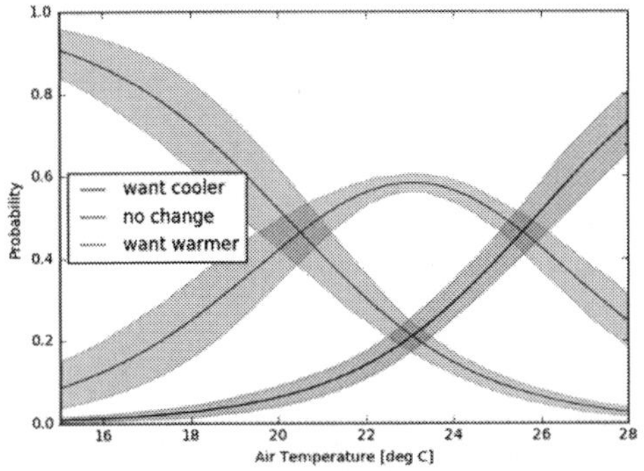

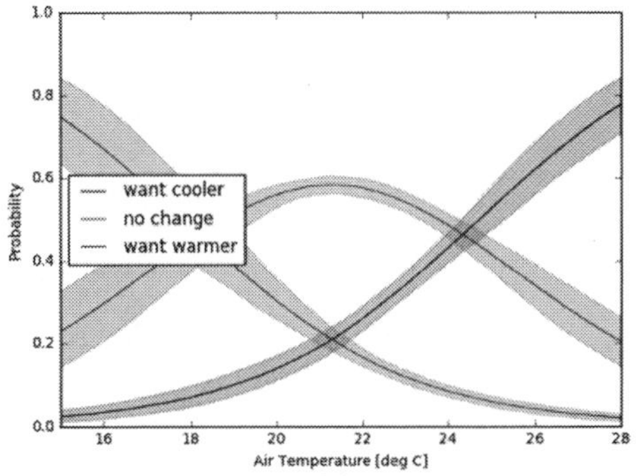

(b) (c)

Figure 6: Probability distribution of an occupant being in each preference class with respect to air temperature change (MRT is equal to air temperature)
(a) air velocity = 0.1m/s, relative humidity = 50%, met = 1.2, clo = 0.65
(b) air velocity = 0.2m/s, relative humidity = 50%, met = 1.2, clo = 0.65
(c) air velocity = 0.1m/s, relative humidity = 50%, met = 1.2, clo = 1.0

3.2 Two Clusters Model

A model which has two clusters (i.e. $K = 2$) was developed. Since developing a full model including PMV equations is computationally expensive, a simpler model was developed for a preliminary study. The simpler model is an ordered logit model of which input is only air temperature. Note that the full model is being developed, and the result will be reported.

Fig. 7 shows MCMC samples for the parameters. The blue histogram and crosses are samples for one of the clusters and the green ones are samples for the other cluster. As shown in the figure, the blue crosses and the green circles for the parameter $\mathbf{w}$ and $\boldsymbol{\theta}$ are clearly distinguished. This provides evidence that there is a high probability of classifying two or more clusters of people with respect to thermal preference characteristic.

Fig. 8 shows the estimated hidden cluster value for each occupant in the dataset. Blue and red pixels denote each cluster respectively, each row of pixels in the figure is a set of samples sampled together from one MCMC iteration and the x-axis denotes each occupant. Therefore, each column means probability of the occupant being in cluster k (i.e. $P(z_d = k \mid \mathbf{x}_{1:D}, y_{1:D})$, $k = 1 \text{ or } 2$). Note that the order of occupants was sorted by the probability of the occupant being in cluster 1, $P(z_d = 1 \mid \mathbf{x}_{1:D}, y_{1:D})$, in order to visualize the results more clearly. As shown in the figure, occupants on the left side of the figure are highly probable to be in the cluster 1, and occupants on the opposite side are highly probable to be in the cluster 2. This provides more evidence that there are two clusters of people having different thermal preference characteristic. If there was only one cluster, the results would show that most of the occupants have the same cluster value z. In case of occupants in the middle, since $P(z_d = k \mid \mathbf{x}_{1:D}, y_{1:D})$ is not highly biased, it can be explained that the occupants are between the two clusters or there are not observations to decide in which cluster the occupants belong too. For the occupants in the dataset, their thermal preference can be predicted by a mixture of sub-models for each cluster with $P(z_d = k \mid \mathbf{x}_{1:D}, y_{1:D})$ visualized in the figure following Eq. 4.

Fig. 9 shows the probability distribution of an occupant being in each preference class with respect to the air temperature variation for each cluster. As shown in the figures, sub-models for a cluster of people preferring cooler condition and another cluster preferring warmer condition were developed respectively. Since the sub-models are developed for each cluster, the shape of distributions becomes sharper compared to that of the single cluster model and the maximum probabilities of an occupant preferring no change are higher than 0.6. These are good characteristics in terms of predicting occupant's thermal preference.

The results from the two clusters model imply that there is a high probability of existence of multi-clusters in people with respect to thermal preference characteristic.

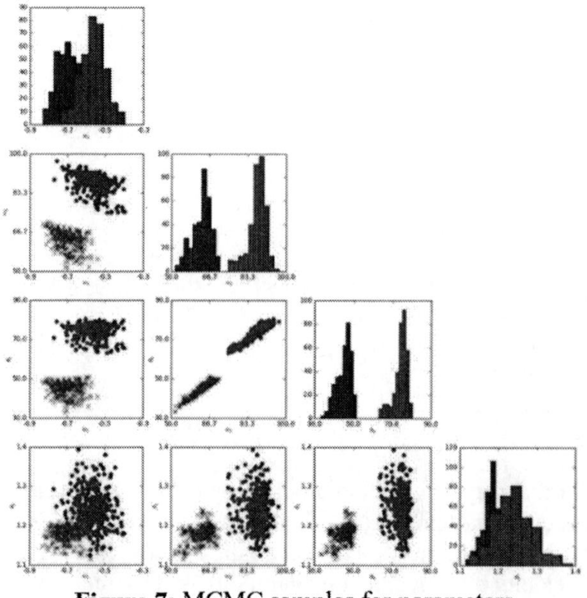

Figure 7: MCMC samples for parameters

Figure 8: Estimated hidden cluster value for each occupant in the dataset

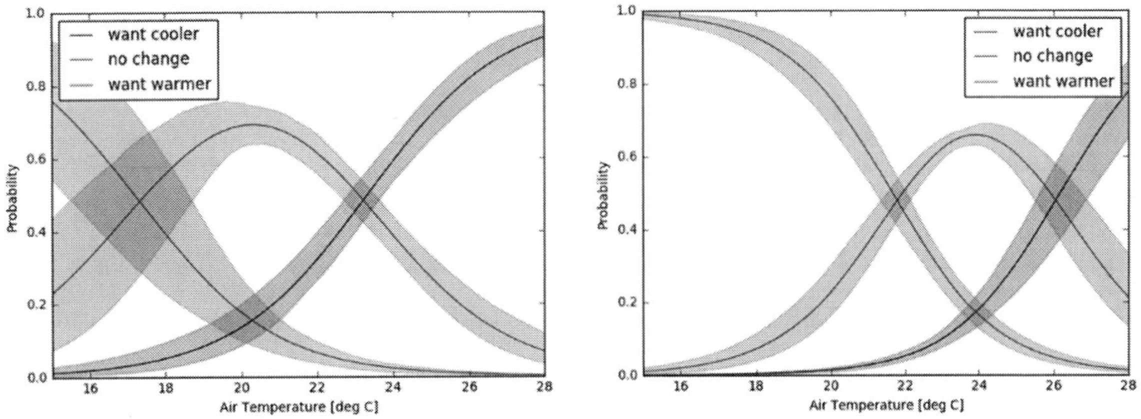

Figure 9: Probability distribution of an occupant being in each preference class with respect to air temperature change for each cluster.

4. DISCUSSION AND CONCLUSIONS

In this paper, a new method for learning and predicting individual occupant's thermal preference was introduced. The results show that the probabilities calculated by the models are in accordance with the general knowledge, and there is a high probability of classifying two or more clusters of people with respect to thermal preference characteristic. As a result, the two cluster model is more promising than the single cluster model considering our modeling purpose, i.e. learning and predicting individual occupant's thermal preference.

In this paper, a single cluster model and a two cluster model were qualitatively evaluated and compared with each other, however, the models should be also evaluated and compared in an objective and quantitative way, which will be presented in forthcoming publications by the authors. Also, the optimal number of clusters will be identified by testing different models and consequently, the hidden cluster value of a new occupant will be predicted.

Since the authors believe that occupant's thermal preference is affected by other factors, additional relationships and variables will be introduced in the general model and tested. Also, carefully designed experiments will be conducted in the future to enrich the dataset for this specific purpose. Since the Bayesian approach can seamlessly combine data from heterogeneous sources, all the data would be exploited for developing a general thermal preference model.

REFERENCES

Auffenberg, F., Stein, S., & Rogers, A. (2015). A personalised thermal comfort model using a Bayesian network.

Barlow, S., & Fiala, D. (2007). Occupant comfort in UK offices—How adaptive comfort theories might influence future low energy office refurbishment strategies. *Energy and Buildings, 39*(7), 837-846.

Brager, G. S., & de Dear, R. J. (1998). Thermal adaptation in the built environment: a literature review. *Energy and Buildings, 27*(1), 83-96.

Corgnati, S. P., Filippi, M., & Viazzo, S. (2007). Perception of the thermal environment in high school and university classrooms: Subjective preferences and thermal comfort. *Building and Environment, 42*(2), 951-959.

Daum, D., Haldi, F., & Morel, N. (2011). A personalized measure of thermal comfort for building controls. *Building and Environment, 46*(1), 3-11.

de Dear, R. J., Brager, G., & Cooper, D. (1997). *Developing an Adaptive Model of Thermal Comfort and Preference.* ASHRAE.

de Dear, R. J., & Brager, G. S. (1998). Developing an adaptive model of thermal comfort and preference. *ASHRAE transactions, 104*(1).

Erickson, V. L., & Cerpa, A. E. (2012). *Thermovote: participatory sensing for efficient building hvac conditioning.* Paper presented at the Proceedings of the Fourth ACM Workshop on Embedded Sensing Systems for Energy-Efficiency in Buildings.

Fanger, P. O. (1967). Calculation of thermal comfort: Introduction of a basic comfort equation. *ASHRAE Transactions, 73.*

Fanger, P. O. (1970). *Thermal comfort-analysis and applications in environmental engineering*: Danish Technical Press, Copenhagen.

Feldmeier, M., & Paradiso, J. A. (2010, Nov. 29 2010-Dec. 1 2010). *Personalized HVAC control system.* Paper presented at the Internet of Things (IOT), 2010.

Fonnesbeck, C., Patil, A., Huard, D., & Salvatier, J. (2015). PyMC (Version 2.3.6).

Ghahramani, A., Jazizadeh, F., & Becerik-Gerber, B. (2014). A knowledge based approach for selecting energy-aware and comfort-driven HVAC temperature set points. *Energy and Buildings, 85*, 536-548.

Ghahramani, A., Tang, C., & Becerik-Gerber, B. (2015). An online learning approach for quantifying personalized thermal comfort via adaptive stochastic modeling. *Building and Environment, 92*, 86-96.

Guillemin, A., & Molteni, S. (2002). An energy-efficient controller for shading devices self-adapting to the user wishes. *Building and Environment, 37*(11), 1091-1097.

Humphreys, M. A., & Nicol, J. F. (2002). The validity of ISO-PMV for predicting comfort votes in every-day thermal environments. *Energy and Buildings, 34*(6), 667-684.

Hwang, R.-L., Cheng, M.-J., Lin, T.-P., & Ho, M.-C. (2009). Thermal perceptions, general adaptation methods and occupant's idea about the trade-off between thermal comfort and energy saving in hot–humid regions. *Building and Environment, 44*(6), 1128-1134.

Jaynes, E. T. (2003). *Probability theory: the logic of science*: Cambridge university press.

Jazizadeh, F., Ghahramani, A., Becerik-Gerber, B., Kichkaylo, T., & Orosz, M. (2014). Human-Building Interaction Framework for Personalized Thermal Comfort-Driven Systems in Office Buildings. *Journal of Computing in Civil Engineering, 28*(1), 2-16.

La Gennusa, M., Nucara, A., Pietrafesa, M., & Rizzo, G. (2007). A model for managing and evaluating solar radiation for indoor thermal comfort. *Solar Energy, 81*(5), 594-606.

McCullagh, P. (1980). Regression Models for Ordinal Data. *Journal of the Royal Statistical Society. Series B (Methodological), 42*(2), 109-142.

Murakami, Y., Terano, M., Mizutani, K., Harada, M., & Kuno, S. (2007). Field experiments on energy consumption and thermal comfort in the office environment controlled by occupants' requirements from PC terminal. *Building and Environment, 42*(12), 4022-4027.

Nicol, J. F., & Humphreys, M. A. (2009). New standards for comfort and energy use in buildings. *Building Research & Information, 37*(1), 68-73.

Van Hoof, J. (2008). Forty years of Fanger's model of thermal comfort: comfort for all? *Indoor Air, 18*(3), 182-201.

Wagner, A., Gossauer, E., Moosmann, C., Gropp, T., & Leonhart, R. (2007). Thermal comfort and workplace occupant satisfaction—Results of field studies in German low energy office buildings. *Energy and Buildings, 39*(7), 758-769.

ACKNOWLEDGEMENT

This work was funded by the National Science Foundation under Grant No. 1539527. Any opinions, findings, and conclusions or recommendations expressed in this material are those of the authors and do not necessarily reflect the views of the National Science Foundation.

Design and Application of Distributed Economic Model Predictive Control for Large-Scale Building Temperature Regulation

Nishith R. PATEL[1*], James B. RAWLINGS[1], Michael J. WENZEL[2], Robert D. TURNEY[2]

[1] University of Wisconsin–Madison, Department of Chemical and Biological Engineering,
Madison, WI, United States of America
nishith.patel@wisc.edu*, james.rawlings@wisc.edu

[2] Johnson Controls Inc., Technology and Advanced Development, Building Efficiency,
Milwaukee, WI, United States of America
mike.wenzel@jci.com, robert.d.turney@jci.com

* Corresponding Author

ABSTRACT

Although recent research has suggested model predictive control as a promising solution for minimizing energy costs of commercial buildings, advanced control systems have not been widely deployed in practice. Large-scale implementations, including industrial complexes and university campuses, may contain thousands of air handler units each serving a multiplicity of zones. A single centralized control system for these applications is not desirable. In this paper, we propose a distributed control system to economically optimize temperature regulation for large-scale commercial building applications. The decomposition strategy considers the complexities of thermal energy storage, zone interactions, and chiller plant equipment while remaining computationally tractable. One of the primary benefits of the proposed formulation is that the low-level airside problem can be decoupled and solved in a distributed manner; hence, it can be easily extended to handle large applications. Peak demand charges, a major source of coupling, are included. The interactions of the airside system with the waterside system are also considered, including discrete decisions, such as turning chillers on and off. To deploy such a control scheme, a system model is required. Since using physical knowledge about building models can greatly reduce the number of parameters that must be identified, grey-box models are recommended to reduce the length of expensive identification testing. We demonstrate the effectiveness of this control system architecture and identification procedure via simulation studies.

1. INTRODUCTION

1.1 Motivation

Commercial buildings account for 20% of the total U.S. energy consumption, corresponding to roughly $200 billion a year in primary energy expenditures (Department of Energy, 2012). The Energy Information Administration projects that commercial floor space and primary energy consumption will continue to grow between 2009 and 2035, at 28% and 22%, respectively. Average energy prices, on the other hand, have been, and are expected to remain, relatively stable; therefore, the amount spent on energy in commercial buildings will continue to increase significantly. Given the significance of these energy cost values and their projected growth, buildings have become a clear target for control strategies designed to reduce costs and improve efficiency, particularly in the area of temperature control.

Almost all heating and cooling systems in commercial buildings and educational facilities use on/off and proportional-integral-derivative (PID) controllers for control of their heating, ventilation, and air conditioning (HVAC) systems (Afram & Janabi-Sharifi, 2014). These systems rely on controllers whose goal is to converge to the desired temperature set point. However, a better goal is to minimize the total energy cost. In a utility market with time-varying prices, the potential exists for cost savings by temporally shifting heating or cooling loads using thermal energy storage. To achieve these savings, predictive optimization is required using a model of the system for forecasting. Load shifting decreases the burden on power plants during peak hours, allowing them to operate more efficiently. Furthermore, chillers can operate more efficiently at night than during the day.

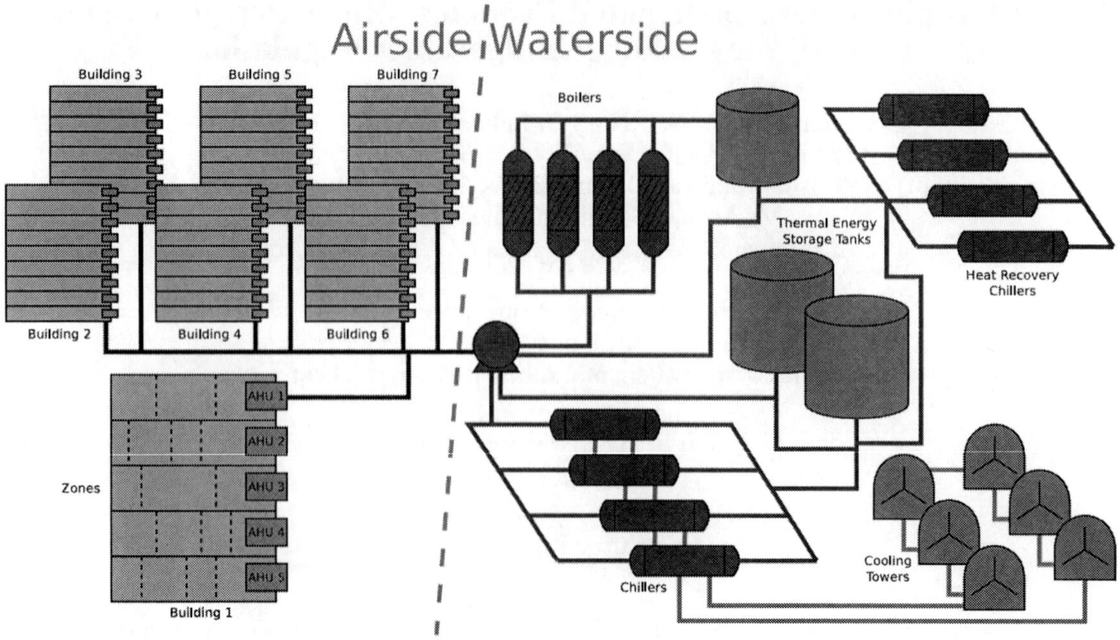

Figure 1: Schematic of large-scale commercial application with airside system shown on the left and waterside system shown on the right.

1.2 Model Predictive Control

Model predictive control (MPC) is an advanced control method that has been highly successful over the past two decades with thousands of applications in the chemical and petroleum industries alone (Qin & Badgwell, 2003). MPC relies on a model of the system to predict the process variables based on the actions taken by the controller (Rawlings & Mayne, 2009). At each step, an online optimization problem is solved. In economic MPC, the objective for this optimization is to minimize total cost (Rawlings & Amrit, 2009). Numerous studies have shown that MPC outperforms existing control systems due to its ability to forecast into the future and anticipate events before they occur (Ma *et al.*, 2012; Mendoza-Serrano & Chmielewski, 2012). MPC enables shifting of the energy load from peak hours to off-peak hours by using the mass of the building for passive thermal energy storage (TES) (Oldewurtel *et al.*, 2010; Avci *et al.*, 2013). Active TES, e.g., chilled water tanks, can also be used to further facilitate load shifting. Through a combination of active and passive storage systems, energy costs can be decreased by concentrating equipment usage to times of low resource prices while maintaining comfort limits within the building.

1.3 Large-Scale Commercial Applications

Economically optimal control systems have not been deployed widely in the HVAC industry. One fundamental obstacle to the successful deployment of MPC is the large number of zones. To implement MPC in HVAC systems, an optimization problem must be solved in a reasonably short time (on the order of a few minutes). Campus wide implementations may contain hundreds of buildings and thousands of air handler units each serving tens of zones. A schematic of such a large-scale application is shown in Figure 1. The airside system consists of all zones and the air-handler units (AHUs) that are responsible for temperature regulation. The waterside system consists of the equipment, such as chillers, that are required to provide cooling and/or heating. The control objective is to determine temperature setpoints for all zones in the airside system and to operate the waterside equipment to meet the corresponding load. A single, combined control system for these applications is impractical and undesirable as the number of zones and equipment increases because the resulting single optimization problem is difficult to maintain and also too large to solve in real time.

Distributed MPC can be implemented so that the large combined problem can be divided into smaller problems, which can be solved in parallel to reduce the overall computation time (Rawlings & Mayne, 2009; Rawlings & Stewart, 2008). Iterative methods are typically used in distributed MPC (Scherer *et al.*, 2013; Lamoudi *et al.*, 2011). However, iterative methods with many information exchanges between the distributed controllers are undesirable in practice because the controllers must wait on other controllers at each iteration, which can result in a greater computational time than that of solving the combined problem directly. Additionally, there are potential practical limitations to information exchanges based on existing HVAC communication protocols.

Power companies not only charge customers according to their total energy usage with time varying electricity costs, but they often also levy a peak demand charge based on the peak power usage during the month. Since peak demand charges comprise a significant portion of total cooling costs in buildings, excluding these charges leads to suboptimal performance with respect to minimizing energy costs. Existing distributed methods do not address these peak demand charges. Coordination is required; otherwise, zone controllers may cool at similar times leading to a high peak power usage. Furthermore, these existing methods often do not consider active TES or include detailed waterside equipment models, where significant cost savings can be achieved.

In this paper, we design a distributed economic MPC architecture for building temperature regulation to address these issues. The formulation is general enough to handle a large class of HVAC systems including interactions between the airside and waterside systems. The overall single optimization problem is decomposed into smaller subproblems for large-scale applications. Aggregate models are used in the high-level problem to reduce the computational burden. Iterations and communication between the low-level airside subsystems are not required. Demand response is addressed for optimal performance in the presence of peak demand charges. Regulatory controller dynamics are modeled to account for the slow closed-loop responses to temperature setpoint changes in commercial buildings.

2. DECOMPOSITION

2.1 Problem Statement

The goal is to develop a solution to the overall problem: given a forecast of electricity prices, ambient conditions, and disturbance loads, decide how to operate the HVAC system to minimize the total operating cost. The decision variables include temperature setpoints sent to the zone controllers, chiller operation schedule and storage tank operation schedule. In making these decisions, comfort bounds on the zone temperatures, capacity constraints on the equipment (air-handlers, chillers, and storage tank), and switching constraints for chillers must be respected.

Given this problem statement, we can formulate a single overall optimization problem for the combined airside and waterside system. As discussed in Section 1, solving this single optimization problem becomes difficult for large-scale applications. The airside system can consist of thousands of zones, and the waterside system can contain multiple pieces of equipment. Inclusion of the discrete decisions (turning equipment on or off) results in a mixed-integer optimization problem, which further increases the problem complexity. Since the single problem cannot be solved in real-time implementations, a decomposition is required to formulate smaller, manageable optimization problems.

2.2 Control Architecture Design

Our decomposition of the overall problem for HVAC control is shown in Figure 2. The MPC layer sits on top of the regulatory layer, which consists of the building automation system (BAS) and PID controllers for the airside and waterside systems. The design is expected to work with any existing BAS, allowing one to retrofit an existing HVAC control architecture with optimization. Only setpoint and measurement communication between the layers is required. In addition to a high- and low-level separation for the MPC problem, we split the low-level problem into *airside* and *waterside* following industrial convention.

For large-scale applications with hundreds or thousands of zones, the low-level airside problem can be further decomposed into separate distributed MPC controllers with each controlling a subsystem of the entire airside system as shown in Figure 2. The information exchanges between the layers are identical for all of the distributed airside controllers. The high-level problem contains an aggregate model of each airside subsystem and allocates cooling to each subsystem. The subsystem MPC controllers contain more detailed zone-level models. The choice of dividing the airside system into separate subsystems is made for computational reasons so that the subsystem MPC problems can be solved within a few minutes. Each subsystem can be anything from a single zone to a collection of zones to an entire building or even multiple buildings. If there is significant coupling between the subsystems (e.g., heat exchange between zones in separate subsystems), then performance may deteriorate as the controllers do not coordinate their

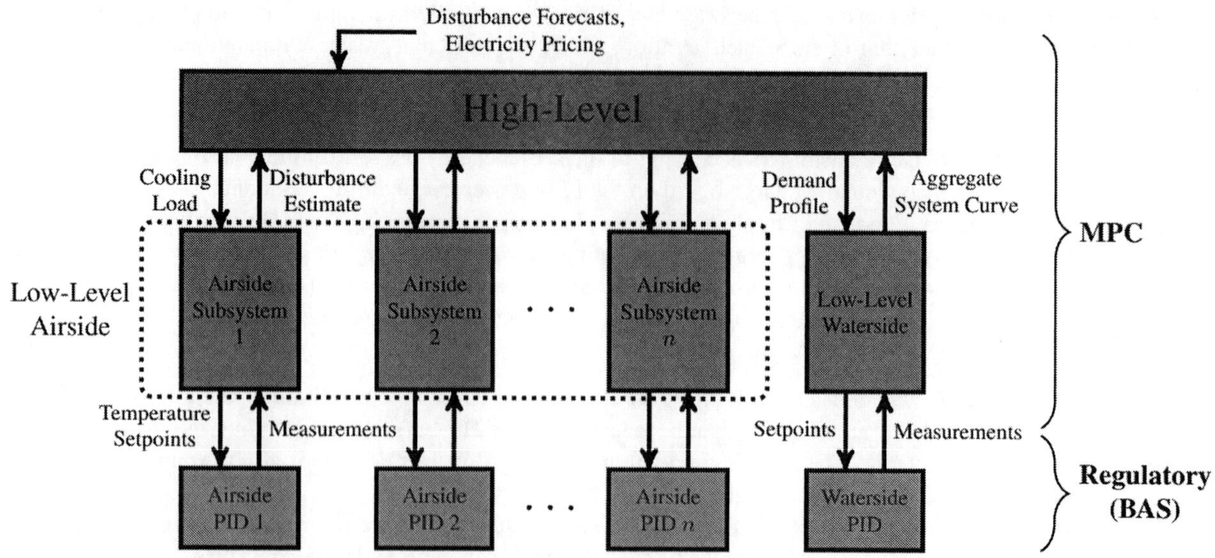

Figure 2: Control architecture with distributed MPC on the low-level airside problem (dotted box) for large-scale applications.

solutions. One way to decompose the airside system to guarantee that there is no coupling between the subsystems is to decompose by building because buildings do not exchange heat with one another. For this reason, we choose each subsystem to be a separate building in the discussion that follows.

The high-level problem determines how much cooling to allocate to each building to optimize total operational cost using an active TES model as well as aggregate airside and waterside models to reduce computational complexity. Each building has a separate low-level airside controller that computes the zone temperature setpoints that minimize energy usage, while maintaining the zone temperatures in a prespecified comfort region and not exceeding the cooling load from the high-level problem. Note that an alternative objective for the low-level airside problem is tracking the average building temperature from the high-level problem. Based on simulations, there is not a significant difference in performance between the two objectives, hence both are suitable. The low-level waterside problem is a mixed-integer linear program that minimizes cost while meeting the load from the high-level problem; its decision variables include equipment operation, thermal loads for chillers, flows for pumps, setpoints for other auxiliary equipment, and storage utilization. Previously, heuristics have been used to make these decisions. Risbeck *et al.* (2015) discuss the details of the low-level waterside problem. This paper primarily addresses the high-level and low-level airside problems. The setpoints from the two low-level problems are sent to the regulatory controllers of the existing building automation system (BAS). Feedback is employed to communicate measurements (temperature, power, equipment state) from the regulatory layer to the low-level MPC layer. Then, updated aggregate models and disturbance forecasts for both airside and waterside systems are communicated to the high-level problem for the next MPC execution step.

2.3 Advantages over Alternative Strategies

There are several advantages of this decomposition. The primary sources of coupling (peak demand charge and total equipment capacity constraint) are handled in the high-level problem, which uses aggregate airside and waterside models to reduce computational burden. Subsystems in the low-level airside problem are decoupled. Hence, iterations between the subsystems are not required, greatly reducing the internal communication burden. The general framework can handle various types of applications: central chiller plants, air handler units (AHUs), roof top units (RTUs), variable refrigerant flow (VRF) systems, and any BAS that can receive temperature setpoints and send measurements. Most building temperature regulation methods do not consider detailed models of chiller plant equipment or integer decision variables, which decreases the fidelity of energy cost calculations. Including integer variables in the waterside problem allows the optimizer to determine the equipment operation schedule rather than using heuristics. We consider active TES, where the greatest potential for load shifting and cost savings lies. However, the control architecture can be applied generally, whether or not active TES is available.

3. PROBLEM FORMULATIONS

3.1 System Model

The dynamics of cooling a single zone or building are described by the energy balance

$$C\frac{dT}{dt} = -H(T - T_a) - \dot{Q}_c + \dot{Q}_{\text{other}} \tag{1}$$

This model lumps all zone mass and air properties into a single zone temperature. Other models can also be used to describe the airside system dynamics. Differential equations can be written to model the zone air and mass temperatures separately, and the mass temperature can be further separated into a shallow mass temperature and deep mass temperature. However, for simplicity, we considered the lumped model in the simulation studies.

The system model given in Equation (1) can be extended to handle buildings with multiple zones. To this end, the coupling between the air spaces needs to be included. As energy is transferred due to a temperature gradient, the coupled model is given by

$$C_i\frac{dT_i}{dt} = -H_i(T_i - T_a) - \sum_{j \neq i} \beta_{ij}(T_i - T_j) - \dot{Q}_{c,i} + \dot{Q}_{\text{other},i} \tag{2}$$

in which β_{ij} characterizes the degree of coupling between zones i and j. If zones i and j are not adjacent, then $\beta_{ij} = 0$. Since buildings are separate and do not directly exchange energy, $\beta = 0$ in the high-level problem. Active TES is modeled using first order linear dynamics

$$\frac{ds}{dt} = -\sigma s + \dot{Q}_{\text{storage}} \tag{3}$$

For convenience, we convert all models to state-space form for use in the MPC optimization problem. Equation (2) is combined with Equation (3) and discretized according to the sample time of the controller to yield a discrete-time state-space model for the high-level problem. Similarly, using Equation (2) and the cooling duty model, which is discussed later, a discrete-time state-space model is obtained for the low-level airside problem.

3.2 High-Level Optimization Problem

As shown in Figure 2, the data for the high-level problem includes electricity pricing, weather forecast, building use model, and disturbance estimates from low-level problem. Since most pricing structures include a peak demand charge based on the peak power usage during the month in addition to the time varying electricity costs, both components are included in the economic cost objective. The decision variables are building temperatures, storage utilization, and thermal loads for waterside system. Dynamics models for the building temperature and active TES are required. The aggregate waterside system model is represented by the coefficient of performance (COP) in the high-level objective. Constraints include comfort bounds on the air temperature and bounds on the cooling duty. The comfort zone constraints are implemented as soft constraints. Mathematically, the high-level MPC problem is formulated as

$$\begin{aligned}
\min_{\mathbf{x},\mathbf{u},\dot{Q}_{\text{peak}}} \quad & \sum_{k=0}^{N-1} c_k \eta \dot{Q}_{\text{HVAC},k}\Delta + c_{\text{peak}}\eta\dot{Q}_{\text{peak}} \\
\text{s.t.} \quad & C_b\frac{dT_b}{dt} = -H_b(T_b - T_a) - \dot{Q}_{c,b} + \dot{Q}_{\text{other},b} \\
& \frac{ds}{dt} = -\sigma s + \dot{Q}_{\text{storage}} \\
& \dot{Q}_{\text{HVAC},k} = \sum_b \dot{Q}_{c,b,k} + \dot{Q}_{\text{storage},k} \\
& 0 \leq \dot{Q}_{\text{HVAC},k} \leq \dot{Q}_{\text{peak}} \\
& \dot{Q}_{\text{peak,past}} \leq \dot{Q}_{\text{peak}} \leq \dot{Q}_{\text{HVAC,max}} \\
& 0 \leq s_k \leq s_{\max} \\
& T_{\min} \leq T_b \leq T_{\max}
\end{aligned} \tag{4}$$

The states in the high-level MPC problem are the building temperatures and storage tank level. The inputs are the cooling duties to each building and the amount of storage charged or discharged. A rate of change penalty can be added to the inputs in the MPC problem. The disturbances are the ambient temperature and external loads placed on each building. A forecast of this disturbance vector is obtained and provided to the optimization problem at each MPC step. In the simulation study, the states are assumed to be measured. In real applications, state estimation is required to reconstruct the states from the measurements. To obtain building models from the zone models that are typically available or identified, model aggregation is employed. The average building temperature is defined by summing over all zones in that building using

$$T_b = \frac{\sum_i C_i T_i}{\sum_i C_i} \tag{5}$$

With this definition, the building parameters for the high-level airside model are obtained by summing parameters from the low-level airside models using

$$C_b = \sum_i C_i \qquad H_b = \sum_i H_i \qquad \dot{Q}_{\text{other},b} = \sum_i \dot{Q}_{\text{other},i} \tag{6}$$

Model aggregation also helps to reduce the amount of information that needs to be exchanged, thus, addressing the potential practical limitations with existing HVAC communication protocols.

3.3 Low-Level Airside Optimization Problem

As shown in Figure 2, the data for the low-level airside problem includes the cooling duty allocated from the high-level problem. Its objective is to minimize energy usage over the horizon by varying zone temperature setpoints. The zone air temperature and temperature setpoint to cooling duty model is required ($\dot{Q}_{c,i} = f(T_i, T_{\text{sp},i})$), which is discussed in the following section. The constraints include comfort bounds on the zone air temperatures and bounds on the cooling duty obtained from the high-level problem; they are all implemented as soft constraints with comfort zone violations penalized more than high-level cooling duty violations.

In each building, the low-level airside MPC problem is formulated as

$$\begin{aligned}
\min_{\mathbf{x},\mathbf{u}} \quad & Q_{\text{total},N} \\
\text{s.t.} \quad & C_i \frac{dT_i}{dt} = -H_i(T_i - T_a) - \sum_{j \neq i} \beta_{ij}(T_i - T_j) - \dot{Q}_{c,i} + \dot{Q}_{\text{other},i} \\
& \frac{dQ_{\text{total}}}{dt} = \sum_i \dot{Q}_{c,i} \\
& T_{\min} \leq T_i \leq T_{\max} \\
& \frac{Q_{\text{total},k+1} - Q_{\text{total},k}}{\Delta} \leq \dot{Q}_{\text{HighLevel},k} \\
& Q_{\text{total},k+1} - Q_{\text{total},k} \geq 0 \\
& \dot{Q}_{c,i} = f(T_i, T_{\text{sp},i})
\end{aligned} \tag{7}$$

To obtain the cooling duty as a function of the temperature setpoint and temperature, two separate models are required:

1). A model of the regulatory zone temperature controller to determine its control action in the air handler unit (AHU) as a function of the zone temperature and temperature setpoint: $v_{\text{air},i} = f_1(T_i, T_{\text{sp},i})$

2). An energy balance relating the control action to the actual cooling duty: $\dot{Q}_{c,i} = f_2(v_{\text{air},i})$

Assuming an ideal proportional-integral (PI) controller and a linear relationship between $\dot{Q}_{c,i}$ and $v_{\text{air},i}$, the simplified linear cooling duty controller model is

$$\begin{aligned}
\dot{Q}_{c,i} &= \dot{Q}_{\text{ss},i} + K_{c,i}\left[\varepsilon_i + \frac{1}{\tau_{I,i}}\int_0^t \varepsilon_i(t')dt'\right] \\
\varepsilon_i &= T_{\text{sp},i} - T_i
\end{aligned} \tag{8}$$

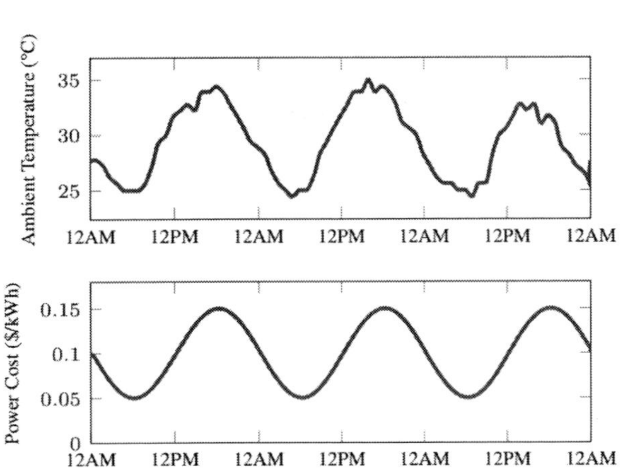

Figure 3: Ambient Temperature and Power Cost.

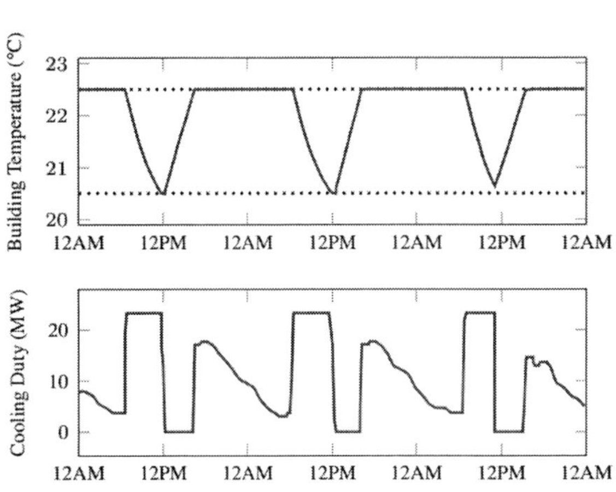

Figure 4: High-Level Optimization Result.

Saturation is represented as constraints on $\dot{Q}_{c,i}$. If a linear model is not sufficiently accurate to model the PI controller and heat transfer in air handler units, a nonlinear cooling duty model can be used instead. In this case, (7) becomes a nonlinear optimization problem. While it may seem atypical to model the regulatory controllers in the MPC optimization problem, the dynamics of zone PI controllers cannot be ignored as their responses are sluggish. In building temperature regulation, it may take zones up to an hour to reach a new temperature setpoint, which does not permit a time-scale separation in the decomposition. Since electricity prices are time-varying, ignoring these dynamics leads to suboptimal operation.

For the low-level airside problem, the states are the zone temperatures, the integrals of the zone tracking errors, and the total amount of cooling delivered. The inputs are the zone temperature setpoints. A rate of change penalty can be added to these inputs as well in the MPC problem to discourage rapid setpoint changes. The disturbances are the ambient temperature, external loads, and steady-state cooling rates.

3.4 Alternative Formulation: Temperature Tracking

An alternative strategy for the low-level airside problem is to track the average building temperature from the high-level problem rather than to minimize energy consumption. To track the temperature computed from the high-level problem, the high-level cooling duty constraint from (7) is removed, and the objective is replaced with

$$\sum_{k=1}^{N} \frac{1}{2} \| T_{b,k} - T_{\text{HighLevel},k} \|^2 + \mu Q_{\text{total},N} \tag{9}$$

in which the average building temperature is defined by Equation (5) and μ is a small penalty placed on energy usage. This penalty on energy usage ensures uniqueness of the solution.

4. SIMULATION RESULTS

A simulation study was performed for a three-zone building with active TES. The weather and electricity pricing data used in the simulation are shown in Figure 3. The weather data were obtained from Atlanta, Georgia for a three-day period in the summer (July 25-27, 2012). The time-varying electricity price and peak demand charge data are representative of data from Johnson Controls. Figure 4 shows the results of the high-level optimization problem, and Figure 5 shows the results of the low-level optimization airside problem. The top graph in each figure shows the temperatures with the comfort zone denoted by dotted lines. The bottom graph in each figure shows the inputs computed from the corresponding optimizations. With passive TES available, the economically optimal strategy is to precool the building before the peak period when electricity prices are highest. The flatness of the cooling duty profile from the high-level problem is a result of the peak demand charge. The low-level airside problem determines the order of precooling based on the dynamics of the individual zones in the building to minimize energy usage. Since Zone 2 is not as well insulated as Zone 3, the low-level airside controller chooses to precool Zone 3 before the Zone 2.

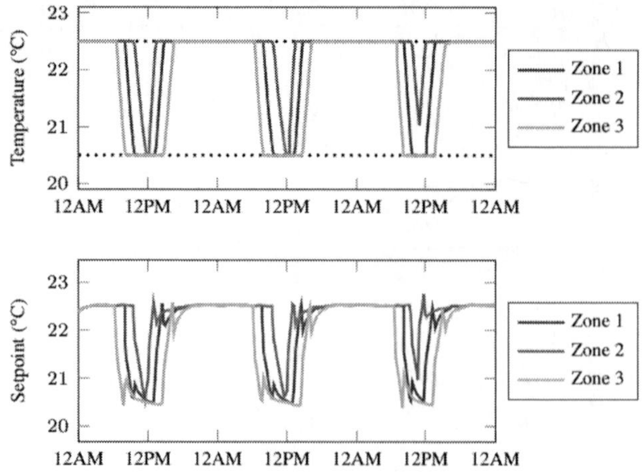

Figure 5: Low-Level Airside Optimization Result.

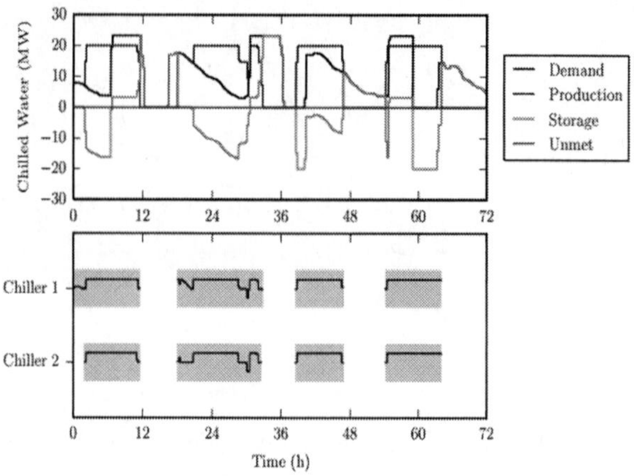

Figure 6: Low-Level Waterside Optimization Result.

Figure 6 shows the results of the low-level waterside optimization problem. The top graph shows how the demand profile from the high-level problem is met using a combination of production from chillers and charging/discharging of the storage tank. The bottom graph shows the corresponding equipment operation schedule for the two chillers. The shaded region depicts when the equipment is turned on with the approximate loading fraction between 0 and 1 depicted by the solid line. With active TES available, chillers produce more chilled water during the night to charge the storage tank so that they can be turned off during the peak period when electricity is more expensive.

The concept can be extended to larger systems with more buildings and more zones. This decomposition strategy has also been applied to a campus with 25 buildings, each with 20 zones (for a total of 500 zones). The resulting optimization problems at each MPC step can be solved within a few minutes when distributed controllers are implemented for the low-level airside problem; however, the results are not shown due to space limitations.

5. AIRSIDE SYSTEM IDENTIFICATION

To implement this control system, an airside model is required. A few issues must be considered in the building problem that are not present in typical chemical process control applications. During testing, exciting inputs using large pseudo-random, binary signals (PRBS) can result in frequent saturation of the PID controllers in air handler units, hence small PRBS or integrated PRBS are used instead. Moreover, one of the primary disturbances affecting buildings, ambient temperature, can be easily measured and predicted. In order to use weather forecasts during online optimization, the effect of ambient temperature on zone temperature dynamics is identified by adding ambient temperature to the input vector before passing the data to the identification routine. The model between other disturbances, such as solar radiation, and the zone temperature can also be identified if measurements and predictions of these disturbances are available. Furthermore, temperatures must be kept in the comfort zone if the building has occupants during testing period. Sample identification results are presented in the next section that account for these issues.

5.1 Black-Box and Grey-Box Identification

In a linear black-box model, no knowledge about the internal structure of the system is assumed. While these techniques can be applied to any system using some knowledge of the system to add structure to the model can greatly increase the accuracy of the identified model, while decreasing data requirements. Models with this added structure are called grey-box models. To determine whether black-box and grey-box identification perform equally well for the building problem, both types of identification were performed on the same dataset using MATLAB's System Identification Toolbox. The free parameters in the grey-box approach include the thermal capacitances of the zones and heat transfer coefficients from Equation (2) as well as the PI controller parameters from Equation (8). The results are shown in Figure 7. Based on the figure, both methods fit the validation data well. However, the two methods do not scale equivalently for large-scale systems.

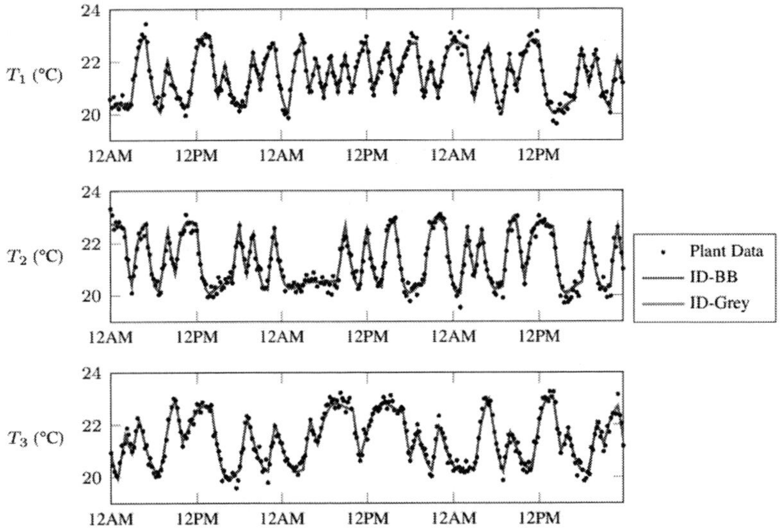

Figure 7: Identification Results.

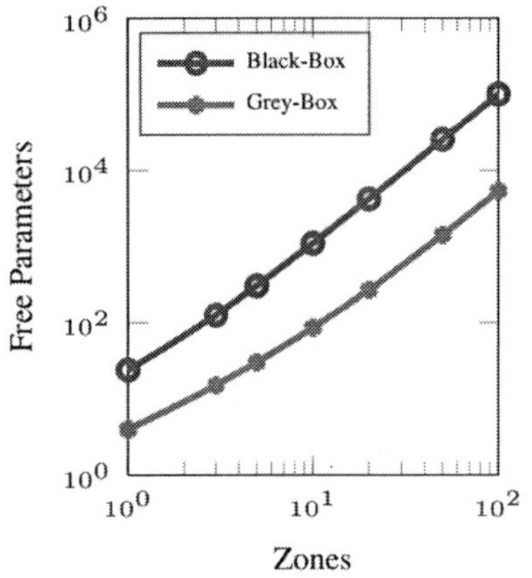

Figure 8: Scaling Analysis.

5.2 Scaling Analysis

Figure 8 shows how the number of free parameters to be identified from data increases as the number of zones increases. The figure shows the worst case scenario for grey-box identification when all heat transfer coefficients between zones need to be determined. Typically, many of these coupling coefficients can be set to zero if zones are not adjacent, which leads to a more linear scaling for the grey-box case. As large-scale commercial building applications may have thousands of zones, identifying black-box models for these systems involves determining many parameters. More data is required when there are more free parameters to identify. Hence, grey-box identification requires less data than black-box identification. In practice, less data means a shorter testing period and faster implementation of the MPC controller, which makes grey-box identification more suitable for large-scale commercial applications.

6. CONCLUSIONS

An economically optimal and practically implementable method of temperature regulation and equipment operation for large-scale commercial building applications was presented. The control system architecture considers both airside and waterside optimization as well as storage, and the architecture can be easily extended to handle more complex models. Since the time scale of regulatory controllers is significant, their dynamics are modeled. Grey-box system identification is preferred over black-box identification for large systems to decrease the testing period. We show the effectiveness of the proposed method through the simulation of a three-zone building.

NOMENCLATURE

Indices

b	building index (high-level)
i	zone index (low-level)
k	time index

Variables		**Parameters**	
T	temperature	Δ	sample time of controller
T_a	ambient temperature	N	horizon length
$T_{\text{sp},i}$	zone temperature setpoint	c_k	cost of electricity at time k
$\dot{Q}_{\text{other}}$	external load, radiation, or disturbance	c_{peak}	peak demand charge
$\dot{Q}_c$	cooling rate delivered	C	thermal capacitance
Q_{total}	total amount of cooling delivered	H	scaled ambient heat transfer coefficient
$\dot{Q}_{\text{HVAC}}$	cooling rate from HVAC system	β_{ij}	scaled inter-zone heat transfer coefficient
$\dot{Q}_{\text{peak}}$	peak HVAC system cooling rate	$K_{\text{c},i}$	scaled zone PI controller gain
s	amount of cooling potential in storage tank	$\tau_{\text{I},i}$	integral time constant for zone PI controller
$\dot{Q}_{\text{storage}}$	cooling rate delivered to storage tank	$\dot{Q}_{\text{ss},i}$	steady-state rate of cooling
$\dot{Q}_{\text{HighLevel}}$	cooling rate from high-level problem	η	inverse of the aggregate COP
$T_{\text{HighLevel}}$	building temperature from high-level	$\dot{Q}_{\text{HVAC,max}}$	max cooling capacity of HVAC system
ε_i	tracking error	$\dot{Q}_{\text{peak,past}}$	peak cooling rate previously achieved
$v_{\text{air},i}$	air flow rate in AHU	σ	decay constant for storage tank
		s_{max}	max cooling capacity of storage tank
		T_{min}	lower bound of comfort region
		T_{max}	upper bound of comfort region
		μ	penalty on energy usage

REFERENCES

Afram, A., & Janabi-Sharifi, F. (2014). Theory and applications of HVAC control systems—a review of model predictive control (MPC). *Building and Environment*, *72*, 343–355.

Avci, M., Erkoc, M., Rahmani, A., & Asfour, S. (2013). Model predictive HVAC load control in buildings using real-time electricity pricing. *Energy and Buildings*, *60*, 199–209.

Department of Energy. (2012, March). *2011 Buildings Energy Data Book.* `http://buildingsdatabook.eren.doe.gov/ChapterIntro3.aspx`.

Lamoudi, M. Y., Alamir, M., & Béguery, P. (2011). Distributed constrained model predictive control based on bundle method for building energy management. In *50th IEEE Conference on Decision and Control and European Control Conference* (pp. 8118–8124). Orlando, FL, USA.

Ma, J., Qin, J., Salsbury, T., & Xu, P. (2012). Demand reduction in building energy systems based on economic model predictive control. *Chemical Engineering Science*, *67*, 92–100.

Mendoza-Serrano, D. I., & Chmielewski, D. J. (2012). HVAC control using infinite-horizon economic MPC. In *51st IEEE Conference on Decision and Control* (pp. 6963–6968). Maui, Hawaii, USA.

Oldewurtel, F., Parisio, A., Jones, C. N., Morari, M., Gyalistras, D., Gwerder, M., ... Wirth, K. (2010). Energy efficient building climate control using stochastic model predictive control and weather predictions. In *American Control Conference* (pp. 5100–5105). Baltimore, MD, USA.

Qin, S. J., & Badgwell, T. A. (2003). A survey of industrial model predictive control technology. *Control Engineering Practice*, *11*(7), 733-764.

Rawlings, J. B., & Amrit, R. (2009). Optimizing process economic performance using model predictive control. In *Nonlinear Model Predictive Control* (Vol. 384, pp. 119–138). Berlin: Springer.

Rawlings, J. B., & Mayne, D. Q. (2009). *Model Predictive Control: Theory and Design.* Madison, WI: Nob Hill Publishing.

Rawlings, J. B., & Stewart, B. T. (2008). Coordinating multiple optimization-based controllers: New opportunities and challenges. *Journal of Process Control*, *18*, 839-845.

Risbeck, M. J., Maravelias, C. T., Rawlings, J. B., & Turney, R. D. (2015). Cost optimization of combined building heating/cooling equipment via mixed-integer linear programming. In *American Control Conference* (p. 1689-1694). Chicago, IL.

Scherer, H., Pasamontes, M., Guzmán, J., Álvarez, J., Camponogara, E., & Normey-Rico, J. (2013). Efficient building energy management using distributed model predictive control. *Journal of Process Control.*

Personalized Shading Control Strategy to Maximize Occupant Satisfaction while Minimizing Lighting Energy Use by Multi-Objective Optimization

Jie XIONG[1, 2]*, Seungjae LEE[1, 2], Athanasios TZEMPELIKOS[1, 2], Panagiota KARAVA[1, 2]

[1] Purdue University, Lyles School of Civil Engineering,
West Lafayette, Indiana, USA
Email: xiong29@purdue.edu, lee1904@purdue.edu, ttzempel@purdue.edu, pkarava@purdue.edu

[2] Ray W. Herrick Laboratories, Purdue University, Center for High Performance Buildings,
West Lafayette, Indiana, USA
Email: xiong29@purdue.edu, lee1904@purdue.edu, ttzempel@purdue.edu, pkarava@purdue.edu

* Corresponding Author

ABSTRACT

In this study, a personalized shading control framework is developed to maximize occupant satisfaction while minimizing lighting energy use using a multi-objective optimization scheme. A personalized satisfaction model was developed based on specially-designed experiments in private offices, to quantify the occupant satisfaction level with motorized roller shades by predicting the override probability of occupants considering different variables. Then, a multi-objective optimization algorithm was constructed, considering the shading override probability and predicted lighting energy use as objectives, where the occupants are the decision makers in the final balancing between their personalized comfort limits and energy use considerations. The developed method serves as a prototype study on adaptive shading controls with learned personalized comfort profiles and parallel energy use considerations.

1. INTRODUCTION

Occupants play a significant role in energy use of buildings, while they have a strong preference for customized indoor environment. Their perception of controlled environment affects their productivity and well-being. In perimeter building zones with glass facades, shading systems are used as comfort delivery systems under dynamic conditions, affecting daylight provision, view and lighting energy use.

Previous studies (Van Moeseke *et al.* 2007; Tzempelikos and Athienitis 2007; Nielsen *et al.* 2011; Shen and Tzempelikos 2012; Grynning 2014) have proven the energy benefits of automated shading control strategies, while occupant dissatisfaction issues with respect to visual comfort could arise with such systems (Reinhart and Voss 2003; Lee *et al.* 2013). Preventing direct sunlight associated with visual discomfort has been the topic of previous studies, using blind cut-off angles (Tzempelikos, 2013) and sun glare considerations (Guillemin and Morel 2001; Zhang and Birru 2012; Din and Kim 2013), including blind optical properties and overall glare evaluation (Chan and Tzempelikos, 2013). Existing glare indices such as DGI have been used as constraints in optimization studies aiming at reducing energy use (Chaiwiwatworakul *et al.* 2009; Oh *et al.* 2012), while Wienold *et al.* (2011) compared different shading devices and simple rule-based controls through simulation with respect to energy performance and visual comfort. More recently, simple and more advanced model-based shading control strategies for venetian blinds (Chan *et al.*, 2015) and roller shades (Xiong and Tzempelikos, 2016) have been proposed considering overall daylight discomfort glare using detailed optical properties and validated daylight-glare models. Several studies developed probability models of occupant actions of shading and/or lighting with a single variable (Reinhart 2004; Gunay *et al.* 2014a; Zarkadis *et al.* 2015; Wang *et al.* 2016) or multivariate (Haldi and Robinson 2010; Fabi *et al.* 2015), and incorporated the models in shading and/or lighting control (Gunay *et al.* 2014b). Gunay *et al.* (2015) compared the effects on energy consumption of existing behavioral models by simulation, but validation and verification of compared models were not investigated.

There are three main challenges with the current and previous approaches related to "optimal" shading and lighting controls, and considerations of visual comfort, occupant satisfaction, and energy use. First, with very few exceptions that considered occupant preferences of some kind (Guillemin and Molteni, 2002), the standard approach has been to try to satisfy visual comfort criteria using heuristic, model-based or simple occupant-based lighting/shading controls, based on existing indices. However, dissatisfaction with the visual environment, especially in perimeter offices, could still exist even if specific criteria are met, since it depends on several factors, personal and subjective preferences. Moreover, development of shading interaction models are challenging by default, since visual conditions in perimeter zones are dynamic and multivariate, and there is no agreement between different studies . But even when advanced probabilistic models are developed based on short or long-term studies, the studies are specific to the shading type used and space functionality and layout, and there is no standard procedure to quantify the uncertainty involved in this type of models or to guide the selection of the right set of variables (including non-physical variables). Therefore, the development of personalized visual comfort profiles related to shading operation, including a generic set of physical and human variables, are required, with quantified implications on energy use.
Second, existing models were developed for the case of manual shading controls (no automation or motorization), which cannot reflect occupant satisfaction and corresponding interactions with motorized shading or fully automated systems. Moreover, the effects of different control interfaces and human variables play a significant role as proven recently (Sadeghi *et al.*, 2016).
Finally, the common approach has been to consider visual comfort and energy use in optimization schemes, with different levels of complexity and formulation types. Typically, multi-objective optimization is used that involves more than one objective function, combining a measure of energy use and a measure or metric related to comfort. Most studies related to multi-objective optimization in this field applied the method to building envelope parameters (Manzan and Padovan 2015; Futrell *et al.* 2015; Carlucci *et al.* 2015), as MOO usually provides a set of non-dominated alternatives (Pareto front) instead of a deterministic solution. Villa and Labayrade (2012) tried to apply MOO in real-time lighting control – controlling the dimming levels of two luminaires to minimize lighting power demand while maximizing modeled satisfaction level – but the Pareto front of MOO solution was not fully taken advantage of. Cassol *et al.* (2011) formulated a general optimization problem by assigning weights to two objectives – light provision deviation and energy consumption. However, one could argue that the rationality of constructing the optimization problem and choosing weights that connect energy and comfort are debatable.
A more realistic approach to deal with these problems, including occupant visual comfort and interactions with shading/lighting systems, would be to consider occupant satisfaction with the visual environment as an objective, quantified by specially-designed experiments with occupant interactions and surveys, in order to develop adaptive and personalized shading or lighting control models. If this is combined with energy use considerations for overall optimization, instead of assigning standard weights, variable weight "coefficients" determined by the occupants themselves could be used for realistic and flexible shading control optimization.
In this study, a personalized shading control framework is developed to maximize occupant satisfaction while minimizing lighting energy use using such a multi-objective optimization scheme. A personalized satisfaction model was developed based on specially-designed experiments in private offices, to quantify the occupant satisfaction level with motorized roller shades by predicting the override probability of occupants considering different variables. Then, a multi-objective optimization algorithm was constructed, considering the shading override probability and predicted lighting energy use as objectives, where the occupants are the decision makers in the final balancing between their personalized comfort limits and energy use considerations. The developed method serves as a prototype study on adaptive shading control strategies with learned personalized comfort profiles and parallel energy use considerations.

2. MODELING

2.1 Experiment and Data

A set of experiments were conducted to gather data for developing personal satisfaction models. The experiments were conducted in four private offices of the Herrick Laboratories building at Purdue University from 09/29/2015 – 10/23/2015. The offices are equipped with controllable roller shades and electric lighting systems, as well as comprehensive indoor environment sensing system, for investigating the impact of façade design and control strategies on occupant comfort perception and corresponding behavior. The offices are 3.17m wide by 4.03m deep by 3.1m high, and have a south facing façade with 60% vision area (window-to-wall ratio) and a high performance glazing with 70% normal visible transmittance.

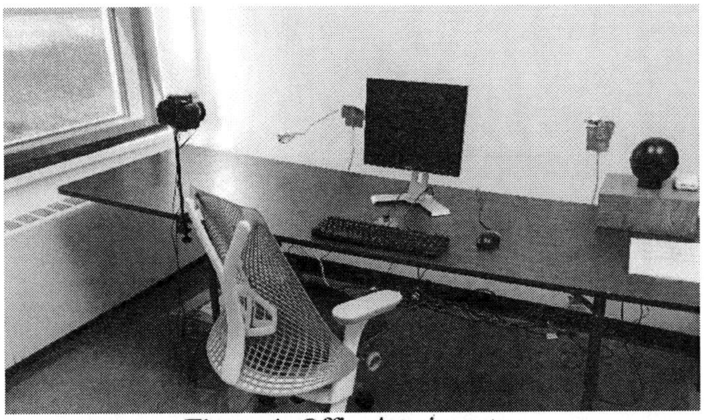

Figure 1: Office interior setup

In the experiments, environmental conditions (e.g. vertical and horizontal indoor illuminances, window transmitted illuminance, etc.), states of the shading system (i.e., shading position), and occupancy status were recorded every one minute. Test-subjects were asked to do general work in the office during working hours. For the purpose of learning occupant preferences, the programmed shading control intentionally raises the shades gradually every 30 minutes in order to increase the data collection rate and improve the learning efficiency. The occupants could control the shading system by overriding the system through a web-based control interface, while the override actions and resulting shade positions were continuously monitored. The test subject of each room remained the same during the experiment, and the data used in modeling study is the data of 14-days (non-consecutive) in one office, which consists of 1735 training data points and 2601 test observations.

2.2 Integrated Daylight, Electric-light and Lighting Energy Model

A validated hybrid ray-tracing and radiosity daylight model (Chan and Tzempelikos, 2012) was used for daylight simulation. The model combines the accuracy of forward ray tracing for direct light with the computational efficiency of radiosity for diffuse light entering the space and interior reflections. Angular direct-direct and direct-diffuse light transmission through roller shades is calculated using the semi-empirical model by Kotey *et al.* (2009), which has been validated using integrated sphere measurements and full-scale experiments (Chan *et al.*, 2015).
Beam and diffuse (sky and ground-reflected) transmitted illuminance through the window (calculated or measured by sensors for real-time control, or TMY3 data for an annual analysis) are required as inputs to the daylight model. At each time step, these quantities are obtained by processing the readings of one exterior sensor (direct-diffuse ratio of horizontal solar radiation) and two interior sensors (total transmitted radiation and total transmitted illuminance).
The daylight simulation results are then used as required inputs in electric lighting model and developed personal satisfaction model. Electric lights dimming levels were determined in electric lighting model based on a work plane illuminance set point (500 lx in this study) and the simulated work plane illuminance (on the calculation grid), by mapping dimming levels and work plane illuminance provided by lights. The lighting power and energy consumption can then be computed based on determined dimming levels.
The integrated daylight, electric-light and energy model based on models above can be mathematically expressed as:

$$[Power, Y] = E(SP, X) \tag{1}$$

where $Power$ is the predicted lighting power density consumption (W/m^2); Y is the simulated results from daylight model which are also required inputs for the developed personal satisfaction model; SP is the shade position; X is the measurement inputs used for daylight model.

2.3 Personal Satisfaction (Override) model

2.3.1 Model Development: The personal satisfaction model developed in this study aims at predicting the satisfaction level of occupant by quantifying the override probability distribution with different shading positions given a set of measured physical conditions. The override model is developed by logistic regression method. Considering that the same physical variable could have reverse effects on shade lowering and raising actions (e.g. high illuminance triggers lowering action while low illuminance leads to raising), two separate models were first developed for predicting lowering and raising probabilities respectively, and then the overall override model was inferred by probability theory:

$$P(lowering) = \frac{\exp(\boldsymbol{\beta}_l \boldsymbol{x}_l + \alpha_l)}{1 + \exp(\boldsymbol{\beta}_l \boldsymbol{x}_l + \alpha_l)} \quad (2)$$

$$P(raising) = \frac{\exp(\boldsymbol{\beta}_r \boldsymbol{x}_r + \alpha_r)}{1 + \exp(\boldsymbol{\beta}_r \boldsymbol{x}_r + \alpha_r)} \quad (3)$$

$$P(override) = 1 - (1 - P(lowering))(1 - P(raising)) \quad (4)$$

where $\boldsymbol{x}_l$ and $\boldsymbol{x}_r$ are input variables; $\boldsymbol{\beta}_l$, α_l, $\boldsymbol{\beta}_r$ and α_r are parameters trained based on experimental data, and shall be different with different occupants.

The input variables were selected based on prior knowledge and sensitivity analysis. To assist in reliable regression results and comparative studies, all the input variables are normalized (e.g. vertical illuminance is normalized by constant 5000 lx, and all values higher than 5000 is set to 1).

The first prior belief is that shade position (SP) should be included in the model, based on the common sense that when shades are already fully closed/open, the probability of lowering/raising action should be 0, and the probability of raising/lowering action would be higher than other positions. Second, at least one illuminance value (transmitted illuminance TI, vertical illuminance VI or work plane illuminance WI) should be selected as they indicate the amount of light perceived by occupants.

Global sensitivity analysis was conducted to investigate the sensitivities of different illuminance values with override action. Sobol' indices (Sobol' 1990; Saltelli *et al.* 2000) were computed with six alternative models (lowering and raising models using normalized SP with VI, SP with WI and SP with TI) to quantify the influence of variables on override prediction, as shown in Figure 2.

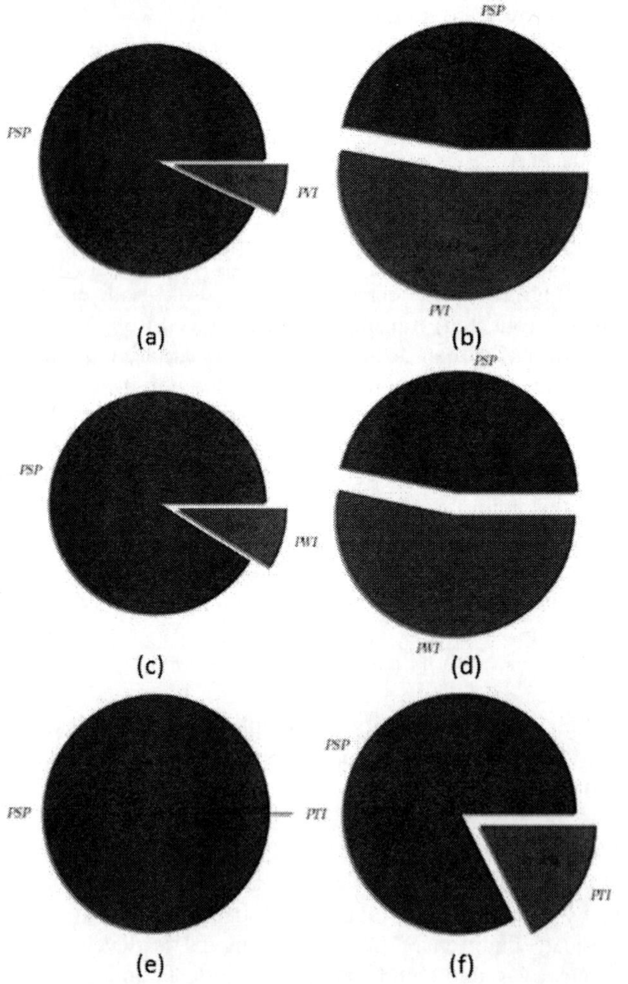

Figure 2: Total Sobol' indices of (a) lowering model with PSP and PVI, (b) raising model with PSP and PVI, (c) lowering model with PSP and PWI, (d) raising model with PSP and PWI, (e) lowering model with PSP and PTI, (f) raising model with PSP and PTI

It can be seen that vertical illuminance and work plane illuminance have similar effect on both lowering and raising models, while transmitted illuminance has little impact on lowering and smaller effect on raising. As vertical illuminance on the eye can better represent the light incident on the occupant (and is used in glare indices such as DGP), it was selected as the input variable in both lowering and raising models.

2.3.2 Modeling Results and Evaluation: Logistic regression was implemented with experiment data based on selected variables and model structure.

$$P(lowering) = \frac{\exp(2.9735SP + 0.21VI - 5.3485)}{1 + \exp(2.9735SP + 0.21VI - 5.3485)} \quad (5)$$

$$P(raising) = \frac{\exp(-2.1882SP - 24.9413VI - 2.9023)}{1 + \exp(-2.1882SP - 24.9413VI - 2.9023)} \quad (6)$$

$$P(override) = 1 - (1 - P(lowering))(1 - P(raising)) \quad (7)$$

The lowering and raising models are visualized in Figure 3.

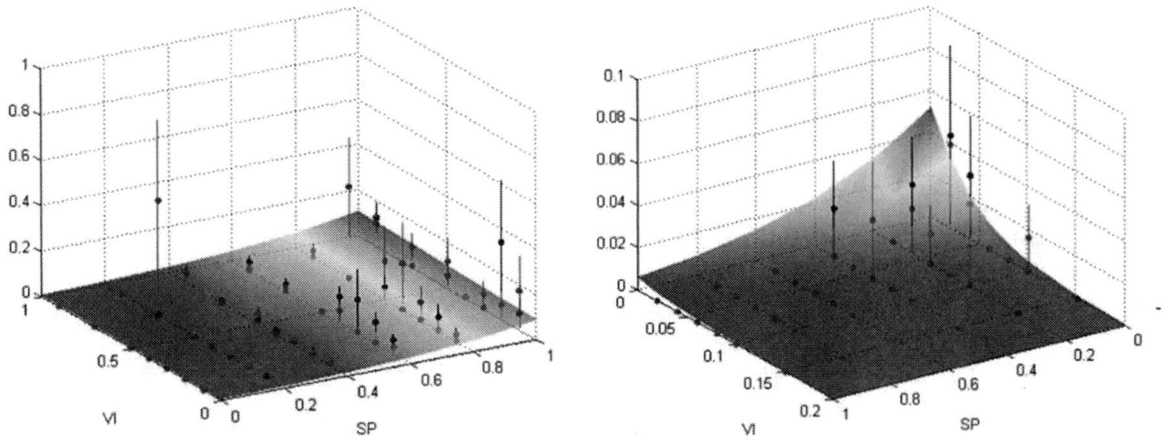

Figure 3: Lowering (left) and raising (right) models

The red and black dots represent the mean frequencies of training and test data, respectively, and the vertical lines indicate the corresponding uncertainty (standard deviation) from data noise and number of observations. From the left figure it can be seen that the lowering action is mainly determined by SP, but not really sensitive with VI. The raising model shows curved surface as expected in the right figure. The observations (red and black dots and lines) are accordant with the models in general, with a few noticeable deviations due to data sparsity (long uncertainty bars).

3. MULTI-OBJECTIVE OPTIMIZATION

3.1 Formulation of the Optimization Problem

3.1.1 Variables and Objectives: The controlled variables of the multi-objective optimization problem is simply one-dimensional shading position, ranged from 0 to 1 (0 – fully open; 1 – fully closed).

The objective of the optimization problem is to maximize the satisfaction of occupant by minimizing the probability of override action, while minimizing lighting energy use. Therefore the two objective functions are derived based on the energy model and developed override model.

As the override model predicts the probability of override every one minute (given the experimental data), the magnitude of the prediction is very low in most of the cases – due to the fact that an occupant cannot be dissatisfied and operate shades every minute. Therefore, a longer analysis horizon is preferred for optimization, to enlarge the output value of the "satisfaction" objective function for more reliable optimization distinction, by inferring the joint probability of any override in the time horizon (15 minute in this study) based on assumptions of independency of the action probabilities at serial time steps and fixed outdoor conditions in the time horizon:

$$P(override\ in\ t{\sim}t+T-1) = 1-\prod_{i=1}^{T}[1-P(override\ at\ t_i)] = 1-(1-P(override\ at\ t))^T \quad (8)$$

where $T = 15$.
The energy objective function should also utilize the 15-minute horizon to predict the average power density consumed. By assuming constant outdoor conditions in 15 minutes, the average power density is the power density at current time step:

$$Power\ in\ 15min = \sum_{i=1}^{15} Power(t_i)\Big/15 = Power(t) \quad (9)$$

Then the optimization problem can be mathematically expressed as:

$$\min_{SP\in[0,1]} \{Power(SP,X), P(SP,Y)\} \quad (10)$$

where $Power(SP,X)$ is the average power consumption in 15 minutes simulated by energy model given shade position SP and inputs X for integrated model; $P(SP,Y)$ is the joint probability of override in 15 minute predicted by personal satisfaction model with shade position SP and inputs Y, which is calculated from the integrated model.

3.1.2 Optimization Strategy and Algorithm: The integrated model and personal satisfaction model, as well as the multi-objective optimization were implemented in Matlab R2014a.
Several toolbox options of different algorithms are provided by Matlab aiming at multi-objective optimization. However the optimization objectives in this study are dependent – personal satisfaction model requires inputs simulated from integrated model – and dependent objectives are not supported by most of algorithms. To address the dependency problem, daylight model needs to run in both the energy model and the personal satisfaction model separately, which is inefficient.
To compromise between computation efficiency and accuracy, a straightforward search strategy was adopted to solve the multi-objective optimization problem. As continuous shade position can be approximated by dividing the full range into a specific number of discrete positions, and very small changes in roller shade position makes no difference in application with respect to energy consumption and occupant comfort perception, the controlled variable is pre-defined with discrete options as a feasible set (e.g. 21 positions with every 5% increments). In the optimization process, two objective models run with all the shade position options, and objective values are provided through the whole feasible region. The optimization solutions are determined by comparing all the feasible objectives in the way of Pareto optimality (if there exists no feasible solution that minimizes two objective simultaneously):

$$\begin{gathered} x\ is\ not\ a\ Pareto\ solution\ if\ there\ exist\ x^*\ such\ that \\ f_1(x^*) \le f_1(x)\ and \\ f_2(x^*) < f_2(x). \end{gathered} \quad (11)$$

where x and x^* are controlled variable in feasible region; f_1 and f_2 are objectives of the optimization, average power density and override probability in this study. The set of Pareto optimal objectives form the Pareto front (or Pareto boundary), which implies the trade-off or different weights of objectives.
The discretization of shade position depends on the continuity requirement of shading control and efficiency demand of optimization algorithm. For example, 11 shade positions (10% increments) result in short computation time, while 101 positions (1% increments) approximate continuous change of roller shades very well. In this study, 11 shade positions are pre-defined with 10% increments from 0 to 1:

$$SP \in \{0, 0.1, 0.2, \dots, 1\} \quad (12)$$

3.2 Results and Discussions

The developed optimization algorithm was implemented on the experiment training data (7 days) for preliminary theoretical evaluation of the performance. A 15-minute interval was set for the data in accordance with the time horizon of the optimization algorithm, which resulted in optimization results of 115 time steps. The outdoor conditions (measured transmitted illuminance values) of these time steps are shown in Figure 4 for analysis reference. The illuminance values can roughly indicate different types of weather conditions from cloudy and partly cloudy (day 1, 2, 3, 7) to sunny (day 4, 5, 6). Transmitted illuminances around step 90 are close to 0 corresponding to data collected at sunset.

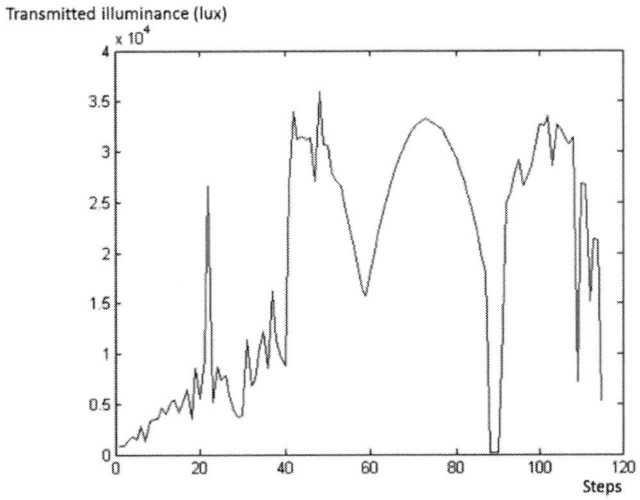

Figure 4: Measured transmitted illuminances

3.2.1 Representative Pareto Fronts: Figure 5 shows the Pareto fronts of optimization results of three representative time steps – step 2 (day 1), step 36 (day 3) and step 86 (day 5). The numbers next to the Pareto optimal or feasible points correspond to the values of controlled variable – shade position (0 – fully open, 1 – fully closed). The number of feasible points, including Pareto optimal, is equal to the number of pre-defined shade positions (11 in this study) in the optimization algorithm, and is kept constant.
As can be seen, the number of Pareto solutions and corresponding optimal objectives vary with different time steps, depending on the shape of feasible region (boundary) in objective space. Figure 5 (b) represents the most general cases, where several Pareto optimal points are located at the left side of the "U-shape", corresponding to partly open shade positions. Figure 5 (a) represents the cases with strictly convex feasible boundary, where the complete left "valley" is the Pareto front (from fully to partly open positions). Figure 5 (c) shows the case when a single global optimal could be achieved.

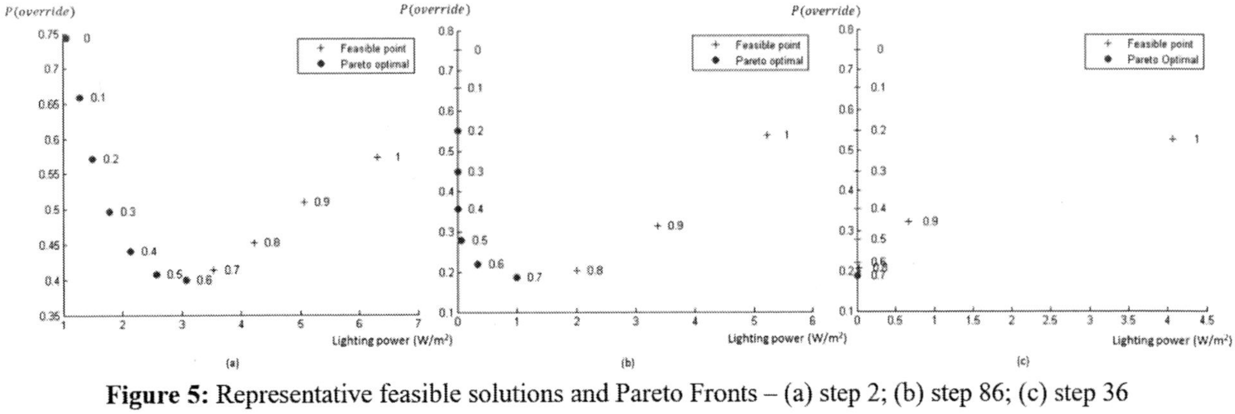

Figure 5: Representative feasible solutions and Pareto Fronts – (a) step 2; (b) step 86; (c) step 36

3.2.2 Tolerance in Objective: Due to large window-to-wall ratio and south-facing facade of the test offices, the lighting power consumed with different shade positions sometimes has very slight difference, which can be seen from Figure 5 (b), where the Pareto points close to the axis have differences that can hardly be noticed. To address this problem, a tolerance could be set in comparison of power objectives while searching for the Pareto optimal:

$$x \textit{ is not a Pareto solution if there exist } x^* \textit{ such that } \\ f_1(x^*) \le f_1(x) + tolerance \\ and\ f_2(x^*) < f_2(x). \tag{13}$$

In this way, points with negligible differences in power consumption but higher override probability would be filtered out as not a Pareto solution. In this study, a tolerance of 0.1 W/m^2 in power objective is investigated as such difference is insignificant in lighting control. Comparison of Pareto fronts with and without tolerance is shown in Figure 6, using data from the same time step. As can be seen, three Pareto optimal points without significant difference in power consumption are eliminated.

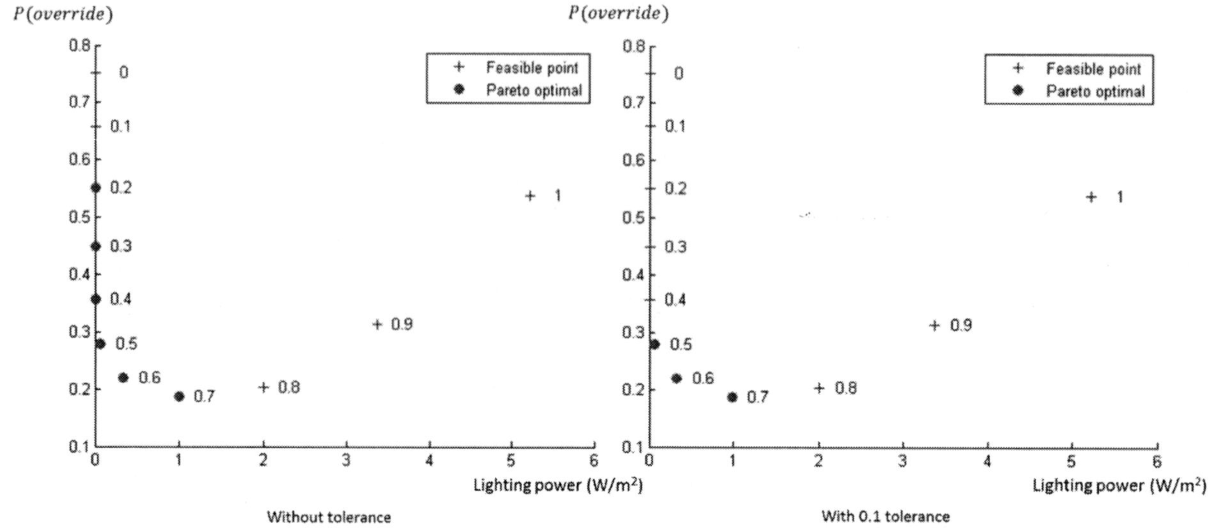

Figure 6: Comparison of Pareto Fronts with and without tolerance

3.2.3 Analysis of Pareto Solutions: The numbers of Pareto solutions of every time step are shown in Figure 7. As can be seen, cases with single optimal (one solution) are rare (only once) without tolerance but more often with tolerance, due to stricter definition of Pareto optimal. The maximum number of Pareto solutions is 7 in this study, appearing with cloudy conditions (low transmitted illuminances). By comparing outdoor conditions in Figure 4 with Figure 7, the trend that sunny condition leads to less Pareto solutions can be concluded.

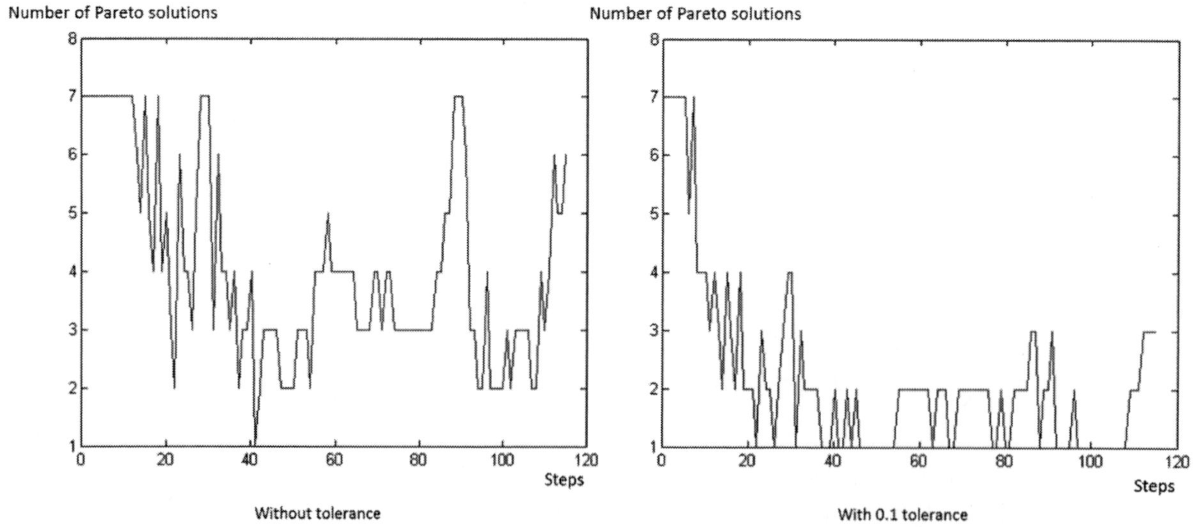

Figure 7: Number of Pareto solutions with and without tolerance

Figure 8 shows the Pareto solution range of every time step with and without tolerance. As can be seen, the Pareto optimal shade positions range from 0 to 0.8 in both scenarios, which means that the shades are never fully closed (SP=1) even in sunny conditions, indirectly indicating the energy saving potential of the developed optimization algorithm. Intermediate shade positions (0.3 – 0.6) are determined as Pareto solutions more frequently than extreme shade positions. The range of Pareto solutions is narrowed with tolerance as expected.

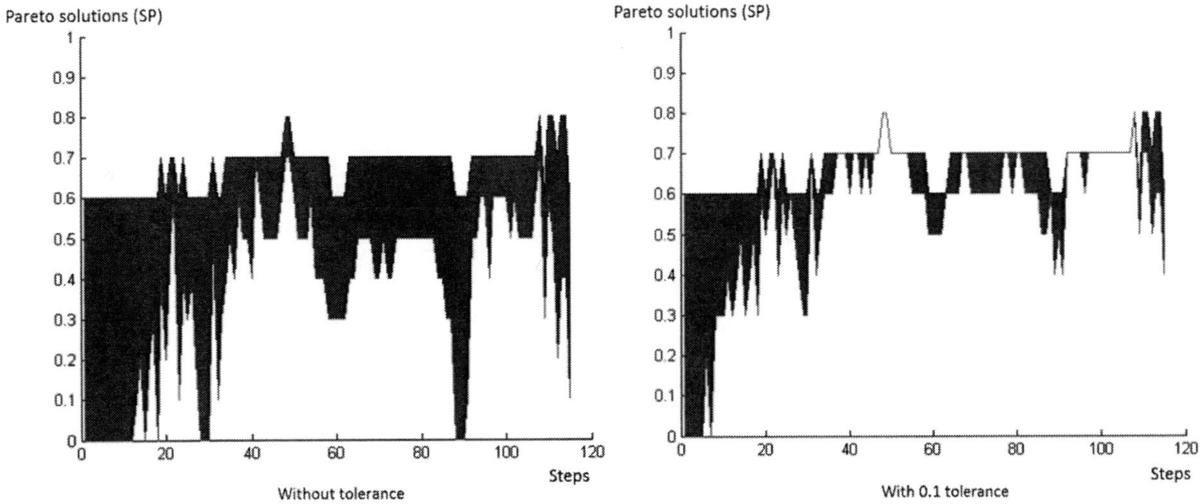

Figure 8: Pareto solution range with and without tolerance

3.2.4 Analysis of Pareto Optimal: The ranges of Pareto optimal objectives of every time step without and with 0.1 W/m^2 tolerance are shown in Figure 9 and Figure 10. The Pareto optimal power density ranges from 0 to approximately 3.5 W/m^2, except for peak power consumed (around 6 W/m^2) when it's nearly dark outside. Average Pareto optimal power consumption is higher in cloudy conditions as expected. The Pareto optimal override probability ranges from 0.2 to 0.75, and the range is larger with cloudy conditions compared to sunny conditions. The Pareto optimal power objective ranges are similar with and without small tolerance, while Pareto optimal override probability range is narrowed with tolerance.

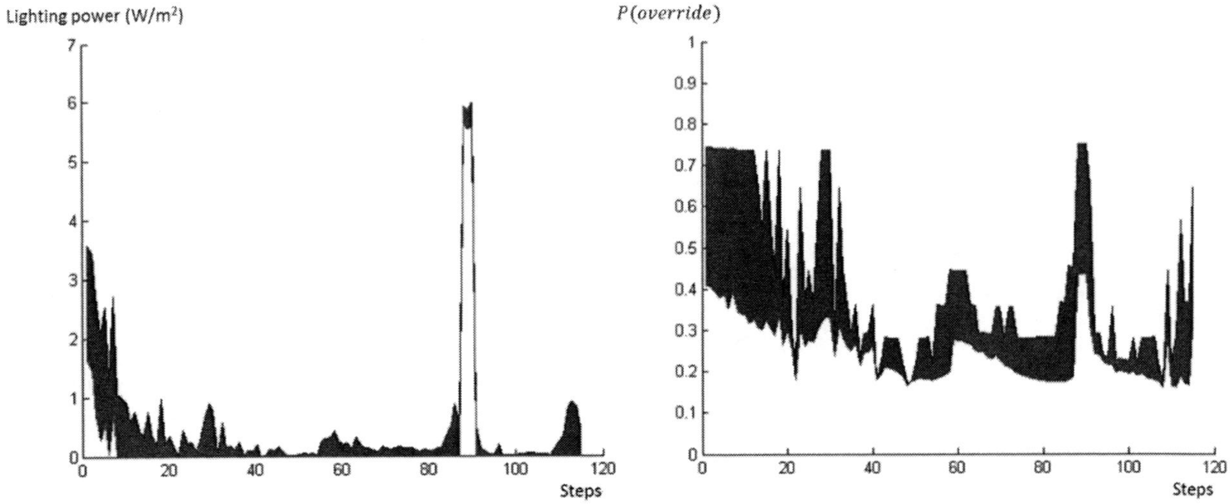

Figure 9: Range of Pareto optimal lighting power (left) and range of Pareto optimal override probability (right) without tolerance

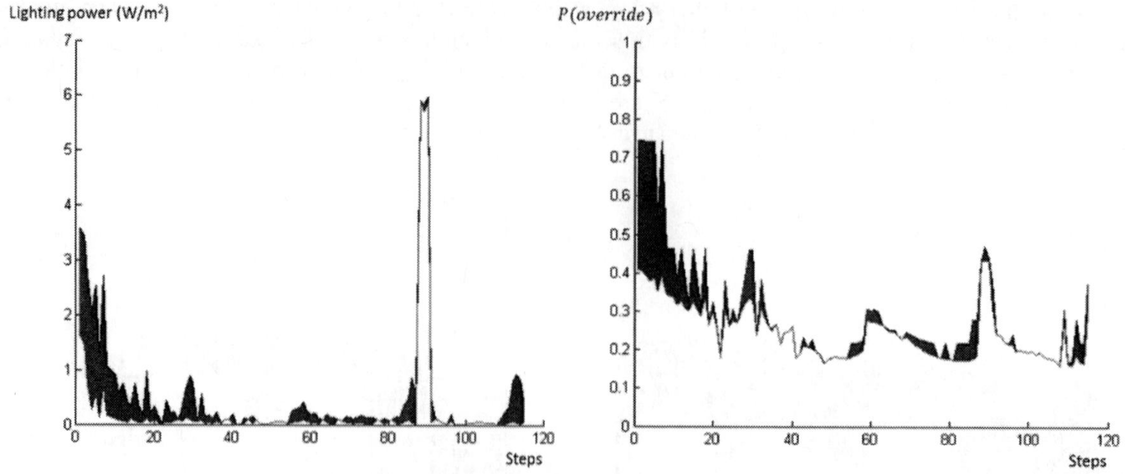

Figure 10: Range of Pareto optimal lighting power (left) and range of Pareto optimal override probability (right) with tolerance

3.3 Framework of Multi-Objective Optimization Application in Shading Control

The optimization process is designed for application of shading control in perimeter zones of buildings with glass facades. As multi-objective optimization usually provides a set of Pareto solutions, it is challenging to select a single "optimal" point for the control system from the solution set. This selection is equivalent to assigning weights w_1, w_2 for objectives in multi-objective optimization so that a normal single-objective optimization problem is formed:

$$\min_{SP\in[0,1]} \; w_1 Energy(SP,X) + w_2 P(SP,Y) \tag{14}$$

where w_1, w_2 will lead to the selected Pareto optimal.

However such weighting of objectives is questionable when considering evaluation of the trade-off between them, especially for personalized control. Therefore, instead of fixing weights for the optimization, variable weight "coefficients" determined by the occupants themselves are introduced.

The idea is to provide various control options which enable the decision maker (occupant or user) to tune the trade-off of different objectives by themselves –but in all cases within the developed personalized satisfaction "limits". For example in this study, a pool of options in Pareto front from most comfort corresponding to minimum probability of override to most energy-saving corresponding to minimum lighting energy use could be provided in every time step, in a way similar to slide bar (Figure 11).

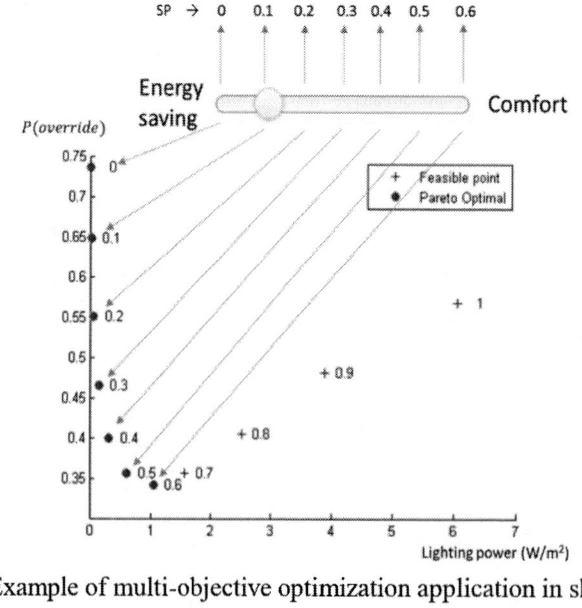

Figure 11: Example of multi-objective optimization application in shading control

Since the number of solutions in Pareto front varies with different time steps, the slide bar consists of a variable number of options, or even only one option when there is a single global minimum for the multi-objective optimization. Further research is required on how to assign changing number of Pareto solutions to the slide bar.

4. CONCLUSION AND FUTURE WORK

This study presents the development of a personalized shading control strategy and the application framework aiming at maximizing personal occupant satisfaction while minimizing lighting energy consumption with automated shading systems, using a multi-objective optimization algorithm. A set of experiments with human subjects in private offices were conducted to develop shading override probability models for quantifying the occupant satisfaction with the visual environment using automated shading. The override probability prediction along with predicted lighting energy use were taken as objectives in the multi-objective optimization algorithm to provide a set of Pareto solutions. Finally, occupants or users are allowed to choose from the Pareto solutions based on subjective preferences.

The override probability model was developed by combining separately-built lowering and raising models trained by multivariate logistic regression, and the input variables were selected based on global sensitivity analysis. Quantitative evaluation method for probability function is required for evaluating the performance of developed models. Further quantitative evaluation of developed models is required such as Jensen–Shannon divergence method.

The multi-objective optimization strategy utilizes enlarged time horizon based on serial-independency and fixed condition assumptions to improve the reliability of optimization. Pre-defined finite controlled variables and straight-forward search strategies were adopted as optimization algorithm for better efficiency without sacrificing accuracy.

The developed optimization algorithm was implemented on part of experimental data for evaluation. Three representative Pareto fronts are shown and compared, and tolerance was introduced in the power objective function in optimization for practical application. Results of Pareto solutions and optimal points show the effectiveness of tolerance, and trends that cloudy conditions lead to more Pareto solutions as well as larger Pareto optimal override probability ranges. Extreme shade positions are less likely to be in Pareto front and energy savings potential can be inferred from that high shaded window fraction (high SP) is not preferred in Pareto solutions.

The application framework of developed personal shading control strategy using multi-objective optimization strategy was introduced. Instead of assigning fixed weights on multiple objectives without specific reasons, complete Pareto solutions are provided as pool of options and decision is left to occupants or users. In this way, in contrast with existing optimization studies, the occupants are the decision makers in the final balancing between their personalized comfort limits and energy use considerations.

The developed method serves as a prototype study on adaptive shading control strategies with learned personalized comfort profiles and parallel energy use considerations. It is a preliminary step for application of multi-objective optimization strategy in integrated building control. Future work includes developing adaptive learning satisfaction or preference models, as well as new strategies for constructing pools of options given varying numbers of Pareto solutions. The developed strategy needs to be investigated and evaluated by application in more buildings and also considering feedback from occupants. Objectives and controlled variables of the control framework are not only limited in lighting energy, visual comfort and shade position, but also extended to energy consumed in building, overall comfort and indoor environment control.

REFERENCES

Athienitis, A. K., & Tzempelikos, A. (2002). A methodology for simulation of daylight room illuminance distribution and light dimming for a room with a controlled shading device. *Solar Energy*, 72(4), 271-281.

Carlucci, S., Cattarin, G., Causone, F., & Pagliano, L. (2015). Multi-objective optimization of a nearly zero-energy building based on thermal and visual discomfort minimization using a non-dominated sorting genetic algorithm (NSGA-II). *Energy and Buildings*, 104, 378-394.

Cassol, F., Schneider, P. S., França, F. H., & Neto, A. J. S. (2011). Multi-objective optimization as a new approach to illumination design of interior spaces. *Building and Environment*, 46(2), 331-338.

Chaiwiwatworakul, P., Chirarattananon, S., & Rakkwamsuk, P. (2009). Application of automated blind for daylighting in tropical region. *Energy Conversion and Management*, 50(12), 2927-2943.

Chan, Y. C., & Tzempelikos, A. (2012). A hybrid ray-tracing and radiosity method for calculating radiation transport and illuminance distribution in spaces with venetian blinds. *Solar energy*, 86(11), 3109-3124.

Chan, Y. C., Tzempelikos, A., & Konstantzos, I. (2015). A systematic method for selecting roller shade properties for glare protection. *Energy and Buildings*, 92, 81-94.

Din, I., & Kim, H. (2013). Joint blind and light control for lighting energy reduction while satisfying light level and anti-glare requirements. *Lighting Research and Technology*, 1477153513497740.

Fabi, V., Andersen, R. K., & Corgnati, S. (2015). Verification of stochastic behavioural models of occupants' interactions with windows in residential buildings. *Building and Environment*, 94, 371-383.

Futrell, B. J., Ozelkan, E. C., & Brentrup, D. (2015). Bi-objective optimization of building enclosure design for thermal and lighting performance. *Building and Environment*, 92, 591-602.

Grynning, S., Time, B., & Matusiak, B. (2014). Solar shading control strategies in cold climates–Heating, cooling demand and daylight availability in office spaces. *Solar Energy*, 107, 182-194.

Guillemin, A., & Molteni, S. (2002). An energy-efficient controller for shading devices self-adapting to the user wishes. *Building and Environment*, 37(11), 1091-1097.

Guillemin, A., & Morel, N. (2001). An innovative lighting controller integrated in a self-adaptive building control system. *Energy and buildings*, 33(5), 477-487.

Gunay, H. B., O'Brien, W., & Beausoleil-Morrison, I. (2015). Implementation and comparison of existing occupant behaviour models in EnergyPlus.*Journal of Building Performance Simulation*, 1-46.

Gunay, H. B., O'Brien, W., Beausoleil-Morrison, I., & Huchuk, B. (2014). On adaptive occupant-learning window blind and lighting controls. *Building Research & Information*, 42(6), 739-756.

Gunay, H. B., O'Brien, W., Beausoleil-Morrison, I., Goldstein, R., Breslav, S., & Khan, A. (2014). Coupling stochastic occupant models to building performance simulation using the discrete event system specification formalism. *Journal of Building Performance Simulation*, 7(6), 457-478.

Haldi, F., & Robinson, D. (2010). Adaptive actions on shading devices in response to local visual stimuli. *Journal of Building Performance Simulation*,3(2), 135-153.

Kotey, N. A., Wright, J. L., & Collins, M. R. (2009). Determining off-normal solar optical properties of roller blinds. *ASHRAE Transactions*, 117(1).

Lee, M., Carswell, C. M., Seidelman, W., & Sublette, M. (2013, September). Green Expectations The Story of a Customizable Lighting Control Panel Designed to Reduce Energy Use. In *Proceedings of the Human Factors and Ergonomics Society Annual Meeting* (Vol. 57, No. 1, pp. 1353-1357). SAGE Publications.

Manzan, M., & Padovan, R. (2015). Multi-criteria energy and daylighting optimization for an office with fixed and moveable shading devices.*Advances in Building Energy Research*, 9(2), 238-252.

Nielsen, M. V., Svendsen, S., & Jensen, L. B. (2011). Quantifying the potential of automated dynamic solar shading in office buildings through integrated simulations of energy and daylight. *Solar Energy*, 85(5), 757-768.

Oh, M. H., Lee, K. H., & Yoon, J. H. (2012). Automated control strategies of inside slat-type blind considering visual comfort and building energy performance. *Energy and Buildings*, 55, 728-737.

Reinhart, C. F. (2004). Lightswitch-2002: a model for manual and automated control of electric lighting and blinds. *Solar Energy*, 77(1), 15-28.

Reinhart, C. F., & Voss, K. (2003). Monitoring manual control of electric lighting and blinds. *Lighting Research and Technology*, 35(3), 243-258.

Saltelli, A., Chan, K., & Scott, E. M. (Eds.). (2000). *Sensitivity analysis*(Vol. 1). New York: Wiley.

Shen, H., & Tzempelikos, A. (2012). Daylighting and energy analysis of private offices with automated interior roller shades. *Solar Energy*, 86(2), 681-704.

Sobol', I. Y. M. (1990). On sensitivity estimation for nonlinear mathematical models. *Matematicheskoe Modelirovanie*, 2(1), 112-118.

Tzempelikos, A., & Athienitis, A. K. (2007). The impact of shading design and control on building cooling and lighting demand. *Solar Energy*, 81(3), 369-382.

Van Moeseke, G., Bruyère, I., & De Herde, A. (2007). Impact of control rules on the efficiency of shading devices and free cooling for office buildings.*Building and environment*, 42(2), 784-793.

Villa, C., & Labayrade, R. (2012). Multi-objective optimisation of lighting installations taking into account user perferences-a pilot study. *Lighting Research and Technology*, 1477153511435629.

Wang, C., Yan, D., Sun, H., & Jiang, Y. (2016). A generalized probabilistic formula relating occupant behavior to environmental conditions. *Building and Environment*, 95, 53-62.

Wienold, J., Frontini, F., Herkel, S., & Mende, S. (2011). Climate based simulation of different shading device systems for comfort and energy demand. In *12th Conference of International Building Performance Simulation Association* (pp. 14-16).

Zarkasdis, N., Morel, N., & Scartezzini, J. L. (2015). A novel occupant-adapted and fuzzy logic-ready visual comfort modelling approach using machine learning algorithms. In *Proceedings of International Conference CISBAT 2015 Future Buildings and Districts Sustainability from Nano to Urban Scale* (No. EPFL-CONF-213360, pp. 419-424). LESO-PB, EPFL.

Zhang, S., & Birru, D. (2012). An open-loop venetian blind control to avoid direct sunlight and enhance daylight utilization. *Solar Energy*, 86(3), 860-866.

ACKNOWLEDGEMENT

This work was funded by the National Science Foundation under Grant No. 1539527. Any opinions, findings, and conclusions or recommendations expressed in this material are those of the authors and do not necessarily reflect the views of the National Science Foundation.

Air Cleaning Performance of a Biowall for Residential Applications

Osama Alraddadi[1], Hannes Leuner[2], Brandon Boor[3],
Bhargav Rajkhowa[4], William Hutzel[5]*, Michael Dana[6]

Purdue University
West Lafayette, IN, U.S.A.
[1]osa@purdue.edu
[2]Hannes.Leuner@gmx.de
[3]bboor@purdue.edu
[4]brajkhow@purdue.edu
[5]hutzelw@purdue.edu
[6]dana@purdue.edu

* Corresponding Author

ABSTRACT

An active botanical air filtration system, called the Biowall, is used to improve indoor air quality (IAQ) and provide the potential for energy savings in residential buildings by reducing the need for air conditioning. The Biowall is an integral part of the HVAC system to actively filter the return air from Volatile Organic Compounds (VOCs). The Biowall was recently evaluated in an environmental chamber with controlled conditions to measure its air cleaning characteristics. A 'pull-down' test method, in which a known amount of contaminant was introduced to the chamber so that its decay could be monitored over time, was used for the evaluation. The decay rate of the contaminants with the Biowall present was then compared to the decay rate of an empty chamber and the growth media alone (without plants). The experiments also evaluated the filter at different airflow rates across the filter and different moisture content inside the growth media.

Based on the experimental data, the clean air delivery rates (CADR) of the Biowall were quantified. The preliminary results showed that the Biowall removed up to 90% of the introduced contaminants within two to three hours inside a sealed environmental chamber. This could potentially translate into energy savings on the ventilation systems of up to 25%. In addition to savings on ventilation, the Biowall could contribute in saving heating and cooling energy by reducing the air temperature during summer months, and reducing the air dryness during the winter months. Beyond these quantifiable benefits, a Biowall provides an intangible benefit by adding a pleasant natural aesthetic to a home.

1. INTRODUCTION

In recent years, increasing attention has been directed to energy consumption patterns in the developed world. Though there has been an increase in energy consumption in all sectors, the increase in building energy consumption has been fairly substantial. Pérez-Lombard, Ortiz, and Pout (2008) mentioned that in developed countries, building energy use alone accounts for up to 40 % of total energy consumption. Fifty percent of this consumption is solely due to the building's HVAC system. This increase in building energy use led to an increase in CO_2 emissions from the built environment. Simultaneously, rising fossil fuel prices meant building operating costs continued to rise (Pérez-Lombard, Ortiz, & Pout, 2008).

To tackle the dual challenge of environmental and financial sustainability, buildings in the early 1970s were made airtight and energy efficient. In addition, ventilation rates were also lowered, all with the aim of minimizing HVAC and building energy use. But this strategy came at a cost, as it was noticed that the quality of air in these buildings declined considerably over time. Unfortunately, the ramifications of this unintended consequence were, and are still being felt, by the occupants of these indoor spaces.

In recent years, indoor air quality (IAQ) has slowly, but steadily, become a problem affecting most Americans. With most of them spending over 90% of their time indoors, the importance of this issue cannot be ignored. Improved insulation and lowered ventilation rates has resulted in less fresh air being brought indoors. Simultaneously, stricter environmental regulations have led to buildings being made more airtight to optimize energy efficiency.

Reduced ventilation and minimal infiltration make the air inside prone to pollution from various sources. Heat loads from electronic devices, CO_2 from humans as well as volatile organic compounds (VOCs) that are off-gassed from various chemical sources, remain trapped indoors. As the levels of these pollutants increases over time, the quality of air indoors is significantly reduced. This results in the occupied space suffering from what is known as sick building syndrome (SBS). SBS, in addition to lowering human productivity, also results in the occupants suffering from nausea, headaches and dizziness amongst other ailments. The effects of poor IAQ on human health have been well documented with the World Health Organization (WHO) naming IAQ as the fourth biggest health threat of the 21st century. This has prompted research into new technologies that filter indoor air and minimize ventilation energy use. With an increasing emphasis on the development of sustainable technology, the best way forward may perhaps be to integrate the natural environment with the built one and use it to mitigate the ill effects of the built environment.

Numerous scientific studies have shown that plants possess the ability to remove contaminants present in soil, water and air. This process, called phytoremediation, has been used to tackle environmental pollution from various human and chemical sources. Furthermore, studies have shown that certain indoor houseplants, and to be more specific, the microbial community associated with the root zone of these plants, can break down pollutants present in indoor air (Wolverton, 1997). Wang and Zhang, studied the air cleaning potential of a botanical bio-filter under both short and long term conditions and found it to be of use in reducing VOC levels (Wang & Zhang, 2011). At Rensselaer Polytechnic Institute, Aydogan (2012) designed, tested and evaluated a full scale BI-AMPS that could improve the quality of indoor air through botanical bio-filtration.

With indoor air pollution being a persistent and pressing concern in the developed world, a hydroponic biofilter, with sufficient plant area to capitalize on the plants' natural phytoremediation capacity, could possibly be used to improve indoor air quality. With plants also taking up CO_2 during photosynthesis, this botanical biofilter can concurrently aid in lowering CO_2 levels indoors, further improving IAQ.

2. Biowall Design

2.1 Design Characteristics

An active botanical air filtration system called the Biowall was designed with a focus on customer convenience, energy efficiency and aesthetics. The Biowall pulls return air from the home, removes pollutants from this air stream, and passes cleaned air back to the HVAC system of the house. The Biowall is optimized to allow easy plant and growth media maintenance with the help of its key design feature, the removable plant trays. Furthermore, space requirements and aesthetic guidelines were taken into account for the design of the Biowall. The Biowall has approximately 1 m^2 of surface area for the plants. Although this paper reports on the performance of the Biowall in a laboratory environment, the Biowall will ultimately be placed in a net zero energy research home that has been constructed.

2.1.1 Components:

The Biowall consists of a plenum, four trays and a back duct as shown in Figure 1. The plenum is designed to fit in a wall of a building and accommodates the plant trays. The back duct connects the Biowall to the return duct of the HVAC-system and combines the individual air streams of the four trays. The trays provide space for the plants and the grow media. For research purposes and reliable maintenance, the trays can be easily removed from the plenum and maintained away from the Biowall or replaced with new trays. The trays were positioned in a horizontal orientation to sufficiently use loose grow media substances and to prevent maldistributions of water.

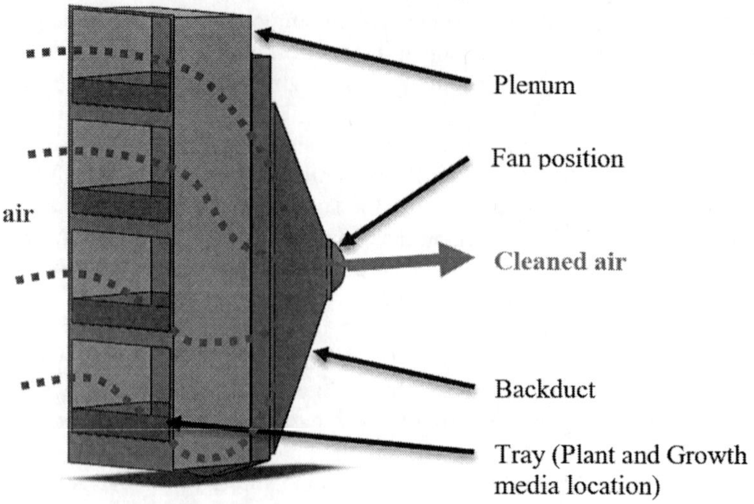

Figure 1: Biowall design sketch.

2.1.2 Airflow:

Figure 1 displays schematically the air flow pathway through the device. Basically, a fan in the back duct of the Biowall passes contaminated air from the inside of a building through the plants and their rhizosphere inside the trays where degradation processes take place. Then, the cleaned air from the trays enters the back duct and is combined to a single air flow outlet which leads to the return duct of the HVAC-system. The air flow has been optimized focusing a low pressure loss and equal air flow treatment across the filter surface in each tray.

2.1.3 Equipment:

The Biowall is planted with Epipremnum aureum (Golden Pothos), Chlorophytum comosum (Spider Plant) and Philodendron scandens (Heartleaf Philodendron), which are equally distributed across all four trays as shown in Figure 2. The growing medium in the trays is a mixture of coconut coir, Growstone® and activated carbon with an elevated air porosity. Furthermore, the plants are watered frequently through soaker hoses embedded in the growing medium. State of the art triple-band LED grow lights are mounted above the trays to supply a photosynthetically active photon flux of minimum 65 μmol m^{-2} sec^{-1}.

Figure 2: Biowall with plants, grow media and grow lights (left side) and detail view of one tray (right side)

3. Contaminant Removal

3.1 Methodology

The Biowall, in this study, was tested in a laboratory environment to understand its removal capacity of air contamination. The challenge gas in this study was Toluene, which is an insoluble and aromatic VOC. Toluene is a commonly found contaminant in homes with concentrations between 2 ppm to 8 ppm and it is usually emitted from different sources such as paints and adhesives (Maroni et al., 1995). An 800-ft^3 (22.6-m^3) environmental chamber with a hydronic cooling and heating system were used to conduct the experiments of this study. The environmental chamber preparation, the data acquisition system, and the toluene injection method are explained in sub-subsection 3.1.1, the testing procedure for toluene removal is explained under sub-subsection 3.1.2.

3.1.1 Experimental setup

The environmental chamber was examined for air-leakage and was also prepared for the experiments before any testing. By using CO_2 as a tracer gas, the air leakage of the chamber was found to be below 0.03 ACH, which is the acceptable limit for testing portable air cleaners according to ANSI/AHAM standard AC-1. The chamber surfaces were cleaned according to ASTM standard D6670-13, which recommends scrubbing the surfaces by a laboratory ionic detergent with phosphate, rinsing with water, and then rinsing with deionized water as a final rinse. The cleaning process was implemented to eliminate the presence of any adsorbed VOCs to the chamber's surfaces as well as to minimize the injected-toluene desorption from these surfaces.

Inside the environmental chamber, multiple variables were monitored and controlled throughout the experiments. The temperature inside the chamber, for instance, was monitored and kept at 73°F±4°F (23°C±2°C) by using the chamber's external hydronic system. The relative humidity and the pressure drop across the filter were monitored but they were not controlled. The water content in the growth media was not monitored nor was it controlled. The plants were given 4 liters of water per tray every 48 hours. During the experiment, a photoionization detector (PID) from RAE systems (MultiRAE gas detector – model: PGM-6628) monitored the bulk concentration of toluene. The manufacturer calibrated the PID to measure toluene concentration from 0.1 ppm to 5,000 ppm.

A gas dispersal system (GDS) was designed and built to inject atomized toluene inside the environmental chamber. The GDS had a 25 ml syringe and a syringe pump (manufacturer: New Era – Model: NE-1010). The syringe had an atomizing nozzle attached to its luer to disperse the liquid toluene as a gas. The GDS was tested for repeatability for different injection amounts, and the system always provided concentration levels within 10% for each injection. For example, a 5 ml injection always provided a concentration of 45 ppm ± 5 ppm.

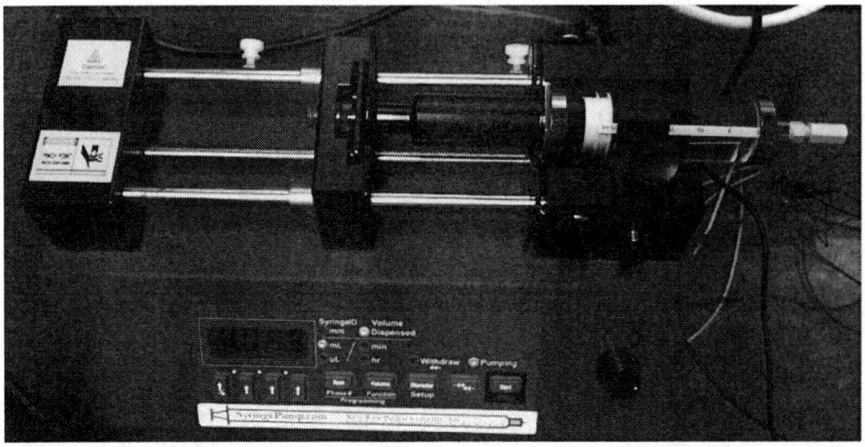

Figure 3: Gas Dispersal System (syringe pump, syringe, and atomizing nozzle)

3.1.2 Testing:
The main approach for testing the removal capacity of the Biowall followed a 'pull-down' testing method. The method requires quantifying the natural decay for toluene inside the chamber (i.e., without the trays of the plants and their growth media) as well as the toluene decay with the presence of the Biowall. The difference between the two decay rates multiplied by the volume of the chamber is the clean air delivery rate (CADR) of the biofilter. The toluene decay with the Biowall was measured in various settings.

The Biowall was tested using two different fan speeds and different initial VOC concentration levels. The high fan speed and the low fan speed of the Biowall had volumetric airflow rates of around 105 cfm (180 m^3 per hour), and 65 cfm (110 m^3 per hour), respectively. On the other hand, each fan speed of the Biowall was tested with three different single injections amount of toluene (i.e., different initial concentration) in addition to a multi-injection of toluene. 5 mL of toluene, 1 ml of toluene, 0.2 ml were the single injection test, and they had initial concentrations of 45 ppm ± 5 ppm, 11 ppm ± 1 ppm, and 1.6 ppm ± 0.2 ppm, respectively. Each single injection test was carried for around 10-hours. The multi-injection test, on the other hand, was for the 5 ml injection in which toluene was injected four times during one test of 48 hours (i.e., one injection every 12 hours).

Table 1 provides a summary of the tests for the Biowall as well as the growth media with both conditions. In addition to testing the Biowall as a biofilter, the growth media of the Biowall (without the plants) was also tested as an absorbent filter in two different conditions. The toluene decay was tested for the growth media mixture a dry condition and in a wet condition. The growth media, which had no previous contact with the plants, was tested under the high fan speed only (i.e., 105 cfm or 180 m^3 per hour). The dry growth media refers to the growth media as it came from the supplier without the addition of any water prior testing. Although the growth media was considered dry, the Coco Coir had some moisture content as they came from the supplier. The wet growth media was watered with 4 L of water prior each test unless it was watered within 48 hours. Both conditions of the growth media were tested in a single injection test and a multi-injection test of 5 ml only.

Table 1: Summary of Tests

Target Test		Fan Speed (cfm) [m^3/hour]	Injection amount [peak concentration in ppm]				Inside the Chamber
			5 ml [45±5]	1 ml [11±1]	0.2 ml [1.6±0.2]	Multi-injection 4 x 5 ml [4 x 45±5]	
Biowall		105 [180]	Yes	Yes	Yes	Yes	Entire Biowall
		65 [110]	Yes	Yes	Yes	Yes	
Growth Media	Dry	105 [180]	Yes	No	No	Yes	Entire Biowall without plants
	Wet		Yes	No	No	Yes	

3.2 Decay Curves
The decay curves for the single injection tests of the Biowall, with both fan speeds, are shown in Figure 3. In general, the Biowall with a low fan speed showed a faster decay rate than the Biowall with a high fan speed. The Biowall with a low fan speed was able to remove up to 90% of the initial concentration within the first two to three hours of the test. In contrast, the Biowall with a high fan speed removed up to 80% of the initial concentration within the first three to four hours of the test. The reason behind the difference between the two speeds is likely due to the empty bed residence time (EBRT), which is the time taken for the contaminated air to pass through the tray. Since the Biowall with a low fan speed has a higher EBRT, the low fan speed provides more time for the contaminated air to be absorbed by the growth media.

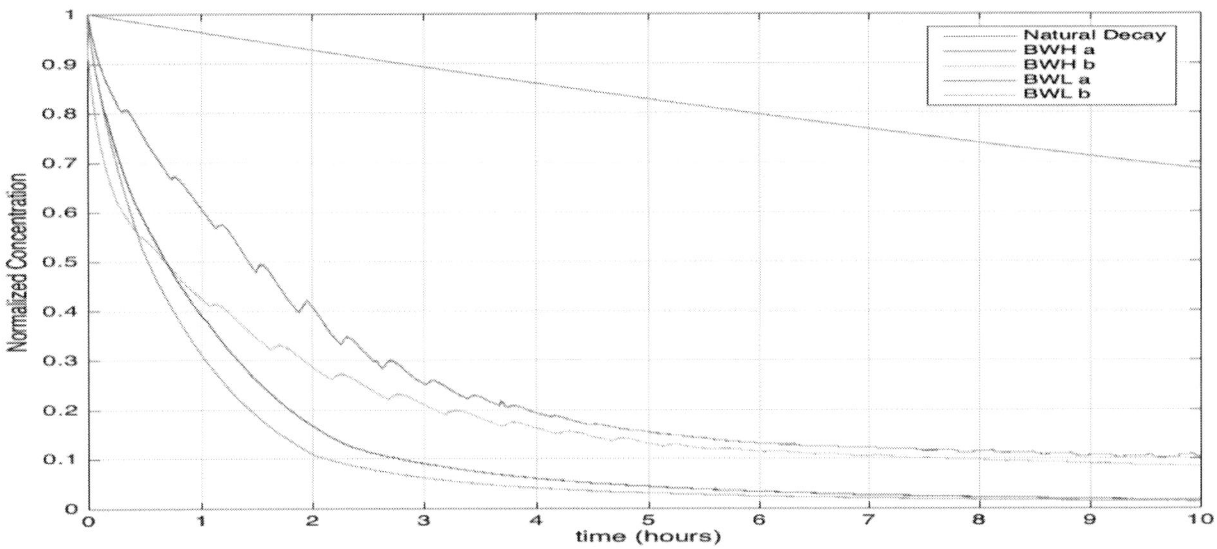

Figure 3: Toluene Decay for Biowall with high speed (BWH) and Low Speed (BWL) for the 5 ml Injection Test

Additionally the water content of the growth media influenced the test results for both fan speeds. In Figure 3, the tests denoted with the letter 'a' indicates that the plants where watered right before the test, while the 'b' letter indicates that the test was performed 24-hours after the first test and without the addition of water. In other words, test 'a' indicates higher water content than test 'b' since the fan was at least running for 10-hours during test 'a' in addition to the plants consumption of water. As shown in Figure 3 for both the Biowall with a low fan speed and the Biowall with a high fan speed, the tests denoted with 'b' had a faster decay rates than the tests denoted with 'a'. The reason is believed to be due to the insolubility of toluene in water. Therefore, the higher the water content, the slower the decay rate of toluene, and vice versa. The 1 ml and 0.2 ml had the same observations.

Figure 4 is the results of multi-injection VOC tests for the Biowall using both fan speeds. Similar to Figure 3, the Biowall with a low fan speed had a faster decay rate than the Biowall with the higher fan speed. The decay curves for the Biowall with a high fan speed also showed some oscillation, and these oscillations became more prominent in each subsequent injection. In addition, the final concentration for each injection in both fan speeds was higher than the previous injection. The differences between the final concentrations and the oscillations are believed to be due diffusion and biodegradation processes, which are time-dependent processes and outside the scope of this study.

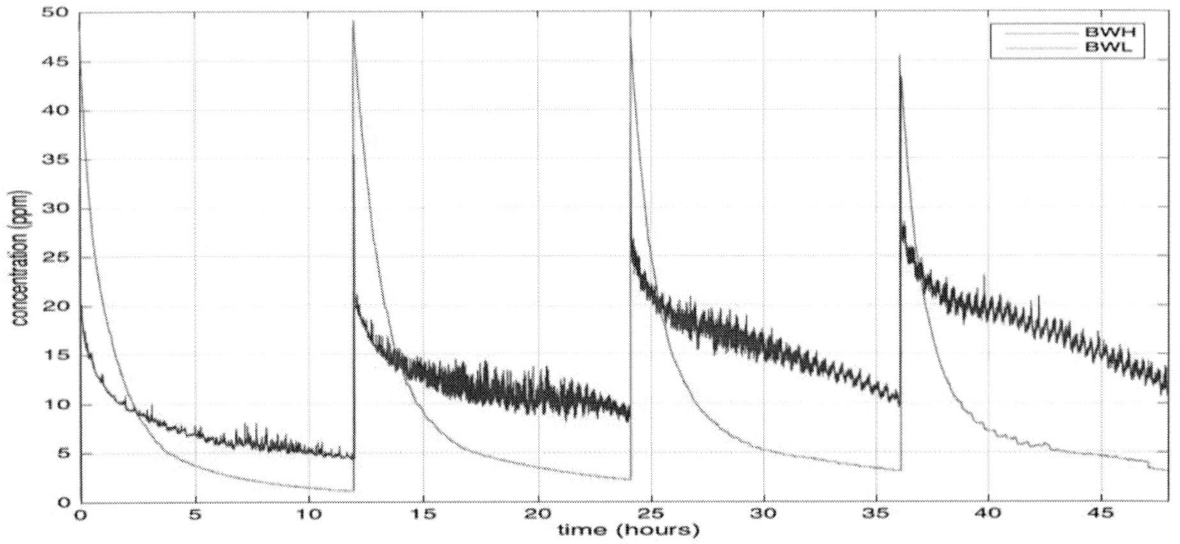

Figure 4: Toluene Decay for Biowall with high speed (BWH) and Low Speed (BWL) for the Multi-Injection Test

The final decay curves are for the growth media in the dry and wet conditions are shown in Figure 5, which shows both the single injection test of 5 ml (left) and the multi-injection test (right). The dry growth media had a faster decay rate than the wet growth media in both test cases. The reason, as noted before, is due to the insolubility of toluene in water. Both dry and wet growth media were tested under the high fan speed, but the oscillation, which also appeared with the Biowall with a high fan speed, only appeared with the wet growth media. The dry growth media had a much smoother curve that is almost similar to the Biowall with a low fan speed. Therefore, despite the EBRT, the water content has a major role in the cleaning ability of the biofilter.

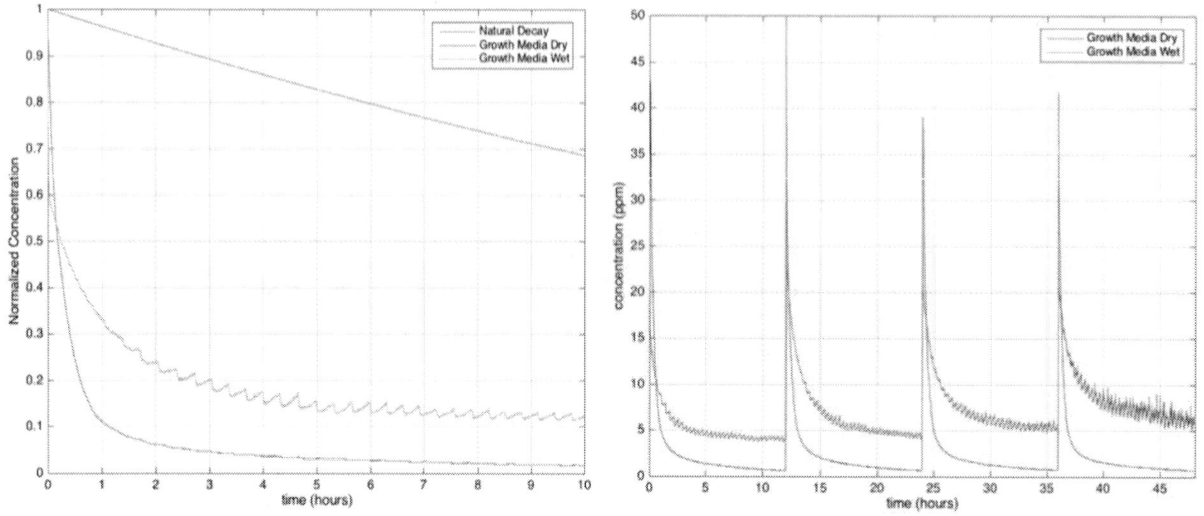

Figure 5: Toluene Decay for Dry and Wet Growth Media with high fan speed for the 5 ml Injection Test (left), and Toluene Decay for Dry and Wet Growth Media with high fan speed for the Multi-Injection Test (right)

3.3 Analysis

The decay curves of the Biowall with both fan speeds and the growth media with both conditions were analyzed to find the Clean Air Delivery Rate (CADR), which is a parameter used by filter manufacturers to quantify an equivalent volumetric flow rate of fresh air developed by a filter. By using the 'Trust region reflective algorithm' (MATLAB built-in function), the curve fit for each the decay curve was optimized. The natural decay rate was found to be -0.038 h^{-1} as an average of several tests. Similarly, the decay rate for the Biowall with a high fan speed, Biowall with low fan speed, dry growth media, and wet growth media were found for each single injection test that was performed. The CADRs were found as discussed in the methodology and the average value for each test category is presented in Figure 7.

The CADRs were different based on the initial concentration, fan speed, and the condition of the growth media. For the 5 ml injection test, the Biowall with a high fan speed and low fan speed had CADRs of approximately 10 m^3 per hour (6-cfm) and 20 m^3 per hour (12-cfm), respectively. For the 1 ml injection test, the CADRs were 23 m^3 per hour (14-cfm) for both fan speeds. For the 0.2 ml injection test, the CADRs were 20 m^3 per hour (12-cfm) for the Biowall with a high fan speed, and 37 m^3 per hour (22-cfm) for the Biowall with a low fan speed. On the other hand, the dry growth media and wet growth media had CADRs of around 61 m^3 per hour (35-cfm), and 35 m^3 per hour (21-cfm), respectively, when both were operated in high fan speed setting with a 5 ml injection of toluene.

The differences between the CADRs have few consistent trends that are obvious from the values or the previously presented decay curves. As shown in Figure 5 and in Figure 7, the addition of water to the growth media significantly lowered the CADR. For the 5 ml injection tests for the Biowall with a high fan speed, and the wet growth media with the same speed, the wet growth media had a higher CADR. The reason could be due to the fact that the Biowall was regularly watered, for weeks, prior to the start of the experiments, while the wet growth media was watered during the experiments timeframe only. Thus, the Biowall, cumulatively, had more water in its own growth media that might cause the activated carbon pellets to be saturated. The saturation of the activated carbon pellets reduces the absorptive capability and the air permeability of the pellets (Wolverton, 1995).

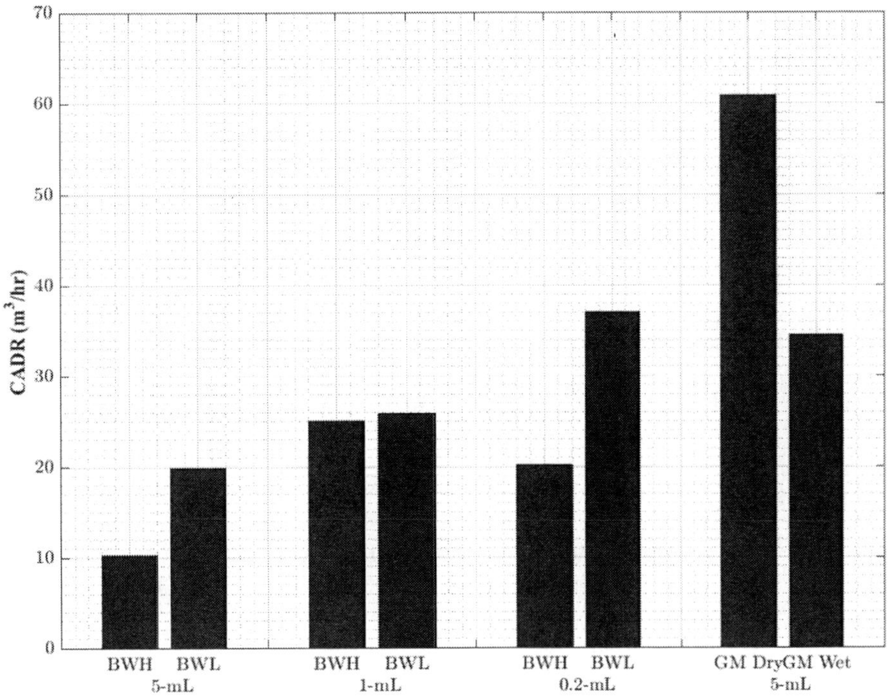

Figure 7: Clean Air Delivery Rates for the Biowall with High Fan Speed (BWH) and Biowall with the Low Fan Speed (BWL), in addition to Dry Growth Media (GM), and Wet growth media with High Fan Speed

3.4 Air Characteristics

During the experiments, temperature, relative humidity and pressure drop were monitored. For the Biowall with both fan speeds, the relative humidity levels downstream the Biowall increased significantly up to 70%, and 80% in some cases. The hydronic system could not overcome the increase in relative humidity, but when a forced air system was conditioning the chamber for different experiments, the relative humidity levels were controllable to acceptable levels. During the first three hours of each of the tests, on the other hand, the temperature downstream of the Biowall dropped by around 3°F (2°C) due to the evaporative cooling, but after the three hours, the temperature dropped by only 1°F (0.5°C) as an average because less evaporative cooling occurred. The pressure drop across the Biowall was approximately 0.15 inches of water column (in w.c.) and 0.08 in w.c. for the Biowall with a high fan speed and the low fan speed, respectively.

4. Preliminary Energy Analysis

In order to understand the potential of the CADR of the Biowall, a preliminary energy analysis was conducted based on ventilation requirement and the results are shown in Table 2. ASHRAE 62.2 standard (2013), which provides the acceptable ventilation rates for residential buildings, was used as a reference for an energy analysis of typical U.S. homes. According to a survey by the U.S. Department of Housing and Urban Development (2010), the median size of a typical occupied home is around 1,800 ft^2 (167 m^2), and the typical numbers of bedrooms is three or more. In ASHRAE 62.2, homes between 1501 ft^2 and 2000 ft^2 need ventilation rates of 90 cfm and 98 cfm for three and four bedrooms, respectively. Since the Biowall with a low fan speed showed a better performance, and a lower pressure drop than the high fan speed, (i.e., 0.08 inches of water column (in w.c.) for the low fan speed, and 0.15 in w.c. for the high fan speed), the low fan speed was chosen for the analysis. The highest CADR for the Biowall with a low fan speed was around about 23 cfm (39 m^3 per hour). Table 2 shows that incorporating a Biowall into the ventilation requirement could provide a savings in ventilation that is up to 25% of the total outside air requirement.

Table 2: Ventilation Requirements and Potential Ventilation Reduction for Typical U.S. Homes

Home Square footage	Number of Bedrooms	Continuous Ventilation Requirement (cfm)	With the Biowall (cfm)	Reduction In Ventilation
1501-2000	3	90	67	25%
	4	98	75	23%

The final part of the preliminarily energy analysis was to estimate the cost savings based on the reduction in ventilation. Osama Alraddadi (2016) created an energy model based on different climate zones. The model estimated the required cooling and heating of air to room temperature based on weather data without taking the humidification effects into account. The model shows the cost of ventilation based on ASHRAE 62.2 requirement alone, the cost with ASHRAE requirements in addition to one Biowall, and the cost of the ventilation requirement based on ASHRAE requirements with two Biowalls installed in the house.

Figure 8 shows the results of the preliminarily energy model for various climates in the U.S. The results show that a Biowall has more cost savings in colder climates than hotter ones. If the humidification effect is taken into account, the benefits may be more for the colder climates. In contrast, the Biowall does not present any notable benefits cities like San Diego, CA where the weather is moderate and less heating and cooling is required. There could be some savings in the hotter climate but the humidification effect could be disadvantageous to these climates.

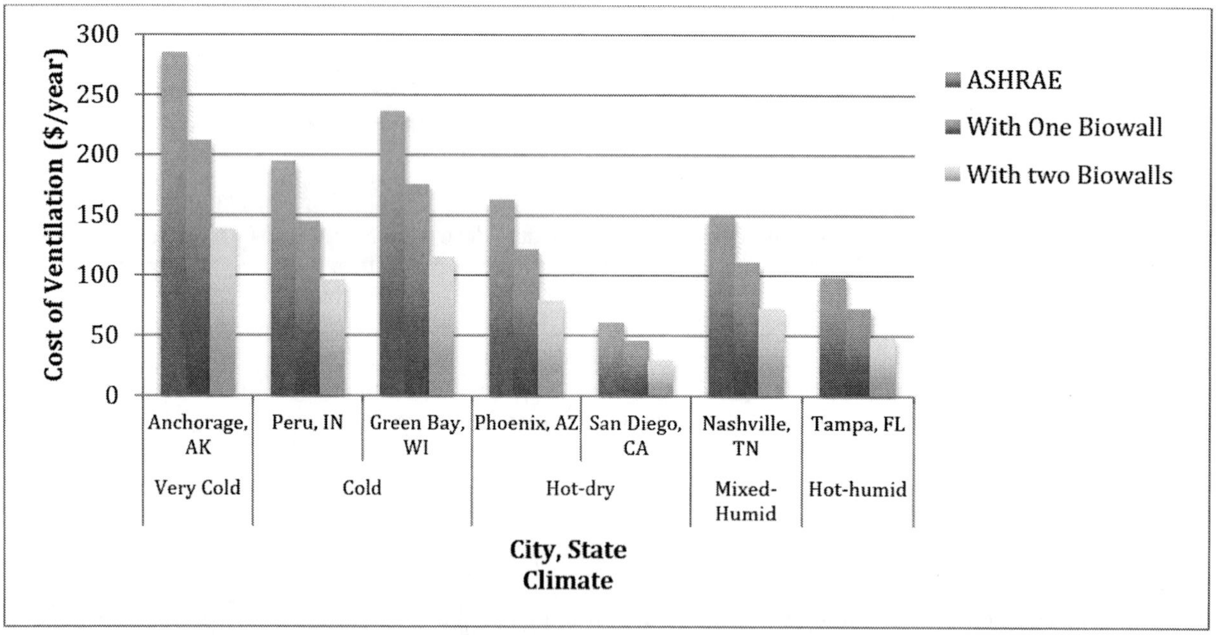

Figure 8: Annual Costs of the Required Ventilation for the Median Size of a Typical Home with Three-Bedrooms at Different Cities in the U.S. based on ASHRAE 62.2 Requirements

5. CONCLUSIONS

There are several outcomes from this research:

- The Biowall was able to remove up to 90% of the initial concentration of toluene (i.e., around 45 ppm) within the first two hours.
- The CADR of the Biowall could be up to 23 cfm (39 m^3 per hour) in the case of toluene.
- The Biowall in a closed chamber increased the relative humidity to 80%, while the temperature differential across the Biowall decreased by up to 3°F (2°C) due to evaporative cooling.
- The estimated savings in ventilation are approximately 25% based on ASHRAE 62.2 requirements for typical U.S. homes for three- and four-bedrooms homes, respectively.
- The Biowall could have more potential energy savings in colder climates than hotter ones.

Finally, the analysis and the results are based on the toluene as a challenge gas, and other VOCs with different physical properties (e.g., soluble in water) could have different results. The dry growth media showed higher CADRs than the Biowall for the short-period tests of this study. The Biowall has the biodegradation advantage for a longer-term bio filtration, while the dry growth media, when used as a filter, would require periodical replacement. Further long-term studies for different VOCs would be beneficial to understand the performance of botanical air filters.

REFERENCES

Alraddadi, O. (2016). *Investigating the Toluene Removal of a Botanical Air Filter with a Loose-Packed Growth Media for Potential Energy Savings.* (Master's thesis). Purdue University, West Lafayette, IN.

ASHRAE. (2013). *ASHRAE Standard 62.2: Ventilation and Acceptable Indoor Air Quality in Low-Rise Residential.* Atlanta: American Society of Heating Refrigeration and Air Conditioning Engineers.

Aydogan, A. (2012). *Building-integrated active modular phytoremediation system* (Doctoral dissertation, Rensselaer Polytechnic Institute).

Maroni, M., Seifert, B., & Lindvall, T. (1995). *Indoor air quality: A comprehensive reference book* (Air quality monographs; vol. 3). Amsterdam; New York: Elsevier.

Pérez-Lombard, L., Ortiz, J., & Pout, C. (2008). A review on buildings energy consumption information. *Energy and buildings*, *40*(3), 394-398.

U.S. Department of Housing and Urban Development. (2010). U.S. homes today are bigger with more bedrooms and bathrooms than 1973 [Press release]. Retrieved from: http://portal.hud.gov/hudportal/HUD?src=/press/press_releases_media_advisories/2010/HUDNo.10-138

Wang, Z., & Zhang, J. S. (2011). Characterization and performance evaluation of a full- scale activated carbon-based dynamic botanical air filtration system for improving indoor air quality. *Building and Environment, 46*, 758-68.

Wolverton, B. C., & Wolverton, J. D. (1995). *U.S. Patent No. 5,217,696.* Washington, D.C: U.S. Patent and Trademark Office

Sensitivity Analysis for the PMV Thermal Comfort Model and the Use of Wearable Devices to Enhance Its Accuracy

Mohammad H. Hasan, Fadi Alsaleem,* Mostafa Rafaie

Wichita State University, Department of Mechanical Engineering,
Wichita, Kansas, USA
Contact Information (316-978-6381, fadi.alsaleem@wichita.edu)

* Corresponding Author

ABSTRACT

This paper studies the sensitivity of the Predicted Mean Vote (PMV) thermal comfort model relative to its environmental and personal parameters. PMV model equations, adapted in the American Society of Heating, Refrigerating, and Air-Conditioning Engineers (ASHRAE) Standard 55–Thermal Environmental Conditions for Human Occupancy, are used in this investigation to generate two-dimensional (2D) and three-dimensional (3D) comfort zone plots for different combinations of parameters. It is found that personal parameters such as clothing and metabolic rate, which are usually ignored or simply assumed to be constant values, have the highest impact. In this work, we demonstrate the use of smart wearable devices to estimate metabolic rate. The metabolic rate for an occupant during normal life activities is recorded using a Fitbit® wearable device. This example is used to do the following: (1) demonstrate the PMV expected error range when personal parameters are ignored, and (2) determine the potential of using a wearable device to enhance PMV comfort model accuracy.

1. INTRODUCTION

Achieving higher performance at work is of high interest. Thermal comfort that is strongly related to productivity has recently received a great deal of attention. Thermal comfort is a subjective matter and may vary from person to person. There have been multiple attempts to develop a unified and widely accepted thermal comfort model that can be received and adopted by large audiences. The most popular model is the Predicted Mean Vote model (PMV model), which was constructed by P. O. Fanger (1970) and later adapted into the American Society of Heating, Refrigerating, and Air-Conditioning Engineers (ASHRAE) Standard 55–Thermal Environmental Conditions for Human Occupancy. This model correlates multiple environmental parameters (air temperature, air velocity, relative humidity, and radiant temperature) and personal parameters (metabolism and clothing) to different levels of comfort based on a rating between –3 and 3, where –3 means the body is very cold and 3 means the body is very hot. Typically, the goal is to control environmental factors in order to keep the PMV value between –0.5 and 0.5, where the body is believed to be thermally satisfied. Since the environmental parameters are relatively easy to measure, they have received a great deal of attention in the literature. Personal parameters, on the other hand, are more difficult to estimate or measure and, therefore, are usually assumed to be constants, thus missing the opportunity to accommodate comfort variation for occupants experiencing the same environmental conditions.

The PMV value can be directly calculated using a system of highly nonlinear and iterative equations, which were later adapted in the ASHARE Standard 55 (ASHRAE, 2013):

$$PMV = (0.028 + 0.3033e^{-0.036M}) \times L \tag{1}$$

$$\begin{aligned} L = (M - W) - 3.05x10^{-3}(5733 - 6.99(M - W) - Pa) - 0.42(M - W - 58.15) \\ - 1.7x10^{-5}(5867 - Pa) - 0.0014M(34 - t_a) - f_{cl}h_c(t_{cl} - t_a) \\ - 3.96x10^{-8}f_{cl}[(t_{cl} + 273)^4 - (\bar{t}_r + 273)^4] \end{aligned} \tag{2}$$

where L defines the overall heat transfer around a single occupant, M is the metabolic rate in Watt/m^2, W is the work excreted by the occupant in Watt/m^2, Pa is the humidity level, $\bar{t}_r$ is the mean radiant temperature, t_a is the air temperature, I_{cl} is the clothing insulation factor defined as the ratio of the area of the clothed body to the area of the nude body, and h_c is given by:

$$h_c = \begin{cases} 2.38(t_{cl} - t_a)^{0.25}, if\ 2.38(t_{cl} - t_a)^{0.25} > 12.1\sqrt{V} \\ 12.1\sqrt{V},\ if\ 2.38(t_{cl} - t_a)^{0.25} > 12.1\sqrt{V} \end{cases} \quad (3)$$

where V is the air velocity in m/s, and t_{cl} is the clothing temperature, which can be calculated based on the conditions of the body using the following simple heat balance equation:

$$t_{cl} = 35.7 - 0.028(M - W) - 0.155I_{cl}\{3.96x10^{-8} \times f_{cl}[(t_{cl} + 273)^4 + (\bar{t}_r + 273)^4] + f_{cl} \times h_c(t_{cl} - t_a)\} \quad (4)$$

A heat balance approach is adapted in the PMV model (Balaras et al., 2007) to infer human thermal comfort. For example, equations (1) and (2) were obtained by balancing the following heat modes: (1) heat generation due to metabolism $(M - W)$, (2) heat transfer by convection $0.0014M(34 - t_a)$, (3) heat transfer through the skin $3.05x10^{-3}(5733 - 6.99(M - W) - Pa)$, (4) heat transfer through latent respiration $1.7x10^{-5}(5867 - Pa)$, (5) heat transfer by dry respiration $0.0014M(34 - t_a)$, and (6) heat transfer by radiation $3.96x10^{-8}f_{cl}[(t_{cl} + 273)^4 - (\bar{t}_r + 273)^4]$. Also in equation (2), clothing temperature is estimated based on the heat generated by metabolism and heat transfer by way of conduction, convection, and radiation. To solve for the clothing temperature t_{cl} in equation (4), ASHERE incorporated an iterative process to continuously update the clothing temperature until the difference between the current and previous iteration is within a predefined margin.

Evaluating PMV equations is computationally intensive and requires iterative processes. Hence, many approximations were made. For example, in the work of Ku et al. (2015), an artificial neural network (ANN) model was employed to capture the dynamics of the PMV model equations and then use them to evaluate the PMV value for any given thermal parameters set. Also, Zhang and You (2014) introduced a sequential approximation to nonlinear equations, not only to simplify the calculation but also to find the "air temperature, relative humidity" pair that leads to maximum energy savings (inverse problem solution).

Most human comfort research work has focused on studying the comfort effect of air temperature, which is widely accepted as the most important parameter in thermal comfort models, coupled with a few other environmental factors, such as air velocity and relative humidity (ASHRAE, 2013). Less work has dealt with the effects of comfort and sensitivity to metabolism and clothing, which are personal parameters. This may reflect the fact that personal parameters are underestimated, or difficult to quantify and measure. In fact, metabolism and clothing thermal resistance play a vital role in defining the optimal thermal comfort conditions. While metabolism increases the rate of heat generation in the human body and decreases the desirable comfort temperature, clothing helps to tolerate colder conditions. Ignoring clothing and metabolism or simply assuming them to be constant values may lead to a false PMV calculation.

The focus in this paper is on metabolism and its direct effect on the PMV model. MET is the unit of metabolism in the PMV model. A single MET is equivalent to the heat a body produces while it inhales 3.5 ml of oxygen per kg of weight each hour $(\frac{ml\ O_2}{Kg.H})$ (Jette et al., 1990). Also, MET can be thought of as multiples of the resting metabolic rate for the occupant while he or she is engaged in a physical (or mental) activity (Institute of Medicine, 2005). It has been found that the body produces 5 Kcal/L of oxygen per inhale, meaning that a man weighing 70 kg would consume 1.2 Kcal/min if his metabolism is equal to 1 MET. Accurate measurement of metabolism requires knowing the amount of oxygen the body inhales or the amount of carbon dioxide and nitrogen waste are produced from the cellular breathing process (Ferrannini, 1990). This task is not trivial, as it involves using devices such as mask calorimeters to measure the gas intake and outtake. Other devices can also be used to estimate the metabolic rate, such as pedometers, load transducers (also known as foot-contact monitors), accelerometers, and heart rate monitors. Those sensors individually provide an indirect estimate of the metabolic rate and often result in numerous errors. Recent advancement in smart wearable devices has made it possible to fit most of these sensors into a single smart band, thus allowing an accurate and continuous estimation of the metabolic rate. In this paper, in order to estimate the metabolic rate, we will use the Fitbit® smart wearable device that is equipped with a pedometer, an accelerometer, and a heart rate sensor.

In this work, we first highlight the PMV model sensitivity to various environmental and personal parameters, and then we study their interaction. Next, we focus on the use of a wearable fitness device to acquire the metabolic rate of an occupant during multiple activities and its impact on his/her comfort level. The organization of this paper is as follows. In section 2, we simulate the PMV parameters interaction and their effect on the comfort zone using 2D and 3D plots. Next, in section 3, we study the variation in MET values during different controlled physical activities. Later in that section we monitor the metabolic rate of two occupants for a day and identify those errors that could result from taking the MET value as a constant value. In section 4, we summarize the paper and provide some conclusions.

2. PMV PARAMETER INTERACTIONS

There has been an increasing interest in studying the combined effect of temperature, humidity, and air velocity on PMV values, but less attention has been paid to the effect of metabolism and clothing. In this section, we follow a general approach in studying the interaction of these factors. First, we plot and discuss multiple areas of comfort under the 10% dissatisfaction criteria, i.e., PMV value is between –0.5 and 0.5, while varying the thermal comfort parameters (including the personal parameters) a pair at a time. Then, we construct 3D plots showing comfort zones as a surface. In the remainder of this paper, unless otherwise stated, the radiant temperature is assumed to be equal air temperature, and any parameter that does not vary in the simulation is assumed to be constant as follows: clothing = 0.65 CLO (clothing condition), relative humidity = 50%, MET = 1.0 (metabolism conditions), and air velocity = 0.5 m/s.

2.1 Interactions of environmental thermal conditions

The comfort area as a function of temperature and humidity is depicted as a trapezoid in Figure 1(a), a popular plot in the literature that can be used to explain the relative humidity effect on comfort. On a dry (low humidity) day, air has extra capacity to hold water compared to moist (high humidity) day. Hence, a body, through evaporative and latent respiration cooling, can lose heat faster, thus making it feel colder at the same ambient temperature. This phenomena explains, for example, why in Figure 1(a) the pair (TA = 28°C, RH = 30%) lies within the comfort zone, while the pair (TA = 28°C, RH = 60%) does not. For the same reason, high relative humidity means high vapor content in the air. Hence, the generated heat from a human body is trapped and cannot be rejected to the surrounding air by evaporative cooling. Figure 1(a) can be used as a guideline for a thermostat logic to maintain occupants' comfort based on temperature and humidity control.

Air velocity helps in maintaining comfort at high temperatures by increasing the heat-rejection rate through force convection. Figure 1(b) shows the comfort zone as a function of air velocity and temperature. It is apparent from this figure that an occupant might tolerate higher temperatures as the air velocity increases. For example, even though the air temperature of 28.5°C is considered to be uncomfortable for all possible relative humidity values above 30%, as shown in Figure 1(a), it is within the comfort zone once the air velocity exceeds 1.5 m/s. This argument necessitates the need for a 3D plot showing the comfort domain as a function of humidity, temperature, and air velocity, as shown in Figure 2. This plot shows the humidity-temperature trapezoid comfort area shift to the left (higher temperature) as air velocity increases.

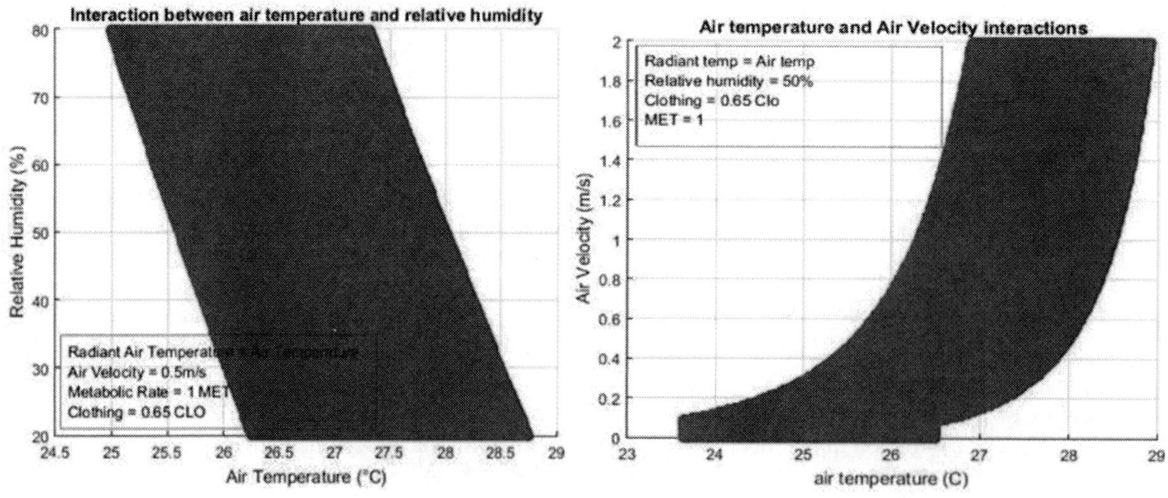

(a) (b)

Figure 1: Interaction of air temperature: (a) with relative humidity, (b) with air velocity, and resultant comfort zones

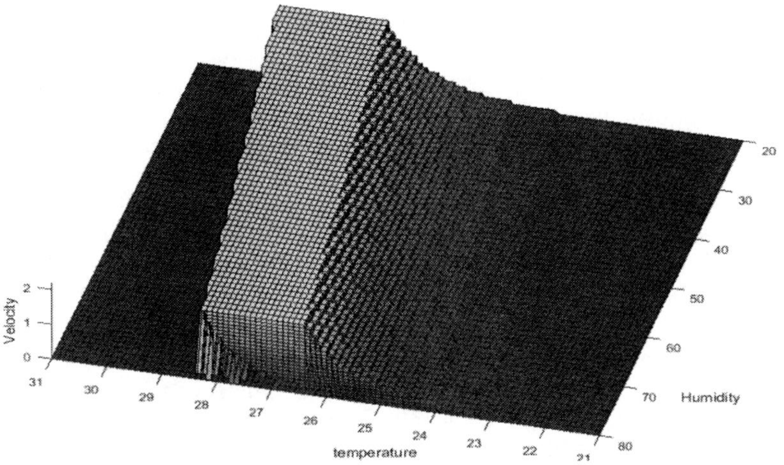

Figure 2: 3D plot of interaction between air temperature, humidity, and air velocity, and resultant comfort zone

2.2 Interactions of environmental thermal and personal parameters

Figure 3(a) shows the comfort zone area as a function of temperature and clothing, indicating that the comfort zone is very sensitive to clothing. An interesting extremely high sensitivity is noticed when CLO is around 0.5. This value lies between 0.36 CLO, the CLO value of wearing a short-sleeved shirt and shorts, and 0.57 CLO, the CLO value of wearing a short-sleeved shirt with trousers. This behavior occurs around a clothing point where a comfortable human body changes from a hot to cold feeling. Figure 3(a) confirms the naive observation that two people under similar thermal conditions can feel differently because of their clothing. More specifically, higher clothing values make the body tolerant to a very wide range of lower temperatures. For example, an occupant at 17°C air temperature and 2 CLO is predicted to be comfortable. This is true because clothing partially isolates the human body and helps to decrease the rate of heat rejection to the outside. On the other hand, high CLO values can quickly move a person out of his/her comfort zone at mild/high temperatures. In that case, body heat becomes trapped inside, creating an extreme feeling of warmth and thermal dissatisfaction. Figure 3(a) also shows that low CLO values result in people tolerating higher ambient temperatures and being very sensitive to mild ambient temperatures. For example, at 0.4 CLO, a human body can tolerate an ambient condition of over 30°C air temperature and humidity ratio of 50%, but the same body can feel cold with ambient temperatures under 28°C. From this discussion, it can be concluded that clothing is very critical to thermal comfort and should not be taken as a constant; otherwise, large errors in calculating the PMV value and the comfort level may be imminent.

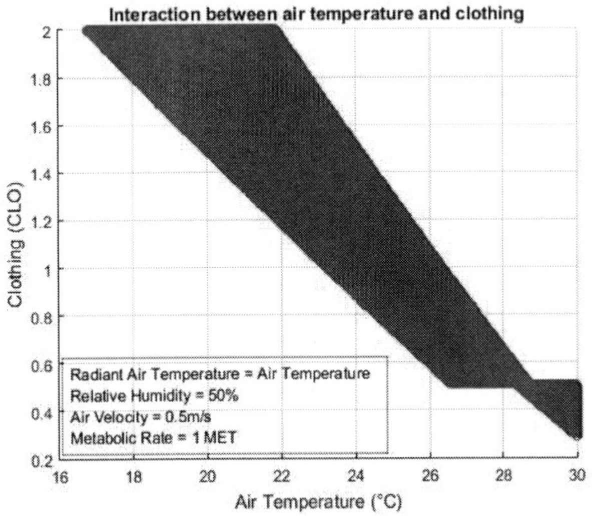

(a)

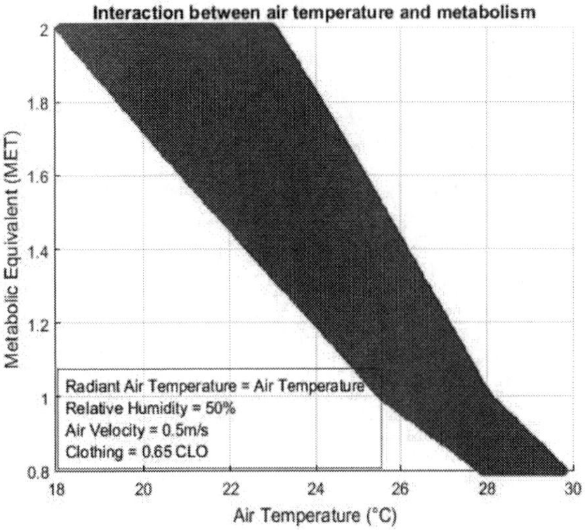

(b)

Figure 3: Interaction of air temperature: (a) with clothing, (b) with air metabolism, and resultant comfort zone

Metabolism is related to heat generation in the human body. As the metabolism (MET) of an occupant increases, the heat generation rate of his/her body increases, leading to a warmness sensation. Metabolism is one of the personal factors in the PMV model that is difficult to estimate and is usually assumed to be constant in the PMV calculation. In a perfect world and if metabolism is measured, a heating, ventilation, and air-conditioning (HVAC) system needs to compensate for an increase in the metabolic rate by controlling the ambient environmental conditions such as decreasing the air temperature, increasing the air velocity, or reducing the relative humidly. Figure 3(b) shows the comfort zone as a function of metabolic rate in METs and air temperature. As expected, the figure shows better comfort results at low air temperature as the metabolic rate increases. This can be explained by the fact that lower ambient temperatures are needed to reject the internal heat through natural or forced conviction. At a high metabolic rate, the body is highly intolerant to high and even mild temperatures. For example, as shown, at a metabolic rate of 1.5 MET, an occupant is thermally uncomfortable at 26°C, a temperature that is well within the comfort zone with a metabolic rate of 1.0 MET. This 1.0 MET, according to ASHARE Standard 55 tables, is the metabolic equivalent of a person sitting at rest without engaging in any physical and mental activities, and is widely accepted to represent metabolism in the PMV equations. In this paper, however, using smart wearable devices data, we show that the metabolic rate may vary for a person even while maintaining minimum activity levels or vary between two persons performing similar activity.

Figure 4(a) shows the metabolism from a different angel, by displaying its interactions with air temperature and relative humidity. As the metabolic rate increases, the body requires a means to reject the heat/gain cooling in order to stay thermally comfortable. One option is to lower the relative humidity (i.e., air has the capacity to absorb water vapor) in order to provide more cooling through latent respiration and/or sweating. Another option is to increase the heat-rejection rate due to radiation, convection, and dry respiration by lowering the ambient temperature. The generated 3D comfort surface shows a very high sensitivity to metabolism compared to relative humidity and temperature (i.e., surface gradient is higher in MET axis than in direction of relative humidity axis). The low-curve gradient in the direction of the relative humidity axis may reflect the fact that sweating and latent respiration alone are not fast enough to reject heat from the body at a higher metabolic rate. As shown in Figure 4(b) for MET = 1.5, Figure 4(a) is sliced to find the temperature-relative humidity comfort zone at a given metabolic rate. Figure 4(b) resembles Figure 1(a) but with a different optimal ambient temperature range.

We close our discussion about clothing and metabolism effect by presenting Figure 5, which indicates the effects of varying air temperature, clothing, and metabolism on the comfort zone. In general, this figure shows that metabolism has a larger effect on an occupant's thermal sensitivity comfort compared to clothing. However, as shown in Figure 3(a), the effect of clothing is more dominate in the area around 0.5 CLO, which makes a quick jump in the comfort zone.

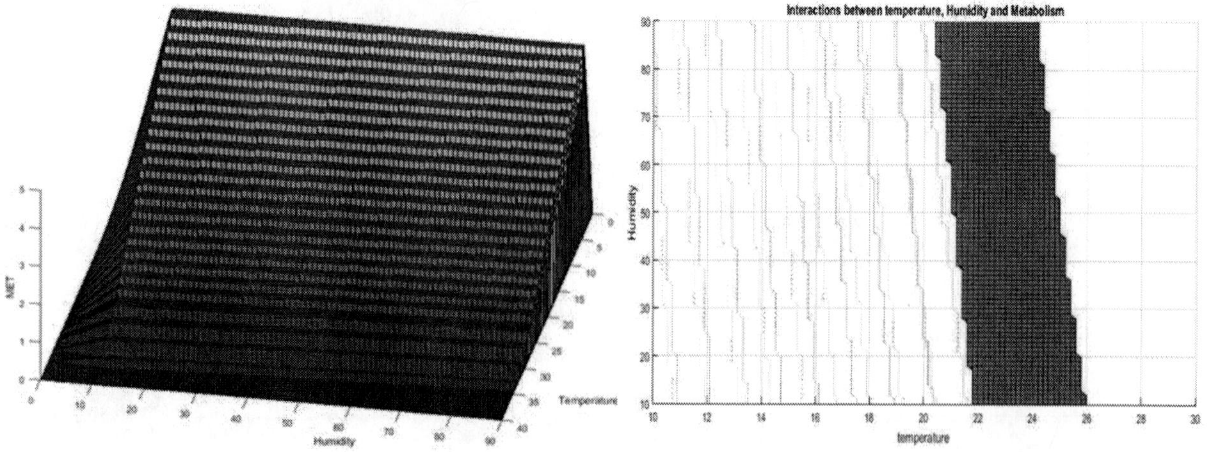

(a) (b)

Figure 4: Comfort region: (a) 3D area relating to interaction between air temperature, clothing, and metabolic rate, (b) slice of 3D region at MET = 1.5

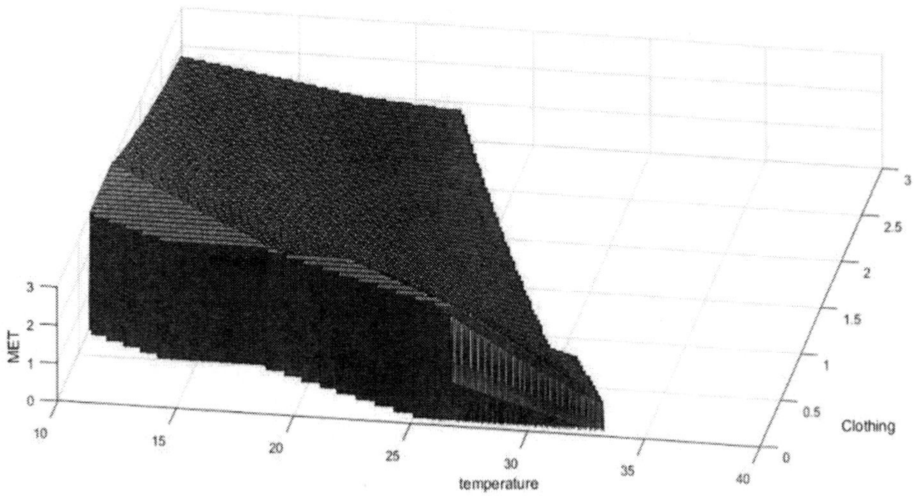

Figure 5: 3D comfort region relating to interaction between air temperature, clothing, and the metabolic rate

3. METABOLISM ESTIMATION USING SMART WEARABLE DEVICE

As mentioned previously, metabolism is difficult to measure and is usually assumed to be constant in the PMV model calculation. However, due to the ever-increasing popularity and advancement of wearable fitness devices, the estimation of metabolism becomes much easier and more convenient. In this paper, we have used Fitbit® band data to estimate metabolism. Fitbit® can be easily configured to share the metabolic rate, heart rate, and activity level to a computing unit in real time. These pieces of information are updated every minute to enable a fast response, if needed.

The metabolism of a Fitbit® user is calculated as a multiple of the basal metabolic rate (BMR), which is defined as the minimum rate of energy expenditure per unit time by an endothermic human at rest (Mifflin et al. 1990):

$$BMR = (\frac{10m}{1\ Kg} + \frac{6.25h}{1\ cm} - \frac{0.5a}{1\ year} + s)\frac{Kcal}{day} \tag{5}$$

where m is the mass of the body (in kilograms), h is the height of the body in cm, a is the age in years, and s is a factor relating to sex, as follows:

$$s = \begin{cases} +5 & for\ males \\ -161 & for\ females \end{cases} \tag{6}$$

Fitbit® also uses a built-in accelerometer to infer the activity level of the wearer (Lam et al., 2013). It uses this information to calculate the estimated the wearer's energy requirement (EER), which is related to the age, sex, weight, height, and physical activity of the user. For males, the EER is

$$\text{EER} = 864 - 9.72\ a\ (years) + PA\left(14.2\ m(kg) + 503\ h\ (meters)\right) \tag{7}$$

and for females, the EER is

$$\text{EER} = 387 - 7.31\ a\ (years) + PA\left(10.9\ m(kg) + 660.7\ h\ (meters)\right) \tag{8}$$

where PA is the physical activity level that is related to the motion of the person and is measured by Fitbit®'s built-in accelerometer as well as the physical properties of the wearer. The PA is calculated for men as (Gerrior et al. 2006)

$$PA = \begin{cases} 1, & 1.0 < PAL < 1.4\ (Sedentary) \\ 1.12, & 1.4 < PAL < 1.6\ (Low\ active) \\ 1.27, & 1.6 < PAL < 1.9\ (Active) \\ 1.54, & 1.9 < PAL < 2.5\ (Very\ active) \end{cases} \quad (9)$$

and for women as

$$PA = \begin{cases} 1, & 1.0 < PAL < 1.4\ (Sedentary) \\ 1.14, & 1.4 < PAL < 1.6\ (Low\ active) \\ 1.27, & 1.6 < PAL < 1.9\ (Active) \\ 1.45, & 1.9 < PAL < 2.5\ (Very\ active) \end{cases} \quad (10)$$

$$PAL = ((I - 1)\ [(1.15/0.9) \times D\ (minutes)]/1440))/\ (BEE/[0.0175 \times 1440 \times w\ (kg)]) \quad (11)$$

where I and D are the activity intensity (inferred from heart rate and accelerometer measurements) and duration, respectively, and BEE is the basal energy expenditure, given by

$$BEE = \begin{cases} 2933.8 \times a\ (years) + 456.4 \times h(meters) + 10.12 \times w\ (kg), & Men \\ 2472.67 \times a\ (years) + 401.5 \times h(meters) + 8.6 \times w(kg), & Women \end{cases} \quad (12)$$

Once EER and EMR are calculated, the metabolic can be calculated as

$$MET = \frac{EER}{BMR} \quad (13)$$

From equation (13), MET is actually the ratio between the energy generated while performing some activity to the rest of the body's metabolic rate. This MET is used to estimate the amount of calories burned by the human within the equation of the PMV model. Next, we show results for MET values monitored over time for an occupant.

3.1 Monitoring metabolic rate during physical activities

In this section, we investigate MET during real physical activities and its effect on the PMV value. Here, we conducted an experiment where a student ran on a treadmill at different speeds while wearing a Fitbit® device. Figure 6 shows the MET and PMV curves for the experiment, as well as the following different periods of the experiment:

- Period 1: started at 9:52 (592 minutes) am and ended at 10:02 am; student ran at a speed of 1 mph
- Period 2: started at 10:02 am and ended at 10:08 am; student rested at that time.
- Period 3: started at 10:08 am and ended at 10:18 am; student ran at a speed of 2 mph.
- Period 4: started at 10:18 am and ended at 10:25 am; student rested
- Period 5: started at 10:25 am and ended at 10:35 am; student ran at a speed of 3 mph.
- Period 6: started at 10:35 am and ended at 10:43 am; student rested
- Period 7: started at 10:43 am and ended at 10:52 am; student started running at a speed of 0.5 mph and increased speed 0.5 mph every minute

The MET value of the student was recorded by the minute, and the actual PMV value was computed for every MET value and then averaged to provide one value per period. To remove noise, a low pass filter was applied over the MET values. Figure 6 reveals some interesting behaviors. For example, looking at the active periods (1, 3, 5, and 7), the MET value increases over time as the student starts to breathe more erratically and his heart rate increases as he starts to show signs of exhaustion. At these times, the MET value rises, due to the increase in the amount of oxygen intake. Period (7) illustrates the relationship between the metabolic rate and the intensity of the exercise. The MET value appears to increase as the speed of the treadmill increases. Moreover and as expected the figure shows that the averaged PMV value is way much higher than the assumed PMV value (MET=1.0), while exercising

As shown in Figure 6, resting periods (2, 4, and 6) show an equally interesting phenomena where the metabolic rate does not immediately match the BMR in the resting condition. Instead, it gradually decreases over time and settles at an MET near 2.0. This indicates that physical exercise has a lingering effect on the metabolic rate, because the body continues to breathe at a higher rate, even after exercise is over, in order to supply sufficient oxygen to the body. This phenomena explains why occupants arriving at their work places by foot often set the air conditioning temperature to a relatively low value, causing discomfort to other employees who arrive by car. As the body performs physical labor, its metabolic rate increases; however, once the body stops, this does not immediately return the heart and breathing rates to normal, causing the body to maintain a higher metabolic rate for a period of time before going back to a MET of around 1.0. Another interesting phenomena happens during rest time is the average PMV value decreases and the metabolism rate converges faster to lower value as the intensity of the exercise increases, before resting. This might be because the more intense is the exercise, the easier is to reject the internal body heat through sweating.

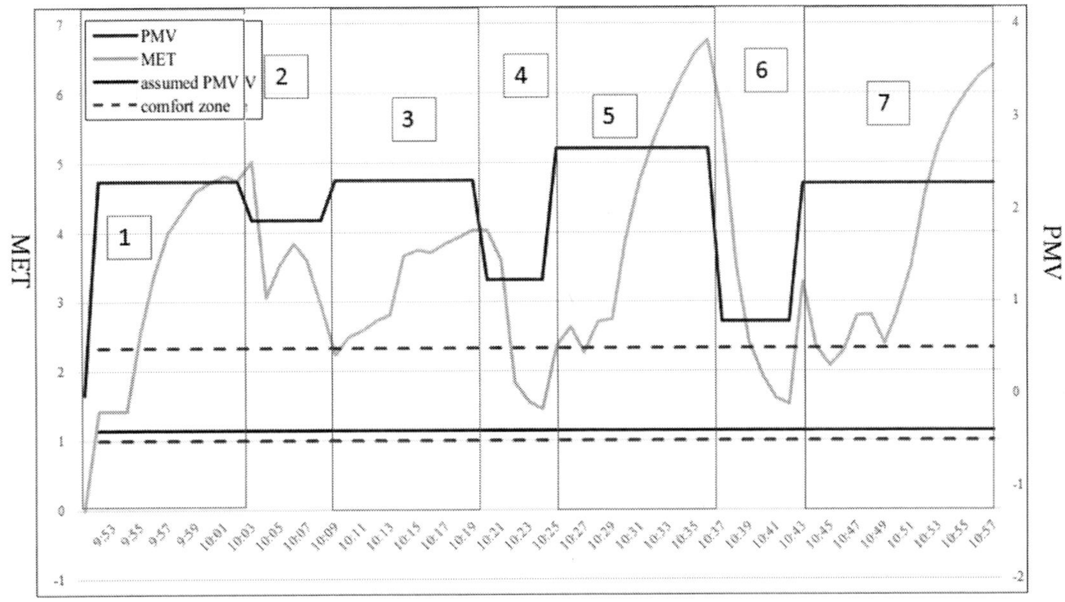

Figure 6: Metabolic rate and PMV values during training periods: periods (1), (3), and (5) are at treadmill speed of 1, 2, and 3 mph; periods (2), (4), and (6) are resting periods; and period (7) is a variable-speed treadmill, starting at 0.5 mph and increasing at a rate of 0.5 mph every minute. The assumed PMV values are calculated with MET=1.0.

3.2 Monitoring metabolic rate during normal day activates

In this section, we investigate the effect of metabolism on the comfort level for building occupants while performing normal life activates. An experiment was conducted on a 22-year-old and 35-year-old male graduate students (the first and last authors of this paper) for more than a half day and a full day, respectively. These students have recorded their indoor environmental conditions and clothing every 30 minutes while wearing a Fitbit® device to monitor their heart rate, activity level, and rate of caloric consumption per minute. The wearable device data were used to determine their MET and PMV values every minute and then averaged for each 30 minutes. In this experiment, the two students were asked to perform similar normal life activities while working at office or at home.

Figures 7(a) and (b) show plots for the measured MET values using the Fitbit® device along with the corresponding PMV value and the assumed PMV value (i.e., metabolic rate of 1.0 MET is assumed). These plots show that the MET value keeps changing throughout the entire day. For example in Figure 7(a), for the younger student, the metabolic rate was consistently over 1.0 MET throughout the entire day, the lowest being 1.09 around 1:00 pm. Even at its lowest value, the metabolic rate was higher than the assumed value of 1.0 MET. The figure also shows a very large increase in the MET value, and consequently PMV value, during the student's study-hours at home between (3:00 pm – 5:00 pm) and (7:00 pm – 9:00pm), while it decreases during the student's relax hours between (5:00 pm and 7:00 pm). Throughout most of the studying period, the student was thermally dissatisfied, feeling hot, with an average PMV value of 2.5, while the assumption is that he should be comfortable with a PMV value of less than -0.2 (assuming a constant MET value of 1.0). Even though the student was not involved in physical labor, the plots show that mental work and simply standing and walking around seem to increase the MET value to an average of greater than 3.0. From our earlier findings about the PMV model's high sensitivity to metabolism, as shown previously in Figures 4 and 5, this should explain the very large error between the assumed and actual PMV calculation.

Figure 7(b) shows similar higher metabolic rate during daily normal activities for the 34-year-old student. These activates include typing (office) and eating (home). The figure shows that the MET value was sometimes less than 1.0 at complete rest (sleeping). Figure 7 confirms that not only the MET value may vary over time for an occupant, but may also vary among occupants preforming similar daily activates.

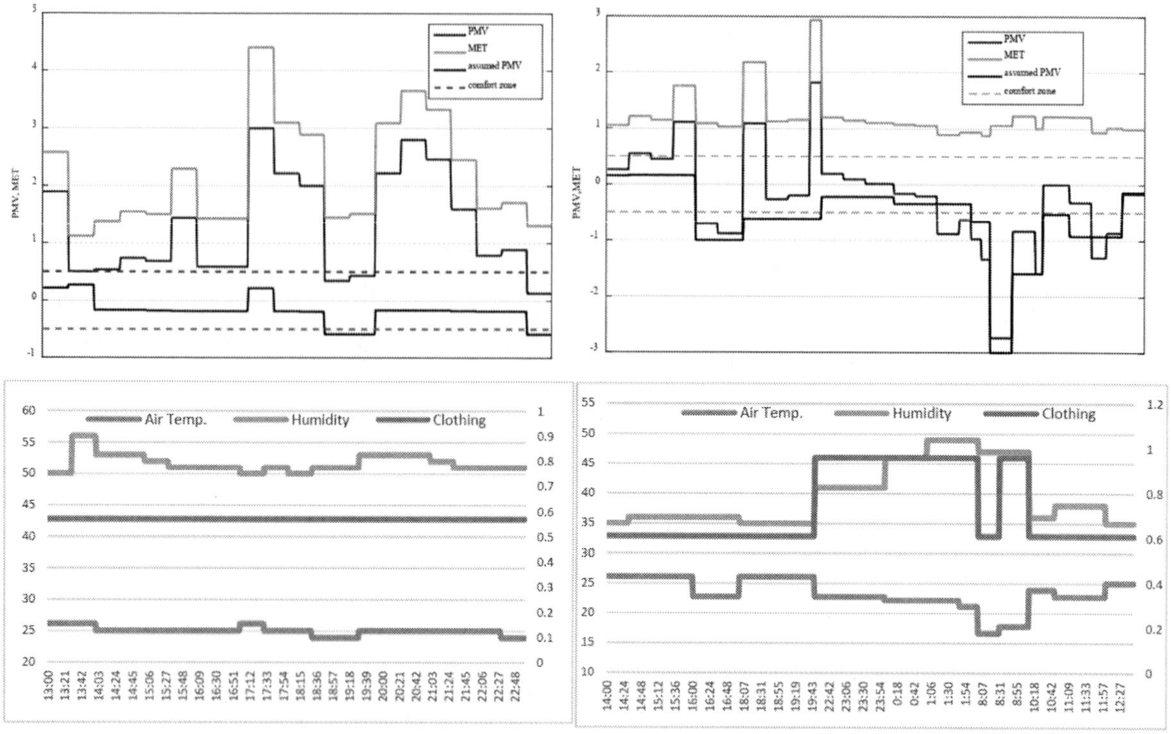

(a) 22-year-old male student (b) 35-year-old male student

Figure 7: MET and PMV values, clothing, and indoor environmental conditions recorded for two students. The figure shows a big difference in the PMV value when the actual metabolism is used in the calculation instead of the assumed value of 1.0 MET. Also this difference is shown to be higher for the younger student. Data while students were not inside a building are ignored.

4. CONCLUSIONS

The PMV value of an occupant is calculated based on multiple thermal environment and personal parameters. While it is not a perfect model, the best case should provide an 80% accuracy level, assuming that all input parameters are accurately measured. In this study, it was found that humidity, compared to other factors, has a minimum effect on thermal comfort, while metabolic rate and clothing have a very profound effect. The metabolic rate, which is assumed to be constant throughout most of the literature, is now easy to estimate using wearable devices. We have shown that a person who is not fully at rest will potentially have a MET value considerably higher than 1.0 MET and hence may experience discomfort during the majority of working hours. This shows the importance of measuring the metabolic rate in buildings to assure the reliability of the PMV model. Accurately measuring the metabolic rate of an occupant can extend the area of application of the PMV model to those who might be engaged in physical activities, such as waiters and waitresses in a restaurant, or people working out in gyms, where the estimated MET value differs greatly from the 1.0 MET.

The PMV model simulation analysis performed in this work can help to prioritize measurements with the highest sensitivity. For example, the relative humidity might cause some "partial discomfort," only if its value exceeds a certain range (less than 30% and more than 60% (Balaras et al. 2007; Wolkoff and Kjaergaad, 2007). Hence, humidity is rarely in need of monitoring, whereas metabolism sensitivity is much higher than relative humidity sensitivity and needs to be closely and continuously measured to assure reliability of the PMV comfort model. The metabolic rate continuously changes over time, even without performing any notable physical activities. This paper shows that simple mental activities and doing regular tasks might lead to some increase in the MET value, which can lead to thermal discomfort.

REFERENCES

Haynes, Barry P. "The impact of office comfort on productivity." *Journal of Facilities Management* 6.1 (2008): 37-51.

Fanger, P.O. Thermal Comfort: Analysis and Applications in Environmental Engineering. McGraw-Hill (1970).

ANSI/ASHRAE. "ASHRAE Standard 55-2013, Thermal Environmental Conditions for Human Occupancy." ASRAE Standards Committee. 2013.

Ferreira, Pedro M., S. M. Silva, A. E. Ruano, A. T. Negrier, and E. Z. E. Conceicao. "Neural network PMV estimation for model-based predictive control of HVAC systems." *2012 International Joint Conference on Neural Networks (IJCNN)*, Brisbane, Australia, June 10–15 (2012): 1-8.

Ku, K. L., et al. "Automatic control system for thermal comfort based on predicted mean vote and energy saving." *IEEE Transactions on Automation Science and Engineering*, 12.1 (2015): 378-383.

Zhang, Tian-hu, and Xue-yi You. "Applying neural networks to solve the inverse problem of indoor environment." *Indoor and Built Environment* 23.8 (2014): 1187-1195.

Xue, Yu, Zhiqiang John Zhai, and Qingyan Chen. "Inverse prediction and optimization of flow control conditions for confined spaces using a CFD-based genetic algorithm." *Building and Environment* 64 (2013): 77-84.

Luo, Xiaonan, Wenbang Hou, Yi Li, ad Zhong Wang. "A fuzzy neural network model for predicting clothing thermal comfort." *Computers & Mathematics with Applications* 53.12 (2007): 1840-1846.

Havenith, George, Ingvar Holmér, and Ken Parsons. "Personal factors in thermal comfort assessment: Clothing properties and metabolic heat production." *Energy and Buildings* 34.6 (2002): 581-591.

Jette, M., K. Sidney, and G. Blümchen. "Metabolic equivalents (METS) in exercise testing, exercise prescription, and evaluation of functional capacity." *Clinical Cardiology* 13.8 (1990): 555-565.

Institute of Medicine of the National Academies. *Dietary Reference Intakes for Energy, Carbohydrate, Fiber, Fat, Fatty Acids, Cholesterol, Protein and Amino Acids*. National Academies Press, Washington, DC, 2005.

Ferrannini E. "The theoretical bases of indirect calorimetry: A review." *Metabolism* 37.3 (1988): 287-301.

Mifflin, M.D., S. T. St. Jeor, L. A. Hill, G. J. Scott, S. A. Daugherty, and Y. O. Koh. "A new predictive equation for resting energy expenditure in healthy individuals." *The American Journal of Clinical Nutrition* 51.1 (1990): 214-247.

Meijer, G. A., K. R. Westerterp, H. Kper, and F. ten Hoor. "Assessment of energy expenditure by recording heart rate and body acceleration." *Medicine and Science in Sports and Exercise* 21.3 (1989): 343-347.

Lam, Abraham Hang-yat, and Dan Wang. "Carrying my environment with me: A participatory-sensing approach to enhance thermal comfort." *Proceedings of the 5th ACM Workshop on Embedded Systems for Energy-Efficient Buildings*, Roma, Italy, November 11-15, 2013, Association for Computing Machinery.

Balaras, Constantinos A., Elena Dascalaki, and Athina Gaglia. "HVAC and indoor thermal conditions in hospital operating rooms." Energy and Buildings 39.4 (2007): 454-470.

Wolkoff, P., and S.K. Kjaergaard. "The dichotomy of relative humidity on indoor air quality." *Environment International* 33.6 (2007): 850-857.

Gerrior, Shirley, WenYen Juan, and Peter Basiotis. "An easy approach to calculating estimated energy requirements." *Preventing Chronic Disease* 3.4 (2006): A129.

Stochastic Modeling of Short-term Occupancy for Energy Efficient Buildings

Jin Dong[1*], Chris Winstead[1], Seddik M. Djouadi[1], James J. Nutaro[2], Teja Kuruganti[2]

[1] University of Tennessee, Department of Electrical Engineering and Computer Science,
Knoxville, TN 37996
{jdong, cwinste2}@vols.utk.edu, djouadi@eecs.utk.edu

[2] Oak Ridge National Laboratory
Oak Ridge, TN 37831.
{nutarojj, kurugantipv}@ornl.gov

* Corresponding Author

ABSTRACT

In this paper, we present a simple yet effective algorithm to automatically assign reference temperature set-points based on the occupancy information. Both the binary and detailed occupancy estimation cases are considered. In the first case study, we assume the schedule involves only binary states (occupied or not occupied), i.e. the room is invariant. With long-term observations occupancy levels can be estimated using statistical tools. In the second case study, two techniques are introduced. Firstly, we propose an identification-based approaches. More precisely, we identify the models via Expectation Maximization (EM) approach. The statistical state space model is built in linear form for the mapping between the occupancy measurements and real occupancy states with noise considered. Secondly, we introduce a novel finite state automata (FSA) which is successfully reconstructed by general systems problem solver (GSPS). As far as we know, no studies have used the finite state machine or general system theory to estimate occupancy in buildings. All above estimates can be used to adaptively update the temperature set-points for Heating, Ventilation, and Air-Conditioning (HVAC) control strategy.

1. INTRODUCTION

The primary energy consumer of smart buildings are HVAC systems, approximately 30% of the building energy use, which usually operate on a fixed schedule. Currently, most modern buildings still condition rooms with a set-point assuming maximum occupancy rather than actual usage. As a result, rooms are often over-conditioned needlessly. Occupancy-based controls can achieve significant energy savings by temporally matching the building energy consumption and building usage, conservative user behavior can save a third of expended energy. Some researchers have proposed the control and use of shading blinds and natural ventilation to make effective use of the available natural resources to reduce building energy consumption such as Tzempelikos and Athienitis (2007); Van Moeseke et al. (2007), while others have proposed optimal control of HVAC schedules Sun et al. (2013).

The above-mentioned methods do not consider the actual number of occupants in the building. Meanwhile, it has been shown that annual energy savings of 10% – 42% can be achieved if a proper HVAC strategy that accounts for actual occupancy levels as reported in Erickson et al. (2011). Consequently, occupancy and the related activities in a building are highly related with the energy consumption of that building. Any off-line strategy for pre-defined control parameters is unable to handle all variations of numbers of humans and their various behaviors in the building.

Despite the large number of potential application scenarios the estimation of buildings' occupancy is still a cumbersome, error-prone, and expensive process Nguyen and Aiello (2013). In residential environments, most of the approaches aim at controlling the HVAC system more efficiently, as heating and cooling account for most of the energy expenditure in an average household. To optimize heating or cooling of the building, the authors typically utilize various sensors to determine the occupancy state of the building or of individual rooms. For instance, Lu et al. instrumented households with passive infrared (PIR) sensors and reed switches on entrance doors to detect when household occupants are at home and when not Lu et al. (2010). They use this data as ground truth information to evaluate the performance of an occupancy prediction algorithm, which is in turn used to drive a smart heating control system.

In this paper, we present a simple yet effective algorithm to automatically assign reference temperature set-points based on the occupancy information. Both the binary and detailed occupancy estimation cases are considered. In the first case study, we assume the schedule involves only binary states (occupied or not occupied), i.e. the room is invariant. With long-term observations, occupancy levels can be estimated using statistical tools. In the second case study, three techniques are introduced. First, we propose an identification-based approach. More precisely, we identify the models via EM approach. The statistical state space model is built in linear form for the mapping between the occupancy measurements and real occupancy states, with noise considered. Secondly, we introduce a novel FSA which is successfully reconstructed by GSPS. As far as we know, no studies have used the finite state machine or general system theory to estimate occupancy in buildings. All above estimates can be used to adaptively update the temperature set-points for HVAC control strategy.

To demonstrate effectiveness of proposed approach, a simulation-based experimental analysis is carried out using occupancy data. We define the estimation accuracy as the total number of correct estimations divided by the total number of estimations, and both Root Mean Squared Error (RMSE) and estimation accuracy analysis are provided. All the proposed estimation techniques could achieve at least 70% accuracy rate. Generally, accuracy for binary states estimation is much higher than that of detailed occupancy. For GSPS model, more training data improves performance of estimation. It should be remarked that although some mismatch exist for non-zero jumps, estimation performance tracks the zero base line (non-occupied status) perfectly. Therefore, the estimation techniques are effective for binary estimation with over 90% accuracy. Finally, the estimated occupancy is applied into a temperature set algorithm to generate reference temperature curve. By adjusting temperature set curve, we can achieve significant energy savings without sacrificing customer's comfort.

This paper is organized as follows: Section 2 defines the mathematical problem under consideration. We also set up a temperature setting algorithm which could adaptively tune the temperature setting point of the room based on the real occupancy information. Section 3 contains details about both the binary and detailed occupancy estimation algorithms. A novel GSPS model is first employed to estimate the occupancy for building rooms. Section 4 presents the simulation results regarding the estimation performance of the aforementioned estimation algorithms. Finally, conclusions are stated in Section 5.

2. PROBLEM FORMULATION

It is now well known that HVAC problems can be approached using Model-based Predictive Control (MPC) strategy. In this section, we describe the model settings for energy efficient control of HVAC systems. The system model was proposed in Gwerder and Tödtli (2005) and employed in Oldewurtel et al. (2010). We briefly described it in the sequel.

2.1 System Model

Using the same model proposed in Ma et al. (2015), we could obtain a discretized system as follows:

$$x_{k+1} = A_d x_k + B_d u_k + C_d \omega_k, \tag{1}$$

where the parameters can be computed from the continuous-time model, and $x_k = [t_{1,k}, t_{2,k}, t_{3,k}]^T$.

We assume the following constraints are imposed on the temperature and control inputs:

$$68°\text{F} \leq t_{1,k} \leq 80.6°\text{F}, \quad -50 \leq u_k \leq 200, \tag{2}$$

where $u_k > 0$ means heating and the opposite means cooling.

2.2 Cost Function

2.2.1 *Cost Function* We consider the problem where the temperature t_1 is required to remain within certain bounds of a constant in the presence of the disturbance vector d. Moreover, we can assign set-points for t_1, t_2 and t_3, but without any other constraint on t_2 and t_3. Thus, we can regulate the output error $e_k := x_k - x_r$ at time k, where x_r is the set-point vector of x. We hope to minimize the error e to keep the temperature t_1 close to the desired value. Meanwhile, we also hope to use as less power as we can to save energy. Ultimately, our objective is to find for the system (1), the M-control sequence $\{u_0, \cdots, u_{M-1}\}$, where $u_i := u(t_i)$, $i = 0, \cdots, M$; M is an integer large enough. Let $t_i = i\Delta T$, where ΔT is the sampling period; and corresponding state sequence $\{x_0, \cdots, x_{M-1}\}$ and error sequence $\{e_0, \cdots, e_{M-1}\}$, we want to minimize the following finite horizon objective function:

$$V_N(e_0, \mathbf{u}, \omega) := \frac{1}{2}[(x_N - x_r)^T P (x_N - x_r) + \sum_{k=1}^{N-1} e_k^T Q e_k + \sum_{k=0}^{N-1} u_k{}^T R u_k], \tag{3}$$

where $P \geq 0$, $Q \geq 0$ (i.e., semi-definite positive matrices), $R > 0$ (i.e., positive definite matrix), N is the prediction horizon, and

$$\mathbf{x} := [x_0^T, ..., x_N^T]^T, \quad \mathbf{u} := [u_0^T, ..., u_{N-1}^T]^T, \quad \omega := [\omega_0^T, ..., \omega_{N-1}^T]^T.$$

Algorithm 1 Temperature Setting Algorithm

Step 1:
  Initialize α
Step 2:
  $n \leftarrow \frac{T_{max} - T_{min}}{k} + 1$
for all hour $h = 1$ to 48 **do**
  $Range \leftarrow max(O_h) - min(O_h)$
  $r_0 \leftarrow min(O_h)$
end for
if $\alpha = 0$ **then**
  Go to step 3
else
  Go to step 4
end if
Step 3:
for all set-point j ($j = 1$ to n) **do**
  $r_j \leftarrow r_{j-1} + \frac{Range}{n}$
  Go to Step 5
end for
Step 4:
for all set-point j ($j = 1$ to n) **do**
  $r_j \leftarrow r_{j-1} + Range * \frac{2^{\alpha(j-1)}(1-2^{\alpha})}{(1-2^{\alpha n})}$
end for
Step 5:
for all hour $h = 1$ to 48 **do**
  $T_h^{set} \leftarrow k[argmin\{j : O_h \leq r_j\} - 1] + T_{min}$
end for

2.3 Temperature Setting Algorithm

In conventional practice, the HVAC operates under a fixed thermostat set-point around which the inside temperature fluctuates. The reference temperature is usually set by the administrator or occupants of the room, and it does not change frequently (if at all) throughout the day - often times, regardless of occupancy fluctuations.

While the above model can accommodate such cases by simply employing the reference temperature values (i.e., $X_r[t]$) as fixed input parameters, it is mainly designed to deploy them as decision variables that are determined as a response to the occupancy fluctuations.

Inspired by the idea from Avci et al. (2013), we employ a simple yet effective algorithm to assign reference temperature set-points for each half hour of the day based on the real-time occupancy. The proposed algorithm first determines the maximum ($\max(O_h)$) and minimum($\min(O_h)$) occupancy values for the next $24h$ period, and from these it identifies a number of occupancy ranges depending on the number of temperature set-point candidates. The temperature set-point candidates are the temperature values between any given maximum and minimum allowed temperature set-points, which can be pre-determined by the users as in (2). Finally, each occupancy range is coupled with a temperature set-point. The temperature set-points are then assigned to each half hour of the day based on the range in which the occupancy number of corresponding half hour falls in.

A discomfort tolerance index α is defined to model consumer choice on thermal comfort, which is used to capture the trade-off between thermal comfort and energy cost. The value of α, in general, varies across individuals. Namely, a high discomfort tolerance (i.e., $\alpha > 0$) represents a consumer who has higher tolerance for higher indoor temperatures in return for cost savings. Such customers prefer that wider price ranges are assigned to higher temperatures. Whereas low tolerance ($\alpha < 0$) customers prefer the opposite as they are less willing to compromise comfort for savings. We refer to the consumer with $\alpha = 0$ as the neutral consumer who has a balanced preference between comfort and savings, leading to uniform price ranges allocated to temperature set-points.

The temperature set-point assignment algorithm is shown in Algorithm 1.

3. OCCUPANCY MODEL

In this section, we first estimate a simple binary occupancy model from collected occupancy data. Then we introduce our algorithm for the detailed occupancy estimation of future occupancy, and we show how we could predict occupancy levels.

In order to study the practical occupancy model, we collected real occupancy data from an occupancy sensor for a room. We randomly pick a segment of data which is dated from "$10/13/2010 \sim 4/5/2011$". The sampling interval is 30 minutes, so any sensor collects 48 occupancy samples each day. Natural questions which arise are:

1. What is the probability for this room to be occupied?
2. How many people will be in the room?

For the first problem, we could compute the probability for the room to be occupied by observing all the historic data. It is noted that this model is only meaningful when we add some assumptions for the usage of the room. Basically, the room is more or less an office or meeting room which is occupied routinely. Thus, whether the room would be occupied or not would exactly follow a fixed schedule. Though slight, this approach still represents a step forward from the traditional control strategies which neglect occupancy information. For the second problem, we need to apply more intelligent techniques to estimate the number of people in the room.

3.1 Binary occupancy state estimation

We use $\gamma(t)$ and $\eta(t)$ (both initiated to be 0) to count the number for the room to be *occupied or not*.
If the number of people coming in the room is > 0 at time step t, we consider the room is occupied, i.e. $\gamma(t) = \gamma(t)+1$. Otherwise, $\eta(t) = \eta(t) + 1$. Therefore, if we count $\gamma(t)$ and$\eta(t)$ for all the 174 days, we will get the probability of occupancy for each time step from 12:00 AM until 11:30 PM. For convenience, make a time map as below:

Table 1: Binary count for $\gamma(j)$ and $\eta(j)$

Actual Time	Time Sequence	Count
12:00 AM	1	$\gamma(1)$ and $\eta(1)$
12:30 AM	2	$\gamma(2)$ and $\eta(2)$
⋮	⋮	⋮
11:00 PM	47	$\gamma(47)$ and $\eta(47)$
11:30 PM	48	$\gamma(48)$ and $\eta(48)$

Then we obtain the occupancy probability for every half hour during a day.

- So the probability to be occupied is denoted as: $P(j) = \frac{\gamma(j)}{\gamma(j)+\eta(j)}$ for $j = 1 \cdots 48$.
- So the probability NOT to be occupied is denoted as: $P(j) = \frac{\eta(j)}{\gamma(j)+\eta(j)}$ for $j = 1 \cdots 48$.

After determining the occupancy probability, we could set the temperature in the room for HVAC according to the *Temperature Setting Algorithm* as described in *Section 2.3*.

Notice that, for the above Temperature Setting Algorithm, it does not make a difference whether we use the occupancy probability or the real occupants for each half hour. Basically, the occupancy probability can be viewed as a normalized occupancy data.

3.2 Detailed occupancy estimation

3.2.1 *Motivation* From the last section, we did gain some energy savings for HVAC by adaptively adjusting the temperature setting point for the controller. However, it's not realistic to desire the real occupancy of the room to exactly follow the given schedule. Therefore, we should be able to estimate and update the occupancy information based on the most recent measurement. Moreover, detailed occupancy estimation not only considers whether the building is occupied or not, but also takes into account the number of occupants in the building. Here we will introduce some background and several alternative techniques that may be adequate for occupancy estimation.

3.2.2 *Expectation Maximization (EM)* The state-space model has been widely employed in control systems and signal processing since it can be used on-line and updated after receiving new observations. It is a purely statistical approach based on collected measurements. It consists of a state (or system) equation (4) and a measurement (or output) equation (5). Denoted by y_k the occupancy measurement at time k, state-space equation in discrete time can be written in the following form:

$$x_{k+1} = A_k x_k + B_k w_k \tag{4}$$

$$y_k = C_k x_k + D_k v_k \tag{5}$$

where $x_{k+1} \in \mathbb{R}^{n\times 1}$ ($\mathbb{R}^{n\times 1}$ denotes the space of real vectors of dimension $n \times 1$) is the state that characterizes the occupancy; it is a variable of the time series $\{x_k\}$ determined by the previous state x_k and the noise term $w_k \in \mathbb{R}^{m\times 1}$ introduced at each k. $A_k \in \mathbb{R}^{n\times n}$ and $B_k \in \mathbb{R}^{n\times m}$ are corresponding coefficients.

The time varying property of the parameters enables the state-space model able to adapt dynamically to a variety of frequencies. The noise terms w_k and v_k can capture small perturbations or uncertainties introduced at each time k, which improves the flexibility of the model. In this paper, the noise terms are assumed to be standard Gaussian variables, and the covariances of them are denoted by coefficients B_k and D_k.

Due to these properties, we propose to use the state-space model to estimate behaviors of the occupancy model. The unknown system parameters $\beta_k = \{A_k, B_k, C_k, D_k\}$ and states $\{x_k\}$ can be estimated through a finite set of received signal measurement data $Y = \{y_1, y_2, ...\}$. Thereafter, the state variables x_k can be reconstructed from the measurements.

Then we employ EM to estimate the parameters in the state-space model. EM is a filter-based iterative numerical scheme to compute maximum likelihood estimates of the parameters given the measurement data. It yields parameter estimates with non-decreasing values of the likelihood function, and converges under mild assumptions. Due to space limitations, the detailed algorithm of EM is not presented here. For more details, interested readers could refer to Dong et al. (2013, 2014) for example. This algorithm yields parameter estimation with non-decreasing values of the likelihood function, and converges under mild assumptions.

3.3 Finite State Automata

Probabilistic FSA Paz (2014) have been introduced to describe distributions over strings. FSA has been used quite successfully to address several complex sequential pattern recognition problems, such as continuous speech recognition, computational biology Lyngso et al. (1999) and linguistics Mohri (1997). General Systems Problem Solver (GSPS) proposed in Klir (1969), provides a novel and highly effective method for reconstructing the input/output behavior of FSA. The detailed algorithm and system formulation can be found in Klir (2013).

Given a system and a sequence of length n generated by that system, we may posit a relation between its variables. This relation takes the form of a function f that maps observations of some variables at times $n, n-1, \cdots, k$ with $k \geq 1$ to observations of other variables at time n.

3.4 Simplified Binary states FSA

In general, the length of the "look back depth" used is decided by actual problem. For our specific occupancy prediction problem, the system of interest has a single variable with two possible values: occupied (b) or not occupied (a). After tail and error check, we posit a rule that looks back three steps and also considers the time of day, i.e. $v_1(n) = f(v_1(n-1), v_1(n-2), v_1(n-3), t(n))$. The time of day $t(n) = t_n$ is used to characterize the different rules for different time period during a day; for the data available to use there are 48 times that can be considered because the sampling interval of the sensor is 30 minutes. Given the data and a rule, we can build a model for forecasting with the simple procedure described by Klir Klir (1969).

To illustrate this procedure, consider the mask and sequence in Table 2. The system that generated this sequence has time variable t_n and a single variable v that can take the value b or a. The rule for this mask built with 3000 data points is shown in Table 3. It should be noted that in order to simplify the explain here, we decouple the time variable from the rule. To be specific, Table 3 represents the rule for $3:00$ pm using 3000 data points. Considering, the sampling interval is every 30 minutes, we could generate 48 such rule tables for the whole day (24 hours).

Table 2: A sequence and mask for a system with uncertain output. Circles are input data for the function f and the square is the output data.

round	time (v_1)	occupancy (v_2)
11	ⓣt_{11} (circled)	a (boxed)
10	t_{10}	ⓑ
9	t_9	ⓐ
8	t_8	ⓑ
7	t_7	a
6	t_6	a
5	t_5	b
4	t_4	a
3	t_3	b
2	t_2	b
1	t_1	a

Table 3: Rule for the sequence and mask using a subset of 3000 points restricted to 3 pm (a-unoccupied, b-occupied).

input	output	count	likelihood
aaa	a	47	0.959
	b	2	0.041
aab	a	0	0
	b	1	1
aba	a	1	1
	b	0	0
abb	a	0	0
	b	1	1
baa	a	1	0.5
	b	1	0.5
bab	a	1	0.33
	b	2	0.67
bba	a	0	0
	b	1	1
bbb	a	0	0
	b	4	1

To illustrate how this model is applied for forecasting, suppose we begin with the latest observation sequence as bab at 3 pm. The next output is a with probability 0.33 or b with probability 0.67. If we were at a different time step other than 3 pm, then the output is selected according to the corresponding rule table at that time step. Once the time step is fixed, then the second output is a with some probability p or b with probability $1-p$. Continuing in this fashion, we can construct a tree of possible futures and use these possible futures to inform the control system.

3.5 Estimating number of occupants

Though the method described in Section 3.4 only involves two states, it is readily extended to estimate the exact number of occupancy by using the number of occupants as a variable rather than the binary occupied/unoccupied. In this occupancy problem, we want to know the exact number of people in the room. So we assume different numbers of occupants as different states in the FSA. As long as finite states are involved, the general form of the rule and the methods for constructing and forecasting are the same. In the following simulation results, we would show that the discussion for the binary model applies directly to the model with multiple states.

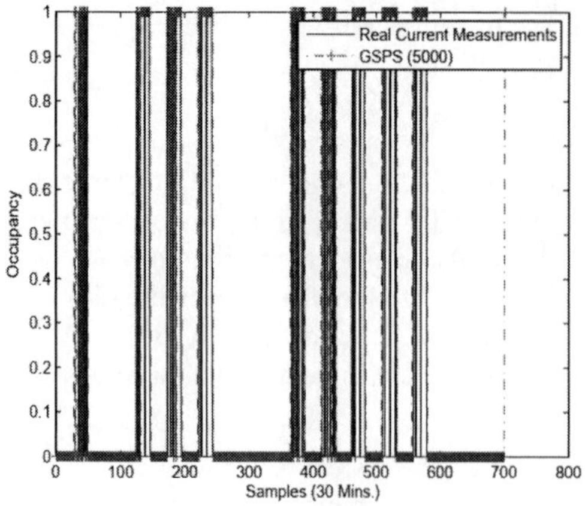

(a) Binary occupancy using GSPS model 3000 points

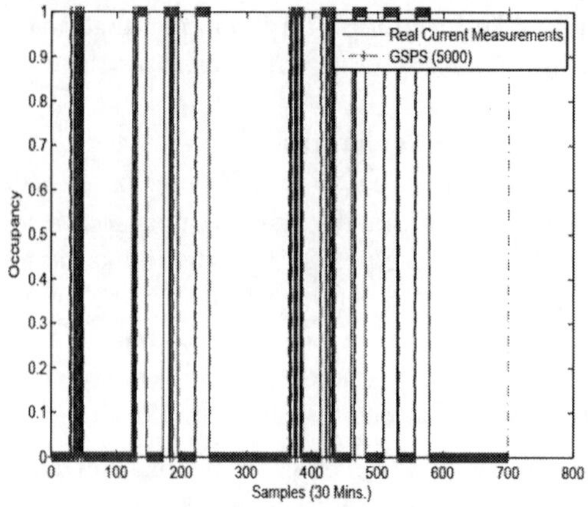

(b) Binary occupancy using GSPS model 5000 points

Figure 1: Binary occupancy using GSPS model

4. SIMULATION RESULT

To illustrate the effectiveness of the proposed occupancy estimation techniques discussed in the last several sections, we assess their performances for different scenarios in this section. In the first scenario, binary occupancy state estimation will be examined where only the state occupied or not occupied will be given. Then a detailed occupancy estimation performance using GSPS model will be explained and compared with some alternative methods. In the second part, the temperature set-points are treated as a decision variable which can be determined using the TSA algorithm.

4.1 Definition of the performance indexes

We define the estimation accuracy as the total number of correct estimations divided by the total number of estimations, and the estimation accuracy by using the probability counting and the GSPS algorithms are listed in Table. 4. The Root Mean Squared Error (RMSE), which characterizes the absolute estimation errors.

More formally, following the mathematical expression in Ebadat et al. (2013), if we let ACC represent the accuracy, then using indicator functions, we obtain the accuracy of the estimator:

$$ACC\left(\hat{O}\right) := \frac{N - \sum_{k=1}^{N} \mathbb{1}\left(O(k) - \hat{O}(k)\right)}{N}, \tag{6}$$

where O and $\hat{O}$ are true and estimated occupancy, respectively; and $\mathbb{1}\left(O(k)\right)$ is given as:

$$\mathbb{1}\left(O(k)\right) := \begin{cases} 1 & \text{if } O(k) > 0, \\ 0 & \text{otherwise.} \end{cases} \tag{7}$$

RMSE is defined as the square root of the mean square error, which is a frequently used measure of the differences between values predicted by a model or an estimator and the values actually observed. Then the MSE associated to the couple $O, \hat{O}$ is denoted by

$$MSE(\hat{O}) := \frac{1}{N}\sum_{k=1}^{N}(\hat{O}(k) - O(k))^2. \tag{8}$$

Therefore, RMSE will be:

$$RMSE(\hat{O}) := \sqrt{MSE(\hat{O})}. \tag{9}$$

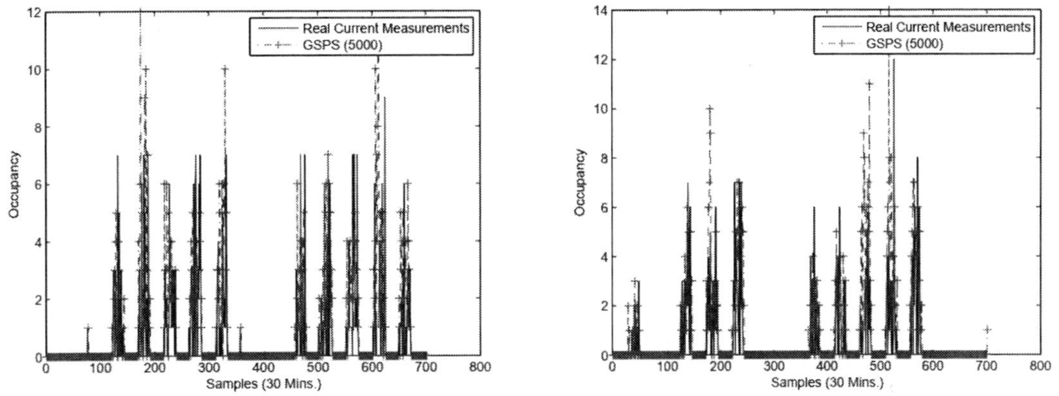

(a) Occupancy estimation using GSPS model 3000 points

(b) Occupancy estimation using GSPS model 5000 points

Figure 2: Detailed occupancy estimation using GSPS model

4.2 Binary occupancy state estimation

4.2.1 GSPS binary model This model predicts a 1, by default, to signify occupation. To compare the results, we build the models with the first 3000 and 5000 input points from the occupancy measurements, respectively. Then we test the obtained models across the last 3000 values as depicted in Fig. 1a and Fig. 1b. In particular, we summarize the binary occupancy estimation performance in Table 4.

Obviously, the GSPS model significantly outperforms the traditional probability counting method. Notice that only two states (occupied or not occupied) in this binary occupancy estimation, both the RMSE and accuracy are very good which is as desired. Moreover, the model built with 5000 data shows better performance than the one with less data. It could be viewed as the an outcome of better model obtained with more data observed by the GSPS model.

4.3 Detailed occupancy estimation

4.3.1 *GSPS Model* Then we test the obtained models across the last 3000 and 5000 values as depicted in Fig. 2a and Fig. 2b. More precisely, in Fig. 2a the model is trained using the last 3000 occupancy data, while it is 5000 for Fig. 2b. As expected, the more data involved for training, the better estimation results we got. This is also confirmed by the estimation error comparison shown in Table 5.

4.3.2 *Temperature set-points* Through the above simulation results, we do get the desired estimation performance, however, our goal is to design temperature set-points based on this occupancy sequences. In order to check the compatibility of these occupancy estimation results, we apply them in the temperature setting algorithms. The one using EM estimation is given in Fig. 3a. Similarly, the one using GSPS model is shown in Fig. 3b.

4.4 Summary of the results

In this section, we compare four estimators, all trained using the same training set described at the beginning of Section 2. Fig 3 shows the realizations of two different estimation strategies when applied to the corresponding test set, while Table 4 and Table 5 summarize the achieved numerical performance. Generally speaking, the accuracy for binary states estimation is much higher than that of detailed occupancy. For GSPS model specifically, more training data points contribute to improved estimation performance.

Table 4: Comparison of binary estimations.

Methods	Estimation RMSE	Accuracy
Probability counting	0.206	0.683
GSPS (3000)	0.094	0.889
GSPS (5000)	0.086	0.912

Table 5: Comparison of estimation algorithms.

Methods	Estimation RMSE	Accuracy
GSPS (3000)	3.078	0.700
GSPS (5000)	2.646	0.715
EM	3.715	0.615

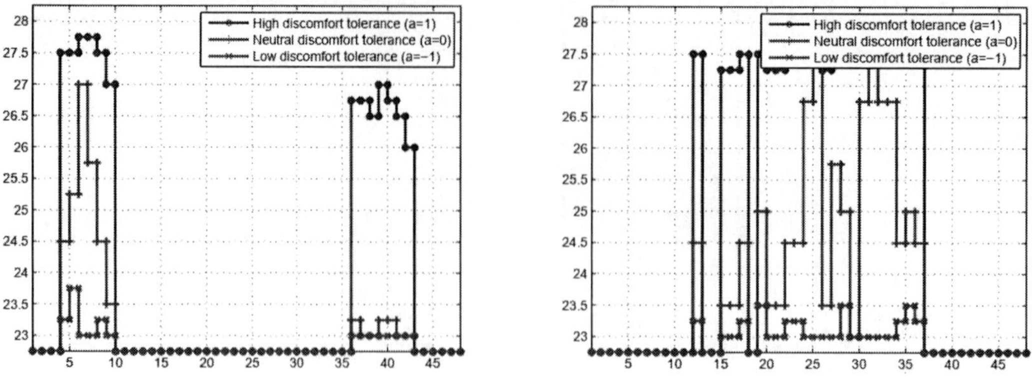

(a) EM

(b) GSPS model 5000 points

Figure 3: Detailed temperature set-points.

5. CONCLUSIONS

In this paper, we present a simple yet effective algorithm to automatically assign reference temperature set-points based on the occupancy information. The proposed strategy aims at reducing the total energy consumption while maintaining a comfortable temperature environment in buildings. Both the binary and detailed occupancy estimation cases are considered. To demonstrate the effectiveness of the proposed approach, a simulation-based experimental analysis is carried out using occupancy data.

For future work, we will apply the proposed occupancy estimation and temperature setting strategy into our control framework to see the exact energy saving results. Moreover, we will investigate more advanced estimation techniques such as particle filters to track the movement of occupants.

REFERENCES

Avci, M., Erkoc, M., Rahmani, A., and Asfour, S. (2013). Model predictive hvac load control in buildings using real-time electricity pricing. *Energy and Buildings*, 60:199–209.

Dong, J., Ma, X., Djouadi, S., Li, H., and Liu, Y. (2014). Frequency prediction of power systems in fnet based on state-space approach and uncertain basis functions. *IEEE Trans. Power Systems*.

Dong, J., Ma, X., Djouadi, S. M., Li, H., and Kuruganti, T. (2013). Real-time prediction of power system frequency in fnet: A state space approach. In *Smart Grid Communications (SmartGridComm), 2013 IEEE International Conference on*, pages 109–114. IEEE.

Ebadat, A., Bottegal, G., Varagnolo, D., Wahlberg, B., and Johansson, K. H. (2013). Estimation of building occupancy levels through environmental signals deconvolution. In *Proceedings of the 5th ACM Workshop on Embedded Systems For Energy-Efficient Buildings*, pages 1–8. ACM.

Erickson, V. L., Carreira-Perpiñán, M. Á., and Cerpa, A. E. (2011). Observe: Occupancy-based system for efficient reduction of hvac energy. In *Information Processing in Sensor Networks (IPSN), 2011 10th International Conference on*, pages 258–269. IEEE.

Gwerder, M. and Tödtli, J. (2005). Predictive control for integrated room automation. In *8th REHVA World Congress Clima*.

Klir, G. (1969). *An Approach to General Systems Theory*. Van Nostrand Reinhold.

Klir, G. (2013). *Architecture of systems problem solving*. Springer Science & Business Media.

Lu, J., Sookoor, T., Srinivasan, V., Gao, G., Holben, B., Stankovic, J., Field, E., and Whitehouse, K. (2010). The smart thermostat: using occupancy sensors to save energy in homes. In *Proceedings of the 8th ACM Conference on Embedded Networked Sensor Systems*, pages 211–224. ACM.

Lyngso, R. B., Pedersen, C., and Nielsen, H. (1999). Metrics and similarity measures for hidden markov models. In *Proc. Int. Conf. Intell. Syst. Mol. Biol*, pages 178–186.

Ma, X., Dong, J., Djouadi, S. M., Nutaro, J. J., and Kuruganti, T. (2015). Stochastic control of energy e cient buildings: A semide nite programming approach. In *Smart Grid Communications (SmartGridComm), 2015 IEEE International Conference on*. IEEE.

Mohri, M. (1997). Finite-state transducers in language and speech processing. *Computational linguistics*, 23(2):269–311.

Nguyen, T. A. and Aiello, M. (2013). Energy intelligent buildings based on user activity: A survey. *Energy and buildings*, 56:244–257.

Oldewurtel, F., Parisio, A., Jones, C. N., Morari, M., Gyalistras, D., Gwerder, M., Stauch, V., Lehmann, B., and Wirth, K. (2010). Energy efficient building climate control using stochastic model predictive control and weather predictions. In *American Control Conference (ACC), 2010*, pages 5100–5105. IEEE.

Paz, A. (2014). *Introduction to probabilistic automata*. Academic Press.

Sun, B., Luh, P. B., Jia, Q.-S., Jiang, Z., Wang, F., and Song, C. (2013). Building energy management: Integrated control of active and passive heating, cooling, lighting, shading, and ventilation systems. *IEEE Trans. Automation Science and Engineering*, 10(3):588–602.

Tzempelikos, A. and Athienitis, A. K. (2007). The impact of shading design and control on building cooling and lighting demand. *Solar Energy*, 81(3):369–382.

Van Moeseke, G., Bruyère, I., and De Herde, A. (2007). Impact of control rules on the efficiency of shading devices and free cooling for office buildings. *Building and environment*, 42(2):784–793.

ACKNOWLEDGEMENT

This material is based upon research sponsored by the Laboratory Directed Research and Development Program of Oak Ridge National Laboratory (ORNL), managed by UT-Battelle, LLC for the U.S. Department of Energy under Contract No. DE-AC05-00OR22725.
The submitted manuscript has been authored by a contractor of the U.S. Government under Contract DE-AC05-00OR22725.
Accordingly, the U.S. Government retains a nonexclusive, royalty-free license to publish or reproduce the published form of this contribution, or allow others to do so, for U.S. Government purposes.

A Bayesian Approach for Modeling Occupants' Use of Window Shades

Seyed Amir SADEGHI[1,3]*, Nimish M AWALGAONKAR[2], Panagiota KARAVA[1,3], Ilias BILIONIS[2]

[1]Lyles School of Civil Engineering, Purdue University, West Lafayette, IN, U.S
(sadeghis@purdue.edu, pkarava@purdue.edu)

[2]Department of Mechanical Engineering, Purdue University, West Lafayette, IN, U.S
(ibilion@purdue.edu, nawalgao@purdue.edu)

[3] Center for High Performance Buildings, Purdue University, West Lafayette, IN, U.S

* Corresponding Author

ABSTRACT

Despite the significant impact of occupant interactions with window shading systems on visual comfort and building energy consumption, there are still significant gaps in understanding and predicting these complex phenomena. This paper presents a Bayesian modeling approach for the prediction of states of motorized roller shades operated by occupants. It is based on a field study with a large number of human test-subjects, conducted in a high performance building with advanced technology and easy-to-access user interfaces for environmental controls. Unlike the Frequentist methods used in previous studies, the Bayesian approach allows for uncertainty quantification which provides further insight on parameter estimates in models. This information is important when dealing with small-sized datasets which is often the case in real applications of human data collection. In addition, this study contributes to the body of knowledge by: (1) expanding the investigation of human-building interactions to motorized interior roller shades; (2) incorporating human attributes and personal characteristics as important underlying variables in occupant-shading interaction models; (3) enabling the prediction of continuous intermediate states of the shading system.

1. INTRODUCTION

In recent years, there has been a growing interest in studying behaviors of building occupants in regards to their interactions with building systems such as window shades. Following the chronological progress in the literature, one can note the two main motivations for conducting these studies. First, to attain an understanding of the reasoning behind the human-building interactions which can provide useful insights to the development of optimal control algorithms for building systems. Second, to discover a mapping path between human-building interactions and environmental conditions, which can be used to predict the operating status of building systems within the framework of Building Performance Simulation (BPS).

Several indoor variables have been monitored in previous studies to investigate the triggers of occupant interactions with shading systems. These include indoor air temperature (Mahdavi et al., 2008; Inkarojrit, 2005), work plane illuminance (Sutter et al., 2006; Mahdavi et al., 2008; Haldi and Robinson, 2010), vertical illuminance on VDU screens (Sutter et al., 2006), daylight glare index and probability (da Silva et al., 2013), daylight work plane illuminance (Love, 1998), solar penetration length (Inoue et al., 1988), and transmitted solar radiation (Inoue et al., 1988; Sutter et al., 2006; Inkarojrit, 2005). Horizontal and vertical global illuminance and irradiance (Sutter et al., 2006; Mahdavi et al., 2008), outdoor temperature, solar altitude are among the outdoor variables which were investigated. Seasonal effects (Mahdavi et al., 2008), façade orientation Oreszczyn (2001), and sky conditions were considered as well. Rubin (1978) considered sunny, cloudy, and hazy conditions and found that blind position seemed to be independent of those. As a counter example, Rea (1984) noticed that blind occlusion was significantly different between different sky conditions. This study also concluded that occupants have a long term perception of solar irradiances that may affect the use of blinds. Overall, findings (Haldi and Robinson, 2010; O'Brien et al.,

2012) suggest that observed variations can be explained by variables such as solar intensity, daylight levels, and geometry. That is, considering the right triggering variables, human-shading interactions throughout the year can be described regardless of façade orientation.

Personal characteristics and human attributes, i.e. non-physical variables, have also been reported to describe occupant interactions with shading and electric lighting systems. These are variables which are not measurable with typical sensors, such as view and connection to the outside or privacy or daylight-health perception. Inoue et al. (1988) reported that most occupants preferred to have seats close to the windows, although these were known to be the most possible locations to have glare and direct sunlight. This finding implies that occupants may tolerate some discomfort in order to have a better quality of view and connection to the outdoors. Veitch et al. (1993) found that people prefer daylight to artificial lighting due to their beliefs regarding health issues. Inkarojrit (2005) reported visual privacy as second important reason for choosing the blind positions. Moreover, Foster and Oreszczyn (2001) unexpectedly observed higher rate of blind lowering in the north facade than the west facade. They believe this was due to the fact that north facade of the building was facing another office building and occupants deployed their blinds to preserve their privacy. Similarly, Reinhart and Voss (2003) tried to correct the privacy-related bias in their observations and suggested that if blinds were lowered at ambient horizontal illuminance less than 1000 lux, it would have occurred due to occupants' desire to maintain privacy.

This paper presents a field study that focuses on motorized roller shades as studies on occupant interactions with such systems are rather limited (Bakker et al., 2014; Meerbeek et al., 2014). The contributions of this work include a Bayesian modeling framework that enables: (a) Uncertainty quantification in model parameter estimates which is important when dealing with small-sized datasets; a ubiquitous issue when collecting human data. (b) Incorporation of underlying personal characteristics and human attributes of human-shading interactions. (c) Prediction of continuous intermediate states of the shading system.

2. FIELD STUDY

This section presents a brief overview of the experimental study that provided the basis for the development of human–shading interaction model. For more detailed information on the experimental setup the reader may refer to Sadeghi et al. (2016). Four south-facing private offices (3.3m×3.7m×3.2m high) in the Herrick Laboratories, a high performance building located in West Lafayette, Indiana, was selected for the purpose of this study. Figure 1 shows a general view of the building and monitored offices. A Building Management System (BMS) is available through the installed Tridium JACE controllers and Niagara/AX software framework, which provides the ability to monitor and control all the building systems. Each office has one exterior curtain wall façade with 54% window-to-wall ratio, and a high-performance glazing unit with a selective low-emissivity coating (visible transmittance: 70%, solar transmittance: 33%). The window is equipped with a dark-colored motorized interior roller shade that has a total visible transmittance equal to 2.53% (measured with an integrating sphere) and an openness factor of 2.18%. There are two electric lighting fixtures with two 32-watt T5 fluorescent lamps (total of 128 watts) in the office which can provide a maximum of 500 lx on the work plane. During the field study, the temperature in the office was well kept within ±0.5 °C of the set point to ensure that there were no thermal effects on occupant interactions with shading.

The field study was conducted in two rounds in order to cover a wide range of sky conditions and solar paths. First, over a period of 40 days between April 1st and June 15th 2015 including 22 sunny days, 10 cloudy days, and 8 mixed sky days. Second, over 38 days between October 19 and December 10 2015 covering 21 sunny days, 11 cloudy days, and 6 mixed sky conditions. Overall, 208 office occupants participated in the study (131 males and 77 females). Participants were students and staff (between 20 and 40 years old) not familiar with this research. Each office was occupied by one participant every day between 9:00 am and 4:00 pm. All participants were asked to perform their usual workload (computer-related work, reading, writing, etc.) during the day and answer four short web-based questionnaires, which were sent by e-mail and combined with phone alarm reminders at specific times during the day. They were free to take breaks or leave the office if they needed to (e.g. attend meetings, classes etc.) to create realistic dynamics of occupation. Participants were advised to interact with electric lights, shading system, and thermostat as they usually would, and to avoid any direct contact with the monitoring instrumentation. The instrumentation was installed so there was no interference with the occupant regular position and task. To eliminate any bias in the results, each person participated in the monitoring campaign only for a single day in one office setup. This sampling method enabled a large number of participants, which is necessary for the purpose of this study, and did not require the installation of experimental equipment in a large number of offices.

Three different control setups were deployed in the offices. In setup 1, participants used commercially available wall switches (Figure 2, right) to control motorized roller shades and electric lights. Participants could open/close roller shades or turn on/off electric lights with a single button push (top and bottom), or they could choose intermediate shade positions or light dimming levels (both in 25% increments) by pressing middle increase/decrease buttons respectively. In setup 2, participants used a modular web-based graphical interface (designed by the authors) to control shade position and electric lighting levels (Figure 2, middle). Participants could use sliders or click on buttons to control roller shade position and electric light levels in 25% increments. Other important features were designed on the graphical interface, including comfort sliders for capturing the level of comfort with the amount of light and visual conditions, as well as a four-scale reasoning slider in the middle to capture non-physical motives of human-shading interactions. In setup 3, roller shades were controlled automatically but occupants could override the shade position using the same interface as in setup 2. The automatic controller followed an algorithm to prevent direct sunlight on the occupant/work plane, but allowed direct light on the floor, up to 1 m from the window. In addition, there were adjustments for low light and high brightness conditions. Once overridden, the automatic controller would remain disabled for one hour. Electric lights were controlled manually using the same graphical interface as in setup 2. Side-by-side comparisons of occupant interactions in different control setups are presented in Sadeghi et al. (2016). For this paper, data collected from all three setups are used to develop occupant-shading interaction models.

The following variables were monitored every five minutes during the field study: window shade position, electric light level, room temperature set point, occupancy state, work plane illuminance, vertical illuminance near eye level (30 cm from occupant's head), transmitted global solar radiation, transmitted illuminance through window, indoor air temperature, black-globe temperature and relative humidity. Figure 2 (left) depicts the office layout, seating position, and locations of sensors and control devices. Two web-based survey questionnaires were designed. Survey type-A included questions about motives of human-building interactions and was completed four times a day. Survey type-B referred to personal characteristics and attributes and was completed once at the end of the day. Survey questionnaires were sent to participants at specific times during the day along with reminders by phone alarms.

Figure 1: The four monitored offices: exterior view (left); Interior view (right)

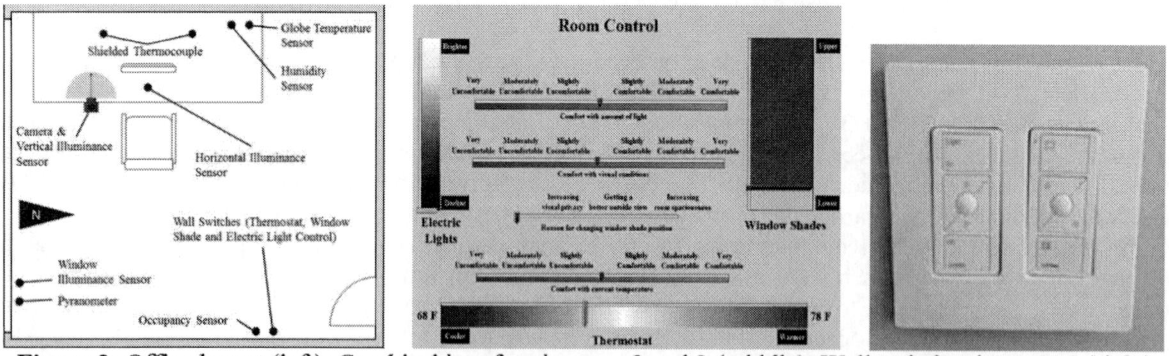

Figure 2: Office layout (left); Graphical interface in setup 2 and 3 (middle); Wall switches in setup 1 (right)

3. EXPERIMENTAL RESULTS

Important observations from the experimental data are presented in this section as they are relevant to the development of human-shading interaction models. On average, 1.57 shading actions were observed in control setup 1 while the occupants' daily average shade movement rate was 2.12 and 2.75 respectively in setups 2 and 3. These values are higher than what is observed in different studies with non-motorized manual shading devices, operated by turning a rod, pulling a chain or cord (Inoue et al., 1988; 1993; Inkarojrit, 2005; da Silva et al., 2013) proving that easy-to-control shading devices reduce the effort required to control/improve indoor environmental conditions and result in more human-shading interactions. It was observed that 53.2% of times, the selected position of motorized window roller shade was different from extreme positions of 0% and 100% (0% being fully lowered and 100% being fully raised). This finding indicates that a good behavioral model for occupant interactions with window shades should predict the intermediate operating states of the system as well as the fully lowered/raised positions.

The occupation dynamics was found to have an impact on the way participants would interact with roller shades. The first ten minutes after arrival in the morning and the last ten minutes before the departure (at the end of the day) were selected as threshold limits for arrival and departure time intervals. The same threshold was used for determining events before and after absences during the intermediate time interval. As shown in Table 1, a considerable portion of shading interactions (55.3%) occurs outside the intermediate time interval with continuous occupation. A high frequency of shading interactions is observed during the arrival time, which is in agreement with findings of previous studies (Inoue et al., 1988; Reinhart and Voss, 2003; Haldi and Robinson, 2010; da Silva et al., 2013). Compared to actions throughout the day, the number of shading actions per unit time is significantly higher upon arrival. Therefore, only shading actions in arrival time has been considered for modeling in this study. A similar approach has been undertaken elsewhere (Reinhart and Voss, 2003; Inkarojrit, 2005).

Table 1: Summary of survey questionnaires

	Arrival	Departure	Intermediate, before absence	Intermediate after absence	Intermediate, continuous occupation
Shading interactions	36.6%	1.7%	0%	17%	44.7%

Among the reasons for human-shading interactions reported in survey type-A, reducing the overall brightness of workspace and glare in workspace were found to be the main motives for lowering the motorized roller shade (respectively reported with frequency of 27% and 45%). Increasing the amount of daylight was also reported to be the main reason for raising the roller shade (with frequency of 47%). Actions governed by these motives can be described by physical variables (e.g. work plane illuminance, vertical illuminance etc.). Figure 3 presents the boxplots of environmental variables during the whole course of the experiment as well as the moments right before raising and lowering actions took place during the arrival time. As shown in the figure, variables such as indoor illuminances, window un-shaded fraction, solar penetration length, transmitted global solar radiation and ratio of diffuse over total solar radiation explain lowering actions to some extent as at lowering moments, they have quite different distribution than the distribution over the course of the experiment. The same can be stated about work plane and vertical illuminance, window un-shaded fraction, work plane daylight illuminance, solar penetration length, solar azimuth and altitude for describing shade raising actions.

Based on the survey questionnaires, achieving a better outside view was a significant motive for raising/opening the window roller shade (with frequency of 37%). The desire to increase visual privacy was also reported by participants as an important reason for window shade lowering/closing actions (with frequency of 26%). Figure 4 illustrates boxplots of the selected shade position versus occupants' self-reported level of importance for having a clear outside view and visual privacy (importance level of one is excluded for visual privacy since there was no vote on it). Higher un-shaded portions are selected by participants to whom having a clear view is more important; and lower shade positions correspond to participants who reported visual privacy to be of high level of importance. Figure 5 illustrates the distribution of all human variables collected from survey type-B in overall and at the moments when actions occurred. The significance of human variables on shade raising and lowering actions is clear from the figure and therefore, need to be considered, when developing models for human-shading interactions.

4. MODELING METHODOLOGY

Four sub-models have been considered to construct a probabilistic model for human-shading interactions while allowing for predictions of intermediate shade positions (Figure 6). Model S1 predicts binary actions of shade raising (raising events versus non-raising events) and model S2 predicts binary actions of shade lowering (lowering events versus non-lowering events). The states of the shade between each two consecutive time steps (five-minute intervals) are compared and if a change is detected, the observed event is coded as "1" and "0" (existence and non-existence of event respectively) for the first time step among the two, corresponding to the row of all explanatory variables (human and environmental state variables) for which occupants decided to undertake the action. Models S3 and S4 respectively predict the magnitude of shade raising and lowering. Figure 6 indicates the hierarchy within which models S1, S2, S3, and S4 work together to predict human-shading interactions given all the explanatory variables.

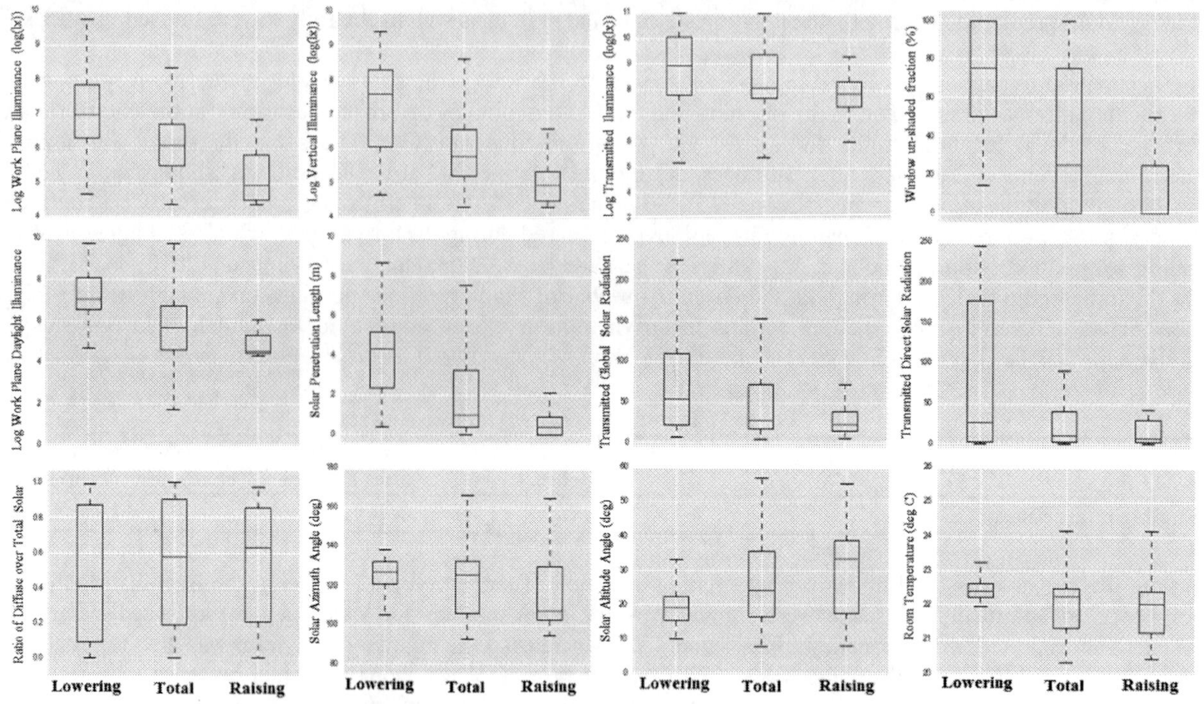

Figure 3: Distribution of environmental variables

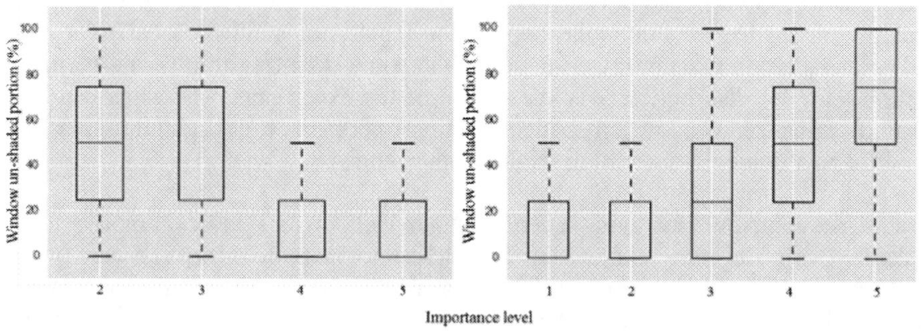

Figure 4: Impact of visual privacy (left) and window view (right) on selected shade positions

The multivariate binary-choice logit model along Bayesian parameter estimation has been used for the set of models in higher level of hierarchy (S1, S2). In addition, Bayesian linear regression has been used for magnitude models, or the models in lower level of hierarchy (S3, S4). Bayesian inference is a method of statistical inference in which the Bayes' theorem is used to update the probability of a hypothesis (beliefs) as more and more evidence or information (data) becomes available. Bayesian inference differs from the traditional Frequentist approach in the sense that it preserves the uncertainty in predictions.

In what follows, $\mathbf{x} = (1, x_1, \ldots, x_d)$ denotes the vector of the d features (explanatory variables) that define the state of the environment as well as any human attributes. Notice that we have prepended a constant unit feature to $\mathbf{x}$. The purpose of this constant feature is to simplify the notation of the regression models presented below. We will be referring to $\mathbf{x}$ as the *feature vector*.

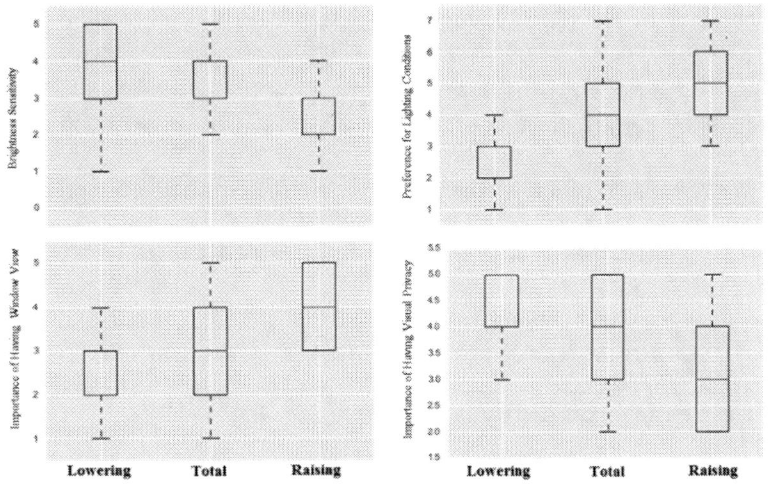

Figure 5: Distribution of human variables

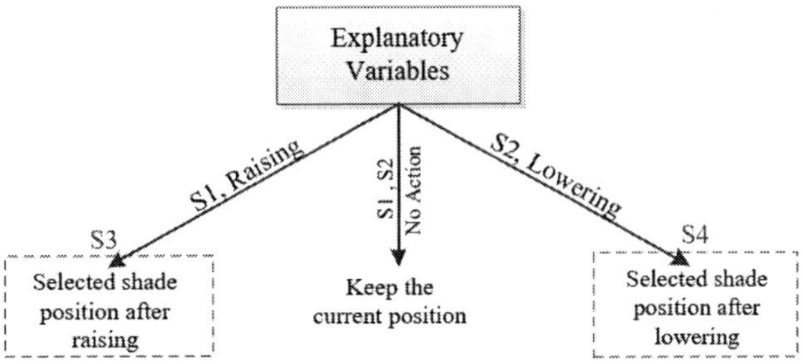

Figure 6: Structure of human-shading interaction model

4.1 Bayesian Hierarchical Modeling – Multivariate Logistic Regression (S1 and S2)

In the higher level models of hierarchy (S1 and S2), we are interested in addressing the following question:
"Given the environment inside the private office, what is the probability that the occupant will lower or raise the shades?"

The model form we develop is identical for S1 and S2. Let z be a binary random variable such that $z = 1$ corresponds to "action" and $z = 0$ to "non-action". The probability of "non-action" conditioned on the observed features is modeled by:

$$p(z = 0|\mathbf{x}, \mathbf{b}) = \mathrm{sigm}(\mathbf{b}^T\mathbf{x}), \tag{1}$$

Where $sigm(x) = \frac{1}{1+e^{-x}}$ denotes the sigmoid function, $\mathbf{b} = (b_0, b_1, \ldots, b_d)$ is a vector of regression coefficients to be inferred from the data, and $\boldsymbol{b}^T\boldsymbol{x}$ is the dot product between $\mathbf{b}$ and $\mathbf{x}$. Using the standard rules of probability, the probability of "action" conditioned on the observed features is:

$$p(z = 1|\mathbf{x}, \mathbf{b}) = 1 - p(z = 0|\mathbf{x}, \mathbf{b}) \tag{2}$$

We will be calling z the *action target*.

Now, let $\mathbf{x}_{1:N} = \{\mathbf{x}_1, \ldots, \mathbf{x}_N\}$ and $\mathbf{z}_{1:N} = \{z_1, \ldots z_N\}$ be the observed features and action targets, respectively. Assuming that the measurements are conditionally independent given the features, the *likelihood* of the observed data set is:

$$p(\mathbf{z}_{1:N}|\mathbf{x}_{1:N}, \mathbf{b}) = \prod_{i=1}^{N} p(z_i|\mathbf{x}_i, \mathbf{b}). \tag{3}$$

To proceed, we need to specify our *prior* state of knowledge about the coefficients $\mathbf{b}$. Since we do not have much prior information about it, we will construct a vague hierarchical prior distribution. Specifically, we assign to $\mathbf{b}$ a zero mean Gaussian prior,

$$p(\mathbf{b}|\alpha) = \mathcal{N}(\mathbf{b}|\mathbf{0}, \alpha^{-1}\mathbf{I}_{d+1}), \tag{4}$$

where $\mathcal{N}(\cdot|\boldsymbol{\mu},\boldsymbol{\Sigma})$ is the probability density function of a multivariate Gaussian distribution with mean $\boldsymbol{\mu}$ and covariance matrix $\boldsymbol{\Sigma}$ and $\mathbf{I}_{d+1}$ is the $(d+1)$-dimensional unit matrix, and α is a priori unknown precision. Completing the model specification, we assign an exponential prior to α:

$$p(\alpha|\lambda) \sim \mathcal{E}(\alpha|\lambda), \tag{5}$$

where $\mathcal{E}(\cdot|\lambda)$ is the probability density function of an exponential random variable with rate parameter λ. Here, we fix $\lambda = 10000$.

Using Bayes rule, our *posterior* state of knowledge about the coefficients $\mathbf{b}$ and the precision parameter α is given by:

$$p(\alpha,\mathbf{b}|\mathbf{x}_{1:N}, z_{1:N}, \lambda) \propto p(z_{1:N}|\mathbf{x}_{1:N}, \mathbf{b})p(\mathbf{b}|\alpha)p(\alpha|\lambda). \tag{6}$$

Using the sum rule of probability theory, the *predictive distribution* at a new set of features $\mathbf{x}^*$ is given by:

$$p(z^*|\mathbf{x}^*,\lambda) = \int p(z^*|\mathbf{x}^*,\mathbf{b})p(\alpha,\mathbf{b}|\mathbf{x}_{1:N}, z_{1:N}, \lambda)d\mathbf{b}d\alpha. \tag{7}$$

This is intractable, but it can be approximated by sampling (see Sec. 4.3).

4.2 Bayesian Hierarchical Modeling – Multivariate Linear Regression (S3 and S4)

In the lower level models of hierarchy (S3 and S4), we are interested in addressing the following question - *"Given that we already know whether the person is going to raise/lower the shades and the current environment inside the office, what is the amount by which the person is going to raise or lower the shades?"*

The model form we develop is identical for S3 and S4. Let y be a random variable indicating the position of the shades selected by the human after an action (either lowering or raising). We will be referring to y as the *target* variable. The probability of a shade position y conditioned on the observed features is modeled by:

$$p(y|\mathbf{x},\boldsymbol{\beta},\sigma) = \mathcal{N}(y|\boldsymbol{\beta}^T\mathbf{x},\sigma^2), \tag{8}$$

Where $\boldsymbol{\beta} = (\beta_0,\beta_1,\dots,\beta_d)$ is the vector of regression coefficients to be determined from the data, and σ^2 is the noise variance. The role of the latter is to capture as random effects everything that cannot be explained by the observed features.

As in the previous section, let $\mathbf{x}_{1:N} = \{\mathbf{x}_1,\dots,\mathbf{x}_N\}$ and $\mathbf{y}_{1:N} = \{y_1,\dots y_N\}$ be the observed features and targets, respectively. Assuming that the measurements are conditionally independent given the features, the *likelihood* of the observed data set is:

$$p(\mathbf{y}_{1:N}|\mathbf{x}_{1:N},\boldsymbol{\beta},\sigma^2) = \prod_{i=1}^{N} p(y_i|\mathbf{x}_i,\boldsymbol{\beta},\sigma^2). \tag{9}$$

Similarly, we assign a hierarchical prior to $\boldsymbol{\beta}$:

$$p(\boldsymbol{\beta}|\psi) = \mathcal{N}(\boldsymbol{\beta}|\mathbf{0},\psi^{-1}\mathbf{I}_{d+1}), \tag{10}$$

where, as before, ψ is an unknown precision parameter distributed exponentially,

$$p(\psi|r) = \mathcal{E}(\psi|r), \tag{11}$$

with rate parameter $r = 10000$, and

$$p(\sigma|\gamma) = \mathcal{E}(\sigma|\gamma), \tag{12}$$

with $\gamma = \frac{1}{25}$. Using Bayes rule, our posterior state of knowledge about the parameters is:

$$p(\psi,\boldsymbol{\beta},\sigma|\mathbf{x}_{1:N},\mathbf{y}_{1:N},r,\gamma) \propto p(\mathbf{y}_{1:N}|\mathbf{x}_{1:N},\boldsymbol{\beta},\sigma)p(\boldsymbol{\beta}|\psi)p(\psi|r)p(\sigma|\gamma). \tag{13}$$

Using the sum rule of probability theory, our *predictive distribution* at a new set of features $\mathbf{x}^*$ is given by:

$$p(y^*|\mathbf{x}^*,\lambda) = \int p(y^*|\mathbf{x}^*,\mathbf{b})p(\psi,\boldsymbol{\beta},\sigma|\mathbf{x}_{1:N},\mathbf{y}_{1:N},r,\gamma)d\psi d\boldsymbol{\beta}d\sigma. \tag{14}$$

This is also intractable, but it can be approximated by sampling (see Sec. 4.3).

4.3 Training and Sampling Approximation to the Predictive Distributions

We implemented our models using the Python package PyMC 2.3.0 (Patil et al. 2010) and used Markov chain Monte Carlo (MCMC) to sample from the posterior of $\mathbf{b}$ ($\boldsymbol{\beta}$). After monitoring the traces and the autocorrelations of the chain, we decided to burn the first 1,000,000 samples, and gather 200 samples by keeping one MCMC sample out of every 1,000. Using these samples, we can approximate the predictive distributions at new sets of features. For example, if $\mathbf{b}_{1:S} = \{\mathbf{b}_1,\dots,\mathbf{b}_S\}$ are samples from Eq. (6), then the predictive distribution of Eq. (7) is approximated by:

$$p(z^*|\mathbf{x}^*,\lambda) = \frac{1}{S}\sum_{i=1}^{S} p(z^*|\mathbf{x}^*,\mathbf{b}_i). \quad (15)$$

In our numerical results, we summarize the predictive distribution by computing its median 5- and 95-percent predictive quantiles. This is accomplished numerically by computing the empirical percentiles of a $z_i^*, i = 1, ..., S$, sampled from $p(z^*|x^*, \mathbf{b}_i)$.

5. ESTIMATION RESULTS AND DISCUSSION

All the variables in Figures 3 and 5 were considered within the multivariate structure of models. Python 2.3.0 was used for programming all models and the forward selection method was applied to determine the most statistically significant variables. As mentioned earlier, only the arrival time is considered for model development. Table 2 presents descriptive statistics of variables selected in the models used to normalize the data. Table 3 presents the estimate parameters for two different forms for model S1. Form 1 represents a shade raising model only including physical variables. Form 2 represents the model constructed using significant human variables. Variable "SV" in this model accounts for the interactive impacts of outside view and shade position. The variable is created by dividing the shade position by the importance level of clear view to outside reported by occupants in survey type-B. An increase in probability of shade raising is expected as this variable decreases, which is confirmed by the negative sign of parameter estimate. Lighting preference is also included as an individual characteristic reported by occupants in survey type-B. The positive sign for this variable indicates that the probability of shade raising increases for occupants who prefer to have brighter conditions.

Table 2: Descriptive statistics of explanatory variables (features)

Variable	Minimum	Mean	Max	Std.
Work plane ill. (lx)	78.11	801.39	7052.8	1526.49
Vertical ill. (lx)	75.78	982.34	12331.6	1056.66
Shade position (%)	0	44.05	100	37.08
SV	0	17.91	100	20.36
SP	0	22.39	100	26.01

Table 3 also presents estimation results for shade lowering actions. Among the environmental variables, vertical illuminance at eye level was found to be the best predictor for shade lowering actions. "SP" is included in model S2 to represent the interaction between the shade position and need to have visual privacy. To create variable "SP", the votes on importance of visual privacy are manipulated to reflect symmetry. That is, "the most important" votes were assigned to level 1 and "the least important" votes to level 5 (1. The most important ... 5. The least important). Variable "SP" is created by dividing the shade position by the new representation of votes on visual privacy need. It is expected for probability of shade lowering to increase as this variable increases which is confirmed by the positive sign of the parameter estimate for this variable in Table 3.

Table 3: Shade raising and lowering models (S1 and S2)

	Parameter Estimates (β | 95% HPD interval)			
	Shade raising models		**Shade lowering models**	
	Form 1	**Form 2**	**Form 1**	**Form 2**
Work plane ill.	-11.112|(-10.635,-5.622)	-6.665|(-9.627,-3.855)	N/A	N/A
Vertical ill.	N/A	N/A	1.046|(0.692,1.288)	1.153|(0.766,1.431)
Lighting preference	N/A	1.343|(0.752,1.635)	N/A	-0.701|(-1.110,-0.201)
SV	N/A	-2.662|(-3.748,-1.042)	N/A	N/A
SP	N/A	N/A	N/A	0.955|(0.635,1.311)
Constant	-4.501|(-5.544,-3.328)	-5.834|(-7.232,-4.846)	-2.000|(-2.373,-1.654)	-2.631|(-3.098,-1.979)
Observation No.	569		397	

The data is divided into training and testing sets and a separation plot for the testing set (10% of data) is shown in Figure 7. The separation plot for binary classifiers is resulted from sorting the data points based on estimated probabilities. Black lines show the estimated probability for each data point while the vertical blue lines indicate occurrence of raising/lowering actions (1s in dependent variable). It is expected that estimated probability increases as density of vertical lines does. It is clear from figure 7 (left) that unlike model form 2 (based on physical and human variables), model form 1 (based on physical variables) mostly fails to estimate high probabilities corresponding to events where raising actions have occurred. Better performance of the lowering model in form 2 is also evident from figure 7 (right). That is, including human variables adds to the power of the model in describing the phenomenon and improves its performance.

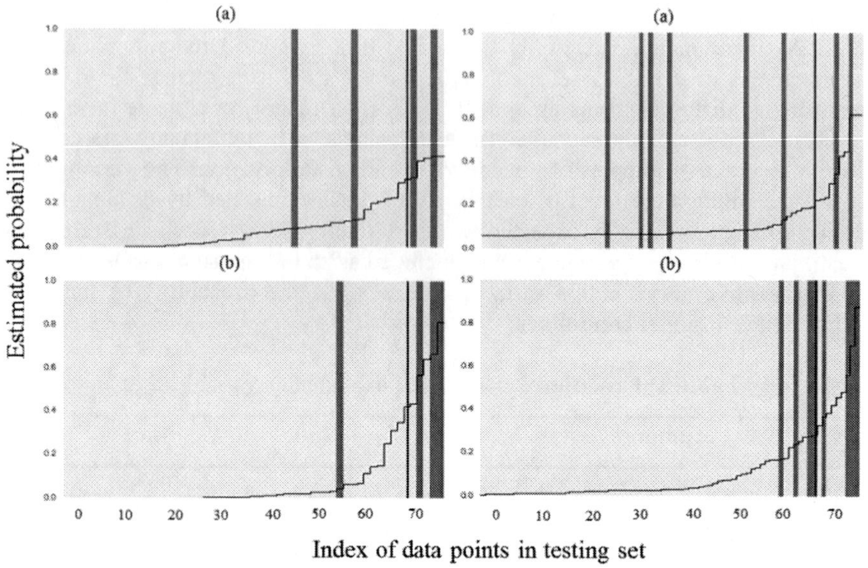

Figure 7: Separation plots for shade raising (left) and lowering (right) models; (a): model with physical variable, (b): model with both physical and human variables

Figure 8 presents the median of estimated probabilities resulted from models S1 and S2 for shade raising and lowering actions versus the change in explanatory variables. Outcome probabilities have been filtered based on lighting preference so that three plots are presented for each model; preferences for dark, moderate, and bright lighting conditions (used vote values of 2, 4, and 6 for lighting preference variable in the models). It is clear from the top plots that shade raising probability increases as the need for window view (SV) increases and the level of work plane illuminance decreases. However, this variation is also influenced by the lighting condition preference conditions. It can be seen that when dark conditions are preferred, shade raising action is of a low probability after the threshold value of 500 lx for the work plane illuminance while this happens around 1200 lx when there is a preference for bright lighting conditions. The same concept can be seen for shade lowering actions in bottom plots. That is, a shade lowering action is more likely to happen when dark conditions are preferred. It is evident from the plots that increase in vertical illuminance and need for visual privacy increases the likelihood of shade lowering actions.

Table 4 presents the estimated linear regression models for the magnitude of shading actions (models S3 and S4) while their performance is evaluated on the testing set (10% of the data) and shown in Figure 9. As a feature of Bayesian estimation, the uncertainty of predicted values is quantified and shown as a box plot for each data point. In this plot, we are only looking at the mean of the response with the Gaussian noise excluded and the uncertainty is the epistemic uncertainty induced by the finite amount of data. It is worth mentioning that the number of data points in the testing set is rather small, which might be the reason for off predictions and wide distributions of predicted values in some data points.

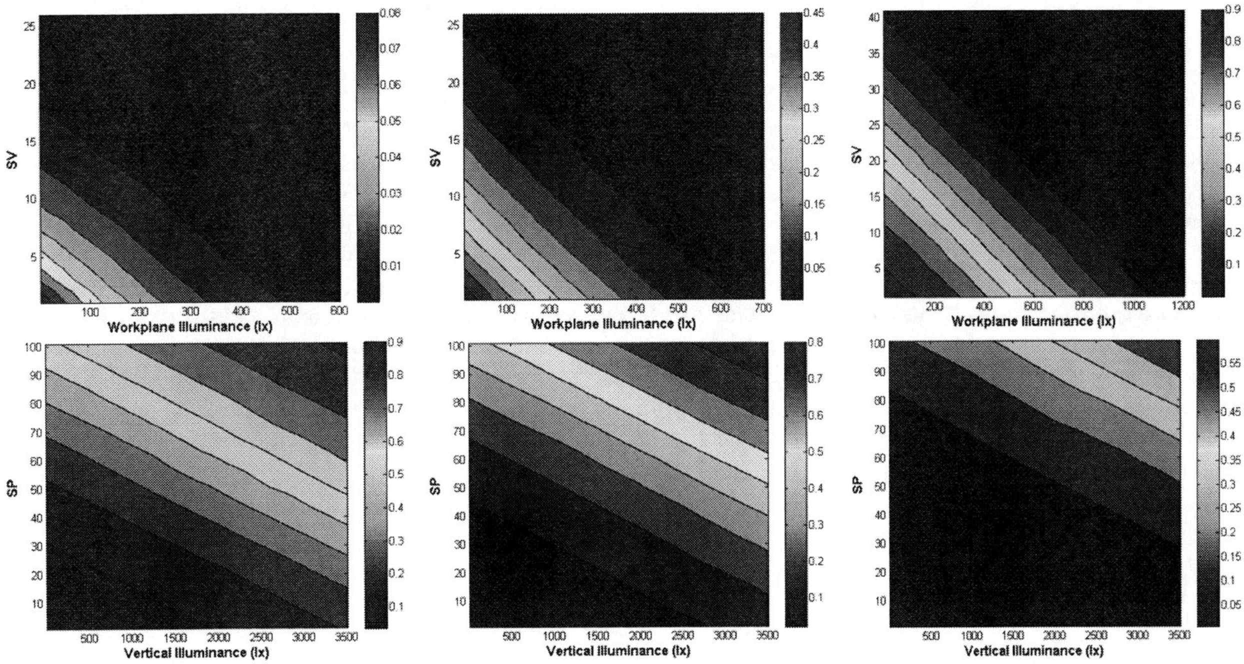

Figure 8: Level plots of estimated outcome probabilities for shade raising (top) and lowering (bottom) filtered based on lighting condition preferences (left: preference for dark conditions, middle: preference for moderate conditions, right: preference for bright conditions)

Table 4: Models for magnitude of shade raising and lowering (S3 and S4)

	Parameter Estimates (β | 95% HPD interval)	
	Raising magnitude	**Lowering magnitude**
Work plane ill.	-3.153|(-9.241,3.413)	N/A
Vertical ill.	N/A	6.234|(1.356,11.040)
Shade position	-3.710|(-8.956,3.537)	17.101|(11.363,22.096)
Constant	40.353|(33.824,45.113)	45.332|(40.234,50.543)
Observation No.	71	69

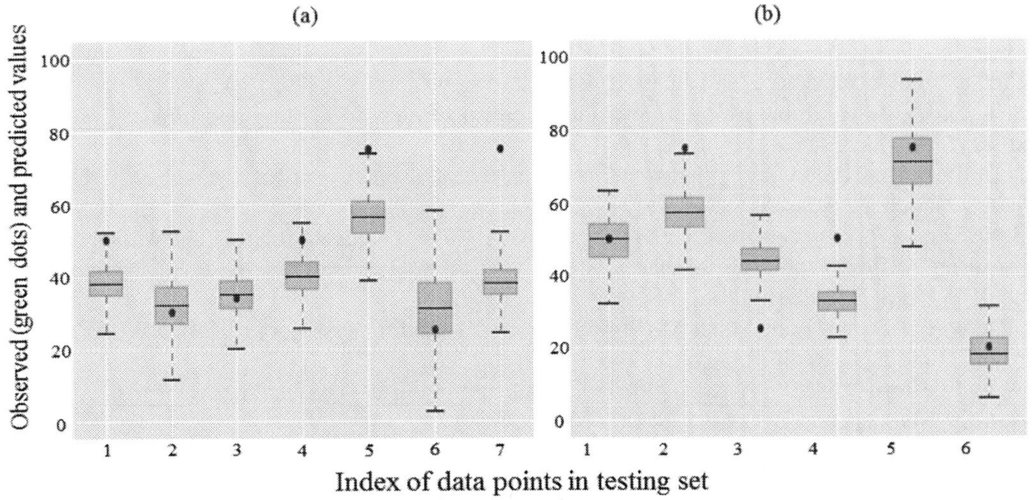

Figure 9: Observed and estimated magnitude of raising (a) and lowering (b) actions

6. CONCLUSIONS

This paper presented a contribution to characterization of human-shading interactions in private offices. Bayesian multivariate binary-choice logit models have been developed to predict shade raising/lowering along with Bayesian linear regression models to estimate the magnitude of shading actions, thus, allowing for prediction of continuous intermediate operating states of the shading system. Based on the findings of our field study, intermediate positions are frequently selected by the occupants and therefore, the new modeling framework is expected to increase the prediction accuracy of human-shading interactions in BPS tools. Human variables proved to have a significant impact on the operation of shading systems. Our findings prove that it is a combination of physical variables along with human attributes and individual characteristics that underlie human interactions with building systems such as window shades and both should be considered in modeling structures and BPS tools. It should be noted that the experimental dataset corresponds to a private office with a south façade orientation and the predicted probabilities to 5 min intervals. Occupant interactions with electric lighting will be incorporated in the modeling framework in future work by the authors.

REFERENCES

Bakker L.G., Hoes-van Oeffelen E.C.M., Loonen R.C.G.M. and Hensen J.L.M., 2014. User satisfaction and interaction with automated dynamic facades: A pilot study. Building and Environment 78: 44-52.

da Silva P.C., Leal V. and Andersen M., 2013. Occupant's interaction with electric lighting and shading systems in real single-occupied offices: Results from a monitoring campaign. Building and Environment, 64: 152-168

Foster M., and Oreszcyn T., 2001. Occupant control of passive systems: the use of venetian blinds. Building and Environment, 36 (2): 149-155.

Haldi F, Robinson D. Adaptive actions on shading devices in response to local visual stimuli. J Build Perform Simul 2010; 3:135-53.

Inkarojrit V., 2005. Balancing Comfort: Occupants' control of window blinds in private offices. PhD dissertation, University of California, Berkeley.

Inoue T, Kawase T, Ibamoto T, Takakusa S, Matsuo Y. The development of an optimal control system for window shading devices based on investigations in office buildings. ASHRAE Transactions 1988; 94:1034-49.

Love J. Manual switching patterns in private offices. Lighting Research and Technology 1998; 30:45-50.

Mahdavi A, Mohammadi A, Kabir E, and Lambeva L., 2008. Occupants' operation of lighting and shading systems in office buildings. Journal of Building Performance Simulation, 1:57-65.

Meerbeek B., Kulve M., Gritti T., Aarts M., Loenen E. and Aarts E., 2014. Building automation and perceived control: A field study on motorized exterior blinds in Dutch offices. Building and Environment 79: 66-77

O'Brien W, Kapsis K, Athienitis AK. Manually-operated window shade patterns in office buildings: a critical review. Building and Environment 2012; 60(0):319-38.

Patil A., Huard D., Fonnesbeck C.J., 2010. PyMC: Bayesian stochastic modeling in Python. J. Stat. Softw. 1:81

Rea M. Window blind occlusion: a pilot study. Building and Environment 1984; 19:133-7.

Reinhart C.F. and Voss K., 2003. Monitoring manual control of electric lighting and blinds. Lighting Research & Technology, 35 (3): 243-260.

Rubin AI, Collins BL, Tibbott RL. Window blinds as a potential energy saver: a case study. US Department of Commerce, National Bureau of Standards; 1978.

Sadeghi S.A., Karava P., Konstantzos I. and Tzempelikos A. 2016. Occupant interactions with shading and lighting systems using different control interfaces: a pilot field study. Building and Environment 97: 177-195. doi:10.1016/j.buildenv.2015.12.008

Sutter Y., Dumortier D. and Fontoynont M., 2006. The use of shading systems in VDU task offices: a pilot study. Energy and Buildings, 38:780-9.

Veitch J.A. and Gifford R., 1996. Assessing beliefs about lighting effects on health, performance, mood, and social behavior. Environment and Behavior, 28: 446-70.

ACKNOWLEDGEMENT

This work was partially funded by Lutron Electronics, Co. Inc. and by the National Science Foundation under Grant No. 1539527. Any opinions, findings, and conclusions or recommendations expressed in this material are those of the authors and do not necessarily reflect the views of the National Science Foundation.

The Rebound Effect after the Energy Refurbishment of Residential Buildings towards High Performances

Vincenzo CORRADO[1]*, Ilaria BALLARINI[1], Simona PADUOS[1], Elisa PRIMO[1]

[1]Politecnico di Torino, Department of Energy, Torino, Italy
Tel.: +39 011 0904456, Fax: +39 011 0904499,
E-mail addresses: vincenzo.corrado@polito.it (V. Corrado),
ilaria.ballarini@polito.it (I. Ballarini),
simona.paduos@polito.it (S. Paduos),
elisa.primo@polito.it (E. Primo).

* Corresponding Author

ABSTRACT

In the last decades, governments worldwide have promoted energy efficiency improvements in order to reduce the building energy consumptions and CO_2 emissions.
In many geographical contexts, the largest and most cost-effective energy saving potential is in the existing housing stock renovation. For this reason, many energy efficiency policies are focused on the energy refurbishment of the existing buildings.
Though the energy consumption of buildings is supposed to decrease after a retrofit, several studies show that the energy performance does not increase as much as it would be expected and that a significant gap between the estimated and the real energy savings - called "rebound effect" - occurs.
This deviation can partly depend on the model used for the energy performance evaluation, but changes in the occupants' behavior after a thermal retrofit of an existing dwelling are also revealed.
The present paper examines the influences of occupant behavior on the energy savings for some retrofitted Italian residential representative buildings. Some typical use patterns are defined according to statistics of the national census of population and dwellings and the expectations of the building occupants are specified before and after the retrofit. The main objective is to investigate the elasticity of the energy consumption in function of the occupancy parameters before and after the application of high energy efficiency improvements.
The analysis is carried out by means of a simplified energy performance assessment model. The outcomes of this study consist in defining synthetic indicators for the quantitative assessment of the rebound effect.

1 INTRODUCTION

In the last decades, governments worldwide have promoted energy efficiency improvements in order to reduce the building energy consumptions and CO_2 emissions.
In many geographical contexts, as in Italy, the largest and most cost-effective energy saving potential is in the existing housing stock renovation. For this reason, many energy efficiency policies are focused on the energy refurbishment of the existing buildings.
Though the energy consumption of buildings is supposed to decrease after a retrofit, several studies show that the energy performance does not increase as much as it would be expected and that a significant gap between the estimated and the real energy savings - called "rebound effect" – occurs (Haas and Biermayr, 2000).
This deviation can partly depend on the model used for the energy performance evaluation (Tronchin and Fabbri, 2008), but changes in the occupants' behavior after a thermal retrofit of an existing dwelling are also revealed (Oreszczyn *et al.*, 2006).
In the nineteenth century the rebound effect was identified in the so called "Jevons paradox", that pointed out the higher consumption of coal resulting from an increase of the efficiency of the steam engines. After the first oil shock in 1973-1974 and the resulting energy efficiency targets this notion revived and both the economists Khazzoom and Brookes, working independently, observed that in fact energy efficiency improvements, which were aimed to

produce savings in energy consumption, had the opposite effect and led to increase the consumption of energy services.
In the midst of the intense discussion among economists over the last two decades of the twentieth century the terms "rebound effect" and "backfire" became mainstream in academic literature. "Backfire" was the preferred term to describe the cases where energy efficiency improvements led to an increase in energy consumption, whereas "rebound effect" came to be used to describe and quantify the shortfalls in expected energy savings.
By the mid-2000s a lot of definitions of the rebound effect and approaches to quantify this phenomenon were used in literature ever since, and some relevant differences in the values for the rebound effect were observed (Galvin, 2015). Among the others, Hans e Biermayr (2000) defined the rebound effect as the increase in energy demand which results from an increase in service demand due to decreasing prices for energy service. They quantified it for some retrofitted dwellings in Austria as the shortfall in energy savings as percentage of the expected energy savings.
According to the previous analysis of Greening *et al.* (2000), Sorrell and Dimitropoulos (2008) delineated the effects of the rebound in three categories, in economics terms: direct, indirect and economy-wide effects.
The present paper concerns the direct rebound effect according to which improved energy efficiency for a particular energy service will decrease the effective price of that service and should therefore lead to an increase in consumption of that service. In terms of occupants' behavior, the direct rebound effect can be exemplified as follows: occupants take back the benefits of energy savings after a thermal retrofit as improved thermal comfort (e.g. higher indoor temperature, more hours of heating).
In the same way Herring and Roy (2007) used the term rebound effect to describe the effect that the lower costs in energy services has on consumer behavior, due to the increased energy efficiency. They also called this phenomenon "take-back effect" because of consumers take back some of the energy savings due to the improved energy efficiency in the form of higher levels of energy services.
Hens *et al.* (2010) estimated the rebound effect in some residential buildings by dividing the difference between the calculated reference consumption and the normalized measured consumption by that reference. References were estimated using the methodology imposed by the energy performance regulation, based on EN ISO 13790.
Sunikka-Blank and Galvin (2012) clarified the different impact of the occupant behavior on the energy consumption of buildings before and after the refurbishment and highlighted that the rebound effect occurs when a proportion of the energy savings after a retrofit is consumed by additional energy use.
Furthermore, Galvin (2014) identified the classic definition of the rebound effect as the energy efficiency elasticity of energy consumption and two further definition of the rebound effect, which can be useful when there is not enough information to calculate an elasticity rebound effect. Galvin named these two further definitions as follows:

- energy performance gap: the ratio between over-consumption and expected consumption;
- energy savings deficit: shortfalls in energy savings as a proportion of expected energy savings.

In the present research, the rebound effect is defined as the difference between expected and actual post-retrofit energy savings taking account of the occupants' behavior change after a retrofit, in absolute terms.
The influence of the occupant behavior on the energy savings for some retrofitted Italian residential representative buildings is examined. Some typical use patterns are defined according to statistics of the national census of population and dwellings and the expectations of the building occupants are specified before and after the retrofit. The main objective is to investigate the elasticity of the energy consumption in function of the occupancy parameters before and after the application of high energy efficiency measures.

2 METHODOLOGY

It is generally recognized in literature that user behavior is related to some building properties, for instance period of construction, roof insulation, wall type and required energy consumption to maintain a steady state temperature (Oreszczyn *et al.*, 2006). Moreover, it is observed that occupant behavior changes after retrofitting and that this change is related to the adoption of energy efficiency measures (Oreszczyn *et al.*, 2006).
For this reason, in the present research it is defined a rebound correlation matrix among energy efficiency measures, energy services and user parameters as shown in Figure 1.
Each energy efficiency measure has an influence on the consumptions related to one or more energy services and at the same time each energy service is related to one or more parameters that the user can control according to his preferences (e.g. temperature set point for heating and cooling, ventilation rate). These parameters can change after retrofitting if a rebound effect occurs.

Heating	Cooling	Domestic Hot Water	Ventilation	Lighting	ENERGY EFFICIENCY MEASURES	
X	X				External wall insulation	Thermal envelope insulation
X	X				Wall vs. unconditioned spaces insulation	Thermal envelope insulation
X	X				Roof insulation	Thermal envelope insulation
X	X				Floor insulation	Thermal envelope insulation
X	X				Window replacement	Thermal envelope insulation
	X			X	Solar shading devices	Solar control
X					Heat recovery unit	Ventilation heat recovery
X					Room temperature PI-controller	Improving heating control
	X				Chiller	Installation of high efficiency generator
X					Space heating generator	Installation of high efficiency generator
		X			DHW generator	Installation of high efficiency generator
X		X			Combined heating and DHW generator	Installation of high efficiency generator
X	X	X			Heat pump for heating, DHW and cooling	Installation of high efficiency generator
		X			Thermal solar system	Energy form renewable sources
(*)	X	(*)		X	PV system	Energy form renewable sources
				X	High efficiency luminaries	Improving lighting efficiency
				X	Automatic lighting control	Improving lighting efficiency

Heating	Cooling	Domestic Hot Water	Ventilation	Lighting	USER PARAMETERS
X					Temperature set point for heating
X					Temperature set back for heating
X					Daily heating period
X			X		Ventilation rate
	X				Temperature set point for cooling
	X				Daily cooling period
		X			Daily water consumption
				X	Daily lighting period

(*) in presence of heat pumps

Figure 1: Rebound correlation matrix among energy efficiency measures, energy services and user parameters

In order to investigate the influence of occupant behavior changes, the building energy consumption was evaluated after the implementation of the the same energy efficiency measure by considering the variables of the user behavior both before and after the refurbishment, when the occupant behavior is affected by rebound.

In the present research, the rebound effect is defined as the difference between expected (ΔEP) and actual post-retrofit energy savings (ΔEP_{RB}) taking account of the occupants' behavior change after a retrofit, in absolute terms as listed in equation (1).

$$RB = \Delta EP - \Delta EP_{RB} = EP_{EB} - EP_{REF} - (EP_{EB} - EP_{REF,RB}) = EP_{REF,RB} - EP_{REF} \tag{1}$$

Inspired by Galvin's analysis (2014), the follow indicators, based on rebound effect and estimated energy savings, were determined:

- $I_{RB,1}$ represents the size of the rebound effect as proportion of the building energy performance after retrofit without changes in occupant behavior, as expressed in equation (2):

$$I_{RB,1} = \frac{RB}{EP_{REF}} = \frac{EP_{REF,RB} - EP_{REF}}{EP_{REF}} \tag{2}$$

- $I_{RB,2}$ estimates the percentage of the expected energy savings, which are not frustrated by the phenomenon of the rebound effect, as formulated in equation (3):

$$I_{RB,2} = \frac{\Delta EP_{RB}}{\Delta EP} = \frac{EP_{EB} - EP_{REF,RB}}{EP_{EB} - EP_{REF}} \tag{3}$$

3 CALCULATION

3.1 Case studies

According to academic literature, the occupant behavior is influenced by a whole string of variables such as user characteristics, lifestyle, personal background and perception of comfort (Guerra Santin, 2011).

In accordance with previous studies, the user characteristics that could affect the building energy use are: household size, age of occupants, education, income, main occupation (Guerra Santin, 2011).

In order to define some typical occupants, the statistics of the last Italian census of population and dwellings were analyzed. Some of the resultant typical householders are presented in Table 1. Each user was then related with a specific building typology for his characteristics. The pictures and the main geometrical and technological data of the building type are shown in Table 2.

Table 1: Household characteristics

# User	Description of family	N. of components	Age	Income	Education level	Ecological awareness	Building typology
U1	Elderly single	1	70	Low	Low	Low	Apartment blocks
U2	Young couple with children	3 or 4	42, 38, 5, (2)	Middle	High	High	Multi-family house
U3	Adult couple with teenagers	3 or 4	53, 48, 18, (16)	High	Middle	Middle	Single-family house

Table 2: Main data of the buildings

				Building type		
				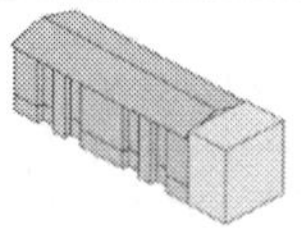 Apartment blocks (AB)	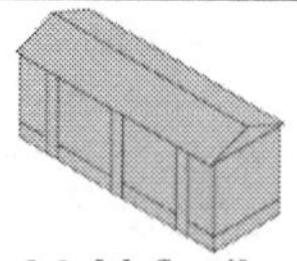 Multi-family house (MF)	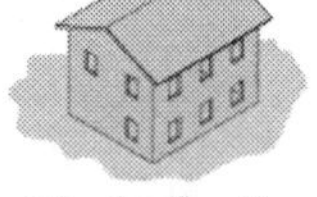 Single-family house (SF)
Geometry	Whole building	V_g	(m^3)	5949	3076	584
		$A_{f,n}$	(m^2)	1552	827	162
		A_{env}/V_g	(m^{-1})	0.46	0.51	0.73
		A_w	(m^2)	217	150	20.26
		no. of floors	(-)	4	3	2
		no. of units	(-)	24	12	1
Construction	Opaque envelope	U_{wl}	($W\,m^{-2}K^{-1}$)	1.15	1.48	1.48
		$U_{wl,u}$	($W\,m^{-2}K^{-1}$)	2.32	1.70	-
		$U_{fl,up}$	($W\,m^{-2}K^{-1}$)	1.65	1.65	1.65
		$U_{fl,lw}$	($W\,m^{-2}K^{-1}$)	1.30	1.30	2.00
	Windows	U_w	($W\,m^{-2}K^{-1}$)	4.90	4.90	4.90
		$g_{gl,n}$	(-)	0.85	0.85	0.85
Technical systems efficiency (mean yearly/seasonal values)	Radiators	$\eta_{H,e}$	(-)	0.90	0.90	0.88
	Room temperature control	$\eta_{H,c}$	(-)	0.85	0.85	0.85
	Central distribution	$\eta_{H,d}$	(-)	0.901	0.889	0.876
	Natural gas standard generator	$\eta_{H,gn}$	(-)	0.85	0.85	0.85
	Electric water heater	$\eta_{W,gn}$	(-)	0.75	0.75	0.80
	Indoor units split systems	$\eta_{C,e}$	(-)	0.97	0.97	0.97

The reference existing buildings were selected among those included in the Italian National Building Typology of the IEE-TABULA research project (Ballarini *et al.*, 2014), as belonging to the age class ranging from 1946 to 1960. The location of the buildings is Milan (Italy, 2404 HDD): this city belongs to the climatic zone (E), which includes most of the municipalities and of the Italian population.

The energy efficiency measures applied to the reference buildings were gathered in four packages: thermal envelope insulation, installation of solar shading devices, replacement of the generator for space heating and domestic hot water, and the building transformation into nZEB, as defined by the national and European legislation (European Directive 2010/31/EU).

The above-mentioned packages include the following energy efficiency measures:

a) insulation of the opaque envelope enclosing the heated space, replacement of the existent windows with new ones with double low-e glass and PVC frame, improvement of heating control with room temperature PI-controllers,
b) installation of external movable solar shading devices,
c) replacement of natural gas standard generator for space heating and electric water heater with a condensing boiler and installation of room control temperature PI-controllers,
d) insulation of the opaque envelope, replacement of the existent windows with new ones with double low-e glass and PVC frame, improvement of the heating control with new room temperature PI-controllers, replacement of the natural gas generator for space heating and electric water heater with a heat pump for heating, DHW and cooling (and associated radiant panels), installation of PV panels.

Each energy efficiency measure was identified by an appropriate parameter (e.g. *U*-value of the building envelope component, efficiency of the heat generator) and its value meets the requirements fixed by the Italian legislation in force (Ministerial Decree of 26 June 2015). The packages of the energy efficiency measures applied to the case studies are listed in Table 3.

Table 3: Energy efficiency measures and options

Retrofit measures	*Existing building*			Thermal envelope insulation*			Solar shading devices			Thermal system*			Major renovation		
Building type	AB	MF	SF	AB	MF	SF	AB	MF	SF	AB	MF	SF	AB	MF	SF
U_{wl} (W m^{-2}K^{-1})	*1.15*	*1.48*	*1.48*	0.28	0.28	0.28	*1.15*	*1.48*	*1.48*	*1.15*	*1.48*	*1.48*	0.26	0.26	0.26
$U_{wl,u}$ (W m^{-2}K^{-1})	*2.32*	*1.70*		0.56	0.47		*2.32*	*1.70*		*2.32*	*1.70*		0.58	0.43	
$U_{fl,up}$ (W m^{-2}K^{-1})	*1.65*	*1.65*	*1.65*	0.27	0.27	0.27	*1.65*	*1.65*	*1.65*	*1.65*	*1.65*	*1.65*	0.24	0.24	0.24
$U_{fl,lw}$ (W m^{-2}K^{-1})	*1.30*	*1.30*	*2.00*	0.58	0.58	0.29	*1.30*	*1.30*	*2.00*	*1.30*	*1.30*	*2.00*	0.52	0.52	0.26
U_w (W m^{-2}K^{-1})	*4.90*	*4.90*	*4.90*	1.40	1.40	1.40	*4.90*	*4.90*	*4.90*	*4.90*	*4.90*	*4.90*	1.40	1.40	1.40
τ_{sol} (–)	-	-	-	-	-	-	0.57	0.55	0.55	-	-	-	0.30	0.20	0.30
$\eta_{H,c}$ (–)	*0.85*	*0.85*	*0.85*	0.995	0.995	0.995	*0.85*	*0.85*	*0.85*	0.995	0.995	0.995	0.995	0.995	0.995
EER (–)	*2.35*	*2.35*	*2.35*	*2.35*	*2.35*	*2.35*	*2.35*	*2.35*	*2.35*	*2.35*	*2.35*	*2.35*			
$\eta_{H,gn}$, COP_H (–)	*0.85*	*0.85*		*0.85*	*0.85*		*0.85*	*0.85*		1.10	1.10				
$\eta_{W,gn}$, COP_W (–)	*0.75*	*0.75*		*0.75*	*0.75*		*0.75*	*0.75*		0.95	0.95				
$\eta_{H+W,gn}$, COP_{H+W} (–)			*0.80*			*0.80*			*0.80*			1.10			
EER; COP_{H+W} (–)													2.50; 3.00	2.50; 3.00	2.50; 3.00
$W_{PV,p}$ (kW)													25	10	3

*plus improved heating system control

The user behavior is characterized by the parameters that influence the energy uses and that can be controlled by the occupants. The value of these parameters was defined in accordance with the level of occupants' expectation and the household characteristics. The European Standard EN 15251 (2007) specifies the values of the indoor environmental parameters for three categories of expectation according to the building properties (new construction, renovation, existing building) and the occupants conditions (sick, children, elderly people).
Combining the household characteristics with the typical level of expectation of each category, the user behavior's variables were identified as listed in Table 4.

Table 4: User parameters in connection with retrofit measures

Retrofit measures	*Existing building*			Thermal envelope insulation*			Solar shading devices			Thermal system*			Major renovation		
User ID	U1	U2	U3	U1	U2	U3	U1	U2	U3	U1	U2	U3	U1	U2	U3
$\theta_{H,set\ point}$ (°C)	*21*	*18*	*20*	22	20	21	21	18	20	22	20	21	22	20	21
$\theta_{H,set\ back}$ (°C)	*16*	*16*	*16*	16	16	16	16	16	16	16	16	16	16	16	16
d_H (h/d)	*14*	*13*	*14*	14	13	14	14	13	14	14	13	14	14	13	14
n (h^{-1})	*0.42*	*0.50*	*0.59*	0.50	0.59	0.70	0.42	0.50	0.59	0.50	0.59	0.70	0.50	0.59	0.70
$\theta_{C,set\ point}$ (°C)	*27*	*27*	*26*	26	26	25.5	26	26	25.5	27	27	26	26	26	25.5
d_C (h/d)	*14*	*13*	*14*	14	13	14	14	13	14	14	13	14	14	13	14
V_{DHW} (l/d unit)	*30*	*175*	*245*	30	175	245	30	175	245	50	245	438	50	245	438
*plus improved heating system control															

3.2 Calculation options and boundary conditions

The energy performance of the case studies was calculated by means of the quasi-steady-state calculation method, as specified by the Italian technical specification UNI/TS 11300, based on the standard EN ISO 13790. To determine the net energy need for space heating and cooling, it analyses the steady state balance of heat losses (transmission and ventilation) and of heat gains (solar and internal) evaluated in average monthly conditions. The dynamic effects on the net heating and cooling energy needs are considered by introducing dynamic parameters, such as the utilization factors, that accounts for the mismatch between transmission plus ventilation heat losses and solar plus internal heat gains; and an adjustment of the set point temperature for intermittent heating/cooling or set-back.
According to the calculation model, the following assumptions and simplifications were applied:

- climatic data of Milan obtained from the national technical standard UNI 10349-1 (2016),
- simplified approach to calculate internal heat gains, building internal heat capacity, temperature of unconditioned spaces, according to the assumptions defined in the Italian technical specifications UNI/TS 11300-1 (2014),
- value of shading reduction factor for external obstacles fixed at 0.8,
- simplified calculation of the mean monthly values of the technical subsystem efficiencies,
- primary energy factors fixed at 2.42 (1.95 non-renewable plus 0.47 renewable) for electricity, at 1.05 for fossil fuels (entirely non-renewable), at 1.00 for both electricity from photovoltaic system and heat energy from outdoor with heat pump (completely renewable in both cases).

3.3 Results and discussion

From Figure 2 to Figure 7 the net energy need for space heating and cooling and the non-renewable primary energy normalized by the heated floor area are shown for the considered case studies. In each figure the existing building is compared with the different retrofit options, both with and without the rebound effect.
Results show that the major renovation toward the nearly zero-energy target reach the lowest non-renewable primary energy consumption, but also that the thermal insulation of the envelope is a very efficient retrofit measure when the building stock is characterized by high U-values.
Considering the rebound effect, the net energy need for space heating decreases of about 70-75% either after the thermal envelope insulation or the building transformation into nZEB, for all the analyzed users; anyway, the energy savings would be 5-10% higher without considering a change in the user behavior after the retrofit. The replacement of the heat generator for space heating and domestic hot water causes 10-30% increase of the net energy need for

space heating, due to a different user behavior in the set-point temperature control and in the ventilation of the building. As regards the net energy need for cooling and considering the rebound effect, the results are not so uniform among the different users and the considered packages of measures: for user 1 and user 2, the thermal envelope insulation makes worse the cooling energy performance on the order of 25-40%; for user 3, the same retrofit measures provide energy savings of about 10% in the cooling period. The reduction of the net energy need for cooling would be significant (about 25-30%) after the installation of solar shading devices, but the changes in the user behavior after the retrofit frustrate it almost completely. The major renovation determines the decrease of the net energy need for cooling for all the users but especially for the user 2 and the user 3 on the order of 25-45%.
As concerns the non-renewable global primary energy, the energy savings result considerable for all the users after the thermal envelope insulation (45-70%) and the building upgrading into nZEB (80-85%), whereas they are limited after the adoption of thermal system retrofit measures singularly (10-20%). On the contrary after the installation of solar shading devices, the non-renewable global primary energy slightly increases (2%). As far as the energy services are concerned, the non-renewable primary energy for space heating is the most sensitive at the analyzed retrofit actions.

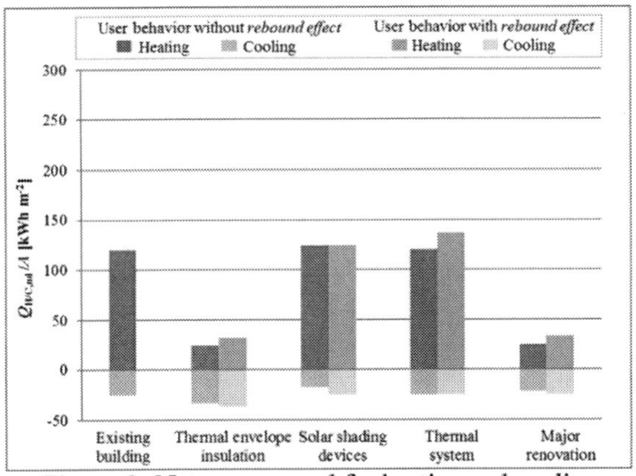

Figure 2: Net energy need for heating and cooling normalized by the heated floor area – User 1

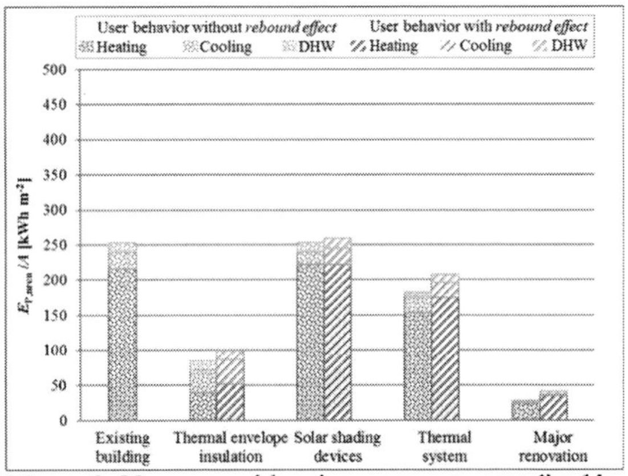

Figure 3: Non-renewable primary energy normalized by the heated floor area – User 1

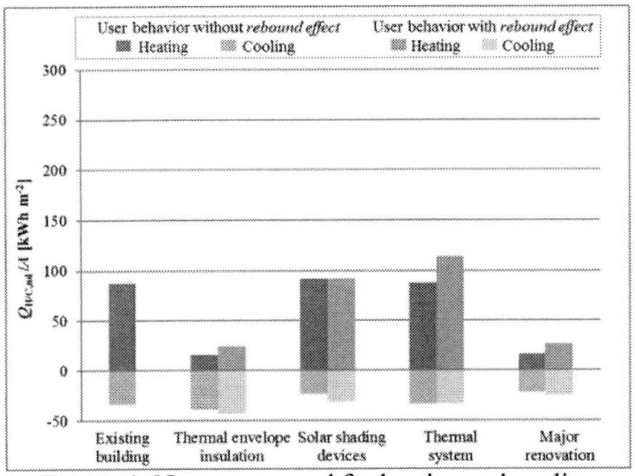

Figure 4: Net energy need for heating and cooling normalized by the heated floor area – User 2

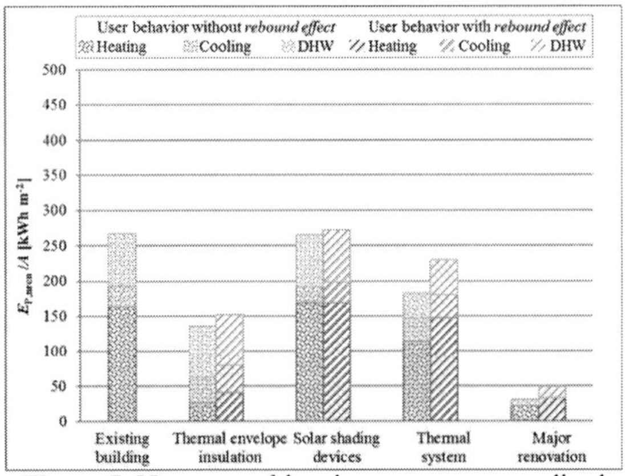

Figure 5: Non-renewable primary energy normalized by the heated floor area – User 2

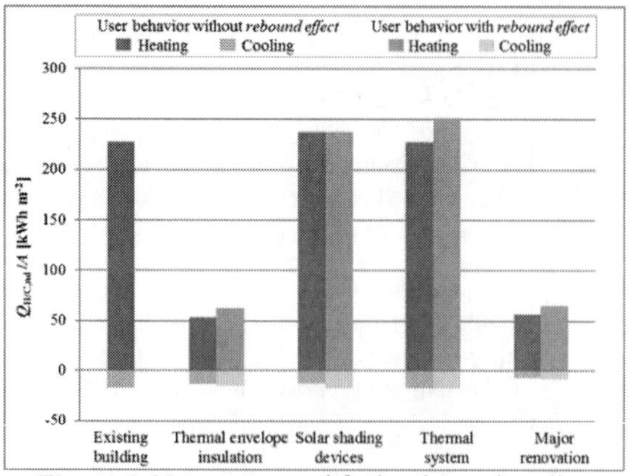

Figure 6: Net energy need for heating and cooling normalized by the heated floor area – User 3

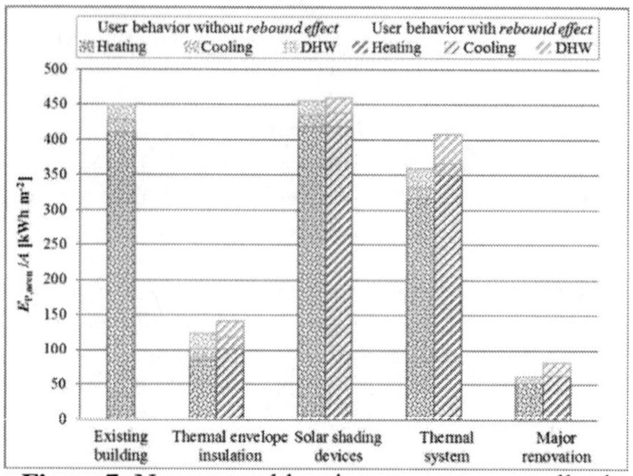

Figure 7: Non-renewable primary energy normalized by the heated floor area – User 3

The rebound effect and the rebound indicators of the case studies are presented in Table 5.
The highest values of the index $I_{RB,1}$ stress that significant energy savings only occur when a major renovation into a nearly zero-energy building is considered.
Looking to the values of the index $I_{RB,2}$, the gap between the real and the expected energy saving is higher for retrofit measures concerning the thermal systems (35-55%). On the opposite, when the energy efficiency improvements concern the thermal envelope insulation or both the envelope and the thermal systems, approximately 10% of expected energy savings are frustrated by the rebound effect.
Moreover, the single installation of solar shading devices is always inefficient: the energy savings thanks to solar shading devices are little or don't come true given that the energy savings due to lower solar gains during cooling period are frustrated by the higher heating demand due to lower solar gains during heating period. In this context, the higher expectations of the occupant after the retrofit can determine negative values of ΔEP_{RB}.

Table 5: Energy efficiency measures and options

Retrofit measures	Thermal envelope insulation*			Solar shading devices			Thermal system*			Major renovation		
User ID	U1	U2	U3	U1	U2	U3	U1	U2	U3	U1	U2	U3
RB (kWh m^{-2})	14.79	16.52	17.36	6.03	6.59	4.13	25.05	47.21	48.87	12.57	17.59	19.18
ΔEP_{RB} (kWh m^{-2})	154.16	114.89	309.51	*-5.60*	*-4.67*	*-9.43*	45.91	37.87	42.30	212.79	218.94	369.06
$I_{RB,1}$ (%)	17.5%	12.2%	14.0%	2.4%	2.5%	0.9%	13.7%	25.9%	13.6%	44.9%	57.1%	30.8%
$I_{RB,2}$ (%)	91.2%	87.4%	94.7%	*-1297.4%*	*-243.7%*	*177.8%*	64.7%	44.5%	46.4%	94.4%	92.6%	95.1%
*plus improved heating system control												

4 CONCLUSIONS

The present work aims to investigate the "rebound effect", defined as the gap between the real and the expected energy saving due to the influence of the occupant behavior on the energy consumption of retrofitted buildings.
For this purpose, three typical users and associated Italian building types were considered, and the expectations of the building occupants were modeled before and after the retrofit.
A correlation matrix among the energy efficiency measures adopted in the refurbishment process, the energy services and the user parameters affected the building energy consumption (e.g. temperature set point for heating and cooling, ventilation rate) was defined. Finally, two indexes ($I_{RB,1}$ and $I_{RB,2}$) were used in order to estimate the real energy saving due to the application of some retrofit measures, and the influence of the modified behavior of the occupants on the energy consumption after the building refurbishment.
The results show that the energy efficiency measures lead to the highest benefits when these are mutually reinforced, that is the case of the major renovation into a nZEB. In contrast, single energy efficiency measures may lead to the

opposite goal of increasing the energy consumption of the building, that is the case of the solar shading devices installation.

Moreover, the results point out that the rebound effect does exist: it especially occurs when the users modify their behavior in favor of a higher thermal comfort as a result of the thermal system renovation (35-55%), while in case of the major renovation of the building toward the nearly zero-energy target, the rebound effect decreases to the 10%.

NOMENCLATURE

A	area	(m^2)
COP	coefficient of performance	(–)
E	energy	(kWh)
EER	energy efficiency ratio	(–)
EP	energy performance	(kWh m^{-2})
g	total solar energy transmittance	(–)
HDD	heating degree days	(°C d))
n	ventilation rate	(h^{-1})
I	indicator	(-)
Q	quantity of heat	(kWh)
RB	absolute rebound effect	(kWh m^{-2})
U	thermal transmittance	(W $m^{-2}K^{-1}$)
V	volume	(m^3)
W	power	(kW)
Δ	variation	(–)
η	efficiency	(–)

Subscript

C	cooling
c	control (subsystem)
d	distribution (subsystem)
e	emission (subsystem)
env	envelope
EX	existing building
f	floor
fl	slab
g	gross
gl	glass
gn	generation (subsystem)
H	heating
lw	lower
n	normal
nd	need (energy)
nren	non-renewable (energy)
P	primary (energy)
p	peak (PV system)
PV	photovoltaic (system)
RB	rebound effect
REF	refurbished
sh	solar shading
u	unheated
up	upper
W	domestic hot water
w	window
wl	wall

ACKNOWLEDGEMENT

This work is based on the results of the research project financed by RSE (Research on Energy System) concerning the "Study of the elasticity of the buildings' energy consumptions".

REFERENCES

1-Article from a periodical

Ballarini, I., Corgnati S.P., Corrado. V. (2014). Use of reference buildings to assess the energy saving potentials of the residential building stock: The experience of TABULA project. *Energy Policy, 68, 273-284.*

Haas, R., Biermayr, P. (2000). The rebound effect for space heating. Empirical evidence from Austria. *Energy Policy, 28*(6-7), 403–410.

Galvin, R. (2014). Making the 'rebound effect' more useful for performance evaluation of thermal retrofits of existing homes: Defining the 'energy savings deficit' and the 'energy performance gap'. *Energy and Buildings, 69*, 515–524.

Greening, L., Greene, L., Difiglio, C. (2000). Energy efficiency and consumption - the rebound effect - a survey, *Energy Policy, 28*(6-7), 389–401.

Guerra Santin, O. (2011). Behavioural Patterns and User Profiles related to energy consumption for heating. *Energy and Buildings, 43*(10), 2662–2672.

Guerra Santin, O., Itard, L., Visscher, H. (2009). The effect of occupancy and building characteristics on energy use for space and water heating in Dutch residential stock. *Energy and Buildings, 41*(11), 1223–1232.

Herring, H., Roy, R. (2007). Technological innovation, energy efficient design and the rebound effect. *Technovation, 27*(4), 194–203.

Oreszczyn, T., Hong, S., Ridley, I., Wilkinson, P. (2006). Determinants of winter indoor temperature in low income households in England. *Energy and Buildings, 38*(3), 245–252.

Sorrell, S., Dimitropoulos, J. (2008). The rebound effect: Microeconomic definitions, limitations and extensions. *Ecological Economics, 65*(3), 636–649.

Sunikka-Blank, M., Galvin, R. (2012). Introducing the prebound effect: the gap between performance and actual energy consumption, *Building Research & Information, 40*(3), 260–273.

Tronchin L., Fabbri K. (2008). Energy performance building evaluation in Mediterranean countries: Comparison between software simulations and operating rating simulation. *Energy and Buildings, 40*(7), 1176–1187.

2-Book

Galvin, R. (2015). *The rebound effect in home heating: A guide for policymakers and practitioners*. London: Routledge.

3-Legislation

European Committee for Standardization (CEN). (2007). EN 15251. Indoor environmental input parameters for design and assessment of energy performance of buildings addressing indoor air quality, thermal environment, lighting and acoustics.

European Committee for Standardization (CEN). (2008). EN ISO 13790. Energy performance of buildings - Calculation of energy use for space heating and cooling.

European Union. (2010). Directive 2010/31/EU of the European Parliament and of the Council of 19 May 2010 on the energy performance of buildings (recast). *Official Journal of the European Union; L153/13*(18 June 2010).

Italian Ministry of Economic Development. (2015). Ministerial Decree of 26 June 2015. Application of energy performance calculation methods and definition of minimum requirements. *Official Journal of Italian Republic, 162, Ordinary Addition 39*(15 July 2015).

Italian Organization for Standardization (UNI). (2016). UNI 10349-1. Building space heating and cooling – Climatic data.

Italian Organization for Standardization (UNI). (2010-2014). UNI/TS 11300 (series). Energy performance of buildings.

Evacuation Hazards in Crowded Subway Stations

Mei-ki HO[1], Chung-yee KU[2], Wan-ki CHOW[3]*

[1] Research Centre for Fire Engineering, Department of Building Services Engineering,
The Hong Kong Polytechnic University, Hong Kong, China
Email: 13018484d@connect.polyu.hk

[2] Research Centre for Fire Engineering, Department of Building Services Engineering,
The Hong Kong Polytechnic University, Hong Kong, China
Email: 09902325r@connect.polyu.hk

[3] Research Centre for Fire Engineering, Department of Building Services Engineering,
The Hong Kong Polytechnic University, Hong Kong, China
Phone: +852 2766 5843; Fax: +852 2765 7198, Email: bewkchow@polyu.edu.hk; beelize@polyu.edu.hk

* Corresponding Author

ABSTRACT

Evacuation hazards of crowded subway stations in the Asia-Oceania region were studied in this presentation. Two examples in Hong Kong, with one being an interchange station and a deep underground station were used. The evacuation time was estimated by using hydraulic models. Different scenarios were assigned on off peak train time and rush hour. Different exit widths and blockage of some exit routes were assumed. Fire safety management is identified to be a key part in keeping evacuation time short.

1. INTRODUCTION

Underground subway system is the key transportation means in dense urban areas in the Asian-Oceania region. Subway stations are crowded with passengers waiting observed to squeeze into the carriages during rush hours (Lee, 2014; Cheung, 2014). As reported in Hong Kong (Lee, 2014), after following the change in maximum capacity from six passengers per meter square to four, the capacity is only 70% full at rush hour. However, the capacity can be over 90% of full loading as observed under emergency or suspension of operation due to whatever reasons (Chow et al., 2011). Subway stations become more crowded with an average weekday patronage of nearly 5.3 million passengers (Cheung, 2014).

Subway stations are mostly located in the basement or ground levels connecting the shopping mall, commercial or residential building in downtown areas. The occupancy density of passengers can be much higher than expected during special events (The Sun, 2016; Apple Daily, 2014) such as fireworks. Therefore, evacuation time in emergency situation will be prolonged. Putting in platform screen doors (Qu and Chow, 2012) as in Figure 1 made the evacuation more difficult. To have better understanding of the safety issue in the subway station, evacuation time in the emergency situation will be studied in this paper.

Two subway stations in Hong Kong, Station I and Station II are selected to study the evacuation hazard in crowded stations when a fire occurs. Station I is an interchange between two railway lines, being one of the most crowded stations with high occupancy density. Station II is the first station (Tam, 2014) in the local rail network to feature a special design - "Lift-only Entrances". This is a deep underground station which lies under 70 m of ground level and, the passengers have to be evacuated by lift. The occupancy density in Station II is built relatively much lower than Station I under normal conditions at the moment, though the subway stations were observed to be very crowded whenever there were delays on train services (Chow et al., 2011), such as signal failure.

(a) Empty (b) Rush hour (c) Very crowded

Figure 1: Crowded subway station with platform screen door

The evacuation of Station I and Station II are studied with three scenarios. Evacuation time in these three different scenarios will be considered by using hydraulic model calculation (Proulx, 2002; Nelson and Mowrer, 2002) in this presentation. Moreover, the special evacuation feature of "Lift-only Entrances" in Station II (Tam 2014) and the fire safety management strategies for emergency evacuation are also discussed.

2. LOCAL EVACUATION REQUIREMENTS

The Stations in railway systems are considered to have a primary purpose for passenger transit as specified in the local codes (Buildings Department, 2011; Fire Services Department, 2011, 2013). Occupants normally stay within a building for a period of time not longer than that necessary to wait for and board a departing vehicle/ship/aircraft or exit the terminal after arrival in an incoming vehicle/ship/aircraft.

The subway station population should be complied with local codes (Buildings Department, 2011; Fire Services Department, 2011,2013). The use classification for subway station is 5c (Transport facilities like passenger terminals, railway station), the occupancy factor is based on actual design and layout. Therefore, the layout of the two stations should follow requirement.

3. EVACUATION TIME

Considering the evacuation effectiveness from crowded station and also the evacuation time when the fire is occurred, the sequence of occupant response to fire should be known as reviewed with a time chart summarized before (Ng and Chow, 2006). The movement time is only a small part of the evacuation time as passengers are not expected to react effectively in emergency situation without appropriate notification by fire safety management (Proulx, 2002). The hydraulic model calculation can be applied to study egress time under different population densities.

The evacuation time with various passenger loadings can be calculated by hydraulic model (Nelson and Mowrer, 2002). The total evacuation time (TET) is calculated by human response time t_{resp}, travel time t_{tra} and waiting time t_{wait}:

$$\text{TET} = t_{resp} + t_{tra} + t_{wait} \tag{1}$$

Time delay (Proulx et al., 1996; Proulx and Fahy, 1997) to start evacuation is a concern. The waiting time (Ng and Chow, 2006) should be watched in the TET as some passenger may be jammed when all the passengers resist at the same time. It is impossible for passengers or small groups of passengers to move under high population density. The waiting time for passengers to escape will be extended. However, there is basically no systematic study with clinical psychology on human behavior in Hong Kong (Chow, 2012, 2015). Therefore, values adopted elsewhere have to be used.

The speed along the line of travel S (in m/s) can be expressed in terms of density D (in persons per unit area) and a constant k (Proulx, 2002):

$$\text{S} = \text{k} - \text{akD} \tag{2}$$

The actual flow rate F_c can be expressed in terms of specific flow rate F_s (in person/s/effective width) and effective width W_e (in m) as:

$$F_c = F_s W_e = SDW_e \tag{3}$$

The total egress time t_p can be expressed in terms of population P (in number of persons):

$$t_p = \frac{P}{F_c} \tag{4}$$

4. IDENTIFIED SCENARIOS TO STUDY

Three identified scenarios will be studied in each station:
Scenario 1: Assuming that the passengers are evenly distributed in different exits in emergency situation. All the possible factors such as passenger behaviors and conditions are eliminated.
Scenario 2: Passengers have higher tendency to evacuate at the larger exit. This is one of the passenger behaviors in emergency situation. Therefore, the passenger distribution depending on exit width will be studied.
Scenario 3: Assuming that some of the exit routes were blocked.

5. INTERCHANGE SUBWAY STATION : STATION I

The layout of the Station I is shown in Figure 2a. There are nine exits in the Station I and all the exits are located at the concourse. There are two major concourses which the locations are circled in Figure 2a. The door width and evacuation distance of different exits are measured in order to calculate the evacuation effectiveness. The concourse area in station I is 1800 m^2 (Buildings Department, 2011) and 5 passengers were observed for 1.53 m^2 in peak hour as in Figure 2b.
Associated key data in studying evacuation about the passenger density in concourse is 0.31m^2 per passenger. Total passenger in the Station I concourse will be 1800/ 0.31 or 5800 persons.

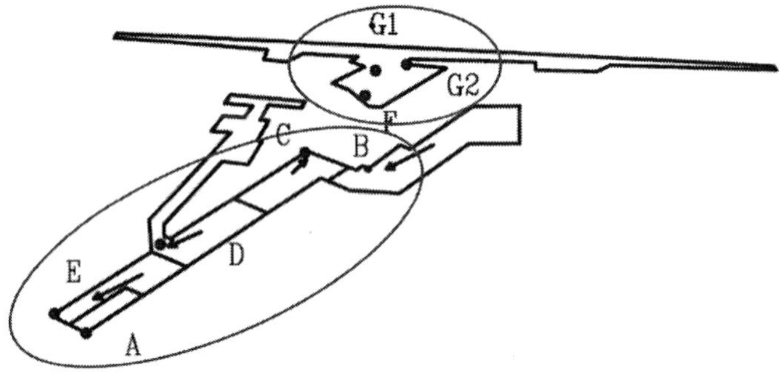

(a) Geometry

(b) Passenger Density

Figure 2: Station I

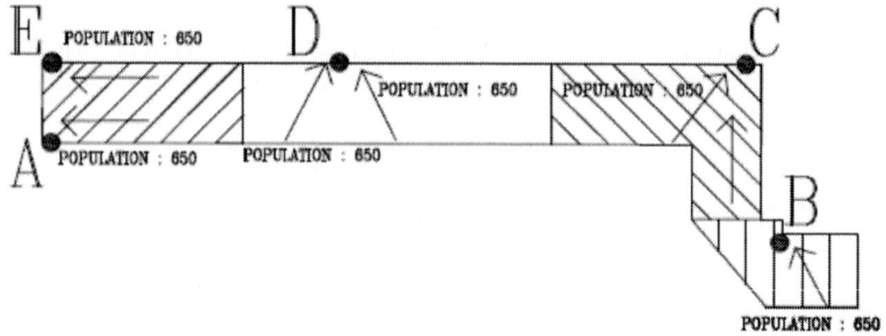

(a) Concourse for scenarios 1 and 2

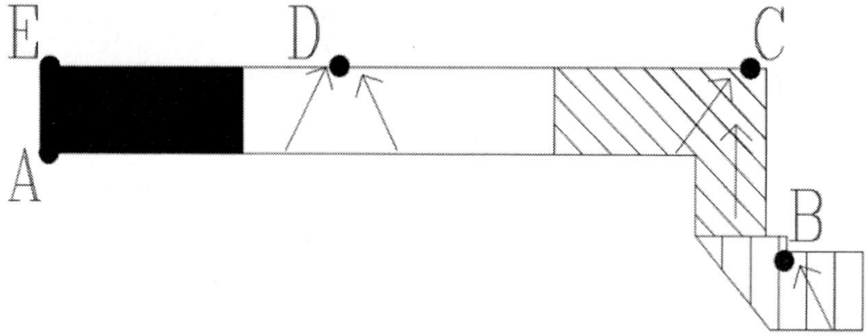

(b) Concourse for scenario 3

Figure 3: Passenger flow in Station I

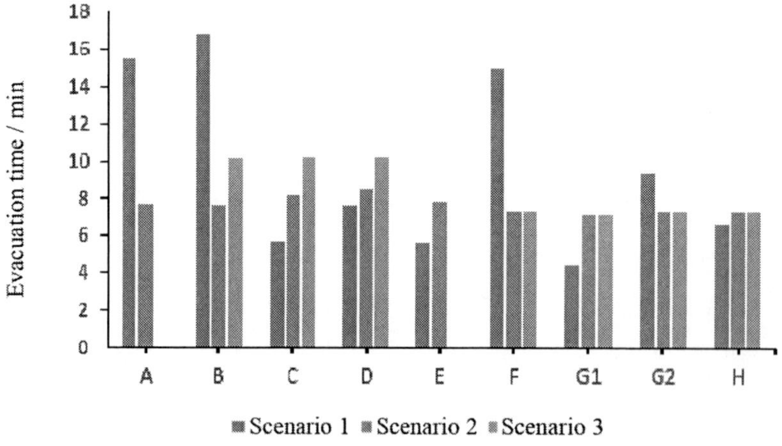

Figure 4: TET in different scenarios for Station I

For scenario I:
The population density in each exit is assumed to be the same, and thus, there are 645 persons (5800/9, i.e. 11%) in each exit. The passenger flow in different exits from one of the concourse is shown in Figure 3a. The evacuation time for scenario I is calculated by using the hydraulic model with results shown in Figure 4.
For scenario 2:
The population density in each exit depends on exit width, the passengers in the basement using Exit A, B, C, D and E. The passengers on the ground floor will be using Exit F, G1, G2 and H. The passenger flow is same as scenario 1 and is shown in Figure 3a, but the density in each exit is different. In order to calculate the evacuation time for scenario 2, the effective width for different exits should be measured and shown in Table 1. The result of evacuation time for scenario 2 is shown in Figure 4.
For scenario 3:
Assuming the exit A and E in the main concourse are blocked. Therefore, only exit B, C and D can be used. In this situation, the number of passenger assuming to evacuate in each exit will depend on exit width and the passenger flow is shown in Figure 3b. The result of evacuation time for scenario C is shown in Figure 4.
From the Figure 4, it shows that the TET in scenario 1 in different exits was not evenly distributed. The TET in exit G1 and B is 4.4 minutes and 16.8 minutes respectively which the difference is very large. That was not reliable as TET in different exits should be similar as the passenger would try their best to evacuate as fast as they could. Thus, they would find the nearest exit or the less crowed exit to evacuate. The result for the scenario 2 was more reliable as the TET was depending on the exit width, and passengers would choose the exit to evacuate immediately, and thus, the TET in different exit should be similar. For the scenario C, the exit A and E are blocked in the main concourse, thus, the evacuation time in exit B, C and D would be longer.

Table 1: Effective width for different exit in Station I

Exit	A	B	C	D	E	F	G1	G2	H
Effective width (m)	1.8	1.65	5.68	4.25	5.42	1.8	6.2	2.9	4.12

6. DEEP UNDERGROUND SUBWAY STATION : STATION II

The layout of station II is studied with schematic shown in Figure 5. There are five exits and one concourse in station II which the concourse location is circled. The passenger density in station II is different from station I as the density should be built much lower than the station I. The station II is not an interchange station and the concourse area is much smaller than station I, and thus, the passenger density is assumed to be 600 only.
For scenario 1:
The population density in each exit is assumed to be the same, and thus, there are 600/3 = 200 persons in each exit. The evacuation time for scenario A is calculated by using hydraulic model with results shown in Figure 6.
For scenario 2:
Same as the scenario 1, as the all the exit width in station II are the same. Thus, the evacuation time is the same as scenario 1.
For scenario 3:
Assuming the exit A in the main concourse is blocked. Therefore, only exit B and C can be used. The result of evacuation time for scenario 3 is shown in Figure 6.
For the passengers who evacuate at exit A, the TET calculated from the Figure 6 is the evacuation time from the exit A to the lift lobby. During emergency situation, the lift will stop serving the lift lobby and the fire curtains installed in front of the lift door will automatically close in order to stop the smoke and fire spread into the lift shaft. The passengers then will be directed to the refuge floor through the emergency exit in Station II. The passengers will wait for the lift in the pressurized refuge lift lobby. The fire resistant material is used in the refuge area to ensure the passengers are safe in the refuge floor.
From the Figure 6, it shows that the TET for all exits in scenarios 1 and 2 are similar, except exit C2. The reason why the TET in C2 is much longer than the other exit as the length of exit route is nearly 320 m, and thus, the TET is double of C1. Although the TET in exit C2 is much longer than the other, the passengers can egress safely as the exit route is pressurized. For the scenario 3, Figure 6 shows that when the exit A is blocked, the TET for the other

exits slightly increases. It is because the population density is not very high in station II, therefore, when one exit is being blocked, the TET of the other exit will not increase sharply.

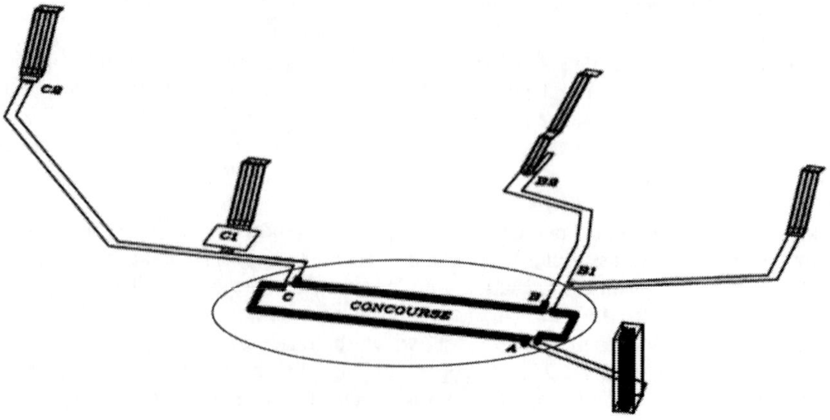

Figure 5: Exits location in Station II

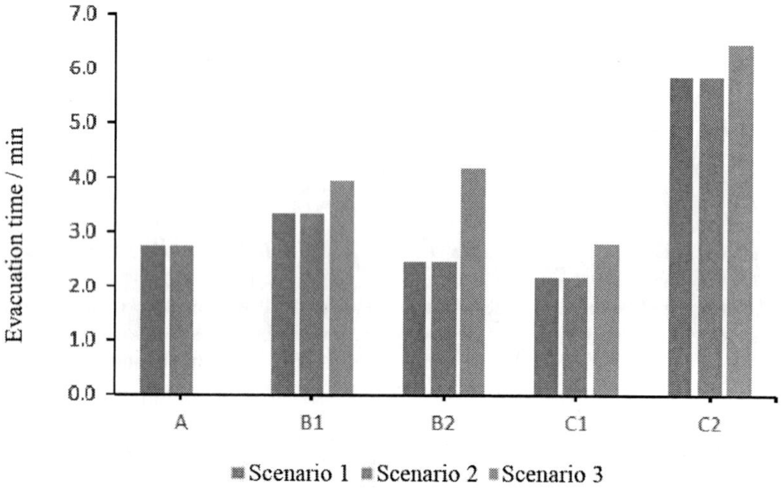

Figure 6: TET in different scenarios for Station II

7. FIRE SAFETY MANAGEMENT

Comparing the TET in station I and station II, the TET in station I is relatively longer than station II as the number of passengers in station I is very high at peak hour. Based on the result, appropriate fire safety management (Malhotra, 1987; Lui and Chow, 2000) should be worked out properly in station I.

For example, crowd control measures during the peak hour should be provided, closing one third to half of the ticket gates can avoid having excessive number of passengers getting into the concourse and the platform.

From the TET for scenario 1 of station I in Figure 3b, the population density in exit A, B and F is relatively higher than the other exit. Management staff should be allocated to guide the passengers at the peak hour.

Besides, the fire alarm signal is the basic warning signal in subway station. However, many passengers would misunderstand the signal to be the false alarms and the passenger response time will be longer.

In order to have better awareness of passengers safety and improvement on the response time in evacuation, public address system should be designed properly in both stations I and II. Passengers can take the immediate action when live broadcasting message is provided with the emergency information. Real-time broadcasting message should be clear, simple and no delay. Moreover, clear instruction that could direct the passengers to a specific exit route should be provided. Such communication method can provide reliable information, and thus, the passengers are more alerted with the live broadcasting message as it can reflect whether the situation is emergency or not.

In the staff training plan under the fire safety management scheme, staff must be trained to take appropriate and immediate actions for crowd control under emergency situation. Fire hazard scenarios should be identified through the use of fire dynamics evaluation. Management staff are trained to be familiar with all the exit routes in the subway station. Besides, they are expected to arrange to announce broadcasting message through the public address system in the first instance during the emergency situation so that passengers can be evacuated in appropriate time.

8. CONCLUSIONS

From the above study on evacuation times of two railway stations I and II in Hong Kong, it is concluded that the evacuation time depends on different factors, such as population, movement speed and exit width. Behaviour of the passengers would affect the evacuation time as observed in scenario 2. Therefore, appropriate fire management system (Malhotra, 1987; Lui and Chow, 2000) should be worked out for crowd control.

REFERENCES

Apple Daily (2014). Over 80 management staff under emergency operation of subway system. 5 September 2014. http://hk.apple.nextmedia.com/news/art/20140905/18856576

Buildings Department (2011). *Code of Practice for Fire Safety in Buildings,* pp. 26-35, Hong Kong Special Administrative Region.

Cheung, M. (2014). *MTR Corporation Limited,* pp. 1-5.

Chow, W.K. (2012). Experience on implementing performance-based design in Hong Kong. Welcome speech, *The 9th Asia-Oceania Symposium on Fire Science and Technology*, 17-20 October 2012, Hefei, China.

Chow, W.K. (2015). Performance-based approach to determining fire safety provisions for buildings in the Asia-Oceania regions. *Building and Environment, 91*, 127-137.

Chow, W.K., Qu, L. & Pang, E.C.L. (2011). Incidents on fire and ventilation provision in subway systems in Hong Kong. *International Journal on Engineering Performance-Based Fire Codes, 10*(3), 41-47.

Fire Services Department (2011). *Code of Practice for Minimum Fire Installations and Equipment and Inspection, Testing and Maintenance of Installations and Equipment.* Hong Kong Special Administrative Region.

Fire Services Department (2013). *Guidelines on Formulation of Fire Safety Requirements for New Railway Infrastructures.* Hong Kong Special Administrative Region.

Lee, A. (2014). MTR passengers say 'ease squeeze with early-bird fares'. *South China Morning Post,* 27 February 2014.

Lui. G.C.H., & Chow, W.K. (2000). A demonstration on working out fire safety management schemes for existing karaoke establishments in Hong Kong. *International Journal on Engineering Performance-Based Fire Codes, 2*(3), 104-123.

Malhotra, H.L. (1987). Fire safety in buildings. Building Research Establishment Report, Department of the Environment, Building Research Establishment, Fire Research Station, Borehamwood, Herts, WD6 2BL, UK.

Nelson, H.E., & Mowrer, F.W. (2002). Emergency movement, In: DiNenno, P.J., Drysdale, D., Beyler, C.L., Walton, W.D., Custer, R.L.P., Hall, J.R., & Watts, J.M. (Eds.) *SFPE Handbook of Fire Protection Engineering,* 3rd edition, Section 3/Chapter 14, pp. 375-380. Quincy, Mass.: National Fire Protection Association; Bethesda, Md.: Society of Fire Protection Engineers.

Ng, C.M.Y., & Chow, W.K. (2006). A brief review on the time line concept in evacuation. *International Journal on Architectural Science, 7*(1), 1-13.

Proulx, G. (2002). Movement of people: The evacuation timing. In: DiNenno, P.J., Drysdale, D., Beyler, C.L., Walton, W.D., Custer, R.L.P., Hall, J.R., & Watts, J.M. (Eds.) *SFPE Handbook of Fire Protection Engineering,* 3rd edition, Chapter 13, pp. 342-363. Quincy, Mass.: National Fire Protection Association; Bethesda, Md.: Society of Fire Protection Engineers.

Proulx, G., Fahy R.F. (1997). The time delay to start evacuation: Review of fire case studies. In: *Proceedings of the Fifth International Symposium on Fire Safety Science, Melbourne, Australia* (783-794). Boston, MA: International Association for Fire Safety Science.

Proulx, G., Kaufman, A., & Pineau, J. (1996). *Evacuation Time and Movement in Office Building*, Chapter 4.1 (Occupant Behavior). Internal Report No. 711, NRC-CNRC, National Research Council Canada, Ottawa.
Qu, L. & Chow, W.K. (2012). Platform screen doors on emergency evacuation in underground railway stations. *Tunnelling and Underground Space Technology*, *30*, 1-9.
Tam, A. (2014). Engineering for fire safety on a modern railway network. *Hong Kong Engineer – The Journal of the Hong Kong Institution of Engineers*, *42*(3), 8-12.
The Sun (2016). Fireworks at Hong Kong. 10 February 2016. http://the-sun.on.cc/cnt/news/20160210/00407_075.html

ACKNOWLEDGEMENT

The work described in this paper was supported by a grant from the Research Grants Council of the Hong Kong Special Administrative Region, China for the Theme-Based Research Scheme Project "Safety, Reliability, and Disruption Management of High Speed Rail and Metro Systems" (Project Number: T32-101/15-R).

PurduePaper2016a

Performance Evaluation of a HP/ORC (Heat Pump/Organic Rankine Cycle) System with Optimal Control of Sensible Thermal Storage

Carolina Carmo[1*], Mads P.Nielsen[1], Brian Elmegaard[2], Olivier Dumont[3]

[1] Aalborg Univ., Department of Energy Tech.,
Aalborg, Denmark
cca@et.aau.dk, mpn@et.aau.dk

[2] DTU, Department Mechanical Eng.,
Lyngby, Denmark
be@mek.dtu.dk

[3] Univ. Liège, Thermodynamics Lab.,
Liège, Belgium
olivier.dumont@ulg.ac.be

* Corresponding Author

ABSTRACT

In energy systems with high share of renewable energy sources, like wind and solar power, it is paramount to deal with their intrinsic variability. The interaction between electric and thermal energy (heating and cooling) demands represent a potential area for balancing supply and demand that could come to contribute to the integration of intermittent renewables. This paper describes an innovative concept that consists of the addition of an Organic Rankine Cycle (ORC) to a combined solar system coupled to a ground-source heat pump (HP) in a single-family building. The ORC enables the use of solar energy in periods of no thermal energy demand and reverses the heat pump cycle to supply electrical power. A dynamic model based on empirical data of this system is used to determine the annual performance. Furthermore, this work assesses the benefits of different control strategies that address the critical concern that thermal supply should always operate considering the actual load conditions in the overall system performance. The performance is assessed in terms of energy efficiency of the system, the integration of renewable energy in the energy supply of the system and the level of comfort of the users. Results show that real load control logic can lessen the adverse effects of cycling in the compressor of the system as well as increase the thermal demand (up to 33%) and the electrical demand (max. 8.4%) covered by renewable energy (solar). However, the extension of these improvements is highly dependent on the thermal mass of the building and the volume of the sensible storage.

1. INTRODUCTION

In the future energy systems scenery, with high share of intermittent renewable sources, both energy efficiency and the time of energy use are considered imperative to ensure security of supply. In order to achieve these objectives, new control strategies have been assessed over the last couple of decades to operate HVAC systems in buildings, as they represent 35% of the world's energy use (IEA, 2013).

Recently, Shaikh *et al.* (2014) presented a comprehensive and extensive review on the state-of-the-art of, so-called, *intelligent control systems* in residential, commercial and office buildings. Perspectives for the future of building automation were given based on 21 different controllers, which objective parameters include running cost reduction, increase of flexibility and thermal comfort represented by humidity, air quality and indoor temperature. Overall, it is considered that buildings occupants' behaviour, activities and preferences are the most important feedbacks for the success of building automation.

In the same study it is shown that Model-based Predictive Control (MPC), multi-agent system technology (MAST), Fuzzy and ON/OFF control methods are the most studied and promising approaches reaching up to 40% energy savings. However, the efforts needed to implement MPC, MAST and Fuzzy - such as lengthy model development and challenges to obtain reliable information for each case- are significant and often considered prohibitive. Yet, other studies have proven it is possible to achieve comparable energy savings with controllers based on traditional reactive control (ON/OFF) methods if enhanced by adaptive setbacks using occupancy presence and domestic hot water consumption (DHW) information (Tammmaro et al., 2016; Rosiek et al., 2013). In practice, this information is easily accessed through real-time occupancy sensors.

In this work, we intend to study the effects of enhanced reactive controllers in a residential building equipped with an innovative co-generation technology based on solar energy, known as HP/ORC unit. The novel control logics intend to minimize the total energy use by optimizing the energy use schedules according to fluctuating occupant presence while maintaining the comfort at suitable levels. The three novel operation strategies investigated are described below:

1. **Occupancy setback (OccSB):** this control minds occupancy information to adapt the indoor comfort temperatures from 20 °C, when occupied to 16.5 °C, when non-occupied.
2. **DHW priority (DHWp):** in this case the space heating distribution pump is turned off when the HP/ORC unit is supplying DHW. This control logic intends to ensure higher energy efficiency by minimizing the mixing inside the storage water tank when it delivers high temperature water for domestic end-uses.
3. **Combined Occupancy setback and DHW priority (OccSB+DHWp):** this variant terminates the space heating supply both when people are not at home or in case they are taking a shower.

These operation strategies are compared with an unbounded control strategy based exclusively on the variation of temperature in a certain height of the tank (0.6 m). A deadband (db) of 20K is used, so that the unit turns OFF when the temperature in tank at 0.6 m exceeds a T_{set} 60°C and then back ON when the temperature is equal to T_{set} - db. This large *db* is used to ensure the hot water is delivered at suitable temperatures to the user ($T_{shower} \geq 38°C$)(Kordana et al., 2014) adjusted by a tempering valve and to protect the compressor from adverse cycling effects (Green, 2012).

2. METHODS

The annual performance of a simulated ground source based HP/ORC system integrated in a house subject to four different control strategies - previously introduced - is assessed. Conclusions are made based on a comparison of the annual performance between the new control logics and the reference case. All the simulation results presented were performed in Modelica with simulation time steps of maximum 5 minutes, using the boundary conditions for a passive house with 2 residents under Danish climate conditions. The space heating load and the domestic hot water are summarized in table 1 for the reference scenario (simulations results).

In Figure 1 a simplified hydraulic scheme of the simulated system is shown. The basic principle of this innovative concept is the modification of the compressor to operate as an ORC (Dumont et al., 2015). Combined with a thermal solar collector the system supplies a single-family residential building. The HP and direct heating from the solar collector (DH) modes cover the thermal demand. The ORC enables the use of solar energy in periods of no thermal energy demand and reverses the heat pump cycle to supply electrical power.

Figure 2 illustrates the reference control strategy that is then enhanced with the three different control logics. T_{roof} is the roof temperature in the outlet, T_{sto} is the hot water tank control temperature at height 0.6m – middle of the tank -,$T_{sto,low}$ is the lower limit of temperature allow in the tank (40°C), $T_{sto,high}$ is the higher limit of temperature allow in the tank (60°C), $W_{ORC,min}$ is the lower limit of the power to start the ORC (2 kW) Note: $W_{ORC,min}$ is used to avoid chattering issues during simulation. A value lower than 2 kW induces a large number of mode changes. If the storage control temperature (T_{sto}) is below a fixed low temperature threshold ($T_{sto,low}$) and the temperature in the roof solar collector (T_{roof}) , the heat pump mode is activated to guarantee the space heating (SH) and the domestic hot water (DHW). The

Table 1: Space heating (SH), domestic hot water(DHW) and light and appliances(L & App) demand for reference scenario

SH [kWh]	DHW[kWh]	L & App[kWh]
3951	845	1570

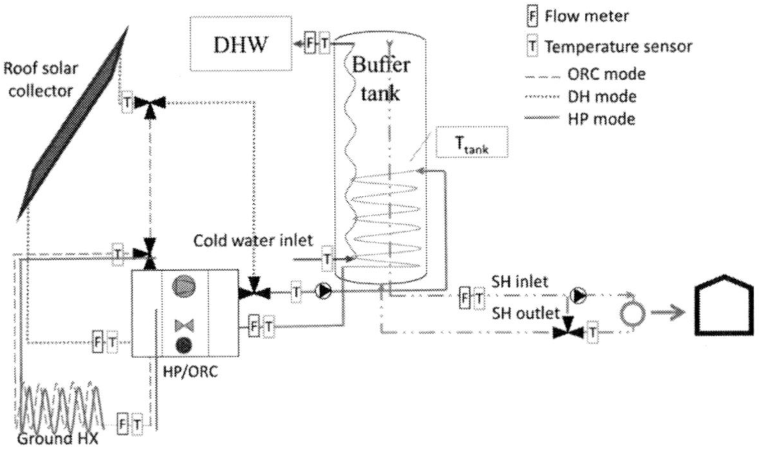

Figure 1: Simplified hydraulic schematic of the HP/ORC system

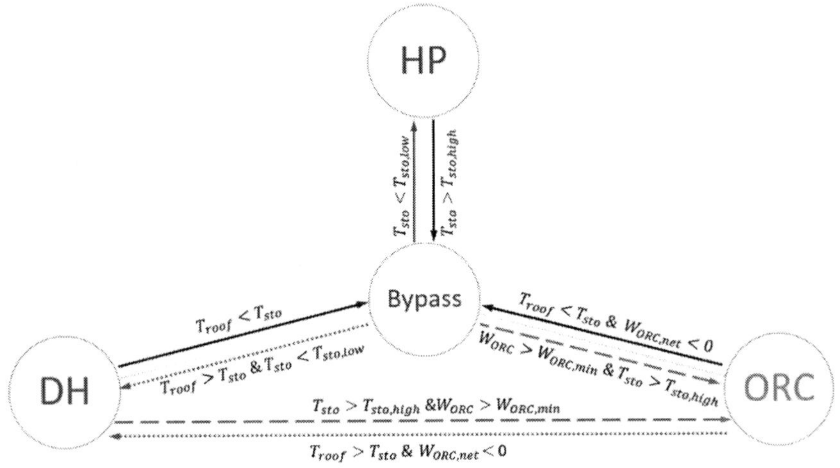

Figure 2: State diagram representing the reference case control of the HP/ORC unit.

heat pump cold source is the ground heat exchanger or the solar roof depending on which one is the warmest. If the storage control temperature is above the high temperature threshold ($T_{sto,high}$), the ORC mode is activated if the roof temperature is above the minimum power to get a net electrical production ($W_{ORC,min}$). If the storage control temperature lies between the high and low threshold, the former mode is kept to avoid excessive chattering between modes reducing performance and reliability in practice. Finally, the by-pass mode is used if the storage temperature cannot be increased and if the ORC mode is not able to produce a positive net electricity production. In the by-pass mode, only the roof pump is running to homogenize the roof temperature. This control strategy is relatively simple.

2.1 Residential buildings and users behaviour

The building model is based on the geometry and the construction characteristics of a real building 140m^2 located in Herning, Denmark.

The building model is split into 5 zones, according to its orientation, air volume and use: dining room, kitchen, main bedroom, bathroom and guest rooms. The boundary conditions of each zone are given by solar irradiance, ambient temperature, heat exchange with adjacent zones, lighting and appliances load, and occupancy state.

The walls and roof of each zone are modelled using a simplified lumped-capacitance methods introduced in Masy (2007). They are modelled with two thermal resistances and heat capacity. The simplicity of this model is appropriate for annual performance assessment. The air volume and floor slab of each zone is modelled with basic components from Buildings library (M. Wetter, 2014). Due to the negligible thermal inertia of the windows, they are considered as pure thermal resistances. These components are standard models available in the basic Modelica library. An extra heat capacity is included to model the thermal inertia of the furniture as well as dissimilarity in the indoor temperature of each zone. Furthermore, to control the solar heat absorbed through the windows an artificial shadow factor is included to resemble the effect of louvers.

Finally, the users' load profile corresponds to hourly lighting and appliances and the DHW consumption according to probability density functions of the time of energy use of lighting and appliances in 1300 households across Europe and their specific energy use in EU-countries (Georges et al., 2013).

2.2 Ground source HP/ORC system

The energy supply system consists of a modified scroll compressor ground source heat pump with a thermal capacity of 13 kW_{th}, which in reversed mode operates as an ORC (5.3 kW_{el}). The different components of the HP/ORC unit are simulated with parametric curves based on an experimental campaign at different operational conditions (Dumont et al., 2015).

The model of the ground source horizontal heat exchanger consists in a 1D finite-volume model with three layers. Each heat exchanger layer is modelled with a heat capacity and two thermal resistances. These represent the pipe and the soil thermal resistance, respectively.

2.3 Solar collector

The building is equipped with a thermal collector incorporated in the whole roof (138.8 m^2) which is and is modelled using Klein's equation (Klein, 1975).

2.4 Stratified storage tank

The water tank (250 *L*) consists of a stainless steel cylinder with two built-in spiral heat exchangers (HXs), 3 *L* and 9.6 *L* respectively – one going from mid-height to bottom of the tank and another going from bottom to the top of the tank. When charging the store, the working fluid in the HP/ORC unit is circulated through the mid-height helical heat exchanger. During discharging, the cold water from the grid is circulated through the all-through heat exchanger to supply DHW. Furthermore, a direct inlet (bottom) and outlet (top) to the tank form part of the space heating loop. In periods of space heating demand the water from the house floor slab enters the inlet and supplies hot water from the outlet in the top.

In the current work, this stratified sensible thermal storage is modelled by a one-dimensional finite-volume discretization comprising i=20 isothermal segments with equal volume. The dynamic temperature profile of the tank is represented by a set of i ordinary differential equations that approximately represent the energy balance of the tank:

$$A_i \Delta x_{seg} \rho_i c_p \frac{dT_i}{dt} = \dot{m}(h_{ex,i} - h_{su,i}) + A_{hx,unit,i} \dot{q}_{hx,unit} + A_{hx,DHW,i} \dot{q}_{hx,DHW} + \alpha_i A_{i+1} \dot{q}_{i+1} + \beta_i A_{i-1} \dot{q}_{i-1} - UA_{amb,i}(T_i - T_{amb}) \quad (1)$$

where

$$\dot{q}_{hx,unit,i} = U_{hx}(T_{i,hx} - T_i) \quad (2)$$

and

$$\dot{q}_{i-1} = \sigma \frac{(T_{i-1} - T_i)}{\Delta x_{seg}} \quad (3)$$

In Eq. (1) α is 0 if the i-th node is the top of the tank and 1 otherwise and β is 0 if the i-th node is the bottom node and 1 otherwise. In addition, the model (Eq. 3) includes the mixing parameter (σ, with units of thermal conductivity) in order to represent the combined effect of the three modes of heat transfer that can occur between tank nodes (conduction, diffusion or axial mixing due to turbulent flow) as well as the mixing that occurs due to temperature inversion. This parameter is issued whenever there is temperature inversion in the tank. This parameter is estimated by minization of errors between simulated and measured temperatures in the tank from experimental tests at operation conditions identical to the conditions in which the storage tank is operated in HP/ORC system (Carmo et al., 2015). In practice, a much higher mixing parameter is used when this situation arises.

3. RESULTS AND DISCUSSION

The HP/ORC system coupled with a passive house($U_{walls} = 0.09W/m^2K$ and $U_{windows} = 0.63W/m^2K$) model results are presented under three different control strategies in comparison with the reference control. The influence of controls on a building which envelope characteristics ($U_{walls} = 0.258W/m^2K$ and $U_{windows} = 1.2W/m^2K$) are similar to current building stock is also assessed. Additionally, a case with a larger water tank (500 l) is considered. The purpose of the study is to evaluate the performance of the system in terms of energy performance, renewable energy integration and comfort levels and how the different controls can enhance its performance. The ratio of heating demand covered by HP (Q_{HP}/Q_{dem}), the annual electricity demand of the HP/ORC unit ($W_{dem,unit}$) and production (W_{ORC}), the number of HP cycles (# cycles), the average duty cycle (**adc**), the seasonal performance factor (**SPF**), the demand cover factor (γ_d), the level of discomfort in DHW (**DL**$_{DHW}$) and indoor temperature (**DL**$_{in}$) are considered to evaluate the performance of the system.

The average duration of an ON period of the HP or average duty cycle (adc) in minutes, is defined in eq.4.

$$adc = \frac{1}{N}\sum_{1}^{N} \int_{t=1}^{t=8760*60} t_{ON} \quad (4)$$

The **SPF** is calculated as shown in eq. 5.

$$SPF = \frac{Q_{unit}}{W_{unit}} \quad (5)$$

The demand cover factor (γ_d), indicates the ratio of electricity demand ($W_{unit+L\&App}$) and the supply that has been produced by the building's energy supply system, in this case by the ORC (W_{ORC}) and is here used as an indicator of renewable energy integration (Equation 6).

$$\gamma_d = \frac{\int_1^N min(W_{dem,unit+L\&App}, W_{ORC})}{\int_1^N W_{dem,unit+L\&App}} \quad (6)$$

Finally, other two parameters are considered to quantify the level of comfort,DL_{DHW} and DL_{in}. These are measured by the percentage of time of DHW (t_{DHW}) or SH demand (t_{SH}) when the DHW or room temperatures are not delivered at the minimum desired levels, 40°C (Kordana et al., 2014) and 19°C (ISO7730, 2004).

$$DL_{DHW} = 100\frac{\int_1^N t_{T_{DHW}<40}}{\int_1^N t_{DHW}} \quad (7)$$

$$DL_{in} = 100\frac{\int_1^N t_{T_{in}<19}}{\int_1^N t_{SH}} \quad (8)$$

3.1 Annual performance results

The results of the simulations are summarized in table 2. The table is divided horizontally in three main cases : original passive house with 250 L, poorly insulated house with 250 L and passive house equipped with a 500 L tank, each presenting the results with the four control logics, except the 500L case which only presents the results for the two most relevant cases for compactness.

From the results summarized it can be noted how the consumption of the compressor of the HP/ORC unit ($W_{dem,unit}$) is not reduced with any of the controls in the passive house case, per contra $W_{dem,unit}$ increases up to 4.3% compared

Table 2: Results for the control comparison

Case	Q_{HP}/Q_{dem} [%]	$W_{dem,unit}$ [kWh]	W_{ORC} [kWh]	**adc** [min]	**# cycles**	**SPF**	γ_d	DL_{DHW} [%]	DL_{in} [%]
Ref.	75	1975.5	3105.8	19.8	861	2.7	6.7	1.6	0.0
OccSB	76	2061.4	3111.7	46.3	377	2.6	6.5	1.0	5.5
DHW_p	75	1975.9	3105.8	19.8	862	2.7	6.7	0.8	0.0
$OccSB + DHW_p$	76	2052.9	3112.2	46.5	375	2.6	6.5	0.3	5.7
Ref.poor	85	3959.9	3100.5	22.2	1544	2	4.3	0.7	0.0
OccSBpoor	84	3775.7	3105.0	49.2	661	2	4.4	1.0	10.0
DHW_ppoor	85	3963.4	3099.9	22.0	1553	2	4.3	0.3	0.0
$OccSB + DHW_p$poor	84	3759.0	3101.8	49.9	651	2	4.4	0.9	10.4
Ref.500L	67	1511.3	3186.3	37.6	450	3.9	8.2	0.0	0.0
$OccSB + DHW_p$500L	69	1464.1	3190.9	72.5	240	4.1	8.4	0.0	5.2

to the reference case. None of the controls promotes significantly the power production in ORC mode (W_{ORC}) and the demand cover factor (γ_d). The maximum improvement (0.2%) of W_{ORC} in comparison with the ref. case occurs when $OccSB + DHW_p$ is implemented. Also, when OccSB controls adc and #cycles are doubled and halved, respectively. Yet, this does not seems to affect the SPF. This is mainly due to the fact that adc is higher than the critical duty cycle time (<18 min), where the compressor efficiency is lessen (see Figure 3).

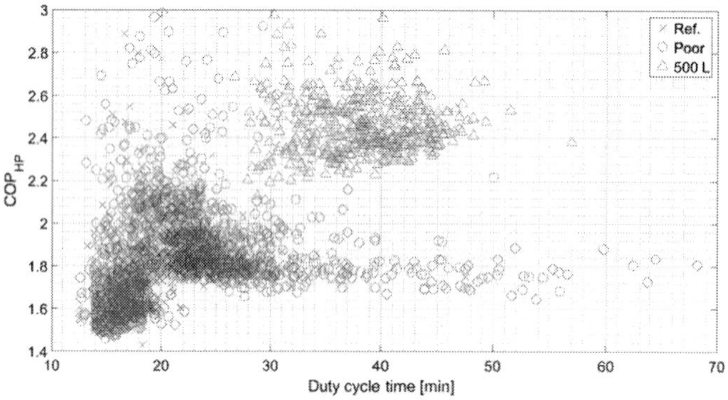

Figure 3: Results showing the effect of short duty cycle on the coefficient of performance (COP) of the compressor for the ref., poor insulation and 500 L cases in HP mode

Furthermore, it should be noted that the discomforts levels of DHW temperatures are reduced up to 80%, $OccSB$+DHW_p case.

On the other hand, in the case of house with poor insulation - $W_{dem,unit}$ of the Ref.poor is two times higher than the value of Ref.case - the new control logics achieve up to 5.1% reduction of the unit power consumption when compared to the ref. poor case, except the $DHW_p poor$. However, this is not reflected in an improvement of the SPF, the reason is the lower thermal mass of the building. While in the reference case, the large thermal mass of the building enables higher integration of solar renewable energy expressed by smaller values of Q_{HP}/Q_{dem} (75%), in poorly insulated buildings 85% of the Q_{dem} is covered by the HP. The comfort levels are also influenced considerably with the new control logics. The DL_{in}, for example, reaches values of 10%, which trespasses the tolerable value (5%).

Finally, the case including a larger water storage tank is considered. It is shown that the introduction of a larger sensible storage in a passive house enables a large integration of renewable sources both in the thermal demand (Q_{HP}/Q_{dem} = 67%) and power (γ_d = 8.2) when compared to the reference case (Ref.) with a tank of 250 L. Additionally, new control strategies, $OccSB + DHW_p$, improve in all remaining performance parameters. The $W_{dem,unit}$ reduces 3%, the #cycles is halved and the SPF improved 5% in comparision with Ref.500L. These values increase to 25%, 72% and 50% when compared to Ref. case.

3.2 Control strategies effects: one-week operation

Figures 4-8 present the weekly (from Sunday to Saturday) profiles of (1) Temperatures in the storage ($T_{sto,top}$,$T_{sto,h=0.6m}$, $T_{sto,bottom}$), outdoor (T_{out}) and indoor (T_{in}) (2) Mass flow rates in floor heating loop ($\dot{m}_{FH}$) and DHW ($\dot{m}_{DHW}$) and occupancy state (3) Heating supply ($\dot{q}_{unit}$), power demand ($\dot{W}_{HP}$ and $\dot{W}_{L\&App}$) and supply ($\dot{W}_{ORC}$) under different control strategies.

It can be seen how the space heating demand is modulated (Fig.5) in comparison with the reference case (Fig.4). In the first 60 hours the number of HP cycles is reduced from 7 to 2. The $\dot{m}_{FH}$ controlled by a PI reaches values three times higher than in the Ref. case due to the larger heat demand after a long unoccupied period where T_{in} reaches a value of 17°C. This induces a large drop and mixing of the temperatures in the storage, which leads to a longer duty cycle (Fig. 5 Time=32h). Thus, no major power savings are achieved. The operation under DHW_p (Fig. 6) is similar to the Ref. case. Likewise, is the operation with $OccSB + DHW_p$ (Fig.7) in comparison with $OccSB$. The space heating demand is only slightly (order of seconds) delayed. Finally, the profiles with 500 L storage tank case are presented in Fig. 8 showing the advantages. Lower power demand from the HP is required. Only one cycle (Fig. 7) compared to the same house with same control (Fig.4) and larger amount of heating demand covered by the solar collector (DH mode) at hours 58, 80 and 128.

Furthermore, the Figures 4-8 (sub-figures 2) show that the occupancy schedule, in the cases studied, does not match the availability of power supply from the ORC. Ultimately, this limits the increase of renewable power integration since the energy system studied does not include an electric battery.

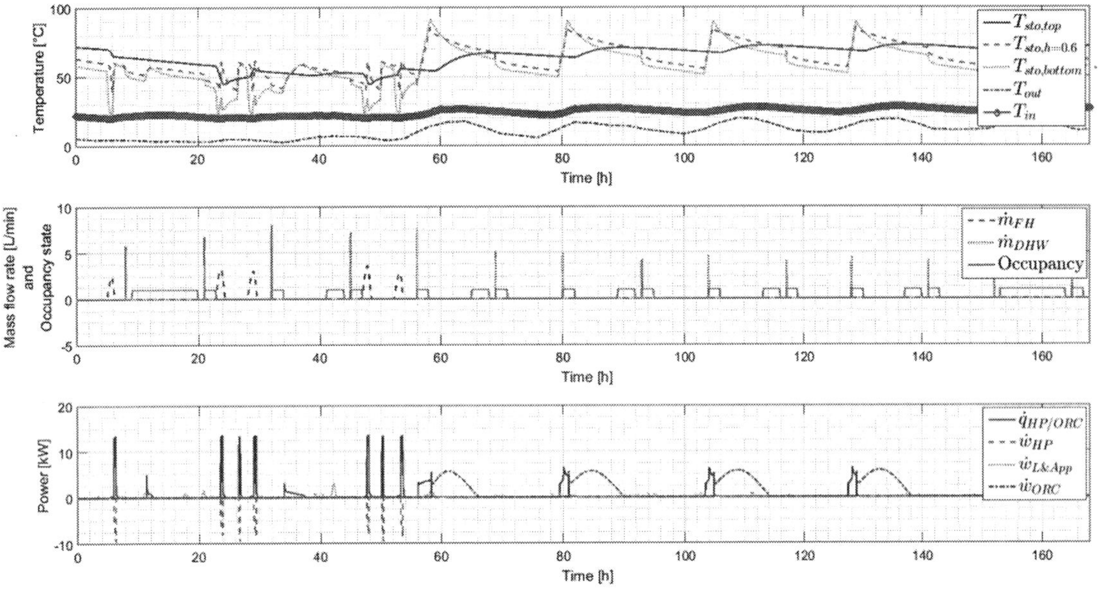

Figure 4: Ref. case operation variables as function of time during a week in May on a ref. year in Danish climate

4. CONCLUSIONS

In the present study, the effect of four different control strategies in the performance of an innovative co-generation unit, known as HP/ORC unit, are presented based on simulation results. These control logics take into account real

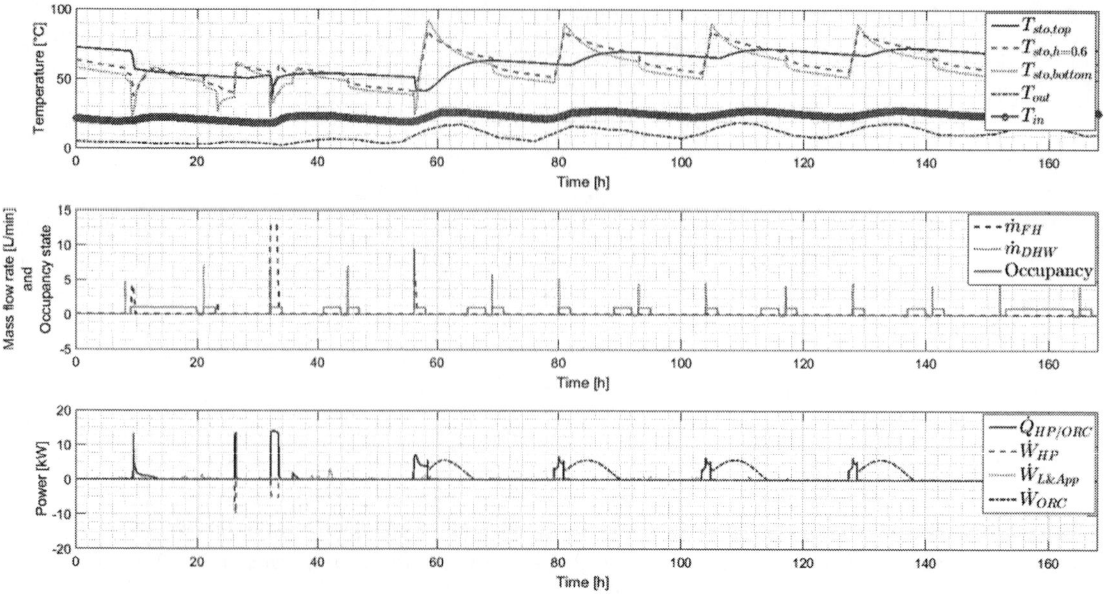

Figure 5: OccSB case operation variables as function of time during a week in May on a ref. year in Danish climate

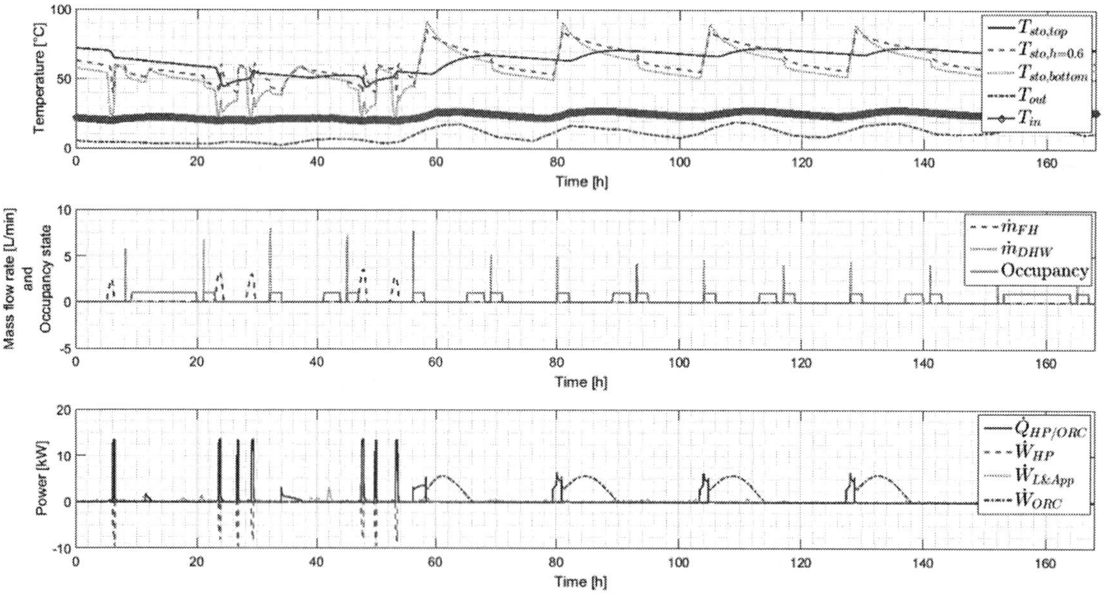

Figure 6: DHW$_p$ case operation variables as function of time during a week in May on a ref. year in Danish climate

load information from occupancy and thermal energy end-use.

- It is shown that, in all cases studied, occupancy driven control logics can avoid adverse cycling effects in terms of compressor lifetime.
- The effects of these controls in improving the performance of the HP/ORC system depend both on the thermal mass of the building and the sensible storage volume as well as the occupancy schedule match with renewable energy availability.
- The controls do not increase the power demand covered by renewable sources but can improve the thermal demand covered by renewable energy.
- In the current building stock, occupancy controls should be implemented with deference as they might disturb the indoor comfort levels significantly.

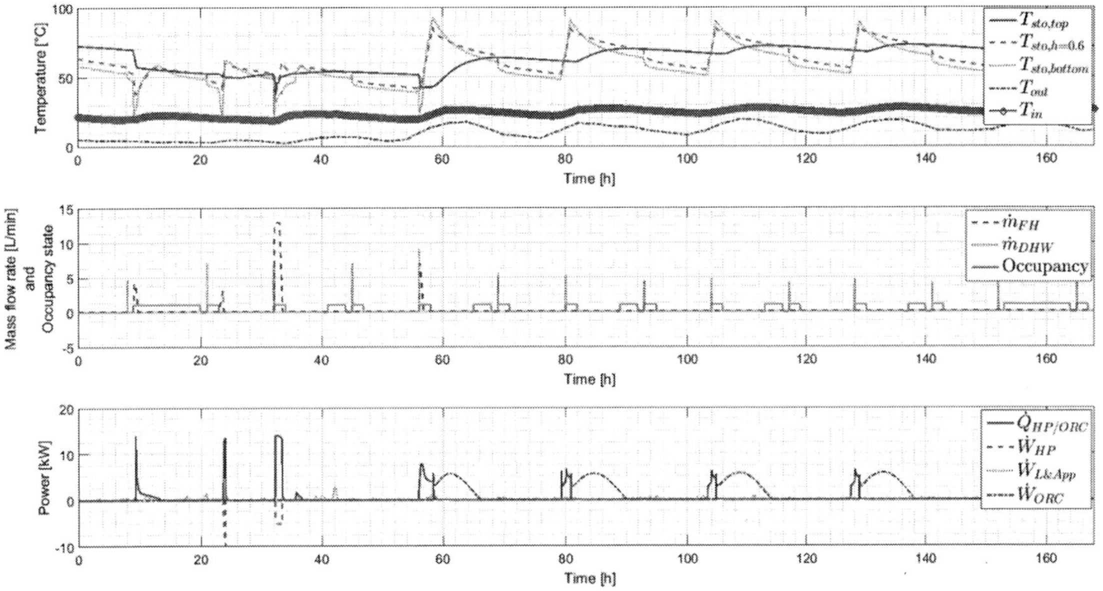

Figure 7: OccSB+DHW$_p$ case operation variables as function of time during a week in May on a ref. year in Danish climate

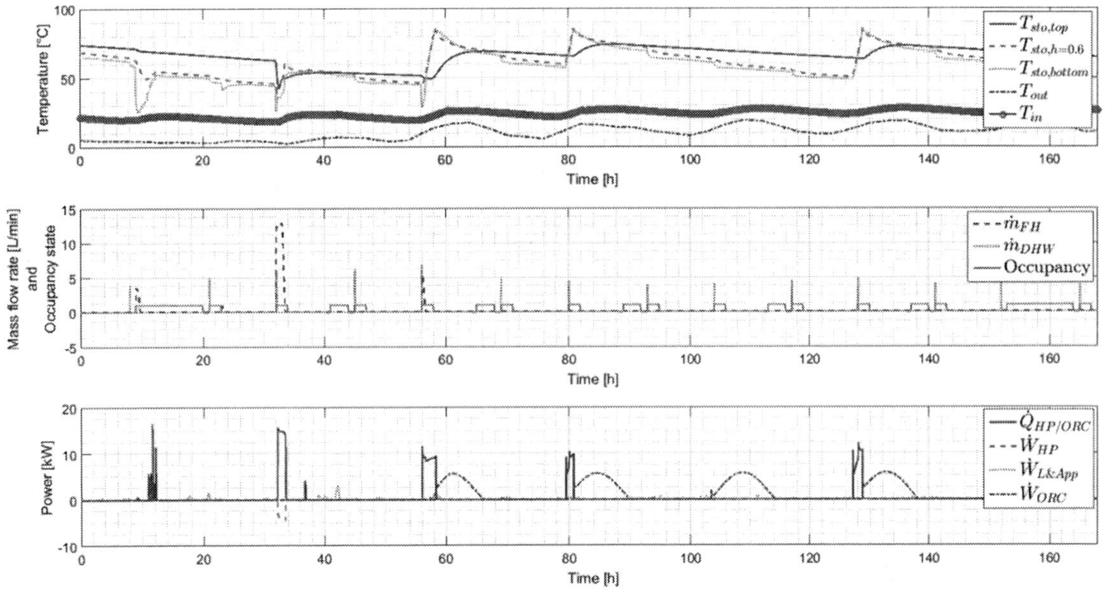

Figure 8: OccSB+DHW$_p$500L case operation variables as function of time during a week in May on a ref. year in Danish climate

NOMENCLATURE

A	area	m^2	C_p	specific heat	J/gK
Δx	height segment	m	h	specific entalphy	J/kg
$\dot{m}$	mass flow rate	kg/s	t	time	s
Q	heat demand	kWh	$\dot{q}$	heat flow	W

T	Temperature	°C	t	time	s
U	heat transfer coef.	W/m^2K	Q	heat	kWh
W	Power	kWh			

Greek letters

α	binary paramater	(-)	β	binary paramater	(-)
σ	artificial mixing term	W/mK			

Subscript

amb	ambient	*seg*	tank segment
dem	demand	*el*	electrical
ex	exhaust	*hx*	heat exchanger
i	tank segment number	*in*	indoor
min	minimum	*p*	priority
out	outdoor	*ORC*	Organic Rankine Cycle
su	supply	*sto*	storage
th	thermal	*unit*	HP/ORC unit

REFERENCES

Carmo, C., et al. (2015). Experimental validation of a domestic stratifed hot water tank model in modelica for annual performance assessment. In *Proceedings of asme-orc2015.*

Dumont, O., et al. (2015). Experimental investigation of a reversible heat pump / organic rankine cycle unit designed to be coupled with a passive house (net zero energy building). *International Journal of Refrigeration*, *54*, 190–203. (http://dx.doi.org/10.1016/j.ijrefrig.2015.03.008.)

Georges, E., et al. (2013). Modeling and simulation of the domestic energy use in belgium. In *Proceedings of clima 2013 11th rehva world congress and 8th international conference on iaqvec.* Praha, Czech Republic.

Green, R. (2012). *The effects of cycling on heat pump performance* (Tech. Rep.). EATechnology for Depart. of Energy and Climate Change.

IEA. (2013). *Transition to sustainable buildings- executive summary* (Tech. Rep.). OECD/IEA.

ISO7730. (2004). Ashrae standard 55 or iso7730 thermal environmental conditions for human occupancy.

Klein, S. (1975). Calculation of flat-plate loss coefficient. *Solar Energy*, *77*, 79–80.

Kordana, S., et al. (2014). Rationalization of water and energy consumption in shower systems of single-family dwelling houses. *Journal of Cleaner Production*, *82*, 58–69. (http://dx.doi.org/10.1016/j.jclepro.2014.06.078)

Masy, G. (2007). Definition and validation of a simplified multizone dynamic building model connected to heating system and hvac unit. *Univ. Liège.*

M. Wetter, e. a., W. Zuo. (2014). Modelica buildings library. *Journal of Building Performance Simulation*, *7*(4), 253–270.

Rosiek, S., et al. (2013). Reducing a solar-assisted air-conditions system's energy consumption by applying real-time occupancy sensors and chilled water storage tanks throughout the summer: A case study. *Energy Conversion and Management*, *76*, 1029–1042. (http://dx.doi.org/10.1016/j.enconman.2013.08.060)

Shaikh, P., et al. (2014). A review on optimized control systems for building energy and comfort management of smart sustainable buildings. *Renewable and Sustainable Energy Reviews*, *34*, 409–429.

Tammmaro, A., et al. (2016). Hot sanitary water production with co_2 heat pumps: Effect of control strategy on system peformance and stratification inside the storage tank. *Applied Thermal Engineering.* (http://dx.doi.org/10.1016/j.applthermaleng.2016.01.094)

ACKNOWLEDGMENT

This work is supported by the Danish Ministry of Science, Technology and Innovation and Insero Energy. We thank the continuous support given by ThermoCycle developers from Thermodynamic Lab., Univ.Liège.

Operation of Cool Thermal Energy Storage to Increase Renewable Energy Utilization

Amy VAN ASSELT*, Douglas T. REINDL, Gregory F. NELLIS

University of Wisconsin – Madison, Department of Mechanical Engineering,
Madison, WI, USA
avanasselt@wisc.edu

* Corresponding Author

ABSTRACT

Recent international agreements on climate change aim to increase the production of electricity derived from renewable energy resources. Renewable energy generation can be pursued on both an individual building and utility scale. Due to the intermittent nature of renewables, some form of energy storage is essential to bridge diurnal mismatches between generation and demand. Air-conditioning loads associated with commercial buildings dominate peak electricity demand on the utility grid in some areas and climates. Therefore, Cool Thermal Energy Storage (CTES) is a relatively technically mature and inexpensive means of providing this "storage" and balancing supply/demand mismatches, thereby enabling the success of increased renewable energy penetration. Electrical energy generated by renewables during periods of higher availability can be used to run chillers that charge CTES systems. The stored thermal energy can subsequently be used to meet air-conditioning demand during periods of low renewable energy resource availability.

In this work, the U.S. Department of Energy Commercial Reference Building Model for a secondary school is used to obtain simulated cooling loads that are met by a combination of two chillers and a stratified chilled water thermal storage system. Control strategies are designed to charge the thermal storage system when renewable resources are available and discharge storage to meet building cooling loads during periods with low or no renewable energy resource. One optimization target is the fraction of the chiller energy consumption met by renewable power. This metric is one that may be of interest to electric utilities trying to manage a grid with increasing renewable penetration. An alternative optimization target is the net economic benefit to the building owner assuming on-site, small scale renewable generation and thermal storage. This metric is based on equipment costs, net electric demand after wind and/or solar generation offsets the chiller electric demand, and time-of-use electricity rate structures.

The results show that there is a trade-off between maximizing the use of renewable power and life-cycle cost, but a storage system designed to optimize either variable will be more cost effective and utilize the renewable resource better than a system without storage. The analysis is carried out for locations in Texas and California. These results suggest that CTES may be a technology enabling utilities to reach higher penetration of renewables while avoiding the so-called "duck curve" generation ramp caused by the time mismatch between the renewable generation and demand peaks.

1. INTRODUCTION

The demand for electricity varies significantly on both a seasonal and diurnal basis. This variation is driven by cooling load differences between winter and summer as well as the daily peaks that occur as a result of typical work schedules. Myers, et al (2010) evaluated the impact of varying penetration rates of solar energy technology to meet the aggregate load in Wisconsin. In 2002, the aggregate peak utility demand in Wisconsin was 13,200 MW. The analysis showed that the deployment of 6,600 MW of photovoltaic capacity (50% of the utility aggregate peak) only reduced the grid demand for electricity from traditional generation sources by 4.3%. The minimal reduction in peak electrical demand even with such a large renewable energy deployment is due to the mismatch between the peak solar energy electricity production (which occurs around noon) and the utility load peak (which occurs at approximately 4 p.m.). The authors of this work noted that the future success of large-scale penetration of renewable energy technology, such as photovoltaics, hinges on cost-effective and reliable energy storage technologies.

At a utility level, the intermittent nature of renewable energy production, coupled with the obligation to provide firm supply with a high degree of reliability, has created both operations and capital challenges. From an operations perspective, utilities have been challenged with significant grid dynamics created by the rapidly changing electricity production rates from renewable energy sources that occur coincidently with changing customer demand. Evidence of such dynamics includes utilities experiencing very high electricity production ramp rates and base load generation encroachment (over-generation). The California Independent System Operator (CAISO) region has been observing these specific issues due to increased solar electricity production as shown on the popular "duck curve" shown in Figure 1 (CAISO, 2013). From a capital perspective, utilities have not experienced a reduction in fixed asset generation or transmission/distribution requirements due to the growing deployment of renewable energy generation. On the contrary, utilities have had to increase investments in transmission, distribution, and grid controls in order to accommodate the increased renewable energy generation. They have not been able to forego investments in traditional generation assets due to the renewables' inability to provide firm generation capacity.

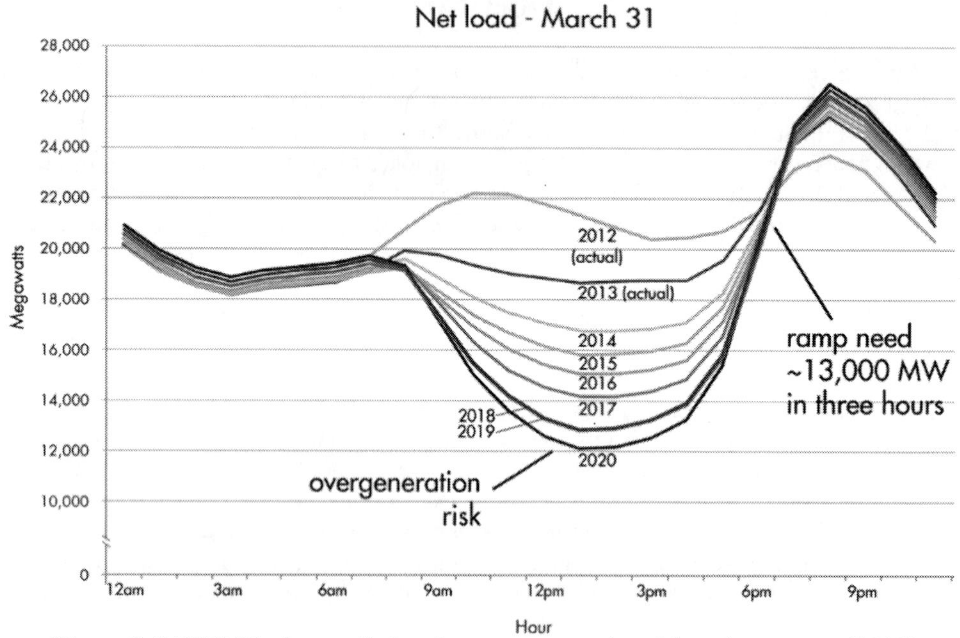

Figure 1 CAISO "duck curve" showing over-generation risk and ramp rate (2013)

Cool thermal energy storage technology is a cost-effective, mature, and high efficiency energy storage technology that has the potential to bridge mismatches between renewable energy production and utility aggregate demands. By operating energy-intensive building chilling systems to charge thermal storage systems during periods when renewable energy resources are available and then discharging the stored thermal energy with chillers idle to meet building air-conditioning loads during periods of resource unavailability, CTES offers the potential to enable a more effective utilization of electricity produced from renewable energy resources. Our hypothesis is that, beyond electricity cost savings, CTES can cost-effectively enable a significantly greater penetration of renewable energy production than currently exists today. The overall objective of this work is to evaluate the potential benefits of thermal energy storage as an enabling technology that will allow a more effective penetration of renewable energy generation.

2. MODELING

2.1 System Models

A chiller plant that directly meets the facility load without storage is used as a baseline comparison for CTES system options. This system is referred to as the no-storage case throughout this work. The chiller for the non-storage system operates anytime the building experiences a cooling load. The fluid flow rate is variable and adjusted to maintain a chilled water temperature differential of 10°F (5.5°C). The installed chiller capacity is sized to meet the annual peak cooling load. The model consists of two equally sized chillers for both redundancy and improved efficiency at part-load operation.

Figure 2 shows the CTES system modeled in this work with a storage tank decoupling the cooling load from the air-cooled chillers. In a stratified chilled water storage system, a single tank stores thermal energy by utilizing water's natural tendency to stratify – warmer, less dense fluid gathers at the top and cooler, more dense fluid moves to the bottom. During a charge cycle, cool water from the chiller is supplied to the bottom of a tank where it is available to subsequently be withdrawn from the bottom of the tank and supplied to meet building loads during its discharge cycle. The discharge cycle can proceed with or without chiller operation. The CTES-based chilling system is designed to operate with a 20°F (11°C) water temperature differential.

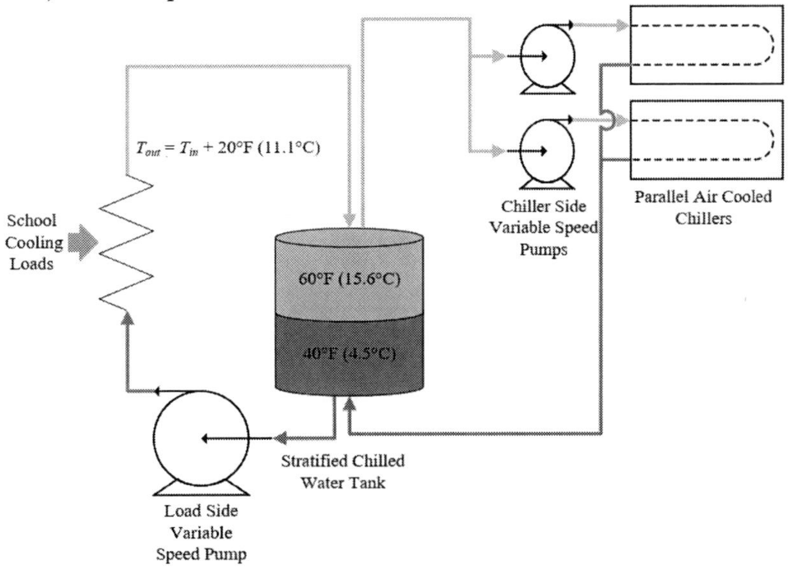

Figure 2 Stratified chiller water storage system schematic

The air-cooled chillers modeled in this work use screw compressors and therefore exhibit coefficients of performance (COP) of 3.0 or better at full-load conditions. However, their part-load performance is poor and leads to a COP of less than 1 at extremely low part-load ratios (PLR). In order to operate less frequently at these poor performance conditions, two equally sized chillers are assumed to be installed in parallel. In this case, each chiller is run at equal part-load ratios down to half of the overall building part-load ratio. Once the building part-load ratio dips below half, only one chiller is in operation and the system performance follows the single chiller performance curve. The performance curves for the overall chilling system are shown in Figure 3. The system is technically capable of running at as little as 7.5% of the two-chiller full-load capacity which corresponds to one chiller operating at 15% of its full-load capacity limit.

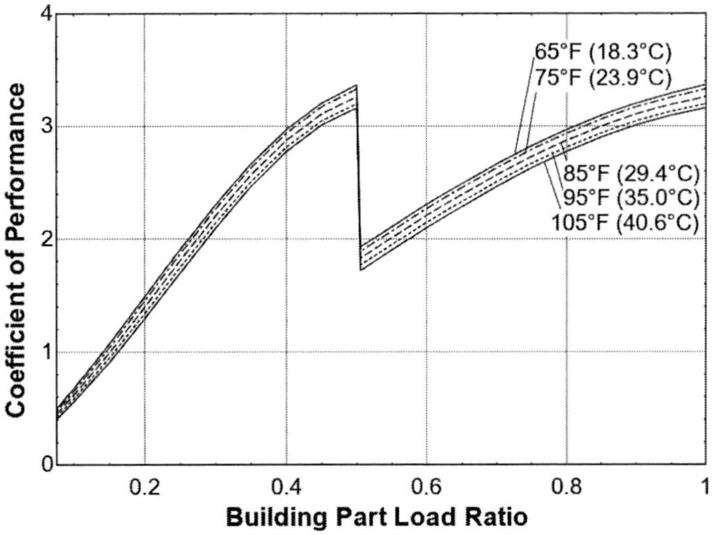

Figure 3 Air-cooled chiller performance curves for two equal sized chillers (CUSCST, 2011)

2.2 Cooling Loads and Equipment Sizing

Cooling loads met by the systems are derived from the Commercial Reference Building model developed for a secondary school by the US Department of Energy (2011). Energy simulations are run for both Amarillo, TX and Los Angeles, CA, each with weather data and building construction parameters that are representative of these climates. The simulations provide cooling loads at 10-minute intervals which then must be met either using chillers directly or the CTES system.

For this work, the sizing of the chiller plant system and stratified chiller water storage tank are based on a partial storage strategy. This strategy determines to the minimum capacity or size of each piece of installed equipment. The integrated design day cooling load, or load on the day with the largest cooling load, is divided by 24-hours in order to determine the minimum required chiller capacity. This implies that the chiller will be running at full-load throughout this day. The resulting chiller capacity for Amarillo is 630 kW (180 tons) and for Los Angeles it is 470 kW (134 tons). The storage tank is the minimum capacity tank that will allow the chiller system to meet the cooling load, 5,200 kWh (1,480 ton-hrs) for Amarillo and 4,500 kWh (1,280 ton-hrs) for Los Angeles.

2.3 Weather, Generation, and Cost Data

The weather data utilized for the Commercial Reference Building energy simulation is Typical Meteorological Year version 3 (TMY3) data (NREL, 2008). The energy model requires temperature, relative humidity, and solar radiation data for solar gain calculations, this data is referred to here as the cooling load weather data. TMY3 weather is also used to estimate the solar resource. A four parameter photovoltaic model uses the total radiation on a tilted surface to predict the solar power data using only parameters provided by panel manufacturers (Duffie and Beckman, 2013). The wind speed data in TMY3 is collected at a height of 33 ft (10 m) and, although it is possible to estimate the equivalent wind speed at other heights using a wind shear exponent that is a function of terrain, data gathered at heights of 164 ft (50 m) or higher are more accurate for estimating the wind resource. The hub-height wind data used in this work comes from White Deer, TX and Kern County, CA (AEI, 2012; NREL, 2015b). Wind turbine power output is then modeled using a characteristic power curve from NREL's System Advisor Model (2015a). The geographic locations for the three weather data types are shown in Figure 4 for the Los Angeles region (a) and the Amarillo region (b).

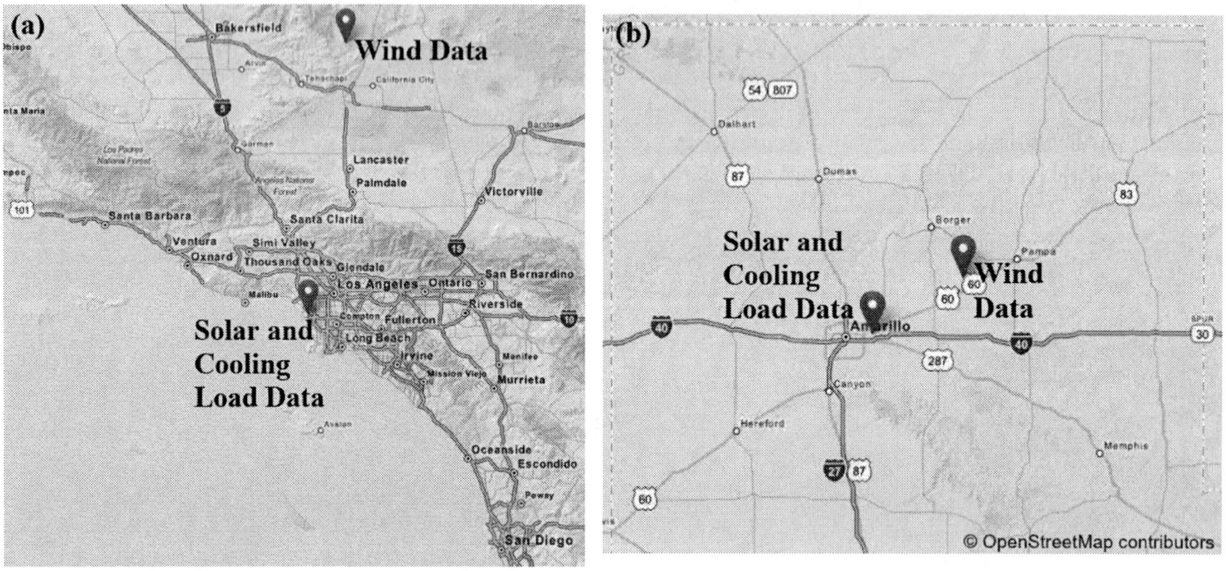

Figure 4 Cooling load, solar, and wind data locations for (a) Los Angeles and (b) Amarillo

Cost data for stratified chilled water storage tanks have been obtained from a consulting firm that specializes in thermal storage systems (J. Schuett, personal communication, April 12, 2016). The air-cooled chiller cost data is obtained from RS Means Mechanical Cost Data for rotary-screw type water chillers with integrated condensers (2015). Capital cost for installed wind turbines is based on the US DOE's Wind Technology Report and cost for solar photovoltaics is reported by the US DOE's SunShot Photovoltaic System Pricing Trends for non-residential systems (2015a, 2015b). A single representative commercial time-of-use electricity rate is used for both locations in order to enable consistent life-cycle cost comparisons. During the summer months, the on-peak rate between the hours of 3 p.m. and 8 p.m. is 16.5 ¢/kWh and the off-peak rate during all other hours is 7.16 ¢/kWh. During the winter months the rates are the same, but there is an additional on-peak period between the hours of 6 a.m. and 8 a.m. (CoServ, 2015). No demand charges are considered. When calculating the present value of the electricity costs over a 20-year period, a discount rate of 8% and an inflation rate of 5% are taken into account.

3. CONTROL STRATEGIES

Two CTES control systems are compared: "Renewable Control" and "Cost Control." Renewable Control runs the chillers whenever there is sufficient renewable power available to operate the chillers as well as a need for the cooling that is produced (either to directly meet building cooling loads or for charging storage). The level of renewable power required to run the chillers is an independent variable that can range from the full load chiller power down to the minimum individual chiller part-load ratio of 15%. Figure 5 illustrates a flow chart of the Renewable Control strategy for the case where wind power is the renewable energy source that is being exploited. The first branch, activated when there is insufficient wind power to run the chiller(s), demonstrates one constraint that requires that the storage tank be fully recharged by the beginning of the next day. This decision assumes perfect knowledge of the facility cooling load for the remainder of the day.

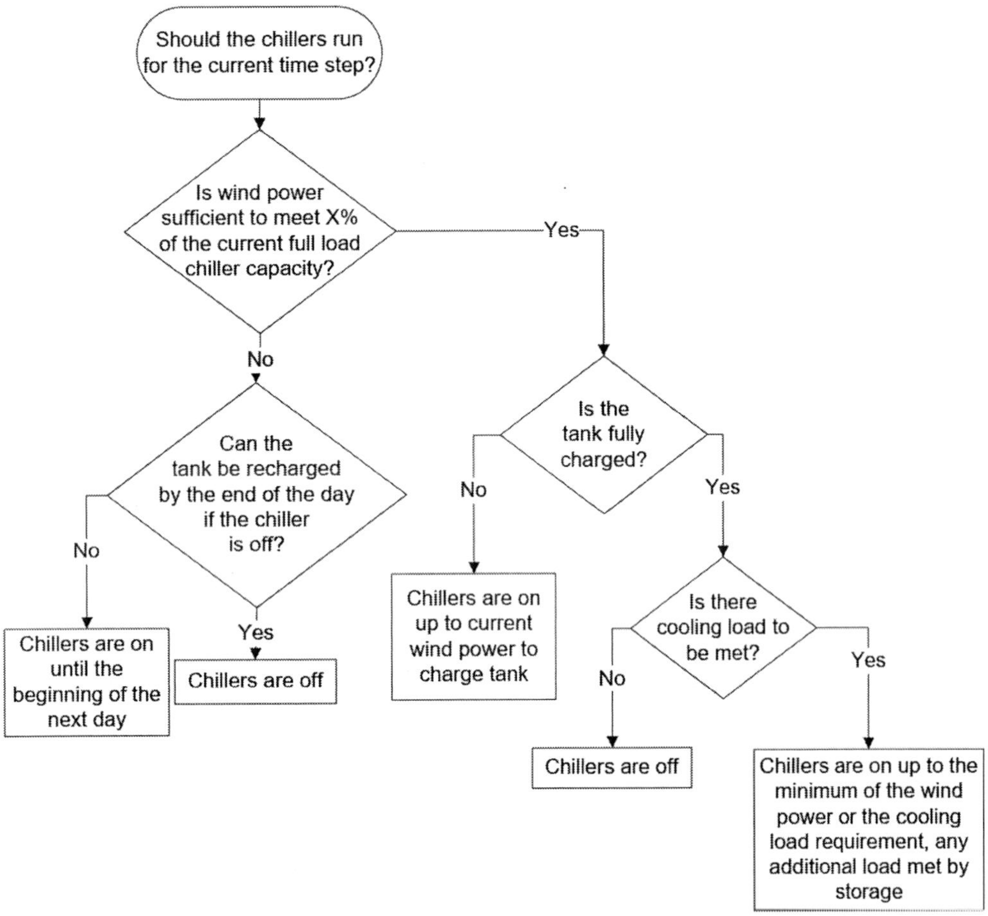

Figure 5 Flow chart for the Renewable Control strategy based on wind

The Cost Control strategy is similar to Renewable Control except that the trigger for operating the chillers is related to off-peak time-of-use electricity rate rather than the level of renewable energy that is available. The chillers will run during the on-peak period only if necessary in order to fully charge the tank by the beginning of the next day.

4. DESIGN PARAMETERS AND METRICS

In order to accomplish the system optimization, the two independent variables (or design parameters) examined are the hour of the day by which the storage tank must be recharged and the chiller system minimum part-load ratio. For each of the strategies utilizing the solar resource, the optimal storage tank recharge hour was found to be noon while for those strategies utilizing wind, the optimal recharge hours was found to be 8 a.m.

Two metrics have been developed for carrying out optimization of the control strategies. The first is the fraction of the chiller power that is met by the renewable resource, defined as the chiller electrical consumption met by renewable divided by the total chiller electric consumption over the simulation year. The second metric is the cost of owning and operating the system for a period of 20 years. This cost includes capital costs for chillers, storage tanks, renewable generation equipment, and the discounted operating cost, as described previously in Section 2.3. Figure 6 shows the method for calculating the daily operating cost for 250 kW of installed solar power with (a) no storage and (b) storage. The top-most plot shows the chiller power consumption over 24 hours as a negative quantity, the second to the top is the positive solar power produced, and the third is the sum of the two, corresponding to the net utility impact. The discontinuities apparent in the chiller power consumption plot are due to the transition between one and two chiller operation which is governed by the performance curve in Figure 3. At each timestep, the time-of-use electricity rate in effect is applied to the net utility impact to get the net electricity value that is shown in the bottom plot. The shaded region represents the time of on-peak rates and discontinuities in the net electricity value are apparent at the beginning and end of this time period. The integration of the net electricity value over a year gives the operating cost. The operating cost does not account for on-going maintenance costs or solar panel performance degradation over time.

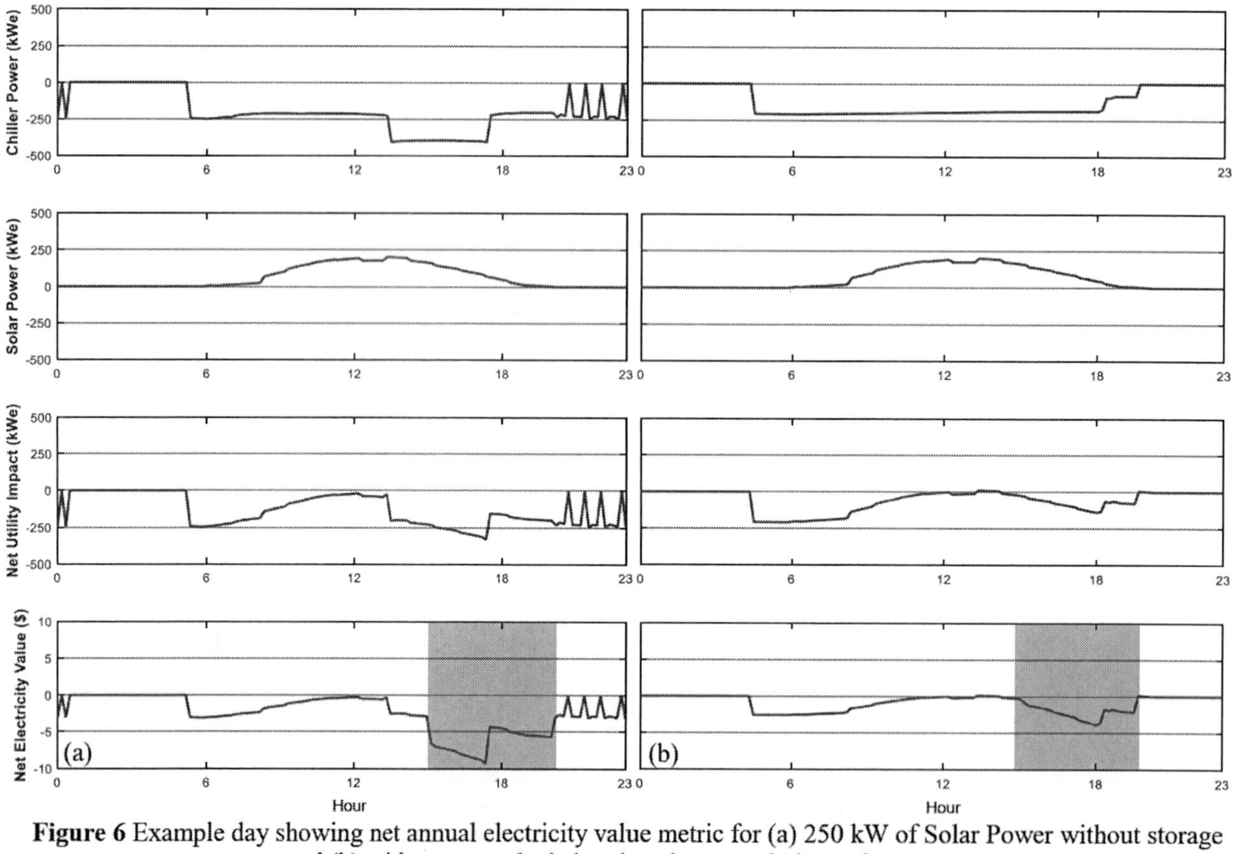

Figure 6 Example day showing net annual electricity value metric for (a) 250 kW of Solar Power without storage and (b) with storage, shaded region shows peak time-of-use rates

5. RESULTS

5.1 Individual Facility Results

Figure 7 shows several results, all as a function of chiller system part-load ratio. Both the fraction of chiller energy consumption met by renewable and life-cycle cost (LCC) are shown in Amarillo, TX and Los Angeles, CA using both the Renewable and Cost Control strategies with the wind resource utilized as the renewable. The installed nameplate renewable capacity is fixed at 250 kW. LCC is calculated using the sum of the present value of 20 years of net operating cost (as described in section 4) as well as the chiller, storage tank, and wind turbine capital costs. Negative LCC indicates that more wind energy was produced than chiller energy consumed, and the operating costs show a gain that outweighs the capital costs. The results in Amarillo show that if the chiller system is restricted to run at a part-load ratio of not less than 0.3, then 84% of the chilling system energy can be met by wind using the Renewable Control strategy as shown by point (1) in Figure 7a. However, the trade-off is that over a 20-year span, the system costs $120,000 more to operate using Renewable Control than it does with the Cost Control strategy if the optimal part-load ratio of 1 is used. This is shown by the difference between point (2) in Figure 7d and point (1) in Figure 7c. On the other hand, point (2), which optimizes cost, results in only 57% of the chiller energy consumption being met by wind. As a point of comparison, to meet the same cooling load without a storage system, the cost is $291,000 more than Renewable Control (-$49,000 for no-storage minus -$340,000 in point (1)) and $411,000 more than Cost Control (-$49,000 minus -$460,000 in point (2)). The no-storage system results in just 50% of the chiller energy met by wind (shown as point (3) in Figure 7). Similarly, for the Los Angeles system without storage, the cost is $254,000 more than Renewable Control (-$86,000 for no-storage minus -$340,000) and $374,000 more than Cost Control (-$86,000 minus -$460,000) with just 56% of the chiller energy met by wind as shown by point (4).

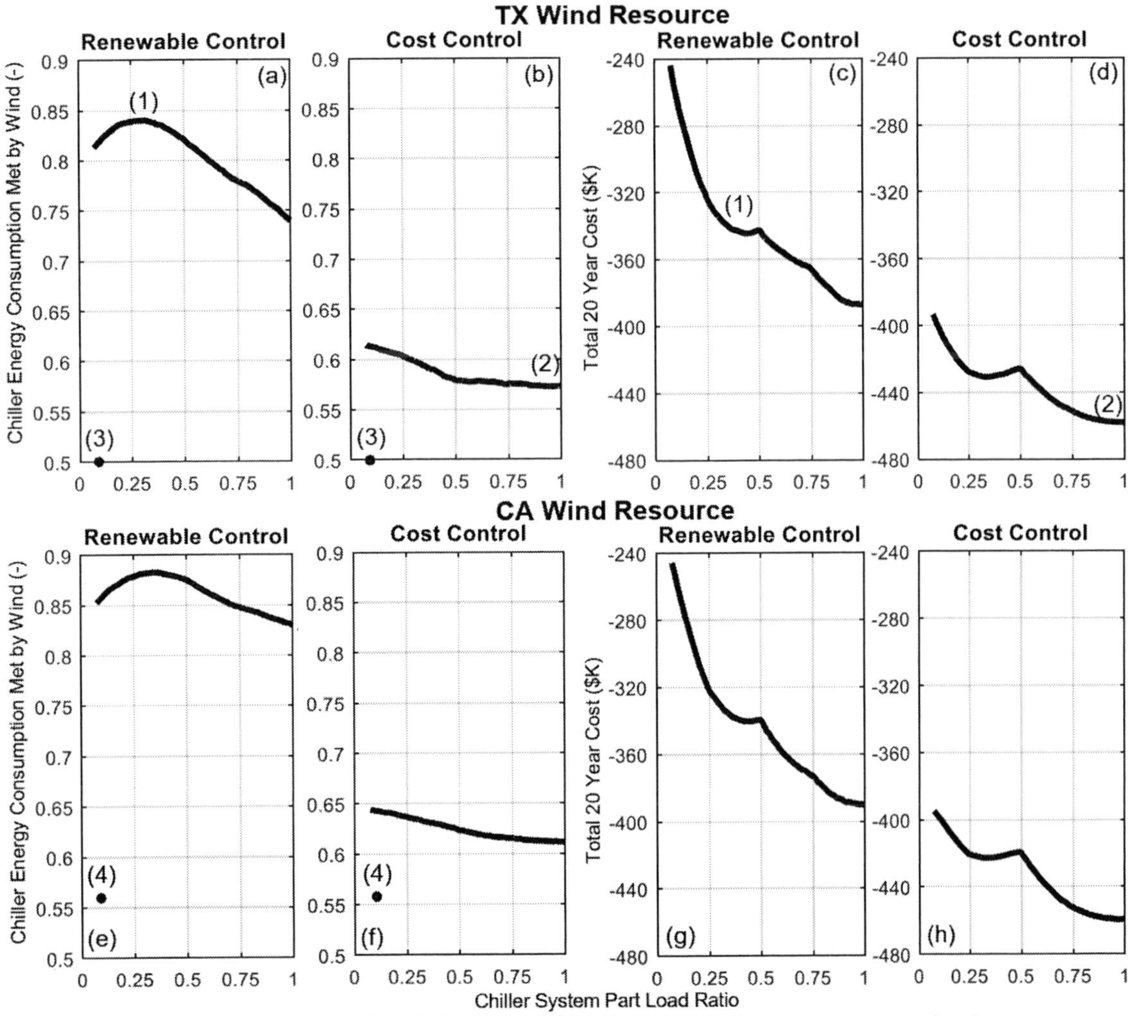

Figure 7 Results for wind resource (single points indicate no-storage values)

Figure 8 displays the same results but using the solar resource. In Amarillo, the capacity factor for the 250 kW of solar PV is less than that for the 250 kW of installed wind power, and therefore the maximum achievable fraction of the chiller energy met by solar is less than that for wind. Looking at Los Angeles in plots (e) through (h), restricting the minimum chiller part-load ratio to 0.3 results in the best utilization of the renewable resource with 82% of the chiller energy consumption being met by solar (see point 1 in Figure 8e). This strategy costs $80,000 more over 20 years than operating with the Cost Control strategy at a part-load ratio of not less than 1 which is shown by the difference between point (2) in Figure 8h and point (1) in Figure 8g. The Cost Control strategy only results in 54% of the chiller energy consumption met by solar (shown as point (2) in Figure 8f). For the Los Angeles system without storage, the 20-year cost is $420,000 more than Renewable Control ($1,100,000 for no-storage minus $680,000 for point (1)) and $500,000 more than Cost Control ($1,100,000 minus $600,000 for point (2)). The no-storage system results in 54% of the chiller energy consumption met by solar (shown as point (3) in Figure 8). In Amarillo, the no-storage 20-year cost is $330,000 more than Renewable Control ($970,000 for no-storage minus $640,000 for Renewable Control) and $405,000 more than Cost Control ($970,000 minus $565,000 for Cost Control) with 44% of the chiller energy consumption met by solar (shown as point (4) in Figure 8). In both locations, the no-storage system has at least as much chiller energy consumption met by solar as the Cost Control system. This is because the Cost Control strategy aims to keep the chiller system off when the on-peak rates are in effect from 3 p.m. to 8 p.m. Solar power is still significant in this timeframe and goes unused by the chiller system. All methods of operating the storage system are effective from a cost and renewable utilization standpoint as compared to a system without storage.

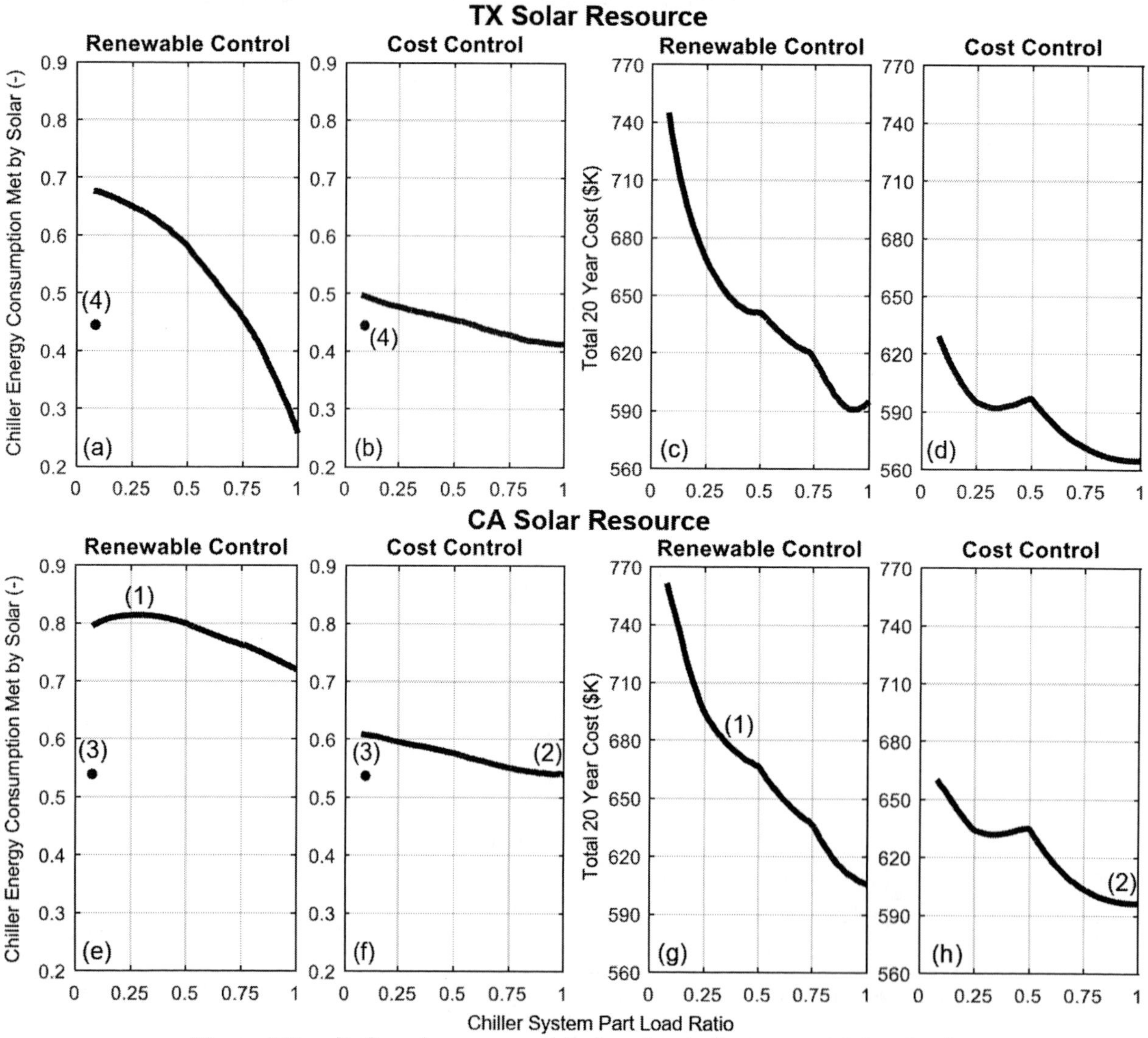

Figure 8 Results for solar resource (single points indicate no-storage values)

5.2 Aggregate Results

The results for an individual secondary school have been aggregated in order to begin to consider how large scale implementation of CTES systems might impact the utility aggregate demand (i.e. mitigate the "duck curve" problem). There are 4,495 secondary schools in California and each is assumed to use an air-cooled chiller system (High-Schools.com, 2015). The total installed solar PV capacity is 250 kW per school or approximately 1,125 MW. Figure 9 illustrates various impacts on the CAISO system on March 31st, the day used for the classic duck curve figure. Each plot shows the total CAISO load as a solid line. Plot (a) shows the impact of the 1,125 MW of solar capacity or the "net load" referred to in the duck curve in Figure 1. Plot (b) shows the net impact of 1,125 MW of solar PV offset by the addition of no-storage chiller systems for each of the 4,495 schools. Plots (c) and (d) show the net impact of exchanging the no-storage chiller systems for CTES systems operating using Renewable Control and Cost Control, respectively. The no-storage case in (b) shows much of the solar power being utilized, but adding the school chiller systems increases the peak evening load by nearly 700 MW as shown by point (1). The secondary schools equipped with CTES and operating with a Renewable Control storage system in (c) shows that some solar power is still left in the middle of the day, encroaching on the base load and fattening the "belly" of the duck. On the other hand, this strategy adds no load during the peak periods and meets the same cooling load with less energy overall. Lastly, the schools equipped with CTES and operating with Cost Control in (d) also use less energy than the no-storage systems and utilize much of the solar power. The time-of-use rate structure used causes some solar power to go unused around 7 a.m. during the on-peak period as shown by point (2). Point (3) shows where it also causes a short peak at 8 p.m. when the off-peak period begins. This shows that by shifting on-peak and off-peak schedules, this rate structure could be better optimized for utilization of the solar resource as well as reduction of peak demand.

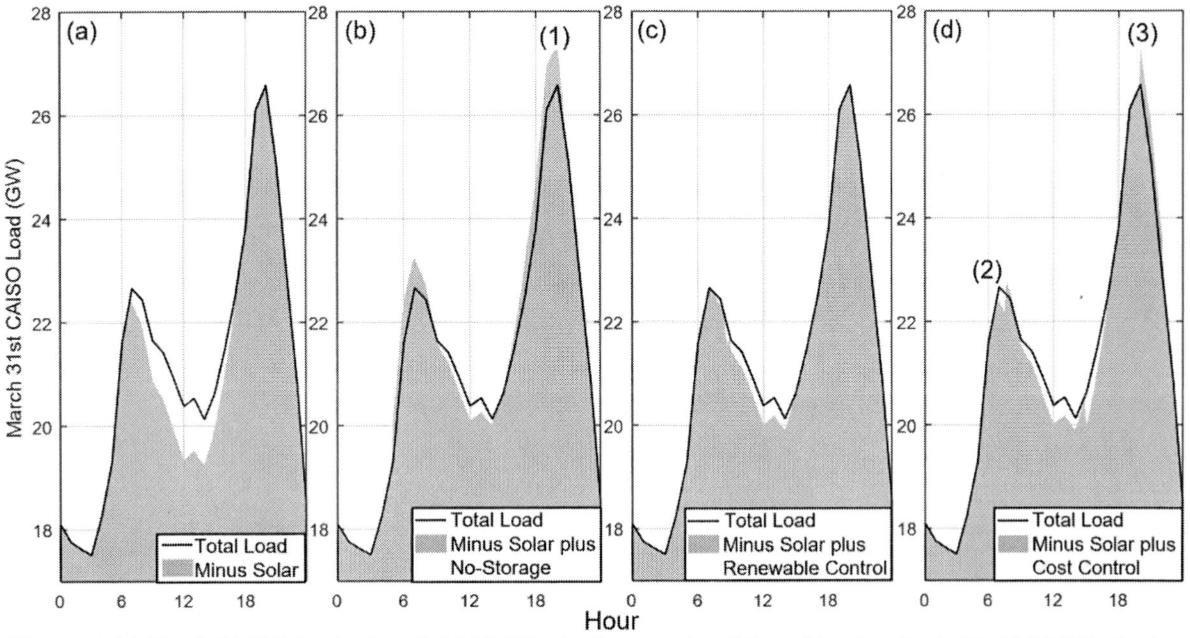

Figure 9 (a) Total CAISO load minus 1,125 MW of solar capacity, (b) total load and net of 1,125 MW of solar capacity and 4,495 schools with no storage, (c) Renewable Control storage, and (d) Cost Control storage

6. CONCLUSIONS

- A CTES system utilizing the Cost Control strategy has a positive externality of increasing the portion of the chiller energy consumption met by wind (by at least 10%), but this is not the case for the solar resource because the system without storage consumes more energy during the day to meet the same cooling load
- Either control strategy implemented with a storage system will be more cost effective than a no-storage system
- Renewable Control of a storage system results in at least 40% more of the chiller energy consumption being met by wind and at least 50% more met by solar as compared to the Cost Control strategy
- Implementing CTES systems with Renewable Control has the potential to reduce the duck curve peak load by approximately 700 MW if implemented solely on the secondary schools in California

NOMENCLATURE

CAISO	California Independent System Operator
COP	Coefficient of performance
CTES	Cool thermal energy storage
ERCOT	Electric Reliability Council of Texas
LCC	Life-cycle cost
NREL	National Renewable Energy Laboratory
PLR	Part-load Ratio
TMY3	Typical Meteorological Year, version 3
USDOE	United States Department of Energy

REFERENCES

Alternative Energy Institute (AEI). (2012). *Wind Data: White Deer, TX.* Retrieved from http://www.windenergy.org/datasites/. Last accessed June 2015.

California Independent System Operator (CAISO). (2013). *What the Duck Curve Tells us about Managing a Green Grid* (CommPR/HS/10.2013). Folsom, CA.

California Utilities Statewide Codes and Standards Team (CUSCST). (2011). *2013 California Building Energy Efficiency Standards: Final Measure Information Template – Chiller Minimum Efficiency.*

Duffie, J. A. & Beckman, W.A. (2013). *Solar Engineering of Thermal Processes* (4th ed.). New York: John Wiley & Sons, Inc.

High-Schools.com. (2015). *California High Schools.* Retrieved from http://high-schools.com/directory/ca/. Last accessed April 2016.

Myers, K.S., Klein, S.A., & Reindl, D.T. (2010). Assessment of high penetration of solar photovoltaics in Wisconsin. *Energy Policy, Volume 38*(11), 7338-7345.

National Renewable Energy Laboratory (NREL). (2008). *User's Manual for TMY3 Data Sets*. Retrieved from http://www.nrel.gov/docs/fy08osti/43156.pdf. Last accessed April 2016.

National Renewable Energy Laboratory (NREL). (2015a). System Advisor Model (Version 2015.6.30) [Software].

National Renewable Energy Laboratory (NREL). (2015b). *Western Wind Dataset.* Retrieved from http://www.nrel.gov/electricity/transmission/western_wind_methodology.html. Last accessed April 2016.

RSMeans. (2015). *RSMeans 2016 Mechanical Cost Data*. Rockland, MA: RSMeans Construction Publishers.

United States Department of Energy (USDOE). (2011). *U.S. Department of Energy Commercial Reference Building Models of the National Building Stock* (NREL/TP-5500-46861). Golden, CO.

United States Department of Energy (USDOE) (2015a). *2014 Wind Technologies Market Report* (DOE/GO-102015-4702).

United States Department of Energy (USDOE) (2015b). *SunShot Photovoltaic System Pricing Trends: Historical, Recent, and Near-Term Projections* (NREL/PR-6A20-64898).

ACKNOWLEDGEMENT

This work was supported by the American Society of Heating, Refrigerating, and Air-Conditioning Engineers (ASHRAE).

Dynamic Modeling and Performance Analysis of Sensible Thermal Energy Storage Systems

Austin NASH*, Neera JAIN

Purdue University,
West Lafayette, IN, USA
nash@purdue.edu

* Corresponding Author

ABSTRACT

In this paper we consider the problem of dynamic performance evaluation for sensible thermal energy storage (TES), with a specific focus on hot water storage tanks. We derive transient performance metrics, from second law principles, that can be used to guide real-time decision-making aimed toward improving demand response. We show how the transient nature of the metrics can be used not only to influence control variables within the system, but also to mitigate adverse effects of disturbances during operation. To evaluate these metrics in the context of hot water storage tanks, a thermal stratification model is needed. We derive a reduced-order model which allows the simulation of tank thermal stratification during all modes of system operation. The proposed performance metrics are analyzed in simulation using the dynamic tank model. The results highlight key trade-offs captured by the metrics that can be incorporated into future optimal control design for sensible TES systems.

1. INTRODUCTION

Motivation and Problem Definition: Within thermodynamic processes, waste heat losses occur on a large scale due to equipment inefficiencies and limitations. It has been estimated that anywhere from 20-50% of energy consumption in U.S. industrial applications alone is ultimately released as waste heat (*Waste Heat Recovery - Technology and Opportunities in U.S. Industry*, 2008; Semkov, Mooney, Connolly, & Adley, 2014). As such, there exists a great opportunity to develop and improve strategies for waste heat recovery. Waste heat recovery is most commonly carried out in centralized power plants through heat recovery steam generators where the thermal energy is immediately used to generate additional electricity (Giaccone & Canova, 2009). However, with increasing penetration of distributed power generation through micro-combined heat and power (micro-CHP) systems, opportunities for waste heat recovery at the residential scale are growing (Barbieri, Melino, & Morini, 2012). In these systems, the recovered heat is typically used to heat water that is stored in a hot water storage tank for domestic use. The use of a thermal energy storage (TES) system enables the recovered energy to meet future thermal demand. However, in order to design optimal control strategies to achieve demand response, dynamic performance metrics for TES systems are needed.

Gaps in Literature: Many researchers have studied the development of efficiency metrics to quantify the performance of hot water storage tanks using second law-based analyses. Most efforts, however, use finite time exergy-based metrics, rather than rate-based, metrics (Han, Wang, & Dai, 2009; Li, 2016; Fernández-Seara, Uhía, & Sieres, 2007a, 2007b). The disadvantage is that, while these types of metrics give a global perspective of performance which can be used in making offline design decisions, they fail to give an idea of how well energy is being used at any given time for online decision-making. Few researchers to date have performed thermodynamic optimization of equipment involving highly transient processes (Pizzolato, Verda, & Sciacovelli, 2015; Sciacovelli, Verda, & Sciubba, 2015). In order to optimize performance of a sensible TES in real-time, we desire transient second-law based performance metrics which can be used to embed advanced control strategies into system operation.

As previously mentioned, a common type of sensible TES system is a hot water storage tank. Dynamic modeling of hot water storage tanks has been studied by numerous researchers (Kleinbach, Beckman, & Klein, 1993; Han et al., 2009). Recently, researchers have also developed control-oriented dynamic models for hot water storage tanks with simultaneous sink and source mass flow loops (Powell & Edgar, 2013). Many hot water storage tanks utilize an immersed coil heat exchanger as a means of heat absorption, and dynamic models incorporating this feature are available within the literature (Cadafalch, Carbonell, Consul, & Ruiz, 2015). However, these models are typically computationally expensive due to the increase in dynamic states that results from the dynamic modeling of the heat exchanger. For eventual control design, we require a reduced-order model of a hot water storage tank with an immersed coil heat exchanger.

Contribution: We define a general set of dynamic performance metrics for the different operation modes of sensible TES systems. These metrics are then analyzed in simulation using a control-oriented dynamic hot water storage tank model; key benefits are demonstrated with respect to real-time decision-making and demand response for hot water storage tank operation.

Outline: In Section 2 we present a background on performance analysis of TES systems and thermal stratification in hot water storage tanks. Section 3 describes the sensible TES model proposed in this work, while Section 4 details transient performance metrics for evaluating system performance. Simulation results demonstrating the dynamic behavior of a hot water storage tank are presented in Section 5, followed by a set of case studies wherein we analyze the performance metrics developed in Section 4. This is followed by concluding remarks in Section 6.

2. BACKGROUND

In this section, we present an overview on the importance of performance metrics for improving demand response in sensible TES systems. We then introduce thermal stratification and its impact on the performance of sensible TES in hot water storage tanks.

2.1 Performance Metrics for Sensible Thermal Storage

In electro-chemical batteries, state-of-charge (SOC) is an important concept that enables the control engineer to design algorithms that effectively utilize the stored electrical energy (Cheng, Divakar, Wu, Ding, & Ho, 2011). In the context of thermal energy storage, little attention is paid to quantifying SOC; instead, performance and efficiency metrics typically offer a steady-state or aggregate perspective of the behavior of the system (Han et al., 2009; Pizzolato et al., 2015). These steady-state metrics are often used with the intent of influencing offline design factors such as storage sizing. From the perspective of real-time control, we require metrics that not only quantify the current SOC but also the charging and/or discharging behavior of the system, so that energy management strategies may be designed and implemented.

2.2 Thermal Stratification in Hot Water Storage Tanks

Thermal stratification in storage tanks is a phenomenon that results when a density gradient is present within the tank. The gradient causes the warmer, less dense water to rise to the top of the tank while the cooler, higher density water sinks to the bottom of the tank. The result is the presence of a high temperature section of water toward the top of the tank and a low temperature section of water at the bottom, as depicted in Fig. 1.

In the mixing region between the hot and cold sections, there exists a sharp temperature gradient called a *thermocline*. Thermal stratification in storage tanks is a very desirable characteristic. It is estimated that the efficiency of the heat storage can be increased by up to 20% when the tank is kept highly stratified with a thin thermocline region (Han et al., 2009). Therefore, modeling the dynamic evolution of the thermocline region has important implications on improving system performance. In order to most accurately quantify the thermal dynamics in a hot water storage tank, partial differential equations are needed in multiple dimensions. Alternatively, a lumped parameter assumption can be combined with a spatial discretization in order to derive a one-dimensional (1D) model of the system. A 1D model was introduced in the 1990s (Kleinbach et al., 1993). More recently, Cadafalch et. al. presented a method for incorporating an immersed coil heat exchanger into the analysis (Cadafalch et al., 2015). In the following section, we present a control-oriented, reduced-order 1D model for a thermally stratified hot water storage tank with an immersed coil heat exchanger.

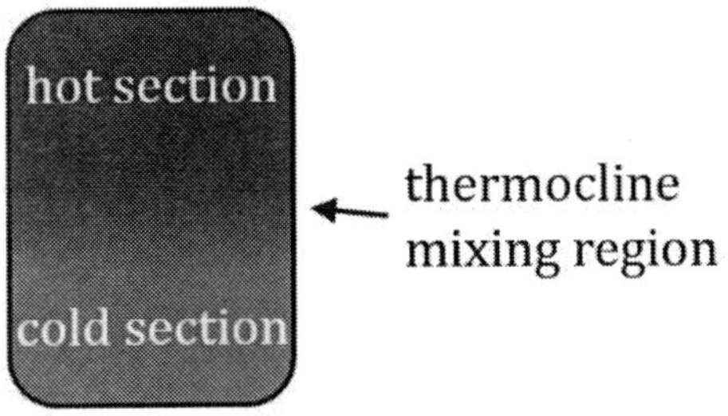

Figure 1: A schematic of a thermally stratified hot water storage tank.

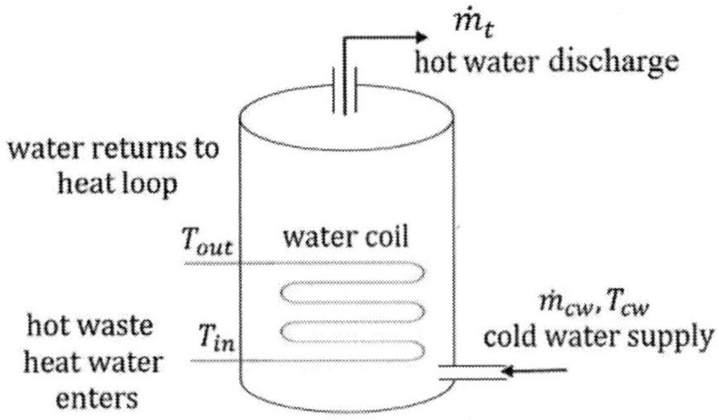

Figure 2: A cylindrical hot water storage tank with an immersed coil heat exchanger.

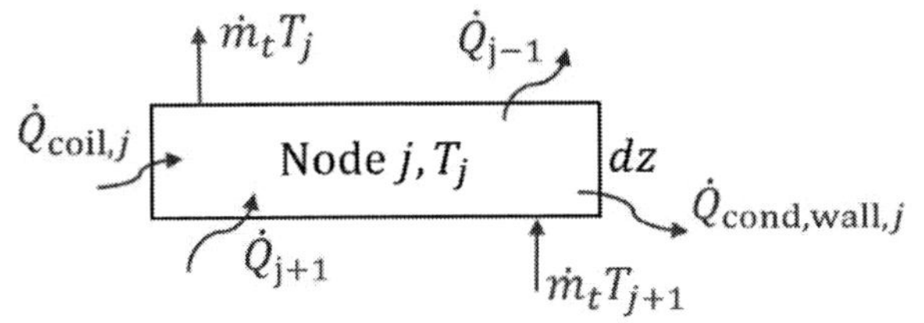

Figure 3: A discretized control volume.

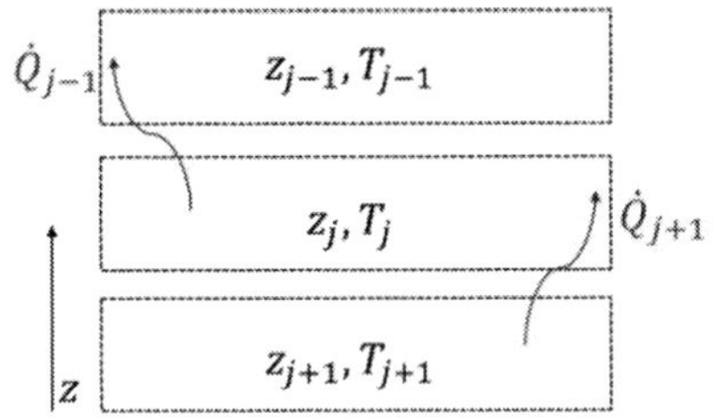

Figure 4: Finite difference scheme for discretized node.

3. HOT WATER STORAGE TANK MODEL

In this section we present the governing equations for modeling the thermal stratification in a cylindrical hot water storage tank. A schematic of the system is shown in Fig. 2. Hot waste heat water at a temperature T_{in} enters an immersed hot water coil situated at the lower portion of the tank. The waste heat water is then pumped vertically through the coil until it exits the tank at T_{out}. Heat is transferred from the warm water flowing through the coil to the colder water in the storage tank. We define three different modes of operation for the system. We define charging to be the heat addition mode, or the mode during which waste heat water is being pumped through the coil in order to heat the water in the tank. Conversely, we define discharging to be the heat rejection mode, or the mode during which the stored hot water is being pumped out of the tank and replaced with cold water. The hot water is removed from the tank at a flow rate $\dot{m}_t$ which acts as a disturbance on the system. Moreover, the water in the tank is replenished by an equal flow of water defined as $\dot{m}_{cw}$ which enters at the bottom of the tank. Finally, we define a third mode of operation to be simultaneous charging and discharging of the tank.

3.1 Governing Equations

In this section we derive a control-oriented model of the thermal storage tank dynamics that can be used for model-based control design. The model presented here builds upon work presented by previous researchers (Kleinbach et al., 1993; Powell & Edgar, 2013; Cadafalch et al., 2015). The tank is discretized vertically into n nodes (Kleinbach et al., 1993), with the top node defined as node one. An illustrative view of a discretized control volume is shown in Fig. 3.

Each node is allowed to exchange heat with its surroundings in several ways. Nodes can absorb heat from the coil, $\dot{Q}_{coil}$, and lose heat to the ambient by conduction through the tank walls, $\dot{Q}_{cond,wall}$. Additionally, each node is allowed to exchange heat with its bordering nodes, modeled as $\dot{Q}_{j+1}$ and $\dot{Q}_{j-1}$. While in discharge mode, water flows upward through each control volume. Each node is assumed to remain in steady-state with respect to mass flow. The bottom node is sized such that it contains the cold water flow inlet valve. Conservation of energy is used to derive a system of

n ordinary differential equations that can be solved numerically, yielding the temperature stratification in the storage tank as a function of time. For the jth node, the energy balance equation is

$$m_j c_v \frac{dT_j}{dt} = \dot{Q}_{coil,j} - \dot{Q}_{cond,wall,j} + \dot{Q}_{j+1} - \dot{Q}_{j-1} + \dot{m}_t c_v (T_{j+1} - T_j) \ . \tag{1}$$

One by one, nodal temperature differentials $\frac{dT_j}{dt}$ can be solved independently. Values for heat loss through the walls are given by $\dot{Q}_{cond,wall,j} = \frac{k_{wall}A}{w_t}\left(T_j - T_0\right)$, where k_{wall} is a tuning parameter, w_t is the approximate wall thickness, and T_0 is the temperature of the ambient environment. Logic is used to determine whether each node is within the coil region which in turn dictates whether or not there is heat transfer between the coil and that particular node. The internal heat transfer rate terms $\dot{Q}_{j-1}$ and $\dot{Q}_{j+1}$ are solved using a finite difference scheme.

3.2 Calculation of Internal Heat Transfer Between Nodes

To quantify the internal heat transfer rates, $\dot{Q}_{j-1}$ and $\dot{Q}_{j+1}$, a finite difference scheme is used (Powell & Edgar, 2013). Fig. 4 helps to illustrate the process. The scheme yields the following expressions for the internal heat transfer rates,

$$\dot{Q}_{j-1} = -k_{j-1} A \frac{T_{j-1} - T_j}{z_{j-1} - z_j} \ , \tag{2}$$

$$\dot{Q}_{j+1} = -k_{j+1} A \frac{T_j - T_{j+1}}{z_j - z_{j+1}} \ . \tag{3}$$

The z terms represent heights of the vertical midpoints of the nodes with respect to a zero datum at the tank's bottom. In Eqns. (2)-(3), the k terms are tuning parameters in the model and are used to handle temperature inversion within the model (Powell & Edgar, 2013). Temperature inversion is a phenomenon which occurs when warmer, less dense water exists below cooler, heavier water. In this work, temperature inversion exists due to the presence of the immersed heat coil in the lower portion of the tank. The coils heat the water at the bottom of the tank; this hot water then rises toward the top of the tank. One-dimensional models are incapable of naturally accounting for temperature inversion. Hence, an alternative way of incorporating the phenomenon is necessary.

At any point in simulation, if the temperature of node j is either lower than the node below or higher than the node above, the k_{j-1} and k_{j+1} terms are increased by several orders of magnitude in order to force heat to be transferred upward, thereby resolving the temperature inversion problem. The algorithm for the variation of the coefficients is given by

$$k_{j-1} = \begin{cases} k_{j-1}\Delta\left|T_j - T_{j-1}\right|, & \text{if } T_j > T_{j-1} \\ k_{j-1}, & \text{otherwise} \end{cases} \quad \text{and} \quad k_{j+1} = \begin{cases} k_{j+1}\Delta\left|T_j - T_{j+1}\right|, & \text{if } T_j < T_{j+1} \\ k_{j+1}, & \text{otherwise} \end{cases} \ , \tag{4}$$

where Δ is a tuning parameter whose magnitude is several orders higher than the magnitude of the k term itself.

3.3 Calculation of Coil Heat Transfer

In previous work, Cadafalch et. al. present a method of incorporating a transient analysis of the immersed heat exchanger (Cadafalch et al., 2015). In this work, we are interested in a reduced-order model of the interaction between the tank water and the coil in order to obtain a computationally efficient control-oriented model. To calculate the heat transfer rate between the coil and the tank nodes within the coil region, we treat the coil as a separate system acting at steady-state and assume a linear temperature profile from the tank coil inlet to the tank coil outlet as shown in Fig. 5.

In Fig. 5, the value of T_{in} is known. At any time t, the value of T_{out} is assumed to be equal to the temperature of the water of the surrounding node. This assumption is based upon the fact that 1) the mass flow rate inside the coil is slow (3-7 liters per minute) and 2) the amount of water in the coil is small in comparison to the overall amount of water in the tank. Assuming the coil to be at steady-state, the value for the heat transfer rate of the jth coil slice is calculated

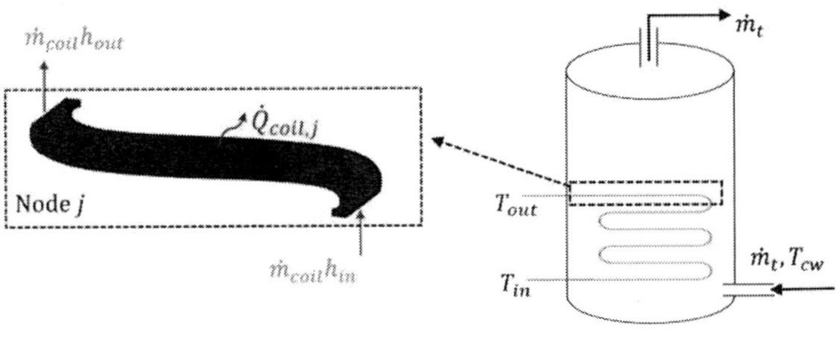

Figure 5: Schematic for calculating heat transfer rates due to immersed coil.

as $\dot{Q}_{coil,j} = \dot{m}_{coil}c_v\left(T_{coil,in,j} - T_{coil,out,j}\right)$. The values for $T_{coil,in,j}$ and $T_{coil,out,j}$ are determined using a linear reduction in temperature from T_{in} at the coil inlet to T_{out} at the coil outlet.

4. TRANSIENT PERFORMANCE METRICS

In this section, we present transient second-law based performance metrics for the modes of operation of the hot water storage tank.

4.1 Charging Performance Measure

To quantify transient performance during charging mode, we require a metric to compare the rate at which useful work (i.e. exergy) is supplied to the system to the rate at which useful work is lost. Overall, the metric should inform how efficiently exergy is being *stored* within the system. Therefore, we believe the best way to characterize performance during a charging mode is to use the standard exergetic efficiency metric, given by

$$\psi_c = 1 - \frac{\dot{X}_{dest}}{\dot{X}_{sup}} , \tag{5}$$

where $\dot{X}_{dest}$ represents the total rate of exergy destruction in the tank, $\dot{X}_{sup}$ represents the total rate of exergy being supplied to the tank, and the subscript c denotes charging. The supplied exergy rate is calculated based on the rate at which exergy is transferred via heat transfer from the immersed coil to the tank water. Within each node in the coil region, there is some amount of exergy transfer via heat transfer. The total rate of supplied exergy to the system is defined as the summation of all nodal supplied exergy rates as shown in Eqn. (6).

$$\dot{X}_{sup} = \sum_j \dot{X}_{sup,j} = \sum_j \left(1 - \frac{T_o}{T_j}\right)\dot{Q}_{coil,j} \tag{6}$$

To determine the rate of exergy destruction in the tank, an entropy balance is applied to each node. For the jth node, the entropy accounting equation is given by

$$\frac{d}{dt}S_j = \sum_i \frac{\dot{Q}_{i,j}}{T_{b,i}} + \dot{m}_t(s_{j+1} - s_j) + \dot{S}_{gen,j} . \tag{7}$$

For the heat transfer terms, boundary temperatures are taken to be as given in Table 1. The left hand side of Eqn. (7) is rearranged to yield

$$\frac{dS_j}{dt} = \frac{m_j c_v}{T_j} \cdot \frac{dT_j}{dt} . \tag{8}$$

Table 1: Boundary temperatures for heat transfer rates in entropy accounting equation

Heat transfer term	Boundary temperature
$\dot{Q}_{coil,j}$	T_j
$\dot{Q}_{j+1}$	$\frac{1}{2}\left(T_j + T_{j+1}\right)$
$\dot{Q}_{j-1}$	$\frac{1}{2}\left(T_j + T_{j-1}\right)$
$\dot{Q}_{cond,wall,j}$	T_o

Equation (7) can be solved for $\dot{S}_{gen,j}$. Then the corresponding exergy destruction rate for the jth node is $\dot{X}_{dest,j} = T_o\dot{S}_{gen,j}$. The total rate of exergy destruction in the tank is found by summing all of the nodal exergy destruction rates: $\dot{X}_{dest} = \sum_j \dot{X}_{dest,j}$.

4.2 Discharging Performance Measure

For any mode involving a discharge of thermal energy, we seek to quantify performance by analyzing the rate at which useful work is being recovered relative to the rate at which the stored thermal energy is changing. We define the disharge performance metric to be

$$\psi_d = \frac{\dot{X}_{rec}}{\dot{X}_{sto}} \,, \tag{9}$$

where $\dot{X}_{rec}$ is the rate at which exergy is being recovered from the TES, $\dot{X}_{sto}$ is the rate at which exergy is being stored, and the subscript d denotes a discharging mode. In order to determine the rate at which exergy is stored, exergy accounting equations are used on each node. For the jth node, the rate of change of exergy is given by

$$\frac{d}{dt}X_j = \sum_i \left(1 - \frac{T_o}{T_{b,i}}\right)\dot{Q}_{i,j} + \dot{m}_t(x_{j+1} - x_j) - T_o\dot{S}_{gen,j} \,, \tag{10}$$

where the boundary temperatures correspond to those given in Table 1 and $\dot{S}_{gen,j}$ is calculated according to Eqn. (7). We define the total rate of exergy stored in the system to be

$$\dot{X}_{sto} = \left|\sum_j \frac{dX_j}{dt}\right| \,. \tag{11}$$

We define the rate of exergy recovered during discharge to be the rate of flow exergy exiting the top of the tank with respect to a setpoint state whose temperature is T_{set}. The setpoint state acts as a threshold for undesirable system performance during a discharging mode. We take the rate of recovered exergy to be

$$\dot{X}_{rec} = \dot{m}_t\,(x_1 - x_{set}) = \dot{m}_t\left[c_v\,(T_1 - T_{set}) - T_o c_v \ln\left(\frac{T_1}{T_{set}}\right)\right] \,. \tag{12}$$

The value of T_{set} is a user-defined parameter which serves to quantify the instant at which useful work is no longer being recovered from the system. Once the temperature of the water at the top of the storage, T_1, is less than T_{set}, the recovery rate will become a negative quantity, thereby changing the sign of the performance metric.

5. SIMULATION RESULTS

In this section, we first present the dynamic evolution of tank thermal stratification during the simultaneous charging and discharging operation mode. Next, we demonstrate how the dynamic performance metrics capture the relative effect of control input variables and disturbance input variables on system performance using the dynamic storage tank model. The physical parameters used in these simulation studies are based on an experimental micro-CHP system at Purdue University; values are provided in Table 2. A discretization of 30 nodes is used in all simulation studies, yielding 30 dynamic states.

Table 2: Baseline values used for hot water storage tank model simulations

Parameter	Value	Parameter	Value
tank height	1.3 m	coil volumetric flow rate	2 lpm
tank diameter	0.4 m	tank volumetric flow rate	2 gpm
coil inlet temperature	45° C	tank wall thickness	0.051 m
dead state temperature	20° C	height of coil inlet	0.15 m
take initial temperature	20° C	height of coil outlet	0.58 m
cold water supply temperature	20° C	set point temperature	30° C

5.1 Dynamic Thermal Stratification Behavior

The tank is initially charged for two hours. During this time, hot waste heat water from an integrated power generation system flows through the coil, thereby exchanging heat with the tank. At t = 120 minutes, the tank enters the simultaneous charging/discharging mode in which hot water is removed from the tank at a specified flow rate $\dot{m}_d$ while heat continues to be added to the system by the coil. The temperature evolution of the thermal stratification for this mode, over a period of 30 minutes, is shown in Fig. 6.

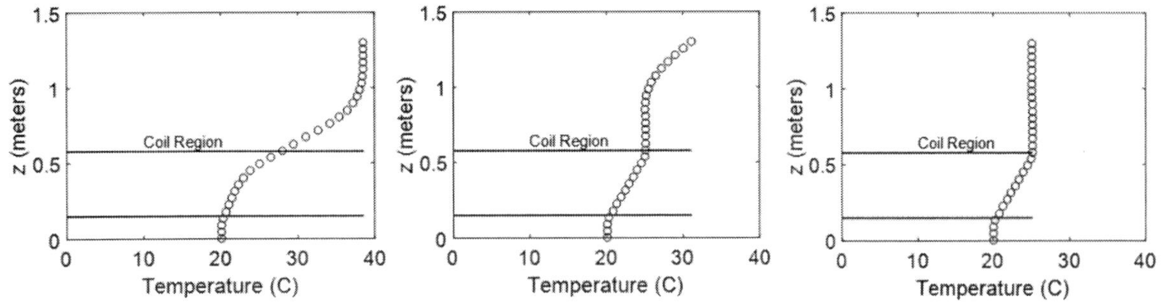

(a) minute 130 (b) minute 140 (c) minute 150

Figure 6: Illustration of thermal stratification during 30 minute simultaneous charge/discharge.

In Fig. 6a, the charge/discharge has been ongoing for ten minutes. A portion of the hot water above the coil region has exited the tank and been replenished with the cold water supply. However, the coil waste heat water continues to heat the cold water as it is drawn upwards through the tank. In Fig. 6b, it is clear that the hottest water in the tank, which was located above the coil at a temperature of 40°C, has been removed from the tank. Finally, in Fig. 6c, all of the water above the coil region is at a temperature below 30°C. At this point, the tank has reached its steady-state thermal gradient; water being removed from the tank is heated from the cold water supply temperature of 20°C to about 27°C before it leaves the tank. The mixing of the water being heated by the coil, and the cold supply water flowing into the tank, results in a widened thermocline.

5.2 Analyzing System Performance

In this section, we present results for performance characterization of the different modes of system operation. Performance is analyzed with respect to two types of variables: (1) control variables and (2) disturbance variables.

5.2.1 *Control Variables and Charging Performance* Within the system, one variable over which we have control authority is the volumetric flow rate of hot waste heat water through the immersed coil. In order to examine the charging performance of the system, a three hour charge is simulated for coil flow rates varying from one to five liters per minute (lpm). In each case, charging performance is quantified according to Eqn. (5) and the results are compared in Fig. 7.

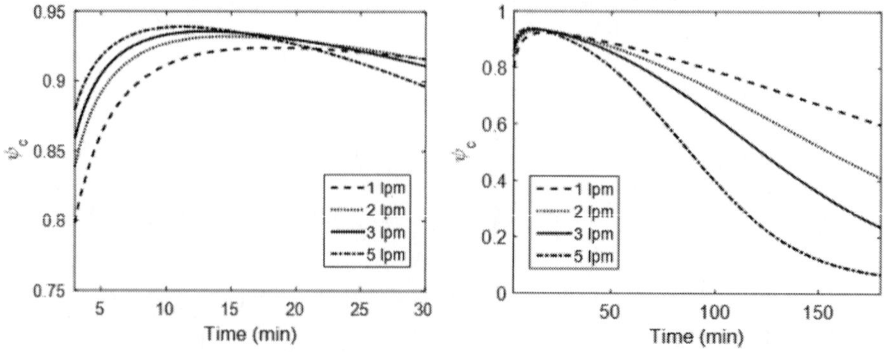

(a) first 30 minutes

(b) full 3 hours

Figure 7: Simulation results depicting effect of varying coil flow rate on charge performance.

Figure 7(a) shows that initially, higher flow rates yield better performance. However, as times progresses, the higher flow rates actually have an adverse effect on system performance. As the tank heats up, there is a decrease in the temperature differential between the waste water in the coil and the water in the tank. This leads to a decrease in the heat transfer rate between the coil and the tank, and consequently, there is less useful work being supplied to the tank. This aspect of system performance can be seen more clearly through Fig. 8 which shows how the coil outlet temperature varies as a function of time. Based on the assumptions presented in Section 3, the coil outlet temperature at any time is equal to that of the surrounding water in the tank. In Fig. 8, we see that for high coil flow rates, the tank heats up, or charges, more quickly. However, the tank then reaches its maximum state of charge more quickly as well, and system performance subsequently degrades as the coil cannot add any additional charge to the system. Once the maximum charge is reached, a higher amount of useful work is lost relative to the amount being supplied via the immersed heat exchanger coil.

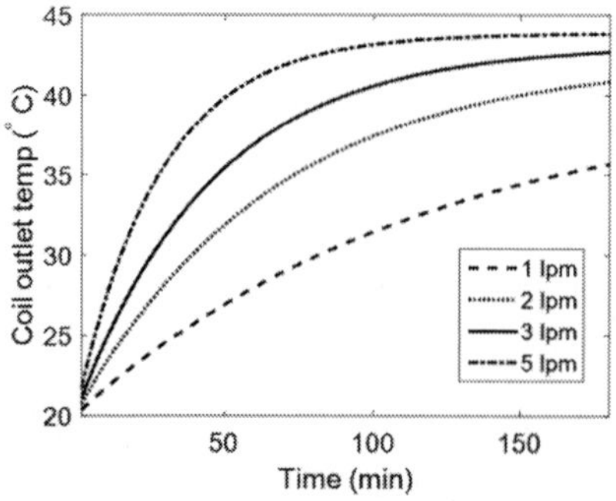

Figure 8: Simulation results depicting effect of varying coil flow rate on coil outlet temperature.

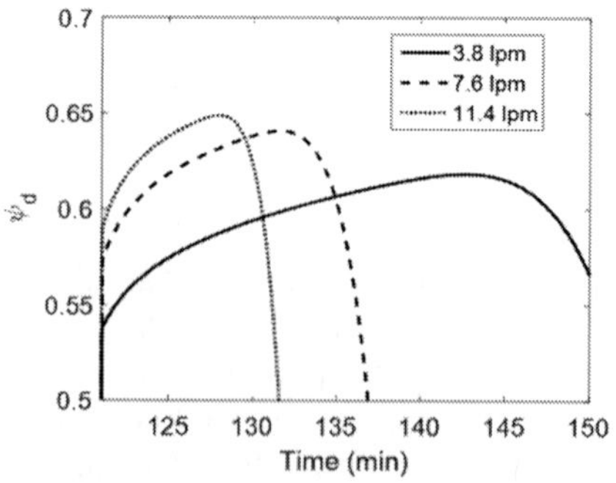

Figure 9: Simulation results depicting effect of varying tank flow rate on discharge performance.

These results suggest that in order to maximize the charging rate while also maximizing charging efficiency, a variable coil flow rate could be used. For example, the coil flow rate could be set high toward the beginning of the charging period but later decreased as the derivative of the efficiency curve becomes negative. In this way, the performance metric for charging modes can be used in real-time to make control decisions to help improve global system performance.

5.2.2 *Disturbance Variables and Discharging Performance* From a controls perspective, we are also interested in understanding the effect of measured disturbances on system performance. The primary disturbance input is the volumetric flow rate of water exiting the tank during a discharge mode. While we cannot control the timing or duration of discharge modes, we can use the performance metric in Eqn. (9) to inform real-time control decisions in order to limit adverse effects from an energy utilization standpoint.

To investigate the effects of different discharge flow rates on the system, we charge the tank from its cold state for two hours. The hot waste heat water is then turned off and a discharge mode is allowed to commence for 30 minutes at a specified tank flow rate. Fig. 9 compares the discharge performance for tank flow rates ranging from one to five gallons per minute, or 3.8-11.4 lpm. The figure shows that higher tank flow rates allow better performance at first. However, as the state-of-charge decreases, performance at higher flow rates degrades quickly. This makes intuitive sense; as hot water is removed from the tank, cold supply water is being added to the tank thereby reducing the overall state-of-charge. This is further illustrated by Fig. 10.

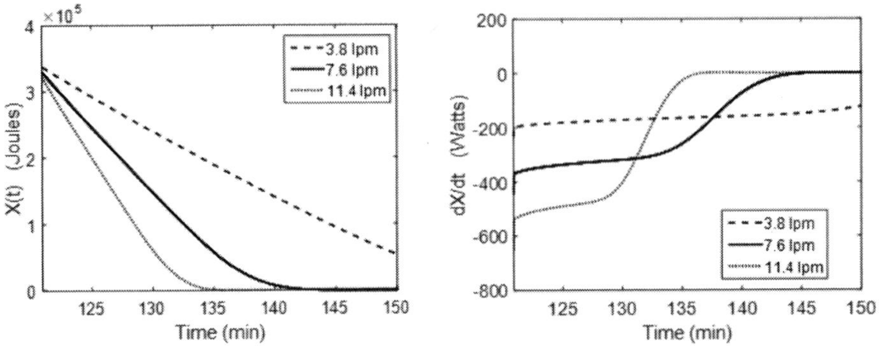

(a) exergy available in tank

(b) rate of change of available exergy

Figure 10: Simulation results depicting effect of varying tank flow rate on exergy storage rate and available storage.

In Fig. 10b, we see that for higher flow rates, the absolute value of the rate of change of exergy, $\left|\frac{dX}{dt}\right|$, or the exergy storage rate, quickly decreases. The result is that the available exergy within the system decreases rapidly, seen in Fig. 10a. As the high flow rates deplete the system of its charge, cold water is pumped into the system and the available thermal energy decreases further. The system rapidly approaches steady state in which cold water exists uniformly throughout the tank and useful work is unavailable. For smaller tank flow rates, the exergy storage rate in the system decreases much less rapidly, signaling that a higher amount of charge remains available within the storage. The simulations presented in this section allow real-time insight into system performance for various modes of operation. Performance metrics can be used to design control algorithms aimed at improving demand response and to mitigate adverse effects on performance with respect to disturbances.

6. CONCLUSION

In this paper we defined a set of dynamic performance metrics that are generalizable to a range of thermal energy storage systems. These metrics were then analyzed in the context of a hot water storage tank with an immersed coil heat exchanger, a common type of sensible TES system. We derived a reduced-order model for simulating thermal stratification within the hot water storage tank. More specifically, a simplified model for the interaction between the tank's immersed coil heat exchanger and the surrounding water reduced the overall number of dynamic states needed to describe the transient behavior of the system. Using the dynamic hot water storage tank model, we demonstrated the sensitivity of the performance metrics to changes in both control input variables and disturbance input variables. The performance metrics presented for each mode of operation will enable future work aimed at designing advanced control algorithms for improved system performance and demand response.

NOMENCLATURE

				Subscript			
A	area	s	entropy			j	node j
c_v	specific heat	T	temperature	b	boundary	$j-1$	node above j
k	heat transfer coefficient	w_t	tank wall thickness	c	charge	$j+1$	node below j
$\dot{m}$	mass flow rate	x	specific exergy	d	discharge	sup	supplied
ψ	performance metric	z	height in tank	$dest$	destroyed	t	tank
$\dot{Q}$	heat transfer rate			gen	generated	0	dead state

REFERENCES

Barbieri, E. S., Melino, F., & Morini, M. (2012, September). Influence of the Thermal Energy Storage on the Profitability of Micro-CHP Systems for Residential Building Applications. *Applied Energy*, *97*, 714–722.

Cadafalch, J., Carbonell, D., Consul, R., & Ruiz, R. (2015, February). Modelling of Storage Tanks with Immersed Heat Exchangers. *Solar Energy*, *112*, 154–162.

Cheng, K. W. E., Divakar, B. P., Wu, H., Ding, K., & Ho, H. F. (2011, January). Battery-Management System (BMS) and SOC Development for Electrical Vehicles. *IEEE Transactions on Vehicular Technology*, *60*(1), 76–88.

Fernández-Seara, J., Uhía, F. J., & Sieres, J. (2007a, January). Experimental Analysis of a Domestic Electric Hot Water Storage Tank. Part II: Dynamic Mode of Operation. *Applied Thermal Engineering*, *27*(1), 137–144.

Fernández-Seara, J., Uhía, F. J., & Sieres, J. (2007b, January). Experimental Analysis of a Domestic Electric Hot Water Storage Tank. Part I: Static Mode of Operation. *Applied Thermal Engineering*, *27*(1), 129–136.

Giaccone, L., & Canova, A. (2009, June). Economical Comparison of CHP Systems for Industrial User with Large Steam Demand. *Applied Energy*, *86*(6), 904–914.

Han, Y., Wang, R., & Dai, Y. (2009, June). Thermal Stratification Within the Water Tank. *Renewable and Sustainable Energy Reviews*, *13*(5), 1014–1026.

Kleinbach, E., Beckman, W., & Klein, S. (1993). Performance Study of One-dimensional Models for Stratified Thermal Storage Tanks. *Solar Energy*, *50*(2), 155–166.

Li, G. (2016, January). Sensible Heat Thermal Storage Energy and Exergy Performance Evaluations. *Renewable and Sustainable Energy Reviews*, *53*, 897–923.

Pizzolato, A., Verda, V., & Sciacovelli, A. (2015). Local Entropy Generation Analysis of Transient Processes: An Innovative Approach for the Design Improvement of a Thermal Energy Storage with Integrated Steam Generator. In *Constructal Law & Second Law Conference*. Parma (Italy).

Powell, K. M., & Edgar, T. F. (2013, December). An Adaptive-Grid Model for Dynamic Simulation of Thermocline Thermal Energy Storage Systems. *Energy Conversion and Management*, *76*, 865–873.

Sciacovelli, A., Verda, V., & Sciubba, E. (2015, March). Entropy Generation Analysis as a Design Tool: A Review. *Renewable and Sustainable Energy Reviews*, *43*, 1167–1181.

Semkov, K., Mooney, E., Connolly, M., & Adley, C. (2014, September). Efficiency Improvement Through Waste Heat Reduction. *Applied Thermal Engineering*, *70*(1), 716–722.

Waste Heat Recovery - Technology and Opportunities in U.S. Industry (Tech. Rep.). (2008, March). BCS, Incorporated.

Optimization of Air-Source Heat Pump Systems over the Heating Season through the Use of Renewable Energy Sources

Elena BEE[1]*, Alessandro PRADA[1], Paolo BAGGIO[1]

[1]Dept. of Civil, Environmental and Mechanical Engineering – Univ. of Trento – Trento (Italy)
elena.bee@unitn.it
alessandro.prada@unitn.it
paolo.baggio@unitn.it

* Corresponding Author

ABSTRACT

The drive to cover with renewable energy sources an increasing quota of the energy use of a building is boosting the installation of heat pumps coupled with PV panels for residential heating applications. This HVAC solution is especially advantageous in high performance buildings when low temperature hydronic systems (e.g. radiant panels) are adopted. Nonetheless, the seasonal performance of the heating system is strongly dependent on HVAC design choices, such as the system sizes (heat pump, water storage tank, PV battery), and on the system control strategies adopted. Besides, if too many heat pumps are installed an unbearable burden may be imposed to the national power grid. Thus, an optimal design and control of the heating systems installed in high performance building is essential to ensure, along the entire heating season, not only a low energy use but also the appropriate exploitation of in-situ available renewable energy sources. This study presents a possible approach to the optimization of different system configurations, storage and control strategies for high performance buildings. A massive building located in the northern part of Italy has been considered to perform a preliminary application of this approach. The dynamic simulation of the building and its heating system was set up using the TRNSYS simulation suite and the optimization of the HVAC systems was performed coupling the simulation code with a Genetic Algorithm, to minimize the total energy consumption. Results highlight the role of thermal and electrical storage to optimize self-consumption and enhance the energy performance of air source heat pump. Besides, the study points out the issues of coupling systems with high response time in high performance buildings.

1. INTRODUCTION

This work aims to optimize the design of a heating system that includes PV panels and an air-source heat pump, two currently widespread technologies used to optimize the exploitation of renewable energy. The risk of covering heating needs with renewable energy is to overlook the objective of limiting energy use, the latter being not proportional to operating costs as in the case of fossil fuels. On the other hand, using renewable and zero emissions energy sources, the mere minimization of the energy needs of a building is not necessarily the only goal. Consequently, a multi-objective evaluation can be more appropriate in order to define optimal design solutions, although it is difficult to establish the right objective functions. Since the overall performance of such a system is strongly dependent on the weather and on the building load, each optimization has been made setting these two pre-conditions. In this preliminary application of the approach, the weather conditions of Trento, as representative of a climate of Northern Italy, have been considered. Regarding the building load, four different cases have been analyzed, based on two reference building types. The final purpose is to investigate the difference in the optimization results for different reference buildings and, in addition, the extent to which variations of some features of the envelope affects the optimal configuration of the HVAC system.

2. METHODS

2.1 General Approach

The work deals with the challenging of correct design and operation of air source heat pump (ASHP) coupled with photovoltaic (PV) systems. The choice of optimal trade-off between low energy consumption of ASHP and high PV integration without oversizing the PV system is based on the domination of a general solution X over a solution Y. According to Pareto, a solution X is said to dominate the other solution Y if both the following conditions are true:

1. The solution X is no worse than Y in all objectives;
2. The solution X is strictly better than Y in at least one objective.

Thus passing from Y to X an improvement for all the objectives is supposed, or an improvement for some, without the other ones be harmed. In other words, the concept of "Optimum of Pareto" simply indicates a situation in which it is impossible to improve the situation of an objective without making worse the others. The two steps of the optimization procedure are therefore the definition of the Pareto front and the selection of a trade-off solution among the Pareto front. For the multi-objective optimization (MOO) analysis of the building retrofit, an Elitist Non-dominated sorting genetic algorithms (NSGA-II) (Deb 2002) was implemented in MatLab (Penna *et al.*, 2014). The fitness function used in the analysis is a Matlab code that launch Trnsys model for the building energy simulation. After the model execution, the function reads the TRNSYS output file and post-processes the results of the simulation in order to compute the objective functions. In particular, the energy demand for heating, the fraction covered by PV power and the PV overproduction are chosen as goals for the MOO. The choice of these objectives aims at achieving the enhancement of the integration among HVAC components. In fact, by decreasing the energy consumption of the heat pump it becomes more difficult the achievement of a minimum value or renewable coverage factor avoiding the PV oversizing.

2.2 Genetic Algorithm Implementation

The implemented GA is an Elitist Non-dominated sorting GA algorithms (NSGA-II) (Deb, 2002). Nonetheless several customization of the code are used such as sampling, crossover, mutation and selection procedure. These adjustments coupled with the selection of mutation rate, population size, crossover fraction are adopted with the purpose of increasing the GA performances.The first step in the GA procedure is the selection of the initial population. The GA optimization is closely related to the initial population. In this regard, random sampling could lead in oversampling of same region, whereas a uniform random number generation produces uniform sample when the population size is high (Saltelli *et al.*, 2004). For this reason, the code was equipped with a Sobol's sequence sampling in order to overcome the clustering which can occur with sample random sampling or quasi random generator (Saltelli *et al.*, 2004). Sobol' sequence is a low-discrepancy sequence, which aim to give a uniform distribution of values; in higher dimensions (Burhenne *et al.*, 2011). The random starting point in the Sobol' sequence was obtained through a pseudo-random generator (Matsumoto and Nishimura, 1988). Once the fitness function is evaluated, the GA proceeds with the selection of the best individuals. Selection is the procedure by which GA chooses parents for the next generation. In this study we adopted the tournament selection without replacement (TSWOR) (Goldberg *et al.*, 1989, Golberg and Deb, 1991). In this method a short list of 4 eligible parents are randomly chosen and the best individual out of that set to be a parent. Following on from this point, the code combines the genetic characteristics of both parents, giving rise to the new generation. The recombination procedures implemented is based on arithmetic weighting of parents' genes to create children. Children are a random (Matsumoto and Nishimura, 1988) arithmetic mean of two parents, uniformly on the line between the parents. (Burjorjee, 2013). The adopted crossover fraction, i.e. the fraction of the next generation that crossover produces is set equal to 0.8. The remaining individuals in the next generation becomes from mutation of population. Mutation is applied at a random point in a random individual. In particular, by means of Mersenne-Twister pseudo random generator (Matsumoto and Nishimura, 1988), a randomly selected gene is replaced by a uniformly distributed random value that meet the gene range.

2.3 Test Cases

Two typologies of building are investigated (Figure 1) with the purpose of extending the representativeness of the analysis. In particular, a penthouse-like house (S/V = 0.63) and an intermediate flat in a multi-story building (S/V=0.3) have been considered. Also the orientation of the glazed façade was varied: south and east oriented windows have been considered. Overall, four test cases were selected; these buildings were developed starting from a reference building module, which is a typical flat having 100 m^2 floor surface, 3 m internal height and façades oriented towards the main cardinal directions (Penna *et al.*, 2015).

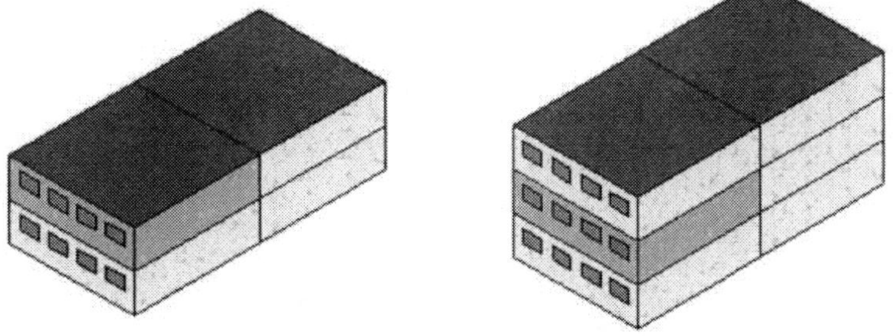

Figure 1: Building typologies used in the GA procedure (Penna *et al.*, 2015).

2.4 Simulation of building-plant system

The dynamic behavior of the system was calculated by means of TRNSYS simulation (Figure 4). The reference heating system is based on an ASHP with variable speed compressor, coupled with radiant floor panels. A water storage tank separates the circuit of the water heated by heat pump and radiant panels loops, on which a circulation is installed. Therefore, the total electric load is the sum of power needed by heat pump (compressor) and circulation pump. The system is powered both by grid and photovoltaic electric power. Two different approaches for handling battery storage were implemented: the parameter "*mode*", defined internally in the simulation project, states which one is applied. In the first strategy (*mode*=0) the PV array is operated in parallel with the grid and the priority is to cover the load with direct photovoltaic power. The excess power is delivered to the batteries until they are fully charged, then is passed to the grid. When the PV are not generating power anymore, the available energy stored in the batteries is used. The second operation mode (*mode*=1) is an UPS like mode that avoids operation of the batteries in parallel with the grid but favors complete battery charge and discharge cycles. The system is connected to the grid only when the battery has been completely discharged and is disconnected when the battery has been fully charged. Figures 2 and 3 better explain these two approaches showing power output management for three representative days within the simulation period (Dec, 30th – Jan, 1st).

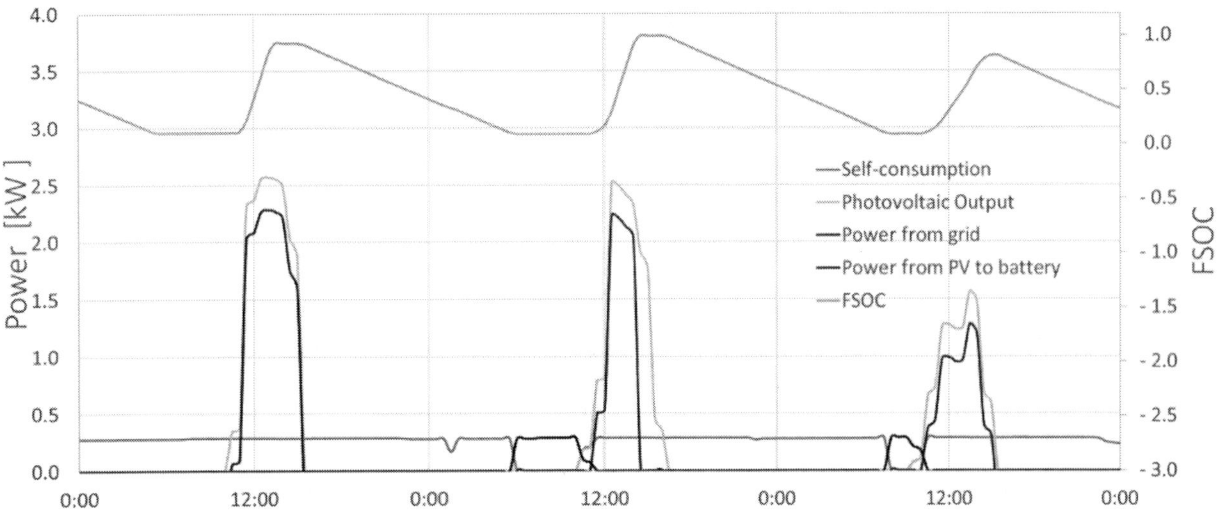

Figure 2: Power managing strategy corresponding to *mode*=0 (PV in parallel with the grid)

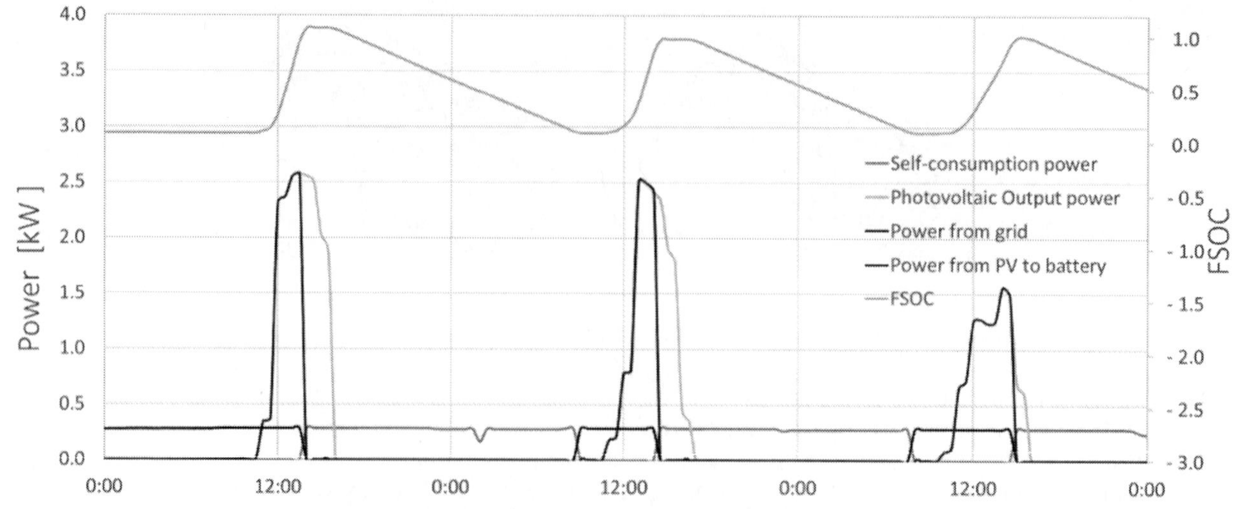

Figure 3: Power managing strategy corresponding to *mode*=1 (UPS mode)

The variable speed heat pump model used in the simulation program TRNSYS is described in (Bee *et al.*, 2016). It takes into account the typical function of performance (f_{corr}) of commercial products for part load working conditions. The model allows to modify that function by means of two factors, that have been set as optimization parameters in the present work (see Table 1). The first parameter is the CR value where the COP begins to degrade; the second is the CR value where the maximum of f_{corr} occurs (this last set to 1.2, as can be the case with the last generation of variable speed heat pumps commercially available). Also the rated thermal power and the rated COP (both referred to 7-35°C) are variable parameters. Thermal power and compressor power at different reference temperatures are given in input to the model as fractions of respective rated values. The photovoltaic array has been described with type 194 in the TESS library. This model is based on the calculation method presented by De Soto *et al.*, 2006. Performance specifications of photovoltaic modules are representative of high efficient products currently used in residential applications. The output of interest from this component is the electrical power, that, at every simulation time-step, can be used in different ways, depending on the described approaches. The number of modules is an optimization parameter and each module has an area of 1.6 m^2. The battery was defined with type 47, a model of a lead-acid storage battery that operates in conjunction with solar cell array and power conditioning components. It specifies how the battery fractional state of charge (FSOC) varies over time, given the rate of charge or discharge. A minimum FSOC (0.1) has been set so that the model is consistent with real applications, being the complete discharge of the battery not recommended. The capacity depends on the total numbers of cells, each having a capacity of 1.2 kWh. Radiant floor panels are defined within the building model as an active layer (Solar Energy Laboratory, 2005). Radiant floor design refers to a typical commercial configuration: pipe spacing 0.12 m, pipe outside diameter 0.016 m, pipe wall thickness 0.002 m and pipe wall conductivity 0.44 W $m^{-1}K^{-1}$. The system is equipped both with climatic and zone control. Climatic control changes the supply water temperature as a linear function of the outdoor air temperature. The curve can be modified throughout the optimization process, by means of two parameters: the supply water temperatures corresponding to outdoor air temperatures of 18 °C and -5 °C (plant operation limits for the considered reference year). If the heat pump reaches his lower modulating limit and water is warmed up more than necessary, an on/off controller turns it off. This controller works with a dead-band, which width was optimized. The zone control is handled by a proportional band thermostat (type 1669) with a variable set-point and a band width of 0.5 °C or 1°C (optimization parameter). Set-point can be adjusted between 19 °C and 22 °C during the day and between 15 °C and 19 °C during the night. Also the hours when the set-point changes are parameters to be optimized. Thermostat output signal controls the flow rate circulating in the radiant floor loops, that can vary from 0 kg/h to the maximum design flow rate (400 kg/h). Finally, the building has been modeled by means of the TRNSYS Multi-zone Building subroutine, type 56. Envelope features and considered test cases are described in the previous paragraph. The complete list of optimization parameters previously mentioned is reported in Table 1. Figure 4 represents a schematic view of the system, as in the Simulation Studio TRNSYS tool.

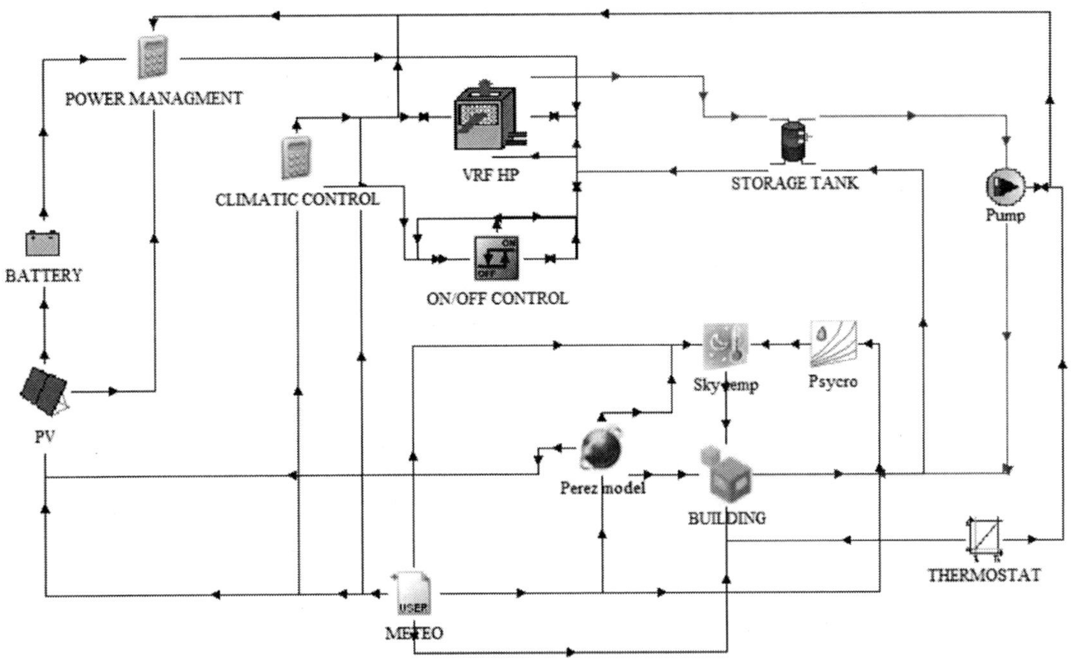

Figure 4: Building and heating system model (TRNSYS)

Table 1: Optimization parameters

Description	**Units**	**Minimum value**	**Maximum value**	**Variation step**
Tank volume	m^3	0.05	2	0.05
Climatic regulation curve - parameter 1	°C	25	30	1
Climatic regulation curve - parameter 2	°C	30	40	1
Dead band half width for HP on-off controller	°C	0.5	2.5	0.5
Dead band half width for thermostat	°C	0.25	0.5	0.25
HP rated thermal power at 7-35 °C	kW	4	10	0.05
HP rated COP at 7-35 °C	-	3.5	5	0.1
HP part load function – parameter 1	-	0.2	0.4	0.05
HP part load function – parameter 2	-	0.45	0.6	0.05
Number of photovoltaic modules	-	2	12	2
Number of battery cells	-	4	16	4
Mode (battery charge management)	-	0	1	1
Set-point temperature	°C	19	22	1
Set-back temperature	°C	15	19	1
Set-back start time	h	19	23	1
Set-point start time	h	5	8	1

3. RESULTS

The results of the calculations are shown in Figure 5, where each point represents a simulation run. The energy consumption and the PV production are normalized with respect to the building floor area. Obviously, all the results lie in the lower left triangular part of the diagram (below the bisector line) because self-consumption of PV generated power cannot be higher than total energy use. As can be expected, the optimal solutions (i.e the Pareto front) are, for the most part, found close to the border of the solutions domain (see Figure 6 and 7). It is immediately clear that some engineering judgment should be applied when evaluating such optimal solutions. In fact, in this preliminary application, the multi-objective optimization of total energy use, PV energy self-consumption and PV energy surplus are pursued with the same weight. From a practical point of view, the most interesting solutions are possibly the ones that ensure low total energy consumption (around 5 kWh m^{-2}), and, at the same time, high PV self-consumption (around 3-4 kWh m^{-2}). These can be found on the lower left part of the graphs. Looking at the "optimal" values assigned to the adjustable parameters for the points some results are obvious, for example the best results are obtained for higher values of the COP (HP rated COP = 4-5) and for operation of the batteries in parallel with the grid (*battery charge management* = 0). A bit more remarkable are the results obtained for the selection of the optimal value for some other parameters such as the capacity of the hot water storage tank (Figure 6) installed in the hydronic heating system and the total capacity of the installed batteries (Figure 7). As can be clearly inferred looking at the color of the dots in figures 6 and 7, the optimal combination of water and electricity storage is noticeably affected by the characteristics of the building. The buildings with the larger value of S/V = 0.63 seem to perform better with larger tanks and smaller batteries capacity with respect to the ones with smaller S/V = 0.3. This is probably due to the higher heat demand of the building that causes a high number of charge/discharge of the storage tank. On the contrary, in S/V=0.3 the storage tank should be smaller in order to decrease the residence time of energy in the storage tank and thus the heat losses.

4. CONCLUSIONS

The HVAC system of a building is becoming more and more complex in order to decrease energy use, increase the renewable energy quota exploited to cover such energy needs and, more recently, to maximize the self-consumption of in-situ generated renewable energy. In order to reach these targets, the role of local energy storage is becoming of fundamental importance. Such energy storage can be pursued at the building level with different methods: thermal energy can be stored exploiting the thermal capacity of the building structure and/or water storage tanks and electric energy using batteries. And it is quite possible that a balanced combination of these methods is the optimal solution. All this is making the task of the HVAC designer increasingly complex and the dynamic interaction among building, energy systems and occupant behavior should be taken into account especially in high performance buildings. The multi-objective optimization approach presented here can be a useful tool in order to steer the predesign and early design phase of a high performance HVAC system. Some preliminary results obtained with this approach are presented here for the heating system of a small building. The study points out the issues of coupling systems with high response time in high performance buildings. The optimal mix of HVAC configurations and managements are strictly related to the building characteristics. The buildings with low dispersing surface (S/V 0.30) and with South oriented windows required a higher battery capacitance, since the PV production is concurrence with the high solar gains and thus with the minimum in heating demand. Moreover, the low heating demand guides to the choice of small storage tank in order to limit the energy waste trough heat losses. On the other hand, the increasing dispersing surface (S/V 0.63) and the East oriented glazings ensure a high number of cycle of charge/discharge of the storage tank, allowing a reduction of the heat losses hence bigger storage thanks can be adopted. The authors recognize that there are still some shortcomings and are currently performing further research work in order to: a) shorten the computing time (now several hours of computer time are required to complete the run for each building) by means of high efficiency algorithms, b) investigate the effectiveness of the MOO objectives (to favor the solutions having lower total energy needs) and c) extend the application to a much larger variety of buildings having different size and structural characteristics and climates.

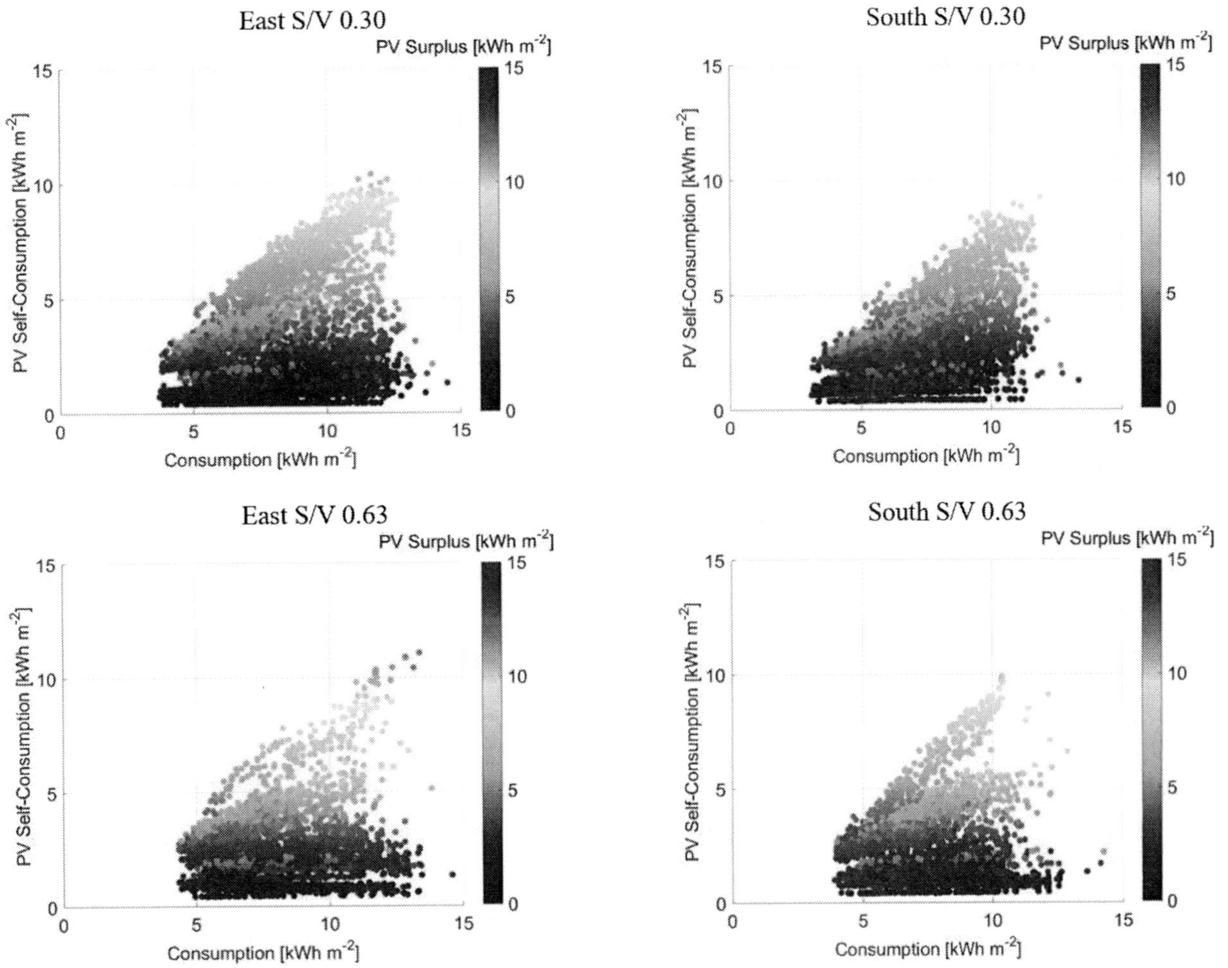

Figure 5: Objective functions for all the explored simulation field and for the four test cases. Each dot represents one of the simulation runs and together they show the spanned solution space. The colored scale shows the surplus energy generated by the photovoltaic (PV) panels delivered to the grid (i.e. the third MOO target parameter)

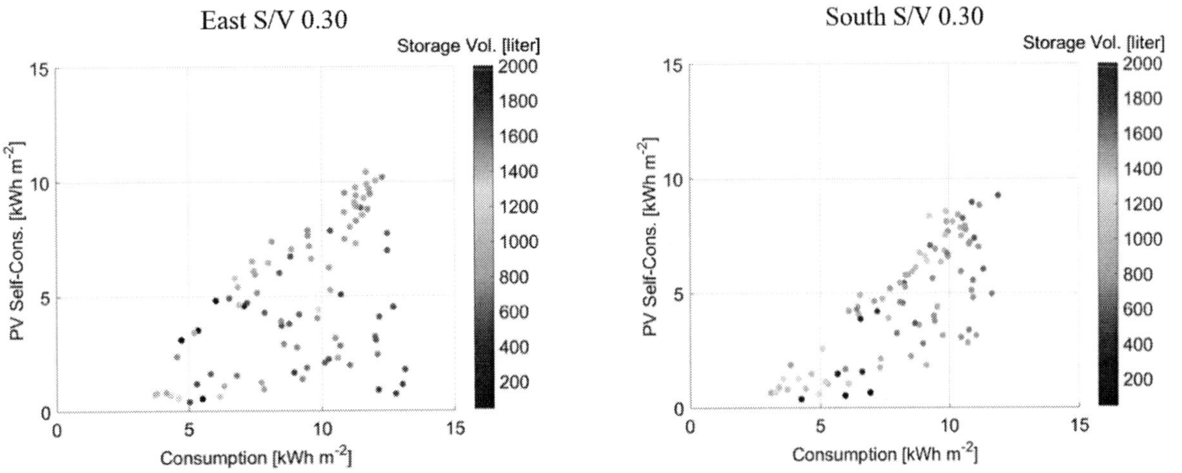

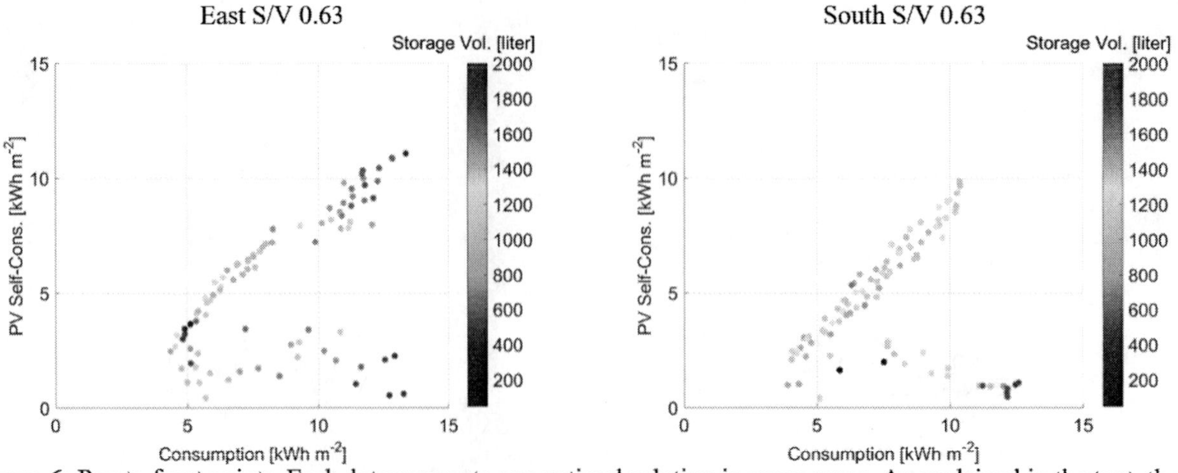

Figure 6: Pareto front points. Each dot represents one optimal solution in some sense. As explained in the text, the "really" optimal solutions are represented by the dots close to the 5 kWh m^{-2} energy consumption and 3-4 kWh m^{-2} PV generated energy consumption values. The colored scale shows the capacity (in liters) of the hot water storage tank installed in the hydronic heating system, calculated by the optimization process for each different solution.

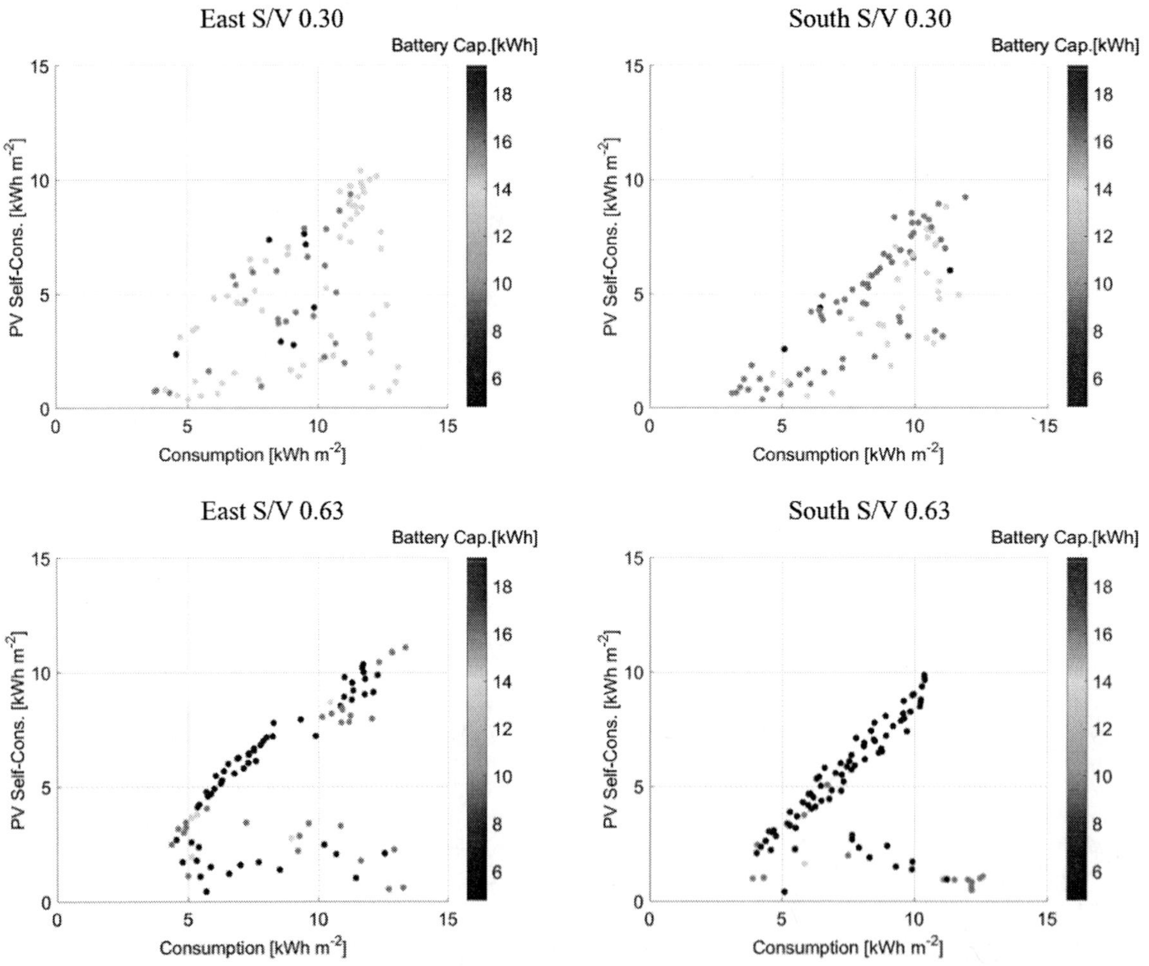

Figure 7: Pareto front points. Each dot represents one optimal solution in some sense. As explained in the text, the "really" optimal solutions are represented by the dots close to the 5 kWh m^{-2} energy consumption and 3-4 kWh m^{-2} PV generated energy consumption values. The colored scale shows the total capacity of the installed batteries (in kWh), calculated by the optimization process for each different solution.

NOMENCLATURE

COP	coefficient of performance	(–)
CR	capacity ratio	(–)
FSOC	fractional state of charge	(–)
f_{corr}	correction function	(–)

REFERENCES

Bee, E., Prada, A. & Baggio, P. (2016). Variable-speed air-to-water heat pumps for residential buildings: evaluation of the performance in northern Italian climate. *Accepted for 12th Rehva World Congress.*

Burjorjee, K. M. (2013). Explaining optimization in genetic algorithms with uniform crossover., *Proceedings of Foundations of Genetic Algorithms*, 37–50.

Burhenne, S. Jacob, D. & Henze, G. P. (2011). Sampling based on Sobol's sequences for Monte Carlo techniques applied to building simulations, *Proceedings of 12th Building Simulation*, p. 1816–1823.

Deb, K. (2002). A fast and elitist multi-objective genetic algorithm: NSGA-II, *IEEE transactions on evolutionary computation*, 6(2), 182–197.

De Soto, W., Klein, S.A., Beckman, W.A. (2006). Improvement and validation of a model for photovoltaic array performance. *Solar Energy,* 80, 78–88

Goldberg, D. E., Korb, B., & Deb, K. (1989). Messy genetic algorithms: motivation, analysis, and first results. *Complex Systems*, 3(5), 493–530.

Goldberg D.E. & Deb, K. (1991). A comparative analysis of selection schemes used in genetic algorithms, *Proceedings of Foundations of Genetic Algorithms*, 69–93

Klein S.A. (2009). TRNSYS, A Transient Systems Simulation Program. Version 17. 1, Solar Energy Laboratory, University of Wisconsin–Madison.

Matsumoto, M. & Nishimura, T. (1998). Mersenne Twister: a 623-dimensionally equidistributed uniform pseudo-random number generator, *ACM Transactions on Modeling and Computer Simulation,* 8(1), 3-30.

Penna, P., Prada, A., Cappelletti, F. & Gasparella, A. (2014). Multi-objectives optimization of Energy Efficiency Measures in existing buildings. Energy and Buildings, 95, 57–69.

Saltelli, A., Tarantola, S., Campolongo, F., and Ratto, M. (2004). Sample generation, in *Sensitivity analysis in practice. A guide to assessing scientific models*. Chichester (UK): John Wiley & Sons, 193– 204.

Solar Energy Laboratory (2005) TRNBuild, TRNSYS 17 Manual, 5 Multi-zone Building modeling with Type56 and TRNBuild, University of Wisconsin-Madison

ACKNOWLEDGEMENT

The authors want to thank the GIS unit of the Edmund Mach Foundation for the raw weather data of the Trento meteorological station.

Investigation on A Ground Source Heat Pump System Integrated with Renewable Sources

Defeng Qian*, Fuxin Niu, Steve Kavanaugh, PhD, Zheng O'Neill, PhD, PE

The University of Alabama,
Tuscaloosa, AL, United States
Phone: 978-328-9009
Email: dqian@crimson.ua.edu

*Corresponding Author

ABSTRACT

This paper presents a laboratory test facility for a solar powered ground source heat pump (GSHP) system. A comprehensive data acquisition system for performance monitoring is introduced as well. The preliminary data collection includes temperatures for water and air loops, solar heat radiation flux, solar panel output power, and power consumptions of GSHP. In addition, the feasibility of the system is discussed in the paper. The feasibility analysis includes the utilization of the solar system, energy balance of the geothermal system and performance of the heat pump. The analysis is based on a 29-day performance test in the cooling mode. The analysis results show that the performance of this solar powered geothermal heat pump system was acceptable during the test.

1. INTRODUCTION

According to the recent research (Berardi, 2015), from 2007 to 2035, the global demand for oil will increase by 30%, while the demand for coal and natural gas will increase by 50%. Those data are telling us that the energy-related carbon emissions will increase significantly if there are no radical changes of the energy structure. Meanwhile, the U.S. consumed 17% of the energy in the world, and buildings consumed 40% of the energy and represented 40% of the carbon emissions in the U.S. (DOE, 2012). This is more than any other sector of the U.S. economy, including transportation and industry. About 24% of all energy used in the nation was for space heating, cooling and water heating in buildings (DOE, 2012). Improving energy efficiency and reducing energy consumption in buildings is one of the most important priorities during the design and operation stages of buildings.

Enhancing building efficiency is one of the simplest, most immediate and most cost-effective ways to reduce the carbon emissions. If an efficient Heating, Ventilation and Air-Conditioning (HVAC) system is combined with the renewable energy source, the goal of net-zero building could be achieved. Theoretically, the goal of net-zero energy building (NZEB) is to reduce the energy consumption (and demand) through efficient designs and operations, and utilize the renewable energy as a major energy source. In a NZEB, the conventional energy source plays a backup role in the building. Therefore, the key points of the goal of NZEB include: 1) a high efficient building envelope and HVAC system and 2) a renewable energy source.

As one of the most efficient system in the market, the Ground Source Heat Pump (GSHP) is considered as one of the best solutions for the building HVAC system for a wide variety of geology conditions. The GSHP utilizes the relatively constant temperature of the ground, ground water, and surface water, and it is about 45% more efficient than conventional heat pumps (ENERGY STAR, 2015).

Generally speaking, renewable energy is defined as a kind of energy can be replenished with the passage of time. Solar energy is one type of renewable energy and is considered as an unlimited and an environment friendly energy source, which has been widely used for solar thermal and solar power applications. Currently, the primary source of electricity is petroleum and coal, which leads a huge amount of carbon emission. Meanwhile, energy arrived at the earth from the sunlight in one hour is more than energy consumed by human beings through one year (Lewis, 2007). If renewable energy can be utilized in the buildings for the energy (e.g., electricity) source, then, there will be a great reduction of carbon emissions due to less electricity generation needed.

This paper presents the design and the setup of a solar powered GSHP system with a full data acquisition system in a laboratory. Some preliminary data is being collected in the cooling mode and will be presented, followed by the feasibility analysis of the system performance using such data and future work.

2. SYSTEM DESCRIPTIONS

In this test rig, a ¾-ton water-to-air GSHP is connected to two 60-feet deep wells (as shown in Figure 1). A group of solar PV panels of 1.12 kW (Note: 16 panels were installed about 20 years ago and 4 panels are new) is connected to two 800 Ah battery banks, which are used to power the GSHP system and a 0.27 kW DC powered well pump. During the daytime, solar PV panels convert solar photons into electrical energy, which will be stored into the battery bank. Whenever the system is on demand, the battery banks will provide the electricity power. Figure 1 shows the design and setup of this GSHP system. Figure 2 shows the design schematics of a water-to-air GSHP system used in this study. Due to the low pH of the groundwater while the water quality at this site is good, a heat exchanger was introduced between the heat pump condenser and the ground water loop. Figure 3 depicts the GSHP with accessories including a circulation water pump, various sensors, a heat exchanger, etc. Table 1 lists the major sensors deployed in this test rig.

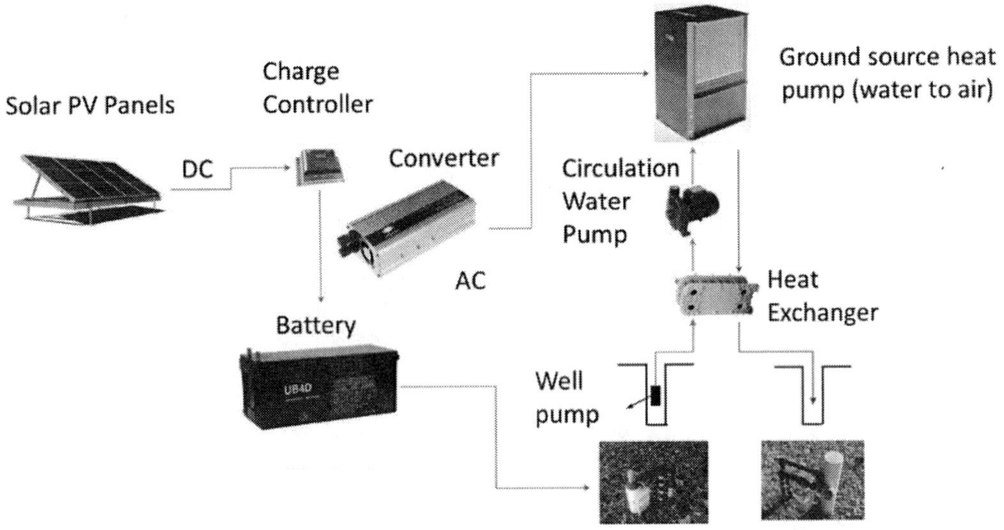

Figure 1. Design and setup of the solar powered GSHP system.

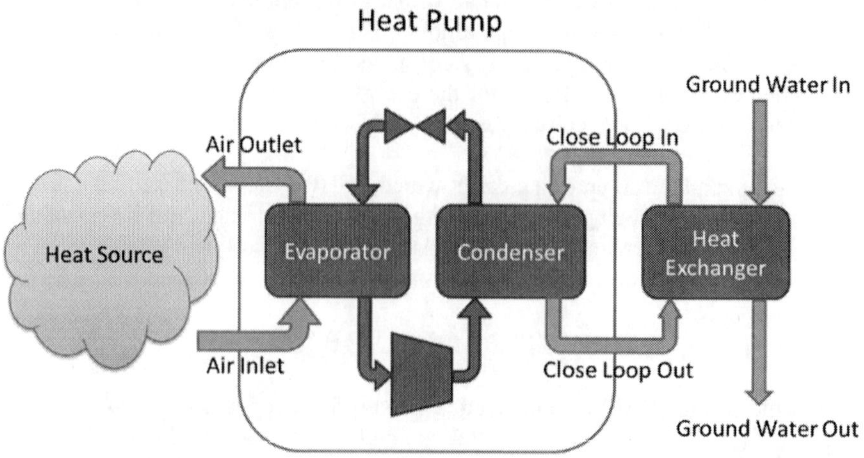

Figure 2. Schematic design of the water-to-air GSHP System (cooling mode).

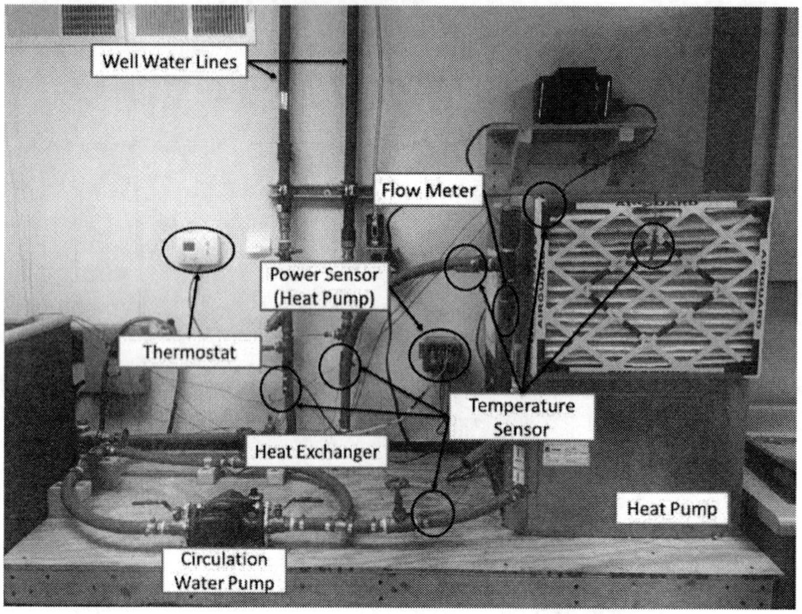

Figure 3. The setup of GSHP in the lab.

Table 1. Major Sensors Deployed in The Test Rig

Position	Sensor*
Ground Water Inlet (from the well)	Temperature Sensor 1
Ground Water Outlet (to the well)	Temperature Sensor 2
Circulate Water Inlet (to the GSHP)	Temperature Sensor 3
Circulate Water Outlet (from the GSHP)	Temperature Sensor 4
Air Intake (to the GSHP)	Temperature Sensor 5
Air Outlet (from the GSHP)	Temperature Sensor 6
Power Sensor (GSHP)	Power Transducer
Solar Panel Input Voltage	Voltage Divider
Solar Panel Input Current	Current Shunt
Circulation Loop (closed water loop)	Flow Meter

*The temperature sensor is a T-type thermocouple.

For the setup of the solar panels, although the solar panel exposes to the sun for a relatively longer time in the summer than that in the winter, there is not much difference on the electricity demand generated by the solar panel between the summer and the winter. To balance the solar energy collection, the angle of the solar panel is fixed at 30°.

In addition, as mentioned before, a comprehensive performance monitoring and data acquisition system was installed. The data acquisition is combined with a chassis system which has 5 hybrid slots, 3 PXI Express slots (up to 250MB/s per-slot bandwidth and 1.75 GB/s system bandwidth) and two input modules. The thermocouple input module has 32-channel, 8 built-in cold junction compensation channels and 0.3 °C accuracy. The voltage module has 16 analog inputs, two analog outputs, 16-bit resolution and a range of $\pm$10V. The sampling rate is once per second for both temperature and power generation. In order to match the sampling rate of the weather station (once per two minutes), the average of the solar radiation flux over two minutes will be used for our analysis.

By using the data acquisition system, well ground water temperatures, refrigerant temperatures, air temperatures, solar power generation, etc. are all real-time monitored, trended and stored in a database. In addition, an on-site weather station is installed to measure outside air temperature, relative humidity, wind speed and direction, and solar radiation intensity (note: total global horizontal solar flux only). Figure 4 shows the data acquisition system and Figure 5 shows the on-site weather station.

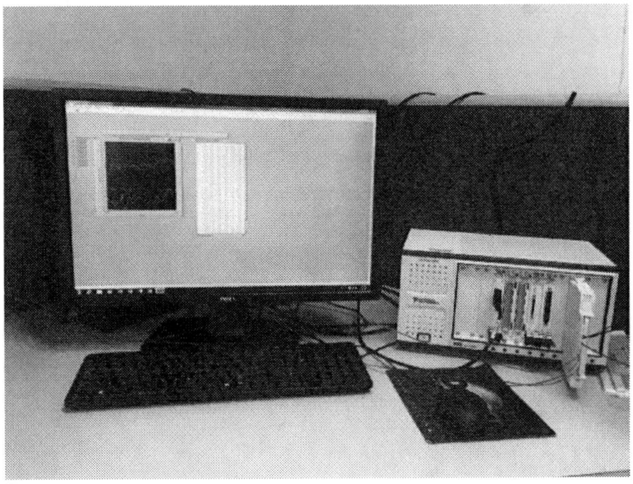

Figure 4. Data acquisition system

Figure 5. On-site weather station

To analyze the system performance, three data analysis were conducted: 1) solar energy collection/power generation, 2) geothermal energy balance and 3) heat pump energy balance.

3. SYSTEM PERFORMANCE ANALYSIS

3.1 Solar Energy Collection/Power Generation

On the supply side, solar panels will only be able to generate the electric energy during the daytime, which means the battery bank needs to store enough energy for the potential usage after the sunset and bad weather such as rainy days. The way to monitor the solar energy collection/power generation in this study is to measure the power voltage generated by the solar panels and current flow between the solar panels and the battery bank. Based on the measured voltage and current, the power generation can be calculated based on the following equation:

$$P = V \times I \tag{1}$$

where

P is the power generation from the solar panels (W),
V is the voltage of the solar panels (V),
I is the current flow between the solar panels and the battery bank (A).

Some of solar panels that are connected to the heat pump were installed about 20 years ago, therefore, it is expected that there are some performance degradations. In addition, the solar angle changes through the years. This will cause the fluctuation of electricity energy generation from the solar panels. The following assumptions are used to simplify the calculation:

1) solar energy collection/power generation can be maintained at the peak,
2) the variations of weather impact are neglected,
3) the batteries can continuously store energy from the solar panels,
4) although the manufacture specifies the efficiency of the inverter is up to 93%, an 85% efficiency will be taken during the calculation, and
5) the discharge capacity is at a 70% level of the total capacity of the batteries.

For the charge and discharge time, the following equations are used respectively:

$$t_{charge} = \frac{C_{battery} \times V_{battery}}{P_{charge}} \tag{2}$$

where

t_{charge} is the time to charge the battery (hr),
$C_{battery}$ is the capacity of the batteries (Ah),
$V_{battery}$ is the battery voltage (V),
P_{charge} is the charge rate (W).

$$t_{discharge} = \frac{C_{discharge} \times \eta \times V_{battery}}{P_{discharge}} \tag{3}$$

where

$t_{discharge}$ is the time to discharge the battery (hr),
$C_{discharge}$ is the discharge capacity of the battery (Ah),
η is the inverter efficiency,
$P_{discharge}$ is the discharge rate (W).

In general, the battery cannot be fully discharged. Usually, the discharge capacity is about 70% to 80% of the capacity of the battery. In this study, a 70% level of the battery capacity is assumed to be used.

3.2 Geothermal Energy Balance

The energy balance between the ground water and the closed-loop water is tested using the following equation:

$$\dot{m}_{GW} \times C_{p_{water}} \times \Delta T_{GW} = \dot{m}_{CL} \times C_{p_{water}} \times \Delta T_{CL} \tag{4}$$

where

$\dot{m}_{GW}$ is the mass flow rate of ground water ($\frac{kg}{s}$),
$\dot{m}_{CL}$ is the mass flow rate of closed-loop water ($\frac{kg}{s}$),
$C_{p_{water}}$ is the specific heat of water ($\frac{J}{kg*K}$),
ΔT_{GW} is the temperature difference between the ground water inlet and outlet (K),
ΔT_{CL} is the temperature difference between the water inlet and outlet of the heat pump (K).

3.3 Heat Pump Energy Balance

The total energy balance represents the system performance in either the cooling or heating mode. The total energy balance is based on the energy consumption of the heat pump, the heat exchange-rate between the water-side and air-side of the heat pump, and the cooling or heating rate of the heat pump. In this paper, all the tests are conducted for the cooling mode. The following equation is used to calculate the cooling rate and check the total energy balance.

$$P_{heat\ pump} + q_{rej} + q_{cooling} = 0 \tag{5}$$

where

$P_{heat\ pump}$ is the power input of the heat pump (W),
q_{rej} is the heat rate rejected to the water-side of the heat pump (W),
$q_{cooling}$ is the cooling rate of the heat pump (W).

The energy consumption of the heat pump is monitored by the data acquisition system and energy exchanged between the water-side and the air-side will be calculated using the measured data including temperatures and flow rates. The actual cooling rate will be calculated using Equation (5).

3.4 The Coefficient of Performance (COP)

The COP is a dimensionless parameter which is used to measure the efficiency of the heat pump. It is calculated by using the cooling rate divided by the power consumption. A higher value of COP reflects a higher efficiency of a heat pump. In this paper, the COPs for both GSHP and the system are calculated using the following equations:

$$COP_{heat\ pump} = \frac{q_{cooling}}{P_{heat\ pump}} \tag{6}$$

$$COP_{system} = \frac{q_{cooling}}{P_{heat\ pump} + P_{water\ pumps}} \tag{7}$$

where

$q_{cooling}$ is the cooling rate of the heat pump (W),
$P_{heat\ pump}$ is the power input of the heat pump (W),
$P_{water\ pump}$ is the total power consumption of the water pumps in the system (W).

4. RESULT AND DISCUSSIONS

4.1 Solar Energy Collection/Power Generation

In this paper, a four-week long data between January 30th, 2016 to February 27th, 2016was collected and analyzed. Figure 6 shows the solar energy collection/power generation results over the time in February. The area under the curve represents the collection of solar energy, high wave crest means relatively high solar power generation (W) and vice versa. In addition, the fluctuation of the curve reflects the change of the available solar irradiance. According to the plot, for some days, such as February 8th and 9th, the highest charging power during the experiment period was around 750 Watts. Meanwhile, for some days, such as February 15th and 22nd, the weather could be the possible reason for a lower peak charging power of around 100 Watts.

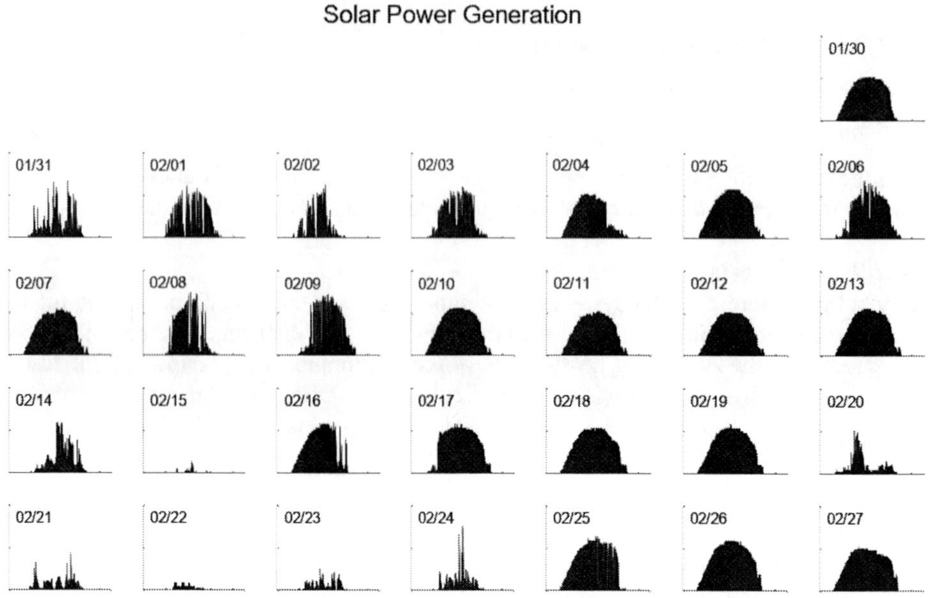

Figure 6. Solar energy collection (in Watts) over the time in February.

In addition, a further analysis was conducted between the theoretical power generation and actual power generation (as shown in Figure 7), the red solid line represents the theoretical power generation from the solar panels, and the blue dot-dash line represents the actual values based on measurements using Equation (1). The theoretical results were calculated using solar radiation heat flux from the on-site weather station and manufacturer's specification (SIEMENS, 2016) of the solar panels. According to the specification, when the cell temperature is 25.0°C and the solar irradiance is at 1000 W/m^2 level, the power generation is 50 Watts for each panel. In this test, 16 panels were used to charge the batteries. This means an 800 Watts charging power in total.

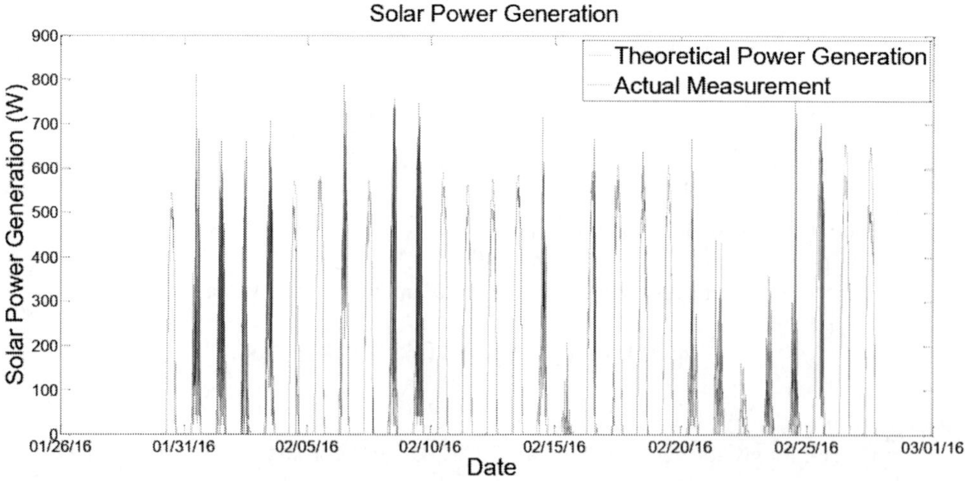

Figure 7. Comparison between the theoretical and actual power generation from the solar panel.

The further analysis shows that the theoretical generated energy was about 297 MJ, while the actual measured generated energy from the solar panels was 242 MJ. The percentage difference between the theoretical value and actual measurement was about 18.6%. There are several reasons that could contribute such difference:

1. The actual location of the on-site weather station is higher than those solar panels, which means a solar panel at the location of the weather station is able to collect more solar radiation compared with the actual solar panels.
2. The actual charging rate depends on the state of charge (SOC) of the batteries. Full charging power was applied only when the SOC was relative low, and with the increase of the SOC, the charging rate could be reduced, which caused the sensors detect a low charging power. Meanwhile, the theoretical result was based on the on-site weather station, and the applied algorithm assumed the batteries were charged with full power. Therefore, the actual charged energy could be less than the theoretical value at the time when SOC was relatively higher.
3. Some of solar panels were installed about 20 years ago, the degradation on the performance could be another factor that may cause the actual reading is less than the theoretical result.

4.2 Charge and Discharge

There is a need to know the charge and discharge time of the batteries. Calculations of the charge and discharge time were based on the assumptions listed in section 4.1.

From the measured data, the peak power generation from the solar panels was about 754 Watts. If all the batteries were empty and the electricity generation from the solar panel was assumed to be at the peak value always, it would take 12.7 hours for the batteries to be fully charged by using Equation (2). At the same time, the test also showed that the heat pump energy consumption was about 600 Watts during the operation, which means the batteries could be drained in about 9.5 hours if the heat pump kept running by only using power provided by the batteries. The experiment was conducted in February with a relatively short daylight time and a low solar radiation level. If the same test was done in the summer, the peak solar radiation flux and the solar panel power generation would be higher than the current result, which leads to a higher charging power and a less charging time. Figure 8 shows the solar radiation flux through the year of 2015, and there was about 25% more solar radiation in the summer than in February. By combining the monthly plots together, it is easy to tell that the solar radiation peak appeared between May to September of the year (as shown in Figure 9).

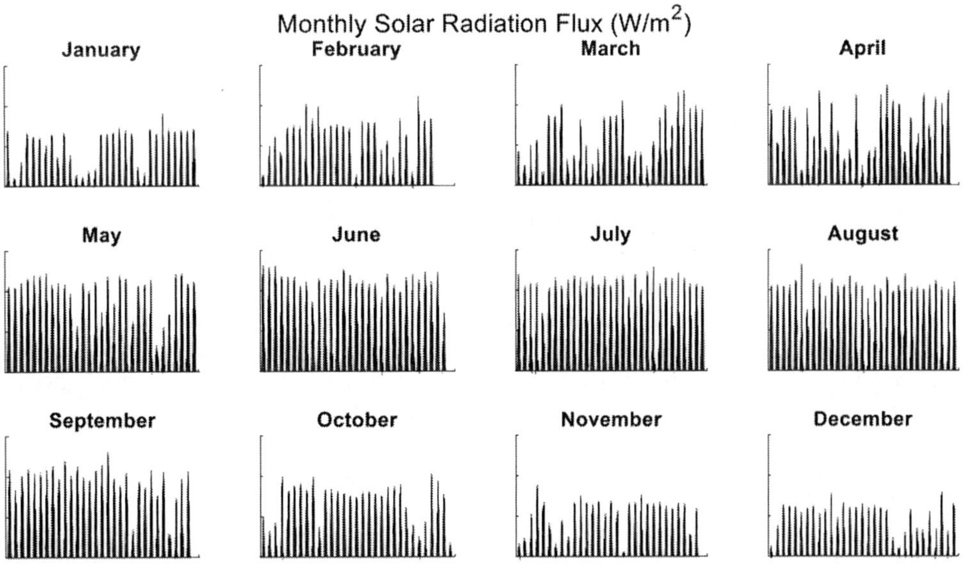

Figure 8. Monthly plots of solar radiation flux (W/m^2).

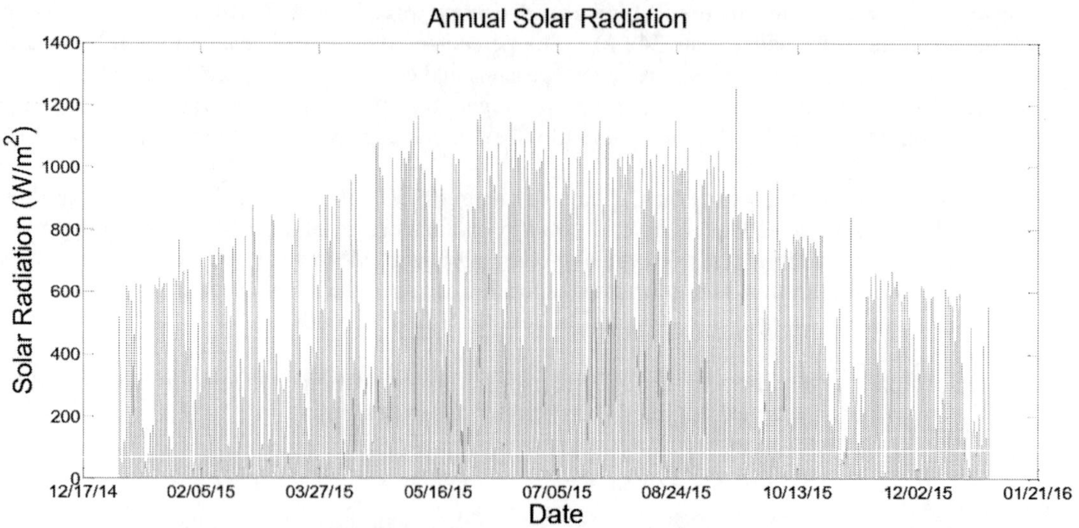

Figure 9. Annual solar radiation flux.

4.3 Heat Pump Energy Balance

In the cooling mode, the energy balance of the heat pump was maintained among the electric power consumption, rejected energy and cooling rate. Since the heat pump power consumption was constant during the test, and the heat rejection from the air-side to the water-side could be calculated by using measured temperatures and flow rate data. The cooling rate of the heat pump was calculated to be 1,938 Watts using Equation (5). However, the manufacturer specified the rated cooling rate of this heat pump is 2,638 Watts. The actual cooling rate is 73.5% of the rated value. This difference might cause by the fact that the rated value was tested in the lab at the standard rating condition. According to the standard rating condition under a cooling mode (AHRI 2009), the air temperatures entering indoor portion of unit are 26.7°C dry-bulb and 19.4°C wet-bulb. Dry-bulb air temperatures surrounding the unit are 26.7°C. Refrigerant temperatures from the test water coil (liquid line) are 25.0°C. However, for the actual operation in this study, all these conditions are different. The test was tested at a different environment.

The COP was calculated using the actual cooling rate from the total energy balance calculation, and the power consumption from the actual measurement. In the standard rating condition, the COP of the tested heat pump was 3.8 listed in the manufacturer catalog. By using Equation (6), the average COP of the heat pump was 3.23 during the test period. For the system COP, more components need to be considered, which means the value of COP would become lower compared with the COP of the heat pump itself only. In this test, the average COP of the system was 2.67 by using Equation (7).

5. CONCLUSIONS AND FUTURE WORK

This paper introduces a preliminary study of a solar powered ground source heat pump (GSHP) test bench in a laboratory. Actual sensor data collected from the system was used to do the feasibility analysis of the energy balance. The preliminary data analysis showed that the solar panel system conditionally satisfied the demand from the heat pump and other equipment. In addition, the heat pump energy balance showed that the performance of the geothermal heat pump system was acceptable during the test.

Some future works are listed as follows:

1. The capacity of the battery banks will be extended, a larger battery capacity will offer relatively longer operating time. Also, some synchronizing relays will be added which will help better control the well pump and the circulation water pump. Currently, both the well pump and the water pump were kept running during the test. The proposed synchronizing relays help control the running time of the water pump and well pump, the pumps would work only when the heat pump is on demand.
2. The air flow meter will be added to measure the air flow rate, which allows to measure the cooling rate more accurately.

3. By considering the feasibility of the applications of the system, different types of heat pumps with different refrigerant will be tested in the same system.
4. A dynamic model of this solar powered GSHP system will be developed and validated using the data collected from this test rig. Such model will be used for model-based control of this system for a better integration of buildings and the power grid.

NOMENCLATURE

P	power generation	(W)
V	voltage	(V)
I	current	(Amp)
t	time	(hour)
C	capacity	(Ah)
η	inverter efficiency	(%)
$\dot{m}$	mass flow rate	$(\frac{kg}{s})$
C_p	specific heat	$(\frac{J}{kg*K})$
ΔT	temperature difference	(K)
q	energy transfer rate	(W)

Subscript

charge	charge
discharge	discharge
battery	battery
GW	ground water
CL	closed-loop water
heat pump	heat pump
rej	rejection
cooling	cooling
water	water
water pump	water pump

REFERENCES

AHRI (2009). *2005 Standard for Performance Rating of Direct GeoExchange Heat Pumps, ANSI/SHRI Standard 870.*

Berardi, U. (2015). Building energy consumption in US, EU, and BRIC countries.

DOE (2012). *2011 Buildings Energy Data Book.*

ENERGY STAR (2015). *ENERGY STAR certified geothermal heat pumps*. http://www.energystar.gov [Accessed: 27 December 2015].

Lewis, N. S. (2007). Toward cost-effective solar energy use. *science,315*(5813), 798-801.

SIEMENS (2016). *Solar module SM50-H.* Retrieved from http://www.a1solar.co.uk/sm50-h.pdf

ACKNOWLEDGEMENT

Part of this project is sponsored through ASHRAE undergraduate senior design project. Danfoss donated the well pump. Supports from ASHRAE and Danfoss are greatly appreciated.

Heat Pump Assisted Solar Thermal System

David Lotz[1], William Hutzel[2*], William Arnett[3], Duane Dunlap[4], Christopher Foreman[5]

School of Engineering Technology, Purdue University
West Lafayette, IN, U.S.A.

[1]dlotz@purdue.edu
[2]hutzelw@purdue.edu
[3]arnettw@purdue.edu
[4]ddunlap@purdue.edu
[5]jchrisf@purdue.edu

* Corresponding Author

ABSTRACT

According to the U.S. Department of Energy, 40% of the energy consumed in the U.S. is used in buildings and almost half of that amount is used for heating or cooling. Current technologies allow for efficient thermal management, but most utilize energy harvested from fossil fuels or convert electricity directly into thermal energy. Alternative heating technologies such as heat pumps and solar thermal collectors can greatly reduce the energy used for heating while producing reliable heating performance.

This paper documents how an off-the-shelf residential heat pump water heater was integrated with a solar thermal system to improve overall heating performance. Solar thermal is the primary source of heat and the heat pump is used as a back-up when the sun is not shining with enough intensity. A primary/secondary pumping system is used to allow multiple modes of operation, including both hot water and hydronic heating. Since the heat pump water heater is located entirely inside the conditioned space, a secondary benefit/detriment for space cooling is noted based on the weather conditions.

An energy dashboard that evaluates the heating performance of this hybrid solar heat pump system in real time was developed. The primary efficiency metric is the Energy Factor (EF), which is the ratio of thermal energy delivered to electrical energy consumed. As a point of reference, heat pump water heaters earn an "Energy Star" rating for an Energy Factor of 2.2 under test conditions. During two-weeks of rigorous 24 hours a day testing in March of 2016, the instantaneous EF for the solar heat pump system varied from 1 to greater than 40 depending on the weather conditions. The overall average EF for the two weeks of continuous testing was 2.29, indicating an efficiency significantly higher than traditional direct electrical heating systems.

1. INTRODUCTION

According to the U.S. Department of Energy, 40% of the energy consumed in the U.S. is used in buildings and almost half of that amount is used for heating or cooling. Current technologies allow for efficient thermal management, but most utilize energy harvested from fossil fuels or convert electricity directly into thermal energy. Alternative heating technologies such as heat pumps and solar thermal collectors can greatly reduce the energy used for heating while producing reliable heating performance.

Because of the increasing trend towards net zero energy buildings, there are many ongoing projects to develop and deploy solar thermal heat pump systems (SHPS). One interesting system that is comparable to this research was recently developed and built at the National Institute of Standards and Technology (NIST) in Gaithersburg, MD. Their Net Zero Energy Residential Test Facility is a full-size home that is evaluating a number of promising new residential technologies, including a SHPS. The NIST system used a heat pump with a rated COP of 2.6, which was coupled with flat plate solar thermal collectors on the roof of the home. Three important differences from this research are that the NIST system 1) included substantial solar storage, 2) was designed for domestic hot water only, no space heating, and 3) the air source heat pump was located in an unconditioned space. The NIST SHPS system achieved an annual Energy Factor (EF) of 2.39. An EF is a performance measure that is the ratio of useful heat delivered to the electricity used for operating the system. Equally important, the NIST research suggests that daily time of use patterns for residential hot water have an impact on the performance results that are achieved.

2. SYSTEM OVERVIEW

Figure 1 shows the main components of a newly constructed solar heat pump system (SHPS) located in the Applied Energy Laboratory at Purdue University in West Lafayette, IN. The primary components are solar thermal collectors, primary and secondary circulating pumps, a heat exchanger for generating domestic hot water, and a heat pump with a storage tank. The SHPS is monitored and controlled by a web-based building automation system that includes comprehensive monitoring of outdoor weather conditions and component-level energy use. The entire system circulates a 50% glycol/water solution for freeze protection during cold winter weather.

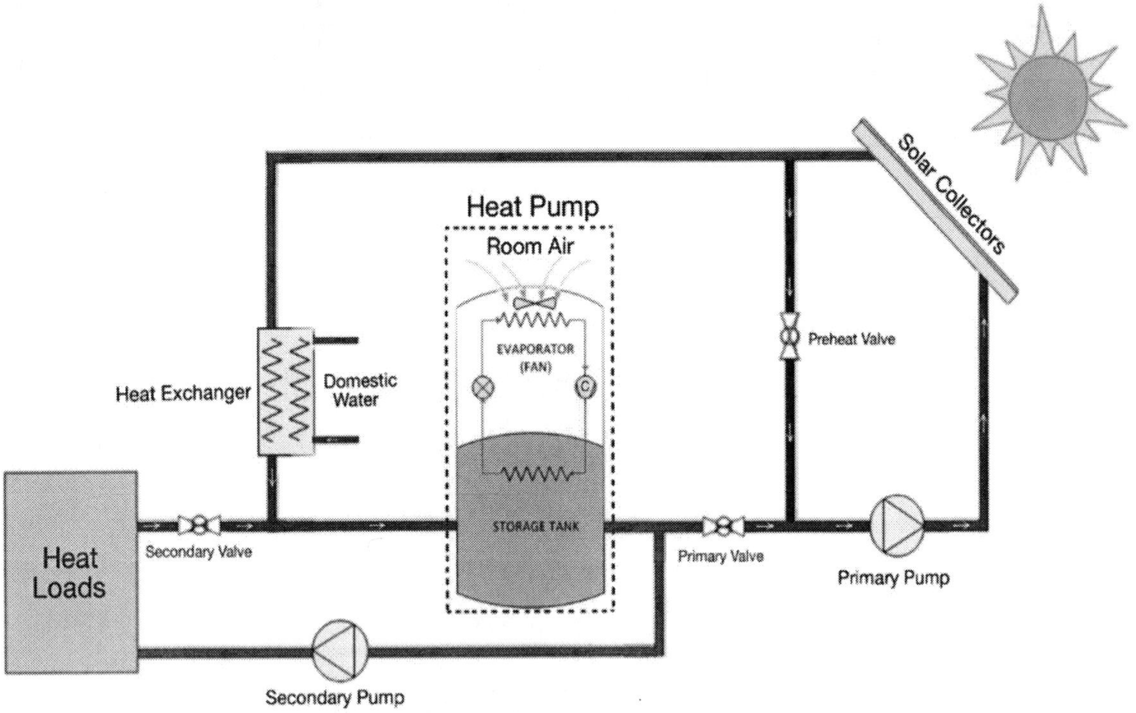

Figure 1: The SHPS has primary and secondary loops.

The primary loop of the SHPS consists of the primary pump, solar collectors, heat exchanger, and heat pump. The primary valve controls the glycol flow through the primary loop. When sufficient solar energy is available the primary pump circulates glycol through the solar collectors, heat exchanger, and heat pump. The collected heat is then used to provide domestic hot water via the heat exchanger or is transferred to the heat pump storage tank for later use. The heat pump supplements the solar heat as needed to maintain a storage tank temperature of 120°F.

The secondary loop of the SHPS consists of the heat pump, secondary pump, and heat loads. The secondary pump circulates the heated glycol to deliver space heating or domestic hot water through the heat exchanger. The secondary valve closes when there is no call for heating to isolate that loop.

The primary and secondary loops can operate independent of each other. This is necessary when but there is call for heating, but not enough solar energy is to maintain temperature. This mode is enabled by closing the primary valve, opening the preheat valve, and opening the secondary valve. The primary pump circulates flow to the solar collectors to gradually increase the temperature in the primary loop. The secondary pump circulates glycol through the heat pump and delivers it to the heat loads.

Figure 2 shows the solar thermal collector array located on the roof of the building including one evacuated heat tubes panel (left) and four flat-plate solar thermal panels (right). Various panel designs are used, however, each design operates in the same basic way. As sunlight hits the surface of the panels, the glycol solution pumped through the panels is heated by the solar energy. The heated glycol then circulates through the primary loop where the heat is transferred to either the heat pump storage tank or directly to the domestic water.

Figure 2: Solar thermal collectors for the SHPS.

Figure 3 shows the residential air source heat pump with an integrated 50-gallon storage tank that is used in the SHPS. The heat pump is necessary so that heating can be provided at night or any other time when solar energy is not available. In heat pump mode, the compressor transfers heat from the room air into the storage tank. Additionally, a resistive heating element provides rapid or backup heating for the system.

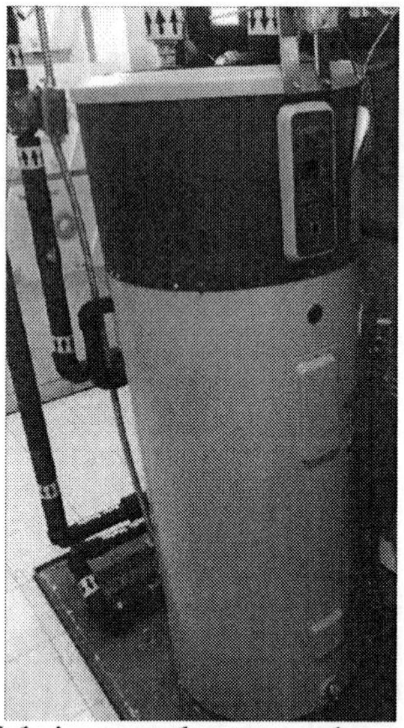

Figure 3: The residential air source heat pump has an integral storage tank.

3. ENERGY ANALYSIS

The SHPS is monitored and controlled by a sophisticated web-based building automation system that includes a large number of temperature, pressure, flow, and power monitoring sensors in addition to actuators for controlling the pumps and valves. An energy dashboard was created to coordinate and simplify this large amount of data into a few key metrics that quantify the performance of the SHPS in real-time. The graphical output of the energy dashboard provides a comprehensive visual summary of the heating system by displaying key metrics in a user-friendly format. The data includes not only current performance metrics, but trends over time so that users can identify equipment malfunctions or weather-related anomalies that effect the performance of the SHPS.

The primary efficiency metric is the Energy Factor (EF), which is the standard used by the U.S Environmental Protection Agency and the U.S. Department of Energy to rate water heating systems. The Energy Factor is the ratio of thermal energy delivered to electrical energy consumed. As a point of reference, a generic air source heat pump water heater can earn an EPA "Energy Star" rating for an Energy Factor of 2.2 under test conditions.

The EF computation for this research project is different from the computations for many other SHPS systems because the heat pump water heater is located inside the conditioned space. This arrangement has a secondary benefit/detriment for space cooling, based on the outdoor weather conditions, that must be accounted for in the EF computations.

3.1 Energy Factor Computations

Energy Factor (EF) is a key metric for evaluating the system's performance. As shown in Equation 1, EF is the ratio of energy collected (Qtotal) to energy consumed (Wtotal). More specifically, Qtotal is the amount of thermal energy collected by the SHPS and Wtotal is the amount of electrical energy used by the SHPS. In order to simplify the calculations, both Qtotal and Wtotal are calculated in Watt-hours (Wh). Furthermore, both Qtotal and Wtotal are comprised of multiple terms that are discussed in greater detail below.

$$EF = \frac{Q_{total}}{W_{total}} = \frac{Q_{solar}+Q_{hp}\pm Q_{air}}{W_{pump}+W_{hp}} \quad (1)$$

The first step in calculating the EF of the system is to determine the total electrical energy consumption of the system (Wtotal). As the denominator of equation 1 shows, Wtotal includes the electrical energy consumed by the primary circulating pump (Wpump) and the heat pump (Whp). In both cases the power of each device is measured in Watts and then integrated over time to determine the actual energy consumption in Watt-hours.

The next step in calculating the EF of the system is to determine the total amount of thermal energy collected by the SHPS (Qtotal). The total energy collected is made up of three parts. Solar energy (Qsolar), heat pump energy (Qhp), and air energy (Qair) all contribute to the thermal energy.

The first energy collection term, Qsolar, represents the amount of thermal energy collected by the solar panels. The heat transfer rate of the panels is calculated using real-time measured flow rates and temperature changes. Then the actual energy collection is determined by integrating the heat transfer rate over time.

The second energy collection term, Qhp, represents the amount of thermal energy collected by the heat pump. The heat pump has two modes of heating: heat pump and resistive. The resistive heating, which directly converts electrical energy to heat, is assumed to be 100% efficient. Therefore, the amount of energy provided by the resistive heating element is equal to Whp. In heat pump mode, however, the heat pump extracts energy from the ambient air and transfers that heat to the glycol solution in addition to the heat produced by the electrical energy used to run the heat pump compressor. Therefore, the total amount of output energy from the heat pump has the potential to be greater than 100% of the electrical energy consumption. The actual energy collected by the heat pump is determined by multiplying Whp by the coefficient of performance (COP) as determined by the specifications of the heat pump in use. In this case, the COP is set to 1.8 meaning that for every 1Wh of electrical energy consumption the heat pump provides 1.8Wh of thermal energy while operating in heat pump mode. The energy produced by the resistive heating and heat pump modes are then added together to determine the total energy collected by the heat pump (Qhp).

The third term, Qair, was included in order to account for the heat removed from the conditioned air of the lab. As previously explained, the heat pump is designed to extract heat directly from the conditioned air of the lab. During warm periods, when the outdoor air temperature is above 60°F, the cooling of the room air is an added benefit to the building. However, when the outdoor air temperature is below 60°F, the building must replace any energy extracted from the room air. The Qair term is included to account for the added benefit or penalty of the heat loss from air. Qair is positive when the measured outdoor air temperature is above 60°F and negative when it is below 60°F.
Equation (2) shows how Qair was calculated

$$Q_{air} = Q_{hp} - W_{hp} = W_{hp}(COP - 1) \quad (2)$$

4. TEST CONDITIONS

The solar heat pump system was evaluated over a two-week period from March 15th to March 28th, 2016. During the test period, the average day length (duration of daylight) was roughly12 hours and 14 minutes. On March 15th, the sunrise and sunset were 7:58AM and 7:55PM respectively. By March 28th, sunrise and sunset had shifted to 7:37AM and 8:09PM respectively.

The heating system was operated at close to 100% capacity for 24 hours a day in all 14 days of the test period. The load was applied by removing heat from the SHPS using finned tube radiators in an environmental chamber that was maintained at a constant temperature. Figure 4 shows the daily heating (Qload) cycles starting at 12:00AM on March 15th and ending at 11:55PM on March 28th. The daily load was 19 kWh each day, roughly the amount that the heat pump could supply if it operated continuously with a COP of 1.8 and without using electric reheat or solar energy.

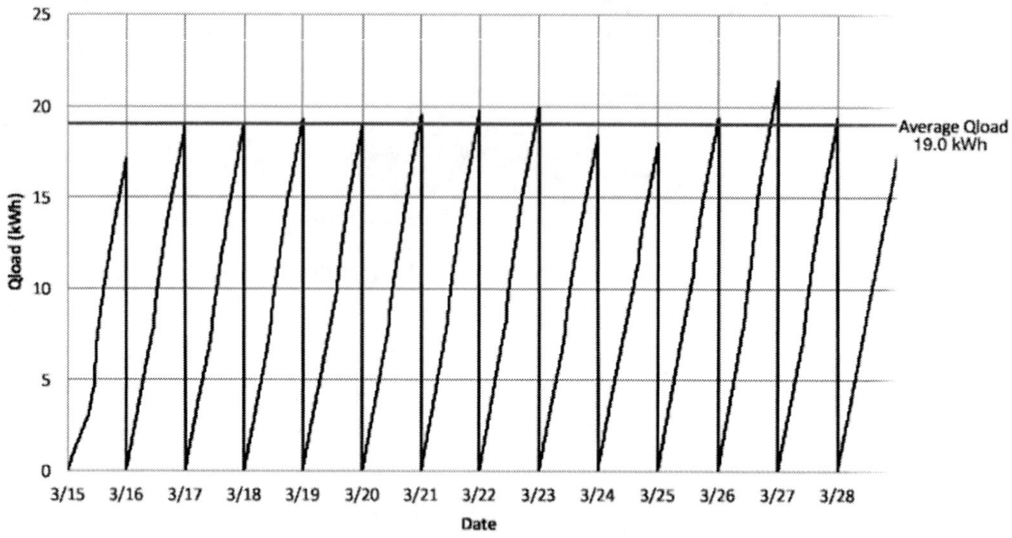

Figure 4: Two weeks of continuous testing was conducted.

As Figure 4 shows, the rate of heat demand was relatively constant for all 14 days. Furthermore, the demand was constant throughout each 24-hour period meaning that the system was rigorously and consistently tested during periods of both daytime and nighttime. Solar energy was used when available but at night or on cloudy days the heat pump was used to meet the load demand. This meant that regardless of the solar energy collection during the day, the heat pump was still loaded at night. At midnight Qload was reset to zero so that the total heat demand for each day, represented by the peaks, could be determined.

The average daily heat demand during the test period was 19.0 kWh, which is 2.6kWh higher than the average hot water energy demand for Midwestern homes (United States Department of Energy's Office of Energy Efficiency and Renewable Energy, March 2012). The higher demand put more stress on the heating system, but was deemed acceptable because the system is designed to provide both domestic hot water and space heating. Furthermore, the graph above shows that the daily heat demand was relatively consistent throughout the test period, which was important for comparing the performance values of each day.

The two-week test period also included a variety of weather patterns. Both solar intensity and outdoor air temperature were measured throughout the test. Each test day was categorized by its average measured solar intensity. A sunny day was defined as any day with an average solar intensity greater than 200 W/m2 whereas a cloudy day was any day with and average less than 100 W/m2. A partly cloudy day was any day with an average solar intensity between the two limits. The test period included eight sunny days, three partly cloudy days, and three cloudy days.

Similarly, each test day was also categorized by the amount of time during the day at which the outdoor temperature was above 60°F. The 60°F limit was used because it is also the point at which the Qair term in Equation (1) begins to be treated as a benefit to the building. Therefore, a cold day was defined as any day in which the outdoor temperature never reached 60°F. A warm day was any day in which the outdoor temperature reached 60°F but for less than 50% of the time. And a hot day was defined as any day in which the outdoor temperature was above 60°F for more than 50% of the time. The test period included one hot day, eight warm days, and five cold days.
Furthermore, the test period occurred during an ideal time of the year.

Figure 5 shows the average daily global horizontal irradiance (blue) and outdoor temperature (orange) for West Lafayette, Indiana (40°24'36.0"N 86°54'00.0"W) based on values from 2010 to 2014 (National Renewable Energy Laboratory, 2016). The vertical shaded area designates the test period and the horizontal shaded area demarcates the average outdoor temperatures measured during that time.

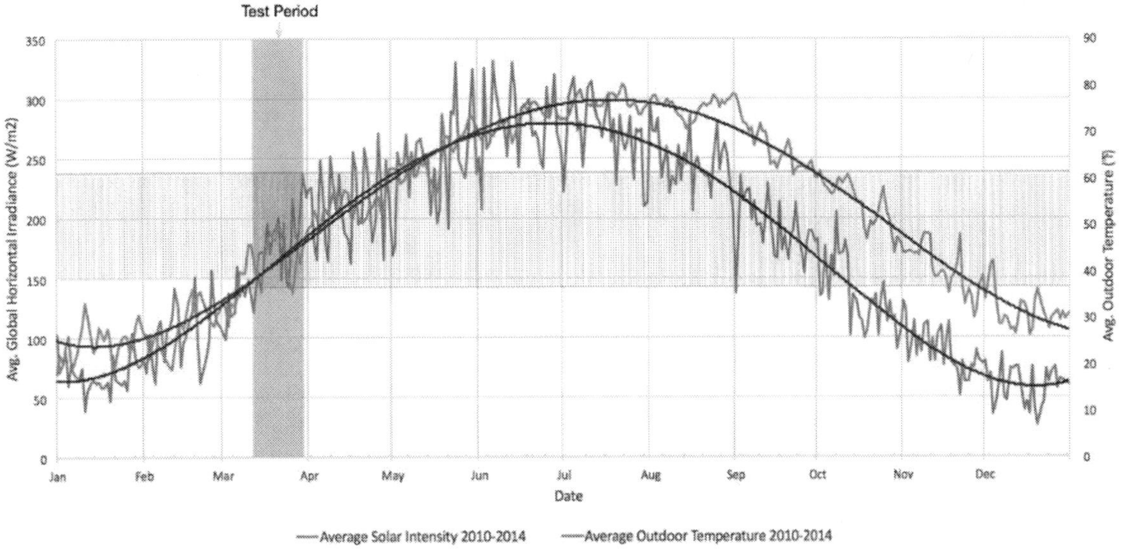

Figure 5: Solar Intensity and Outdoor Air Temperature Averages for 2010 to 2014

As Figure 5 indicates, the test period occurred at an intermediate time of the year. Specifically, the median of the measured solar irradiance during the test period was nearly identical to the median of the values shown in the graph. Figure 5 also shows that the range of measured outdoor temperatures during the test period encompasses a large, central portion of the outdoor temperature averages shown in the graph. All of this suggests that the measured performance of the heating system during the two-week test period is a good estimate of the annual performance of the system. As previously mentioned, both the energy dashboard program and the solar heat pump system were evaluated during the test period. The dashboard needed to provide a comprehensive and user-friendly summary of the system's energy performance. The goal of the solar heat pump system evaluation was to achieve an overall average energy factor of 2.0 or greater.

5. RESULTS

As previously discussed, the actual performance of the solar heat pump system was evaluated during the test period. The system was loaded 24 hours a day for 14 days. Of those 14 days there were three cloudy, three partly cloudy, and eight sunny days. Similarly, there were five cold days, eight warm days, and 1 hot day.

Table 1 shows the two-week averages of the solar EF, heat pump EF, and overall EF values, which are the key metrics used to evaluate the SHPS. As previously discussed, EF is the ratio of collected to consumed energy. The values shown in the table were determined by averaging the fourteen daily EF values for each of the systems. The system achieved an average EF of 2.29 during the two-week test period.

Table 1: Average Measured Energy Factors

# of Days	Solar EF	Heat Pump EF	Overall EF
14	26.95	1.25	2.29

Table 2 shows the average EF values for the cloudy, partly cloudy, and sunny days. Instead of determining the average daily EF values for all fourteen days, as shown in the previous table, the respective daily EF values were averaged separately for each type of day. As expected, there was a significant difference in the solar EF values between the cloudy, partly cloudy, and sunny days. The average solar EF value was 35.68 for the eight sunny days and 2.77 for the three cloudy days. The average value during the partly cloudy days predictably fell between the two ranges at 18.67. The results also show a correlation between the average solar intensity and the overall EF. The driving factor of this correlation is obviously the amount of energy collected by the solar panels, which was directly effected by the amount of sunlight and is reflected in the solar EF values.

Table 2: Average Measured EF Values for Sunny, Partly Cloudy, and Cloudy Days

Type of Day	# of Days	Solar EF	Heat Pump EF	Overall EF
Cloudy	3	2.77	1.20	1.23
Partly Cloudy	3	18.67	1.27	1.74
Sunny	8	35.68	1.24	2.75

Similarly, Table 3 compares the average measured EF values for the cold, warm, and hot days of the test period. The respective daily EF values were averaged separately for each type of day. As previously discussed, the outdoor temperature directly effects the heat pump EF value due to the inclusion of the Qair term in the calculation. Specifically, on cold days the expected heat pump EF values is 1.00, which matched the measured average exactly. Similarly, the measured averages for both the warm and hot days fell within their respective expected ranges of 1.0-1.8 and 1.8-2.6. Again, there appears to be a correlation between the overall EF and the outdoor air temperature with the heat pump EF being the driving variable. However, there also appears to be a correlation between the outdoor air temperature and the solar EF value. In both cases, further testing is required to verify the significance of the correlations.

Table 3: Average Measured EF Values for Cold, Warm, and Hot Days

Type of Day	# of Days	Solar EF	Heat Pump EF	Overall EF
Cold	5	18.21	1.00	1.66
Warm	8	27.76	1.32	2.38
Hot	1	36.65	1.81	3.53

During the test period energy consumption and collection values for both the solar and heat pump sub-systems were measured. Figure 8 contrasts the two forms of energy in this system. The pie chart to the left compares the average electrical energy consumption of the solar and heat pump systems. The pie chart to the right shows the average contributions of the solar and heat pump systems. As the charts show, the heat pump consumed the majority of the electric energy, roughly 9.13 kWh per day as compared to the 0.32 kWh per day consumed by the solar energy system. On the other hand, the heat pump only contributed 11.24 kWh per day compared to the 8.91 kWh per day collected by the solar panels.

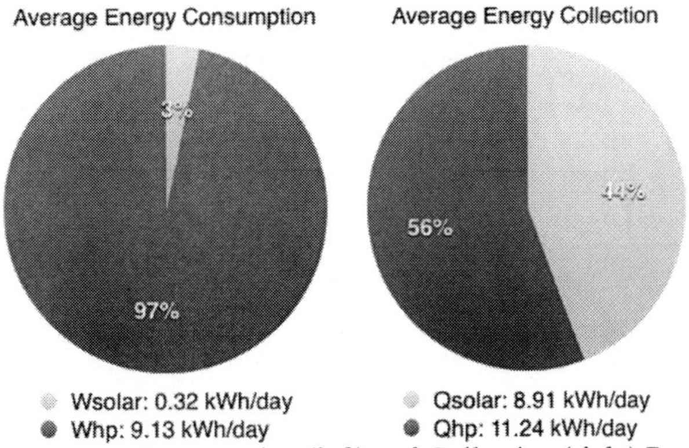

Figure 8: Average Consumption (left) and Collection (right) Percentages

6. CONCLUSIONS

During the two-week test period the SHPS achieved an average EF of 2.29. The average solar and heat pump EF ratings were also as expected. Specifically, the average solar EF rating was roughly 27 and the average heat pump EF rating was roughly 1.3. The solar EF was significantly effected by the amount solar intensity throughout the day. Similarly, the outdoor temperature had a significant effect on the heat pump EF. In particular, the percentage of each day spent over 60°F closely correlated to the EF rating of the heat pump. The outdoor temperature also appeared to have an effect on the solar EF. The solar and heat pump EF values directly effected the overall EF rating. Therefore, solar intensity and outdoor temperature also had similar effects on the overall EF value.

The fact that the heat pump was responsible for the majority (97%) of the electrical energy consumption but provided just over half (55%) of the energy collection was a concern. On the other hand, the solar collection system was extremely efficient when solar energy was available. Despite only accounting for 3% of the total electrical energy consumption the solar energy provided just under half of the total heating (44%). All of this indicates that the overall efficiency of the system could potentially be improved by minimizing the use of the heat pump and maximizing the use of solar energy. However, some heat pump electrical consumption is necessary for the system to adequately provide heat even when solar energy is not available.

The solar heat pump system was very successful in meeting the heating demand placed on the system. The average daily load on the SHPS was 19.0 kWh per day and the system was loaded day and night meaning that it could not rely solely on solar energy to meet the demand. During the night and during cloudy days, the heat pump was necessary. Despite the heavy demand the heat pump rarely resorted to resistive heating, which helped the system to achieve a higher EF rating. The fact that the SHPS was able to meet the high demand, even in the absence of sunlight, was very important because it showed that the SHPS is a viable heating solution. Many of the features of the system are meant for educational purposes, however, the results of the energy dashboard test revealed that it also has potential as a commercial heating system.

NOMENCLATURE

EF	Energy Factor	
Q	Energy (heat) collected	(Wh)
SHPS	Solar Heat Pump System	
W	Electrical energy consumed	(Wh)

Subscript

air	Conditioned lab air
hp	Heat Pump
load	Heat load
pump	Primary circulating pump
solar	Solar collection system

REFERENCES

Fanney, A., Payne, V., Ullah, Ng, Boyd, Omar, Davis, Skye, Dougherty, Polidoro, Healy, Kneifel, and Pettit. (2015). Net-zero and beyond! Design and performance of NIST's net-zero energy residential test facility. Energy & Buildings, 101, pp 95-109.

Krockenberger, Kyle, Degrove, John, Hutzel, William, and Foreman, Christopher (2014), *Design of a heat pump assisted solar thermal system*, Proceedings of the 3rd International High Performance Buildings Conference at Purdue, West Lafayette, IN, July 14-17, 8 p.

Lotz, D. (2016). *Energy Dashboard for Real-time Evaluation of a Heat Pump Assisted Solar Thermal System,* College of Technology Masters Theses.

National Renewable Energy Laboratory. (2016). National Solar Radiation Database (V2.0.1) [Datafile]. Retrieved from https://nsrdb.nrel.gov/nsrdb-viewer

United States. Dept. of Energy. Office of Energy Efficiency and Renewable Energy. (March 2012). 2005 Delivered Energy End-Uses for and Average Household, by Region [Data File]. Retrieved March 20, 2016, from http://buildingsdatabook.eren.doe.gov/docs/xls_pdf/2.1.9.pdf

ACKNOWLEDGEMENT

This project could not have been completed without generous support from GE, Grundfos, Belimo, and ASHRAE. These organizations provided equipment and funding for constructing the solar heat pump system.

Energy and Economic Performance Analysis of Heat Recovery Devices Under Different Climate Conditions

Stefanie TAFELMEIER[1]*, Giovanni A. LONGO[2], Andrea GASPARELLA[1]

[1]Free University of Bolzano, Faculty of Science and Technology,
Bolzano, South Tirol, Italy
(stefanie.tafelmeier@natec.unibz.it)

[2]University of Padova, Department of Management and Engineering,
Vicenza, Veneto, Italy
(giovanniantonio.longo@unipd.it)

* Corresponding Author

ABSTRACT

High performance buildings typically rely on extreme envelope performances in terms of insulation level, solar gains valorization and airtightness. The energy needs are limited and the system sizing is reduced to facilitate the recourse to renewable energy sources. The only exception to this trend is related to the ventilation and the associated energy needs, which is even more crucial because of the tight infiltration control. Mechanical ventilation is a common option to ensure an appropriate ventilation rate and a suitable indoor air quality. Heat recovery devices are widely used to save energy in buildings mechanical ventilation systems. Their contribution in the design or operating phases is typically assessed in terms of effectiveness, which expresses the ratio between the amount of energy – sensible and/or total – actually recovered and the maximum recoverable under ideal configuration. However, the effectiveness of recovery devices is not a constant value and depends on the operating – outdoor and indoor –conditions and control strategy. In order to evaluate the energy and cost savings allowed by heat recovery systems, these aspects have to be considered, and their relative impact on different heat recovery technologies accounted for. In this work, some of those aspects are analyzed in detail and their contribution to the device effectiveness is discussed. The objective is to understand the influence of the air conditions (fresh and exhaust) on the energy savings considering different sensible and total heat recovery devices and operative constraints which could affect the control strategy. The analysis starts from hourly weather data, representative of different climate types, to quantify the actual effectiveness of different heat recovery systems. The impact of indoor conditions on the sensible and latent recovery is analyzed, defining appropriate control strategies to prevent excessive moisture recovery and discomfort issues when the indoor humidity is considered. Moreover, the effect of condensation and frosting on the hourly and annual performance of each kind of device is analyzed. Three different climatic conditions have been considered. Sensible fixed-plate cross flow heat recovery system has been compared with enthalpy wheels. By means of a parameterization of the operating constraints and the expression of the main quantities in specific terms, depending only on the building use, it has been possible to generalize the main results, making them independent of a specific case study, and to compare different technologies and climatic contexts. Seasonal and annual energy and economic performances have been quantified, providing a synthetic overview of the technical and economic aspects of the considered technologies.

1. INTRODUCTION

The movement toward better and better insulated buildings is accompanied by not only a reduction in energy consumptions but also the necessity of better ventilation. The energy consumption of buildings equals 41% of the final energy uses in Europe. More than 60% of it is due to space heating (Bosseboeuf et at.; 2012). Space heating is more efficient if the building is equipped with an airtight envelope, which reduces heat losses by unwanted infiltration. In contrast, the lack of natural ventilation and of fresh air, compromises indoor air quality, and thermal comfort can hardly be reached without additional devices (Rasouli, Simonson, & Basant, 2010). To overcome these problems it is possible to install a mechanical ventilation cycle that supplies fresh air for achiving indoor air quality. The need of a suitable ventilation rate, and the use of those devices, contributes to an additional energy demand and raises the total energy consumption of the building. The most effective solution to reduce energy need due to mechanical ventilation are heat recovery devices.

The effectiveness of an air-to-air heat recovery device can limit the energy demand that is necessary to take the supply air to the indoor comfort conditions, which is known as the building ventilation load. To contain these needs while providing appropriate comfort conditions, a control strategy is required aimed at maintaining suitable indoor air conditions, namely temperature and particularly relative humidity. The influence of these conditions was investigated though field tests and experiments. Generally, a high tolerance in the sensation of relative humidity is often reported by field studies (Hwang, Lin, & Kuo, 2006) (Mishra & Ramgopal, 2014). Nevertheless, some further works discuss an influence of the relative humidity not only on the comfort sensation but also on the occupants' performance. A theoretical approach by Kosonen and Tan (2004) based on reviewing results – using the Predicted Mean Vote (PMV) as a tool to estimate the productivity correlations – led to the conclusion that a rising relative humidity is expected to increase productivity losses. In addition, they showed that occupations requiring a higher activity level, and therefore leading to a higher specific latent load simply due to a rising metabolic rate, could result in a more significant performance loss if the humidity is not controlled. Tsutsumi et al. (2007) investigated experimentally the correlation between a step change in humidity level and occupants' comfort as well as productivity. Even though there were no significant correlations between the sensation or performance and the humidity variation following a step-change, it could be seen that the total number of complaints for relative humidity of 70% were higher than for any lower humidity value. The authors stated that longer exposure time to higher humidity in their climate test chamber possibly leads to a performance loss. In a previous paper similar observation were made, within a similar study with college-aged test persons, which had more complaints about fatigue, but for low humidity (Y. Chen *et al.*, 2003). Apart from the thermal comfort sensation and the productivity of occupants, there are also effects on the building materials and occupants' health, which can occur when humidity is not properly controlled, as for instance fungi and mold growth or surface or interstitial condensation (Sterling, Arundel, & Sterling, 1985). Thus, the effect of ventilation device on not only the temperature, but also the humidity, should be considered in the control strategy to avoid the described inconveniences.

Not only the control strategy of heat recovery devices has to adapt to the dynamic control of the indoor conditions, but it has also to take into account the effects of the fresh air conditions on the device's actual effectiveness, especially under the so-called wet operating conditions. These occur in case of condensation within the recovery device, which may happen if the temperature of the separation interface falls below the dew point of the return or supply air. As the return air temperature from the room is quite high, it would only induce condensation when very humid outside conditions corresponding to high dew points. In contrast, saturation on the returning air in case of cold climates is a quite common situation. In theory, condensation of vapor on a plate brings along a release of the latent heat. That additional energy can contribute to the increment of the heat recovery effectiveness. This has been quantified by Nam and Han (2016) who studied a plate type heat exchanger for broiler houses modelled using computational fluid dynamics. The analysis showed an effectiveness increase with increasing return air humidity. In their example it is shown numerically and experimentally a rise up to 15% of effectiveness from dry return air to very humid (rH=90%) condition for outside air temperature between 0°C and 10°C. The numerical simulation of Anisimov et al. (2015) of plate heat exchanger (crossflow) reports an effectiveness increase of 20% for a relative humidity of 90% and an outside air temperature of -20°C. By reaching such low outside temperatures the separation interface approaches the freezing/frosting limit. Broad overview of the findings on frosting in air-to-air heat and energy exchangers is given in the review of Rafati Nasr et al. (2014), showing clearly that frost must be avoided in order to prevent additional pressure drop and especially to avoid mechanical damages to the devices.

The onset of frosting depends largely on the humidity of the return air, and on the outside temperature ((Deshko, Ya, & Sukhodub, 2016) (Fisk et al., 1985)). In the case of total heat recovery, the danger of frosting is reduced by the fact of mass transfer. Nevertheless, studies based on numerical modelling (Simonson & Besant, 1998) showed that for outside temperature of -20°C blocking starts also within an enthalpy wheel for a relative humidity of the return air of 30% and a total effectiveness from ca. 65% to 75% . Fisk et al. (1985) observed for an enthalpy-type cross flow heat exchanger a frosting onset of -8 to -12°C and Liu et al. (2016) a limit of -10°C (return air condition: T≈22°C, rH=30%; outside air rH≈55%) for a crossflow plate enthalpy exchanger. To give some examples regarding the actual applications, suppliers suggests a frost protection starting from -11°C for enthalpy wheel (for return air temperature of 21°C & relative humidity of 50%) and from -5°C for cross-flow plate total heat exchanger THR. A similar threshold temperature is proposed for sensible heat recovery SHR. This value is in line with the experimental and simulated results of Fisk et al. (1985), that are in the range of -7°C to -3°C for a cross-flow plate heat exchanger and ca. -6°C for a countercurrent one. Liu et al. (2016) reports a frosting limit of – 5°C for a return air relative humidity of 30% (return air: T≈22°C; outside air rH≈55%) for a cross flow heat exchanger. Summarizing it can be stated that for cold climate the frosting limit should be considered for both sensible and total heat recovery allowing for a dependency on the relative humidity of the return air. Once that limit is reached the use of the recovery device requires a frost protection strategy.

Frost protection can be realized in multiple ways and contributes to the annual energy consumption. Its impact was investigated by Kassai et al.(2015). Within their study, the pre-heating frost protection appears to be more convenient than a supply side closing strategy both for the climate of Helsinki and Krakow. The work of Zhang and Fung (2015) includes the influence of frost protection (defrost cycles) for heat and energy recovery in Toronto, reporting that the heat recovery annual defrost period is 3,5 time than for total heat recovery. Consequently, for a comparison between different recovery devices, its necessary to consider defrosting needs

The aim of this research is the evaluation of sensible and total heat recovery devices for ventilation purposes. The analysis has been generalized with respect to any specific applications. To do so the specific latent load was introduced (Lazzarin *et al.*, 2000), which is calculated by considering the latent load based on a person's activity and the necessary air flow rate per person in an air conditioned space. Showing that the ventilation load reduction can be analyzed taking into account only the specific latent load of the air-conditioned space and the impact of the outside air conditions. The impact of condensation and freezing/frosting within a recovery device have been estimated, both for sensible and total heat recovery, and considered in the analysis. A control strategy for total heat recovery has been proposed and investigated in order to avoid excessive humidity due to excessive humidity recovery when total heat recovery devices are adopted. All considerations have been repeated for 7 different cities (representing 3 climate classes after Köppen) and evaluated by assessing the annual savings. To allow a comparison not only from an energy point but also from a financial point of view, an economic analysis is included based on currents year's price for fuel (natural gas) and electricity and on the investment related to the heat recovery system.

2. METHODS

2.1 Air Handling Unit and Heat Recovery System

For the determination of the energy savings due to sensible and total heat recovery, a simple air handling unit (AHU) has been considered, in which the Return Air (RA) from the conditioned space is partially recirculated (CA) and partially exhausted (EA). An equivalent amount of outside air (OA) and the recirculated air are combined into a Mixed Air (MA) which is then taken to the supply conditions (SA) in a set of heating/cooling and dehumidification, humidification and reheating coils. Heat recovery is typically operated between EA and OA, can be characterized by the values of the sensible ε_s, latent ε_l and enthalpy or total ε_t effectiveness, according to the following equations:

$$\varepsilon_s = \frac{\dot{m}_{OA}\,(T_R - T_{OA})}{\dot{m}_{min}\,(T_{RA} - T_{OA})} \quad (1)$$

$$\varepsilon_l = \frac{\dot{m}_{OA}\,(x_R - x_{OA})}{\dot{m}_{min}\,(x_{RA} - x_{OA})} \quad (2)$$

$$\varepsilon_t = \frac{\dot{m}_{OA}\,(h_R - h_{OA})}{\dot{m}_{min}\,(h_{RA} - h_{OA})} \quad (3)$$

Where $\dot{m}_{OA}$ and $\dot{m}_{min}$ are the OA mass flow rate, and the minimum between the OA and the EA mass flow rates, T is the temperature, x the humidity ratio, h the enthalpy and the remaining subscripts refer to the positions in in Fig. 1. In the following, the ratio of OA and EA mass flow rates is considered as unitary. Moreover, sensible heat exchangers have typically null latent heat effectiveness, while it is quite common to assume an equal value for sensible, latent and total effectiveness of total heat recovery devices.

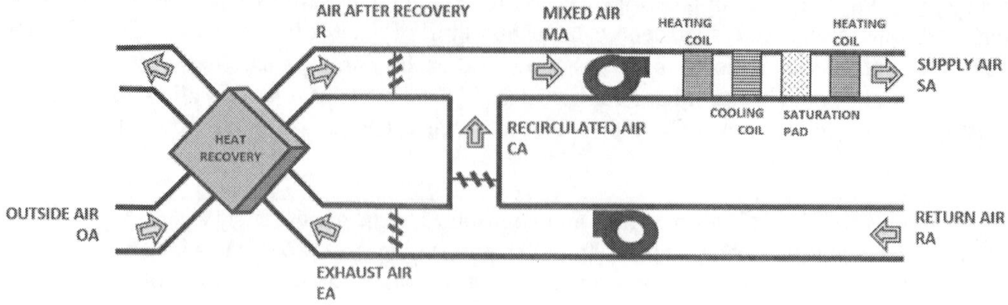

Figure 1: Ventilation and air conditioning cycle with heat recovery.

2.2 Relative Humidity Control and Heat Recovery Limitation

In winter mode operation, humidification of the MA is often necessary to the supply conditions, because of the quite low winter outside humidity ratio. Nevertheless, SA should be provided with a humidity ratio lower than the internal one, in order to compensate for the internal latent load:

$$x_{SA} = x_{RA} - \Delta x_{load} \tag{4}$$

where

$$\Delta x_{load} = \dot{m}_l / \dot{m}_{MA} \tag{5}$$

is the ratio between the mass rate of moisture internal production $\dot{m}_l$ and the dry supply air flow rate MA or SA. While for $x_{MA} \leq x_{SA}$ a quite energy inexpensive process of humidification by water injection can fill the gap, when the $x_{MA} > x_{SA}$ a very expensive dehumidification by cooling process is required.

Mixing conditions depends on the EA and OA conditions, but also on the heat recovery when total heat exchangers are operated. Excessive moisture recovery, can compromise the benefits of ventilation heat recovery because of the need of dehumidification of the MA.

It can be shown (Lazzarin et al., 2000) that in case of humidity recovery an equivalent condition to $x_{MA} \leq x_{SA}$ can be expressed just as a function of the OA mass flow rate, the moisture production and the outside humidity ratio:

$$x_R \leq x_{RA} - \Delta x_{spec,load} \tag{6}$$

$$\Delta x_{spec,load} = \dot{m}_l / \dot{m}_{OA} \tag{7}$$

In case of no heat recovery or sensible heat recovery only, x_R in equation (6) corresponds to x_{OA}. The potential of the equations (6) and (7) is clear when considering that both terms in the expression of the specific latent load can be referred just to the activity level in a given environment, neglecting any consideration about the actual building characteristics, as reported in table 1.

Table 1: Thermal loads gain depending on the activity and building type

		Thermal load per person (1)					
Activity	**Type of Building**	**Total W**	**Sensible W**	**Latent W**	**Latent g/h**	**Air exchange(2) l/(s person)**	**Specific latent load g_v/kg_{da}**
Seated, relaxed	Theatre, Cinema	100	60	40	57.6	5.5	2.4
Seated, writing	Offices, Hotel, Apartments	120	65	55	79.2	11.0	1.7
Eating	Restaurants	170	75	95	136.7	10.0	3.2
Seated, light activity, typing	Offices, Hotel, Apartments	150	75	75	108.0	11.0	2.3
Slowly walking	Retail store, Bank	185	90	95	136.7	11.5	2.8
Moderate dancing	Dance Hall	375	120	255	367.1	16.5	5.1
Heavy activity	Gymnasium	525	185	340	489.4	16.5	6.9

(1) ISO 7730; (2) UNI 10339

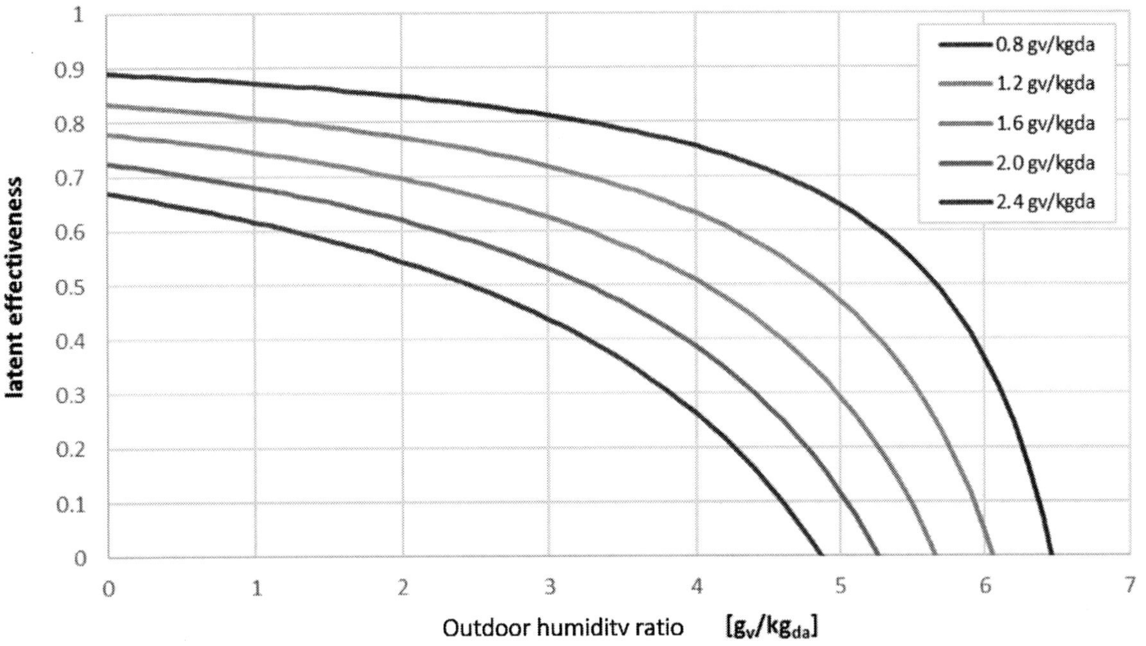

Figure 2: Correlation between the maximum allowed effectiveness and the outside air in case of humidity control & sketch of heat exchanger nomenclature (ASHRAE, 2004)

There, the thermal load and suggested air changes per person are listed, depending on the person's activity, together with the resulting specific latent load. Reference values may vary from country to country because of different standards and regulations, but the specific latent load for a given combination of values stays quite constant along the occupation time, which allows conducting the analysis from a more general perspective.
To prevent excessive humidity, a control strategy can be applied to limit the recovery effectiveness depending on the outside humidity ratio. In practice the control can be realized by changing the rotational speed of an enthalpy wheel or by introducing a bypass for a plate heat exchanger. Following the THR with controlled effectiveness is called THRc. The limitation in (6) can be expressed in terms of a maximum allowed latent effectiveness - and in practice any effectiveness - and the specific latent load:

$$\varepsilon_l \leq 1 - \frac{\Delta x_{spec,load}}{x_{RA} - x_{OA}} \quad (8)$$

Once the level of activity in a building is known, the effectiveness control depends only on the outside humidity. It can be seen (Fig. 2) that for specific latent loads higher than 2, the maximum operative effectiveness should be lower than 0.73, which in other words means that total heat recovery with every device with higher latent effectiveness has to be partialized. When the outside humidity ratio overcomes the limit in (6) the recovery device has to be totally bypassed. In similar cases, heat recovery has to be excluded also for SHR (controlled SHRc).
For values of the supply humidity ratio higher than the return air, such as in summer operation, dehumidification is always required and there is no reason to limit to the effectiveness, provided heat recovery reduces enthalpy.
For the investigation a slightly higher value of 60 % besides the nominal 50 % relative humidity has been taken into consideration for RA, to analyse the sensitivity of the saving potential on the recovery constraints, while maintaining still acceptable humidity levels.

2.3 Condensation and Frosting

The onset of saturation of vapor in the exhaust air outlet has been assumed to occur as soon as the air temperature falls below the dew point. Furthermore, an estimation of the latent heat gain contributing to heat recovery can be based on the amount of condensed vapor, calculated from the difference between the humidity ratio of RA and of its value at saturation. For total heat recovery, the mass transfer tends to reduce the RA humidity within the process and condensation is less likely.

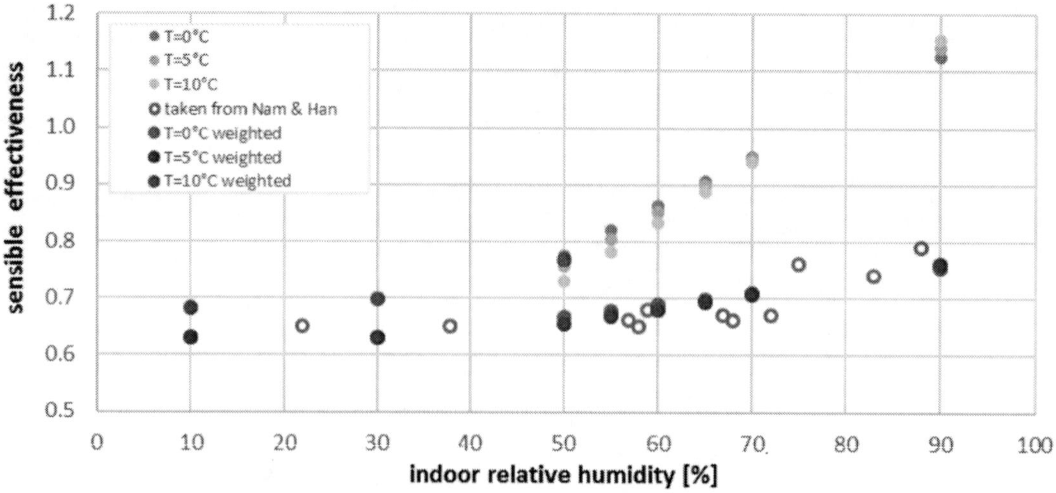

Figure 3: Sensible effectiveness on the indoor relative humidity: calculated vs of Nam and Han (2016)

In real applications, part of the air stream does not reach thermal equilibrium with the cold surfaces of the device to (bypass effect) (Keller, 2009). Consequently, the contribution from the latent heat is reduced. For that reason, a weighting factor has been introduced to consider the actual fraction of latent heat actually recovered. To validate this value, the results were compared with the experiments of (Nam & Han, 2016) for a plate-type heat exchanger, for different relative humidity and temperature (Figure 3) showing an good agreement with a factor of 0.25.
The impact of frosting conditions was considered in terms of amount of time when defrost must be applied and accounts for an additional energy demand. A pre-heating system was assumed for frost avoiding. As freezing onset depends as mentioned on many factors, an outside temperature limit for the start of the pre-heating has been chosen in line with threshold limits suggested by the manufacturers. For sensible heat recovery the onset of pre-heat is at -5°C and for an enthalpy wheel at -11°C for a relative humidity of 50%, and -8°C for a relative humidity of 60%.
The calculations on savings were repeated with and without consideration condensation respectively frosting.

2.4 Energy Savings and Economic Analysis

The savings have been determined considering hourly weather data (outside temperature and relative humidity) for a whole representative year, as provided by the EnergyPlus weather data base (energyplus, 2016). The savings in heating and cooling demand have been calculated in terms of recovered enthalpy, and expressed per unit of outside air flow rate [l/s]. This way the results are independent of the set actual size of the ventilation or air conditioning system and can be generalized for buildings of different dimensions.The following assumptions have been made for the analysis:

- a nominal effectiveness for sensible and total heat recovery of 70%
- a seasonal efficiency of 0.9 for the gas water heater [onset heating for T_{OA}<20°C for SHR, h_{OA}<38.55 kJ/kg (20°C;50%) for THR]
- a Seasonal Energy Efficiency Ratio (SEER) of 2.2 for the chiller [onset cooling for T_{OA}>26°C for SHR, h_{OA}<52.91 kJ/kg (20°C;50%) for THR]
- a pressure drop of 200 Pa for the heat exchanger and a fan mechanical efficiency of 0.3
- costs charged per MWh (Table 2) for natural gas referred to final users with annual gas consumption between 20 and 200 GJ (Eurostat, 2016) and electricity between 2500 and 5000 kWh (StatisticsFinland, 2016).

Investment costs per flow rate are based on the data given by the price list for performance of public works and maintenance of the municipality of Milan (Milano, 2015) and are for SHR 0.7 €/(l/s) and for THR 2.3 €/(l/s).
Savings for smaller sizing of the air conditioning equipment have been neglected.
The performed economic analysis includes the payback time of the recovery devices and its present worth value. The payback helps to understand after what time period the savings overcome the investment. The higher the payback time is the lower the savings. The present worth value represents the value an investment would have in the future considering the discount. For the calculation a time period of 10 years and a discount rate of 5% have been assumed.

Table 2: Natural gas and electricity costs MWh for the considered cities

costs [€/MWh]	Berlin	Helsinki	Lisbon	London	Madrid	Rome	Warsaw
natural gas	68	38	98	64	73	77	50
electricity	295	155	228	213	231	245	144

3. RESULTS AND DISCUSSION

3.1 Specific Latent Load and Recovery Limitation

The percentage of time in heating mode (T_{OA} lower than T_{RA}) with outside air humidity not overcoming the limit given by Equation (6) is reported in fig. 4 (i.e. with conditions lie in the area A). Under these conditions introducing fresh outside air or sensible heat recovery for winter operation would need dehumidification to meet the supply air condition. Obviously, the percentage is decreasing with increasing specific latent load, but it also strongly depends on the climate condition. In London, Lisbon and Madrid the annual outside humidity often reaches an uncomfortable condition if no air dehumidification is available, whereas for Helsinki it is comparably less problematic.

Figure 5 (solid color bars) shows the percentage of time in heating mode (h_{OA} lower than h_{RA}) with outdoor humidity leading to excessive humidity in the MA, when using THR with 70 % effectiveness if not properly controlled (i.e. with conditions in the area A). For all cities, assuming a specific latent load of 1.6 g_v/kg_{da}, which is comparable to the load in an office, only in less than 10% of the heating time the use of such THR does not lead to excessive humidification. In dashed it is shown the additional percentage of time (i.e. with conditions in area B) for heating mode in which controlled THR effectiveness allows preventing excessive humidity. The percentage of operative time without additional dehumidification is almost always more than doubled.

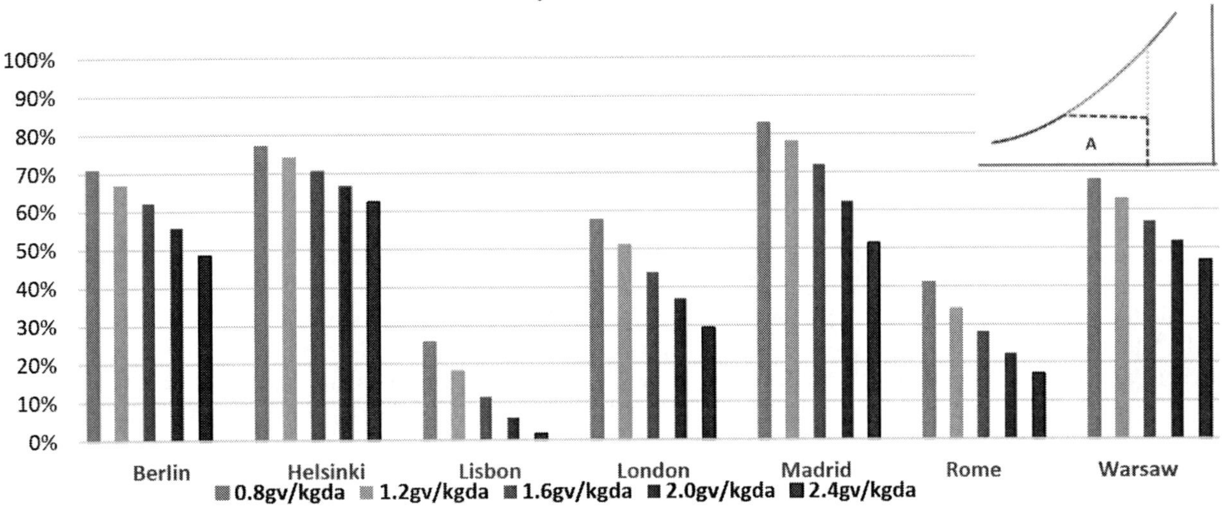

Figure 4: Percentage of time within heating conditions with outside air under excessive humidity limit, RH_{RA}=50%. Top right: corresponding conditions in the psychrometric chart.

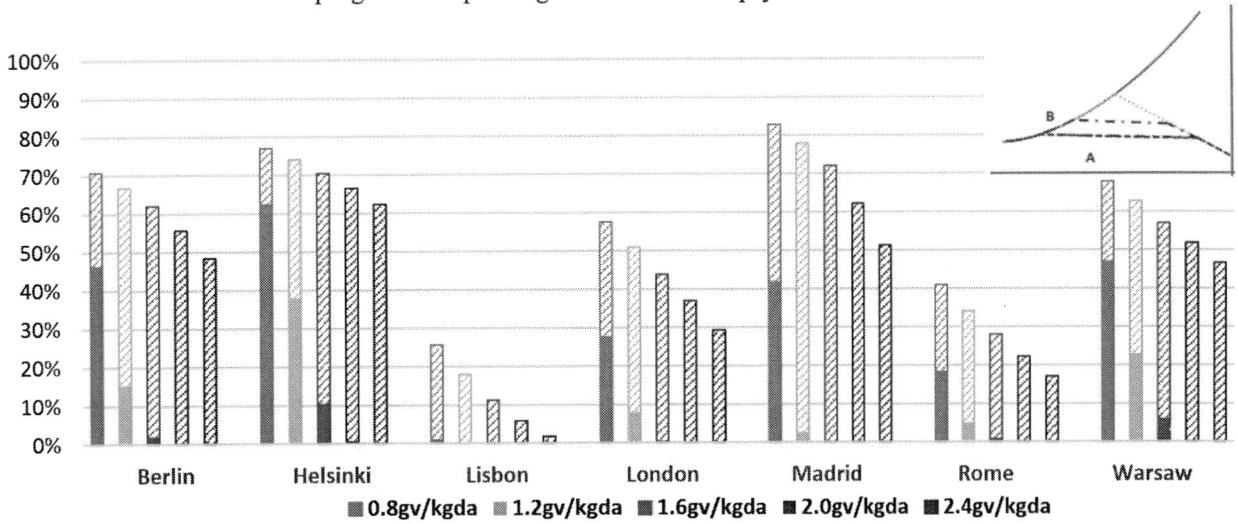

Figure 5: Percentage of time within heating conditions with outside air under excessive humidity limit without (A solid color) and with effectiveness control (B dashed), RH_{RA}=50% for a THR with 70 % effectiveness. Top right: corresponding conditions in the psychrometric chart.

3.2 Condensation and Frosting

The impact of condensation, is significant for SHR while it takes place for THR only for outside temperature far below 0 °C and mainly under the defrost limit excluding gains related to the latent heat. In SHR the additional heat transfers due to phase change results in an average annual effectiveness rise from 1 % in Berlin/Warsaw to 4 % for Helsinki. The maximum rise of annual costs due to frost protection occurs for Helsinki. The presumed frost limits for SHR, lead to an annual additional pre-heating cost per flow rate of 0.20 €/(l/s). The range of frost limits provided by the suppliers depending on the return air humidity leads to a pre-heating cost from 0.06 €/(l/s) to 0.10 €/(l/s) when using THR. In the case of a higher RA humidity, the subsequent rise in the frost limit strongly influences the additional costs: moving from 50 % to 60 %, the pre-heating costs nearly double. It is worth mentioning, that just because the outside temperature doesn't reach the frosting limit so often it does not mean it is not necessary a pre-heating system, as damage should be avoided.

3.3 Energy Savings and Economic Analysis

The energy and cost savings per flow rate of annual demand are listed in Table 3 for both SHRc and THRc. As expected the savings in heating are the lowest for the warm climate (Csa), and due to the highly humid conditions in the cold seasons the THRc for Lisbon and Rome does not bring along a significant benefit for heating conditions. The high costs for natural gas in Germany compared to Finland and Poland lead to cost savings that are higher than for Helsinki or Warsaw, even though the reduction in energy demand is lower. The humidity controlled THRc reduces the savings with respect to uncontrolled recovery, but avoids excessive indoor humidity and dehumidification costs. The actual operative effectiveness could be much lower than the nominal one which could be misleading to designers. Benefits can be not significantly larger than, and in some cases even lower than those provided by SHRc.
If an indoor relative humidity of 60 % is assumed, the benefits in cost savings are higher (Table 4). Depending on the climate, the payback time halves or nearly halves as for Rome and Berlin/Madrid respectively. For Lisbon the impact of the humidity limit is the most significant. This is due to the fact that, in heating season many conditions occur with humidity close to the return air humidity. In general, if it is feasible in terms of thermal comfort the allowance of a higher indoor humidity could be taken into account for reducing expenses.
For the following observations it is distinguished, additional to THRc and THR, between SHRc and SHR. The SHR and THR are not controlled depending on the outside humidity. Hence, this results determine the economic aspects without taking into account the dehumidification needed. The difference in the present worth value is illustrated in Figure 6 (left). The lower the value the less beneficial is the investment. The THRc allows a high present values for cold climates for which heating is responsible for the main part of the annual costs. One exception is Lisbon.

Table 3: Annual reduction in energy for heating and cooling demand [kWh/(l/s)] and cost savings [€/(l/s)] per air flow rate (assumed specific latent load of 1.6 g_v/kg_{da})

	SHRc				THRc				
City	heating	costs	cooling	costs	heating	costs	cooling	costs	Climate class
Berlin	74	4.97	0.072	0.02	79	5.32	0.10	0.03	Dfb
Helsinki	110	4.21	0.002	0.00	140	5.39	0.02	0.00	Dfb
Lisbon	7	0.72	0.571	0.13	3	0.29	2.07	0.39	Csa
London	52	3.28	0.019	0.00	49	3.11	0.01	0.00	Cfb
Madrid	49	3.58	0.000	0.00	49	3.62	0.00	0.00	Csa
Rome	21	1.62	0.339	0.08	19	4.56	7.22	1.47	Csa
Warsaw	79	3.98	0.057	0.01	79	5.32	0.23	0.03	Dfb

Table 4: Payback time [a] for different indoor humidity values for THRc (assumed specific latent load of 1.6 g_v/kg_{da})

Indoor relative humidity	Berlin	Helsinki	Lisbon	London	Madrid	Rome	Warsaw
50%	0.61	0.61	11.17	1.05	0.90	2.27	0.72
60%	0.38	0.45	1.57	0.55	0.47	1.00	0.48

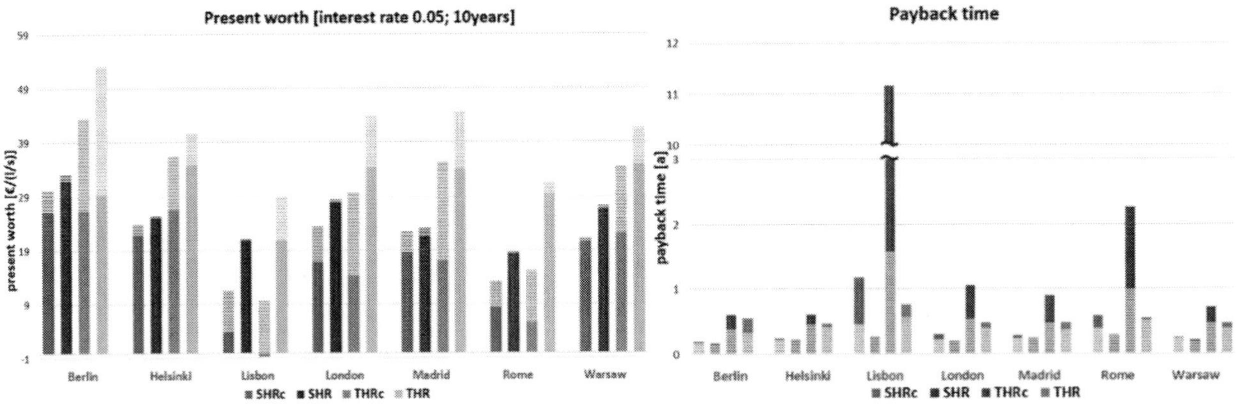

Figure 6: Present worth – interest rate 0,05 10 years – (left) and payback time (right) for RH_{RA}=50% (dark color) and RH_{RA}=60% (light color)

For that climate the humidity control leads to a very high payback time and a negative present worth value, due to that fact that the outside air is in cold condition comparable humid and in order to avoid the excessive humidification the effectiveness is lower, but also the resulting savings. Generally, energy savings are depending on the climate class, while the economic savings are independent of the climate due to different natural gas and electricity costs.

4. CONCLUSIONS

Heat recovery through sensible and total heat exchanger represents an opportunity for the improvement of energy efficiency in existing buildings and an important resource to move towards nearly zero energy buildings. However, the actual saving potential is often overestimated when only the driving gradient is considered, i.e. temperature difference for SHR and enthalpy difference for THR. In fact, in winter mode operation, when the outside air humidity ratio produces excessive mixed air humidity for SHR or even more when latent recovery with THR leads to a similar effect even with lower external humidity, a control strategy (SHRc and THRc) has to be introduced to limitation or exclude any heat recovery and avoid the need for an energy expensive dehumidification by cooling process. This appears particularly challenging with tighter internal humidity control and higher internal latent loads.

Considering the energy savings for SHRc and THRc, the saving potential is higher for heating than for cooling. Only cities in warm climates (Csa) and with relatively humid summer benefit significantly also for cooling. The energy savings of THR is not much higher than that of SHR, when the control strategy is assumed.

As for the economic performance, the payback time for SHRc is much lower than a year in most of the cases. Since the recovery limitation is stronger with lower internal setpoint humidity ratios, the analysis has been repeated with a higher internal relative humidity (60 % instead of 50 %), finding significant improvements of the economic performance (present worth and payback time in particular), but leaving THRc quite far from SHRc.

The contribution of condensation of the exhaust air in the recovery system appears quite limited, with a maximum increase of the annual effectiveness of 4% for cold climates (Helsinki) and SHR. Defrosting has been shown to impact the savings up to 0.2 €/(l/s) in Helsinki for SHR.

NOMENCLATURE

ε	effectiveness	
h	spesific enthalpy	(kJ/kg)
$\dot{m}$	mass flow rate	(kg/s)
RH	relative humidity	(%)
T	temperature	(°C)
x	humidity ratio	(kg_v/kg_{da})

Subscripts

CA	recirculated air
da	dry air
EA	exhaust air
l	latent
MA	mixed air
min	minimum
OA	outside
R	air after recovery
RA	return air
SA	supply air
s	sensible
t	total
W	water
V	vapour

REFERENCES

Anisimov, S., Jedlikowski, A., & Pandelidis, D. (2015). Frost formation in the cross-flow plate heat exchanger for energy recovery. *International Journal of Heat and Mass Transfer*, *90*(February 2016), 201–217. http://doi.org/10.1016/j.ijheatmasstransfer.2015.06.056

Deshko, V. I., Ya, A., & Sukhodub, I. O. (2016). Heat and mass transfer in cross-flow air-to-air membrane heat exchanger in heating mode, *100*, 133–145. http://doi.org/10.1016/j.applthermaleng.2016.01.139

energyplus. (2016). Retrieved from https://energyplus.net/weather

Eurostat. (2016). (s. o. Union, Producer) Retrieved from http://ec.europa.eu/eurostat/about/overview

Fisk, W. J., Chant, R. E., Archer, K. M., Hekmat, D., Offermann, F. J., & Pedersen, B. S. (1985). Onset of Freezing in Residential Air-To-Air Heat Exchangers. *ASHRAE Transactions*, *91*(pt 1B), 145–158.

Hwang, R. L., Lin, T. P., & Kuo, N. J. (2006). Field experiments on thermal comfort in campus classrooms in Taiwan. *Energy and Buildings*, *38*(1), 53–62. http://doi.org/10.1016/j.enbuild.2005.05.001

Kassai, M., Nasr, M. R., & Simonson, C. J. (2015). A developed procedure to predict annual heating energy by heat- and energy recovery technologies in different climate European countries. *Energy and Buildings*, *109*, 267–273. http://doi.org/10.1016/j.enbuild.2015.10.008

Keller, L. (2009). Leitfaden für Lüftungs- und Klimaanlagen. München: Oldenbourg Industrieverlag.

Kosonen, R., & Tan, F. (2004). Assessment of productivity loss in air-conditioned buildings using PMV index. *Energy and Buildings*, *36*(10 SPEC. ISS.), 987–993. http://doi.org/10.1016/j.enbuild.2004.06.021

Lazzarin, R. M., Gasparella, A., Longo, G. a., & Perbellini, M. (2000). Gli scambia tori di calore aria aria: potenzialità applicative nella riduzione del carico di ventilazione degli edifici. *Atti Del 55°. Congr. Naz. ATI, Matera 16.*

Liu, P., Rafati Nasr, M., Ge, G., Justo Alonso, M., Mathisen, H. M., Fathieh, F., & Simonson, C. (2016). A theoretical model to predict frosting limits in cross-flow air-to-air flat plate heat/energy exchangers. *Energy and Buildings*, *110*, 404–414. http://doi.org/10.1016/j.enbuild.2015.11.007

Milano, C. d. (2015). Listino prezzi per l'esecuzione di opere pubbliche e manutenzioni.

Mishra, A. K., & Ramgopal, M. (2014). Thermal comfort field study in undergraduate laboratories - An analysis of occupant perceptions. *Building and Environment*, *76*, 62–72. http://doi.org/10.1016/j.buildenv.2014.03.005

Nam, S. H., & Han, H. (2016). Computational modeling and experimental validation of heat recovery ventilator under partially wet conditions. *Applied Thermal Engineering*, *95*, 229–235. http://doi.org/10.1016/j.applthermaleng.2015.11.023

Rafati Nasr, M., Fauchoux, M., Besant, R. W., & Simonson, C. J. (2014). A review of frosting in air-to-air energy exchangers. *Renewable and Sustainable Energy Reviews*, *30*, 538–554. http://doi.org/10.1016/j.rser.2013.10.038

Simonson, C. J., & Besant, R. W. (1998). Heat and Moisture Transfer in Energy Wh.eels During Sorption, Condensation, and Frosting Conditions, *120*(August).

StatisticsFinland. (2016). Retrieved from http://www.stat.fi/til/ehi/2015/04/ehi_2015_04_2016-03-09_tau_002_en.html

Sterling, E. M., Arundel, a., & Sterling, T. D. (1985). Criteria for Human Exposure To Humidity in Occupied Buildings. *ASHRAE Transactions*, *91*(pt 1B), 611–622.

Tsutsumi, H., Tanabe, S. I., Harigaya, J., Iguchi, Y., & Nakamura, G. (2007). Effect of humidity on human comfort and productivity after step changes from warm and humid environment. *Building and Environment*, *42*(12), 4034–4042. http://doi.org/10.1016/j.buildenv.2006.06.037

Y. Chen, H. Tsutsumi, T. Akimotob, S. Tanabe, T. T. (2003). Subjective experiments on the effects of relative humidity and humidity ratio during summer season. *Proceedings of ISIAQ 7th Lnternational Conference*, 293–298.

Zhang, J., & Fung, A. S. (2015). Experimental and numerical investigation of the thermal impact of defrost cycle of residential heat and energy recovery ventilators. *Energy and Buildings*, *97*, 129–136. http://doi.org/10.1016/j.enbuild.2015.03.034

Gradient-Based Estimation of Air Flow and Geometry Configurations in a Building Using Fluid Dynamic Adjoint Equations

Runxin HE[1*], Humberto GONZALEZ[2]

Department of Electrical & Systems Engineering, Washington University in St. Louis, St. Louis, MO, USA

[1] runxinhe@email.wustl.edu

[2] hgonzale@wustl.edu

* Corresponding Author

ABSTRACT

Real-time estimations of temperature distributions and geometric configurations are important to energy efficient buildings and the development of smarter cities. In this paper we formulate a gradient-based estimation algorithm capable of reconstructing the states of doors in a building, as well as its temperature distribution, based on a floor plan and a set of thermostats. Our algorithm solves in real time a convection-diffusion Computer Fluid Dynamics (CFD) model for the air flow in the building as a function of its geometric configuration. We formulate the estimation algorithm as an optimization problem, and we solve it by computing the adjoint equations of our CFD model, which we then use to obtain the gradients of the cost function with respect to the flow's temperature and door states. We evaluate the performance of our method using simulations of an apartment in the St. Louis area. Our results show that the estimation method is both efficient and accurate, establishing its potential for the design of smarter control schemes in the operation of high-performance buildings.

1. INTRODUCTION

Buildings currently account for more than 40% of the total energy consumption in the U.S. (U.S. Department of Energy, 2012), and they cost $130 billion in energy leakage and inefficiency (U.S. Green Building Council, 2014). For this reason many research groups have developed new control algorithms to improve the performance and efficiency of Heating, Ventilation, and Air Conditioning (HVAC) systems in buildings (Wang & Ma, 2008; Dounis & Caraiscos, 2009; Afram & Janabi-Sharifi, 2014; Shaikh, Nor, Nallagownden, Elamvazuthi, & Ibrahim, 2014).

Among the many control algorithms used in smart building applications, Model Predictive Control (MPC) stands out thanks to its flexible mathematical formulation, as well as its accurate and robust responses in real-world implementations (Huang, 2011; Xi, Poo, & Chou, 2007; Moroşan, Bourdais, Dumur, & Buisson, 2010). Moreover, MPC has become the standard for solving complex constrained multivariate control problems in process control applications (Afram & Janabi-Sharifi, 2014). MPC has been used in HVAC control applications such as zoned temperature control (J. Ma, Qin, Salsbury, & Xu, 2012; He & Gonzalez, 2016) and overall temperature regularization (Prívara, Široký, Ferkl, & Cigler, 2011) among others, experimentally showing significant increments in energy efficiency.

MPC algorithms require the use of dynamical models, which are used as predictors. Compared to concentrated-parameter models such as the Resistance-Capacitance (RC) networks (Y. Ma, Kelman, Daly, & Borrelli, 2012), Computational Fluid Dynamics (CFD) models for building temperature control have a significant advantage since they naturally incorporate geometric and air flow information. Moreover, CFD models can accurately describe short time scales, allowing us to reflect indoor climate changes in minutes or even seconds (Jones & Whittle, 1992; Chow, 1996). Hence, using MPC with CFD models as predictors enables the study of control and estimation strategies beyond the standard temperature control, such as controlling fan speeds or finding the optimal location for thermostats.

However, due to the predictive nature of its formulation, MPC is sensitive to mismatches in the dynamic prediction model and inaccurate initial state estimations, which might lead to steady-state offsets or even system instability (Faanes & Skogestad, 2005). Therefore MPC algorithms are usually implemented in coordination with estimation algorithms capable of inferring relevant parameters and initial conditions from sensor data, such as Kalman filters. Other optimization-based estimation algorithms have been developed, such as the results by Isakov (2006) in Partial Differential Equation (PDE) estimation. Banks et al. viewed the parameters for the inverse problem as random variables, and used probabilistic inference methods to estimate the desired parameters (Banks & Bihari, 2001; Banks, Holm, & Kappel, 2011). Gutman (1990) fully discretized a weak form of the Stokes Equations in time and space and identified the system's discontinuous parameters.

In this paper, we develop a gradient-based optimization method to estimate the doors' state and temperature distribution in the apartment. In particular, our contribution is twofold. First, we mathematically formulate a gradient-based estimation method to identify real-time indoor climate distribution and the apartment doors' states based on only thermostatic data. Second, we show the accuracy of our estimation method under a limited number of thermostats by simulating an apartment in the St. Louis area. Our results show that thermostatic information, when used together with CFD models, provide enough information to estimate most of the variables relevant for building climate control. In other words, a handful of thermostats can provide information, such as the configuration of doors, without the need to physically install extra sensors in a building.

The paper is organized as follows. The fluid dynamic model and finite element method are formulated in Section 2. We present the theoretical basis for our gradient-based estimation algorithm in Section 3. Finally, our simulation results are presented in Section 4.

2. COMPUTATIONAL FLUID DYNAMIC MODEL AND OPTIMIZATION PROBLEM

The kernel of our model is the incompressible Navier-Stokes equation, which is a good approximation for the coupling of temperature with free flow convection at atmospheric conditions (Awbi, 1989; Dobrzynski, Pironneau, & Frey, 2004). Throughout the paper we make two major simplifications to this model. First, we assume that the air flow behaves as a laminar fluid which reaches steady-state behavior much faster than the temperature in the building. Theoretical (Awbi, 2003) and experimental (Sun, Stowell, Keener, & Michel Jr., 2002) results have shown that turbulent flows are present in residential building, such as in the area around HVAC vents, yet their overall effect in the temperature distribution is negligible. Hence, we consider a stationary laminar Navier-Stokes equation to describe the fluid behavior, and a time-dependent equation to describe the temperature behavior. Second, we consider only two-dimensional air flows moving parallel to the ground. These assumptions reduce the accuracy of our model to some extent (van der Poel, Stevens, & Lohse, 2013), yet they allow us to significantly simplify the computational complexity of our CFD-based control design.

Let $\Omega \subset \mathbb{R}^2$ be the area of interest, assumed to be bounded and connected, and let $\partial\Omega$ be its boundary. Let $u: \Omega \to \mathbb{R}^2$ be the *stationary air flow velocity*, and $p: \Omega \to \mathbb{R}$ be the *stationary air pressure* in Ω. Also, given $T > 0$, let $T_e: \Omega \times [0, T] \to \mathbb{R}$ be the *temperature* in Ω. Then, following Landau and Lifshitz (1959), the non-dimensional temperature convection-diffusion model in Ω can be described by the following PDE:

$$\frac{\partial T_e}{\partial t}(x,t) - \nabla_x \cdot \big(\kappa(x)\, \nabla_x T_e(x,t)\big) + u(x) \cdot \nabla_x T_e(x,t) = g_{T_e}(x,t), \tag{1}$$

where $g_{T_e}: \Omega \times [0, T] \to \mathbb{R}$ represents the heat source in the room, $\kappa: \Omega \to \mathbb{R}$ is the *thermal diffusivity*, Pr is the *Prandtl number* of the air, Re is the *Reynolds number* of the air, and $\nabla_x = \left(\frac{\partial}{\partial x_1}, \frac{\partial}{\partial x_2}\right)^T$ is the *gradient operator*. The initial condition of the temperature is:

$$T_e(x,0) = \pi_0(x), \quad \text{for } x \in \Omega. \tag{2}$$

Similarly, the non-dimensional stationary air flow in Ω is governed by the following incompressible Navier-Stokes stationary PDE:

$$-\frac{1}{Re}\, \Delta_x u(x) + \big(u(x) \cdot \nabla_x\big) u(x) + \nabla_x p(x) + \alpha(x)\, u(x) = g_u(x), \quad \text{and,} \tag{3}$$

$$\nabla_x \cdot u(x) = 0, \tag{4}$$

where $g_u: \Omega \to \mathbb{R}^2$ represents all the external forces applied to the air (such as fans), and $\triangle_x = \frac{\partial^2}{\partial x_1^2} + \frac{\partial^2}{\partial x_2^2}$ is the *Laplacian operator*. The we introduce the *viscous friction coefficient* $\alpha: \Omega \to \mathbb{R}$, following the technique in Gersborg-Hansen, Sigmund, and Haber (2005), to model different materials in Ω. Indeed, when the point x corresponds to a material that blocks air, we choose $\alpha(x) \gg u(x)$, which results in $u(x) \approx 0$. When the point x corresponds to air, then we choose $\alpha(x) = 0$.

The building's exterior wall are denoted by $\Gamma_w \subset \partial\Omega$, and the air inlet of the HVAC system is modeled as a gap in the wall, denoted $\Gamma_i \subset \partial\Omega$. Hence, $\Gamma_w \cup \Gamma_i = \partial\Omega$. The boundary condition for the temperature is:

$$T_e(x) \equiv T_A, \quad \text{for } x \in \partial\Omega, \tag{5}$$

where T_A is the *atmospheric temperature*. We only apply a boundary condition for the pressure equation at the inlet, setting $p(x) \equiv p_A$ for each $x \in \Gamma_i$, where p_A is the *atmospheric pressure*. We do not define a boundary condition for the air flow at the inlet Γ_i, and we set the air flow at the exterior wall as follows:

$$u(x) \equiv 0, \quad \text{for } x \in \Gamma_w. \tag{6}$$

We assume that there are n_t thermostats in the building. The i-th thermostat is located at $x_i \in \Omega$, and samples the temperature in a neighborhood averaged using the *bump weight function* $\Phi_i(x) = \sigma \exp\left(-\left(r^2 - \|x - x_i\|^2\right)^{-1}\right)$ for $\|x - x_i\| < r$, and $\Phi_i(x) = 0$ otherwise, where $\sigma > 0$ is a normalization factor such that $\int_\Omega \Phi_i(x)\,\mathrm{d}x = 1$.

We also assume that there are n_d doors in the building. We define $\theta_i \in \{0, 1\}$ as the configuration of the i-th door, i.e., $\theta_i = 1$ when the i-th door is open, and $\theta_i = 0$ when is closed. Let $\Omega_{\theta_i} \subset \Omega$ be the area occupied by the i-th door when it is closed, and let $\mathbb{I}_i$ be the indicator function of Ω_{θ_i}, i.e., $\mathbb{I}_i(x) = 1$ for $x \in \Omega_{\theta_i}$, and $\mathbb{I}_i(x) = 0$ otherwise.

When the door configuration changes, so does the prediction generated by our CFD model in equations (1) and (3). In particular, the parameters α and κ change for each $x \in \Omega_{\theta_i}$ as a function of θ_i. We model this relation by defining $\alpha: \Omega \times \{0, 1\}^{n_d} \to \mathbb{R}$ and $\kappa: \Omega \times \{0, 1\}^{n_d} \to \mathbb{R}$ as follows:

$$\alpha(x, \theta) = \alpha_0 + \sum_{i=0}^{n_d} (1 - \theta_i)\,(\alpha_w - \alpha_0)\,\mathbb{I}_i(x), \quad \text{and} \quad \kappa(x, \theta) = \kappa_0 + \sum_{i=0}^{n_d} (1 - \theta_i)\,(\kappa_w - \kappa_0)\,\mathbb{I}_i(x), \tag{7}$$

where α_0 and κ_0 are the parameters for open air, while α_w and κ_w are the parameters for solid walls. Note that both α and κ are affine functions of $\theta \in \mathbb{R}^{n_d}$.

Now, using binary values for each θ_i means that our estimation algorithm will have to use combinatorial methods, which tend to scale poorly in both computation time and computational resources. To avoid this problem we relax the binary parameters $\theta_i \in \{0, 1\}$, instead allowing them to belong to the unit interval $[0, 1]$. Although for each θ_i only the extreme values have meaningful physical interpretations, non-integer values can theoretically be interpreted as averaged observations over the optimization horizon, as explained in Vasudevan, Gonzalez, Bajcsy, and Sastry (2013a, 2013b). For example, if throughout the optimization horizon a door is open half the time, and closed half the time, it is likely that we will observe $\theta_i \approx 0.5$. The relaxation of each θ_i is also important in our numerical calculations, since it transforms the optimization program from a mixed-integer program to a more convenient nonlinear format (Gersborg-Hansen et al., 2005).

Now we can formulate our main estimation algorithm to compute the door configuration θ and the initial temperature π_0 using the information from the n_t thermostats in the building. Given an arbitrary estimation time horizon, say $[0, T]$, we write our optimal estimation problem as follows:

$$\begin{aligned}
\min_{\pi_0:\Omega\to\mathbb{R},\, \theta\in\mathbb{R}^{n_d}} \; & J(\pi_0, \theta) = \sum_{i=1}^{n_t} \int_0^T \left(\int_{\Omega_i} \Phi_i\, T_e(x, t; \pi_0, \theta)\,\mathrm{d}x - T^*_{e,i} \right)^2 \mathrm{d}t + \eta_0 \sum_{i=1}^{n_t} \left(\int_{\Omega_i} \Phi_i\, \pi_0\,\mathrm{d}x - \pi^*_{0,i} \right)^2 + \eta_1\, \|\pi_0\|^2_\Omega, \\
\text{subject to:} \; & \text{partial differential equations (1), (3), and (4),} \\
& \text{boundary and initial conditions (2), (5), and (6),} \\
& 0 \le \theta_i \le 1, \quad \forall i \in \{1, \ldots, n_d\},
\end{aligned} \tag{8}$$

were, $\eta_0, \eta_1 > 0$ are weight parameters, $T_e(x, t; \pi_0, \theta)$ is the unique solution of equation (1) with initial condition π_0 and configuration θ, $T^*_{e,i}(t)$ is the time signal obtained from the i-th thermostat over the horizon $[0, T]$, and $\pi^*_{0,i}$ is just notation for the initial thermostat temperature, i.e., $\pi^*_{0,i} = T^*_{e,i}(0)$.

3. ADJOINT-BASED GRADIENT COMPUTATION

In this section we develop a numerical algorithm to solve the optimization problem defined in equation (8). We use a gradient-based optimization algorithm to find local minimizers of our optimization problem, where the gradients are computed using the adjoint equations of the CFD model, similar to the techniques in Gunzburger (2000) and Yang and Tomlin (2014). We then discretize the adjoint equations using the Finite Element Method (FEM), resulting in a practical algorithm which we test in Section 4.

3.1 Adjoint Equations and Fréchet Derivatives

In order to derive our CFD model's adjoint equations, first we need to write the *Lagrangian function* of the optimization problem (Giles & Pierce, 1997; Gunzburger, 2000). Let $\{\lambda_i\}_{i=1}^{6}$ be the set of *Lagrange multipliers*, or *adjoint variables*, each associated to one of the equations (1) to (6) and defined in its respective dual space. Then, the Lagrangian function of our optimal estimation problem is:

$$L\left(T_e, u, p, \pi_0, \theta, \{\lambda_i\}_{i=1}^{6}\right) = J(\pi_0, \theta) + \left\langle \lambda_1, \frac{\partial T_e}{\partial t} - \nabla_x \cdot (\kappa(x)\, \nabla_x T_e) + u \cdot \nabla_x T_e - g_{T_e} \right\rangle_{\Omega\times[0,T]} + \langle \lambda_4, T_e \rangle_{\partial\Omega\times[0,T]} + \\ + \left\langle \lambda_2, -\frac{1}{Re}\, \Delta_x u + (u\cdot\nabla_x)\, u + \nabla_x p + \alpha u - g_u \right\rangle_{\Omega} + \langle \lambda_3, \nabla_x \cdot u \rangle_{\Omega} + \langle \lambda_5, u \rangle_{\Gamma_w} + \langle \lambda_6, T_e(0,\cdot) - \pi_0 \rangle_{\Omega}, \quad (9)$$

where $\langle f_1, f_2 \rangle_S = \int_S f_1(z) f_2(z)\, \mathrm{d}z$ is the inner product of the Hilbert space of square integrable functions $\mathcal{L}^2(S)$. We write the necessary conditions for optimality using Galerkin methods (Girault & Raviart, 1979), i.e., by setting the inner product of the partial derivatives of L with respect to all the dual directions equal to zero. That is, we look for solutions such that $\left\langle \frac{\partial L}{\partial T_e}, w \right\rangle_{\Omega\times[0,T]} = 0$, $\left\langle \frac{\partial L}{\partial u}, v \right\rangle_{\Omega} = 0$, and $\left\langle \frac{\partial L}{\partial p}, q \right\rangle_{\Omega} = 0$ for each set of functions (w, v, q) in the respective dual spaces, and sufficiently weakly differentiable. As detailed in Appendix A, the conditions above are satisfied when the dual variables satisfy:

$$-2\sum_{i=1}^{n_t}\left(\int_{\Omega_i} \Phi_i(z)\, T_e(z,t)\, \mathrm{d}z - T^*_{e,i}(t)\right) + \frac{\partial \lambda_1}{\partial t}(x,t) + \nabla_x \cdot (\kappa(x)\, \nabla_x \lambda_1(x,t)) + u(x)\cdot\nabla_x \lambda_1(x,t) = 0, \quad (10)$$

$$\lambda_6(x) = \lambda_1(x, 0), \quad (11)$$

$$\int_0^T \lambda_1(x,t)\, \nabla_x T_e(x,t)\, \mathrm{d}t + \alpha(x)\, \lambda_2(x) - \frac{1}{Re}\, \Delta_x \lambda_2(x) - u(x)\cdot\nabla_x\lambda_2(x) + \lambda_2(x)\cdot\nabla_x u(x) - \nabla_x \lambda_3(x) = 0, \quad \text{and,} \quad (12)$$

$$\nabla_x \cdot \lambda_2(x) = 0, \quad (13)$$

with boundary conditions $\lambda_1(x,t) = 0$ and $\lambda_2(x,t) = 0$ for each $x \in \partial\Omega$ and $t \in [0,T]$, together with final condition $\lambda_1(x,T) = 0$ for each $x \in \Omega$. The adjoint functions λ_4 and λ_5 are irrelevant to our Fréchet derivative calculation, therefore we omit them from this presentation.

Now we can compute the Fréchet derivatives of the cost function with respect to θ and π_0. Consider a parameter change from (θ, π_0) to $(\theta + \delta\theta, \pi_0 + \delta\pi_0)$. Since both α and κ are affine in θ, these variations will result in changes from (α, κ) to $(\alpha + \delta\alpha, \kappa + \delta\kappa)$, which will also imply changes from (T_e, u, p) to $(T_e + \delta T_e, u + \delta u, p + \delta p)$. As detailed in Appendix B, these variations allow us to compute a first-order approximation of the cost function J, which result in:

$$\left\langle \mathcal{D}_\alpha J, \delta\alpha \right\rangle_{\Omega} = \langle \lambda_2 \cdot u, \delta\alpha \rangle_{\Omega}, \quad \text{and} \quad \left\langle \mathcal{D}_\kappa J, \delta\kappa \right\rangle_{\Omega} = \int_0^T \langle \nabla_x \lambda_1 \cdot \nabla_x T_e, \delta\kappa \rangle_{\Omega}\, \mathrm{d}t, \quad (14)$$

and using the chain rule and the formulas in equation (14) we get the desired directional derivatives for J:

$$\left\langle \mathcal{D}_{\pi_0} J, \delta\pi_0 \right\rangle_{\Omega} = \langle \nabla_{\pi_0} J - \lambda_6, \delta\pi_0 \rangle_{\Omega}, \quad \text{and} \quad \mathcal{D}_\theta J \cdot \delta\theta = \sum_{i=1}^{n_d}\left(\left\langle \mathcal{D}_\alpha J, \frac{\partial \alpha}{\partial \theta_i}\right\rangle_{\Omega} + \left\langle \mathcal{D}_\kappa J, \frac{\partial \kappa}{\partial \theta}\right\rangle\right) \delta\theta_i. \quad (15)$$

Note that both directional derivatives are linear bounded operators, hence they are also Fréchet derivatives as desired.

Algorithm 1 Gradient-based estimation algorithm

Require: Initial values for θ and π_0.

1: **loop**
2: Compute T_e, u, and p by solving the CFD model in equations (1) to (6).
3: Compute λ_1, λ_2, λ_3, and λ_6 by solving the adjoint equations (10) to (13).
4: Compute the gradients $\mathcal{D}_{\pi_0}$ and $\mathcal{D}_\theta$ in equation (15).
5: Compute the projected-gradient descent directions $(\delta\pi_0, \delta\theta)$ by solving the QP in equation (16), with value V.
6: **if** $V = 0$ **then**
7: Stop.
8: **end if**
9: Compute the step size β using the Armijo line search method in equation (17).
10: Update $\pi_0 \leftarrow \pi_0 + \delta\pi_0$ and $\theta \leftarrow \theta + \beta\,\delta\theta$.
11: **end loop**

3.2 Gradient-Based Optimization Algorithm

Using the closed-form formulas for the Fréchet derivatives of J with respect to π_0 and θ, we build a gradient-based optimization algorithm to solve the problem in equation (8) using a projected-gradient method (Nocedal & Wright, 2006, Chapter 18.6).

First, we find descent directions $\delta\pi_0$ and $\delta\theta$ as solutions of the following Quadratic Program (QP) with value V:

$$\begin{aligned} V = \min_{\delta\pi_0:\Omega\to\mathbb{R},\ \delta\theta\in\mathbb{R}^{n_d}} & \langle \mathcal{D}_{\pi_0}J, \delta\pi_0\rangle_\Omega + \mathcal{D}_\theta J\cdot\delta\theta + \frac{\gamma}{2}\|\delta\pi_0\|_\Omega^2 + \frac{\gamma}{2}\|\delta\theta\|^2, \\ \text{subject to: } & 0 \le \theta_i + \delta\theta_i \le 1,\ \forall i \in \{1,\ldots,n_d\}, \end{aligned} \tag{16}$$

where $\gamma > 0$ is a parameter. The QP in equation (16) is derived using first-order approximations for the cost function using the derivatives in equation (15), together with a condition to guarantee the feasibility of the desired direction. Note that $V \le 0$, since $\delta\pi_0 = 0$ and $\delta\theta = 0$ always belong to the feasible set. Hence, if $V = 0$ then our method cannot find further descent directions, and it thus terminates.

Second, a step size is computed using the following Armijo line search method:

$$\begin{aligned} \beta = \arg\max_{j\in\mathbb{N}} & \bar{\beta}^j, \\ \text{subject to: } & J\big(\pi_0 + \bar{\beta}^j\,\delta\pi_0, \theta + \bar{\beta}^j\,\delta\theta\big) - J(T_e, \pi_0, \theta) \le \bar{\alpha}\bar{\beta}^j\,V. \end{aligned} \tag{17}$$

where $\bar{\alpha}, \bar{\beta} \in (0, 1)$ are parameters.

Our gradient-based optimization method is detailed in Algorithm 1. Steps 2 and 3, are numerically solved using FEM discretizations, implemented using the FEniCS package (Logg, Mardal, & Wells, 2012).

4. SIMULATION RESULTS

We applied our estimation algorithm to a simulated St. Louis area apartment with $n_d = 4$ doors, labeled $\{d_i\}_{i=1}^4$, and $n_t = 3$ thermostats, labeled $\{s_i\}_{i=1}^3$. The floor plan of the apartment is shown in Figure 1, with dimensions $7.6 \times 16.8\,\text{m}^2$ (approx. 1375 sq ft). The apartment is equipped with four HVAC vents, labeled $\{h_i\}_{i=1}^4$. We assume that each vent is endowed with a fan acting on a $1 \times 0.5\,\text{m}^2$ area, and oriented in a fixed direction.

The CFD model is governed by the constants $Re = 10^2$, $\alpha_0 = 0$, and $\kappa_0 = 10^{-2}$ when $x \in \Omega$ corresponds to free air, while $\alpha_w = 10^3$ and $\kappa_w = 10^{-4}$ when $x \in \Omega$ corresponds to a wall. The atmospheric pressure is $p_A = 101.3\,\text{kPa}$, and the atmospheric temperature is $T_A = 23.83\,°\text{C}$. We assume that h_1 and h_2 work at a low output setting, producing 0.1 kW of heat and an air flow speed of 0.1 m/s. On the other hand, h_3 and h_4 work at a normal setting, producing 4 kW of heat and and an air flow speed of 0.5 m/s. The time horizon is 300 s, sampled uniformly at 10 s steps. The sensors' observation radius is $r = 1.0\,\text{m}$. The parameters in (8) are set to $\eta_0 = 1.0$, $\eta_1 = 0.1$. The parameter in (16) is set to $\gamma = 1.0$. The parameters in (17) are set to $\bar{\alpha} = 0.01$ and $\bar{\beta} = 0.7$. We wrote our code in Python, the FEM discretization

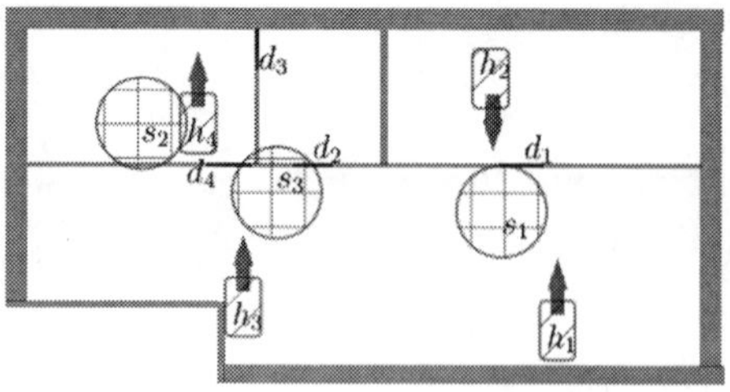

Figure 1: Floor plan of the apartment simulated in Sec. 4.

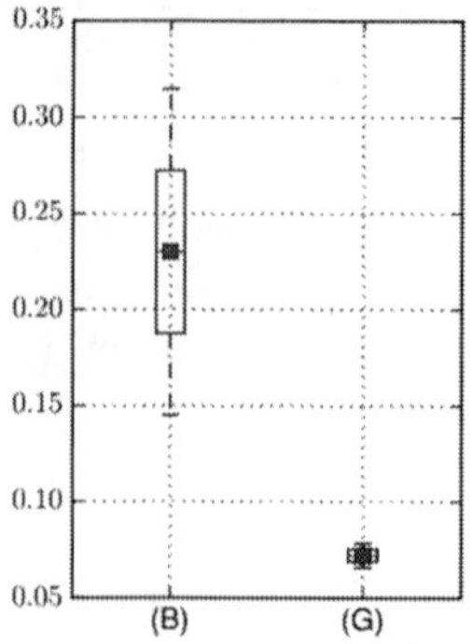

(a) Average estimation error of θ.

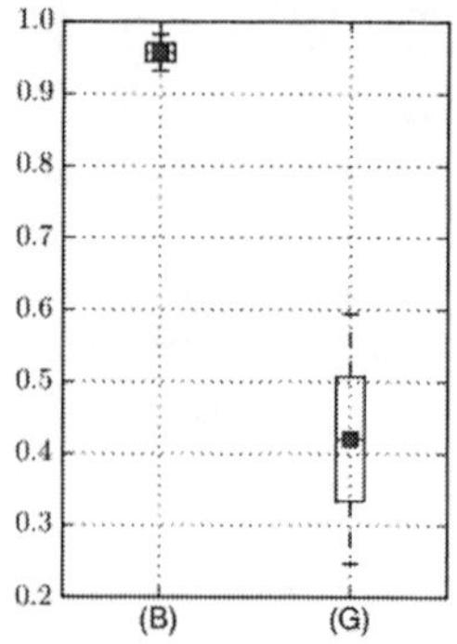

(b) Relative estimation error of π_0.

Figure 2: Results of the experiments in Sec. 4.2. Columns: (B) Banks and Bihari's method, (G) Gradient-based method, Algorithm 1.

was computed using tools from the FEniCS Project (Logg et al., 2012), and the building plan was discretized into $n_{\text{elem}} = 6276$ elements.

4.1 Probabilistic Estimation Method

In our simulations below we compare our estimation method with a probabilistic-based estimation algorithm formulated by Banks and Bihari (2001), and applied to problems involving parameter estimation of differential equations (Banks et al., 2011). Under Banks and Bihari's framework, π_0 and θ are random variables with unknown probability distributions, thus the estimation problem is formulated to find the optimal distributions that would most likely produce the acquired sensor data in expectation. Due to space constraints we omit a detailed description of this method, and we refer interested readers to (Banks & Bihari, 2001).

Let $\pi_{0,\Delta}$ be the FEM discretization of π_0, hence $\pi_{0,\Delta} \in \mathbb{R}^{n_{\text{elem}}}$. We assume that θ and $\pi_{0,\Delta}$ follow probability distributions $\mathbb{P}(\theta)$ and $\mathbb{P}(\pi_0)$. In the particular case of θ, since it is a vector of independent binary variables, its distribution is $\mathbb{P}(\theta) = \prod_{i=1}^{n_d} p_i^{\theta_i} (1 - p_i)^{1-\theta_i}$. We assume that θ and $\pi_{0,\Delta}$ are independent.

Banks and Bihari's estimation algorithm relies on closed-form formulas of the expected values of each of the random variables in the cost function. Using the cost function in equation (8), the only nontrivial expected value is that of $T_{e,\Delta}(x, t; \pi_{0,\Delta}, \theta)$, the FEM discretization of T_e. Note that given $x \in \Omega$, $t \in [0, T]$, and $\theta \in \{0, 1\}^{n_d}$, then $T_{e,\Delta}(x, t; \pi_{0,\Delta}, \theta)$ is a linear function of $\pi_{0,\Delta}$; hence, as shown by Kumar and Varaiya (2015, Chapter 3), the conditional expected value of $T_{e,\Delta}$ is $\mathbb{E}[T_e(x, t; \pi_{0,\Delta}, \theta) \mid \theta] = T_{e,\Delta}(x, t; \mathbb{E}[\pi_{0,\Delta}], \theta)$ for each pair (x, t). Then, using Bayes' rule,

$$\mathbb{E}[T_{e,\Delta}(x, t; \pi_{0,\Delta}, \theta)] = \sum_{\theta \in \{0,1\}^{n_d}} T_{e,\Delta}(x, t; \mathbb{E}[\pi_{0,\Delta}], \theta)\, \mathbb{P}(\theta), \quad \forall x \in \Omega,\ t \in [0, T]. \tag{18}$$

It is worth noting that the cardinality of $\{0, 1\}^{n_d}$ is 2^{n_d}, hence each evaluation of equation (18) involves solving a PDE an exponentially growing number of times as a function of n_d.

4.2 Estimation Using Three Thermostats

We run both estimation algorithms, Banks and Bihari's method and gradient-based method, Algorithm 1, under six different combinations for $\theta \in \{(0,0,0,0), (0,0,1,1), (0,1,1,0), (0,1,1,1), (1,0,0,0), (1,1,1,1)\}$ and two different initial temperatures π_0. Since Algorithm 1 converges to local minimizers, we also run five estimations for each pair (π_0, θ) initializing the algorithm with different values.

Figure 2a is a bar plot of the average estimation errors of θ, calculated as $e_\theta = \frac{1}{n_d} \sum_{i=1}^{n_d} |\theta_i - \hat{\theta}_i|$, where $\hat{\theta}$ is either the estimated probability distribution from Banks and Bihari's method, or the estimated relaxed configuration from gradient-based method, Algorithm 1. Figure 2b shows a similar bar plot for the relative estimation error of π_0, calculated as $e_{\pi_0} = \frac{\|\pi_0 - \hat{\pi}_0\|_\Omega}{\|\pi_0\|_\Omega}$, where $\hat{\pi}_0$ is either the estimated expected value of π_0 from Banks and Bihari's method, or the estimated

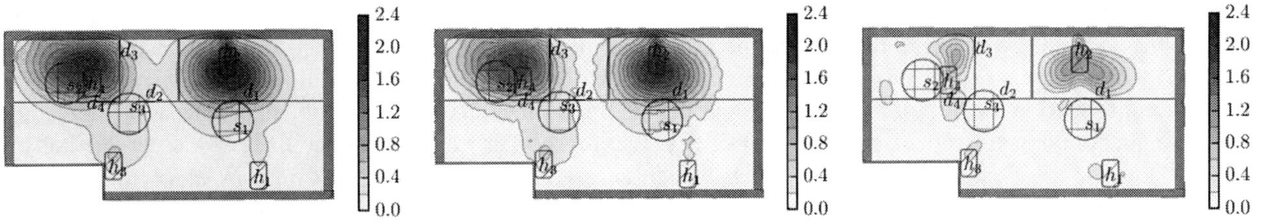

(a) Initial temperature π_0 in °C with respect to T_A.

(b) Estimation error of π_0 by Banks and Bihari's method.

(c) Estimation error of π_0 by gradient-based method, Algorithm 1.

Figure 3: Results of the experiments in Sec. 4.2. All doors are closed.

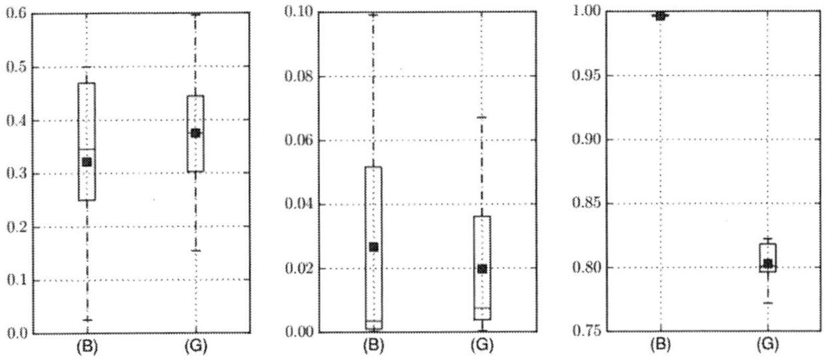

(a) Average estimation error of θ.

(b) Estimation error of θ_1.

(c) Relative estimation error of π_0.

Figure 4: Results of the experiments in Sec. 4.3. Columns: (B) Banks and Bihari's method, (G) Gradient-based method, Algorithm 1.

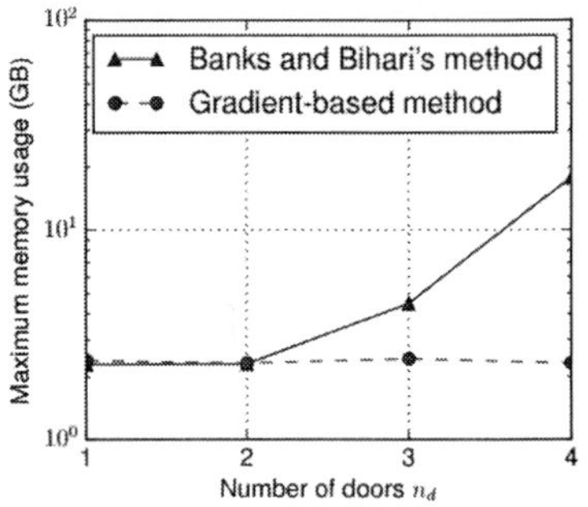

Figure 5: Memory usage in GB of the experiments in Sec. 4.4.

initial distribution from gradient-based method, Algorithm 1. From these results we can observe that Algorithm 1 is significantly more accurate than the probabilistic method in estimating both doors' states and initial temperature distribution. It is worth noting that the average error of gradient-based method, Algorithm 1, in Figure 2a is small enough so that one can use a constant threshold to convert from relaxed values of θ to binary values. Also, the accuracy of our results enables further smart applications, such as the locating the residents in a building by using thermostat data and further behavioural assumptions. In Figure 3 we show the actual initial temperature distribution for one configuration θ, and the estimation errors by both algorithms. These results show that even with the temperature of three points, the gradient-based method, Algorithm 1, can accurately reconstruct the initial temperature distribution in the building, thus enabling advanced control methods such as MPC to significantly improve the energy efficiency of the HVAC system (He & Gonzalez, 2016).

4.3 Estimation Using One Thermostat

Now we only assume that only one thermostat, s_1, is functional. The motivation is to show the performance of both estimation algorithms in a realistic scenario, since most residential buildings' HVAC systems operate using a single thermostat. We simulated the same scenarios as in Section 4.2.

Figures 4a and 4c are analogous to those in Figure 2, while Figure 4b shows the estimation error just for door d_1, which is located very close to thermostat s_1. As shown in these figures, both estimation algorithms do an almost equally poor job at estimating the doors' states, and the gradient-based method, Algorithm 1, is marginally better at estimating the initial temperature distribution. Yet, both algorithms are capable of accurately estimating the configuration of the door closest to the thermostat.

4.4 Memory usage comparison

A significant advantage of the gradient-based method, Algorithm 1, when compared to probabilistic estimation algorithms is that our method does not need to compute numerical solutions of the set of differential equations for each possible configuration $\theta \in \{0,1\}^{n_d}$. Figure 5 shows the maximum memory usage of both algorithm implementations as the number of doors to estimate increases from 1 to 4. Our results show that the probabilistic estimation method can be used only for small values of n_d, quickly outgrowing the amount of memory in standard computers (for $n_d = 4$ the usage was 16 GB approx.), while Algorithm 1 memory usage remains almost constant (at 2.4 GB approx.).

5. CONCLUSION

Our gradient-based estimation method and simulation results show the potential for reconstructing indoor climate and building configuration by using only thermostat sensor data, thus reducing the need for extra sensors to monitor a smart buildings. Also, since the method can accurately estimate the indoor climate and configuration with acceptable memory usage, it can be used in coordination to advanced MPC control strategies, significantly increasing the efficiency of HVAC units without a decrease in human comfort. Our method has the potential to enable interesting new applications. For example, since it is able to identify a building's configuration in real-time, it can potentially be applied to monitor an unexpected break-in.

A. DERIVATION OF ADJOINT EQUATIONS

Consider the Lagrangian function in equation (9). For each set of functions (w, v, q) in the respective dual space of the tuple (Te, u, p), we can write $\left\langle \frac{\partial L}{\partial T_e}, w \right\rangle_{\Omega\times[0,T]} = 0$ as follows:

$$\left\langle \frac{\partial L}{\partial T_e}, w \right\rangle_{\Omega\times[0,T]} = \left\langle \frac{\partial J}{\partial T_e}, w \right\rangle_{\Omega\times[0,T]} + \left\langle \lambda_1, \frac{\partial w}{\partial t} - \nabla_x \cdot (\kappa \nabla_x w) + u \cdot \nabla_x w \right\rangle_{\Omega\times[0,T]} + \langle \lambda_4, w \rangle_{\partial\Omega\times[0,T]} + \langle \lambda_6, w(0,\cdot) \rangle_{\Omega} = 0. \quad (19)$$

Similarly, we can write $\left\langle \frac{\partial L}{\partial u}, v \right\rangle_{\Omega} = 0$ as

$$\left\langle \frac{\partial L}{\partial u}, v \right\rangle_{\Omega} = \left\langle \lambda_1, v \cdot \nabla_x T_e \right\rangle_{\Omega\times[0,T]} + \left\langle \lambda_2, -\frac{1}{Re} \Delta_x v + (u \cdot \nabla_x) v + (v \cdot \nabla_x) u + \alpha v \right\rangle_{\Omega} + \langle \lambda_3, \nabla_x \cdot v \rangle_{\Omega} + \langle \lambda_5, v \rangle_{\Gamma_w} = 0, \quad (20)$$

and we can write $\left\langle \frac{\partial L}{\partial p}, q \right\rangle_{\Omega} = 0$ as $\left\langle \lambda_2, \nabla_x q \right\rangle_{\Omega} = 0$.

Applying integration by parts and Green's formula, equations (19) and (20) become

$$\begin{aligned} &\left\langle \frac{\partial J}{\partial T_e}, w \right\rangle_{\Omega\times[0,T]} - \left\langle \frac{\partial \lambda_1}{\partial t} + \nabla_x \cdot (\kappa \nabla_x \lambda_1) + u \cdot \nabla_x \lambda_1, w \right\rangle_{\Omega\times[0,T]} + \big\langle \lambda_1(\cdot,T), w(\cdot,T) \big\rangle_{\Omega} + \\ &\quad + \big\langle \lambda_6 - \lambda_1(\cdot,0), w(\cdot,0) \big\rangle_{\Omega} + \left\langle \kappa \frac{\partial \lambda_1}{\partial \vec{n}} + \lambda_4 + \vec{n} \cdot u \lambda_1, w \right\rangle_{\partial\Omega\times[0,T]} - \left\langle \kappa \lambda_1, \frac{\partial w}{\partial \vec{n}} \right\rangle_{\partial\Omega\times[0,T]} = 0, \quad (21) \end{aligned}$$

$$\begin{aligned} &\left\langle \lambda_1, v \cdot \nabla_x T_e \right\rangle_{\Omega\times[0,T]} + \left\langle \alpha \lambda_2 - \frac{1}{Re} \Delta_x \lambda_2 + \nabla_x u \cdot \lambda_2 - u \cdot \nabla_x \lambda_2 - \nabla_x \lambda_3, v \right\rangle_{\Omega} + \\ &\quad - \frac{1}{Re} \left\langle \lambda_2, \frac{\partial v}{\partial \vec{n}} \right\rangle_{\partial\Omega} + \left\langle \frac{1}{Re} \frac{\partial \lambda_2}{\partial \vec{n}} + \lambda_3 \vec{n} + (u \cdot \vec{n}) \lambda_2, v \right\rangle_{\partial\Omega} + \langle \lambda_5, v \rangle_{\Gamma_w} = 0, \quad (22) \end{aligned}$$

where $\vec{n}$ is the vector normal to the boundary at $x \in \partial\Omega$.

From the identities above it follows that, in order to make $\left\langle \frac{\partial L}{\partial T_e}, w \right\rangle_{\Omega\times[0,T]}$, $\left\langle \frac{\partial L}{\partial u}, v \right\rangle_{\Omega}$, and $\left\langle \frac{\partial L}{\partial p}, q \right\rangle_{\Omega}$ all equal to 0 for any set of functions (w, v, q), a sufficient condition for the dual variables $\lambda_{1,2,3,6}$ is to satisfy the differential equations (10) to (13) and their boundary conditions.

B. DERIVATION OF FRÉCHET DERIVATIVES

As explained in Section 3.1, if we take variations $(\delta\pi_0, \delta\theta)$ of our optimization variables, they will induce variations $\delta\alpha$, $\delta\kappa$, δT_e, δu, and δp. Then, from equations (1), (3), and (4), it follows that the variations satisfy the following differential equations:

$$\frac{\partial \delta T_e}{\partial t} - \nabla_x \cdot (\delta\kappa \nabla_x T_e) - \nabla_x \cdot (\kappa \nabla_x \delta T_e) + \delta u \cdot \nabla_x T_e + u \cdot \nabla_x \delta T_e = 0, \tag{23}$$

$$\delta\alpha\, u + \alpha\, \delta u - \frac{1}{Re} \triangle_x \delta u + \delta u \cdot \nabla_x u + u \cdot \nabla_x \delta u + \nabla \delta p = 0, \tag{24}$$

$$\nabla_x \cdot \delta p = 0, \tag{25}$$

with the following boundary and initial conditions: $\delta T_e(x,t) = 0$ for each $x \in \partial\Omega$ and $t \in [0,T]$, $\delta T_e(x,0) = \delta\pi_0(x)$ for each $x \in \Omega$, and $\delta u(x) = 0$ for each $x \in \Gamma_w$.

Now, using equations (8), (23), (24) and (25) and their boundary and initial conditions, we get

$$\begin{aligned} J(\theta + \delta\theta, \pi_0 + \delta\pi_0) - J(\theta, \pi_0) &= \left\langle \frac{\partial J}{\partial T_e}, \delta T_e \right\rangle_{\Omega\times[0,T]} + \left\langle \frac{\partial J}{\partial T_e}, \delta\pi_0 \right\rangle_{\Omega} + \\ &+ \left\langle \lambda_1, \frac{\partial \delta T_e}{\partial t} - \nabla_x \cdot (\delta\kappa \nabla_x T_e) - \nabla_x \cdot (\kappa \nabla_x \delta T_e) + \delta u \cdot \nabla_x T_e + u \cdot \nabla_x \delta T_e \right\rangle_{\Omega\times[0,T]} + \\ &+ \left\langle \lambda_2, \delta\alpha\, u + \alpha\, \delta u - \frac{1}{Re} \triangle_x \delta u + \delta u \cdot \nabla_x u + u \cdot \nabla_x \delta u + \nabla_x \delta p \right\rangle_{\Omega} + \left\langle \lambda_3, \nabla_x \cdot \delta u \right\rangle_{\Omega} + \\ &+ \left\langle \lambda_4, \delta T_e \right\rangle_{\partial\Omega\times[0,T]} + \left\langle \lambda_5, \delta u \right\rangle_{\Gamma_w} + \left\langle \lambda_6, \delta Te(\cdot,0) - \delta\pi_0 \right\rangle_{\Omega}, \end{aligned} \tag{26}$$

where $\{\lambda_i\}_{i=1}^{6}$ are the adjoint variables defined in Section 3.1. Then, applying integration by parts and Green's formula to equation (26), and after canceling terms using the identities in equations (10) to (13), we can get

$$J(\theta + \delta\theta, \pi_0 + \delta\pi_0) - J(\theta, \pi_0) = \left\langle \nabla_{\pi_0} J - \lambda_6, \delta\pi_0 \right\rangle_{\Omega} + \left\langle \nabla_x \lambda_1 \cdot \nabla_x T_e, \delta\kappa \right\rangle_{\Omega\times[0,T]} + \left\langle \lambda_2 \cdot u, \delta\alpha \right\rangle_{\Omega}, \tag{27}$$

which are equivalent to the directional derivatives in equations (14) and (15), as desired.

REFERENCES

Afram, A., & Janabi-Sharifi, F. (2014). Theory and applications of HVAC control systems—a review of model predictive control (MPC). *Building and Environment*, *72*, 343–355.

Awbi, H. B. (1989). Application of computational fluid dynamics in room ventilation. *Building and Environment*, *24*(1), 73–84.

Awbi, H. B. (2003). *Ventilation of buildings*. Taylor & Francis.

Banks, H. T., & Bihari, K. L. (2001). Modelling and estimating uncertainty in parameter estimation. *Inverse Problems*, *17*(1).

Banks, H. T., Holm, K., & Kappel, F. (2011). Comparison of optimal design methods in inverse problems. *Inverse Problems*, *27*(7).

Chow, W. (1996). Application of computational fluid dynamics in building services engineering. *Building and Environment*, *31*(5), 425 - 436.

Dobrzynski, C., Pironneau, O., & Frey, P. (2004). Numerical Coupling for Air Flow Computations in Complex Architectures. In *Proceedings of the european congress on computational methods in applied sciences and engineering (eccomas)*.

Dounis, A. I., & Caraiscos, C. (2009). Advanced control systems engineering for energy and comfort management in a building environment—A review. *Renewable and Sustainable Energy Reviews*, *13*(6), 1246–1261.

Faanes, A., & Skogestad, S. (2005). Offset-free tracking of model predictive control with model mismatch: Experimental results. *Industrial & Engineering Chemistry Research*, *44*(11), 3966–3972.

Gersborg-Hansen, A., Sigmund, O., & Haber, R. B. (2005). Topology optimization of channel flow problems. *Structural and Multidisciplinary Optimization*, *30*(3), 181–192.

Giles, M. B., & Pierce, N. A. (1997). Adjoint equations in CFD: duality, boundary conditions and solution behaviour. In *Proceedings of the 13th computational fluid dynamics conference.*

Girault, V., & Raviart, P.-A. (1979). *Finite element approximation of the Navier-Stokes equations*. Springer-Verlag.

Gunzburger, M. (2000). Adjoint equation-based methods for control problems in incompressible, viscous flows. *Flow, Turbulence and Combustion, 65*(3-4), 249–272.

Gutman, S. (1990). Identification of discontinuous parameters in flow equations. *SIAM Journal on Control and Optimization, 28*(5), 1049–1060.

He, R., & Gonzalez, H. (2016). *Zoned HVAC control via PDE-constrained optimization.* (To appear in Proceedings of the 2016 American Control Conference. arXiv: 1504.04680)

Huang, G. (2011). Model predictive control of VAV zone thermal systems concerning bi-linearity and gain nonlinearity. *Control Engineering Practice, 19*(7), 700–710.

Isakov, V. (2006). *Inverse problems for partial differential equations*. Springer.

Jones, P. J., & Whittle, G. E. (1992). Computational fluid dynamics for building air flow prediction—Current status and capabilities. *Building and Environment, 27*(3), 321–338.

Kumar, P. R., & Varaiya, P. (2015). *Stochastic systems: Estimation, identification, and adaptive control.* SIAM.

Landau, L. D., & Lifshitz, E. M. (1959). *Course of theoretical physics, volume 6: Fluid mechanics* (1st ed.). Pergamon Press.

Logg, A., Mardal, K.-A., & Wells, G. (2012). *Automated solution of differential equations by the finite element method: The FEniCS book.* Springer.

Ma, J., Qin, J., Salsbury, T., & Xu, P. (2012). Demand reduction in building energy systems based on economic model predictive control. *Chemical Engineering Science, 67*(1), 92–100.

Ma, Y., Kelman, A., Daly, A., & Borrelli, F. (2012). Predictive control for energy efficient buildings with thermal storage. *IEEE Control System Magazine, 32*(1), 44–64.

Moroşan, P.-D., Bourdais, R., Dumur, D., & Buisson, J. (2010). Building temperature regulation using a distributed model predictive control. *Energy and Buildings, 42*(9), 1445–1452.

Nocedal, J., & Wright, S. (2006). *Numerical optimization.* Springer.

Prívara, S., Široký, J., Ferkl, L., & Cigler, J. (2011). Model predictive control of a building heating system: The first experience. *Energy and Buildings, 43*(2), 564–572.

Shaikh, P. H., Nor, N. B. M., Nallagownden, P., Elamvazuthi, I., & Ibrahim, T. (2014). A review on optimized control systems for building energy and comfort management of smart sustainable buildings. *Renewable and Sustainable Energy Reviews, 34*, 409 - 429.

Sun, H., Stowell, R. R., Keener, H. M., & Michel Jr., F. C. (2002). Two–dimensional computational fluid dynamics (CFD) modeling of air velocity and ammonia distribution in a high-rise hog building. *Transactions of the American Society of Agricultural and Biological Engineers, 45*(5), 1559–1568.

U.S. Department of Energy. (2012, March). *2011 Buildings Energy Data Book.* Retrieved from `http://buildingsdatabook.eren.doe.gov/docs/DataBooks/2011_BEDB.pdf`

U.S. Green Building Council. (2014). *Better Buildings, Better Policy: A compilation of green building policy adoptions in the United States, 2011-2014.* Retrieved from `http://www.usgbc.org/resources/better-buildings-better-policy-compilation-green-building-policy-adoptions-united-states-2`

van der Poel, E. P., Stevens, R. J., & Lohse, D. (2013). Comparison between two-and three-dimensional Rayleigh-Bénard convection. *Journal of Fluid Mechanics, 736*, 177–194.

Vasudevan, R., Gonzalez, H., Bajcsy, R., & Sastry, S. S. (2013a). Consistent approximations for the optimal control of constrained switched systems—Part 1: A conceptual algorithm. *SIAM Journal on Control and Optimization, 51*(6), 4463–4483.

Vasudevan, R., Gonzalez, H., Bajcsy, R., & Sastry, S. S. (2013b). Consistent approximations for the optimal control of constrained switched systems—Part 2: An implementable algorithm. *SIAM Journal on Control and Optimization, 51*(6), 4484–4503.

Wang, S., & Ma, Z. (2008). Supervisory and optimal control of building HVAC systems: A review. *HVAC&R Research, 14*(1), 3–32.

Xi, X.-C., Poo, A.-N., & Chou, S.-K. (2007). Support vector regression model predictive control on a HVAC plant. *Control Engineering Practice, 15*(8), 897–908.

Yang, I., & Tomlin, C. J. (2014). Reaction–diffusion systems in protein networks: Global existence and identification. *Systems & Control Letters, 74*, 50–57.

Bayesian Calibration – What, Why And How

Ralph T Muehleisen*, Joshua Bergerson

Argonne National Laboratory, Energy Systems Division,
Lemont, IL, United States
630-252-2547, rmuehleisen@anl.gov
630-252-9333, jbergerson@anl.gov

* Corresponding Author

ABSTRACT

Calibration of building energy models is important to ensure accurate modeling of existing buildings. Typically this calibration is done manually by modeling experts, which can be both expensive and time consuming. Additionally, biases of the individual modelers will creep into the process. Many methods of automated calibration have been developed which reduce costs, time and biases, including optimization using genetic or swarming algorithms, machine learning methods, and Bayesian methods. Bayesian methods differ significantly from the other optimization and machine learning methods in that inputs are assumed to be uncertain and main goal is not to match the prediction to the measured data as closely as possible, but to reduce the uncertainty in the inputs in a manner consistent with the measured data. Bayesian methods are particularly useful when there are model inputs that have high sensitivity and high uncertainty and where there is limited measured data to use for calibration. In this paper, the basic concepts of Bayesian calibration are explained and a typical application and results are demonstrated.

1. INTRODUCTION

With the desire for higher performance building designs, the use of building energy models (BEMs) for retrofit analysis has been growing and the accuracy of building energy models has become critically important. When models are used for designing high performing retrofits or improving building operation, models should be calibrated to measured performance to ensure an accurate model. Conventional calibration methods have several shortcomings that may counteract the goal of the calibration procedure, including overfitting and amplification of model errors. Bayesian calibration is a method of calibration that overcomes the majority of the disadvantages of conventional calibration methods.

2. BUILDING ENERGY MODELS

Over the past several decades, building energy models have dramatically evolved into the dynamic, highly detailed BEMs of today. Modern full fluid dynamic BEMs, such as Energy Plus, contain several hundred parameters. Each of these parameters has an associated uncertainty, and thus the uncertainty inherent in the total system can be enormous. Errors also exist in the model itself as the complex heat and moisture balance equations need to be approximated at time intervals in order to improve model run time as these balances are approximating a steady state solution in a non-steady state condition. While modern BEMs offer a significant improvement over hand calculations and early generation BEMs in the sense that they more accurately model the thermodynamics and loads of buildings systems, they present a new challenge in the level of information needed to inform the hundreds of model parameters. The uncertainties in model parameter quantities and errors in modeling must be considered at a minimum and ideally quantified such that a modeler can have a certain level of confidence in the predictive power of a BEM.

Estimating BEM parameter values is a time consuming and challenging process. For new construction, parameters generally fall into one of three categories. The first category is BEM parameters that are known with little-to-no uncertainty when the modeler begins assembling the BEM, such as building dimensions and climate parameters. The second category is parameters that have a high level of uncertainty when the modeler begins creating a BEM but become known with little uncertainty by the end of the design phase. BEM parameters in this category include parameters related to building systems (lighting, plumbing, structural) whose design is incomplete (or not started) when the modeler begins assembling the BEM and become known with little uncertainty as the building systems' designs progress and are finalized. Finally, a third category of BEM parameters exists which have a high level of uncertainty throughout the entire design process such as BEM parameters related to occupant behavior.

For existing buildings, BEMs are often utilized for exploring the energy savings potential of various energy efficiency measures (EEMs). As the building being modeled in retrofit building energy modeling is an existing building, most of the BEM parameters could be determined with little uncertainty. However, obtaining these parameters through field measurements and testing can be prohibitively expensive, disruptive, and time consuming, and thus these parameters are often estimated with a moderate to high level of uncertainty.

2.1 Need for Calibration

When the building energy model is for retrofit of existing construction, the model should undergo calibration in order to improve the predictive power of the model (reducing the error between model outputs and observed data), presumably by decreasing the error in model parameters. This calibration is critically important for accurately assessing the energy savings potential of various EEMs. Observed data is readily available in the forms of natural gas, electricity, and water bills showing energy and water consumption. New construction buildings are not often calibrated, as the design is finalized prior to construction and system performance data will not become available until after construction is complete. This means that calibrating new construction BEMs will not provide better model predictions at a time when they matter (i.e. during design phase), and thus is rare in practice. When new construction BEMs are calibrated, this calibration comes during the initial commissioning of building system. Building commissioning is the process of ensuring all building automation, sensors, and controls systems are properly functioning and of tuning building systems for peak performance. This process is especially important for high performance buildings as even small deviations from peak performance or improper operation of building automation systems can lead to significant shortcomings in expected building efficiency. This is particularly true when the building has a performance based contract in which threshold energy efficiencies must be satisfied to avoid significant economic penalties. Buildings trying to achieve certain green building certifications, such as from the Leadership in Energy and Environmental Design (LEED), often are required to validate BEM designs using commissioning.

3. CONVENTIONAL CALIBRATION

Conventional calibration generally involves an expert modeler iteratively adjusting BEM parameters to minimize the difference between observed data and model outputs (Reddy *et al.*, 2007). This process can be very labor intensive and inherently introduces modeler biases into the calibrated model. More recently, optimization schemes have been implemented to automate calibration (New and Chandler, 2013; Raftery *et al.*, 2011). These schemes determine the combination of model parameters that minimize the error between data and model outputs.

3.1 Advantages

Conventional calibration methods offer certain advantages. Expert based conventional calibration draws on the wealth of experience and success of expert modelers to reduce the error between model outputs and data. Optimization based calibration methods explore parameter domains to find local minimum error and can find multiple parameter combinations that provide local minimum errors. Optimization based calibration often results in lower error than expert based calibration. Optimization calibration methods are generally less labor intensive but are much more computationally expensive, especially when there are numerous calibration parameters.

3.2 Disadvantages

It is an underlying assumption of conventional calibration that minimizing the difference between real world data and model outputs minimizes the error in the model outputs, and thus the calibrated model is more predictive of future observable data. This assumption implies that no uncertainty exists in real world data and that all uncertainty exists in model output due to uncertainties in model parameters that are reduced during the calibration process. However, uncertainty exists in measured data itself and neglecting to account for this uncertainty can introduce biases into the calibrated model. Additionally, conventional calibration methods cannot address modeling errors. In fact, by neither considering data uncertainty nor modeling error, conventional calibration methods may actually amplify model output uncertainties if the data is biased. Another problem with conditional calibration is that if multiple combinations of calibration parameters yield nearly identical minimal error, there is no way to determine which of these calibrated parameter combinations is the most probable. Finally, as most conventional calibration methods do not attempt to quantify uncertainties in model parameters or in the model itself, they are unable to estimate how much better predictive the calibrated model is relative to the uncalibrated model. Even when uncertainties are quantified in conventional calibration, total model output uncertainty may not be reduced during calibration as the goal of calibration is not to reduce uncertainty but rather to make the model more aggregable with observed data. This limitation means that conventional calibration is unable to quantify how much more likely the calibrated model parameters and outputs are than the uncalibrated model parameters and outputs. Modelers are unable to express any quantitative confidence level in the accuracy of the calibrated model.

3.2.1 Overfitting: Overfitting is a problem in regression analysis, multi-objective optimization, and machine learning when the fitting/optimization routine fits the model to the noise in measured data (Dietterich, 1995). This most commonly occurs when a model has more degrees of freedom than the underlying physics (Cawley and Talbot, 2010). One indicator of overfitting would be when the when the error in the fit is less than the uncertainty in the underlying measured data.

4. BAYESIAN CALIBRATION

Bayesian calibration is a method of calibration that is fundamentally different than conventional calibration methods. Bayesian calibration has a long history within computer modeling in general (Dunsmore, 1968; Racine-Poon, 1988; Kennedy and O'Hagan, 2001) and has been adapted to building energy models by Heo *et al.* (2011). Bayesian calibration is an iterative process of updating uncertainty distributions on the BEM parameters in a way that is consistent with the observed data. This process does not minimize the difference between observed data and model outputs, but rather determines the most likely uncertainties for input parameters that yield an output uncertainty in which the observed data is most likely. Bayesian calibration is an application of Bayes Theorem, which relates prior information with uncertainty to future information based on the likelihood of observed outputs from the model (Bergerson and Muehleisen, 2015).

4.1 Advantages

As each iteration of the Bayesian calibration updates the probability density functions (PDFs) of BEM parameters in a way that is most likely to yield the observed data, the PDFs determined after a sufficiently large number of iterations are the most probable calibration of the system. This means that a BEM calibrated with Bayesian calibration is best able to predict future real world observations by determining the best estimate of parameter uncertainties. One of the primary advantages of Bayesian calibration is that as uncertainty is integral to the model definition and calibration procedure, a modeler can quantify a confidence level in the calibrated model. Additionally, a probabilistic risk analysis can be performed and competing retrofits can be ranked according to a desired risk level, based on mean values, variances, etc. Finally, Bayesian calibration reduces the tendency to overfit the model to the observed data. Bayesian methods tend to reduce this potential problem as they are not trying to minimize the difference between the model output and the measured data but rather are trying to maximize the likelihood that the model output is statistically consistent with the measured data. This consistency naturally takes into account the uncertainty in the measured data (Heckerman, 1998).

4.2 Disadvantages

One of the disadvantages of all calibration methods, both conventional and Bayesian, is that they require observed data from the physical system being modeled against which to compare model outputs. The collection of this data is typically very challenging, and as such is generally the most time consuming and expensive step of model calibration. Another disadvantage of Bayesian calibration is that as the calibration process is iterative, calibration may require a significant number of iterations to converge to the most likely PDFs. This is especially true if the prior probability density functions are poorly chosen (and thus require significant updating during the calibration process). Additionally, due to the high computational demand of Bayesian calibration, Bayesian calibration often requires a parameter screening and selection process in order to select the most significant parameters to use in the calibration process in large models with numerous parameters to improve model run time.

4.3 Methodology

The general methodology of Bayesian calibration is 1) define uncertainty PDFs for uncertain model parameters, 2) collect real system performance data for known input parameter values, and 3) calibrate (assumed) prior parameter PDFs based on the observed data by iteratively using Bayes Theorem until iterations converge to an acceptable level (Kennedy and O'Hagan, 2001). Establishing prior uncertainty PDFs for input parameters generally involves both expert opinion solicitation and literature review. When only a range of appropriate values can be determined for a parameter, a uniform distribution may be assumed (Riddle and Muehleisen, 2014). When a range of values and a most likely value can be determined, a triangular distribution may be used. As previously mentioned, when the model is large, the calibration is done on a subset of parameters in order to dramatically decrease both the data collection and computational demands of the calibration. Collecting data is usually the most financially expensive step in calibration. As the goal of the calibration is to improve the predictive power of the model for future model outputs by arriving at the most likely calibration parameter uncertainty distributions, it is critical to sample data from across calibration parameters' uncertainty ranges. Intelligent sampling strategies can significantly reduce the quantity of data required for calibration and thus reduce the expense of data acquisition. The final step of Bayesian calibration is iteratively updating the selected calibration parameters prior uncertainty PDFs based on the collected data using Bayes Theorem. Usually a random subset of the observed data is utilized in each iteration of the calibration.

5. NUMERICAL EXAMPLE

A case study was undertaken at the National Renewable Energy Laboratory (NREL) in Golden, CO. The study developed a building energy model for an existing 850ft^2 one story building, built in 1994. The building is a site entrance building, and thus is occupied 24 hr/day by one to four people. The space conditioning is provided by a split system direct expansion (DX) cooler with a natural gas furnace. Auxiliary baseboard electric heaters are utilized by the operable badging window. Figure 1 shows a screen shot of the BEM and plan view of the building.

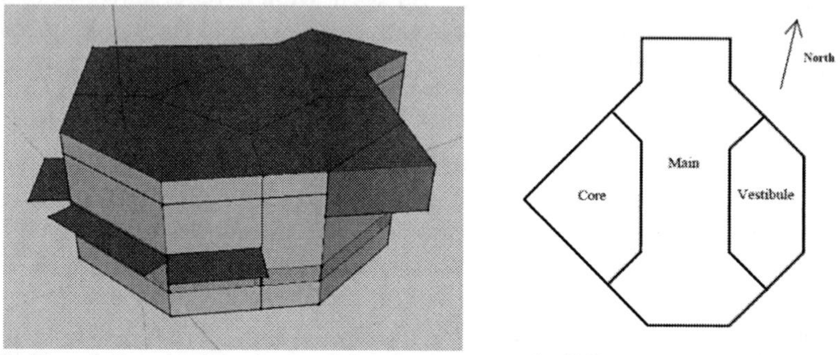

Figure 1: BEM (left) and plan (right) views of the site entrance building at NREL examined in the numerical example

5.1 Building Energy Model and Parameter Screening

A building energy model was developed for the building using OpenStudio. Bayesian calibration was utilized to improve the predictive power of the BEM. In order to reduce the computational demand of the calibration procedure, a sensitivity analysis was undertaken using the enhanced Morris method (Campolongo *et al.*, 2005). The sensitivity analysis evaluated the BEM output electricity and natural gas consumption sensitivity to a large subset BEM parameters considered to be uncertain. The four most significant parameters determined from the parameter screening are infiltration rate, plugload power density, lighting power density, and occupant density, and thus these four parameters were selected as calibration parameters.

5.2 Results

The four calibration parameters were calibrated using Bayesian calibration using monthly electricity and natural gas consumption. The prior probability density functions and calibrated PDFs are shown in Figure 2 by dashed lines and vertical bars respectively. It is clear from Figure 2 that the uncertainty distributions for both the infiltration rate and plugload power density BEM parameters calibrated well using Bayesian calibration, with both having significantly reduced uncertainty (variance) at the end of the calibration process. The lighting power density and occupant density parameters were less sensitive to calibration, with both the mean and variance of the prior and final PDFs being relatively constant for both parameters.

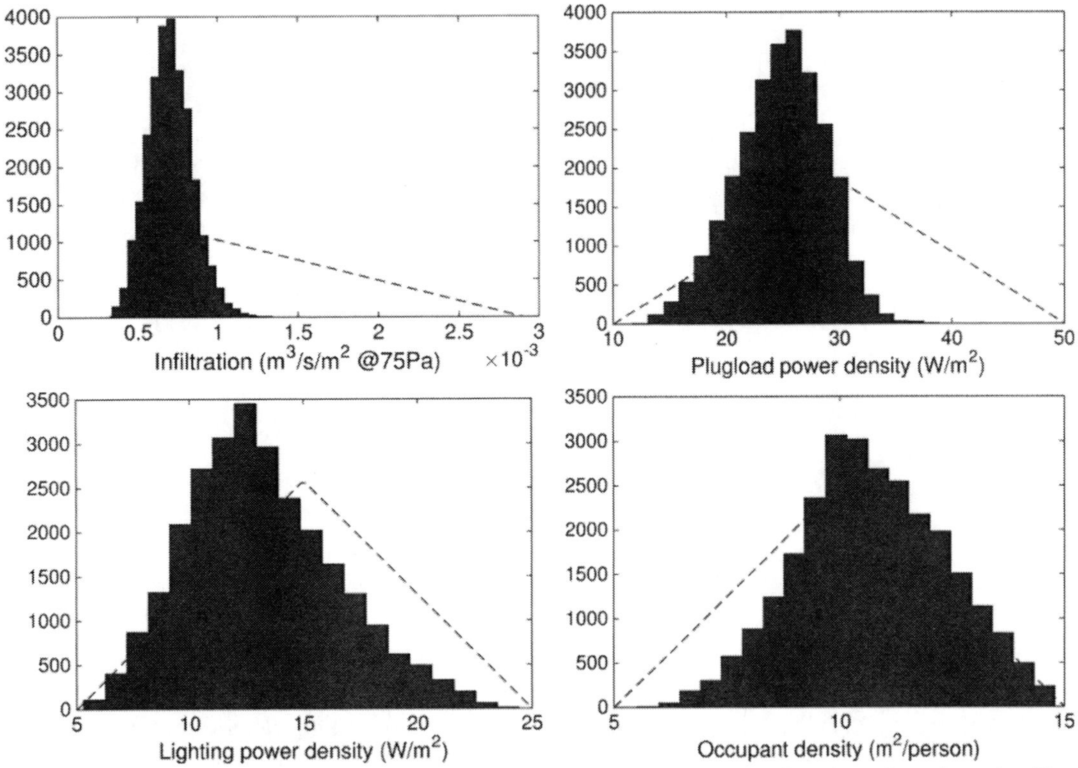

Figure 2: Prior (dashed line) and posterior (vertical bars) probability density functions for selected calibration parameters

The results of the calibration on the BEM projected electricity and natural gas consumption are given in Figure 3, which shows the monthly pre-calibrated and calibrated BEM output consumption against the monthly metered data. It is clear from Figure 3 that Bayesian calibration dramatically reduces the error between metered and BEM projected monthly electricity consumption. The calibration also reduced the error between metered and BEM projected monthly natural gas consumption, but moderate error still exists. This is primarily due to the fact that a nearly linear relationship exists between the plugload power density and lighting power density calibration parameters and the BEM projected electricity consumption.

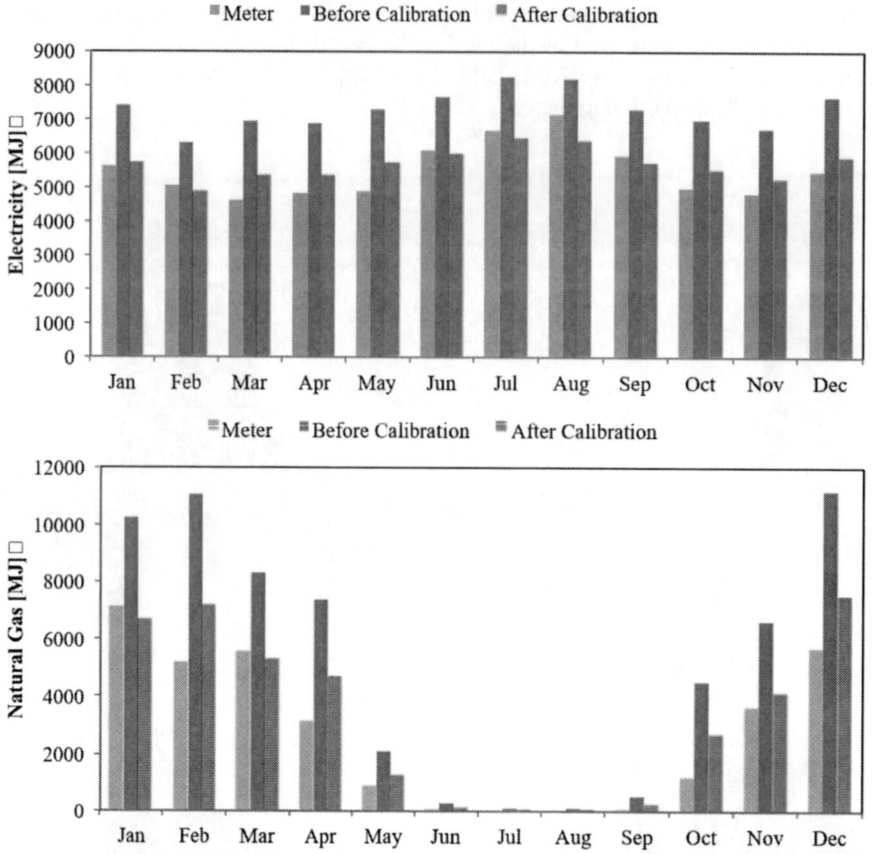

Figure 3: Pre-calibrated and calibrated building energy model projected monthly electricity (above) and natural gas (below) consumption relative to monthly metered consumption

6. CONCLUSION

Bayesian calibration is a method of calibration that is fundamentally different than conventional expert based, multivariate optimization, or machine learning calibration methods which overcomes a majority of the shortcomings of conventional calibration schemes. The primary advantage of Bayesian calibration is that it provides a most probable quantification of the uncertainty inherent in the calibration parameters and thus in the model outputs, allowing for probabilistic analysis, risk assessment, and confidence interval determination. This provides a greater level of information than conventional deterministic calibration procedures. Additionally, a model calibrated with Bayesian calibration yields the most probable uncertainty PDFs for the calibration parameters that is consistent with the observed data and prevents overfitting. This means that future model outputs from the calibrated model should have minimal error with respect to future observed data, given that the future combination of parameters for observed data are within the uncertainty bounds of the calibration parameters in the model. Parameter screening and efficient sampling methods help reduce the cost of data collection and computational demand of Bayesian calibration. A numerical example was included that highlights the application of Bayesian calibration to building energy modeling.

NOMENCLATURE

The nomenclature should be located at the end of the text using the following format:

BEM	building energy model
DX	direct expansion
EEM	energy efficiency measure
LEED	Leadership in Energy and Environmental Design

NREL	National Renewable Energy Laboratory
PDF	probability density function

REFERENCES

Bergerson, J., & Muehleisen, R. T. (2015). Bayesian Large Model Calibration Using Simulation and Measured Data for Improved Predictions. *SAE Int. J. Passeng. Cars - Mech. Syst., 8*(2). doi:10.4271/2015-01-0481

Campolongo, F., Cariboni, J., Saltelli, A., & Schoutens, W. (2005). Enhancing the Morris Method. *Proceedings of the 4th International Conference on Sensitivity of Model Output*, (pp. 369-379).

Cawley, G. C., & Talbot, N. L. (2010). On Over-Fitting in Model Selection and Subsequent Selection Bias in Performance Evaluation. *J. Mach. Learn. Res., 11*(August), 2079-2107.

Dietterich, T. (1995). Overfitting and Undercomputing in Machine Learning. *ACM Comput. Surv., 27*(3), 326-327. doi:10.1145/212094.212114

Dunsmore, I. R. (1968). A Bayesian Approach to Calibration. *Journal of the Royal Statistical Society, Series B (Methodological), 30*(2), 396-405.

Heckerman, D. (1998). A Tutorial on Learning with Bayesian Networks. In M. I. Jordan (Ed.), *Learning in Graphical Models* (Vol. NATO ASI Series 89, pp. 301-354). Netherlands: Springer. Retrieved from http://link.springer.com/chapter/10.1007/978-94-011-5014-9_11

Heo, Y., Augenbroe, G. A., & Choudhary, R. (2011). Risk Analysis of Energy-Efficiency Projects Based on Bayesian Calibration of Building Energy Model. *Building Simulation 2011.* Retrieved from http://www.ibpsa.org/proceedings/BS2011/P_1799.pdf

Kennedy, M. C., & O'Hagan, A. (2001). Bayesian Calibration of Computer Models. *Journal of the Royal Statistical Society: Series B, 63*(3), 425-464.

New, J., & Chandler, T. (2013). Evolutionary Tuning of Building Models to Monthly Electrical Consumption. *ASHRAE Transactions, 119*(89).

Racine-Poon, A. (1988). A Bayesian Approach to Nonlinear Calibration Problems. *Journal of the American Statistical Association, 83*(403), 650-656. doi:10.2307/2289287

Raftery, P., Keane, M., & O'Donnell, J. (2011). Calibrating Whole Building Energy Models: An Evidence-Based Methodology. *Energy and Buildings, 43*(9), 2356-2364. doi:10.1016/j.enbuild.2011.05.020

Reddy, T., Maor, I., & Panjapornpon, C. (2007). Calibrating Detailed Building Energy Simulation Programs with Measured Data - Part I: General Methodology (RP-1051). *HVAC&R Research, 13*(2), 221-241.

Riddle, M., & Muehleisen, R. T. (2014). A Guide to Bayesian Calibration of Building Energy Models. *Building Simulation Conference.* Atlanta, GA: ASHRAE.

ACKNOWLEDGEMENT

The authors would like to acknowledge Yuming Sun for assistance in Bayesian calibration example used in this study and Brian Ball for providing the basic building energy model and measured data used for the example. This work was supported by the U. S. Department of Energy under Contract No. DE-AC02-06CH11357 with Argonne National Laboratory.

An Efficient and Accurate Building Optimization Strategy Using Singular Value Decomposition

Yeonjin BAE[1]*, W. Travis HORTON [1]

[1]Purdue University, School of Civil Engineering,
West Lafayette, IN, U.S.A
ybae@purdue.edu, wthorton@purdue.edu

* Corresponding Author

ABSTRACT

Optimizing the life cycle cost of a building typically involves a large number of variables due to the many options that exist at the time that a building is being designed. Such large-scale optimization problems are often prohibitive within the building industry because of the excessive computational time required by the building energy modeling software; therefore, any optimization studies that are performed during a building design are typically only completed using a small number of variables. To achieve the goal of performing a life cycle building optimization in an acceptable time frame, this paper proposes an accurate and efficient method using singular value decomposition on the design variables. Through the use of singular value decomposition a large number of design variables can be reduced to a smaller subset of design variables that can be solved more quickly by the optimization algorithm. In this paper the authors apply this methodology to a case study of a typical residential building in six separate locations across the U.S. and compare the results with those of the full optimization process over the entire design space.

1. INTRODUCTION

As energy consumption in the U.S. continues to increase, building energy modeling and simulation software is being incorporated more regularly into the building design process to simulate thermal loads and energy consumption, and to predict the energy performance characteristics of the building prior to construction. Energy simulation programs typically require hundreds of design parameters as inputs, which means they are not often employed at a very early design stage because a large number of the design parameters are still uncertain (Augenbroe, 2002). Since any given building project can include a large number of design variables, and each variable may take on many different values, it is impossible to do an exhaustive search of the entire design space to find the variables that optimize the building design subject to various project constraints and objectives, such as construction cost, energy cost, and environmental impact.

The current study proposes an accurate and efficient method to perform the optimization, using detailed energy calculation with existing energy simulation software, and using actual construction costs. This optimization is normally prohibitive because of calculation time and therefore it is processed with only a limited number of variables. The main reason for the long running process is that there are too many input variables and the building energy simulation software takes a long time to predict energy consumption. To perform the optimization within an acceptable time frame, several approaches can be made. The focus of this study is on reducing the number of design variables that are considered during the optimization process. Many approaches to reduce the number of design variables have been studied in the past. Bettonvil and Klenijnen (Bettonvil & Klenijnen, 1997) applied a group screening method by detecting the important factors using sequential bifurcation to the building energy model. The author asserts that important factors need to be further explored in a later phase, for example in the optimization stage. Rahni (Rahni, et al., 1997) partitioned factors into groups and tested which group demonstrates a significant effect. Only 6% of the input variables were identified as important variables for analyzing the output model.

However, not many studies have been performed to further examine the trade-offs between accuracy of the optimization study and the efficiency of the optimization algorithm with a reduced number of design variables.
To achieve the goal of performing a life cycle building optimization in an acceptable time frame, this paper proposes an accurate and efficient method using singular value decomposition on the design variables. Through the use of singular value decomposition a large number of design variables can be reduced to a smaller subset of design variables that can be solved more quickly by the optimization algorithm. In this paper the authors apply this methodology to a case study of a typical residential building in six separate locations across the U.S. and compare the results with those of the full optimization process over the entire design space.

2. METHODOLOGY

To have an accurate methodology for determining the optimized building system, a detailed energy consumption model using energy simulation software needs to be utilized to evaluate the energy cost along with a detailed cost model of the building. This study is focused on reducing the variables in the design space by eliminating insignificant design variables through a variable selection method. Significant variables are considered to be those that demonstrate the strongest contribution to the output results. Using this approach, significant variables are identified for use in the optimization process while redundant variables and insignificant variables are discarded prior to optimization. To validate the accuracy of this methodology, two optimization processes are set. The processes are described in Figure 1. The first is the original optimization process, which is performed using all available design variables. The second is a reduced optimization process, which is performed with a smaller subset of only the significant design variables based on the variable selection approach. In this study, singular value decomposition is used to explore the design space and determine the most significant variables for use in the optimization study.

To perform an optimization study on the buildings and compare the alternatives, minimizing life cycle cost (LCC) is used as the objective function. To have an accurate analysis, the first step is to develop a detailed building energy simulation model and cost database. To develop a detailed building model, the energy simulation software EnergyPlus is selected and used. For the cost database, RSMeans (RSMeans, 2011) is used for construction cost, while HVAC equipment cost data is taken from online equipment suppliers.

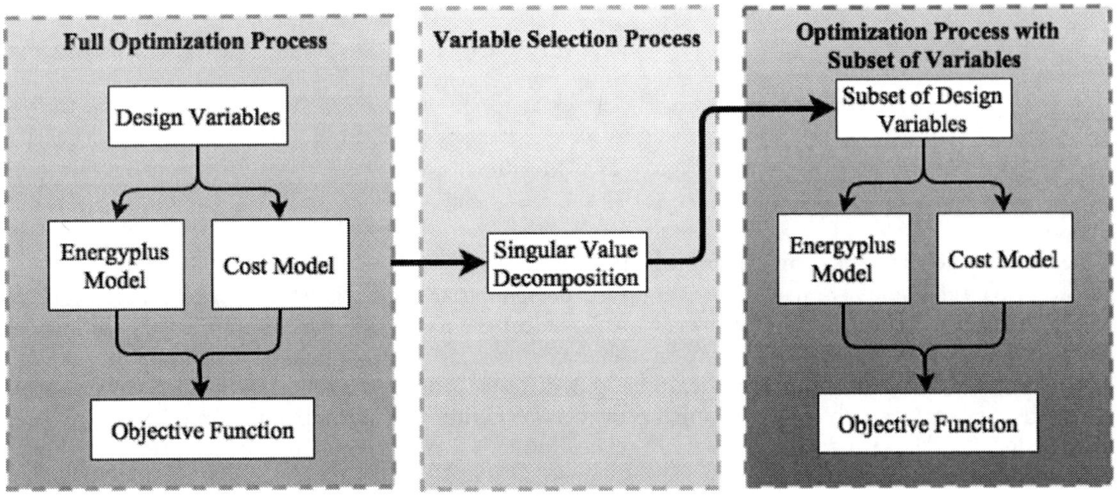

Figure 1: Description of methodology

3. CASE STUDY

3.1 Case study building characteristics

The case study building is assumed to be a typical U.S residential single family home (Cooper, 2011) with a crawl space and attached garage. Gross floor area of the house is 139.1m^2 (1497 ft^2) and detailed dimensions are shown in Figure 2. To investigate the impact of different climates on building energy consumption and optimization results, six different locations are considered; including, Denver, CO, Indianapolis, IN, Minneapolis, MN, Phoenix, AZ, Seattle, WA, Tampa, FL. The house has four different thermal zones, which include a living space, garage, crawl space, and attic. The living space, which is 111m^2 (1194 ft^2) is the only actively conditioned zone in the home while the remaining zones are unconditioned.

All of the major building construction elements are defined on a layer-by-layer basis to emulate typical residential house construction. The primary HVAC system for the house is modeled as a single-zone unitary system and the heating and cooling capacities are sized based on calculated building thermal loads. The heating set point temperature is fixed at 21.1°C (70°F) and the cooling set point temperature is set to 23.3°C (74°F). For realistic seasonal control of the HVAC system, a seasonal on-off schedule is set for all locations. Since each location has a different heating/cooling season, a generalized approach to cover all location is assumed as follows: a) the heating system is on from August 22 to May 1 and b) cooling system is on from March 21 to October 21. The overlapping dates allow both the heating and cooling systems to be active during the shoulder months when any given day may require heating during one portion of the day and cooling during another.

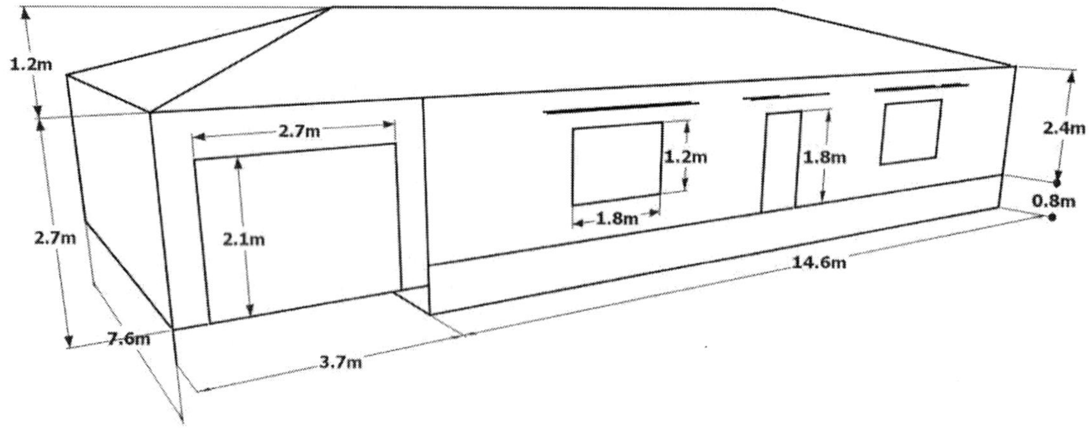

Figure 2: Model dimension

3.2 Selected design variables for optimization

The following considerations are made in selecting the design variables for the study: 1) select elements that may have a high impact on house energy consumption, 2) select elements that may have a strong influence in the construction cost, 3) select elements that may have energy saving or cost saving but whose influence is not strictly known beforehand. Based on these considerations, 13 different construction design variables were selected to investigate their impact on life cycle cost of the residential building. The first nine variables are related to the building envelope and the remaining variables are related to the HVAC system. Only those design variables relative to the envelope and the HVAC system that are commercially available have been selected for inclusion in the current study. As a result, the design variables can only take on discrete values or categorical values, which are listed in Table 1. Each variable can take anywhere from two to 16 values shown in the second column. For example, roofing material can take on a value of asphalt shingles, metal surface, or concrete tile roof. Design variables of the wall core can be a given stud dimension and filled with any type of insulation shown in the table. Structural insulated panels or insulated concrete forms with various thicknesses are also considered as wall core material for this study.

Table 1: Design variables and values

<table>
<tr><th>Design variables (Number of options)</th><th colspan="2">Values</th></tr>
<tr><td>Roofing Material (3)</td><td colspan="2">Asphalt shingles, Metal surface, Concrete tile roof</td></tr>
<tr><td>Roof Eave overhang Depth (3)</td><td colspan="2">305mm, 457mm, 610mm</td></tr>
<tr><td rowspan="2">Attic Insulation Material (12)</td><td colspan="2">Loose fill cellulose: SI-R3.3, 4.4, 5.3, 6.7, 8.6, 10.6</td></tr>
<tr><td colspan="2">Fiberglass batting: SI-R3.3, 4.4, 5.3, 6.7, 8.6, 10.6</td></tr>
<tr><td>External Wall Siding Material (4)</td><td colspan="2">Vinyl siding, Wood siding, Fiber cement siding, Brick</td></tr>
<tr><td rowspan="3">Wall Core (16)</td><td>Stud: 38mmx89mm (2"X4" nominal) studs at 400mm (16") on center, 38mmx140mm (2"X6" nominal) studs at 600mm (24") on center, 38mmx184mm (2"X8" nominal) studs at 600mm (24") on center</td><td>Insulation: Filled with fiberglass batting insulation, Sprayed on foam insulation, Loose fill cellulose insulation</td></tr>
<tr><td colspan="2">Structural Insulated Panels (SIPS) [mm]: 114, 165, 210, 260</td></tr>
<tr><td colspan="2">Insulated Concrete Forms (ICF) [mm]: 228, 278, 328</td></tr>
<tr><td>External Foam Board Insulation (6)</td><td colspan="2">Board insulation [mm]: 12.7, 25.4, 38.1, 50.8, 63.5, 76.2</td></tr>
<tr><td rowspan="2">Foundation Wall Insulation (9)</td><td colspan="2">Foundation board insulation [mm]: 12.7, 25.4, 38.1, 50.8, 63.5, 76.2</td></tr>
<tr><td colspan="2">Foundation spray foam [mm]: 25.4, 50.8, 76.2</td></tr>
<tr><td rowspan="2">Under Floor Insulation (7)</td><td colspan="2">Spray on polyurethane foam [mm]: 25.4, 50.8, 76.2</td></tr>
<tr><td colspan="2">Fiberglass batting: SI-R2.3, 3.3, 5.3, 6.7</td></tr>
<tr><td>Window Type (2)</td><td colspan="2">Double pane window, Triple pane window</td></tr>
<tr><td>Garage Door (4)</td><td colspan="2">1 Layer with no insulation, 2 Layer with polystyrene, 3 Layer with polystyrene, 3 Layer with polyurethane</td></tr>
<tr><td>Heat Recovery Type (2)</td><td colspan="2">None, Sensible heat recovery</td></tr>
<tr><td>Air Conditioner Seasonal COP (6)</td><td colspan="2">3.81, 4.10, 4.40, 4.69, 4.98, 5.28</td></tr>
<tr><td>Natural Gas Furnace Efficiency (4)</td><td colspan="2">80%, 85%, 90%, 95%</td></tr>
</table>

3.3 Life cycle cost

To assess the total cost associated with any given building construction permutation, a life cycle cost analysis (LCCA) approach is performed using a 20-year time horizon. The true life cycle cost of a building includes the initial construction cost, annual energy utility cost, and ongoing maintenance cost. However, in the current study only construction materials and annual energy costs are considered since it is anticipated that these will have the strongest influence on the LCCA. The impact of construction labor and maintenance costs will be considered in a later study. Equation (1) shows the method that is used to calculate life cycle cost in this study.

$$LCC = C_{Mat} + C_{Equip} + C_{Elec} + C_{Ng} \tag{1}$$

C_{Mat}: *Material cost*

C_{Equip}: *HVAC equipment cost*

C_{Elec}: *Electricity cost*

C_{Ng}: *Natural gas cost*

A database of material costs for the envelope was developed on a regional basis across the U.S. to accurately capture not only the physical structure and energy consumption of the home but also the material cost of the various components. Several available sources were used in developing the material cost database. The primary tool utilized for estimating cost is RSMeans residential cost data. HVAC equipment costs are modeled using multiple linear regression to fit cost data taken from online equipment suppliers. To determine an appropriate model for the cost of the air conditioner and furnace, several factors are taken into account. In this study the system cost is correlated to both its heating/cooling capacity as well as its efficiency.

3.4 Singular Value Decomposition

To reduce the dimensionality of a design space, the focus is on trying to find only those significant variables that need to be considered as optimization input variables, which means they are the variables that most strongly affect the life cycle cost. The next thing to be considered is that in many cases involving nonlinear systems, including categorical variables and discrete variables, a linear approach cannot be applied or will not have an accurate result. To overcome these problems, a new index is introduced. Each material is converted into an insulation value per unit cost, and the materials are then ordered from least to greatest and assigned an index. Using the methodology described in the previous section, 100 data samples with a different combination of design variables for each location is generated to explore the design space and identify the most significant variables to be used later in the optimization study. After generating the initial 100 data samples, singular value decomposition (SVD) is applied to the data set. Singular value decomposition is one method that can identify the most significant input variables by transforming correlated variables to uncorrelated variables. By selecting the first few dimensions or variables that explain most of the variance within the data set, the input variables can be reduced from a higher dimensional space to a lower dimensional space (Baker, 2005). After applying SVD to matrix A, which is the data sample, matrix A is broken into an orthogonal matrix U, and a diagonal matrix S with the singular values in decreasing order; and the transpose of an orthogonal matrix V. By selecting the first k dimensions on the diagonal matrix S, matrix A can be approximated as follows.

$$A = USV^T = U_1S_1V_1^T + U_2S_2V_2^T \approx U_1S_1V_1^T \tag{2}$$

3.5 Optimization

The optimization of the building design based on energy simulations may require large scale computing resources to properly explore the numerous variables (both discrete and continuous) and nonlinear functions, which also leads to discontinuous outputs. To deal with these characteristics, a discrete binary version of the particle swarm optimization methodology (Kennedy & Eberhart, 1997) has been selected and used for this study based on its ability to efficiently explore the design space and arrive at an optimal solution. To prevent fast convergence to a local optimum, the inertial version (Eberhart, et al., 2001) is used.

The first optimization study is a full optimization in which the life cycle cost is optimized using all 13 of the initial variables. The second optimization study is performed with a subset of design variables based on the results of the singular value decomposition. After finding the significant variables for the optimization, the remaining variables can be fixed to any values since they have only a minor affect on the life cycle cost. In this study, the cheapest material is chosen as the value of the insignificant variables during the optimization process.

According to Parsopoulos and Vrahatis (Parsopoulos & Vrahatis, 2002) when there are up to 15 variables under consideration, the recommended swarm size is the number of variables multiplied by 5. Based on this suggestion, the swarm sized is set to 64 and a von Neumann topology is selected having a neighborhood size of eight for the first study. For the second study, a swarm size of 36 and neighborhood size of six is used. The total number of generations are set to 500 and 300 respectively to make sure the particles are converged fully.

4. RESULTS

4.1 Singular Value decomposition result by location

After applying singular value decomposition on the sample data, the most meaningful components are identified based on the singular values. Figure 3 shows the singular values in descending order in Indianapolis. Engineering judgement is used to determine how many components need to be selected. There is a trade-off between the speed of the optimization process and the accuracy of the optimization result depending on the number of components that are selected. If only the first few components are selected, the optimization process will be considerably faster; however, the optimum point that is determined using a reduced number of design variables could be significantly further away from the true optimal point. In this study, the first four components are chosen as significant components to explain the most of the data samples and rest of the components are considered insignificant for all six locations.

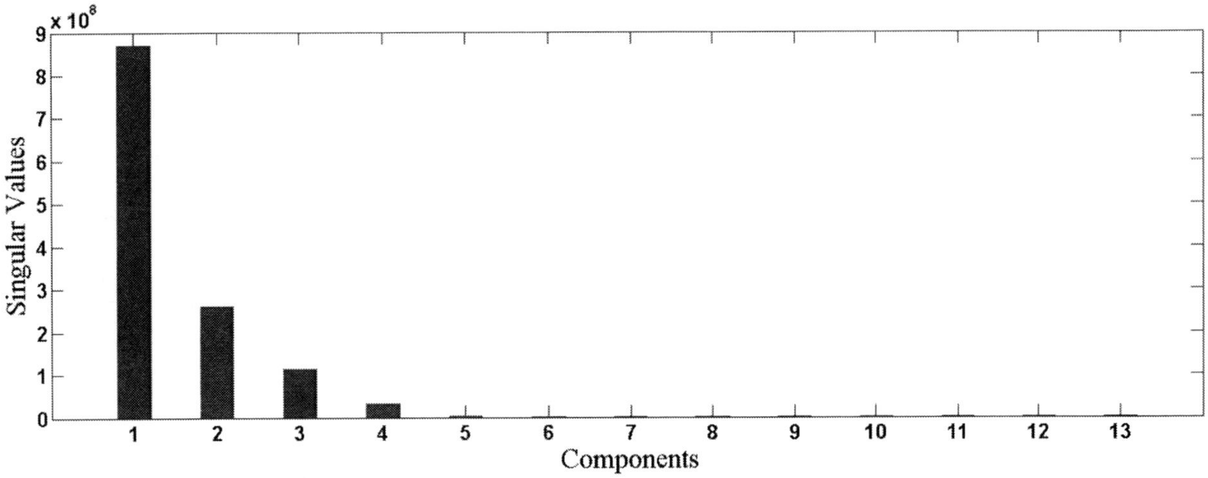

Figure 3: Result of singular values in Indianapolis

The first four singular vectors of the matrix are examined to identify the significant variables for the optimization study. As shown in Table 2, external foam board insulation is a strong contributor to the first singular vector, while a linear combination of roofing materials, window type, and heat recovery type are significant variables for the second singular vector. Wall core and under floor insulation are chosen as the main contributors for the third and fourth vectors.

Table 2: Indianapolis result of first four singular vectors

	1	2	3	4
Roofing Materials	-0.009	**0.238**	0.007	0.001
Roof Eave Over Hang Depth	-0.001	0.002	-0.001	0.002
Attic Insulation Material	0.006	-0.004	0.025	0.033
External Wall Siding Material	**-0.999**	-0.033	-0.019	-0.017
External Foam Board	0.004	-0.054	-0.019	-0.011
Wall Core	-0.018	-0.032	**0.999**	-0.008
Foundation Wall Insulation	0.000	-0.021	-0.012	0.006
Under Floor Insulation	-0.018	0.003	0.007	**0.999**
Window Type	-0.029	**0.875**	0.025	-0.003
Garage Door	0.004	-0.061	-0.010	-0.002
Heat Recovery Type	0.013	**-0.401**	-0.012	0.004
Air Conditioner Seasonal COP	0.002	-0.085	-0.005	0.006
Natural Gas Furnace Efficiency	0.002	-0.020	-0.004	-0.015

The same approach is applied to all six locations, and the selected significant design variables for each location are summarized in Table 3. Some items such as wall core, window type, heat recovery, external wall siding material, and under floor insulation are chosen for all locations, which indicates that LCC will be affected greatly by these variables, which means they should be included in the optimization process. The result also shows variation of the significant variables by location. In Denver and Seattle, the air conditioner seasonal COP is significant, while natural gas furnace efficiency is significant in Phoenix and Tampa. This occurs because there is no significant energy savings to annual utility cost by increasing the natural gas furnace efficiency in these warmer climates, and as a result the LCC is greatly increased as the efficiency goes up while air conditioner seasonal COP is not selected because there is a significant trade-off between air conditioner seasonal COP and LCC.

Table 3: Selected design variables for each location

Location	Denver	Indianapolis	Minneapolis
Selected design variables	Roofing Materials	Roofing Materials	Roofing Materials
	External Wall Siding Material	External Wall Siding Material	External Wall Siding Material
	Wall Core	Wall Core	Wall Core
	Under Floor Insulation	Under Floor Insulation	Under Floor Insulation
	Window Type	Window Type	Window Type
	Heat Recovery Type	Heat Recovery Type	Heat Recovery Type
	Air Conditioner Seasonal COP		
Location	**Phoenix**	**Seattle**	**Tampa**
Selected design variables	External Wall Siding Material	Roofing Materials	External Wall Siding Material
	Wall Core	External Wall Siding Material	Wall Core
	Under Floor Insulation	Wall Core	Under Floor Insulation
	Window Type	Under Floor Insulation	Window Type
	Heat Recovery Type	Window Type	Heat Recovery Type
	Natural Gas Furnace Efficiency	Heat Recovery Type	Natural Gas Furnace Efficiency
		Air Conditioner Seasonal COP	

4.2 Optimization result comparison

With the original 13 variables, or the selected six to seven variables from the previous section, the two optimization studies are performed. The optimized LCC for a 20-year horizon are compared in Table 4. Results show that the overall optimized LCC using only a subset of significant design variables are close to the original optimal point by 3.2 to 6.2%.

Table 4: comparison of optimized LCC

	Optimized LCC [$]	Optimized LCC with subset of variables [$]	Absolute Cost Difference [$]	Percentage Difference [%]
Denver	17,850	18,722	873	4.9
Indianapolis	20,097	21,025	928	4.6
Minneapolis	22,046	23,206	1,160	5.3
Phoenix	17,892	18,999	1,107	6.2
Seattle	16,543	17,182	639	3.9
Tampa	16,655	17,181	527	3.2

Using a 6-core processor and 6 gigabytes of RAM, the first (13 variable) optimization process requires approximately 36 hours per each location to complete the study while the second (reduced number of design variables using SVD) optimization process requires an average of 4 hours and 20 min per each location, which includes one hour for data generation that is used in the variable selection procedure. This significant reduction in computational calculations is mainly caused by the reduced design space and as a result, the number of evaluations requested by the optimization algorithm is significantly reduced. Also, the reduced swarm size due to a smaller number of design variables allows the optimizer to more quickly search the design space. Table 5 shows the comparison of time requirements and evaluation time between original full optimization and reduced optimization process. The total number of evaluations is decreased by 60% and the time requirement is decreased by 88% of the original computational time.

Table 5: comparison of two optimization methodologies

	Full optimization	**Optimization with subset of variables**
Number of evaluations	32,000	12,200
Time requirement[hrs]	36	4.35
Average LCC	18,514	19,386

5. CONCLUSION

The primary focus of this paper was to develop a methodology to perform an efficient and accurate life cycle building optimization in an acceptable time frame and apply the methodology to a case study of typical residential building and compare the results with those of the full optimization process over the entire design space. The result of the optimization process with a subset of design variables after performing singular value decomposition on the data sample for the design space shows a significantly shortened time requirement for the optimization process of 88%, while the optimized life cycle cost is close enough to the original optimum point by 3.2 to 6.2 %.

In the future, a more detailed study will be carried out to determine the relationship between accuracy and computational time based on the criteria selection as well as the optimization methodology to improve the efficiency and accuracy proposed in this paper.

REFERENCES

Augenbroe. (2002). Trends in building simulation. *Building and environment*, 891-902.

Baker, K. (2005). Singular value decomposition tutorial. *The ohio state University*, 24.

Bettonvil, B., & Klenijnen, J. (1997). Searching for important factors in simulation models with many factors: Sequential bitfurcation. *European Journal of Operational Research, 96*(1), 180-194.

Cooper, N. S. (2011). Optimization of energy efficient housing for the lower income demographic utilizing a generalized pattern search particle swarm optimization algorithm. *Purdue University*.

Eberhart, R. C., Shi, Y., & Kennedy, J. (2001). *Swarm intelligence.*

Kennedy, J., & Eberhart, R. C. (1997). A discrete binary version of the particle swarm algorithm. *IEEE International conference on systems, Man, and cyvernetics, 5*, 4104-4108.

Parsopoulos, K., & Vrahatis, M. (2002). REcent approaches to global optimization problem through particle swarm optimization. *Naturla Computing, 1*, 235-306.

Rahni, N., Ramdani, N., Candau, Y., & Dalicieux, P. (1997). *Application of Group Screening to Dynamic Building Energy Simulation Model.* 2Research Centre, Electricité de France.

RSMeans. (2011). *RSMeans Residential Cost Data 2011. 30th ed.* R.S.Means.

State-Space Modeling of Thermal Spaces in a Multi-Zone Building

Vahid R. DEHKORDI[1]*, Jose A. CANDANEDO[2]

[1]CanmetENERGY, Natural Resources Canada,
Varennes, Quebec, Canada
vahid.raissidehkordi@canada.ca

[2]CanmetENERGY, Natural Resources Canada,
Varennes, Quebec, Canada
jose.candanedoibarra@canada.ca

* Corresponding Author

ABSTRACT

In this paper, a study on system identification and modeling of thermal spaces in a large institutional building is presented. Thermal models are essential in predictive control since they are required to predict the thermal load of a single building zone, a collection of various thermal spaces, or a whole building. The main goal of this paper is to show how the optimum model order associated which each thermal zone depends on factors such as the location of the zone within the building, its orientation and its exposure to outdoor space. The results of this study will serve as a guideline for choosing the appropriate order of linear models in similar buildings.

1. INTRODUCTION

Control-oriented modeling of heating, ventilation, and air conditioning (HVAC) systems has a significant potential to improve the operation of HVAC systems, as it is an essential step in the optimization of energy consumption and indoor environment quality. Thermal space models, specifically, can be used to calculate building energy needs over a prediction horizon of up to several days, when a reasonably accurate weather forecast and occupancy data are available. This is a valuable asset for the formulation and optimization of a model predictive control (MPC) setup (Prívara *et al.*, 2012; Candanedo *et al.*, 2015).

HVAC systems are modeled using various approaches, which result in quite diverse types of models (Janczura *et al.*, 2013). The modeling approaches most commonly used can be categorized roughly in three types (Afram and Janabi-Sharifi, 2014). The data-driven "black box" approach consists of collecting system performance data under different conditions and then deriving a mathematical relationship between the input and output variables. Data-driven models can be obtained in numerous ways, for example, by using statistical regression or artificial neural network (ANN) techniques. Conversely, the "white box" approach involves deriving the model based on the knowledge of physical dynamics governing the plant to be modeled. Finally, an intermediate "grey box" approach is also available which uses physical characteristics of the system to form the basic structure of the model. Subsequently, measured data of the system are processed using parameter estimation algorithms to tune the model parameters. Data-driven RC models used for characterization of thermal spaces are a typical example of grey box models, as the RC circuit structure is determined first via physical considerations; the value of the RC parameters can then be determined by using recorded weather data and indoor temperature (Saberi-Derakhtenjani *et al.*, 2015; Candanedo *et al.*, 2013). This classification in "black box", "grey box" and "white box" is of course only a very coarse approximation of a continuum of model types, as all models are really some form of grey-box model.

White box models require detailed knowledge of the system and its processes. On the other hand, data driven gray box and black box models can be obtained with little or no knowledge of the system. White box models can be generalized easily since they follow the 'true' physical notions and relations between different parts of a system, but at the same time, on account of their reliance on abundant *a priori* information, they are prone to function poorly when evaluated numerically against actual processes. On the other hand, data driven models function well under situations closely mimicking a defined set of training data, but their performance varies considerably when different datasets are applied as input data. Grey box models benefit from the advantages of the other two types, providing good generalization capabilities in comparison with data driven models, as well as better accuracy in comparison with the physics-based models.

The above mentioned approaches to the modeling of building dynamics and HVAC systems result not only in diverse types of models, but also in different levels of complexity. For example, the building simulation software EnergyPlus creates and uses a highly detailed, white box RC network model; this model is a valuable tool for design, but it is arguably too complex for the formulation and later optimization of a MPC setup. Conversely, a low-order model can provide a satisfactory description of large-scale dynamic phenomena in the building (Kim and Braun, 2012; Cole *et al.*, 2014). For instance, a low-order thermal space model of the whole building with very few inputs/outputs to a coarse estimation of thermal load of a commercial building during the next 48 hours. Other examples are low-order thermal networks which are used to estimate the thermal response of thermal zones from office spaces to whole communities. The simplicity of low-order models makes them more manageable and flexible. Moreover, they provide clarity and insight about the effect of different inputs and can facilitate the management of uncertainty. However, the simplicity of low-order models also means that they fail to represent the smaller-scale details of their subject. Representing the whole building with a single model provides a good example of such limitation; while the building model predicts well the overall load, it does not contain enough information to show how much cooling or heating is required by each individual zone.

This paper explores system identification and modeling of thermal spaces in a large institutional building as nearly "black box" models with very few input signals. These data-driven models are in the form of state-space representations which do not take the physical characteristics of the thermal zones into account. The main goal is to find how the optimum state-space model order associated which each thermal zone depends on factors such as the location of the zone within the building, its orientation and its exposure to outdoor space, *without* explicitly incorporating the physical structure of each thermal space into the model.

2. PRELIMINARIES

This section provides information regarding the simulation building considered for this study, as well as some fundamental definitions regarding state-space models.

2.1 Case Study Building

The case study building is a model of a two-story school building, with a floor area of 24,000m^2 (258,000ft^2) with 46 thermal zones. A detailed thermal model of the building was created in EnergyPlus. The advantages of having such a model will be explained in Section 3.

The zones considered include a large diversity of spaces: small offices, classrooms, long hallways and two gymnasia. Because of this diversity, some zones span both the main and the upper floor. Also, as it can be seen in Figure 1, the western side of the building is divided into three wings. All these factors cause the number of zones adjacent to each zone varies greatly.

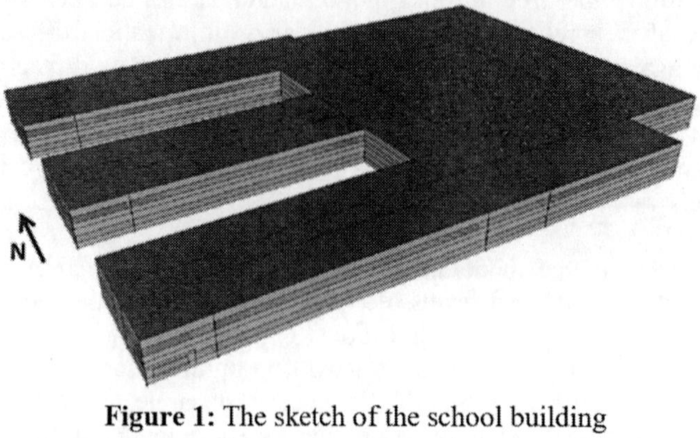

Figure 1: The sketch of the school building

2.2 State-space Representation

A state-space representation is a mathematical model of any physical system in which inputs, outputs and state variables are related using first-order differential equations. Consider the LTI discrete-time multi-input multi-output (MIMO) model of a system, expressed in state-space form:

$$\begin{aligned} \mathbf{x}[i+1] &= \mathbf{A}_{n\times n}\mathbf{x}[i] + \mathbf{B}_{n\times p}\mathbf{u}[i] \\ \mathbf{y}[i] &= \mathbf{C}_{q\times n}\mathbf{x}[i] + \mathbf{D}_{q\times p}\mathbf{u}[i] \end{aligned} \tag{1}$$

where matrices **A**, **B**, **C** and **D** relate the input and output signals; *i* indicates the discrete time sample and vectors $\mathbf{u} \in \mathbb{R}^p$, $\mathbf{x} \in \mathbb{R}^n$ and $\mathbf{y} \in \mathbb{R}^q$ represent input signal, state vector and output signal, respectively. In case of a multi-input single-output (MISO) system, **C** and **D** will become row vectors.

2.3 State-space models: determination through RC networks vs. direct derivation from data

According to the definition of state-space models, the RC models of thermal spaces can also be represented in state-space format. In this case, the order of the state-space model or *n*, as in equation (1), will be equal to the number of capacitors included in the RC model, which is the same as the number of differential equations required to express the dynamics of the RC circuit. A standardized method for the transformation from RC network to state-space for thermal networks was proposed by (Candanedo et al., 2013).

Since the heat transfer phenomena in the building are mainly described by linear differential equations, it is reasonable to assume that, as a whole, the thermal space behaves a linear dynamical system (Putta *et al.*, 2014). Given this important assumption, it is also possible to obtain a state-space model *directly from recorded data.* Directly identifying a state space model from measured data provides a significantly higher number of degrees of freedom as compared to a derivation from a calibrated RC model. This is because every individual coefficient in the matrices **A**, **B**, **C** and **D** can be considered as a degree of freedom, whereas in the RC model the degree of freedom is equal to the total number of resistors and capacitors. The main drawback of obtaining such "free-form" state-space model is that the state vector **x** cannot be readily associated with physical quantities or features of the thermal space. In contrast, in a state-space representation derived from an RC thermal network model, **x** (the vector of states) corresponds to temperatures in the building. Figure 2 summarizes the above discussion.

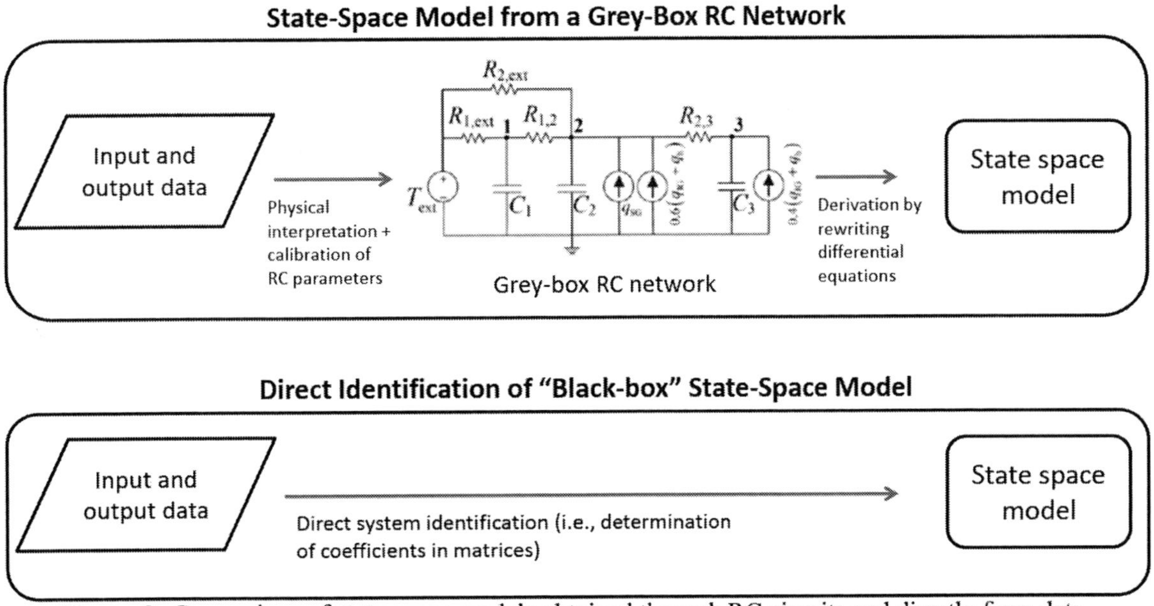

Figure 2: Comparison of state-space models obtained through RC circuits and directly from data.

3. MODELING

As mentioned in Section 2.1, a detailed model of the school building was developed on the EnergyPlus platform, which allows running long-term simulation scenarios or "virtual experiments" and recording practically any required performance index. The weather data file used for simulations contains various weather indices, such as solar radiation, which determines the solar gains (q_{SG}) into the building, and outdoor temperature (T_{ext}). Internal heat gains (q_{IG}), corresponding to the heat released by lights, equipment and occupants, are assumed to follow semi-rectangular, uniform pulses according to the typical operation of a school building (Dehkordi and Candanedo, 2014).

Each thermal zone is individually modeled as a MISO LTI discrete-time state-space system. The output of each model is the indoor temperature (T_{in}) of its corresponding thermal zone (EnergyPlus output), while T_{ext}, q_{SG} and q_{IG} are considered to be the input signals. The heating/cooling power (q) injected to each thermal zone is also treated as an input.

The structure of the EnergyPlus thermal zones is assumed not to be adiabatic, meaning that there will be an exchange of heat between different thermal spaces. In other words, the thermal dynamics of each thermal zone will affect the thermal dynamics of neighboring zones. Knowing this, four modeling scenarios are considered:

1) The mutual effects between zones are ignored. In other words, the thermal response of each zone is considered to be solely affected by the input factors directly affecting it, i.e., T_{ext}, q_{SG}, q_{IG} and q.

2) The heating/cooling power delivered to a certain thermal zone is considered to have an effect on thermal dynamics of its adjacent thermal zones. It means that beside T_{ext}, q_{SG}, q_{IG} and q, a fifth input is considered which is the sum of the q delivered to adjacent zones (q_{adj}).

3) The indoor temperature of a certain thermal zone is considered to have an effect on thermal dynamics of its adjacent thermal zones. Similar to the previous case, beside T_{ext}, q_{SG}, q_{IG} and q, the indoor temperature of the adjacent zones, T_{adj} are also individually considered as inputs. In other words, the number of inputs in this case will vary depending on the number of adjacent zones.

4) Similar to 3, the indoor temperature is considered to have an effect on thermal dynamics of its adjacent thermal zones. In this case, however, the *average* indoor temperature of the adjacent zones, $\overline{T}_{adj}$, is considered as the fifth input.

Note that using models generated following scenarios 2 to 4 in a building control algorithm requires for each zonal model to have access to real-time data from its adjacent zones. Figure 3 shows the model structures considered:

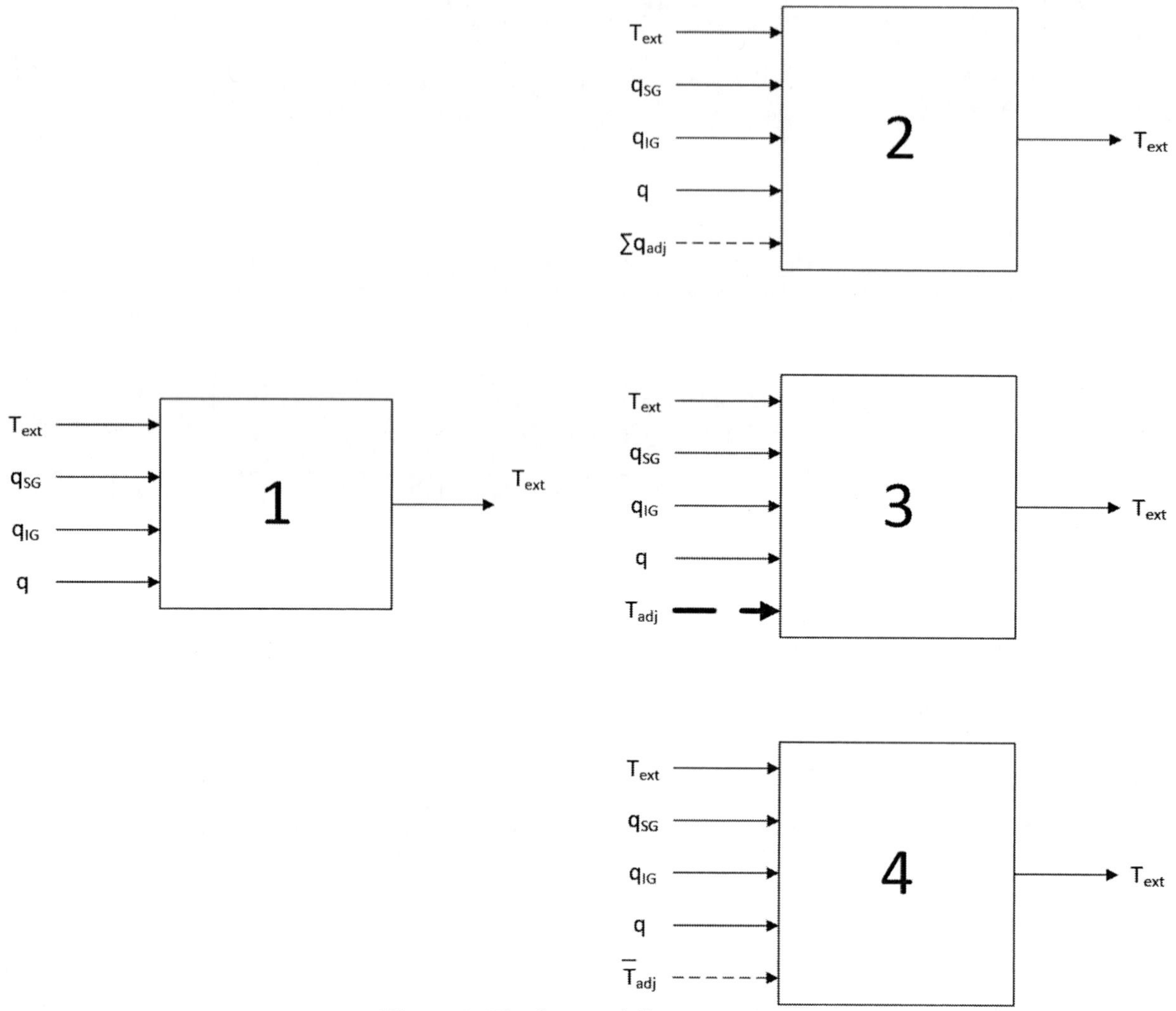

Figure 3: The four modeling scenarios

The EnergyPlus simulation data is reported at 15-minute intervals for each thermal zone. It provides the portion of total solar gains received by the building applicable to each thermal zone, which in practice can be measured on-site. Also, the internal gains follow a typical school profile and the outdoor temperature is the same for all thermal zones. The q for each thermal zone is provided for a typical temperature setpoint of 23°C.

For every zone, the adjacent zones are defined as the ones who share a wall or floor, i.e., corners are not considered as connection between any two zones. Figure 4 shows the floor map as well as the number of adjacent zones for each zone. Note that the empty spaces in the second floor are mostly the zones which extend to both floors (e.g., gymnasium with a high ceiling). Also, the color map is chosen in such a way that the darker color, the smaller the value (here the number of adjacent zones) assigned value is.

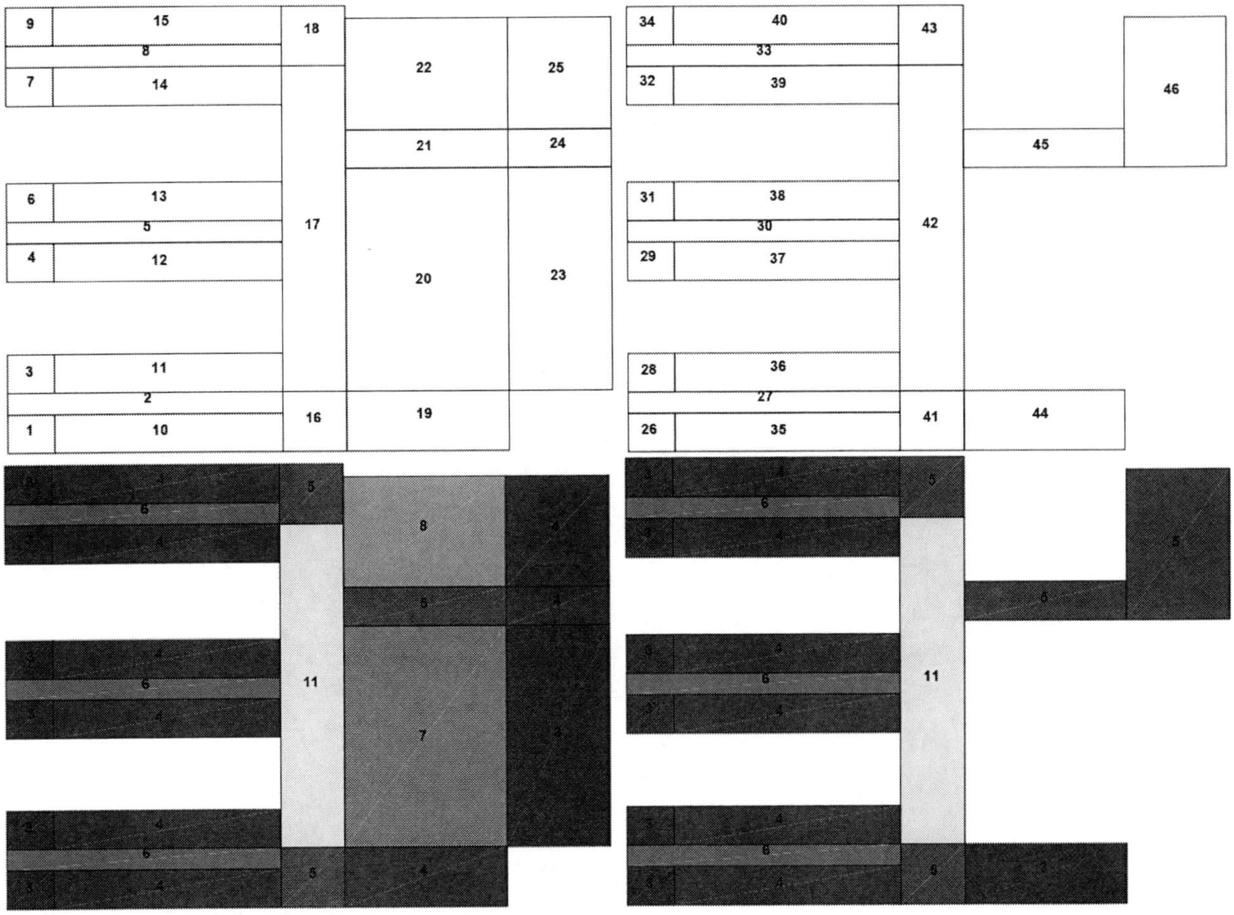

Figure 4: The floor map of first (left) and second (right) floors with zone indices (top) and the color-coded number of adjacent zones (bottom)

4. SIMULATION RESULTS

As discussed in Section 3, state-space models are developed for each thermal zone, following all four scenarios depicted in Figure 3. The order of the models is set to be varying between 1 and 10, therefore putting a hard cap on the maximum order of the state-space models. The system identification is performed using *ssest* command in MATLAB software environment, which uses a state-space subspace system identification (4SID) approach. Following Equation (1), no noise model is taken into account. Figure 5 shows the best modeling RMSE achieved for each zone, for all four scenarios.

As depicted in Figure 5, the performance of scenario 2 is not that much different from the base scenario 1 since the least RMSE achieved for individual thermal zones are generally very close. Also, the number of cases indicating otherwise is far less than being meaningful. The performance of scenario 3, however, shows a substantial improvement of the performance as the average RMSE value drops by a factor of 3.

Figure 6 shows the order of the model resulting in the least RMSE for each thermal zone following scenarios 1, 3 and 4. As it can be clearly seen, the location and the orientation of the zones have an important impact on the modeling performance when adiabatic conditions are assumed (scenario 1). For example, at the first floor, the zones facing southwest and north west – zones 1, 3, 4, 6, 7 and 9 (see Figure 4) - have higher orders compared to south facing zones (10, 12 and 14). Observing the numerical results in Figure 5 reveals that using the same modeling setup, the error level is more or less in the same range for zones 35 to 40. Zones 36, 38 and 40 require only a second-order model to reach the lowest error level. This becomes more interesting since at the same time, the south-facing zones 35, 37 and 39 require have optimal modeling orders of 6 and 7. Therefore, the orientation of thermal zones definitely plays a role in their thermal response and subsequently, the performance of modeling.

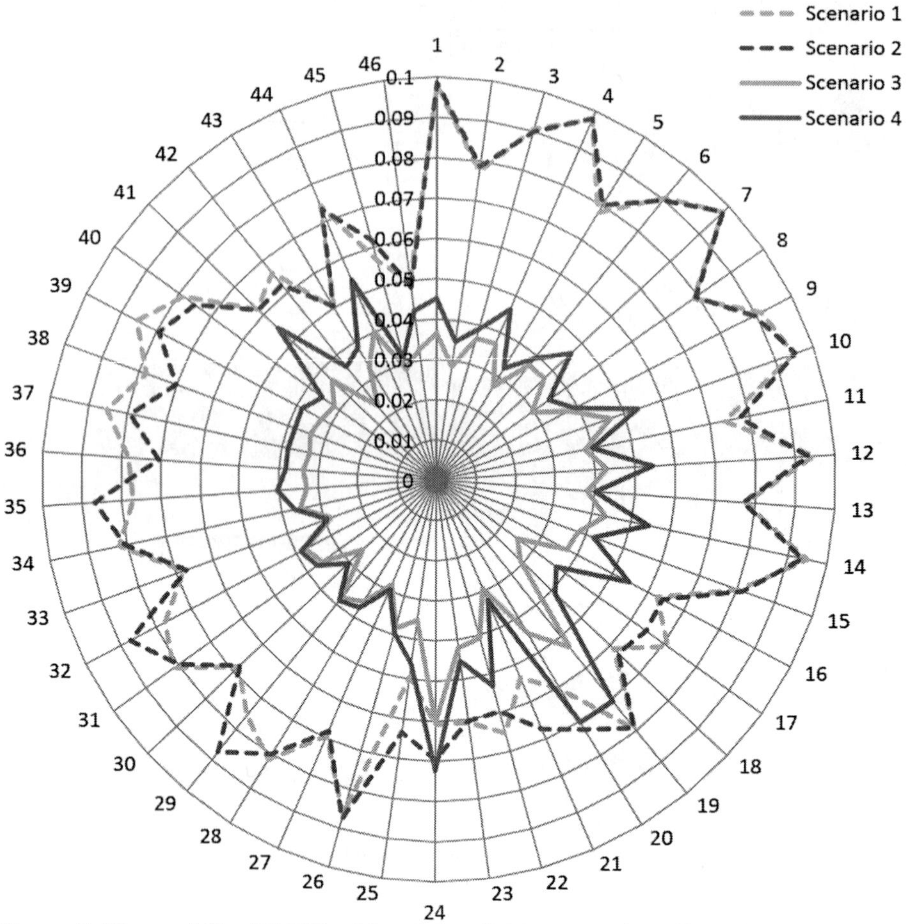

Figure 5: The modeling RMSE achieved for all 46 zones using all modeling scenarios

Focusing now on the results of scenario 3, the optimum model orders for both floors seem to be quite high for all thermal zones. However, it should be noted that upon observing all RMSE values obtained from all ten models developed for each zone (not shown in this paper), the variation in modeling performance for each individual thermal zone in this scenario is relatively low, often negligible – sometimes less than 5% of the best value. Therefore, one can safely reduce the order to much lower values, even 2 to 4, and still benefit from a low energy of modeling error. In other words, relatively high order of the models depicted in figure 6 (right) is purely because of numerical optimization and does not have anything to do with the geometrical characteristics of the thermal zones.

Scenario 4 plays the role of an alternative between scenarios 2 and 3: similar to the model structure in scenario 2, it has only one additional input compared to that of scenario 1, but its performance is more or less similar to that of scenario 3, as it can be seen in Figure 5. By observing Figure 6, it is even more interesting to see that (unlike scenario 2) the model orders can be somewhat associated with certain orientations and indoor locations. For example, the north facing zones have relatively smaller orders while the corner zones facing west, somehow similar to scenario 1, have higher orders. Interior thermal zones have relatively low orders; the ones with high orders exhibit the same numerical situation of scenario 3: the variation in modeling performance is often negligible which means that, for example, zones 2, 27 and 33 can safely be modeled with orders as low as 4 or 5, similar to the zones 5, 8 and 30.

Back to Figure 5, it can be seen that in all modeling scenarios, in general, the least RMSE associated with a certain thermal zone on the second floor is lower than that of its equivalent on the first floor. Also, given the fact that the optimum model order for some zones in scenario 1, e.g. zone 1 or 4, is low while their corresponding RMSE is relatively high, one can conclude that there is a very little chance that increasing the order of model beyond the maximum 10 would do any better. At the same time, if the modeling performance is crucial, i.e. very low error level/RMSE is needed, scenarios 3 and 4 are the most prone to leading to better results if the model order increased, although by a small extent.

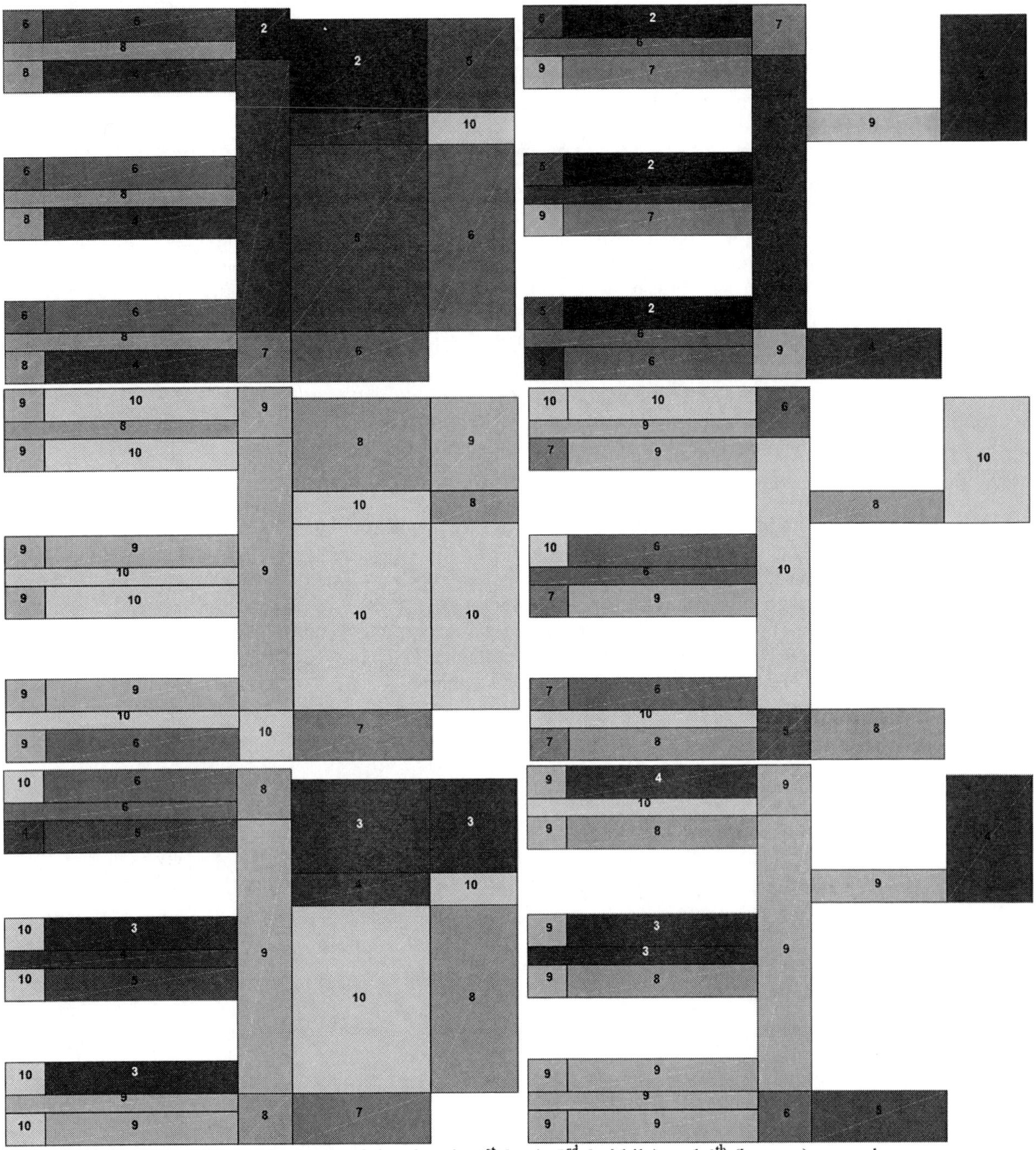

Figure 6: Optimum model orders in 1st (top), 3rd (middle) and 4th (bottom) scenarios

5. CONCLUSIONS

Four modeling scenarios have been presented in this paper to assess the performance of linear state-space models for the representation of thermal response of the thermal spaces in a model school building. Simulation data provided by EnergyPlus (the building simulation environment) has been used directly to create the state-space models (black-box modeling). When adiabatic conditions are assumed, the optimum order of the models can be associated with the geographical location and orientation of the thermal zones. Depending on which parameters are considered as input variables, taking the indoor temperature of the adjacent zones into account proves to be useful in reducing the error levels while still maintaining a ≤ 10 model order. Also, taking the average temperature of the adjacent zones proves to be useful in reducing the error while leading to even lower model orders and keeping the number of input variables low. Future works will include the comparison of the performance of different types of RC models with that of state-space models with the same order.

REFERENCES

Afram, A., Janabi-Sharifi, F., 2014. Review of modeling methods for HVAC systems. Applied Thermal Engineering In Press.

Candanedo, J. A., Dehkordi, V. R., Lopez, P., A control-oriented simplified modelling strategy. 13th International Building Simulation Conference (IBPSA 2013), August 26-28 2013 Chambéry, France.

Candanedo, J. A., Dehkordi, V. R., Tamasauskas, J., Impact of heat pumps and predictive control on residential energy use, load and grid interaction: a Canadian perspective. 14th International Building Simulation Conference (IBPSA 2015), December 7th-9th 2015 Hyderabad, India.

Cole, W. J., Powell, K. M., Hale, E. T., Edgar, T. F., 2014. Reduced-order residential home modeling for model predictive control. Energy and Buildings 74, pp. 69-77.

Dehkordi, V. R., Candanedo, J., Multi-level model predictive control of a building. ASME 2014 Dynamic Systems and Control Conference (DSCC 2014), October 22-24 2014 San Antonio, Texas.

Janczura, J., Maciejewska, M., Szczurek, A., Wyłomanska, A., 2013. Stochastic modeling of indoor air temperature. Journal of Statistical Physics.

Kim, D., Braun, J. E., Reduced-order building modeling for application to Model-Based Predictive Control. SimBuild 2012 August 1-3 2012 Madison, Wisconsin.

Prívara, S., Cigler, J., Vana, Z., Oldewurtel, F., Sagerschnig, C., Zacekova, E., 2012. Building modeling as a crucial part for building predictive control. Energy and Buildings In Press.

Putta, V., Kim, D., Cai, J., Hu, J., Braun, J. E., Distributed model predictive control for building HVAC systems: a case study. Third High Performance Buildings Conference, July 14-17 2014 Purdue University, Indiana.

Saberi-Derakhtenjani, A., Candanedo, J. A., Chen, Y., Dehkordi, V. R., Athienitis, A. K., 2015. Modeling approaches for the characterization of building thermal dynamics and model-based control: A case study. Science and Technology for the Built Environment 21, pp. 824-836.

Parallel Object-Oriented Algorithms for Building Performance Simulation. Application to an existing dwelling.

López JOAN[1]*, Capdevila ROSER[1], Souaihi OUSSAMA[1], Rigola JOAQUIM[1], Oliva ASSENSI[1]

[1] *Heat and Mass Transfer Technological Center (CTTC), Universitat Politècnica de Catalunya- BarcelonaTech (UPC)*
ESEIAAT, Carrer Colom 11, 08222 Terrassa (Barcelona), Spain
joanl@cttc.upc.edu

ABSTRACT

In the present work an existing dwelling, situated in the Netherlands, has been modeled by means of a parallel object-oriented simulation tool called NEST-Buildings. The model is based on a pre-defined collection of elements (e.g., walls, rooms, openings, outdoors, occupants, ventilation tubes and boxes, solar radiation distributors, HVAC equipment, etc.) that are connected to each other composing a dynamic thermal system. New configurations can be easily handled by adding or removing elements. Moreover, the building elements can be modeled at distinct levels of accuracy ranging from lumped volumes mixed with one-dimensional to detailed CFD&HT models. This approach makes possible the assessment of general-type buildings (residential, services, old, modern, etc.) using the appropriate modeling level at each component. The work is focused on the global resolution algorithm currently implemented in this computer simulation tool.

1. INTRODUCTION

New policies are progressively converging to nearly zero energy standards. Zero net energy buildings, commonly known as ZEB (Marszal 20011, ACEEE 2015), combine high levels of energy efficiency with renewable energy systems (solar, biomass, windmill, etc.) to offset the energy delivered to the building over the year. The EISA 2007 specifies a zero-energy target for 50% of US commercial buildings by 2040 and for 100% of them by 2050 (US Congress 2007). In Europe, the new regulations seeks nearly ZEB target for all public buildings by 2018 and from 2020 for all new public buildings (EU 2010, 2012). In this regard, architects and engineers should be able to use appropriate tools for assessing the energy consumption in buildings. Experimental methods based on in situ measurements are not only challenging and expensive but they are normally not feasible because of the space usage (residential, hospitals, hazardous environments, etc.). Alternatively, computer simulation software is much more affordable. Buildings performance can be evaluated by means of virtual designs providing a lot more flexibility.

The main challenge in building simulation is the non-linearity of the mathematical model that rules the physical phenomena. Transient heat transfer by conduction, convection and radiation is mixed with moisture and air flows between the outdoor environment and the inside of the building and between the distinct zones, too. All these phenomena are driven by the effect of the weather conditions (sun irradiation, rain, wind, etc.) and the surrounding elements (orientation, other buildings, trees, etc.) on the building through the envelope (walls, panes, roof, soil, etc.), the openings and the infiltrations. In addition, there is the action of the non-natural events such as the HVAC systems or the occupants themselves. Hence, the simulation of buildings at this level requires the use of *i)* highly flexible software tools and *ii)* well though-out system resolution approach in order to succeed within a reasonable time.

The present work has been developed within the "Dwelling Climate Control System (DCCS)" project (DCCS 2013-2016), one of its main objectives is to develop an affordable, easy to use and apply, integral dwelling climate control system in order to reduce the energy use of buildings and improve the air quality by optimizing the parameters of the heating and ventilation systems. For the achievement of such objective, a real dwelling (a semi-detached house in the Netherlands) has been chosen and has been modeled by means of a parallel object-oriented simulation tool called NEST-Buildings. That *virtual building model* has been used as a test bench for the testing of the developed control system. The model is based on a pre-defined collection of elements (e.g., walls, rooms, openings, outdoors, occupants, ventilation tubes and boxes, solar radiation distributors, HVAC equipment, etc.) that are connected to each other composing a dynamic thermal system. New configurations can be easily handled by adding or removing elements. Moreover, the building elements can be modeled at distinct levels of accuracy ranging from lumped volumes mixed with one-dimensional to detailed CFD&HT models. This approach makes possible the assessment of general-type buildings (residential, services, old, modern, etc.) using the appropriate modeling level at each component.

2. NEST-BUILDINGS

2.1 Code description

NEST-Buildings (Damle 2011, López 2016) is a parallel object-oriented code for building performance simulation. The overall model is based on several predefined classes that provide abstraction of the involved physical phenomena, equipment, boundary conditions and events. For illustrating this, the sketch of a house (walls, windows, outdoor, radiation element, rooms, ground, etc.) and its corresponding NEST-Buildings conceptual graph view is shown in Figure 1. A description of each of these concepts is provided in Table 1 as well as the all possible connections among them.

Therefore, the elements can be linked to each other to form a specific building or room configuration. This makes easy to create new configurations by adding or removing elements to/from the overall model. Each element can be represented in any form as long as it can exchange the necessary boundary information from the rest of the elements in the system. Such an approach gives flexibility of choosing a model for each element and having different levels of modeling (lumped volumes, 1D, CFD&HT...) for different elements in the system without changing the basic program structure. Furthermore, the numerical calculations can be easily done in several parallel processes as the models are uncoupled to each other at code level.

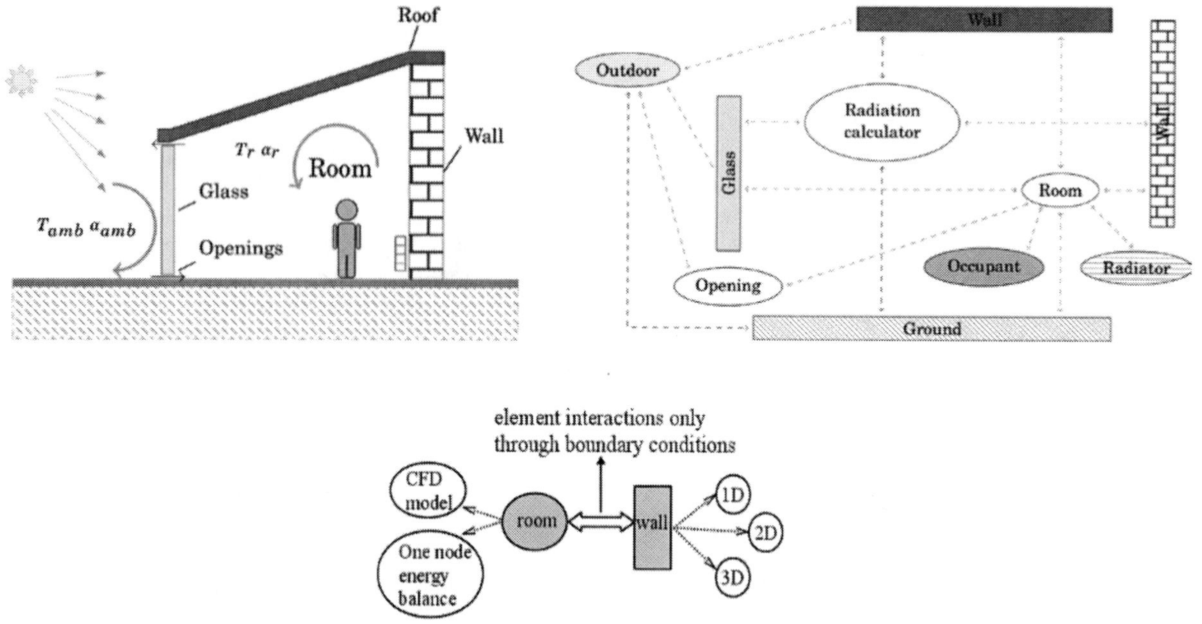

Figure 1. (top-left) Scheme of a room. (top-right) NEST views the system as a vertex-edge graph. (bottom) Flexibility in choosing individual elements models.

Table 1. Summary of the elements (classes) developed for NEST-Buildings.

Elements/Classes	Elements Description	Possible Connections
Outdoor	Weather database manager depending on building location.	Wall, Glass, Opening
Wall	1D Heat conduction and moisture transport is solved. Heat convection, thermal radiation and solar gains can be considered at the boundaries.	Wall, Outdoor, Room, RadBox, SolarDistributor
CompWall	Sub-system composed of multiple Wall layers of different materials.	Wall, Outdoor, Room, RadBox, SolarDistributor
Glass	Wall-like object with optical properties (e.g. solar radiation transparency).	Glass, Outdoor, Room, RadBox, SolarDistributor
MultiGlass	Sub-system composed of N Glass objects and N-1 Rooms (air chambers), e.g. double glazing	Outdoor, Room, RadBox, SolarDistributor
Room	Lumped volume with mass, moisture, energy and CO_2 balance.	Wall, Opening, HVAC, SolarDistributor
RadBox	Solves thermal radiation based on view factors of a given number of walls.	Wall, CompWall ,1)Glass, MultiGlass,
SolarDistributor	Distributes the transmitted solar radiation uniformly to room walls.	Wall, CompWall ,Glass, MultiGlass, Outdoor
HVAC (radiator, TRV, mechanical ventilation)	Controls the heating and ventilation of rooms.	Room
Opening	Air movement between rooms and outdoor due to density differences. Infiltrations are also managed ~~here~~ small Openings.	Room, Outdoor
Occupant	This includes people behavior (heat, moisture and CO_2 generation) into the simulation. People events are managed by means of Schedule objects.	Room
Schedule	Manages events that produce the Occupants or the controller devices.	Not a system element

2.2 Algorithm for solving the overall model

The resolution scheme of the overall model is based on a block-Jacobi like method. During this procedure, every element of the system is visited to call the resolution routines of their particular sub-models. After this, each element changes its state (e.g. room temperature, CO_2 concentration, doors and windows mass flows, radiators heat output, etc.). Once all the elements have been iterated, the program synchronizes their input/output data in order to maintain them up to date with the state of their neighboring elements (i.e. new mass flows, room temperatures, ambient temperatures, solar irradiation, etc.). This is repeated several times until the elements are well coupled so that they do not change their state. At this point, the whole set of elements is converged to an instantaneous building state. Then, the program checks for changes in the schedules of the openings and the occupants. Finally, the simulation ends or it continues towards the next time step.

This scheme is summarized in Figure 2. It is important to highlight that the distinct elements composing the system are computed in several parallel processes (i.e. the different branches in the scheme). This is an important property of the block-Jacobi like approach. The fixed-point iterations approach requires that the system components can be resolved apart from any other, which in turn enables parallelization. In particular, the code is programmed for running in distributed memory systems. This means that the elements that belong to different processes cannot directly share their data. For managing this problem, during the synchronization operation the processes communicate the appropriate data with each other using the Message Passing Interface standard (MPI).

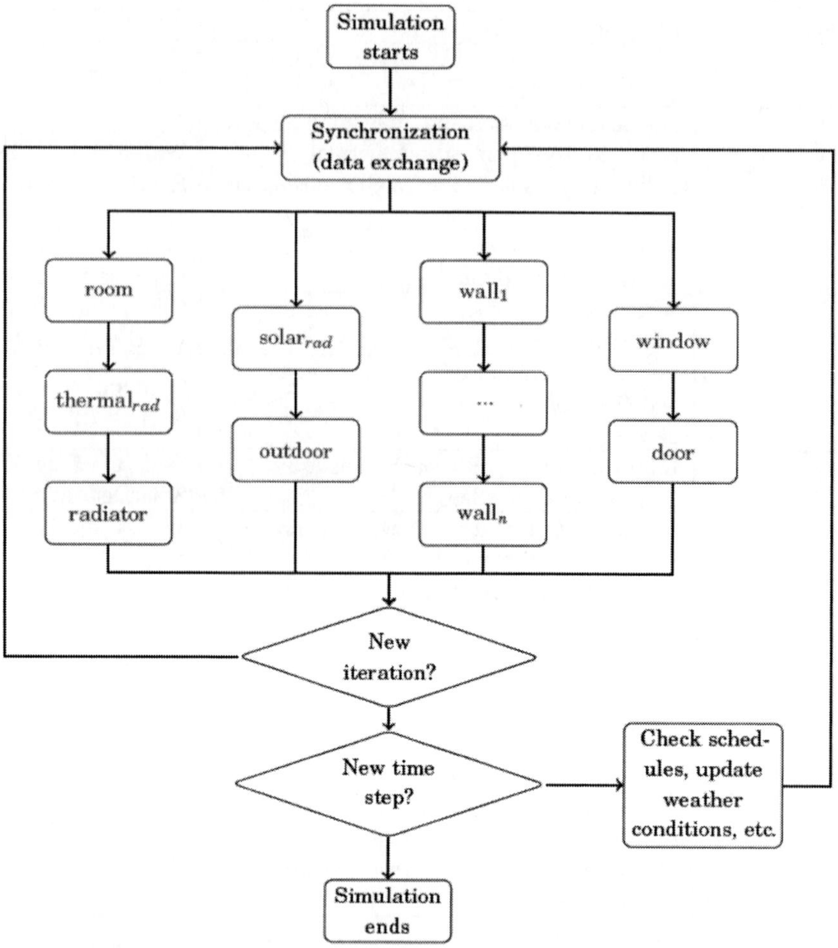

Figure 2. Block-Jacobi like algorithm implemented in NEST-Buildings.

2. CASE STUDIES

A real semi-detached house located in Lemmer (Netherlands), with two floors and 12 rooms (see planes of the house in Figure 3), has been virtually modeled by linking elements like walls, composite walls, glass walls, rooms, radiation calculators, outdoors, openings, ventilation tubes, radiators, boiler, ventilation box, etc. and simulated in a transient mode. The front of the house is oriented 20° azimuth. The walls are assumed to be made of common bricks and the floors of concrete. The mechanical ventilation extracts air from the livingroom-kitchen (room3), the toilet located in the ground floor (room 5) and the bathroom (room 10). There are radiators in all rooms except the garage (room 1), the hallway with stairs (rooms 4 and 11), the toilet (room 5) and the attic (room 12).

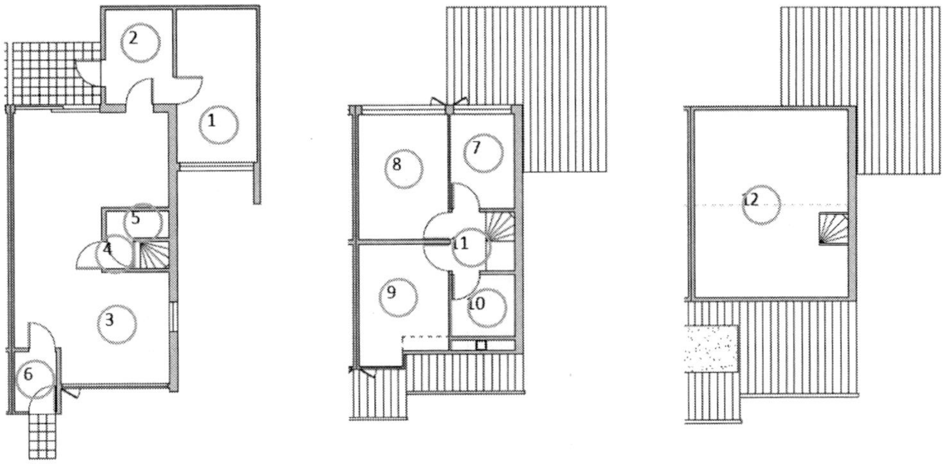

Figure 3. Planes of the simulated semi-detached house. From left to right: ground, first and attic floors.

That *virtual building model* has been used as a test bench for the testing of the developed control system. The developed control system controls two systems in the building: the boiler, through the water *setpoint* temperature (which also acts as the on/off mechanism), and the ventilation system, through the whole-house extract fan speed.

In present work three simple control strategies have been studied (see Table 2). In all them the control of the boiler is the same: Boiler turns on when temperature of the room where thermostat is located, i.e. the living room (room 3) drops below *Tsetpoint-threshold*, but does not turn off until the temperature rises above *Tsetpoint+threshold*. The boiler control is a simple timer with room *setpoint* temperature settings according to the occupancy schedule.

In the first case the ventilation fan does not work, rooms are closed (just infiltrations are considered). In the second case, the control of the ventilation system is assumed to run at the lowest fixed speed setting, which is the default manual setting for most households. In the third case the controller controls the fan speed according to the occupancy schedule: the fan speed is the minimum (1) when the house is not occupied and to a higher speed (2) when it is occupied.

In the three situations the rooms are kept closed and just infiltrations through the doors are considered. Infiltration is set to assure minimum required ventilation rates in mechanical ventilated houses at minimum fan speed.

Table 2. Studied ventilation strategies.

case	Ventilation control
1	No – fan does not function
2	No – ventilation speed is constant
3	Yes – fixed schedule

An occupancy schedule which is the same during weekdays and different for the weekend has been used and shown in Figure 4. There is an adult couple living in the house. On Saturday two guests visit the house in the evening. Mainly the living room, bedroom and sporadically the bathroom are used. All doors are closed, however infiltrations through doors and windows are considered. Three days have been simulated: from Friday to Sunday. Nevertheless, simulations start 24h hours before in order to have a more reliable initial map of temperatures, velocities and pressures at the beginning of the two or week analysed days. So, the total simulated time is 96h for these 72h tests.

Simulations were performed in February, using the weather for a typical year in Eelde, Netherlands (closest meteorological station to Lemmer, where the real test bench is located (METEO 1999)). A typical winter week has been chosen: temperatures around 0°C, 2 sunny days and 2 cloudy days, and not windy.

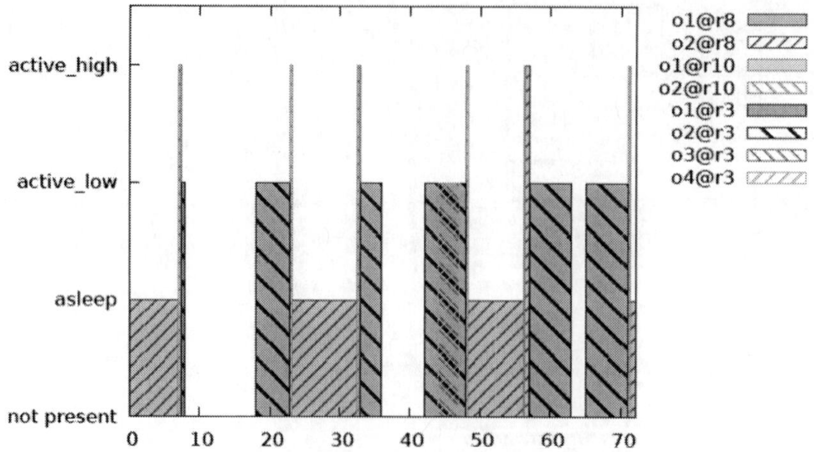

Figure 4. Occupancy patterns and activity levels for 2 occupants, o1 and o2, plus two guests, o3 and o4, over the 72hr Friday - Sunday period.

The thermal performance of the house without air flow model was previously analysed in detail in [4].

3. RESULTS

The left side of Figure 5 shows the thermal characteristics for the three studied cases. It includes the room setpoint temperature from the controller and the resulting temperatures in the various rooms. The room 3 is the living room and kitchen, 5 is the downstairs toilet, 10 is the bathroom and 8 the master bedroom. Solar irradiation on two external vertical façades is also shown, along with the external air temperature. Thermal behaviour of the whole house in the three situations is very similar, especially in heated rooms. However, it can be observed that in case 1 at the beginning of the first day, Friday, the house is at higher temperature than in the other two cases. This is due to the fact that in this case there is not ventilation, so the house does not cool down by the infiltrated fresh air.

The ventilation characteristics are depicted in the right side of Figure 5 and show the resulting CO_2 levels in the rooms, along with the exhaust duct and ventilation set point. The CO_2 levels are a function of occupancy and changes in ventilation rates are due to temperature and pressure driven ventilation effects.

CO_2 profiles follow the occupancy schedule: at night the CO_2 concentration increases in the bedroom. The air quality in this room would be regarded as poor. On Saturday, when there are two guests in the living room, CO_2 concentration in that room increases noticeably. On Sunday, the occupants spend almost all day in the house, except for a couple of hours at lunchtime, CO_2 profile shows two peaks according to these occupancy periods. When occupants leave a room, an exponential decay of CO_2 is observed. The higher is the fan speed, the greater is the decay.

Case 2 shows higher CO_2 levels than case 3, mainly in those periods where in case 3 fan speed is set to 2. CO_2 concentrations in bedroom are similar in these two cases since in case 3 for night period (for noise reasons) fan speed has been set to the minimum speed, which is the same as in case 2. Anyhow it can be clearly seen the beneficial effect of having mechanical ventilation since the CO_2 concentration is much higher in case 1.

For comparison purposes a set of performance indicators have been defined. They measure the differences of the different cases in terms of energy consumption and air quality and are defined as follows:

Energy consumption performance indicator ($ECPI_{room\_number}$): Sum of the heat output from all the radiators of room during a simulation period. ECPI is in kW·h.

Air quality performance indicator ($AQPI_{room\_number}$): The duration that a room when occupied (one or more people present) has a CO_2 level above the threshold of 1200ppm (which is given as air quality limit in Dutch regulations). AQPI is in ppm·min, thus gives an indication of the duration and magnitude that the threshold CO_2 level has been exceeded.

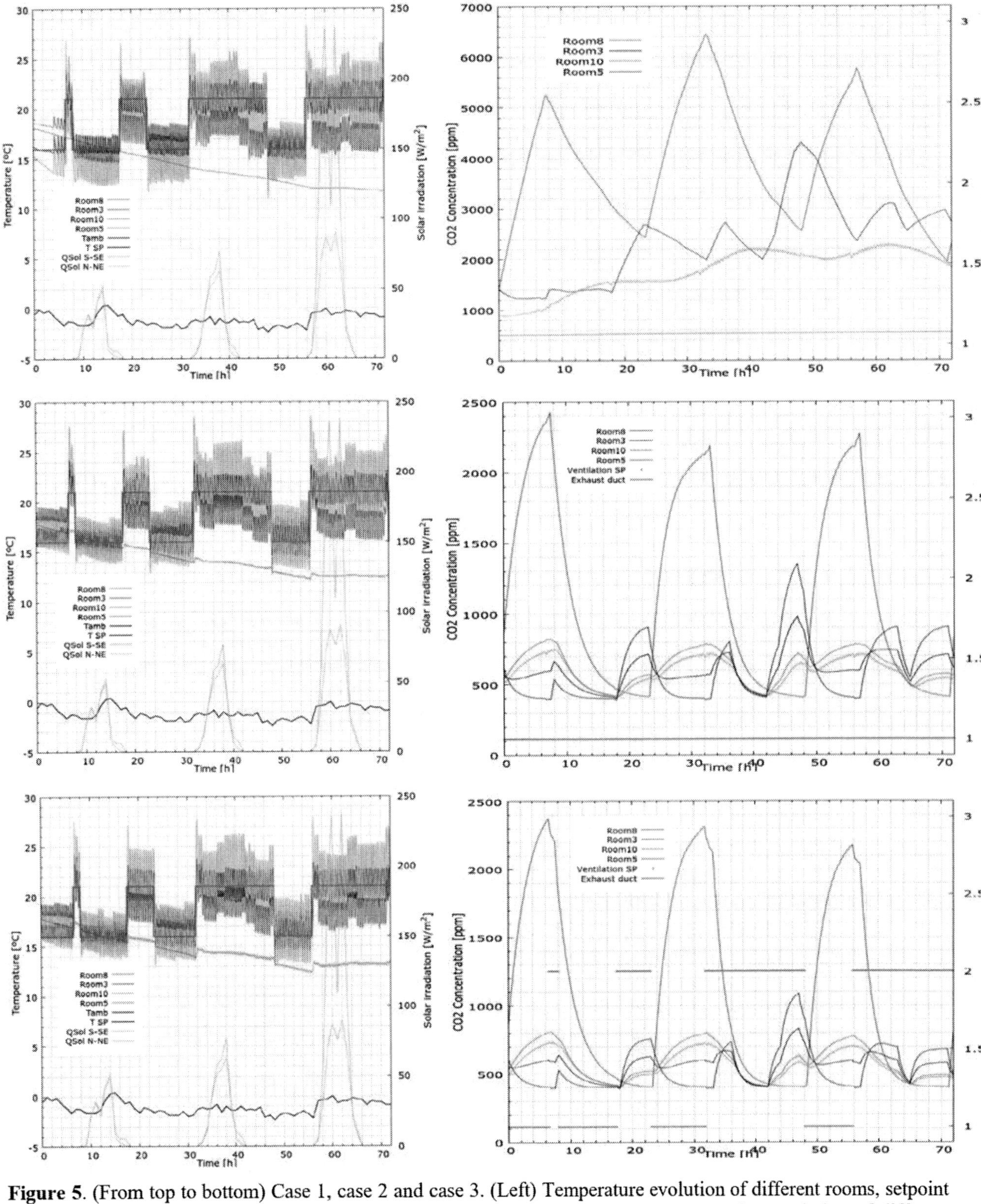

Figure 5. (From top to bottom) Case 1, case 2 and case 3. (Left) Temperature evolution of different rooms, setpoint and ambient and solar irradiation on south and north façades. (Right) CO_2 concentration evolution of different rooms and at exhaust duct, and ventilation setpoint.

Table 3 shows performance indicators at the three studied cases. In general it can be observed that on one hand the ECPI indicator is higher for those cases where ventilation is working, and on the other hand AQPI values are smaller. From a thermal and fluid-dynamic point of view it responds to a logical behaviour: when ventilation is working means that the air sucked by the ventilation system is replaced by air coming from the outside, that it is much colder that the inner (heated) air. For that reason the heating demand is greater, which is translated directly to a higher ECPI. This is why in case 3, where higher ventilation fan speed is used, the amount of cold air that enters the house is higher, thus, the heating demand is greater. However, since the house is more ventilated, the CO_2 concentration is lower, meaning that the quality of the air is better, so the AQPI is lower.

Table 3. Performance indicator of the three studied cases

	Case 1	**Case 2**	**Case 3**
ECPI3	62.7	86.1	93.3
ECPI2	19.3	25.2	27.0
ECPI7	16.7	22.0	23.7
ECPI8	17.5	23.0	24.9
ECPI9	21.9	28.5	30.6
ECPI10	19.7	26.0	28.2
TOTAL	157.9	210.9	227.7

AQPI3	$2.33\cdot10^6$	$1.02\cdot10^4$	$0.00\cdot10^0$
AQPI8	$4.77\cdot10^6$	$9.71\cdot10^5$	$1.03\cdot10^6$
TOTAL	$7.09\cdot10^6$	$9.82\cdot10^5$	$1.03\cdot10^6$

6. CONCLUSIONS

A virtual building model of a real dwelling located in the Netherlands has been built-up and tested at three different control situations. The effect of the ventilation control on the air quality and the energy performance of the house has been analysed. It has been observed that the higher ventilation fan speed is used, higher is the amount of cold air that enters the house, thus, the heating demand is higher. However, the quality of the air, measured as CO_2 concentration in the occupied rooms, is better.

So far, the full simulation of the overall building model is based on block-Jacobi and/or Gauss-Seidel algorithms. With the current implementation, the computational time for performing practical simulations may become an important impediment as the size of the building and events probability increases (e.g. block of flats simulation, a hospital, a mall center, etc.). Further research is necessary for attaining fast simulations. Better overall algorithms, the use of optimum time stepping, proper mesh sizes, convergence criteria, loop control strategies and the use of other non-linear solvers (i.e. Newton like methods), will be assessed in future works. The advances in this direction will help first to better understand the behavior of the already available algorithms and later to speed up the simulations. The second is important in the attainment of optimal designs of dwellings and buildings in general.

REFERENCES

American Council for an Energy-Efficient Economy - ACEEE (2015). Energy Efficiency in the US: 35 Years and Continuing.

Damle, R., Lehmkuhl, O., Colomer, G., (2011). Modular simulation of buildings with an object-oriented tool, *23rd Int. Congress of Refrigeration*, Prague, Czech Republic.

Dwelling Climate Control System (DCCS). http://www.kic-innoenergy.com/innovationproject/our-innovation-projects/dwelling-climate-control-system-dccs, 2013-2016. EIT-KIC InnoEnergy.

European Commission. Energy strategy for competitive, secure and sustainable energy (2010). Retrieved from http://eur-lex.europa.eu/legal-content/EN/TXT/PDF/?uri=CELEX:52010DC0639&from=EN (2016-04-28).

European Parliament. Energy Efficiency Directive (2012). Retrieved from http://eur-lex.europa.eu/legal-content/EN/TXT/PDF/?uri=CELEX:32012L0027&from=EN (2016-04-28).

López, J. (2016). *Parallel object-oriented algorithms for simulation of multiphysics. Application to thermal systems.* Universitat Politècnica de Catalunya. Thesis.

Marszal, A. J. et al. (2011). Zero energy building-a review of definitions and calculation methodologies. Energy and Buildings. 43,970–979.

Message-Passing Interface-MPI. http://www.mpi-forum.org (2016-04-28).

METEONROM 4.0 (1999). Meteotest Fabrikstrasse 14, CH-3012 Bern, Switzerland.

US Congress. Energy Independence and Security Act (2007). Retrieved from http://www.gpo.gov/fdsys/pkg/STATUTE-121/pdf/STATUTE-121-Pg1492.pdf (2016-04-28).

ACKNOWLEDGEMENT

This work has been performed within the "DCCS Dwelling Climate Control System" collaborative project, supported by the EIT-KIC InnoEnergy (project ref. 15-2013_IP65).

The air distribution around nozzles based on active chilled beam system

Bingjie Wu[1,2], Wenjian Cai[2,*], Qinguo Wang[3],Can Chen[2], Chen Lin[2], Haoran Chen[2]

[1]*Energy Research Institute @ NTU (ERI@N), Interdisciplinary Graduate School, Nanyang Technological University, Singapore*
[2]*School of Electrical and Electronic Engineering, Nanyang Technological University, Singapore*
[3]*Institute for Intelligent Systems,the University of Johannesburg,South Africa*
* *Corresponding author. W Cai, Email: EWJCAI@ntu.edu.sg*

Abstract

During the past two decades, the utilization of Active Chilled Beam (ACB) systems as promising air-conditioning systems has becoming increasingly prevalent in Europe, North America and Asia. However, the studies on air distribution of ACB systems are still inadequate. The air distribution has a great impact on thermal comfort. The un-uniformity of air distribution will easily lead to turbulent flow which can cause unpleased feeling such as draught in the occupied zone. ACB terminal unit is the source of the air flow entering into the occupied place, which plays a crucial role on the air distribution inside the room. Therefore, it's of great importance to evaluate the air distribution in the vicinity of ACB nozzles.

In order to fulfil the gap, air distribution for a two-way discharge ACB terminal unit is investigated in this study. The air velocities around the nozzles under different conditions are tested in a 7.3m*3.3m*2.5m thermal isolated room and simulated by a three dimensional Computational Fluid Dynamics (CFD). After being verified, the CFD model is utilized to examine the effects of nozzle diameter and inlet pressure. From the results of experiments and simulation, it is found out that the air flow is discharged in an asymmetric way from nozzles, which is ascribed to un-uniformity of pressure distribution inside ACB caused by the layout of the duct. Moreover, the un-uniformity is significant when the nozzle diameter is large and the elevation of the inlet pressure would aggravate this un-uniformity. Therefore, as we design the ACB systems, high attention on the nozzle diameter should be paid to prevent the un-uniformity air flow when the nozzles and the inlet pressure are large. Eventually, a proper strategy to solve this problem is also proposed and validated by CFD simulation.

Key words: Air distribution, Uniformity, Active chilled beam, CFD, Thermal comfort

1 Introduction

The chilled beam system has been proved to be a reliable Heating, Ventilation and Air Conditioning (HVAC) solution ever since it was introduced in 1980s. It can be classified into Active Chilled Beam (ACB) system and passive chilled beam system according to whether there is ventilation of fresh air. Owing to the advantages of providing good energy efficiency, thermal comfort and noise control with less space requirement, ACB system has been increasingly popular in not only Europe, but also North America and Asia [1]. It's wildly utilized in a variety of commercial buildings, schools, laboratories and hospitals[2].

ACB is the terminal part of ACB system where the conditioned air is discharged into the occupied place. It is normally mounted inside or below ceiling and consists of a primary air plenum, nozzles, a mixing chamber and a heat exchanger. The working principle is depicted in Figure 1. After being processed by air handling unit and dehumidifier, the conditioned air is sent to terminal unit of ACB system. Firstly, it is forced into dozens of small nozzles by high pressure from the primary air plenum and ends up with high speed primary air. An entrainment effect is generated by this high speed fluid. The indoor air is induced into the beam through a water-air cooling heat exchanger, namely secondary air. The primary air and secondary air mix together in the mixing chamber and serve as the supply air for the occupied zone as a whole.

In modern society, people spend most of their time in the artificial environment. Increasing diseases such as sick building syndrome make people suffer so much because of the poor thermal environment and air quality[3]. Studies show that many factors including air distribution, temperature, heat sources and human activities all have significant effects on the thermal comfort. Air distribution is one of the most important factors. Kosonen et al. [4] discussed the influence of thermal load strength and position on airflow pattern of chilled beams in a test room. The experiment result showed that the maximum velocity increased as the rising of heat load. However, the location of the thermal load did not significantly affect the maximum velocity. The un-uniform air distribution is easily to cause turbulence, which has high risk to induce draught sensation. Jan et al. [5] examined the velocities around the chilled beams based on a thermistor anemometer system. It was demonstrated that the airflow generated by a chilled beams exhibits strong fluctuations which may result in discomfort draught. The airflow pattern was extremely unstable and sensitive to the presence of heat sources. Cao et al. [6, 7] utilized an original method, particle image velocimetry (PIV), to trace the path of the airflow. The attached plane jet moved in a turbulent way regardless of the Reynolds number.

The previous studies for air distribution are mainly focused on the occupied places. However, little attention is paid to air distribution in the vicinity of ACB terminal unit, which is exactly the air source for the indoor room. The non-uniformity of the airflow has a huge potential to cause uncomfortable draught for occupants. In order to fill the gap, air distribution for a two-way discharge ACB terminal unit is investigated in this study.

Computational Fluid Dynamics (CFD) technology has been widely used in field of fluid analysis as a useful tool to predict the air movement since 1970s. Besides, it also plays an important part in visualizing the flows in 3D spaces [8-10].The airflow in the room is described mathematically by a set of couplet differential equations known as the Navier-Stokes equations and the procedure of solving these equations by a numerical method is called computational fluid dynamics [11]. With the rapid development of ACB, CFD is also applied as powerful tool by some scientists studying chilled beams. Hannu et al. [12, 13] characterized the air distribution in office environment that was ventilated by ACB under summer, spring, autumn and winter cases. The layout of ACB was in an asymmetric way and the corresponding CFD models were set up to simulate the flow pattern. The conclusions were drawn that the heat sources provide a notable impact on the air distribution. More importantly, a large circulation occurred because of asymmetric layout of chilled beams. Einberg et al. [14] predicted the velocity close to an air diffuser by means of CFD simulations and laboratory experiments. The conclusion was that the airflow near the diffuser was a complex combination of jet flow and flow from a displacement diffuser. As the flow pattern near the nozzle duct were too complicated to be modelled in practical CFD simulations, a simplified CFD model was developed for nozzle duct diffuser by Hannu. However, he found that the simplified model could not be extensively applicable[15]. Therefore, a CFD model with original scale was set up in this paper to investigate the air distribution of ACB.

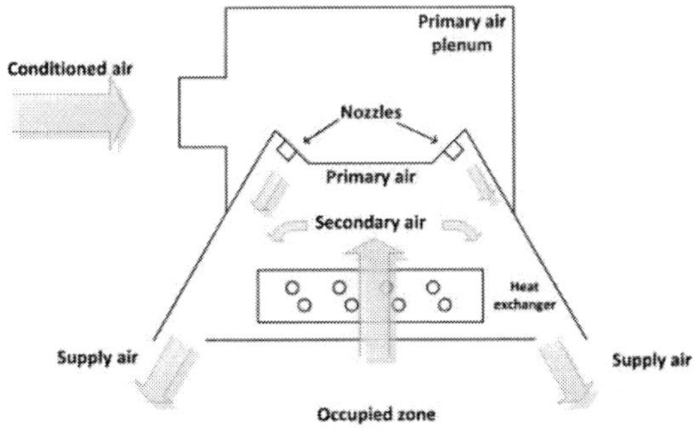

Figure 1 Schematic diagram of ACB

2 CFD modeling of ACB

2.1 CFD modeling of ACB

2.1.1 Computational grid

A three dimensional ACB model was set up by Computational Fluid Dynamics (CFD) of Ansys edition 12.0. As ACB is symmetry from left to right side, only half of ACB was built to simplify the calculation. The computational grid is formed by tetrahedron cells (Figure 2) because of the complex structrue of ACB. The minimum element quality is 0.213 and the elements number is 2595478.

Figure 2: CFD model mesh diagram of ACB

2.1.2 Boundary conditions

The Reynolds number for inlet is estimated from the equation:

$$Re = vL/\nu = \rho vL/\mu$$

Where v is the velocity of fluid, L is the characteristic length, ν is the kinematic viscosity, μ is the dynamic viscosity of the fluid, ρ is the density of the fluid.

For fully developed pipe flow the turbulence intensity at the core, the turbulence intensity can be estimated from the equation:

$$I = 0.16 * (Re)^{-\frac{1}{8}}$$

where Re is the Reynolds number based on the pipe hydraulic diameter.

As the Reynolds number is around 1.93×10^4, the air flow is considered to be constant and incompressible. The turbulence intensity is estimated as 5%.The inlet pressure is set to be 30Pa. The outlet pressure and secondary air inlet pressure is 0 Pa. Other prameters are all default values.

2.1.3 Solver settings

The standard k-ε model, which is widely used in fluid dynamics, is adoped as the turbulence model. The SIMPLEC algorithm is used for pressure-velocity coupling. Second order upwind disccretization scheme is applied in the solving momenterm, turbulence energy and turbulent dissipation rate terms.

The basic governing equations for incompressible fluid are:

(1) Mass conservation equation(Continuity equation)

$$\frac{\partial u}{\partial x}+\frac{\partial v}{\partial y}+\frac{\partial w}{\partial z}=0 \quad (1)$$

where, u, v and w are the velocities of x, y and z direction.

(2) Momentum conservation equation

$$\begin{cases} \frac{\partial u}{\partial t}+\text{div}(u\,\text{U})=\text{div}(\upsilon\,\text{grad}u)-\frac{1}{\rho}\frac{\partial p}{\partial x} \\ \frac{\partial v}{\partial t}+\text{div}(v\,\text{U})=\text{div}(\upsilon\,\text{graduv})-\frac{1}{\rho}\frac{\partial p}{\partial y} \\ \frac{\partial w}{\partial t}+\text{div}(w\,\text{U})=\text{div}(\upsilon\,\text{gradw})-\frac{1}{\rho}\frac{\partial p}{\partial z} \end{cases} \quad (2)$$

where p is the pressure of the air; υ denotes the kinematic viscosity; ρ is the density.

(3)The k-ε model equations:

$$\rho\frac{\text{dk}}{dt}=\frac{\partial}{\partial x_i}\left[\left(\mu+\frac{\mu_t}{\sigma_k}\right)\frac{\partial k}{\partial x_i}\right]+G_k+G_b-\rho\varepsilon-Y_M \quad (3)$$

$$\rho\frac{\text{d}\varepsilon}{dt}=\frac{\partial}{\partial x_i}\left[\left(\mu+\frac{\mu_t}{\sigma_\varepsilon}\right)\frac{\partial \varepsilon}{\partial x_i}\right]+G_{1\varepsilon}\frac{\varepsilon}{\text{k}}(\text{G}_k+G_{3\varepsilon}G_b)-G_{2\varepsilon}\rho\frac{\varepsilon^2}{\text{k}} \quad (4)$$

where k is turbulence energy; ε is turbulent dissipation rate; G_k is turbulent kinetic energy caused by mean velocity gradient; G_b is the turbulent kinetic energy caused by buoyancy; Y_M denotes effects of compressible turbulent flow pulsation to the total turbulent dissipation.; μ is kinematic viscosity; others are considered constants in CFD.

2.2 CFD model calibration by experiments

This CFD model is verified by experiments in a 7.3m*3.3m*2.5m mock up room (Figure 3). An ACB (Figure 4) with a dimension of 1200mm*400mm*158mm is equipped latitudinal in the middle of the room. 30 nozzles with diameter of 21.5mm are distributed on each side of this ACB. The velocities around the nozzles are tested by a hot wire air velocity transducer. The full scale is 10m/s with accuracy of ±0.5%.

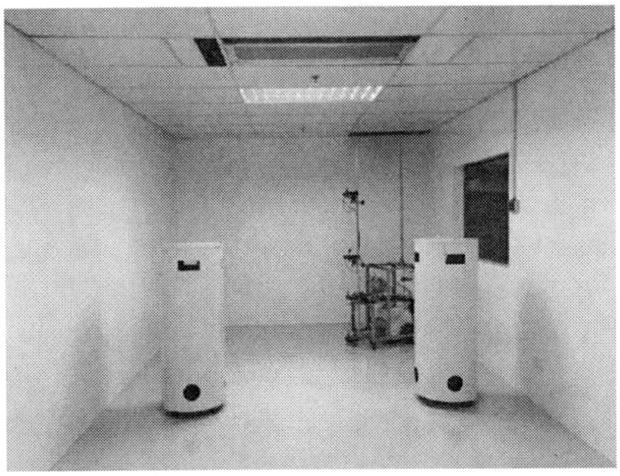

Figure 3: Thermal room

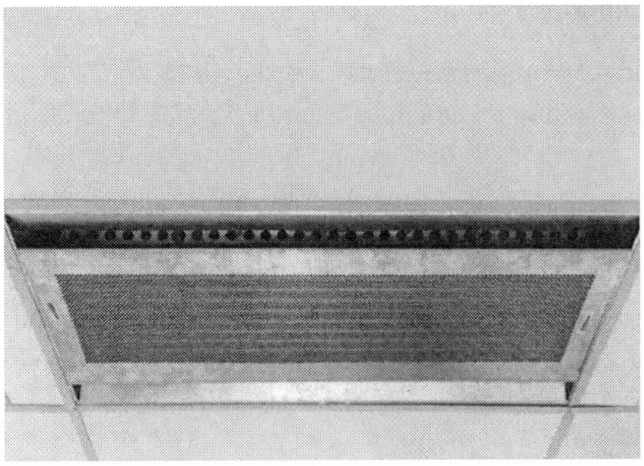

Figure 4: ACB terminal unit

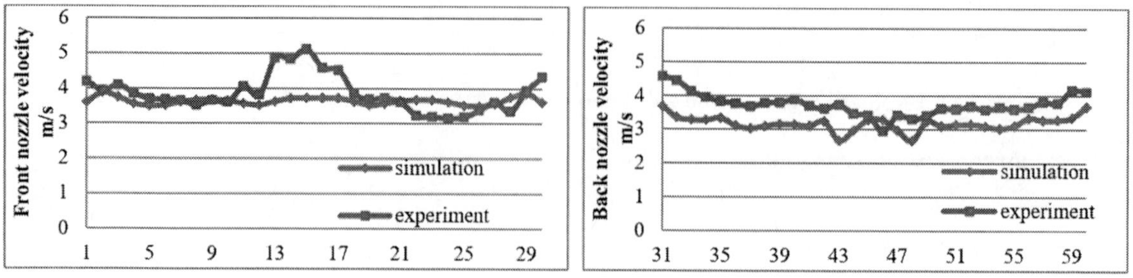

Figure 5: Nozzle velocity comparison between simulation and experiment

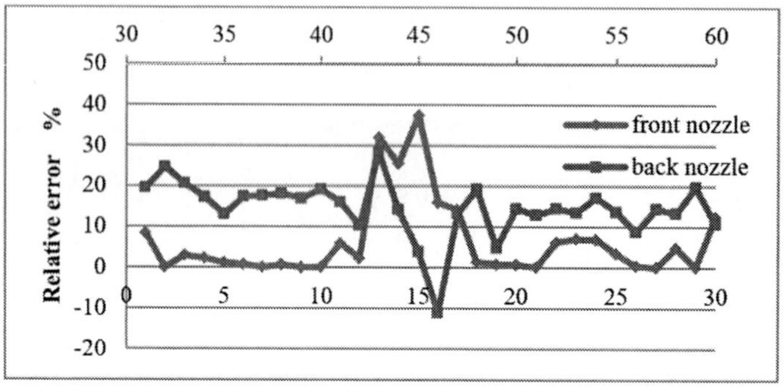

Figure 6: Relative error between experiment and simulation

Assume the inlet side is the back side of ACB, where nozzles are sequenced from 31 to 60. On the contrary, points 1-30 are nozzles located in the opposite side of inlet. The nozzle velocity comparison between simulation and experiment is show in Figure 5. For front nozzles, the velocities match quite well with the simulation velocities except for a few middle points. The overall trend for experiment is similar for back nozzles compared with the simulation results. The velocity magnitude difference between experiment and simulations may be caused by simulation and testing errors. The relative error between experiment and simulation are also calculated (Figure 6). Most relative errors are within 20%, which can be acceptable for CFD simulation.

3 Results and analysis

3.1Simulation results

The simulation air streamlines are shown in Figure 7, from which we can see the air velocity distribution is quite fluctuant in the primary air plenum. It can be drawn from the line charts (Figure 8) that, a huge velocity difference exists between back nozzles and front nozzles. For back nozzles (nozzles 31-60), the velocities reach the bottom at the middle area which is close to inlet area. However, the velocities on the two edges are slightly large. In terms of front nozzles (nozzles 1-30), the velocities have higher magnitudes at the middle and sides area. The maximum velocity difference is 0.957m/s. As a result of the velocity un-uniformity between front and back nozzles, the air flow rates discharging from two outlets are also different, which may have an effect on the ventilation in the room.

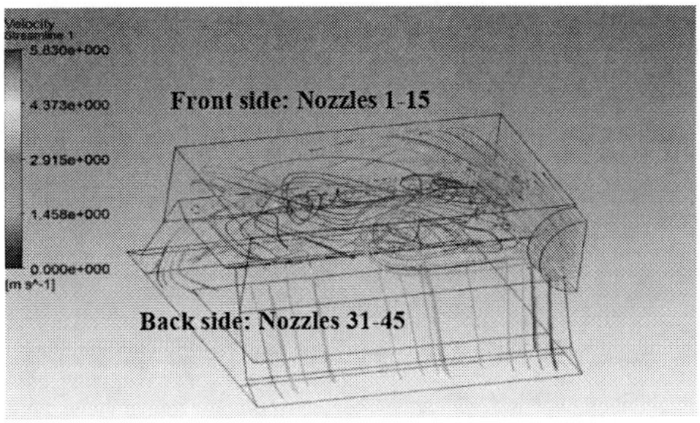

Figure 7: ACB simulation streamlines

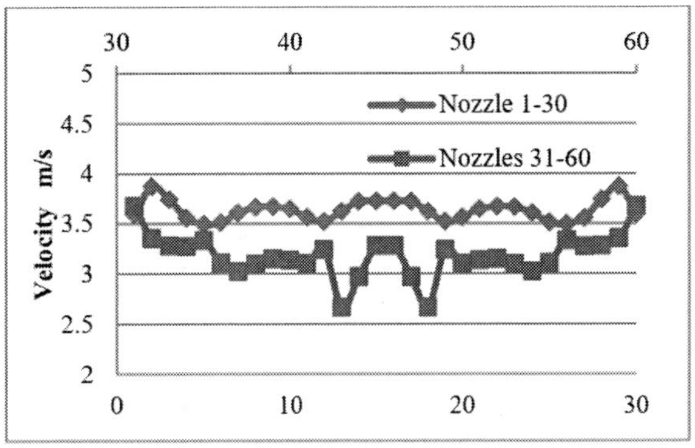

Figure 8: Simulation velocities for nozzle diameter 21.5mm under 30Pa

In order to investigate the effects of nozzle diameter and inlet pressure on this velocity un-uniformity between front and back nozzles, the CFD model is applied to simulate the velocities for nozzle diameter21.5mm, 15mm and 9mm under inlet pressure 30 Pa, 60 Pa and 90 Pa, respectively. The simulation results are depicted in Figure 9, 10 and 11.

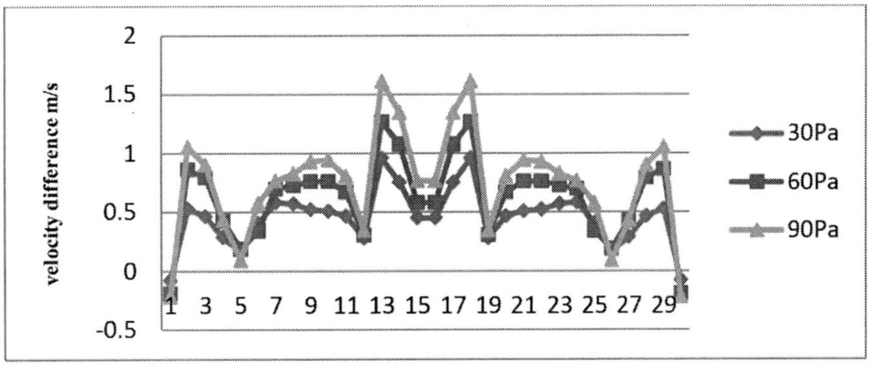

Figure 9: Velocities difference for nozzle diameter 21.5mm under 30Pa, 60Spa, 90Pa

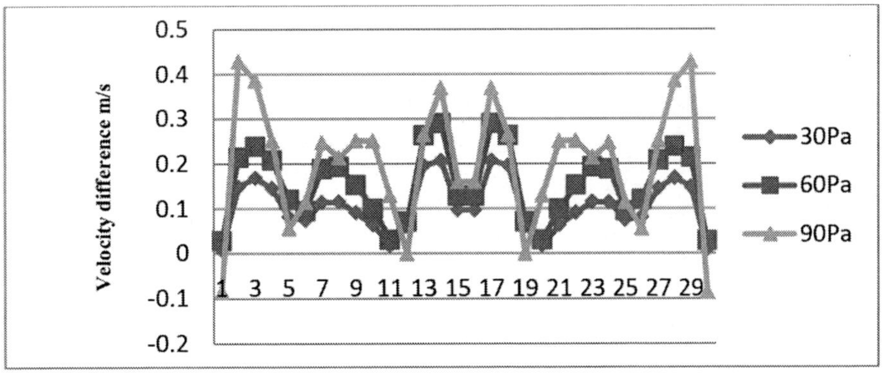

Figure 10: Velocities difference for nozzle diameter 15mm under 30Pa, 60Spa, 90Pa

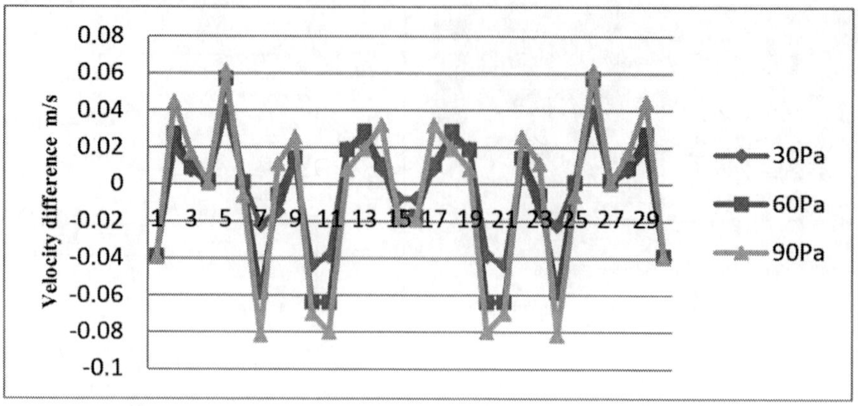

Figure 11: Velocities difference for nozzle diameter 9mm under 30Pa, 60Spa, 90Pa

As is shown in the Figure 9, 10 and 11, with a decrease of nozzle diameter, the velocity difference reduces. When the nozzle is small enough, the velocity differences are small enough to be neglected. Moreover, for the same nozzle diameter, the higher the inlet pressure, the bigger of the gap between two lines nozzles. The maximum velocity difference magnitude is depicted in Table 1.

Table 1: maximum velocity difference magnitude

Inlet Pressure / Diameter	30 Pa	60 Pa	90 Pa
9mm	0.044m/s	0.064 m/s	0.082 m/s
15mm	0.207 m/s	0.290 m/s	0.428m/s
21.5mm	0.957 m/s	1.265m/s	1.609 m/s

3.2A improvement for un-uniformity of nozzle velocity

In order to solve this unpleasant un-uniformity, several solutions have been proposed. Only few have been proved to be feasible through validation of CFD. One of the strategies is to add a board (200*70*2mm) in the middle of the primary air plenum as is shown in Figure 12.

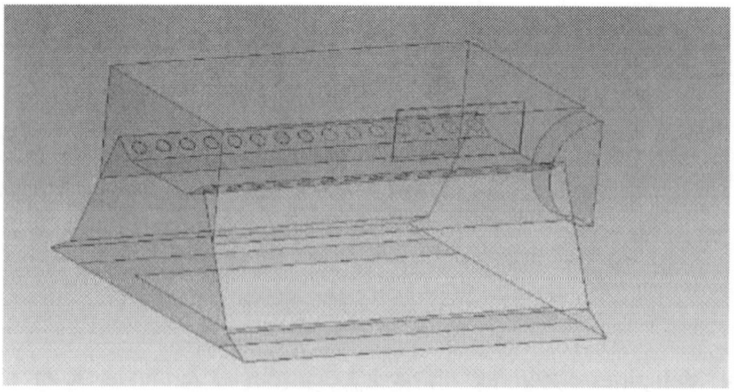

Figure 1: Modified ACB

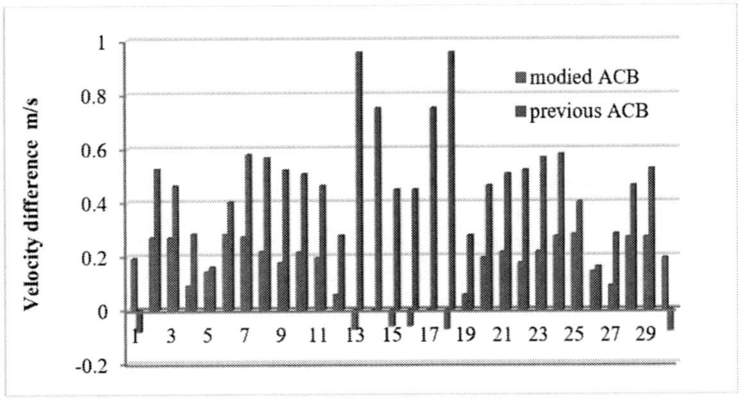

Figure 2: Velocity differences between front and back nozzles

As is depicted in the **Error! Reference source not found.**, the velocities differences between front side and back nozzles are smaller after being modified. The maximum velocities between front and back nozzles are 0.378m/s, which is obviously smaller than maximum difference 0.957 m/s of the original ACB. It can be concluded that adding a board in the middle will significantly improve the uniformity of velocities.

4 Conclusions

From the results of simulation and experiments of air distribution for ACB nozzles, the conclusion can be drawn that the un-uniformity is significant when the nozzle diameter is large. Moreover, the rise of the inlet pressure will aggravate this un-uniformity. The smallest velocities are located close to inlet of back side, while the largest velocities are distributed in the nozzles of front side and the two verges. The pressure around the opposite side of inlet and edge area is comparatively larger which may leads to the high speed in those areas. However, this troublesome un-uniformity will be eliminated when the nozzle is small enough. When nozzles are small, the air flow velocity is primarily restricted by the pore size. However, for the nozzles owning a large pore size, the air velocity will mainly depend on the pressure for the nozzle. Therefore, as we design the ACB systems, high attention on the nozzle diameter should be paid to prevent the un-uniformity air flow when the nozzles and the inlet pressure are large.

In addition, a solution of adding a board in the middle is proposed and validated by CFD simulation. With the help of this board, the air will be discharged evenly from the nozzles. Through a complex process, the air will be distributed to all of the nozzles in a more balanced way. The corresponding experiments will be conducted later. The study of air distribution inside ACB is of great significance to reduce the un-uniformity of air flow form the source and provide occupants a pleasant indoor environment with little chance of draught problem.

Reference

[1] C. Chen, W. Cai, Y. Wang, and C. Lin, "Performance comparison of heat exchangers with different circuitry arrangements for active chilled beam applications," *Energy and Buildings,* vol. 79, pp. 164-172, 8// 2014.

[2] ASHRAE Standard, "113 (2005)"Method of testing for room air diffusion" American Society of Heating, Refrigerating and Air-Conditioning Engineers," ed: Inc.

[3] G. Cao, H. Awbi, R. Yao, Y. Fan, K. Sirén, R. Kosonen, *et al.*, "A review of the performance of different ventilation and airflow distribution systems in buildings," *Building and Environment,* vol. 73, pp. 171-186, 3// 2014.

[4] R. Kosonen, A. Melikov, L. Bozkhov, and B. Yordanova, "Impact of heat load distribution and strength on airflow pattern in rooms with exposed chilled beams," in *Proceedings of Roomvent*, 2007.

[5] J. Fredriksson, M. Sandberg, and B. Moshfegh, "Experimental investigation of the velocity field and airflow pattern generated by cooling ceiling beams," *Building and Environment,* vol. 36, pp. 891-899, 8// 2001.

[6] G. Cao, "Modelling the attached plane jet in a room," 2009.

[7] G. Cao, M. Sivukari, J. Kurnitski, M. Ruponen, and O. Seppänen, "Particle Image Velocimetry (PIV) application in the measurement of indoor air distribution by an active chilled beam," *Building and Environment,* vol. 45, pp. 1932-1940, 9// 2010.

[8] J. J. Martinez-Almansa, A. Fernandez-Gutierrez, L. Parras, and C. del Pino, "Numerical and experimental study of a HVAC wall diffuser," *Building and Environment,* vol. 80, pp. 1-10, 10// 2014.

[9] J. D. Posner, C. R. Buchanan, and D. Dunn-Rankin, "Measurement and prediction of indoor air flow in a model room," *Energy and Buildings,* vol. 35, pp. 515-526, 6// 2003.

[10] G. Zhang, S. Morsing, B. Bjerg, K. Svidt, and J. S. Strøm, "Test Room for Validation of Airflow Patterns estimated by Computational Fluid Dynamics," *Journal of Agricultural Engineering Research,* vol. 76, pp. 141-148, 6// 2000.

[11] P. V. Nielsen, "Fifty years of CFD for room air distribution," *Building and Environment,* vol. 91, pp. 78-90, 9// 2015.

[12] H. Koskela, H. Häggblom, R. Kosonen, and M. Ruponen, "Air distribution in office environment with asymmetric workstation layout using chilled beams," *Building and Environment,* vol. 45, pp. 1923-1931, 9// 2010.

[13] H. Koskela, H. Häggblom, R. Kosonen, and M. Ruponen, "Flow pattern and thermal comfort in office environment with active chilled beams," *HVAC&R Research,* vol. 18, pp. 723-736, 2012.

[14] H. Awbi, "Application of computational fluid dynamics in room ventilation," *Building and environment,* vol. 24, pp. 73-84, 1989.

[15] M. Cehlin and B. Moshfegh, "Numerical and experimental investigations of air flow and temperature patterns of a low velocity diffuser," in *Proceedings of the ninth international conference on indoor air quality and climate*, 2002, pp. 765-70.

Performance Evaluation of a Passive Chilled Beam System and Comparison with a Conventional Air System

Janghyun KIM[1*], James E. BRAUN[1,2], Athanasios TZEMPELIKOS[1,2], W. Travis HORTON[1,2]

[1]Ray W. Herrick Laboratories, School of Mechanical Engineering, Purdue University, 140 S. Martin Jischke Dr., West Lafayette, IN 47907, USA
kim1302@purdue.edu, jbraun@purdue.edu, ttzempel@purdue.edu, wthorton@purdue.edu

[2]School of Civil Engineering, Purdue University, 550 Stadium Mall Dr., West Lafayette, IN 47907, USA

* Corresponding Author

ABSTRACT

This study compares the performance of a system employing passive chilled beams with a conventional air system in a typical open plan office setting. Both energy efficiency and thermal comfort are taken into account as performance indicators. A typical single duct variable air volume system, widely used in large office buildings, was considered as a conventional system; a combination of the air system and passive chilled beams was considered as a typical passive chilled beam system. Measurements from full-scale experiments conducted in open plan offices (Living Labs of the Center for High Performance Buildings at Purdue) are used to develop a data-driven passive chilled beam model that can predict the total cooling capacity as well as surface temperatures under different operating conditions. In simulating the performance of the two systems, the same operative temperature was used in order to have a fair comparison. Additionally, the effect of radiation cooling of the passive chilled beam was investigated. The overall results showed about maximum 10-21% total electrical energy savings (5-23% reduction in chiller and 55% reduction in supply fan electrical energy) and thermal comfort improvement of 0.3-0.4 on a PMV-scale (12-15% on a PPD scale) associated with the passive chilled beam system, depending on the system configuration. The radiation cooling of the passive chilled beam is not significant, since the effective surface area for radiant exchange with the room is much less than the surface area for convective heat transfer and temperature differences are relatively low.

1. INTRODUCTION

Reducing energy consumption in buildings is an important part of reducing global energy usage and environmental impact. It has been reported by the International Energy Agency (*Energy technology perspectives 2010*, 2012) that the global energy demand in buildings will approximately double by 2050 due to the "rising number of residential and commercial buildings ....". More than two-thirds of existing buildings are expected to remain until 2050. Therefore, significantly reducing energy use in the building sector requires retrofitting existing buildings with advanced technologies. Furthermore, modeling of these technologies is a critical part of assessing their economics towards realizing success in the marketplace.

Passive radiant cooling systems can lead to reduced cooling requirements, improved comfort, less fan energy and downsized ductwork, compared to conventional all-air systems. Passive chilled beams, that typically use square vertical fins along a water pipe and a horizontal perforated panel at the bottom to increase radiation effects, are a promising technology. However, passive chilled beam modeling is challenging because (i) they have a complex geometry and (ii) the primary heat transfer mechanisms of radiation and natural convection are strongly coupled to the space characteristics and thermal conditions. The only available model, currently implemented in EnergyPlus (Livchak & Lowell, 2012), is not only limited to calculation of the convective heat extraction rate (where it neglects radiation heat transfer at the bottom surface of the chilled beam and the air circulation driven by the natural convection), but it is also an iterative model that needs to compute the mean water temperature at each time step by iteration. Experimental and simulation studies of passive chilled beams are both useful in evaluating their overall performance. One of the benefits of passive chilled beams is that since they only provide sensible cooling they operate at higher surface temperatures (above the dew point of the air) compared to conventional cooling systems leading to improved energy efficiency for chillers. It is also possible to couple the radiant cooling system with energy storage within the building thermal mass, to allow load shifting and reductions in peak load.

Energy modeling tools are important for designing building systems and for assessing new technologies such as passive ceiling cooling systems (Kim, Tzempelikos, & Braun, 2015). To evaluate both thermal comfort and energy saving aspects of passive chilled beams, models should consider both the chilled beam and its surroundings since the performance of passive chilled beams are strongly coupled to the specific characteristics and conditions within the spaces. There have not been enough studies that focus on the potential of energy savings and thermal comfort improvement using passive chilled beam systems. Some studies have focused on radiant ceiling panel systems rather than passive chilled beams (J. D. Feng, Bauman, & Schiavon, 2013; J. (Dove) Feng, Schiavon, & Bauman, 2013; Imanari, Omori, & Bogaki, 1999; Mustakallio, Bolashikov, Kostov, Melikov, & Kosonen, 2016; Niu, Zhang, & Zuo, 2002; Sastry & Rumsey, 2014; Stetiu, 1999); in the few studies on passive chilled beams, the focus has been on the performance of the passive chilled beam itself (Fredriksson & Sandberg, 2009; Kosonen, Saarinen, Koskela, & Hole, 2010; Nelson, Culp, Rimmer, & Tully, 2016) and not the overall system performance or comparison with other system types. In this context, this paper presents a comparison of the performance of a passive chilled beam system versus a conventional air system, based on system simulation for a thermal zone. Measurements from full-scale experiments conducted in open plan offices (Living Labs of the Center for High Performance Buildings at Purdue) are used to develop a data-driven passive chilled beam model that can predict the total cooling capacity as well as the chilled surface temperature of a passive chilled beam under different operating conditions. The variable air volume system model represents a conventional air system. The performance of the passive chilled beam and air system is evaluated in terms of energy consumption in each part of the HVAC system and in terms of thermal comfort, using the same operative temperature as a basis. The radiation cooling effect of passive chilled beams is also taken into account to verify whether it is significant in terms of performance evaluation.

2. BUILDING ENERGY MODEL DESCRIPTION

The details and parameters of the simulation model are described in the following sections. The model was implemented in TRNSYS software (A Transient System Simulation Program, Version 17, 2015).

2.1 Building Envelope Parameters and Climatic Data

The building envelope model is developed based on construction drawings of one of the Living Lab offices in the Center for High Performance Buildings at Purdue University (West Lafayette, Indiana) as shown in Figure 1. The construction specifications of this envelope model are listed in Table 1. Typical Meteorological Year weather for West Lafayette were used as inputs in the energy model. As shown in Figure 1, the thermal zone model is divided into three separate zones, including a double façade feature, acting as a buffer zone between external and internal windows, and a plenum space, where the return air is gathered and flows back to either the HVAC system or to the passive chilled beams. The detailed radiation model is selected in TRNSYS to properly reflect features of this thermal zone such as the double façade, the estimation of human comfort in the space and the radiation heat transfer between indoor surfaces and passive chilled beam surfaces. The surface between the plenum and the office area is partially covered with nine cloud surfaces as shown in the office photo in Figure 1. Passive chilled beams are installed on these clouds, thus, when the passive chilled beam is in operation, the relatively warm air in the plenum is pulled up between the passive chilled beams by natural convection, and the cooled air is pushed towards the office area. To approximate the temperature difference between the office area and the plenum due to thermal stratification, several assumptions were made as shown in Figure 2. First, the lighting gain in the space is applied in the plenum rather than applying it in the office area. The only thermal coupling considered between the office zone and plenum is due to forced convection. The total supply air flow rate from the diffusers is assumed to flow to the plenum but at the temperature and humidity of the office area. Heat transfer from the office zone to the cloud bottom surfaces was forced to be zero by setting the underside convection coefficient and solar absorption to zero. Solar heat gain to the top surface from the south facing windows is considered along with convective heat transfer between the plenum and clouds. However, the temperature within the clouds is considered to be uniform which was implemented by utilizing a very high thermal conductivity. All the other necessary assumptions are shown in Figure 2.

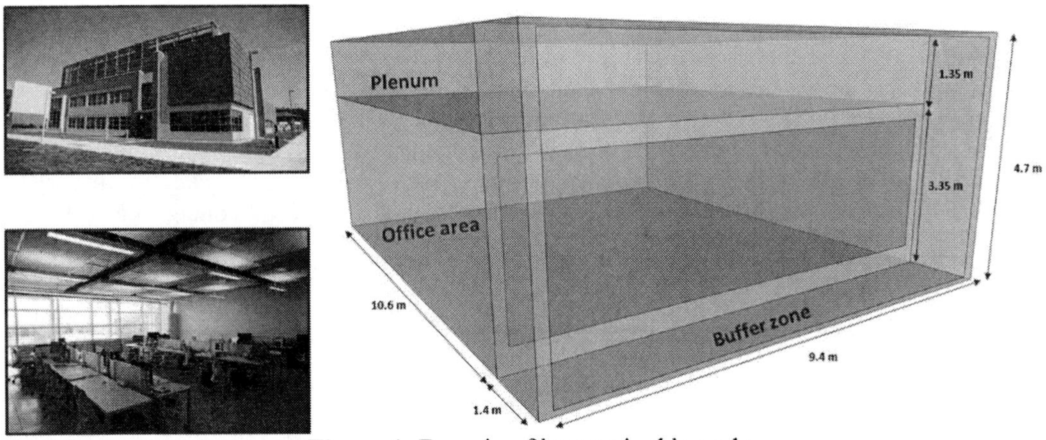

Figure 1: Domain of interest in this study

Table 1: Specification of envelope components

Type	Layer Description	Thickness	U-Value
		m	W/m^{2}K
Roof	Insulation	0.1	0.221
	Concrete	0.038	
Raised Floor	Concrete	0.102	0.086
	Air	0.3	
External Wall	Gypsum Board	0.013	0.358
	Insulation	0.105	
	Stucco	0.025	
Internal Wall	Gypsum Board	0.016	0.136
	Air	0.184	
	Gypsum Board	0.016	
External Window	Double Glazing, ID: 7041	0.0178	3.2
Internal Window	Double Glazing, ID: 7002	0.0188	1.58
Cloud	Virtual surface	0	5.76
Plenum south	Steel panel and Insulation	0.08	0.598

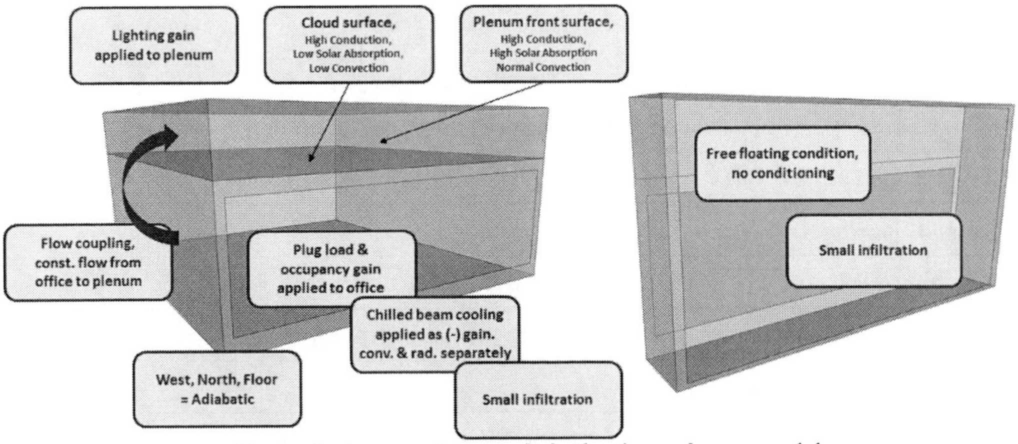

Figure 2: Assumptions made in the thermal zone model

2.2 Internal Gains and Thermal Comfort Parameters

Internal gains schedules for occupancy, lighting and plug loads were adopted from ASHRAE Standard 90.1 (ASHRAE, American Society of Heating, Refrigerating and Air-conditioning Engineers, 2010, p. 1) as shown in Figure 3. These schedules represent typical load profiles of commercial buildings. Parameters related to thermal comfort calculations are listed in Table 2. The operative temperature calculation is adopted from the ASHRAE Standard 55 (ASHRAE, American Society of Heating, Refrigerating and Air-conditioning Engineers, 2013, p. 55) where there is a different weighting factor between dry-bulb temperature and mean radiant temperature based on the relative air speed. Mean radiant temperature is calculated as shown below.

$$T_{MRT}^4 = T_1^4 F_{p-1} + T_2^4 F_{p-2} + \cdots + T_N^4 F_{p-N}$$

T_1 to T_N represent interior surface temperatures in the office and F represents view factors between occupants and the interior surfaces. The view factor in the equation is adopted from the literature (*Thermal Radiation Heat Transfer, 5th Edition*, 2011). Parameters that are necessary for the calculation of the Predicted Mean Vote (PMV) and Predicted Percentage Dissatisfied (PPD) are also listed in Table 2. All the parameters for PMV and PPD calculation are assumed to be the same except the relative air speed between the conventional and the passive chilled beam system. According to ASHRAE Standard 55, a conventional air system's maximum relative air speed can be limited to 0.2 m/s when the operative temperature is maintained below 23°C. A passive chilled beam's relative air speed can vary based on the cooling power of each passive chilled beam installed in the space. Based on an experimental study that focused on the air velocity field in terms of passive chilled beam's cooling power, the maximum relative air velocity was limited to 0.1 m/s in this study (Fredriksson, Sandberg, & Moshfegh, 2001). These two maximum relative air speeds were applied during the simulation to estimate the PMV and PPD differences between the two systems.

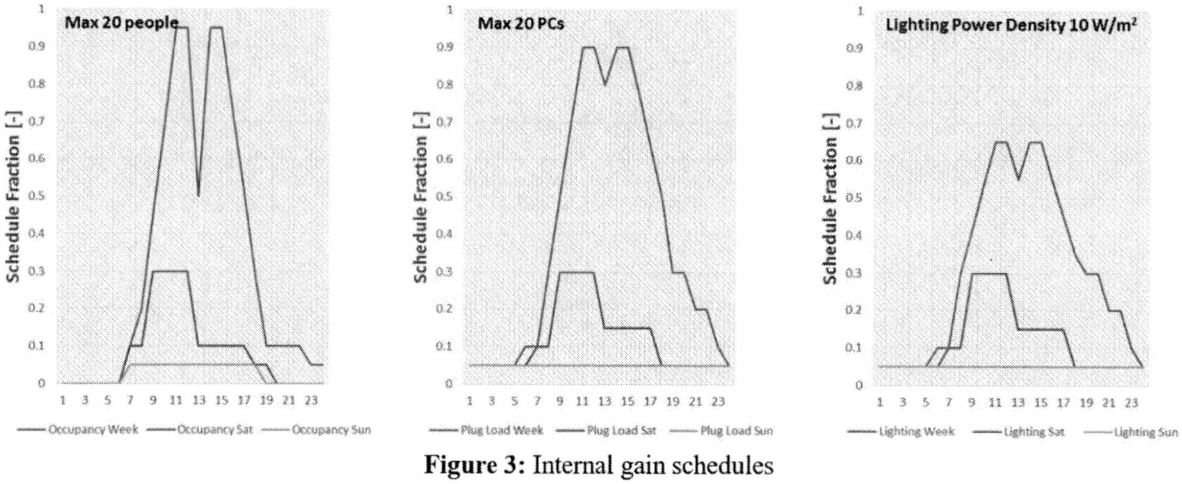

Figure 3: Internal gain schedules

Table 2: Thermal comfort parameters

<table>
<tr><th colspan="3">Specification for Thermal Condition Calculation</th></tr>
<tr><td>Operative Temperature, PMV, PPD</td><td colspan="2">ASHRAE Standard 55</td></tr>
<tr><td>Occupant location</td><td colspan="2">Center of the office</td></tr>
<tr><td>Clothing level</td><td colspan="2">0.57 clo : Trousers, short-sleeve shirt</td></tr>
<tr><td>Metabolic rate</td><td colspan="2">1.1 met : Office activities (typing)</td></tr>
<tr><td>External work</td><td colspan="2">0 met</td></tr>
<tr><td>Relative air speed</td><td>PCB System (<0.1 m/s)</td><td>Air System (<0.2 m/s)</td></tr>
</table>

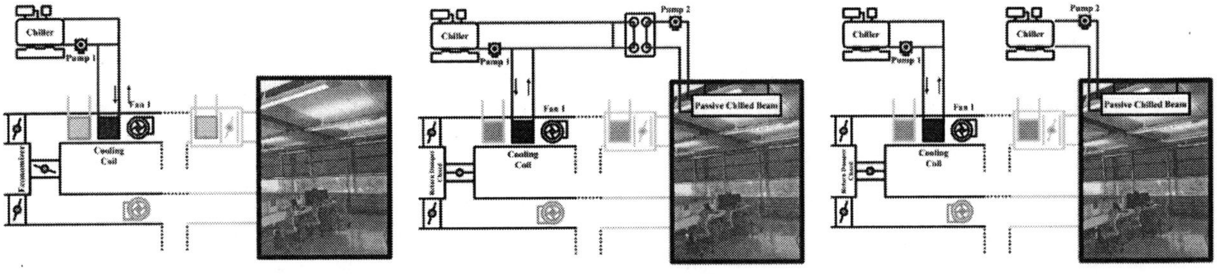

(a) Conventional air system (b) Passive chilled beam system (shared chiller) (c) Passive chilled beam system (separate chillers)

Figure 4: HVAC system configurations

2.3 Configuration of HVAC Systems

Figure 4 shows configurations of the conventional air system and two different passive chilled beam systems. There are ways to improve the energy efficiency of the passive chilled beam system by configuring the system with different methods, such as adding a dedicated chiller for passive chilled beams with relatively higher operating temperature of chilled water, and implementing a dedicated outdoor air system with desiccant and enthalpy wheels for efficient humidity control. In this study, the two different passive chilled beam system configurations are considered depending on whether a separate chiller is used or not used for passive chilled beams. The differences between these two cases will show the effect of using a chiller with higher chilled water operating temperatures.

2.3.1 Conventional Variable Air Volume System

A typical single duct variable air volume (VAV) system, which is widely used in large office buildings (Thornton, B.A. et al., 2011), is considered as the conventional system. As shown in Figure 4 (a), the system includes a cooling coil, supply fan and an economizer. The return fan, preheating coil, reheating coil and VAV box were not included in this study to simplify the approach for the consideration of a summer cooling season. The chilled water source for the cooling coil is provided from an air-cooled electric chiller and a constant speed pump (pump 1) was used to provide chilled water from the chiller to the cooling coil. A bypass cooling coil is used in this model to modulate the amount of chilled water to the coil to maintain a supply air temperature setpoint.

The air-cooled chiller, fan, and pump models are incorporated in the building energy model. The air-cooled chiller model is adopted from the TRNSYS library (Type 655). A normalized performance map was used to calculate the power of the chiller from different conditions of rated capacity, rated COP, water mass flow rate, chilled water leaving temperature and ambient air temperature. Performance characteristics of the fan and pump were estimated from actual measurements. Supply flow rate and power consumption of the fan and the pump were directly coupled by using a 2nd order polynomial. Performance characteristics of these components are shown in Figure 5. The same fan and pump models are used in the passive chilled beam system as well.

A minimum outdoor air requirement suggested from the ASHRAE Standard 90.1 (ASHRAE, American Society of Heating, Refrigerating and Air-conditioning Engineers, 2010, p. 1) is applied in both systems. For the conventional air system, the minimum outdoor air requirement is 173 CFM and an economizer control opens up the outdoor air damper when the outdoor conditions allow free cooling.

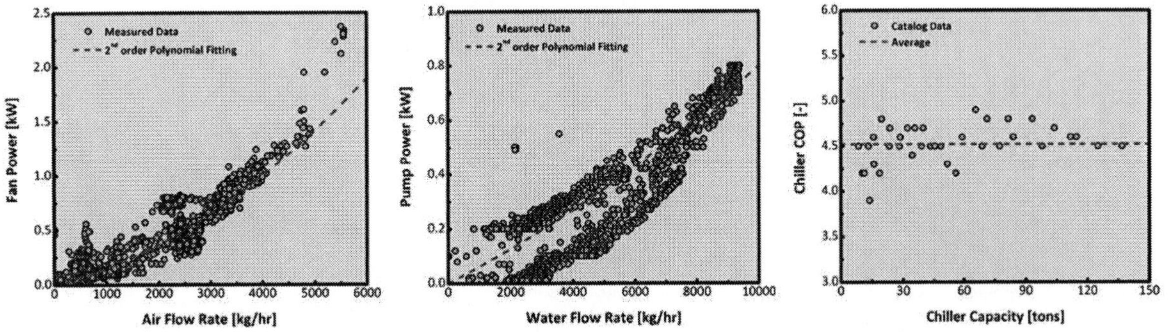

Figure 5: Performance fitting of fan, pump and chiller

2.3.2 Passive Chilled Beam System

The configurations of passive chilled beam systems considered in this study are shown in Figure 4 (b) and (c). The parallel air system in these configurations always provides the minimum outdoor air requirement and primarily handles ventilation load (sensible and latent) requirements along with small latent load requirements associated with the space. The primary space sensible load requirements are met by the passive chilled beams. In Figure 4 (b), a separate water loop for passive chilled beams is coupled to the air system's cooling coil water loop with a bypass heat exchanger. The use of the heat exchanger allows variation of the pump speed (Pump 2) at the passive chilled beam's water loop to meet the sensible load requirement while the bypass valve can modulate the valve position to maintain water supply temperature to passive chilled beams. In Figure 4 (c), a separate chiller is connected to passive chilled beams. Chilled water is provided to each passive chilled beam unit individually, providing the same inlet temperature to each unit, as it is usually applied in practical design. The economizer in the passive chilled beam system is turned off to simulate a 100% dedicated outdoor air system. Thus, the return damper in the economizer is completely closed at all times as shown in Figure 4 (b) and (c) and the minimum outdoor air requirement is only drawn from the outdoor air inlet.

Figure 6 shows the passive chilled beam model that is used in this study. This model is an extension of a previous model (Kim, Braun, & Tzempelikos, 2015). Actual field measurements from the Living Lab were used to estimate the total cooling capacity and chilled surface temperature of the passive chilled beam under different operating conditions in developing this model. The model estimates the total cooling capacity as a function of water supply flow rate, water supply temperature and air temperature in the plenum. The accuracy of the total cooling capacity predictions is depicted in the upper left graph in Figure 6. The chilled surface temperature of the passive chilled beam is assumed to be between the air temperature in the plenum and the water supply temperature to the passive chilled beam. Thus, it is estimated with the air temperature in the plenum, water supply temperature, water supply flow rate and area-weighted uncooled surface temperature (AUST). The accuracy of the chilled surface temperature predictions is illustrated in the lower left graph in Figure 6. The total cooling capacity is the sum of convection and radiation cooling rates. The predicted surface temperature of the passive chilled beam is used with a simplified radiation heat transfer rate calculation shown in Figure 6 in order to estimate the fraction of convective and radiative cooling provided by the chilled beams. The actual passive chilled beam installed in the Living Lab has $0.52m^2$ effective surface area per each unit and this value was used to calculate the total effective surface area for the entire passive chilled beams according to the number of units installed.

Although the TYPE 56 thermal zone model in TRNSYS provides an easier way to calculate the mean radiant temperature, operative temperature and PMV, it is impossible to reflect the effect of chilled surface in the space to calculate these parameters. It also has to be noted that the resultant indoor surface temperatures take this effect into account, since the radiation cooling power of passive chilled beams is directly included in the energy balance equation in TYPE 56. Due to this limitation in the software, an external calculation of the mean radiant temperature, operative temperature and PMV is done at each time step and used as references in the simulation.

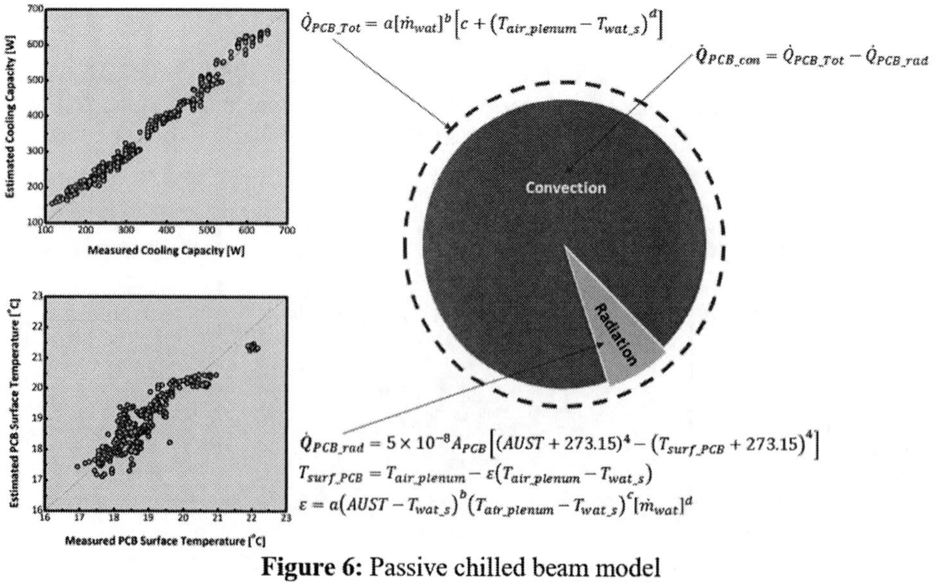

Figure 6: Passive chilled beam model

2.4 Sizing and Control of HVAC Systems

Table 3 summarizes the HVAC sizing and control information for the three systems shown in Figure 4. HVAC component sizing is based on the peak weather conditions. Cooling coil air outlet temperature is set to provide enough dehumidification in the space. Two simulation cases are considered in the conventional system where the cooling coil air outlet temperature setpoint is varied between 10 and 15°C. These cases are necessary to include the range of different dehumidification levels. For passive chilled beam systems, the dehumidification setpoint in the space is set to 0.01 kg_{H2O}/kg_{air} with a deadband of ±0.002 kg_{H2O}/kg_{air}. Constant water flow rate of pump 1 and a chiller water supply temperature are set to maintain a cooling coil air outlet temperature. The chiller capacity is set to maintain a chiller water supply temperature.

Supply air flow rate of the fan in the conventional air system is set to modulate to meet the indoor temperature setpoint. However, since minimum outdoor air requirement is forced at all times, the minimum fan speed is fixed to this requirement. An economizer in the conventional system will take advantage of outdoor air whenever there is a chance to deliver the air with the same condition that would be delivered by the cooling coil. Supply air flow rate of the fan in the passive chilled beam system is set to modulate to meet an indoor humidity ratio setpoint. As it is set in the conventional system, this supply fan's minimum speed is also limited to the minimum outdoor air requirement.

There is an additional pump (pump 2) in the passive chilled beam water loop to modulate the cooling capacity of passive chilled beams based on indoor operative temperature setpoint. In this simulation, ten, twenty and thirty passive chilled beams are considered which corresponds to 5, 10 and 15 m^2 effective chilled surface area (5, 10 and 15% of total ceiling area). The effective chilled surface area is positioned in the center cloud surface to estimate radiation heat transfer between chilled surfaces and other indoor surfaces, and chilled surfaces and the occupant.

PID controllers used in this simulation perform like the real world controllers based on output responses at simulation time steps. Controls of supply air fan and water flow rate in passive chilled beams are used with the PID controllers and a deadband of ±1°C is applied for on/off control.

Table 3: Sizing and control of systems

Sizing Component	Conventional Air System	Passive Chilled Beam System (shared chiller)	Passive Chilled Beam System (separate chillers)
Cooling Coil Outlet Temperature	10°C (15°C), setback to 15°C in unoccupied hrs	Indoor RH < 0.012 kg_{H2O}/kg_{air} → 8°C, setback to 15°C in unoccupied hrs	Indoor RH < 0.012 kg_{H2O}/kg_{air} → 8°C, setback to 15°C in unoccupied hrs
Chiller Water Outlet Temperature	4.44°C	2°C	2°C
Pump 1 Flow Rate (cooling coil)	Cooling coil outlet temp. → 3300 (1200) kg/hr	Cooling coil outlet temp. → 3000 kg/hr	Cooling coil outlet temp. → 3000 kg/hr
Chiller Capacity	Chiller outlet temperature setpoint → 5 (6) tons	Chiller outlet temperature setpoint → 6 tons	Chiller outlet temperature setpoint → 6 tons
Chiller_PCB Water Outlet Temperature	-	-	PCB setpoint → 14°C
Chiller_PCB Capacity	-	-	PCB setpoint → 3 tons
Supply Fan Flow Rate	Temp setpoint or Min OA 350kg/hr → modulate	Ind HR = 0.012 kg_{H2O}/kg_{air} or Min OA 350kg/hr → modulate	Ind HR = 0.012 kg_{H2O}/kg_{air} or Min OA 350kg/hr → modulate
Pump 2 Flow Rate (PCB)	-	Operative temp setpoint → modulate	Operative temp setpoint → modulate
Number of Total PCBs	-	10 (20, 30) units, 5 m^2 (10 m^2, 15 m^2) surface area	10 units, 5 m^2 surface area
PCB Water Supply Temperature	-	14°C	14°C

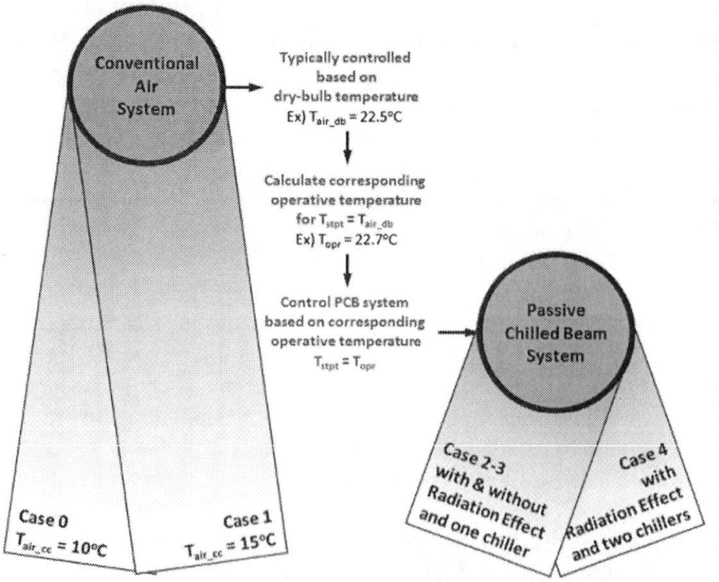

Case Number		Energy Saving	Thermal Comfort	Case Description
0				Air System T_{air_cc} = 10°C
1				Air System T_{air_cc} = 15°C
2				PCB System with **shared** chiller & 0% Effective Chilled Surface Area
3	3-1			PCB System with **shared** chiller & 5% Effective Chilled Surface Area
	3-2			PCB System with **shared** chiller & 10% Effective Chilled Surface Area
	3-3			PCB System with **shared** chiller & 15% Effective Chilled Surface Area
4				PCB System with **separate** chillers & 5% Effective Chilled Surface Area

Figure 7: Comparison criterion for both systems

3. BASIS OF PERFORMANCE COMPARISON

The components described in the previous section are incorporated in TRNSYS to conduct transient simulations of both the conventional air system and the passive chilled beam system over the cooling season. Since both convection and radiation cooling are considered in the passive chilled beam system, it is important to set a basis of a fair comparison between both systems. Energy and comfort should be compared based on the same the thermal condition in the space. The operative temperature, which accounts for both dry-bulb temperature and mean radiant temperature, is used in these simulations as an office temperature setpoint in order to achieve similar thermal comfort conditions. Figure 7 shows the process of simulation sequence and the entire simulation cases for energy savings and thermal comfort improvement comparisons. The air system is first simulated with a typical dry-bulb air temperature setpoint (22.5°C), and based on post-processing, the average operative temperature (22.7°C) was found during the simulation period. This operative temperature was used as the setpoint for the simulation of the passive chilled beam system. A 0.6-min time step for twelve days in July weather data was used in the simulations.

The right hand side table in Figure 7 shows all simulation cases considered in this study. Case 0, 1, 2, 3-1 and 4 are useful to verify the "energy savings" potential of passive chilled beams where two different passive chilled beam system configurations are included in this scenario. Case 0, 1, 2, 3-1, 3-2 and 3-3 are useful in verifying the "thermal comfort" improvement potential of passive chilled beams where four different radiation cooling levels are considered in this scenario. Case 2 uses total cooling capacity as the convection cooling capacity and the chilled surface temperature effect of passive chilled beams is completely removed from the calculation. Case 3-1 to 3-3 represents the case where 10, 20 and 30 passive chilled beams (5, 10 and 15% effective chilled surface area) are installed in the space.

4. RESULTS

Figure 8 shows the average thermal conditions for every case during the entire simulation period but only in occupied hours (9am to 5pm). The dry-bulb temperature in the conventional system in the office area is maintained at 22.5°C for most of the time during the working hours as shown in the graph in the middle. After the average operative temperature was found to be 22.7°C, simulations for the passive chilled beam system were performed based on this operative temperature setpoint. The thermal conditions in terms of the operative temperature setpoint are achieved in the passive chilled beam simulations as shown in the left hand side graph in Figure 8. Since the passive chilled beam system with radiation effect (Case 3 and 4) has an additional chilled surface of passive chilled beams which can reduce the mean radiant temperature in the space, the average mean radiant temperature in these cases are slightly lower (≤22.9°C) than the other simulation cases as shown in the right hand side graph in Figure 8. And this also results in relatively higher average dry-bulb temperature (≥22.5°C) for those cases.

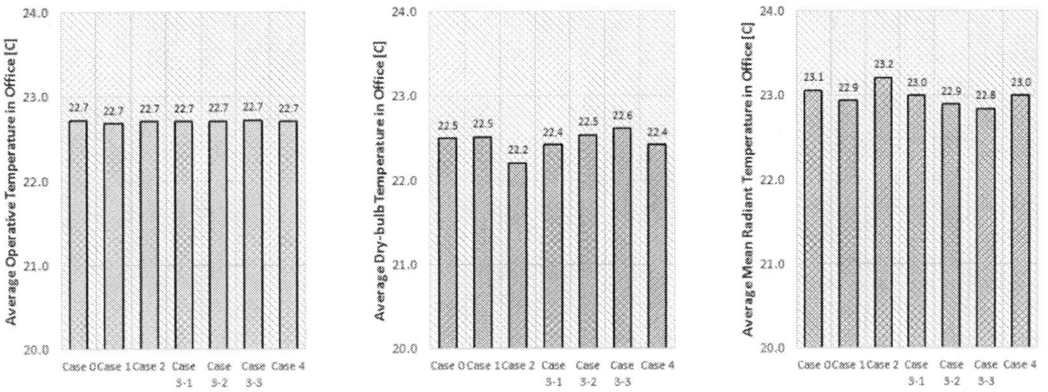

Figure 8: Thermal conditions in office area during simulation period (9am to 5pm)

4.1 Energy Savings Potential

Figure 9 shows the total power consumption for different cases, relative energy savings from the chiller and fan against case 0 and the sensible and latent loads of the cooling coil and passive chilled beams in each case. Compared to case 0, the passive chilled beam system can save 10-12% of total energy (5-6% from the chiller and 60-76% from the fan) with shared chiller configuration (Figure 4 (b)) while using separate chillers (Figure 4 (c)) can increase this savings to 23% (25% from the chiller and 60-76% from the fan). The passive chilled beam system is not favorable in terms of energy savings compared to case 1. This is mostly because of more efficient use of the economizer by increasing the cooling coil outlet temperature to 15°C. It has to be noted again that case 0 and 1 represent the minimum and maximum dehumidification levels in the office space.

There are several useful observations from these results. First is the effect of decoupling of sensible and latent load. Supply air fan changes its role from handling sensible load to latent load in the passive chilled beam system. Since latent load is typically much smaller than sensible load, supply fan energy savings is always guaranteed. As it is also covered in a previous study (Kim, Braun, & Tzempelikos, 2014), a conventional system which is operated based on the sensible load requirement can dehumidify the space excessively in some cases. This excessive dehumidification usually stands out when the climatic condition is "dry" during the cooling season. In this study where the climate is relatively "humid", savings from the latent load reduction were insignificant. The last graph in Figure 9 shows the sensible and latent loads for the cooling coil and passive chilled beams between cases during the entire simulation period. Case 1 shows less latent load than all the other cases. This is where the economizer takes advantage of the situation. Since the cooling coil air outlet temperature has been increased to 15°C, the economizer recovers more favorable energy from the return air stream and can reduce the latent load on the cooling coil. Although a smaller size pump can be used in case 1, increased cooling coil air outlet temperature results in greater fan use. As shown in Figure 9, using a dedicated chiller for passive chilled beams can increase energy savings on chiller by 11%. This is the advantage of using a chiller with higher operating temperature (14°C rather than 4.44°C), which makes the chiller run at higher efficiency. Differences in radiation cooling do not affect the energy consumption estimates.

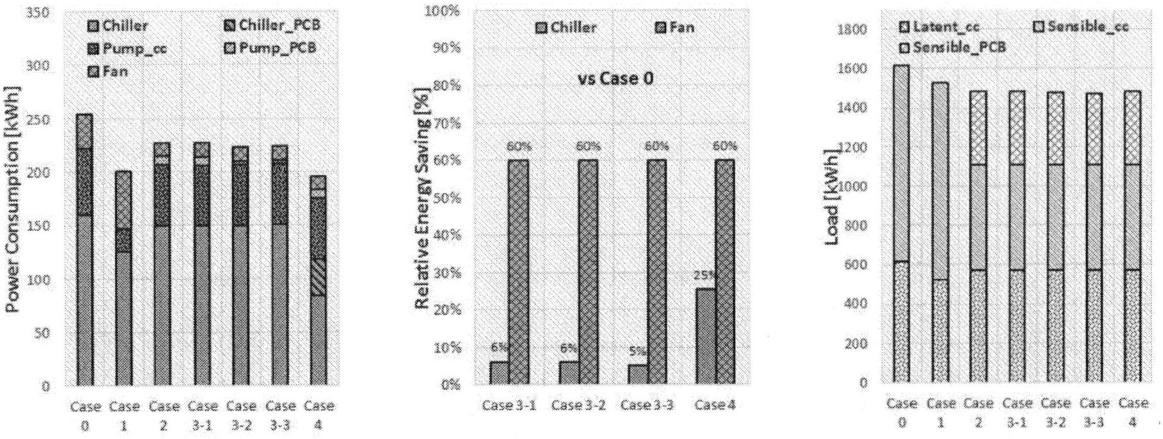

Figure 9: Energy savings of passive chilled beam system

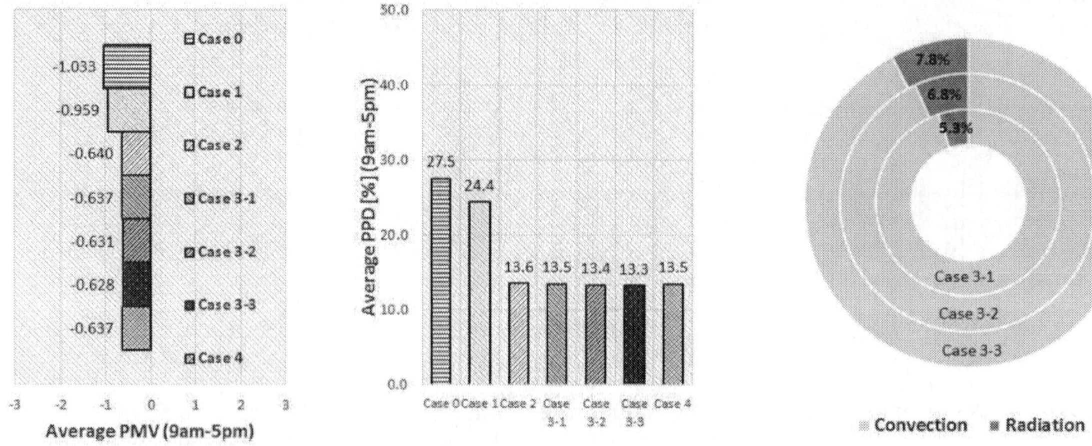

Figure 10: Radiation cooling of passive chilled beams

4.2 Thermal Comfort Improvement Potential

In this study, the effect of the chilled surface of passive chilled beams is taken into account for more precise thermal comfort estimation. However, the radiation portion of passive chilled beams cooling turned out very small in terms of total cooling capacity. Figure 10 shows the average PMV, PPD and average portion of radiation cooling, all during the occupied hours.

The first and second graphs in Figure 10 show that there is a thermal comfort improvement with the passive chilled beam system, by 0.3-0.4 on a PMV scale, which corresponds to 11-14% improvement in PPD. This improvement is mostly due to the different relative air speed that is defined between the two systems in Table 2 rather than due to the radiation cooling effect. Results of case 2 and 3, which represents 0% to 15% effective chilled surface area, show that the improvement in thermal comfort is almost insignificant with increased effective surface area. The reason behind this is shown in the last graph in Figure 10, which shows the radiation cooling portion only varying between 5.3 to 7.8% for 5 to 15% effective chilled surface area. Based on these results, the radiation cooling of passive chilled beams can be neglected in terms of designing a passive chilled beam system and analyzing its performance.

5. CONCLUSIONS

The following conclusions were found in this study:

- Energy can be saved up to 12% by using a passive chilled beam system under Midwest weather condition compared to a typical conventional air system that only controls the space temperature and not space humidity.
- Using a separate chiller with higher chiller water operating temperature (14°C) for passive chilled beams can provide an additional 11% energy savings.
- Thermal comfort can be improved with the passive chilled beam system by 0.3-0.4 on a PMV scale, which corresponds to 11-14% improvement in PPD. The improvement mostly comes from the reduced relative air speed rather than the increased radiation cooling effect.
- Radiation cooling of passive chilled beams is calculated as 5-7% of total cooling capacity for the configuration considered in this study and it is almost insignificant in terms of thermal comfort estimation.

REFERENCES

ASHRAE, American Society of Heating, Refrigerating and Air-conditioning Engineers. (2010). *Standard 90.1-Energy Standard for Buildings except Low-rise Residential Buildings*.

ASHRAE, American Society of Heating, Refrigerating and Air-conditioning Engineers. (2013). *Standard 55-Thermal Environmental Conditions for Human Occupancy*.

A Transient System Simulation Program, Version 17. (2015). TRNSYS (Version 17). University of Wisconsin. Retrieved from http://sel.me.wisc.edu/trnsys/

Energy technology perspectives 2012, Pathways to a Clean Energy System. (2012) (SECOND EDITION). Paris: International Energy Agency (IEA).

Feng, J. D., Bauman, F., & Schiavon, S. (2013). Experimental comparison of zone cooling load between radiant and air systems. *Energy and Buildings*. Retrieved from https://escholarship.org/uc/item/9dq6p2j7#page-1

Feng, J. (Dove), Schiavon, S., & Bauman, F. (2013). Cooling load differences between radiant and air systems. *Energy and Buildings*, *65*, 310–321. http://doi.org/10.1016/j.enbuild.2013.06.009

Fredriksson, J., & Sandberg, M. (2009). The effect of false ceiling on the cooling capacity of passive chilled beams. *Building and Environment*, *44*(7), 1426–1430. http://doi.org/10.1016/j.buildenv.2008.06.017

Fredriksson, J., Sandberg, M., & Moshfegh, B. (2001). Experimental investigation of the velocity field and airflow pattern generated by cooling ceiling beams. *Building and Environment*, *36*(7), 891–899.

Imanari, T., Omori, T., & Bogaki, K. (1999). Thermal comfort and energy consumption of the radiant ceiling panel system.: Comparison with the conventional all-air system. *Energy and Buildings*, *30*, 167–175. http://doi.org/10.1016/S0378-7788(98)00084-X

Kim, J., Braun, J. E., & Tzempelikos, A. (2014). *Energy Savings Potential of Passive Chilled Beam System as a Retrofit Option for Commercial Buildings in Different Climates*. Presented at the International High Performance Buildings Conference. Retrieved from http://docs.lib.purdue.edu/cgi/viewcontent.cgi?article=1106&context=ihpbc

Kim, J., Braun, J. E., & Tzempelikos, A. (2015). Characterizing the In-Situ Performance of Passive Chilled Beams. *ASHRAE Transactions*, *121*, 1N,2N,3N,4N,5N,6N,7N,8N.

Kim, J., Tzempelikos, A., & Braun, J. E. (2015). Review of modelling approaches for passive ceiling cooling systems. *Journal of Building Performance Simulation*, *8*(3), 145–172. http://doi.org/10.1080/19401493.2014.899394

Kosonen, R., Saarinen, P., Koskela, H., & Hole, A. (2010). Impact of heat load location and strength on air flow pattern with a passive chilled beam system. *Energy and Buildings*, *42*(1), 34–42. http://doi.org/10.1016/j.enbuild.2009.07.008

Livchak, A., & Lowell, C. (2012). Don't turn active beams into expensive diffusers. *ASHRAE Journal*, 52–60.

Mustakallio, P., Bolashikov, Z., Kostov, K., Melikov, A., & Kosonen, R. (2016). Thermal environment in simulated offices with convective and radiant cooling systems under cooling (summer) mode of operation. *Building and Environment*, *100*, 82–91. http://doi.org/10.1016/j.buildenv.2016.02.001

Nelson, I. C., Culp, C. H., Rimmer, J., & Tully, B. (2016). The effect of thermal load configuration on the performance of passive chilled beams. *Building and Environment*, *96*, 188–197. http://doi.org/10.1016/j.buildenv.2015.11.017

Niu, J. L., Zhang, L. Z., & Zuo, H. G. (2002). Energy savings potential of chilled-ceiling combined with desiccant cooling in hot and humid climates. *Energy and Buildings*, *34*(5), 487–495. http://doi.org/10.1016/S0378-7788(01)00132-3

Sastry, G., & Rumsey, P. (2014). VAV vs. Radiant, Side-by-Side Comparison. *ASHRAE Journal*, 16–24.

Stetiu, C. (1999). Energy and peak power savings potential of radiant cooling systems in US commercial buildings. *Energy and Buildings*, *30*(2), 127–138.

Thermal Radiation Heat Transfer, 5th Edition. (2011). Retrieved from https://www.crcpress.com/Thermal-Radiation-Heat-Transfer-5th-Edition/Howell-Menguc-Siegel/9781439866689

Thornton, B.A. et al. (2011). *Achieving the 30% Goal: Energy and Cost Savings Analysis of ASHRAE Standard 90.1-2010* (No. PNNL-20405). Pacific Northwest National Laboratory. Retrieved from http://www.pnl.gov/main/publications/external/technical_reports/PNNL-20405.pdf

ACKNOWLEDGEMENT

The authors would like to acknowledge Donghun Kim and Jie Cai for their preliminary work on the Living Lab envelope model and experimental data measurements.

Demand Controlled Heat Recovery for Residential Applications

Zhen Li[1] and William Hutzel[2]*

[1]School of Engineering Technology, Purdue University
West Lafayette, IN, USA
[1]li2215@purdue.edu
[2]hutzelw@purdue.edu

*Corresponding Author

ABSTRACT

Heating and cooling for residential buildings consumes a lot of energy in the United States every year, which leads to Energy Recovery Ventilator (ERV) systems becoming more common for residential applications. An ERV is a relatively simple air to air heat exchanger with supply and exhaust fans. The supply fan pulls outdoor air into the house while the exhaust fan pulls stale air from the house. The ERV preheats outside air during cold weather and cools/dehumidifies outside air during warm weather. These systems are a good investment because they typically achieve a simple payback of 2 years or less based on energy savings. ERV's also have an important role for Indoor Air Quality (IAQ) in modern energy efficient homes that are carefully sealed to minimize the infiltration. One challenge is determining an optimal strategy for when to turn on the fans for the ERV. Unnecessary run time wastes energy by running equipment when it is not necessarily needed for IAQ.

This paper explores a demand controlled ventilation strategy that has been deployed in a net zero energy home which was designed and built by Purdue University for the 2011 Solar Decathlon. The control strategy only runs the ERV when direct measurements for IAQ indicate that fresh outside air is needed. CO_2, VOC, and relative humidity levels are monitored in the home on a continuous basis. When any of these measurements rise above a specified threshold, the heat recovery system turns on and continues to run until the IAQ levels return to healthy levels. This paper summarizes the results of several years of demand controlled ERV performance in terms of IAQ and energy savings. The results clearly show that this optimized control strategy should be considered in more ERV installations.

1. INTRODUCTION

Data from the U.S. Energy Information Administration (EIA) shows that residential buildings consume approximately 20% of the primary energy used in the United States. Roughly 1/2 of that total is for Heating, Ventilating, and Air Conditioning (HVAC) and the amount of energy for HVAC is increasing. The unit cost of energy is also increasing and that, along with concerns about climate change, have led to increased attention for energy saving opportunities. Residential houses are increasingly being built airtight in order to decrease infiltration and HVAC energy consumption. An unintended sided consequence is that a well-insulated and tightly sealed building envelope can have a negative effect on Indoor Air Quality (IAQ). To counter this trend, an HVAC system must bring in a certain amount of fresh air while taking out exhaust air to meet IAQ and residential health. Both the demands of indoor air quality and energy savings led to Energy Recovery Ventilation (ERV) systems becoming more common in residential applications.

An ERV is a heat exchanger that transfers heat between exhaust air and fresh air in an HVAC system. It is designed to provide fresh air into the building while exhausting an equal amount of stale air to the outside. Energy, sometimes including latent loads due to moisture, is transferred between the airstreams. This reduces the loads on the heating and cooling equipment while delivering the air comfortably and quietly into the space. ERV systems also have an important role for indoor air quality in modern energy efficient homes that are carefully sealed to minimize the infiltration by providing a consistent amount of fresh outside air to maintain healthy living conditions. Although ERV's are not common in residential applications in the U.S., they have been widely used in many European countries, especially in the high latitude nations such as Germany and Sweden, where ERV systems have become a requirement in residential buildings.

Figure 1 is the schematic of an ERV system that highlights the energy transfers taking place. Suppose the season is winter and outside air is cold. One duct of the ERV system carries cool (blue: from outside to inside) outside air (T_1) into the home and through the system, where the air will be pre-heated to state point T_2. The warm (red: from inside to outside) air from the home at state point T_3 is exhausted through the ERV and is cooled to T_4 before it is delivered outside. The red and blue streams exchange energy such that the outside air is heated before entering the home.

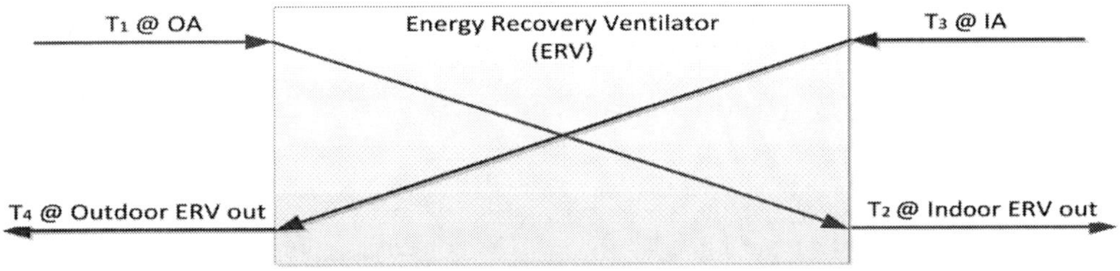

Figure 1. ERV schematic of heating outdoor air in the winter.

During the summer, the process works in reverse. Warm air from outdoors enters at T_1 and is cooled by the air T_3 that is being exhausted. This process saves energy by using less cooling from the HVAC system. At the same time the contaminated indoor air can be moved out while the fresh outdoor air can supply to the house to maintain high levels of IAQ and a comfortable environment.

Effectiveness is one parameter to evaluate the performance of an ERV system. The effectiveness (ε) of an ERV system is the ratio of the sensible energy transferred to the outside air as compared to the total energy transported through the heat exchanger. From the temperatures shown in Figure 1, the effectiveness of an ERV system can be calculated as shown in equation 1:

$$\varepsilon = \frac{T_2 - T_1}{T_3 - T_1} \tag{1}$$

For this equation, T means temperature, and T_1 is the temperature of outside air, T_2 is for supply air to indoors, T_3 is for return air and T_4 is for the exhaust air. In this paper, effectiveness is used to quantify the sensible energy performance of the ERV in sample house. Effectiveness values for common residential ERV systems range from 60% to 80%.

Some ERVs also transfer moisture (latent heat) between the supply and exhaust streams. The latent performance of ERV can be significant in humid climates because the ERV reduces the humidity of the fresh air entering the building. When latent energy is considered, other indices like Coefficient of Performance (COP) or Recovery Efficiency Ratio (RER) can be used to evaluate the performance of ERV system. The determination of whether a sensible or total ERV is appropriate is largely dependent on the weather conditions as a specific geographic location.

A number of researchers have modelled and evaluated the performance of residential ERV's. Zhang*et al* (2006) analyzed ERV efficiency with different factor and developed an equation for Coefficient of Performance (COP). Liu *et al* (2007) simulated the performance of an ERV in different weather conditions and compared the energy savings. Some studies have shown that ERV systems can decrease the annual heating energy consumption significantly. Apart from heating and cooling, an ERV also provides moisture control. Also, some studies have compared the performance of ERV systems in different locations and given suggestions about choosing the type of ERV (sensible or total energy) based on climate conditions.

There is also one practical challenge for ERV system operation in terms of determining when to turn on the fans. Most ERVs use a simple timed schedule that runs the supply and exhaust fans on a 25%, 50%, 75% or 100% duty cycle, which wastes energy by running equipment when it is not necessarily needed for IAQ. It is also possible that the room

needs fresh air while the ERV is not running according to its time cycle. Although some research has been conducted on optimal ERV systems, it has been mostly limited to theoretical modelling. The purpose of this paper is to propose a potential solution for improving the efficiency of an ERV system using a smart controller that makes direct measurements of CO_2, VOC and relative humidity. This control strategy can improve ERV operation by determining when the system is on or off, which means that it only works
when it is necessary to improve IAQ.

2. SMART HOME LAYOUT

A student team from Purdue University designed and built a home for the 2011 Solar Decathlon. Called the INhome (short for Indiana Home) at that time, it earned a 2nd place finish in this international competition. After the competition, the home was moved to Lafayette, IN and became a private residence. The home includes many smart features, including an ERV that has operated on a smart demand controlled strategy. It was also designed to be net zero in terms of its demand for energy all from the electric grid using a 9 kW solar array located on its roof.

The INhome was designed with the intention of proving that attractive, practical, and efficient housing is available at an affordable price. Figure 2 shows that the house was designed for a typical Midwestern consumer, with a contemporary style that would fit in a typical neighborhood. The home was competitive in terms of in today's cost expectations for a new home. The total cost of construction was $250,000, including the solar array that provides all the electricity for the home. Based on the size of the home, the cost translates to $250/ft^2, which includes the cost of the solar array that provides all the electricity. The goal was to show people that the transition to solar powered living could be a smooth and financially realistic one, while maintaining the familiar comforts of home.

Figure 2. Front view of INhome.

The INhome is a new generation of smart house. It is connected to local electric grid but can provide all of its electricity from solar panels. Apart from solar energy, the INhome has passive and active features to minimize energy consumption. It uses high levels of insulation and air sealing and also has large south-facing windows that make daylighting possible. A web-based controller monitors HVAC and security in this house. All of these features combine to fulfill the goal of a net zero energy house.

3. ERV SYSTEM FOR INHOME

Figure 3 is a three-dimensional view that highlights the HVAC system of the INhome that was designed to provide high thermal comfort and exceptional indoor air quality. The green ducts show supply air from outside the house and the orange ducts show return air moving inside the house. The key components of the HVAC system are numbered ① through ④ in Figure 2. ① is a biofilter that was located in the living room to provide natural filtration for house but was recently decommissioned. ② is the Air Handling Unit that modulates air flow throughout the house through carefully placed ductwork within the INhome's conditioned space. The highly insulated duct also conserves the heated or cooled air as it is distributed throughout the home.

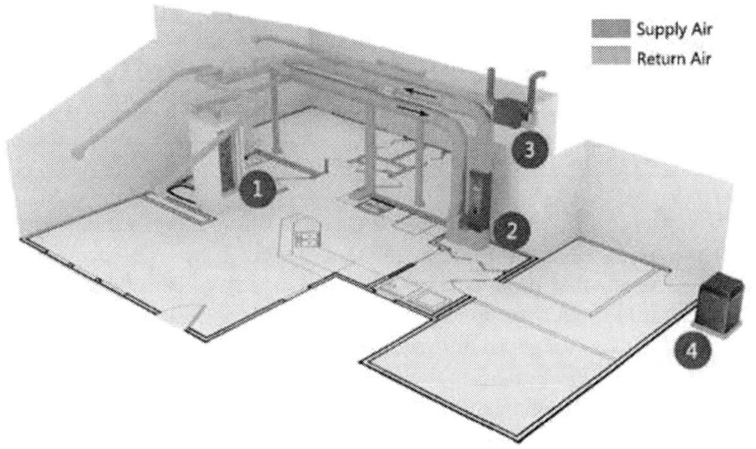

Figure 3. Routing of supply and return ductwork in the INhome.

An ERV was incorporated into the mechanical design to regulate the ventilation air brought into the home. The total cost of the ERV was approximately $1,000, which includes $500 for the ERV itself and $500 for installation. In Figure 3, ③ is the ERV that is located in the utility closet. The ERV is ducted to draw return air from the bathroom and also works as a bathroom vent. Outside air is routed through the ERV and delivered to the space through a grill located high on a wall of the kitchen. ④ is an air source heat pump, which serves as the primary source of heating and cooling for the INhome.

The default operation for a typical ERV is a simple timed cycle. Under this mode of operation, an ERV operates at least 30 minutes out of every hour, regardless of occupancy or the need for fresh air. This is potentially wasteful because the HVAC system has to operate to provide conditioning for the outside air.

Instead of a timed cycle, the INhome used a more sophisticated demand controlled ventilation strategy. This approach was made possible by the automation and control used in the home. An advanced web-based building automation system monitored and controlled the performance of the ERV. It allowed researchers to remotely monitor VOC, CO_2 and relative humidity levels in the home and turn on/off the ERV accordingly. Despite all the automation, the homeowners also had a local over-ride in the bathroom to manually turn on the ERV. In addition to the building automation system, a separate web-based system for monitoring electricity use provided component level energy data, including data on the energy consumption of the HVAC system and the ERV. Taken as a whole, the level of automation in the home provided a basis for this research into demand controlled ventilation for residential applications.

Equation 1 in this paper shows how to compute effectiveness (ε) based on temperature data. The effectiveness of an ERV is similar to efficiency because it measures how well the device is pre-heating and pre-cooling air. Using data from the web-based control system in the home, the outside air, supply air, and return air temperature were monitored while the ERV was running and used to compute effectiveness. An average effectiveness of 87% was then computed from these individual readings. The ERV in the INhome has a typical effectiveness of 60-80% which is better than the typical ERV that is used in a residence. The effectiveness is essentially constant over the whole year, so this 87% value was used as the basis for further computation and analysis.

Table 1 lists the sensors that the ERV uses as triggers for the demand controlled ventilation along with the range and accuracy of each sensor. All of these devices are commercial-grade sensors for general HVAC use and are not as accurate as research-grade sensors. The VOC sensor detects Volatile Organic Compounds (VOCs) from human respiration, building materials, perfume, colognes and furniture off-gassing and is calibrated in ppm of CO_2 equivalent. The CO_2 sensor primarily measures human respiration, so the reading tends to increase with more people or higher activity levels in the home. The range for both the VOC and CO_2 are 0-2000ppm. The relative humidity sensor records moisture levels in the home from 5% to 95%.

Table 1. Sensors for smart control of ERV.

Sensor	Accuracy	Range	On/Off thresholds
Volatile Organic Compounds (VOC)	±5% ppm	0-2000 ppm	1000/900 ppm
Carbon Dioxide (CO_2)	±3% ppm	0-2000 ppm	900/800 ppm
Relative humidity	±5% RH	0%-95%	60/55 %

The building automation system turns on the ERV when direct measurements from one of the three sensors in Table 1 indicates that fresh outside air is needed. These thresholds were based on recommendations from ASHRAE Standard 62.2, Ventilation for Acceptable Indoor Air Quality in Residential Buildings. To minimize short cycling of the ERV, each sensor has an initial higher set point to initially turn on and then a second lower set point to turn off. Table 1 shows that a CO_2 level above 1000 ppm turned on the ERV and it continued to run until the level went below 900 ppm. VOC readings above 900 ppm turned on the ERV and it continued to run until the level went down to 800 ppm. Sensors in both the restroom and living room would trigger the ERV if the relative humidity level rose above 60% and remain on until the level dropped to 55%.

The ERV system in the INhome also has manual control. A timer switch located in bathroom allows the house owner to turn it on whenever it is needed. Apart from maintaining and improving indoor air quality in the whole house, the ERV also works as the bathroom exhaust fan. When the residents take a shower or want fresh air of bathroom they can operate the ERV system manually. This means even though the level of CO_2, VOC and relative humidity all may be below thresholds, the ERV system may run based on the requirements of the homeowners. So the manual control needs to be considered when analyzing the performance of ERV system.

An analysis of the data collected for the ERV shows that the demand control algorithm turned "on" the ERV for about 3000 hours out of 8760 total hours in a year. This means that the ERV had a 34% duty cycle to maintain IAQ in the home. This estimate does not include a small amount of time when data was inadvertently lost due to disruptions in internet connectivity. The 34% ERV duty cycle is significantly less than the 50% to 100% duty cycle that is common for time-operated ERVs. The reduction in runtime also translates to significant energy savings due to decreased loads on the HVAC system in the house and decreased ERV fan run time.

Figure 4 summarizes the energy savings of the ERV system during its run time. The horizontal axis (independent variable) is the cumulative number of hours in one year when the ERV operated. The vertical axis (dependent variable) is the temperature of the outdoor air before and after the ERV in degrees Fahrenheit. The horizontal boundary shows the constant 70 °F room air temperature that is essentially constant. The lower boundary of the curve (green) shows outdoor air temperature on an hourly basis when ERV was running. The middle curve, at the boundary of the two color zones, shows the beneficial impact on air temperature from a heat recovery system with an effectiveness of 87%.

In Figure 4, the colored zones illustrate heat recovery performance. The area of the green zone is proportional to the energy recovered. When the temperature is cold, the green zone is pre- heating energy. During warmer weather it means pre-cooling of outside air. The area of red zone is the additional energy required from heating or cooling the space using the air source heat pump. So the building benefits from pre cooling and heating with the ERV system. Figure 4 illustrates that heat recovery is more beneficial during cold winter weather, but it also helps reduce the cooling loads during hot summer weather by a small amount.

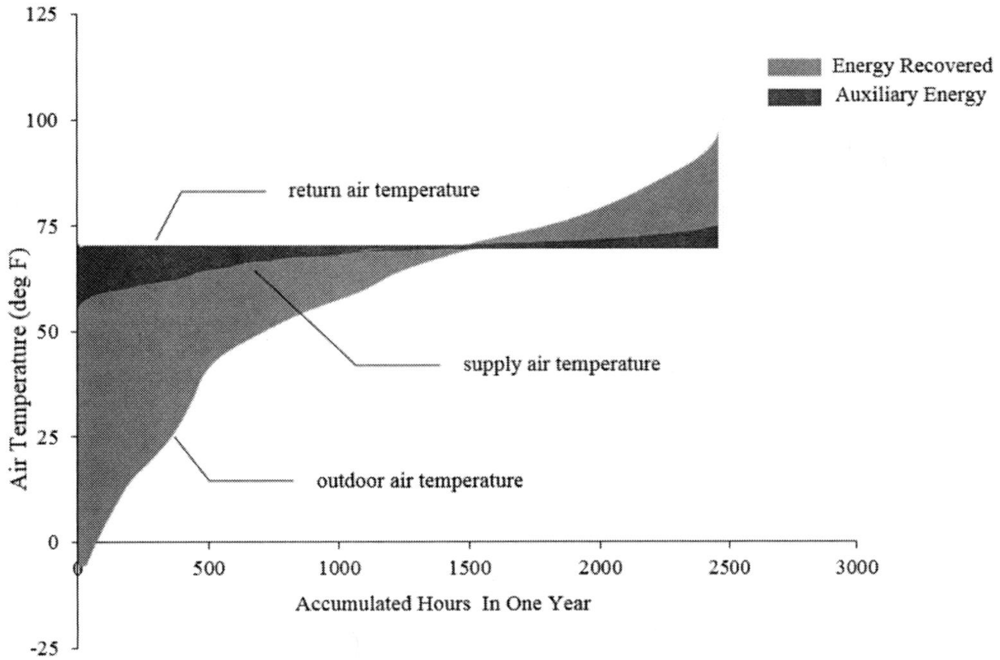

Figure 4. Annual benefit of the ERV system.

4. RESULTS

4.1 Indoor Air Quality Result

Although the INhome has been occupied since 2013, that first year had many disruptions. During 2013, the house was painted and custom furniture was installed. Considering all the decoration work in 2013, the CO_2 and VOC data for 2013 does not accurately reflect IAQ levels for a common residential house. This paper analyzes data from 2014 and 2015. In addition, there were several disruptions in data collection in 2014 and 2015 so that data is not 100% complete for any one year. The data disruptions occurred when the web-based control system turned itself off while performing maintenance or due to disruptions in internet service. As the data for these two years covered most of the time, the data can accurately represent the annual indoor air quality from the demand controlled ERV system.

Table 2 is a summary of the percentage of time that the data collected from IAQ sensors were above the limits for acceptable IAQ using the thresholds described above. Table 2 shows that the VOC exceed the limitation about 21% of the time, while CO_2 and relative humidity exceeded only about 5%. Most of the time these parameters were below the preset threshold levels, indicating acceptable IAQ in the INhome. Table 2 also shows that the zone VOC exceeded its threshold value most often, indicating that VOC is the main index that influenced IAQ.

Table 2. Percentage of parameters above limitation

Sensor	Zone VOC	Zone CO_2	Bathroom RH
Percentage of Time Above IAQ Threshold	21.1%	6.2%	4.6%

The author chose a long time period without interruption and a constant sampling rate when analyzing the relationship between different signals and ERV operation. However, a software update inadvertently took the data acquisition system offline and prevented data logging for a period of time. To counter this gap the paper will present the weighted average of available data to represent the annual performance of the signals and sensors.

Figure 5 shows the percentage of time that multiple triggers turned on the ERV. There are four cases to turn on the ERV. One option has no sensor above the threshold, indicating that the homeowner turned on the ERV manually. This case accounts for about 25% of the measuring time. Another case is when one sensor is above its threshold, which takes more than 50% of the time. Two sensors above their threshold occurred about 20% of the time. All of three sensors were above their limits only about 3% of the time, which is the least common option. Figure 5 shows that just one sensor energized the ERV on and about 1/2 the time. About 25% of the time, the ERV was operated by hand.

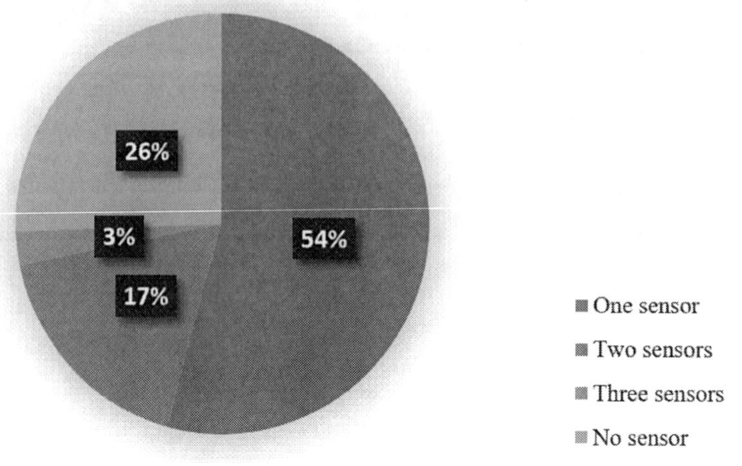

Figure 5. Number of sensors above limitation when ERV works

Through Figure 5, it can be seen that this system can fulfill smart control. The ERV in the INhome was always controlled by indoor air quality sensors. As the percentage when more than one signal energized ERV is 20%, so there was always one or no poor signal to influence air quality. When combined with smart control strategies, the indoor air quality is acceptable for the INhome. Notice that the system can operate manually, so the ERV system may save more energy if it is just controlled by VOC, CO_2 and RH signals in this house.

Figure 6 is the weighted percentage of measured time when a parameter energized the ERV system. The ERV operates about 35% of the time each year. As mentioned above, the number of sensors that drive the ERV in INhome is varied. For this figure, it counts for all cases of ERV worked without manual. It takes indices of VOC, CO_2 and relative humidity of bathroom. It is easy to see that VOC is main source that energized ERV system, since it took 65% of possibility to trigger ERV while RH is about 20% and CO_2 is 15%. This result is similar to Table 2.

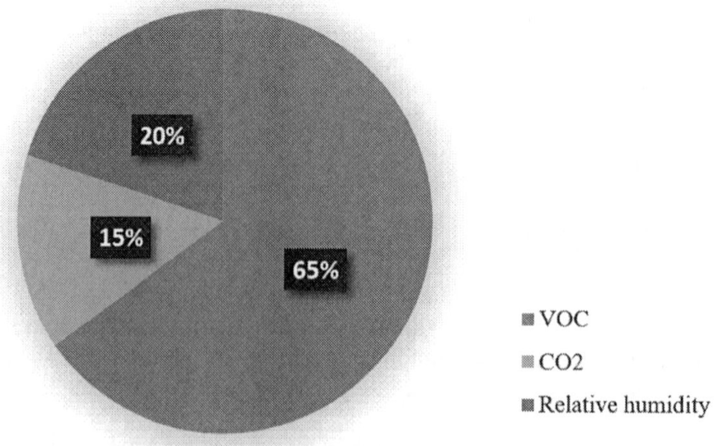

Figure 6. Sensor which energized ERV system.

4.2 Savings for Smart Control ERV

A demand controlled ERV can save energy cost compared with a common ERV. From Emonitor, a website to record energy consumption of the INhome, the annual energy usage of the ERV system is 110kWh. This is based on an instantaneous power consumption of 30 to 40 Watts by the fans. It takes a very small part of the whole house annual energy consumption and less than 1% of the total energy consumption for the home. The residential electricity rate in Lafayette is about 0.12 \$/kWh and suppose the ERV system is used in a residential house without smart control. In this instance, the ERV works the whole year for a total of 8760 hours. The annual energy cost can be calculated using equation (2). In this equation, η represents percentage of annual work time of ERV system, and t is working time with unit of hour, here is 8760 hours. The cost of electrical energy is around \$42.

$$C = P \times t \times electic\ rate * \eta \tag{2}$$

For the smart control ERV system used in the INhome, the annual cost was about \$13, which means residents can save \$30 each year on fan energy. A detailed comparison is shown below.

Figure 7 shows the monthly electrical consumption of the ERV system in the INhome compared with a traditional house that has an ERV system running all the time. The traditional house used the same model of ERV equipment as the INhome but without smart control. This means that the ERV was not controlled by IAQ and operated 24 hours per day. In this figure, a traditional house using an ERV for 24 hours will have a constant energy consumption from month to month, while electric usage of ERV in INhome varied with time and more less than traditional houses.

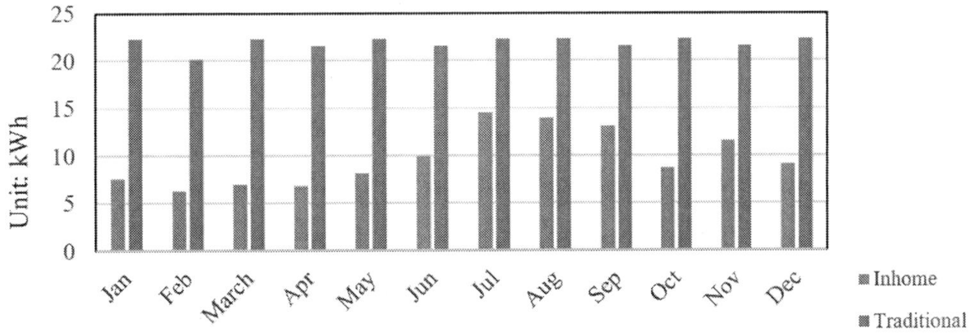

Figure 7. Monthly energy consumption of ERV system.

Figure 7 shows how the ERV energy consumption varied through time for the INhome, which also can reflect smart control of ERV system. During in summer and winter, energy consumption is higher, which means there is a need for more air exchange at that time. One potential reason for this is the residents used heat pump and air handling unit at that time which leads to increased use of the ERV. Consider that the ERV in the INhome also works the vent in the bathroom which can save money on the installation of a vent.

This analysis does not account for the HVAC energy savings avoided from not having to heat or cool outside air unnecessarily. The potential savings from avoided heat and cooling should be substantially larger than the fan electricity savings discussed here. The computations to quantify this additional savings potential is ongoing.

5. CONCLUSIONS

The INhome is a new generation of net zero energy residential house because the cost of ownership is affordable for a typical US family. One challenge with this type of home is that it is tightly sealed and has the potential for poor indoor air quality. The INhome uses an ERV system to save energy and maintain comfort while providing fresh outdoor air. A smart control strategy to make the ERV system more efficient was developed. By setting thresholds for VOC, CO_2 and Relative Humidity, the ERV system only operated when it was needed for IAQ. An analysis of several years' worth of data showed that indoor the air quality in INhome was acceptable most of the time. VOC's were the main contaminant that energized the ERV in the INhome. Through smart control in the INhome, the homeowners saved \$30 per year in fan electricity as compared to a traditional ERV that operates continuously. So this optimized control strategy should be considered in more heat recovery installations in residential houses.

NOMENCLATURE

ERV	Energy Recovery Ventilation	
INhome	Indiana Home	
ε	Effectiveness of ERV	
C	Electric cost	$
P	Power	kW

REFERENCES

ASHRAE (2010). ASHRAE Standard 62.2: Ventilation and Acceptable Indoor Air Quality in Low-Rise Residential Buildings: Atlanta, GA.

Besant R.W & Simonson C.J.. (2010) Air-to-air Energy Recovery, *ASHRAE Journal*, *5*, 31-38

Causey, L. (2012). *Evaluating home controls for the 2011 Solar Decathlon*. Master Degree. Purdue University.

Dunham B., (2013) Effects *of Botanical Air Filtration on Energy Efficient Homes,* Master Degree, Purdue University

Fauchoux, M.T., Simonson, C.J. & Torvi, D.A. (2007). The effect of energy recovery on per- received air quality, energy consumption, and the economics of an office building, *ASHRAE Transactions 113 (2)* 437–449.

Hughes, B., Chaudhry, H., & Calautit, J. (2014). Passive energy recovery from natural ventilation air streams. *Applied Energy, 113,* 127-140.

Juodis, E. (2006). Extracted ventilation air heat recovery efficiency as a function of a building's thermal properties. *Energy and Buildings, 38(6),* 568-573.

Liu, J., Li, W., Liu, J., & Wang, B. (2010). Efficiency of energy recovery ventilator with various weathers and its energy saving performance in a residential apartment. *Energy and Buildings, 42(1),* 43-49.

Purdue Solar Decathlon. (2016). *Purdue.edu.* Retrieved 29 April 2016, from http://www.purdue.edu/inhome/

Zhang, J., Fung, A., &Jhingan, S. (2014). Analysis and feasibility study of residential integrated heat and energy recovery ventilator with built-in economizer using an excel spreadsheet program. *Energy and Buildings, 75,* 430-438.

Zhang, L. & Niu, J., (2001). Energy requirements for conditioning fresh air and the long term savings with a membrane-based energy recovery ventilator in Hong Kong, *Energy and Buildings 26* 119–135.

Zhang, J., Fung, A. and Jhingan, S. (2014). Analysis and feasibility study of residential integrated heat and energy recovery ventilator with built-in economizer using an excel spreadsheet program. *Energy and Buildings, 75,* pp.430-438.

Zhong, Y. K., (2009) Applicability of air-to-air heat recovery ventilators in China, *Applied Thermal Engineering 29 (5–6)* 830–840.

Zhou, Y., Wu, J., & Wang, R. (2007). Performance of energy recovery ventilator with various weathers and temperature set-points. *Energy and Buildings, 39(12),* 1202-1210.

Simonson, C.J., Shang W. & Besant, R.W. (2000). 'Part load performance of energy wheels: Part I. Wheel speed control, Part II. By-pass control and correlations.' *ASHRAE Transactions*, 106(1).

ACKNOWLEDGEMENT

We would like to show our gratitude to Marifran and Jim Matson, the house owners of INhome, for assistance during data collection, participation in the survey, and for sharing their experiences of living in the INhome.

Temperature, Relative Humidity, and Carbon-Dioxide Modulation in a Near-Zero Energy Efficient Retrofit House

Saurabh SUDHAKARAN[1], Mark SHAURETTE[2]

[1]Purdue University, School of Construction Management,
West Lafayette, Indiana, U.S.A
765-476-6964, sudhakas@purdue.edu

[2]Purdue University, School of Construction Management,
West Lafayette, Indiana, U.S.A
765-494-6812, 765-496-2246 mshauret@purdue.edu

ABSTRACT

The concept of Net Zero Energy Buildings (NZEB) has reached a phase where countries all around the world are encouraging its implementation into mainstream construction. In the United States, both private and public sector buildings are incorporating energy efficient technologies to reduce their environmental impact, while increasing the productivity and comfort of its occupants. A Net Zero Energy Building (NZEB) performs as expected only when the building's envelope, HVAC and other mechanical/electrical systems work in unison. Subsequently, once these buildings are occupied, the behavior of its occupants significantly influences the buildings energy performance. The authors have captured the modulation of temperature, relative humidity, and carbon-dioxide within a home retrofit to near Net Zero Energy Building conditions during the heating season. This house is a Deep-Energy Retrofit Home completed as a marketing and demonstration home for a joint neighborhood stabilization project and U.S. Department of Energy funded community-wide retrofit grant program. The house includes an internet based real-time home energy monitoring system, which facilitates reviewing the changes in the houses energy consumption as a consequence of fluctuating internal temperature settings and external climate conditions. Post retrofit blower-door test result conform that the house has been made fairly air-tight during the retrofit. Hence ventilation of the house is achieved via an Energy Recovery Ventilator (ERV) with multiple stages of operation. To this end, the paper is an exploratory examination of the inter-relationships between occupancy, interior temperature, relative humidity, carbon-dioxide levels and energy consumption within a Near-Zero Energy Building.

1. INTRODUCTION

Development in building technology has helped in reducing air infiltration within modern homes, hence reducing unwanted energy loss through uncontrolled air exchange. Energy star rated equipment installed in these modern homes are highly energy efficient, and contribute in reducing the overall energy consumption RECS (Residential Energy Consumption Survey, 2011). EIA's 2011 survey found that over 40 million householders in America (35 percent) have used caulking or weather-stripping to seal cracks and air leakages around their house and 26 million (23 percent) had added some form of insulation to their existing homes.

Houses with lower air infiltration consume considerably less energy for space heating and air-conditioning. However, as a result of reducing the infiltration of fresh outside air to very low levels, residents of modern residential building may run a risk of less than recommended ventilation rates or Air Change Rates (ACR) within their homes. For any building a ventilation rate of 20cfm per person is recommended for maintaining optimum level of Indoor Air Quality (IAQ) (ANSI/ASHRAE Standard 62-1999, Ventilation for Acceptable Indoor Air Quality). Occupant health and perceived IAQ will usually be improved by avoiding ventilation rates below 20 cfm (9 L/s) per occupant and further improvements in health and perceived IAQ will sometimes result from higher ventilation rates up to 40 cfm (18 L/s) per person. (O. Seppänen, W.J. Fisk, M.J. Mendell, 2002)

The primary reason behind making a house airtight is to reduce energy consumption with the additional benefit of increased occupant comfort. Comparing the average energy consumption of modern homes, with the ones built before 2000 reveals that the way residential buildings in the United States consume energy has changed drastically over the past few decades. Homes built in the 2000s accounted for 14% of the total occupied housing units in 2009 and these new homes consumed 21% less energy than older homes for space heating. Energy use units in 2009 and these new homes consumed 21% less energy than older homes for space heating. Energy use per household for the year 2005 was 95 million British Thermal Units (BTUs) compared with 138 million BTUs per household in the year 1978 (RECS, 2009), a drop of 31 percent.

Increasing energy costs, strict building codes and reduced cost of energy efficient alternatives are some explanations for this decrease in the energy consumption of residential buildings. The RECS 2009 survey has identified space heating and household appliances as the two main end-uses that consume the most energy within residential buildings in the United States. With this knowledge, making a house airtight appears to be a logical step for homeowners who want to reduce their energy bills.

2. DEEP ENERGY RETROFIT HOME

Taking the concept of reducing energy consumption to the next level, some modern buildings are designed and constructed to be a Net Zero Energy Building (NZEB). A NZEB is defined as a building with very high energy performance, when the primary energy that it produces and feeds-in to the grid or any other network is equal to total energy that is delivered to the building by the grid and all other sources. It is important that the annual energy balance of energy exchanged must be 0 BTU/sf.

These modern buildings are capable of providing more comfortable interior environments than traditional homes at a fraction of its energy cost. The operation and maintenance of such energy efficient building is just as important as their design and construction. This is especially important in a residential setting, where the home owners largely do not have sufficient knowledge of all the installed systems. These buildings perform as predicted only if all the components of the house work in unison.

Providing a high quality and comfortable interior environment is desirable for all residential buildings. Over the past two decades, the scientific community has shown increased concern for indoor air quality (IAQ) and the effect of interior air on the health of occupants. As mentioned earlier, modern homes are more airtight than older homes, these modern homes use more synthetic materials that generate Volatile Organic Compounds (VOCs) into indoor air. Whilst these improvements have led to more comfortable buildings with lower running costs, they also provide indoor environments in which contaminants are readily produced and may build up to a much higher concentration than are found outside. (Jones, 1999)

Purdue University participated in the Lafayette Energy Assistance Program (LEAP) and guided the technical development of the program as well as the design and retrofit of a Near-Zero Energy demonstration retrofit home. As a retrofit of an existing home it was not able to achieve the full goal of a NZEB, but far exceeded the energy performance of both new and existing homes in the community in which it is located. The demonstration home was completed to provide visibility to attract LEAP program participation and to provide opportunities for homeowner education within the community where it was located. The Energy Conservation Measures (ECM's) for the retrofit house listed in Table 1 were carefully chosen and recommended to the builder competing the retrofit and the Neighborhood Stabilization Program (NSP) program manager. Alternative energy production through roof-mounted solar panels was used on the demonstration home to attract attention. All other ECMs were selected to reduce energy consumption in addition to the following decision criteria:

- Appropriate for homes found in the project communities
- Easy for local building trades to understand and install
- Available through traditional supply channels without delay
- Assessed from a whole-building energy performance viewpoint
- Expected to provide near-term potential for positive payback, but with no specific cut-off
- Appropriate to or related to energy retrofits that could be funded by LEAP program grants
- A contributor to energy conservation first with introduction of alternative energy sources only when energy consumption had been minimized

Table 1: Energy Conservation Measures

ECM	Description
Windows	R-5.56 triple glazed casement
Sun Tube	One in each bath with dimmer to provide daylight illumination and control
Exterior Doors	Insulated steel, thermal break frame, magnetic weather-strip, polyurethane core R-8.3
Crawl Space	Damp Proof w/ sealed 20 mil poly floor cover
Attic Access	R-40 insulated, weather-stripped attic closure system
Air Seal Attic	Air seal all top plates and ceiling penetrations with closed cell foam
Air Seal Walls	Expanding foam seal all exterior wall penetrations
Insulate Attic	R-60 Loose Fill Cellulose with 3" closed cell foam - 3' wide at roof edge (R-20+)
Insl. Crawl Space	Conditioned crawl, 2" closed cell foam on interior of crawl wall and band joist (R-13+)
Insulate Ext. Walls	R-11 batts @ 2x3 wall cavity plus 4" (R-20) extruded polystyrene sheathing (2 layers of 2" foam with lapped and taped joints)
South Overhang	Extend to 16" for summer shading and add continuous vent
Hot Water	Heat Pump Water Heater with min. COP rating of 2.0 or greater
Solar Energy	Nominal 4 KW Solar PV System
Furnace & AC	Multi-speed air handler, min. 25,000 BTU gas furnace, 1 ton AC
Ductwork - Supply	Within conditioned space, Mastic seal all ductwork
Ductwork - Return	Within conditioned space were possible, Attic runs min. R- 20 insulation, Mastic seal all duct
ERV	Energy Recovery Ventilator min. 60% heat recovery, installed in conditioned space
Thermostat	7-Day Setback
Washer	Front Load – Energy Star Rated
Dryer	Electric – Energy Star Rated
Refrigerator	Top Freezer, No water and ice through door, Energy Star Rated
Dishwasher	Energy Star Rated
Electrical	44 circuit eMonitor® energy monitor, real-time internet energy use dashboard
Lighting	All lighting CFL or T-8 florescent except LED kitchen task lighting
Window Coverings	Living Room Insulating Cellular Shades with air sealing tracks
Air-Tightness	Infiltration post-retrofit 2.28 air changes/hour @ 50 pcal pressure (blower door)

Significant care was exercised in use of caulk, air barriers and spray foam to provide an airtight envelope. After the completion of retrofits, a blower door test was conducted to measure the infiltration post-retrofit. The test results indicated that the house was highly air tight, with 2.28 air changes per hour under 50 pcal of pressure. Achieving this level of air tightness was an important step towards creating the highly energy efficient home reflected in the HERS certification rating of 17 (Figure 1).

3. METHODOLOGY

For the purpose of this study, the authors started with identifying some measurable parameters within the retrofit house which provide basic information about the house's HVAC system performance. Temperature and relative humidity were the two obvious parameters that were chosen to be monitored for this purpose. It was predicted that these readings would allow the authors to study the occupants comfort preference and also let the authors monitor how the house performed in meeting these needs. Indoor carbon-dioxide levels correlate reasonably well with perceived odor and levels of human bio effluents. Thus, interior carbon-dioxide concentrations are often used to judge the acceptability of indoor air. Its concentration is an indicator that may be used to identify pollutant sources, determine building and heating, ventilating, and air conditioning (HVAC) system deficiencies, verify corrective actions, and correlate occupant symptoms to IAQ. (Batterman &Peng, 1995). Therefore, the third parameter chosen to be monitored for this study was in-door concentration of carbon-dioxide as it would provide the authors valuable information about effect of air-tightness of the retrofit house on the indoor air quality.

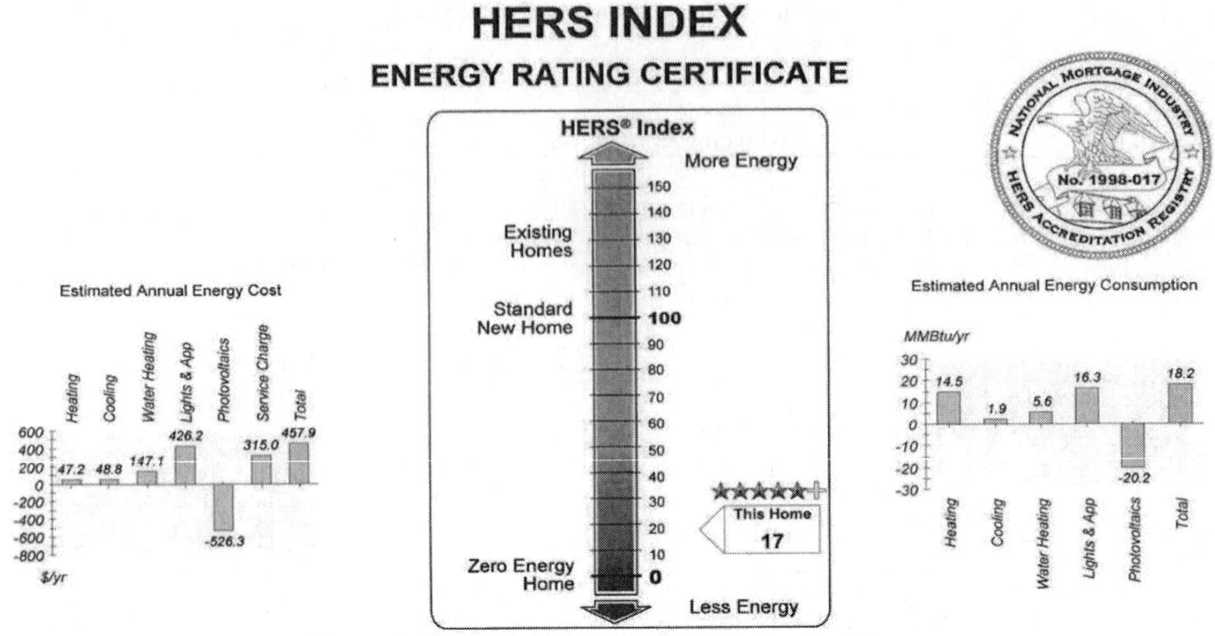

Figure 1: Demonstration Home HERS Rating Certificate

ASHRAE recommends carbon-dioxide concentration between 1000ppm – 1200ppm for spaces housing sedentary people, in this case the retrofit house. As per guidelines in ASHRAE Standard 62.1-2013, "Ventilation for Acceptable Indoor Air Quality" carbon-dioxide concentration greater than 5000 ppm within indoor environments can pose health risks.

For this experiment the authors used Fluke 975 AirMeter, this device was capable of recording and recording the concentration of carbon-dioxide, relative humidity and interior temperature continuously for 99 hours within the retrofit house. Three sets of 99 hour readings were collected between 7th February 2016 and 2nd March 2016 for this study. The first set of readings was recorded in the living room of the house. During this time the measuring device was placed at the floor level (6 inches above the floor level). The second set was recorded in the master bedroom of the retrofit house. This time the measuring device was placed at a height of 3 feet above the floor level. The third set was recorded in the kitchen and the meter was placed just 1 foot below the ceiling level. The authors placed the meter at different levels for all the three readings to study the stratification of indoor air-temperature within the retrofit house. The retrofit house is occupied by a single resident. The authors interviewed the occupant before starting data collection to learn about the occupant's comfort preferences and knowledge of the installed systems. Since the occupant did not know the advantages and procedure of operation of the setback thermostat, the authors informed the owner about some advantages of using this system, and how it can be operated. The authors also encouraged the resident to use the setback thermostat of house during the course of this study. The use of setback thermostat caused the indoor temperatures to fluctuate, indoor carbon-dioxide concentrations and relative humidity levels also reacted to the changing indoor temperatures. The next section presents these findings in detail.

4. ANALYSIS OF DATA

Comfort criteria due to odors with respect to human bio effluents are likely to be satisfied if the ventilation results in indoor carbon-dioxide concentrations are less than 700 PPM above the outdoor air concentration. (ASHRAE 62-2001). This reference to a carbon-dioxide concentration of 700 PPM over outdoor air is not an upper limit for acceptable IAQ, rather a recommendation for comfort. The concentration of carbon-dioxide within the retrofit house measured during the study was between 200 PPM -3333 PPM. Although typical outdoor CO2 concentrations are approximately 380 ppm, outdoor levels in urban areas as high as 500 ppm have been reported (Persily 1997). Hence indoor carbon-dioxide concentrations recorded during the study were higher than ASHRAE recommendations. Prior research has documented direct health effects of carbon-dioxide on humans, but only at concentrations much higher than those found in normal indoor settings. Carbon-dioxide concentrations greater than 20,000 ppm cause deepened breathing; levels of 40,000 ppm increases respiration markedly; 100,000 ppm levels cause visual disturbances and tremors and has been associated with loss of consciousness; and at 250,000 ppm carbon-dioxide (a 25% concentration) can cause death (Lipsett et al. 1994). Maximum recommended occupational exposure limits for an 8-hr workday are 5,000 ppm as a time weighted average, for the Occupational Safety and Health Administration (OSHA 2012) and the American Conference of Government Industrial Hygienists (ACGIH 2011). The ASHRAE 62-2001 standard recommends maintaining a maximum 60% humidity during the summer and at least 25–45% during the winter. ASHRAE Standard 62.1-2013 recommends that relative humidity in occupied spaces be controlled to less than 65% to reduce the likelihood of conditions that can lead to microbial growth. The recorded humidity readings within the retrofit house were within acceptable range, with a low of 40 %RH and a high of 58.8%RH.

Through an internet connected energy monitor the authors have access to minute-by-minute, real-time and past electricity usage of all major appliances and circuits within the retrofit house. The retrofit house uses natural gas furnace for space heating, but the air handler runs on electricity. The house is billed monthly for gas with no other gas consumption monitoring so its real-time consumption was not available. The authors used the real-time electricity consumption from the air handler unit to monitor the operation of the furnace.

Figure 2, Figure 3 and Figure 4 exhibit the modulation of outside temperature, and indoor temperature of the retrofit house in the living room (7th Feb- 11th Feb), bedroom (14th Feb – 18th Feb) and kitchen (26th Feb – 2nd March) as well as energy used by the air handler unit circuit which includes the ERV. All three graphs present 99 data points representing each hour the Fluke AirMeter was recording. The indoor temperature reading was collected by the Fluke meter, while outdoor temperature and electricity used by the air hander was recorded by the real-time energy monitoring system.

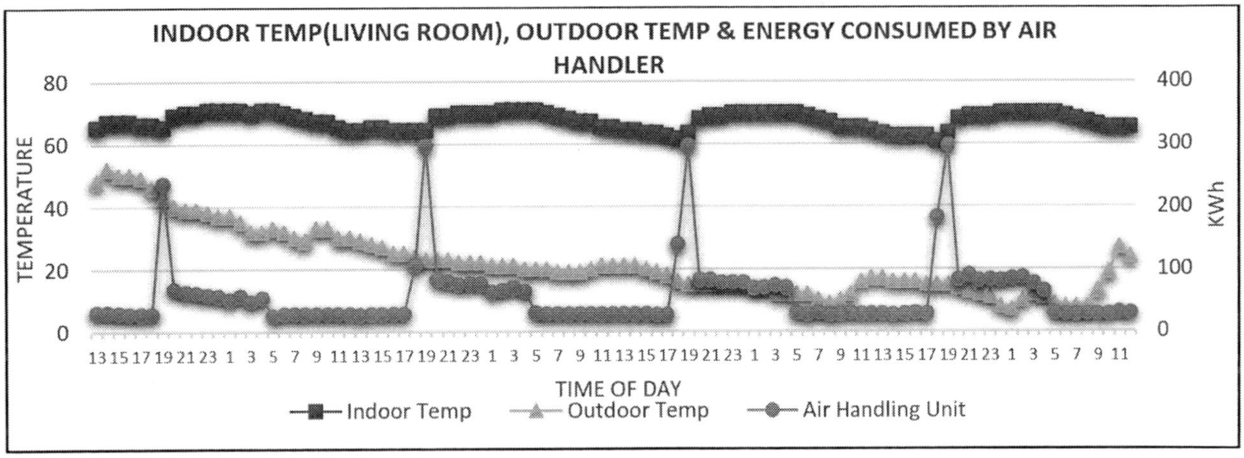

Figure 2: Indoor Temperature in the Living Room, Outdoor Temperature & Energy Consumed by Air Handler

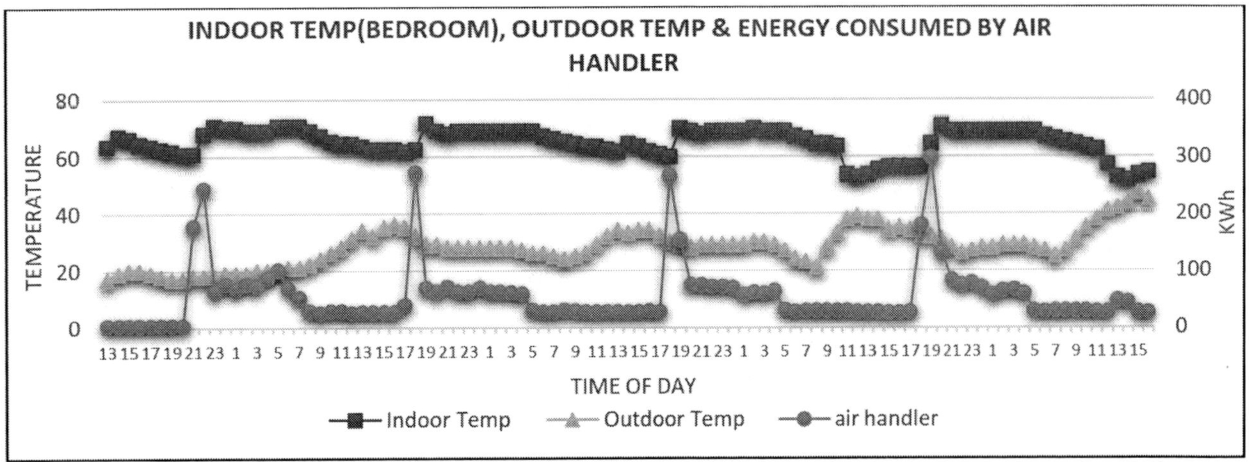

Figure 3: Indoor Temperature in the Bedroom, Outdoor Temperature & Energy Consumed by Air Handler

In the brief interview with the occupant prior to beginning the indoor air measurements they mentioned the setback capable thermostat was programmed to maintain the interior temperature of the house at 70 degree Fahrenheit from 7PM to 5 AM. Figure 1 and Figure 2 shows that every day at 7PM the house starts consuming energy and brings the interior temperature back to 70 degree Fahrenheit. ASHRAE Standard 55-2013, Thermal Environmental Conditions for Human Occupancy, notes that for thermal comfort purposes, indoor temperature could range from between approximately 67 and 82 °F. The data indicates that the HVAC system as controlled by the occupant's settings of the setback thermostat successfully controlled indoor temperature and humidity at acceptable comfort levels". During this study, the outdoor temperature was an average of 22.19 °F in the first set of readings (7th Feb- 11th Feb) in Figure 1 and an average of 28.38 °F in the second set of readings (14th Feb – 18th Feb) as seen in Figure 2. The occupant also mentioned that the house was allowed to cool down to a minimum of 50 °F when it was un-occupied between 5AM and 7PM which allowed maximum energy savings without any compromises in comfort.

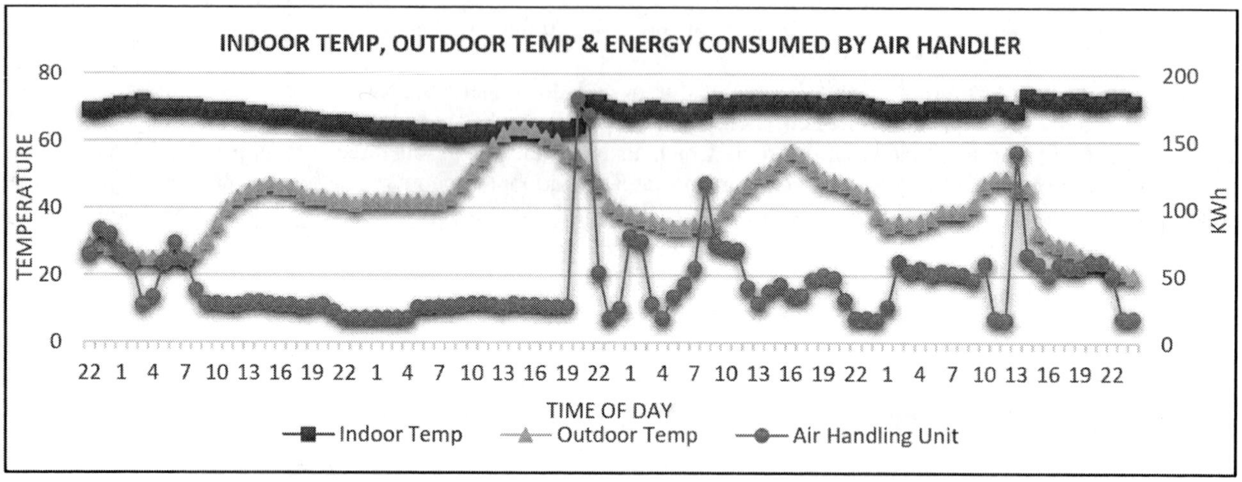

Figure 4: Indoor Temperature in the Kitchen, Outdoor Temperature & Energy Consumed by Air Handler

Figure 4 shows the modulation of indoor temperature, outdoor temperature and energy consumed by air handler in the last set of recorded data (26th Feb – 2nd March). In this data set the average outdoor temperature was 40.9 °F, which was highest in all the three sets. Also, in this data set it was observed that the house did not use any heat between 3PM on 27th Feb. to 10PM on the 28th of February. During this time the house was unoccupied and the occupant had turned off the heating system.

Most people spend 70-90 % of their time indoors, with a large percentage of this time spent within one's own residence. Some pollutants found in outside air find their way indoors, and indoors the home's occupants release toxins further creating the possibility of higher indoor pollutant concentrations. As a result, it is acknowledged that inhalation of indoor air is the major determinant of human exposure to many pollutants (Samet et al., 1987, 1988). Although not considered a pollutant itself, carbon-dioxide concentration is often used as a simple and inexpensive method to indicate effective ventilation levels.

Figure 5, Figure 6 and Figure 7 show that the carbon-dioxide concentrations within the retrofit house varied between 200 PPM to 3333 PPM. The average carbon-dioxide concentration recorded in the living room was 1915 PPM, in the bedroom the average concentration of carbon-dioxide was 1934 PPM and in the kitchen the concentration of carbon-dioxide was an average of 1979 PPM. Since average outdoor concentration of carbon-dioxide varies ranges from 380 PPM to 500 PPM, the difference in concentration between indoor and outdoor concentration of carbon-dioxide suggested the building needs additional ventilation.

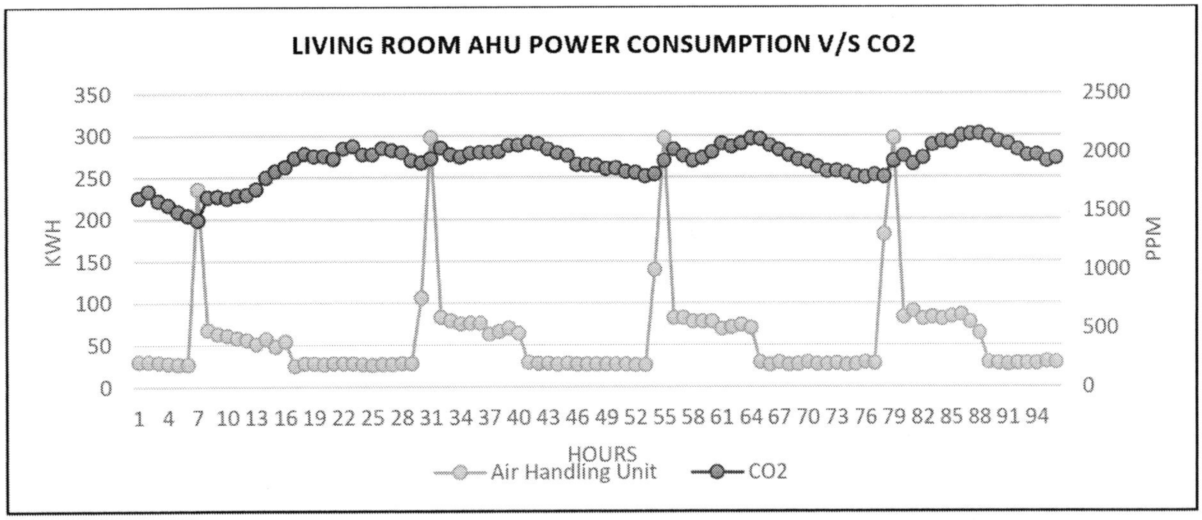

Figure 5: Modulation of Carbon-Dioxide in the Living Room

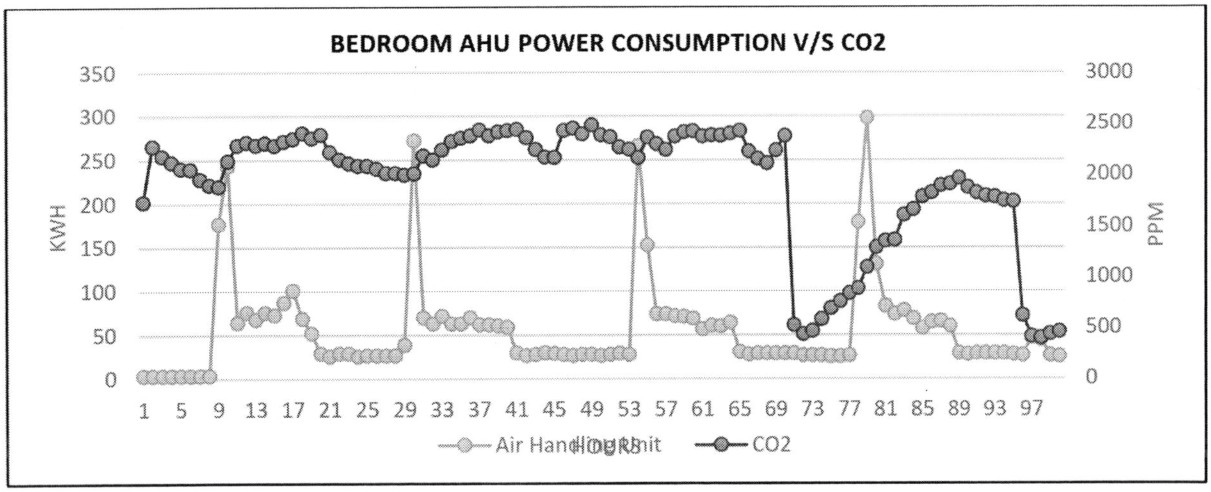

Figure 6: Modulation of Carbon-Dioxide in the Bed-Room

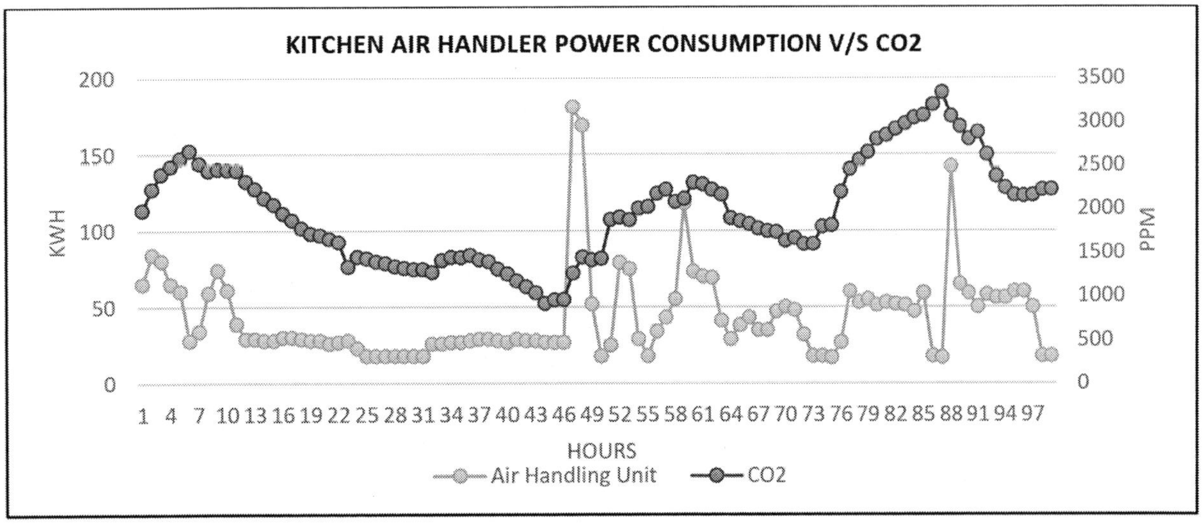

Figure 7: Modulation of Carbon-Dioxide in the Kitchen

A closer look at the modulation of carbon-dioxide appears to show that the concentration of carbon-dioxide increases with when AHU is in operation. The authors hypothesize that this increase in the concentration of carbon-dioxide is not an actual increase but an apparent increase, captured by the meter as a result of increased air flow through the registers which creates air drafts within the house. To study the correlation between the concentration of carbon-dioxide and AHU operation, the authors performed a Pearson Product Moment correlation (Pearson's r test). In this case the tests measured the linear correlation between two variables X (carbon-dioxide concentration) and Y (AHU energy usage), giving a value between +1 and −1 inclusive, where 1 is total positive correlation, 0 is no correlation, and −1 is total negative correlation.

Overall, the Pearson's coefficient of correlation between carbon-dioxide concentration and AHU operation for the first and second set of data were -0.04 and 0.00 respectively, which meant that there was absolutely no correlation. However, for the third data set, the coefficient of correlation was 0.27, showing a slight positive correlation and some possibility that the concentration of carbon-dioxide increased with AHU operation.

Figure 8, Figure 9 and Figure 10 are graphs that show the modulation of relative humidity with operation of AHU. To study the correlation of relative humidity with AHU operation, the authors again performed the Pearson's r test. In this case the tests measured the linear correlation between two variables X (Relative humidity) and Y (AHU energy usage). In the first set of data (Figure 8), Pearson's coefficient was 0.20 which suggested that there was slight correlation, and there was a slight chance that RH increased with AHU operation. In the second set of data (Figure 9), the coefficient was – 0.24, which suggested that there was a slight chance that RH decreased with AHU operation. Finally, for the third data set (Figure 10), Pearson's coefficient was 0.31, suggesting a slight correlation of increased RH with AHU operation.

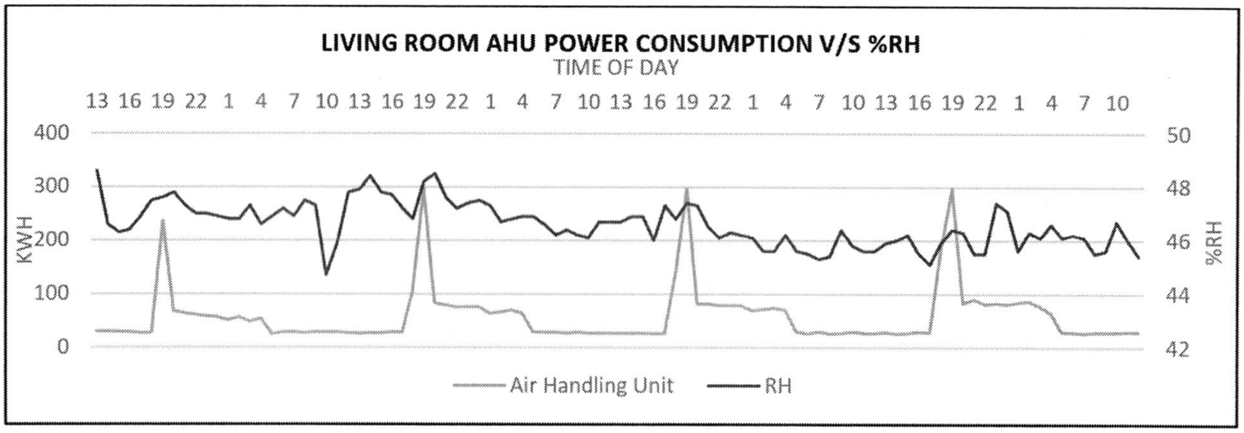

Figure 8: Modulation of Relative Humidity in the Living Room

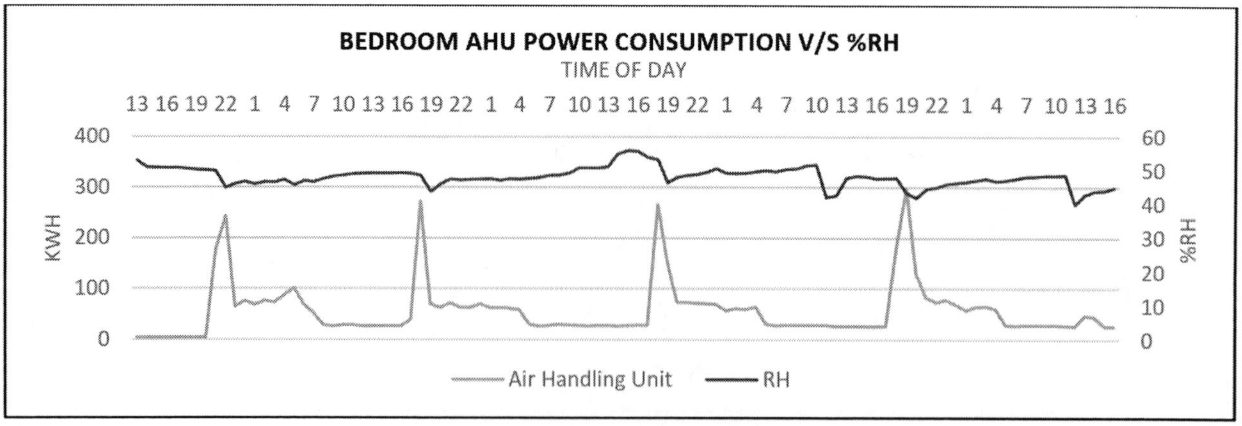

Figure 9: Modulation of Relative Humidity in the Bedroom

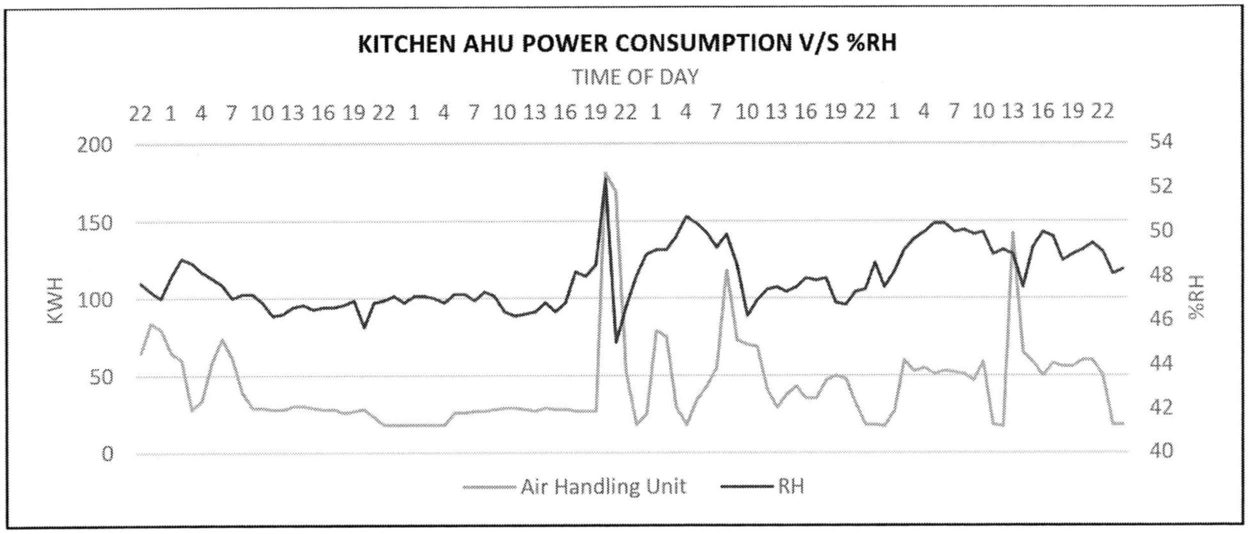

Figure 10: Modulation of Relative Humidity in the Kitchen

5. CONCLUSION

This study compared the effect of operation and control of the HVAC system of a home retrofit for low air infiltration on various interior air parameters. During the course of this study, authors measured interior concentrations of carbon-dioxide higher than ASHRAE's recommendation for the retrofit house. The house being highly airtight and the occupant's preference for having all windows and doors closed are some primary explanations for this phenomenon. In a questionnaire the occupant revealed that the ERV was at its highest setting throughout the heating season. Hence it can be concluded that the ERV is not able to providing adequate ventilation for the retrofit house under the present conditions. While the elevated carbon-dioxide levels are not an indication of unhealthy air quality, they do indicate that adequate ventilation may not be available for odor control. The relative humidity and indoor temperature readings from the retrofit house did not show any unusual patterns, and overall it can be concluded that the operation of furnace with the setback capable thermostat allows the occupant to reduce energy consumption, while maintaining a comfortable interior environment during hours of occupancy.

LIST OF REFERENCES

ACGIH (American Conference of Governmental Industrial Hygienists). 2011. TLVs and BEIs. Cincinnati, OH: American Conference of Governmental Industrial Hygienists.

ASHRAE Fundamentals Handbook, Ventilation and infiltration, American Society of Heating, Refrigerating and Air-Conditioning Engineers, Inc., Atlanta, GA, 1997 (Chapter 25).

ASHRAE, ASHRAE Standard 55-2013, Thermal Environmental Conditions for Human Occupancy; 2013.

ASHRAE Standard 62–1999, Ventilation for acceptable indoor air quality, American Society of Heating, Refrigerating and Air-Conditioning Engineers, Inc., Atlanta, GA (1999).

Batterman, S., & Peng, C. (1995). TVOC and CO2 Concentrations as Indicators in Indoor Air Quality Studies. *Aihaj, 56*(1), 55-65.

U.S. energy information administration (EIA) 2003 commercial building energy consumption survey (CBECS). 2013. http://www.eia.gov/consumption/commercial/index.cfm, accesses on 03/14/2015

EIA 2009 residential energy consumption survey (RECS). 2009. Accessed at: http://www.eia.gov/consumption/residential/data/2009/, accesses 03/21/2015

Energy Information Administration. (2009). *EIA Residential Energy Consumption Survey* (RECS). //www.eia.gov/consumption/residential/index.cfm, accessed on 10/27/15

Jones, A. (1999). Indoor air quality and health. *Atmospheric Environment, 33*(28), 4535-4564.

Lipsett MJ. Shusterman DJ, Beard RR. 1994. Inorganic compounds of carbon, nitrogen, and oxygen. *In: Patty's Industrial Hygiene and Toxicology (Clayton GD, Clayton FD, eds)*. New YorUohn Wiley & Sons, 4523-4554

O. Seppänen, W.J. Fisk, M.J. Mendell, Ventilation rates and health, ASHRAE Journal (August) (2002), pp. 56–58

OSHA (Occupational Safety and Health Administration). 2012. Sampling and Analytical Methods: Carbon Dioxide in Workplace Atmospheres. Available: http://www.osha.gov/ dts/sltc/methods/inorganic/idl72/id 172.html

Persily, A.K. 1993. Ventilation, carbon dioxide, and ASHRAE Standard. *ASHRAE Journal* 35(7): 40-44.

Persily AK. 1997 Evaluating building IAQ and ventilation with carbon dioxide. *ASHRAE Transactions* 103(2): 193-204.

Samet, J.M., Marbury, M.C. and Spengler, J.D. (1987) "Health effects and sources of indoor air pollution. Part I," *American Review of Respiratory Disorders,* 136, 14861508.

Samet, J.M., Marbury, M.C. and Spengler, J.D. (1988)" Health effects and sources of indoor air pollution. Part 11," *American Review of Respiratory Disorders,* 137, 221-242.

Satish, U., Mendell, M. J., Shekhar, K., Hotchi, T., Sullivan, D., Streufert, S., and Fisk, W. B. J. 2012. Is CO2 an indoor pollutant? Direct effects of low-to-moderate CO2 concentrations on human decision-making performance. *Environ. Health Perspect.*120: 1671–1677.

Study of Performance Characteristics of a Triple Duty Valve using CFD

Sagar Dilip Jadhav[1], Shyam Narayan Shukla[2], Ravindra Sharanappa Birajadar[3],

[1]Kirloskar Brothers Limited, Corporate Research and Engineering Development,
Pune, Maharashtra, India
(+91-9422373771, +91-20-27211060, sagar.jadhav@kbl.co.in)

[2]Kirloskar Brothers Limited, Corporate Research and Engineering Development,
Pune, Maharashtra, India
(+91-9922934629, +91-20-27211060, shyam.shukla@kbl.co.in)

[3]Kirloskar Brothers Limited, Corporate Research and Engineering Development,
Pune, Maharashtra, India
(+91-9850965670, +91-20-27211060, ravindra.birajdar@kbl.co.in)

ABSTRACT

A triple duty valve is used to control the functions such as on-off, throttling and non-return on the discharge side of hydronic pump systems. In this paper, computational fluid dynamic techniques are used to simulate the 2 inch triple duty valve performance characteristics for various stroke lengths with uniform outgoing velocity. The performance characteristics such as flow coefficient and pressure loss coefficient for triple duty valve are calculated for different stroke lengths (8 mm, 12 mm and 20 mm) through simulation. Numerical simulation is conducted to study effect of various Reynolds number and valve openings on the performance characteristics of triple duty valve. The permanent pressure loss coefficient and its dependence on Reynolds number is computed and analyzed respectively so that the recovery of pressure after passage through valve could be quantitatively understood. The comparison of flow pattern at different stroke length is investigated. Pressure drop across the triple duty valve and formation of vortex flow after the valve section have been discussed. Magnitude of flow coefficient, pressure loss coefficient and permanent pressure loss coefficient calculated through numerical analysis are compared with the experimental results. Furthermore, this study can be used for the evaluation of performance characteristics of triple duty valve at different sizes.

1. INTRODUCTION

1.1 Overview

A valve plays a vital role in process industries. When valve is used for flow control, the dynamics of the valve has to match the dynamics of the flow system. The relationship between valve position and the pipe system would create the pressure drop and flow becomes highly turbulent. This implies difficulties in predicting the flow properties. While designing a new kind of valve, simple structures along with optimum energy consumption is common concern. It is very important to understand fluid flow inside the valve to improve the performance at different operating conditions. Therefore, it is important to involve other techniques, such as numerical methods, to compliment experimental studies and provide additional information regarding valve flow patterns. In hydronic systems, three different types of valve being used to control the functions normally required on the discharge side such as throttling, on-off and non-return. Triple Duty Valve (TDV) is a combination of all these valves used on the discharge side of the hydronic pump systems. TDV features are design to balance system accurately and throttle excess flow. It includes a spring loaded non-slam check valve. This helps to prevent damage of pump and unwanted heat transfer through the system. It also provides positive shut-off allowing a person to isolate the pump from the rest of the system. It features a disc style design with a soft seat which provides complete shut-off and prevents valve chatter. The soft seat design of the disc prevents noise and potential damage that can be associated with valve chatter. TDV have stopper mechanism which allows returning the disc to its set position after shut-off. While designing the TDV, the main governing parameter which is responsible for throttling is stroke length. Manual operation through hand wheel is used to adjust the stroke length for required flow and complete shut-off.

1.2 Literature Survey

The behavior of valve in flow field has been studied by several researchers. Some of them are discussed here. Henderson A. D. et.al (2007), showed the hydrodynamic torque versus opening angle characteristics during constant head test. It was observed that flow separation increases with angle of closing i.e. from opening to closing travel of the disc. The strong vortical flow implies lift force and cause fluctuations in pressure. Yang Qin et.al (2011), observed flow patterns through stop valve with numerical simulation. They observed valve flow coefficient and fluctuation at different flow with uniform incoming velocity. Shirazi Tavakoli Nima et.al (2012), showed comparison of flow pattern with various opening angle. It was observed that for larger recirculation length, more pressure drop occurred. Tamizharasi G. and Kathiresan S., (2012), observed flow characteristics of butterfly valve under various operating conditions with different disc shape. It was observed that pressure-drop in case of 40° and 60° opening increases significantly. Azad et.al (2014), determined valve characteristics such as flow coefficient and head loss coefficient using CFX 12 for different valve opening angle for incompressible fluid. They observed that the downstream stream line flow seems to be uniform with increase in opening angle which indicates the disturbances are reduced for higher valve opening angle. C Kumar et.al (2015), observed the effect of Reynolds number and the various valve openings on pressure loss coefficient of traditional butterfly valve. The value of pressure loss coefficient rapidly increases with increase in valve opening. Reynolds number has a significant effect on the value of pressure loss coefficient. Jadhav S. D. et.al (2015), showed the flow behavior across the triple duty valve. It was observed that at initial openings (close to open) recirculation is more which leads to energy loss in valve. At higher openings reduction of recirculation was observed which consequently improves more uniform pressure and velocity field.

Several studies had been done and reported in the literature by researcher using CFD to analyze performance characteristics of different type of valves. However, for a triple duty valve these studies have not been covered for different stroke length. Hence the present study is aimed at understanding some of these aspects. In this paper in continuation to prior work of Jadhav S. D. et.al (2015), study has been carried out to calculate the performance characteristics of TDV at different stroke lengths. Here, more emphasis is given on the effect of various Reynolds number and valve openings on characteristics of triple duty valve. While studying these aspects at different Reynolds number, numerically choking is simulated to identify limiting velocity below which it is eradicated.

1.3 Flow Performance Calculations

It is important for fluid system designer to understand performance characteristics of the valve, which helps to achieving desired act without physical damage. Performance characteristics include the pressure drop, flow coefficient, pressure loss coefficient and permanent pressure loss coefficient. They are defined below,

1.3.1 Flow Coefficient (Cv): Cv is defined as one U.S. gallon (3.8 liters) of 60°F (16°C) water that flows through an opening, such as a valve, during 1 min with a 1 psi (0.07 bar) pressure drop. As specified by Society of America (ANSI/ISA Standard S 75.01), the simplified equation for Cv is,

$$C_v = Q \times \sqrt{\frac{S_g}{\Delta p}} \tag{1}$$

1.3.2 Pressure Loss Coefficient (k) and Permanent Pressure Loss Coefficient (k_{perm}): The value of pressure loss coefficient (k) is calculated by considering the difference in pressure between two locations, one at upstream and other at downstream of the pipeline. Thus, k is defined as,

$$\mathrm{k} = \frac{\Delta p}{\frac{1}{2} \times \rho \times v^2} \tag{2}$$

The valve disc introduces certain permanent pressure loss in the duct. In order to evaluate its magnitude, the pressure loss in the duct with and without valve is computed at identical flow condition. Pressure drops namely, Δp_1 and Δp_2 are calculated between the inlet and outlet of the valve and obviously,

$$\Delta p_1 > \Delta p_2$$

Thus, permanent pressure loss is defined as,

$$\Delta p_{perm} = \Delta p_1 - \Delta p_2$$

Permanent pressure loss coefficient is defined as,

$$k_{perm} = \frac{\Delta p_{perm}}{\frac{1}{2} \times \rho \times v^2} \quad (3)$$

The parameter represents the extent of the energy loss in the duct due to the introduction of the valve. It has been evaluated for three case studies.

2. NUMERICAL MODELLING

CFD provides numerical approximation to the equation that governs fluid motion. Application of the CFD to analyze a fluid problem requires the following steps.

2.1 Model Description for CFD

In the present case, experimental investigation is followed by numerical simulation on three-dimensional geometry using commercially available CFD software to understand the fluid flow physics and validation of software.

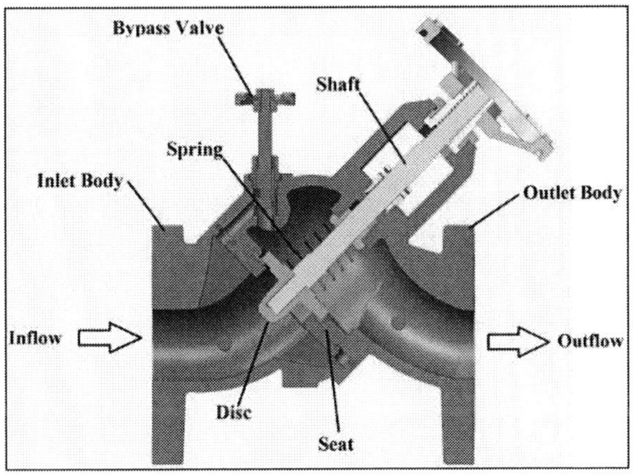

Figure 1: Sectional view of Triple Duty Valve assembly

Fig. 1 shows the cross sectional view of 2" TDV model developed using Creo Parametric 2.0. It mainly consists of inlet body, outlet body, disc, spring, shaft and hand wheel. Subsequent to experimental work, the same geometry is taken for numerical simulation. The CFD analysis is carried out at three different stroke lengths. The geometry of TDV includes inlet pipe, inlet body, disc, shaft, outlet body and outlet pipe as shown in Fig. 2.

Precise geometry of 3D models is prepared using Creo Parametric 2.0. ANSYS ICEM CFD 16.1 is used for extraction of 3D fluid domain (Fig. 2) and for creation of good quality meshing of TDV. The solution and post-processing is done in ANSYS CFX-Solver and ANSYS CFX-Post respectively. The computational mesh is generated using mostly tetrahedral elements in the fluid volume and generating prismatic elements near walls region. The purpose of creating a hybrid prismatic mesh is to substantially increase mesh resolution in boundary layers, where high Reynolds number flows exhibit large gradients in the wall-normal direction. The grid distribution is optimized keeping a good balance between computational time and better computational results. Number of generated elements and nodes in the passage are 1.94 million and 0.44 million respectively. Three prismatic layers are generated at the wall boundaries of the TDV. Grid independence test is performed for the geometry to check whether the results are independent of the mesh sizes. The deviations in results are within 0.05%. The final grid points are considered to be 0.44 million.

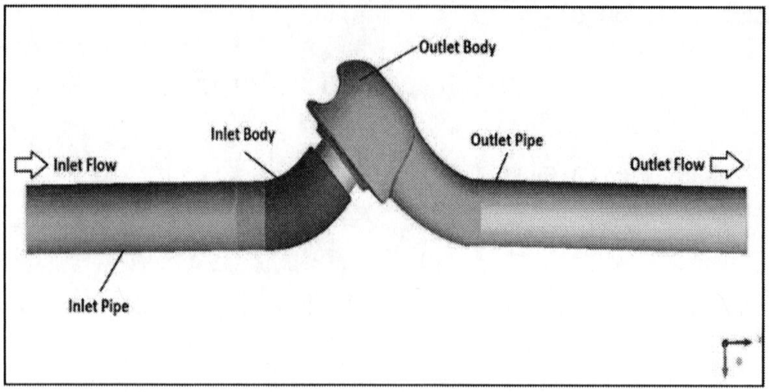

Figure 2: 3D fluid domain

2.2 Boundary Conditions

In CFD analysis, solution depends upon the appropriate boundary conditions. The problem is solved considering steady state simulation. Appropriate boundary conditions are given at the inlet and outlet of the TDV geometry. Numerically chocking is considered to identify the limiting flow at each opening. The domain type is fluid domain with water as the working fluid. The standard SST turbulence model is used for analysis. The advection scheme used is high resolution. This scheme is a higher order scheme and gives good results especially in the case of recirculation of flows. The physical properties of water are considered at 25°C. The turbulent kinetic energy and turbulence eddy dissipation are assigned as automatic. As the experimental study was conducted on cast iron valve, the value of surface roughness on walls is considered as 0.26×10^{-3} m.

3. RESULTS AND DISCUSSION

The simulation is carried out for different openings and with different Reynolds number. The values of performance characteristics for 8 mm, 12 mm and 20 mm stroke lengths are calculated through experimentation and CFD simulation. These values are discussed here.

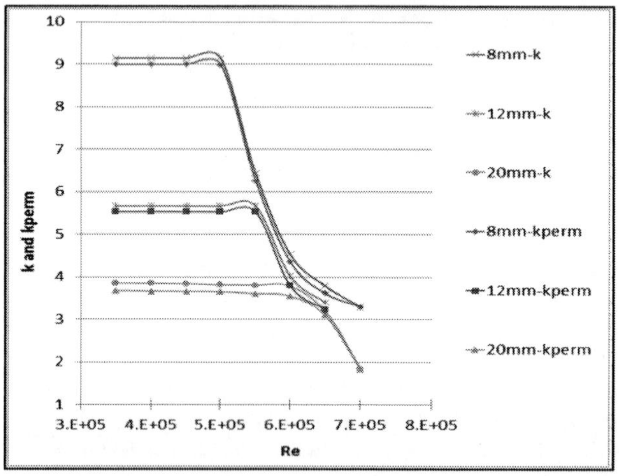

Figure 3: k and k_{perm} comparison at different stroke length with different Reynolds number

Fig. 3 shows that initially the value of k and kperm are almost constant with increase in R_e at each stroke length. Further, in each case it goes up initially and drops down. As R_e reaches to 5.0×10^5 and above, k and k_{perm} start decreasing. This is because, at this R_e choking flinches. As stroke length increases choking get shifted towards the highest R_e. For a given R_e, k and k_{perm} decreases with increase in stroke length (i.e. from 8 mm to 20 mm). This is obviously due to the reduction in the obstruction caused by the disc to the flow and consequently decreases in the energy dissipation. The behavior of k_{perm} and k are similar. The value of k_{perm} is always lower than k at any given stroke length and R_e, although the difference is not very large. This is because, partial recovery of the pressure after valve passage.

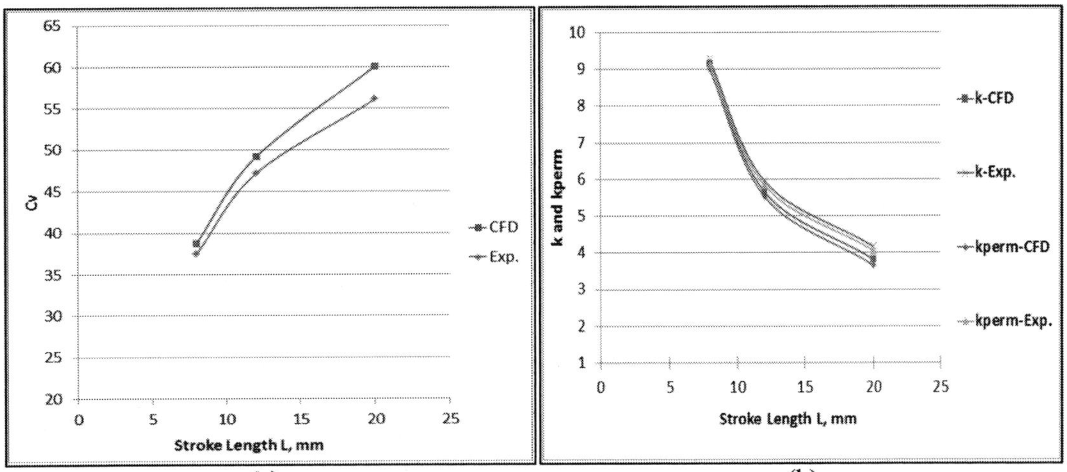

(a) (b)

Figure 4: CFD and Experimental: Cv, k and k_{perm} values comparison at different stroke length

Fig. 4 shows comparison of CFD analysis values with experimental results for performance characteristics. Fig. 4(a) shows that the value of Cv from CFD results are higher than the experimental values. While the values of k and k_{perm} from CFD results are less than experimental as shows in fig. 4(b). The values of performance characteristics computed through CFD simulations are well matched with experimentation. In CFD and experimental values maximum deviation (7%) is observed at higher stroke length due to fluctuations in higher openings.

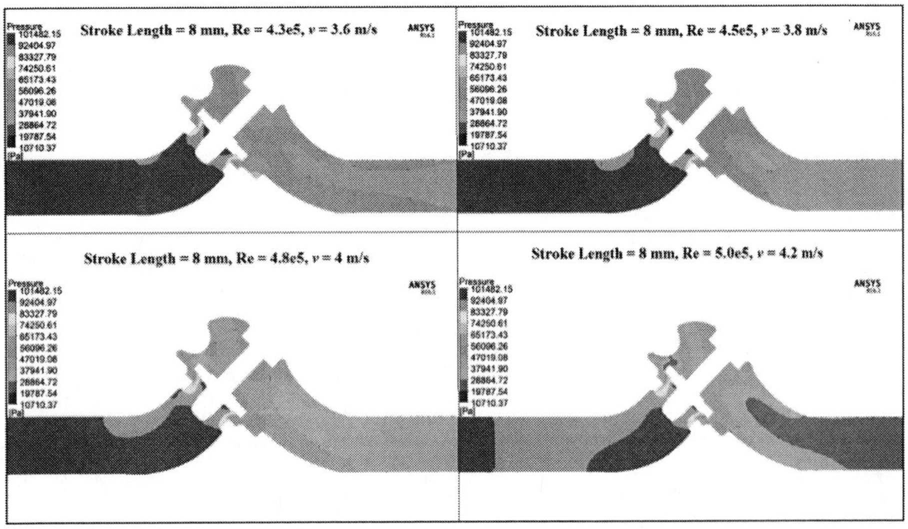

Figure 5: Pressure contour plots at 8 mm stroke length with different R_e

Fig. 5 to fig. 10 shows the flow behavior patterns inside the valve at various Reynolds number in the form of pressure plots and velocity plots at mid-section. These plots are taken for different stroke lengths such as 8 mm, 12 mm, and 20 mm.

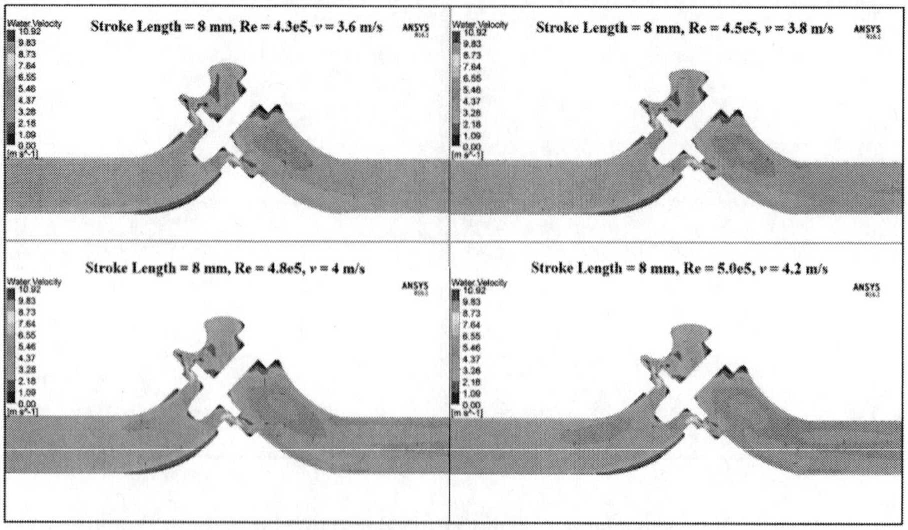

Figure 6: Velocity contour plots at 8 mm stroke length with different R_e

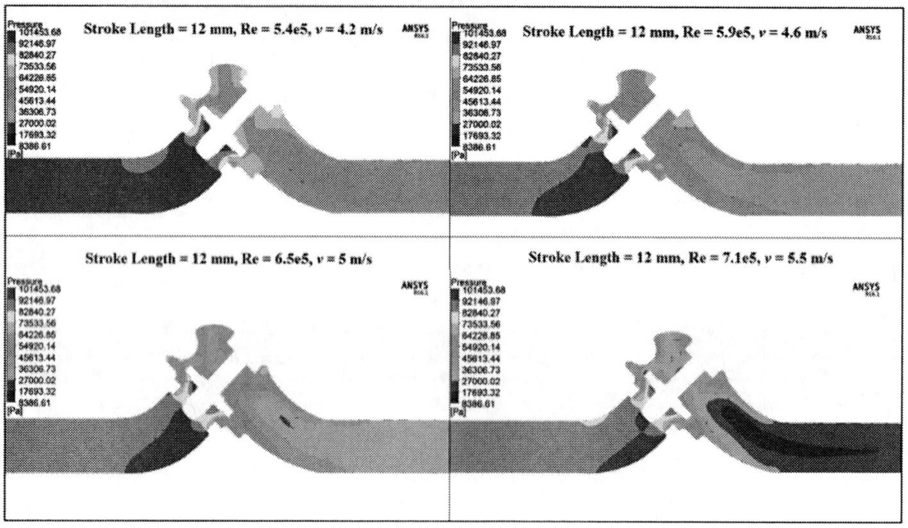

Figure 7: Pressure contour plots at 12 mm stroke length with different R_e

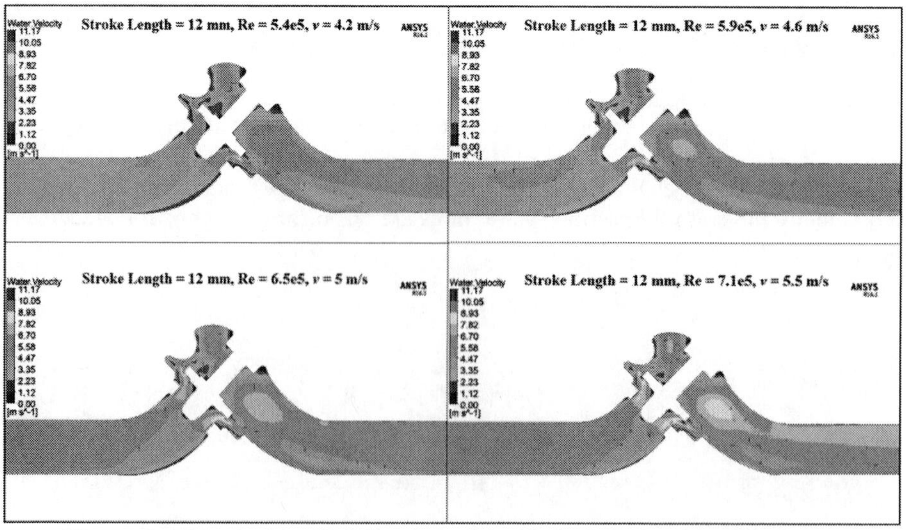

Figure 8: Velocity contour plots at 12 mm stroke length with different R_e

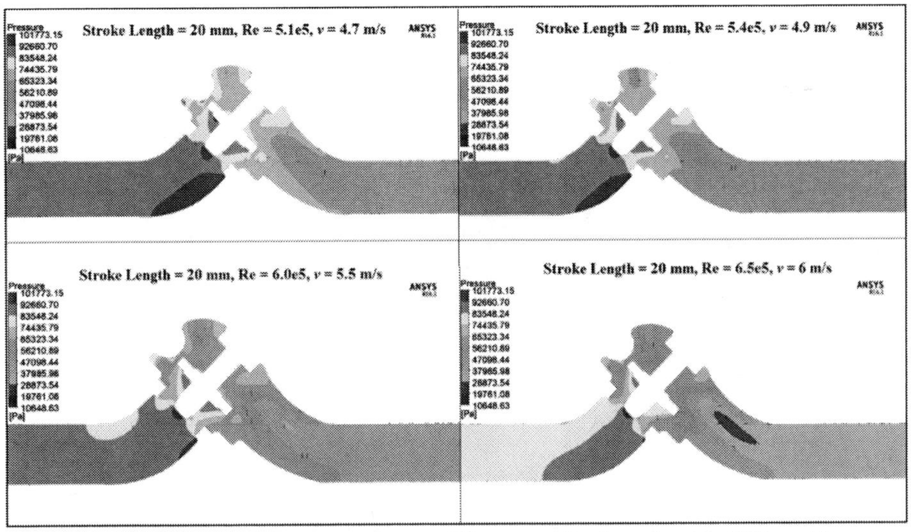

Figure 9: Pressure contour plots at 20 mm stroke length with different R_e

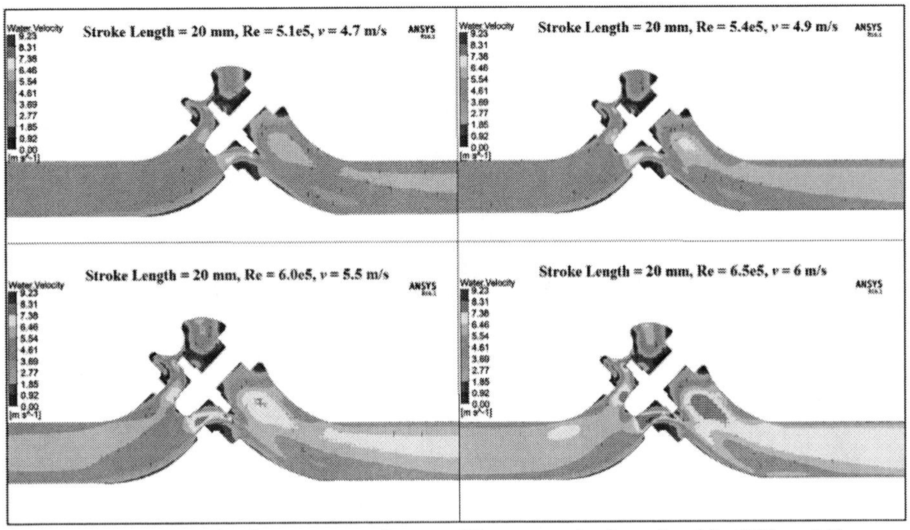

Figure 10: Velocity contour plots at 20 mm stroke length with different R_e

It is observed that at lower R_e, the value of pressure drop across the valve disc is minimal. The pressure and velocity development with different R_e are similar. The only difference observed is in magnitude. The backflow is observed behind the disc region. The recirculation at downstream side observed more in case of higher R_e and lower stroke length. These causes large amount of drag and total loss. From lower to higher R_e, inlet pressure drops down marginally at upstream side. It drops more in higher openings. Above this R_e value, the pressure drops at downstream drastically while there is not much difference in inlet pressure drop. Further, pressure drop in outlet initiates choking. At higher R_e ($R_e \geq 5.0 \times 10^5$) the pressure imbalance is more which creates higher forces on the valve disc. The pressure zone near the inlet and across the valve disc dome is similar in all R_e and in respective stroke lengths. For same R_e, the pressure and velocity distribution is more uniform with higher stroke length. This is because, the reduction in obstruction in flow passage. Flow separation is observed in all R_e and stroke lengths but, it is observed more in case of lower stroke length. The value of k and k_{perm} decreases with increase in head loss as valve stroke length increases. This indicates pressure loss coefficient is important function of valve stroke length.

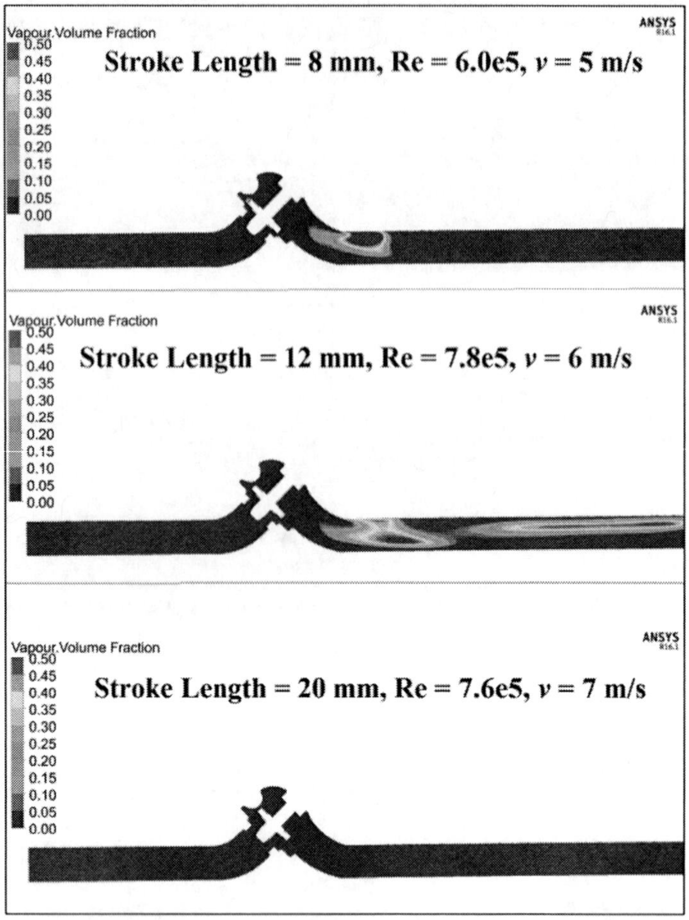

Figure 11: Vapour-volume fraction contour plots at different stroke length with $R_e \geq 5 \times 10^5$

Fig. 11 shows the initiation of chocking at different R_e and stroke length. In case of 8 mm stroke length the choking zone observed at outlet of the TDV. Further, this region shifted to the downstream of the pipeline in case of 12 mm stroke length. This choking zone is shifted at the pipe outlet in case of 20 mm stroke length. It is observed that at 8 mm, 12 mm and 20 mm stroke lengths the choking initiates approximately at $R_e \geq 5.0 \times 10^5$, $R_e \geq 7.8 \times 10^5$ and $R_e \geq 7.6 \times 10^5$ respectively. This indicates that value of the R_e needs to be kept below 5×10^5 for 2" TDV to avoid choking.

4. CONCLUSION

A validated numerical technique for analyzing performance characteristics through triple duty valve for different stroke length with various R_e has been presented. For the same stroke length R_e (turbulent zone) does not have significant effect on k and k_{perm} but, it decreases rapidly in higher stroke length (i.e. from full close to open). The pressure and velocity growth with different R_e are same. The only difference observed is in magnitude. This analysis has been extended further to analyze the initiation of choking. This implies that for lower stroke lengths the choking initiates at $R_e \geq 5.0 \times 10^5$.

5. FUTURE SCOPE

This technique will be applied on higher sizes of triple duty valve to analyze performance characteristics and limiting values of R_e at different stroke length to avoid choking.

NOMENCLATURE

Cv	Flow coefficient or flow capacity rating of valve	(–)
k	Pressure loss coefficient	(–)
k_{perm}	Permanent pressure loss coefficient	(–)
Q	Flow rate	(USGPM)
R_e	Reynolds Number	(–)
S_g	Specific gravity of fluid	(–)
v	Velocity	(m/s)
ρ	Density of fluid	(kg/m^3)
Δp	Pressure drop across the valve	(psi)
Δp_{perm}	Permanent pressure drop across the valve	(Pa)
Δp_1	Pressure drop between inlet and outlet of the pipe with valve	(Pa)
Δp_2	Pressure drop between inlet and outlet of the pipe without the valve	(Pa)

REFERENCES

1-Paper published in conference proceedings

Azad Arun, Baranwal Deepak, Arya Rajeev, Diwakar Nilesh (2014), Flow Analysis of Butterfly Valve Using CFD, IJMER, ISSN 2249-6645 (pp.50), Vol.4, Iss.6, TIEIT-Bhopal, India.

C Kumar Naveen, Seshadri V, K Kumar Yogesh (2015), Parametric Study of Flow Characteristics of Butterfly Valve Using CFD, IJETE, ISSN (2250-2459), Volume 5,Iss. 11, MIT-Mysore, India.

Henderson A. D., Sargison J. E., Walker G. J. and Haynes J. (2007), A Numerical Study of the Flow Through a Safety Butterfly Valve in a Hydro-Electric Power Scheme, 16th Australasian Fluid Mechanic Conference, (pp.1116-1122), Australia.

Jadhav S. D., Shukla S. N., Birajdar R.S. (2015), Numerical Analysis of Triple Duty Valve, ISBN-978-1-910643-24-2, NAFEMS World Congress, San Diego, California, USA.

Shirazi Tavakoli Nima, Azizyan Reza Gholam, Akbari Hossein Gholam (2012), CFD Analysis of the Ball Valve Performance in Presence of Cavitation, Life Sci. J, ISSN: 1097-8135 (pp.1460-1467), Chabahar, Iran.

Tamizharasi G., Kathiresan S. (2012), CFD Analysis of a Butterfly Valve in a Compressible Fluid, International Journal of Engineering Trends and Technology, Volume 3 Iss. 2 No3, Bharat University-Chennai, India.

Yang Qin, Zhang Zhiguo, Liu Mingyue, Hu Jing (2011), Numerical Simulation of Fluid Flow Inside the Valve, ELSEVIER, Procedia Engineering 23 (pp.543-550), Wuhan, China.

2-Book

Skousen P.L.(2014), Valve handbook, Second Edition, McGraw-Hill.

3-Book chapter

White F M, "Fluid Mechanics", McGraw-Hill, 2012, Seventh Edition, Chapter 6.

ACKNOWLEDGEMENT

The authors would like to thank the Management of Kirloskar Brothers Ltd. Pune, India for the encouraging attitude towards Research & Engineering development activities, which led to this paper. We are also very thankful to our colleagues for their invaluable suggestions in research work.

Ray W. Herrick Laboratories
Center for High Performance Buildings

Author Index

By Paper ID